2nd
EDITION

PHYSICAL GEOLOGY

THE SCIENCE OF EARTH

WILEY

2nd EDITION

PHYSICAL GEOLOGY

THE SCIENCE OF EARTH

Charles Fletcher
University of Hawai'i

WILEY

VP and PUBLISHER	Petra Recter
EXECUTIVE EDITOR	Ryan Flahive
SENIOR PROJECT EDITOR	Jennifer Yee
ASSISTANT EDITOR	Julia Nollen
SENIOR CONTENT MANAGER	Micheline Frederick
SENIOR PRODUCTION EDITOR	Sandra Rigby
MARKETING MANAGER	Suzanne Bochet
DESIGN DIRECTOR	Harry Nolan
SENIOR DESIGNER	Wendy Lai
SENIOR PHOTO EDITOR	Billy Ray
COVER PHOTO	MDN/Shutterstock

This book was set in 10/12 Times by codeMantra and printed and bound by Quad/Graphics. The cover was printed by Quad/Graphics.

This book is printed on acid-free paper. ∞

Founded in 1807, John Wiley & Sons, Inc. has been a valued source of knowledge and understanding for more than 200 years, helping people around the world meet their needs and fulfill their aspirations. Our company is built on a foundation of principles that include responsibility to the communities we serve and where we live and work. In 2008, we launched a Corporate Citizenship Initiative, a global effort to address the environmental, social, economic, and ethical challenges we face in our business. Among the issues we are addressing are carbon impact, paper specifications and procurement, ethical conduct within our business and among our vendors, and community and charitable support. For more information, please visit our website: www.wiley.com/go/citizenship.

ISBN: 978-1-118-55968-0
ISBN: 978-1-118-73642-5 (BRV)

Printed in the United States of America

10 9 8 7 6 5 4 3 2 1

Hello, my name is Charles Fletcher. I am a geologist and I work as the Associate Dean for Academic Affairs in the School of Ocean and Earth Science and Technology at the University of Hawai'i.

Even after 30 years I look forward to the start of each new semester. As a teacher, my goal is to help students to understand and appreciate how geology intersects our lives every day. My hope is that this understanding will lead students to make good decisions about their own lives, become exemplary citizens, and help lead us to a positive future.

Are you a student? If so, after you graduate, you will inherit an Earth with a number of global challenges. Climate change, pollution, shrinking resources, and growing vulnerability to natural hazards are unfortunate consequences of the rising human population on our planet.

I believe the discussion that takes place in the physical geology classroom is an important part of your training to be Earth's next generation of caretakers. What you learn in this course can influence your future decision-making—for instance, how you vote, and how you use natural resources. This course may lead you to become more effective Earth citizens.

This is the second edition of *Physical Geology: The Science of Earth*. It provides training to think critically and to acquire an enduring understanding of geologic principles. Physical geology is the science of Earth, and the preparation of tomorrow's decision makers begins in the physical geology classroom.

Let me know what you think—I want to hear from you. You can email me at fletcher@soest.hawaii.edu. Together let's discuss what you are learning, what is important to you, and how to create a better planet for your future and for your kids.

To my family

Kellen, Katie, Chase, Estevan, Javier,

and most especially, Ruth.

The world has changed

Global warming brings new intensity to natural hazards. The human population has reached record levels, and geologic resources have grown more expensive and scarce. These are global challenges requiring our best thinking and a need to be knowledgeable about how Earth works.

In 2012, Hurricane Sandy caused over $65 billion in damage and took 253 lives. The effects of global warming and enormous population growth along the U.S. east coast amplified the amount of damage caused by the storm. Geologists can be important sources of information in building community resiliency to natural hazards.

The physical geology classroom has evolved as well. Across the U.S., students are taking geoscience classes in increasing numbers and more students are graduating with geoscience degrees than ever before. The U.S. Bureau of Labor Statistics predicts that employment of geoscientists in the U.S. will grow disproportionately, by 21% between 2010 and 2020; the number of qualified scientists worldwide is unlikely to meet demand.

Instructors, faced with burgeoning classrooms, need new tools to relate to modern students on their own terms, and to continue the work of assessing learning among large class sections. Interactive lectures, online resources, blogs, and affordable *eBooks* increasingly characterize the physical geology classroom. *Physical Geology: The Science of Earth 2e* is fully prepared to meet these needs with a comprehensive course management system, online grading, new animations, an *eBook*, interactive critical thinking exercises, and more.

Why Physical Geology: The Science of Earth 2e?

The study of physical geology has never been more relevant, timely, and fast-paced. However, three things have not changed:

1. Instructors want to enrich the lives of their students and help them leave the course as more responsible and informed Earth citizens. This requires students to:

The Tohoku, Japan earthquake and tsunami in 2011 caused the death of 20,000 people and destroyed over 1 million buildings.

a. Develop a basic understanding of geologic processes

b. Cultivate strong critical thinking skills

2. Students need a book that enhances learning. It must be easy to read, well illustrated, relevant, and engaging.

3. Instructors need a text with a *course management* system that focuses on basic principles with the most up-to-date information and also provides *assessment tools* for large class sections.

What sets *Physical Geology; The Science of Earth* apart from the other physical geology books on the market? The second edition of this text continues our tradition of emphasizing **student learning, critical thinking**, and **basic geologic principles**. It achieves this with important innovations while maintaining traditional strengths.

Innovations

- **Ease of Reading:** We deliver the fundamental principles of geology arranged around specific learning objectives. Students find the book approachable, succinct, easy to read, and extremely well organized.

- **Critical Thinking:** Woven into each page are critical-thinking exercises (similar to lab exercises, homework, and class discussion sessions) linked to illustrations and photos. These exercises are guided by Bloom's Taxonomy, a classic study that identifies six sequential levels of building intellectual skill (see the box below, "Bloom's Taxonomy of Learning Objectives").

- **On-line Grading:** The hundreds of critical thinking questions and exercises in this text are all graded automatically on line with the WileyPLUS Learning Space course management system. The WileyPLUS Learning Space system automatically records grades in a spreadsheet. Thus, the instructor's workload, especially in large classrooms, is significantly reduced. For example, each of the 21 chapters has 20 multiple-choice questions—these can be assigned as homework that is automatically graded on-line. I do this in my class every semester. This keeps the students focused day by day on the course content, encourages them to keep up with reading, and provides additional grading opportunities (augmenting grades on mid-terms and final exams).

Although my class isn't particularly large, I do have about 100 students. The WileyPLUS Learning Space system allows me to give 21 homeworks and three exams each semester and grade them all automatically—this is a HUGE relief!

- **Current Topics:** In this edition I discuss relevant and provocative global issues such as peak oil, global warming, drought and water stress, population growth, natural hazards, and environmental management. Chapters on these socially relevant topics are not buried at the end of the book, as in so many other texts. Rather, they are integrated into the main body of the text because I feel these issues characterize the practice of modern-day geology.

Physical Geology is a learning system rich with critical thinking questions and activities that drive students to challenge their understanding of the world. What is critical thinking? We identify it as the use of reasoning to define the world around us by developing hypotheses and theories that can be rigorously tested.

© Dleonis/iStockphoto

In 2012 North America experienced the worst drought in its history. Covering at least 80% of the contiguous United States, parts of Mexico, and central and eastern Canada, the drought was responsible for over $150 billion in losses.

Physical Geology **treats the art and photo program as a critical thinking opportunity**. It is common for texts to be illustrated with colorful, three-dimensional art and timely, revealing photographs. We achieve this and take it a step further: we make almost every art piece a critical thinking exercise that works through the six learning objectives in Bloom's Taxonomy. Our illustrations are part of problem-solving exercises that students can do in place of lecture, as homework, in teams or solo, or simply as another learning opportunity.

Good questions are the driving force of critical thinking. This integration of art and critical thinking promotes the growth of problem-solving skills in the context of a lifelong understanding of Earth. Students are reconnected with the natural resources and landscapes that sustain us all, and they exercise the thinking tools needed to address global challenges.

Our critical thinking approach is guided by Bloom's Taxonomy (1956), a classic study that identified six sequential levels of cognitive skill. Bloom's Taxonomy is considered a foundational and essential element within the education community.

Bloom's Taxonomy of Learning Objectives
(updated by Anderson et al., 1994[1])

1. **Knowledge (or remembering)**—This is the basic level. Bloom found that over 80% of test questions simply test knowledge. Students are asked to arrange, list, recall, recognize, and relate information.

   Example: How are loose sediments lithified? What is Bowen's reaction series?

2. **Comprehension (or understanding)**—Knowledge should build comprehension. Students must classify, describe, and explain concepts. For instance, students are asked to state a problem in their own words.

---

[1]Anderson, Lorin W., and Lauren A. Sosniak, eds. (1994), *Bloom's Taxonomy: A Forty-Year Retrospective*. Chicago National Society for the Study of Education.

Example: Explain why waves change shape as they enter shallow water. Why do corners and edges of rock weather faster than a flat surface? Describe the crystallization of minerals in magma of average composition.

We use **knowledge** and **comprehension** questioning often throughout the book, typically with every art piece and as a warm-up to more complex activities.

3. **Application**—Students are asked to apply what they learned in the classroom and their reading in novel problem-solving situations.

Example: Using your understanding of mass wasting and stream behavior, define three general guidelines for safely locating a new road in a hilly region.

4. **Analysis**—This skill level requires students to calculate, analyze, categorize, and compare/contrast information.

Example: What evidence would you look for to test the hypothesis that humans are contributing to or are the cause of global warming?

Application and **analysis** are higher levels of cognition. Students are asked to practice these skills in our two-page "Expand Your Thinking" exercises.

5. **Evaluation**—This requires students to make judgments about the value of ideas or materials. We ask them to appraise, assess, choose, and predict outcomes.

Example: What criteria would you use to assess the hazards associated with a volcano that no one had previously studied?

6. **Synthesis**—Students must put parts together to form a new meaning or structure. This involves arranging information to construct new knowledge.

Example: In some areas, climate change is causing less overall rainfall but more intense rain events. As a community leader, what new challenges do these trends offer you and how will you manage them?

Evaluation and **synthesis** are the highest levels of learning. Students are asked to use these skills in our "Expand Your Thinking" exercises and our two-page "Critical Thinking" exercises.

Physical Geology is carefully organized into two to four page segments tied to Learning Objectives. Students find that a reading assignment consisting of "bite-size pieces" is easily accomplished, the content is easier to understand, and thus the assignment is more readily started and more likely to be completed. As instructors, we want students to do the reading and the work we assign. This book advances that goal with a purposeful design to facilitate student learning and keep their attention. Every segment concludes with an "Expand Your Thinking" question that provides further critical thinking opportunities for students to check their conceptual understanding of the material before moving forward.

Don Becker/USGS

Floods claim nearly 100 lives and cost billions in property damage in the United States annually. Spring is peak flood season in many parts of the country, but floods can happen anywhere, at any time of the year.

Strengths

Emphasizing the basic principles of geology, *Physical Geology* is easy to adopt into an existing syllabus. For example, plate tectonics is introduced early, and the entire text teaches to tectonics as a unifying theory of geology.

The writing is clear and jargon free. Throughout the review process, as dozens of faculty from around the nation were asked to critically review each chapter, one resounding comment surfaced again and again—"This text is well-written and easy to read."

Every page has several pieces of accurate, detailed, beautiful illustrations and photos.

Nearly every page discusses relevant and provocative aspects of geologic principles and relates them to the modern world around us.

Physical Geology contains chapters on each rock group and surface environments plus detailed coverage not normally seen in other texts: *Climate Change*, and *Glaciers and Paleoclimatology*, as well as separate chapters on *Coastal Geology* and *Marine Geology*. There are also chapters on *Geologic Time* and *Earth's History*. Because the book has 21 chapters, many instructors will want to choose a subset of chapters that best fits their learning objectives.

Pedagogical Help for the Student

Each student in the classroom is capable of learning, yet none think or process information in the same ways. Some are social and learn when interacting with other people. Some are analytical and learn best by problem solving. Others are spatial learners and enjoy graphics. Teachers, parents, and modern brain-based researchers understand that many different types of learners populate the classroom.

As teachers, we are challenged to reach all of these learners. Hence, *Physical Geology* delivers information in a variety of ways.

- **Specific learning objectives**—There are 21 chapters, each one with 10 to 15 learning objectives. The learning objectives identify what the students should know, understand, or be able to accomplish as a result of their reading. Achieving each learning objective will help students study and help the instructor assess and evaluate student progress. This design facilitates learning in a busy world where our attention is challenged by texting, cell phones, social networking, and other distractions.

- **A diversity of assessment**—Every chapter has several types of review, critical thinking, and problem-solving exercises; students are asked to contemplate geological problems by focusing on art, calculations, inferring outcomes, recognizing patterns, connecting concepts, and other tasks.

- **Relevant and provocative**—Four themes recur throughout the book to remind us that dynamic and vital natural processes needing careful management surround us. *Critical Thinking, Geology in Our Lives,* *Geology in Action,* and *Earth Citizenship* provide examples, applications, and activities for students to further their understanding.

- **Easy to read, easy to teach from**—Students will love reading this book because of its simple organization and direct language. Instructors will find this book easy to adopt because it presents the basic principles of geology in a flexible array of chapters.

In Closing

The role of teaching is to educate the next generation of leaders. Frankly, I think physical geology should be a required course for every student on the university campus. So many geology-related problems have surfaced in the past decade that adults with even minimal training in the basic principles of geology are capable of offering important and critically needed help. Indeed, I feel this is one of the major reasons to teach physical geology.

Climate is changing, resources are growing in expense, droughts and storminess both are increasing, environments are threatened, communities are expanding into hazardous areas; how we respond to these challenges will write the future and fate of humankind on Earth. These grand challenges are on our doorstep now. Our greatest hope lies with the best critical thinkers who understand how Earth works.

Supplements

Geo Media Library

This easy-to-use multimedia web site helps reinforce and illustrate key concepts from the text through the use of animations, videos, and interactive exercises. Students can use the resources for tutorials as well as self-quizzing to complement the textbook and enhance understanding of physical geology. Easy integration of this content into course management systems and homework assignments gives instructors the opportunity to incorporate multimedia with their syllabi and with more traditional reading and writing assignments. Resources include:

- **Animations**—Key diagrams and drawings from our rich signature art program have been animated to provide a virtual experience of difficult concepts. These animations have proven influential in helping visual learners understand this content.

their understanding of key concepts and explore additional visual resources.

- **Concept Caching**—an online database of photographs explores what a physical feature looks like. Photographs and GPS coordinates are "cached" and categorized along core concepts of geography. Professors can access the images or submit their own by visiting www.ConceptCaching.com.

Book Companion Site (www.wiley.com/college/fletcher)

In addition to our multimedia content, our book companion site offers a wealth of study and practice materials, including:

Student Online Resources

- *Self-quizzes*—chapter-based multiple-choice and fill-in-the-blank questions.

- *Web Resources*—useful web sites to explore more about geology.

- *Geo Media Library*—link to the media library for students to explore key concepts in greater depth using videos, animations, and interactive exercises.

- *Google Earth™ Tours*—virtual field trips allow students to discover and view geological landscapes around the world. Tours are available as kmz files for use in Google Earth™ or other virtual earth programs.

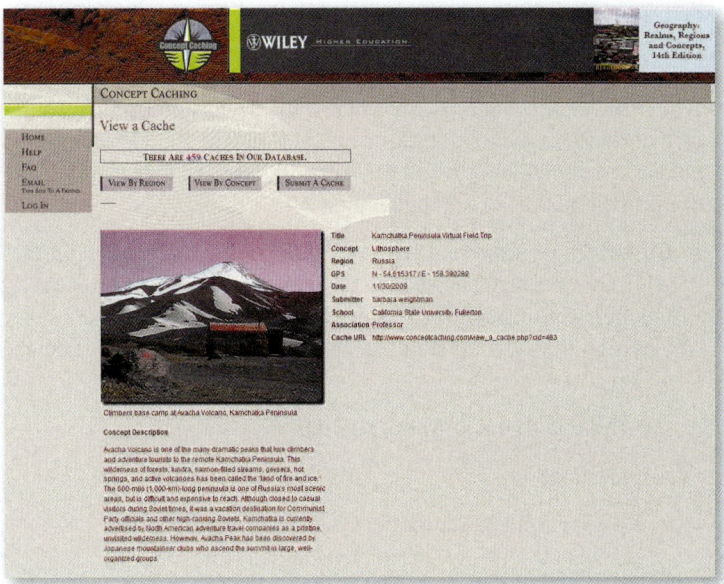

Instructor Online Resources include all student resources, plus the following:

- *PowerPoint Lecture Slides*—chapter-oriented slides including lecture notes and text art.

- *Image Gallery*—images containing both line art and photos from the text.

- **Videos**— Brief video clips provide real-world examples of geographic features and processes, and put these examples into context with the concepts covered in the text.

- **Simulations**—Computer-based models of geographic processes let students manipulate data and variables to explore and interact with virtual environments.

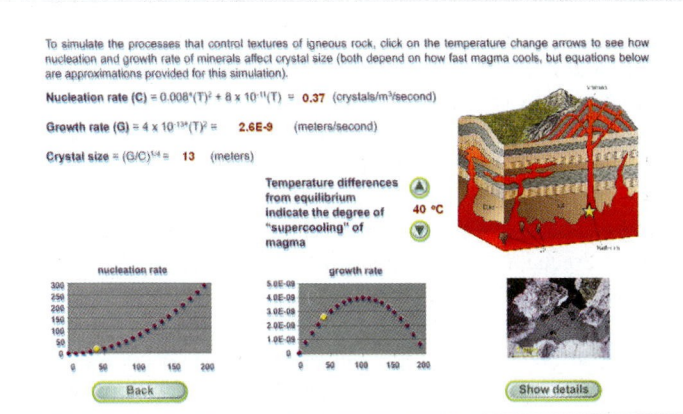

- **Interactive Exercises**—Learning activities and games built from our presentation material give students an opportunity to test

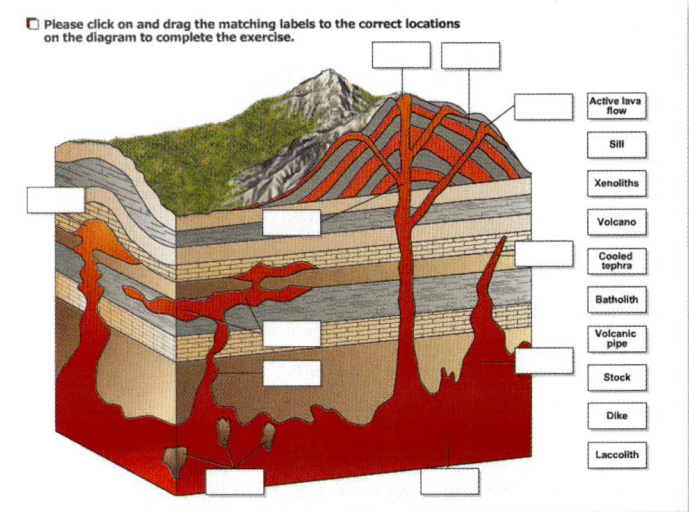

WileyPLUS Learning Space

WileyPLUS Learning Space is an innovative, research-based, online environment for *effective* teaching and learning.

What do students receive with WileyPLUS Learning Space?

A Research-Based Design. *WileyPLUS Learning Space* provides an online environment that integrates relevant resources, including the entire digital textbook, in an easy-to-navigate framework that helps students study more effectively.

- *WileyPLUS Learning Space* adds structure by organizing textbook content into smaller, more manageable "chunks."

- Related media, examples, and sample practice items reinforce the learning objectives.

- Innovative features such as calendars, visual progress tracking, and self-evaluation tools improve time management and strengthen areas of weakness.

One-on-one Engagement. With *WileyPLUS Learning Space* for Fletcher/Physical Geology, students receive 24/7 access to resources that promote positive learning outcomes. Students engage with related examples in various media and practice sample items.

Measurable Outcomes. Throughout each study session, students can assess their progress and gain immediate feedback. *WileyPLUS Learning Space* provides precise reporting of strengths and weaknesses, as well as individualized quizzes, so that students are confident they are spending their time on the right things. With *Wiley PLUS Learning Space*, students always know the exact outcome of their efforts.

What do instructors receive with WileyPLUS Learning Space?

WileyPLUS Learning Space provides reliable, customizable resources that reinforce course goals inside and outside the classroom and help assess individual student progress. Pre-created materials and activities help instructors optimize their time.

- *Customizable Course Plan:* WileyPLUS Learning Space comes with a pre-created Course Plan designed by a subject-matter expert uniquely for this course. Simple drag-and-drop tools make it easy to assign the course plan as is or modify it to reflect your course syllabus.

- *Pre-Created Activity Types Include:*
  - Questions
  - Readings and Resources
  - Presentations
  - Concept Mastery
  - Projects

- *Course Materials and Assessment Content Include:*
  - Art PowerPoint Slides
  - Image Gallery
  - Testbank

- *Gradebook:* *WileyPLUS Learning Space* provides instant access to reports on trends in class performance, student use of course materials, and progress toward learning objectives, helping inform decisions and drive classroom discussions.

WileyPLUS Learning Space. Learn more at www.wileyplus.com

Powered by proven technology and built on a foundation of cognitive research, *WileyPLUS Learning Space* has enriched the education of millions of students in over 20 countries around the world.

Acknowledgments

Thank you to the University of Hawai'i at Manoa, School of Ocean and Earth Science and Technology, Department of Geology and Geophysics for support and encouragement. An effort like this is not possible without the work of many people. I want to thank Gil Wiswall of West Chester University for early inspiration and mentoring. Special thanks to Raghu Reddy, who made it happen. To the amazing team at Wiley, I love working with you guys. In particular, Billy Ray, senior photo editor, for his patience and artistic knowledge to secure the most beautiful and rich photo program; Sandra Rigby, senior production editor, for her keen eye and attention to detail throughout the process; and Ryan Flahive, executive editor, special thanks, you were always there, always smart, and always encouraging.

Thanks also to the many learners who have traveled through my classroom: you have been teachers and I have been the student in ways you may not realize. An incredible art team has made this book a thing of beauty: Nancy, Barbara, Brooks, Byron … you guys are the best! Precision Graphics, you have done great work and should be proud.

Many reviewers and focus groups provided valuable input, and I thank you all for your time and energy. Several colleagues with expertise in various fields answered my questions with patience and friendship; thank you.

Reviewers, Class Testers, and Focus Group Participants

John All, *Western Kentucky University*
Laura Sue Allen-Long, *Indiana-Purdue University*
Robert Altamura, *Florida Community College–Jacksonville*
Laurie Anderson, *Louisiana State University*
Jake Armour, *University of North Carolina–Charlotte*
Abbed Babaei, *Cleveland State University*
Cathy Baker, *Arkansas Tech University*
Leslie Baker, *University of Idaho*
James Barrick, *Texas Tech University*
Jerry Bartholomew, *University of Memphis*
Michael Barton, *Ohio State University*
Jay Bass, *University of Illinois at Urbana-Champaign*
Raymond Beiersdorfer, *Youngstown State University*
Barbara Bekken, *Virginia Tech University*

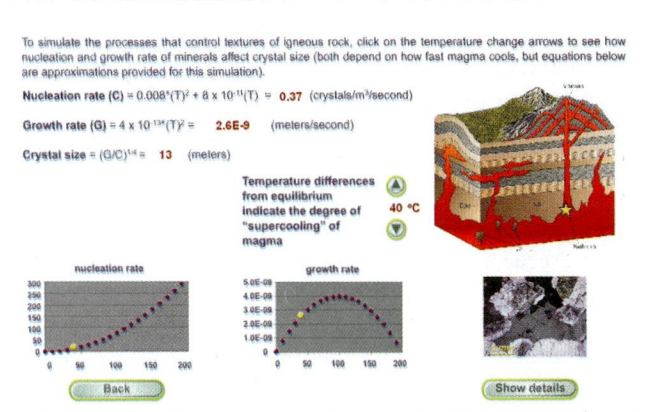

their understanding of key concepts and explore additional visual resources.

- **Concept Caching**—an online database of photographs explores what a physical feature looks like. Photographs and GPS coordinates are "cached" and categorized along core concepts of geography. Professors can access the images or submit their own by visiting www.ConceptCaching.com.

Book Companion Site (www.wiley.com/college/fletcher)

In addition to our multimedia content, our book companion site offers a wealth of study and practice materials, including:

Student Online Resources

- *Self-quizzes*—chapter-based multiple-choice and fill-in-the-blank questions.

- *Web Resources*—useful web sites to explore more about geology.

- *Geo Media Library*—link to the media library for students to explore key concepts in greater depth using videos, animations, and interactive exercises.

- *Google Earth™ Tours*—virtual field trips allow students to discover and view geological landscapes around the world. Tours are available as kmz files for use in Google Earth™ or other virtual earth programs.

- **Videos**—
  Brief video clips
  provide real-world examples
  of geographic features and processes,
  and put these examples into context with the
  concepts covered in the text.

- **Simulations**—Computer-based models of geographic processes let students manipulate data and variables to explore and interact with virtual environments.

- **Interactive Exercises**—Learning activities and games built from our presentation material give students an opportunity to test

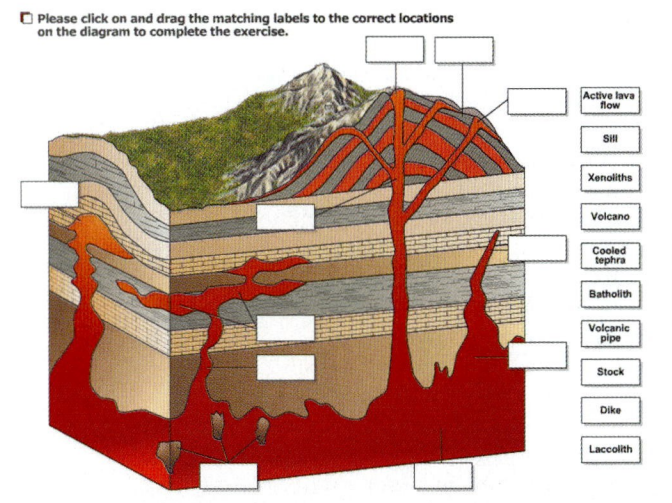

Instructor Online Resources include all student resources, plus the following:

- *PowerPoint Lecture Slides*—chapter-oriented slides including lecture notes and text art.

- *Image Gallery*—images containing both line art and photos from the text.

WileyPLUS Learning Space

WileyPLUS Learning Space is an innovative, research-based, online environment for *effective* teaching and learning.

What do students receive with WileyPLUS Learning Space?

A Research-Based Design. *WileyPLUS Learning Space* provides an online environment that integrates relevant resources, including the entire digital textbook, in an easy-to-navigate framework that helps students study more effectively.

- *WileyPLUS Learning Space* adds structure by organizing textbook content into smaller, more manageable "chunks."

- Related media, examples, and sample practice items reinforce the learning objectives.

- Innovative features such as calendars, visual progress tracking, and self-evaluation tools improve time management and strengthen areas of weakness.

One-on-one Engagement. With *WileyPLUS Learning Space* for Fletcher/Physical Geology, students receive 24/7 access to resources that promote positive learning outcomes. Students engage with related examples in various media and practice sample items.

Measurable Outcomes. Throughout each study session, students can assess their progress and gain immediate feedback. *WileyPLUS Learning Space* provides precise reporting of strengths and weaknesses, as well as individualized quizzes, so that students are confident they are spending their time on the right things. With *Wiley PLUS Learning Space*, students always know the exact outcome of their efforts.

What do instructors receive with WileyPLUS Learning Space?

WileyPLUS Learning Space provides reliable, customizable resources that reinforce course goals inside and outside the classroom and help assess individual student progress. Pre-created materials and activities help instructors optimize their time.

- *Customizable Course Plan:* WileyPLUS Learning Space comes with a pre-created Course Plan designed by a subject-matter expert uniquely for this course. Simple drag-and-drop tools make it easy to assign the course plan as is or modify it to reflect your course syllabus.

- *Pre-Created Activity Types Include:*
  - Questions
  - Readings and Resources
  - Presentations
  - Concept Mastery
  - Projects

- *Course Materials and Assessment Content Include:*
  - Art PowerPoint Slides
  - Image Gallery
  - Testbank

- *Gradebook:* *WileyPLUS Learning Space* provides instant access to reports on trends in class performance, student use of course materials, and progress toward learning objectives, helping inform decisions and drive classroom discussions.

WileyPLUS Learning Space. Learn more at www.wileyplus. com

Powered by proven technology and built on a foundation of cognitive research, *WileyPLUS Learning Space* has enriched the education of millions of students in over 20 countries around the world.

Acknowledgments

Thank you to the University of Hawai'i at Manoa, School of Ocean and Earth Science and Technology, Department of Geology and Geophysics for support and encouragement. An effort like this is not possible without the work of many people. I want to thank Gil Wiswall of West Chester University for early inspiration and mentoring. Special thanks to Raghu Reddy, who made it happen. To the amazing team at Wiley, I love working with you guys. In particular, Billy Ray, senior photo editor, for his patience and artistic knowledge to secure the most beautiful and rich photo program; Sandra Rigby, senior production editor, for her keen eye and attention to detail throughout the process; and Ryan Flahive, executive editor, special thanks, you were always there, always smart, and always encouraging.

Thanks also to the many learners who have traveled through my classroom: you have been teachers and I have been the student in ways you may not realize. An incredible art team has made this book a thing of beauty: Nancy, Barbara, Brooks, Byron … you guys are the best! Precision Graphics, you have done great work and should be proud.

Many reviewers and focus groups provided valuable input, and I thank you all for your time and energy. Several colleagues with expertise in various fields answered my questions with patience and friendship; thank you.

Reviewers, Class Testers, and Focus Group Participants

John All, *Western Kentucky University*
Laura Sue Allen-Long, *Indiana-Purdue University*
Robert Altamura, *Florida Community College–Jacksonville*
Laurie Anderson, *Louisiana State University*
Jake Armour, *University of North Carolina–Charlotte*
Abbed Babaei, *Cleveland State University*
Cathy Baker, *Arkansas Tech University*
Leslie Baker, *University of Idaho*
James Barrick, *Texas Tech University*
Jerry Bartholomew, *University of Memphis*
Michael Barton, *Ohio State University*
Jay Bass, *University of Illinois at Urbana-Champaign*
Raymond Beiersdorfer, *Youngstown State University*
Barbara Bekken, *Virginia Tech University*

Gene Bender, *Minot State University*
David Berner, *Normandale Community College*
Pier Binda, *University of Regina*
Gregory Bishop, *Orange Coast College*
Ross Black, *University of Kansas*
Claude Bolze, *Tulsa Community College*
Theodore Bornhorst, *Michigan Tech University*
Robert Brenner, *University of Iowa*
John Breyer, *California State University, Los Angeles*
Douglas Britton, *Long Beach City College*
Art Busbey, *Texas Christian University*

Phyllis Camilleri, *Austin Peay State University*
Patricia Campbell, *Slippery Rock University*
Michael Canestaro, *Sinclair Community College*
Richard Carlson, *Texas A&M University*
Roseann Carlson, *Tidewater Community College*
Victor Cavaroc, *North Carolina State University*
Christina Chan, *University of Houston*
Marshall Chapman, *Coastline Community College*
Chu-Yung Chen, *University of Idaho*
Yu-Ping Chin, *Ohio State University*
Lindgren Chyi, *University of Akron*
Russell Clark, *Albion College*
Mark Colberg, *Southern Utah University*
Jennifer Cole, *Northeastern University*
Melissa Connely, *Casper College*
Susan Conrad, *Dutchess Community College*
Jennifer Coombs, *Northeastern University*
Dee Cooper, *University of Texas*
Constantin Cranganu, *CUNY Brooklyn College*
Juliet Crider, *Western Washington University*

Michael Dalman, *Blinn College–Brenham*
John Dassinger, *Chandler Gilbert Community College*
René De Hon, *University of Louisiana–Monroe*
Jeffrey Dick, *Youngstown State University*
Bruce Dod, *Mercer University*
Jennifer Duncan, *Grossmont College*
James Duvall, *Eastern Washington University*
David Elliott, *Northern Arizona University*
W. Crawford Elliott, *Georgia State University*
Robert Eves, *Southern Utah University*

Mike Farabee, *Georgia College & StateUniversity*
Mark Feigenson, *Rutgers University*
Colin Ferguson, *Butte College*
David Foster, *University of Florida*
Bill Frazier, *Columbus State University*
Todd Fritch, *Northeastern University*
Tracy Furutani, *North Seattle Community College*

William Garcia, *University of NorthCarolina–Charlotte*
Shemin Ge, *University of Colorado*
Richard Graus, *University of Rochester*
Jeff Greenberg, *Wheaton College*
Jay Gregg, *Missouri University of Science and Technology*
James Grenda, *Angelo State University*
Harold Gurrola, *Texas Tech University*
Erich Guy, *Ohio University*

Daniel Habib, *Purdue University*
Duane Hampton, *Western Michigan University*

Michael Harrison, *Tennessee Technological University*
Nicole Harvey, *Nashville State Tech Community College*
Kevin Hefferan, *University of Wisconsin–Stevens Point*
Don Hellstern, *Dallas Community College*
Barry Hibbs, *California State University, Los Angeles*
John Hickey, *Inver Hills Community College*
Eric Holdener, *Kenyon College*
Paul Howell, *University of Kentucky*
William Hoyt, *University of Northern Colorado*
Paul Hudak, *University of North Texas–Denton*
Melinda Hutson, *Portland Community College*

Solomon Isiorho, *Indiana University-Purdue University Fort Wayne*

Eric Jerde, *Morehead State University*
Megan Jones, *North Hennepin Community College*
Richard Josephs, *University of North Carolina–Wilmington*

Michael Katuna, *College of Charleston*
David King, *Auburn University*
John Kocurko, *Midwestern State University*

David Lageson, *Montana Stage University*
Cynthia Lawry, *Blinn College*
Peter Leavens, *University of Delaware*
Patty Lee, *University of Hawai'i at Manoa*

Denise Lemaire, *Rowan University*
Steven Lower, *Ohio State University*
James Lowry, *Stephen F. Austin State University*

Jaime Marso, *Grossmont College*
James Martin-Hayden, *University of Toledo*
Stephen Mattox, *Grand Valley State University*
Dan McNally, *Bryant University*
Brendan McNulty, *Cosumnes River College*
Joseph Meert, *University of Florida*
Katherine Milla, *Florida A&M University*
Tim Millen, *Elgin Community College*
Basil Miller, *Henderson State University*
David Miller, *Clark State Community College*
James Miller, *Southwest Missouri State University*
Ken Miller, *Rutgers University*
Keith Montgomery, *University of Wisconsin–Marathon County*
Daniel Moore, *Brigham Young University–Idaho*
Greg Moore, *University of Hawaii*
Gregory Moore, *University of Georgia*
Sadredin Moosavi, *Tulane University*
David Morris, *Eastern Arizona College*
Stephen Moshier, *Wheaton College*

Clive Neal, *University of Notre Dame*
Hanna Nekvasil, *Stony Brook University*

Poorna Pal, *Glendale Community College*
Chris Parkinson, *University of New Orleans*
Terri Plake, *Northwest Indian College*
Harry Pylypiw, *Quinnipiac University*

Usha Rao, *St. Joseph's University*
Kenneth Rasmussen, *Northern Virginia Community College–Annandale*
Max Reams, *Olivet Nazarene University*
Robert Regis, *Northern Michigan University*
Brady Rhodes, *University of California, Santa Barbara*
Eliza Richardson, *Penn State University*

BRIEF CONTENTS

CONTENTS

HOW TO GET THE MOST FROM YOUR TEXT

Human impacts on Earth are now global in extent. Managing these challenges requires our best thinking and knowledge of how Earth works. For many of you, this course in physical geology may be your only opportunity to learn about the planet you call home. We have designed the book to help you develop a lifelong understanding about Earth and its processes. This will enhance your role as a decision maker and a steward of Earth. Let's walk through the book and examine the special features that will help make your learning more enjoyable and successful.

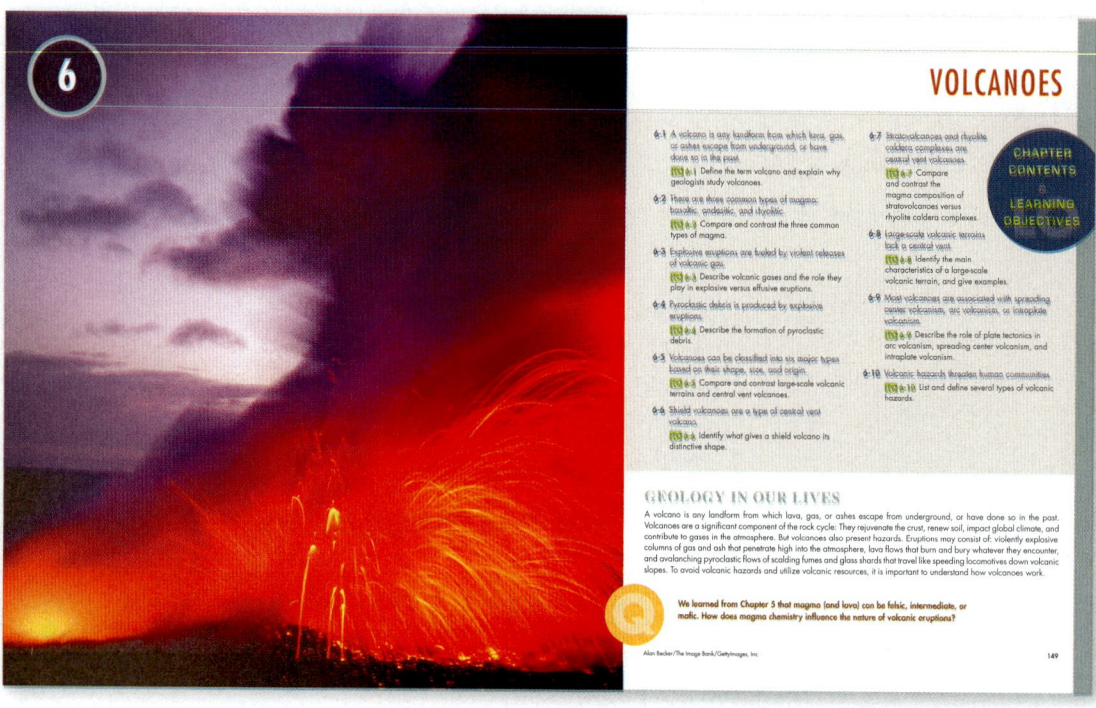

Chapter Opener

Every chapter opens with a critical thinking question to keep in mind as you read the chapter and a brief section called *Geology in Our Lives*. These introduce you to the overall theme and relevance of the chapter.

Chapter Contents and Learning Objectives

Each chapter is divided into sections of two to four pages with specific Learning Objectives (LO) to help you understand what you should know or be able to accomplish after studying that section. Each LO is designed to guide and reinforce what you learn by reading the text. This presentation allows a full discussion of a topic in just the right amount of detail and allows you, the student, to study in succinct, digestible segments. Short, easy to read segments mean your reading assignments are approachable and fit in with your busy life.

Visual Learning

This text contains hundreds of extraordinary photos, illustrations, graphs, and tables, all of which were selected to enhance your learning of physical geology. But instead of just being "pretty pictures," we have added review and critical thinking questions to many of the visuals. These questions are marked with a and ask you to really look at and think about the image being presented. By answering these questions, you will be more involved in the subject matter and have a better understanding of what you are seeing. Your instructor may use these questions as lecture starters and class discussion topics, or may assign them as quiz items.

FIGURE 6.25 A stack of several Columbia River basalt lava flows in the Columbia River Gorge, Washington State. Each flow is 15 to 100 meters thick.

Why is it unlikely for lava plateaus to emerge above a rhyolite magma source?

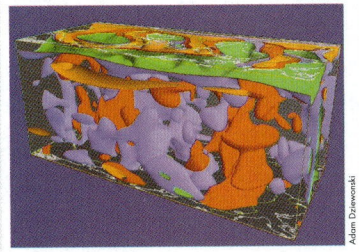

FIGURE 3.5 Earth's interior as viewed from the north. This complicated model of the mantle is based on records of the behavior of seismic waves passing through warm (red) and cool (blue) rock, indicating the structure of the planet is more complex than a simple sequence of uniform layers.

How are the red areas and blue areas going to change over time?

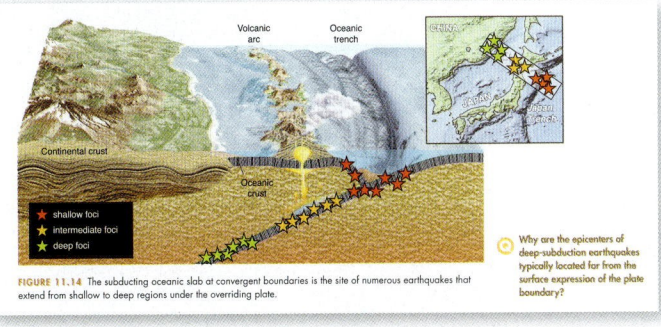

FIGURE 11.14 The subducting oceanic slab at convergent boundaries is the site of numerous earthquakes that extend from shallow to deep regions under the overriding plate.

- shallow foci
- intermediate foci
- deep foci

Why are the epicenters of deep-subduction earthquakes typically located far from the surface expression of the plate boundary?

Geology in Action, Geology in Our Lives, Earth Citizenship, and Critical Thinking

The text supports four major themes to help you understand the importance and relevance of physical geology. Embedded within chapters are icons that bring these physical geology themes to life and provide examples of what you are learning.

Geology in Action presents topics that show how Earth is in constant change and demonstrate the relationships among the many forces of nature.

Geology in Our Lives offers examples of how geology is relevant to everyone in day-to-day life and encourages you to appreciate the impact we all have on Earth.

Earth Citizenship encourages important understanding of the role we play in the community and the world. We all have the power to influence and manage natural resources locally and globally to meet the needs of present and future generations. An important goal of this course is to encourage you to become a geologically informed global citizen.

GEOLOGY IN ACTION

The Sumatra–Andaman Earthquake: The Quake That Moved the Entire Planet

The Sumatra–Andaman earthquake of 2004 set many records: It was the most powerful quake in more than 40 years; it triggered a devastating tsunami that killed more people than any other natural catastrophe in modern history; it lasted an unbelievable 500 to 600 seconds (most quakes last only a few seconds); and it shook the ground everywhere on Earth's surface. Weeks later, the planet was still trembling. The quake was produced by the longest fault rupture ever observed: 1,255 kilometers—the distance from Los Angeles, California, to Portland, Oregon. The fault shifted continuously for 10 minutes, also a record.

The earthquake and the resulting tsunami (FIGURE 11.10), which swept across the Indian Ocean and was recorded on instruments around the world, killed approximately 230,000 people in 11 countries and left over 1 million people homeless. Ground movement of as much as 1 centimeter occurred everywhere on Earth's surface, and in some areas it was felt...

FIGURE 11.10 The tsunami (shown in Banda Aceh, Indonesia) triggered by the 2004 Sumatra–Andaman earthquake devastated the shores of Indonesia, Sri Lanka, India, Thailand.

GEOLOGY IN OUR LIVES

Bringing European Air Travel to a Standstill

On April 15, 2010, British civil aviation authorities ordered the country's airspace closed when it became too dangerous to allow air traffic through the area due to a cloud of ash drifting from the erupting Eyjafjallajökull (Icelandic for "island-mountain glacier") volcano in Iceland (FIGURE 6.11).

Flying through volcanic ash can abrade the cockpit windshield and make it impossible for the pilots to see outside; it can also damage communication and navigation instruments on the outside of the aircraft; worst of all, it can coat the inside of jet engines with concretelike solidified ash, which can force the engine to shut down.

Within 48 hours the ash cloud had spread across northern Europe. Three hundred and thirteen airports in England, France, Germany, and other European nations closed for about one week, with more than 100,000 flights cancelled. Airplane schedules around the world were affected, and some 10 million air travelers were stranded, at a cost estimated at $3.3 billion.

As a consequence of this event, brought on by what volcanologists describe as a rather modest eruption, European authorities are considering bringing all European air traffic under the control of a single agency, a move designed to increase flexibility in response to future...

FIGURE 6.11 The April 14, 2010, eruption of Eyjafjallajök brought air travel throughout Europe to a standstill for nearly cancellation of over 100,000 flights stranded some 10 milli around the world.

EARTH CITIZENSHIP

Fracking: An Energy Revolution with an Environmental Cost

According to the International Energy Association, the United States will become the largest oil producer in the world by 2017, surpassing Saudi Arabia, Russia, and other traditional oil giants in the Middle East and South America. Oil imports to America are at their lowest levels in two decades, and the country may become a net oil exporter by 2020 and energy self-sufficient by 2035.

As the United States grows less and less dependent on oil from the Middle East, the energy revolution in the country is being spurred by new energy policies initiated early this century and continuing under the present administration. According to the White House, the United States holds a 100-year supply of natural gas, and the current administration has pushed to increase production of both gas and oil.

Natural gas (known as *methane*) production jumped 15% between 2008 and 2012. Oil and petroleum imports have fallen, on average, more than 1.5 million barrels per day; and since 2008, domestic crude oil production has increased by an average of more than 720,000 barrels per day. All the drilling required to speed this production has important and positive consequences: It has provided jobs for a weak economy—job growth in these industries has risen 25% since January 2010—and the reduced dependence of the United States on external sources of energy has improved the security of the nation.

The increase in U.S. oil and natural gas production is the result of a controversial technology known as *hydraulic fracturing*, or "fracking." Fracking is a drilling process whereby millions of gallons of chemically enhanced water and sand are injected at high pressure into shale formations that hold large volumes of fossil fuels (FIGURE 1.10). The chemical mixture fractures the shale, allowing for the release of previously unobtainable oil and methane. Fracking typically takes place thousands of feet underground, where it is not harmful; but closer to the surface, escaping methane from poorly constructed wells can pollute freshwater aquifers, seep through the ground, and cause extensive damage to ecosystems and drinking water supplies.

Fracking presents several challenges: 1) The process requires 2 to 4 million gallons of chemically enhanced water, which can put a strain on local water supplies. 2) Oil and natural gas can escape from poorly constructed wellheads (the tops of wells) and pollute local aquatic ecosystems and aquifers. 3) The process produces millions of gallons of wastewater, which must be managed by a combination of recycling and disposal. 4) The drilling and fracking process requires heavy machinery that produces both noxious exhaust and heavy traffic.

As is the case with the extraction of many natural resources, the fracking industry can minimize these environmental impacts by instituting conscientious and powerful safeguards and best management practices. To date, however, the fracking industry is exempted from the federal Clean Air and Clean Water Acts, which means there is a lack of federal oversight. Consequently, there is growing concern regarding the potential for environmental and community damage related to fracking. To address this concern, a broader conversation is in order, to address the development process and the steps being taken to ensure safe operations.

FIGURE 1.10 A typical hydraulic fracturing operation needs to institute powerful safeguards to minimize impacts on the environment and nearby community.

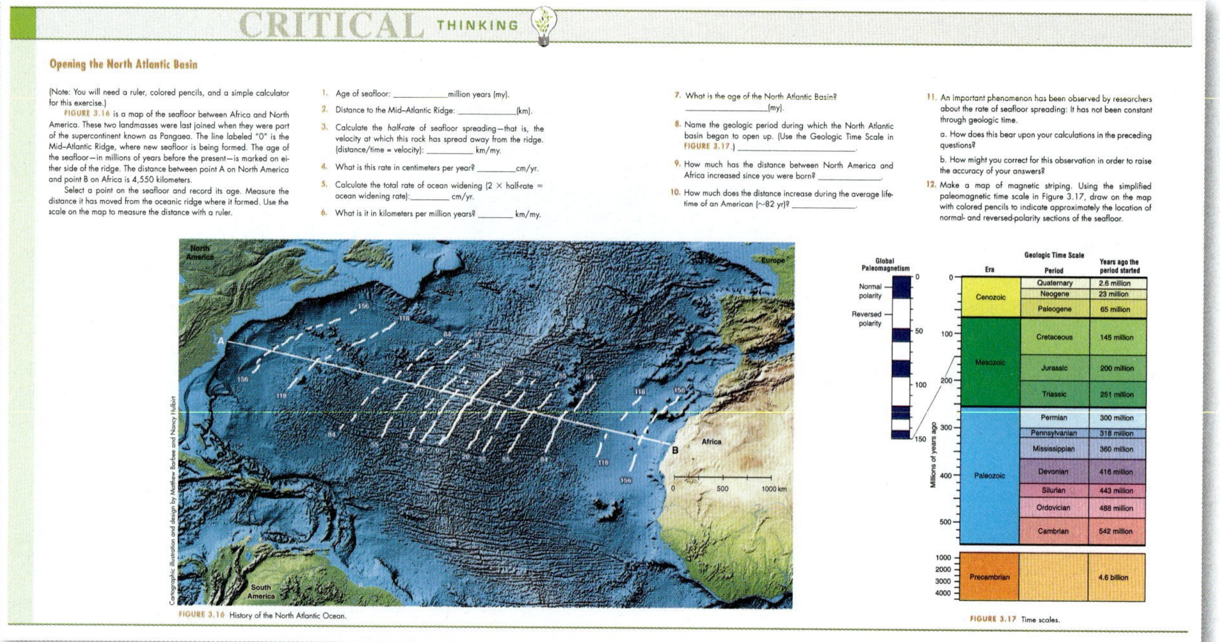

CRITICAL THINKING

Opening the North Atlantic Basin

(Note: You will need a ruler, colored pencils, and a simple calculator for this exercise.)

FIGURE 3.16 is a map of the seafloor between Africa and North America. These two landmasses were last joined when they were part of the supercontinent known as Pangaea. The line labeled "0" is the Mid-Atlantic Ridge, where new seafloor is being formed. The age of the seafloor—in millions of years before the present—is marked on either side of the ridge. The distance between point A on North America and point B on Africa is 4,550 kilometers.

Select a point on the seafloor and record its age. Measure the distance it has moved from the oceanic ridge where it formed. Use the scale on the map to measure the distance with a ruler.

1. Age of seafloor: _____ million years (my).

2. Distance to the Mid-Atlantic Ridge: _____ (km).

3. Calculate the half-rate of seafloor spreading—that is, the velocity at which this rock has spread away from the ridge. (distance/time = velocity): _____ km/my.

4. What is this rate in centimeters per year? _____ cm/yr.

5. Calculate the total rate of ocean widening (2 × half-rate = ocean widening rate): _____ cm/yr.

6. What is it in kilometers per million years? _____ km/my.

7. What is the age of the North Atlantic Basin? _____ (my).

8. Name the geologic period during which the North Atlantic basin began to open up. (Use the Geologic Time Scale in FIGURE 3.17.)

9. How much has the distance between North America and Africa increased since you were born? _____

10. How much does the distance increase during the average lifetime of an American (~82 yr)? _____

11. An important phenomenon has been observed by researchers about the rate of seafloor spreading: It has not been constant through geologic time.
   a. How does this bear upon your calculations in the preceding questions?
   b. How might you correct for this observation in order to raise the accuracy of your answers?

12. Make a map of magnetic striping. Using the simplified paleomagnetic time scale in Figure 3.17, draw on the map with colored pencils to indicate approximately the location of normal- and reversed-polarity sections of the seafloor.

FIGURE 3.16 History of the North Atlantic Ocean.

FIGURE 3.17 Time scales.

Critical Thinking features give you skills you need to fully understand the science and develop problemsolving abilities that can be applied in many aspects of your studies and future careers. In addition to the embedded boxes, critical thinking is further developed with expanded two-page showcase activities that allow you to further improve your knowledge and skills.

Expand Your Thinking

The theme of critical thinking is continued through *Expand Your Thinking* questions posed at the end of every two-page section. These questions provide a check for you to build self-confidence by making sure you have mastered the ideas in the discussion and enable you to further your understanding.

EXPAND YOUR THINKING

What human activities have added to the amount of greenhouse gases in the atmosphere?

Study Materials

Every chapter ends with a variety of study materials for you to review and help you prepare for quizzes and exams.

Let's Review "Geology in Our Lives" and Study Guide

The *Let's Review* section revisits the *Geology in Our Lives* from the chapter opening page and reviews it in terms of the knowledge you have gained after reading the chapter. The *Study Guide* provides an outline of the major sections in the chapter and summarizes the key points, while providing additional review of the learning objectives for each chapter.

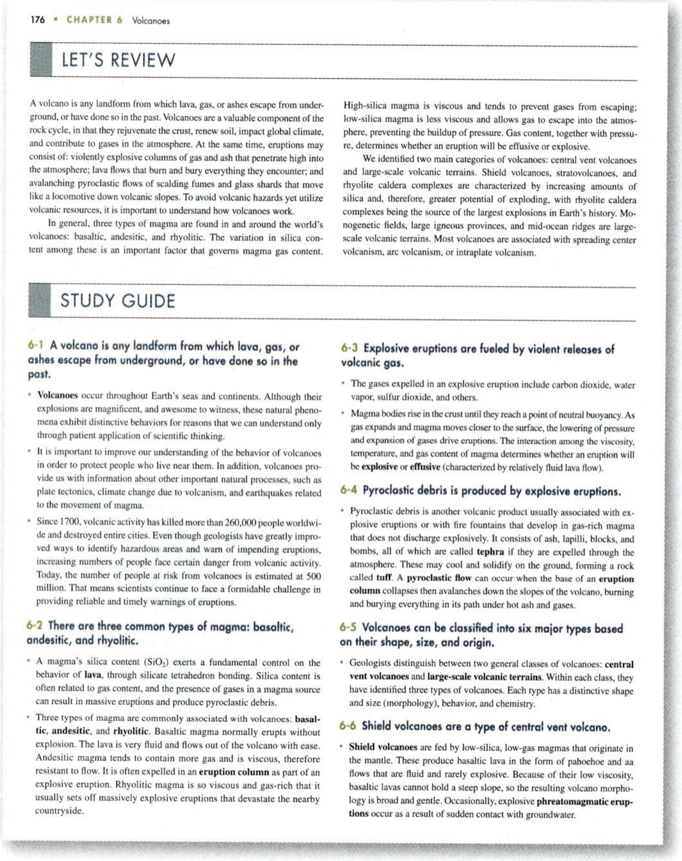

176 • CHAPTER 6 Volcanoes

LET'S REVIEW

A volcano is any landform from which lava, gas, or ashes escape from underground, or have done so in the past. Volcanoes are a valuable component of the rock cycle, in that they rejuvenate the crust, renew soil, impact global climate, and contribute to gases in the atmosphere. At the same time, eruptions may consist of: violently explosive columns of gas and ash that penetrate high into the atmosphere; lava flows that burn and bury everything they encounter; and avalanching pyroclastic flows of scalding fumes and glass shards that move like a locomotive down volcanic slopes. To avoid volcanic hazards yet utilize volcanic resources, it is important to understand how volcanoes work.

In general, three types of magma are found in and around the world's volcanoes: basaltic, andesitic, and rhyolitic. The variation in silica content among these is an important factor that governs magma gas content.

High-silica magma is viscous and tends to prevent gases from escaping; low-silica magma is less viscous and allows gas to escape into the atmosphere, preventing the buildup of pressure. Gas content, together with pressure, determines whether an eruption will be effusive or explosive.

We identified two main categories of volcanoes: central vent volcanoes and large-scale volcanic terrains. Shield volcanoes, stratovolcanoes, and rhyolite caldera complexes are characterized by increasing amounts of silica and, therefore, greater potential of exploding, with rhyolite caldera complexes being the source of the largest explosions in Earth's history. Monogenetic fields, large igneous provinces, and mid-ocean ridges are large-scale volcanic terrains. Most volcanoes are associated with spreading center volcanism, arc volcanism, or intraplate volcanism.

STUDY GUIDE

6-1 A volcano is any landform from which lava, gas, or ashes escape from underground, or have done so in the past.

- **Volcanoes** occur throughout Earth's seas and continents. Although their explosions are magnificent, and awesome to witness, these natural phenomena exhibit distinctive behaviors for reasons that we can understand only through patient application of scientific thinking.

- It is important to improve our understanding of the behavior of volcanoes in order to protect people who live near them. In addition, volcanoes provide us with information about other important natural processes, such as plate tectonics, climate change due to volcanism, and earthquakes related to the movement of magma.

- Since 1700, volcanic activity has killed more than 260,000 people worldwide and destroyed entire cities. Even though geologists have greatly improved ways to identify hazardous areas and warn of impending eruptions, increasing numbers of people face certain danger from volcanic activity. Today, the number of people at risk from volcanoes is estimated at 500 million. That means scientists continue to face a formidable challenge in providing reliable and timely warnings of eruptions.

6-2 There are three common types of magma: basaltic, andesitic, and rhyolitic.

- A magma's silica content (SiO_2) exerts a fundamental control on the behavior of lava, through silicate tetrahedron bonding. Silica content is often related to gas content, and the presence of gases in a magma source can result in massive eruptions and produce pyroclastic debris.

- Three types of magma are commonly associated with volcanoes: **basaltic, andesitic,** and **rhyolitic**. Basaltic magma normally erupts without explosion. The lava is very fluid and flows out of the volcano with ease. Andesitic magma tends to contain more gas and is viscous, resistant to flow. It is often expelled in an **eruption column** as part of an explosive eruption. Rhyolitic magma is so viscous and gas-rich that it usually sets off massively explosive eruptions that devastate the nearby countryside.

6-3 Explosive eruptions are fueled by violent releases of volcanic gas.

- The gases expelled in an explosive eruption include carbon dioxide, water vapor, sulfur dioxide, and others.

- Magma bodies rise in the crust until they reach a point of neutral buoyancy. As gas expands and magma moves closer to the surface, the lowering of pressure and expansion of gases drive eruptions. The interaction among the viscosity, temperature, and gas content of magma determines whether an eruption will be **explosive** or **effusive** (characterized by relatively fluid lava flow).

6-4 Pyroclastic debris is produced by explosive eruptions.

- Pyroclastic debris is another volcanic product usually associated with explosive eruptions or with fire fountains that develop in gas-rich magma that does not discharge explosively. It consists of ash, lapilli, blocks, and bombs, all of which are called **tephra** if they are expelled through the atmosphere. These may cool and accumulate on the ground, forming a rock called **tuff**. A **pyroclastic flow** can occur when the base of an **eruption column** collapses then avalanches down the slopes of the volcano, burning and burying everything in its path under hot ash and gases.

6-5 Volcanoes can be classified into six major types based on their shape, size, and origin.

- Geologists distinguish between two general classes of volcanoes: **central vent volcanoes** and **large-scale volcanic terrains**. Within each class, they have identified three types of volcanoes. Each type has a distinctive shape and size (morphology), behavior, and chemistry.

6-6 Shield volcanoes are a type of central vent volcano.

- **Shield volcanoes** are fed by low-silica, low-gas magmas that originate in the mantle. These produce basaltic lava in the form of pahoehoe and aa flows that are fluid and rarely explosive. Because of their low viscosity, basaltic lavas cannot hold a steep slope, so the resulting volcano morphology is broad and gentle. Occasionally, explosive **phreatomagmatic eruptions** occur as a result of sudden contact with groundwater.

KEY TERMS

andesitic magma (p. 155)
basaltic magma (p. 155)
calderas (p. 167)
central vent volcanoes (p. 152)
cinder cones (p. 164)
craters (p. 167)
effusive eruption (p. 153)
eruption column (p. 159)
explosive eruption (p. 153)
flood basalt (p. 170)

large igneous provinces (LIPs) (p. 170)
large-scale volcanic terrains (p. 152)
lava (p. 155)
monogenetic volcanic fields (p. 170)
phreatomagmatic eruption (p. 164)
Plinian eruption (p. 159)
pyroclastic debris (p. 155)
pyroclastic flows (p. 160)
rhyolite caldera complexes (p. 167)
rhyolitic magma (p. 155)

shield volcanoes (p. 163)
stratovolcanoes (p. 156)
tephra (p. 158)
tuff (p. 160)
vent (p. 152)
volcano (p. 152)
volcanologists (p. 152)
welded tuff (p. 160)

Key Terms

A list of the *Key Terms* (boldfaced terms in the text), along with page references, provide a quick resource to highlight the most important terms in the chapter. These terms are again defined in the *Glossary*. This is a helpful tool to use as you prepare for quizzes and exams.

Assessing Your Knowledge, Further Learning, and Online Resources

Multiple-choice questions at the end of each chapter review and assess your knowledge of the main concepts and offer you another way to check your mastery of the chapter. By answering these questions, you will know whether you have accomplished the learning objectives for the chapter. These questions may serve as a self-check or be assigned by your instructor as homework.

Further Learning assignments offer a higher level of questions or assignments to encourage additional critical thinking and ask you to bring together multiple concepts.

Helpful web sites are listed with ideas for investigating additional information of interest and relevance if you would like further information on a topic. These resources are useful for preparing class papers or projects.

In addition, the Book Companion Site (www.wiley.com/college/fletcher) and *WileyPLUS Learning Space* offer you a variety of ways to enhance your study and help you succeed in the course, including Self Quizzes, automatically graded quizzes for immediate feedback; Lecture Notes, art from the text and select lecture slides to use in taking notes in class; and Google Earth Tours, where you can take a trip to see first-hand what real-world geological processes look like.

---

212 • CHAPTER 7 Weathering

ASSESSING YOUR KNOWLEDGE

Please complete this exercise before coming to class. Identify the one best answer to each question.

1. Weathering consists of:
   a. Erosion, tectonics, and uplift.
   b. Chemical, biological, and physical degradation.
   c. Crust age, chemistry, and sedimentary minerals.
   d. Sedimentary quartz, hematite, and alumina.
   e. None of the above.

2. The pile of rocks at the base of this cliff is called _____ (talus, humus, vertisol); the rocks weathered off the cliff face by _____ (salt weathering, hydraulic action, abrasion, ice wedging).

©Marli Bryant Miller

3. Spheroidal weathering is caused by:
   a. Sand abrasion in running water.
   b. Crystal growth in cold climates.
   c. Chemical weathering of angular rocks.
   d. A combination of slaking and mass wasting.
   e. None of the above.

4. The chemical interaction of oxygen with other substances is known as:
   a. Dissolution.
   b. Hydrolysis.
   c. Saturation.
   d. Oxidation.
   e. None of the above.

5. The most important form of chemical weathering of silicate minerals is:
   a. Crystal growth.
   b. Slaking.
   c. Hydrolysis.
   d. Dissolution.
   e. Frost wedging.

6. _____ (Oxidation, Hydrolysis, Dissolution) causes feldspar to break down and produce clay.

Courtesy of Chip Fletcher

7. Insoluble residues are:
   a. Minerals produced by weathering.
   b. Dissolved compounds resulting from chemical weathering.
   c. Soils rich in organics.
   d. All of the above.
   e. Typically dissolved in hydraulic acid.

8. The tendency of silicates to weather on Earth's surface is predicted by:
   a. Mineral texture.
   b. Rock color and environment of deposition.
   c. Crystallization temperature.
   d. Tectonic setting.
   e. Their roundness.

9. In _____, rock breaks down into solid fragments by processes that do not change the rock's chemical composition.
   a. Chemical weathering
   b. Physical weathering
   c. Insolation weathering
   d. Space weathering

Gregory G. Dimijian/Photo Researchers

10. Which of the following statements about carbon dioxide is true?
    a. It is an important gas that regulates Earth's climate and influences groundwater chemistry.
    b. The amount present in the atmosphere is affected by the rate of crustal weathering.
    c. Its decrease has caused net global cooling over recent geologic history.
    d. It is a greenhouse gas.
    e. All of the above.

11. The variables that most affect the weathering process are rock composition and _____.
    a. Topography
    b. Surface area
    c. Living things
    d. Climate
    e. None of the above.

12. Which of the following statements about soil erosion is true?
    a. It is a form of pollution that affects biological communities.
    b. It is a major problem affecting million of acres of croplands.
    c. It threatens to impact food production.
    d. It takes centuries to make soil and only minutes to erode it.
    e. All of the above.

FURTHER RESEARCH

Go to the Web and research one of these volcanoes. Then answer the questions below as part of a two-page report that you turn in to your instructor.

Vesuvius, Italy
Kilauea, Hawaii
Etna, Sicily
Mount St. Helens, Washington
Krakatau, Indonesia
Tambora, Indonesia

Novarupta and Amak, Alaska
Unzen, Japan
Rotorua, New Zealand
Yellowstone, Wyoming
Juan de Fuca Ridge
Eastern Pacific

1. What is the name of the volcano or eruption that you selected?
2. Describe the volcanic processes occurring in your volcano, and classify it using the classification system presented in this chapter.

3. Is the volcano currently active? If not, when did it last erupt? What is its eruption history?
4. What is its eruption style: explosive, effusive, or some other type? Please elaborate.
5. What type of magma is present? How does its chemistry influence the volcanic processes occurring in this volcano?
6. What is the tectonic setting?
7. Is the volcano near a human settlement? How is this significant?
8. Describe unique and important features of the activity or eruption.
9. Use the Web to discover additional information about your volcano.
10. Turn in your typed (not handwritten) report, complete with a title page, illustrations, reference sources, and other research information. Your instructor will detail the proper format for you.

ONLINE RESOURCES

Explore more about volcanoes on the following Web sites:

NASA Earth Observatory contains up-to-date information on the latest eruptions:
http://earthobservatory.nasa.gov/NaturalHazards

USGS Cascades Volcano Observatory is full of activity updates, event reports, and an excellent photo glossary:
http://vulcan.wr.usgs.gov

USGS Alaska Volcano Observatory is a storehouse of information about volcanoes:
www.avo.alaska.edu

USGS Hawaii Volcano Observatory contains excellent information about the Hawaiian hotspot, shield volcanoes, and the ongoing eruption of Kilauea Volcano:
http://hvo.wr.usgs.gov

Additional animations, videos, and other online resources are available at this book's companion Web site:
www.wiley.com/college/fletcher

This companion Web site also has more information about WileyPLUS and other Wiley teaching and learning resources.

AN INTRODUCTION TO GEOLOGY

CHAPTER CONTENTS & LEARNING OBJECTIVES

GEOLOGY IN OUR LIVES

The human population has dramatically increased over the past century—with consequences. Our expanding activities are damaging the natural environment. We are depleting natural resources that we depend on, and we are increasingly exposed to natural hazards. Global climate is changing; in many areas water resources are in short supply; energy costs have risen; and some mineral resources are dwindling. Communities are expanding now into areas formerly considered too dangerous to live in, and the annual cost of natural disasters is at an all-time high. These are global challenges that must be met and managed by critical thinkers with an understanding of Earth. Critical thinking is the use of reasoning to explain the world around us.

What are the grand global challenges that the next generation of Earth's citizens will face?

1-1 Geology Is the Scientific Study of Earth and the Other Planets

LO 1-1 Describe why the science of geology is important in our daily lives.

Geology literally means "study of Earth," but geologists also study other planets in the solar system and beyond.

Geology Is an Integrative Science

Physical geology is the study of the materials that compose Earth; the chemical, biological, and physical processes that create them; and the ways in which they are organized and distributed throughout the planet. Earth is our resource-limited, sometimes hazardous, always-changing home in space. By understanding our restless planet, we improve our ability to conserve geologic resources, avoid geologic hazards, and manage critical environments. This makes us better Earth citizens.

Physical geology is one of two broad branches of geology. The other is **historical geology,** the study of Earth's history. We introduce you to Earth's history so you will understand when and how Earth (and the solar system) came into existence. We also explain how geologists interpret Earth's past and how they collect evidence revealing the evolution of life.

Geology is an integrative science. To figure out how Earth works, geologists must *integrate*, or combine, elements of chemistry, physics, mathematics, and biology (**FIGURE 1.1**). For example, when examining a piece of shale – common type of rock, a geologist must think like a biologist to describe the fossils it contains, analyze its chemistry to identify the minerals within it, and assess its material properties using principles of physics and mathematics.

The information produced by this analysis enables the geologist to determine the rock's history, how it might be economically useful, and whether it represents a **sustainable resource** for humanity. A sustainable resource is any natural product used by humans that meets the needs of the present without compromising the ability of subsequent generations to meet the needs of the future. In doing so, geologists rely on **critical thinking,** the use of reasoning to explain the world around us by developing hypotheses and theories that can be rigorously tested.

In recent years, climate change, natural hazards, human impacts on the environment, and the search for new energy sources have all emerged as overriding global challenges. These concerns do not fall exclusively within the fields of chemistry, or geology, or biology. Addressing these issues requires integrating knowledge from many fields; thus, the natural sciences have evolved to become more integrated. We highlight these integrated relationships throughout this course.

Jeff Greenberg/Photolibrary/GettyImages

FIGURE 1.1 Geology students make a map of a rock formation along the coast.

Maps are widely used by geologists to identify natural resources and geologic hazards. Why would maps be effective for this purpose?

Geology Enters Our Daily Lives

Geology is relevant to everyone's day-to-day life, a point we reiterate at the start of each chapter. For example, the protection provided by Earth's *geomagnetic field* (magnetism that emanates from Earth's core and surrounds the planet) shields us from harmful solar radiation, making life on the planet possible; metals and energy sources are geologic products that build and power modern society; and water, used in industry, agriculture, and to slake our thirst, is also a geologic product. These examples comprise a small sample of literally thousands of geologic products and processes that are indispensable in our lives.

Geology affects our lives hundreds of times a day, usually without our being aware of it. To appreciate its presence in your life, it is important to become conscious of where geologic resources (water, soil, fossil fuels, metals, and others) come from, how they are obtained, and how they can be conserved. It is also a good idea to become aware of the many geologic hazards (**FIGURE 1.2**) that threaten our safety, including tsunamis, landslides, volcanic eruptions, flash floods, and others.

Geologists know the time scales necessary to create geologic resources such as fossil fuels and groundwater, and they recognize that rates of human consumption of those resources greatly exceed their natural rates of renewal. It is, therefore, incumbent on everyone to learn this lesson and to practice lifelong Earth citizenship—for instance, through conservation, informed voting, and participation in community affairs.

Geologists are uniquely trained to understand both the potentially destructive and potentially beneficial role of humankind in the future of Earth's environments and geologic resources. This understanding is embodied in the concept of *sustainability,* meeting the needs of the present without compromising the ability of future generations to meet their own needs. Humans have the power to manage environments and ensure that geologic resources remain abundant for the use of subsequent generations.

ASSOCIATED PRESS

FIGURE 1.2 In the early afternoon of January 10, 2005, a mudslide in the coastal California town of La Conchita took 10 lives, destroyed 15 houses, and damaged another dozen.

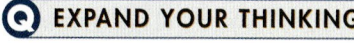 **EXPAND YOUR THINKING**

Where have you encountered geology in your life today?

 How could critical thinking have helped to prevent this tragedy?

1-2 Critical Thinking Is the Use of Reasoning to Explain the World around Us

LO 1-2 Explain the scientific method.

The purpose of science is to find laws or theories that explain the nature of the world we observe (**FIGURE 1.3**). This is achieved using critical thinking.

The Scientific Method

The **scientific method** comprises a series of steps for conducting an objective investigation. The number of steps isn't standard but the process always invokes critical thinking. Critical thinking begins with the use of *reasoning* (the process of drawing logical inferences).

The reasoning process goes like this: A scientist observes a phenomenon that is not explained well by current knowledge. To improve understanding of that phenomenon, the scientist uses *inductive reasoning*. He or she makes specific observations, gathers measurements of the phenomenon, begins to detect patterns and relationships, and, finally, formulates a tentative, testable **hypothesis** that attempts to explain these observations. For instance, *Observations:* A robin is a bird that flies; a crow is a bird that flies. *Hypothesis:* All birds fly.

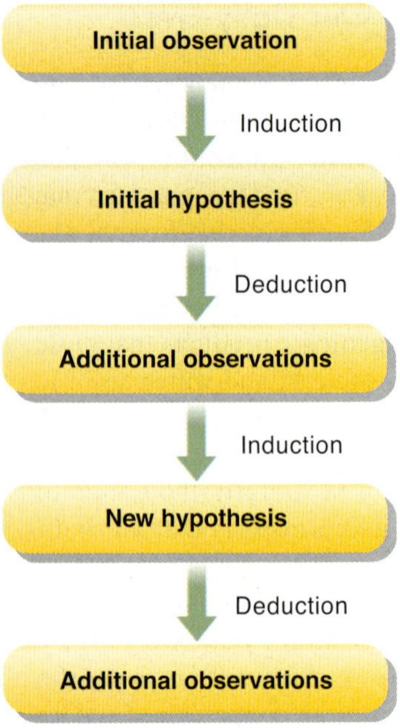

Initial observation
↓ Induction
Initial hypothesis
↓ Deduction
Additional observations
↓ Induction
New hypothesis
↓ Deduction
Additional observations

FIGURE 1.4 Critical thinking relies on inductive and deductive reasoning.

How does inductive reasoning differ from deductive reasoning?

A hypothesis is a *testable* educated guess that attempts to explain a phenomenon. A true hypothesis lends itself to being tested and to being revised or rejected if it fails the test. A hypothesis can be tested using *deductive reasoning*. Deductive reasoning uses generalizations to predict specific occurrences. For example, our hypothesis about birds, while it has a high success rate, ultimately will be proven false by deductive reasoning, as follows: Sparrows fly (true); ducks fly (true); penguins fly (false).

Inductive reasoning can lead to deductive reasoning (**FIGURE 1.4**). That is, observations can be used to build a hypothesis (induction), and the hypothesis can be used to generate testable predictions (deduction).

If a hypothesis passes repeated and rigorous tests, it may become a **theory**. A theory must be as simple and self-contained as possible and must account for all the known facts that are relevant to a subject. A theory is a hypothesis that has been aggressively tested and is generally accepted as true. Examples of successful theories include *Darwin's theory of evolution*, which explains how life has developed into diverse forms, and the *theory of plate tectonics*, which explains the organization and history of many geologic phenomena.

FIGURE 1.3 Scientists study natural phenomena such as volcanic eruptions using the scientific method.

What are the products of volcanic eruption? How would you test your hypothesis?

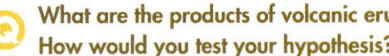
Ammit/iStockphoto

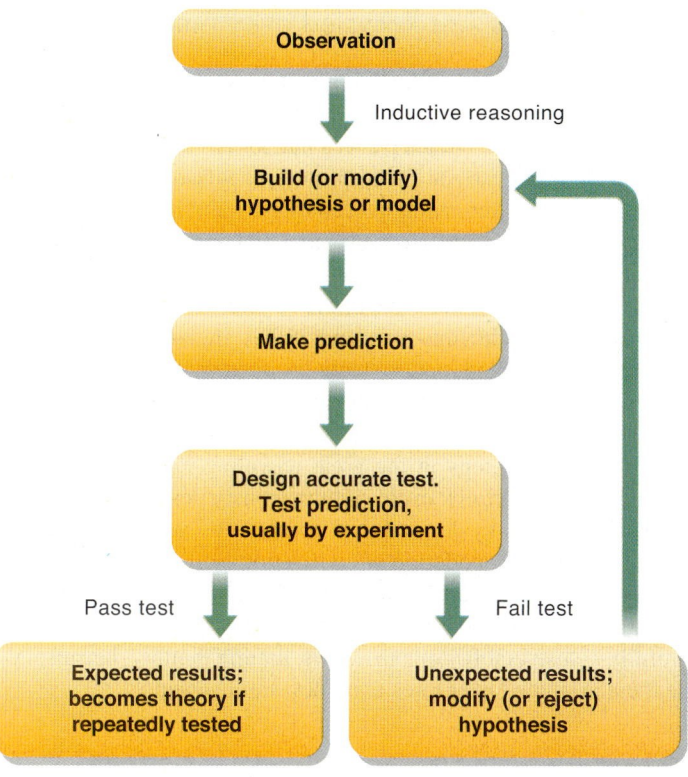

FIGURE 1.5 The scientific method.

🔍 **Why is the scientific method a useful tool for making predictions?**

Critical thinking works only if carefully constructed tests are conducted to examine the hypothesis (or theory). A poorly designed test can lead to false conclusions. Eventually, a successful theory that over time has been shown always to be true and if it is useful in predicting a wide range of phenomena can come to be considered a **natural law**. The *laws of thermodynamics* are examples of natural laws that describe the behavior of energy in all types of settings. The theory of evolution, proven highly useful in combating diseases of various types, has been proposed as a natural law.

Another important type of reasoning is the *principle of parsimony*. This principle states that among competing hypotheses, the simplest and most direct one is preferable. Parsimony assumes that an inefficient and complicated explanation of a phenomenon is not desirable if a simpler explanation would serve the same purpose. Scientists use parsimony to select the simplest explanations of nature.

We all use inductive, deductive, and parsimonious reasoning several times a day. Next time you are standing on a busy street corner, see if you can identify when and how you employ all three types of reasoning in deciding when to cross the street. (Be careful!) By recognizing and naming these thinking tools, we improve the ways in which we implement them to make new discoveries.

Scientists use these reasoning tools in the scientific method. The method has multiple steps (**FIGURE 1.5**), and is used in many professions to improve our lives (**FIGURE 1.6**).

🔍 **EXPAND YOUR THINKING**

What might the outcome be if a hypothesis is not testable?

ChameleonsEye/Shutterstock

FIGURE 1.6 Scientists and engineers use the scientific method to develop and test hypotheses about why and how buildings fail during an earthquake. This leads to designing better buildings.

🔍 **How might the engineers in this photo use reasoning in their hypothesis building?**

1-3 Global Challenges Will Be Met by Critical Thinkers with an Enduring Understanding of Earth

LO **1-3** Describe some characteristics of the science of geology.

The goal of this geology course is to embed an enduring understanding of Earth in your thinking.

The Study of Geology Encompasses a Vast Range of Time and Space

Years from now, when you recall this geology class, we hope you remember vividly the concepts you learned here and put them to good use when making decisions about how you live on our planet. The global challenges that come with living on a crowded planet will be met most effectively by critical thinkers with training in geology—like you.

Natural phenomena studied by geologists cover an immense span of time and space. They range from the megascopic (the length of the solar system is measured in trillions of kilometers; about 8.9 trillion, to be exact) to the microscopic (the bonding of atoms occupies an infinitesimally small space, on the order of 0.000000000001 kilometers). Amazingly, these phenomena exist side by side (**FIGURE 1.7**).

Massive planets are constructed from the tiniest molecules. Eons of geologic time (4.6 billion years since the birth of the solar system and the origin of Earth) are made up of long periods of slow and gradual change punctuated by short, violent convulsions, such as earthquakes, floods, and landslides.

NASA

FIGURE 1.7 The solar system.

How old is the solar system? Where did it come from?

Earth Materials Are Recycled over Time

You will learn in Chapter 2 that shortly after it was formed, Earth became molten. The outer layers cooled and hardened (forming the **crust**, Earths rocky surface), but the planet's interior is still hot enough today to melt rock. Since early in Earth's history, chemical and physical processes, part of a never-ending recycling machine that incessantly destroys and renews Earth materials, have continuously attacked the crust.

Although rocks may seem indestructible, in fact they are perpetually degraded by chemicals in the air and water, baked and contorted by forces within the crust, and, if they are conveyed to Earth's interior, recrystallized and potentially melted. But molten rock will solidify once it returns to the crust, whereupon the process of recycling rock will start over again (**FIGURE 1.8**).

These are all steps in the great recycling machine known as the **rock cycle.** The rock cycle simultaneously destroys and renews Earth's crust. This cycle produces many of the geologic resources that constitute the foundation of civilization: soil, water, petroleum, coal, natural gas, metals, and others. These resources form on geologic time scales of thousands to millions of years. Humans must carefully—sustainably—manage this natural bounty because we use them at much faster rates than they are renewed.

Paul Cloutier/Getty Images, Inc.

FIGURE 1.8 Energy from the Sun streams onto individual grains of quartz sand composed of bonded oxygen and silicon atoms. These accumulated in an ancient sand dune many millions of years ago (revealed by the fossilized sand layers), which today are being eroded by flash floods caused by heavy winter thunderstorms in the desert.

What is the evidence in this photo that the rocks are being eroded?

Plate Tectonics Controls the Geology of Earth's Surface

Earth's crust is in constant motion. Right now the ground under your feet is moving at about the same pace that your fingernails grow. Geologists did not understand this phenomenon until only a few decades ago when your parents were in school. The hypothesis of **plate tectonics** was first proposed in the 1960s, and it has since evolved into a theory.

According to this theory, Earth's crust is organized into a dozen or so pieces, like those of a jigsaw puzzle; these pieces are called *plates*. At their edges, some plates are recycled into Earth's interior, where some collide and crumple, and others separate, allowing molten rock to rise from below. This behavior is called *tectonics* (from the Greek word *tekton*, which means "builder"), and it has far-reaching implications.

As plates move, they change the way Earth looks. Mountain ranges rise when plates collide, only to be worn down by *weathering* and *erosion* (the carrying away of bits of weathered rock by running water, gravity, and wind). Ocean basins open and close as continents separate and converge again. Nearly every aspect of Earth's surface is related to how plates interact and change over time. The theory of plate tectonics, which is invoked to help explain the content of every chapter in this text, is described further in section 1–4 of this chapter and it is the subject of Chapter 3.

Geologic Systems Are the Product of Interactions among the Solid Earth, Water, Atmosphere, and Living Organisms

Earth is organized into overlapping geologic systems that influence and react to one another. *Geologic systems* consist of interdependent materials (such as rocks, sediments, living organisms, various gases, and water) that interact with physical and chemical processes. In a broad sense, these interactions occur because nuclear energy (heat from the Sun and from Earth's interior) and gravitational energy are at work mixing the air, water, and solid Earth.

For instance (**FIGURE 1.9**), volcanoes erupt gases, including carbon dioxide (CO_2), into the atmosphere. Carbon dioxide is used for respiration by plants. Plants feed animals, yield oxygen (O_2) back to the atmosphere (by photosynthesis), help to stabilize hillsides, and contribute to organic soil when they decay. Soil is composed of these organics, water, atmospheric gases, and minerals that crystallize in Earth's interior and on the surface. Heat in Earth's interior rises and breaks the surface into immense rocky plates that interact at their edges. When plates collide, one plate may recycle back into the interior of the planet. Recycling plates feed molten rock to volcanoes that erupt and produce CO_2 needed by plants. This is only one of thousands of examples of the interdependent nature of geologic systems that characterize our complex and diverse planet.

Earth Continuously Changes

We live on an ancient and restless landscape, one that is changing under our feet. All forms of life have evolved partially in response to this geologic revision over time. Today's Earth is the result of gradual, rapid, and catastrophic processes whose products (including minerals, rocks, topography, ecosystems, plants and animals, and others) have been accumulating for 4.6 billion years. Hence, our planet looked very different in the past than it does today, and it will look very different in the future. We humans, with our technological abilities to extract materials from the crust, alter the landscape, pollute the air and water, and transform the climate, may rival, and often exceed, the pace and extent of natural processes that change Earth.

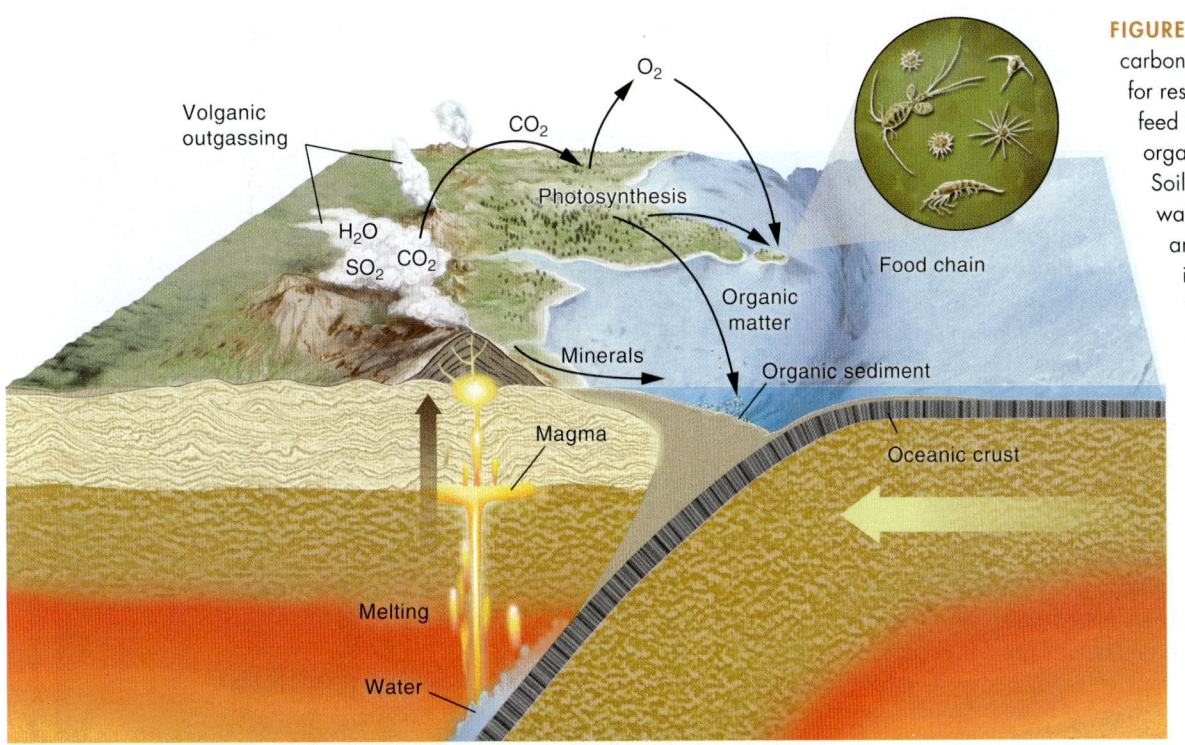

Volganic outgassing
O_2
CO_2
Photosynthesis
H_2O
SO_2 CO_2
Organic matter
Minerals
Organic sediment
Food chain
Magma
Oceanic crust
Melting
Water

FIGURE 1.9 Volcanoes erupt carbon dioxide (CO_2) that is used for respiration by plants. Plants feed animals and contribute to organic soil when they decay. Soil is composed of organics, water, atmospheric gases, and minerals that crystallize in Earth's interior. Heat from Earth's interior rises and breaks the crust into rocky plates. When plates collide, one plate may recycle back into the interior of the planet and feed molten rock to volcanoes that erupt and produce CO_2 needed by plants.

Why is the term "system" used to describe the concept in Figure 1.9?

Rocks and Sediments Are Pages in the Book of Earth's History

By "reading" the evidence of past events in Earth's crust, geologists piece together the history of our restless planet. This evidence shows that: Earth is very old; evolution is responsible for life's incredible diversity; continuous change is a characteristic of geologic systems; and geologic processes operate on an immense stage of time and space.

Rocks and sediments also hold the valuable fossil fuels that power our modern society. One consequence of this fact is discussed in the Earth Citizenship feature titled "Fracking: An Energy Revolution with an Environmental Cost."

EXPAND YOUR THINKING

Why is understanding geologic systems crucial to managing geologic resources sustainably?

EARTH CITIZENSHIP

Fracking: An Energy Revolution with an Environmental Cost

According to the International Energy Association, the United States will become the largest oil producer in the world by 2017, surpassing Saudi Arabia, Russia, and other traditional oil giants in the Middle East and South America. Oil imports to America are at their lowest levels in two decades, and the country may become a net oil exporter by 2020 and energy self-sufficient by 2035.

As the United States grows less and less dependent on oil from the Middle East, the energy revolution in the country is being spurred by new energy policies initiated early this century and continuing under the present administration. According to the White House, the United States holds a 100-year supply of natural gas, and the current administration has pushed to increase production of both gas and oil.

Natural gas (known as *methane*) production jumped 15% between 2008 and 2012. Oil and petroleum imports have fallen, on average, more than 1.5 million barrels per day; and since 2008, domestic crude oil production has increased by an average of more than 720,000 barrels per day. All the drilling required to speed this production has important and positive consequences: It has provided jobs for a weak economy—job growth in these industries has risen 25% since January 2010—and the reduced dependence of the United States on external sources of energy has improved the security of the nation.

The increase in U.S. oil and natural gas production is the result of a controversial technology known as *hydraulic fracturing,* or "fracking." Fracking is a drilling process whereby millions of gallons of chemically enhanced water and sand are injected at high pressure into shale formations that hold large volumes of fossil fuels (**FIGURE 1.10**). The chemical mixture fractures the shale, allowing for the release of previously unobtainable oil and methane. Fracking typically takes place thousands of feet underground, where it is not harmful; but closer to the surface, escaping methane from poorly constructed wells can pollute freshwater aquifers, seep through the ground, and cause extensive damage to ecosystems and drinking water supplies.

Fracking presents several challenges: 1) The process requires 2 to 4 million gallons of chemically enhanced water, which can put a strain on local water supplies. 2) Oil and natural gas can escape from poorly constructed wellheads (the tops of wells) and pollute local aquatic ecosystems and aquifers. 3) The process produces millions of gallons of wastewater, which must be managed by a combination of recycling and disposal. 4) The drilling and fracking process requires heavy machinery that produces both noxious exhaust and heavy traffic.

As is the case with the extraction of many natural resources, the fracking industry can minimize these environmental impacts by instituting conscientious and powerful safeguards and best management practices. To date, however, the fracking industry is exempted from the federal Clean Air and Clean Water Acts, which means there is a lack of federal oversight. Consequently, there is growing concern regarding the potential for environmental and community damage related to fracking. To address this concern, a broader conversation is in order, to address the development process and the steps being taken to ensure safe operations.

FIGURE 1.10 A typical hydraulic fracturing operation needs to institute powerful safeguards to minimize impacts on the environment and nearby community.

1-4 The Theory of Plate Tectonics Is a Product of Critical Thinking

LO 1-4 List three observations that support the theory of plate tectonics.

No one person is responsible for developing the theory of plate tectonics. It is still evolving, as thousands of critical thinkers all try to make sense of the patterns of geology that surround us.

Alfred Wegener and Continental Drift

Why are there continents? Why are there ocean basins and islands? Why are earthquakes common in some areas and rare in others? Why are most volcanoes found in chains? Many scientists have worked hard to answer these questions, and a handful of researchers stand out as particularly important in developing the early components of the theory of plate tectonics. *Alfred Wegener* is preeminent among them.

Like many observers of nature before him, Wegener, an early twentieth-century German meteorologist, noticed the natural "fit" of the continents. Conducting his own research, as well as drawing on that of others, Wegener employed inductive reasoning to propose the hypothesis known as *continental drift*. He cited observations that the edges of the continents seem to fit together like pieces of a puzzle, that continents separated by wide oceans have similar topographic and geologic features along opposite shores, and that fossil evidence suggests that continents were formerly connected.

Based on these observations, Wegener proposed that continents move, or "drift," across Earth's surface; but he was unable to describe how this movement occurs. He hypothesized that Europe, North America, Africa, and South America, formerly were joined together in a "supercontinent," which he named **Pangaea**.

Evidence for Wegener's continental drift can be found in the similar deposits of rock that occur across Europe, Africa, and North America (**FIGURE 1.11**) and in the glacial deposits from a single ice source found in the southern portions of South America, Africa, India, Australia, and Antarctica. Wegener also recognized that the presence of fossils of similar extinct plants and animals on different continents could be best explained by his proposition that those continents were formerly joined together.

Seafloor Spreading

At the same time, the basic laws of physics suggest that continents simply cannot plow through Earth's crust like ships through seawater; and despite strong supportive evidence, Wegener's continental drift hypothesis failed to gain acceptance by most scientists. Then, after World War II, new observations led to a modification of the hypothesis and the emergence of **seafloor spreading** as an explanation of continental drift.

During the war, a geologist from Princeton University, *Captain Harry Hess* (1906–1969), commanded a ship that mapped portions of the Pacific Ocean seafloor. He was able to show that the ocean floor is neither flat nor featureless, as was commonly thought. At the time, scientists assumed that ocean basins were unchanging and that the ocean floor was covered in thick layers of sediment that buried the topography beneath it. Instead, Hess found sharp, rugged ridges on the seafloor. He inferred from this evidence that the sediment cover was thin and, therefore, that the seafloor was young. But how could the crust on an old planet be young?

In 1962, Hess published his answer to that question in a paper titled "History of Ocean Basins." Using deductive reasoning to establish a robust, predictive hypothesis that improved on the idea of continental drift, he proposed the important concept of seafloor spreading. He predicted that new seafloor is formed at places he called *spreading centers*, where Earth's crust tears open and molten rock from below rises

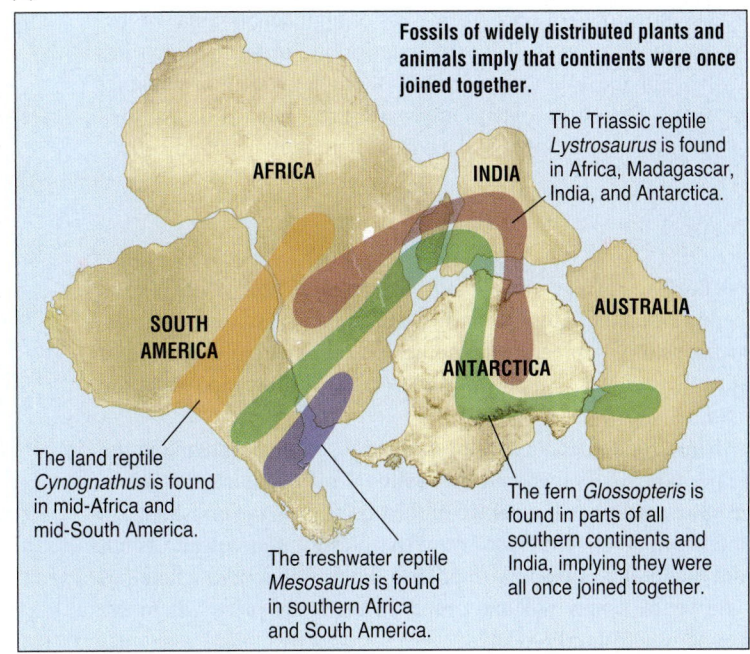

(a)

Older mountain belts   Younger mountain belts

(b)

Fossils of widely distributed plants and animals imply that continents were once joined together.

The Triassic reptile *Lystrosaurus* is found in Africa, Madagascar, India, and Antarctica.

AFRICA

INDIA

AUSTRALIA

SOUTH AMERICA

ANTARCTICA

NORTH AMERICA

EUROPE

Close match in mountain ages

The land reptile *Cynognathus* is found in mid-Africa and mid-South America.

The freshwater reptile *Mesosaurus* is found in southern Africa and South America.

The fern *Glossopteris* is found in parts of all southern continents and India, implying they were all once joined together.

FIGURE 1.11 Wegener's evidence included (a) the "fit" between continental outlines, the similar nature of their rocks, and (b) the landbound fossils they share.

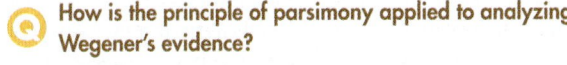

How is the principle of parsimony applied to analyzing Wegener's evidence?

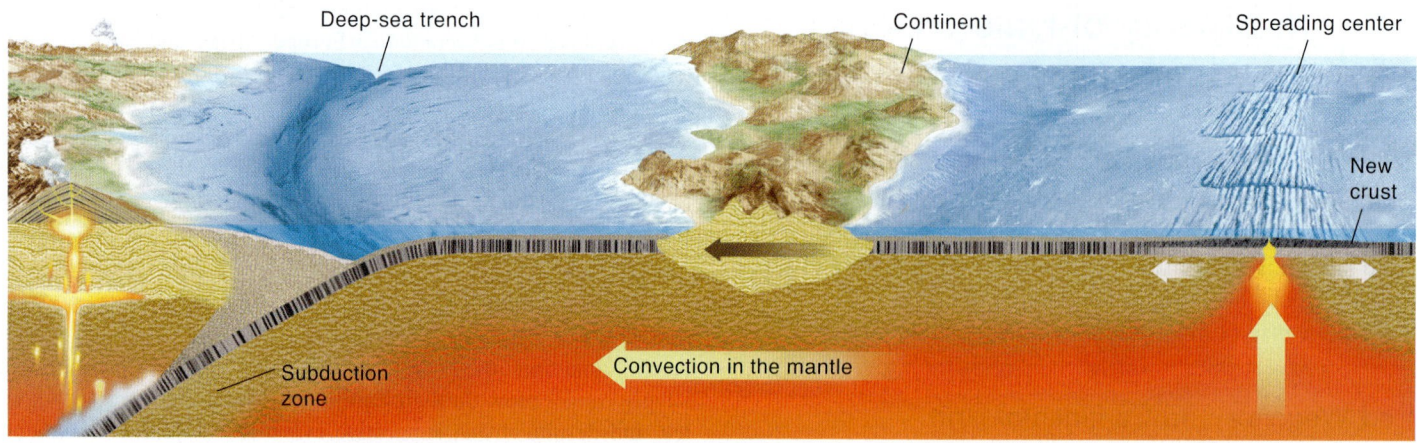

FIGURE 1.12 Spreading centers are long, narrow openings in the crust where molten rock rises from below to cool and form new seafloor. Like a conveyor belt, the seafloor moves away from these areas toward subduction zones, where the seafloor is recycled into the mantle.

Q **What do you think will happen when the continent arrives at the subduction zone?**

and cools to form new crust. Spreading centers, he said, are marked by submerged mountainous ridges because they are regions of high heat flow and crust that is hot and buoyant.

At other locations where ocean basins have long and narrow *deep-sea trenches*, the seafloor is recycled into Earth's interior at **subduction zones.** Hence, like a conveyor belt, new oceanic crust is created at a spreading center and moves across Earth's surface until it reaches a subduction zone, where it is recycled back into Earth's interior (**FIGURE 1.12**). Rather than plowing through Earth's crust, then, continents, ride embedded within these moving plates of rock.

Seafloor spreading differs from continental drift in that the movement of continents is driven by *motion of the seafloor*. Also different from the continental drift hypothesis is that Hess's concept of seafloor spreading proposed an explanation for seafloor motion, namely that plates ride on currents of hot rock circulating in Earth's interior. These currents are the product of **convection,** the transfer of heat by the movement of rock from areas of high heat to areas of low heat. In the thickest layer of Earth's interior, the **mantle** (which lies directly beneath the crust), convection is thought to occur in a manner similar to the motion in a lava lamp, with great islands of hot rock migrating upward and downward as they alternately heat and cool.

Hotspots

In 1963, the Canadian geophysicist *J. Tuzo Wilson* (1908–1993) seized on the significance of Hess's ideas. He used deductive reasoning based on the seafloor-spreading hypothesis to predict the existence of **hotspots** that would explain midplate chains of volcanoes like the Hawaiian Islands (**FIGURE 1.13**).

Wilson deduced that a single, stationary source of magma in the mantle periodically erupts onto the seafloor, forming an active volcano that emerges above the surface of the ocean, creating an island. As the plate shifts a volcano away from the hotspot, the volcano is cut off from the molten rock and becomes extinct. Within a few hundred thousand years, a new volcano erupts in the same place where the extinct one was formerly located.

No new land is formed on the extinct volcano, so it is eroded by rainfall, streams, landslides, and ocean waves. As it travels away from the hotspot, the plate beneath it cools and thins. This, along with persistent erosion of the

land surface, eventually causes the island to subside below the sea and form a topographic feature called a *seamount*. This mechanism is responsible for chains of volcanic islands and seamounts found around the world.

The hotspot hypothesis predicts that islands farther from a hotspot are older than those closer to it, a prediction that has been repeatedly supported with geologic data. For instance, the Hawaiian island of Kauai (the farthest of the main islands from the Hawaiian hotspot) is the most highly weathered and the oldest (approximately 1.4 to 5.7 million years).

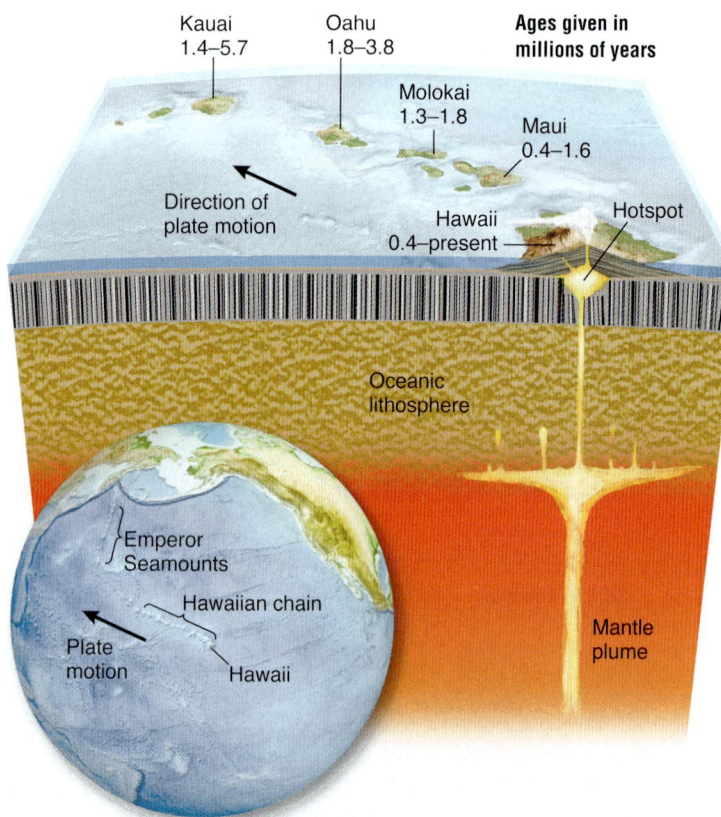

FIGURE 1.13 Chains of midplate volcanoes can be explained by the existence of a single stationary source of magma, a "hotspot" in the mantle.

Q **What characteristic of the island ages supports hotspot theory?**

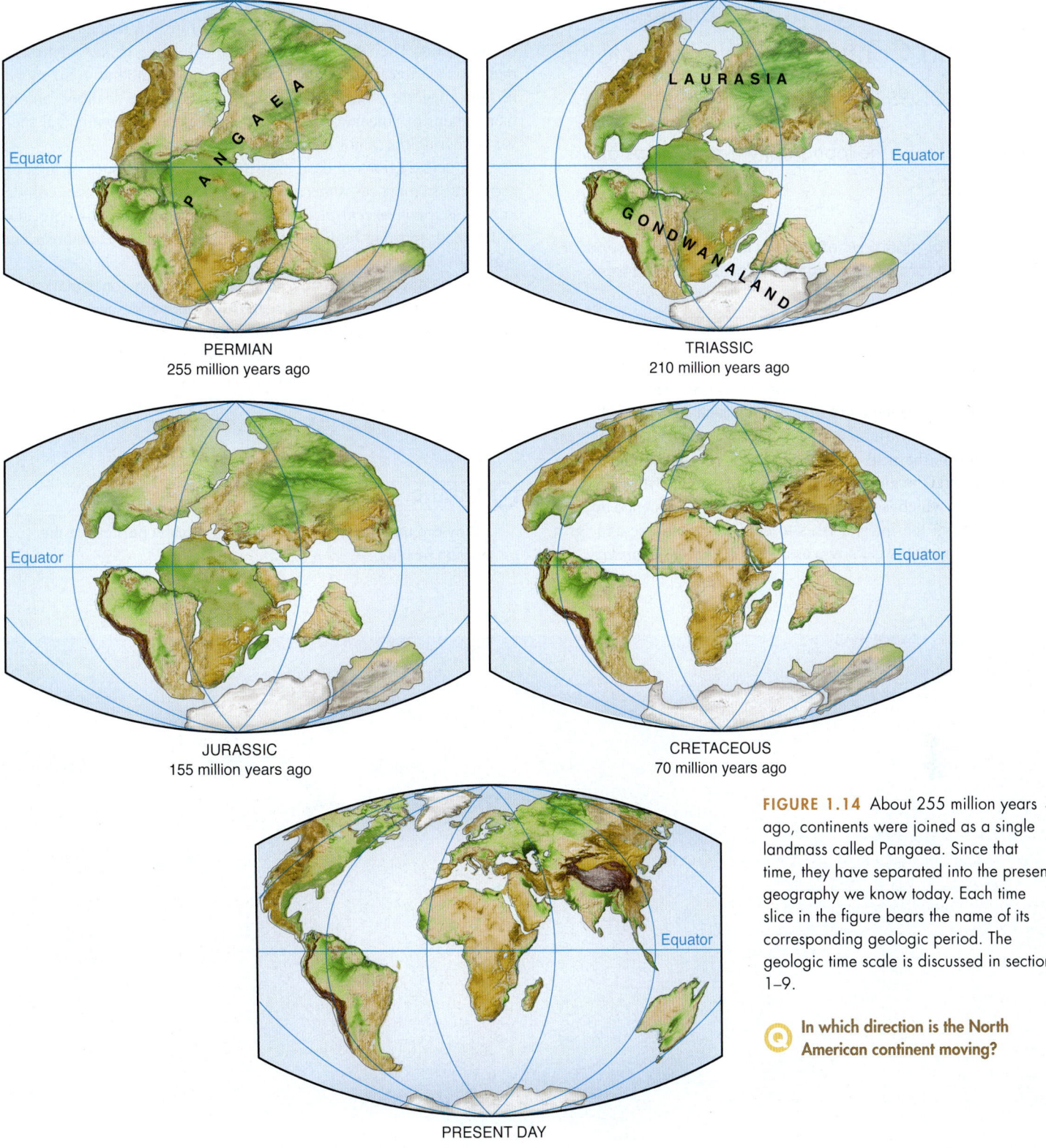

PERMIAN
255 million years ago

TRIASSIC
210 million years ago

JURASSIC
155 million years ago

CRETACEOUS
70 million years ago

PRESENT DAY

FIGURE 1.14 About 255 million years ago, continents were joined as a single landmass called Pangaea. Since that time, they have separated into the present geography we know today. Each time slice in the figure bears the name of its corresponding geologic period. The geologic time scale is discussed in section 1–9.

Q In which direction is the North American continent moving?

Pangaea

Modern research has verified Wegener's hypothesis regarding the former existence of a supercontinent. About 255 million years ago, plates carried continents in such a way that they all merged together and formed a single supercontinent called Pangaea (from the ancient Greek words for "all lands"). Since then plates have shifted causing Eurasia, Antarctica, Africa, South America, North America, and Australia to break away into separate pieces, creating the continental pattern we know today (**FIGURE 1.14**).

Q **EXPAND YOUR THINKING**

Drawing geologic features can help us identify what we do not understand. Draw a simple cross section of Earth's crust and upper mantle showing a spreading center, a subduction zone, and mantle convection. Label as many aspects of the diagram as possible.

The Topography of the Oceans and Continents Reflects Tectonic Processes

1-5

LO **1-5** Describe the topography that develops at plate boundaries.

The interaction of plates, whether they are colliding or separating, leads to the formation of some of Earth's greatest topographic features.

Global Topography

Because spreading centers experience high heat flow from Earth's interior, the crust is warm and sits higher than the surrounding seafloor. This forms the **mid-ocean ridge** that runs for thousands of kilometers around the globe (**FIGURE 1.15**). The top of this ridge has a narrow valley, which is where the seafloor is *rifted* (opened up) and molten rock from the interior rises to cool at the surface and form new oceanic crust. This is known as a *divergent plate boundary*.

The map of global topography in Figure 1.15 shows the long broad mid-ocean ridge (in light blue) that runs through all the major ocean basins. The ridge is the site of a spreading center where two plates (sometimes three) move away from one another. Also shown in light blue are other topographic features on the seafloor, among them chains of volcanic islands and seamounts and continental shelves bordering the major landmasses.

On land, regions in dark brown on the map are sites of high topography. In most cases these mark the location where plates collide, known as a *convergent plate boundary*. Here the crust is deformed and folded, forming high mountain ranges (e.g., the Himalayas and Alps).

Other convergent boundaries form subduction zones marked by deep-sea trenches. Thus, mid-ocean ridges, trenches, and continental mountain ranges usually mark the location of past or present plate boundaries where the crust is either separating or colliding. Plate boundaries are discussed in more detail in Chapter 3.

The Geology in Our Lives feature titled "Earth Science Literacy" introduces the "Nine Big Ideas" of Earth science.

Q **EXPAND YOUR THINKING**

Why are mid-ocean ridges relatively shallow portions of the ocean basins?

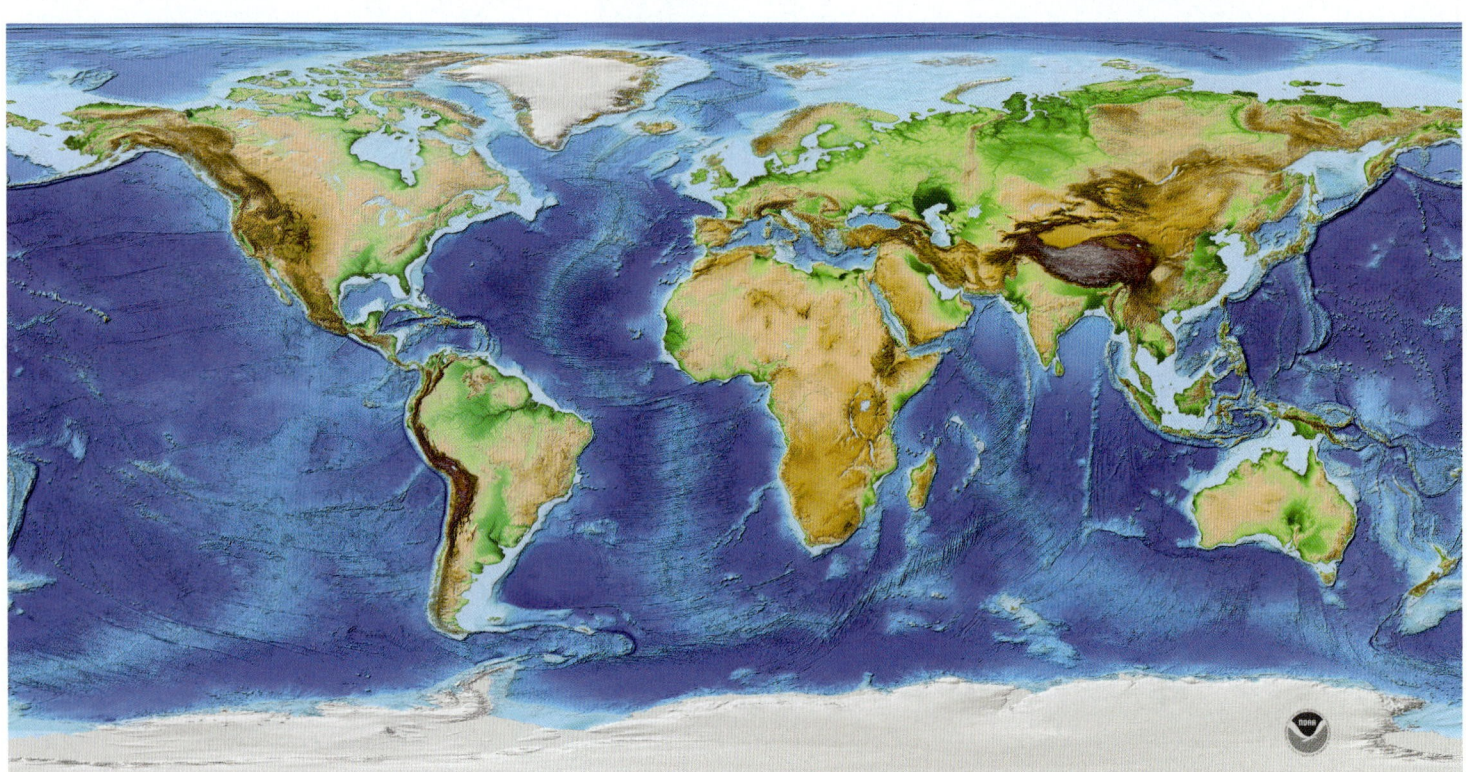

NOAA

FIGURE 1.15 Topographic map of the world. Running through every ocean basin is a continuous mountain range on the seafloor (light blue) that marks the location of a spreading center on a mid-ocean ridge. Brown regions mark mountain ranges on land.

Q Using the map in Figure 1.14, draw arrows on each continent in Figure 1.15 indicating its direction of movement.

GEOLOGY IN OUR LIVES

Earth Science Literacy

Educators and researchers have summarized "Nine Big Ideas" of Earth science, concepts that capture the complexities of how our planet works. (For more on these concepts, go to www.earth-scienceliteracy.org). An Earth-science-literate person understands the fundamental concepts of Earth's many systems and is able to find and access scientifically credible information about Earth, and then communicate these concepts in meaningful ways. Here are the big ideas of Earth science:

1. *Earth scientists use repeatable observations and testable ideas to understand and explain our planet.* This was discussed in section 1–2 as the scientific method. Scientists develop testable hypotheses as a way of separating opinion from true understanding of how Earth works.

2. *Earth is 4.6 billion years old.* Over Earth's vast history, both gradual and catastrophic processes have produced enormous changes. Earth's age is supported by multiple independent observations of Earth phenomena. This is explained in more detail in section 12–10.

3. *Earth is a complex system of interacting rock, water, and life.* The four major systems of Earth are the geosphere (rock and sediment), biosphere (living organisms and their ecosystems), hydrosphere (water in its many forms), and atmosphere (the envelope of gases surrounding Earth). These interact with, and change, each other on a continual basis.

4. *Earth is continuously changing.* The weathering of Earth's surface, movement of lithospheric plates, release of heat from the interior, pull of gravity, changing climate, and other phenomena all work continuously to alter the planet.

5. *Earth is the water planet.* Water is found everywhere on Earth, from the top of the atmosphere to the bottom of the mantle. Water is involved in multiple geological processes that drive Earth's many complex systems.

6. *Life evolves on a dynamic Earth and continuously modifies it.* Evolution is a natural and ongoing process that is responsible for the great diversity of life, both past and present. Life is found in a wide range of Earth environments, changing the physical and chemical properties of the geosphere, hydrosphere, and atmosphere.

7. *Humans depend on Earth for resources.* Earth is our home, and its resources mold civilization and all types of human endeavor. Geologic resources are limited and unevenly distributed around the planet. Geologists help society move toward greater sustainability of these resources.

8. *Natural hazards pose risks to humans.* Hazards shape human history and, conversely, humans can contribute to the frequency and intensity of some natural hazards. An earth-science-literate public is essential for reducing risks from natural hazards.

9. *Humans significantly alter Earth.* The human population has expanded to the point where human activities markedly alter the rates of many of Earth's surface processes and the availability of many types of natural resources. For instance, humans cause global climate change, affect the quality and distribution of freshwater, modify the land surface, and impact global ecosystems.

1-6 Rock Is a Solid Aggregate of Minerals

LO 1-6 Compare and contrast the three families of rock.

It is commonly said that geology is "the study of rocks." In fact, geology involves much more than the study of rocks. Nevertheless, "What is rock?" is a fundamental question that we need to answer.

What Is Rock?

Rock is a solid aggregation of minerals. *Minerals* (the subject of Chapter 4) are solid chemical compounds that can be seen in rocks as crystals or grains. If you hold loose sand in your hand, you are not holding a rock. If you cement those sand grains together, you have a rock called *sandstone* (studied in Chapter 8). Look closely at the speckled surface of a rock sample (**FIGURE 1.16**). Each speck of white, black, gray, brown, red, or other color is a separate mineral grain.

As we will learn in later chapters, there are many types of minerals. Hence, there are also many kinds of rock. Despite this diversity, all rocks can be grouped into three distinct families, depending on how they were formed. These three families are **igneous rock, sedimentary rock,** and **metamorphic rock.** These rock families do not exist independently of one another. Rather, they interact and mix together and are recycled through the rock cycle.

Granite

Courtesy of Chip Fletcher

FIGURE 1.16 Rock is a solid aggregation of minerals. Each speck of black, brown, red, gray, and white is a separate mineral in this igneous rock sample of *granite*.

If this rock came from one, uniform type of magma, how can it have many different types of minerals?

Igneous Rock

More than four-fifths of Earth's crust is made of igneous rock (**FIGURE 1.17**). Igneous rock (Chapter 5) is produced by the development and movement of molten rock, called *magma*. Igneous rock is formed in two ways:

1. Cooled magma within Earth's crust forms *intrusive* igneous rock.

2. Cooled *lava*—magma located on Earth's surface (after erupting from a volcano)—forms *extrusive* igneous rock.

Sedimentary Rock

Fragments of mineral and rock (known as *sediment*) are constantly dissolving or breaking off the crust and being moved (by gravity, wind, and water) from one place to another on Earth's surface. The process of degrading and breaking the crust through chemical reactions or mechanical processes is called *weathering* (Chapter 7).

Weathering is usually followed by *erosion*, which is the process of moving sediment from its place of origin, where it was formed by weathering, to its final resting place, where it is *deposited*. You can see sediment as dust blowing in the wind or in the mud and sand carried along in a stream or river (Chapter 17).

Over time, huge quantities of eroded sediment are carried by wind and water from the land into the oceans, where they are eventually deposited. Sedimentary rock is made of sediment that has

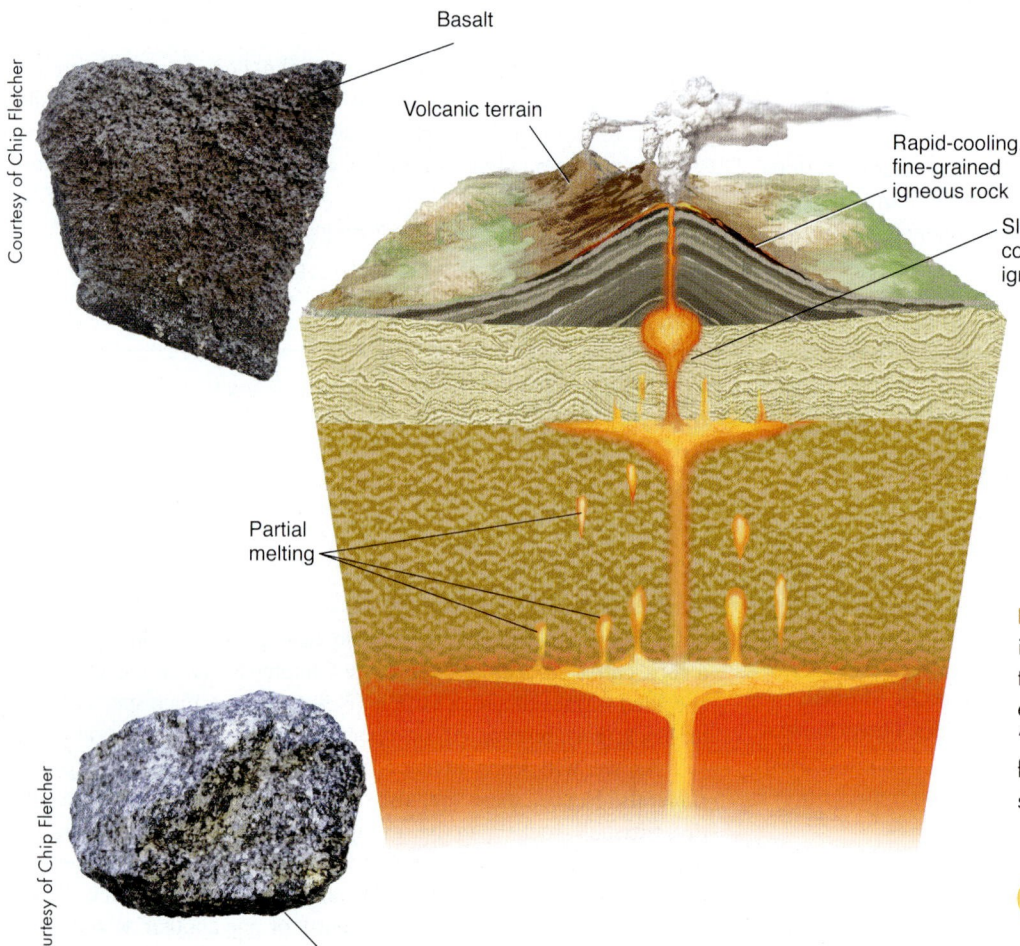

Courtesy of Chip Fletcher

Basalt

Volcanic terrain

Rapid-cooling, fine-grained igneous rock

Slow-cooling, coarse-grained igneous rock

Partial melting

Courtesy of Chip Fletcher

Gabbro

FIGURE 1.17 Molten rock, or magma, forms in Earth's interior. Magma in contact with the ocean or atmosphere is lava and forms extrusive igneous rock (in this case called "basalt"); cooled magma in Earth's interior forms intrusive igneous rock ("gabbro" is shown here).

Why would intrusive rocks typically have large minerals and extrusive rocks typically have very small crystals?

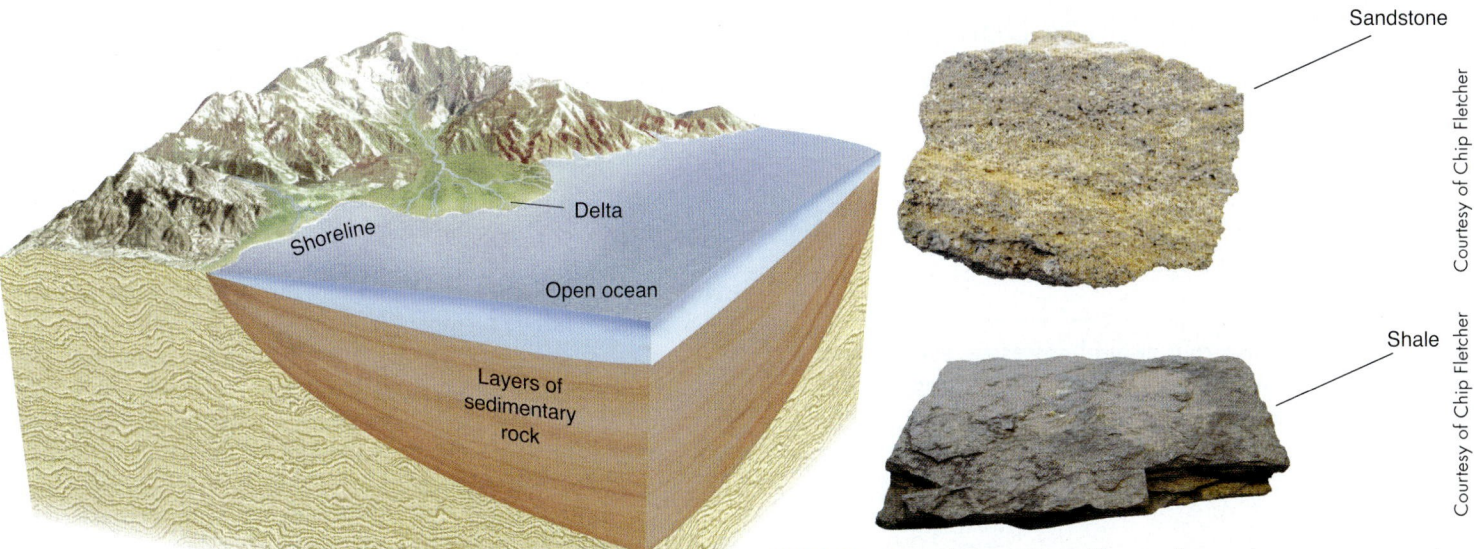

Courtesy of Chip Fletcher

Sandstone

Shale

FIGURE 1.18 Sediments, formed by weathering the crust, are moved by erosion and, ultimately, deposited, where they become sedimentary rock. One of the rocks shown is *sandstone*. It is possible to see layers of sand as they originally collected near the shoreline preserved in the rock. The other rock shown is *shale,* made of mud accumulating on the deeper ocean floor.

Ⓠ Why would mud collect on the deep seafloor, and sand collect in shallow water?

collected and become solidified into rock by *compaction* or by *cementation* of grains. Although sedimentary rock (**FIGURE 1.18**) is less plentiful than igneous rock in Earth's interior, it makes up about 75% of the exposed rock on Earth's surface.

Metamorphic Rock

The third family of rock is metamorphic rock (Chapter 9; **FIGURE 1.19**). It forms beneath Earth's surface in several ways. When rock is buried at great depths—for example, beneath mountain ranges—it is subjected to enormous heat and pressure. This causes minerals to change, and allows for new minerals to grow out of

elements that formed the old ones. *Metamorphism* (change to a rock caused by heat and pressure) also occurs when magma bakes the rock surrounding it.

Both the pressure of deep burial and exposure to the heat of Earth's interior produce new environmental conditions that lead to the growth of *metamorphic minerals*. It is important to note that rock does not melt during metamorphism. It usually becomes harder as the minerals *recrystallize* into new types. Metamorphic rock is made of new minerals that are formed under relatively high-temperature and high-pressure conditions, while the rock remains solid. Very often, metamorphic rock is folded or contorted by the pressure that forms it.

Ⓠ **EXPAND YOUR THINKING**

Why is sedimentary rock most common upon Earth's surface, and igneous rock most common below the surface?

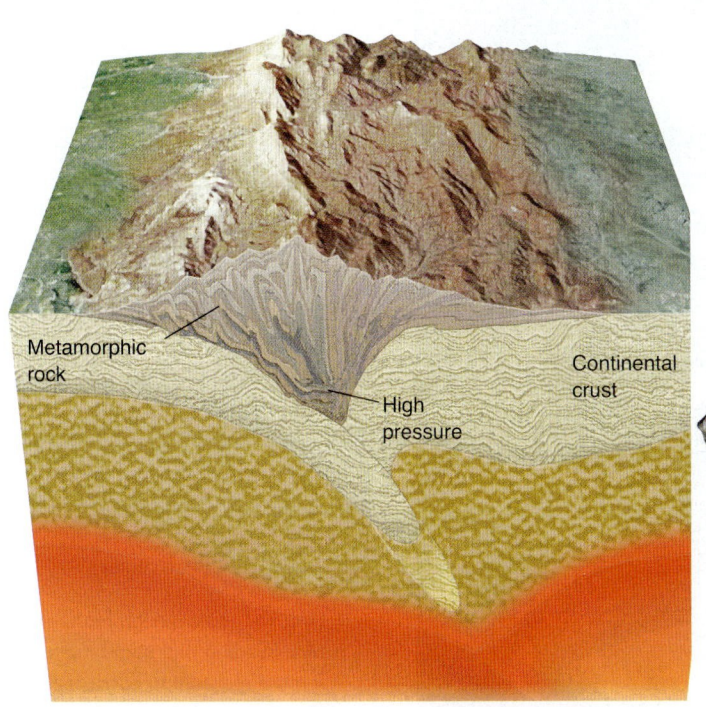

Metamorphic rock

High pressure

Continental crust

Gneiss

Courtesy of Chip Fletcher

Ⓠ What is the significance of the contortions in the sample of gneiss?

FIGURE 1.19 Contorted rock is a sign of metamorphism.

1-7 Geologists Study Dangerous Natural Processes Known as Geologic Hazards

LO **1-7** Summarize geologic hazards that are capable of causing severe damage.

Geologists study dangerous natural processes, such as landslides, floods, erosion, volcanic eruptions, tsunamis, earthquakes, hurricanes, and other **geologic hazards**.

Floods

In most cases, when geologic hazards occur in remote areas, floods, eruptions, and other dangerous processes do not affect humans. However, too often we have built our neighborhoods in potentially dangerous areas, and, with population growth, geologic hazards are affecting human communities with growing frequency. Geologists monitor geologic hazards to help communities avoid casualties and reduce damage. Hazard avoidance is one type of *mitigation*; mitigation is achieved by lowering community vulnerability to a geologic hazard.

Floods (Chapter 17) are part of the natural life cycle of streams. Stream channels and their *floodplains* depend on regular flooding to maintain healthy ecosystems and to clear accumulated sediment. Nevertheless, in the United States, stream flooding causes about $5.5 billion in losses each year; and in the last decade, nearly 1,000 people lost their lives to floods. One reason for these losses is the human tendency to build houses, farms, and even entire towns in areas that are vulnerable to flooding. Scientists predict that flooding will increase as a result of global warming, as rainfall events in many areas becomes more intense due to increased water vapor in the atmosphere.

Hurricanes

Hurricane Sandy devastated portions of the Caribbean and the U.S. Atlantic in late October 2012 (**FIGURE 1.20**). The storm became the largest and second costliest (behind hurricane Katrina, in 2005) Atlantic hurricane on record, with losses exceeding $74 billion and a death toll of 253 people in 7 countries. Its enormous size and the widespread damage it caused led the media to nickname the hurricane "Superstorm Sandy."

Hurricanes (Chapter 20) are tropical cyclones characterized by winds that reach 119 kilometers per hour, or higher. High winds are a primary cause of injuries and property damage from hurricanes. In addition, flooding resulting from storm waves, high water levels, and torrential rains also cause property damage and loss of life.

Volcanic Eruptions

In 1991, thousands of lives were saved when government geologists evacuated towns and villages on the slopes of Mount Pinatubo in the Philippines. Soon afterward, the volcano exploded in a massive eruption that devastated vast tracts of land in the region.

There are several types of volcanoes (Chapter 6), and learning how and why they differ is a key to understanding the threat they pose to local populations (**FIGURE 1.21**).

Associated Press

FIGURE 1.20 In late October 2012, Hurricane Sandy, with a combination of high winds, storm surge, coastal erosion, and intense rainfall, severely damaged dozens of communities along the Atlantic shoreline of the United States.

Q **Describe some design features of a building that might make it more resistant to hurricane damage.**

FIGURE 1.21 Stromboli volcano erupts in Italy.

 What drives the lava eruption into the air?

FIGURE 1.22 In March 2011, the most powerful earthquake in Japan's history spawned a massive tsunami that killed more than 20,000 people and destroyed dozens of cities and villages.

 Based on what you have learned about plate tectonics, where are earthquakes and tsunamis most likely to occur?

Earthquakes

When rock in Earth's crust breaks, it may cause an *earthquake* (Chapter 11). Shock waves radiating outward from the fracture shake buildings and jolt bridges, often causing massive damage.

In Haiti, more than 230,000 people died and over 1 million were left homeless when an earthquake struck there on January 12, 2010. Injuries and damage caused by earthquakes are due not only to poorly constructed buildings that collapse and trap people, but also to landslides that bury cars, buildings, and entire neighborhoods, fires ignited by ruptured gas lines and downed electrical lines, and lack of adequate rescue and medical facilities.

Tsunamis

The most damaging and deadly *tsunami* (a series of huge waves caused by sudden movement of the seafloor) in recorded history occurred on December 26, 2004, when coastlines around the Indian Ocean were inundated without warning. Known as the *Sumatra-Andaman tsunami*, it was caused by an undersea earthquake in Indonesia (the second-largest ever recorded) and resulted in the deaths of more than 230,000 people. On some shorelines, the tsunami produced water levels over 30 meters high, and many coasts were swept clear of all human development.

In 2011, a magnitude 9.0 earthquake, the fourth largest ever measured, occurred offshore of Tohoku, Japan, and generated a tsunami with a maximum run-up elevation of 39 meters. The tsunami and earthquake caused the death or disappearance of more than 20,000 people and cost over $30 billion in damages (**FIGURE 1.22**).

Mass Wasting

Mass wasting (Chapter 16) is the process that forms valleys; it is the movement of soil and surface materials down a slope under the force of gravity. *Landslides* and other types of mass wasting are ways by which land erodes.

As mentioned earlier, erosion is important in recycling sediments and nutrients through the environment. At the same time, excessive mass wasting caused by heavy rains, earthquakes, or other triggering events constitute a major geologic hazard. Mass wasting occurs in all 50 states, costing $1 to $2 billion in damage and causing more than 25 deaths each year.

Q EXPAND YOUR THINKING

Describe the geologic hazards that are prevalent where you live.

1-8 Geologic Resources Are Not Limitless, and So Must Be Managed

LO 1-8 Explain the management of geologic resources.

Although Earth is large, its **geologic resources** are not limitless and the expanding human population uses many critical resources faster than natural processes can replace them; this is not sustainable.

Water

Geologic resources are materials that occur in nature and are essential and/or useful to humans, such as water, air, building stone, topsoil, minerals, and many others. To ensure that future generations will have the resources they need, geologists work to understand how geologic resources are formed and distributed, and how they can be conserved; they also determine how abundant they are.

Freshwater is relatively scarce, representing only about 3% of all water on Earth. Of this total, almost 70% is unusable, as it is stored in icecaps and glaciers. Although water is a renewable resource, it takes decades to centuries for rain and snow to replenish the amounts we pump out of the ground in only a few years (Chapter 18). Moreover, water-rich rock and sediment layers can be damaged by overpumping, rendering them no longer useful.

Agriculture is by far the largest water consumer, accounting for some 70% of total water withdrawals. (Industry accounts for 20%; household use for 10%.) While the daily drinking water needs of humans are relatively small—about 4 liters per person—the water required to produce the food a person consumes each day is much higher, ranging between 2,000 and 5,000 liters (**FIGURE 1.23**).

Soil

Soil (Chapter 7) is the layer of minerals and organic matter that covers the land. Soil ranges in thickness from centimeters to several meters or more. Its main components are bits of rock and minerals, organic matter, water, and air.

Researchers report that 65% of the soil on Earth is degraded by erosion (**FIGURE 1.24**), *desertification* (Chapter 19), and *salinization* (high salt content). Soil quality must be protected because soil provides us with food and sustains natural ecosystems. Soil is a renewable resource, but only if it is treated with natural fertilizer (such as manure) several times each year and carefully protected from erosion.

Under natural conditions, it takes 200 to 1,000 years to form only 2.5 centimeters of topsoil, and this can be swept away in a single rainstorm if it is carelessly exposed to erosion. Worldwide, soil is being lost 16 to 300 times faster than it can be replaced naturally; and in the United States, 90% of cropland is losing its soil faster than it can be replaced.

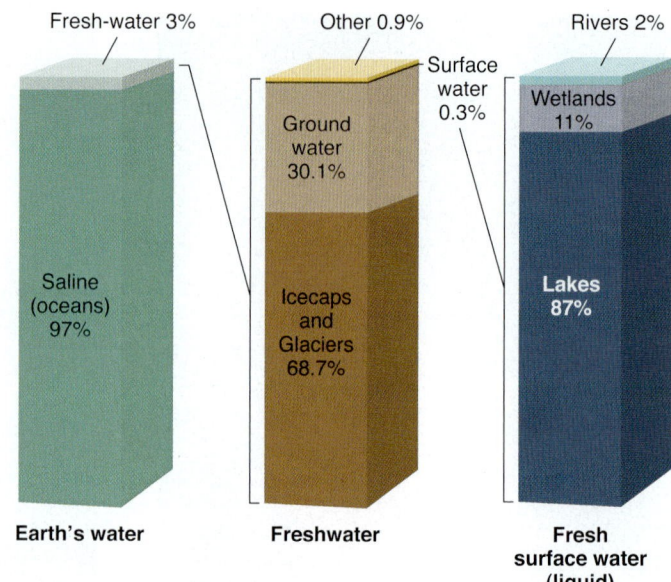

FIGURE 1.23 About 97% of all water on Earth is in the oceans, and only 3% is freshwater. The majority of freshwater, about 69%, is locked up in glaciers and icecaps (mainly in Greenland and Antarctica). Of the remaining freshwater, almost all of it lies below our feet, as groundwater. Of all the freshwater on Earth, only about 0.3% is contained in rivers, wetlands, and lakes.

Q Does the water you drink come from surface environments or from groundwater?

Mint Images - Frans Lanting/MintImages/Getty Images

FIGURE 1.24 Researchers report that 65% of the soil on Earth is degraded by erosion, desertification, and salinization.

Q How would you manage soil erosion?

Fossil Fuels

Our major means of generating electricity, by burning fossil fuels such as coal, oil, and natural gas (**FIGURE 1.25**), pollutes air and water. In addition, carbon dioxide released while burning these fuels is changing Earth's climate (Chapter 14).

Since the 1970s, far more oil has been consumed than has been discovered. Humans now burn 1,000 barrels of oil per second. Yet we are finding only 1 barrel of new oil for every 4 barrels that we consume. Many geologists believe that all the significant oil fields have already been discovered, and they predict that a permanent downturn in oil production will occur in less than 20 years.

Oil prices are very sensitive to small changes in supply, so when the rate of production begins to decline, the combination of growing demand and falling production will bring about a marked increase in oil prices. As a result, gasoline, which has long been cheaper than bottled water, may become unaffordable. Indeed, some economists indicate that this pattern has been set in motion.

The global energy market has thus far reacted to reductions in oil availability by flexibly shifting to natural gas or coal as sources of energy. In fact, from one year to another, as natural disasters or economic impacts cause the production or shipping of one type of energy source to decrease, other types of energy are used to make up for the deficit. The flexibility and integrated nature of the global energy market is such that impacts to one type of energy source are compensated by the scaling up of another type. Nonetheless, it is widely recognized that because of limitations and restrictions in fossil fuel availability, the cost of fossil fuel power will likely increase in the future.

Minerals

Every year each person in the United States uses, on average, 18,000 kilograms of rocks and minerals, including stone, gravel and sand, clays, limestone for cement, salt, and various metals, such as iron, zinc, aluminum, copper, silver, and many others.

These mineral resources are used to build, among many other things, highways, cars, computers, medical equipment, and buildings (**FIGURE 1.26**). But mineral resources are finite, and increased demand can, over time, make them unaffordable. Geologists are responsible for finding and assessing the precious minerals we depend on, and for identifying alternatives when necessary.

The feature "Applying Geology in Our Lives" contains an exercise intended to prompt thought about geologic resources and geologic hazards.

EXPAND YOUR THINKING

Describe the geologic resources that are prevalent where you live."

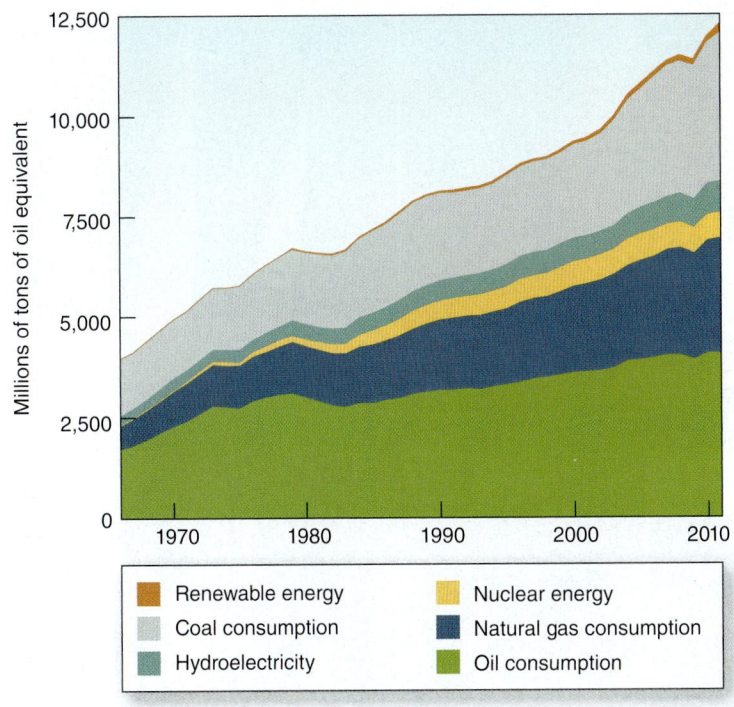

FIGURE 1.25 The global demand for energy—measured in millions of tons of oil equivalent (MTOE)—increases every year by about 2.5%. Oil is the world's predominant energy source, followed by coal and natural gas. These fossil fuels account for approximately 87% of all energy use.

Is the current mix of energy sources sustainable for the next 100 years? Explain your answer.

Michael Melford/National Geographic Stock

FIGURE 1.26 Southwest of Salt Lake City, Utah, the Bingham Canyon Mine is an open-pit mining operation extracting a large mineral deposit. In production since 1906, mining here has resulted in the creation of a pit over 0.97 kilometers deep, 4 kilometers wide, and covering 770 hectares. It was designated a National Historic Landmark in 1966.

What is one effective way to conserve mineral resources?

CRITICAL THINKING

Applying Geology in Our Lives

Throughout this chapter, we have emphasized the fact that geology is relevant in our lives. The images in **FIGURE 1.27** depict a number of common geologic resources and geologic hazards. Working with a partner, complete the following exercise:

1. Classify each image in Figure 1.27 as either "geologic hazard" or "geologic resource."

2. What criteria would you use to rank the *value* of each geologic resource in our lives?

3. Describe how answering question 2 led you to think of new ways to value resources.

4. For each hazard shown in Figure 1.27, describe how humans might decrease the damage they cause in our communities.

5. Write a short description of how each image in the figure depicts the role of geology in our lives.

6. What additional geologic hazards and geologic resources can you think of that are not depicted here?

tristan tan/Shutterstock

jordache/Shutterstock

STEPHEN ALVAREZ/National Geographic Stock

NOAA

© Tetra Images/Corbis

© Jeff Greenberg / Alamy

FIGURE 1.27 The role of geology in our lives.

Nightman1965/Shutterstock

Don Becker/USGS

Aaron Amat/Shutterstock

© acilo/iStockphoto

JAMES TAVARES/AFP/GettyImages

© Dleonis/iStockphoto

Lester Lefkowitz/Iconica/GettyImages

Rainer Albiez/Shutterstock

1-9 The Geologic Time Scale Summarizes Earth's History

LO 1-9 Discuss the geologic time scale.

Everyone has heard of dinosaurs, but *Tyrannosaurus rex* and its cousins are only one group of animals among hundreds of thousands (and millions of insects) that have come and gone in the course of Earth's 4.6 billion-year history.

Uniformitarianism

Until the late 1700s, Earth was believed to be only a few thousand years old, and all the species on it were thought to have been created simultaneously. The fact that Earth is much, much older was recognized by, among others, geologist *James Hutton* (1726–1797), who based his claim on a close analysis of fossils (**FIGURE 1.28**) and rock strata.

Based on his studies, Hutton proposed one of the most important concepts in geology: the **principle of uniformitarianism.** This principle states that Earth is very old, that natural processes have been essentially uniform through time, and that the study of modern geologic processes is useful for understanding past geologic events.

Uniformitarianism is often summarized by the statement, "The present is a key to the past." In Hutton's day, this idea was a direct rejection of the belief that Earth's landscape and history were the product of the great flood described in the Bible. Today, geologists hold uniformitarianism to be true, and recognize natural disasters such as earthquakes, asteroid impacts, volcanic eruptions, and floods as normal Earth processes.

The Geologic Time Scale

Geologists have created a calendar of events that covers Earth's entire history, called the **geologic time scale**. Like any other calendar, it is divided into periods of time. However, since days, weeks, and months have very little meaning over millions of years, the geologic time scale is divided into longer units of time, which vary in length: *epochs* (tens of thousands to millions of years), *periods* (millions to tens of millions of years), *eras* (tens of millions to hundreds of millions of years), and *eons* (hundreds of millions to billions of years). **TABLE 1.1** provides an overview of how these time units are organized.

By the end of the semester, you will know a great deal about how Earth works. This means you will then be able to use your knowledge to make better decisions about conserving geologic resources, avoiding geologic hazards, and managing natural environments.

Q EXPAND YOUR THINKING

The earliest human settlements were built, and intensive agriculture undertaken, about 10,000 years ago. Calculate the percentage of total geologic time this represents.

FIGURE 1.28 *Archaeopteryx,* considered to be the first bird, was a transitional form between birds and reptiles that lived approximately 150 million years ago.

Q What evidence supports the conclusion that this was a birdlike creature?

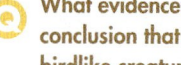

Francois Gohier/Photo Researchers,Inc.

TABLE 1.1 Overview of the Geologic Time Scale

Began Years Ago	Notable Events	Epochs	Periods	Eras	Eons
10 thousand	Modern epoch	Holocene	Quaternary	Cenozoic	Phanerozoic
2.6 million	Global cooling	Pleistocene			
23 million	Earliest Hominids, first apes		Neogene		
65 million	Early modern mammals		Paleogene		
	Extinction of dinosaurs Flowering plants abundant Modern sharks		Cretaceous	Mesozoic	
	First birds and lizards Dinosaurs abundant		Jurassic		
251 million	First dinosaurs First mammals and crocodilians		Triassic		
	Major extinction of many life forms Beetles and flies evolve Reef ecosystems flourish Pangaea forms		Permian	Paleozoic	
	First reptiles and coal forests		Pennsylvanian		
	First land vertebrates Early sharks		Mississippian		
	First insects and amphibians First ferns and seed-bearing plants		Devonian		
	First land plants First jawed fishes and vascular plants		Silurian		
	First green plants and fungi on land		Ordovician		
542 million	First abundant fossils		Cambrian		
2.5 billion	Oxygenated atmosphere				Proterozoic
3.8 billion	Earliest life (single-celled algae), oldest crust				Archean
4.6 billion	Oldest mineral grain, oldest asteroids and Moon rocks				Hadean

LET'S REVIEW

By the end of the semester, you will know a great deal about how Earth works. This means you will then be able to use your knowledge to make better decisions about conserving geologic resources, avoiding geologic hazards, and managing natural environments.

After you graduate, you may become a businessperson, a teacher, an artist, an engineer, or any of a multitude of other professionals. But regardless of the career you choose, you will take with you the knowledge that we have only one planet, and the realization that humanity is its caretaker.

We only have one Earth, and the best tool to keep it clean, safe, and livable is to use *critical thinking*. By applying the skills of critical thinking, we stand the best chance of meeting the global challenges of our time and overcoming difficulties that may stand between this important goal and us.

STUDY GUIDE

1-1 Geology is the scientific study of Earth and the other planets.

- **Geology** is the scientific study of Earth and the other planets: the materials they are made of, the processes that act on those materials, and the products that are formed as a result. Geology includes the study of Earth's history and the life-forms that have emerged on Earth since its origin.

- Geological knowledge contributes to the production of the materials that are used in building modern society and constructing the everyday conveniences that we rely on in our lives.

- Dynamic interactions occur among rocks, water, gases, sediments, and living plants and animals. These interactions are powered by solar energy, gravity, and heat from Earth's interior. That means Earth's surface is actively evolving and that natural materials are being recycled and renewed across a broad range of time and space.

- Geology fosters recognition of the importance of conserving geologic resources, as well as the human need to use those resources in producing material goods. Learning to carefully use and manage resources, and planning for their eventual depletion and replacement by alternatives, are important aspects of the practice of modern geology.

1-2 Critical thinking is the use of reasoning to explain the world around us.

- Science is conducted with the help of **critical thinking,** which requires that any explanation for a process or a phenomenon must be testable. This **scientific method** consists of the use of observations and experiments to build and refine **hypotheses** that explain phenomena.

- Scientists use several types of reasoning to solve problems, including inductive reasoning, deductive reasoning, and parsimony.

1-3 Global challenges will be met by critical thinkers with an enduring understanding of Earth.

- The goal of this geology class is to instill in your mind an enduring understanding of Earth. By doing so, we hope to encourage you to make decisions that take into account the need to live in sustainable ways on our planet, whose resources are limited.

- By becoming trained in geology, you can make better decisions about the use of precious **geologic resources,** and know how to avoid **geologic hazards.** Global challenges faced by a population with dwindling geologic resources, overwhelming geologic hazards, and worsening pollution will be solved by critical thinkers who have an enduring understanding of Earth.

- Six concepts provide an overarching, big picture of Earth: 1) The study of geology encompasses a vast range of time and space. 2) Earth materials are recycled over time. 3) **Plate tectonics** control the geology of Earth's surface. 4) Geologic systems are the product of interactions among the solid Earth, water, the atmosphere, and living organisms. 5) Earth continuously changes. 6) Rocks and sediments are pages in the book of Earth's history.

1-4 The theory of plate tectonics is a product of critical thinking.

- Alfred Wegener deduced that the continents have shifted positions over Earth's history. His evidence included similar rock formations and fossils on continents separated by wide ocean basins. His hypothesis of continental drift was never widely accepted, however, because it failed to provide a physical process describing how continents move across Earth's surface.

- The theory of plate tectonics describes **spreading centers** (where new crust is manufactured), **subduction zones** (where old crust is recycled), and moving plates of crust that act like conveyor belts carrying continents and the seafloor across Earth's surface. Plate movement was originally described by Harry Hess as the result of convection in the **mantle.**

- Plate tectonics provide an explanation for chains of volcanic islands **(hotspots),** the past existence of the supercontinent **Pangaea,** and the topography of **rift valleys** at spreading centers.

1-5 The topography of the oceans and continents reflects tectonic processes.

- Mid-ocean ridges, trenches, and continental mountain ranges usually mark the location of past or present plate boundaries where the crust is either separating (divergent boundary), or colliding (convergent boundary).

1-6 Rock is a solid aggregation of minerals.

- **Rock** is a solid aggregation of minerals. All types of rock can be grouped into three distinct families according to how they were formed: **igneous, sedimentary,** and **metamorphic.** These families do not coexist independently; rather, they interact and mix and are recycled into new forms through the **rock cycle.**

- Igneous rock is composed of cooled molten rock. Sedimentary rock is made of sediment that collects and becomes solidified into rock. Metamorphic rock is made of recrystallized minerals that are formed under relatively high-temperature and high-pressure conditions while the rock remains solid.

1-7 Geologists study dangerous natural processes known as geologic hazards.

- Geologists study and monitor landslides, floods, erosion, volcanic eruptions, storms, earthquakes, hurricanes, and other **geologic hazards.**

- When human land use interferes with natural processes, the results may damage property and harm the environment.

- Population growth has exposed more people than ever before to dangerous geologic hazards.

1-8 Geologic resources are not limitless and so must be managed.

- **Geologic resources** are materials that occur in nature and are essential or useful to humans, such as water, air, building stone, topsoil, minerals, and others.

- Although water is a renewable resource, it takes time for rain and snow to replenish the water we pump out of the ground. Moreover, water-rich rock and sediment layers can be damaged by overpumping, such that they are no longer useful.

- Researchers report that 65% of the soil on Earth is degraded by erosion, desertification, and salinization. Soil quality must be protected because soil provides us with food and sustains natural ecosystems.

- Our major means of generating electricity, by burning fossil fuels such as coal, oil, and natural gas, pollutes air and water. In addition, carbon dioxide released while burning these fuels is changing Earth's climate.

- Every year each person in the United States uses, on average, 18,000 kilograms of rocks and minerals, including stone, gravel and sand, clays, limestone for cement, salt, and various metals, such as iron, zinc, aluminum, copper, silver, and many others.

1-9 The geologic time scale summarizes Earth's history.

- Geologists have developed a **geologic time scale** that divides time into eons, eras, periods, and epochs. Each time interval has unique characteristics in terms of living organisms and geologic events that are often related to plate tectonics.

- The **principle of uniformitarianism** is a unifying concept in the geosciences that states that Earth is very old, that natural processes have been uniform through time, and that the study of modern geologic processes is useful in understanding past geologic events. Uniformitarianism is often summarized by the statement, "The present is a key to the past."

KEY TERMS

convection (p. 12)
critical thinking (p. 4)
crust (p. 8)
geologic hazards (p. 18)
geologic resources (p. 20)
geologic time scale (p. 24)
geology (p. 4)
historical geology (p. 4)
hotspots (p. 12)
hypothesis (p. 6)
igneous rock (p. 15)

mantle (p. 12)
mass wasting (p. 19)
metamorphic rock (p. 15)
mid-ocean ridge (p. 14)
natural law (p. 7)
Pangaea (p. 11)
physical geology (p. 4)
plate tectonics (p. 9)
principle of uniformitarianism
  (p. 24)
rock (p. 15)

rock cycle (p. 8)
scientific method (p. 6)
seafloor spreading (p. 11)
sedimentary rock (p. 15)
subduction zone (p. 12)
sustainable resource (p. 4)
theory (p. 6)

ASSESSING YOUR KNOWLEDGE

Please complete this exercise before coming to class. Identify the one best answer to each question.

1. We emphasize learning the most important ideas of geology in this text because:
   a. Your knowledge of geology can guide your decision making throughout your life.
   b. As an adult, it is important that you understand resource sustainability and hazard mitigation.
   c. An understanding of Earth can help you make good decisions in the voting booth.
   d. Geology is important in our daily lives.
   e. All of the above.

2. Study the activities shown in the following images. In the blanks next to each description, use one of three options to describe whether it exploits geologic resources: 1) at the same rate, 2) faster, or 3) slower than they are naturally renewed.

Fresh water use _____

Pumping oil _____

Soil erosion _____

Gold mining _____

3. Physical geology is the study of:
   a. The materials that compose Earth and the ways in which they are organized and distributed throughout the planet.
   b. The history of life on Earth.
   c. The nature of how humans and animals interact and influence one another.
   d. Fossils and past environments.
   e. None of the above.

4. A hypothesis must be testable so that:
   a. It can be objectively evaluated.
   b. Its accuracy can be independently assessed.
   c. Its viability to make accurate predictions can be appraised.
   d. It is possible to revise the hypothesis if it does not provide accurate predictions.
   e. All of the above.

5. Geologists "read" the evidence of past events in earth's history by analyzing rocks and sediments. What event does the outcrop in this image reveal? _____

6. Which of the following statements is most likely to be proven using the scientific method?
   a. My dog is better than your dog.
   b. Bacteria cause tooth decay.
   c. That is a beautiful flower.
   d. Apples taste better than bananas.
   e. All of the above.

7. The rock cycle is:
   a. A concept describing how rocks are moved in running water.
   b. A concept describing the fact that rocks roll downhill.
   c. A concept proposing that rocks are naturally recycled.
   d. A concept predicting that rocks eventually return to their place of origin.
   e. None of the above.

8. Subduction occurs:
   a. When one plate slides beneath another.
   b. When a plate is recycled into Earth's interior.
   c. When a plate enters the mantle at a deep-sea trench.
   d. When seafloor is recycled beneath an overriding plate.
   e. All of the above.

9. Hess's mechanism for moving plates was:
   a. Continental drift.
   b. Ocean currents.
   c. Mantle convection.
   d. Volcanic eruptions.
   e. None of the above.

10. Place the following labels where they belong on this cross section of plate tectonics. subduction zone, deep-sea trench, seafloor spreading, mantle, rift valley, magma.

11. Mantle convection:
    a. Was proposed by Hess as the driver of seafloor movement.
    b. Is the product of heat flow within Earth's interior.
    c. Is thought to resemble the action in a lava lamp.
    d. Causes rifting and seafloor movement between spreading centers and subduction zones.
    e. All of the above.

12. The process of seafloor spreading predicts that:
    a. The oldest seafloor will be found at subduction zones.
    b. Seafloor moves like a conveyor belt from spreading centers toward subduction zones.
    c. There is high heat flow from Earth's interior into the crust at spreading centers.
    d. Continents shift locations because they are embedded in moving plates.
    e. All of the above.

13. Pangaea is:
    a. The name for ancient Greece.
    b. The name of an ancient supercontinent.
    c. The Greek name for plate tectonics.
    d. The concept that rocks are recycled at subduction zones.
    e. The portion of crust that is dry land.

14. Plates influence Earth's topography because:
    a. Mountain belts form where plates collide.
    b. Mid-ocean ridges develop at spreading centers.
    c. Trenches develop at subduction zones.
    d. Rift valleys mark sites of plate divergence.
    e. All of the above.

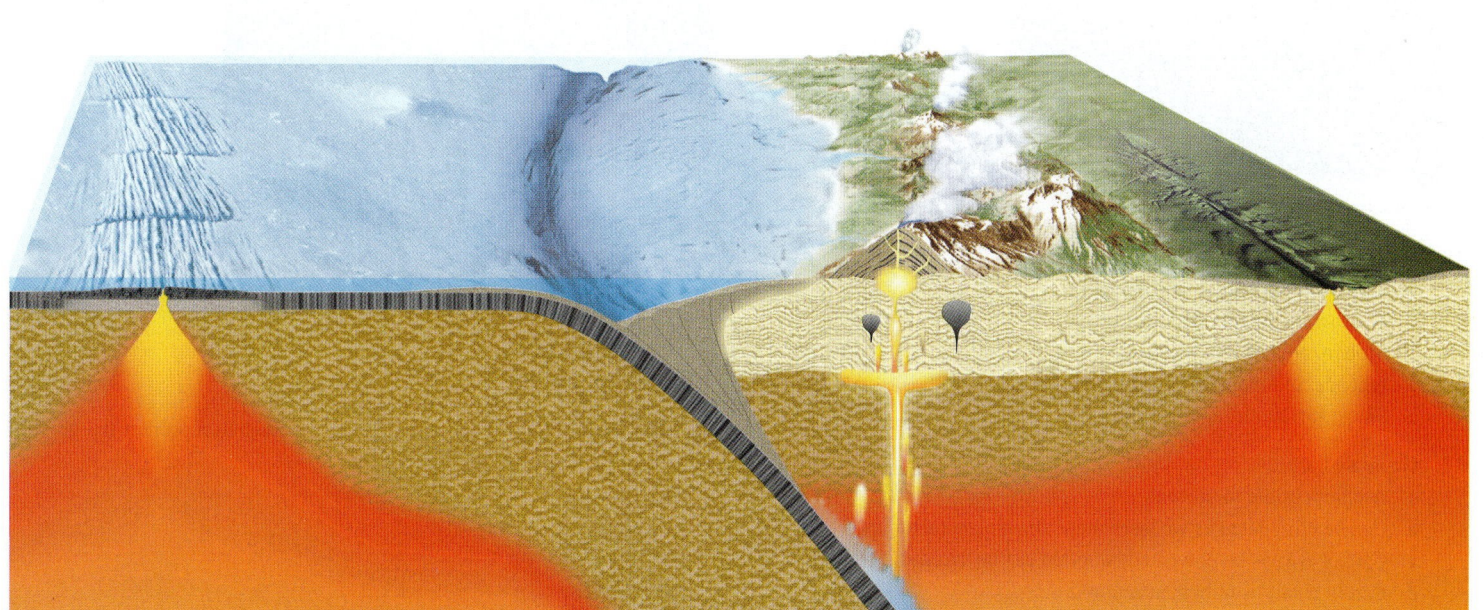

15. Label the following rocks as igneous, sedimentary, or metamorphic.

_____

_____

Mineral images courtesy of Chip Fletcher

_____

16. The three rock groups are:
  a. Oceanic, continental, and Pangaea.
  b. Weathered, eroded, and contorted.
  c. Molten, crustal, and recycled.
  d. Igneous, sedimentary, and metamorphic.
  e. Shale, sandstone, and basalt.

17. Geologic hazards:
  a. Are impossible to prevent.
  b. Can be eliminated as we learn more about plate tectonics.
  c. Cannot always be prevented but often can be avoided.
  d. Rarely result from human actions.
  e. Tend to cluster in distant areas.

18. Describe the geologic hazards that threaten development in each of the following images.

Brett R. Henry/Alamy

_____

David Greedy/Getty Images, Inc.

_____

Reuters/Landov LLC

_____

19. Common geologic resources studied by geologists include:
  a. Fossil fuels.
  b. Freshwater.
  c. Soil.
  d. Building materials.
  e. All of the above.

20. The subdivisions of the geologic time scale that represent the greatest expanse of time are called:
  a. Epochs.
  b. Eras.
  c. Eons.
  d. Periods.
  e. Eternity.

FURTHER LEARNING

1. Search the Web for a description or story about a geologist. What does he or she do? Make contact by email and ask why he or she became a geologist, and what he or she likes about their job.

2. Why do geologists study geologic hazards if we usually cannot prevent them?

3. From your perspective, what is the reason for learning about science?

4. Why are fossils important markers for understanding geologic time?

5. Give an example of how you solved a problem using inductive reasoning. Give an example of how you used deductive reasoning.

6. Describe what it means when scientists say that a hypothesis must be testable.

7. List two geologic resources. What happens when they are poorly managed? How can they be managed effectively?

ONLINE RESOURCES

Explore more about geology on the following Web sites:

USGS Undergraduate education page:
education.usgs.gov/common/undergraduate.htm

ScienCentral stories and videos:
sciencentral.com/index2.php3?nav=phy

See the latest scientific findings announced at:
Science Daily Web site:
www.sciencedaily.com/news/earth_climate/geology

BP Annual Energy Statistical Review...
http://www.bp.com/sectionbodycopy.do?categoryId=7500&content Id=7068481...and video: http://clients.world-television.com/bp_media_ coverage/GB4415/index.html

Additional animations, videos, and other online resources are available at this book's companion Web site:
www.wiley.com/college/fletcher

This companion Web site also has more information about WileyPLUS and other Wiley teaching and learning resources.

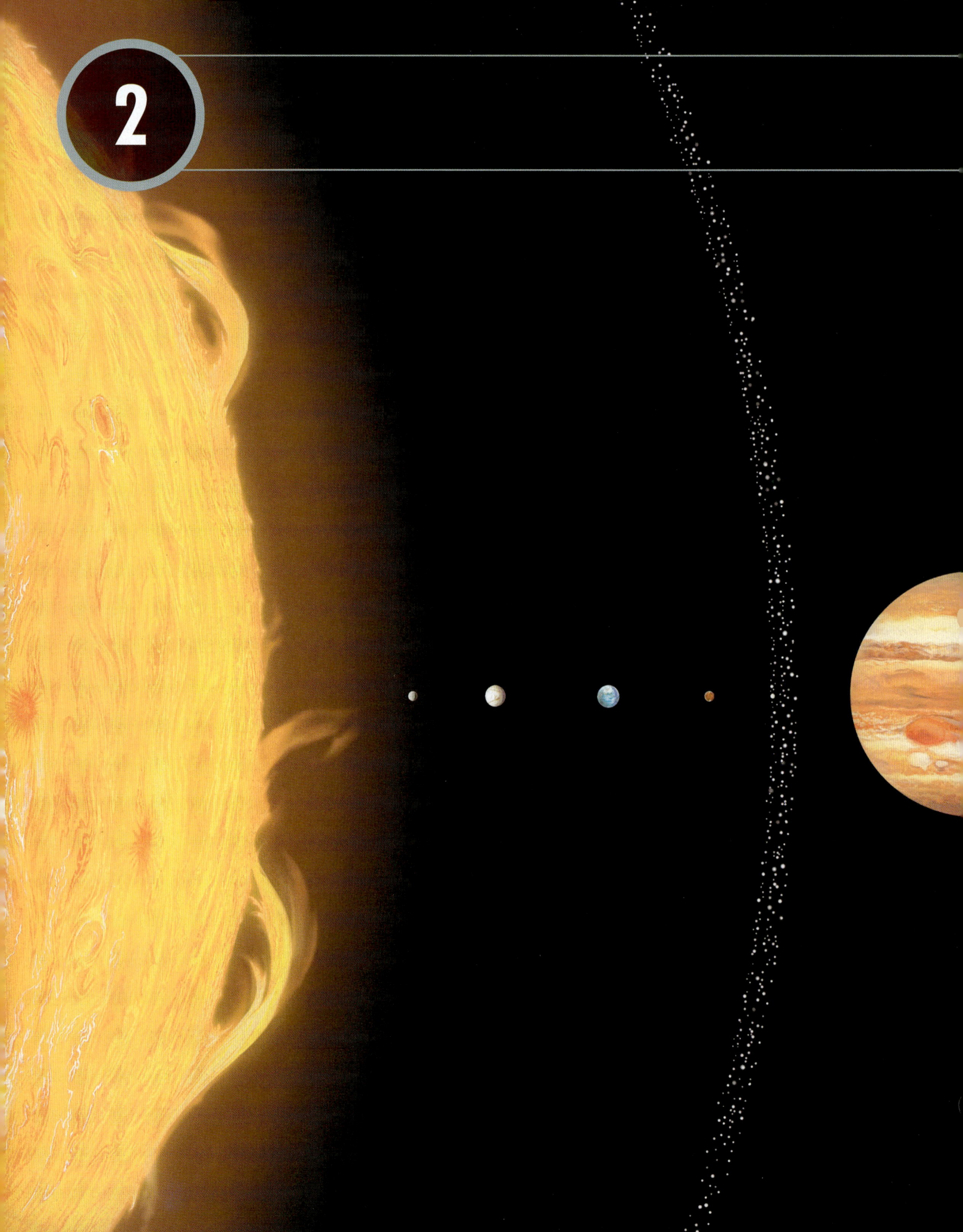

SOLAR SYSTEM

GEOLOGY IN OUR LIVES

Earth, the Sun, and other objects in the solar system originated at the same time and from the same source, and have evolved in varying ways since then. The solar system began with the collapse and condensation of a planetary nebula. Understanding Earth's place in the cosmos gives us a perspective on and appreciation for the scale of nature. Our exploration of the solar system is one of humanity's grand accomplishments. Someday our planetary neighbors may provide us with useful resources.

Why do planets close to the Sun, such as Earth, have thin atmospheres, and those far from the Sun, such as Jupiter, have thick atmospheres?

2-1 Earth's Origin Is Described by the Solar Nebula Hypothesis

LO 2-1 Explain the solar nebula hypothesis.

Earth and the other planets have a common origin that is described by the **solar nebula hypothesis.** The solar nebula hypothesis states that the Sun, the planets, and other objects orbiting the Sun (**FIGURE 2.1**) originated at the same time and from the same source through the collapse and condensation of a *planetary nebula* (a great cloud of gas made by an exploding star), and have evolved in varying ways since that time.

The Solar Nebula Hypothesis

Support for the solar nebula hypothesis is found in the observation that the planets orbit on nearly the same plane and in the same direction around a common focus, the Sun. Differences among the planets and other solar system objects that are not explained by the solar nebula hypothesis are attributable to events that have happened since the origin of these objects.

The **solar system** consists of these objects:

- The Sun

- Eight "classical" planets: Mercury, Venus, Earth, Mars, Jupiter, Saturn, Uranus, and Neptune

- Five "dwarf" planets (Pluto, Ceres, Haumea, Makemake, and Eris)

- Small solar system bodies (including asteroids, comets, and objects in the Kuiper belt, scattered disk region, and Oort cloud)

- Two hundred forty known satellites (moons), including 162 orbiting the classical planets

- Countless particles and interplanetary space

A star is a celestial body of hot gases that radiates energy generated by nuclear reactions in the interior. The Sun is a star located at the center of the solar system. It is huge; if the Sun were the size of a basketball, Earth would be the size of the head of a pin—a mere speck.

The **Milky Way galaxy** contains billions of other stars besides the Sun, most smaller than the Sun, which is in the top 10% of stars by mass—it contains over 99% of the solar system's mass.

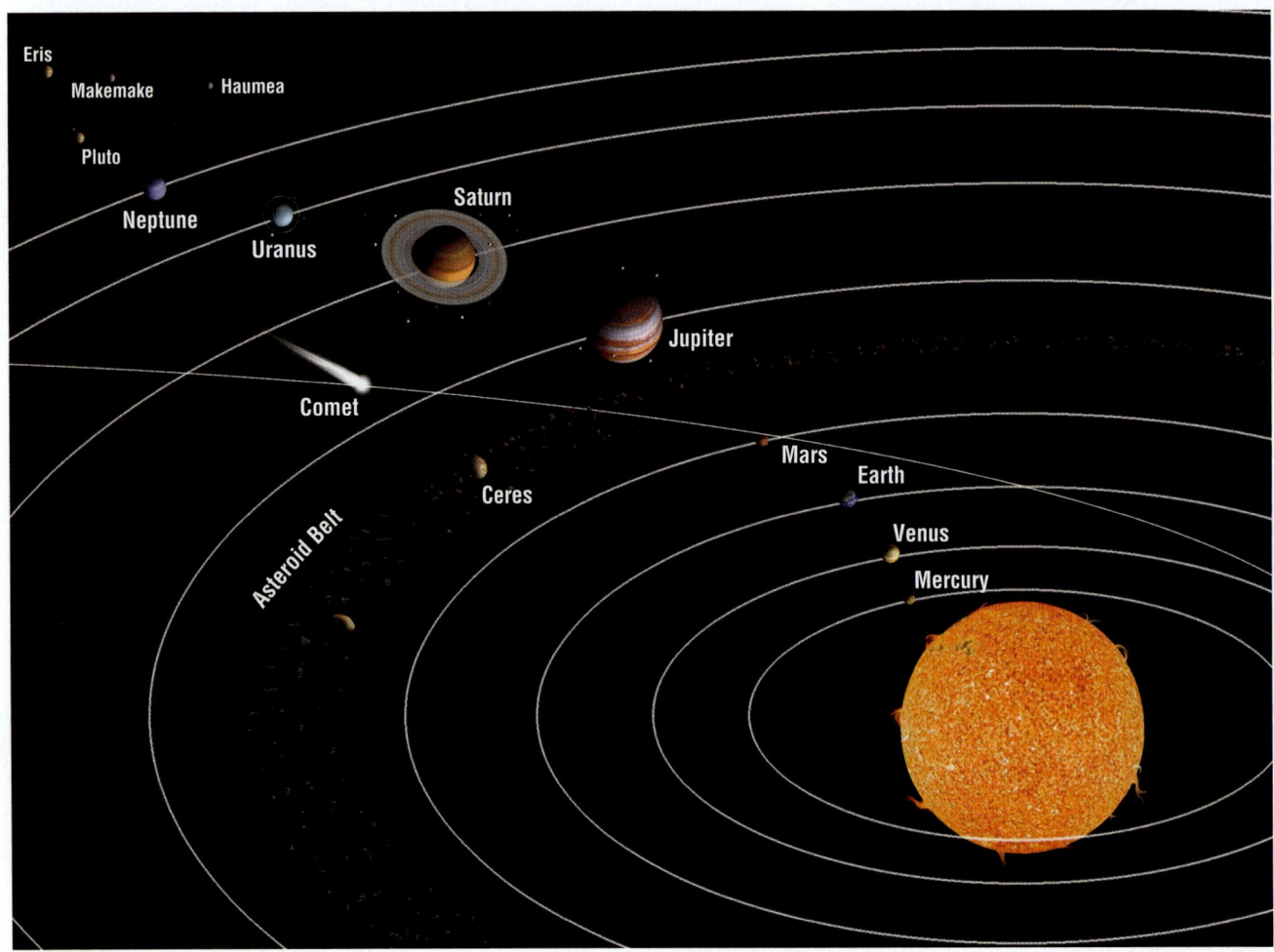

FIGURE 2.1 The solar system.

Q What is the source of the elements and compounds that compose the solar system?

The median size of stars in our galaxy is probably less than half that of the Sun.

Everything in our solar system orbits around the Sun, on a path that balances the Sun's gravity with the *centrifugal force* (an outward tug such as you feel when you're in a car going around a curve) acting on the object. As shown in Figure 2.1, the classical planets lie nearly on a single plane, called the *ecliptic*, with the Sun at the center. All the planets except Venus rotate, or spin on their axes, in the same direction.

Venus rotates in the opposite direction, for reasons about which scientists are unsure. Perhaps it suffered a huge collision that tipped it nearly upside down. Uranus also has a steeply tilted axis, thought to be the result of a collision with a large object early in its history. All the classical planets orbit the Sun in the same direction.

According to the solar nebula hypothesis, about 5 billion years ago a swirling nebula (**FIGURE 2.2**), a cloud of interstellar gas and dust, began to collapse inward under the pull of gravity. Shock waves from the explosion of a nearby star, or some other disturbance, may have initiated this collapse.

As the volume of the cloud decreased, its density and rate of rotation increased—a phenomenon similar to that which occurs when ice skaters pull in their arms in order to spin faster. As the spinning accelerated, the cloud flattened and developed a hot core, giving birth to an infant star, our future Sun.

The cloud continued to collapse as material streamed into its center. Climbing temperatures vaporized dust, and the resulting gases joined the other gases feeding the growing core. Astrophysicists calculate that it took less than 100,000 years for the nebula to evolve into a hot "protostar," fed by gases flowing inward. Simultaneously, centrifugal force prevented some gas and dust from reaching the core of the nebula. This material formed a cooler outer region where rocky

© Don Dixon

FIGURE 2.3 Planetesimal accretion.

Q **Write a short description of the planetesimal accretion process.**

particles and gases coalesced to become our system of planets. This process of planet growth is termed **planetesimal accretion.**

The original nebula was composed mostly of simple atoms of hydrogen (H) and helium (He), the two simplest and lightest elements, in gaseous form. It also contained heavier atoms, such as oxygen (O), carbon (C), silicon (Si), iron (Fe), and additional metals and other elements, some of them in the form of dust particles.

As the new star took shape in the center of the disklike formation, matter also coalesced in the outer regions of the disk. This was a critical phase in the development of the solar system. Earth and other planets would soon be formed in an area of the disk that was enriched with many of the elements we rely on: iron for steel, calcium (Ca) for cement, oxygen to form important types of compounds, silicon for glass and computer chips, and others.

The outer region of the disk cooled rapidly, with the result that metal, rock, and ice condensed as tiny particles. These particles grew by colliding with one another, forming larger masses of rock that eventually developed gravitational fields. Gravity gave larger particles an advantage over smaller ones, and their growth accelerated as they pulled in additional solid and gaseous matter. These objects grew still further, becoming *planetesimals,* small, solid celestial bodies. Planetesimals, in turn, coalesced to form *planetary embryos,* a process that unfolded over a period of hundreds of thousands of years. Finally, over the course of another few million years, planetary embryos combined to form true planets (**FIGURE 2.3**).

Today, there are two major groups of planets in the solar system: the four **terrestrial planets** (Mercury, Venus, Earth, and Mars) and the four **gas giants** (Jupiter, Saturn, Uranus, and Neptune). The asteroids orbit the Sun in a belt beyond the orbit of Mars; made mostly of rock and iron, they are probably debris from the planetary nebula that never coalesced.

NASA, ESA, J. Hester and A. Loll (Arizona State University)

FIGURE 2.2 The solar nebula hypothesis proposes that the solar system originated from a cloud of interstellar gas and dust (the solar nebula). This is a photo of the Crab Nebula, a supernova remnant, all that remains of a tremendous stellar explosion. Observers in China and Japan recorded the supernova nearly 1000 years ago, in 1054.

Q **EXPAND YOUR THINKING**

Describe how the solar system would be different if the original solar nebula did not rotate.

2-2 The Sun Is a Star That Releases Energy and Builds Elements Through Nuclear Fusion

LO 2-2 Describe the Sun and how it works.

The Sun is our star, and it dominates the characteristics and history of the solar system.

Nuclear Fusion

When the universe first came into existence some 14 billion years ago, it was composed only of hydrogen, helium, and small amounts of lithium (Li), beryllium (Be), and boron (B), the first five elements in the periodic table. Today, it consists of 90 naturally occurring elements, all of which have been discovered here on Earth. If there were only five elements to begin with, where did the remaining 85 come from? They were created through **nuclear fusion** in stars. In other words, stars are the source of most of the elements that make up our world.

To visualize the process of nuclear fusion, think of a car: The car's internal combustion engine converts gasoline into a mixture of gases that explodes to release energy and exhaust fumes. The Sun resembles an engine; but instead of being powered by internal combustion, it runs on nuclear fusion. Hydrogen is the fuel used in the process of nuclear fusion that powers the Sun and other stars. Over time, a star's supply of hydrogen will be used up. When that happens, other fuels must be used to power the fusion process.

The fusion process produces an explosive burst of energy that adds protons and neutrons to the nucleus of a hydrogen atom, thereby forming a heavier element: helium. When the concentration of hydrogen runs low, helium is used as the fuel for fusion. Helium fusion produces lithium, an even heavier element.

In the billions of years since the origin of the universe, nuclear fusion within stars has formed the rest of the heavier elements (elements with more protons in the nucleus) found in nature. Thus, when the concentration of helium fuel runs low, lithium is used, producing carbon and oxygen—which, in turn, produce still heavier elements. This process can be likened to burning the ashes of one fire as fuel for another. This is not the end of the story, however.

Planetary Nebulas

Stars die, or burn out, when they use up their nuclear fuel. When a star's concentration of hydrogen is too low to support the fusion process, it begins fusing helium in its core. At this point, it will be 1,000 to 10,000 times brighter than previously, becoming what is termed a *red giant*, a large, bright, unstable star whose hydrogen has run out.

Our Sun will eventually become a red giant that is likely to engulf Mercury, Venus, and, perhaps, Earth. Astrophysicists hypothesize that the Sun, which has now completed approximately half its life, will enter the red giant phase in 4 to 5 billion years. Red giants, like all other stars, will eventually burn out. If it is large enough, a red giant ends its life in a process in which its core collapses. The intense fusion taking place in the collapsing core

FIGURE 2.4 The Hubble Space Telescope took this image of hot gas fleeing a dying star 3800 light-years away in the Scorpius constellation. This planetary nebula is known as the Butterfly Nebula. What resemble butterfly wings are actually roiling cauldrons of gas heated to more than 20,000°C. The star itself, once about five times as massive as the Sun, is over 200,000°C, making it one of the hottest objects in the galaxy.

NASA, ESA, and the Hubble SM4 ERO Team

ⓔ Describe the source of heat in a star.

causes the outer gaseous layers to be ejected explosively into space. An exploding star is called a *nova*.

When more massive stars explode, they are called *supernovae*. The material ejected by a nova or supernova is composed of newly made elements, and the resulting field of debris—the stuff of which planets are made—is a *planetary nebula* (**FIGURE 2.4**). As described earlier, it was a planetary nebula that evolved into our solar system.

Our Sun: A Massive Hydrogen Bomb

The Sun is the most prominent feature of our solar system and the most massive, yet its interior structure is still somewhat of a mystery. The diameter of one hundred and nine Earths would be required to equal that of the Sun, and its interior could hold more than 1.3 million Earths. Like Earth, the Sun is composed of several layers that define its internal structure (**FIGURE 2.5**). Unlike Earth, however, it is completely gaseous and lacks a solid surface.

FIGURE 2.5 Energy is generated in the Sun's core, where temperatures reach 16,000,000°C.

NASA Goddard Space Flight Center

ⓔ Why is the Sun getting lighter with time?

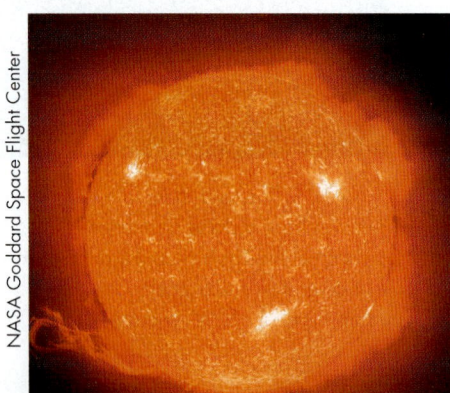

The Sun's extremely dense core accounts for about 50% of its total mass but only about 1.5% of its total volume. Physical conditions inside the core are extreme. The temperature is thought to be around 16,000,000°C and the pressure perhaps 250 billion times the air pressure at Earth's surface.

Under these intense conditions, atoms are stripped of their electrons. As a result, the Sun's core is a mixture of protons, neutrons, nuclei, and free electrons. Energy deep within the core is so intense that nuclear reactions take place there, causing the fusion of hydrogen nuclei to form helium nuclei.

The helium is about 0.7% less massive than the hydrogen used to manufacture it. The difference in mass is expelled as energy, which is carried to the surface, where it is released as light and heat. Every second, 700 million tons of hydrogen are converted into helium. In the process, 5 million tons of pure energy is released; over time, therefore, the Sun weighs less.

If we could see into the Sun's core, it would appear black, since none of the energy produced there is visible to our eyes. It mainly produces X-rays, which work their way slowly to the surface. As they move upward, the X-rays collide with subatomic particles and change direction in random ways.

Each X-ray may travel only a few millimeters before it experiences another collision and sets off in a different direction. The time to complete this journey out of the Sun's interior is measured in millions of years—an incredible fact, given that X-rays travel at the speed of light. To put it another way, sunlight leaving the Sun's surface came from a nuclear reaction that took place perhaps 1 million years ago deep within the core. Before we can see this radiation, we have to wait another 8 minutes, because that is how long it takes for it to travel the 150 million kilometers from the Sun to Earth.

The Sun is essentially an enormous hydrogen bomb in a constant state of explosion. But it does not explode outward. Instead, its immense gravity, which pulls its mass inward toward the core, keeps the explosion in check; at the same time, the outward pressure generated by the continuous nuclear reaction prevents it from collapsing under its own gravity. The Sun thus is in a state of *dynamic equilibrium* between exploding and collapsing, a balance of nature to which we owe our existence.

The Sun is, clearly, fundamental to life on Earth. If we wish to learn the potential for life elsewhere in the universe, we first need to know many stars there are in the universe. For the answer, read the feature titled "Critical Thinking: How Many Stars Are There?"

Ⓠ EXPAND YOUR THINKING

What would the result be if the force of gravity in the Sun were greater than the release of energy by nuclear fusion?

CRITICAL THINKING

How Many Stars Are There?

It is impossible to physically count every star in the universe (**FIGURE 2.6**). But what if you wanted to know how many stars actually exist? How could you find out? You could construct an estimate based on **extrapolation.** Extrapolation is a form of critical thinking that is commonly used to estimate things that cannot be directly observed. It consists of inferring or estimating an answer by projecting or extending a known value. Questions such as

How many cells are in the human body?

How many tuna are in the sea?

How many grains of sand are on the world's beaches?

can be answered using *extrapolation.*

So, how many stars are there? The answer to this question has been estimated using extrapolation.

FIGURE 2.6 How would you count the stars in the universe? This is a view into one of the more crowded places in the universe, a small region inside the globular cluster Omega Centauri, which has nearly 10 million stars. Globular clusters are ancient swarms of stars united by gravity. The stars in Omega Centauri are 10 to 12 billion years old.

Astronomers at Australia National University used two telescopes to count the stars in a single strip of sky. Within that one region, some 10,000 galaxies were pinpointed, and detailed measurements of their brightness were taken. These measurements were then used to calculate how many stars they contained. That number was multiplied by the number of similar-size strips it would take to cover the entire sky, then multiplied again out to the edge of the visible universe. The result was 70 sextillion, or 70,000,000,000,000,000,000,000. And as it turns out, this is about 10 times more than the total number (also calculated using extrapolation) of grains of sand on all the beaches in the world.

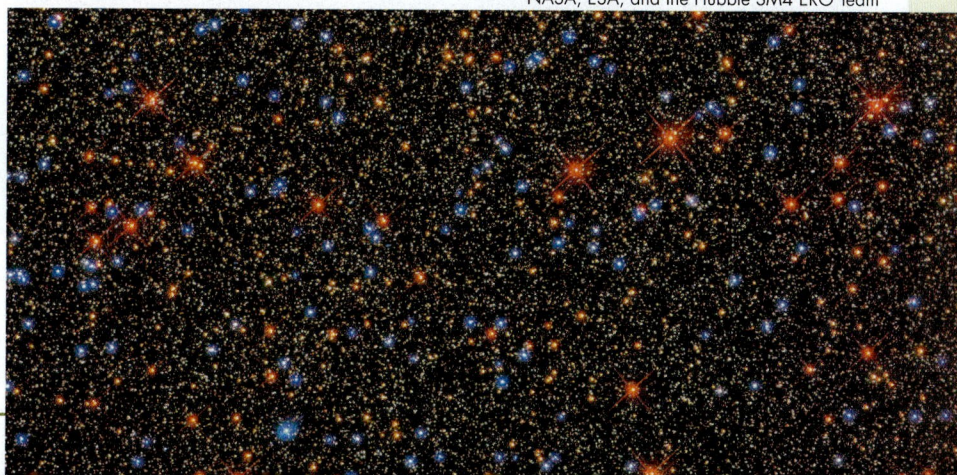

NASA, ESA, and the Hubble SM4 ERO Team

2-3 Terrestrial Planets Are Small and Rocky, with Thin Atmospheres

LO 2-3 State the ways that Mercury, Venus, and Mars are different from Earth.

The terrestrial planets include Mercury, Venus, Earth, and Mars. They are relatively small and consist primarily of compounds of silicon, oxygen, iron, magnesium (Mg), and other metals— potassium (K), calcium, sodium (Na), and aluminum (Al).

The Terrestrial Planets

The terrestrial planets were formed close to the developing Sun during the first 100 million years after its birth, when the heat was too great to allow them to accumulate much water (H_2O) or other *volatile gases* (easily evaporated gases).

These gases, which accumulated as thin primitive atmospheres on the terrestrial planets but are found in much larger quantities farther from the Sun, include **carbon dioxide** (CO_2), *methane* (CH_4), and *ammonia* (NH_3), all of which were abundant in the planetary nebula but too unstable to survive in great abundance close to the Sun.

At the same time, in the cooler outer portion of the solar system, the original nebula gases froze on some planets, and formed thick gaseous atmospheres on others (**FIGURE 2.7**). The accumulation of gases

allowed the outer planets to acquire massive gravitational power. This gravitational force enabled them to hold onto huge quantities of helium and hydrogen, which were still present in the outer region of the condensing nebula. They thus became gas giants.

Each gas giant has several moons, covered with thick icy surfaces made largely of water, ammonia, and carbon dioxide. Many scientists think that conditions on some of these moons might be conducive to the formation of life.

While planetesimals were forming in the outer region of the condensing solar nebula, nuclear fusion began in the nebula's core. With fusion came a release of energy that stopped the contraction of the nebula. This event marked the birth of our Sun.

Fusion let loose a fiery storm of high-energy subatomic particles and radiation called the **solar wind.** The solar wind, still present today, swept away the gases left within the nebula and stripped the inner planets of their thin, primitive atmospheres. Mercury, Venus, Earth, and Mars likely became barren rocky spheres with no atmospheres and no surface water. Beyond Mars, the huge gravitational mass of the gas giants stabilized them and enabled them to retain their immense atmospheres.

Earth

Earth's atmosphere: nitrogen (N_2), 78%; oxygen (O_2), 21%; argon (Ar), 1%

In many ways, Earth (**FIGURE 2.8**) is unlike its neighbors in the solar system. Its hot interior supplies the energy to fuel volcanoes and drive internal convection, yet its surface is cool enough to support life.

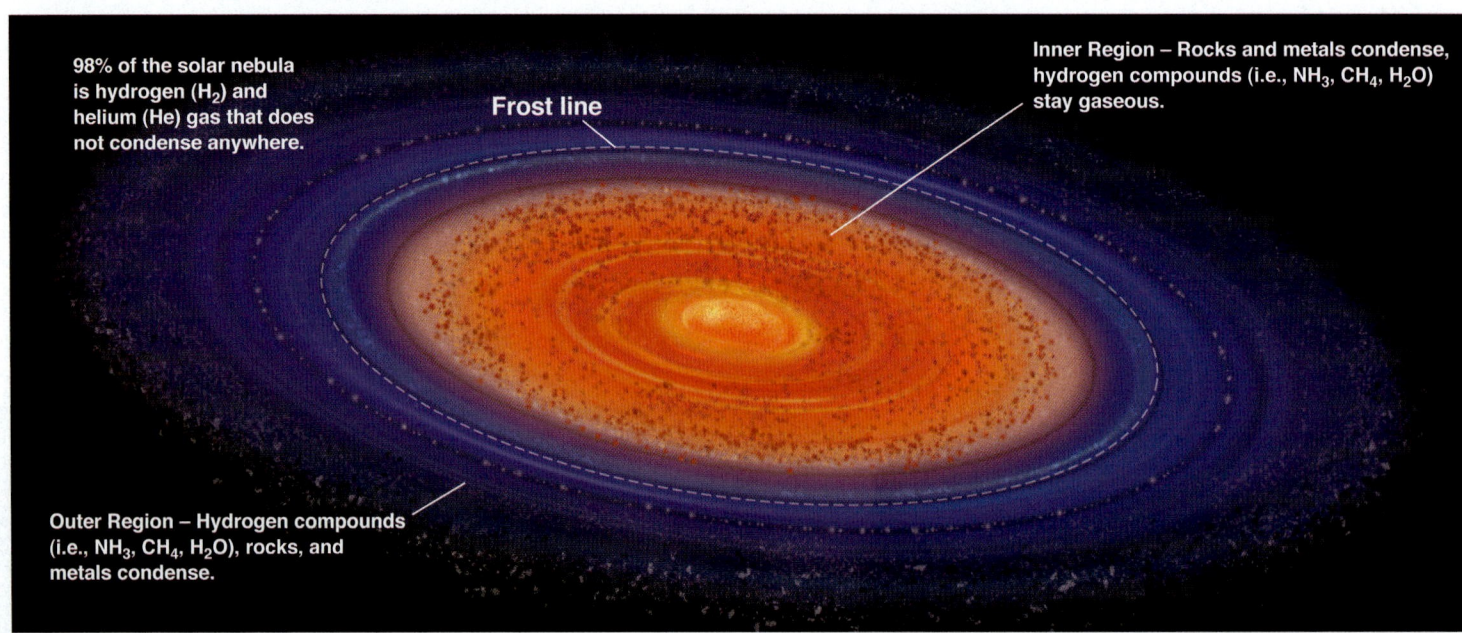

98% of the solar nebula is hydrogen (H_2) and helium (He) gas that does not condense anywhere.

Frost line

Inner Region – Rocks and metals condense, hydrogen compounds (i.e., NH_3, CH_4, H_2O) stay gaseous.

Outer Region – Hydrogen compounds (i.e., NH_3, CH_4, H_2O), rocks, and metals condense.

FIGURE 2.7 Rocky particles and metallic compounds formed solids in the inner portion of the condensing solar nebula. But it was too hot for hydrogen compounds to solidify. In the cooler outer region, hydrogen compounds, metals, and rocks all condensed into a solid state. The transition zone between the two regions is known informally as the "frost line".

Q Why do scientists use the term "frost line"?

FIGURE 2.8 Calling this "the most detailed true-color image of the entire Earth to date," NASA put together this mosaic using satellite-based data that shows every square kilometer of our planet.

This image clearly reveals one of Earth's most distinguishing features. What is it?

Earth's position relative to the Sun provides sufficient warmth to maintain liquid water, yet it still experiences freezing seasons. Earth's winds are relatively calm; the atmosphere is thin and transparent, so it is easily penetrated by light; and daily extremes in temperature are comparatively modest. All these conditions promote the existence of life, which are rare in the rest of the solar system. By taking a tour of Earth's "neighborhood" in space, we can find out just how unique our planet really is.

Mercury

Mercury's atmosphere: oxygen (O_2), 42%; sodium (Na), 29%; hydrogen (H_2), 22%; helium (He), 6%; and potassium (K), 0.5%

Five elements have been detected in minute quantities in Mercury's unstable atmosphere. Oxygen, sodium, and hydrogen are present at all times, whereas helium and potassium are found only at night, as during the day they are absorbed by rocks on the surface.

Mercury is the closest planet to the Sun, hence is dominated by unrelenting and intense heat during the day. But because it lacks a robust atmosphere to trap that heat, during the night temperatures plummet. Thus, surface temperatures may exceed 227°C on the sunlit side then chill to −137°C at night.

Mercury's surface resembles that of Earth's Moon in that it is heavily marked by impact craters and includes areas that appear to be *lava plains* (**FIGURE 2.9**). On one side of the planet is a large impact crater called the Caloris Basin. Exactly opposite, on the other side, is an area of unusual topography suitably called the "weird terrain," where the shock waves from the impact that created the Caloris Basin apparently converged and disturbed the surface of Mercury.

FIGURE 2.9 Mercury, the closest planet to the Sun, and the smallest, completes an orbit of the Sun once in 88 days.

This image clearly reveals Mercury's most distinguishing features. What are they?

Mercury's axis of rotation does not tilt as does Earth's. Therefore, craters at Mercury's poles may be in perpetual shadow and escape the planet's staggering temperature extremes. Radar mapping of Mercury's surface in 1991 revealed unusual signals from within the polar craters that may indicate the presence of ice. The possibility of a polar icecap so close to the Sun is certainly a surprise.

Venus

Venus's atmosphere: carbon dioxide (CO_2), 96.5%; nitrogen (N_2), 3.5%

Venus (**FIGURE 2.10**), the second planet from the Sun, is often referred to as Earth's sister because it is our immediate neighbor, and of all the planets is closest to Earth in size. Its atmosphere, however, makes Venus a radically different place.

Carbon dioxide, which is very efficient at trapping heat, makes up most of Venus's atmosphere. As a result, surface temperatures there reach a hellish 477°C—even hotter than Mercury's daytime temperatures. Global winds distribute heat around the planet and, combined with the thick atmosphere, ensure that nighttime and daytime temperatures are about the same. Deep, white clouds composed of sulfuric acid (H_2SO_4) fill the atmosphere, causing acid rain to fall continually on the planet's surface.

As Venus travels around the Sun, it rotates very slowly on its axis, only once every 243 Earth days. Venus's axis is very nearly perpendicular, although it may be upside down. This is because, unlike Earth, Venus does not rotate in the same direction it travels around the Sun. Rather, it rotates in the retrograde (opposite) direction—the only planet to do so. One explanation for this is that, perhaps, it was knocked over by a collision with another planetesimal during the accretion phase early in its history.

Mars

Mars's atmosphere: carbon dioxide (CO_2), 95.3%; nitrogen (N_2), 2.7%; argon (Ar), 1.6%; oxygen (O_2), 0.13%

Mars is the most Earth-like of the planets in the solar system. Although it is only half the size of Earth, it has many similar features, such as a nearly 24-hour day, clouds, layers of sediment, sand dunes, volcanoes, and polar icecaps that grow and recede with the changing seasons.

Mars has dry river channels, along with floodplains and watersheds, indicating that water flowed across its surface in the past, perhaps from springs. Physical features closely resembling shorelines, gorges, riverbeds, and islands suggest that the planet formerly possessed great rivers, lakes, and even seas (**FIGURE 2.11**).

NASA

FIGURE 2.10 Venus is the second planet from the Sun. It has a dense carbon dioxide atmosphere that traps heat so effectively that the surface is a scorching 477°C.

Q Why is Venus hotter than Mercury, even though Mercury is closer to the Sun?

NASA

FIGURE 2.11 Mars is the fourth planet from the Sun. This image reveals the *Valles Marineris*, a system of canyons that run along nearly a quarter of the planet's surface. Researchers have suggested that the feature is a large tectonic crack in the Martian crust that has been widened by erosion.

Q Mars has several features that are similar to Earth; can you name three?

Scientists hypothesize that after its formation, Mars cooled enough so that large volumes of water vapor collected as thick clouds in the atmosphere. The resulting rain fell onto the planet's surface, forming the numerous water-carved features we observe today. The lack of a protective ozone layer meant that water molecules were easily destroyed by sunlight, and the constituent elements probably escaped the atmosphere because of Mars's weak gravity field. Mars today may have liquid water on a seasonal basis, although frozen water is found in the polar icecaps and in the form of a shallow layer of permafrost under the ground.

The Geology in Action feature, "The Surface of Mars," describes the exciting new discovery that Mars may develop flowing water during its warm season.

EXPAND YOUR THINKING

Why is an atmosphere important in uniformly heating a planet's surface?

GEOLOGY IN ACTION

The Surface of Mars

NASA scientists have discovered what may be the first evidence of flowing water during the warmest months on Mars by comparing images from warm and cold seasons. The images, collected by the Mars Reconnaissance Orbiter mission, show dark, fingerlike features (**FIGURE 2.12**) that appear and extend down some Martian slopes during late spring through summer, fade in winter, and return during the next spring.

The best explanation for these features is that flowing saltwater makes them. Why saltwater? Because saltiness lowers the freezing point of water, and the ground temperature on Mars would allow liquid saltwater, whereas freshwater would freeze.

These results are the closest scientists have come to finding evidence of liquid water on the planet's surface today. Previously, frozen water has been detected near the surface in many regions with cooler climates. At other sites, fresh-looking gullies suggest that slope movements have occurred in geologically recent times, perhaps aided by water.

FIGURE 2.12 NASA scientists have discovered what may be the first evidence of water flowing on modern Mars. Narrow flow features, 1/2 to 5 meters wide, appear as relatively dark markings on steep slopes at several locations. Repeat imaging by the Mars Reconnaissance Orbiter shows that the features appear and grow incrementally during warm seasons and then fade in cold seasons.

NASA

2-4 Gas Giants Are Massive Planets with Thick Atmospheres

LO 2-4 Describe each of the gas giants.

The gaseous planets, or gas giants, include Jupiter, Saturn, Uranus, and Neptune.

Jupiter

Jupiter's atmosphere: hydrogen (H_2), 89.8%; helium (He), 10.2%

Jupiter (**FIGURE 2.13**) is the largest planet in the solar system, more than twice as massive as all the other planets combined. Its enormous size is due to its huge gaseous atmosphere, a holdover from the early days of the solar nebula.

Jupiter has 67 confirmed moons, which have attracted scientific interest in recent decades. Organic molecules that are believed to be necessary for the formation of life have been detected on some of those moons. Scientists study such conditions in the hope of learning more about the development of life on Earth. Jupiter's four largest moons, known as the Galilean satellites—*Io, Europa, Ganymede*, and *Callisto*—can readily be seen from Earth.

By reading the Geology in Action feature, "Jupiter's Top," (p. 47) you can learn about the famous banding in Jupiter's atmosphere.

Saturn

Saturn's atmosphere: hydrogen (H_2), 96.3%; helium (He), 3.25%

You have probably seen photos of Saturn, the planet with the famous rings (**FIGURE 2.14**). These rings are thought to be the remains of satellites that were pulled apart by Saturn's gravity. They are composed of myriad particles of ice and rock. The ring system is approximately 100,000 kilometers wide and 200 meters thick. Data show that Saturn has a mass about 95 times that of Earth, as well as a very rapid spin rate, which has the effect of flattening the planet at the poles and bulging it at the equator.

Saturn's exterior is composed primarily of frozen ammonia. The interior consists mostly of hydrogen, with lesser amounts of helium and methane. Saturn is the only planet that is less dense than water (about 30% less). Amazingly, if you were able to place it in a freshwater bath, Saturn would float!

NASA

FIGURE 2.13 The gas giant Jupiter.

Q Why does Jupiter display color banding in its atmosphere?

NASA

FIGURE 2.14 This ultraviolet image of Saturn, taken by the Hubble Space Telescope, shows its ring structure and southern hemisphere when the planet was at its maximum tilt of 27 degrees toward Earth. During the course of its 29.5-year orbit, Saturn experiences seasonal tilts away from and toward the Sun, much the same way Earth does.

Q Saturn's spin axis is tilted. How could this have happened?

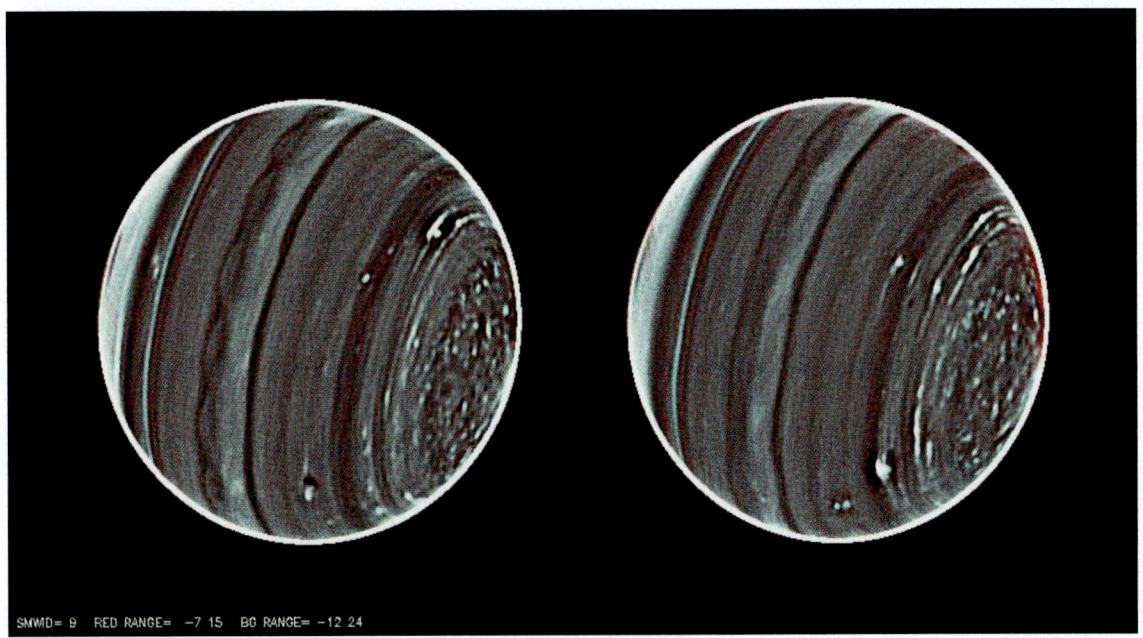

FIGURE 2.15 Two specially processed sets of images of Uranus taken from the Keck II telescope on July 25 and 26, 2012, reveal an astounding level of detail in Uranus's clouds. In each image, the north pole is at the right and slightly below center. The broad bright band just to the left of the disk center covers the latitude range from the equator to about 10 degrees north. Just south of the equator, and never before seen on Uranus, is a scalloped wave pattern, which is similar to instabilities that develop in regions of horizontal wind shear. Winds blow mainly east to west at speeds reaching 900 kilometers per hour Near the bottom of the left image is a small dark spot with bright companion clouds. Uranus's atmosphere consists mostly of hydrogen and helium; but compared to Jupiter and Saturn, it has a greater abundance of "ices" of water, ammonia, and methane.

What gives Uranus its blue color?

Uranus

Uranus's atmosphere: hydrogen (H_2), 82.5%; helium (He), 15.2%; methane (CH_4), 2.3%

Uranus (**FIGURE 2.15**) has a unique feature: Its axis of spin is approximately parallel to the ecliptic. In other words, Uranus lies on its side, with its north pole pointing toward the Sun for half of a Uranian year. Its south pole points toward the Sun for the other half of the year. This strange tilt produces some extreme seasonal effects. One is that the polar regions get the greatest amount of sunlight, whereas the rest of the planet fails to experience the daily heating and cooling experienced by other planets. Researchers hypothesize that collision with another object altered the planet's spin axis, or "tipped it over."

Neptune

Neptune's atmosphere: hydrogen (H_2), 80%; helium (He), 18.5%; methane (CH_4), 1.5%

Neptune (**FIGURE 2.16**) is the outermost classical planet in our solar system. Scientists hypothesize that the inner two-thirds of the planet are composed of a mixture of molten rock, water, liquid ammonia, and methane. The outer third, they believe, is a mixture of heated gases comprising hydrogen, helium, water, and methane. It is methane that gives Neptune its blue cloud color.

The strongest winds on any planet have been measured on Neptune. Most of them blow westward, in the direction opposite that of the rotation of the planet. Near one spot, wind speed reaches 2000

FIGURE 2.16 Neptune is the farthest classical planet from the Sun. It is similar in composition to Uranus in that both have compositions that differ from Jupiter and Saturn. Neptune's atmosphere is composed primarily of hydrogen and helium, along with a higher proportion of "ices," such as water, ammonia, and methane. Astronomers sometimes categorize Uranus and Neptune as "ice giants" to emphasize these distinctions.

How does the chemical composition of Neptune and Uranus differ from that of Jupiter and Saturn?

kilometers per hour. Neptune has four rings, which are narrow and very faint. The rings are made up of dust particles originating from tiny meteorites smashing into Neptune's 13 known moons.

The solar system is part of the *Milky Way galaxy*, one of billions of galaxies in the universe. For more on this, read the Critical Thinking feature, "The Nature of the Universe."

EXPAND YOUR THINKING

What are the five most abundant gases in the solar system? How do the atmospheres of the gas giants differ from the atmospheres of the terrestrial planets?

CRITICAL THINKING

The Nature of the Universe

Have you ever wondered whether the universe is endless? If it is not, and you were able to travel to its edge, what might you see? And if there is an edge, there must be a center; but what is in the center of the universe? We can begin to answer these questions by thinking critically about a simple observation: It gets dark at night.

If the universe were infinite and uniformly filled with stars, every point in the sky would be filled with a star radiating heat, and it would not get dark at night. Indeed, the glare would be overpowering, and the combined radiation of all that light would heat Earth and all other objects in the universe to the temperature of the Sun's surface. This observation is called *Olbers' paradox* (after astronomer Heinrich Olbers, 1758–1840) because although in his day it was assumed (largely for religious reasons) that the universe must be infinite, his observations suggested that it was finite. The solution to the paradox was proposed in 1848 by the poet Edgar Allen Poe, who stated that the night sky is dark (**FIGURE 2.17**) because the *universe is not infinite.*

Does the universe have an edge and, therefore, a center? The *Doppler effect* (named for physicist Christian Doppler, 1803–1853) helps answer this question. This term refers to the change in frequency of energy as the source of that energy approaches or recedes. To illustrate: You can determine the direction in which an ambulance is traveling based on the pitch of its siren. The pitch rises when the ambulance is coming toward you, and falls as it travels away from you. Inside the vehicle, however, the sound is constant; the driver does not notice a change in pitch. Light from a distant galaxy has the same effect. Wavelengths of light produce certain colors, lower-frequency wavelengths produce reddish color and higher-frequency wave lengths produce bluish color. The wavelengths of light from a retreating galaxy appear to decrease in frequency (the light is "red-shifted"). The wavelengths of light from an approaching galaxy appear to increase in frequency (the light is "blue-shifted").

In 1929 astronomer Edwin P. Hubble (1889–1953) astounded the scientific community by announcing that other galaxies in the universe are red-shifted, and that the amount of the shift increases the farther away a galaxy is. That is, galaxies are moving away from Earth, and more distant galaxies are moving away faster. The explanation for this phenomenon is simple but revolutionary: *The universe is expanding.*

Hubble's finding, which has been tested many times by other astronomers, indicates that the universe is expanding. What does that

NASA and the Hubble Heritage Team (STScI/AURA)

FIGURE 2.17 If the universe were infinite and filled with stars, the sky should be brilliantly lit at all times and the radiation would heat Earth thousands of degrees. It gets dark at night because the universe is *not* infinite.

mean, actually? In part it means that no matter which galaxy we happen to be in, virtually all the other galaxies are moving away from us. In other words, it's not as if Earth (or any other location) is at the center of the universe and everything else is receding from it. Thus, the universe has no center and, therefore, no edge, as such.

GEOLOGY IN ACTION

Jupiter's Top

The gas giant Jupiter (**FIGURE 2.18**) is famous for its colorful bands, which appear to be on the planet surface. But they are not; the bands are at the top of its massive atmosphere. In fact, although no one knows for sure, it is thought that Jupiter has no hard surface. Instead, the atmosphere just gets denser toward the planet center, until it turns to liquid.

It is ammonia clouds, organized into a dozen bands that parallel the equator, that impose the banding on Jupiter. The bands are bounded by powerful winds, known as *jets*, that blow at different speeds and thus make the turbulence visible in the clouds. Dark bands are called *belts*, and light bands are called *zones*. Zones are colder than belts and are sites of upwelling air. Belts are composed of descending air; the source of their dark color is not known.

The bright blue emissions above the North and South Poles are auroral lights, similar to those seen above Earth's polar regions. These are curtains of light resulting from electrons following the planet's magnetic field into the upper atmosphere, where collisions with atmospheric atoms and molecules produce the light.

Unlike the terrestrial planets, which lost their early atmospheres to the heat of the developing Sun, Jupiter did not. Jupiter's most famous feature is the *Great Red Spot*. Still there after more than 300 years, the spot is the largest storm in the solar system, with winds reaching 434 kilometers per hour. Fed by heat generated by the huge planet, it presumably continues to exist because it never runs over land, where it would lose its source of energy.

FIGURE 2.18 Jupiter is characterized by the unusual banding at the top of its atmosphere.

NASA

Objects in the Solar System Include the Dwarf Planets, Comets, and Asteroids

2-5

LO 2-5 Define a dwarf planet and name the five that are currently recognized.

In addition to the so-called classical planets, the solar system includes five named dwarf planets and numerous comets and asteroids.

A dwarf planet is defined as an object in the solar system that orbits the Sun and is not a satellite of a planet or other celestial body; and it must be spherical (or nearly so) in shape (**FIGURE 2.19**). This means that it is large enough so that its own gravitational field has overcome the internal rigidity of rock to shape it into a round object. Dwarf planets are not massive enough to clear their neighborhood region of planetesimals. Under these strict requirements, the International Astronomical Union recognizes five dwarf planets: Eris, Makemake, Haumea, Pluto, and Ceres.

Eris, Makemake, Haumea, Pluto, and Ceres

Pluto's atmosphere: nitrogen (N_2), 98%; water (H_2O), < 1%; methane (CH_4), < 1%; carbon monoxide (CO), < 1%

Pluto, Haumea, and Makemake reside in the region of the *Kuiper belt*, a sea of icy bodies extending beyond Neptune that is one source of comets. Eris is thought to be a member of the *scattered disk* region. The scattered disk region overlaps with the Kuiper belt, but objects within

it have highly eccentric and inclined orbits, and its outer limits extend much farther from the Sun. Although there are just five dwarf planets now, their number is expected to grow. Scientists estimate there may be many dozens of dwarf planets in the outer solar system. But since they don't know the actual sizes or shapes of many of these faraway objects, they can't yet determine which are dwarf planets and which are not. More observations through more powerful telescopes are likely to result in additions to the list in coming years.

Pluto was discovered in 1930 and considered a planet until it was reclassified as a dwarf planet in 2006. Haumea, Makemake, Eris, and Ceres have been classified since then. Pluto's mass is thought to be only 0.0025 that of Earth. Eris is slightly larger than Pluto. The mass of these bodies is closer to that of a moon than to that of a planet. Indeed, Pluto, together with its moon, Charon, is so small that together they could fit within the borders of the continental United States. And because Charon and Pluto are similar in size, they are really considered a double dwarf planet system, rather than the more typical combination of a planet with a much smaller moon.

The dwarf planets do not resemble the inner terrestrial planets or the outer gaseous planets in their makeup. Instead, they more closely resemble the ice moons of the outer planets, a resemblance that has led some researchers to suspect that they are large icy chunks of debris left over from the formation of the solar system.

Ceres is a dwarf planet in the asteroid belt. Although it is the largest of the asteroids (about 950 kilometers in diameter), it is less than half the size of Pluto and Eris. Still, Ceres accounts for approximately one-third of the belt's total mass. The physical aspects of Ceres are not well understood. Researchers hypothesize that its surface is warm, and that it has a tenuous atmosphere that includes frost. One study concluded that the dwarf planet has a rocky core overlain by an icy mantle, possibly containing 200 million cubic kilometers of water, more than the total amount of freshwater on Earth.

FIGURE 2.19 The dwarf planets include Eris, Makemake, Haumea, Pluto (with its moon Charon), and Ceres.

 What is a dwarf planet?

Asteroids and Comets

Not all of the planetesimals ended up becoming planets. Some were made up primarily of rocky and metallic substances, and they became *asteroids*. Other planetesimals were made up primarily of water ice and other frozen liquids or gases—hydrogen, helium, methane, carbon dioxide, and others—and are known as *comets* (**FIGURE 2.20**).

Because they are made of ice, comets would not last long if they inhabited the asteroid belt. Continuous exposure to the Sun's heat would cause them to waste away. Instead, most of the comets in the solar system are found in two distant regions: the *Oort cloud*, a vast spherical cloud of comets at the extreme outer edge of the solar system, and the closer Kuiper belt, a wide band of comets located beyond the orbit of Neptune.

Comets range from 1 to 10 kilometers in diameter. In addition to water ice, the nucleus, or center, of a comet may contain frozen carbon dioxide, carbon monoxide, methane, and other easily evaporated compounds. Interspersed with this ice are tiny mineral grains (dust); together, the ice and dust in the nucleus of a comet form a "dirty snowball," an idea first put forth by astronomer Fred Whipple in 1950 and verified in 1986 by the spacecraft *Giotto*'s flyby of Halley's comet.

When the solar wind encounters the material in the coma (a diffuse cloud of gas and dust surrounding the nucleus of a comet), it blows it back behind the nucleus, thus creating the comet's tail (which usually extends behind the comet, away from the Sun). Early in the

NASA Jet Propulsion Laboratory (NASA-JPL)

FIGURE 2.21 Two hemispheres of the asteroid Eros. Most asteroids reside in a belt of rocky debris between Earth and Jupiter, which may be left over from the original nebular disk. The total mass of all the asteroids is less than that of our Moon.

Where are most asteroids found?

history of the solar system, large numbers of comets bombarded the terrestrial planets, delivering some of the first water and important atmospheric components. These impacts also added heat, contributing to the creation of the planets' internal structure.

Comets actually have two tails: the dust tail and the gas, or ion, tail. The dust tail is formed of solid particles escaping from the comet along its flight path, and may be slightly curved. Its whitish color is due to reflected sunlight. The gas tail is pushed straight back from the comet by the solar wind (it always points directly away from the Sun) and has a bluish glow due to the fluorescence of ionized molecules of carbon dioxide, carbon monoxide, nitrogen, and methane.

The asteroid belt is a zone of rocky debris (**FIGURE 2.21**) between Mars and Jupiter. Asteroids are rocky, metallic objects that orbit the Sun but are too small to be considered planets. Asteroids range across the central solar system from inside Earth's orbit to beyond Saturn's. Most, however, are contained in the region between Mars and Jupiter. To learn more about asteroids, see the Geology in Action feature in Chapter 5 titled "The Oldest Rocks."

To expand your understanding of this discussion, study Figure 2.22 in the Critical Thinking feature titled "Summarizing the Solar System," and then complete the exercise there.

Jerry Lodriguss/Photo Researchers, Inc.

FIGURE 2.20 Comets orbiting the Sun exhibit a coma (a diffuse cloud of gas and dust surrounding the nucleus of a comet) and tail.

What causes a comet to have a blue tail and a white tail?

EXPAND YOUR THINKING

If a comet had the composition of an asteroid, how would it look different when observed from Earth?

CRITICAL THINKING

Summarizing the Solar System

Work with a partner to complete the following exercise, based on **FIGURE 2.22**.

1. Label each of the objects in this image. Identify the classical planets and dwarf planets.

2. Identify differences in the chemistry of the terrestrial planets and the gas giants.

3. What was the source of heat for the solar nebula? What role does heat play in organizing the solar system?

4. How did the planetesimal accretion process shape the appearance of the modern solar system?

5. What is unique about Earth compared to the other planets? Describe differences and similarities in the chemical composition of the planets.

6. Itemize a timeline of events in the formation and future of the solar system.

7. How would the solar system be different if the entire system rotated at a slower rate?

8. Scientists classify the planets as either classical or dwarf. Use the chemical information provided in this chapter to suggest a new classification system, one based on their chemistry.

FIGURE 2.22 The solar system.

2-6 Earth's Interior Accumulated Heat During the Planet's Early History

LO **2-6** Describe the sources of heat to early Earth and the consequences of heat buildup.

Earth's formation included several steps: planetesimal accretion, condensation, heating, and chemical differentiation.

Early History

Most scientists agree that Earth and the solar system were formed by planetesimal accretion and gravitational contraction of a massive nebula. Driven by gravity, the nebula contracted, and its temperature rose until it achieved nuclear fusion, forming the Sun. Energy from the Sun constitutes the solar wind, which stripped the inner planets of their primitive atmospheres of helium and hydrogen (**FIGURE 2.23**).

The solar nebula hypothesis describes the formation of the early solar system approximately 4.6 billion years ago. This period of Earth's history is informally known among geologists as the "Hadean eon" (named after Hades, another word for hell) because shortly after its formation, Earth's surface turned into a sea of molten rock. The Hadean eon lasted about 800 million years, from approximately 4.6 to 3.8 billion years ago. Changes that occurred during that time set the stage for plate tectonics and the origin of life on Earth.

Condensing the Planet

Earth formed in four basic steps:

1. It began to accrete from the nebular cloud as particles smashed into each other, forming planetesimals.

2. As Earth's mass grew, so did its gravitational field, causing it to condense into a smaller, denser spherical body.

3. This condensation generated heat in Earth's interior; heat was also produced by the decay of *radioactive elements* in the interior, as well as by the impacts of many thousands of extraterrestrial objects. Earth's interior began to melt. Of the common elements, iron is the heaviest and droplets of liquid iron sank toward the planet's center and eventually accumulated to form the core.

4. Melting iron moving through the planet generated friction that contributed more heat and raised Earth's temperature. A deep magma ocean formed on the planet surface and the planet developed the basic internal structure of an inner and outer core, mantle, and crust – a process known as *chemical differentiation*.

Extraterrestrial Bombardment

Early in its history, Earth (and other planets) endured heavy impacts by asteroids and comets (**FIGURE 2.24**). Evidence in the form of overlapping craters on other planets and moons suggests that these impacts occurred in two waves, the first consisting of small to medium-size rocky asteroids, most likely from the nearby asteroid belt, and the second (known as the "late heavy bombardment period"—ca. 4 to 3.8 billion years ago) probably consisting of icy comets from the distant Kuiper belt and scattered disk region that surround the solar system. Some meteorite debris on Earth is even thought to have come from the planet Mars when it sustained large impacts that hurtled Martian rock into space.

As you can imagine, when an asteroid the size of several city blocks slams into the planet at thousands of kilometers per hour, the collision releases a great deal of heat. In fact, just one such collision would release more heat than if all the current nuclear bombs on Earth detonated at once.

During the Hadean eon, from tens of thousands to millions of such violent collisions produced enormous quantities of heat, much of which was stored in the solid rock of the planet. With more impacts came more heat. Large impacts, such as the one that may have formed our Moon, were especially important in this process. In time, Earth was transformed from a cold planet into one that was warmed by the energy of extraterrestrial collisions.

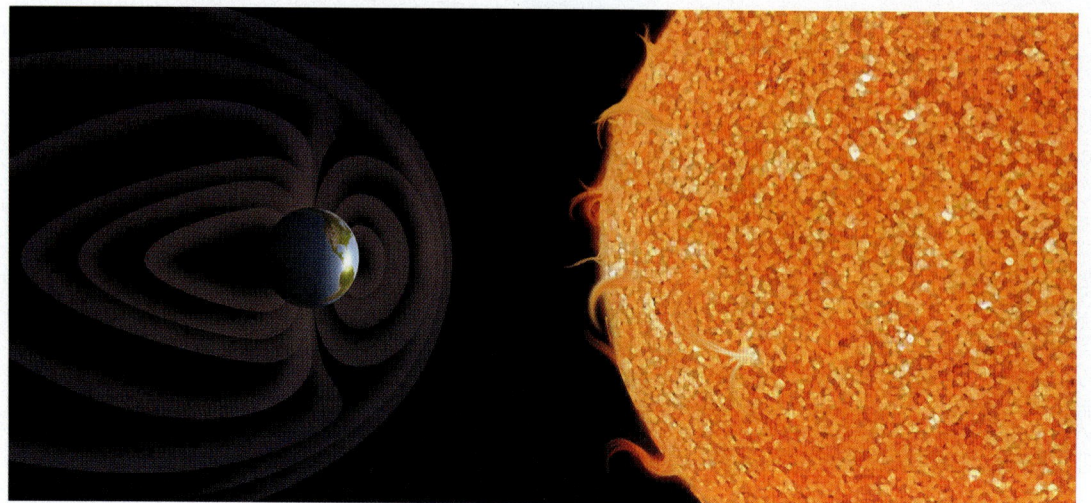

FIGURE 2.23 Early in the history of the solar system, the solar wind stripped the inner planets of their primitive atmospheres. In this image, the modern solar wind deforms the geomagnetic field into a streamlined shape, with the blunt end facing into the solar wind and the tail extending downwind.

Q Earth lost its early atmosphere to the solar wind. What is the source of Earth's present atmosphere?

FIGURE 2.24 With our Moon rising in the background, a young, crater-scarred Earth sustains yet another impact. Comets streak over primitive Earth, bringing water to fill impact basins. These impacts created the first shallow seas and added new gases to the young planet.

🅠 **A thin atmosphere is shown. Where did it come from? What is it made of?**

Planetary scientists hypothesize that icy comets delivered substantial amounts of water to Earth. This water accumulated as pools of liquid and as clouds of gaseous water vapor in a newly forming atmosphere. Much of Earth's water also came from **volcanic outgassing** of moisture trapped in minerals and released by melting within Earth's interior. An important question debated by scientists in recent years is the relative significance of these two sources of water.

According to one school of thought, comets may have supplied the majority of Earth's water during the late heavy bombardment phase of the solar system. If this is true, the chance that organic matter, which is also found in comets, played an important part in the origin of life on Earth is increased. In rebuttal, one study suggests that most of Earth's water may not have come from comets. Researchers found that Comet Hale–Bopp (visible from Earth in 1997) contains substantial amounts of "heavy water," water rich in deuterium, an isotope of hydrogen. If Comet Hale–Bopp is typical in this respect, and if cometary collisions were a major source of terrestrial oceans, Earth's ocean water should be similarly rich in deuterium, but it is not.

Heat stored in Earth's interior led to the creation of the first volcanoes. After it lost its hydrogen, helium, and other hydrogen-containing gases due to the Sun's radiation, Earth lacked an atmosphere. In time, delivery of icy gases by comets, along with volcanic outgassing of volatile molecules such as water, methane, ammonia, hydrogen, nitrogen, and carbon dioxide, produced Earth's second atmosphere. After approximately 4 billion years ago, rain fell to form the first shallow oceans. Geologists hypothesize that the other terrestrial planets experienced a similar process in their early histories. On Earth, plant photosynthesis and other surface processes have also contributed to the chemical makeup of our atmosphere.

What About Our Moon?

Where did our Moon come from? Several hypotheses describe how it was formed, but many of them have trouble accounting for certain aspects of the Earth–Moon system.

- The *capture hypothesis* proposes that Earth's gravity captured a passing planetesimal, which then became the Moon. This theory thus predicts that Earth and the Moon were formed at separate locations. However, the chemistry of Moon rocks is similar to that of Earth rocks, suggesting that both developed either from a common source or at least at the same location in the planetary nebula.

- According to the *double planet hypothesis*, Earth and the Moon were formed concurrently from a local cloud of gas and dust. But this hypothesis fails to account for several aspects of our Moon's history, such as the unusual tilt of its axis, the melting of its surface rocks, and the fact that it is less than half as dense as Earth.

- The *fission hypothesis* holds that centrifugal force associated with Earth's spin caused a bulge of material to separate from Earth in the area of the equator. Unfortunately, this theory requires Earth to rotate once every 2.5 hours in order to develop the necessary force.

- The *impact hypothesis* states that at some point during planetesimal accretion, Earth suffered a massive collision with a huge object the size of Mars, and this collision led to the formation of our Moon (**FIGURE 2.25**).

Currently, the impact hypothesis is the most widely accepted account of the Moon's formation.

FIGURE 2.25 The impact hypothesis suggests that Earth suffered a massive collision that led to formation of our Moon. This artist's recreation shows Earth and a Mars-size object, each peppered by hundreds of smaller impacts, colliding with one another in the early solar system.

🅠 **EXPAND YOUR THINKING**

Why is the impact hypothesis for the moons origin currently favored?

LET'S REVIEW

Earth, the Sun, and other objects in the solar system originated at the same time from the same source and have evolved in varying ways since then. The solar system began with the collapse and condensation of a planetary nebula. Understanding Earth's place in the cosmos gives us a perspective on and appreciation for the scale of nature. The ability to explore the solar system is one of humanity's grand accomplishments. Someday our planetary neighbors may provide us with useful resources. As far as we know, Earth is the only habitable planet in the universe. Certainly it is the only habitable planet in the solar system.

You now know the context of Earth's origin and why it is habitable (and why other planets are not). The stage is set for us to return to Earth and investigate its earliest history, how and why it became layered, and the processes of plate tectonics that govern Earth's evolution through time. These are the subjects of Chapter 3, "Plate Tectonics."

STUDY GUIDE

2-1 Earth's origin is described by the solar nebula hypothesis.

- The **solar system** consists of our Sun, eight classical planets (in order from the Sun: Mercury, Venus, Earth, Mars, Jupiter, Saturn, Uranus, and Neptune); five dwarf planets (Pluto, Ceres, Haumea, Makemake, and Eris); small solar system bodies (including asteroids, comets, and objects in the Kuiper belt, the scattered disk region, and the Oort cloud); 240 known satellites (moons), including 162 orbiting the classical planets; countless particles; and interplanetary space. The Sun is a star located at the center of the solar system; all other components orbit around it.

- The origin of the solar system is described by the **solar nebula hypothesis.** This hypothesis holds that the components of the solar system were all formed together approximately 4.6 billion years ago and have evolved in different ways since then.

2-2 The Sun is a star that releases energy and builds elements through nuclear fusion.

- Stars burn by means of **nuclear fusion,** which creates new elements. When stars use up their hydrogen fuel, they become brighter. A star the size of our Sun will become a red giant, engulf its inner planets, and eventually burn out. A field of mineral particles and gas, known as a planetary nebula, is produced when a star explodes at the end of its "lifetime." The planetary nebula provides the raw material for building new planets, such as Earth, as well as new stars, such as the Sun.

- The solar core is the site of nuclear fusion. There, hydrogen is converted into helium, which has less mass. The difference in mass is expelled as energy that is carried to the surface of the Sun by convection, where it is released as light and heat. Our Sun has been active for approximately 4.6 billion years and has enough fuel to last another 4 to 5 billion years.

2-3 Terrestrial planets are small and rocky, with thin atmospheres.

- In many ways, Earth is unlike the other planets. Earth's hot interior generates the energy to drive plate tectonics, yet its surface is cool enough to support life. Earth's position relative to the Sun provides enough warmth to maintain liquid water, yet it still experiences freezing seasons. Earth's winds are relatively calm; the atmosphere is thin and transparent, so it is easily penetrated by light; and daily extremes in temperature are comparatively modest. All these conditions promote the existence of life on Earth.

- The planet Mercury lies closest to the Sun and is dominated by unrelenting and intense heat during the day. During the night, because it lacks an atmosphere to trap that heat, temperatures plummet to extreme cold.

- Venus is radically different from Earth. Carbon dioxide, which is very efficient at trapping heat, makes up 96% of its atmosphere. As a result, temperatures on the surface reach 477°C, making Venus even hotter than Mercury.

- Mars is the most Earth-like of the other planets. It has dry river channels that indicate that water once flowed across its surface.

2-4 Gas giants are massive planets with thick atmospheres.

- Jupiter is the third-brightest object in the night sky, after the Moon and Venus. It is also the first of the **gas giants.** It is composed mainly of hydrogen and helium, with small amounts of methane, ammonia, water vapor, and other compounds.

- Saturn is flattened at the poles, a condition that is caused by an outward push in the equatorial region due to a high rate of rotation. The atmosphere consists primarily of hydrogen, with lesser amounts of helium and methane.

- Uranus is the only planet that does not rotate perpendicular (or nearly so) to the ecliptic. That is, Uranus lies on its side—or, is tipped over—and its north pole points toward the Sun for half a Uranian year.

- The inner two-thirds of Neptune are likely composed of a mixture of molten rock, water, liquid ammonia, and methane. The outer third is a mixture of heated gases composed of hydrogen, helium, water, and methane.

2-5 Objects in the solar system include the dwarf planets, comets, and asteroids.

- Pluto and its moon, Charon, are similar in size. This duo is, therefore, more of a "double dwarf planet" system than a planet and moon system. Pluto and Charon, Haumea, and Makemake are objects in the Kuiper belt.

- Eris is an object in the scattered disk region, which overlaps with the Kuiper belt, but extends farther into space and is not confined to the plane of the ecliptic.

- Ceres is the fifth dwarf planet. It is a large asteroid located in the asteroid belt.

- Comets are "dirty snowballs" of frozen volatiles that are locked in gravitational orbit around the Sun.

- The asteroid belt is a zone of rocky debris between Mars and the giant gaseous planet Jupiter.

2-6 Earth's interior accumulated heat during the planet's early history.

- The earliest phase of Earth's history, the Hadean eon, was characterized by extraterrestrial impacts and heating of the planet's interior. Earth gained a new atmosphere as a result of **volcanic outgassing,** together with the delivery of gases and water by ice-covered comets.

KEY TERMS

carbon dioxide (p. 38)
extrapolation (p. 37)
gas giants (p. 35)
Milky Way galaxy (p. 34)

nuclear fusion (p. 36)
planetesimal accretion (p. 35)
solar nebula hypothesis (p. 34)
solar system (p. 34)

solar wind (p. 38)
terrestrial planets (p. 35)
volcanic outgassing (p. 51)

ASSESSING YOUR KNOWLEDGE

Please complete this exercise before coming to class. Identify the best answer to each question.

1. The classical planets in order from the Sun are:
   a. Mercury, Venus, Mars, Earth, Jupiter, Saturn, Neptune, Uranus.
   b. Jupiter, Venus, Mars, Earth, Mercury, Saturn, Neptune, Uranus.
   c. Mars, Venus, Earth, Jupiter, Saturn, Mercury, Neptune, Uranus.
   d. Venus, Mars, Mercury, Earth, Jupiter, Saturn, Neptune, Uranus.
   e. None of the above.

2. What is responsible for deforming Earth's geomagnetic field in this image? _____

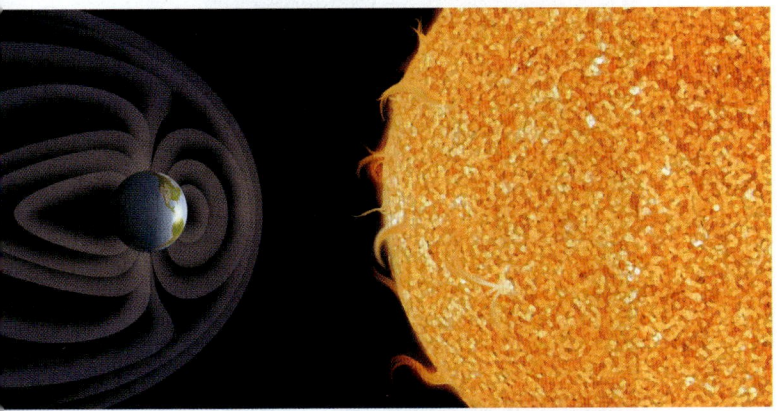

3. The energy that is destined to become sunlight may take _____ years to travel from the Sun's core to its surface.
   a. 10
   b. 100
   c. 100,000
   d. 1,000,000
   e. 1000

4. Nuclear fusion:
   a. Is the source of the Sun's energy.
   b. Occurs when the nucleus of an atom fissions and releases energy.
   c. Radiates throughout the universe.
   d. Is the reason that Jupiter has no solid surface.
   e. Has not yet occurred in the Sun.

5. Nuclear fusion is causing the Sun to weigh less over time.

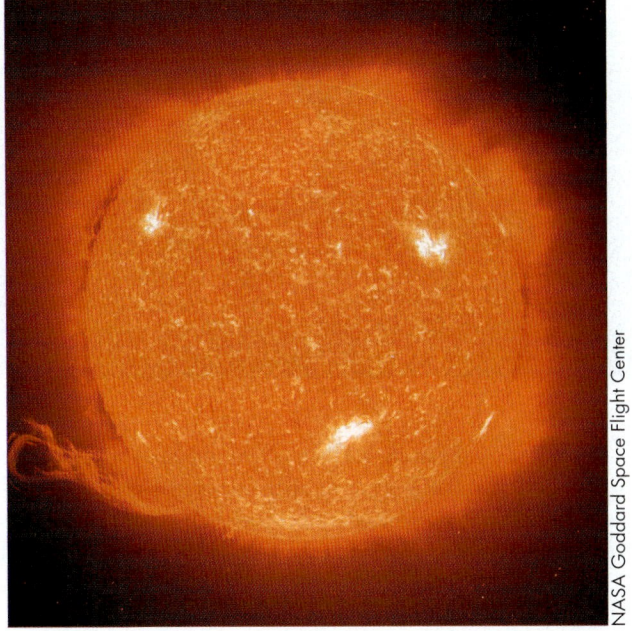

NASA Goddard Space Flight Center

True _____

False _____

6. Mercury, Venus, and Mars are different from Earth because:
   a. They are closer to the Sun.
   b. Earth has volcanoes and they do not.
   c. Earth currently has liquid water and they do not.
   d. Earth is the only planet with ice this close to the Sun.
   e. They are all in retrograde orbit.

7. The largest storm in the solar system is found on which planet?
   a Uranus
   b. Earth
   c. Mars
   d. Jupiter
   e. Pluto

8. The fastest winds in the solar system are found on which planet?
   a. Neptune
   b. Jupiter
   c. Venus
   d. Earth
   e. Mars

9. The basic structure of the solar system is described as:
   a. The ecliptic
   b. The Oort cloud
   c. Inner terrestrial and outer gaseous planets
   d. The asteroid filter
   e. Rotating nuclear fission

10. The outer gaseous giants are characterized by abundant (silica, hydrogen, metals, solid rock) _____ while the inner terrestrial planets are characterized by abundant (silica, helium, hydrogen, methane) _____.

NASA

11. Why do the outer planets and their moons consist mostly of ice and gas while the inner planets are made up mostly of rock and metal?
    a. The solar wind stripped the inner planets of volatile compounds.
    b. The outer gas giants had greater volcanism, which produced large quantities of gases.
    c. Gravity sucked the gases from the inner planets into the Sun.
    d. Solar heat is so limited in the outer portion of the solar system that solids turn into gas.
    e. Far from the Sun, the outer gas giants are made of ice.

12. The dwarf planets are:
    a. Mercury, Earth, and Mars
    b. Ceres, Pluto, Haumea, and Mercury
    c. Eris, Ceres, Pluto, Haumea, and Makemake
    d. There are no dwarf planets, only moons
    e. None of the above

13. What is planetesimal accretion?
    a. The collapse of the Kuiper belt into the core region.
    b. Collisions of bits of ice, gas, and dust grew into planetesimals, planetary embryos, and, eventually, planets.
    c. Jupiter, with its huge mass, broke into pieces that eventually became the major planets.
    d. The solar wind tore the young planets into smaller pieces called planetesimals, and these later grew together to form the present planets.
    e. The solar wind kept the planets in their positions.

14. Which of the following is the name of a hypothesis explaining the origin of the solar system?
    a. Planetesimal collision
    b. Nebular expansion
    c. Solar nebula
    d. Nuclear fusion
    e. Solar objects

15. The major gases in the solar system include:
    a. Ice, argon, methane, and carbon.
    b. Water, carbon dioxide, ammonia, helium, hydrogen, and carbon monoxide.
    c. Lithium, beryllium, carbon, hydrogen, and carbon monoxide.
    d. Ammonia, oxygen, helium, hydrogen, carbon monoxide, and water.
    e. Water, hydrogen sulfide, ammonia, helium, lithium, and carbon monoxide.

16. Comets have two tails. One points (away from, toward) _____ the (Sun, Earth, asteroid belt) _____ and the other points (away from, toward) _____ the direction of travel.

Jerry Lodriguss/Photo Researchers, Inc.

17. Comets are made of:
    a. Molten rock.
    b. Ice and mineral grains.
    c. Gas and ice.
    d. Rock and a thin atmosphere of argon.
    e. Mostly potassium and oxygen.

18. Extraterrestrial impacts:
    a. Probably occurred in two waves.
    b. May have delivered water to Earth and an early atmosphere.
    c. May have originated at the Oort cloud and Kuiper belt regions.
    d. Produced the scars on the Moon's surface.
    e. All of the above.

19. The source of Earth's heat is a combination of:
    a. Extraterrestrial impacts, gravitational energy, and radioactivity.
    b. Nuclear fusion, volcanism, and compression.
    c. Compression, volcanism, and solar wind.
    d. Solar wind, radioactivity, and gravitational energy.
    e. None of the above.

20. Thus far, there has been no direct observation of flowing water on Mars.
    True _____
    False _____

NASA

FURTHER LEARNING

1. If you were in charge of a scientific mission to Mars, what goals for the mission would you establish?

2. Do you think plate tectonics exists on other planets? How would you be able to tell from here, on Earth?

3. In this chapter we stated that Jupiter's Great Red Spot has been in existence for hundreds of years. How do we know this?

4. When you look at any of the gas giants, what are you seeing: the land or the top of the clouds?

5. When you look at the terrestrial planets, what are you seeing: the land or the top of the clouds?

6. What is the difference in general chemistry between the gas giants and the terrestrial planets?

ONLINE RESOURCES

Explore more about planetary geology on these Web sites:

NASA images of the planets and other objects:
http://photojournal.jpl.nasa.gov/index.html

Images and discussion of the solar system:
www.solarviews.com/eng/homepage.htm

USGS Astrogeology:
http://astrogeology.usgs.gov/Projects/BrowseTheSolarSystem/index.html

Additional animations, videos, and other online resources are available at this book's companion Web site:
www.wiley.com/college/fletcher

This companion Web site also has more information about WileyPLUS and other Wiley teaching and learning resources.

PLATE TECTONICS

CHAPTER CONTENTS & LEARNING OBJECTIVES

GEOLOGY IN OUR LIVES

The theory of plate tectonics predicts that Earth's surface is organized into massive slabs of lithosphere called plates in which are embedded continents and ocean basins. Plates move across Earth's surface, on average, at about the same speed that your fingernails grow. At their edges, where plates collide, separate, and slide past one another, rock is recycled, mountains are built, new crust is formed, and numerous geologic hazards and geologic resources develop. The history of Earth's surface is largely interpreted by recreating the past positions of plates. The theory of plate tectonics has advanced our understanding of managing geologic hazards and geologic resources.

Q Thingvellir National Park, Iceland, lies on the boundary of two plates that are pulling away from each other. What geologic hazards are likely to occur here?

3-1 Earth Has Three Major Layers: Core, Mantle, and Crust

LO **3-1** Describe Earth's history during the Hadean eon.

As heat accumulated in early Earth, a deep magma ocean formed on the surface; volcanic outgassing released gaseous compounds that were formerly bound in the rocks; and the chemical layering of Earth's interior was set in motion.

Recall from Chapter 2 that during the Hadean eon, Earth's interior accumulated and stored heat produced by several processes: the impact of extraterrestrial objects, decay of radioactive elements within the planet, gravitational energy associated with condensation of the rocky sphere, and friction generated by flowing liquid iron (Fe).

Chemical Differentiation

Perhaps the most significant single event in Earth's history occurred when the planet's temperature passed the melting point of iron (1538°C). Based on critical thinking including laboratory experiments, calculations, and evidence of Earth's chemistry, scientists hypothesize that it was this event, known informally as the **iron catastrophe**, that resulted in the internal layers that characterize Earth today.

The iron catastrophe began when molten iron, one of Earth's most abundant elements, flowed toward the planet's interior under the pull of gravity. Compounds of lighter elements (e.g., silicon [Si], oxygen [O], and other light elements that were also molten) were displaced toward the surface. Iron and nickel (Ni) accumulated in Earth's deep interior, and less dense compounds accumulated near the surface. Ultimately this process, called **chemical differentiation**, created Earth's internal structure.

As the first great tide of molten iron moved slowly through Earth, the friction it generated raised the temperature another 2000°C (**FIGURE 3.1**). This had a dramatic effect, causing the surface to

FIGURE 3.1 During the Hadean eon, Earth's surface developed a magma ocean that extended to a depth of several kilometers.

Q How did the iron catastrophe change the distribution of compounds in Earth's interior?

develop a deep ocean of molten rock. Later, as Earth cooled, this ocean solidified to form a solid _primordial crust_.

Scientists date this molten period based on two observations: 1) The oldest meteorites and lunar rocks are about 4.4 to 4.6 billion years old; and 2) the oldest known Earth rocks are about 3.8 to 4.1 billion years old (although some mineral grains from Australia are 4.4 billion years old). The difference in these two ages, roughly half a billion years, may represent the molten phase.

As Earth cooled and the rate of extraterrestrial bombardment waned, the surface formed a solid volcanic crust. The period of geologic time that followed the Hadean eon, beginning 3.8 billion years ago, is called the **Archean eon**. This was an important time in Earth's history because the planet could now support life.

Earth's Interior

During the iron catastrophe, about one-third of the primitive planet's mass sank toward the center. It was through this process that Earth was transformed from a _homogeneous_ body, with roughly the same kind of material at all depths, into a _heterogeneous_, or layered, body. It developed a dense iron **core,** with a solid inner layer and a liquid outer layer, a brittle crust composed of less dense compounds with _lower melting points_, and between them a solid **mantle** of intermediate density. The mantle is Earth's thickest layer; it is composed of solid rock that is capable of flow because of the high temperature and pressure in the mantle.

Compare the abundance of elements in the crust and in Earth as a whole (**FIGURE 3.2**). Because most of the iron sank to the core, iron drops to fourth place in the crust in terms of abundance. Silicon, aluminum (Al), calcium (Ca), potassium (K), and sodium (Na) are, in contrast, more abundant in the crust than in the planet as a whole. The reason for this difference is that the more abundant elements in the crust form lightweight chemical compounds that tend to melt at lower temperatures than compounds in the mantle. Materials such as these melted early during the period of chemical differentiation, then rose to the surface through _convection_ (whereby hot materials rise) in the mantle, and accumulated to form the crust.

Much of the way Earth works is based on the structure of its interior. Using _inductive reasoning_ (see Chapter 1) derived from observations of how seismic waves from earthquakes behave as they pass through Earth's interior, along with laboratory reconstructions of pressure and temperature conditions, scientists construct hypotheses describing conditions inside the planet.

Earth is thought to be composed of four zones—although more recent research indicates it may have a more complex interior (**FIGURE 3.3**). At the center is the **inner core**, a rock body made up of solid metal alloy consisting mostly of iron and nickel (two dense elements), that is very hot (about 5000°C). Normally at this temperature the rock would be molten, but deep inside the planet, high pressure prevents it from melting. The inner core therefore is solid, perhaps even crystalline.

The inner core is surrounded by the **outer core**, which is also extremely hot, but because it is under less pressure, the rock there is melted. The outer core is also rich in iron and nickel, probably mixed by convection. The mantle is massive (averaging 2,900 kilometers thick) and consists of hot solid rock that moves based on differences in density. Geologists hypothesize that hot rock rises through the mantle very slowly, perhaps only a few centimeters in hundreds of years. Rock moves through the mantle by convection, wherein hot rock rises while cool rock sinks.

© Don Dixon

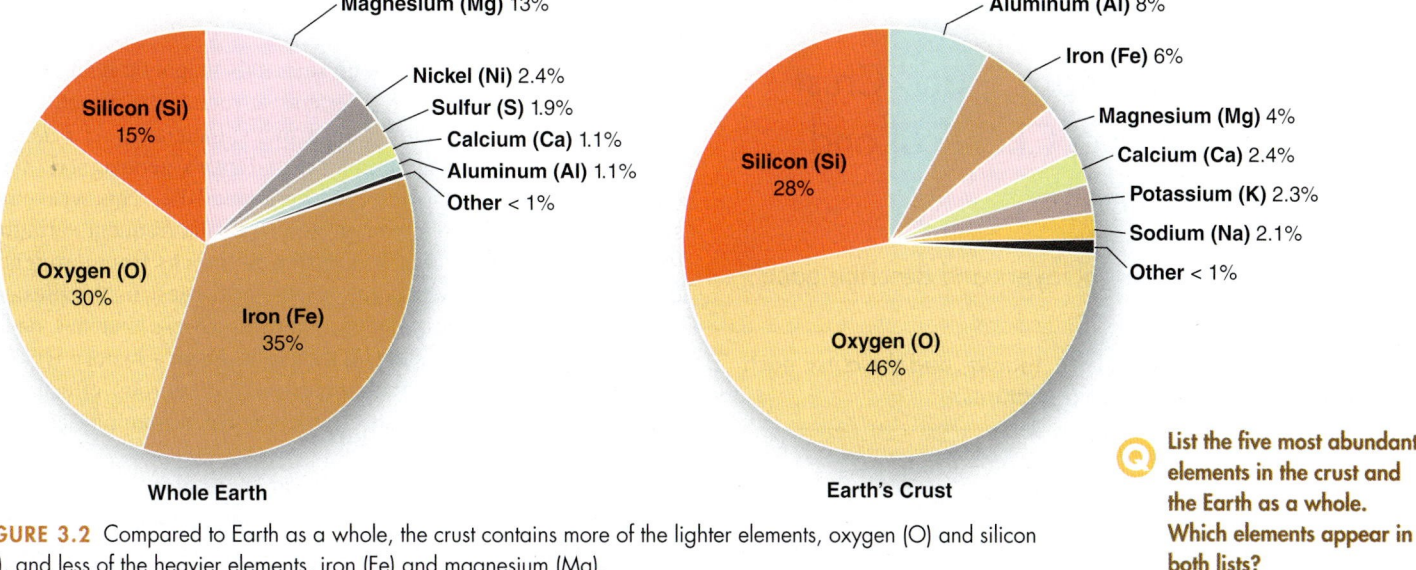

Whole Earth

Earth's Crust

FIGURE 3.2 Compared to Earth as a whole, the crust contains more of the lighter elements, oxygen (O) and silicon (Si), and less of the heavier elements, iron (Fe) and magnesium (Mg).

List the five most abundant elements in the crust and the Earth as a whole. Which elements appear in both lists?

Above the mantle is the solid outer layer of rock that we live on: Earth's **crust**. Relative to the planet's radius, the crust is thinner than an eggshell. The crust is the outer portion of the **lithosphere**, which consists of the upper mantle (lower lithosphere) and the crust (upper lithosphere). Below the lithosphere lies the **asthenosphere**, a weak and ductile layer in the upper mantle at depths between 150 and 300 kilometers, but perhaps extending as deep as 400 kilometers.

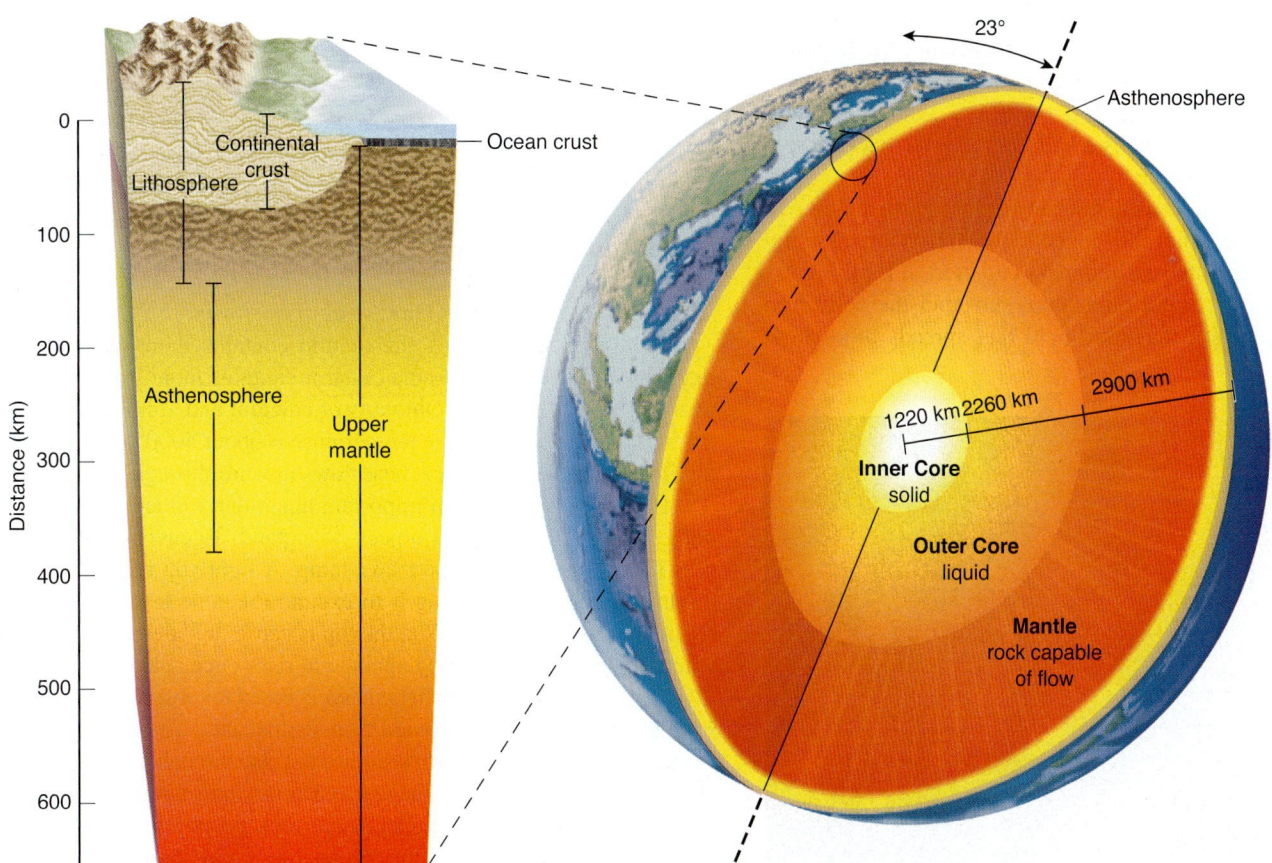

FIGURE 3.3 Evidence indicates that Earth's interior consists of the inner core, the outer core, the mantle, and the crust. The lithosphere consists of the crust (upper lithosphere) and the lower lithosphere, which is part of the upper mantle. The asthenosphere, also in the upper mantle, lies below the lithosphere.

In general, in which direction does heat move in Earth's interior?

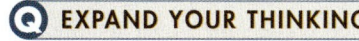 **EXPAND YOUR THINKING**

Because the crust is composed of lighter compounds than the mantle, is it appropriate to refer to it as "floating" on the mantle? What effect does this have? Explain.

3-2 The Core, Mantle, and Crust Have Distinct Chemical and Physical Features

LO 3-2 List Earth's internal layers and describe each.

Earth's interior consists of the core, mantle, and the crust; each has distinct chemical and physical features.

The Core

The inner part of Earth is the core, which is found about 2,900 kilometers below the surface. The core is a dense ball estimated to be composed of 90% iron and about 10% nickel, oxygen, and sulfur. The inner core—Earth's center—is solid and about 1,220 kilometers thick, while the outer core is about 2,260 kilometers thick. Pressure in the inner core is so great that it cannot melt, even though its temperature may exceed 5000°C.

The outer core, which experiences less pressure, is molten. Research indicates that because Earth rotates, the motion of the liquid outer core around the solid iron inner core creates the planet's **geomagnetic field** that surrounds Earth (**FIGURE 3.4**). Some scientists hypothesize that the core is actually a solid iron crystal that may spin faster than the planet as a whole.

Elementary physics tells us that when an electric current moves around a magnet, it intensifies the magnetic force. Iron is a naturally magnetic metal that, as noted earlier, makes up most of Earth's core. Electric currents in the liquid outer core surround the solid iron inner core and enhance a magnetic field that spreads throughout

FIGURE 3.4 The geomagnetic field emanates from the core, surrounds Earth (right) as a protective shield, and shelters sensitive living tissue from damaging solar radiation.

Why is the geomagnetic field not symmetrical around Earth in this image?

the planet. Although it cannot be seen, this magnetic field emerges through the surface near the poles and surrounds the planet.

Since 1845, when the German mathematician *Carl Friedrich Gauss* made the first measurements of Earth's geomagnetic field, the field's intensity has weakened by about 10%. This weakening suggests that the field is undergoing one of the frequent variations in its strength that are recorded in iron-rich rocks from throughout Earth's history. However, some scientists suggest that it presages a *reversal in magnetic polarity*, another process that has occurred at irregular, but frequent, intervals in Earth's history (for largely unknown reasons), with potentially disastrous effects. For one, the impact on our electricity-dependent technology could be devastating.

The magnetic field acts as a shield against solar radiation, which is very damaging to living tissue. The sunburn you feel after spending an hour in direct sunlight is a reminder of the power of solar radiation. Without the geomagnetic field, living tissue on Earth's surface would be fried to a crisp. Humans (and, in fact, all life on Earth) exist today only because the magnetic field protects us from the Sun's high-energy radiation. It does this by preventing high-energy particles from entering Earth's atmosphere. Thus, in a powerful way, life was made possible by the iron catastrophe.

The Mantle

The layer above the core is the mantle. Its upper surface is about 5 to 10 kilometers below the **oceanic crust** and about 20 to 80 kilometers below the **continental crust** (Earth's crust can be divided into an oceanic type and a continental type). The mantle accounts for nearly 80% of Earth's total volume. A unique property of the mantle is its *plasticity*; even though the rock is solid, high temperature and pressure enable it to deform and flow extremely slowly.

Compared to the core, the mantle contains more silicon and oxygen; while compared to the crust, it contains more iron and magnesium, as well as larger amounts of calcium and aluminum. Near the bottom of the mantle the temperature is about 4000°C, but at the top it is about 870°C. Hence, heat flows upward through the mantle.

Like gravity, heat is an important agent that influences our planet's structure. Heat causes objects to expand. To envision this process in the mantle, think of a lava lamp. A lightbulb in the base of the lamp heats wax, causing it to expand (become less dense) and rise to the surface. Cool wax at the top is denser and descends until it reaches the bottom, is heated again, and rises once again.

This mantle convection hypothesis is based on observations of the seismic waves that travel outward from earthquakes. These energy waves travel not only along Earth's surface but also through its interior. By observing how these waves change speed as they pass through rising and descending columns, or *plumes*, of mantle rock, geophysicists can calculate the shape of those convection columns and create three-dimensional models (**FIGURE 3.5**); these models reveal that Earth's interior is more complex than the simple layered model shown in Figure 3.3.

The Crust

As noted earlier, the upper mantle forms the lower portion of the lithosphere. The upper lithosphere is the solid outer shell that we live on, Earth's crust. The crust is composed of brittle rock that fractures easily. It varies in thickness and chemical composition, depending

Courtesy NASA Mashall Space Flight Center (NASA-MSFC)

on whether it is oceanic or continental. Entirely different geologic processes form each of the two types of crust.

• Oceanic crust is formed at places where iron- and magnesium-enriched magma emerges from the mantle onto the seafloor. The rock is dense and relatively lacking in lighter compounds compared to continental crust.

• Continental crust is generally composed of rock that is relatively enriched with respect to lighter compounds, and depleted with respect to iron and magnesium. Continental crust essentially "floats" on the denser mantle.

Most of the heat moving upward through the mantle comes out through the seafloor. However, typically under the continents, cold rock sinks slowly toward the base of the mantle, where it can recycle. Earth's interior is thought to experience convection similar to that in a lava lamp, only at much slower rates. Check out the Critical Thinking exercise "Mantle Plumes" to gain a better understanding of how mantle convection might work.

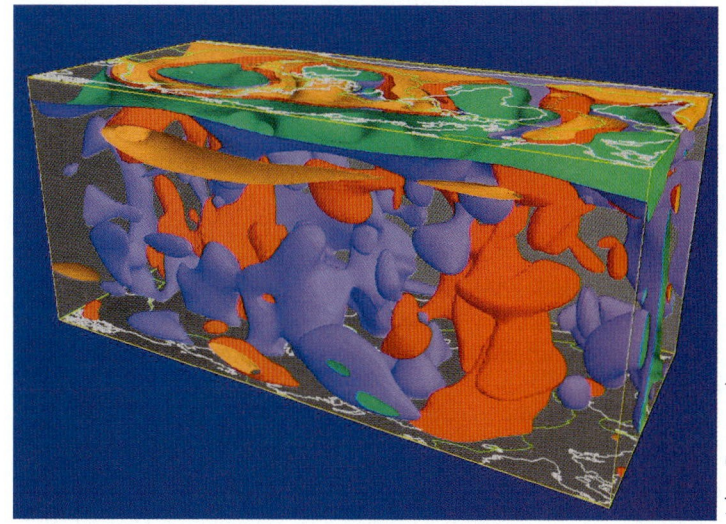

FIGURE 3.5 Earth's interior as viewed from the north. This complicated model of the mantle is based on records of the behavior of seismic waves passing through warm (red) and cool (blue) rock, indicating the structure of the planet is more complex than a simple sequence of uniform layers.

Adam Dziewonski

EXPAND YOUR THINKING

Imagine there had been no iron catastrophe. How would Earth be different today?

How are the red areas and blue areas going to change over time?

CRITICAL THINKING

Mantle Plumes

Fill a clear, tall glass container four-fifths with water and one-fifth with vegetable oil. Drop a pinch of salt at the top. The salt will form an irregular-shaped ball and sink to the bottom of the glass. Then watch as plumes of vegetable oil begin to rise through the water toward the surface. Add more salt so that multiple plumes develop at once.

1. What is the composition of the plumes? Why do they rise through the water?

2. What fundamental physical property causes the plumes?

3. What is the shape of the plumes? Draw them. Geologists hypothesize that plumes of hot rock many kilometers wide move upward through Earth's interior and come to a stop against the underside of the lithosphere, causing it to arch upward and crack in a zone that can be hundreds of kilometers across.

Your experiment creates an effect similar to that which appears in a lava lamp, and may simulate how rock migrates in the mantle (**FIGURE 3.6**).

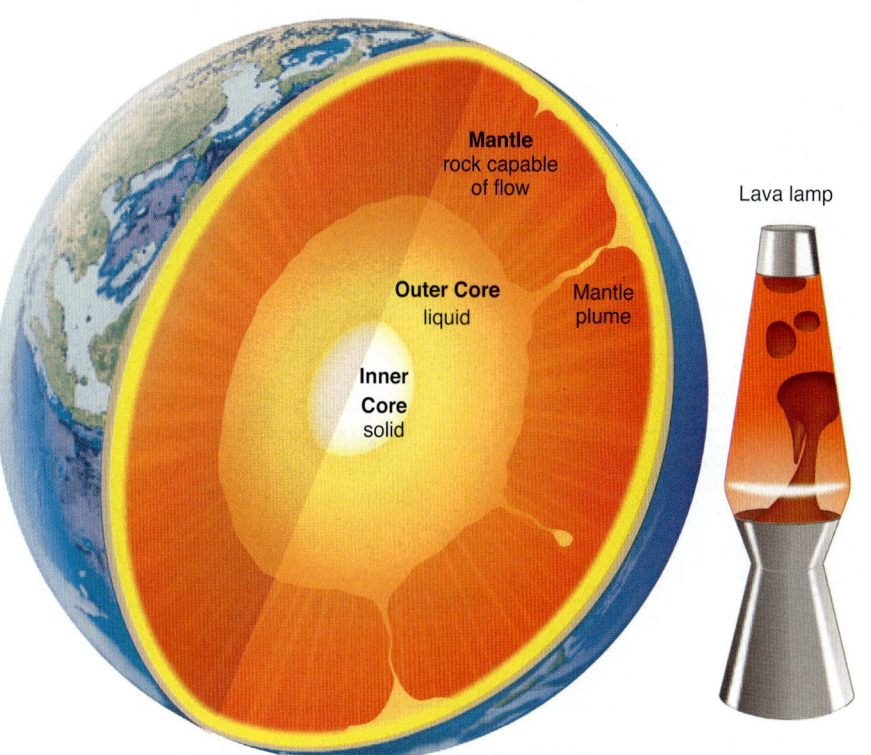

FIGURE 3.6 What is the nature of convection in the mantle?

3-3 Lithospheric "Plates" Carry Continents and Oceans

LO 3-3 Describe the origin and recycling of oceanic crust.

Much of our understanding of Earth's geology is explained by the remarkable theory of plate tectonics, which has revolutionized and unified the science of geology over the last few decades.

Lithospheric Plates

We live on the surface of a planet that is actively shaped by plate tectonics. Scientists generally agree that plate tectonics is possible only because Earth's interior is a source of heat that rises to the surface, and because the mantle experiences convection, as Harry Hess (introduced in Chapter 1) originally hypothesized. This convection contributes to the movement of lithospheric plates, thus keeping Earth's surface in constant motion.

Mountain ranges, ocean basins, continents, chains of oceanic islands, and other features of Earth's surface all owe their origin to, and can be explained in part by, the movement of massive lithospheric plates. Right now you are riding on a lithospheric plate. Which plate are you on? In which direction are you moving? How many different plates have you visited in your life? Before you finish reading this chapter, you will have the answers to these questions.

Tectonic theory predicts that Earth's lithosphere is broken up into plates that move and interact with one another (**FIGURE 3.7**). These plates move in response to forces in the mantle; as a result, **plate boundaries**, where plates bump into and grind against one another, are locations of great geologic change.

Plates come in a wide range of sizes; they may be hundreds or thousands of kilometers across. And because plates are portions of the lithosphere, a single plate may carry both continental and oceanic crust. Hence, plate movement by mantle convection is responsible for the drifting of continents; this fact also solves Alfred Wegener's problem of how this movement occurred and the supercontinent Pangaea came into being.

Twelve major plates have been identified. These are known as the Nubian, Arabian, Eurasian, Antarctic, Indian–Australian, Pacific, Philippine, South American, North American, Cocos, Nazca, and Caribbean Plates. (Note that some geologists consider the Indian and Australian Plates to be separate.) There are several minor plates as well, such as the Bismarck and Caroline Plates.

In places where two plates run into each other, mountains, volcanoes, and deep-sea trenches may be formed and earthquakes may occur. In places where two plates pull apart, long, deep **rift valleys** are formed along the tops of high mid-ocean ridges on the seafloor. Because plates move, the continents and oceans riding on them shift over time. Consider that the Atlantic Ocean exists today because the North American, European, Nubian, and South American Plates are moving away from one another, opening up an ocean basin between them.

Subduction Zones and Spreading Centers

A significant feature of many plates is that they slide or dive into the mantle along one edge and are renewed with magma welling

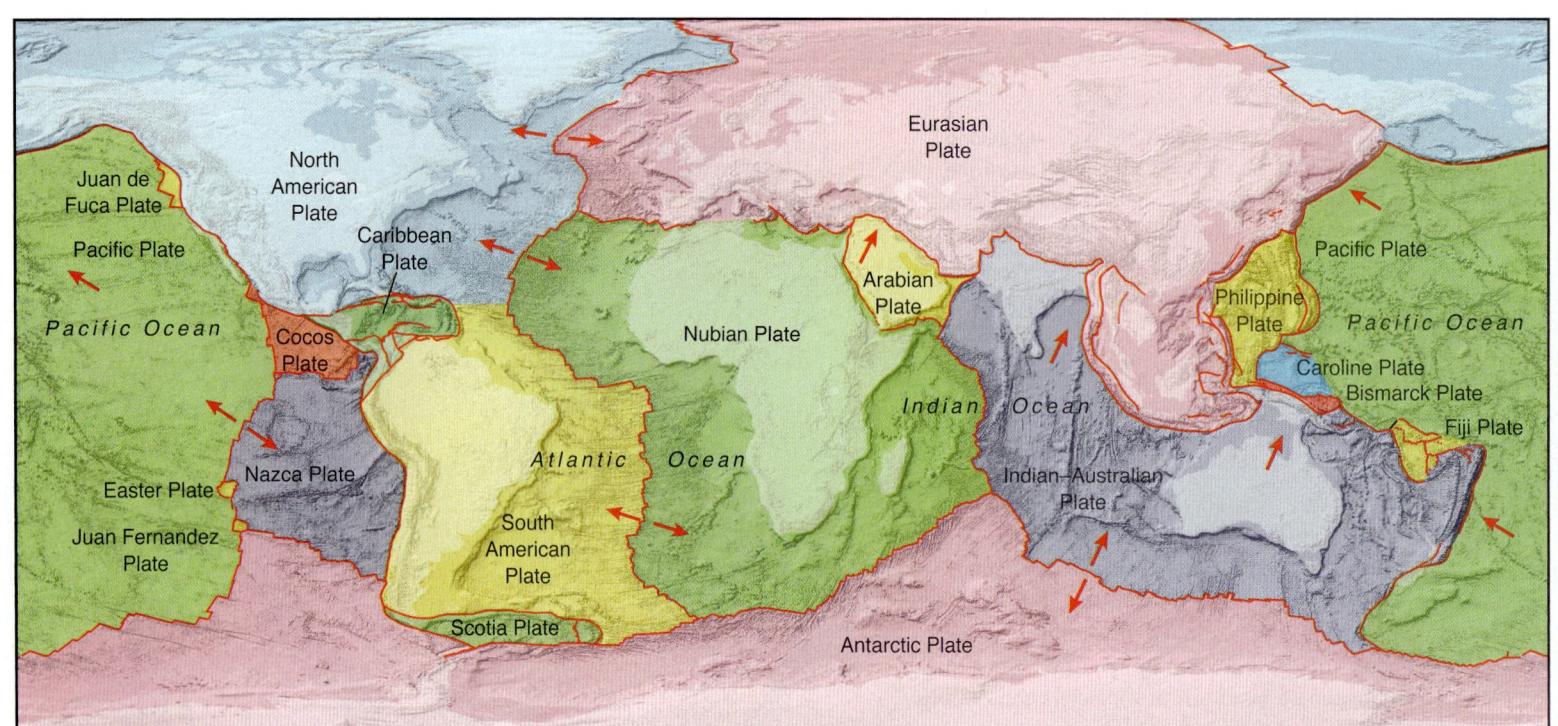

FIGURE 3.7 There are 12 major plates and several smaller ones. Plates are not static; they move several centimeters per year, on average about as fast as your fingernails grow, carrying continents and ocean basins across the planet's surface.

Using the directions of plate movement as your guide, circle places on this map where heat can most effectively escape from Earth's interior to the surface.

up from the mantle along another edge. Places where plates are recycled into the mantle—subduction zones—are characterized by deep-sea trenches where the ocean above the downward-curving surface of a plate is extremely deep.

At these locations, a denser plate dives, or subducts, beneath a less dense plate. If, say, the edge of one plate is characterized by continental crust (relatively enriched with the low-density molecule silica [SiO_2]) and is pushed into another plate whose edge is characterized by oceanic crust (which is relatively enriched in molecules composed of the dense elements iron and magnesium), the oceanic crust will subduct beneath the continental crust because it is denser (**FIGURE 3.8**).

While subduction is occurring along one edge of a plate, new lithosphere may be added to another edge of the same plate at a spreading center. Spreading centers are places where magma wells up from beneath the lithosphere and solidifies, adding fresh rock to the edge of the plate. Some geologists hypothesize that spreading centers are located above upward-moving convection currents in the mantle. Others disagree, proposing instead that spreading centers are complex cracks in the mantle that react to stress in the crust and, by lowering pressure, cause melting in the upper mantle.

Geologists are unsure exactly why spreading centers are formed. Are they the result of a massive upwelling of hot, buoyant rock pushing its way up through the mantle? Or are they cracks in the crust where lowered pressure causes rock in the upper mantle to melt? We know that at a spreading center the lithosphere arcs up, developing a wide bulge on Earth's surface at a mid-ocean ridge. But is it pushed up from below or swollen from the heat escaping from Earth's interior? These and other questions regarding how plate tectonics actually "works" are widely debated as part of the scientific process.

We do know that a spreading center is characterized by fresh crust, delivered from the mantle as magma enriched in Fe and Mg and depleted in Si and O. An opening, or *rift zone*, develops at the surface and is filled with this young rock. Rifting of a continent leads to the formation of a low valley, which may evolve into a narrow seaway (like the Red Sea) or a continental rift valley (like the East African Rift Valley). In time—tens of millions of years—rifted lithosphere may widen into an ocean basin with a mid-ocean ridge, the birthplace of young oceanic crust.

Between their birth at spreading centers and their recycling at subduction zones, plates move like conveyor belts on a one-way journey, taking tens of millions of years to get from one zone to the other. The formation and movement of lithospheric plates has shaped and moved our continents and ocean basins.

To match your understanding of world geography with your new understanding of plate tectonics, try the Critical Thinking exercise on "Global Tectonics."

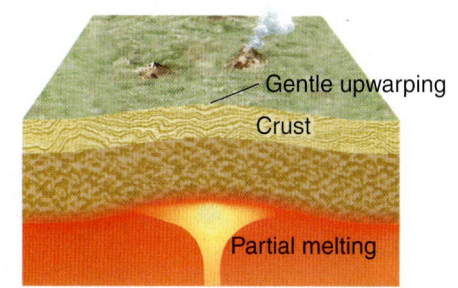

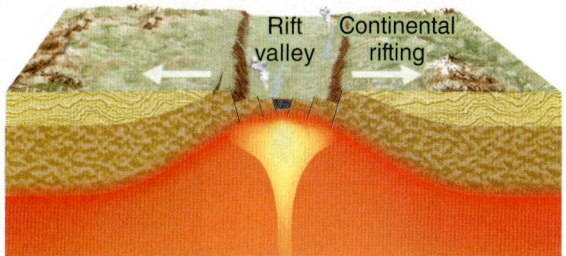

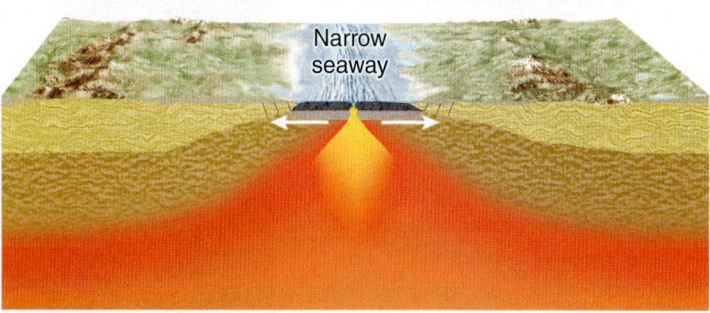

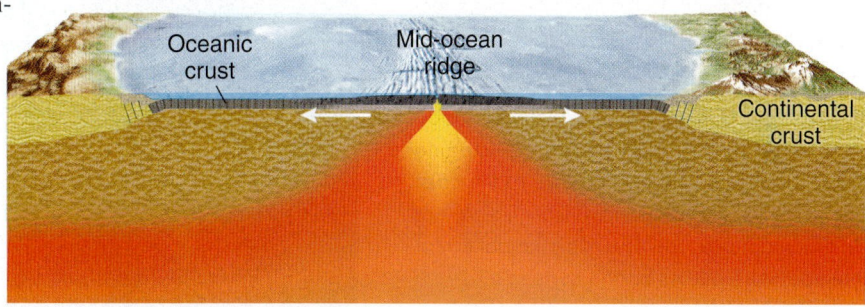

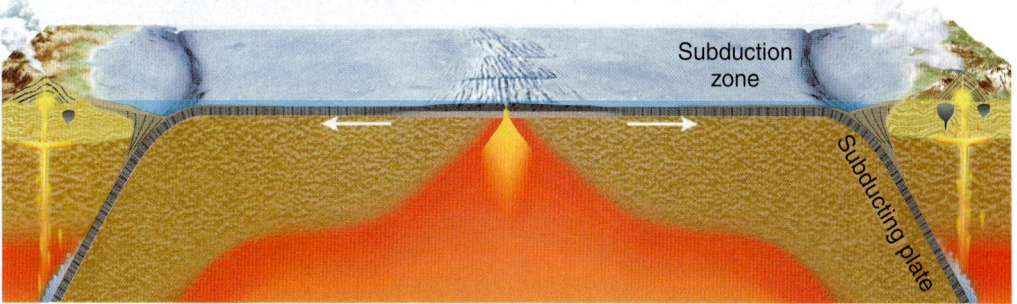

FIGURE 3.8 Lithosphere is produced at spreading centers and carried to subduction zones, where it is recycled into the mantle.

Ⓠ On Earth's surface what is an early sign that a rift is developing?

Ⓠ **EXPAND YOUR THINKING**

What is the significance of rock density in the movement of lithospheric plates?

CRITICAL THINKING

Global Tectonics

The map in **FIGURE 3.9** shows the topography of the continents and seafloor. Working with a partner, follow instructions 1 through 3 and then attempt to answer the following questions. (Note: You will need a pen.)

1. Draw the plate boundaries on the map and label each plate. Include arrows indicating the direction of plate movement.

2. Circle three subduction zones and three spreading centers; describe the topography at each.

3. Hypothesize a type of plate boundary motion where plates are neither converging nor diverging. What evidence would you look for to support your hypothesis?

4. Which plate do you live on? In which direction is it moving? Describe the processes acting along its boundaries, the types of boundaries it has, and where the boundaries are located.

5. If you had to substantiate the theory of plate tectonics, what evidence (direct observations) would you be able to draw from this map?

6. Heat from Earth's interior and plate boundaries are closely related. In what way? Imagine you made a map of the temperature of the crust (including the seafloor) using a satellite. What would that map look like? Describe the pattern of crust temperature across Earth's surface.

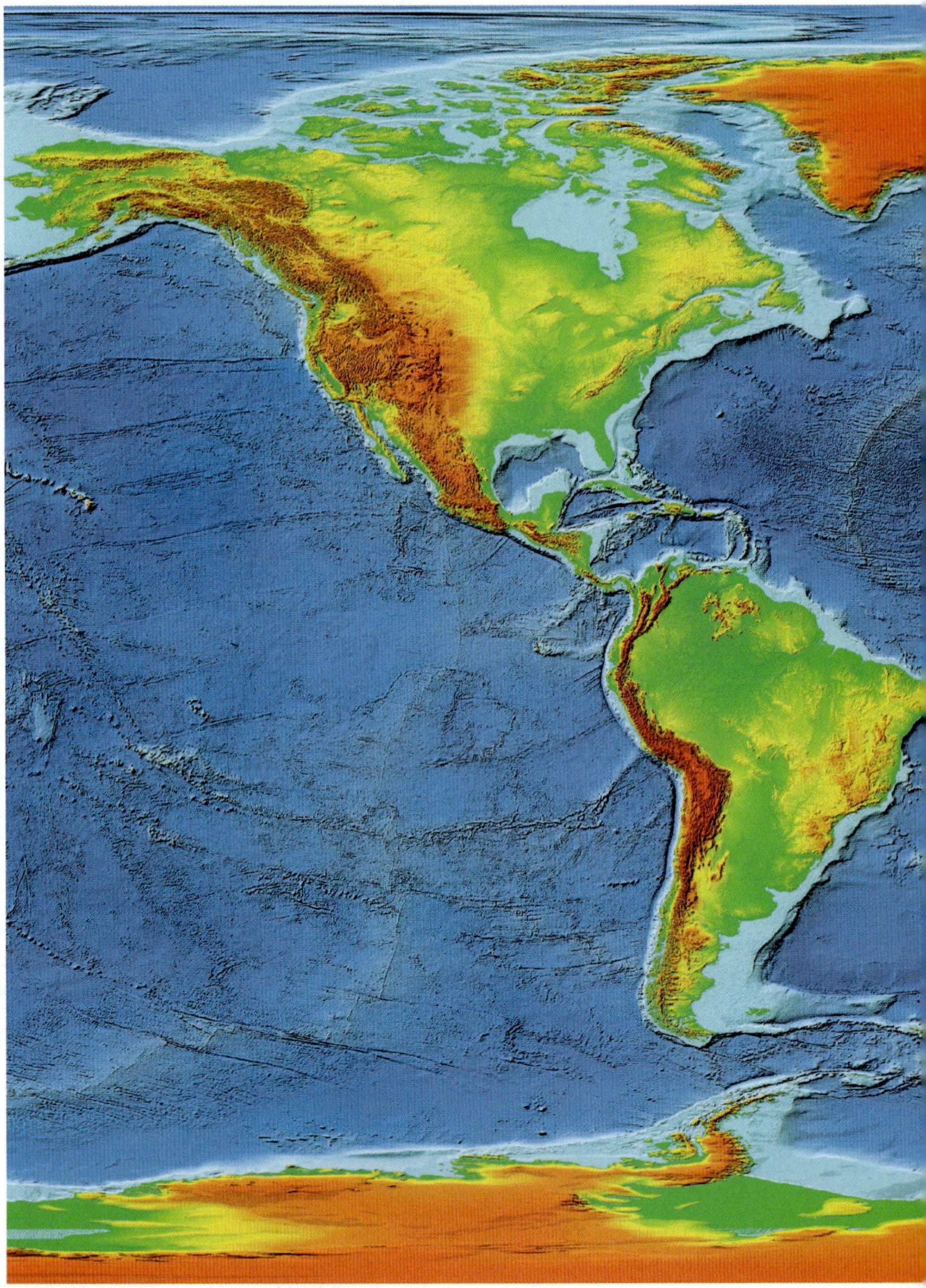

FIGURE 3.9 Global topographic map.

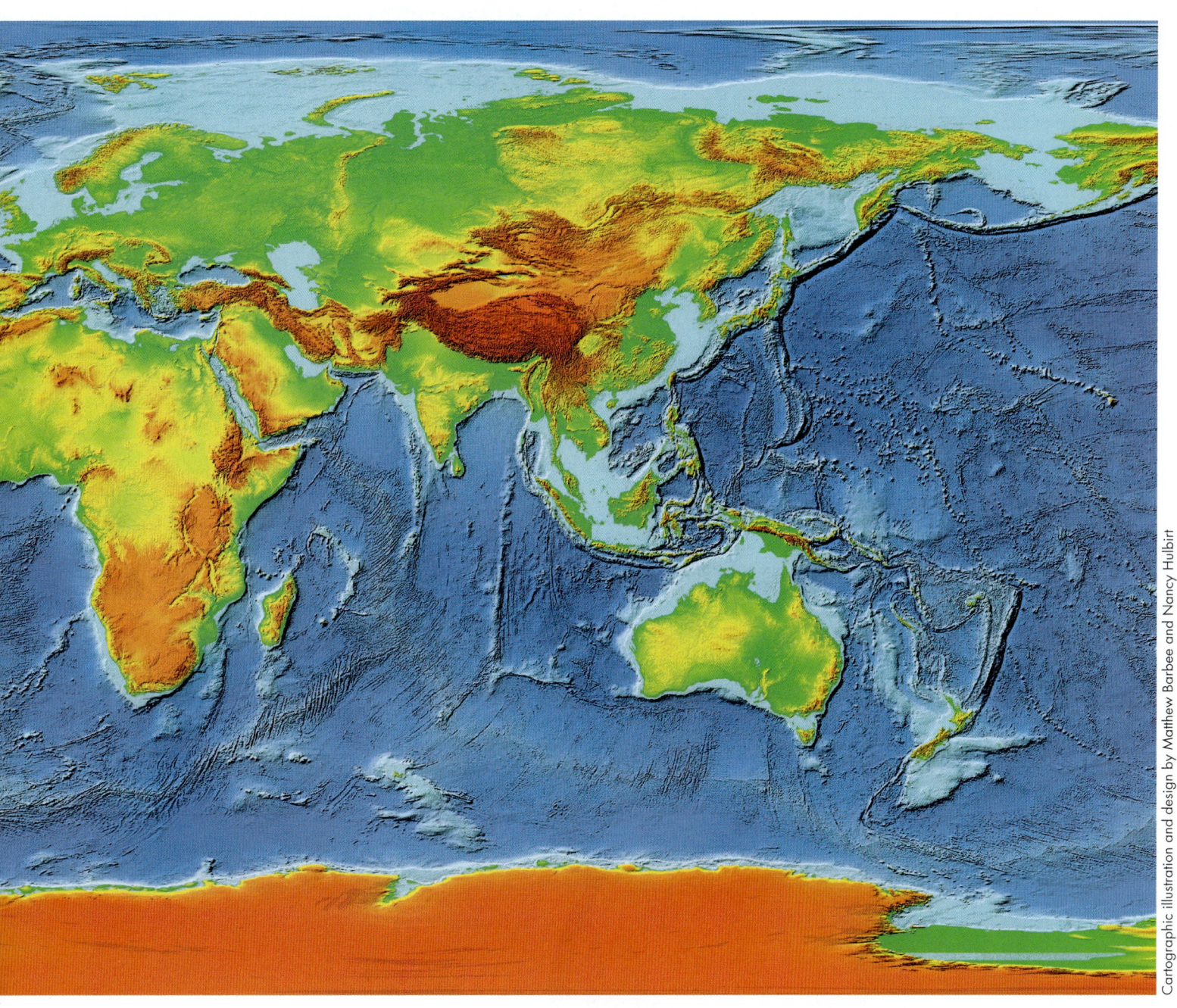

Cartographic illustration and design by Matthew Barbee and Nancy Hulbirt

Paleomagnetism Confirms the Seafloor-Spreading Hypothesis

3-4

LO 3-4 Identify the evidence that the polarity of Earth's geomagnetic field has reversed in the past.

Paleomagnetism, the study of past geomagnetic events, provides proof that seafloor spreading is a real process and that plate tectonics is actively changing the shape of Earth's surface.

A magnetic field is *polarized,* meaning that it is characterized by two equal but opposite states, or "poles." Earth's magnetic poles are named the "north magnetic pole" and the "south magnetic pole." They emerge from the planet's surface close to, but not exactly, where its rotational (geographic) axis is located (**FIGURE 3.10**).

Paleomagnetism

Notably, Earth's magnetic poles apparently have switched at various times in the past. Specifically, a compass needle that today points to the north magnetic pole would have in past times of reversed polarity pointed to the south magnetic pole. How do we know this? Because of **paleomagnetism** (*paleo-* is Greek for "old").

Iron-rich minerals are sensitive to the character of Earth's magnetic field and provide geologic evidence of its polarity. As magnetic minerals crystallize in molten rock, they incorporate into their atomic structure the magnetic orientation of Earth's field. Likewise, iron-containing sediment

particles settling through water physically rotate and align themselves to the geomagnetic field, like compass needles. When these particles are transformed into hard rock, their magnetic orientation is permanently recorded in it. The orientation of ancient geomagnetic fields preserved in rocks can be measured by instruments called *magnetometers.*

An example of a place where paleomagnetism is recorded is in cooling magma at spreading centers. Iron-rich minerals crystallizing in newly erupted oceanic crust acquire Earth's prevailing magnetic polarity. When new crust cools, this magnetic "signature" is preserved. Geophysicists' measurements of these minerals reveal characteristics of the geomagnetic field at the time during which the rock formed.

Much to the amazement of the scientific community at the time, paleomagnetic studies conducted in the 1950s indicated that the polarity of the geomagnetic field has reversed numerous times in the past. Most scientists believe that these changes in polarity result from random, chaotic instabilities in the way in which liquid iron moves in the core. In other words, every once in a while (at intervals ranging from hundreds of thousands to millions of years), Earth's magnetic field changes because (presumably) the movement of liquid iron in the core changes in some significant way. Exactly why it changes, or exactly when it will change again, is unknown. What is known is that the magnetic poles have reversed hundreds of times throughout Earth history—and likely will again.

Geologists studying this phenomenon recognize that rocks can be divided into two groups based on their magnetic properties. One group has *normal polarity,* characterized by magnetic minerals with the same polarity as today's magnetic field. The other group has *reversed polarity.* The age of a group of rocks showing a particular pattern of reversals can be determined by comparing it to a paleomagnetic reference sequence whose age has been determined by other methods of dating using geochemical techniques (**FIGURE 3.11**).

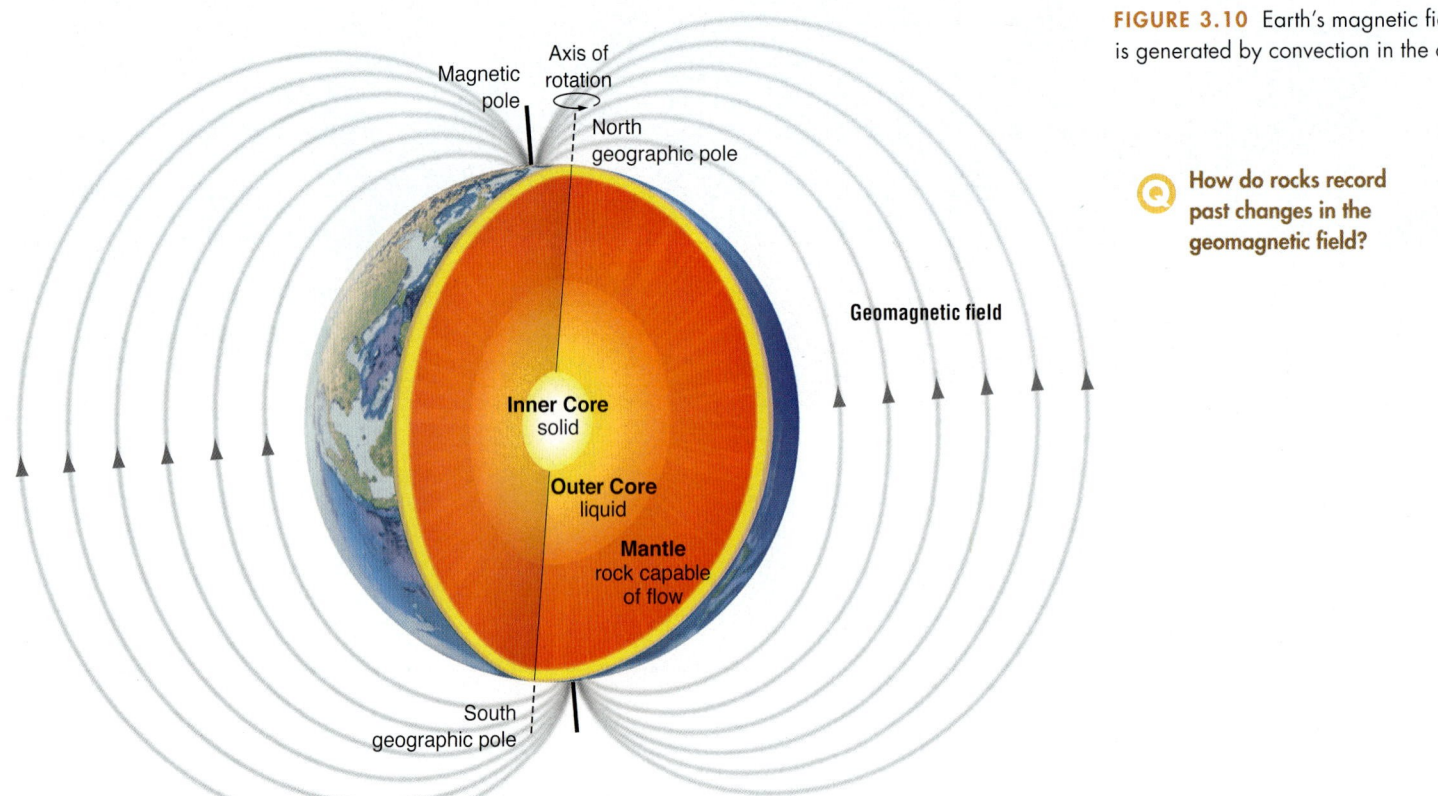

FIGURE 3.10 Earth's magnetic field is generated by convection in the core.

Q How do rocks record past changes in the geomagnetic field?

Axis of rotation

Magnetic pole

North geographic pole

Geomagnetic field

Inner Core solid

Outer Core liquid

Mantle rock capable of flow

South geographic pole

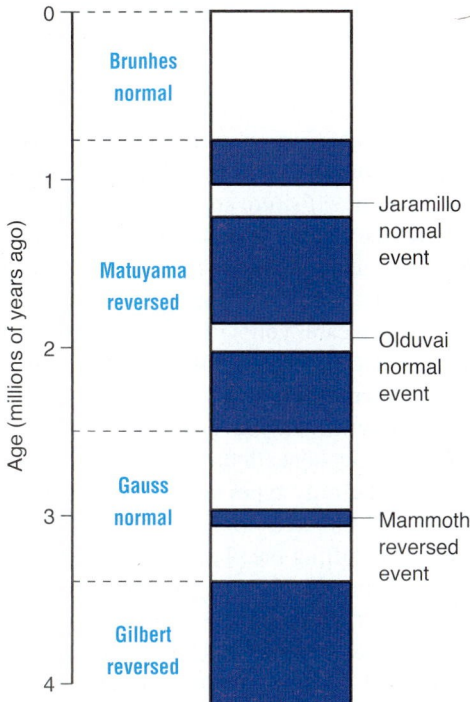

FIGURE 3.11 The history of global polarity for the past 4 million years consists of major "polarity reversals" that last for long periods, and minor "polarity events" that are short-lived. Minor polarity events are named for geographic locations where these paleomagnetic events were discovered (Jaramillo Creek, New Mexico; Olduvai Gorge, Tanzania; Mammoth, California). Major polarity reversals are named for scientists who made important contributions in the field of paleomagnetism: Bernard Brunhes was a French geophysicist who discovered the phenomenon of geomagnetic reversal; Motomori Matuyama was the first to surmise that the geomagnetic field had undergone reversals in the past; Carl Friedrich Gauss made numerous contributions to understanding magnetism; Grove Karl Gilbert was an American geologist who made fundamental contributions to the science of geomorphology (the study of landscape processes).

Q What kind of plate boundary produces a record of seafloor magnetism?

Seafloor Magnetism

Paleomagnetism is a valuable tool for understanding Earth's history, especially at **divergent plate boundaries**, where two or more plates pull away from each other. Because new rock, in which iron-rich minerals are abundant, is formed at oceanic spreading centers, the seafloor records changes in polarity through time. As it moves away from the spreading center, the new rock carries with it a distinct magnetic signature. In this way, the moving oceanic crust preserves a record of switching polarity (**FIGURE 3.12**).

Oceanic crust formed during a period of normal polarity preserves a record of that polarity. When global polarity switches, crust forming at that time will record the change. Hence, a nearly symmetrical pattern of magnetic polarity is found on either side of an oceanic ridge as the seafloor continues to spread over time. Mapping the paleomagnetism of the ocean floor adjacent to spreading centers is an important test of seafloor spreading, one that has confirmed the basic elements of the theory.

Geologists make maps of seafloor magnetism that depict a series of stripes on the seafloor, known as *magnetic striping*. Magnetic striping of oceanic crust usually produces nearly symmetrical patterns with an oceanic ridge in the center. This process provides strong support for two basic elements of plate tectonic theory: 1) that new lithosphere is created from rock emerging from an upwelling source in the mantle, and 2) that the crust migrates away from these plate boundaries.

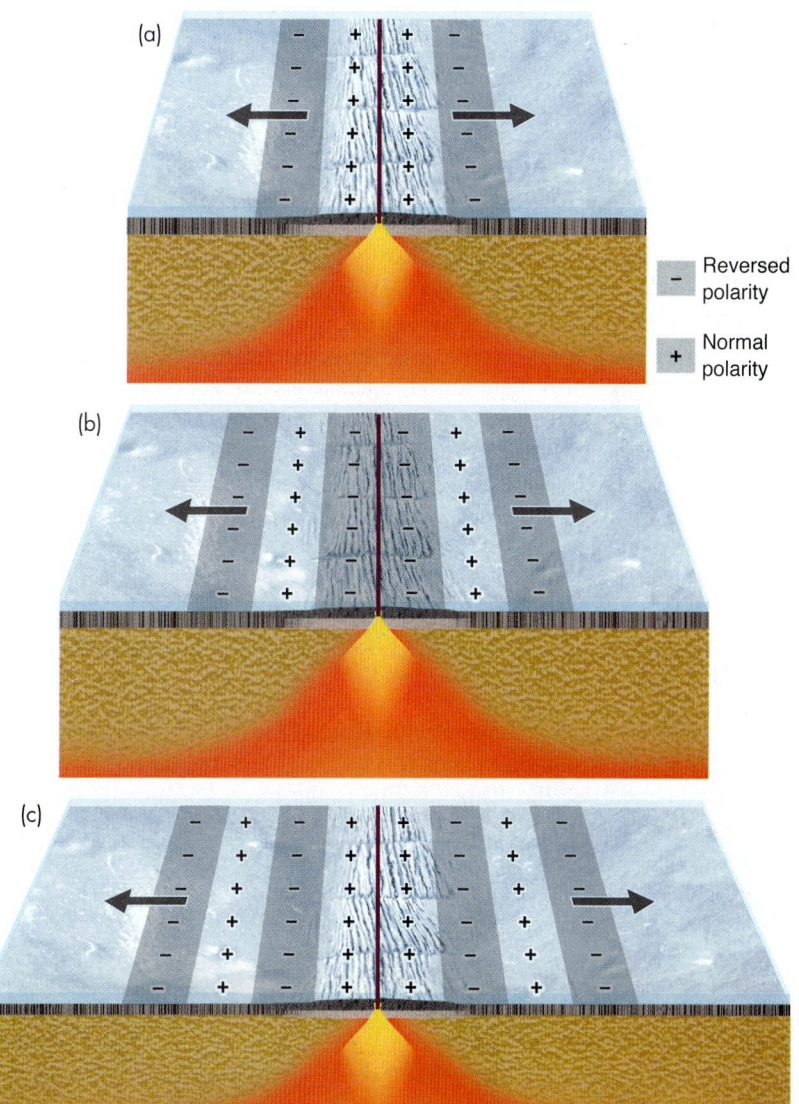

FIGURE 3.12 New oceanic crust is formed continuously at the crest of an oceanic ridge. (a) A spreading ridge during a period of normal magnetism, (b) followed by a period of reversed magnetism, (c) followed by a period of normal magnetism.

Q Why is seafloor magnetism considered strong evidence of plate tectonics?

Q EXPAND YOUR THINKING

Based on your reading of the first two chapters and so far in this one, describe the role of critical thinking in the development of the theory of plate tectonics.

3-5 Plates Have Divergent, Convergent, and Transform Boundaries

LO 3-5 List the three types of plate boundaries.

Plate tectonics is far more than just an academic theory of how Earth's surface works. It is a critical tool that is used in the hunt for metals, groundwater, fossil fuels and other geologic resources that we depend on. Plate tectonics is also the basis for our understanding of the most dangerous geologic hazards such as tsunamis, earthquakes, volcanic eruptions, and others.

Since it was first proposed, the concept of seafloor spreading has developed from a hypothesis into the modern theory of plate tectonics. It is used by exploration geologists (geologists who search for geologic resources) to assist them in their hunt for oil, coal, and natural gas deposits. The theory can be used to successfully predict in which areas valuable minerals may be found. The metals and fossil fuels you use today are cheaper and easier to find because of the predictive power of tectonic theory. The theory also can be used to successfully predict the location of important geologic hazards, such as volcanoes and earthquakes. Earthquakes and volcanic eruptions do not strike randomly, but occur most often (though not exclusively) along the boundaries where plates meet.

The Ring of Fire

One such area is the circum-Pacific **Ring of Fire**, where the Pacific Plate meets surrounding plates (**FIGURE 3.13**). Bordered by active plate boundaries, the Ring of Fire is the most seismically and volcanically active zone in the world. People living in the Pacific region are, generally speaking, at greater risk of experiencing geologic hazards than are people living anywhere else in the world.

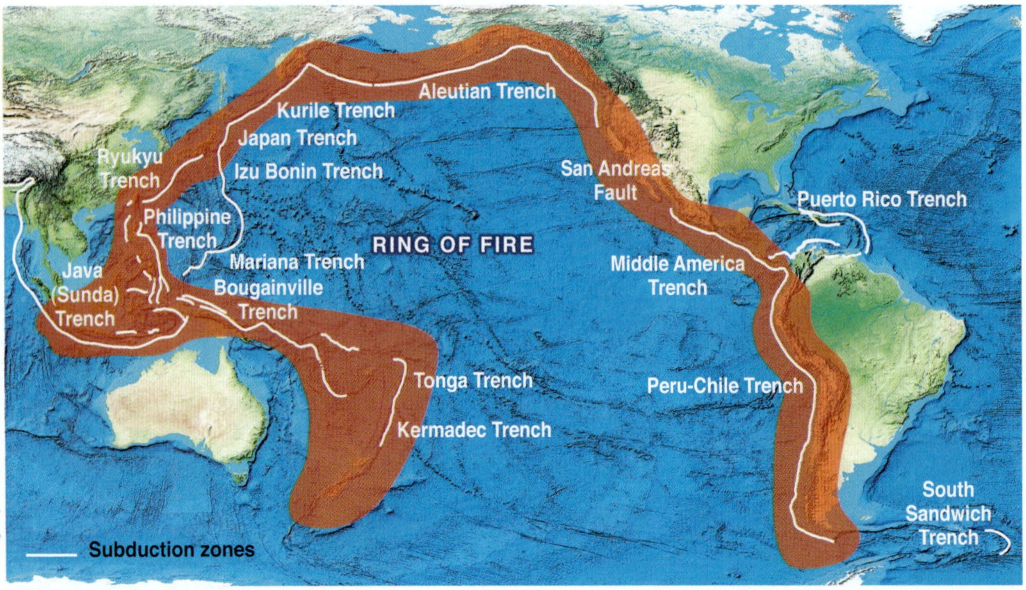

Plate Boundaries

To understand the dramatic processes that occur at plate boundaries, it helps to use **reasoning by analogy**, another tool of scientific thinking. Here's how it works: Picture an orange. Peel the orange so that the skin is in, say, six pieces. Then fit the pieces back together on the orange. Now try sliding one of the pieces across the surface and observe how it interacts with the others.

If you do this with care, you will see three types of interaction: pieces *spreading apart* (**divergent boundaries**), pieces *sliding past one another* (**transform boundaries**), and pieces *pushing together* (**convergent boundaries**). If you push two pieces together hard enough, one will subduct beneath the other, or the two will crumple together. These are the only types of interaction that are possible when solid pieces are forced to move around a sphere. It is one of these motions that is shifting the plate beneath you.

Because Earth is nearly a sphere, the same processes you observed in the experiment with the orange peel segments apply to lithospheric plates. As plates move across Earth's surface, three types of boundaries are formed. These are defined by their relative motion, as follows and as shown in **FIGURE 3.14**:

1. *Divergent boundaries*: New lithosphere is formed as plates pull away from each other.

2. *Convergent boundaries*: Lithosphere is subducted as one plate dives beneath another, or two plates collide head-on without either one subducting.

3. *Transform boundaries*: Lithosphere is neither formed nor recycled; plates simply grind past each other.

Divergent Boundaries

Divergent boundaries occur in areas where two plates are moving away from one another due to seafloor spreading. Seafloor spreading creates new oceanic crust from iron-rich magma that wells up out of the mantle. How does this happen?

Rifting (tearing open) of the lithosphere at a divergent boundary exposes rock in the mantle; as it rises under the rift zone, there is a dramatic decrease in pressure. The decreased pressure causes the rock to melt. The sudden release of pressure that causes melting is similar to what happens when you suddenly open a can of soda and the carbon dioxide contained within bubbles forth. The carbon dioxide is converted from its dissolved state into gas bubbles, and rock in the mantle is converted from solid to liquid (melted). *Decompression melting* creates magma,

Cartographic illustration and design by Mathew Barbee and Nancy Hulbirt

FIGURE 3.13 The Pacific Ocean is surrounded by plate boundaries where volcanoes, earthquakes, and tsunamis have the highest probability of occurring.

Describe the role of plate tectonics in making the Ring of Fire.

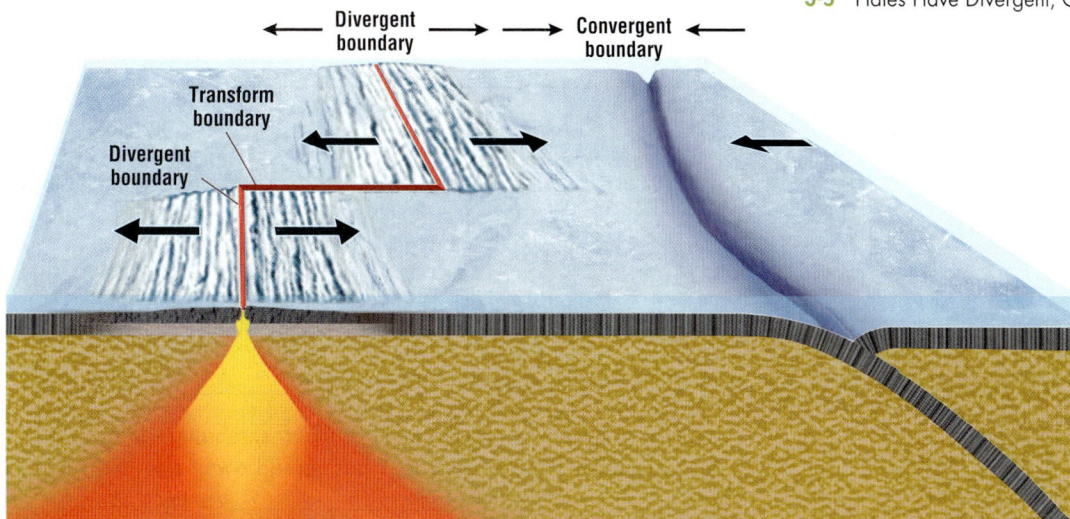

FIGURE 3.14 Geologists recognize three types of plate boundaries: divergent, convergent, and transform.

Describe why the crust is younger at divergent plate boundaries than at convergent plate boundaries.

which fills the opening in the lithosphere made by rifting. This rock becomes new lithosphere forming high-density, iron-rich oceanic crust (**FIGURE 3.15**).

As newly formed seafloor is transported away from the rift zone on a moving plate, it is replaced by mantle upwelling from below. In this way, the seafloor acts like a conveyor belt carrying new lithosphere away from the rift zone. Earlier we discussed how this process can open up space between continents and lead to the formation of a narrow shal-low sea. Rifting in the Red Sea is one such example; it is widening as Saudi Arabia moves away from Africa (refer back to Figure 3.9).

Seafloor spreading leads to the formation of new crust that, compared to continental crust, is relatively enriched in iron and magnesium and depleted in silica (SiO_2), because it reflects the chemistry of the mantle. As two plates continue to move apart, the rock in the seafloor grows older as its distance from the rift zone increases; and as it ages, it cools and becomes denser and is buried under marine sediments that are deposited on the seafloor.

The crust is brittle, meaning that it has a tendency to fracture. Hence, during rifting, the edges of the plates typically break into blocks that slide downward along parallel fracture surfaces. Earthquakes occur when the crust breaks and the rift valley widens. In Figures 3.8 and 3.15 you can see that this breakage occurs both at the edges of the continent, where the rifting originated, and along the boundary between the oceanic plates in the active rift zone.

The motion of opening and the downward settling of the blocks form a central rift valley marking the plate boundary at the spreading center. These blocks tilt slightly outward, away from the center of the valley. Magma migrates upward into the rift valley along fracture surfaces among the blocks. Submarine volcanic action is common in the rift zone.

Divergent boundaries on land form rift valleys that are typically 30 to 50 kilometers wide. The East Africa rift in Kenya and Ethiopia and the Thingvellir rift in Iceland (see the opening photo of this chapter) are examples. Rift valleys on the seafloor tend to be much narrower, and they run along the tops of mid-ocean ridges. To learn more about seafloor spreading, go to the Critical Thinking feature titled "Opening the North Atlantic Basin" and work through the exercise of measuring the history of the Atlantic Ocean basin as it opened up following the breakup of Pangaea.

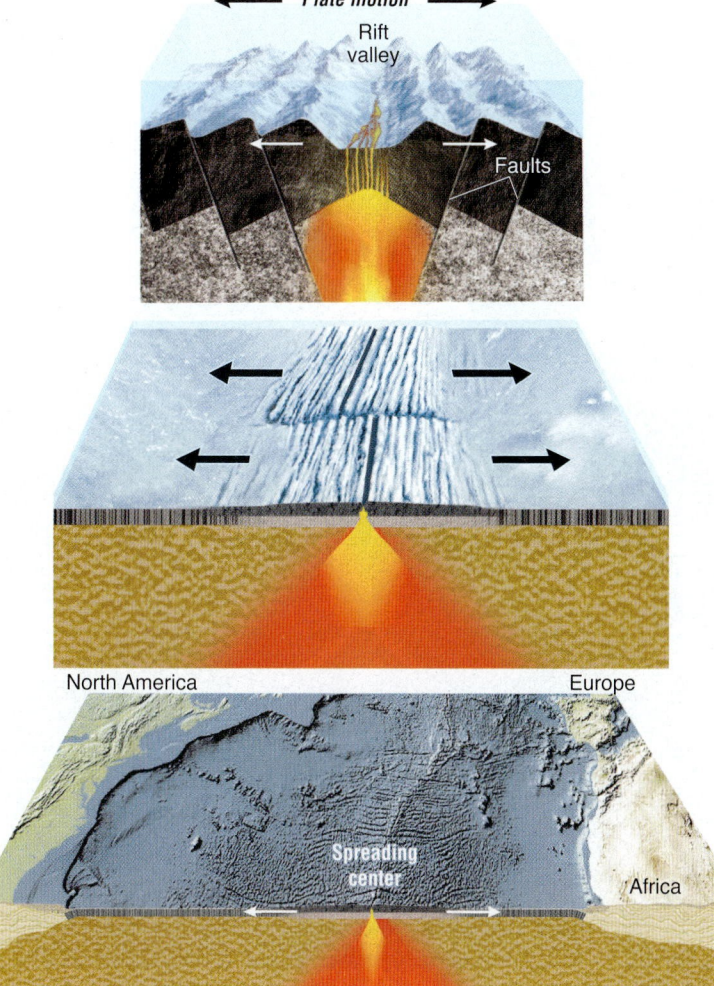

FIGURE 3.15 In places where two plates are moving apart, a central rift valley is formed, characterized by earthquakes and submarine volcanism.

Predict what geologists working in a research submersible would discover if they dove into an active submarine rift valley.

Q **EXPAND YOUR THINKING**

How would you test the hypothesis that there are three types of plate boundaries?

CRITICAL THINKING

Opening the North Atlantic Basin

(Note: You will need a ruler, colored pencils, and a simple calculator for this exercise.)

FIGURE 3.16 is a map of the seafloor between Africa and North America. These two landmasses were last joined when they were part of the supercontinent known as Pangaea. The line labeled "0" is the Mid–Atlantic Ridge, where new seafloor is being formed. The age of the seafloor—in millions of years before the present—is marked on either side of the ridge. The distance between point A on North America and point B on Africa is 4,550 kilometers.

Select a point on the seafloor and record its age. Measure the distance it has moved from the oceanic ridge where it formed. Use the scale on the map to measure the distance with a ruler.

1. Age of seafloor: _____million years (my).

2. Distance to the Mid–Atlantic Ridge: _____(km).

3. Calculate the *half-rate* of seafloor spreading—that is, the velocity at which this rock has spread away from the ridge. (distance/time = velocity): _____ km/my.

4. What is this rate in centimeters per year? _____cm/yr.

5. Calculate the total rate of ocean widening (2 × half-rate = ocean widening rate):_____ cm/yr.

6. What is it in kilometers per million years? _____ km/my.

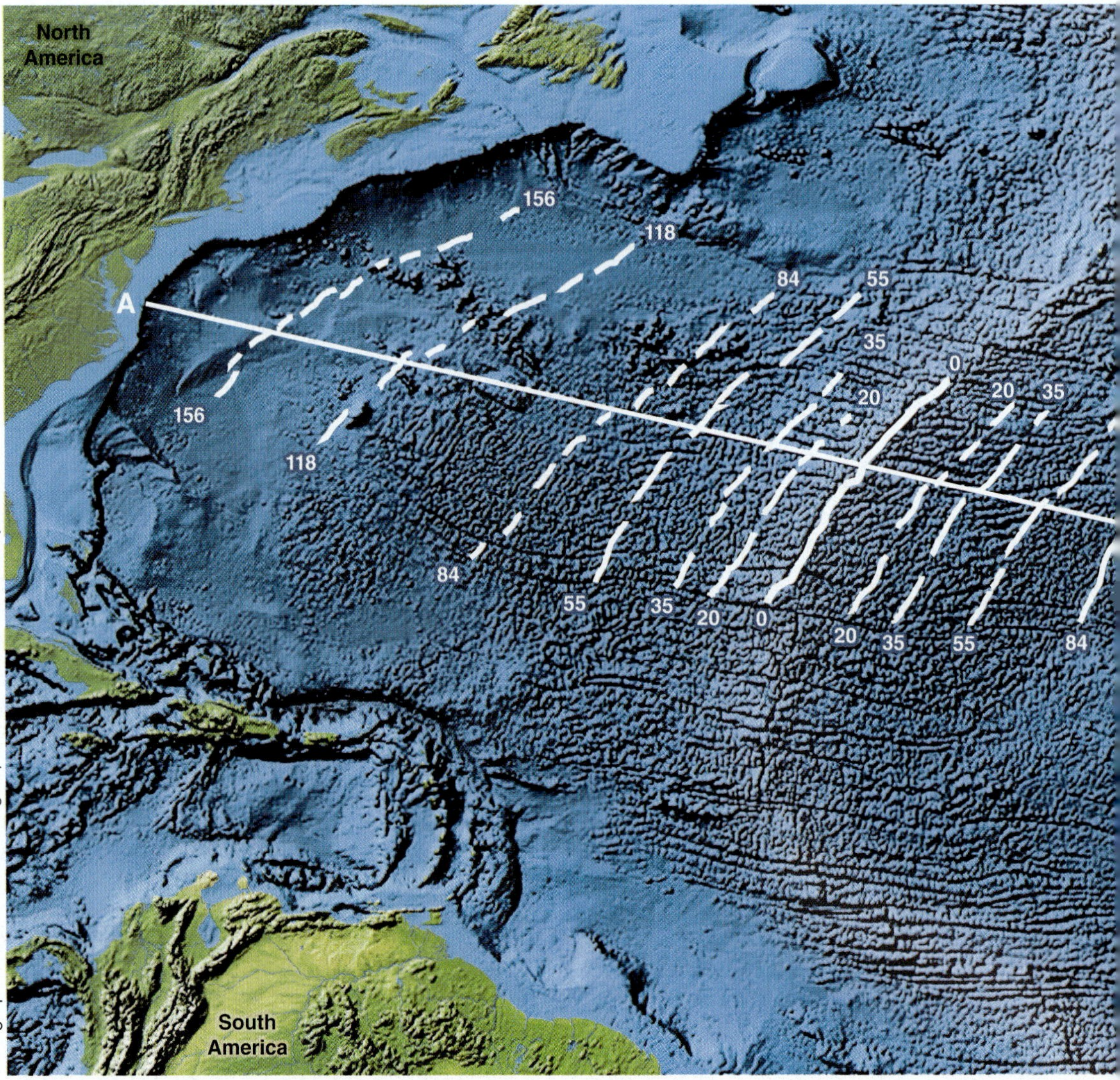

FIGURE 3.16 History of the North Atlantic Ocean.

Cartographic illustration and design by Matthew Barbee and Nancy Hulbirt

7. What is the age of the North Atlantic Basin?
_____(my).

8. Name the geologic period during which the North Atlantic basin began to open up. (Use the Geologic Time Scale in **FIGURE 3.17**.) _____.

9. How much has the distance between North America and Africa increased since you were born? _____.

10. How much does the distance increase during the average lifetime of an American (~82 yr)? _____.

11. An important phenomenon has been observed by researchers about the rate of seafloor spreading: It has not been constant through geologic time.

   a. How does this bear upon your calculations in the preceding questions?

   b. How might you correct for this observation in order to raise the accuracy of your answers?

12. Make a map of magnetic striping. Using the simplified paleomagnetic time scale in Figure 3.17, draw on the map with colored pencils to indicate approximately the location of normal- and reversed-polarity sections of the seafloor.

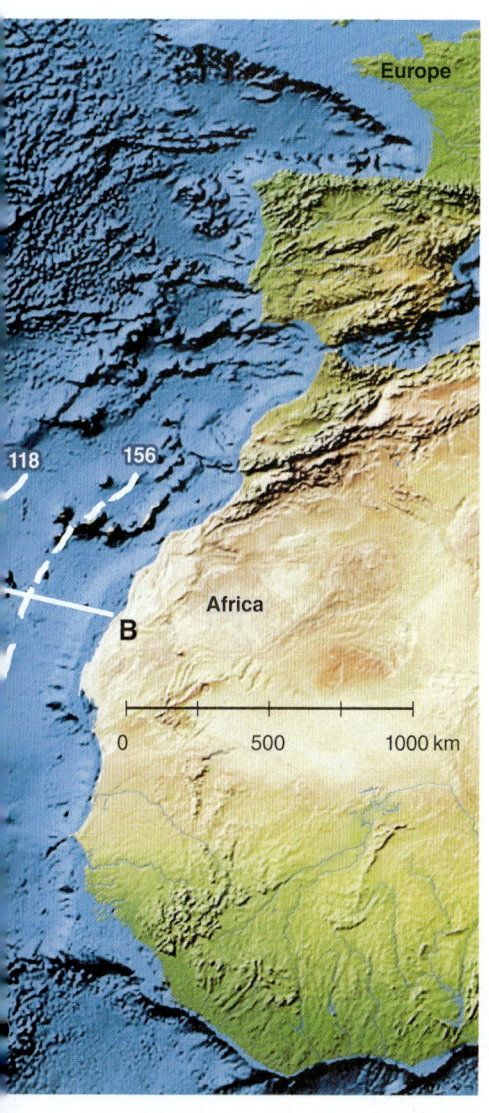

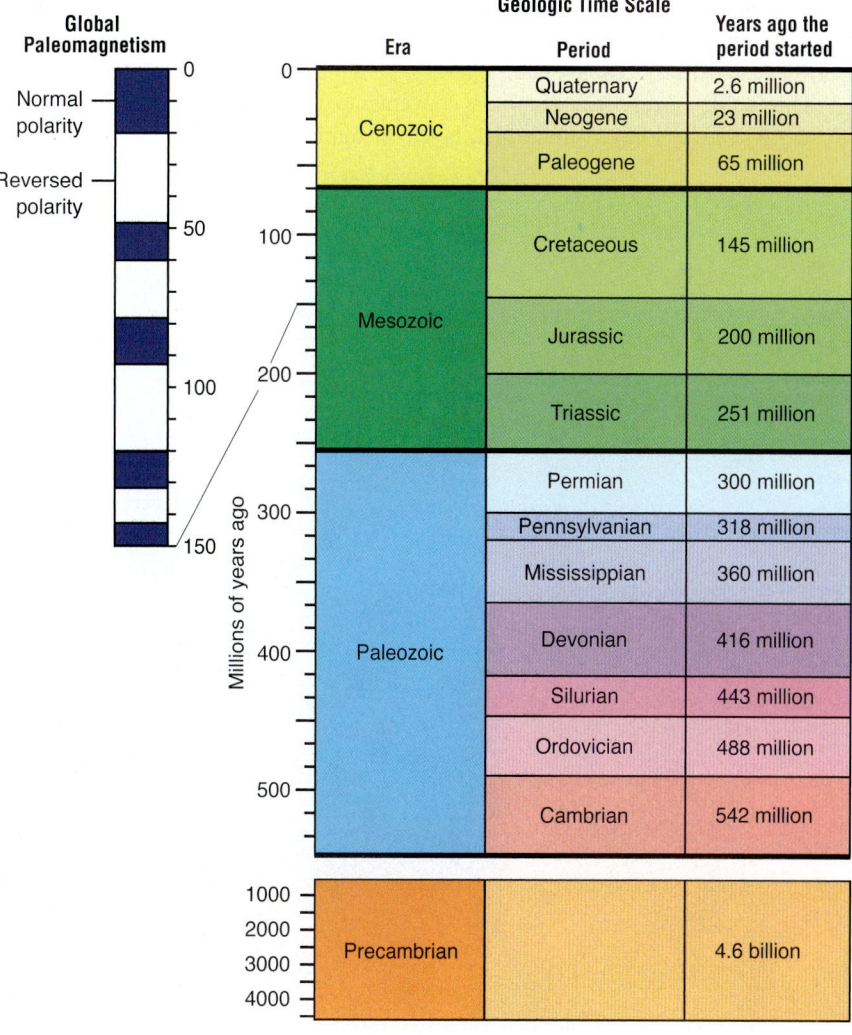

Global Paleomagnetism

Normal polarity

Reversed polarity

	Geologic Time Scale		
	Era	Period	Years ago the period started
	Cenozoic	Quaternary	2.6 million
		Neogene	23 million
		Paleogene	65 million
	Mesozoic	Cretaceous	145 million
		Jurassic	200 million
		Triassic	251 million
	Paleozoic	Permian	300 million
		Pennsylvanian	318 million
		Mississippian	360 million
		Devonian	416 million
		Silurian	443 million
		Ordovician	488 million
		Cambrian	542 million
	Precambrian		4.6 billion

FIGURE 3.17 Time scales.

3-6 Oceanic Crust Subducts at Convergent Boundaries

LO 3-6 Describe the processes occurring at ocean-continent and ocean-ocean convergent boundaries.

The type of rock found at two converging plates is an important factor controlling what happens at the boundary between them. There are three types of convergent boundaries, each named for the type of crust present: ocean-continent, ocean-ocean, and continent-continent.

Ocean-Continent Convergence

FIGURE 3.18 illustrates an **ocean-continent convergent boundary**. Here, a plate composed of oceanic crust has collided with one composed of continental crust. Keep in mind that oceanic crust is relatively high in iron and magnesium, and is dense and thin, so it subducts, or goes under, the continental plate. The plate composed of continental crust is built of rock that is relatively enriched in low-density compounds. It "floats" on the denser mantle and naturally overrides the denser oceanic plate.

The geology at an ocean-continent convergent zone is quite complicated. Geologists have found thick layers of marine sediment on the oceanic plate. This sediment, consisting of clay particles washed off the continents and *microskeletal sediment* from plankton living in the ocean, has collected on the oceanic plate since it first formed at the spreading center. In general, the older a plate is, the thicker the layer of sediment covering it. (Sediment thickness can reach hundreds of meters or more.)

As a plate subducts, some of this sediment is scraped off and collects, or "accretes," as a series of angular rock slabs, forming a wedge along the front of the overriding plate. This process is similar to what happens when snow accumulates in front of a snowplow. The wedge is known as an **accretionary prism** because of its three-dimensional shape.

Continued accretion pushes the prism into a ridge called a *fore-arc ridge*; on the landward side of this ridge, a depression known as a *fore-arc basin* can develop; such a basin naturally will collect sand and mud coming off the continent. These features, the ridge and the sediment-filled basin, form the *continental shelf* that marks the front edge of the continental lithosphere. Some of the sediment is subducted along with the oceanic crust.

Our knowledge of the geologic processes occurring at convergence zones is complicated by the difficulty of studying these large-scale features. At some locations, it appears that ocean-continent convergence leads to accretion, as just described. At other locations, researchers have proposed that *tectonic erosion* actually scrapes rock off the underside of the overriding plate. Rather than building a continental shelf by forming an accretionary prism, tectonic erosion removes rock from the underside and leading edge of the overriding plate. Geologists hypothesize that thick layers of sediment on a subducting plate tend to produce accretion, whereas thin layers of sediment lead to erosion.

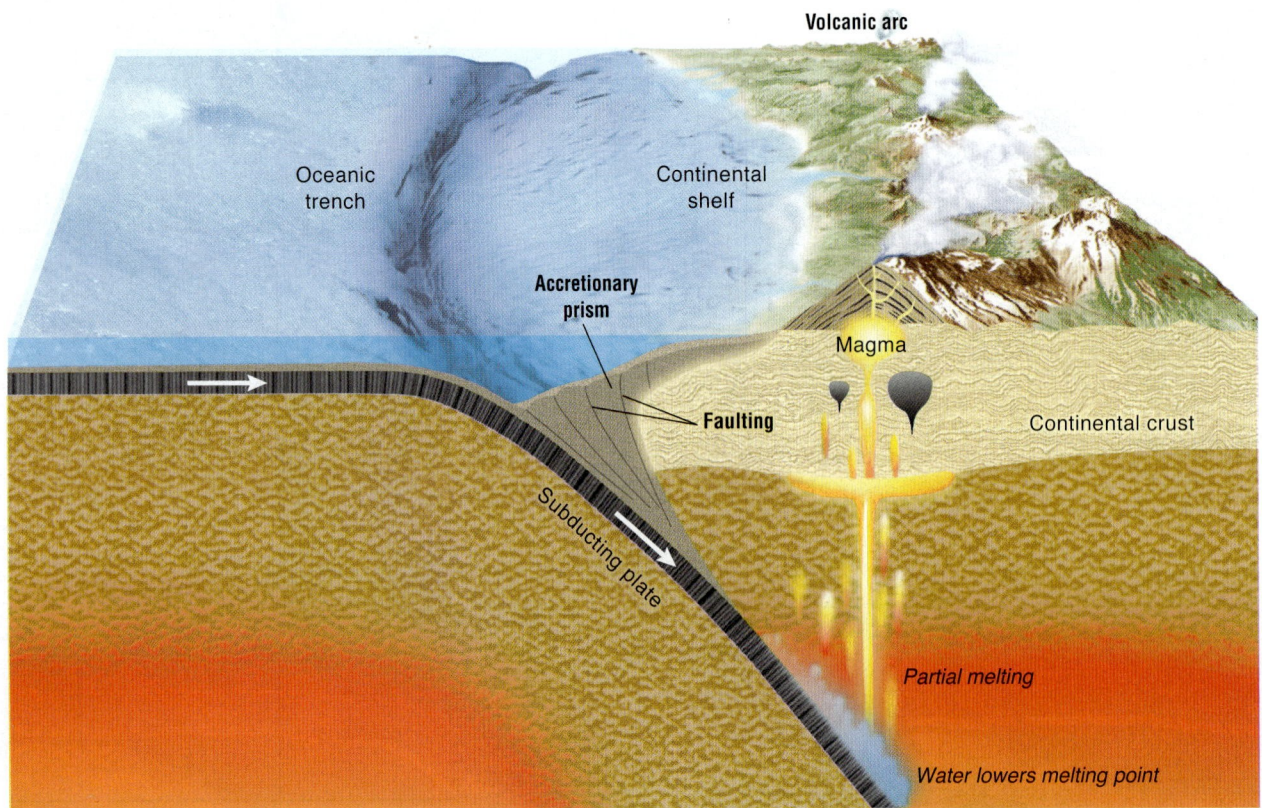

FIGURE 3.18 At many sites, ocean-continent convergence is characterized by the growth of an accretionary prism and a line of volcanoes called a volcanic arc.

What is the source of rock in an accretionary prism?

Releasing Water

As an oceanic plate subducts into the mantle, it experiences an increase in temperature and pressure. These increases result in the release of large amounts of water and other volatile fluids and gases trapped within the marine sediments and oceanic crust.

Released water floods upward into the hot rocks of the upper mantle and causes minerals there to recrystallize into new minerals that melt more easily than the original ones. This widespread melting is caused by the lower melting point of the recrystallized minerals, similar to the way by which salt causes ice to melt prematurely; it lowers the melting point of the rock. Hence, as this water rises into the mantle, the rocks there melt, forming magma.

Magma generated above a subducting plate pushes its way upward into the overlying plate, taking advantage of fractures and zones of weakness in the rock, and partially melts the crust to form yet more magma. Eventually, it may erupt onto the surface as a volcano. An entire line of highly explosive volcanoes, called a **volcanic arc**, will grow along the edge of the continental plate on the landward side of the subduction zone. This volcanic range marks the presence of the convergent boundary.

Partial melting produces magma that is high in Si and O (and other low-density elements) compared to the parent rock. This means that the chemistry of the rock in the line of volcanoes differs from that of the rock in the upper mantle—in relative terms, the volcanoes are enriched in Si and O and depleted in Fe and Mg compared to the upper mantle. This knowledge will prove valuable in Chapter 6, where we discuss why some volcanoes are violently explosive while others are not.

Oceanic Trenches

The deepest seafloor in the world is the *Challenger Deep*, located in the oceanic trench where the Pacific Plate subducts beneath the Philippine Plate at a steep angle. An oceanic trench is a deep, curved valley thousands of kilometers long and 8 to 11 kilometers deep that cuts into the ocean floor. Trenches are caused by subduction of one plate under another. They are curved because it is not possible to form a straight depression in Earth's curved surface. (Push in the surface of a Ping-Pong ball with your thumb and note that the edge of the depression is curved.)

Ocean-Ocean Convergence

An **ocean-ocean convergent boundary** (**FIGURE 3.19**) is formed where two plates composed of oceanic crust collide. Usually the older and, therefore, denser of the two plates subducts below the other. As in ocean-continent convergence, water released from the rock and sediment of the subducting plate causes partial melting in the mantle above the subducted plate. As the resulting magma migrates into the overlying oceanic crust, it causes partial melting and the eventual eruption of a line of volcanoes on the seafloor. In time, these volcanoes build a chain of volcanic islands known as an **island arc**.

Ocean-ocean boundaries differ from ocean-continent boundaries in that they typically do not exhibit well-developed accretionary prisms or fore-arc basins. Some geologists think that ocean-ocean margins are erosional, and that, over time, the overriding plate is worn back (by tectonic erosion) toward the island arc.

Read the Geology in Action feature titled "Back-Arc Basins" to learn about another characteristic sometimes seen at oceanic-convergent margins.

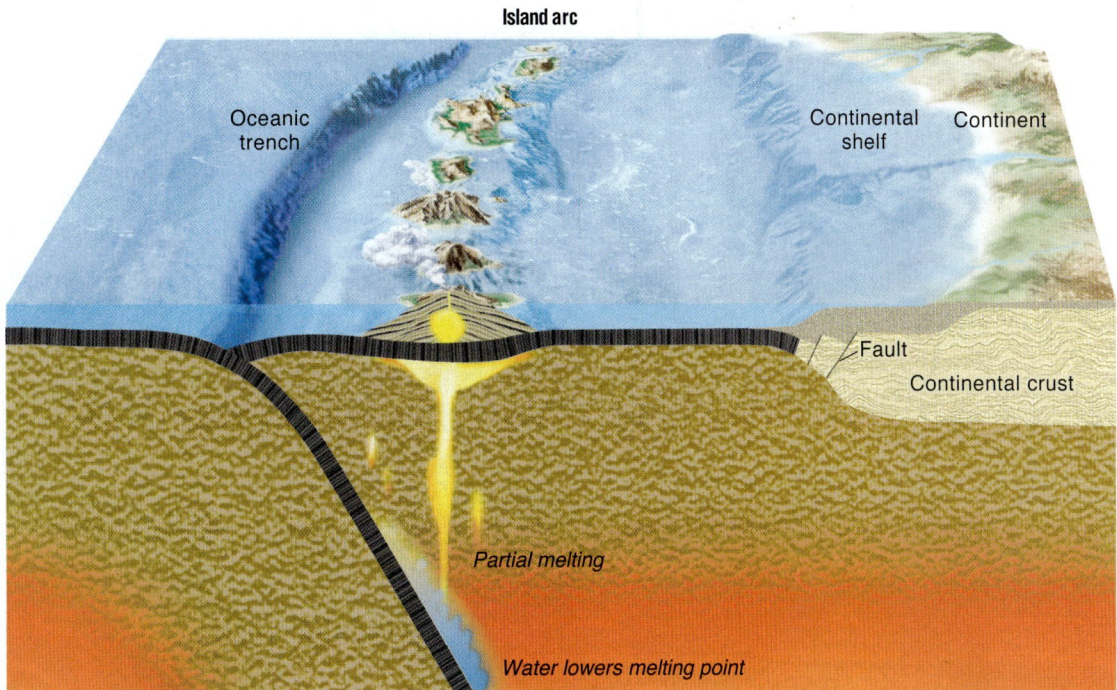

FIGURE 3.19 An ocean-ocean convergent boundary is characterized by a line of volcanic islands called an island arc.

When two plates composed of oceanic crust converge, which one is likely to subduct?

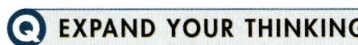

EXPAND YOUR THINKING

What features or parts of a volcanic arc and an island arc are similar; what parts are dissimilar?

Back-Arc Basins

As plate tectonic theory developed in the second half of the twentieth century, most scientists assumed that all convergent boundaries would be dominated by *compression* (squeezing), a form of pressure that lowers the volume of rocks in the crust. Then, in the early 1970s, it was noted at several locations in the western Pacific that rifting had developed in the overriding plate above subduction zones, a phenomenon that was not predicted by early tectonic theory. Rifting is a sign that *extension* (stretching, increasing the volume of rock), the opposite of compression, characterizes the crust.

In these regions (e.g., the Mariana, Okinawa, and Tonga–Kermadec trenches), a long and narrow basin, known as a *back-arc basin,* develops in the overriding plate. Back-arc basins develop where the trench migrates in the direction of the subducting seafloor, a process called *trench rollback.* Trench rollback may develop when the seafloor of the subducting plate sinks into the trench, or when the leading edge of the overriding plate collapses. In either case, the overriding plate experiences extension, and rifting develops behind the arc (**FIGURE 3.20**). These rift zones erupt basaltic magma and form new seafloor; the magma is similar to that erupted from mid-ocean ridges, except that in back-arc basins the magma is very rich in water,

produced by the subducting plate. Mid-ocean ridge basalt magmas are very dry.

Back-arc spreading has several interesting characteristics. For one, rifting may be asymmetrical, meaning that the production of new seafloor may be faster in one direction than in another. Also, researchers have found large rift jumps, where rifting "propagates" by abandoning one rift in order to start another closer to the arc.

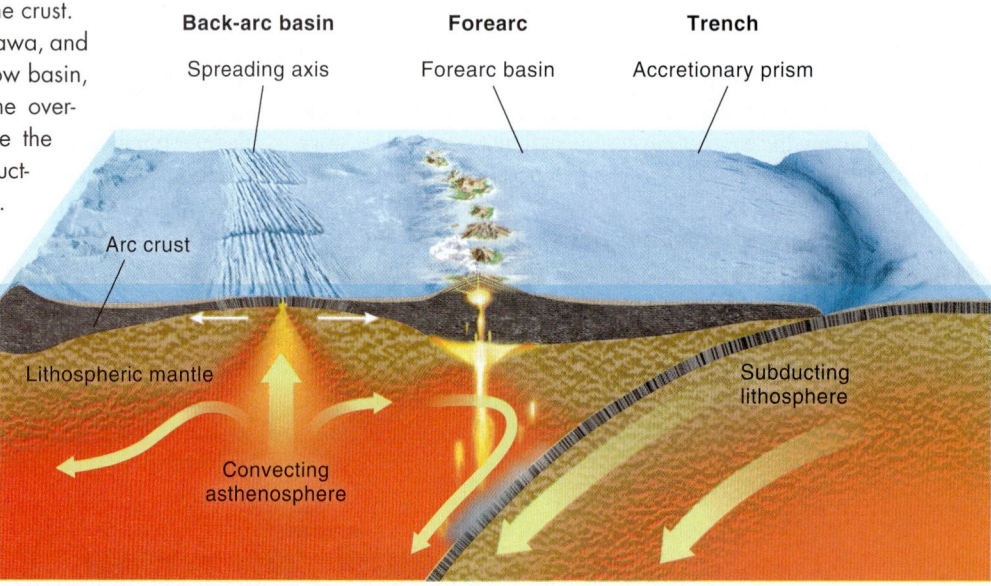

(a)

FIGURE 3.20 Back-arc basins are found at convergent margins where trench rollback causes extension in the overriding plate.
(a) A cross-section through a subduction zone shows the relative positions of back-arc spreading, the arc, forearc, and trench settings. (b) The western Pacific is, presently, the site of most of the world's back-arc basins (BAB).

(b)

Cartographic illustration and design by Matthew Barbee and Nancy

3-7 Orogenesis Occurs in Places Where Two Continents Converge

LO 3-7 Describe the origin of the Himalayas.

Like a slow-motion train wreck, the third type of convergent boundary occurs when an ocean basin closes and two plates composed of continental crust collide. Because neither of the two plates will subduct, a **continent-continent convergent boundary** builds mountains in a dramatic fashion.

Continent-Continent Convergence

Many plates consist of both continental and oceanic crust, and the entire plate moves toward any site where the oceanic portion is being subducted. If the subduction zone is characterized by ocean-continent convergence, eventually any continent riding on the subducting plate will collide with the continent on the overriding plate, pushing up the displaced rock and thus creating mountains (**FIGURE 3.21**). Continent-continent convergence is responsible for the formation of the Himalaya, Ural, and Appalachian mountain systems, as well as many others.

As we have seen, continental crust is composed of relatively buoyant rock that resists subduction. As a result, when two plates experience continent-continent convergence, both undergo compression that leaves their edges broken, contorted, and deformed. This process results in the creation of a high range of mountain peaks with deep continental "roots." Geologists refer to upward movement of rock as *crustal uplift*.

Continent-continent convergence is always preceded by ocean-continent convergence. As subduction of oceanic crust continues, any continent riding on the subducting plate will be drawn toward the overriding plate. The evolving margin of the continent on the overriding plate (known as an **active margin**), with its volcanic arc and accretionary prism, will collide with the tectonically quiet margin (known as a **passive margin**) of the continent riding on the subducting plate.

When the ocean finally closes, a collision zone will be formed, with the forearc ridge, the forearc basin, the accretionary prism, portions of the oceanic crust, the volcanic arc, and back-arc basin all squeezed between the two colliding continents. This results in the formation of a new mountain range with very complex geology.

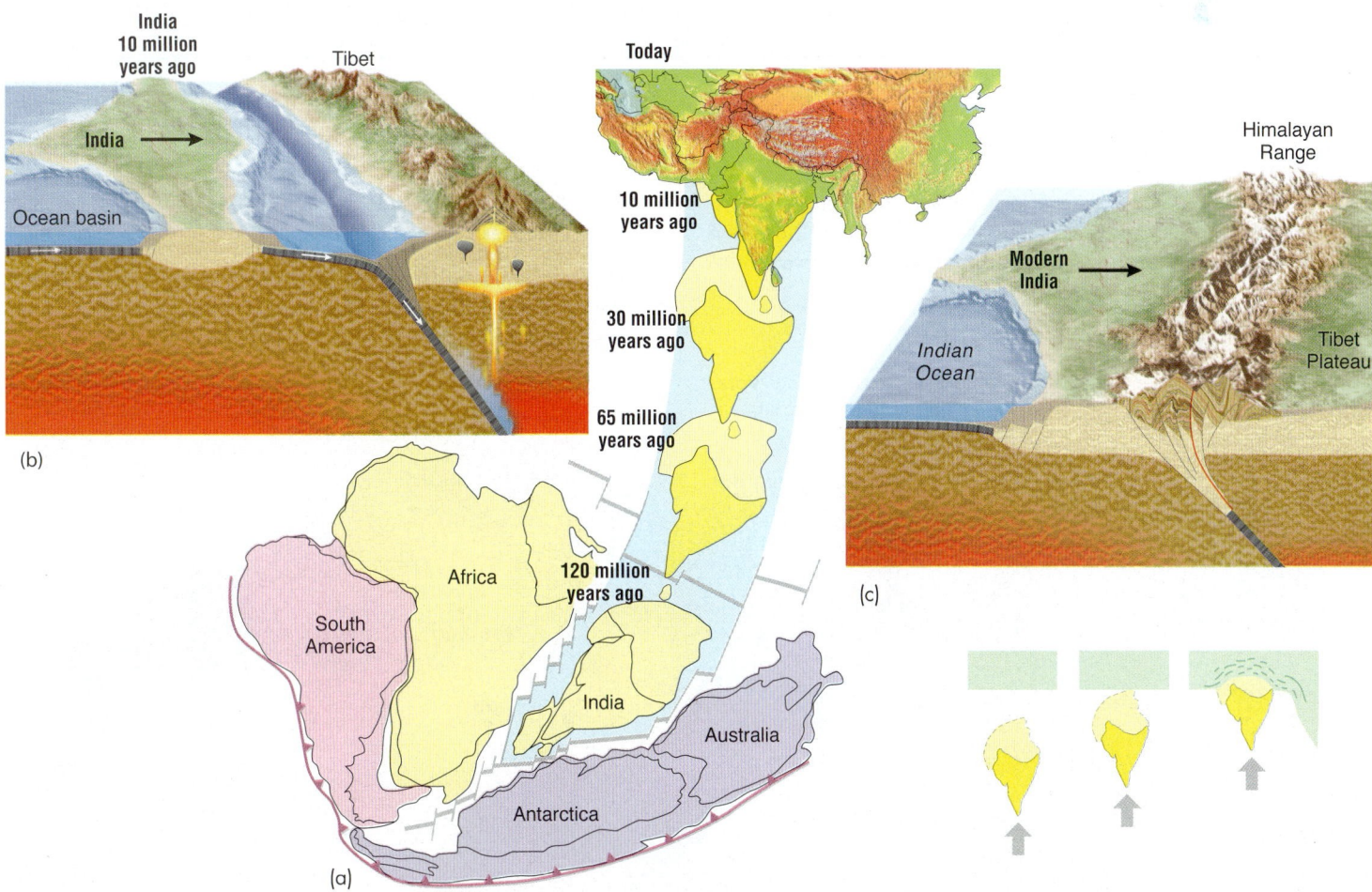

FIGURE 3.21 Two continents separated by a subduction zone will eventually collide, squeezing fore-arc sediments and rocks. The Himalayas, the highest system of mountains in the world, are forming at a continent-continent convergent boundary resulting from the collision of the Indian–Australian and Eurasian Plates. (a) After separating from the southern supercontinent Gondwana approximately 120 million years ago, the continent of India migrated toward the Eurasian Plate. (b) As India approached, the ocean between India and Eurasia narrowed and a deep-sea trench characterized the seafloor. (c) Eventually, India collided with Eurasia and produced the Himalaya range.

What happened to the oceanic crust between India and Eurasia?

Eventually, the last of the oceanic crust will be recycled into the mantle, and subduction will cease.

Mountain building resulting from this kind of continent-continent collision is known as **orogenesis** (Greek for "mountain" and "origin"). Orogenesis is characterized by thickening and deformation of crust at the collision site, as well as by folding, breaking, and crustal uplift. The resulting mountain range is embedded within a new, larger continent composed of two formerly separate continents. The new landmass sits higher in the mantle because of the combined buoyancy of the two continents. In this way continents may grow larger over time.

The Himalayas are an example of orogenesis caused by continent-continent convergence. These mountains are the result of collision between the Indian–Australian Plate and the southern edge of the Eurasian Plate. Over 100 million years ago, India broke away from the huge southern continent, known as *Gondwana*, which was composed of South America, Africa, Madagascar, Antarctica, India, Sri Lanka, and Australia. Gondwana itself was assembled during late Precambrian and Cambrian time and later participated in the formation of Pangaea. The breakup of Pangaea, which initially rifted near the equator, formed a large northern continent called *Laurasia*, and the southern landmass, Gondwana, reemerged in the southern hemisphere.

The Himalayas comprise the highest system of mountains in the world, most notably Mount Everest, the tallest mountain. The mountain-building process in this range is ongoing, as India continues to push its way to the north and the Himalayas deform upward. Mount Everest itself grows by over 6 millimeters and shifts northeast by about 44.5 millimeters each year. Deadly earthquakes in Pakistan and India are evidence of the role of active tectonics in this region.

To test what you have learned about plate boundaries, complete the exercise in the Critical Thinking section titled "Tectonic Features."

Q EXPAND YOUR THINKING

If you were asked to assess the origin of a mountain range about which you had no prior information, what observations would you collect to build your hypothesis? Identify some observations that would point you in the direction of one type of origin versus another.

CRITICAL THINKING

Tectonic Features

1. In **FIGURE 3.22**, label plate tectonic features in the cross section. Label as many features as you can—there are at least 10 different ones—and draw arrows indicating the direction of movement of the crust.

2. Label Earth's layers in the cross section.

3. For each layer, identify its general chemistry and physical state, and indicate important processes and features that occur within it.

4. Two rift zones are shown in Figure 3.22: What are the differences and similarities between them?

5. Describe (or draw) how this scene would look different after a period of time has passed. Identify an area of the world that looks like your drawing.

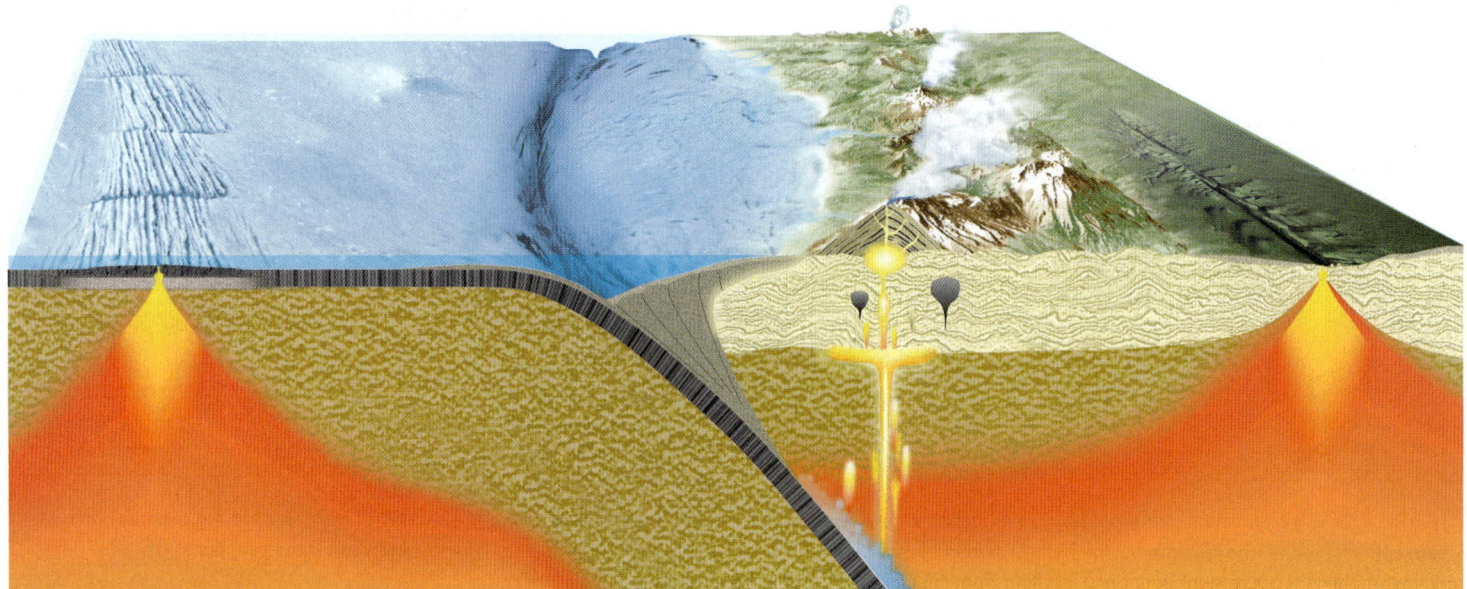

FIGURE 3.22 Major components of the plate tectonic model.

3-8 Transform Boundaries Connect Two Spreading Centers

LO 3-8 Describe the San Andreas transform fault.

Transform boundaries are characterized by side-to-side plate movement. That is, they occur where two plates slide past each other, rather than colliding or separating. You modeled transform boundaries in your experiment with the orange peel, earlier in the chapter.

Transform Boundaries

Typically, transform boundaries produce little direct collision or separation between plates and tend to be marked by linear valleys, offset ridges, long lakes that have been filled by groundwater, and stream valleys that zigzag across the plate boundary (**FIGURE 3.23**). This type of motion is called *shearing*.

Transform boundaries are important for three reasons: 1) They complete our understanding of plate tectonics; 2) they explain significant features on Earth's surface; and 3) they are characterized by frequent earthquakes. Geophysicist J. Tuzo Wilson, whom we introduced in Chapter 1, first proposed the hotspot hypothesis; he also proposed that transform boundaries connect two spreading centers (or, less commonly, two subduction zones). Let us take a closer look at two common types of transform boundaries that connect two spreading centers.

FIGURE 3.24 The San Andreas Fault is a transform boundary between the North American Plate (right side) and the Pacific Plate (left side). *Source:* U.S. Geological Survey, Department of the Interior, USGS.

What evidence would you look for to indicate the direction of displacement associated with movement along the San Andreas Fault?

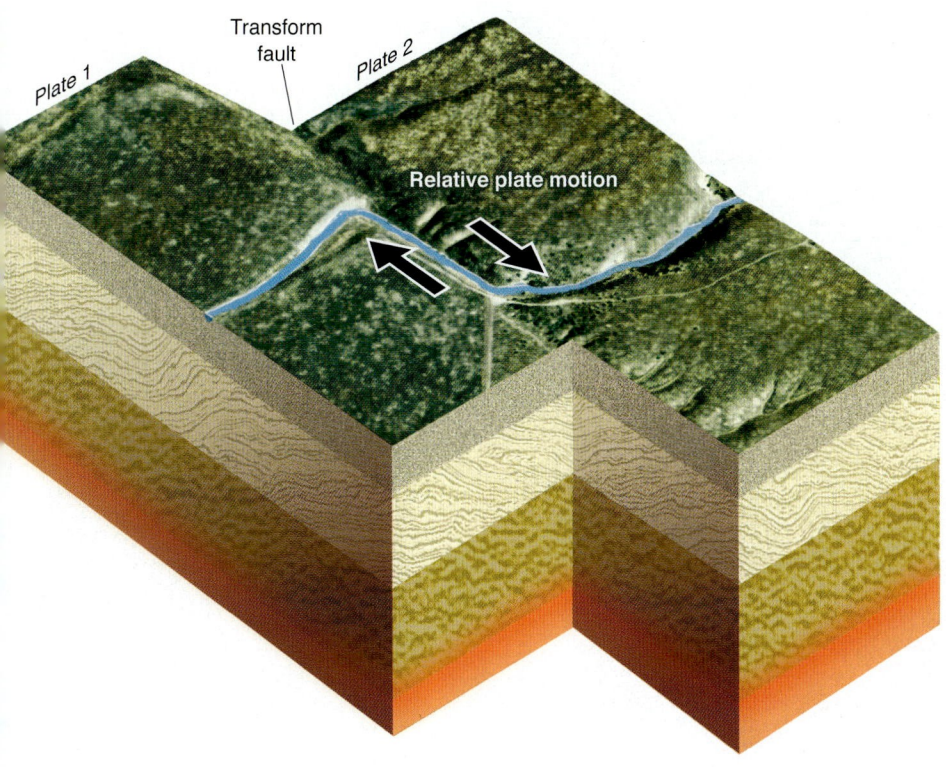

FIGURE 3.23 A transform boundary is characterized by side-to-side plate movement.

What has happened to the stream in this setting?

Probably the most famous transform boundary in the world is the *San Andreas transform fault* (**FIGURE 3.24**). A **fault** is a place where the crust is broken and the broken edges are offset relative to each other, either vertically or horizontally. The San Andreas Fault is a transform boundary between the North American Plate to the east and the Pacific Plate to the west.

The Pacific Plate grinds northward along the edge of the North American Plate at a rate of about 6 centimeters a year. The city of Los Angeles, sitting largely on the Pacific Plate, is slowly moving toward the city of San Francisco, which sits largely on the North American Plate. At the present rate of movement, the two will be joined in about 10 million years (plus or minus a couple of suburbs).

But what about Wilson's prediction? The San Andreas transform fault connects two spreading centers. To the north, offshore of Northern California and Oregon, is the *Gorda Ridge*, where a spreading center separates the *Gorda Plate* from the Pacific Plate; to the south is the *East Pacific Rise* in the Gulf of California, where a spreading center separates the Pacific Plate from the North American Plate. Thus, the San Andreas Fault is a transform boundary that connects two spreading centers, just as Wilson predicted.

A common second type of transform boundary is an **oceanic fracture zone** consisting of inactive (*fracture zone*) and active (*transform fault*) portions. These are ocean-floor valleys that extend horizontally away from spreading centers (**FIGURE 3.25**). Some fracture zones are hundreds to thousands of kilometers long and as much as 6 to 7 kilometers deep. Examples include the Clarion, Molokai, and Pioneer fracture zones in the east Pacific between California and Hawaii. Only small portions of fracture zones, the transform fault portion, are active at any given time.

You may be curious about why transform faults develop. The answer, however, is not clear. Reasoning by analogy, the process that results in the formation of a fracture zone may be similar to what happens when you tear a sheet of paper. If you look closely at one side of the tear, you will notice dozens of microtears at angles (often, right angles) to the main direction of the torn sheet. Each of these microtears is an expression of the tendency of fractured objects (the torn paper, in this case) to splinter. The torn paper is analogous to a spreading center, with oceanic transform faults beginning as splinters of the main rift and having since extended.

Examine Figure 3.25. Notice that a single transform fault has split a spreading center—as Wilson put it, "The transform boundary connects two spreading centers." Along the portion of the fracture between the two spreading centers, opposite sides of the boundary move in opposite directions, creating a shear zone in the transform fault portion. But where the fracture is not located between spreading centers, the two sides of the boundary are locked together and moving in the same direction. This is the inactive fracture zone.

Notice, too, the age of the crust on either side of a fracture zone. Under normal circumstances new seafloor is made at approximately the same rate on either side of a spreading center, suggesting the seafloor at equal distances from an oceanic ridge should be similar in age. But because an oceanic fracture offsets a spreading center, the seafloor on either side of a fracture may differ in age. On the side closer to where it was made, the seafloor is younger; on the side farther from where it was made, it is older.

Q EXPAND YOUR THINKING

On Figure 3.25, write the words "older," "younger," and "same" to indicate the age relationships of the seafloor on both sides of the transform fault and fracture zones.

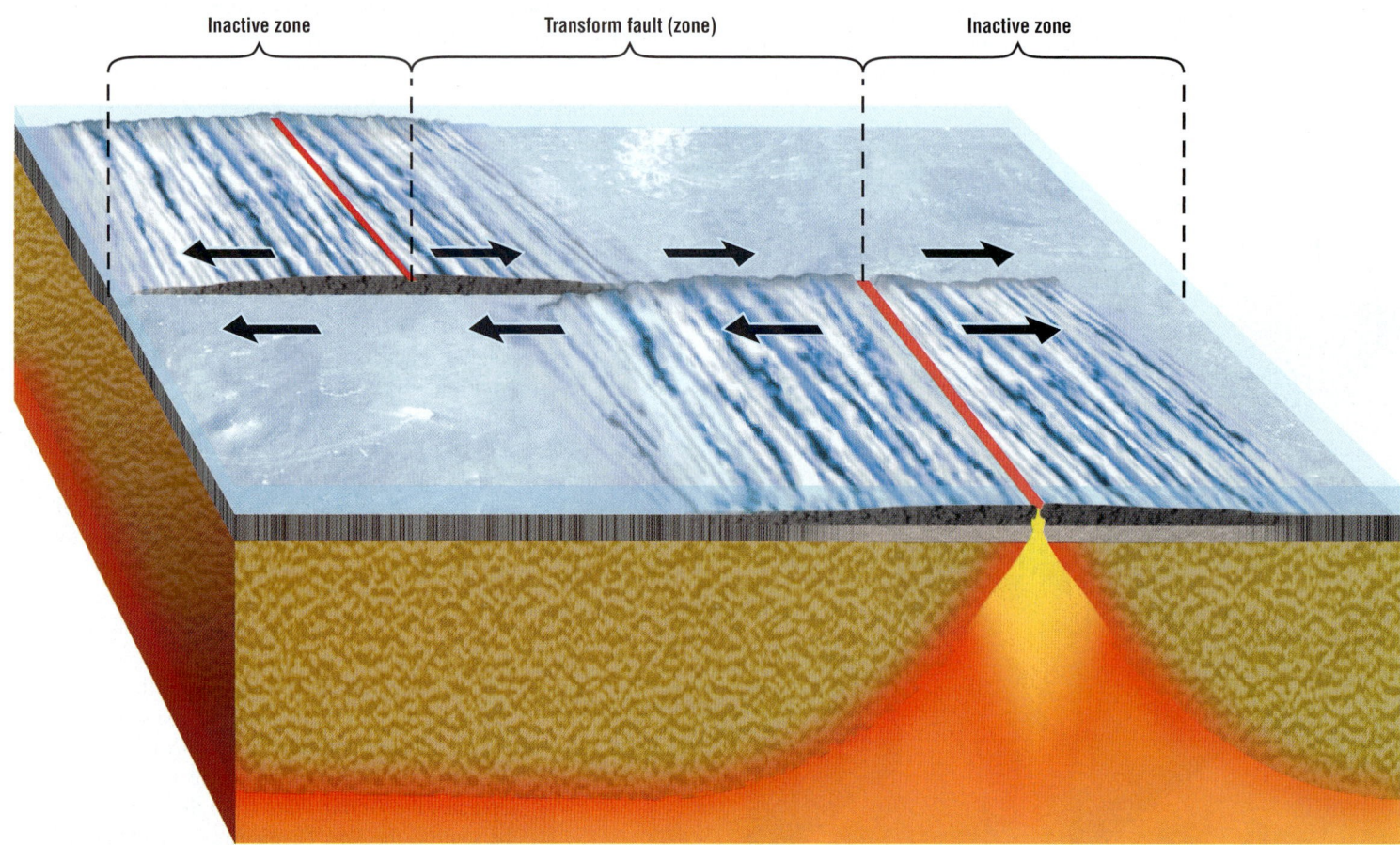

FIGURE 3.25 Oceanic fracture zones consist of transform faults, which are a type of active plate boundary, and fracture zones, which are inactive.

Q Why are fracture zones inactive?

3-9 Earthquakes Tend to Occur at Plate Boundaries

LO 3-9 Describe where earthquakes tend to occur and why.

How do the three boundaries defined by crust type (ocean-ocean, ocean-continent, continent-continent) and the three plate motion settings (divergent, convergent, transform) relate to one another? **TABLE 3.1** summarizes the geologic processes we have discussed and gives examples of the types of crust and styles of plate motion found in various tectonic settings.

Risky Business

In general, plate boundaries are geologically hazardous places. A notable example is the Pacific Ring of Fire, mentioned earlier. This region is hazardous because of the geologic processes occurring at plate boundaries. The forces associated with plate separation in rift valleys commonly lead to earthquakes (Chapter 11) and volcanic eruptions (Chapter 6). Transform boundaries, where one plate shears past another, are characterized by earthquakes that can be very destructive. Convergent boundaries, the site of plate subduction and orogenesis, are subject to numerous earthquakes as well as explosive volcanic eruptions.

The most hazardous earthquakes occur when a section of a transform boundary or a subduction zone becomes "locked." This happens when a local plate motion temporarily comes to a halt because two segments of plate cannot slide past each other, even though overall plate movement continues. As the pressure generated by plate movement builds, the potential for a large-scale earthquake grows as well. Finally, the locked segment can no longer withstand the accumulated stress and the boundary breaks free with a sudden jolt, causing an earthquake (**FIGURE 3.26**). In the case of a locked subduction zone, these are called *megathrust quakes*.

Because the seafloor moves suddenly during a megathrust quake, the possibility that a damaging tsunami will occur is very high. Such were the circumstances that led to the massive Andaman–Sumatra earthquake and tsunami on December 26, 2004, which killed more than 230,000 people, and the Tohoku earthquake and tsunami on

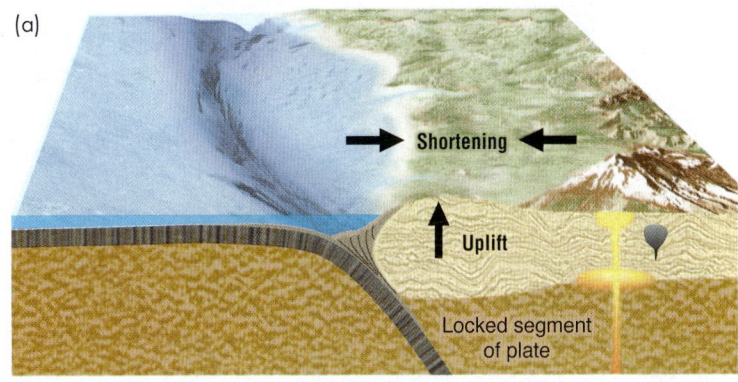
(a)

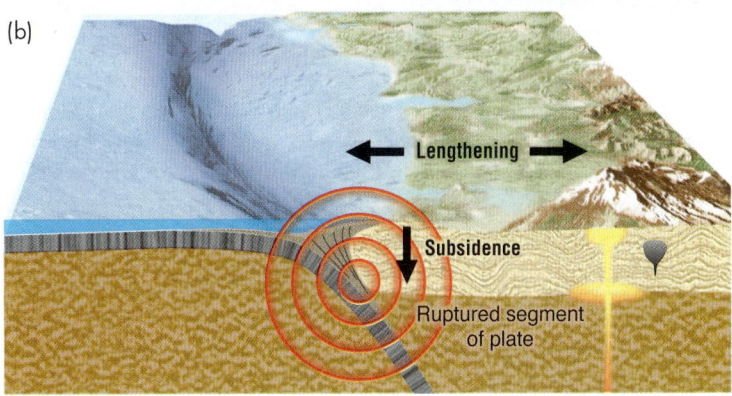
(b)

FIGURE 3.26 When a locked subduction zone ruptures, it sets in motion a megathrust quake that can be very damaging. (a) As strain accumulates, the crust shortens and uplift occurs above the subduction zone. (b) When the plates unlock, the crust extends, the land subsides, and a megathrust earthquake occurs. Sudden movement of the seafloor associated with a megathrust quake can produce damaging tsunamis.

What hazards and environmental changes does the coastal zone on the overriding plate experience during and following a megathrust earthquake?

March 11, 2011, which killed more than 15,000 people and partially or totally damaged over 1 million buildings.

Earthquakes cause the ground to shake because seismic energy released by the breakage of rock passes through the crust. The shaking can lead to disastrous consequences, as when houses, bridges, and

TABLE 3.1 Geologic Processes at Plate Boundaries

Type of Crust	Type of Plate Motion		
	Divergent	**Convergent**	**Transform**
Ocean-ocean	Oceanic ridge, rift valley, spreading center	Island arc, tectonic erosion	Oceanic fracture zone, transform fault
Example	*Mid–Atlantic Ridge*	*Indonesian Islands*	*Molokai fracture zone*
Ocean-continent	Not common	Volcanic arc, subduction zone, oceanic trench	Not common
Example		*Cascade Range, Andes*	
Continent-continent	Continental rift valley	Collision zone, orogenesis, crustal uplift	Transform fault zone
Example	*East African rift*	*Himalayas*	*Portions of San Andreas Fault*

(a)

(b)

© AP/Wide World Photos

Guo Jian She/Redlink/Redlink/Corbis Images

FIGURE 3.27 (a) A major earthquake hit the Caribbean nation of Haiti on January 12, 2010. One month after the event, the government there estimated that more than 230,000 people had died and over 1 million were left homeless. (b) In 2008, a strong earthquake hit China and killed over 80,000 people, caused 300,000 injuries, and left 5 million homeless.

Based on these photos, what is a major cause of injury during an earthquake?

roadways fracture and topple as a result (**FIGURE 3.27**). Earthquakes also loosen unstable hillsides, setting in motion landslides and rock avalanches.

Perhaps you are among the millions of people living in one of the beautiful regions created by plate interactions. If so, you are, unfortunately, also at relatively high risk of experiencing geologic hazards, such as earthquakes. Mountain ranges, volcanic slopes, shorelines—these and other environments located on plate boundaries can be hazardous places to live because of the very forces that formed them. In the course of human history, millions of people have been killed by earthquakes, volcanic eruptions, tsunamis, landslides, and other hazards produced at plate boundaries.

Occasionally, major eruptions or earthquakes kill large numbers of people. Here are a number of examples:

- An earthquake in Shaanxi, China in 1557 killed an estimated 830,000 people living in hillside caves that collapsed.

- In 1883, an eruption of Krakatau volcano in Indonesia, along with the resulting tsunami, killed 37,000 people.

- In 1976, an earthquake in Tangshan, China, killed 250,000 people.

- In 1983, a mudflow on Nevada del Ruiz volcano in Colombia, the result of a volcanic eruption, killed 25,000 people.

- In December 2003, 45,000 people died as a result of a major earthquake in Bam, Iran.

- In Sichuan, China, 80,000 people died in 2008 from a massive earthquake.

The two worst disasters in more recent history happened only six years apart. The Andaman–Sumatra tsunami in 2004 destroyed hundreds of communities on the shores of the Indian Ocean, killing over 230,000 people; and the 2010 Haiti earthquake in the Caribbean Sea killed over 230,000.

All these events had one thing in common: They all occurred on plate boundaries—clearly, dangerous places.

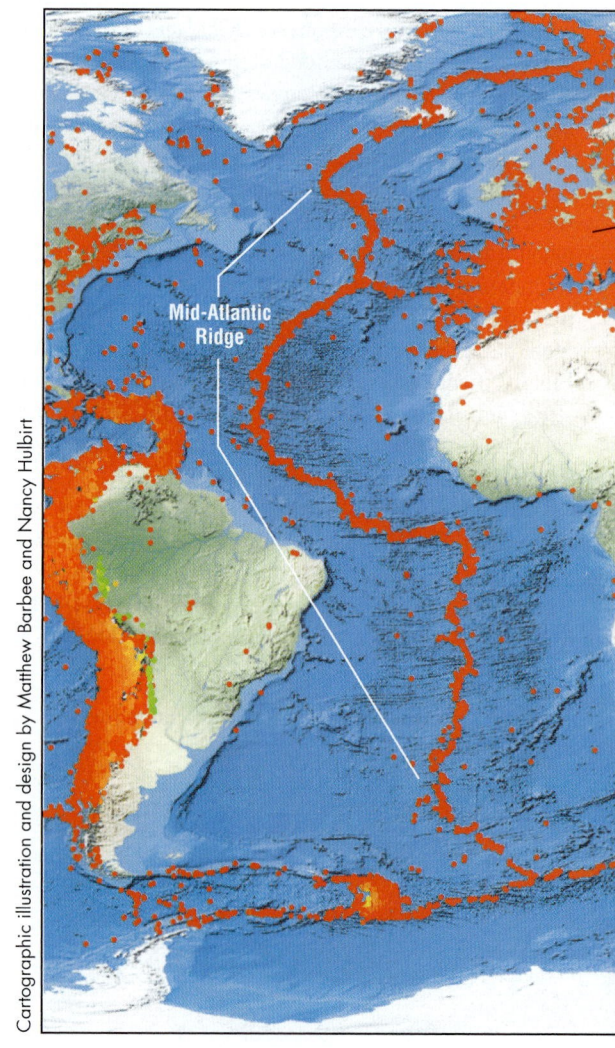

Mid-Atlantic Ridge

Cartographic illustration and design by Matthew Barbee and Nancy Hulbirt

Where Do Earthquakes Occur?

Do you live near a seismic hazard zone? We know that earthquakes are generated by broken crust moving suddenly along a fault surface. We also know that the theory of plate tectonics enables us to predict active faulting at plate boundaries. Therefore, we may deduce that earthquakes are associated with plate boundaries (**FIGURE 3.28**), where crust is known to break and move suddenly. In large part, this is an accurate deduction. But even if you do not live near a plate boundary, you may be at risk. Damaging earthquakes sometimes occur in areas that are not located at plate boundaries. In the United States, these areas include New Madrid, Missouri; Salt Lake City, Utah; and Charleston, South Carolina. That said, many years elapse between major quakes at these sites, and we may yet find that they are associated with former, or future, plate boundaries.

About 81% of the world's largest earthquakes occur in the *circum–Pacific seismic* belt. This belt, which includes the rim of the Pacific Plate and nearby plates, corresponds to the Pacific Ring of Fire and extends from Chile north along the west coast of North America through the Aleutian Islands to Japan, and south through the Philippine Islands, the island groups of the southwest Pacific, and New Zealand.

Eight of the 10 largest earthquakes since 1900 have occurred along this seismic belt.

The second most important earthquake belt, the Alpide belt, extends from Java to Sumatra through the Himalayas and the Mediterranean and out into the Atlantic. This belt accounts for about 17% of the world's largest earthquakes, including some of the most destructive, such as the Andaman–Sumatra quake in 2004, the Iran shock in 2003, and tremors in Turkey in 1999 that killed more than 17,000 people. This belt is marked by largely convergent and transform boundaries between numerous microplates moving from the south against the Eurasian Plate to the north.

The third prominent earthquake belt follows the submerged *Mid–Atlantic Ridge*. Quakes in this belt have not been particularly destructive, as they generally occur far from population centers. Although earthquakes in prominent seismic zones are expected, damaging shocks can and do occur outside these areas.

Ⓠ EXPAND YOUR THINKING

Why do oceanic fracture zones typically experience shallow earthquakes?

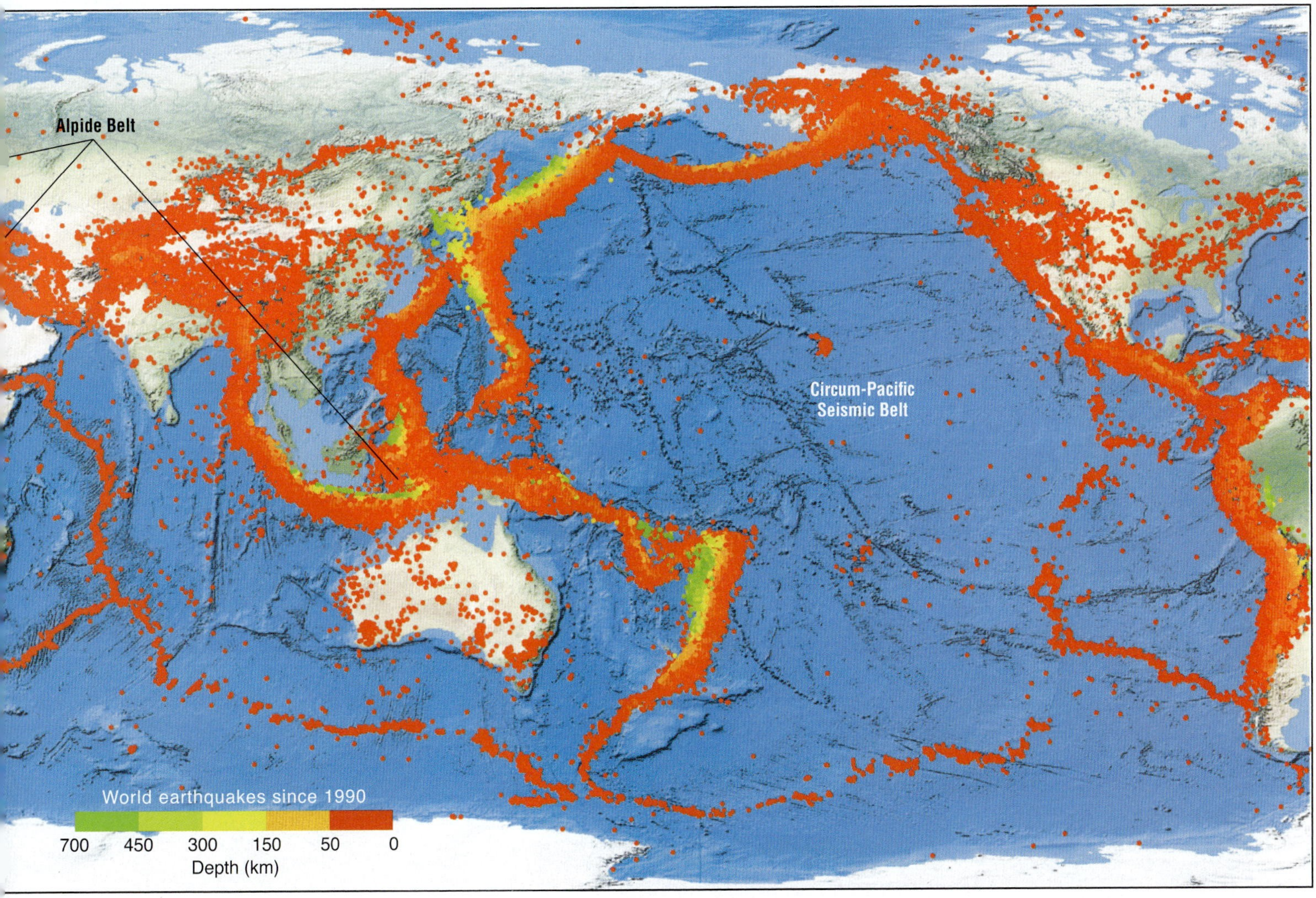

FIGURE 3.28 Earthquakes tend to be aligned with plate boundaries. The circum–Pacific seismic belt, the Alpide belt, and the Mid–Atlantic Ridge are the most seismically active regions.

 Based on the alignment of earthquakes, can you locate and name the 12 major lithospheric plates on this map?

3-10 Plate Movement Powers the Rock Cycle

LO 3-10 Describe the rock cycle and its relationship to plate tectonics.

In time, most types of rocks are exposed to changing environmental conditions that may include chemical reactions on Earth's surface, or crushing temperatures and pressures deep in the crust. Elements that make up these rocks are neither created nor destroyed but, rather, are redistributed and transformed from one natural form to another. This process is known as the rock cycle; it is driven by plate tectonics.

Plate Movement

Scientists are not sure what causes plates to move. Four possible mechanisms have been hypothesized: Plates may be: 1) pushed away from spreading centers, 2) pulled down into trenches, 3) dragged along by the friction created by mantle convection on their under-sides, or 4) driven by gravity to slide down the slope from the ridge to the trench. Conventional wisdom holds that all four mechanisms may contribute somewhat to plate movement, although researchers are undecided as to the proportion of total movement contributed by each mechanism (**FIGURE 3.29**).

Ridge push is the force applied to plates at spreading centers. As plates separate, new, hot magma is extruded between them from the upper mantle. The high temperature in the rift lowers the density of the new rock, causing it to float higher in the mantle; this creates a "push" effect from below. Gravity draws the young seafloor downhill and away from the rift zone, widening the gap and allowing new hot magma to well up and push again.

Slab pull is another potential driving mechanism. Subduction zones typically occur far from spreading centers; therefore, the plates have plenty of time (tens of millions of years) to cool. The lower temperature and changes in the mineral composition of this older seafloor make the plate denser than the hot mantle below it. Hence, the leading edge of the plate, or "slab," sinks under its own weight. As the slab is drawn under, it pulls the rest of the plate behind it. Suggesting that slab pull is an important process based on the observation that plate movement is faster where the age (and thus thickness and density) of subducted crust is greater.

Several models of plate tectonics support the mantle convection hypothesis originally proposed by Harry Hess. Two of them are the *plate drag* model and the *ridge slide mechanism.*

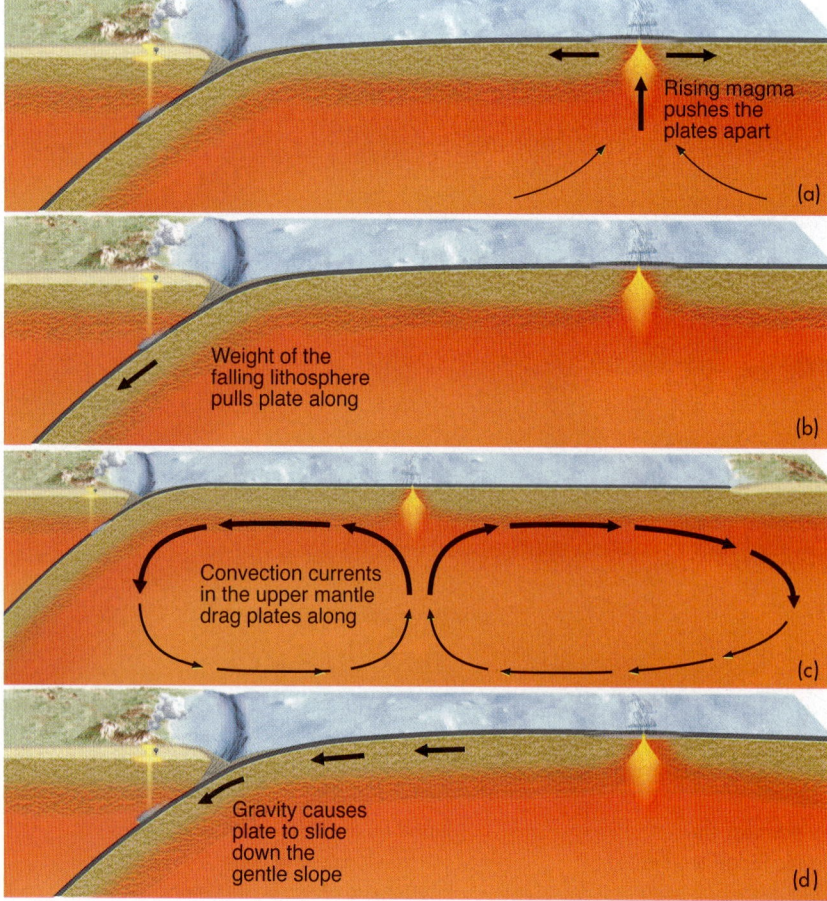

Rising magma pushes the plates apart (a)

Weight of the falling lithosphere pulls plate along (b)

Convection currents in the upper mantle drag plates along (c)

Gravity causes plate to slide down the gentle slope (d)

FIGURE 3.29 Researchers hypothesize four mechanisms of plate movement: (a) the ridge push model, (b) the slab pull model, (c) the plate drag model, (d) the ridge slide model.

Q Each of these models of plate movement is a separate hypothesis. How would you test each one?

In the *plate drag* model, the lithosphere rides passively on currents in the upper mantle. In one version of plate drag, the upper mantle is separate from the lower mantle, with little or no transfer of rock between them. Small, local convection cells in the upper mantle occur beneath the interiors of plates and create frictional drag on the underside, thereby reinforcing plate movement. Plate drag thus is the result of mantle convection currents dragging plates away from ridges. However, many geologists are not convinced that passive drag can generate the energy needed to move massive plates like the North American Plate.

The *ridge slide* mechanism relies on the gravitational slope between the crest of an oceanic ridge and the base of a deep-sea trench to instigate a plate to slide downhill. Such action might lead to ripping of the plate (rifting) and pressure-release melting in the upper mantle.

The Rock Cycle

James Hutton (introduced in Chapter 1), an eighteenth-century gentleman farmer and one of the founders of modern geology, originated the concept of the rock cycle. This concept illustrates the interrelationships among igneous, sedimentary, and metamorphic rocks. Today, we know that it is tectonic processes that drive important aspects of the rock cycle.

The modern rock cycle concept proposes that the mantle, the crust, the atmosphere, the biosphere, and the hydrosphere can be envisioned as a giant recycling machine in which the elements that make up rocks are neither created nor destroyed but, rather, are redistributed and transformed from one natural form to another over time. Because the rock cycle is a continuous process, we can begin to examine it at any step in the cycle.

Heat from the Sun produces chemical reactions among living organisms, fluids, gases, and solid materials on Earth's surface, causing weathering. (Weathering is reviewed in Chapter 7.) Weathering reactions destroy minerals in sedimentary, igneous, and metamorphic rocks that are exposed to the atmosphere. New minerals form as a result of weathering processes.

Mechanical, chemical, and biological weathering produces sediment that consists of new minerals, as well as the broken remains of older weathered rock. Sediments may be eroded (moved) and eventually deposited in the ocean or some other depositional environment. (Sedimentary processes are covered in Chapter 8.) As sediments accumulate, buried particles are turned into solid rock through compaction and cementation.

Sedimentary rock, when exposed to high temperatures and pressures at plate margins or deep in the crust, will metamorphose. (Metamorphic processes are covered in Chapter 9.) If metamorphic rock is exposed to higher temperatures originating within Earth (i.e., geothermal heat), it will melt and turn into magma. Magma that becomes part of oceanic or continental crust will cool and form igneous rock. (Igneous processes are covered in Chapters 5 and 6.) In time, continental crust composed of igneous rock will be uplifted by orogenesis and exposed to weathering—whereupon it turns to sediment and enters a new round of the rock cycle.

The rock cycle can be defined and linked to the tectonic framework (**FIGURE 3.30**). As we saw earlier, tectonics leads to the formation of new igneous rock (oceanic crust) at a spreading center.

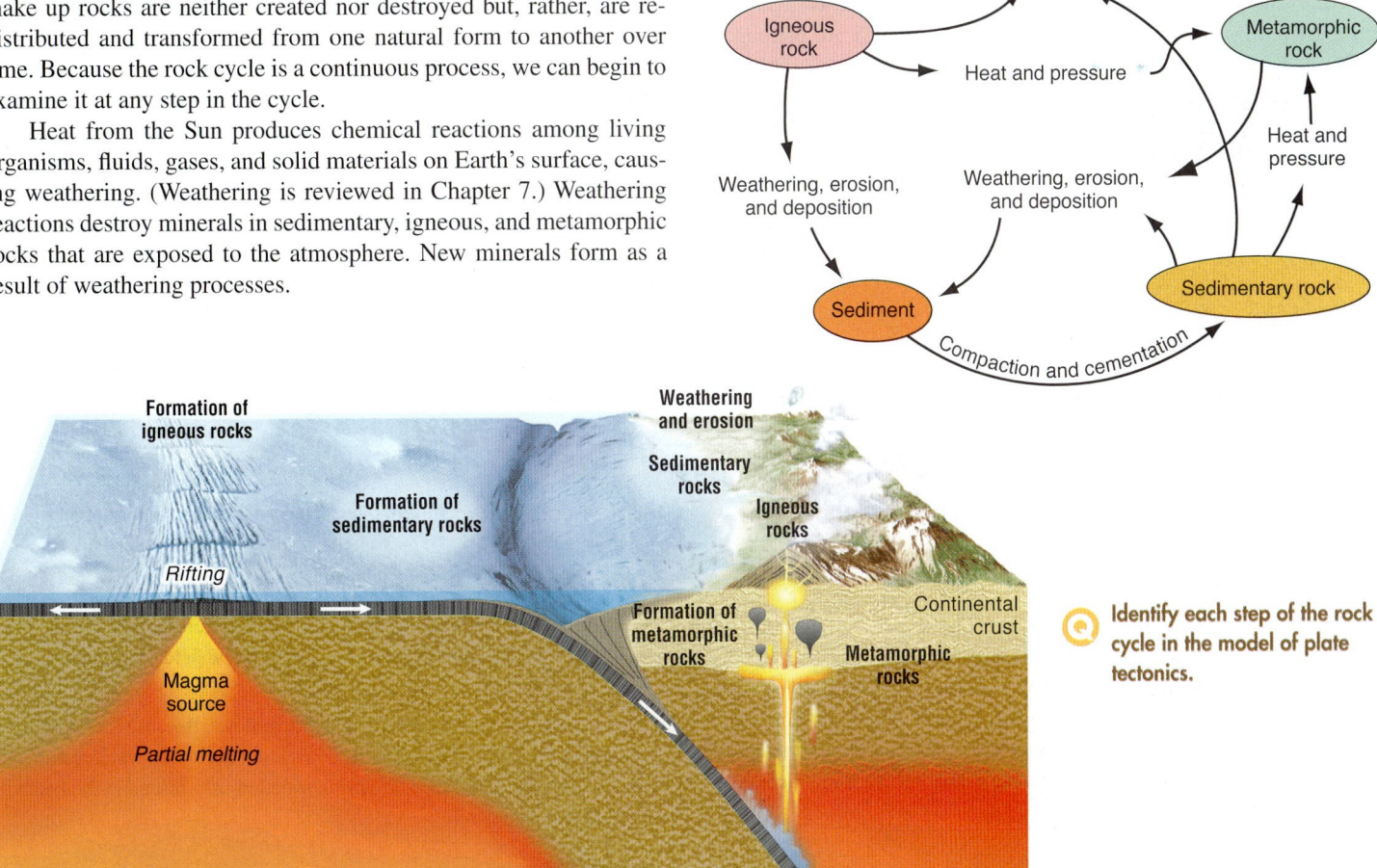

Identify each step of the rock cycle in the model of plate tectonics.

FIGURE 3.30 The rock cycle is closely tied to plate tectonics.

As ocean crust moves slowly toward subduction, it collects thick layers of marine sediment consisting of fine particles from the land and skeletal remains of plankton that live in the ocean. Eventually, the crust subducts at a convergent margin. The heat and pressure of subduction releases water and other volatile compounds from the crust as the plate descends into the mantle.

Sediment may subduct with the plate or collect as an accretionary ridge and prism above the subduction zone. Both subducted and accreting sediments are metamorphosed by pressure at the convergent margin. Water escaping from the subducting plate causes partial melting in the overlying mantle. The resulting magma is high in silicon and oxygen and other easily melted components. It intrudes into the overlying crust, where it generates more magma through partial melting of the crust. Eventually, magma extrudes onto Earth's surface and forms a volcanic arc or island arc composed of igneous rock.

Both oceanic crust and the sediments on it experience metamorphism in the subduction zone due to the extreme pressures and temperatures related to plate convergence. When subduction eventually leads to orogenesis, the level of metamorphism increases due to the enormous pressure generated by colliding continents—in fact, large segments of continental crust may become metamorphosed. Over time, these rocks will weather and the sediments will return to the sea.

Crust in the mountainous arcs experiences weathering and breaks down into particles of sediment. This sediment is eroded by wind and running water and collects on oceanic crust, forming sedimentary rock. This rock is eventually subducted or joins the accretionary prism and ridge, thus completing the rock cycle.

The rock cycle is Earth's great recycling engine, perpetually renewing and destroying rock and ultimately providing humans with critically important natural resources. Simultaneously, it exposes us to dangerous natural hazards, such as earthquakes, volcanism, and landslides.

⊙ EXPAND YOUR THINKING

How would the rock cycle change if the rate of heat flow from the mantle doubled?

LET'S REVIEW

The theory of plate tectonics predicts that Earth's surface is organized into massive slabs of lithosphere called plates in which are embedded continents and ocean basins. Plates move across Earth's surface, on average, at about the same speed that your fingernails grow. At their edges where plates collide, separate, and slide past one another, rock is recycled, mountains are built, new crust is formed, and numerous geologic hazards and geologic resources develop. The history of Earth's surface is largely interpreted by recreating the past position of plates. The theory of plate tectonics has advanced our understanding of geologic hazards and resources.

You have learned that Earth is not an unchanging ball of rock in orbit around the Sun. We hope that you will retain this knowledge throughout your life, and that you will continue to be aware that plate tectonics perpetually renews the crust and drives the rock cycle, that it is a major cause of hazardous earthquakes and volcanoes, and that these processes have been shaping Earth's history for many millions of years.

STUDY GUIDE

3-1 Earth has three major layers: core, mantle, and crust.

- When Earth's temperature reached the melting point of iron, the **iron catastrophe** occurred.

- The iron catastrophe led to the differentiation of Earth into layers. The **inner** and **outer cores** form the core complex, a region that is rich in iron and nickel. The **mantle** is rich in heavy elements compared to the **crust,** but contains more silicon and oxygen than the **core.** The uppermost layer is the **lithosphere,** which consists of the crust and a portion of the upper mantle. The crust is low in iron and magnesium and high in silicon and oxygen relative to Earth as a whole.

3-2 The core, mantle, and crust have distinct chemical and physical features.

- Earth's core is over 2,900 kilometers below the surface. The inner core is thought to be a solid iron crystal, and the outer core a turbulent iron-rich fluid. The core complex generates a **geomagnetic field** that emanates from the planet's surface near the poles.

- Mantle rock has plasticity (the capability to flow). Because the lower mantle is hotter than the upper mantle, convection occurs, in which warm rock rises and cold rock descends. This convection may resemble the behavior of fluid in a lava lamp.

- Earth's crust consists of continental and oceanic types. Together these form the uppermost portion of a thicker layer called the lithosphere. **Oceanic crust** is 5 to 10 kilometers thick; **continental crust** is 20 to 80 kilometers thick, or more. Oceanic crust is relatively rich in iron and magnesium, while continental crust is enriched in low-density compounds.

- The portion of the lithosphere underlying the crust belongs to the upper mantle. The density of continental rocks allows the crust to float on the mantle. The lower lithosphere rests on the asthenosphere.

3-3 Lithospheric "plates" carry continents and oceans.

- Tectonic theory predicts that Earth's lithosphere is broken up into plates that move and interact with one another. These plates move in response to forces in the mantle. As a result, **plate boundaries,** where plates bump into and grind against one another, are locations of great geologic change.

- Twelve major plates have been identified. These are known as the Nubian, Arabian, Eurasian, Antarctic, Indian–Australian, Pacific, Philippine, South American, North American, Cocos, Nazca, and Caribbean Plates. (Note that the Indian and Australian Plates are sometimes treated separately.) There are several minor plates as well.

- Places where plates are recycled into the mantle, called subduction zones, are characterized by deep-sea trenches, where the ocean above the downward-curving surface of a plate is extremely deep. At these locations, a denser plate dives, or subducts, beneath a less dense plate.

3-4 Paleomagnetism confirms the seafloor-spreading hypothesis.

- As magnetic minerals crystallize in molten rock, they incorporate into their structure the magnetic orientation of Earth's field. Likewise, iron-containing sediment particles settling through water physically rotate and align themselves to the geomagnetic field, like compass needles. When these particles are transformed into hard rock, their magnetic orientation is permanently recorded. This is called **paleomagnetism.**

- Paleomagnetism is a valuable tool for understanding Earth's history, especially at **divergent plate boundaries** (where two or more plates pull away from each other).

- Geologists studying this phenomenon recognize that rocks can be divided into two groups based on their magnetic properties. One group has normal polarity, characterized by magnetic minerals with the same polarity as today's magnetic field. The other group has reversed polarity.

3-5 Plates have divergent, convergent, and transform boundaries.

- Bordered by active plate boundaries, the **Ring of Fire** is the most seismically and volcanically active zone in the world. People living in the Pacific region are, generally speaking, at greater risk of experiencing geologic hazards than are people living elsewhere in the world.

- **Divergent boundaries** are where new lithosphere is formed as plates pull away from each other.

- **Convergent boundaries** are where lithosphere is subducted as one plate dives beneath another or two plates collide head-on without either one subducting.

- **Transform boundaries** are where lithosphere is neither formed nor recycled; plates simply grind past each other.

3-6 Oceanic crust subducts at convergent boundaries.

- An **ocean-continent convergent boundary** occurs where a plate composed of oceanic crust has collided with one composed of continental crust. Because oceanic crust is relatively high in iron and magnesium, and is dense and thin, it subducts, or goes under, the continental plate.

- As a plate subducts, some of this sediment is scraped off and collects, or "accretes," as a series of angular rock slabs, forming a wedge along the front of the overriding plate. This process is similar to what happens when snow accumulates in front of a snowplow. The wedge is known as an **accretionary prism** because of its three-dimensional shape.

- An **ocean-ocean convergent boundary** is formed where two plates composed of oceanic crust collide. Usually the older and, therefore, denser of the two plates subducts below the other. These are sites of island arc formation.

3-7 Orogenesis occurs in places where two continents converge.

- The third type of convergent boundary occurs when an ocean basin closes and two plates composed of continental crust collide. Because neither of the two plates will subduct, a **continent-continent convergent boundary** builds mountains in a dramatic fashion.

- Continent-continent convergence is always preceded by ocean-continent convergence. As subduction of oceanic crust continues, any continent riding on the subducting plate will be drawn toward the overriding plate. The evolving margin of the continent on the overriding plate (known as an **active margin**), with its volcanic arc and accretionary prism, will collide with the tectonically quiet margin (known as a **passive margin**) of the continent riding on the subducting plate.

- Mountain building resulting from this kind of continent-continent collision is known as **orogenesis.** The Himalayas are an example of orogenesis, the result of collision between the Indian Plate and the southern edge of the Eurasian Plate.

3-8 Transform boundaries connect two spreading centers.

- Transform boundaries are characterized by side-to-side plate movement. They occur where two plates slide past each other rather than colliding or separating. This type of motion is called shearing.

- Geophysicist J. Tuzo Wilson, who first proposed the hotspot hypothesis, also proposed that transform boundaries connect two spreading centers (or, less commonly, two subduction zones).

- Probably the most famous transform boundary in the world is the San Andreas transform fault. A **fault** is a place where the crust is broken and the broken edges are offset relative to each other either vertically or horizontally. The San Andreas Fault is a transform boundary between the North American Plate to the east and the Pacific Plate to the west.

- A common type of transform boundary is an **oceanic fracture zone** consisting of inactive (fracture zone) and active (transform fault) portions. These are ocean-floor valleys that extend horizontally away from spreading centers.

- About 81% of the world's largest earthquakes occur in the circum–Pacific seismic belt. This belt, which includes the rim of the Pacific Plate and nearby plates, corresponds to the Pacific Ring of Fire. The second most important earthquake belt, the Alpide belt, extends from Java to Sumatra through the Himalayas and the Mediterranean and out into the Atlantic. The third prominent earthquake belt follows the submerged Mid–Atlantic Ridge. Quakes in this belt have not been particularly destructive, as they generally occur far from population centers.

3-9 Earthquakes tend to occur at plate boundaries.

- The forces associated with plate separation in rift valleys lead to earthquakes and volcanic eruptions. Transform boundaries are characterized by earthquakes that can be very destructive. Convergent boundaries are subject to numerous earthquakes, as well as explosive volcanism.

- The most hazardous earthquakes occur when a section of a transform boundary or a subduction zone becomes "locked." This happens when local plate motion comes to a halt because two segments of plate cannot slide past each other even though overall plate movement continues. As the pressure generated by plate movement builds, the potential for a large-scale earthquake grows as well. In the case of a locked subduction zone, these are called megathrust quakes.

- Earthquakes cause the ground to shake because seismic energy released by the breakage of rock passes through the crust. The shaking can have disastrous consequences when houses, bridges, and roadways fracture and topple as a result.

3-10 Plate movement powers the rock cycle.

- Scientists are not sure what causes plates to move. Four possible mechanisms have been hypothesized: Plates may be: 1) pushed away from spreading centers, 2) pulled down into trenches, 3) dragged along by the friction created by mantle convection on their undersides, or 4) driven by gravity to slide down the slope from the ridge to the trench. Conventional wisdom holds that all four mechanisms contribute to plate movement, although researchers are undecided as to the proportion of total movement contributed by each mechanism.

- James Hutton (1727–1797), an eighteenth-century gentleman farmer and one of the founders of modern geology, originated the concept of the rock cycle. The modern rock cycle concept proposes that the mantle, the crust, the atmosphere, the biosphere, and the hydrosphere can be envisioned as a giant recycling machine in which the elements that make up rocks are neither created nor destroyed but, rather, are redistributed and transformed from one natural form to another over time.

KEY TERMS

accretionary prism (p. 72)
active margin (p. 75)
Archean eon (p. 58)
asthenosphere (p. 59)
chemical differentiation (p. 58)
continental crust (p. 60)
continent-continent convergent boundary (p. 75)
convergent boundaries (p. 68)
core (p. 58)
crust (p. 59)
divergent boundaries (p. 68)

fault (p. 77)
geomagnetic field (p. 60)
inner core (p. 58)
iron catastrophe (p. 58)
island arc (p. 73)
lithosphere (p. 59)
mantle (p. 58)
ocean-continent convergent boundary (p. 72)
oceanic crust (p. 60)
oceanic fracture zone (p. 78)
ocean-ocean convergent boundary (p. 73)

orogenesis (p. 76)
outer core (p. 58)
paleomagnetism (p. 66)
passive margin (p. 75)
plate boundaries (p. 62)
reasoning by analogy (p. 68)
rift valleys (p. 62)
Ring of Fire (p. 68)
transform boundaries (p. 68)
volcanic arc (p. 73)

ASSESSING YOUR KNOWLEDGE

Please complete this exercise before coming to class. Identify the one best answer to each question.

1. During the Hadean eon, which of the following is thought to have occurred?
   a. Growth of the modern seas
   b. Formation of modern continents

   c. The iron catastrophe
   d. Origin of life on Earth
   e. All of the above

2. How does the chemical differentiation of Earth today reflect the influence of the iron catastrophe?
   a. There is more iron in the core than in the crust.

b. The lower lithosphere stores most of Earth's iron.

c. Much of Earth's iron has escaped as a result of extraterrestrial impacts.

d. Iron is rare in Earth.

e. None of the above.

3. The mantle consists largely of (solid rock, magma, turbulent gases, liquid iron) _____.

4. What are the principal differences between the average chemistry of the crust and the average chemistry of Earth as a whole?

a. The crust is relatively enriched in less dense compounds and relatively depleted in iron.

b. The crust is relatively enriched in magnesium and relatively depleted in oxygen.

c. Earth as a whole has a greater relative abundance of silicon than does the crust.

d. The crust contains a greater relative abundance of heavier elements than does Earth as a whole.

e. None of the above.

5. How is Earth organized?

a. Earth has an inner and outer core, a mantle, and a crust.

b. Earth has an inner mantle and an outer lithosphere, with a liquid inner core.

c. Earth's crust rests atop the liquid mantle and the solid outer core.

d. The inner core is solid, the mantle is solid, and the crust is solid under the continents and liquid under the oceans.

e. None of the above.

6. Subduction of oceanic lithosphere at a _____ (convergent, divergent, transform) boundary releases (water, gases, magma, sediment)_____ into the overlying plate that causes (occasional, partial, reverse) _____ melting.

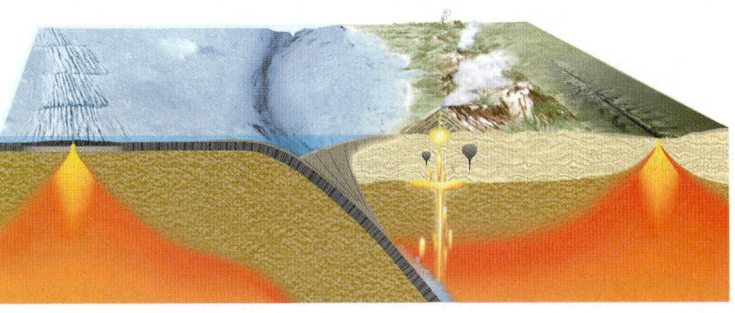

7. Subduction occurs:

a. When one plate crashes into another.

b. When a lithospheric plate is recycled into Earth's interior.

c. When a plate enters the inner core.

d. When a continent is recycled beneath an overriding plate.

e. During orogenesis.

8. Oceanic crust:

a. Is formed by asteroid impact.

b. Is enriched in iron and magnesium, compared to continental crust.

c. Forms from sea salt.

d. Is made of metamorphic rock.

e. None of the above.

9. Magnetic reversals are caused by:

a. Lunar gravitational effects.

b. Changes in the rate at which Earth orbits the Sun.

c. Impacts of extraterrestrial objects.

d. Unknown causes.

e. Faster subduction rates across Earth.

10. This map shows the age of the seafloor across the northern extent of the Atlantic Ocean. The Mid–Atlantic Ridge can be seen extending roughly north-south (in the yellow band) down the middle of the map. Yellow through red colors show rocks of similar age. Number them on the map from 1 (oldest) through 5 (youngest).

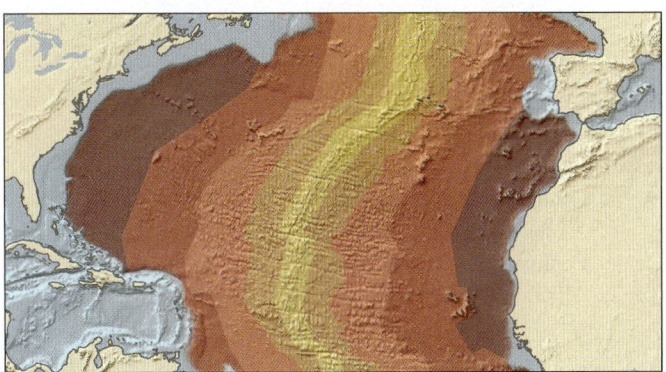

11. Evidence that the polarity of Earth's geomagnetic field has reversed in the past is found:

a. In magnetic striping in volcanic arcs.

b. In magnetic reversals recorded by iron minerals in oceanic crust.

c. In accretionary prisms.

d. Where magma develops above a subducting slab.

e. All of the above.

12. Three plate boundaries, defined by relative motion, are:

a. Converging, diverging, and lateral.

b. Convergent, divergent, and transform.

c. Strike slip, hotspot, and spreading center.

d. Spreading center, transform, and divergent.

e. All of the above.

13. The three types of convergent plate boundaries are:

a. Convergent, divergent, and volcanic.

b. Ocean-ocean, ocean-continent, and continent-continent.

   c. Subducting, divergent, and shearing.

   d. Igneous, sedimentary, and metamorphic.

   e. None of the above.

14. At ocean-ocean convergent boundaries:

   a. Older, denser crust tends to subduct.

   b. Island arcs tend to subduct.

   c. Transform faults typically will develop.

   d. There are rarely earthquakes.

   e. None of the above.

15. At a _____ (convergent, transform, divergent) boundary new oceanic lithosphere forms along a (accretionary, mid-ocean, forearc) _____ ridge. These are locations of seafloor (recycling, spreading, erosion) _____.

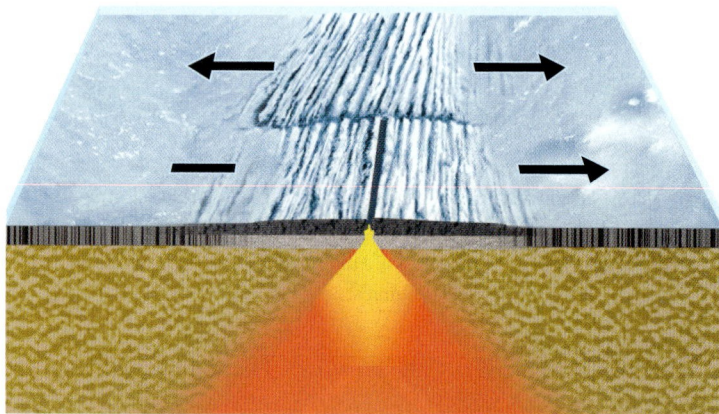

16. The Himalayas are an example of:

   a. Extraterrestrial impact.

   b. Continent-ocean convergence.

   c. A subduction zone.

   d. Continent-continent convergence.

   e. All of the above.

17. At the San Andreas transform fault:

   a. Lithosphere is subducted as one plate dives below another.

   b. New lithosphere is formed as two plates pull away from each other.

   c. Pressure-release melting recycles old crust.

   d. The Pacific Plate moves to the north relative to the North American Plate.

   e. All of the above.

18. Earthquakes occur at:

   a. Divergent plate boundaries.

   b. Ocean-ocean convergent plate boundaries.

   c. Ocean-continent plate boundaries.

   d. Transform boundaries.

   e. All of the above.

19. The rock cycle is a concept that:

   a. Has no relationship to plate tectonics.

   b. Is not widely accepted.

   c. Describes the recycling of rock.

   d. Was first described only two decades ago.

   e. All of the above.

20. Continent-_____ (ocean, continent, transform) convergence is always preceded by (ocean, continent, transform) _____-continent convergence. Mountain building at this type of boundary is known as _____ (volcanism, subduction, orogenesis).

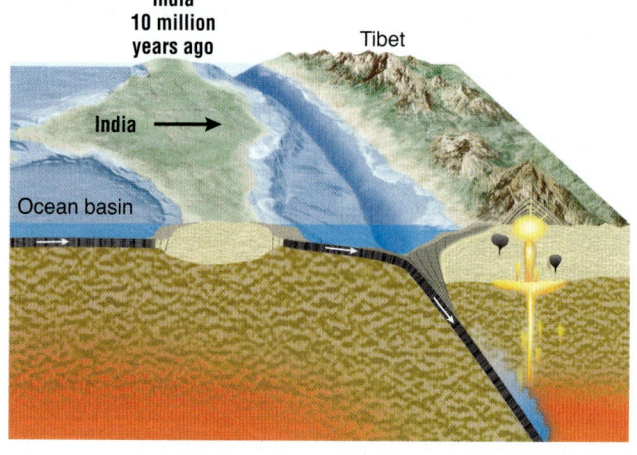

FURTHER RESEARCH

1. Search the Web and find out when Earth's polarity last reversed.

2. What are some of the geologic characteristics of plate boundaries?

3. Why do oceanic ridges have such high elevation in comparison to the surrounding seafloor?

4. What would Earth be like if plate tectonics did not exist?

5. How do geologic processes on an active margin differ from those on a passive margin?

6. How does the chemistry of lava at an island arc differ from that of lava at a volcanic arc?

7. Search the Web for a story about a recent earthquake or volcanic eruption. Did it occur at a plate boundary? Name the plates involved and identify the type of boundary at which the event occurred. From this information, describe the most likely cause of the event.

ONLINE RESOURCES

Explore more about plate tectonics on the following Web sites:

USGS discussion of rock families:
http://wrgis.wr.usgs.gov/docs/parks/rxmin/index.html

USGS story of plate tectonics:
http://pubs.usgs.gov/publications/text/dynamic.html

USGS discussion of Earth's interior:
http://pubs.usgs.gov/gip/interior

USGS building stones of our nation's capital:
http://pubs.usgs.gov/gip/stones

Additional animations, videos, and other online resources are available at this book's companion Web site:
www.wiley.com/college/fletcher

This companion Web site also has more information about WileyPLUS and other Wiley teaching and learning resources.

MINERALS

CHAPTER CONTENTS & LEARNING OBJECTIVES

GEOLOGY IN OUR LIVES

Minerals are critical geologic resources used every day in manufacturing, agriculture, electronics, household items—even in the clothes you wear. Therefore, it is vital that we ensure the sustainability of mineral resources. Minerals are solid crystalline compounds that are found in the crust and mantle and have a definite, but variable, chemical composition.

Q **Can you identify where minerals have entered your life today?**

4-1 Minerals Are Solid Crystalline Compounds with a Definite, but Variable, Chemical Composition

LO 4-1 Define the term *mineral.*

Minerals are crystalline compounds that are mined from the crust. Minerals provide us with metals, nutrients, food, power, building materials, and other critical resources. You use hundreds of mineral-based products every day.

Minerals in Our Lives

Your alarm rings in the morning; you turn it off and switch on the light and head to the bathroom. After washing your face and brushing your teeth, you get dressed, then head to the kitchen where you turn on the radio to listen to the news as you eat breakfast. Your day has hardly begun and already you have used dozens of minerals or mineral-based products. Nearly everything you have done so far, and will do during the rest of the day, will involve the use of minerals in one form or another. Minerals are solid chemical compounds that we use every day.

In the United States alone, each individual, including children, uses about 10 tons of minerals and rocks per year. Minerals are consumed in manufacturing the products we use, growing the food we eat, and building our homes, roads, and office buildings. This great demand for minerals places a heavy burden on mining companies to find mineral reserves. These companies seek minerals in Earth's crust using remote sensing, identifying the chemistry of rock samples, and implementing the theory of plate tectonics to assess the location and size of potential mineral deposits.

Once an economical mineral deposit has been located, geological and mining engineers determine how it can be extracted safely and profitably without irreparably harming surrounding natural ecosystems. Balancing the need for mineral resources against the need to conserve natural environments is a constant challenge to the mineral industry.

Are we in danger of running out of minerals? That is unlikely. However, as we exhaust the more easily mined deposits, the cost of some mineral resources is likely to rise. Read "Will We Run Out of Minerals?" in the Earth Citizenship feature to learn more about this important issue.

Of the thousands of known minerals, only about two dozen are common in crustal rocks. In this chapter, we introduce and discuss the most common minerals in rocks, how and where they are formed, and their importance to humans.

What Is a Mineral?

Perhaps none of nature's phenomena are as beautiful and diverse as the 4,000-plus minerals in Earth's crust. Minerals occur in every hue of the rainbow and in widely varied and fantastic shapes. Some minerals are harder than steel; others are so soft that you can scratch them with your fingernail. Minerals sometimes appear as tiny specks of dust, but when viewed through a microscope, are found to be composed of intricate crystals. Certain minerals can bend light, taste salty or sour, catch fire, dissolve in water, change color, and even shift shape on their own.

A **mineral** is a naturally occurring, inorganic, crystalline solid with a definite, but sometimes variable, chemical composition. This definition is quite a mouthful, and each of its four parts is important, so let us take a closer look at each of them in turn.

1. *Naturally occurring.* A mineral occurs in nature but is not made by humans. Natural substances, like quartz, salt, gold, and gemstones, are minerals. Other substances, such as Formica, plastics, glass, and manufactured gems such as cubic zirconia, may appear similar to minerals, but they do not occur naturally and therefore are not minerals.

2. *Inorganic.* A mineral is inorganic, meaning that it cannot contain compounds composed of organic carbon. Organic carbon, the type of carbon found in all living things, bonds with hydrogen to create compounds that can be represented by multiples of the simple chemical compound *formaldehyde*, CH_2O (e.g., $C_6H_{12}O_6$). All five kingdoms of living organisms (Animals, Plants, Monera, Fungi, and Protista) are composed of organic carbon. Inorganic carbon is formed when carbon bonds with elements other than hydrogen—for instance, *calcium carbonate*, $CaCO_3$, as in the mineral *calcite*. Thus, coal cannot be a mineral because it is composed of plant remains made of organic carbon. But the mother-of-pearl lining of a shell is considered a mineral because it is made of calcite.

3. *Crystalline solid.* The atoms in minerals are arranged in an orderly fashion to create a **crystalline structure.** To have a crystalline structure, a substance must be solid at Earth's surface temperatures, not liquid (like water) or gaseous (like carbon dioxide). It must have an internal network of atoms that forms an orderly, repeating, three-dimensional pattern. So, when a mineral breaks, the shape of its pieces is predictable. For example, the mineral *halite*, NaCl (table salt, **FIGURE 4.1**), contains atoms of sodium (Na) and chloride (Cl) arranged in the shape of a cube. Because of its crystalline structure, if you drop a large block of salt or smash it with a hammer, it breaks into many cubes (and rectangles), large and small.

4. *Chemical composition.* A mineral has a definite, but sometimes variable, chemical composition. Consider the mineral *olivine*: It consists of iron and magnesium atoms in various quantities, plus silicon and oxygen. The formula for olivine is definite: (Mg, Fe)$_2$SiO$_4$, or two parts magnesium (Mg) or iron (Fe) atoms, one part silicon atoms, and four parts oxygen atoms. Because magnesium and iron atoms are similar in size and have identical electrical charges, they are interchangeable in the crystalline structure of olivine. That means they can be present in any combination—all iron and no magnesium, or vice versa, or any ratio in between. The result can be an iron-rich olivine (tends to be brownish), a magnesium-rich olivine (tends to be a brilliant green), or something in between. Hence, the chemical composition is "definite, but sometimes variable" in a mineral.

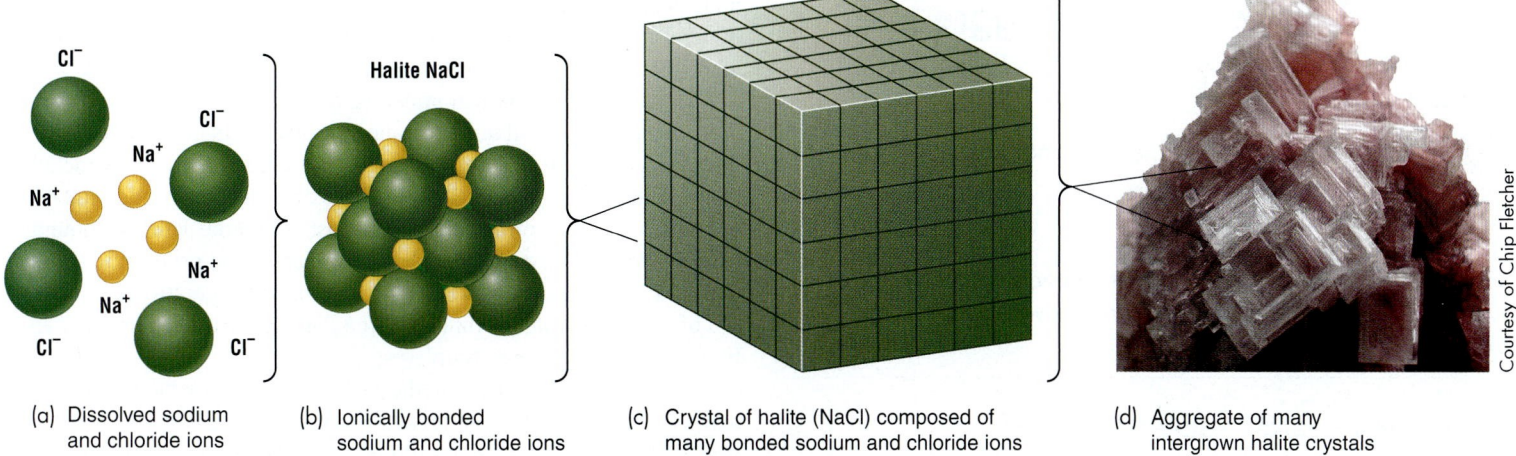

(a) Dissolved sodium and chloride ions

(b) Ionically bonded sodium and chloride ions

(c) Crystal of halite (NaCl) composed of many bonded sodium and chloride ions

(d) Aggregate of many intergrown halite crystals

Courtesy of Chip Fletcher

FIGURE 4.1 Minerals are crystalline. That is, the atoms within them are arranged in an orderly fashion. Halite (common table salt) consists of: (a) sodium (Na) and chloride (Cl) atoms, (b) arranged in the shape of a cube, (c) such that large groups of NaCl, (d) make the mineral in its natural form.

Ⓠ Why does halite conform to the definition of a mineral?

Ⓠ **EXPAND YOUR THINKING**

What environment features dissolved Na and Cl? What process concentrates these elements and promotes their tendency to bond?

EARTH CITIZENSHIP

Will We Run Out of Minerals?

In the 1970s, some scientists predicted that the world would run out of economically important minerals by the end of the century. Fortunately, several factors have led to a revision of that dire scenario:

1. Recycling has greatly reduced the need to mine several metals (**FIGURE 4.2**).

2. New technologies have made it possible to mine low-quality mineral deposits that were previously unprofitable.

3. The rate of mineral consumption has slowed as the primary focus of the U.S. economy has shifted away from manufacturing toward services and technology. It is now widely recognized that availability of minerals is probably based more on cost (can we afford them?) than on depletion (have we run out of them?).

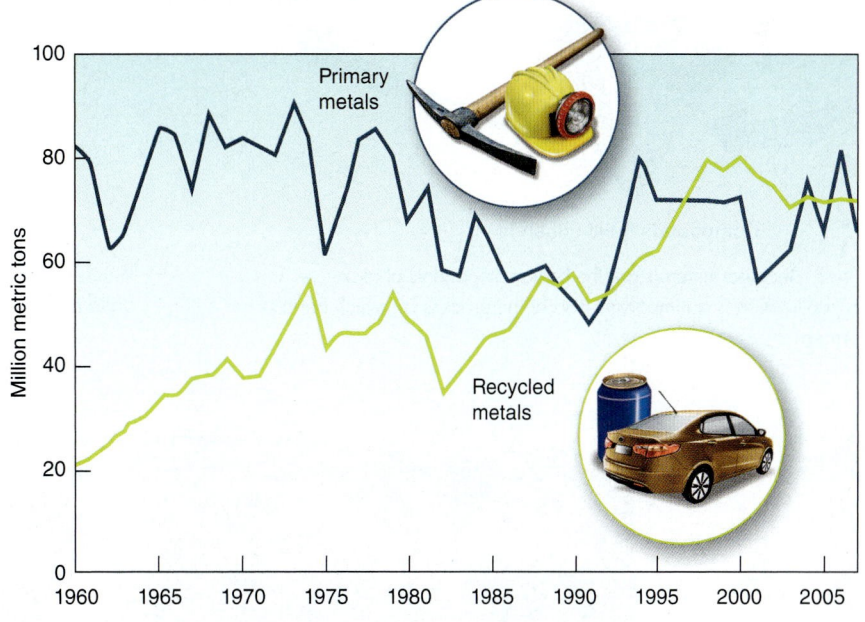

FIGURE 4.2 Improvements in recycling technology have reduced the need for new metals in the United States. Beginning in the 1990s, the amount of recycled metals grew to match the amount of metals mined from the crust. *Source*: U.S. Geological Survey, Department of the Interior, "Metals and Minerals," *Minerals Yearbook*, Vol. 1, 1998–2007.

4-2 A Rock Is a Solid Aggregate of Minerals

LO 4-2 List five useful minerals, and describe their composition and how they are used.

Because minerals are the basic components of rocks, they are an essential part of the process through which Earth's crust is formed (FIGURE 4.3).

How Are Minerals Related to Rocks?

Geologists define rock as a solid aggregate of minerals. Granite, for example, is an aggregate of several minerals—typically, quartz, plagioclase feldspar, orthoclase feldspar, biotite, and others (FIGURE 4.4). To understand rocks, then, we must start with their basic components—minerals. The study of minerals is known as *mineralogy*, and geologists who study minerals are *mineralogists*.

Humans have exploited minerals, especially *gemstones* and *precious metals* (gold, silver, and others), since prehistoric times. The modern economy is heavily dependent on metals, which are extracted from minerals. Iron is an example. We extract iron from the mineral hematite, which is mined from the crust. Hematite is a compound of iron (Fe) and oxygen (O). Like all minerals, the proportion of its chemical elements can describe it: every piece of hematite is composed of two parts iron and three parts oxygen, expressed chemically as Fe_2O_3. Hematite is, essentially, rust.

We can separate or refine the iron in hematite by heating it to a high enough temperature to break the bonds between the oxygen and iron atoms. This process releases pure iron. We use pure iron to make steel by combining the iron with carbon (C) and other elements, such as chromium (Cr) and silicon (Si). To learn more about steel, read "How Is Steel Made?" in the Geology in Our Lives feature.

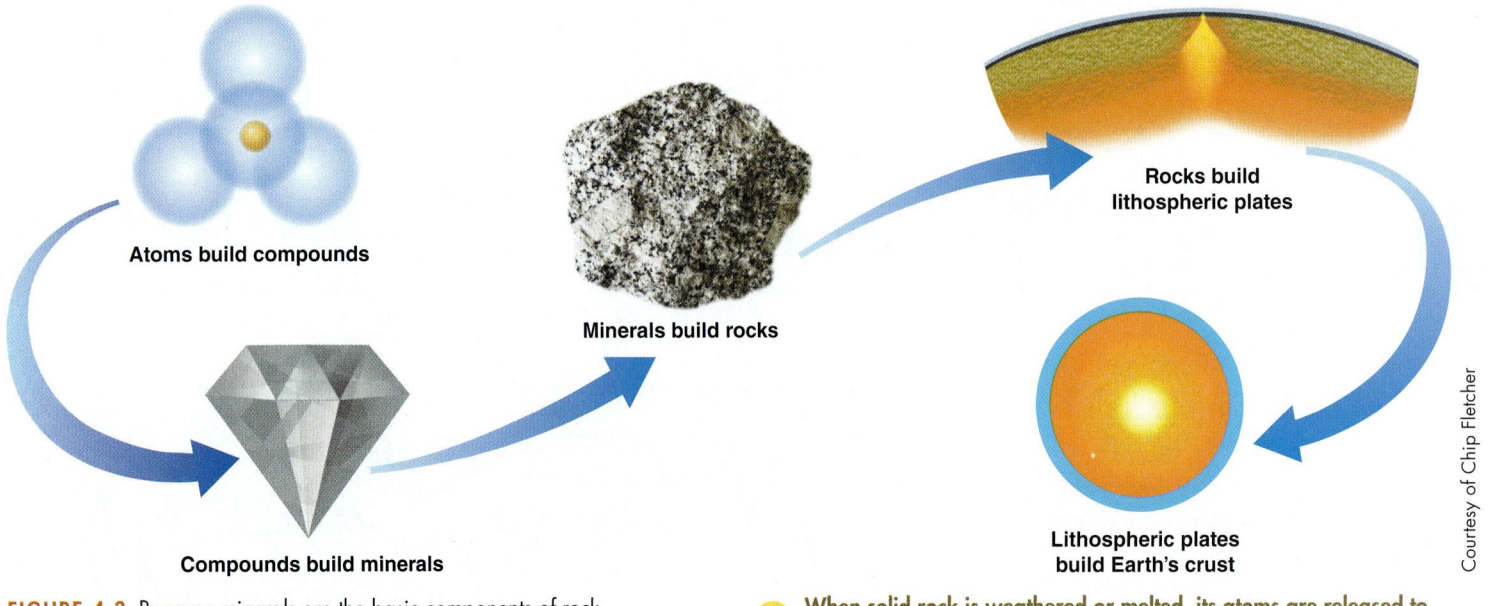

Atoms build compounds

Compounds build minerals

Minerals build rocks

Rocks build lithospheric plates

Lithospheric plates build Earth's crust

Courtesy of Chip Fletcher

FIGURE 4.3 Because minerals are the basic components of rock, mineral crystallization is an important part of the process by which Earth's crust is formed.

When solid rock is weathered or melted, its atoms are released to build new compounds. What is this natural recycling process called?

Mineral images courtesy of Chip Fletcher

(a) (b) (c) (d)

Granite

FIGURE 4.4 Granite is a solid aggregate of several minerals. These minerals include (a) quartz (clear or white), (b) plagioclase feldspar (gray or white), (c) potassium feldspar (pink), and (d) biotite (black).

Which of these minerals dominates the color of the granite sample?

TABLE 4.1 Common Minerals and Their Uses

Mineral	Composition	Uses
Chalcopyrite	Copper-iron-sulfur mineral—$CuFeS_2$	Mined for copper
Feldspar	Large mineral family; aluminum-silicon-oxygen composition—$x(Al,Si)_4O_8$, where x = various elements such as sodium or calcium	Ceramics and porcelain
Fluorite	Calcium-fluorine mineral—CaF_2	Mined for fluorine (its most important ore); steel manufacture
Galena	Lead-sulfur mineral—PbS	Mined for lead
Graphite	Pure carbon—C	Pencil "lead" (replacing the actual lead metal once used in pencils); dry lubricant
Gypsum	Hydrous-calcium-sulfur mineral—$CaSO_4 • 2H_2O$	Drywall, plaster of paris
Halite	Sodium chloride—$NaCl$	Table salt, road salt
Hematite	Iron-oxygen mineral—Fe_2O_3	Mined for iron (to make steel)
Magnetite	Iron-magnesium-oxygen mineral—$(Fe, Mg)Fe_2O_4$	Mined for iron
Pyrite	Iron-sulfur mineral—FeS_2	Mined for sulfur and iron
Quartz	Silicon-oxygen mineral—SiO_2	In pure form, for making glass
Sphalerite	Zinc-iron-sulfur mineral—$(Zn, Fe)S$	Mined for zinc
Talc	Magnesium-silicon-oxygen-hydrogen mineral—$Mg_3Si_4O_{10}(OH)_2$	Used in ceramics, paint, talcum powder, plastics, and lubricants
Rutile	Titanium dioxide—TiO_2	Mined as titanium ore, a valuable metal used in aerospace and other industries
Calcite	Calcium carbonate—$CaCO_3$	Toothpaste, cement, drywall, sheetrock

Other minerals (**TABLE 4.1**) contain valuable metals such as copper, tin, zinc, lead, and titanium. But many minerals besides those that contain metals are useful (**FIGURE 4.5**). Gypsum for one is used for making drywall and wallboard. Gypsum consists of the elements calcium (Ca), sulfur (S), and oxygen (O), which are bound to water molecules (H_2O) in the formula $CaSO_4 • 2H_2O$.

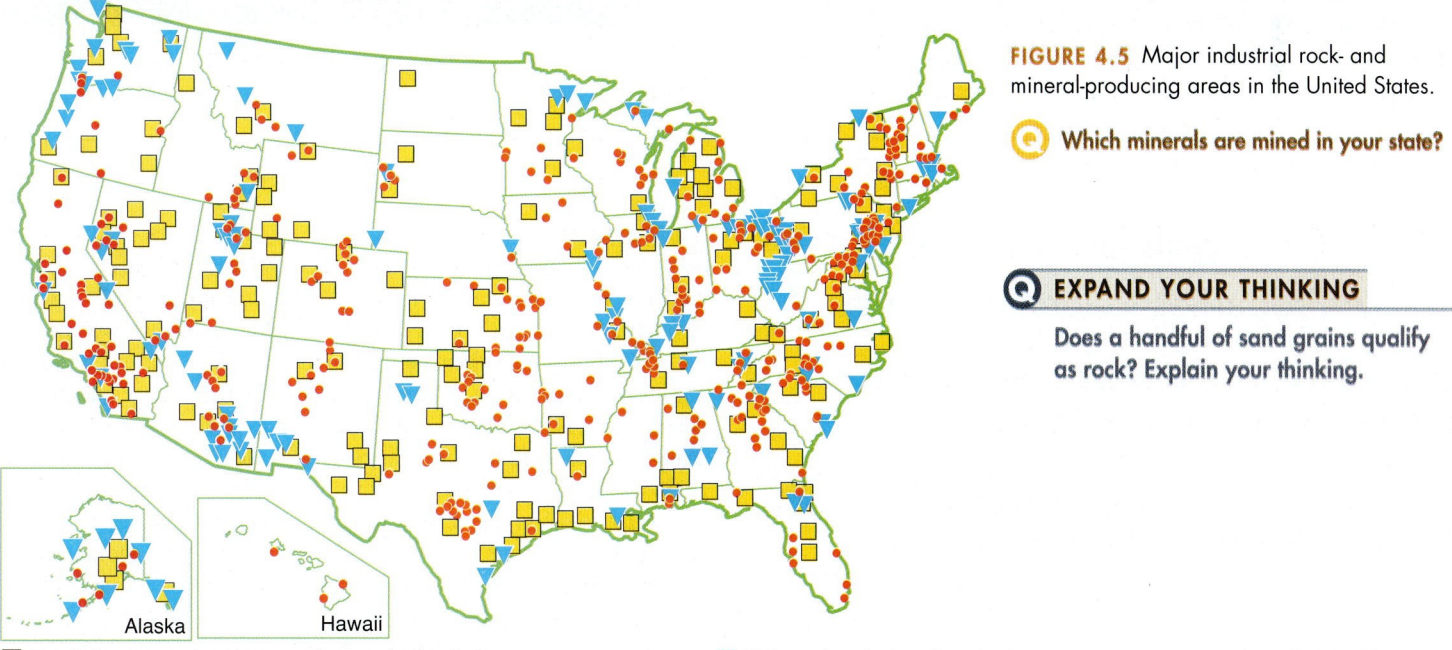

FIGURE 4.5 Major industrial rock- and mineral-producing areas in the United States.

Q Which minerals are mined in your state?

Q **EXPAND YOUR THINKING**

Does a handful of sand grains qualify as rock? Explain your thinking.

Alaska   Hawaii

■ **Metals** (ore processing plants, smelters, and refineries):

Aluminum	Gold	Rhenium	Tungsten
Beryllium	Iron ore	Selenium	Vanadium
Bismuth	Lead	Silicon	Zinc
Cadmium	Magnesium	Silver	Zircon
Chromium	Molybdenum	Tellurium	
Cobalt	Nickel	Thorium	
Columbium/tantalum	Platinum group	Tin	
Copper	Rare earths	Titanium metal	

▼ **Major rock and mineral production:**

Barite	Kyanite	Salt
Bismuth	Magnesium	Soda ash
Bromine	compounds	Sodium sulfate
Diatomite	Mica	Sulfur
Garnet	Olivine	Talc
Gypsum	Peat	Wollastonite
Helium	Phosphate	Zeolite
Indium	Potash	Zircon
Iodine	Rutile	

● **Construction minerals:**

Ball clay
Cement
Dimension stone
Gypsum
Mica
Perlite
Pumice

GEOLOGY IN OUR LIVES

How Is Steel Made?

Steel is essentially iron plus a consistent concentration of carbon throughout (less than 1%) that has had the impurities remove; this, it turns out, is not so easy to do. The process of producing steel from iron ore has several steps. The basic raw materials are coal, iron ore, limestone, and various chemicals. Initially, iron ore is mined from the crust and ground to fine particles. Iron oxide is separated from waste rock using magnets or by floating in a dense liquid that isolates iron by buoyancy. Separated iron oxide is rolled into pellets and transported, usually by train or ship, to steel plants.

At the plant, coal, which does not burn hot enough to melt iron, is turned into coke, which is baked coal that burns at higher temperature. Coal contains impurities such as tar, ammonia, and sulfur. These are removed in the "coking process" in which coal is baked in four-story-tall ovens at 1315°C for 18 hours and quenched using 15,000 liters of water. Fourteen tons of coal yields 11 tons of purified coke. The missing 3 tons is captured gas that is cooled and refined into tar, fertilizer, and oils used to make plastics, perfume, aspirin, nylon, cosmetics, and other products.

Iron pellets are fed into a blast furnace, a tall steel vessel up to 10 stories high that melts the iron using coke, and produces unpurified pig iron or cast iron. In a blast furnace, air is heated to 980°C and blown into the bottom of the furnace at 800 kilometers per hour (the blast). The hot air helps the coke burn, reaching 2100°C. Molten iron ore is mixed with limestone, which combines with impurities and makes "slag" that is skimmed off.

Cast iron is about 4% carbon, and still needs purification to make steel. Molten cast iron (**FIGURE 4.6**) is fed into a basic oxygen furnace (BOF) where the carbon content is taken down to 0.02% (1% carbon or less differentiates steel from cast iron). The BOF is charged with 40 tons of scrap steel followed by 100 tons of hot cast iron poured from giant ladles. High-purity oxygen is blown through the molten bath to lower the carbon, silicon, manganese, and phosphorus content. After about 20 minutes, a sample is taken to verify that the iron is purified to steel.

Molten steel is formed into sheets, rods, or ingots for delivery to a customer, where it is shaped into various products. Higher and lower grades of steel can be produced using additional levels of processing.

Associated Press

FIGURE 4.6 Giant ladles pour molten iron in the basic oxygen furnace to produce steel.

4-3 Geologists Use Physical Properties to Identify Minerals

LO **4-3** List and describe the common physical properties used in mineral identification.

Geologists learn to identify minerals using their physical properties such as cleavage, hardness, luster, and others.

Physical Properties

During the gold rushes of the 1800s, California, Colorado, and Alaska were flooded with thousands of hopeful prospectors. To be successful, these eager amateurs had to be able to distinguish between real gold and so-called fool's gold (pyrite). The two minerals do look alike; both are golden and shiny. However, conducting simple physical tests will quickly betray the difference: If you bite pure gold, it will show tooth marks; if you bite pyrite, you will break a tooth. (Don't try it!) When a knife blade is dragged over real gold, it leaves a gouge; a knife makes no impression on pyrite. When rubbed on broken crockery, real gold leaves a bright golden streak; pyrite leaves a dirty black streak. Truly, only a fool would be unable to distinguish between real gold and its lookalike, pyrite.

Usually, distinguishing between minerals is not so easy. Geologists must learn to identify hundreds of minerals in the field so that they do not have to send samples to a lab for testing—a time-consuming, expensive procedure. Fortunately, many minerals have physical properties that aid in their identification. Each mineral's unique chemistry and crystalline structure determine its physical qualities—cleavage, fracture, hardness, luster, and others (**TABLE 4.2**). Essentially, the field geologist uses critical thinking to observe a mineral's physical properties and develop a hypothesis about it. If it is important to identify a sample exactly, the sample may be tested in a lab using more sophisticated methods.

Cleavage and Fracture

Cleavage is the tendency of a mineral to break, leaving a fairly smooth, flat surface. *Fracture* is the term used to describe breakage that is irregular, meaning that it does not occur along a cleavage plane. A *conchoidal* fracture exhibits seashell-shape curved surfaces like those of broken glass. A *fibrous* fracture looks like splintered wood.

TABLE 4.2 Physical Properties of Minerals

Property	Definition	Testing Method
Cleavage	Breakage along planes of weakness	Examine sample for planar surfaces.
Fracture	Breakage not along cleavage plane	Breakage is either irregular or conchoidal.
Hardness	Resistance to scratching or abrasion	Materials of known hardness are used to determine hardness of sample.
Luster	Character of reflected light	Does sample appear metallic or nonmetallic?
Crystal form	Geometric shape	Describe geometric shape: cubic, hexagonal, etc. Not commonly seen in most samples.
Reaction to HCl	Chemical interaction of weak hydrochloric acid (HCl) and calcium carbonate ($CaCO_3$)	Place drop of HCl on sample and watch for a reaction (bubbles).
Streak	Color when powdered.	Rub sample on porcelain to determine color of streak.

Cleavage occurs in crystals that have specific planes of weakness. These planes, which are produced by weak chemical bonding, are inherent in the crystal's structure. Cleavage is reproducible, which means that a crystal can be cleaved along the same parallel plane an infinite number of times (**FIGURE 4.7**). Any sample of a particular type of crystal will have the same cleavage. Cleavage planes are identifiable because they are usually repeated within the interior of a specimen, producing observable incipient breakage surfaces parallel to an external cleavage surface.

Hardness

Hardness is the ease with which a mineral can be scratched. It is measured by scratching one mineral with another, or by scratching an object of known hardness with a mineral sample (**FIGURE 4.8**). Two centuries ago, the Austrian mineralogist *Friedrich Mohs* proposed a system for measuring hardness, now called the **Mohs hardness scale.** The scale uses numerical values from 1 to 10 as relative measures of hardness based on comparison with specific minerals. Talc, the

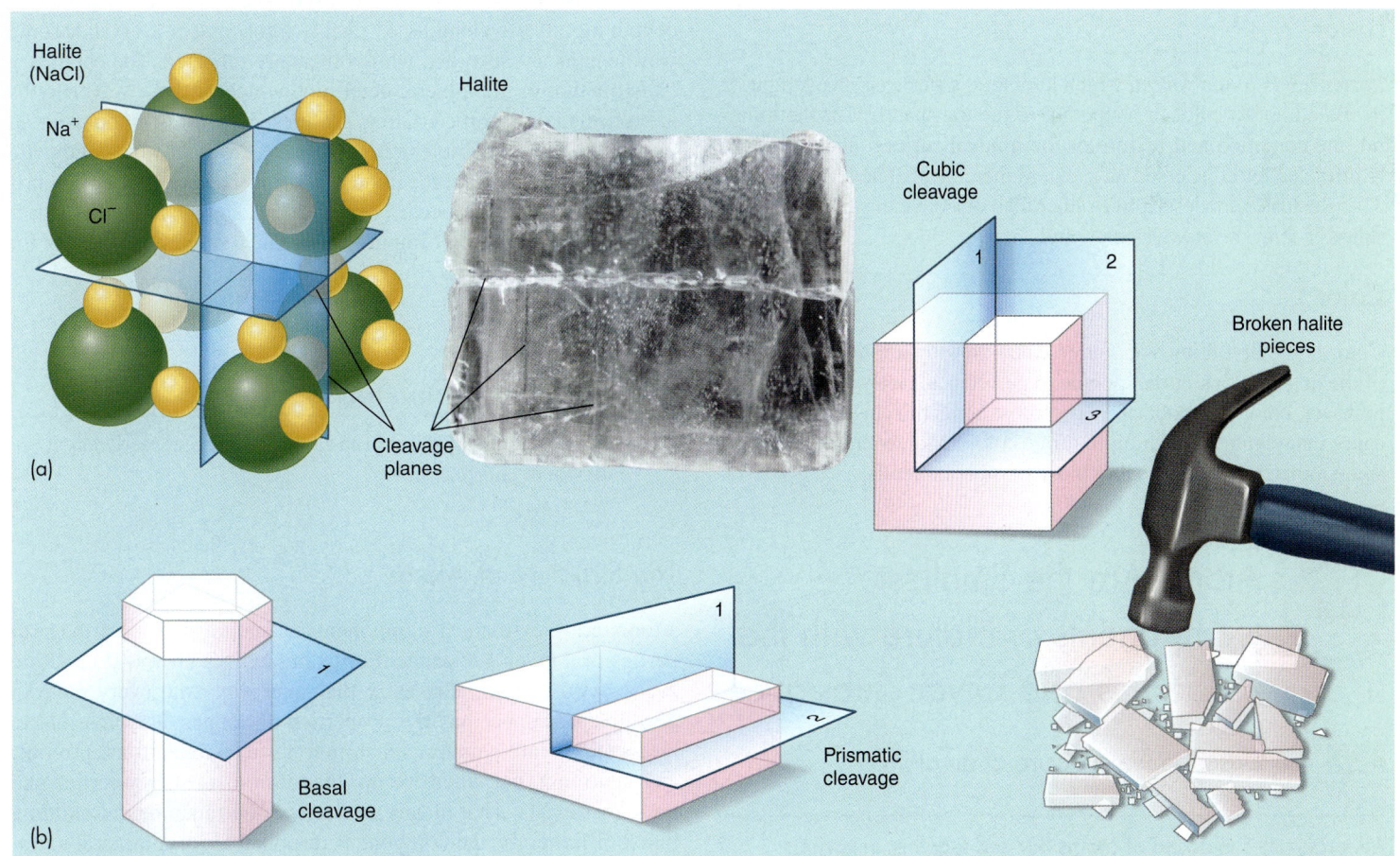

Courtesy of Chip Fletcher

FIGURE 4.7 (a) Halite breaks along three planes of cleavage, which means it has cubic cleavage. (b) Also shown are examples of basal cleavage (cleavage parallel to the base of a mineral) and prismatic cleavage (cleavage forming a six-sided prism).

When a mineral cleaves, which is breaking: atoms or the bond between atoms? Is it easy to break an atom?

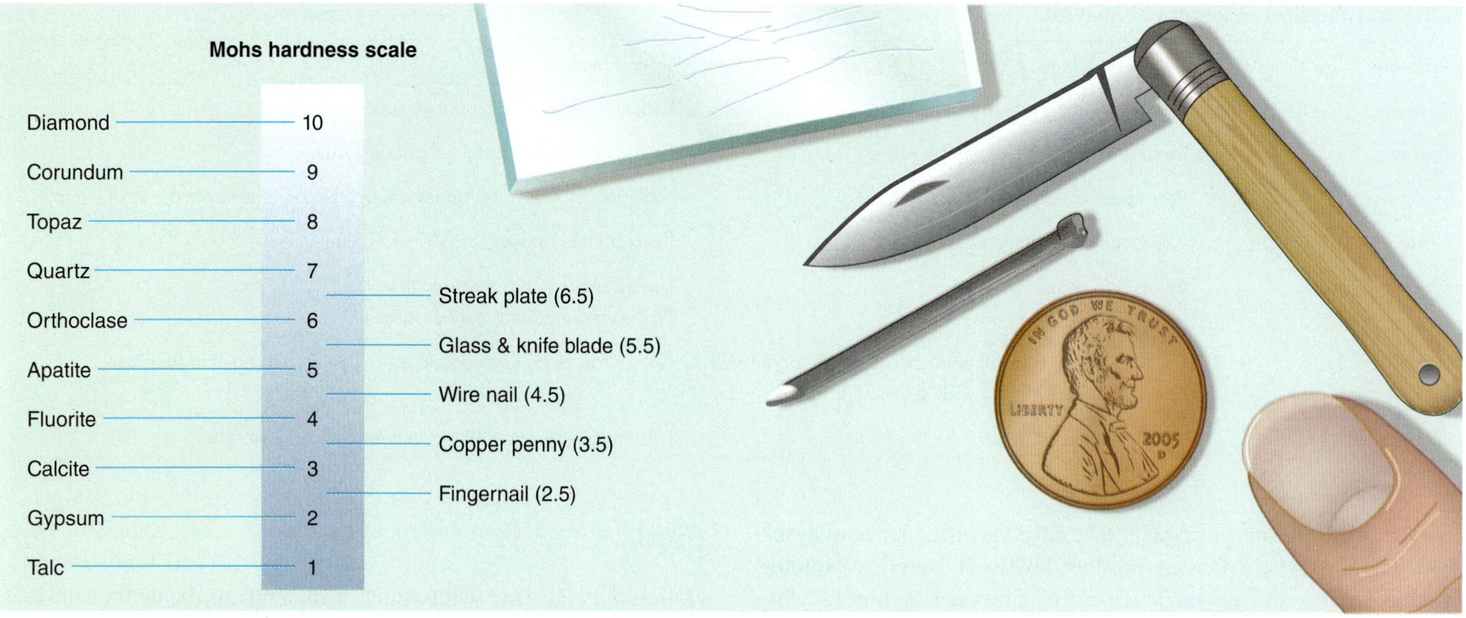

Mohs hardness scale

Diamond	10
Corundum	9
Topaz	8
Quartz	7
Orthoclase	6
Apatite	5
Fluorite	4
Calcite	3
Gypsum	2
Talc	1

Streak plate (6.5)
Glass & knife blade (5.5)
Wire nail (4.5)
Copper penny (3.5)
Fingernail (2.5)

FIGURE 4.8 To test hardness, scratch one mineral across the surface of another. A mineral that leaves a scratch on another is harder.

Q What is the hardness of plastic?

softest mineral, is assigned a hardness value of 1. Diamond, the hardest natural mineral, is assigned a value of 10. All other minerals fall somewhere between these two extremes.

Luster

Luster refers to the way in which a mineral's surface reflects light. It is related not to color or shape but to the transparency of the mineral, the condition and texture of the mineral surface, and how light is refracted and dispersed as it enters the crystal. The terms used to describe luster are descriptive rather than objective: *metallic, vitreous* (glassy), *silky, resinous, pearly,* and *earthy.*

Color

Color is the first thing you notice when viewing a mineral. It is one of the qualities that attract people to gemstones. Generally speaking, however, color is a poor property to use in identifying minerals because many minerals exhibit multiple colors. Moreover, the colors of some minerals are identical to those of other minerals.

There is a scientific explanation for why the colors of minerals are often misleading. Technically, the color of a mineral is caused by the absorption, or lack of absorption, of various wavelengths of light (red is longest, purple is shortest). When pure white light (light containing all wavelengths of visible light) enters a crystal, some wavelengths are absorbed while others are reflected. The chemistry and structure of the mineral determine which of those wavelengths are reflected or absorbed. Often, the presence of tiny amounts of an impurity—some other element—will affect which wavelengths are reflected and, therefore, the mineral's color. The presence of impurities is why quartz can occur in milky, rosy, smoky, and other varieties. In short, because of impurities, color is an unreliable basis for identifying a mineral.

Q EXPAND YOUR THINKING

How is using the physical properties of a mineral to identify it an example of critical thinking?

(4-4) Atoms Are the Smallest Components of Nature with the Properties of a Given Substance

LO 4-4 Describe the structure of an atom.

Minerals are composed of **atoms** bonded together in various configurations. An atom is made of subatomic particles (electrons, protons, and neutrons). The number and relative abundance of subatomic particles in an atom determines its charge and mass, which in turn governs its behavior in the presence of other atoms.

The Structure of Atoms

Atoms are the smallest components in nature that have the properties of a given substance. For example, in a chemical reaction, every oxygen atom behaves in the same way. And every atom in a sample of copper has the properties of copper. It is possible to break atoms into smaller, subatomic particles (electrons, protons, and neutrons), but when we do, the atom loses its properties and no longer exists. That means the atom is the fundamental building block of minerals, the component responsible for a mineral's unique properties.

Every atom in the known universe is a tiny structural unit consisting of *electrons*, *protons*, and (usually) *neutrons* (**FIGURE 4.9**). An atom's center, or *nucleus*, is composed of protons (large, heavy, and having a positive electrical charge (+)) and neutrons (large,

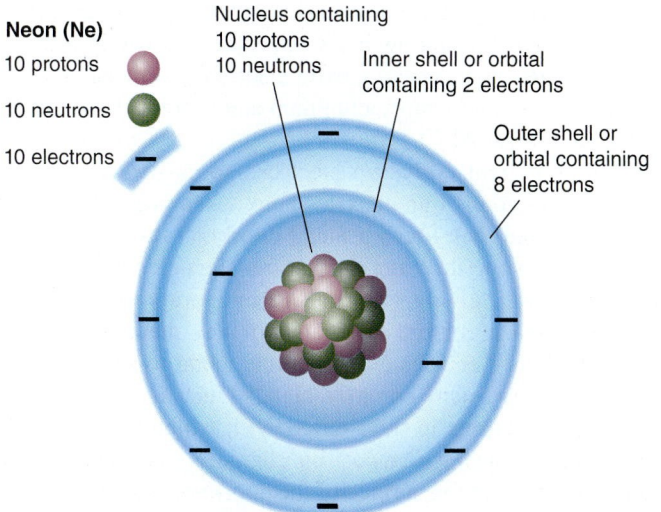

Neon (Ne)
10 protons
10 neutrons
10 electrons

Nucleus containing
10 protons
10 neutrons

Inner shell or orbital
containing 2 electrons

Outer shell or
orbital containing
8 electrons

FIGURE 4.9 Atoms consist of a nucleus constructed of protons and neutrons surrounded by electrons. Hydrogen is the only atom without a neutron in its nucleus.

What makes one element different from another?

heavy, and having no electrical charge). A normal hydrogen (H) atom does not have any neutrons, *but all other atoms do.* Electrons surround the nucleus. They are tiny and light, and have a negative electrical charge (−).

Each element is distinctive because its atoms have a unique number of protons. And opposite electrical charges are attracted to each other, so the negatively charged electrons in an atom form a cloud around its nucleus, with its positively charged protons. For example, hydrogen (H) is distinctive because it has only 1 proton and 1 electron (and no neutrons), making it the lightest element (also the most abundant element in the universe). Iron is distinctive because it has 26 electrons, 26 protons, and 26 neutrons.

The subatomic world is fascinating. If we were to look inside an atom, we would see that the electrons orbit around the nucleus but do not enter it. We would notice that the nucleus remains intact because the neutrons help "glue" the protons within it together, preventing it from flying apart, in most cases. We would also notice that electrons orbit relatively far from the nucleus. If we were to draw a hydrogen atom to scale with the nucleus the diameter of a pencil, the electron's orbit would be about 0.5 kilometers from the nucleus. The whole atom would be the size of a baseball stadium—mostly empty space. Atoms are incredibly tiny. One hydrogen atom is approximately 0.00000005 millimeter in diameter. It would take almost 20 million hydrogen atoms to form a line as long as a hyphen (-).

Protons and neutrons are particles, resembling tiny marbles. Electrons, in contrast, are more like light or radio waves. They are organized in negatively charged energy levels (called shells or *orbitals*) around the nucleus. In most atoms, the innermost shell holds only 2 electrons, and each subsequent shell holds a maximum of 8 electrons. The number of protons in its nucleus defines each element: This is its **atomic number.** An atom whose nucleus contains a different number of protons is a different element. For example, carbon atoms have 6 protons (atomic no. 6), nitrogen atoms have 7 (atomic no. 7), and oxygen atoms have 8 (atomic no. 8). The number of neutrons plus protons in the nucleus is the atom's **mass number;** the most abundant form of carbon, for instance, has 6 protons and 6 neutrons in its nucleus, for a mass number of 12.

Isotopes and Ions

As just noted, carbon atoms normally contain 6 protons and 6 neutrons; some, however, contain 7 or 8 neutrons. Hence, carbon always has an atomic number of 6, but its mass number may be 12, 13, or 14. These variations in mass number create **isotopes** of carbon (**FIGURE 4.10**).

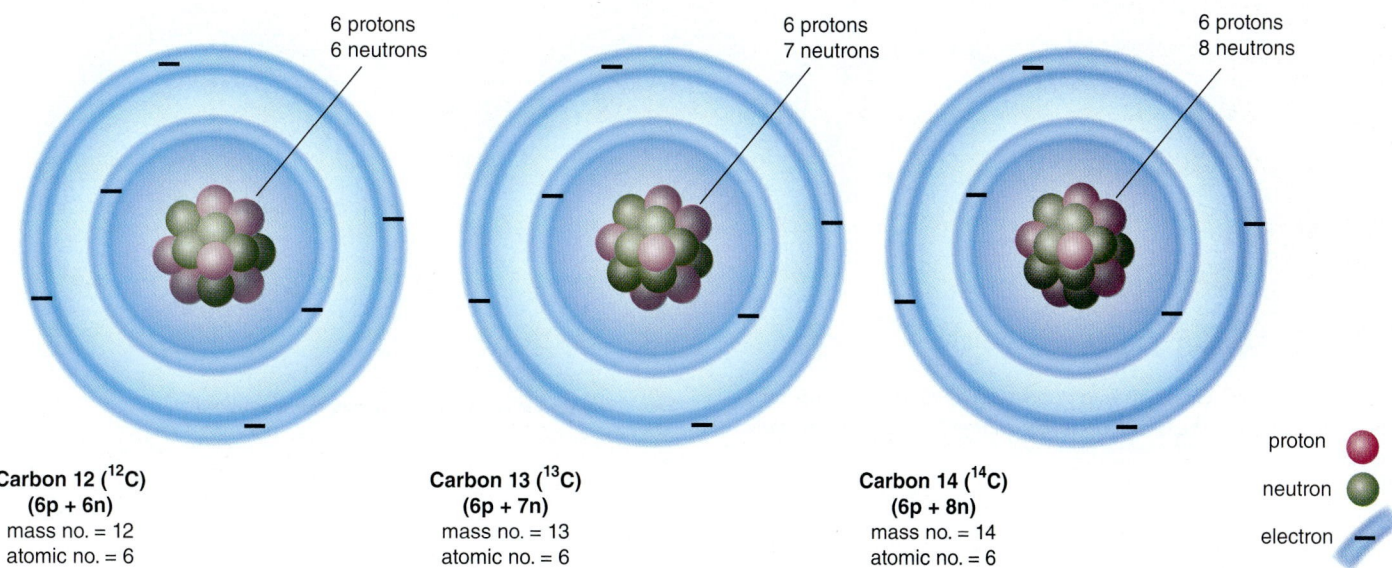

6 protons
6 neutrons

6 protons
7 neutrons

6 protons
8 neutrons

Carbon 12 (^{12}C)
(6p + 6n)
mass no. = 12
atomic no. = 6

Carbon 13 (^{13}C)
(6p + 7n)
mass no. = 13
atomic no. = 6

Carbon 14 (^{14}C)
(6p + 8n)
mass no. = 14
atomic no. = 6

proton
neutron
electron

FIGURE 4.10 Carbon can occur in nature as three isotopes, with 6, 7, or 8 neutrons. All three have the same atomic number (6) but their mass numbers vary: 12, 13, or 14.

What does an atom become if it gains or loses electrons?

All isotopes of carbon have the same atomic number, but each isotope has a different mass number. Isotopes of an element typically have different physical properties, and so behave differently in the physical world. For instance, a water molecule (H_2O) composed of ^{16}O (mass number is written as a superscript to the left of the element) can evaporate more readily than one with ^{18}O, a fact that will become important in our discussion of climate change in Chapters 14 and 15. The term *isotope* is from the Greek for "equal place," meaning that all isotopes of an element belong in the same place in the *periodic table*. Read more about "The Periodic Table of Elements" in the Geology in Action feature.

An atom can either gain or lose electrons and become an **ion** (an atom with a positive or negative charge). If an atom gains electrons, it acquires an overall negative charge and is called an **anion.** If the atom loses electrons, it acquires an overall positive charge and is called a **cation.** The number of protons and neutrons in the atom does not change, so both the atomic number and the mass number remain the same. Generally (but not always), it is the electrons in the outermost shell, called *valence electrons*, that are involved in making ions.

Why do atoms gain and lose electrons? To achieve a *stable electron configuration*, which it does by filling its outer shell with a maximum number of electrons—8—or, alternatively, by losing its outer shell to reveal a filled inner shell. Atoms that gain or lose only a few electrons to become stable do so readily and become ions. Those that give up their outer electrons to reveal a filled inner shell have a positive charge (cations). Those that acquire new electrons to fill their outer shell have a negative charge (anions). Take aluminum (Al), which has 3 electrons in its outer shell. It readily loses these (often to other atoms seeking extra electrons) to reveal a filled inner shell. That is why most aluminum atoms are cations with a +3 charge (written as Al^{3+}).

Do atoms really give up electrons so easily? Yes. And that explains why you get an electric shock when you scuff your feet on a carpet and then touch a metal doorknob (**FIGURE 4.11**). Friction between your shoe and the rug transfers electrons from your body to the rug, giving your body a temporary positive charge. When you reach for the doorknob (or any metal surface), you feel a shock as electrons pass from the metal to your body. This happens because many common metals (such as aluminum) have only a few electrons in their outer shell, which they easily lose. When you approach with a positive charge, 3 electrons from every aluminum atom on the surface of the doorknob leap onto your fingertips. That can amount to a large number of electrons, causing a shock powerful enough to be painful. The metal atoms become ionized (cations), you return to a neutral state, and the shock you feel is the exchange of electrons in what is known as static electricity.

EXPAND YOUR THINKING

The tendency to fill the outer orbital with 8 electrons has been called the "octet rule." Why would this be called a "rule"?

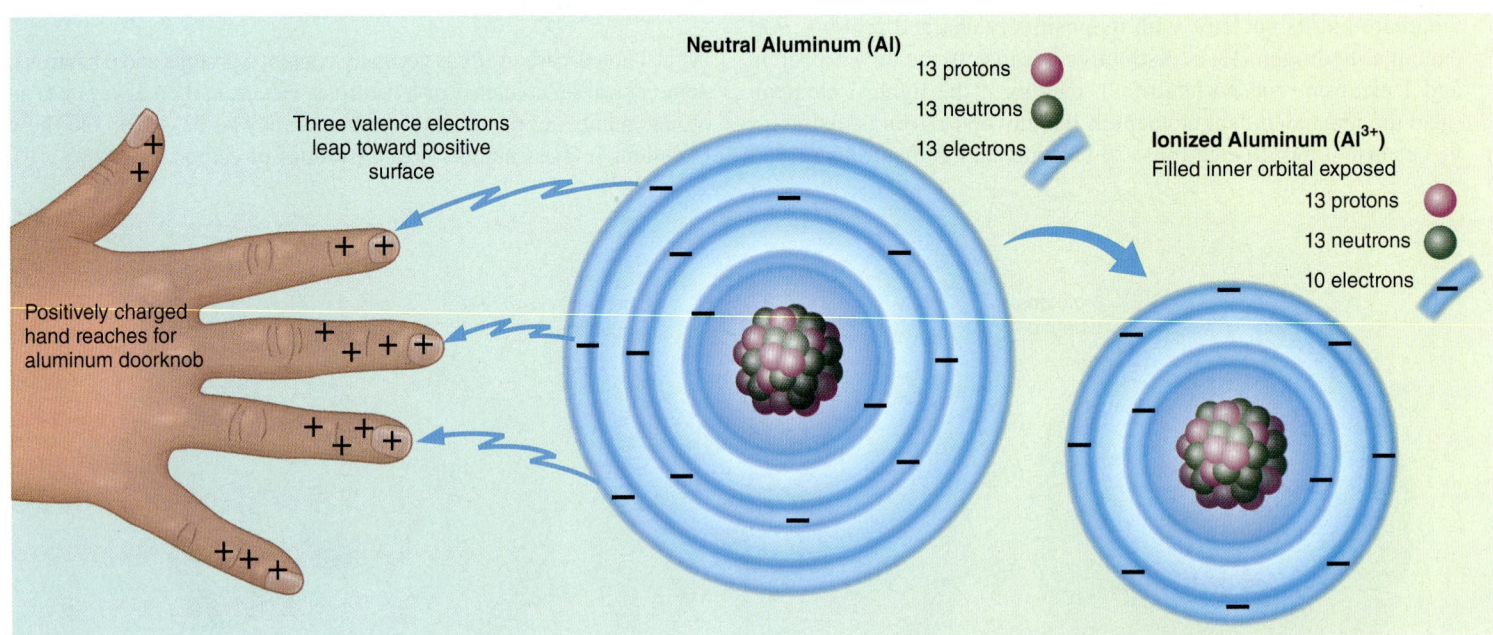

FIGURE 4.11 Aluminum atoms in a doorknob will readily give up the 3 electrons in their outer shells to a positively charged hand, creating static electricity.

Why do electrons readily leave the aluminum?

GEOLOGY IN ACTION

The Periodic Table of Elements

The periodic table of elements organizes the known 112 elements into vertical columns (called *groups*) and horizontal rows (called *periods*) (FIGURE 4.12). Groups are filled with elements that have the same number of electrons in their outermost shells. This means these elements share certain chemical properties. Periods are filled from left to right by elements based on the ordering of their atomic numbers.

Figure 4.12 highlights eight elements that occur in major abundance in Earth's crust: Si, O, Al, K, Ca, Na, Fe, Mg. These eight will become increasingly important as we learn how minerals and rocks are made, in this and the next several chapters.

Elements That Tend to Become Cations

The group to the far left on the periodic table, called the *alkali metals* (column 1), consists of elements that have a single electron in their outermost shell. This is an unstable configuration, and the single electron will be easily lost by each of these elements to form a cation. The *alkaline earth metals* (column 2) consist of elements with 2 electrons in their outermost

shell. These also have an unstable configuration and will lose the outer electrons to become cations.

These two groups include the important crustal elements sodium (NA^{1+}), potassium (K^{1+}), calcium (Ca^{2+}), and magnesium (Mg^{2+}), all of which have a tendency to give up their outermost electrons and become cations.

In light gray are other metals that tend to lose electrons from their outer shells and form ionic bonds. Of these, aluminum (Al^{3+}) is the most abundant metal in Earth's crust, and an important element in the formation of clay minerals.

Elements That Do Not Easily Give Up Electrons

To the far right (column 18) are the inert gases. These elements have filled outermost shells and resist bonding unless under unusual circumstances. As a result they behave as gases, avoiding chemical reactions that form bonds.

Shown in medium gray are the *nonmetals*. These elements have a tendency to gain electrons in their outer shells and form covalent bonds. Two columns headed by fluorine (F^{1-}), which forms a toxic gas, and oxygen (O^{1-}), Earth's most abundant crustal element, are located immediately to the left of the inert gases. These two groups of elements tend to gain electrons in their outer shells and become anions. The elements grouped under fluorine all have 7 electrons in their outer shells and will tend to gain 1 electron to fill the shell. The elements grouped under oxygen tend to gain 2 electrons in addition to the 6 already in their outer shells.

A semimetal, silicon (Si^{4+}), is the second most abundant element in Earth's crust. Silicon commonly bonds covalently with 4 oxygen atoms to build the largest class of minerals in the crust (the *silicates*, discussed in the next section).

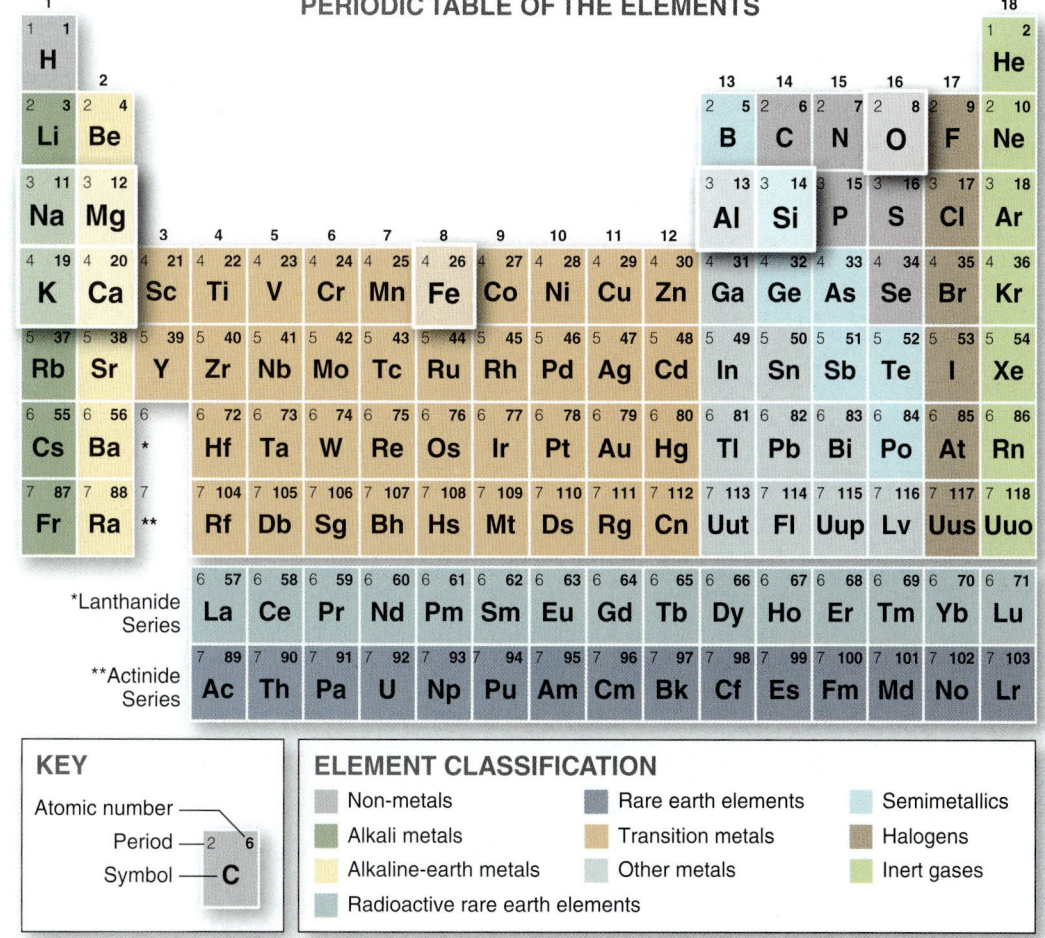

FIGURE 4.12 The periodic table of elements.

Minerals Are Compounds of Atoms Bonded Together

4-5

LO 4-5 Describe the three common types of bonds.

A **compound** is composed of two or more types of elements that are chemically combined with a bond. The elements occur in a fixed ratio that is described by a chemical formula.

Atoms of most elements have the capability to bond with certain atoms of other elements to form compounds. For example, 2 hydrogen atoms and 1 oxygen atom will bond to form the compound water (H_2O). Sodium and chlorine atoms bond to form halite (NaCl), common salt. Most of Earth's 4,000-plus minerals are compounds formed by the bonding of two or more elements.

Bonding

Why do atoms bond together? To achieve a *stable electron configuration*. As we have seen, every atom has a tendency to fill its outermost shell with the maximum number it can hold (2 for the first shell, 8 for the others). A small group of elements have 8 electrons in their outer shells: helium, neon, argon, krypton, xenon, and radon. These are all known as *inert gases* because they will not bond with other elements. **TABLE 4.3** shows the pattern of electrons in the first 20 elements (atomic numbers 1–20).

As described in the example of static electricity, atoms exchange electrons quite easily. This tendency drives the chemical bonding process. A bond between chemical units (atoms, molecules, ions) is formed by the attraction of atoms to one another through the sharing or exchange of electrons. To put it more simply, atoms may share or exchange electrons to attain a full complement of 8 electrons in their outer shells. The tendency of atoms to exchange electrons causes atoms to form one of three common types of bonds: *ionic*, *covalent*, or *metallic*.

Ionic Bonds

Think of ordinary salt, a mineral that is mined from thick layers underground and extracted from seawater to deice roads and season food. Salt is the mineral halite, a compound that is formed when atoms of sodium (a metal) and chlorine (a poisonous gas) are attracted to one another and form ionic bonds.

In ionic bonding, 2 atoms exchange an electron. The atom receiving the electron becomes an anion, while the atom giving up the electron becomes a cation. The two oppositely charged ions are attracted to each other and form a bond. Thus, ionic bonding happens because cations and anions are attracted to one another. In the case of halite, the result of ionic bonding is the compound, sodium chloride

TABLE 4.3 Electron Patterns of the First 20 Elements

| Element | Symbol | Atomic Number | Number of electrons in each shell | | | |
			First (2 is stable)	Second (8 is stable)	Third (8 is stable)	Fourth (8 is stable)
Hydrogen	H	1	1			
Helium	He	2	2			
Lithium	Li	3	2	1		
Beryllium	Be	4	2	2		
Boron	B	5	2	3		
Carbon	C	6	2	4		
Nitrogen	N	7	2	5		
Oxygen	O	8	2	6		
Fluorine	F	9	2	7		
Neon	Ne	10	2	8		
Sodium	Na	11	2	8	1	
Magnesium	Mg	12	2	8	2	
Aluminum	Al	13	2	8	3	
Silicon	Si	14	2	8	4	
Phosphorus	P	15	2	8	5	
Sulfur	S	16	2	8	6	
Chlorine	Cl	17	2	8	7	
Argon	Ar	18	2	8	8	
Potassium	K	19	2	8	8	1
Calcium	Ca	20	2	8	8	2

which has no net charge (**FIGURE 4.13A**). The compound has properties that are vastly different from those of its components (it is tasty; it is nonmetallic; it is not a dangerous gas; and it shatters easily).

Covalent Bonds

Now think of ordinary water, the liquid that makes up about 60% of our bodies, covers 71% of Earth's surface, and is essential to life. Water is a compound that is formed when 2 atoms of hydrogen (a light, explosive gas) and 1 atom of oxygen (a heavier explosive gas) are joined by covalent bonds (**FIGURE 4.13B**).

In covalent bonding, elements "share" electrons in order to fill their outermost shells. The hydrogen atom and the oxygen atom each donate 1 electron to form a chemical bond. Both atoms, resulting in a single covalent bond, share the 2 electrons that form the bond.

Think of covalent bonding in terms of two blocks of wood joined with two nails. The pieces of wood are the atoms, and the nails the electrons that the wood blocks share to form the covalent bond. Each piece of wood shares a portion of the nails.

Metallic Bonds

A third type of bond is a metallic bond. Gold is an example. Pure gold is an element, not a compound, but gold atoms are joined by metallic bonds. Metallic bonding is like covalent bonding because both types involve the sharing of electrons. However, in metallic bonds, all the extra valence electrons of a large group of atoms (millions or billions

of them) are considered together in what we call an "electron sea" that is shared by all the atoms in the group.

Elements That Tend to Become Ions

Many elements tend to become ions. Those that have 1 or 2 electrons in their outermost shell have an unstable configuration, and the electrons are easily lost to form a cation. These include the important crustal elements sodium (Na^+), potassium (K^+), calcium (Ca^{2+}), and magnesium (Mg^{2+}), all of which have a tendency to give up their outermost electrons and become cations. Aluminum (Al^{3+}) has 3 valence electrons and also readily loses them to become a cation.

Other elements have a tendency to gain electrons in their outer shells during the formation of covalent bonds. Fluorine (F^-), which forms a toxic gas, and oxygen (O^-), Earth's most abundant crustal element, both tend to gain electrons in their outer shells and become anions. All elements with a similar valence state will seek to gain one more electron in order to fill their outer shell.

As noted previously, semimetal silicon (Si^{4+}), is the second most abundant element in Earth's crust. Silicon commonly bonds covalently with 4 oxygen atoms to build the largest class of minerals in the crust: the **silicates** (discussed in section 4–6).

> **Q EXPAND YOUR THINKING**
>
> Why do elements with 8 electrons in their outer shells occur as gases and not solids?

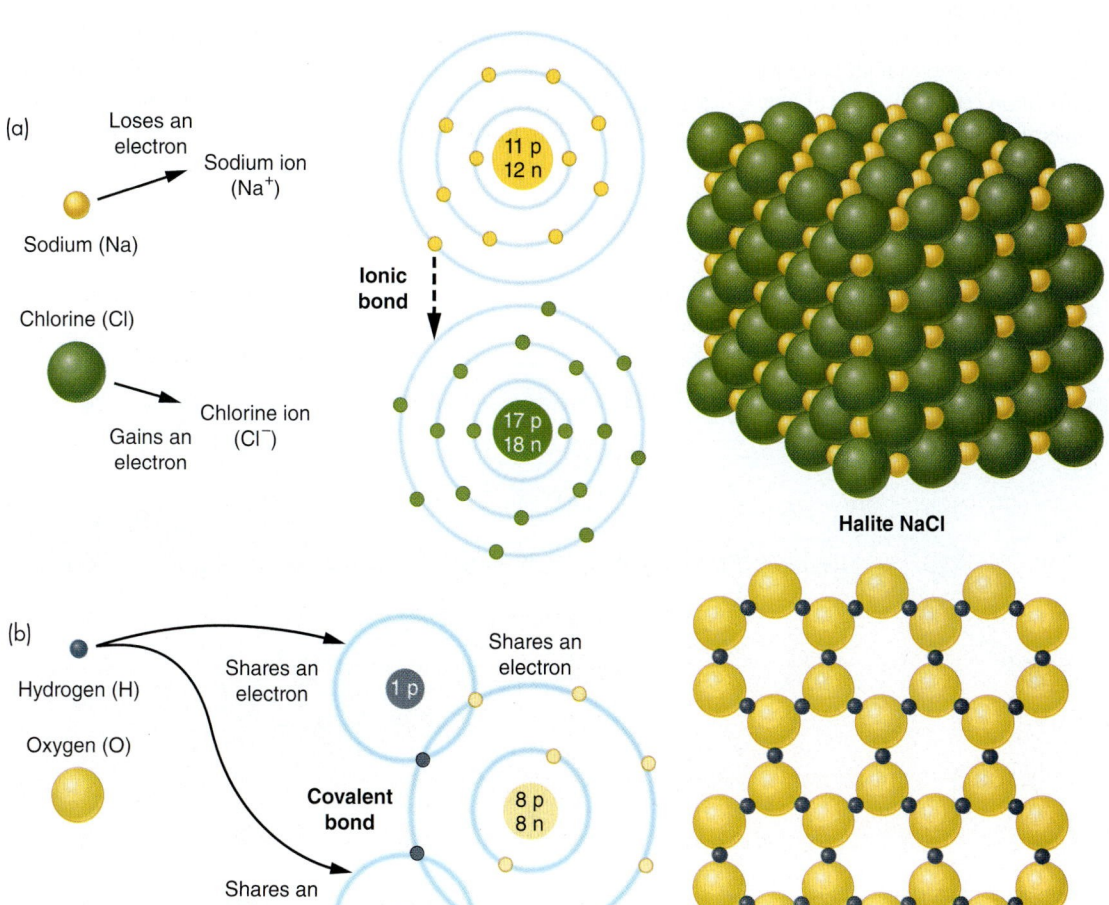

Halite NaCl

Ice H₂O

FIGURE 4.13 (a) Common table salt, sodium chloride (NaCl), is a crystalline solid that is formed by ionic bonding of sodium and chlorine atoms at a 1:1 ratio. (b) The mineral ice is formed by covalent bonding of 2 hydrogen atoms and 1 oxygen atom. Compounds that are formed by covalent bonding simultaneously share electrons in order to fill their outermost shells.

Q **What causes certain atoms, but not others, to bond?**

4-6 Oxygen and Silicon Are the Two Most Abundant Elements in Earth's Crust

LO 4-6 Describe the formation of silicate structures.

Silicon and oxygen are the two most abundant elements in the crust; together they form the compound silica. *Silica combines with metallic cations to build* silicates, *the largest group of minerals.*

Mineral Formation

Magma rising from the mantle and from melted crust cools as it penetrates solid rock. Cooling allows **igneous minerals** to crystallize. Igneous crystallization occurs in three important tectonic settings (**FIGURE 4.14**): subduction zones, spreading centers, and hotspots.

Other types of minerals do not have igneous origins. **Sedimentary minerals** (Chapters 7 and 8) crystallize among sediments from groundwater, as well as in other settings. They include calcite ($CaCO_3$), hematite (Fe_2O_3), and quartz (SiO_2). Some minerals tend to crystallize in the absence of oxygen; pyrite (FeS_2), for one, will develop in iron- and sulfur-rich environments (such as *anoxic* mud). **Metamorphic minerals** (see Chapter 9) also crystallize in particular environments. Where conditions in the crust lead to high heat and pressure, a rock may undergo *recrystallization* (crystallization of new minerals using elements of an old mineral without melting) to form metamorphic minerals.

Crystallization

Crystallization is the process through which atoms or compounds that are in a liquid state (e.g., in magma or dissolved in water) are arranged into an orderly solid state (mineral). This process occurs in two steps: nucleation and crystal growth.

Nucleation is the initial grouping of a few atoms that starts the process of crystal growth. A major barrier to nucleation is heat energy, which keeps magma in a liquid state in which atoms are vibrating too strongly with kinetic energy to bond together. This "energy barrier" can be circumvented if a solid is present in the liquid. The solid acts as a *nucleation seed*. It can be a particle of unmelted rock, another crystal, or even the solid wall of a magma chamber. Crystal growth will occur sooner on a nucleation seed than when no seed is present in the liquid.

Crystals grow as atoms are deposited on the surface of a seed. Atoms arrange themselves to achieve an electrically neutral crystalline compound. After many millions of atoms have bonded together, the result is an orderly crystal with a smooth (and usually) glassy surface—a mineral.

Rapidly growing crystals typically are smaller and less developed than slowly growing crystals. During laboratory experiments, mineralogists have observed that if crystals begin to grow at multiple nucleation sites, the resulting rock will contain many small crystals. Conversely, with fewer nucleation sites, the resulting rock will contain fewer, and larger, crystals.

The Eight Most Abundant Elements in the Crust

On average, iron constitutes over a third of the entire planet but only 6% of the crust. By contrast, oxygen and silicon together form, on average, 74% of the crust. These proportions mean that oxygen and silicon are high in minerals that crystallize from melted crustal rocks, whereas iron is high in minerals that come from the mantle.

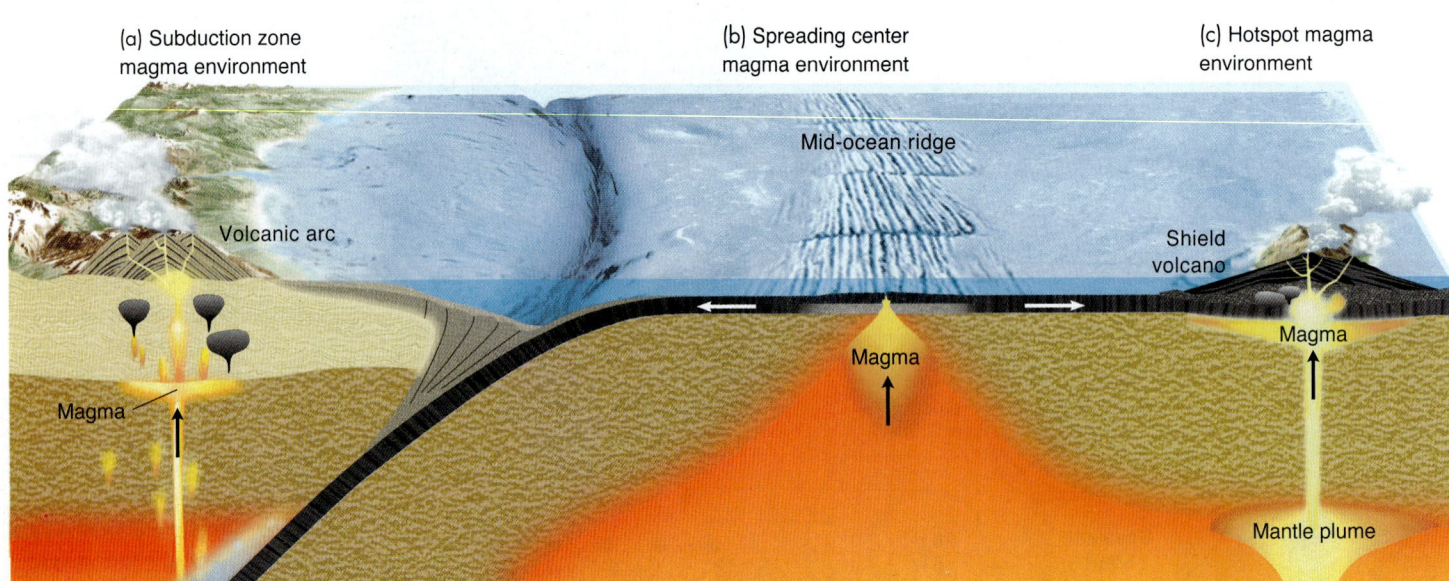

(a) Subduction zone magma environment

(b) Spreading center magma environment

(c) Hotspot magma environment

Volcanic arc

Mid-ocean ridge

Shield volcano

Magma

Magma

Magma

Mantle plume

FIGURE 4.14 Igneous crystallization occurs most commonly in three tectonic settings: (a) subduction zones, (b) spreading centers, and (c) midplate hotspots.

What types of plate boundaries can you identify in this figure?

The eight most common elements in the crust, in order of average abundance, are shown in **TABLE 4.4,** along with their typical ionized states.

Silicates

Silicon and oxygen dominate the composition of Earth's crust. Oxygen, a large, bulky atom, has 6 electrons in its outermost shell, 2 short of the desired 8. It is typically found in an ionized state as O^{2-}. Silicon is the second most abundant element by weight. It is a small atom with 4 electrons in its outermost shell, usually shown as Si^{4+}. Together, silicon and oxygen readily form a covalent bond, creating a *silicate compound*. Minerals made of silicate compounds are the most abundant naturally occurring inorganic substances in Earth's crust. In these compounds, 4 oxygen atoms surround a single silicon atom, forming $(SiO_4)^{4-}$. Each oxygen atom covalently shares 1 electron with the silicon atom, jointly filling its outermost shell.

Silicate compounds have a geometric shape known as a *tetrahedron* (from the Greek for "four faces"). A tetrahedron is a pyramid with 4 large oxygen atoms packed around a much smaller silicon atom (**FIGURE 4.15A**). Note the charged state of the silicate (4^-) resulting from the 4 valence electrons that are picked up by the 4 oxygen atoms.

Because each oxygen atom in a silicate tetrahedron has 7 electrons in its outer shell, it needs 1 more electron to achieve a filled

outer shell. It therefore seeks another silicon atom to bond with. Since there is abundant silicon in most magma, the oxygen atom often finds a second silicon atom by forming a second tetrahedron. This process results in linked pairs of tetrahedra created by the sharing of oxygen atoms. Two isolated tetrahedra ($[SiO_4]^{4-} + [SiO_4]^{4-}$) have a combined charge of 8−. A pair of linked tetrahedra $(Si_2O_7)^{6-}$ sharing a single oxygen atom has a charge of 6−. The formation of these tetrahedra is the first step in magma crystallization.

TABLE 4.4 The Eight Most Common Elements in Earth's Crust

Element	Average Abundance	Typical Ionization
Oxygen	46%	(O^{2-})
Silicon	28%	(Si^{4+})
Aluminum	8%	(Al^{3+})
Iron	6%	$(Fe^{2+}$ or $Fe^{3+})$
Magnesium	4%	(Mg^{2+})
Calcium	2.4%	(Ca^{2+})
Potassium	2.3%	(K^+)
Sodium	2.1%	(Na^+)

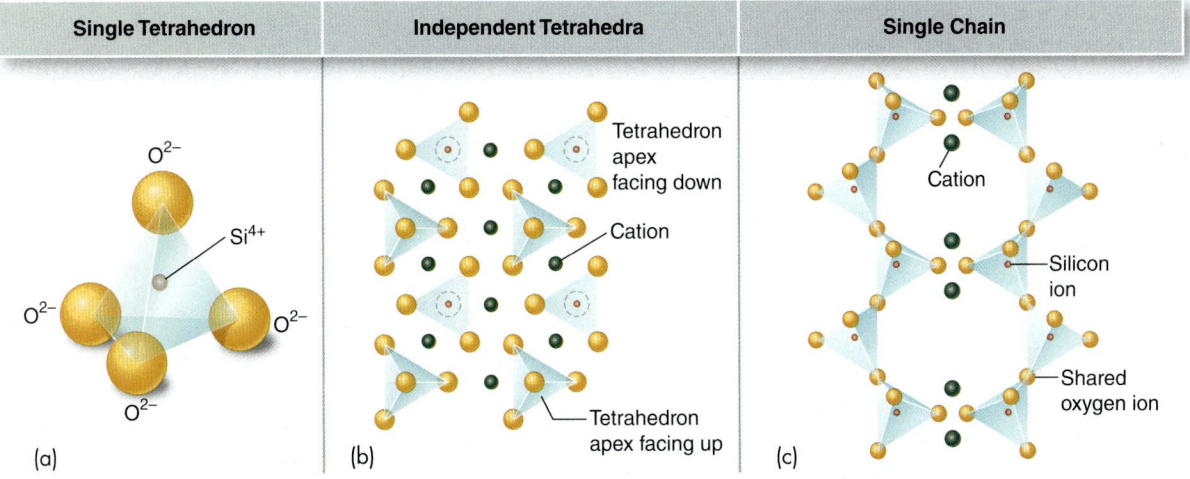

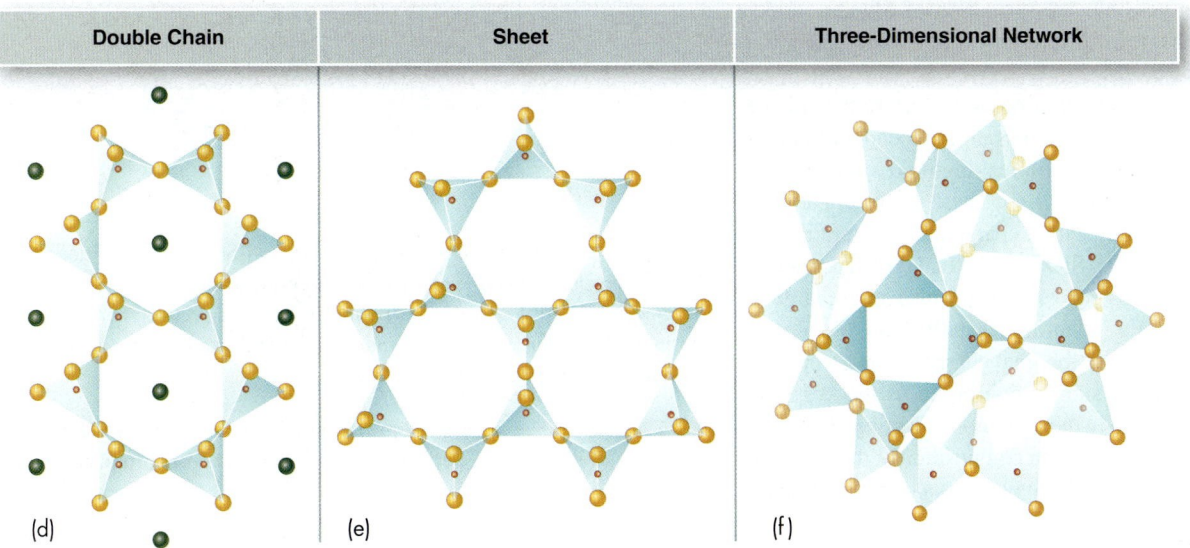

FIGURE 4.15

The silicate structures: (a) single tetrahedron; (b) independent tetrahedra bonded with metallic cations; (c) single chain structure; (d) double chain; (e) sheet silicate; (f) three-dimensional network silicate structure.

What type of bond joins silicate tetrahedra to make silicate structures?

By the time crystallization is completed, a mineral has become a neutral compound; it does not have an electrical charge. So, before a silicate mineral can crystallize, all negative charges associated with silicate tetrahedra must be balanced with cations. How does this happen?

Referring again to Table 4.4, the most abundant cations are aluminum (Al^{3+}), iron, ($Fe^{2+, 3+}$), magnesium (Mg^{2+}), calcium (Ca^{2+}), potassium (K^+), and sodium (Na^+). Most crustal magma contains large quantities of silicon and oxygen, and, therefore, many negatively charged tetrahedra are formed as it cools. There is also an abundance of positively charged metallic cations, so there is tremendous potential for forming many different silicate minerals.

Tetrahedron linkage does not stop at just one pair. Because an oxygen atom achieves stability by bonding with 2 silicon atoms,

4-7 Metallic Cations Join with Silicate Structures to Form Neutral Compounds

LO 4-7 State why single substitution and double substitution occur during crystallization.

Alone, silicate compounds, including single tetrahedra, chains, sheets, and framework silicates carry a negative charge that must be neutralized through bonding with one or more cations.

Cation Substitution

The most abundant cations in magma of average composition are the *metallic cations* we are familiar with: Al^{3+}, $Fe^{2+,3+}$, Mg^{2+}, Ca^{2+}, K^+, Na^+. The metallic cations differ in size and electrical charge. These differences are significant because they control which cations bond with which silicate structure, and thus determine the resulting mineral structure and chemistry.

In magma, cations of similar size and charge jockey like bumper cars for position in the silicate network. Certain cations can substitute for one another because they have a similar size and charge. This cation *substitution* makes possible the formation of a wide variety of minerals. The cations that most often substitute for each other are Na^+/Ca^{2+}, Al^{3+}/Si^{4+}, and Fe^{2+}/Mg^{2+}.

To illustrate the process of substitution, we return to the silicate mineral olivine (**FIGURE 4.16**). Olivine has an independent tetrahedron silicate structure (see Figure 4.15b), making it possible to write a portion of its chemical formula as $(SiO_4)^{4-}$. These tetrahedra are linked to one another by iron cations with a charge of 2^+ in a ratio of 2 cations for each tetrahedron: $Fe_2^{2+}(SiO_4)^{4-}$; the total charge is then 0.

But that is not the end of the story. Magnesium also has a charge of 2^+, and it is similar in size to Fe^{2+}. During crystallization, Mg^{2+} may replace Fe^{2+} (or vice versa) because the total charge of the compound remains at 0. In this way, cation substitution produces the "variable chemical composition" referred to in the definition of a mineral given at the beginning of the chapter. The chemical composition of the resulting mineral, olivine (Fe^{2+}, Mg^{2+})$_2(SiO_4)^{4-}$, may range along a continuum

complex structures of many linked tetrahedra are formed as magma cools. The crystalline structure of silicate minerals ranges from independent tetrahedra that depend entirely on cations of other elements to link them, to single chains, double chains, and sheets of tetrahedra. There are even three-dimensional *framework silicates* in which all the oxygen atoms are shared by adjacent tetrahedra (Figure 4.15b, c, d, e, and f). These are all variations of **silicate structures.**

Q EXPAND YOUR THINKING

Common rust is actually a mineral composed of iron oxide. Is this mineral igneous, sedimentary, or metamorphic?

from an iron-dominated form (known as fayalite, $[Fe_2^{2+}(SiO_4)^{4-}]$) to a magnesium-dominated form (known as forsterite $[Mg_2^{2+}(SiO_4)^{4-}]$).

Single substitution explains the interaction of silicon and oxygen with iron and magnesium. What about the remaining cations? Imagine that two cations of the same size but with different charges are jockeying for the same space between tetrahedra. If one cation gets in, the resulting compound will be neutral, and a mineral will have

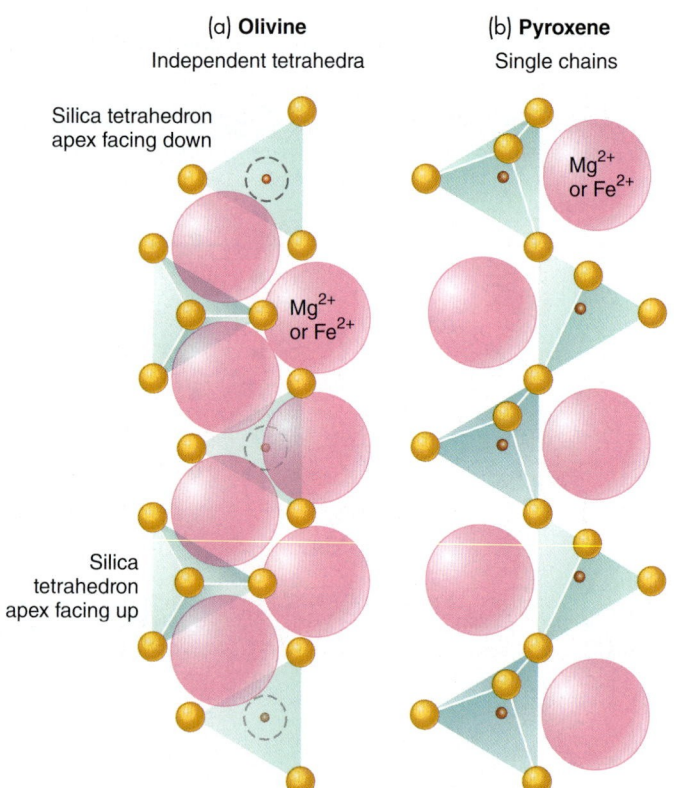

(a) Olivine
Independent tetrahedra

Silica tetrahedron apex facing down

Mg^{2+} or Fe^{2+}

Silica tetrahedron apex facing up

(b) Pyroxene
Single chains

Mg^{2+} or Fe^{2+}

FIGURE 4.16 (a) The mineral olivine, $(Mg, Fe)_2SiO_4$, has an independent silicate structure wherein separate tetrahedra are linked by iron (Fe^{2+}) and magnesium (Mg^{2+}) cations. (b) The mineral pyroxene, $(Mg, Fe)SiO_3$, has a single-chain silicate structure wherein iron and magnesium cations are bonded to a silicate chain.

Q Would pyroxene be the same mineral if the tetrahedra were not linked into a chain but were instead isolated tetrahedra joined by Mg^{2+} or Fe^{2+}? Write the new chemical formula if this were the case.

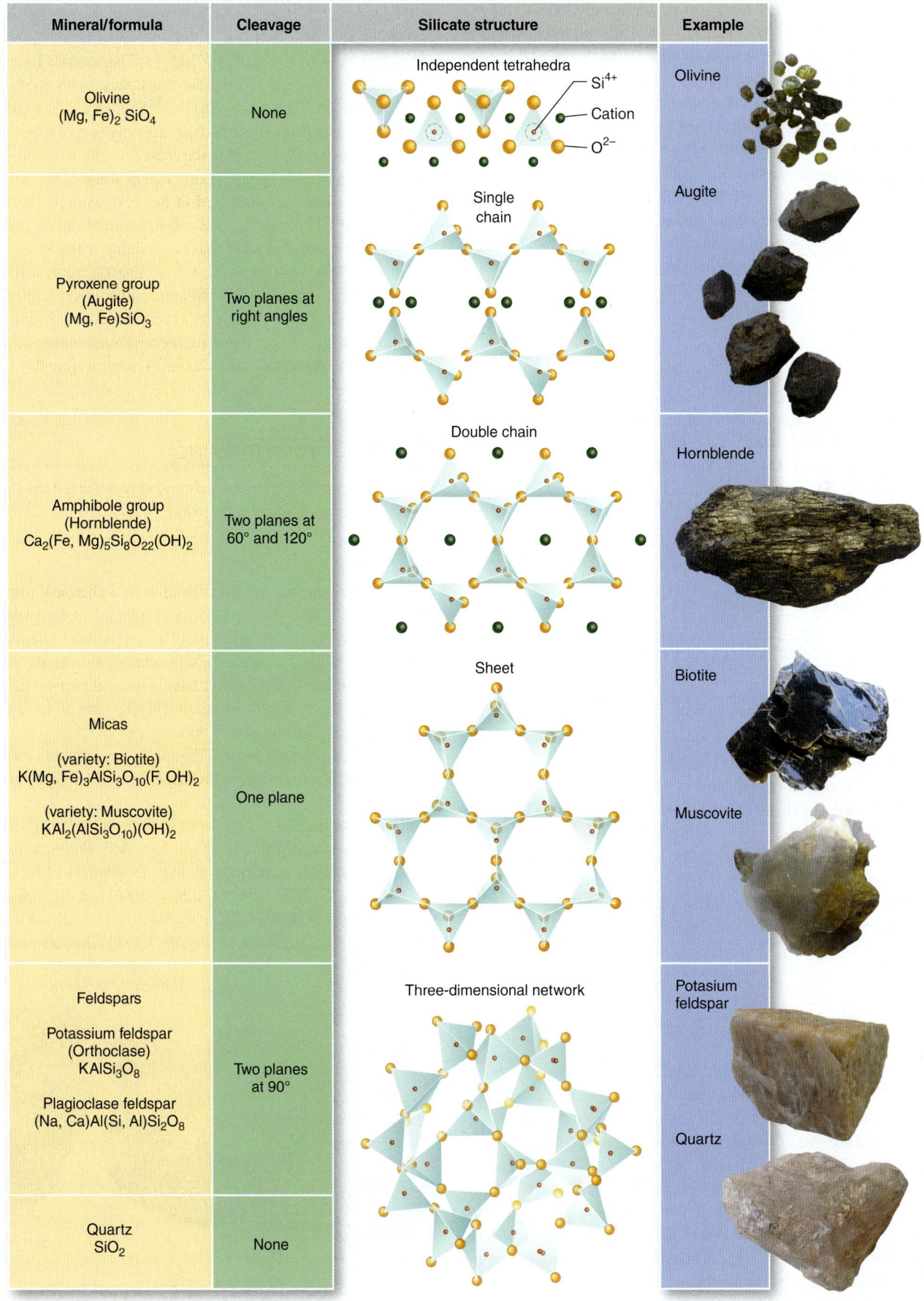

Mineral/formula	Cleavage	Silicate structure	Example
Olivine $(Mg, Fe)_2 SiO_4$	None	**Independent tetrahedra** Si^{4+} Cation O^{2-}	Olivine
Pyroxene group (Augite) $(Mg, Fe)SiO_3$	Two planes at right angles	**Single chain**	Augite
Amphibole group (Hornblende) $Ca_2(Fe, Mg)_5Si_8O_{22}(OH)_2$	Two planes at 60° and 120°	**Double chain**	Hornblende
Micas (variety: Biotite) $K(Mg, Fe)_3AlSi_3O_{10}(F, OH)_2$ (variety: Muscovite) $KAl_2(AlSi_3O_{10})(OH)_2$	One plane	**Sheet**	Biotite Muscovite
Feldspars Potassium feldspar (Orthoclase) $KAlSi_3O_8$ Plagioclase feldspar $(Na, Ca)Al(Si, Al)Si_2O_8$	Two planes at 90°	**Three-dimensional network**	Potasium feldspar
Quartz SiO_2	None		Quartz

FIGURE 4.17 Common silicate minerals and their structures.

Mineral images courtesy of Chip Fletcher

been formed. But if the other cation gets in, the compound will not be neutral. Since minerals must be neutral, one of the silicon or oxygen atoms usually experiences a substitution that results in a neutral compound. This is called *double substitution*: The first cation substitution is followed by a second cation substitution that reestablishes the neutrality of the compound involving Si or O.

An example of double substitution can be seen in the case of plagioclase feldspar. Feldspar, a framework silicate (refer back to Figure 4.15f), incorporates calcium, sodium, and aluminum cations within its structure through double substitution. Calcium and sodium atoms are nearly the same size, so they freely substitute for each other during crystallization. However, since they have different charges, a second substitution must occur to restore neutrality. To do this, an aluminum ion will substitute for a silicon ion, since they, too, are similar in size.

Double Cation Substitution in Plagioclase Feldspar

Double substitution of cations in the mineral plagioclase feldspar occurs in the following manner: Initially, the compound $Na^{1+}Al^{3+}Si_3^{4+}O_8^{2-}$ is formed in cooling magma. This compound is neutral, with the -16 charge of the 8 oxygens (O_8^{2-}) balanced by the $+1$ of the sodium (Na^+), $+3$ of the aluminum (Al^{3+}), and $+12$ of the 3 silicon atoms (Si_3^{4+}).

4-8 There Are Seven Common Rock-Forming Minerals

LO 4-8 List the seven common rock-forming minerals.

Minerals, because of their great chemical diversity, display a wide variety of physical properties.

Some minerals, such as galena (lead ore), are very dense; others, such as corundum, are very hard. Still others are light and soft; gypsum, for example, can be scratched with a fingernail. Individual minerals can also have variations; fluorite, for one, can be found in green, blue, pink, clear, and other varieties. What causes minerals to display such a range of properties?

What Is a Mineral?

Let us return to the definition of a mineral: "a naturally occurring, inorganic, crystalline solid with a definite, but sometimes variable, chemical composition." The crystalline structure of a given mineral cannot vary; every sample of a specific mineral will exhibit the same geometric arrangement of atoms. However, the "definite, but sometimes variable" part of the definition indicates that a mineral's chemical composition can vary, within limits. That means a mineral may exhibit some variation in its chemistry (and its color), provided that its fundamental components are always present. Recall that the proportions of iron and magnesium in olivine differ from one sample to another.

Some variations are so common that they are named. Minerals that have a fixed crystalline structure and whose chemical composition varies are given a *group name*. Within a group, mineral *species* possess a particular and restricted chemical composition. Within a species,

Calcium (Ca^{2+}) atoms are able to substitute for the sodium atoms because they are nearly the same size. However, this upsets the charge balance. (Check this: Is $Ca^{2+}Al^{3+}Si_3^{4+}O_8^{2-}$ balanced?) To restore electrical neutrality, a second substitution takes place, with another aluminum replacing silicon. The result—$Ca^{2+}Al_2^{3+}Si_2^{4+}O_8^{2-}$—is a neutral mineral.

Olivine and plagioclase feldspar are only two of many types of silicate minerals that are formed in the crust. **FIGURE 4.17** presents a guide to the chemistry and crystalline structure of some of the more important rock-forming minerals discussed in the next section. Through the formation of silicate tetrahedra and cation substitution, magma will eventually crystallize into a solid mass containing many types of minerals. In Chapter 5 we look more closely at igneous mineral formation and discover that as magma cools, different minerals crystallize at different temperatures. For now, though, let's complete our study of minerals with an introduction to seven important rock-forming minerals, along with a short survey of the seven major classes of minerals found in Earth's crust.

Q **EXPAND YOUR THINKING**

Why are both the size *and* charge of metallic cations significant in contributing to mineral diversity?

a common irregularity in crystallization or a chemical impurity that is often observed may produce a distinct mineral *variety*. Identifying the mineral's group, species, and variety accommodates chemical variation in a mineral. Consider the example of feldspar shown in (**FIGURE 4.18**).

Group: Feldspar. All members have a three-dimensional silicate framework crystalline structure (Si_3O_8), and all have two cleavage planes forming nearly 90° angles.

Species: Plagioclase feldspar (Figure 4.18a) contains sodium and calcium ions that freely substitute for one another during crystallization.

Varieties: Plagioclase ranges from sodium-rich *albite* ($NaAlSi_3O_8$) to calcium-rich *anorthite* ($CaAl_2Si_2O_8$). Between these two extremes are two varieties with intermediate abundances of Na and Ca, *andesine* ($Na[70\%-50\%] Ca[30\%-50\%] (Al, Si)AlSi_2O_8$) and *labradorite* ($Ca[50\%-70\%] Na[50-30\%] (Al, Si)AlSi_2O_8$).

Species: Orthoclase feldspar (Figure 4.18b) contains potassium ions ($KAlSi_3O_8$).

Courtesy of Chip Fletcher

(a)

(b)

FIGURE 4.18 Two feldspar species: (a) plagioclase feldspar, and (b) orthoclase feldspar.

Q What physical properties would you use to identify plagioclase and orthoclase in the field?

Common Rock-Forming Minerals

There are seven important rock-forming minerals or mineral groups: the **olivine group**, the **pyroxene group**, the **amphibole group**, **biotite**, the **feldspar group**, **quartz**, and **calcite** (FIGURE 4.19).

1. *Olivine group* [$(Fe, Mg)_2 SiO_4$]: Olivines are silicates that crystallize in cooling magmas and are rich in iron or magnesium. Olivine is named for its olive green color. It is an important component of iron/magnesium-rich volcanic rocks.

2. *Pyroxene group* [$(Fe, Mg, Al, Ca, Na) SiO_3$]: Pyroxene, an important single-chain silicate, is a common rock-forming mineral. It crystallizes at a high temperature in the absence of water. If water were present, a double-chained amphibole would most likely be formed instead.

3. *Amphibole group* [$(Ca, Na, Fe, Mg, Al) Si_8O_{22}OH_2$]: Hornblende, the most common species of amphibole, is found in many rocks.

Hornblende is actually the name given to a series of minerals that are rather difficult to distinguish based on ordinary physical properties alone. The iron, magnesium, and aluminum ions can freely substitute for one another and form separate varieties of hornblende, of which there are many (magnesio-hornblende and ferro-hornblende, to name just two).

4. *Biotite* [$K (Fe, Mg)_3 AlSi_3O_{10}(F, OH)_2$]: A member of the *mica group*, biotite is present in most rocks and is formed under a wide variety of conditions. Typically ranging in color from black to brown, it has a layered structure composed of sheets of iron magnesium aluminum silicate weakly bonded together by layers of potassium ions.

5. *Feldspar group* [$(K, Na, Ca)(Al, Si)_4O_8$]: As described earlier, the feldspar group contains both plagioclase and orthoclase feldspar species. The feldspars are the most common mineral group in Earth's crust. Because they crystallize under a wide range of pressures and temperatures, they occur in many rocks.

Mineral images courtesy of Chip Fletcher

FIGURE 4.19 Seven common rock-forming minerals: (a) olivine, (b) pyroxene (augite), (c) amphibole (hornblende), (d) biotite, (e) orthoclase (feldspar group), (f) plagioclase (feldspar group), (g) quartz, and (h) calcite.

Q Which of these minerals are silicates?

6. *Quartz* (SiO_2): Quartz is found in nearly every geologic environment and is a component of almost every type of rock. It is extraordinarily common, and in many rocks is the primary mineral. Quartz is also very diverse in terms of varieties, colors, and forms.

7. *Calcite* ($CaCO_3$): Calcite gets its name from *calx*, a Greek word meaning "lime." It is one of the most common minerals on Earth, comprising about 4% of the crust by weight. It is formed in many different geologic environments but is especially common as marine sediment because several types of marine organisms, such as coral and certain types of very abundant plankton, secrete exoskeletons of calcite. Calcite is used in the steel and glass industries, as ornamental stone, and in the chemical and optical industries.

Q EXPAND YOUR THINKING

Study the color and chemistry of the seven common minerals. There is a strong relationship between color and chemistry. What is it?

Most Minerals Fall into Seven Major Classes

4-9

LO 4-9 Describe each of the seven major classes of minerals.

The seven major mineral classes are: silicates, native elements, oxides, sulfides, sulfates, halides, and carbonates.

Mineral Classes

No chapter on mineralogy would be complete without introducing the seven major classes of minerals (**FIGURE 4.20**), defined by chemical composition. **TABLE 4.5** presents these classes and some of their economic uses.

The mineral classes will remain little more than abstractions unless you can see where each occurs in Earth's crust. We have therefore included an illustration of the environments in which each class of minerals is formed. Please refer to **FIGURE 4.21** in the Critical Thinking feature "Mineral Environments" as you read the following summary of the seven mineral classes.

1. *Silicates* are compounds of silicon, oxygen, and other elements. They are the largest, most complex mineral group. About 30% of all minerals are silicates, and up to 90% of the crust is composed of silicates. This is not surprising, given that oxygen and silicon are the two most abundant elements in the crust. Important silicates include quartz, feldspar, olivine, amphibole, and biotite. Silicates are formed in cooling magma, certain sedimentary environments, and metamorphic rocks. They also crystallize in chemically enriched **hydrothermal fluids** that are formed in the crust when groundwater comes into contact with hot rock.

2. *Native elements* include over 100 known minerals, mostly metals. The native elements are pure masses of a single element, such as gold, silver, lead, copper, and sulfur. They accumulate in several environments, but they are especially concentrated in *vein deposits* (mineral deposits that fill preexisting cracks in rock) formed by hydrothermal fluids (discussed more fully in the next chapter).

(a) (b) (c) (d)

(e) (f) (g)

Mineral images courtesy of Chip Fletcher

FIGURE 4.20 The seven major mineral classes are: (a) silicates (orthoclase feldspar), (b) native elements (copper), (c) oxides (hematite), (d) sulfides (pyrite), (e) sulfates (gypsum), (f) halides (fluorite), and (g) carbonates (malachite).

Q These classes are defined by chemical composition. Describe another criterion by which you could define major classes of minerals.

TABLE 4.5 Seven Major Mineral Classes

Mineral Group	Example	Chemical Formula	Economic Use
Silicates	Quartz	SiO_2 (silicon dioxide)	Silicon source for electronics
	Olivine	$(Mg,Fe)_2SiO_4$ (magnesium-iron silicate)	Abrasives
	Orthoclase	$KAlSi_3O_8$ (potassium-aluminum silicate)	Glass, porcelain
	Plagioclase	$(Na,Ca)Al(Si,Al)Si_2O_8$ (sodium, calcium, aluminum silicate)	Ceramics, building stone
Native Elements	Copper	Cu	Electrical wiring
	Gold	Au	Trade, jewelry, medical
	Lead	Pb	Batteries, metallurgy
	Graphite	C (carbon)	Pencil "lead"; lubricant
Oxides	Ice	H_2O (hydrogen oxide)	Crystalline form of water
	Hematite	Fe_2O_3 (ferric oxide)	Ore of iron, pigment
	Corundum	Al_2O_3 (aluminum oxide)	Sapphire, abrasive
	Magnetite	Fe_3O_4 (ferrous oxide)	Ore of iron
Sulfides	Galena	PbS (lead sulfide)	Ore of lead
	Pyrite	FeS_2 (iron sulfide)	Sulfuric acid production
	Chalcopyrite	$CuFeS_2$ (copper-iron sulfide)	Ore of copper
	Sphalerite	ZnS (zinc sulfide)	Ore of zinc
Sulfates	Barite	$BaSO_4$ (barium sulfate)	Drilling mud
	Gypsum	$CaSO_4 \cdot 2H_2O$ (The dot represents water embedded within the crystalline structure as the water molecule.)	Plaster
	Anhydrite	$CaSO_4$ (calcium sulfate)	Plaster
Halides	Fluorite	CaF_2 (calcium fluoride)	Used in steel making
	Halite	$NaCl$ (sodium chloride)	Common salt
Carbonates	Calcite	$CaCO_3$ (calcium carbonate)	Portland cement, lime
	Dolomite	$CaMg(CO_3)_2$ (calcium-magnesium carbonate)	Cement, lime
	Malachite	$Cu_2CO_3(OH)_2$ (copper carbonate)	Jewelry

3. *Oxides* are compounds of metallic cations (iron, copper, aluminum, and others) bonded to oxygen atoms. Oxides are frequently ores that are mined and refined to extract pure metals for commercial use. Examples include hematite (iron oxide), corundum (aluminum oxide), cuprite (copper oxide), and ice (hydrogen oxide, a mineral only at temperatures below 0°C). Oxides are present everywhere in the crust because of the prevalence of oxygen. They are formed in surface environments where iron, manganese, and other metals and native elements bond with oxygen found in the atmosphere and elsewhere.

4. *Sulfides* are compounds of metallic cations (lead, iron, zinc, etc.) bonded with sulfur (an anion, S^{2-}), such as the mineral pyrite (FeS_2). As with oxides, commercially valuable metals constitute a high proportion of this group and are mined as valuable ores for refinement. Important sulfides include pyrite (iron sulfide), galena (lead sulfide), sphalerite (zinc sulfide), cinnabar (mercury sulfide), and chalcopyrite (copper-iron sulfide). Sulfides tend to be formed as vein deposits from hydrothermal fluids.

5. *Sulfates* are compounds of metallic cations bonded with the sulfate anion (SO_4^{2-}). They include anhydrite (calcium sulfate), which is used to make plaster, and barite (barium sulfate), important in the drilling industry. Sulfates are formed in surface environments.

6. *Halides* are compounds whose anions are halogens, a group of highly reactive gases including fluorine, chlorine, bromine, iodine, and astatine. Halogens usually have a charge of 1− when chemically bonded, so they tend to react with many other elements. Common halides include fluorite (calcium fluoride) and halite (sodium chloride).

7. *Carbonates* are compounds of cations bonded with the carbonate ion (CO_3^{2-}). Important carbonates include calcite (calcium carbonate) and dolomite (calcium-magnesium carbonate). Carbonates are formed in calcium-saturated waters, such as oceans and other complex fluids (e.g., hydrothermal fluids).

Q **EXPAND YOUR THINKING**

Describe where an example of each of the seven major mineral classes might precipitate on Earth's surface as a sedimentary mineral (not an igneous or metamorphic mineral).

CRITICAL THINKING

Mineral Environments

FIGURE 4.21 illustrates some of the geologic environments where the seven major mineral groups are found. Silicates commonly form in cooling magma; native elements tend to concentrate in hydrothermal veins or sedimentary deposits; oxides can be found in ordinary igneous rocks or sedimentary environments; sulfides form in oxygen-depleted environments where metallic cations bond with sulfur; sulfates form in the presence of oxygen and sulfur with metallic cations; halides form in places where evaporation of seawater or other chemical fluids leads to crystallization of dissolved compounds; and carbonates form as skeletal components of marine organisms such as plankton (deep and shallow sea) or corals (shallow sea), or where water evaporates.

Working with a partner, please complete the following exercise:

1. In Figure 4.21, identify the mineral class that is prevalent at each of the seven sites where there is a photo of a mineral. Can you identify the mineral pictured?

2. What compound forms the basic structure of the silicate minerals? Draw it.

3. Name an environment in which sulfates are found.

4. Give three examples of marine organisms that produce carbonate minerals.

5. Which element must be present for oxide minerals to form? Give an example of an oxide mineral that you encounter in your everyday life.

6. Why are native elements economically valuable? Give examples of three valuable native elements and how you use them.

7. What are the differences between the environmental conditions that lead to formation of sulfides and those that lead to sulfates?

8. Describe an environment in which halides form. Give an example of a halide mineral that is economically important.

9. List five economically important minerals that have been used in objects you encountered today.

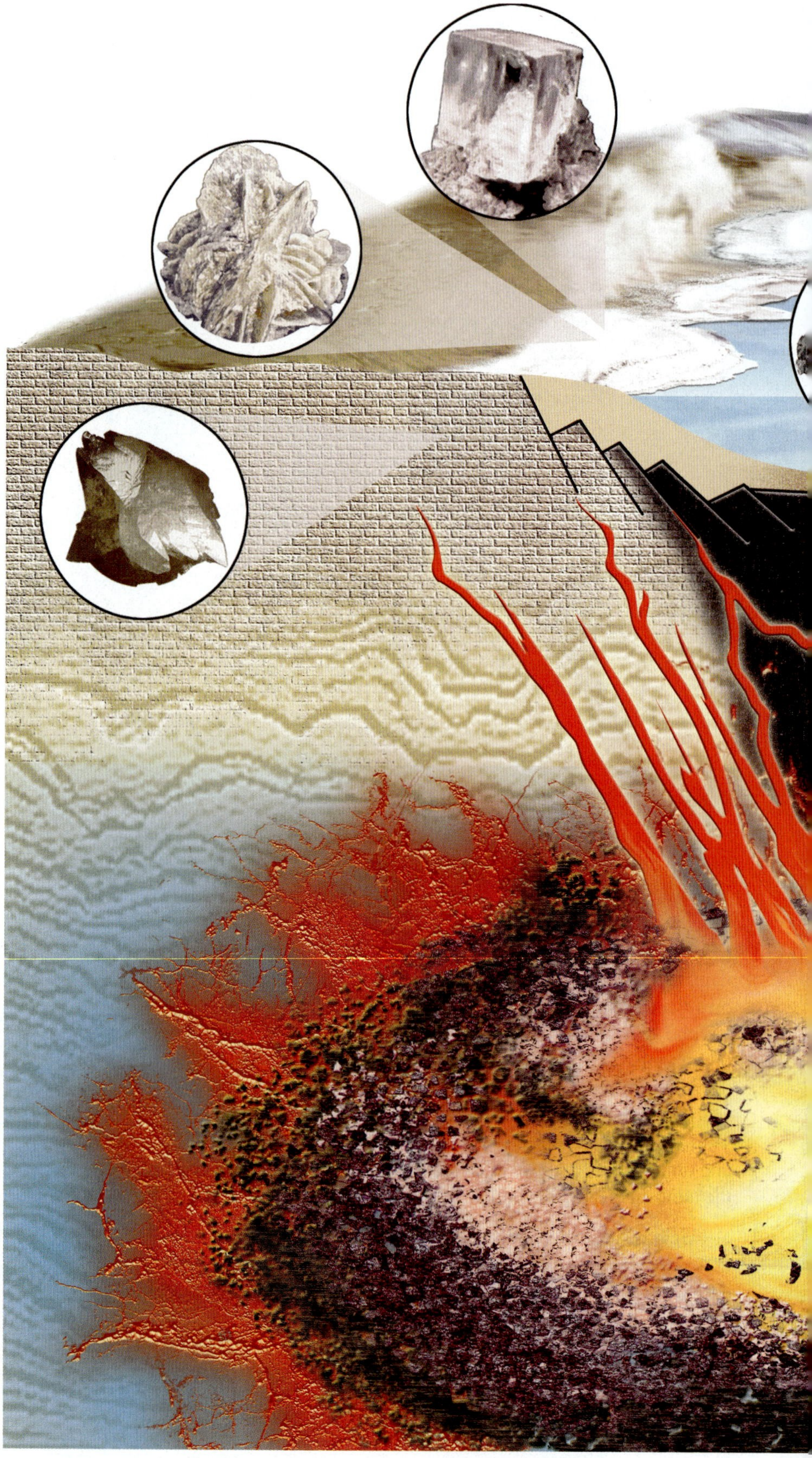

FIGURE 4.21 Environments of the seven common mineral groups.

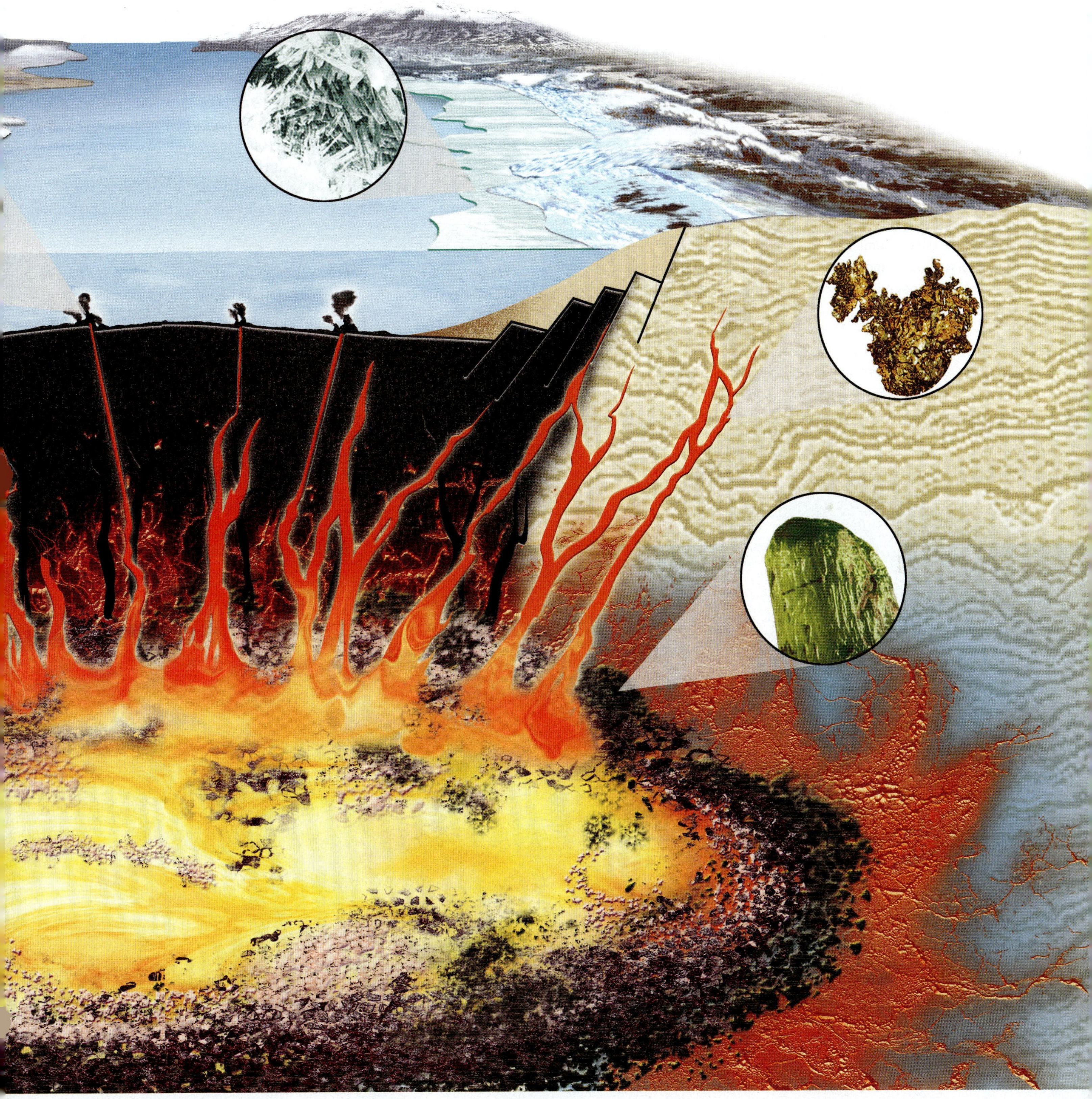

4-10 Mining Is Necessary, and the Resulting Environmental Damage Can Be Minimized

LO **4-10** Identify the ways in which mining can damage the environment

Mineral ores are rarely found in dense concentrations. Most of the time huge quantities of rock must be excavated to extract small quantities of valuable ore. This leads to severe environmental damage. But this can be mitigated with careful planning.

The Bingham Canyon Mine

Minerals rarely occur in pure, concentrated form. Usually, a precious metal such as copper (Cu) is combined with other elements to form a mineral ore such as chalcopyrite ($CuFeS_2$), which accounts for about 50% of copper production.

Mineral ores are often dispersed throughout a rock in sparse quantities called **disseminated ore deposits.** As a result, huge volumes of rock must be mined—at considerable cost and environmental damage—to extract even a little of the desired mineral. This process can be quite destructive to the nearby environment. Various mining methods are used, some underground and some at the surface. One surface-mining technique is *open-pit mining*, whereby a giant hole is dug and rock containing the desired mineral is removed.

Under current U.S. laws, a mine design must protect the environment, reclaiming the land as mining proceeds. **Reclamation** plans must include preserving streams, protecting groundwater, preserving topsoil, handling waste rock, controlling erosion and sediment, controlling dust, and restoring the shape of the land. All this is very expensive, and often dictates whether a mining operation will be profitable or not.

Kennecott Copper Company's Bingham Canyon Mine in Utah (**FIGURE 4.22**) is an open-pit mine. Visible from space, it is the largest excavation on the planet. It has produced more copper than any mine

FIGURE 4.22 Bingham Canyon Mine, in Utah, produces copper, silver, molybdenum, gold, and other minerals.

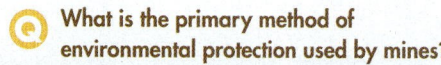
What is the primary method of environmental protection used by mines?

Bettmann/Corbis

in history and so is sometimes called "the richest hole on Earth." The mine produces silver, molybdenum, and gold, but its most important product is copper. The ore deposits are *disseminated,* containing only 0.6% copper when removed from the ground. To extract the copper, the rock must be crushed, ground, chemically washed, heated, and electrically treated until the copper is concentrated in 99.99% pure copper plates called cathodes.

Copper is used in electrical appliances, wiring, plumbing, cars, trains, airplanes, ships, satellites—the list is endless. The average home today contains about 180 kilograms of copper for electrical wiring, water pipes, and appliances, while the automobile you drive contains about 23 kilograms. Each child born today in America will need 680 kilograms of copper in the course of his or her lifetime to enjoy our current standard of living. Total copper consumption in the United States exceeds 3 million tons per year. Of that total, 1.8 million tons comes from domestic mines; the rest comes from imports and recycled scrap. (Copper is the most recycled of all metals.)

Since excavation of Bingham Canyon Mine began in 1906, more than 6 billion tons of rock have been moved, and more than 16 million tons of copper produced, enough to wire every dwelling in Canada, Mexico, and the United States. In the future, this mine will be converted to an underground operation and the existing open pit will be reclaimed as mixed forest and open land.

Ronald Karpilo/Alamy

FIGURE 4.23 Reddish acid mine drainage carries acid and heavy metals throughout the local watershed, threatening plants, animals, and people.

Ⓠ **Describe the environmental damage that can result from mining.**

The Summitville, Colorado, Mine

Gold was discovered at Summitville, Colorado, in 1870. In 1984, the Summitville Consolidated Mining Company, Inc. started open-pit mining, using the *cyanide heap leach* method to accelerate gold production. But less than 10 years later, in 1992, the company had declared bankruptcy and abandoned the site without performing any environmental cleanup.

The heap leach method extracts gold from disseminated ore deposits by spraying a cyanide solution onto a large pile of crushed ore (the "heap") in order to dissolve ("leach") gold that is disseminated through the rock. Ideally, the heap sits on a pad, or impervious liner, to keep the poisonous cyanide from entering the environment. The chemical solution is collected from the base of the heap leach pile, and dissolved gold is chemically extracted from the solution.

At Summitville, environmental problems developed soon after the initiation of open-pit mining, and rapidly worsened after the site was abandoned. During the mining operations, poisonous cyanide leaked from transfer pipes directly into the local watershed. Of additional concern was the release of **acid mine drainage (AMD),** an acidic, metal-rich runoff that drained into the nearby Alamosa River and groundwater system whenever it rained or snowed (**FIGURE 4.23**). The U.S. Geological Survey measured increases in dissolved copper from 25 milligrams/liter in uncontaminated water to over 300 milligrams/liter during the period when the mine was in operation, an important indicator of heavily polluted runoff.

The geology of the Summitville region is characterized by rocks that weather into copper, aluminum, and iron-rich soils. These create naturally acidic and metal-rich runoff. When the rocks are exposed and crushed by mining operations, the weathering process is greatly

accelerated. As a result, the acid and metal content of runoff from the affected area is significantly increased.

In response to concerns about the environmental impacts of these conditions, the State of Colorado requested that the U.S. Environmental Protection Agency (EPA) take control of the site. The EPA decommissioned all mining operations and averted a potentially catastrophic release of cyanide from the heap. Ongoing remediation efforts by the EPA include dismantling the heap leach pad, plugging all mine shafts, backfilling the open pit with mine waste, and capping the backfilled pit to prevent water inflow. The total cost of the cleanup has been estimated at $100 to $120 million in taxpayer dollars.

The Summitville environmental problems are of particular concern because of the extensive use of Alamosa River water downstream from the site for irrigation, livestock, fishing, and wildlife habitat. In 1990, the water's high acid and metal content caused stocked fish along the river to die out. Crops irrigated with the contaminated water included alfalfa (used for livestock feed), barley (used in beer production), wheat, and potatoes; there was concern about potential health hazards due to the increased metal content of these crops. The Alamosa River also feeds wetlands that provide habitat for aquatic life and migratory waterfowl, such as ducks and the endangered whooping crane; acidic and metal-rich runoff may pose hazards to these ecosystems.

The Summitville experience suggests that some types of mineral deposits may be too environmentally dangerous to mine because they contain high concentrations of toxic substances.

Ⓠ **EXPAND YOUR THINKING**

What criteria would you use to decide whether a mining operation should be permitted to open or not?

LET'S REVIEW

Minerals are a critical geologic resource used every day in manufacturing, agriculture, electronics, household items—even in the clothes you wear. It is therefore vital to ensure the sustainability of mineral resources. Minerals are solid crystalline compounds with a definite, but variable, chemical composition that are found in the crust and mantle. From the mineral nutrients in-gested in our food to the metals used to make cars, computers, and buildings, minerals sustain us, our cities and farms, and the world at large. Keep these important ideas in mind, along with the fact that minerals are not limitless resources. Remember to use them sparingly, and recycle them whenever possible.

STUDY GUIDE

4-1 Minerals are solid crystalline compounds with a definite, but variable, chemical composition.

- In the United States, every person consumes, on average, about 10 tons of **minerals** and rocks per year. Minerals make possible the manufacture of almost every product sold today and are essential to maintaining our way of life.

- A mineral is a naturally occurring, inorganic, crystalline solid with a definite, but sometimes variable, chemical composition.

4-2 A rock is a solid aggregate of minerals.

- For example, granite is an aggregate of several minerals—typically, quartz, plagioclase feldspar, orthoclase feldspar, biotite, and others.

4-3 Geologists use physical properties to identify minerals.

- Many minerals have several physical properties that can be used to identify them. Each mineral's unique chemistry and **crystalline structure** give it distinctive physical qualities, such as hardness, cleavage, fracture, luster, color, and streak.

4-4 Atoms are the smallest components of nature with the properties of a given substance.

- **Atoms** are the smallest components in nature that have the properties of a given substance. Each element is defined by its **atomic number**, the number of protons in its nucleus. The number of neutrons and protons in the nucleus is the element's **mass number**. Variations in mass number create **isotopes**. A variation in electrical charge forms an **ion**: A negative charge forms an **anion**; a positive charge forms a **cation**.

- Most atoms seek to achieve a stable electron configuration by bonding with another atom.

4-5 Minerals are compounds of atoms bonded together.

- Atoms of most elements have the capability to bond with atoms of other elements in order to achieve a stable electron configuration. The result is the formation of compounds.

- The tendency of atoms to seek to attain the full complement of electrons in their outermost shells causes them to form one of three common types of bonds: ionic, covalent, or **metallic.**

4-6 Oxygen and silicon are the two most abundant elements in Earth's crust.

- Cooling magma allows igneous minerals to crystallize in three important tectonic settings: subduction zones, spreading centers, and hotspots. Most magma consists of the eight major elements in the crust: aluminum (Al^{3+}), iron ($Fe^{2+,3+}$), magnesium (Mg^{2+}), calcium (Ca^{2+}), potassium (K^+), sodium (Na^+), silicon (Si^{4+}), oxygen (O^{2-}), as well as dozens of trace elements.

- **Crystallization** is the process through which atoms or compounds that are in a liquid state (magma) are arranged into an orderly solid state. This process involves **nucleation** followed by crystal growth. Many nucleation sites produce smaller crystals that are less well developed.

- Silicon and oxygen readily form a covalent bond to create a **silicate** compound. Four oxygen atoms surround a single silicon atom, forming the silicate compound $(SiO_4)^{4-}$. The four-sided geometric shape of silicate crystals is known as a tetrahedron.

- When the process of crystallization is complete, a mineral compound is neutral. All negative charges associated with silicate tetrahedra are balanced by the positive charges of cations.

4-7 Metallic cations join with silicate structures to form neutral compounds.

- In magma, cations of similar size and charge jockey for position within the silicate network in a process called cation substitution. This process leads to the formation of a wide variety of minerals. The cation pairs that most often substitute for one another are Na^+/Ca^{2+}, Al^{3+}/Si^{4+}, and Fe^{2+}/Mg^{2+}. Olivine forms by single substitution; feldspar forms by double substitution.

4-8 There are seven common rock-forming minerals.

- Minerals whose composition varies over a particular range are given a group name. Within the group, each mineral species has a specific composition. Within species, a common irregularity in crystallization or a chemical impurity produces a distinct mineral variety.

- There are seven important rock-forming minerals: the **olivine group**, the **pyroxene group**, the **amphibole group**, **biotite**, the **feldspar group**, **quartz**, and **calcite.**

4-9 Most minerals fall into seven major classes.

- There are seven major mineral classes based on their composition: silicates, native elements, oxides, sulfides, sulfates, halides, and carbonates.

4-10 Mining is necessary, and the resulting environmental damage can be minimized.

- Huge volumes of rock must be mined and refined to extract even small amounts of a desired mineral. This process can be quite destructive to the nearby environment.

- Under current U.S. laws, mining operations must be designed to protect the environment, reclaiming the land as mining proceeds. **Reclamation** plans must include preserving streams, protecting groundwater, preserving topsoil, handling waste rock, controlling erosion and sediment, controlling dust, and restoring the shape of the land.

KEY TERMS

acid mine drainage (AMD) (p. 117)
amphibole group (p. 111)
anion (p. 102)
atomic number (p. 101)
atoms (p. 100)
biotite (p. 111)
calcite (p. 111)
cation (p. 102)
compound (p. 104)
crystalline structure (p. 94)

crystallization (p. 106)
disseminated ore deposits (p. 116)
feldspar group (p. 111)
hydrothermal fluids (p. 112)
igneous minerals (p. 106)
ion (p. 102)
isotopes (p. 101)
mass number (p. 101)
metamorphic minerals (p. 106)
mineral (p. 94)

Mohs hardness scale (p. 99)
nucleation (p. 106)
olivine group (p. 111)
pyroxene group (p. 111)
quartz (p.111)
reclamation (p. 116)
sedimentary minerals (p. 106)
silicate structures (p. 108)
silicates (p. 105)

ASSESSING YOUR KNOWLEDGE

Please complete this exercise before coming to class. Identify the best answer to each question.

1. Which of the following is part of the definition of a mineral?
   a. Liquid
   b. Electrically charged
   c. Inorganic
   d. Synthetic
   e. None of the above

   - positively charged
   - uncharged
   - negatively charged

2. On this diagram, locate and following parts of the atom:
   a. Proton
   b. Nucleus
   c. Electron
   d. First energy-level electron shell
   e. Neutron
   f. Second energy-level electron shell

3. Many minerals are useful in everyday life. Some examples include:
   a. Feldspar and quartz.
   b. Clay and gypsum.
   c. Graphite and chalcopyrite.
   d. Copper and titanium.
   e. All of the above.

4. To identify a mineral sample quickly, geologists use:
   a. Physical size.
   b. Color.
   c. Physical properties.
   d. Laboratory analysis.
   e. None of the above.

5. To be considered a mineral, a substance must _____.
   a. Have a specific chemical composition.
   b. Be formed by inorganic processes.
   c. Be a naturally formed solid.
   d. Have a characteristic crystal structure.
   e. Have all of the characteristics listed above.

Courtesy of Chip Fletcher

6. Fool's gold is _____.
   a. Hematite.
   b. Calcite.
   c. Pyrite.
   d. Native gold.
   e. None of the above.

7. One of the isotopes of the element carbon (atomic no. 6) has a mass number of 13. How many neutrons does this isotope have in its nucleus?
   a. 5
   b. 6
   c. 7
   d. 14
   e. None of the above

8. What are formed when sodium ions and chlorine ions combine to produce NaCl?
   a. Ionic bonds
   b. Covalent bonds
   c. Organic structures
   d. Isotopes
   e. Native elements

9. What property causes the mineral biotite to break into flat sheets?
   a. Its density
   b. Its electrical charge
   c. Its crystalline structure
   d. Its hardness
   e. None of the above

10. What physical property of minerals is illustrated in this photo? _____ (cleavage, hardness, fracture)

11. Silicates are constructed by:
    a. Carbon and hydrogen.
    b. Iron and oxygen.
    c. Silica and feldspar.
    d. Silicon and oxygen.
    e. None of the above.

12. Single substitution occurs during crystallization because:
    a. Neutral compounds attract ions.
    b. The number of leftover ions must be balanced.
    c. A charged compound is formed.
    d. Ions of similar size can substitute for one another.
    e. Forming dense compounds requires single substitution.

13. The two most abundant elements in the crust form:
    a. Oxides.
    b. Sulfates.
    c. Silicates.
    d. Carbonates.
    e. Halides.

14. Copper is an example of a(n) _____; quartz is a _____; and pyrite is a _____.
    a. oxide mineral; sulfide mineral; phosphate mineral
    b. phosphate mineral; sulfide mineral; carbonate mineral
    c. oxide mineral; silicate mineral; phosphate mineral
    d. native element; silicate mineral; sulfide mineral

Courtesy of Chip Fletcher

15. The important rock-forming minerals include:
    a. Feldspars, biotite, and garnet.
    b. Calcite, feldspars, biotite, and amphiboles.
    c. Amphiboles, feldspars, quartz, and rutile.
    d. Rutile, amphibole, calcite, and garnet.
    e. Quartz, feldspar, granite, and basalt.

16. The silica compound takes the shape of:
    a. A rectangle.
    b. A tetrahedron.
    c. A polygon.
    d. A polymer.
    e. Magma.

17. Plagioclase feldspar is a:
    a. Mineral group.
    b. Mineral species.
    c. Mineral variety.
    d. All of the above.
    e. Type of quartz.

18. Which of the following best describes the difference between sulfates and sulfides?
    a. Sulfates include nitrogen; sulfides do not.
    b. Sulfides are metals bonded with sulfur; sulfates are metals bonded with sulfate anion.
    c. Sulfates are metals bonded with inorganic carbon; sulfides are metals bonded with water molecules.
    d. Sulfates are formed only in igneous rocks; sulfides are formed in all types of rock.
    e. All of the above.

© Borislav Dopudja/Alamy

19. Mining can damage the environment by:
    a. Acid mine drainage.
    b. Chemically polluted runoff.
    c. Disturbing the ecosystem and surface environment.
    d. Exposing rocks that are a pollutant source for long time periods.
    e. All of the above.

20. The _____ (silicates, oxides, halides, native elements) are the most abundant class of minerals on Earth; and of these, _____ (copper, olivine, pyroxene, feldspar) is the most abundant group.

Bettmann/Corbis

Courtesy of Chip Fletcher

FURTHER RESEARCH

1. Look around you. Can you identify 10 substances made from minerals? (See Tables 4.1 and 4.4.)

2. Define *mineral* and explain the meaning of each component of the definition.

3. Referring to Table 4.2, draw the oxygen atom. Be sure to include the nucleus and its components, as well as the electrons in their orbitals.

4. Why are cations needed for minerals to be formed? Refer to the charge of a silicate tetrahedron in your answer.

5. Go to the Web and find photos of minerals from each of the seven mineral groups. Identify the minerals and describe their chemistry, their crystalline structure, and the environment in which they are formed.

6. What controls the order in which minerals crystallize from magma?

7. What part of Earth produces the magma that forms minerals with high iron and magnesium content? Where does the magma that forms silicon- and oxygen-enriched minerals come from?

ONLINE RESOURCES

Explore more about minerals on the following Web sites:

Classroom activities and review of key concepts:
http://csmres.jmu.edu/geollab/Fichter/Minerals/index.html

National Park Service site with information about minerals: www2.nature.nps.gov/geology/usgsnps/rxmin/mineral.html

Additional animations, videos, and other online resources are available at this book's companion Web site:
www.wiley.com/college/fletcher

This companion Web site also has more information about WileyPLUS and other Wiley teaching and learning resources.

IGNEOUS ROCK

CHAPTER CONTENTS & LEARNING OBJECTIVES

GEOLOGY IN OUR LIVES

Melting recycles Earth materials, and igneous rock is formed when molten rock (magma) cools and crystallizes. Igneous rocks were the first rocks on Earth, formed over 4 billion years ago, and all other rocks have evolved from it. Igneous rock makes up Earth's interior and provides critical building material and mineral resources.

Q What important natural resources are found in igneous rock?

5-1 Igneous Rock Is Formed When Molten, or Partially Molten, Rock Solidifies

LO 5-1 Describe igneous rock.

Igneous rocks, one of the three main rock groups (igneous, sedimentary, metamorphic), are important because they provide information about the crust and mantle, they reveal Earth's tectonic history, and they host valuable mineral deposits.

What Is Igneous Rock?

At this moment, **igneous rock** (*ignis* is Latin for "fire") lies beneath your feet. Whether visible at the surface (**FIGURE 5.1**) or deeply buried, igneous rock accounts for four-fifths of the material in Earth's crust. It is everywhere, but it is formed only in certain tectonic settings.

These include spreading centers, where decompression melting of the upper mantle produces liquid **magma** that rises and cools to form the igneous seafloor; subduction zones, where hot fluids emerge from subducting oceanic crust and cause recrystallization and melting in the overlying upper mantle; and midplate settings, where a line of volcanoes grow from a magma source at a hotspot.

Igneous rock starts out as liquid magma created by melting, either in the mantle or in the crust. As magma migrates toward the surface, it may cool and crystallize, adding to the mass of the crust. Magma that reaches the surface creates a volcano built up of accumulated **lava** and **pyroclastic debris,** erupted bits of rock, such as ash (see Chapter 6, "Volcanoes").

You may already be familiar with some types of igneous rock (**FIGURE 5.2**). *Granite*, for one, is a handsome building stone used to make floors, countertops, and gravestones. It forms the core of the Rocky Mountains, the Sierra Nevada, and many New England landscapes.

Basalt, another, builds chains of volcanic islands across the Pacific, Indian, and Atlantic oceans and makes up the bedrock of every ocean, covering millions of square kilometers. In short, igneous rock can be found everywhere beneath Earth's surface.

Carsten Peter/NationalGeographic//Getty Images, Inc.

FIGURE 5.1 When volcanoes erupt, they discharge ash, various gases, and magma. Solidified magma, whether within the crust or on it, becomes igneous rock.

Q How would an igneous rock composed of ash look different from one composed of lava?

(a)

(b)

Mineral images Courtesy of Chip Fletcher

FIGURE 5.2 (a) Granite and (b) basalt are two common igneous rocks built of silicate minerals. On average, granite tends to be relatively enriched in silica and depleted in iron and magnesium. Basalt tends to be relatively enriched in iron and magnesium and depleted in silica.

Name two basic differences in the appearance of these two rocks.

Why Study Igneous Rocks?

Igneous rocks were the first to be formed as the young Earth's molten surface cooled and crystallized billions of years ago. Since that time, those original rocks have weathered and eroded, metamorphosed, melted, and crystallized many times over.

This sequence of changes has led to the formation of the myriad sedimentary, metamorphic, and igneous rocks that make up our planet today. This process is known as **igneous evolution.** Igneous rock, thus, was Earth's first geologic generation, and all other rocks have evolved from it.

Although geologists have learned much about igneous rock, questions persist about how magma crystallizes, what conditions lead to the concentration of certain valuable minerals, and exactly how igneous processes have created continental crust. And while important clues to conditions deep in magma chambers have been revealed by laboratory experiments and field observations, questions concerning the origin of igneous rock and its valuable mineral deposits are subjects of ongoing geologic research and critical thinking.

The United States has a diverse geology of both **extrusive** igneous rock (lava and other volcanic products that have been forced onto Earth's surface) and **intrusive** igneous rock (magma that has crystallized within the crust without being exposed to the cool temperatures of the atmosphere and shallow crust).

Some intrusive rock occurs at great depths within the crust or mantle and is referred to as **plutonic** (for Pluto, the Roman god of the underworld).

Igneous rock is found in the Pacific Northwest, where the Cascade Range arose through plate collisions that enlarged the western continent during Neogene time. This mountain range contains large and active **volcanoes** such as Mount Rainier, Mount Hood, Mount St. Helens, and Mount Shasta.

The massive accumulation of volcanic rock in the Northwest created the Columbia Plateau and the Snake River Plain. Hence, igneous rocks are found in two fundamental types, *plutonic* and *volcanic.*

Igneous rock may also be found farther south, in the Sierra Nevada of California, a westward-tilting 560-kilometer-long block of granite containing the spectacular Yosemite, Kings Canyon, and Sequoia National Parks. Massive quantities of granite magma crystallized in the crust to form this region during the Mesozoic era.

Sediment that eroded from Sierra Nevada igneous rocks has filled the Central Valley of California, creating fertile soil, as well as the gold deposits that gave rise to the 1849 Gold Rush. (Read more about this in the Geology in Action feature titled "Geology of Yosemite.")

Occurrences of both extrusive and intrusive igneous rock are found in many U.S. states, including Texas, Missouri, Michigan, North Dakota, Wisconsin, Minnesota, and throughout the Appalachian Mountains and New England, among other places.

EXPAND YOUR THINKING

Igneous rock comes from magma. What plate tectonics environments produce magma?

Geology of Yosemite

Yosemite National Park (**FIGURE 5.3**) is one of the great natural wonders of the world. It is known for its vast exposures of granite, the processes that are weathering and eroding the landscape, and the multistage geologic history that led to its formation.

Granite is a silica-enriched intrusive igneous rock. At Yosemite, the granite exposed at the surface today originated as a large intrusion of magma, known as the *Sierra Nevada Batholith* (a batholith is a massive deep igneous intrusion), approximately 210 to 150 million years ago.

The Sierra Nevada Batholith was located about 10 kilometers deep, but over time the entire region was uplifted and tilted by tectonic forces associated with the westward acceleration of the North American Plate. Uplift subjected the rocks overlying the batholith to erosion, and they were largely removed from the region, exposing the massive granitic batholith at the surface. A series of global ice ages, culminating only 20 to 30 thousand years ago, sculpted the U-shaped Yosemite Valley, and made the dramatic hanging valleys that today display beautiful waterfalls (see Chapter 17).

Videowokart/Shutterstock

FIGURE 5.3 Yosemite Valley.

5-2 Igneous Rock Is Formed Through a Process of Crystallization and Magma Differentiation

LO 5-2 Describe the processes that lead to magma differentiation.

The great diversity of igneous rock types is a result of decompression melting, partial melting, and magma differentiation, which produce different magma types.

Decompression Melting

Although rock in the mantle is very hot, it remains solid yet is capable of flow because of the high pressures there. At the same time, *convection* is thought to cause some "mixing" in the mantle, consisting of hot rock rising and cooler rock subsiding.

As hot rock rises to shallower depths, it experiences lower pressure, and this leads to **decompression melting**. Melting occurs because as pressure is reduced, the rock's melting point is lowered as well.

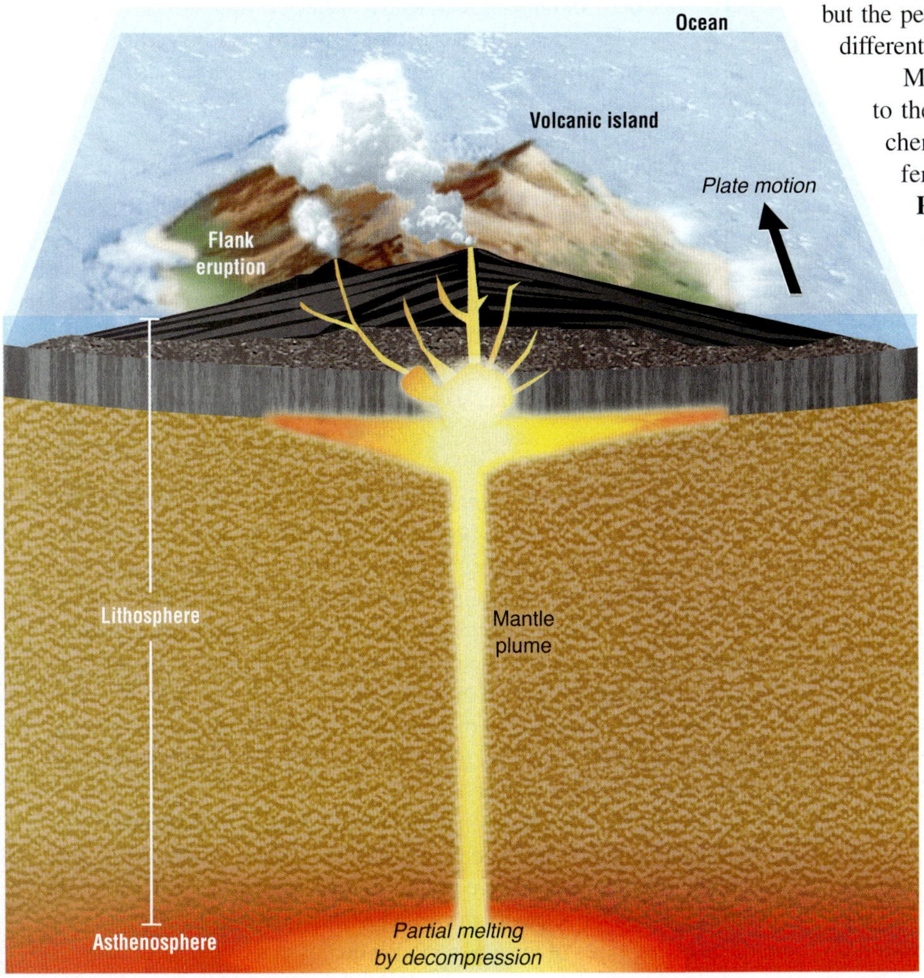

FIGURE 5.4 Magma may form when rock in the mantle experiences a decrease in pressure at shallow depths and undergoes decompression melting.

How would Earth's surface be different if rock in the mantle did not experience decompression melting?

Decompression melting takes place under the crust where the upper mantle is exposed to decreases in pressure caused by *crustal rifting, volcanic eruption*, or upward directed mantle convection.

Formed at temperatures of 600°C to 1300°C, magma accumulates in the upper mantle at depths typically below 60 to 100 kilometers. And because it is less dense than the surrounding solid rock, magma rises (**FIGURE 5.4**) and intrudes into the crust. Some of it cools and crystallizes, forming (intrusive) igneous rock.

Magma crystallization takes thousands of years. Depending on the size of the magma body and its rate of cooling, roughly a million years may be needed to fully solidify a large accumulation. The magma may either crystallize within the crust or erupt at the surface and then crystallize. The depth at which crystallization occurs determines the characteristics of the resulting igneous rock—its chemistry, mineralogy, crystal size, and value as a resource.

Partial Melting

How does rock melt to create magma? Imagine mixing copper pennies, bits of candle wax, and water in a bowl, freezing the mixture, and then thawing it at room temperature. When the mass thaws, the ice melts first, while the wax and copper remain solid.

If you put the remaining mixture in the oven, the wax will melt but the pennies will not. Why? Because the three substances have different melting points.

Magma, made by melting rock, develops in a similar way to the mixture of wax, pennies, and water because different chemical compounds—that is, different minerals—have different melting points. Thus, partial melting forms magma. **Partial melting** occurs when some minerals melt while others remain solid.

As you might imagine, as magma cools, it experiences the reverse of partial melting.

During cooling, different compounds solidify and become minerals at different temperatures, with some portion of the magma remaining fluid until the very end. This process of *crystallization* was described in Chapter 4, "Minerals."

The important point here is that the chemical composition of magma changes as elements are removed from it through crystallization. As the magma cools or moves to places where pressures are lower, the minerals formed vary in their chemical composition because the magma itself has a different chemical composition.

Magma Differentiation

Magma is a hot soup whose ingredients include Earth's major elements—Si, O, Ca, Al, Fe, Na, Mg, K—plus other elements in lesser quantities. As magma migrates into and through the crust, its composition changes. This process, called **magma differentiation,** explains how a single "parent" magma body can produce dozens of different types of igneous rocks (**FIGURE 5.5**).

Since the average body of magma is relatively rich in iron, magnesium, and calcium compounds, and the average continental lithosphere is relatively rich in silica (and depleted in iron, magnesium, and calcium), the effect of magma differentiation is to separate components that are high in iron, magnesium, and calcium from those that are high in silica. Consequently, in most cases, magma differentiation produces one magma body that contains relatively more silica than the parent, and another magma body that contains relatively more iron, magnesium, and calcium.

Magma differentiation can include one or more, or any combination of, the following four processes: *crystal settling, magma migration, magma assimilation,* and *magma mixing.*

Crystal Settling

As magma cools, iron-, magnesium-, and calcium-based compounds crystallize first (at relatively high temperatures) and become solid minerals. As cooling continues, therefore, the magma becomes relatively depleted in iron, magnesium, and calcium and relatively enriched in silicates, which crystallize at relatively lower temperatures.

One way by which this depletion occurs is through crystal settling, whereby the early-forming minerals, which are denser than the surrounding liquid, settle out of the magma body. Crystal settling may produce a zone of highly concentrated minerals in the magma, called a *cumulate.* Cumulates often contain economically valuable ore deposits.

Magma Migration

As we have seen, magma may migrate within the crust. The cumulate then becomes isolated from the magma and forms a separate body of igneous rock with a distinctive mineral composition.

The separated magma continues to crystallize at its new location, but without some of its original components (i.e., those in the early-crystallizing minerals), forming new types of igneous rock. If magma erupts onto Earth's surface, still other types of rocks will be produced because crystallization will occur much more rapidly in the cool atmosphere.

Magma Assimilation

Magma can also melt the surrounding crust, thereby adding (or "assimilating") new compounds to the mixture originally produced in the upper mantle. This crustal melting can result in a variety of minerals and rocks as the magma crystallizes.

Magma Mixing

Sometimes, magma encounters, and becomes mixed with, other bodies of magma. Alternatively, a body of magma may be separated into two or more masses. Each mass will form its own minerals as it cools, with iron-, magnesium-, and calcium-based minerals crystallizing first.

In each of these cases, magma passes through predictable *stages of crystallization,* each of which produces specific minerals. These stages establish the crystallization environment for the specific sets of minerals that make up each type of igneous rock. Clearly, since any body of magma can have a complex history, a wide variety of igneous rocks can form from a single magma body.

During the crystallization of most igneous rocks, oxygen and silicon form silicate tetrahedra bonded with metallic cations, producing mineral compounds. In Chapter 4 we examined how silicates crystallize, but we did not look at the sequence in which particular minerals are formed or at the resulting types of igneous rock. We do this in the next section.

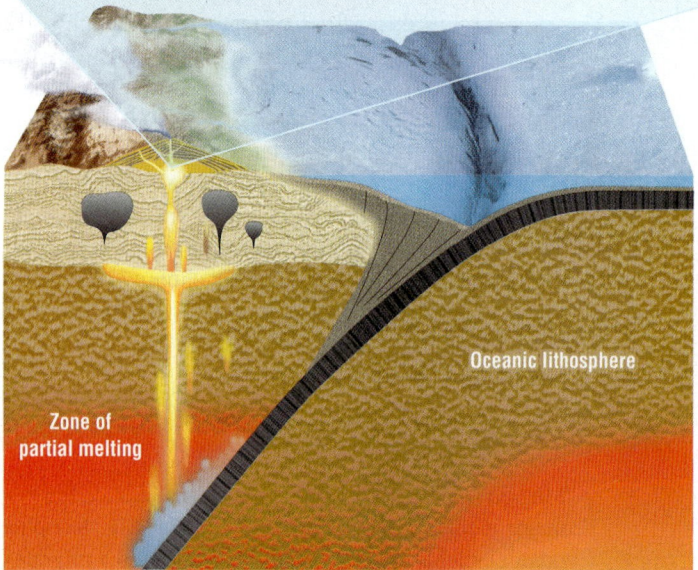

FIGURE 5.5 Magma differentiation can include four processes: crystal settling, magma migration, magma assimilation, and magma mixing. All these processes can change the composition of magma, producing many kinds of igneous rock.

Ⓠ How would magma ready to erupt from a volcano differ from magma just entering the lower magma chamber?

Ⓠ EXPAND YOUR THINKING

Magma is said to "evolve" by differentiation. What does this mean?

5-3 Bowen's Reaction Series Describes the Crystallization of Magma

LO 5-3 Describe the process of mineral crystallization in magma of average composition.

As magma cools, minerals crystallize and separate from the liquid magma in a predictable sequence that depends on the chemistry and temperature of the magma body.

Magma Crystallization

Crystallization is essentially freezing: the change of state in which the random distribution of compounds in a liquid align themselves in the rigid crystal structures of solids—just as water freezes to create ice crystals.

Each compound has its own freezing point, so as magma cools the crystallization process forms different solid minerals in a predictable sequence. The first minerals to crystallize are those that are stable at the highest temperatures, followed by those that are stable at progressively lower temperatures. With the formation of each new crystal, the chemistry of the remaining magma changes to reflect the loss of the elements that went into the solidifying minerals.

In general, as crystallization progresses, magmas first lose their iron, magnesium, and calcium. These elements bind into early-forming minerals such as olivine, pyroxene, and calcium-rich plagioclase feldspar, which are stable in a solid state at high temperatures (they have a high melting point).

Because these minerals have relatively low silica content, the magma experiences no significant reduction in silicon and oxygen. As a result, it evolves into a new chemical state that is relatively enriched in silica (and depleted in Fe, Mg, and Ca). This process is described by **Bowen's reaction series,** explained in the next section and diagramed in **FIGURE 5.6**.

Bowen's Reaction Series

In the early 1900s, mineralogist *N. L. Bowen* (1887-1956) conducted experiments to determine the sequence in which common silicate minerals crystallize from magma. He developed an idealized progression, now known as Bowen's reaction series, which is still widely accepted as a general model of magma crystallization.

Bowen determined that certain minerals are formed at specific temperatures as magma cools. Minerals at the top of his series crystallize early in the cooling process, while the magma is still very hot. These minerals have the highest melting temperatures, are mostly dark in color, and are relatively enriched in iron and magnesium. They also have the highest *specific gravity* (i.e., they are dense and feel noticeably heavy).

We describe igneous rocks composed of these minerals, and others like them, as **mafic** (a contraction of *ma*gnesium and *fer-ic*) in composition. (**Ultramafic** rocks are greatly enriched in iron and magnesium.) The term *mafic* is used to describe any silicate mineral, magma, or rock that is relatively enriched in the heavier elements, including calcium. Common rock-forming mafic minerals include olivine, pyroxene, amphibole, biotite mica, and the calcium-rich plagioclase feldspars.

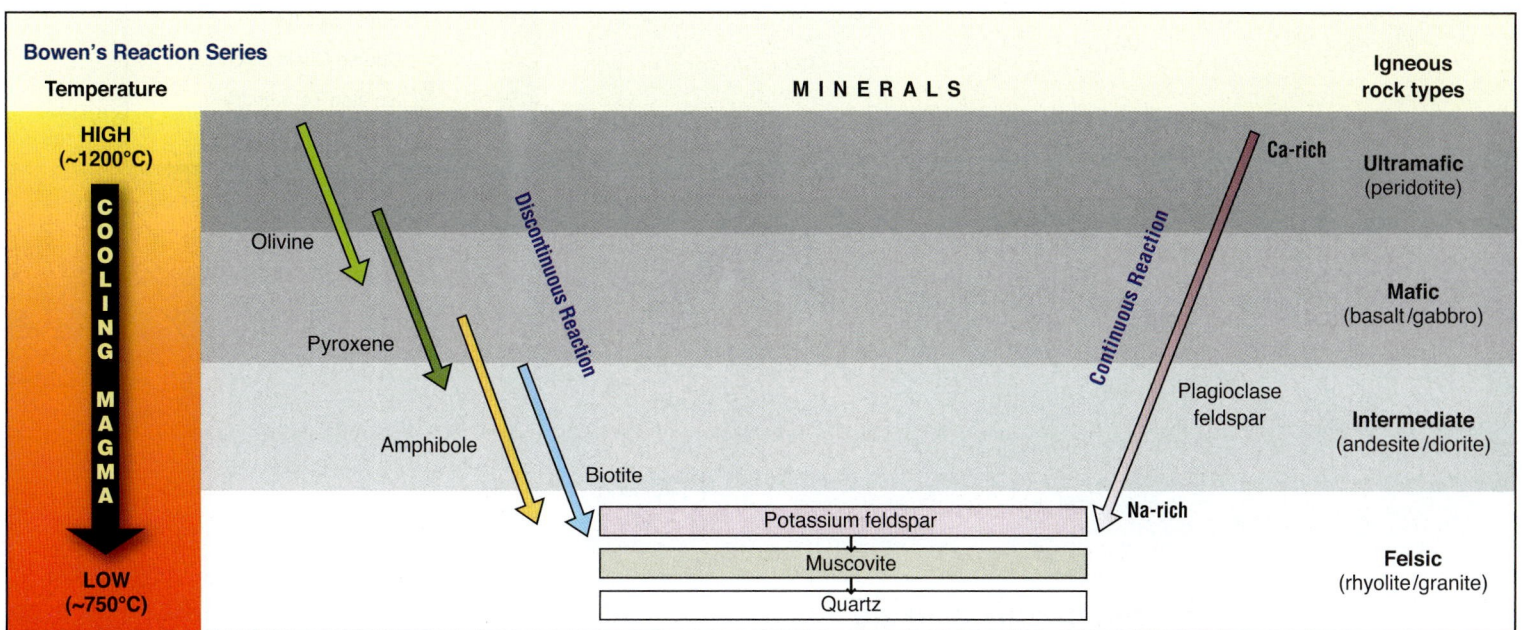

FIGURE 5.6 Mineralogist N. L. Bowen described two crystallization pathways leading to the formation of common silicate minerals in magma of average composition. The horizontal shaded bands highlight the mineral compositions of various types of rocks.

Describe why the chemical composition of magma changes.

Minerals that are formed at the bottom of the series crystallize late in the cooling process, when the magma has cooled significantly. They have the lowest melting temperatures; are light in color and relatively enriched in sodium, potassium, and silica; and have a relatively low specific gravity (i.e., they are less dense than mafic minerals).

We describe igneous rocks composed of these elements as **felsic** (a contraction of *feldspar* and *silica*) in composition. Common felsic minerals include quartz, muscovite mica, the orthoclase feldspars, and the sodium-rich plagioclase feldspars.

Igneous rocks whose composition is **intermediate**—between that of mafic rocks and felsic rocks—are composed of minerals that are somewhat abundant in iron, magnesium, calcium potassium, sodium, and oxygen. Most magma is mafic or intermediate in composition.

Refer back to Figure 5.6 and note that there are two pathways for mineral crystallization: discontinuous, on the left, and continuous, on the right. This reflects Bowen's observation that crystallization at higher temperatures in mafic and intermediate magmas tends to separate into two simultaneous crystallization pathways, known as branches.

The *discontinuous branch* describes the formation of the mafic minerals olivine, pyroxene, amphibole, and biotite mica. It is discontinuous because crystallized minerals formed at higher temperatures react chemically with the surrounding magma and recrystallize at a lower temperature, to create a different mineral with a different crystalline structure.

The *continuous branch* of crystallization describes the evolution of a common type of feldspar—plagioclase feldspar, a framework silicate—as it changes from calcium-rich to sodium-rich. The transition is smooth, as it involves single and double substitutions of cations without changes in the crystalline structure.

At lower temperatures, the two pathways merge, and the minerals common to felsic rocks—potassium feldspar, muscovite mica, and quartz—are formed. **TABLE 5.1** lists the most common igneous rock-forming minerals, and the Geology in Action feature, "The Oldest Rocks," discusses the oldest known rocks—which are igneous.

TABLE 5.1 Common Igneous Rock-Forming Minerals

Plagioclase Feldspar	**Orthoclase Feldspar**	**Quartz**	**Muscovite**	Name
$NaAlSi_3O_8$ (albite)	$KAlSi_3O_8$	SiO_2	$KAl_3Si_3O_{10}(OH)_2$	Composition

Felsic Minerals (lighter in color)

Plagioclase Feldspar	**Amphibole Group**	**Biotite**
$(Ca, Na) AlSi_3O_8$	(Mg, Al, Fe, Ca, Na) $Si_8O_{22} (OH)_2$	(K, Mg, FeF, Al) $Si_3O_{10} (F, OH)_2$

Intermediate Minerals

Plagioclase Feldspar	**Olivine**	**Pyroxene**
$CaAl_2Si_2O_8$ (anorthite)	$(Mg, Fe)_2SiO_4$	(Mg, Fe, Ca, Na, Al) SiO_3

Mafic Minerals (darker in color)

Increasing darkness

Legend:
- Framework Silicate
- Sheet Silicate
- Double Chain Silicate
- Isolated Tetrahedra
- Single Chain Silicate

Mineral images courtesy of Chip Fletcher. Amphibole: © Marli Bryant Miller

EXPAND YOUR THINKING

Bowen's reaction series is a *model* that predicts mineral crystallization in magma of average composition. Explain how this is an example of critical thinking.

GEOLOGY IN ACTION

The Oldest Rocks

The oldest rocks known to scientists do not come from Earth; they are meteorites, and they come from space. Although meteorites are not produced by Earth magma, geologists classify them as igneous in origin because many of them crystallized from molten rock associated with planetesimal accretion. Scientists have collected, classified, and studied over 2,000 meteorites (**FIGURE 5.7**), many of which date from the origin of the solar system and reflect the chemistry of the original solar nebula. At approximately 4.6 billion years old, these are the oldest rocks. Some meteorites are younger and have come to Earth from the Moon or other planets.

The term *meteor* comes from the Greek *meteoron*, meaning "phenomenon in the sky." It describes the streak of light that is emitted when a solid object falls into the atmosphere and is ignited by friction with the air.

Several meteorite samples collected in Antarctica are thought to have come from the Moon because their composition matches that of Moon rocks obtained by Apollo astronauts. Another set of eight meteorites are believed to have come from Mars because gases trapped in the samples match the composition of the Martian atmosphere as measured by Viking landers in 1976. Other types of meteorites probably originated on asteroids or comets; in fact, the majority of meteorites are believed to be fragments of asteroids.

Geologists distinguish among three types or groups of meteorites: *stony meteorites*, *stony-iron meteorites*, and *iron meteorites* (**TABLE 5.2**). Stony meteorites, the most abundant, are most similar to Earth rocks. They often contain the silicate minerals olivine, pyroxene, and plagioclase feldspar. The group is divided into two subgroups based on the presence of chondrules, small, nearly spherical structures in the rock that many scientists believe represent the most primitive material in the solar system: *chondrites*, which contain chondrules, and *achondrites*, which do not.

Achondrites are meteorites that have melted and recrystallized. Scientists hypothesize that they result from magma that has chemically differentiated in a process like that experienced by Earth during the iron catastrophe. The only bodies large enough to heat to this extent

TABLE 5.2 Types and Abundances of Meteorites Collected on Earth

Type	Abundance
Stony meteorites (two types)	92.8%
Chondrites	85.7%
Achondrites	7.1%
Stony iron meteorites (two types)	1.5%
Iron meteorites (two types)	5.7%

must be planetary in size. Therefore, many achondrites are thought to have originated on other planets, including Mars.

Chondrites are meteorites that have not undergone melting or differentiation. Scientists infer from this that they formed very early in the history of the solar system from various types of dust and small grains that accreted to form primitive asteroids. Study of chondrites provides important information for interpreting the origin and age of the solar system.

Stony-iron meteorites consist of almost equal amounts of nickel-iron alloy and silicate minerals. The *pallasite* group is characterized by olivine crystals surrounded by a nickel-iron coating. Pallasites are regarded as samples of core/mantle boundary material from asteroids that experienced chemical differentiation at some point in their history. *Mesosiderites*, on the other hand, consist mainly of plagioclase and pyroxene silicates intermixed with a metal alloy. Their origin is debated, but according to one hypothesis, they were formed by the collision of two differentiated asteroids that allowed the still-liquid core of one asteroid to mix with the solidified crust of the other, thus creating the stony-iron mixture.

Iron meteorites are of two main kinds: *nickel iron* (formed inside the molten core of an asteroid or moon) and *silicated iron* (iron meteorites that also contain silicates). Iron meteorites must have been made as part of the process of chemical differentiation undergone by large bodies that experienced heating. Subsequently, these bodies must have suffered collision and fragmentation into the small bits entering our atmosphere today.

(a)   (b)   (c)

Detlev van Ravensway/Science Source   Walter Geiersperger/Corbis   Detlev van Ravensway/Science Source

FIGURE 5.7 (a) Stony meteorite, (b) stony-iron meteorite, and (c) iron meteorite.

5-4 The Texture of Igneous Rock Records Its Crystallization History

LO 5-4 Identify, in detail, the information revealed by igneous texture.

Igneous rock is formed as a result of magma crystallization. As minerals crystallize, magma becomes more viscous (resists flow) until the entire mass solidifies. In this way, the magma becomes a mass of interlocked crystals that create the texture of the igneous rock.

Texture and Intrusive/Extrusive Rocks

The texture of an igneous rock depends on the size of its minerals, which, in turn, depends partially on how quickly magma crystallizes. Igneous rocks that cool beneath the surface for thousands of years may develop relatively large crystals and a coarse texture (**FIGURE 5.8**). In essence, crystallization is slower, and crystals tend to be larger than those in a rapidly cooled rock.

Additional factors come into play in determining crystal growth, but in general the rate of cooling exerts an important control on crystal size. Rocks that have larger crystals are generally considered to be intrusive or, in the case of deep intrusions, plutonic. Magma that erupts at Earth's surface develops small crystals (or none at all) because it cools rapidly—within minutes to months—allowing far less time for crystallization. These rocks are generally considered to be extrusive or volcanic.

Based on this knowledge of the way crystals are formed, geologists can *infer* (develop a hypothesis about) the crystallization history of magma from the texture of the resulting rock—whether it is coarse- or fine-grained (**FIGURE 5.9**). Within the categories of coarse-grained and fine-grained, rocks can vary widely, ranging from coarse-grained rocks with fist-size or even larger crystals to fine-grained rocks of volcanic glass that cooled so quickly that they contain no true crystals at all.

Common Igneous Textures

The best way to understand textures is to look at them, so while reading the text in this section, please refer to both **TABLE 5.3** and Figure 5.9.

Coarse-grained (intrusive) rocks are *phaneritic* (from the Greek *phaneros*, meaning "visible"): the mineral grains within them are large enough to be seen with the naked eye. Phaneritic rocks are intrusive because they have large crystals, indicating that crystallization proceeded slowly, and the resultant mineral grains are relatively large. Granite and gabbro are examples of phaneritic igneous rocks.

Fine-grained rocks are *aphanitic* (from the Greek *aphanes*, meaning "hidden" or "obscure"); the mineral grains within them are too small to be seen with the naked eye. These rocks are extrusive; basalt is an example. The finest rock texture is *glassy*; volcanic glass (obsidian) contains no crystals at all and, therefore, no minerals. (Recall that a mineral is defined as having a crystalline structure.) Volcanic glass is really an extremely viscous, noncrystalline liquid.

A rock with two distinct textures—mixed large and small grains—is *porphyritic*. The larger crystals are *phenocrysts*, and the finer ones are referred to as the *matrix* or *groundmass*. Rocks with porphyritic texture have undergone two separate stages of cooling: an intrusive stage, during which phenocrysts were formed, and a volcanic stage, during which matrix grains crystallized.

(a)

(b)

Courtesy of Chip Fletcher

Science Photo Library/Photo Researchers, Inc.

FIGURE 5.8 (a) Close-up of the surface of granite showing the interlocking nature of mineral grains. (b) Microphotograph showing the interlocking nature of minerals in an igneous rock.

Which minerals can you identify in Figure 5.8a?

TABLE 5.3 Common Igneous Textures

Texture	Definition	Example
Aphanitic	Minerals too small to see	Rhyolite
Phaneritic	Minerals large enough to see with the naked eye	Granite
Glassy	No obvious minerals	Obsidian
Pyroclastic	Fused, glassy volcanic rock fragments and ash from explosive volcanic eruption	Tuff
Vesicular	Many holes or pits in rock surface caused by escaping gas	Vesicular basalt
Porphyritic	Two distinct mineral sizes	Porphyritic basalt

Pyroclastic texture

Porphyritic texture

Aphanitic texture

Phaneritic texture

Mineral images courtesy of Chip Fletcher

FIGURE 5.9 Igneous rock texture reflects crystallization history.

Q How does porphyritic texture form?

An igneous rock that is full of bubblelike holes is called *vesicular*. Expanding gas bubbles escaping from magma made the holes, or vesicles. This process occurs during volcanic eruptions set off by the rapid expansion of gas as magma moves toward Earth's surface. Basalt can have vesicles, but they are more likely to be found in *pumice* and *scoria*. In fact, pumice contains so much gas that it actually floats on water.

Finally, rocks that solidify when they are expelled into the air during violent volcanic eruptions are termed *pyroclastic* or *fragmental*. Pyroclastic volcanic rocks consist of numerous glassy fragments and shards that have been welded together by the heat of the eruption. Pyroclastic rocks often feel grainy, like sandpaper; and often there are shards of volcanic glass embedded in the rock. Most pyroclastic-textured rocks are referred to as *tuff*.

Q EXPAND YOUR THINKING

Why would volcanically erupted rocks be glassy?

Igneous Rocks Are Named on the Basis of Their Texture and Composition

5-5

LO **5-5** Identify how igneous rock color relates to its chemical composition.

A close look at the size of mineral grains reveals a rock's texture. Similarly, a close look at a rock's color reveals its general chemical composition.

The Color of Igneous Rock

Minerals at the top of Bowen's reaction series, such as pyroxene and amphibole, tend to be dark-colored and mafic to ultramafic in composition. Their dark coloration reflects the fact that iron and magnesium tend to be dark.

Minerals at the bottom of Bowen's reaction series, such as sodium plagioclase, orthoclase, and quartz, tend to be light-colored and felsic in composition. Geologists use these color trends, along with texture, to identify and name igneous rocks.

The chemical content of an igneous rock determines its composition, but to analyze rock chemistry requires extensive training and expensive equipment. Geologists also must use their observations of the physical properties of igneous rocks to identify their components, just as we used physical properties to identify minerals, in Chapter 4.

One component geologists note is color, which serves as a visual aid to estimate the proportion of light minerals to dark ones in a rock sample. Light colors (white, light gray, tan, and pink) most often indicate a felsic composition. Felsic rocks are relatively high in silica, potassium, and sodium.

Dark colors (black, dark gray, and dark brown) most often point to a mafic composition. Mafic rocks are relatively low in silica, potassium, and sodium, and relatively high in iron, magnesium, and calcium. Rocks with intermediate compositions are often gray, or consist of equal parts of dark and light minerals.

To fine-tune these color observations, geologists consult a color index, like the one in **FIGURE 5.10**, to estimate the chemical composition of igneous rocks. The color index is not always an accurate indicator of a rock's composition, however; there are exceptions.

Obsidian, for example, is volcanic glass that has cooled quickly after eruption. Most obsidian is felsic in composition, yet it commonly is dark brown to black in color because it contains volcanic ash impurities.

Dunite has an ultramafic composition (it is composed entirely of olivine, which has a high iron and magnesium content), yet it is apple green to yellowish green in color.

In summary, while the color index can be used to approximate the composition of many igneous rocks, judgments based on it should be viewed just that way: as approximations. Rocks with a felsic composition can, and usually do, contain dark-colored minerals, and mafic rocks can contain light-colored minerals.

Up to this point, we have discussed igneous mineralogy (Bowen's reaction series), texture (phaneritic, aphanitic, porphyritic, etc.), and the composition of igneous rocks (ultramafic, mafic, intermediate, and felsic). When we consider these characteristics together, we can

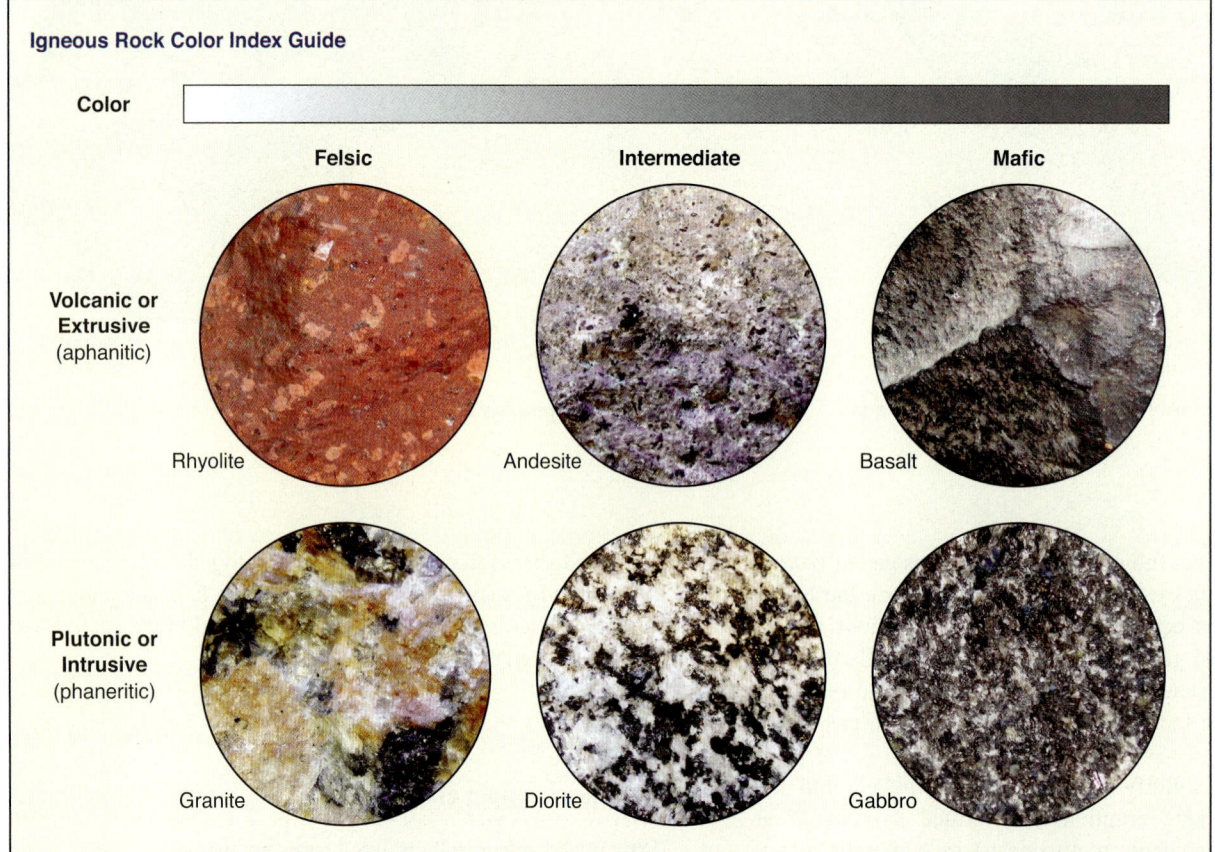

FIGURE 5.10
In the igneous rock color guide, samples are arranged vertically by texture (top to bottom, aphanitic to phaneritic). Notice that the color or shading of the rock generally corresponds to its composition—darker colors to mafic and lighter colors to felsic.

Q Use this information to describe the two rocks shown earlier, in Figure 5.2.

Mineral images Courtesy of Chip Fletcher

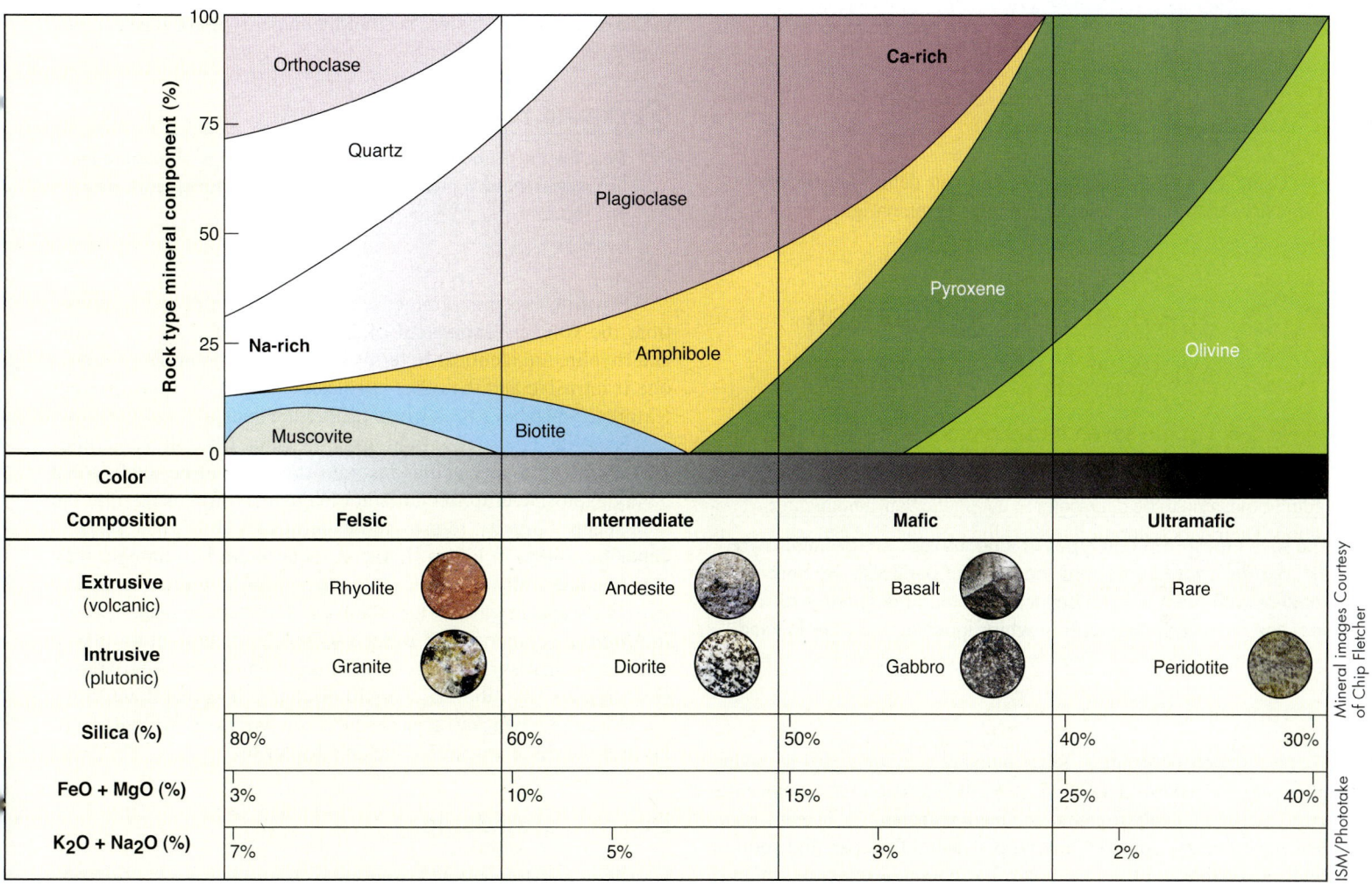

FIGURE 5.11   As explained in the text, igneous rocks are classified using mineralogy (upper portion of chart), color, and chemistry (lower portion).

Ⓠ **What are the texture and composition of, say, andesite?**

use them to name the seven most common types of igneous rock, each of which is described in the next section.

Classifying the Seven Most Common Types of Igneous Rock

Geologists classify igneous rocks based on a combination of the color/texture, mineral/texture, or chemical composition/texture properties of the rock. These three classification systems are shown in **FIGURE 5.11**.

Although at first glance this chart looks complex, it neatly pulls together our knowledge about the minerals, texture, and composition of igneous rocks to make it easier to name them. And why is naming so important? Because names are what geologists around the world use to communicate about the rocks they find and study. By applying the same naming system, geologists everywhere can, for example, identify a particular sample of light-colored, felsic, phaneritic rock of intrusive origin as granite.

Look at Figure 5.11 more closely and examine the top portion. It identifies the eight common igneous rock-forming minerals in Bowen's reaction series: orthoclase, quartz, plagioclase, muscovite, biotite, amphibole, pyroxene, and olivine. These minerals are found in the seven most common igneous rocks: *rhyolite, granite,*

andesite, diorite, basalt, gabbro, and *peridotite* (shown in the lower portion of the figure and described in detail in the next section).

The first column on the left shows the mineral percentage, color, composition, and texture of each type of rock. Column 2 lists the felsic igneous rocks (rhyolite and granite) with their mineral content, color, and chemistry. The lower portion shows the chemistry of the felsic rocks: approximately 6% to 7% compounds of potassium and sodium ($K_2O + Na_2O$), 3% to 9% compounds of iron and magnesium ($FeO + MgO$), and about 62% to 80% silica.

Above the chemistry specifications are the names of the intrusive rock (granite) and the extrusive rock (rhyolite) in the felsic category. Most of these rocks are light in color.

The top portion of the chart shows the minerals in the rock and their abundance in typical samples; in the case of granite and rhyolite: approximately 0% to 25% orthoclase, 25% to 40% quartz, 15% to 40% sodium-rich plagioclase, 0% to 10% amphibole, 5% to 15% biotite mica, and 0% to 15% muscovite mica.

Column 3 lists the intermediate igneous rocks (andesite and diorite) and their characteristics; column 4 lists the mafic rocks (basalt and gabbro); and column 5 lists an ultramafic rock (peridotite).

This chart tells us three important things:

1. Intrusive and extrusive igneous rocks with the same composition contain the same minerals even though they differ in color and texture.

2. Silica becomes less abundant as iron and magnesium become more abundant.
3. The abundance of potassium and sodium decreases as silica content decreases.

Using Figures 5.10 and 5.11, you can define the mineralogy, chemistry, texture, and color of the most common igneous rocks. In the next section we explore each of the major types of igneous rock in more detail.

EXPAND YOUR THINKING

Describe the probable chemistry and mineralogy of magma that has experienced significant differentiation. What igneous rock will result?

5-6 There Are Seven Common Types of Igneous Rock

LO 5-6 List the seven most common igneous rocks, and describe their composition and texture.

The seven most common types of igneous rock are granite, rhyolite, diorite, andesite, gabbro, basalt, and peridotite. By becoming familiar with them we can better understand the *igneous rock system,* the major environments in which igneous rocks are formed.

Granite and Rhyolite

Granite is a coarse-grained, felsic intrusive rock that may be white, gray, pink, or reddish (**FIGURE 5.12A**). It is a rock of the continental crust, most commonly found in mountainous areas. The average granite contains coarse grains of quartz, potassium feldspar, and sodium feldspar (orthoclase), as well as other common minerals, including the mica group (silvery muscovite and black biotite) and small amounts of amphibole.

Granites are the most abundant intrusive rocks of mountain belts and the bedrock of continents. They solidify in great magma chambers and form large intrusive structures called *batholiths* that may extend over hundreds of kilometers and are often associated with diorite and gabbro. Granite is an active intruder of continental crust, frequently forming **dikes** (arms of intrusive rock) that cut across other rocks in the crust.

Rhyolite is a fine-grained extrusive rock of felsic composition, the volcanic equivalent of granite (**FIGURE 5.12B**). Granite and rhyolite are identical in terms of mineral content. But because one is intrusive and the other is volcanic, they have very different textures and so have been given different names.

Rhyolite is usually light gray to pink (though sometimes dark purple). In spectacular fashion, rhyolite can be made when a very explosive eruption hurls hot magma and fragments into the atmosphere. Gravity returns these pyroclasts to Earth's surface, depositing them in layers. If the pyroclasts are hot enough, they may become welded together to form a *rhyolite tuff* (composed of pyroclasts and fragments less than 2 millimeters across). A *rhyolite breccia* (composed of pyroclasts and fragments more than 2 millimeters across) is also formed by welding of coarse pyroclasts. Occasionally rhyolitic magma will discharge lava that flows down the slopes of a volcano and produce *rhyolite lava*. We will learn more about these volcanic rocks in Chapter 6.

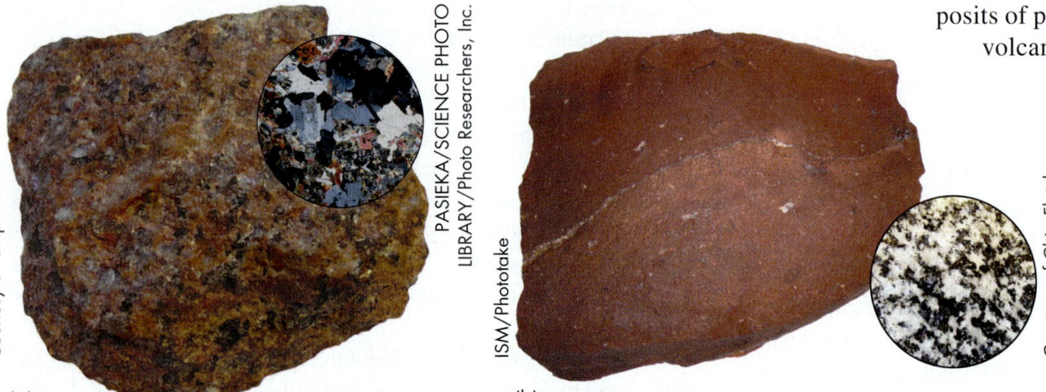

Courtesy of Chip Fletcher

PASIEKA/SCIENCE PHOTO LIBRARY/Photo Researchers, Inc.

ISM/Phototake

Courtesy of Chip Fletcher

(a) (b)

FIGURE 5.12 Granite (a), an intrusive felsic igneous rock, is an important component of continental crust. Rhyolite (b) is its extrusive equivalent, identical chemically but fine-grained, a reflection of its volcanic origin.

How is the appearance of these rocks related to their history?

Diorite and Andesite

Diorite is a coarse-grained, intermediate intrusive rock composed of sodium-rich plagioclase feldspar (light-colored) and amphibole (dark-colored) (**FIGURE 5.13A**). These minerals give diorite its characteristic salt-and-pepper appearance. Small amounts of quartz and biotite mica may also be present. Diorite is the intermediate, intrusive member of the igneous rock family. Although it is coarse-grained (phaneritic), it usually has little or no visible quartz minerals and is darker than granite.

Andesite is the fine-grained volcanic equivalent of diorite (**FIGURE 5.13B**). It is formed by the eruption of intermediate or felsic magma. Such eruptions often begin explosively, making deposits of pyroclast layers common on and around volcanoes. The explosive phase is followed by a flow of lava that cools, hardens, and protects the underlying layer of pyroclasts from erosion. The volcanoes resulting from a series of such eruptions are layered (*stratified*), and are therefore called *stratovolcanoes.* Andesite is an important and frequent component of stratovolcanoes.

Gabbro and Basalt

Gabbro is a coarse-grained, mafic intrusive rock composed mostly of light-colored calcium-rich

(a)

Courtesy of Chip Fletcher

PASIEKA/SCIENCE PHOTOLIBRARY/Photo Researchers, Inc.

(b)

Courtesy of Chip Fletcher

FIGURE 5.13 Diorite (a) is an intrusive igneous rock with an intermediate composition. It has a salt-and-pepper appearance. Andesite (b) has the same composition as diorite but is finer grained due to its volcanic origin.

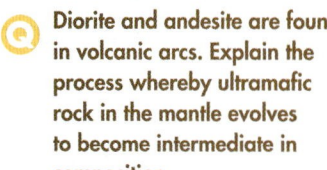 Diorite and andesite are found in volcanic arcs. Explain the process whereby ultramafic rock in the mantle evolves to become intermediate in composition

(a)

PASIEKA/SCIENCE PHOTOLIBRARY/Photo Researchers, Inc.

(b)

PASIEKA/SCIENCE PHOTOLIBRARY/Photo Researchers, Inc.

FIGURE 5.14 Gabbro (a) is a mafic intrusive rock composed of olivine, pyroxene, amphibole, and Ca-plagioclase. Basalt (b) is a mafic volcanic rock of the same composition.

Describe where gabbro and basalt are found in greatest abundance, and explain why mafic rock is predicted at that site

plagioclase feldspar, dark pyroxene, and dark or green olivine (**FIGURE 5.14A**). The dark-colored minerals give gabbro a dark green to black color that clearly indicates its mafic composition. At mid-ocean ridges, gabbro rises from the upper mantle and forms the intrusive component of oceanic crust. This makes gabbro among the most abundant forms of intrusive igneous rock on the planet and a very important component of oceanic crust.

Basalt is the fine-grained, mafic, extrusive equivalent of gabbro. Usually black, it is formed when mafic magma cools at the Earth's surface. **FIGURE 5.14B** shows a typical sample. Basalt erupts on the ocean floor at mid-ocean ridges where mafic magma flows into the rift during the process of seafloor spreading. Basalt is so abundant that it is the single most important component of oceanic crust.

Chains of high volcanic islands that develop at hotspots, such as the Hawaiian Islands and Tahiti, are composed mostly of basalt that erupts from volcanoes on the seafloor and eventually builds up to create islands above the ocean surface. The Big Island of Hawaii is an example; it is composed of five huge volcanoes that have the classic, gentle slopes and rounded, shieldlike shape found in volcanoes composed of basalt, giving them their name, *shield volcanoes*.

Peridotite

Peridotite is the only ultramafic igneous rock that occurs on Earth's surface in any appreciable quantity (**FIGURE 5.15**). It is composed of the ferromagnesian silicate minerals olivine and pyroxene. The magma that forms ultramafic rocks is uniquely differentiated and composed of very dense compounds, and hence it is very rarely found in a volcanic form. Indeed, many geologists consider peridotite to

represent the average composition of the upper mantle, making it one of the most abundant rocks on the planet, even though it is extremely rare on the surface, and thus rarely seen.

Courtesy of Chip Fletcher

Dirk Wiersma/Photo Researchers, Inc.

FIGURE 5.15 Many geologists think that peridotite, although rarely seen in the crust, may represent the average composition of the upper mantle. This sample consists largely of green olivine crystals and black pyroxenes.

Explain why ultramafic rock is rare on Earth's surface.

EXPAND YOUR THINKING

What might happen if mafic magma mixed with felsic magma? What minerals and rocks would result, and in what order?

5-7

All Rocks on Earth Have Evolved from the First Igneous Rocks

LO 5-7 Describe the process of igneous evolution and the role of plate tectonics in it.

Igneous rocks were the first rocks on Earth. All other rocks have evolved from the igneous mafic/ultramafic parent rocks that formed over 4 billion years ago. Today, as a result of igneous evolution, igneous rocks are found in a profusion of types and tectonic environments.

All Igneous Rocks Result from Magma Differentiation

Through the process of igneous evolution, the original, relatively uniform composition of primitive, molten Earth gave rise not only to igneous rocks but also, through the rock cycle, to sedimentary and metamorphic rocks. The idea that igneous rocks evolve over time is one of the most important to emerge in modern geology because it explains the sequence of steps that have produced the great diversity of rocks that exist on Earth today. (To deepen your understanding of the environments in which igneous rocks are formed, work through the exercise in the Critical Thinking feature, "Igneous Environments.")

Plate tectonics theory tells us that the lithosphere is recycled into the mantle at subduction zones. Partial melting of the upper mantle above a subducting slab, coupled with partial melting of the crust due to magma intrusion, produces chemically distinct rocks that are relatively high in silicate compounds. This process, along with magma differentiation, has caused the original composition of Earth's crust to evolve into more chemically complex types.

As just noted, the igneous rock system has evolved in a series of steps. In each step, parent igneous rock differentiates into two portions whose composition differs from that of the parent rock (**FIGURE 5.16**). These portions are:

- A relatively more felsic magma dominated by compounds with lower melting points, typically high in silica relative to the parent.

- A "residue" more mafic fraction that is a solid mineral assemblage; it is high in minerals enriched in iron, magnesium, and calcium, such as olivine and calcium-rich plagioclase, relative to the parent.

Igneous evolution can occur through partial melting of the crust or upper mantle or through crystallization. In either case, two new, different bodies of rock are created from a single magma source.

For example, during partial melting (say, at a spreading center where decompression melting takes place), ultramafic parent rock (peridotite) rises from the mantle and partially melts, producing two

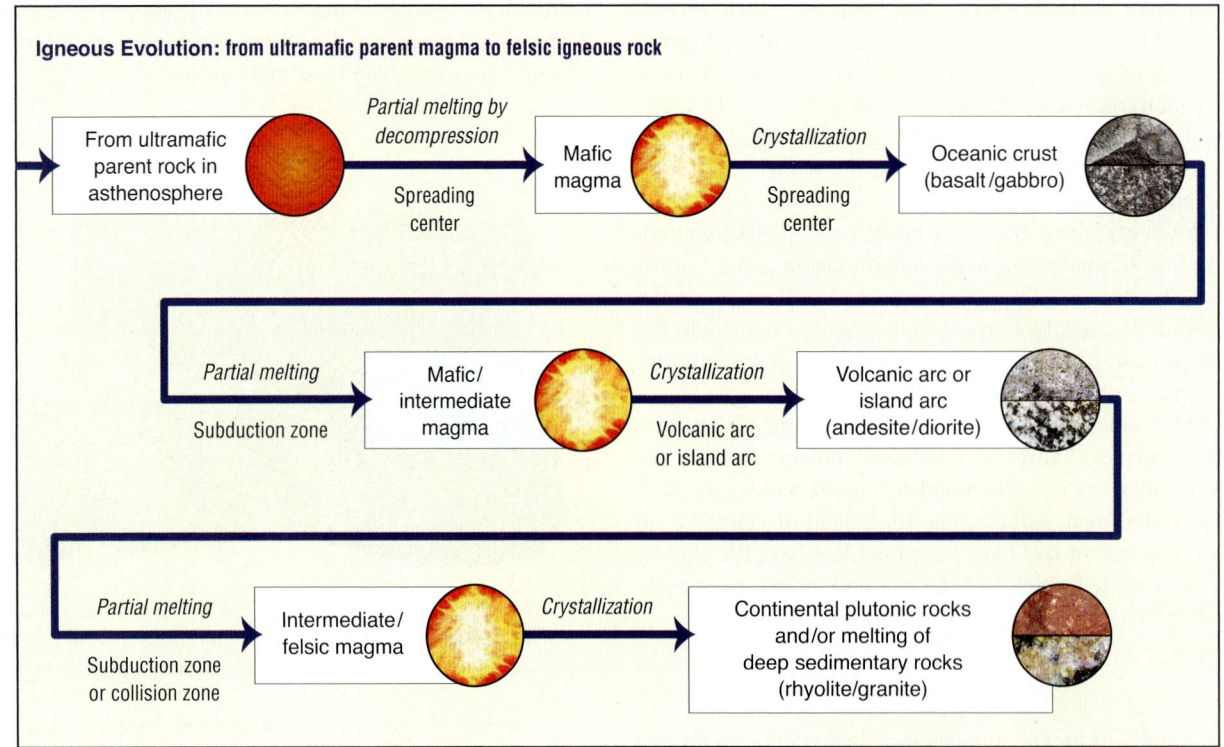

Igneous Evolution: from ultramafic parent magma to felsic igneous rock

From ultramafic parent rock in asthenosphere → *Partial melting by decompression* / Spreading center → Mafic magma → *Crystallization* / Spreading center → Oceanic crust (basalt/gabbro)

Partial melting / Subduction zone → Mafic/intermediate magma → *Crystallization* / Volcanic arc or island arc → Volcanic arc or island arc (andesite/diorite)

Partial melting / Subduction zone or collision zone → Intermediate/felsic magma → *Crystallization* → Continental plutonic rocks and/or melting of deep sedimentary rocks (rhyolite/granite)

FIGURE 5.16 Igneous evolution is driven by magma differentiation, including crystallization and partial melting, which segregates parent material into chemically distinct products.

At which step in this sequence would continental crust begin to develop?

fractions: 1) basalt or gabbro magma that forms new oceanic crust, and 2) ultramafic residue left behind in the upper mantle.

Later, when the same basalt and gabbro oceanic crust is recycled during subduction, hot water and other volatiles escaping from the slab lower the melting point of the overlying upper mantle, leading to partial melting. The resulting magma, now mafic to intermediate in composition, intrudes into the crust. Partial melting of the crust may lead to the formation of intermediate to felsic magma.

If time and conditions allow, magma that has intruded into continental crust can continue to crystallize, and intermediate magma produced above a subduction zone can itself differentiate into felsic magma (granite), leaving behind a crystal residue that is more mafic than the intermediate rock.

Geologists also refer to this process of differentiation as *fractionation*, meaning that chemically distinct fractions of the parent magma are formed. Fractionation continues until everything that can be fractionated from the original magma body has been removed. The process is complete when rocks have been formed. These rocks will not remain solid forever, however; they will enter the rock cycle and be weathered, metamorphosed, or even remelted.

Magma Evolution Is Related to Tectonics

Fractionation occurs mainly at divergent and convergent plate boundaries because both settings have buoyant magma that rises and crystallizes (**FIGURE 5.17**).

At divergent plate boundaries (spreading centers), ultramafic parent magma from the upper mantle fractionates to form mafic rocks—intrusive gabbro and extrusive basalt.

At convergent plate boundaries, partial melting of upper mantle rock above subducting oceanic lithosphere (composed of basalt and gabbro) generates intermediate magmas, such as diorite and andesite, but may eventually create felsic magmas, such as granite and rhyolite.

Igneous fractionation is an extremely important process, as it is responsible for the formation of volcanic and island arcs, the seafloors, the continents, landmasses at hotspots, and other distinct crust types. One implication of this process is that Earth originally had no continents and that the total size of the continents has increased over geologic time as a result of igneous evolution into felsic products. If we imagine an Earth with no continents, it is easy to appreciate the importance of igneous evolution to all Earth's systems.

A final outcome of igneous evolution is that different types of igneous rock occur in different places on Earth, and the distribution of rock types is related to plate tectonics processes and to Earth's history. In general, continents are made of felsic igneous rocks (granite); oceanic crust is made of mafic igneous rocks (basalt and gabbro); and volcanic and island arcs are made of intermediate igneous rocks (diorite and andesite).

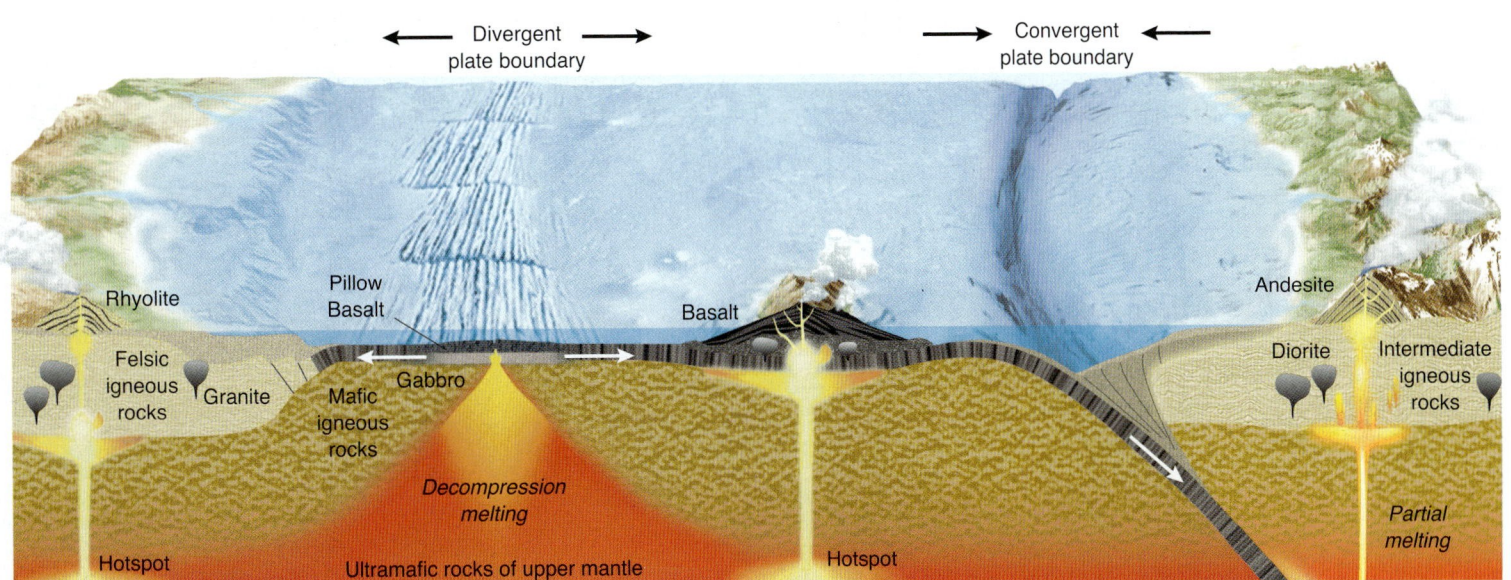

FIGURE 5.17 Igneous rock is a ubiquitous component of Earth's crust because it evolves as a product of tectonic processes.

Ⓠ Identify where partial melting is taking place in this illustration, and describe the chemistry of the parent and the fractionated portions.

Ⓠ **EXPAND YOUR THINKING**

After magma differentiates, explain why the liquid fraction is more felsic than the parent. How would plate tectonics be different if the liquid fraction were more mafic and the solid residue more felsic?

CRITICAL THINKING

Igneous Environments

Working with a partner, and using **FIGURE 5.18**, complete the following exercise:

1. Add labels to the figure identifying all igneous processes and environments and plate tectonics processes and environments. Describe the igneous environments.

2. Summarize tectonic processes that are responsible for each environment you identified.

3. Identify locations where each of the seven most common igneous rocks is likely to be found.

4. Describe the texture and composition of each type of rock, and explain why these characteristics occur in that environment. Specifically, identify which igneous processes in each environment produce the unique texture and composition of each type of rock?

5. What suggestions might you offer to the governor of a state with a volcanic arc for reducing vulnerability to volcanic hazards?

6. A shield volcano at a hotspot is on a moving plate. Describe its long-term fate.

7. Imagine a volcano made of rhyolite above a hotspot. What can you infer about the magma source and the geologic setting?

Stratovolcano

Volcanic arc

Mid-ocea

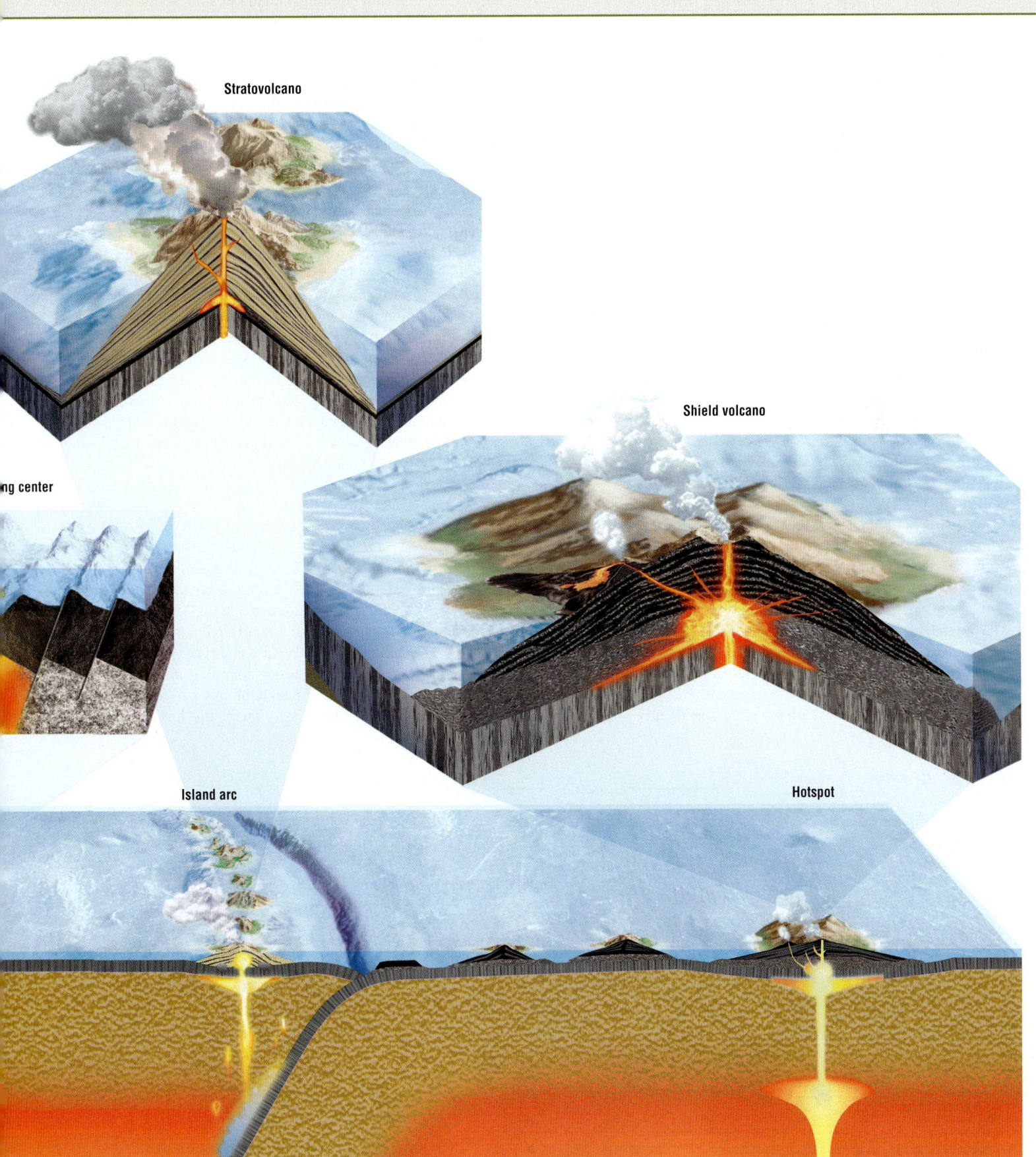

Stratovolcano

Shield volcano

ng center

Island arc

Hotspot

5-8 Basalt Forms at Both Spreading Centers and Hotspots

LO 5-8 Describe the environments where basalt accumulates, and explain why.

Magma actively intrudes Earth's crust throughout the world, forming various kinds of igneous rock. These processes are not random; rather, tectonic activity creates the specific geologic conditions necessary for intrusive and volcanic processes to occur.

Spreading Centers

Magma intrudes into the crust at spreading centers where two plates diverge, forming a rift. As the plates spread apart, hot rock from the asthenosphere flows upward and fills voids in the fractured lithosphere as it separates. At the beginning of this process, mantle rock is solid and ultramafic in composition, probably composed of peridotite.

Upward movement adds thermal buoyancy to the ridge area; in other words, the hot, less-dense rock rises under the cold crust, pushing upward and causing it to bulge up. This is the reason spreading centers in ocean basins are characterized by ridges rising 1000 meters or more above the surrounding seafloor (**FIGURE 5.19**).

As ultramafic rock from the upper mantle rises to shallow levels, it decompresses, and the lighter compounds in it melt to produce mafic magma. Magma that rises beneath a spreading center shapes the intrusive igneous rock gabbro. Magma that rises through fissures in the crust to the surface forms its volcanic equivalent, basalt. Basalt eruptions are typically *fissure eruptions*, whereby lava exits from cracks in the oceanic ridge.

The many pathways that basalt takes in its upward journey to the seafloor form dense dikes, more specifically referred to as *sheeted dikes* to reflect their formation of vertical columns and curtains of intrusive rock in thick sequences, like cards in a deck stood on end. Basalt erupting onto the seafloor is quickly quenched, or "frozen," by cold seawater, and generates submarine deposits of *talus* (in this case, composed of broken glass and basalt ash) and bulbous rocks called **pillow lava** (**FIGURE 5.20**).

The high heat flow at mid-ocean ridges produces numerous **hydrothermal vents** in the seafloor. These vents open when cold seawater seeps downward through cracks in the crust and meets the hot intrusive rock below.

The water warms as it nears hot rock or a magma body and becomes chemically enriched with dissolved metallic cations that it leaches from the surrounding crust. When the water reaches a critical temperature and turns into steam, it is forced back toward the surface and erupts as a hot chemical spring.

Some of these springs are known as **black smokers,** so named for their release of dark, billowing clouds of metal-rich particles. Massive, economically valuable *metallic sulfide deposits* crystallize out of the hot water emerging from the vent.

As this process continues over millions of years, it builds new oceanic lithosphere consisting of intrusive gabbros in the lower part, sheeted vertical dikes, and basaltic lavas of glassy talus and pillows in the upper portions. Interlayered within the basaltic lava are mineral-rich metallic sulfide deposits associated with fields of black smokers (see Chapter 21).

Mantle Plumes (Hotspots)

Several exceptionally active sites of plutonism and volcanism are found at hotspots located far from plate boundaries. One hypothesis suggests that massive plumes of anomalously hot mantle rock underlie active hotspots. These mantle plumes (**FIGURE 5.21**) appear to originate in the lower mantle and rise (presumably, slowly) because they are less dense (hotter) than the surrounding rock.

Research suggests that a mantle plume rises as a plastically deforming mass with a bulbous head. The plume head is fed by a long,

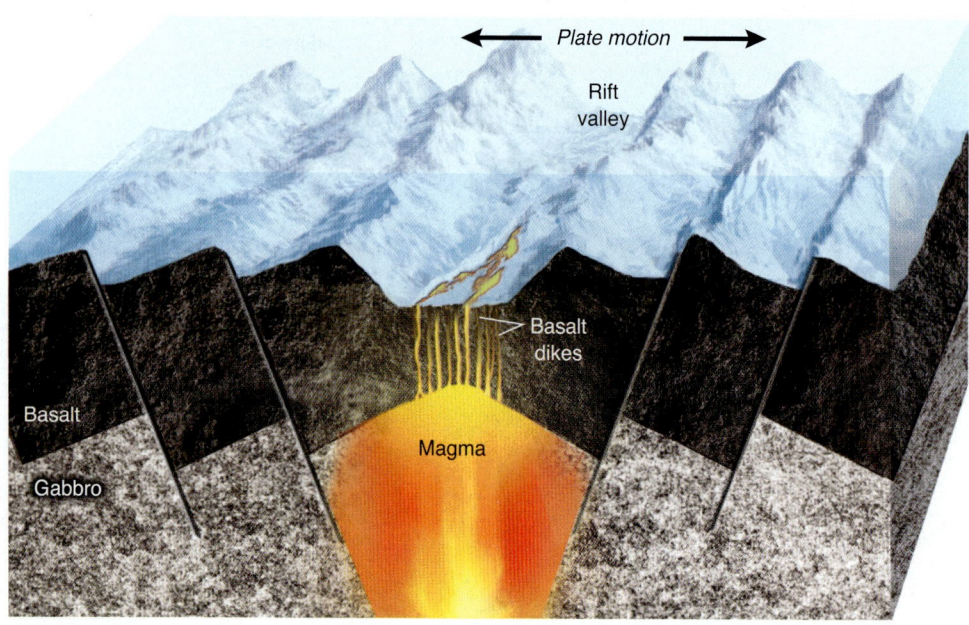

FIGURE 5.19 New seafloor is created at divergent boundaries where peridotite from the upper mantle rises into the crust and partially melts to produce mafic magma. This is the source of the gabbro at the base of the crust and the basalt dikes that make up the shallow seafloor.

Why is the upper crust composed of basalt and the lower crust composed of gabbro?

NOAA

FIGURE 5.20 Pillow lava forms when hot lava meets cold seawater. The outer edge of the lava chills in the water (making glass) while the interior remains hot and pushes forward, stretching the glass layer into a round, bulbous shape.

Ⓠ **Why are pillow lavas made of glass?**

narrow tail, so that the overall shape resembles that of a tadpole. As the head encounters the base of the lithosphere, it spreads outward into a mushroom shape. As a mantle plume rises, it experiences a decrease in pressure that can lead to melting in its upper part. Decompression melting of these hot rocks can generate huge volumes of basalt magma that feed plutons at shallow depths in the crust.

At the surface, massive **large igneous provinces** composed of *flood basalts* take shape above mantle hotspots in either continental or oceanic settings. (These are discussed more fully in Chapter 6, "Volcanoes.") Large igneous provinces contain huge volumes of basalt released over millions of years and may create immense plateaus covering thousands of square kilometers. The Columbia River Plateau in Idaho, Washington, and Oregon is one example.

Geologists still have much to learn about hotspots. Although most of them accept the hotspot concept, the number of hotspots worldwide, and exactly how they are formed, are still subjects of debate and controversy.

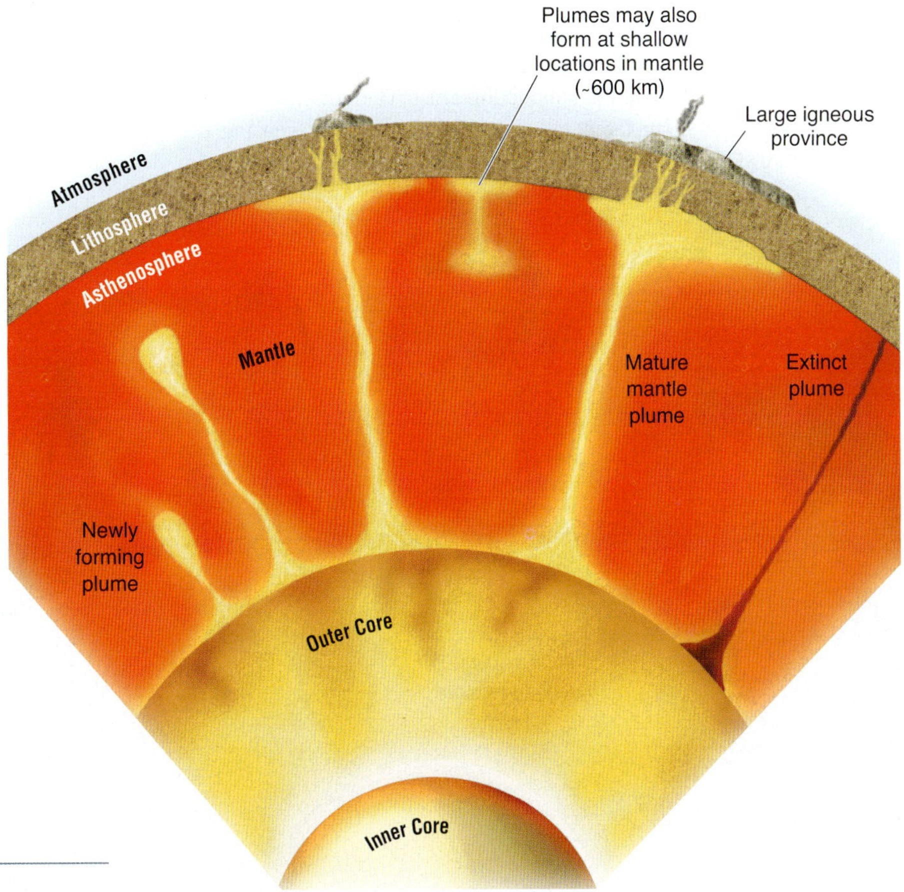

Ⓠ **EXPAND YOUR THINKING**

Would you expect magma at a spreading center to have the same composition as magma at a hotspot? Why or why not? What type of magma will feed a volcano where a hotspot intrudes continental crust?

Plumes may also form at shallow locations in mantle (~600 km)

Large igneous province

Atmosphere

Lithosphere

Asthenosphere

Mantle

Mature mantle plume

Extinct plume

Newly forming plume

Outer Core

Inner Core

FIGURE 5.21 Conceptual model of mantle plumes. Several plumes, at different locations and in various stages of formation, may be active at the same time.

Igneous Intrusions Occur in a Variety of Sizes and Shapes

5-9

LO 5-9 Describe intrusions in a variety of geologic settings.

Magma that intrudes the crust is forced to assume whatever shape the environment dictates: from individual bodies that intrude localized parts of the crust to massive units that occupy hundreds of cubic kilometers and form the cores of mountain ranges.

Magma Production

As a descending plate bends downward during subduction, it creates the long, arcing depression in the seafloor of an *oceanic trench*. These trenches are the deepest topographic features on Earth's surface.

Because continental crust is too buoyant, a plate will subduct only if it is composed of oceanic crust. Deposited on the basalt crust is *marine sediment* that contains large volumes of seawater within its countless pores. Large volumes of water also reside within the crystal structure of certain minerals (called *hydrated minerals*) in the weathered basalt.

As a plate subducts, it encounters progressively higher temperatures and pressures. These force water out of the descending slab and into the mantle lying above the subducting crust. The presence of this water lowers the melting temperature of the mantle rock, causing it to melt (**FIGURE 5.22**). The resulting magma is depleted in iron and magnesium (thus, it is more felsic) compared to the ultramafic rock of the upper mantle; hence, it varies from basalt to andesite in composition. As the magma rises into the overriding plate, differentiation produces intrusive bodies of varying composition that feed a belt of volcanoes aligned parallel to the trench.

If the overriding plate is composed of continental crust, the resulting chain of volcanoes is a *volcanic arc*. Examples include the Cascade Range in the U.S. Pacific Northwest and the Andes in South America.

If the overriding plate is composed of oceanic crust, the resulting chain is an *island arc*. The Aleutian Island chain in the North Pacific is an example of an island arc.

Lava at subduction zones is typically basaltic at oceanic island arcs and andesitic at continental volcanic arcs. This difference occurs because basaltic magma rising through a continental plate to form a volcanic arc is initially surrounded by felsic rocks and grows more felsic (turns into andesite) as a result of differentiation.

Basaltic magma rising into an overriding plate of basalt at an ocean-ocean boundary experiences less chemical change because the surrounding rocks are similar to it in composition.

How Plutons Are Formed

When magma intrudes into cold crustal rocks, several things happen. First, the magma creates space for itself by wedging open the overlying rock layers, breaking off large blocks of rock. If these blocks of crust or mantle rocks are enclosed within the melted magma, they are called *xenoliths* (from the ancient Greek *xenos*, meaning "foreign").

Second, heat from the magma melts the walls of the intruded crustal rock. This melting adds new chemicals to the liquid (magma assimilation).

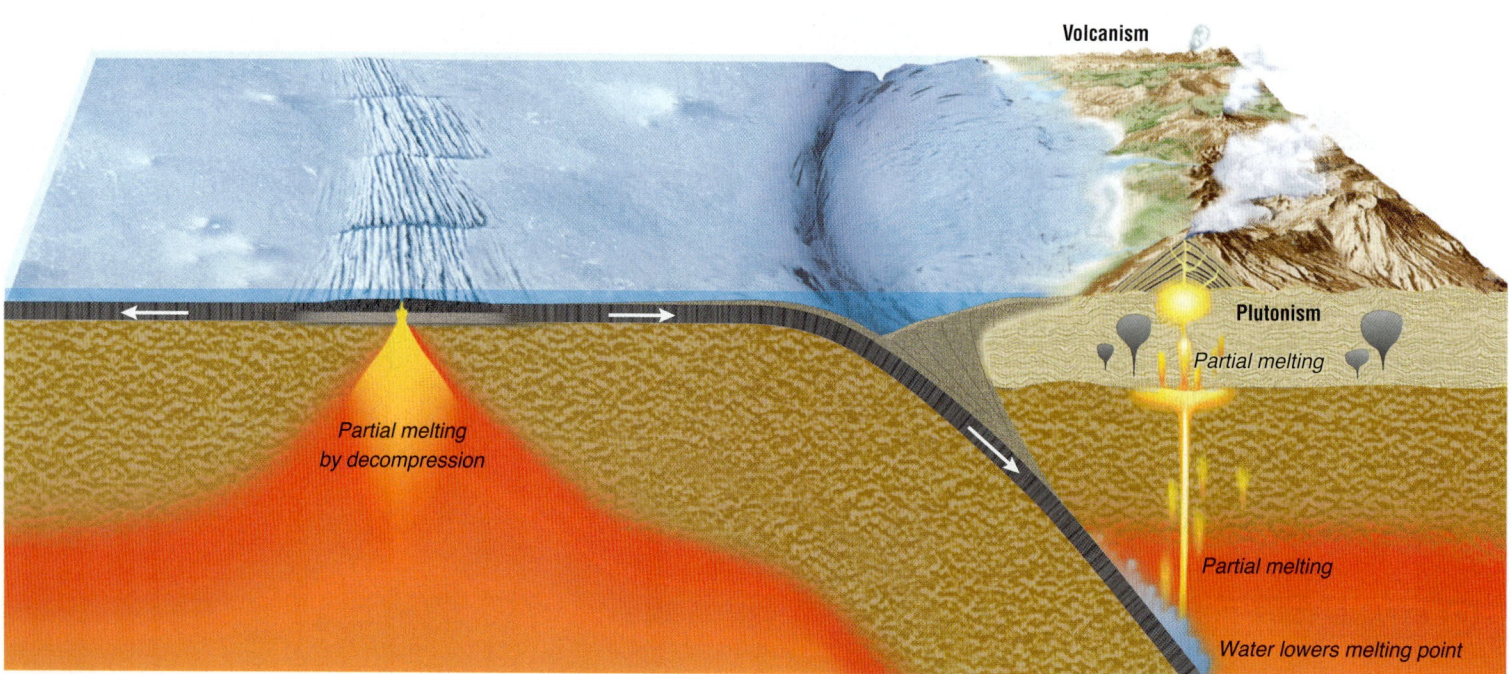

FIGURE 5.22 Subduction zones and spreading centers are areas in which magma is formed as a result of partial melting, plutonism, and volcanism.

Describe the role of water in producing magma.

Third, heat from the magma bakes the surrounding crust for distances that can range from meters to kilometers and alters its mineral composition and texture (a process called *contact metamorphism*).

Plutons occur in a variety of shapes, from individual bodies that intrude localized parts of the crust (such as dikes and sills) to massive units that occupy hundreds of cubic kilometers and form the cores of mountain ranges (recall from earlier that intrusions this large are called batholiths).

Some plutons spend their entire "lives" underground, but others may become exposed at the surface as a result of erosion of the overlying rocks. This can happen during prolonged weathering that removes the overlying rock layers, a process that is enhanced by tectonic forces causing the pluton itself to slowly shove the crust upward. This leads to accelerated erosion, because as the crust lifts to higher elevations, the gradient of the land steepens, providing running water and gravity with more erosive energy.

Miners refer to intruded crust as **country rock,** a term geologists have adopted. As magma intrudes into country rock, it takes whatever shape the environment dictates. (Think of squeezing toothpaste between your clenched teeth.) As the magma cools, the intrusive rock bodies retain that shape. Geologists identify these bodies by their shape and orientation, as summarized in **FIGURE 5.23**.

The massive intrusion at the bottom of the figure is a batholith (note the xenoliths incorporated into the magma); intrusive bodies smaller than batholiths (less than about 100 square kilometers) are *stocks*.

When magma pushes up rock layers into a dome and then cools in that shape, a *laccolith* is formed. Study the dike in the figure. It consists of magma that followed fissures and cracks in the overlying crust, generally vertically (or nearly so), and hardened into igneous rock. Note that the dike cuts across *strata* in the country rock. As the magma encountered different conditions, it intruded horizontally parallel to those strata, taking shape as a *sill* (like a windowsill).

Batholiths, stocks, and dikes are considered *discordant* intrusions because they cut across the layers of country rock into which they intrude. Sills and laccoliths are *concordant* intrusions because their boundaries are essentially parallel to those of the layers of country rock into which they intrude. All are plutons.

Magma that pushes to the surface feeds a volcanic eruption through a *volcanic pipe*. Volcanologists often refer to the "plumbing" of a volcano because magma can flow through complex systems of pipes and fissures to feed surface eruptions.

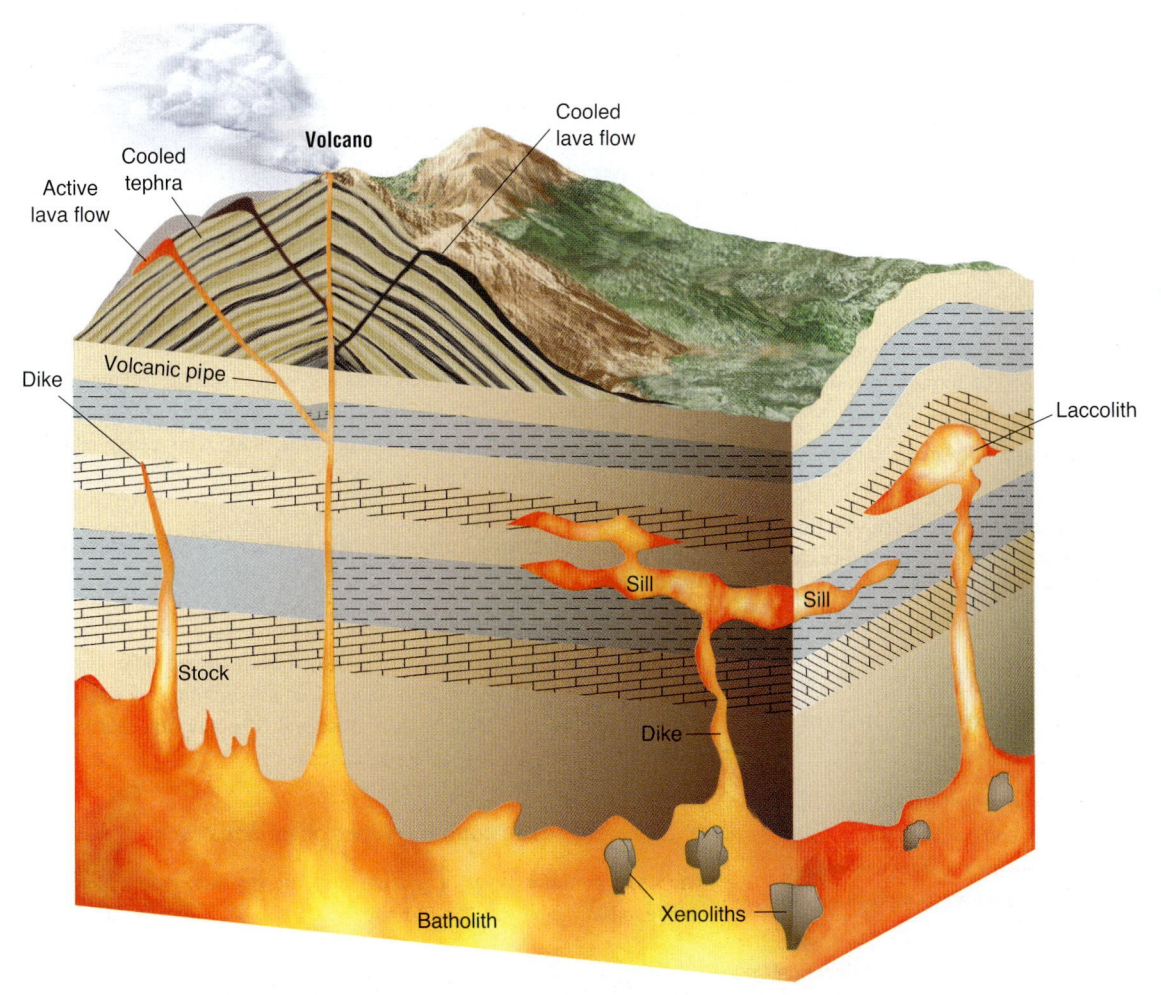

FIGURE 5.23 Plutons can take various shapes, depending on the volume of the magma, the force of the intrusion, and the structure and strength of the surrounding crust.

Describe the influence that xenoliths will have on magma chemistry.

Q EXPAND YOUR THINKING

Do you agree or disagree that, over time, Earth's crust is growing more felsic and the upper mantle is growing more ultramafic? Why or why not?

LET'S REVIEW

Three important concepts emerge from our understanding of igneous rock formation: 1) The rock cycle renews and recycles rocks, sediments, and other materials of the crust; 2) because igneous rock is the most abundant rock on the planet, melting and crystallization are important and widespread processes in the rock cycle; and 3) melting and crystallization occur in various tectonic environments, including spreading centers, subduction zones, and hotspots. As a result, igneous rock occurs in a wide variety of types and is formed in diverse environments.

Despite this diversity, all igneous rock originates from a single type of parent rock, the peridotite of the upper mantle. Through partial melting, this parent rock produces magma that is basaltic in composition, which in turn evolves into other types of magma. In the very earliest phases of Earth's history, igneous rock constituted 100% of the planet. Since then all rock—igneous, sedimentary, and metamorphic—has been produced by igneous evolution. All rock is related to, and derived from, the first molten rock on the ancient planet. Why? Because melting recycles Earth materials.

STUDY GUIDE

5-1 Igneous rock is formed when molten, or partially molten, rock solidifies.

- Four-fifths of Earth's crust is composed of **igneous rock.** Igneous rocks result from melting in the upper mantle and crust.

- Volcanic igneous rock consists of **lava** and other volcanic products that are extruded at the surface of the crust or at very shallow levels below the surface, forming **extrusive** igneous rock. **Intrusive** igneous rock is formed when magma crystallizes within the crust without being exposed to the cool temperatures of the atmosphere.

5-2 Igneous rock is formed through a process of crystallization and magma differentiation.

- Rifting of the crust lowers the pressure in the upper mantle, and the resulting **decompression melting** causes magma to be formed. This magma migrates into the crust and interacts with it as it crystallizes.

- Rocks undergo **partial melting,** which means that silica-rich compounds in the rock melt before other compounds.

- **Magma differentiation** is caused by magma mixing, crystal settling, magma assimilation, and magma migration. Through these processes, magmas of differing composition evolve from single-parent magma. In most cases, differentiation produces magma that is richer in silica than the parent magma.

5-3 Bowen's reaction series describes the crystallization of magma.

- **Bowen's reaction series** describes the order in which the minerals in a magma body crystallize. In magma of average composition, the crystallization process begins with **mafic** minerals (those high in iron, magnesium, and calcium), continues with minerals of **intermediate** composition, and ends with **felsic** (silica-rich) minerals.

- There are two branches of crystallization: 1) Discontinuous crystallization occurs when minerals formed at higher temperatures react chemically with the surrounding magma and recrystallize to form a mineral with a new crystalline structure. 2) Continuous crystallization involves progressive changes in plagioclase feldspar as it evolves from calcium-rich to sodium-rich composition.

5-4 The texture of igneous rock records its crystallization history.

- The texture of an igneous rock depends on the size of the crystals within it, which, in turn, is influenced by the rate at which the rock cooled. Igneous rocks that cool slowly tend to contain large mineral grains and to have a coarser texture (called phaneritic) than those that cool rapidly. These rocks are intrusive or plutonic. Igneous rocks that cool rapidly on Earth's surface usually contain smaller mineral grains and have a finer texture (called aphanitic). These rocks are extrusive or volcanic. Igneous rocks may also be glassy, pyroclastic, vesicular, or porphyritic.

- Minerals at the top of Bowen's reaction series (such as pyroxene and amphibole) tend to be dark in color and mafic to **ultramafic** in composition because they are rich in iron and magnesium, which tend to be dark.

- Minerals at the bottom of Bowen's reaction series (such as sodium plagioclase, potassium plagioclase, and quartz) tend to be light in color and felsic in composition. Geologists use these color trends, along with texture information, to identify and name igneous rocks.

5-5 Igneous rocks are named on the basis of their texture and composition.

- The classification system for igneous rocks tells us three important things: 1) Intrusive and extrusive igneous rocks with the same chemical composition contain the same minerals; 2) silica content decreases as iron and magnesium content increases; and 3) potassium and sodium content decreases as silica content decreases.

5-6 There are seven common types of igneous rock.

- The seven common types of igneous rock are rhyolite, granite, andesite, diorite, basalt, gabbro, and peridotite.

5-7 All rocks on Earth have evolved from the first igneous rocks.

- Igneous rocks were the first rocks on Earth. All other rocks have evolved from parent igneous rocks that were formed over 4 billion years ago. Igneous evolution has produced the great diversity of rock on Earth today.

- The igneous rock system evolves in a series of steps. In each step, the parent igneous rock differentiates into two fractions, each with a different composition from that of the parent. The two fractions are: 1) a felsic fraction of compounds with lower melting points, typically high in silica relative to the parent, and 2) a mafic fraction of solid minerals (i.e., rock that is high in iron, magnesium, and calcium minerals relative to the parent rock).

5-8 Basalt forms at both spreading centers and hotspots.

- Tectonic activity creates the geologic settings and conditions necessary for intrusive and extrusive processes to occur.

- Magma intrudes into the crust at spreading centers where two plates diverge. Oceanic crust consists of several layers of igneous rock. Fed by magma generated by decompression melting in the peridotite upper mantle, gabbro crystallizes at the base of the crust. Above the gabbro are sheeted dikes of basalt; and above these are layers of pillow basalt, glassy talus, and metallic sulfide deposits.

- Several active sites of plutonism and volcanism are located at hotspots far from plate boundaries. It is hypothesized that a massive plume of anomalously hot mantle rock underlies active hotspots. Plumes rise slowly through the mantle due to their positive buoyancy with respect to the surrounding rock.

KEY TERMS

black smokers (p. 142)
Bowen's reaction series (p. 129)
country rock (p. 145)
decompression melting (p. 127)
dikes (p. 136)
extrusive (p. 125)
felsic (p. 130)
hydrothermal vents (p. 142)

igneous evolution (p. 125)
igneous rock (p. 124)
intermediate (p. 130)
intrusive (p. 125)
large igneous provinces (p. 143)
lava (p. 124)
mafic (p. 129)
magma (p. 124)

magma differentiation (p. 127)
partial melting (p. 127)
pillow lava (p. 142)
plutonic (p. 125)
pyroclastic debris (p. 124)
ultramafic (p. 129)
volcanoes (p. 125)

ASSESSING YOUR KNOWLEDGE

Please complete this exercise before coming to class. Identify the best answer to each question.

1. What is igneous rock?
   a. Rock produced by crystallization of magma
   b. Rock composed of sediments
   c. Rock derived from pressure
   d. Rock that mixes the mantle and crust
   e. None of the above

2. These are igneous rocks. Rock (a) is an _____ (intrusive, extrusive) igneous rock and rock (b) is an _____ (intrusive, extrusive) igneous rock.

Mineral images Courtesy of Chip Fletcher

3. Most melting in the mantle is a result of:
   a. High-pressure melting.
   b. Decompression melting.
   c. Sudden increases in temperature.
   d. Turbulent mantle plumes.
   e. None of the above.

4. In most cases, magma differentiation produces magma with higher _____ content than the parent magma.
   a. Aluminum
   b. Silica
   c. Pyroclastic
   d. Mineral
   e. None of the above

5. Magma from the mantle is relatively enriched in _____ compared to magma made from continental rocks.
   a. silica
   b. felsic compounds
   c. quartz
   d. sodium
   e. none of the above

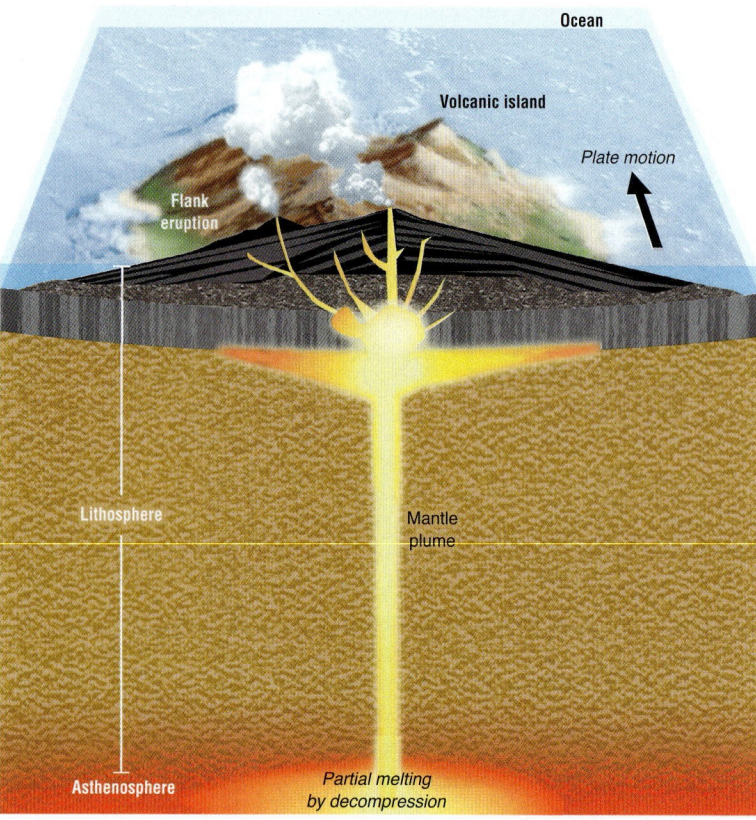

9. Identify the texture on each of the following rocks.
   a. _____ (aphanitic, phaneritic)
   b. _____ (aphanitic, phaneritic)

10. Identify the composition of each of the following rocks.
    a. _____ (mafic, intermediate)
    b. _____ (mafic, intermediate)

(a)   (b)   Courtesy of Chip Fletc

11. Fill in the blanks with the appropriate texture terms:
    a. _____ Two sizes of crystals
    b. _____ Fused, glassy shards
    c. _____ Small crystals
    d. _____ Many small openings produced by escaping gas

6. Magma that is cooling undergoes:
   a. Crystallization.
   b. Recrystallization.
   c. Partial melting.
   d. Refractionation.
   e. Erosion.

12. *Mafic* means_____; *felsic* means_____.
    a. High in iron, magnesium, and calcium; high in silicon and oxygen
    b. High in calcium and magnesium; high in silicon, oxygen, and iron
    c. High in iron and oxygen; high in silicon, calcium, and magnesium
    d. High in silicon, oxygen, and calcium; high in iron and magnesium
    e. Volcanic; plutonic

7. Bowen's reaction series describes:
   a. The sequence in which minerals melt in rapidly heating magma.
   b. The sequence in which plutons are formed in migrating magma.
   c. The sequence in which rocks are formed in average continental crust.
   d. The sequence in which minerals crystallize in cooling magma.
   e. None of the above.

13. The composition of dark igneous rock is likely to be:
    a. Felsic.
    b. Mafic.
    c. Rhyolitic.
    d. Plutonic.
    e. None of the above.

14. List the seven common types of igneous rock.
    a. _____
    b. _____
    c. _____
    d. _____
    e. _____
    f. _____
    g. _____

8. The order of mineral crystallization is, typically:
   a. Felsic, mafic, intermediate, ultramafic.
   b. Felsic, intermediate, mafic, ultramafic.
   c. Ultramafic, mafic, intermediate, felsic.
   d. Mafic, ultramafic, felsic, intermediate.
   e. All of the above.

15. Which of the following best describes igneous evolution?
    a. All rocks evolved as a result of hotspots.
    b. All rocks evolved as a result of spreading-center volcanism.
    c. All rocks evolved as a result of differentiation of early igneous rocks.
    d. All rocks are a result of meteorite impacts.
    e. None of the above.

16. Which of the following statements is/are correct? (There may be more than one correct answer.)
    a. Granite is formed at spreading centers.
    b. Andesite is formed at subduction zones.

Mineral images Courtesy of Chip Fletcher

c. Basalt is formed at hotspots.

d. Gabbro is formed at spreading centers.

e. None of the above.

17. Label this block diagram, depicting various plutonic bodies, using the following terms:

a. Dike

b. Volcanic neck

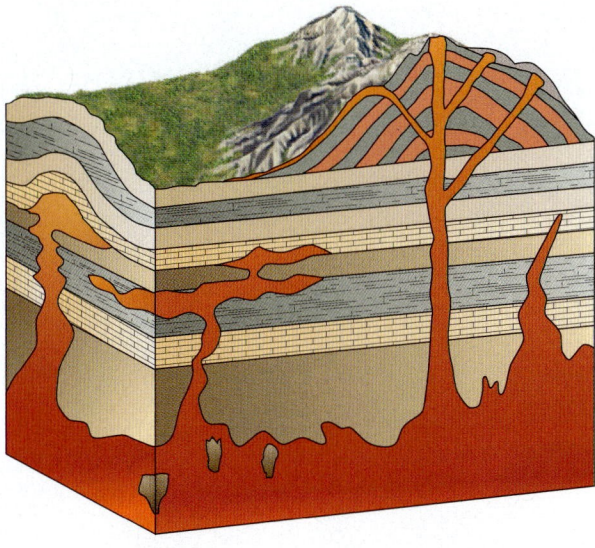

c. Sill

d. Stock

e. Batholith

f. Xenoliths

18. Volcanic arcs are typically composed of:

a. Granite and gabbro.

b. Gabbro and peridotite.

c. Rhyolite and andesite.

d. Andesite and diorite.

e. All of the above.

19. Plutons are:

a. Magma bodies within the deep crust.

b. Intrusive igneous rocks in the core.

c. Magma bodies produced by volcanism.

d. Igneous rocks produced by fissure eruptions.

e. Made by contact metamorphism.

20. Plate tectonics is important to igneous evolution because:

a. Plate tectonics formed the first igneous rocks billions of years ago.

b. Melting does not occur at plate boundaries.

c. Mantle plumes occur only at spreading centers.

d. Plate tectonics provides for many igneous environments.

e. Plate tectonics does not allow for partial melting.

FURTHER RESEARCH

1. Which tectonic environment leads to the production of granite?

2. Which tectonic environment leads to the production of basalt?

3. What is pillow basalt, and how is it formed?

4. On a map of the world identify several locations where you would expect to find hydrothermal vents.

ONLINE RESOURCES

Explore more about igneous rocks on the following Web sites:

Read stories and view photos about igneous rocks at:
www.usgs.gov/science/science.php?term=572&type=theme

Learn about igneous features at Big Bend National Park at:
www.maroon.com/bigbend/ig/index.html

Additional animations, videos, and other online resources are available at this book's companion Web site:

www.wiley.com/colleges/fletcher
This companion Web site also has more information about WileyPLUS and other Wiley teaching and learning resources.

6

VOLCANOES

CHAPTER CONTENTS & LEARNING OBJECTIVES

GEOLOGY IN OUR LIVES

A volcano is any landform from which lava, gas, or ashes escape from underground, or have done so in the past. Volcanoes are a significant component of the rock cycle: They rejuvenate the crust, renew soil, impact global climate, and contribute to gases in the atmosphere. But volcanoes also present hazards. Eruptions may consist of: violently explosive columns of gas and ash that penetrate high into the atmosphere, lava flows that burn and bury whatever they encounter, and avalanching pyroclastic flows of scalding fumes and glass shards that travel like speeding locomotives down volcanic slopes. To avoid volcanic hazards and utilize volcanic resources, it is important to understand how volcanoes work.

Q We learned from Chapter 5 that magma (and lava) can be felsic, intermediate, or mafic. How does magma chemistry influence the nature of volcanic eruptions?

Alan Becker/The Image Bank/Getty Images, Inc.

149

6-1 A Volcano Is Any Landform from Which Lava, Gas, or Ashes Escape from Underground, or Have Done So in the Past

LO 6-1 Define the term volcano and explain why geologists study volcanoes.

Volcanoes, familiar to many of us through popular media, are complex geologic phenomena that impact global climate, threaten major cities, and contribute critical natural resources.

Few geologic processes are more familiar to the average person than a volcanic eruption. We frequently see erupting volcanoes in movies, on TV, and on the Web, and read about them in newspapers and magazines. But despite this familiarity, the average person does not know how volcanoes work, why they erupt, or why they are located where they are. We will learn the answers to these questions in this chapter.

We will also learn that an important clue to a volcano's origin and eruptive processes is found in its shape and size.

Volcanoes

Not many spectacles in nature are as awe-inspiring as an erupting volcano. That is why volcanoes have long figured in mythology, worldwide. In European culture, for example, people once believed that Vulcan, the Roman god of fire and metalworking, worked a subterranean forge whose chimney was the tiny island of Vulcano in the Mediterranean Sea north of Sicily. According to the myth, Vulcano's hot lava fragments and dark ash clouds burst from Vulcan's furnace as he beat out thunderbolts for Jupiter, the king of the gods, and weapons for Mars, the god of war. Our word "volcano" is derived from the name of this island.

Today we know that Vulcano and its sister islands Stromboli and Lipari are fed by partial melting in the upper mantle as the Nubian Plate subducts beneath the southern margin of the Eurasian Plate, creating an island arc of active volcanoes.

In Hawaiian culture, people attribute eruptions to Pele (PEL-lay), the goddess of volcanoes. Hawaii is home to one of the world's longest continuously erupting volcanoes, Kilauea. Since 1952, Kilauea has erupted 34 times; and since January 1983, the eruption has continued without interruption. All told, Kilauea ranks among the world's most active volcanoes, and may even top the list.

What Is a Volcano?

Volcanoes come in many shapes and sizes, but generally speaking a **volcano** is defined as any landform that releases lava, gas, or ashes, or has done so in the past. (In the Geology in Our Lives feature, "Mount St. Helens," you will read about America's most famous volcano.) Hot materials escape from an opening called a **vent,** or fissure. Often, a cone of volcanic rock builds up around the vent. The cone is generally composed of accumulated erupted materials, such as ash, pumice, and lava. The cone may be a few meters high or an entire mountain. The term *volcano* can refer both to the vent (opening) and to the cone.

In this chapter we examine three basic types of **central vent volcanoes** (classic volcanoes built around a central vent) and three varieties of **large-scale volcanic terrains** formed by volcanic action.

GEOLOGY IN ACTION

I will never forget seeing lava for the first time as a young professor at the University of Hawaii. I took a group of students on a field trip to Kilauea in Volcano National Park on the Big Island of Hawaii. We hiked across several kilometers of black crunchy rock only weeks old that broke into glassy fragments underfoot. The landscape was barren and bleak, and the air was rank with the odor of sulfur. Clouds of steam boiled up from cracks in the ground.

Is this what Earth looked like 4 billion years ago?

The lavas of Kilauea flow like syrup. Although they are very hot, by taking great care it is possible to approach small rivulets of molten rock.

Soon we eagerly assembled around a finger of glowing lava that crept forward like taffy. I was ecstatic. "Brand-new rock!" I shouted, and then talked excitedly about the formation of volcanic islands, ancient landscapes, mineral crystallization, plate tectonics, and the Hawaiian hotspot.

Wearing gloves and protecting myself from the heat (**FIGURE 6.1**), I scooped up a dollop of lava with my rock hammer and watched as it chilled to glass.

Volcanoes are indeed spectacular and full of mystery. They are natural phenomena that are continually being explored and studied by **volcanologists.**

FIGURE 6.1 Though the heat is intense, the lava at Kilauea Volcano in Hawaii can be approached cautiously and sampled by scientists.

What kind of lava is Kilauea Volcano made of?

Univertisy of Hawai'i photo by Robert Chinn

We also identify two sorts of lava (shiny, fluid *pahoehoe*, and dark, jagged *aa*), and discuss other volcanic products, a few of which are named in **FIGURE 6.2**. By the end of the chapter, you will understand that some eruptions are **explosive** and rip open entire mountainsides, while others are **effusive** and pour out fluid lava, adding to the land surface with relative calm. These different behaviors can be explained by the chemistry of the magma and its origin as revealed through our understanding of plate tectonics.

Why Study Volcanoes?

The most important human-interest issue related to volcanoes is safety. In many places, Earth's surface—both above and below sea level—is of volcanic origin, so it is little wonder that volcanoes pose such a hazard to life and are so widely studied. Scientists estimate that, today, the total population at risk from volcanoes is at least 500 million worldwide.

Since 1700, volcanic activity has killed more than 260,000 people, destroyed entire cities and forests, and severely disrupted local economies for months, and even years. The 1991 eruption of Mount Pinatubo in the Philippines caused atmospheric cooling of about 0.6°C, temporarily offsetting a decade-long global warming trend.

It is important, therefore, that scientists continue to improve their understanding of volcanoes and how they work. As just noted, millions of people are vulnerable to the effects of dangerous eruptions.

The subject of this chapter is how volcanoes are formed, how they behave, and why they are located where they are. Volcanoes attract attention because they are such dynamic and awe-inspiring displays of power.

Right now, volcanoes are active on every continent—even Antarctica—and on the floor of every major ocean (**FIGURE 6.3**).

© Bryan Lowry/Alamy

FIGURE 6.2 Hot lava flows into the Pacific Ocean after erupting from Kilauea Volcano in Hawaii. Volcanic products include gas, lava, and pyroclastic debris.

Ⓠ **What do think happens to the lava after it enters the ocean?**

Of more than 1,500 volcanoes considered active on Earth's land surface, an average of 10 are erupting each day. More than a dozen active volcanic islands populate the oceans. Many eruptions are small, but even these can be powerful and occasionally cause great damage.

Volcanoes occur not only on Earth, but on other planets and moons. Notably, Venus is highly volcanic; Mars used to be; and one of Jupiter's four big moons, Io, is the most volcanically active body in the solar system.

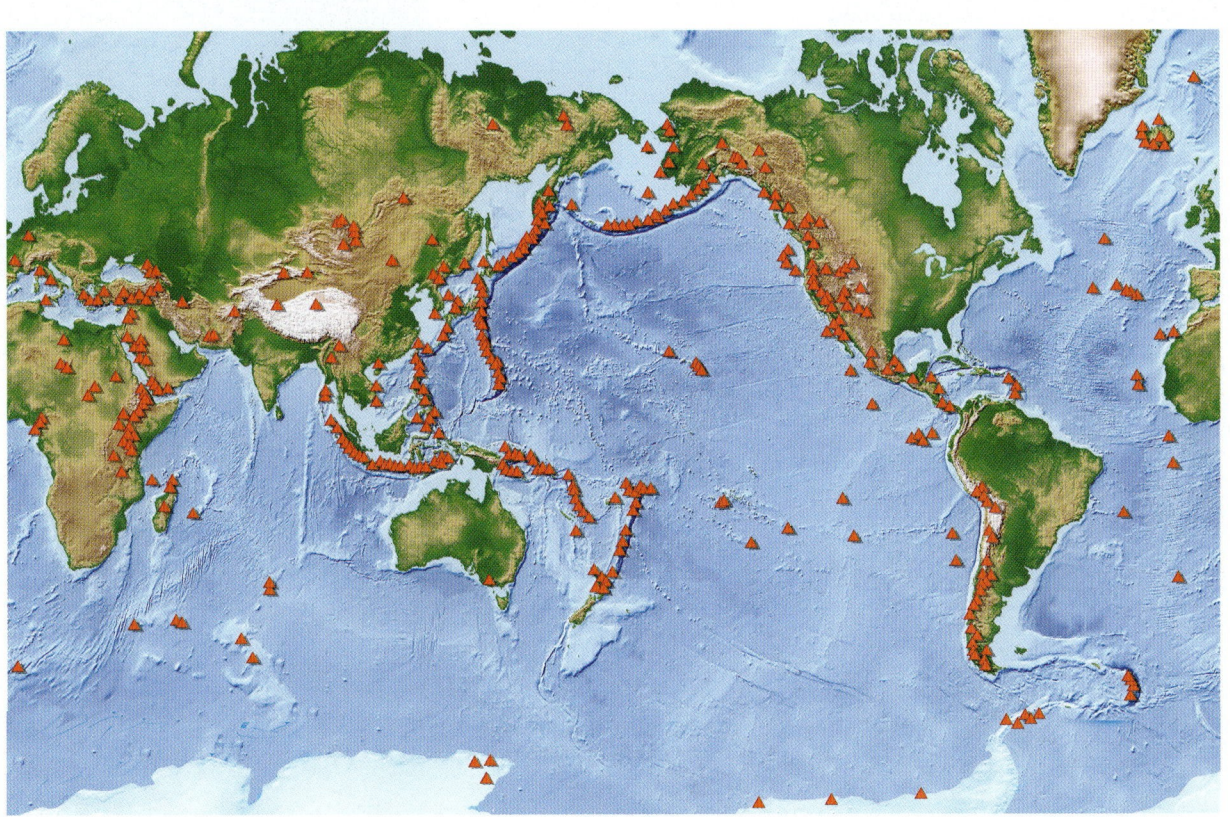

FIGURE 6.3 Global map of active volcanoes. *Source:* U.S. Geological Survey, Department of the Interior, USGS.

Ⓠ **Are the world's active volcanoes arranged in any recognizable pattern?**

Ⓠ **EXPAND YOUR THINKING**

What tectonic environments promote volcanism?

GEOLOGY IN OUR LIVES

Mount St.Helens

Before Mount St. Helens (**FIGURE 6.4**) erupted in 1980, few Americans knew of its existence. Today it is the best-known volcano in the United States. Lying 150 kilometers south of Seattle, Washington, in the Cascade Range, Mount St. Helens is now infamous for its catastrophic eruption, the deadliest and most costly in the history of the United States.

Despite its celebrity, however, Mount St. Helens is similar to other volcanoes in the Cascade Range. It is a large central-vent volcano built of interlayered beds of ash, pumice, and lava made of andesite, basalt, and a type of rock called *dacite* that is more felsic than andesite but more mafic than rhyolite. Thus, the composition of the magma that feeds Mount St. Helens varies across a chemical range.

Over the course of several months in the winter and spring of 1980, the north side of the mountain had been increasingly bulging outward.

Scientists knew that pressure was building and that an eruption was imminent. It occurred on May 18, when a magnitude 5.1 earthquake triggered a massive collapse of the north face of the mountain—the largest known debris slide in recorded history. The sudden collapse of the mountainside rapidly lowered the pressure on a large volume of magma within the volcano, and it exploded laterally in the direction of the debris slide.

The lateral blast marked the start of the eruption, which spawned a *pyroclastic flow* that rolled down the north slope of the volcano and buried over 600 square kilometers of land area, including houses, roads, pristine forest, and entire watersheds.

Soon, an eruption column of gas and ash pulsed skyward and lasted more than 9 hours, reaching a height 20 to 27 kilometers in the atmosphere. High-altitude winds carried ash to the east at an average speed of 100 kilometers per hour. More than 1.5 million metric tons of sulfur dioxide were released into the atmosphere, and in total the eruption ejected almost 3 cubic kilometers of material. The eruption reduced St. Helens' height by 400 meters and left a crater 1.6 kilometers wide and 800 meters deep, with its north end open in a huge breach.

FIGURE 6.4 Mount St. Helens is a large stratovolcano in the Cascade Range of the Northwest United States. The start of the unforgettable May 18, 1980, eruption occurred with a lateral blast that caused the largest debris slide in recorded history. An eruption column of gas and ash that lasted for over nine hours followed the blast.

Science Source/USGS/Getty Images

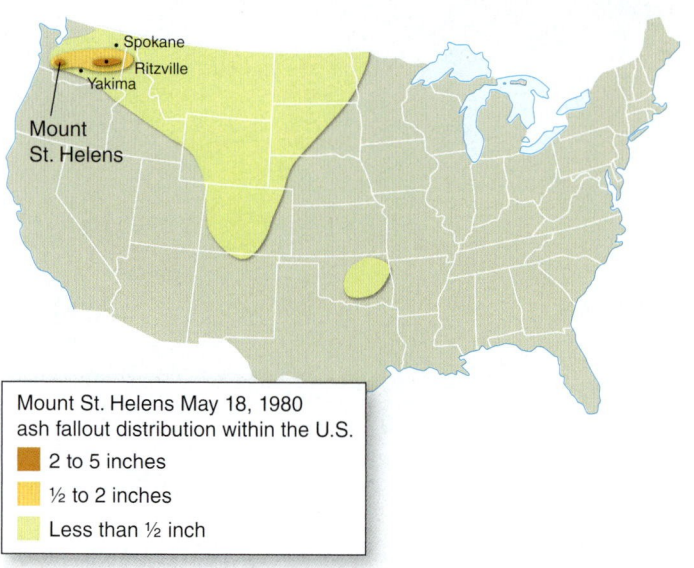

Mount St. Helens May 18, 1980 ash fallout distribution within the U.S.

- 2 to 5 inches
- ½ to 2 inches
- Less than ½ inch

There Are Three Common Types of Magma: Basaltic, Andesitic, and Rhyolitic

6-2

LO 6-2 Compare and contrast the three common types of magma.

The magma that feeds most volcanoes is basaltic, andesitic, or rhyolitic in composition.

Magma Chemistry

What are the products of volcanism? Lava is one, but there are others, of which **pyroclastic** ("fiery pieces") **debris** and volcanic gas are the most important. In this section we review the main types of magma that feed volcanoes; we discuss other volcanic products in later sections.

Lava, a type of magma, is fluid rock that comes from a vent, or fissure. It is also the name of the solid rock formed from cooling lava. Recall from Chapter 5 that we described three types of extrusive rock; these are also the main types of magma: **rhyolitic** (felsic in composition), **andesitic** (intermediate), and **basaltic** (mafic).

In studying volcanoes, researchers have found that three characteristics most influence the behavior of magma—and, therefore, volcanic processes and the shape and size of a volcano. These are: 1) the magma silica content, 2) its temperature, and 3) the amount of dissolved gas it contains.

High-silica magma tends to be rich in dissolved gases. It is viscous (thick and sticky) because it contains many tiny chains of silicate tetrahedra. Consequently, gas does not escape easily and tends to accumulate within the rock. High-silica magma does not flow easily and so will build steep slopes, resulting in tall volcanoes.

Low-silica magma has fewer sturdy silica bonds and therefore is characterized by low viscosity and lower levels of dissolved gas; it is runny and fluid. It is not possible to build a steep slope with such runny material. Consider pancake batter when poured: It quickly assumes the low, flat profile of a pancake. Low-silica magma tends to shape long, low-lying volcanoes with gentle slopes. The characteristics of the three magma types are summarized in **TABLE 6.1**.

Most magma is basaltic in composition. (The oceanic crust is basalt, and it represents over 70% of Earth's surface.) Basaltic magma is mafic; compared to felsic and intermediate rocks, it is high in magnesium and iron and low in silica. Basaltic magma is

hotter and more fluid than andesitic and rhyolitic magma. It contains fewer silica chains, so gases that come out of solution can rise unimpeded through the fluid magma, and there is no significant buildup of gas pressure. The outcome is a relatively gentle, effusive eruption.

Because basalt flows easily, some basalt eruptions release large volumes of lava. A good example of this is the long, sinuous rivers of lava issuing from Kilauea that flow for many kilometers across the Island of Hawaii (**FIGURE 6.5**).

Basaltic lava erupts from magma sources that are mafic or ultramafic in composition. Since most mantle sources are mafic, basaltic lava is found at oceanic hotspots, mid-ocean ridges, and ocean-ocean convergent volcanic arcs (because there is no continental crust, which tends to be relatively high in silica). Basaltic lava is even found at many ocean-continent convergent volcanic arcs where partial melting of the continental crust is not a significant source of magma.

These fluid lava flows can be subdivided into two types, both Hawaiian terms, based primarily on the nature of the lava's surface texture (**FIGURE 6.6**): aa (pronounced *ah-ah*) lava, which has a rough, fragmented surface and a jagged appearance; and pahoehoe (*pah-hoy-hoy*) lava, which has a smooth surface that is shiny and ropy in appearance.

G. Brad Lewis/Stone/Getty Images

FIGURE 6.5 Basaltic lava flows easily because of its low viscosity.

Q What aspect of magma chemistry influences its viscosity?

TABLE 6.1 Magma Types

Magma Type	Composition	Silica Content and Viscosity	Gas Content	Explosivity Potential	Lava Temperature	Examples of Volcanoes
Basaltic	Mafic	Least: ~50% (thin, runny)	0.5%–2%	Least	Hottest: 1100°C–1200°C	Mid-ocean ridges; plateau basalts like the Columbia Plateau; the Hawaiian Islands
Andesitic	Intermediate	Intermediate: ~60%	3%–4%	Intermediate	Cooler: ~900°C–1000°C	Mount St. Helens, Mount Rainier
Rhyolitic	Felsic	Greatest: >70% (thick, stiff)	4%–6%	Greatest	Coolest: ~700°C–800°C	Yellowstone volcano

(a)

USGS, Hawaiian Volcano Observatory

(b)

© Martin Rietze/Westend61/Corbis

FIGURE 6.6 (a) *Aa* is a Hawaiian term for basaltic lava that has a rough, fragmented surface composed of broken lava blocks called clinkers. (b) *Pahoehoe* is a Hawaiian term for basaltic lava that has a smooth, shiny, and ropy surface.

In which tectonic environments are you likely to find aa and pahoehoe lava flows?

Andesitic Magma

Andesitic magma has an intermediate composition. Compared to basaltic magma, it contains more silica and less iron and magnesium (making it lighter in color). Andesite (**FIGURE 6.7**) has many sturdy silicate bonds, making it less fluid than basalt, with higher gas content.

When magma migrates toward the surface, it undergoes a decrease in pressure. This frees gas from its dissolved state, forming bubbles in the ascending molten rock. The released gas tries to escape, but in magma with high silica content it is trapped by the formation of silicate chains during cooling. Gas pressure builds up, eventually generating explosive eruptions that expel great volumes of pyroclastic debris.

If andesitic magma has a chance to depressurize before erupting—which is to say, if conditions in the mantle are such that it

is possible for gases to escape slowly—it can erupt nonviolently. That is not the norm, however; volcanoes containing andesitic magma are prone to violent eruptions because of its high gas content and the fact that it does not flow as easily as basaltic magma.

Unlike basaltic magma, which is found in both aa and pahoehoe lava flows, andesitic magma feeds a different type of flow consisting of smooth-sided, angular blocks that are less porous than those in aa lava flows. Andesitic magma is viscous enough to form immobile plugs, called *lava domes* (**FIGURE 6.8**) that block vents and prevent lava extrusion. Consequently, these lava domes are often forcefully expelled in extremely violent eruptions.

Andesitic magma commonly erupts from high, steep-sided volcanoes known as *composite volcanoes*, or **stratovolcanoes,** that are formed above convergent plate margins. The lava typically emerges

Per-Andre Hoffmann/Look/Getty Images

FIGURE 6.7 An andesitic stratovolcano, Tavurvur is an active volcano in Papua New Guinea in the western Pacific.

Rich Frishman/The Image Bank/Getty Images

FIGURE 6.8 A lava dome is a rounded, steep-sided mound built by highly viscous magma that cannot flow very far from the vent before cooling. These three domes are found in the central vent of Mount St. Helens, Washington

in small-volume flows that advance only short distances (hundreds of meters to a kilometer or so) down the flanks of a volcano. These flows tend to form stiff ridges that rise above the surrounding landscape, a result of their high viscosity.

Rhyolitic Magma

Rhyolitic magma has a felsic composition. Its higher silica content makes it more viscous and slower moving than other types of magma (**FIGURE 6.9**). The high viscosity and gas content of rhyolitic magma means that eruptions of volcanoes composed of this type of magma are usually violently explosive and generate high volumes of pyroclastic debris.

Such eruptions, however, deplete dissolved gases in the magma source. The degassed magma can then rise to the surface and extrude less violently. Thus, after an initial eruption phase marked by enormous and catastrophic explosions, viscous rhyolitic magma generally oozes from a volcano's central vent to form symmetrical lava domes.

Rhyolitic lava flows, like those produced by andesitic volcanoes, produce ridges of viscous lava that rise as much as 10 meters above the surrounding landscape.

FIGURE 6.9 Rhyolitic lava flow, Newberry National Volcanic Monument, Oregon.

(Q) **Describe the eruption style of rhyolitic magma.**

 **EXPAND YOUR THINKING**

Explain or write about the tectonic processes responsible for each of the three primary magma types.

(6-3) # Explosive Eruptions Are Fueled by Violent Releases of Volcanic Gas

LO 6-3 Describe volcanic gases and the role they play in explosive versus effusive eruptions.

Knowledge about the dissolved gases contained in magma is key to understanding why volcanoes erupt.

Magma bodies rise in Earth's crust until they reach a point of neutral buoyancy. As gas expands and magma moves closer to the surface, lessening pressure and gaseous expansion produce an eruption. The interaction among the viscosity, temperature, and gas content of the magma determines whether an eruption will be explosive or effusive.

(a)   Associated press   (b)   Beboy_ltd_iStockphoto

FIGURE 6.10 (a) Explosive eruption of pyroclastic debris, Mount St. Helens, 1980. (b) Effusive eruption of Piton de la Fournaise volcano on Reunion Island, Indian Ocean, 2007.

(Q) **Why do explosive eruptions occur?**

Explosive versus Effusive Eruptions

Most magma is stored in the mantle or crust prior to eruption. During this period it typically reaches, or comes close to, saturation by a number of gases, particularly water vapor. Most explosive eruptions are fueled by the explosive release of these gases at shallow levels in the volcano's "plumbing system."

The potential for any magma to erupt explosively depends on how dissolvable the gases are and the capability of the magma to

retain them during its ascent to the surface. If gases can escape passively, a lava eruption (effusion) will ensue. If gases cannot escape passively—if the magma has high silica content—an explosive eruption will ensue (**FIGURE 6.10**).

TABLE 6.2 Percentage of Volcanic Gas Content at Three Volcanoes

Gas	Kilauea, Hawaii (basaltic magma, 1170°C, hotspot, shield volcano)	Erta Ale, Ethiopia (basaltic magma, 1130°C, divergent margin, shield volcano)	Momotombo, Nicaragua (andesitic magma, 820°C, convergent margin, stratovolcano)
H_2O	37.1	77.2	97.1
CO_2	48.9	11.3	1.44
SO_2	11.8	8.34	0.50
H_2	0.49	1.39	0.70
CO	1.51	0.44	0.01
H_2S	0.04	0.68	0.23
HCl	0.08	0.42	2.89
HF	—	—	0.26

Source: R. B. Symonds, W. I. Rose, G. Bluth, and T. M. Gerlach, "Volcanic Gas Studies: Methods, Results, and Applications," in M. R. Carroll and J. R. Holloway, eds., *Volatiles in Magmas, Mineralogical Society of America Reviews in Mineralogy,* 30 (1994): 1–66.

An explosive eruption occurs for two reasons: First, the solubility (dissolvability) of water vapor and other gases typically is several times greater in silica-rich rhyolite magma than in basalt magma (4 to 6% in weight versus 1%). Second, the high viscosity of rhyolite magma inhibits the rise and escape of gas bubbles. (The buoyant rise rate of a bubble is 106 times faster in basalt than in rhyolite.)

Where does volcanic gas come from? Under high pressure gases are dissolved in molten rock within the crust or upper mantle. Pressure decreases as magma rises toward the surface, and the gases begin to form tiny bubbles. With continued bubble formation, the volume of magma increases, making it less dense and causing it to rise faster.

Near the surface, as bubbles grow in number and size, the volume of gas may actually exceed the volume of molten rock. This creates "magma foam." The large number of rapidly expanding gas bubbles in the foam fragments the magma and produces **tephra** (airborne pyroclasts) when erupted. It is this gas fragmentation, and not just the explosive energy of the eruption, that leads to tephra formation.

The increase in magma volume due to gas fragmentation is truly remarkable. Consider that if 1 cubic meter of 900°C rhyolite magma containing 5% dissolved water were suddenly brought to the surface, it would expand over 600 times in size, occupying 670 cubic meters at atmospheric pressure. Large amounts of gas in the magma can lead to massive eruption columns that spew tephra high into the atmosphere.

Gas Chemistry

Gases spread from an erupting vent primarily in the form of acid aerosols (tiny acid droplets). These compounds attach themselves to tephra particles and microscopic salt particles or other dust in the atmosphere and travel in air currents.

Typically, the most abundant volcanic gas is water vapor (H_2O), followed by carbon dioxide (CO_2) and sulfur dioxide (SO_2). Volcanoes also release smaller amounts of hydrogen sulfide (H_2S), hydrogen (H_2), carbon monoxide (CO), hydrogen chloride (HCl),

hydrogen fluoride (HF), and helium (He). The gas content measured at three different volcanoes is listed in **TABLE 6.2**.

The gases released from a volcano can be as deadly as hot, fiery lava. Volcanic gases (principally sulfur dioxide, carbon dioxide, and hydrogen fluoride) can threaten the health of humans and animals, damage or destroy crops and property, and interfere with air traffic. (The Geology in Our Lives feature, "Bringing European Air Travel to a Standstill," gives a firsthand account of one such incident.)

Sulfur dioxide produces acid rain and air pollution downwind from a volcano, and large explosive eruptions inject massive volumes of sulfur aerosols into the upper atmosphere. These can lower the air temperature by blocking radiation from the Sun. They also damage Earth's ozone layer by forming new molecules with ozone gas.

The water in the Hawaiian city of Kona, located downwind of Kilauea Volcano, used to have high lead concentrations because sulfuric rain dissolved the lead used in the plumbing pipes of the city's buildings. This became a serious health concern when lead was found to damage the human nervous system—particularly in young children, causing learning difficulties and, in extreme cases, brain damage. The lead has since been removed from Kona's plumbing system.

Carbon dioxide is an especially hazardous gas because it is heavier than air and so is capable of flowing into low-lying areas and killing by asphyxiation. Case in point: Late one night in 1986, carbon dioxide seeped out of Lake Nyos, a supposedly dormant, water-filled volcanic crater in Cameroon, Africa. The gas spread into the surrounding valleys and killed more than 1,700 people while they slept.

Ⓠ **EXPAND YOUR THINKING**

Describe how lava chemistry influences the style of volcanic eruptions.

GEOLOGY IN OUR LIVES

Bringing European Air Travel to a Standstill

On April 15, 2010, British civil aviation authorities ordered the country's airspace closed when it became too dangerous to allow air traffic through the area due to a cloud of ash drifting from the erupting Eyjafjallajökull (Icelandic for "island-mountain glacier") volcano in Iceland (**FIGURE 6.11**).

Flying through volcanic ash can abrade the cockpit windshield and make it impossible for the pilots to see outside; it can also damage communication and navigation instruments on the outside of the aircraft; worst of all, it can coat the inside of jet engines with concretelike solidified ash, which can force the engine to shut down.

Within 48 hours the ash cloud had spread across northern Europe. Three hundred and thirteen airports in England, France, Germany, and other European nations closed for about one week, with more than 100,000 flights cancelled. Airplane schedules around the world were affected, and some 10 million air travelers were stranded, at a cost estimated at $3.3 billion.

As a consequence of this event, brought on by what volcanologists describe as a rather modest eruption, European authorities are considering bringing all European air traffic under the control of a single agency, a move designed to increase flexibility in response to future such disruptive episodes.

Scientists familiar with Iceland's volcano system fear that future eruptions at Eyjafjallajökull could activate a nearby volcano named Katla. Katla has been dormant for decades, but it could be reawakened by the activity at Eyjafjallajökull. The two are known to have erupted together in the past.

FIGURE 6.11 The April 14, 2010, eruption of Eyjafjallajökull in Iceland brought air travel throughout Europe to a standstill for nearly a week. The cancellation of over 100,000 flights stranded some 10 million passengers around the world.

©AP/Wide World Photos

Katla, a larger volcano, is buried under the glacier Myrdalsjökull, which is half a kilometer thick. The large amount of ice atop Katla is a major ingredient in a recipe to produce ash clouds of greater magnitude than seen from Eyjafjallajökull, presumably capable of again shutting down airports around the world.

6-4 Pyroclastic Debris Is Produced by Explosive Eruptions

LO 6-4 Describe the formation of pyroclastic debris.

Although lava flows are better known to the general public, it is pyroclastic debris that makes up the largest volume of volcanic products on land.

Most pyroclastic eruptions are explosive in nature and associated with andesitic or rhyolitic magma. That said, if the eruption is gas rich, basaltic magma may form pyroclastic debris as well. Within a volcano, the rapid formation of gas bubbles tends to fragment magma into particles. Explosive eruptions are characterized by the violent expulsion of these fragments in the form of pyroclastic debris. These eruptions also contain high amounts of gas that, upon reaching Earth's surface, expands to many hundreds of times its original volume.

Tephra

Explosive eruptions are typically accompanied by an **eruption column**—a massive, high-velocity, billowing pillar of gas, molten rock, and solid particles that is blasted into the air with tremendous force (**FIGURE 6.12**). *Tephra* is the general term used by volcanologists to characterize airborne pyroclastic debris produced by an eruption column.

Tephra typically includes volcanic *ash* (with particles less than 2.5 millimeters across), larger pyroclastic fragments called *lapilli* (particles 2.5 to 63 millimeters across), *blocks* (more than 63 millimeters across) that are ejected in a solid state, and *bombs* (also more than 63 millimeters across) that are ejected in a semisolid or plastic condition. Together, these materials are known as pyroclastic debris (**FIGURE 6.13**).

In a major explosive eruption, most of the pyroclastic debris consists of lapilli and ash. Bombs are ejected in a semimolten form, so they often take on a rounded, aerodynamic shape as they fly through the air.

Klaus Nigge/NationalGeographic/Getty Images

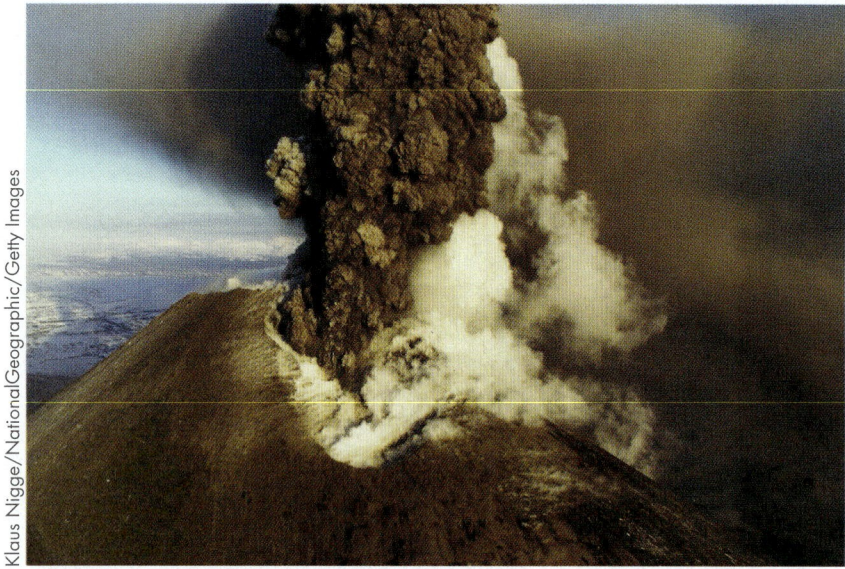

FIGURE 6.12 An eruption column is composed of pyroclastic debris, gas, and lava.

What types of magma tend to produce pyroclastic debris?

When pyroclastic debris accumulates on the ground and solidifies, it forms a rock called **tuff.** Tuff is made up of the various bits of volcanic debris that have been ejected, and usually consists of particles of various sizes contained in an ash matrix.

Pyroclastic Flows

It is common for the base of an eruption column to collapse and form searing **pyroclastic flows** that are driven down the slopes of the volcano by the force of gravity (**FIGURE 6.14**). A pyroclastic flow is a high-speed avalanche of hot ash, rock fragments, and gas that can reach temperatures of up to 800°C and move at speeds of up to 200 to 250 kilometers/hour. A pyroclastic flow is capable of overcoming, knocking down, burying, and/or burning anything in its path. When these materials lose their energy and come to rest, as thick beds of ash and lapilli, they eventually solidify into a welded mass of glass shards. The resulting rock is referred to as a **welded tuff.**

Pyroclastic flows move at incredible speeds and constitute a lethal hazard for communities of people living on the slopes or at the base of explosive volcanoes. In 1902, a pyroclastic flow raced down the side of Mount Pelée on the Caribbean island of Martinique. Despite warnings of an impending eruption, the mayor of a nearby town had convinced the residents not to leave because an election was scheduled within the next few days. Twenty-eight thousand people died in a matter of minutes.

One way by which volcanologists gauge the potential danger of a volcano is to examine rock in the vicinity, searching for welded tuffs. Past pyroclastic flows will have produced a stratigraphic record consisting of sequences of welded tuffs whose thickness

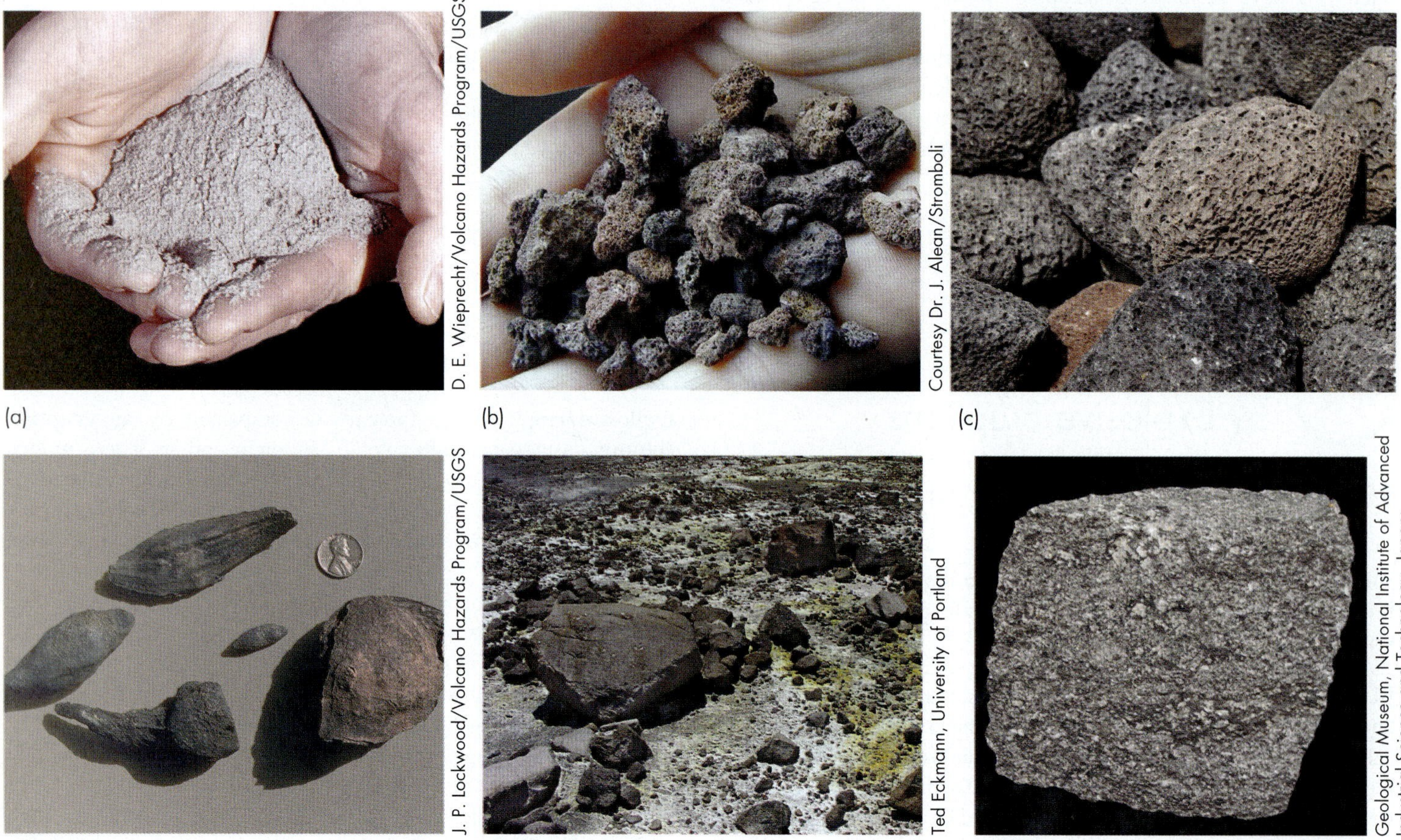

(a) D. E. Wieprecht/Volcano Hazards Program/USGS

(b) Courtesy Dr. J. Alean/Stromboli

(c) Victor Ivanov/Agefotostock

(d) J. P. Lockwood/Volcano Hazards Program/USGS

(e) Ted Eckmann, University of Portland

(f) Geological Museum, National Institute of Advanced Industrial Science and Technology, Japan: http://www.gsj.jp/Muse

FIGURE 6.13 Pyroclastic debris may consist of (a) ash, (b) lapilli, (c) pumice, (d) volcanic bombs, (e) volcanic block, and (f) welded tuff.

What types of magma tend to produce pyroclastic debris?

and frequency offer a guideline for assessing the probability of future flows, their likely direction, and their probable size.

Another important, and curious, product of explosive eruptions is *pumice*. As is well known, pumice is a volcanic rock that floats. Lumps of pumice resemble sponges in that they contain a network of gas bubbles congealed within the fragile volcanic glass and minerals composing the rock.

All types of magma (basalt, andesite, and rhyolite) form pumice, which is created when gases rapidly escape from magma, producing a froth or foam. The foam quickly cools and solidifies to form pumice.

FIGURE 6.14 In 1984, the explosive eruption of Mayon Volcano in the Philippines generated several large pyroclastic flows.

Ⓠ **What is the origin of pyroclastic flows?**

Ⓠ **EXPAND YOUR THINKING**

As a geologist living in a city at the foot of a volcano, what research would you conduct to assess the threat to your community from pyroclastic flows?

6-5 Volcanoes Can Be Classified into Six Major Types Based on Their Shape, Size, and Origin

LO 6-5 Compare and contrast large-scale volcanic terrains and central vent volcanoes.

Volcanoes can be classified by their shape, size, and origin. In this section we define two broad classes of volcanoes: central vent volcanoes and large-scale volcanic terrains (**FIGURE 6.15**).

Volcano Classification

Central vent volcanoes tend to build a volcanic landform from a more or less central vent. These volcanoes most often have a *summit crater*, although they may also produce *flank eruptions* that emerge from the side of the volcano, *fissure eruptions* that originate from an elongated fracture on the side of a volcano (as illustrated in the Critical Thinking exercise, "The Laki Fissure Eruption," at the end of this section), and other types of eruptions.

Central vent volcanoes are the most widespread and best-known type; it is familiar to the public in the image of a volcano as a cone-shape mountain.

Large-scale volcanic terrains lack a central vent; they are formed by eruptive products coming from a network of sources. They also generally (but not always) constitute massive features that have reshaped the land (or seafloor) over an area of hundreds of thousands of square kilometers or more.

Large-scale volcanic terrains are globally significant, as they account for gaps in our understanding of volcano origins that are not explained by the other types and conform to the definition of volcanoes cited at the beginning of the chapter—"A volcano is any landform from which lava, gas, or ashes escape from underground, or have done so in the past."

While there are examples of volcanoes with features of both of these categories, as well as volcanoes that do not fit into either category, this classification system gives a sense of the diversity of scale and origin found among the world's volcanoes. Within each category we find at least three distinct varieties:

1. Central vent volcanoes
   a. Shield volcanoes
   b. Stratovolcanoes (or composite cones)
   c. Rhyolite caldera complexes

2. Large-scale volcanic terrains
   a. Monogenetic fields
   b. Large igneous provinces
   c. Mid-ocean ridges

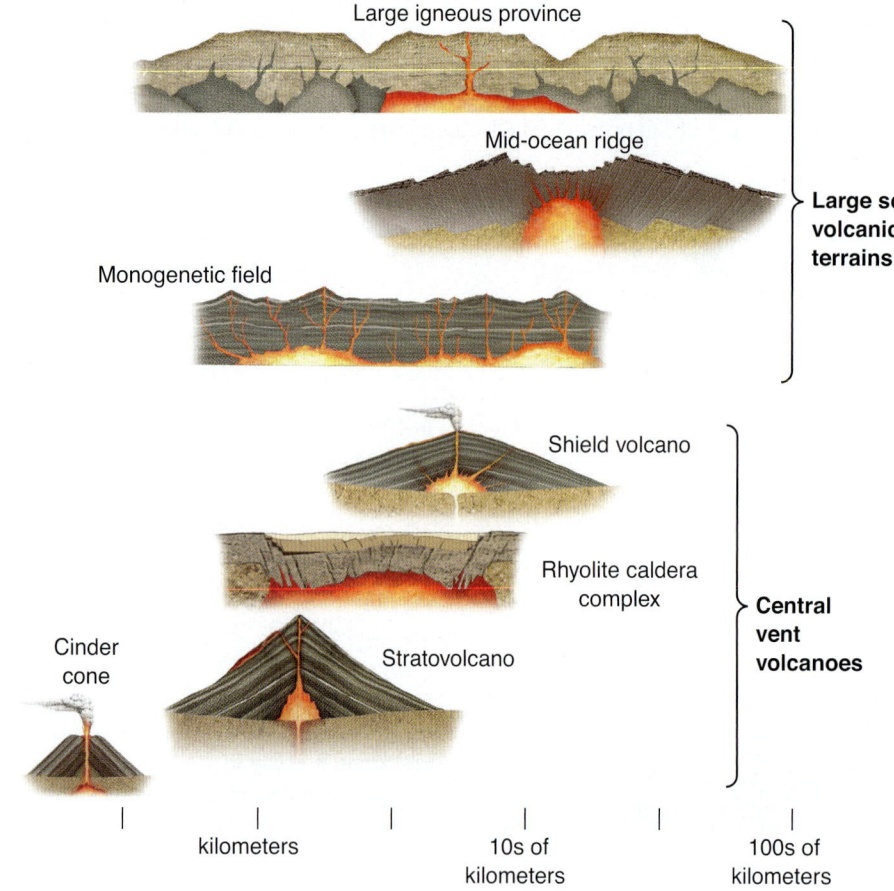

Approximate diameter at base

FIGURE 6.15 Central vent volcanoes consist of shield volcanoes, stratovolcanoes, and rhyolite caldera complexes. Large-scale volcanic terrains consist of flood basalts, monogenetic fields, and mid-ocean ridges. Cinder cones are small-scale volcanoes that are superimposed on other types.

Which volcano type is most dangerous? Why?

The names of these six volcanic types may at first appear complex and technical, but you will find they are easy to remember once you understand the concepts behind the words, which are explained in **TABLE 6.3**.

Q EXPAND YOUR THINKING

The largest types of volcanoes require enormous volumes of magma. Which type of magma is generally available in the largest quantity? Why?

TABLE 6.3 Types of Volcanoes

Type	Shape	Magma Type	Tectonic Setting	Example
Central Vent Volcanoes				
Shield volcano	Large volume, gentle, low-angle slopes	Basaltic, low-silica	Midplate setting or convergent boundary	Mauna Loa Volcano, Hawaii
Composite volcano or stratovolcano	Tall, with steep slopes; often irregular outline from past explosions and rugged dome areas	Andesitic, silica-rich magma at subduction zone	Convergent boundary (usually)	Mount Pinatubo, Mount St. Helens
Rhyolite caldera complex	Low-relief system of collapsed calderas and many small vents	Rhyolitic, silica-rich magma (including melted crustal rock)	Convergent boundary or isolated midplate setting	Yellowstone National Park
Large-Scale Volcanic Terrains				
Monogenetic field	Low-relief system of vents, cones; occasional stratovolcano	Basaltic, low-silica	Convergent boundary, or isolated midplate	San Francisco Volcanic Field, Arizona
Large igneous province	High plateau, massive volume (> 100,000 cubic meters); many layers of lava; no single distinct mountain	Basaltic, low-silica	Variable setting; often midplate or continental margin	Columbia Plateau, Washington
Mid-ocean ridge	Broad slopes on long, linear ridge with central rift valley	Basaltic, low-silica	Divergent boundary; spreading center along mid-ocean ridge	Mid–Atlantic Ridge

CRITICAL THINKING

The Laki Fissure Eruption

On June 8, 1783, the Laki fissure zone in Iceland began to erupt in what was to become the largest basalt eruption in recorded history. The eruption (**FIGURE 6.16**) lasted eight months, during which time 14.73 cubic kilometers of basaltic lava and some tephra were erupted, covering 565 square kilometers of land. Haze from the eruption was reported from Iceland to Syria.

In Iceland, volcanic outgassing led to the loss of most of the island's livestock (poisoned by grass contaminated by fluorine); crop failure (due to acid rain); and the death of one-quarter of the human residents (due to famine). Lava erupted from fissures located 45 kilometers from the coast and flowed toward the ocean at speeds averaging 0.4 kilometers/hour.

As a graduate student working on your master's thesis, you are studying the history of the Laki fissure eruption. Your first task is to graph the history of lava flow to the coast and determine the distance covered during every six-hour period.

1. Design a data table to show the distance traveled by the lava over a five-day period. Calculate the distance covered every six hours.

2. Plot the data on a graph with time on the x-axis and distance on the y-axis.

3. How long did it take the lava to reach the coast?

4. How many kilometers did the lava travel in three days?

FIGURE 6.16 The Laki eruption site is now quiet, but in 1783 it erupted in long lines of lava fountains that discharged massive volumes of lava, forming a small basalt plateau.

Arctic-Images/Corbis

5. If you lived in a village located along the lava path only 10 kilometers from the eruption site, how much time after the first sign of eruption would you have to evacuate? Imagine that you owned a farm with animals and crops. What would you be able to save, and what would most likely be lost to the eruption?

6-6 Shield Volcanoes Are a Type of Central Vent Volcano

LO 6-6 Identify what gives a shield volcano its distinctive shape.

Shield volcanoes are massive, low angle volcanoes with a distinct summit crater. They are fed by low-silica, low-gas basaltic magma from the upper mantle.

Shield Volcanoes

Some of the largest volcanoes in the world are shield volcanoes (**FIGURE 6.17**). They are named for their fanciful resemblance to a warrior's shield laid on the ground; they are also noted for their massiveness and long, low profiles.

Various stages of shield building may be characterized by flank eruptions, fissure eruptions, and the construction of secondary cones on the flanks of the volcano. Made of fluid basaltic lava, most shield volcanoes are built of thousands of thin basaltic layers that accumulate over time.

In northern California and Oregon, many shield volcanoes have diameters of 5 to 8 kilometers and reach heights of 500 to 700 meters. The most famous examples of shield volcanoes are found in the Hawaiian Islands, a linear chain of volcanoes that includes Kilauea and Mauna Loa. Both are among Earth's most active volcanoes, and Mauna Loa is the largest.

Mauna Loa's true size is not obvious at first, however. It projects over 4 kilometers above sea level, but its base on the seafloor lies beneath more than 4.5 kilometers of water. Added together, Mauna Loa's total height is over 8.5 kilometers above the seafloor, making it the second most massive mountain on Earth.

The key to understanding a shield volcano's shape is the chemistry of its magma. Shield volcanoes are fed by a low-silica, low-gas magma source in the upper mantle. Eruptions release repeated flows of pahoehoe and aa that give the volcano its layered structure.

Shield volcanoes also tend to have a relatively constant magma supply rate, which keeps the magma's "plumbing system" hot and allows the lava to maintain a high temperature; thus, it remains fluid longer and flows farther. The chemical composition of magma in a shield volcano probably does not change significantly since its formation in the mantle.

Shield volcanoes are approximately 90% lava and 10% pyroclastic deposits. Explosions do occur on shield volcanoes, often in the form of a **phreatomagmatic eruption** caused by the rapid expansion of steam when magma comes in contact with groundwater. The occurrence of these eruptions is unpredictable, and they can be highly dangerous to life and property in their immediate vicinity.

Shield volcanoes may be formed wherever a basaltic magma source erupts on the land surface or ocean floor. They commonly occur in midplate settings (oceanic hotspots), but may also be found in convergent settings, such as subduction-related volcanic arcs and island arcs.

Midplate shield volcanoes are formed in places where a persistent source in the mantle continuously feeds basaltic lava to the surface, where it erupts and builds a volcano. Most often this occurs in the oceanic lithosphere, presumably because the crust is thin and easily breached and the magma source is low in silica.

As a plate moves over a hotspot, a shield volcano is born and grows, then eventually dies as it moves off the hotspot; thereafter, a new one forms. In this way, a chain of shield volcanoes develops. Many examples are found in the Pacific Ocean basin, with the Hawaiian Islands being the best known.

Cinder Cones

Lava pours out of a shield volcano from vents at the summit or along its flanks. It is common for eruptions from shield volcanoes to include high lava fountains that form **cinder cones** at the vent (**FIGURE 6.18**). This style of eruption is spectacular as it sends hot fragments high into the atmosphere, often for days to months, but it is not explosive in the sense that it is not a sudden, singular, violent release of gas.

Cinder cones, which some scientists consider to be a separate category of volcano, are built up by the ash, lapilli, blocks, and bombs of congealed lava ejected from a single vent. High gas content in the magma source tends to drive the fiery eruption. When the gas-charged magma is released into the air, it accumulates in the form of small fragments that solidify and fall as cinders around the vent, forming a nearly circular cone (or oval, if the wind is more persistent in one direction). Most cinder cones have a bowl-shape summit crater.

Numerous in western North America and other volcanic environments of the world, cinder cones are frequently found as secondary eruption events in association with a longer-lived, historically persistent central vent volcano. They are also found in association with large-scale volcanic terrains and thus are associated with both classes of volcanoes. It is uncommon to find a cinder cone more than 300 meters high.

You can read about the most dangerous volcano in North America, one fed by a hotspot, in the Geology in Action feature, "Supervolcano—Yellowstone National Park."

Ⓠ EXPAND YOUR THINKING

What is cinder? What conditions lead to the production of cinder?

FIGURE 6.17 Shield volcanoes, such as Mauna Loa on the island of Hawaii, are built up by eruptions of fluid basaltic lava.

Stephen & Donna O'Meara/Photo Researchers/Getty Image

Ⓠ Do you agree that it is highly unlikely for rhyolitic magma to build a shield volcano? Why or why not?

G. Brad Lewis/Getty Images

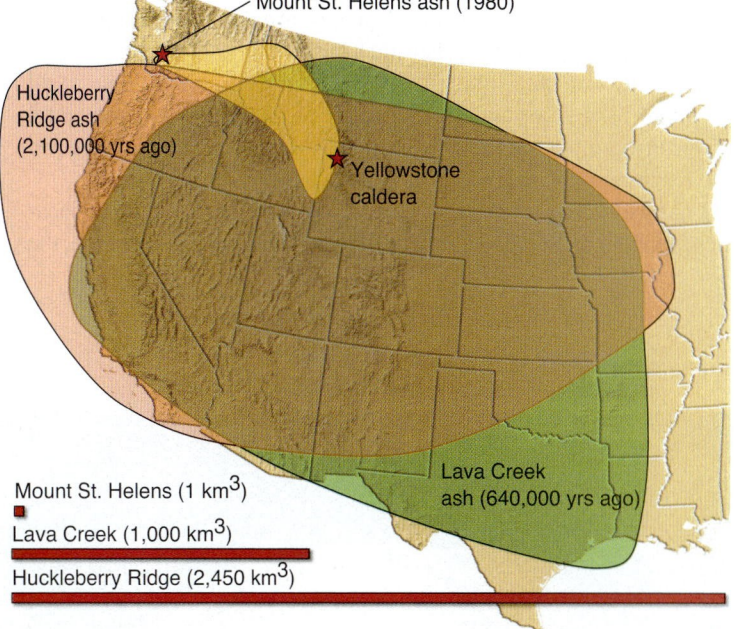

FIGURE 6.18 Cinder cones

GEOLOGY IN ACTION

Supervolcano—Yellowstone National Park

Yellowstone National Park sits on a *supervolcano*. If it were to erupt, as it has in the past, it would cause the largest natural disaster in the history of humankind.

The Yellowstone supervolcano, fed by an 8-kilometer-deep subterranean chamber of molten rock and gases, is so vast that it is arguably one of the largest active volcanoes in the world. Nevertheless, its appearance makes it hard for the average person to accept that it is such a powerful eruptive force. Yellowstone is a *rhyolite caldera complex*, a network of eruptive sites fed by sticky, gas-rich rhyolite magma so violent that the only topographic feature remaining after an eruption is a gigantic hole in the ground.

Compare Yellowstone to Mount St. Helens. The Mount St. Helens crater is about 5 square kilometers (km²), whereas the Yellowstone caldera is over 3,884 square kilometers (km²). When Mount St. Helens erupted in 1980, it blasted a dense column of ash and gas 20 to 27 kilometers into the atmosphere, ejected over 3 cubic kilometers (km³) of ash detectable over 57,000 square kilometers (km²), killed 54 people, and lowered the mountain peak by 400 vertical meters. Impressive, indeed. But 640,000 years ago, the Lava Creek eruption, Yellowstone's last, ejected over 1000 times the ash and lava of Mount St. Helens (**FIGURE 6.19**); and an earlier eruption, called the Huckleberry Ridge eruption, 2.1 million years ago, was more than 2400 times the volume of the Mount St. Helens eruption.

Yellowstone is fed by a hotspot located under the felsic rock of the North American continent. As demonstrated repeatedly over its 16.5-million-year history, this gives it a distinctly explosive character.

As the North American Plate moves to the west, at least six caldera-forming eruptions have produced a trail stretching from the Nevada–Oregon border to the current location of Yellowstone National Park. Calderas are formed by massive explosive eruptions followed

FIGURE 6.19 An eruption of the Yellowstone supervolcano has the potential to blanket the western and midwestern United States with ash. The ash layer produced by the eruption of Mount St. Helens is shown in yellow. The ash deposits of the Huckleberry Ridge eruption (in pink) and the Lava Creek eruption (in green) are many times larger. Red bars illustrate the relative volumes of these erupted products *Source:* U.S. Geological Survey, Department of the Interior, USGS.

by enormous outpourings of lava. As lava is discharged, the lack of support causes the roof of the magma chamber to collapse and the ground above to subside by hundreds of meters, thus forming a caldera.

6-7 Stratovolcanoes and Rhyolite Caldera Complexes Are Central Vent Volcanoes

LO 6-7 Compare and contrast the magma composition of stratovolcanoes versus rhyolite caldera complexes.

Stratovolcanoes are the most common volcano; they characterize the majority of island arcs and volcanic arcs in the world. Rhyolite caldera complexes are relatively rare, as they require a felsic magma source. Both types of volcanoes occur at specific types of volcanic settings.

Magma Genesis

When a plate is recycled into the mantle by subduction, the increased heat causes water and other volatile substances to escape from the subducting oceanic crust. The escaping water causes rocks in the overlying mantle to recrystallize and adopt a lower melting point; the result is partial melting at the base of the overlying plate. The magma produced (typically water-rich basalt) migrates toward the surface, initiating partial melting in the silica-enriched crust and thereby generating molten rock that is also silica enriched.

If the overriding plate is oceanic, the resulting magma could be either basaltic or andesitic. If the overriding plate is continental (typically composed of granite or a metamorphic rock called *gneiss*), the resulting magma could be andesitic, basaltic, or rhyolitic.

As we learned in Chapter 5, section 5–2, during partial melting, silica-rich compounds melt early, and mafic compounds melt later; hence, the

thicker the overlying crust, the greater the potential for magma-crust interaction leading to enhanced production of silica. For this reason, ocean-continent subduction sites tend to have silica-rich eruptions, whereas ocean-ocean sites often do not. The resulting magma feeds the growth of a line of volcanoes on the overriding plate: a *volcanic arc* if the overriding plate is continental; an *island arc* if the overriding plate is oceanic.

Stratovolcanoes

Stratovolcanoes account for the largest proportion (about 60%) of Earth's individual volcanoes, and the great majority of volcanic arcs and island arcs are composed of stratovolcanoes.

As the name implies, stratovolcanoes are composed of alternating layers of stratified lava flows and pyroclastic deposits (**FIGURE 6.20**). There are dozens of famous examples of this volcano type, including Mount St. Helens (Washington State), Mount Pinatubo (Philippines), Mount Fuji (Japan), and Soufrière Hills (West Indies).

Stratovolcanoes have steep slopes and are often explosively eruptive. They may be composed of several types of lava. Normally, they build over many years of eruptions that produce layers of silica-rich lava and ash.

These volcanoes usually are fed by intermediate-composition magma (originating at subduction zones), so the magma source is gas-rich and viscous. Internal gas pressure builds behind the sticky magma, eventually leading to an explosive release that turns out pyroclastic debris, followed by lava flows down the sides of the volcano.

Stratovolcanoes contrast sharply with shield volcanoes, as you'll learn by working through the exercise in the Critical Thinking feature, "The Geology of Shield Volcanoes and Stratovolcanoes." Whereas shield volcanoes are generally characterized by effusive eruptions of fluid basaltic lava, stratovolcanoes are just the opposite: They typically erupt with massively explosive force because the magma is gas-rich and too viscous for volcanic gases to escape easily. This allows tremendous

FIGURE 6.20 Stratovolcano: (a) Mount Fuji, Japan, a classic stratovolcano; and (b) its internal structure.

Andesite lava flow

Pyroclastic flow

Layers of lava and pyroclastic debris

Gavin Hellier/Getty Images

(a)

(b)

How does the chemistry of magma determine the shape and eruption style of stratovolcanoes?

internal pressure to build up. Strong silica chains are formed in the andes-itic magma, making it much less fluid, so gas cannot readily bubble out.

The same geologic conditions that lead to the formation of high-silica magmas also produce a high level of dissolved gas in the magma. Trapped gas generates explosive conditions by adding to the pressure buildup that accompanies the ejection of highly viscous lavas. As a result, lava cannot easily flow from the vent, and so the volcano may develop a plug of cooled lava—a lava dome.

When the pressure becomes high enough, the lava dome is explosively ejected. In the worst case, a **Plinian eruption** may occur. Plinian eruptions are large volcanic explosions that push thick, dark columns of pyroclasts and gas high into the stratosphere.

Such eruptions are named for the Roman scholar Pliny the Younger, who carefully described the disastrous eruption of Mount Vesuvius in AD 79. This eruption formed a huge column of ash that rose high into the sky. Many thousands of people evacuated areas around the volcano; nevertheless, approximately 2000 people were killed and the cities of Pompeii and Herculaneum were destroyed.

Stratovolcanoes cause far more human tragedy than do volcanoes of any other type. One reason is, simply, that stratovolcanoes are more numerous than other types of volcanoes, so more people live on their flanks. Other reasons reflect the inherent characteristics of a steep-sided cone built from highly viscous magma interlayered with pyroclastic deposits:

- Stratovolcanoes are steep piles of ash and lava. Both their steepness and their composition sharply increase the likelihood of slope failure (landslides, avalanches, mudflows).

- Stratovolcanoes have a strong tendency to erupt in mighty explosions that disrupt communities and threaten human life across broad areas.

Rhyolite Caldera Complexes

Rhyolite caldera complexes originate from magma that is so gas-rich and viscous that it almost always generates catastrophic explosions. In fact, the explosions are so violent that the volcano tends to collapse upon itself, to become nothing more than a series of large holes ("caldera complexes") in the ground.

Rhyolite caldera complexes are the most explosive of Earth's volcanoes but, ironically, often do not even look like volcanoes. Their tendency to erupt explosively is due to the high silica and gas content of their rhyolitic magma.

The collapsed depressions are **calderas** (Spanish for "cauldron"), large, usually circular depressions that may be several kilometers wide, indicating that huge magma chambers (10 kilometers or more in width) are associated with the eruptions.

Layers of ash (either ash falls or pyroclastic flows) often extend over thousands of square kilometers in all directions from these calderas. A good example is Yellowstone Volcano, which spread ash over several surrounding states when it last erupted 640,000 years ago.

A caldera (**FIGURE 6.21**) is formed when magma is withdrawn, usually by eruption, from a shallow underground magma reservoir. The rapid depletion of large volumes of magma dislodges structural support for the overlying roof of the reservoir, leaving it to collapse and create the caldera. This phenomenon has led some observers to label these complexes "inverse volcanoes."

Calderas are different from **craters**, which are smaller, circular depressions created primarily by the explosive excavation of rock during eruptions. Calderas can be found on shield volcanoes and stratovolcanoes, as well as in a caldera complex.

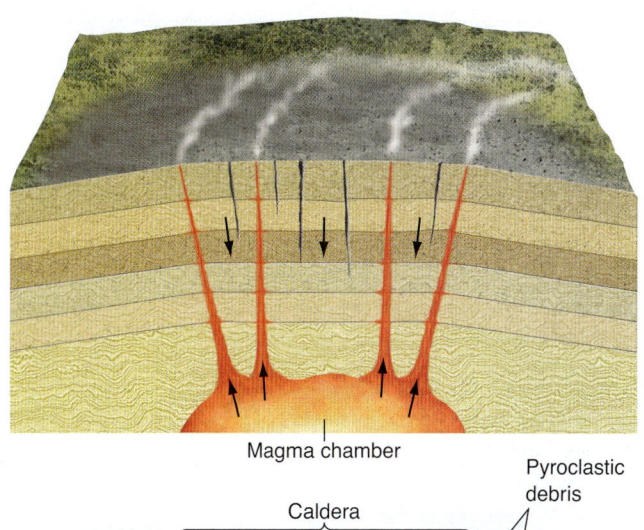

Magma chamber

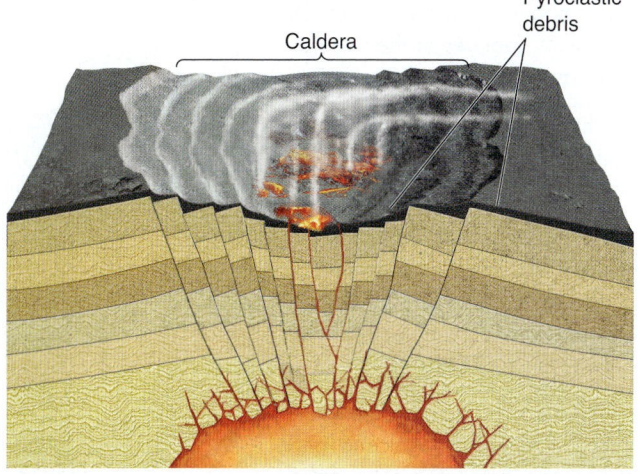

Caldera

Pyroclastic debris

(a)

(b)

M. Williams, National Park Service/USGS

FIGURE 6.21 Rhyolite caldera: (a) Cross sections before (upper) and after (lower) formation of a caldera; and (b) Aniakchak Caldera, Alaska, which was formed during an enormous explosive eruption about 3,450 years ago and expelled more than 50 cubic kilometers of magma. The caldera is 10 kilometers in diameter and about 1 kilometer deep. Later eruptions formed domes, cinder cones, and explosion pits on the floor of the caldera.

Q How does the chemistry of magma determine the shape and eruption style of a rhyolite caldera complex?

Q EXPAND YOUR THINKING

It is likely that someday the Atlantic seafloor will begin to subduct beneath the eastern seaboard of North America. Describe the volcanism that will result.

CRITICAL THINKING

The Geology of Shield Volcanoes and Stratovolcanoes

Working with a partner, complete the following exercise based on **FIGURE 6.22**:

1. Add labels to the figure. Identify the major features associated with shield volcanoes and stratovolcanoes, including magma composition, volcanic products, features of the two volcanoes, and plate tectonics settings.

2. What happens to a midplate shield volcano as it moves away from a hotspot?

3. Summarize the processes acting on a volcanic arc. How do weathering, erosion, explosive volcanism, and effusive volcanism combine to shape the appearance of the arc?

4. Describe the processes that generate magma for the shield volcano and stratovolcano. How are they different? How are they similar?

5. Explain the differences and similarities in the internal structure of the two types of volcanoes.

6. Describe the geologic hazards associated with shield volcanoes, and compare them to the hazards associated with stratovolcanoes. If you were a governor or mayor, how would you mitigate the impact of these hazards on nearby communities?

7. It is not uncommon for a shield volcano to develop at a volcanic arc. How is this possible?

8. You are the first to study a volcano that was previously thought dormant, but has reawakened and started to emit gas. What criteria would you use to assess the danger it poses to local communities?

9. What might happen if basalt magma mixed with rhyolitic magma and erupted? Describe the resulting volcano and the volcanic processes.

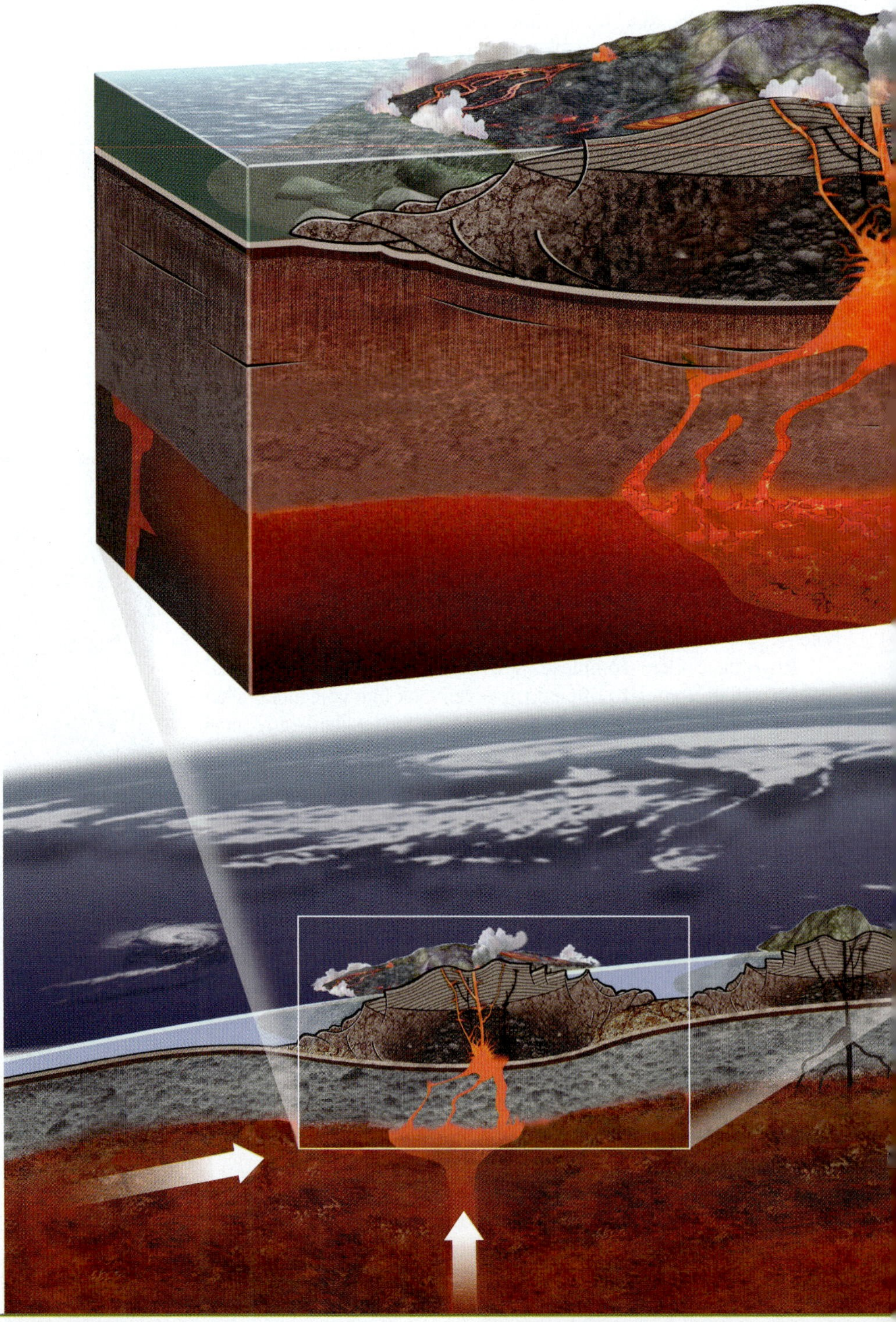

FIGURE 6.22

6-8 Large-Scale Volcanic Terrains Lack a Central Vent

LO 6-8 Identify the main characteristics of a large-scale volcanic terrain, and give examples.

Large-scale volcanic terrains are formed by massive eruptions of, typically, basaltic magma over sustained periods. These types of volcanoes have no central vent and often repave large areas of Earth's surface with volcanic products.

Monogenetic Volcanic Field

A **monogenetic volcanic field** is a collection—a field—of vents and flows, sometimes numbering in the hundreds or thousands. A monogenetic volcanic field (**FIGURE 6.23**) does not look like a classic volcano. Although there may be a single magma source for an entire field (making it *mono*genetic), it sends magma to the surface at a low rate, in a diffuse manner, and through multiple vents.

Studies indicate that in many cases, each vent erupts only once. Furthermore, monogenetic fields grow laterally rather than vertically, and form fields instead of mountains. A monogenetic field can, therefore, be described as the end product of all the separate eruptions and flows of a single large volcano spread across a broad landscape as separate features (**FIGURE 6.24**).

Monogenetic fields are collections of cinder cones and/or maar vents (low-relief, broad volcanic craters formed by shallow explosive eruptions) and the lava flows and pyroclastic deposits associated with them. The explosions are usually set off when groundwater is invaded by magma, provoking a phreatomagmatic eruption. Often, a maar later fills with water to form a lake. Sometimes, a stratovolcano is at the center of the field, as in the San Francisco Volcanic Field in Arizona, which has produced more than 600 volcanoes in its short (geologically speaking) 6-million-year history. In this case, both volcanic categories appear in the same geographic area.

Large Igneous Provinces

Large igneous provinces (LIPs) are formed at locations where massive quantities of basaltic lava pour from systems of long fissures (ground cracks) instead of from central vents. An immense outwelling of lava floods the surrounding countryside or seafloor over a period of millions of years, forming broad plateaus hundreds of meters high. There are places where so much lava has issued from the ground that it is referred to as **flood basalt.**

Layer after layer of this very fluid, silica-depleted basaltic lava takes shape as high plateaus under the sea and on continents, extending over areas of at least 100,000 square kilometers. These lava flows are often visible in the walls of eroded river valleys, measuring a kilometer or more in total thickness (**FIGURE 6.25**).

Lava plateaus composed of flood basalts can be seen in Iceland, Washington and Oregon, India, Siberia, and elsewhere. The massive Ontong Java Plateau on the Pacific seafloor is built up of flood basalts.

Courtesy Zoltan Sylvester

FIGURE 6.23 The San Francisco Volcanic Field in Arizona has produced more than 600 volcanoes during its 6-million-year history. This image shows over two dozen separate volcanoes in one portion of the field.

Q Monogenetic fields consist of dozens to hundreds of separate volcanoes. Are these each fed by a separate magma source or a single magma source? Using the magma source characteristics, what can you infer about the tectonic setting?

LIPs have been formed, some in the oceans and some on land, at various times in the past. Notable examples are the Siberian Traps in Russia and the Deccan Traps in northwestern India. (*Trap* is a Sanskrit word meaning "step," referring to the steplike topography produced by the stacked layers of lava.)

It was originally thought that these flows went whooshing over the countryside at incredible velocities (like a flash flood). Now it is believed that they are more like pahoehoe lava flows, except that they emanate from large magma chambers through fissures. Clearly, such a phenomenon requires a massive magma source.

The most famous example of a LIP in the United States is the Columbia Basalt Plateau, which covers most of southeastern Washington State and northeast Oregon and extends to the Pacific. Active 6 to 17

DaveArnoldPhoto.com/Flickr Open/Getty Images

FIGURE 6.24 Ship Rock is the erosional remnant of the "throat" of a volcano in the Navajo volcanic field of New Mexico.

FIGURE 6.25 A stack of several Columbia River basalt lava flows in the Columbia River Gorge, Washington State. Each flow is 15 to 100 meters thick.

Kaj R. Svensson/Photo Researchers, Inc.

Why is it unlikely for lava plateaus to emerge above a rhyolite magma source?

Image Courtesy of NSF, NOAA, and WHOI Advanced Imaging and Visualization Lab

million years ago, the Columbia Plateau was created by the first eruption of the same hotspot that today feeds Yellowstone Volcano.

Geologists hypothesize that large igneous provinces are shaped by the arrival of an upward-moving mantle plume into the lithosphere (as described in Chapter 5, section 5–8). It is thought that mantle plumes are enriched in lighter elements and hotter than the surrounding rock.

As plumes rise, the pressure they encounter is reduced. This generates magma by partial melting at the top of the plume. As magma reaches the base of the lithosphere, it mushrooms outward, into a large pluton. This feeds magma that erupts into the crust to form huge basalt flows.

Magma appears to be produced in greatest abundance during the first few million years after a mantle plume reaches the surface; therefore, flood basalts develop quickly, in geologic terms. If the plume maintains its position and magma is produced at a consistent rate, it could constitute a hotspot, though this is not always the case.

Mantle plumes are thought to have a deep origin, perhaps at the core-mantle boundary for the larger ones and at a depth of about 600 kilometers for the smaller ones.

Mid-Ocean Ridges

Mid-ocean ridges develop at locations where new oceanic crust is continually being formed by magma rising from the mantle (**FIGURE 6.26**). Because this young crust is still warm and is being pushed up from below, it sits higher than the surrounding seafloor, forming a long ridge that can rise over 2 kilometers above the adjacent ocean bottom.

Eruptions along a central rift valley in the center of the ridge tend to be quiet affairs composed of fluid, silica-depleted basalt. A mid-ocean ridge runs along the floor of every major ocean, tending to run down the middle except in the Pacific, where it runs along the

FIGURE 6.26 This photo, taken in December 2009, shows the first direct observation of submarine volcanism at a spreading center. The photographer was a remotely operated unmanned robot named Jason, controlled from a research ship. Glassy basaltic talus collects around a vent that is spewing sulfur-rich gases into the water. The water depth is approximately 1200 meters.

Why is most submarine volcanism fed by basaltic magma?

south and southeast side of the ocean basin. These ridges are interconnected, linking up as a global underwater mountain chain.

EXPAND YOUR THINKING

Why is it likely that large-scale volcanic terrains are typically (but not always) fed by basaltic magma?

6-9 Most Volcanoes Are Associated with Spreading Center Volcanism, Arc Volcanism, or Intraplate Volcanism

LO 6-9 Describe the role of plate tectonics in arc volcanism, spreading center volcanism, and intraplate volcanism.

Most volcanic activity is linked to plate tectonics processes at convergent and divergent boundaries. There is also considerable volcanic activity at intrplate settings.

Many of the world's active above-sea volcanoes are located near convergent plate boundaries, particularly around the Pacific Basin's Ring of Fire. Much more volcanism—producing about three-quarters of all the lava erupted on Earth—takes place unseen beneath the waters of the world's oceans, mostly along spreading centers, such as the East Pacific Rise and the Mid–Atlantic Ridge.

Some volcanoes erupt in midplate settings. These are probably tied to plumes in the mantle rather than to plate interaction in the lithosphere.

Three tectonic settings are known to foster volcanoes (**FIGURE 6.27**):

1. *Spreading center volcanism*, characterized by fluid basaltic magma that occurs at divergent plate margins forming mid-ocean ridges.

2. *Arc volcanism*, characterized by explosive rhyolitic, andesitic, and basaltic magma that occurs at convergent margins.

3. *Intraplate volcanism* (also known as midplate volcanism), characterized by shield volcanoes, rhyolite caldera complexes, and monogenetic fields.

Spreading Center Volcanism

Spreading center volcanism occurs at divergent plate boundaries. Spreading center magmas are basaltic, with the erupted products reflecting that chemistry. The relatively low silica content of the lavas means that they are characterized by low viscosity and low dissolved gas content. These magmas are least likely to produce violent eruptions.

Spreading center magmas originate in partially molten chambers within the upper mantle. When the magma feeding a spreading center moves toward the surface, it forms an intrusive *dike*.

Dikes are magma-filled cracks in the crust that feed eruptions on the seafloor. Geologists hypothesize that a typical ridge eruption leaves behind a dike from tens of centimeters to a few meters in width and extending between the magma chamber and the eruptive fissure at the surface.

Often, a molten layer of liquid rock develops on the surface of a magma source. This molten rock is periodically tapped by vertical fractures, which provide conduits for the rapid rise of magma to the surface. After volcanism has occurred and magma crystallizes within these fractures, a system of *sheeted dikes* is created that extends upward from the source of the magma (discussed further in Chapter 21). These dikes are an important component of the rock that makes up the oceanic crust.

Lavas extruding from a spreading center fissure chill quickly in contact with cold seawater and form pillow basalts and deposits of glassy *talus* (broken rock). This process of magma ascent and eruption occurs at different times and at different places on the oceanic ridge system.

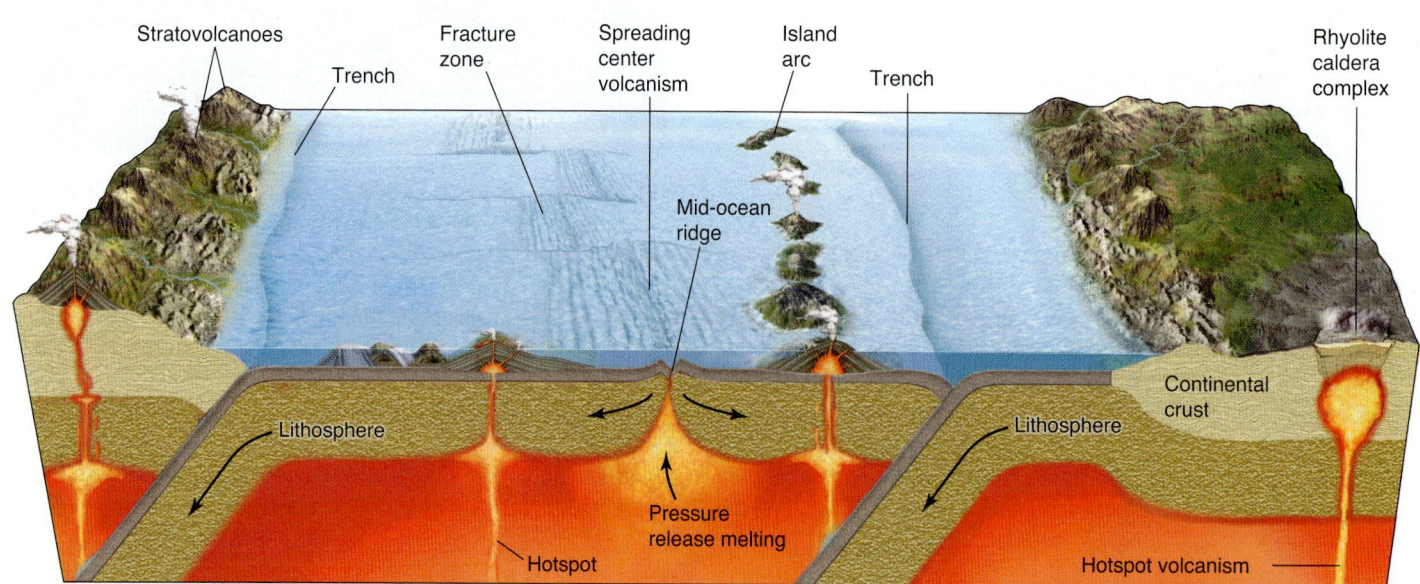

FIGURE 6.27 Most global volcanism is found at spreading centers, convergent arcs, or within plate interiors.

Q Describe the magma chemistry directly beneath each of the active volcanoes in this figure.

The rise of heat from the asthenosphere adds thermal buoyancy to the ridge area. This is the reason that ridges emerge high above the surrounding seafloor. Eruptions produced in this manner are typically quiet fissure eruptions, enabling scientists in research submersibles to study them.

Eruptions may produce pillow basalt interbedded with hydrothermal metallic sulfide mineral deposits. The base of this newly formed seafloor is composed of plutons made of gabbro originating from the same magma source. As a result, the crust consists of a lower layer of gabbro, a layer of sheeted dikes, a layer of basalt pillow deposits and metallic sulfides, and a surface coating of marine sediments that thickens as time passes and the seafloor moves farther from the rift zone.

Arc Volcanism

Arc volcanism occurs at locations where two plates converge. One plate containing oceanic crust descends beneath the adjacent plate, which may be composed of either continental or oceanic crust.

If the overriding plate is oceanic, an island arc will be created. If the overriding plate is continental, a volcanic arc will develop (**FIGURE 6.28**). The Cascade Range of the U.S. Pacific Northwest is an example of a volcanic arc; the islands of Indonesia are an example of an island arc. The majority of volcanoes produced by arc volcanism are violent stratovolcanoes, although shield volcanoes may be formed from a basaltic magma source in an arc setting.

The crust portion of a subducting slab contains a significant amount of water in hydrated minerals (minerals that contain water molecules) and marine sediments. As the subducting slab encounters high temperatures and pressures, the water is released into the overlying mantle.

Water has the effect of lowering the melting temperature of rock, thus causing it to melt more easily than it would if no water were present. The magma produced by this mechanism typically varies from basalt to andesite in composition and is gas rich. The magma's high water content leads to partial melting and then silica-enriched magma.

Plutonic bodies of mafic or intermediate composition may evolve through crystallization and produce felsic magmas. As the silica content increases, so too does the violence of the resulting volcano, as well as the tendency to create pyroclastic products.

If rhyolitic magma is produced through differentiation, the chances of a highly explosive eruption are great. This is most likely to happen with magma that has intruded and partially melted continental crust, producing a high-silica liquid. The high-silica content of continental crust can enhance the differentiation process that produces rhyolite-type magma, leading to the formation of some rhyolite caldera complexes.

Intraplate Settings

Intraplate settings such as hotspots can be exceptionally active sites of volcanism—although the exact relationship between large igneous provinces and hotspots is still somewhat unclear. What is clear is that if you follow some of the longer hotspot tracks from their current eruptions toward their oldest volcanoes, you end up at a LIP.

This is true of the Yellowstone hotspot and the Columbia Basalt Plateau, for example. For this reason, it is thought that some LIPs are produced above mantle hotspots when the plume first arrives.

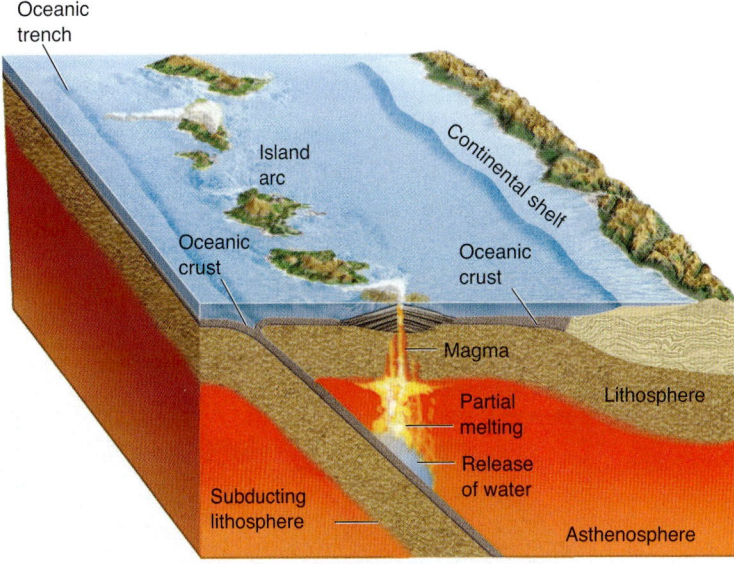

(a)

(b)

FIGURE 6.28 (a) Island arc volcanism occurs when a plate composed of oceanic crust subducts beneath another plate that is also composed of oceanic crust. (b) Volcanic arc volcanism occurs when oceanic crust subducts beneath continental crust.

Q How will magma chemistry differ between an island arc and a volcanic arc?

Intraplate volcanism is also characterized by monogenetic fields and, in some cases, by rhyolite caldera complexes. Indeed, intraplate volcanism, the subject of ongoing research by volcanologists, is marked by a diversity of volcanic types and processes.

FIGURE 6.29 illustrates the three tectonic settings we have just discussed.

Q **EXPAND YOUR THINKING**

Why do hotspots beneath continents tend to produce rhyolite caldera complexes?

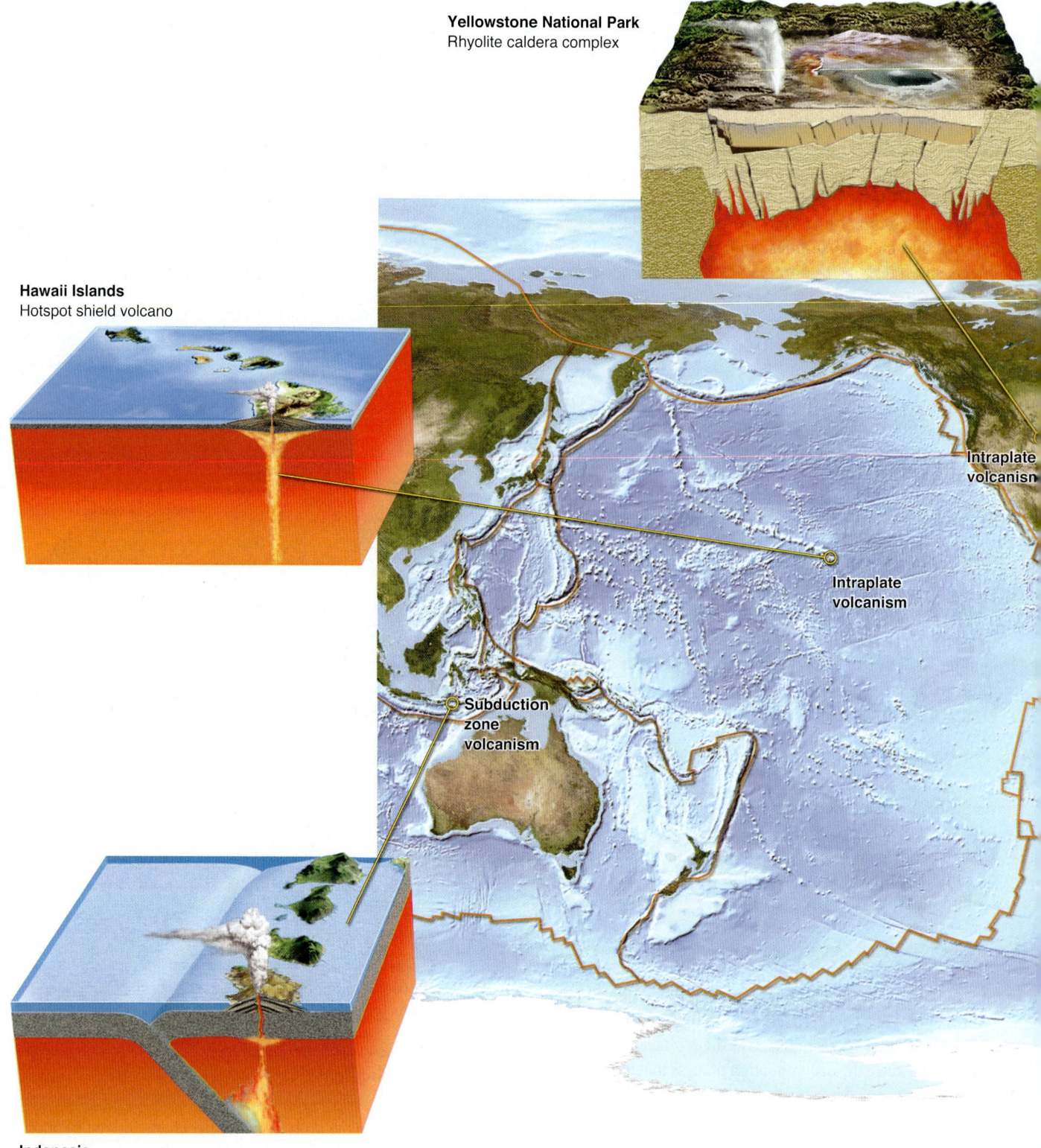

Yellowstone National Park
Rhyolite caldera complex

Hawaii Islands
Hotspot shield volcano

Intraplate
volcanism

Intraplate
volcanism

Subduction
zone
volcanism

Indonesia
Island arc stratovolcano

FIGURE 6.29 Types of volcanoes and tectonic settings.

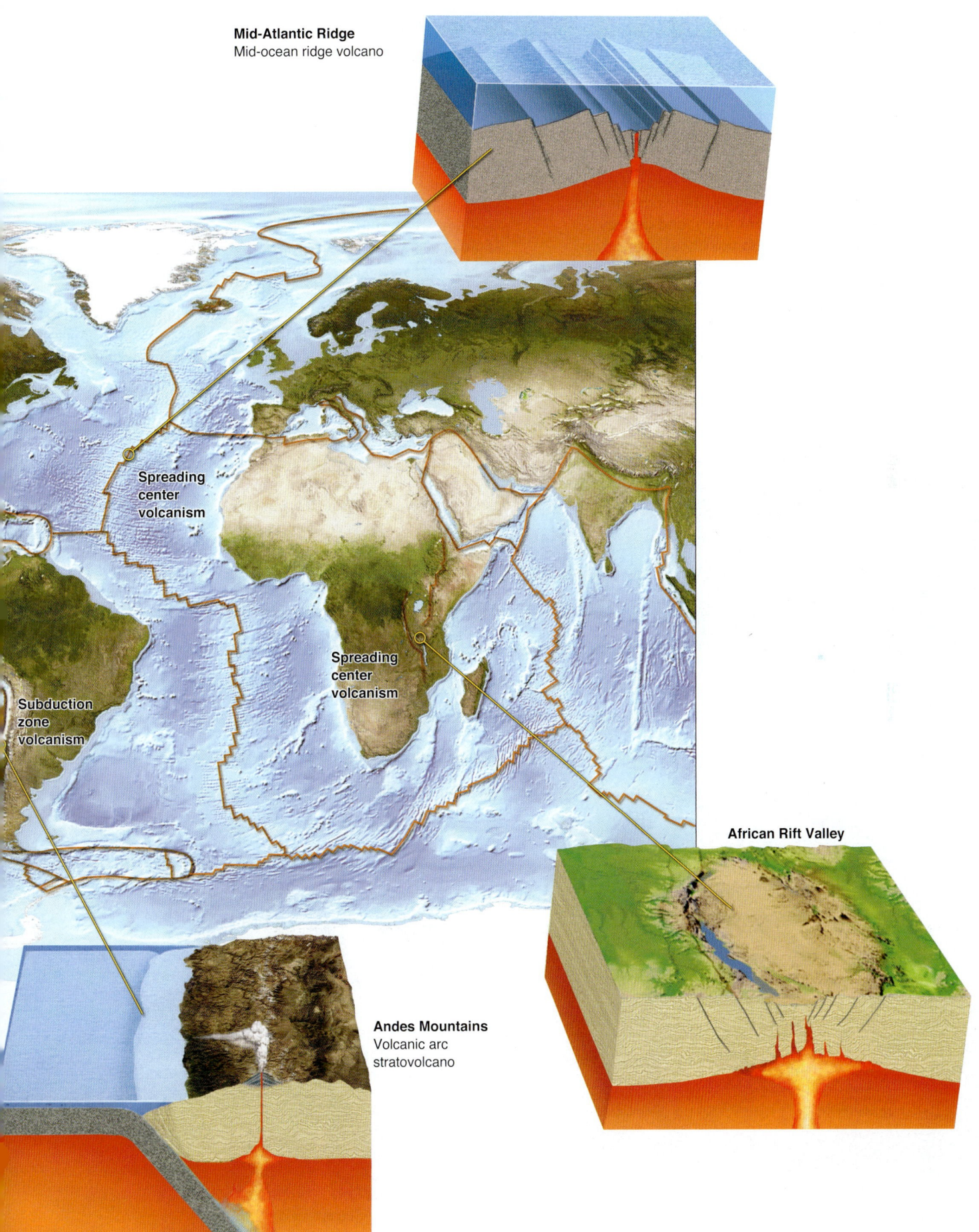

Mid-Atlantic Ridge
Mid-ocean ridge volcano

Spreading
center
volcanism

Spreading
center
volcanism

Subduction
zone
volcanism

Andes Mountains
Volcanic arc
stratovolcano

African Rift Valley

6-10 Volcanic Hazards Threaten Human Communities

LO 6-10 List and define several types of volcanic hazards.

We know that volcanoes are dangerous. They explode and their lava and pyroclastic flows burn, bury, and destroy anything in their path. In addition to volcanoes themselves, there are other kinds of volcanic hazards that can endanger life and threaten property.

FIGURE 6.30 Geologists measure a ground crack and monitor gas chemistry at the base of an active volcano.

Q What kinds of gases are emitted by volcanoes?

FIGURE 6.31 The buoyant plume of gas and ash formed by a Plinian eruption column can collapse around its outer edges and set off an avalanching, gas-charged cloud of ash, a pyroclastic flow that rolls down the slopes of the volcano.

Volcanic Hazards

Volcanoes tend to be seismically active. That is, they are places where rapid movement of magma through the ground breaks Earth's crust and causes earthquakes that can damage buildings, set off landslides, and open cracks and chasms in the ground (**FIGURE 6.30**).

The ground will tilt, swell, and fracture as magma migrates beneath the surface from one location to another. To follow the movement of magma and predict the growth of magma reservoirs, both of which may indicate that an eruption is imminent, volcanologists place networks of tiltmeters on active volcanoes and use GPS technology to measure the swelling and tilting slopes of a volcano.

Many arc volcanoes exhibit catastrophic eruptive styles (because of their viscous, gas-rich andesitic and rhyolitic magma), ejecting hot rock particles, ash, noxious gases, and needle-thin shards of glass. Even volcanoes at mid-ocean ridges, hotspots, and ocean-ocean convergent margins, which are characterized by more fluid, basaltic magma, can erupt explosively when they encounter groundwater, setting off phreatomagmatic eruptions. Also, because so much of the material in massive shield volcanoes is rapidly chilled lava (i.e., glass), the mechanical strength of these volcanoes is poor; they fracture easily and frequently, putting in motion large landslides.

As an eruption starts, gas-charged magma moves through the conduit system inside a volcano toward the surface. As it moves, the confining pressure is reduced, allowing dissolved gases to be released in much the same way that the carbon dioxide in a soda drink is released when its container is opened. Where these conditions occur in basaltic magma, the eruption is characterized by fire fountains hundreds of meters high (like the spray from a shaken soda drink spewing forth). When these conditions occur in the silica-enriched magmas that characterize arc volcanoes, the result will be a cataclysmic Plinian-style eruption.

As mentioned earlier, in a Plinian eruption, the outer edge of the column of erupted gas and ash can collapse downward to instigate an avalanche of hot, gas-charged ash and glass (**FIGURE 6.31**).

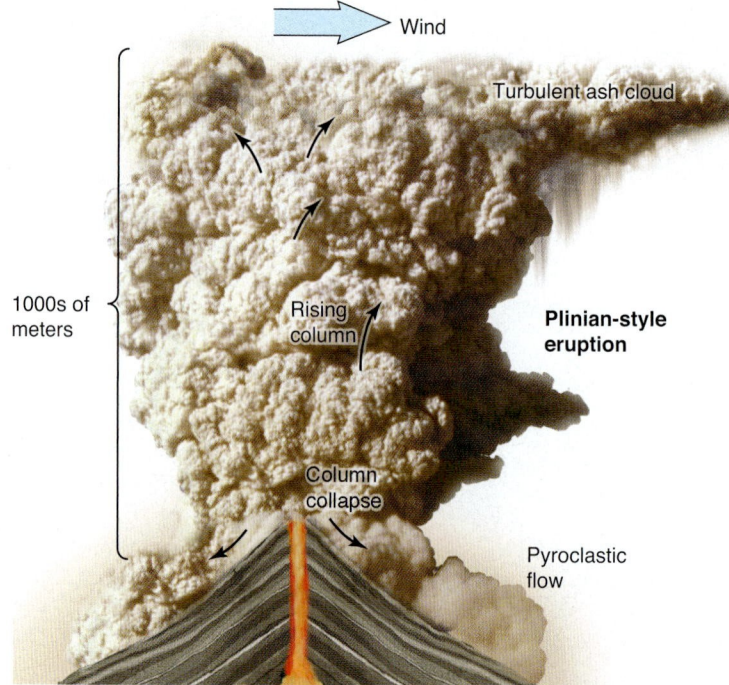

Stratovolcano

Q As intermediate and felsic magma rises inside a volcano during an eruption, it turns to ash. Why does this happen?

FIGURE 6.32 Lahars from the slopes of Mayon Volcano, Philippines, buried buildings and farms in Abay Province on December 2, 2006.

Ⓠ **What volcanic product mixes with water to make a lahar?**

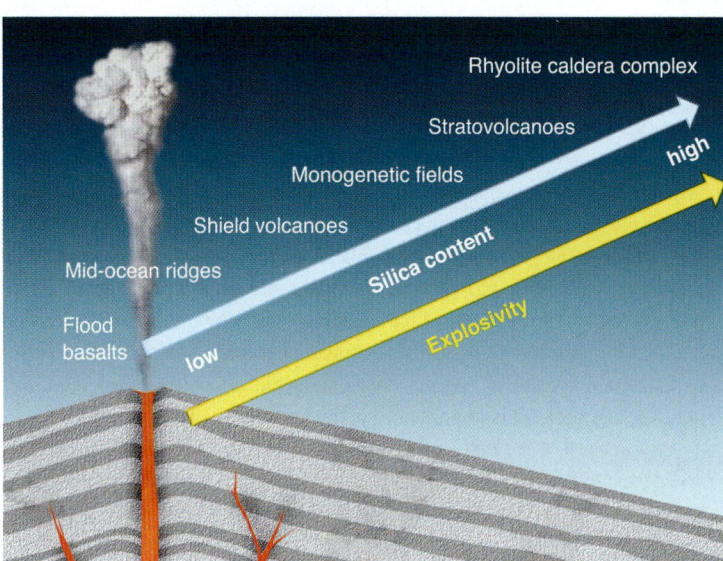

Bullit Marquez/©AP/Wide World Photos

FIGURE 6.33 The higher the silica content of its magma source, the more likely a volcano is to erupt explosively.

Ⓠ **Why is silica content a good guide to the explosivity of a volcano?**

The resulting pyroclastic flow rolls with the speed of a locomotive down the slopes of the volcano. Anything in the way of this turbulent flow will be instantly burned and buried under thick layers of hot ash. As noted earlier, Plinian eruptions are among the most devastating and deadly events in nature.

Heavy rain, tremors precipitated by the movement of magma, or oversteepening of a slope due to internal magma intrusions—all can trigger slope failure. A common deadly hazard on stratovolcanoes is a *lahar*, an Indonesian word denoting any type of flow but used by volcanologists to describe highly liquefied ash flows on volcanoes. Lahars are rapidly flowing mixtures of rock, mud, and water on the slopes of a volcano (**FIGURE 6.32**).

Lahars are formed in a variety of ways, including rapid melting of snow and ice, intense rainfall on loose volcanic deposits, the breakthrough of a lake dammed by volcanic deposits, or as a consequence of avalanches. Lahars are deadly events that bury everything in their path.

Volcanoes and Humans

As mentioned at the beginning of the chapter, scientists estimate that the total population at risk from volcanoes is at least 500 million worldwide. Volcanic activity can cause thousands of deaths, destroy towns and forests, and severely disrupt local economies for periods of months to years.

Clearly, volcanic eruptions are among Earth's most dramatic and violent agents of change. And the damage wreaked by these powerfully explosive events extends far beyond the tens of kilometers around a volcano; tiny volcanic particles erupted into the stratosphere can temporarily alter the climate of our entire planet.

Eruptions often force people living nearby to abandon their land and homes, usually forever. Those living farther away are likely to be spared complete destruction of their property, but their communities, crops, industries, transportation systems, and electrical grids may be damaged by explosively ejected tephra, lahars, earthquakes, pyroclastic flows, and landslides.

Furthermore, despite the improved ability of scientists to identify hazardous areas and to warn of impending eruptions, increasing numbers of people are today exposed to volcanic hazards. Scientists, therefore, continue to face a formidable challenge in providing reliable and timely alerts of eruptions to so many people at risk.

Volcanic Risk

As we have seen, the highest gas content is found in silica-rich magma. Higher gas content associated with explosive eruptions makes it possible to establish a ranking of volcanic explosivity potential, as illustrated in **FIGURE 6.33**. High-silica volcanoes (stratovolcanoes, rhyolite caldera complexes) are the most dangerous. But keep in mind that every volcano is dangerous: Long-quiet volcanoes have been known to release killer gases that seep across the countryside at night; passive shield volcanoes can suddenly explode when magma encounters groundwater; and once lava starts moving downhill, it is an unstoppable force. The point is, we must appreciate volcanoes for their might and majesty, respect them for their immense power and unpredictability, and, finally, avoid them because they are so dangerous.

Ⓠ **EXPAND YOUR THINKING**

Consider the various types of volcanoes in the United States (don't forget Hawaii and Alaska). Which state has the most explosive? Which state has the least explosive?

LET'S REVIEW

A volcano is any landform from which lava, gas, or ashes escape from underground, or have done so in the past. Volcanoes are a valuable component of the rock cycle, in that they rejuvenate the crust, renew soil, impact global climate, and contribute to gases in the atmosphere. At the same time, eruptions may consist of: violently explosive columns of gas and ash that penetrate high into the atmosphere; lava flows that burn and bury everything they encounter; and avalanching pyroclastic flows of scalding fumes and glass shards that move like a locomotive down volcanic slopes. To avoid volcanic hazards yet utilize volcanic resources, it is important to understand how volcanoes work.

In general, three types of magma are found in and around the world's volcanoes: basaltic, andesitic, and rhyolitic. The variation in silica content among these is an important factor that governs magma gas content.

High-silica magma is viscous and tends to prevent gases from escaping; low-silica magma is less viscous and allows gas to escape into the atmosphere, preventing the buildup of pressure. Gas content, together with pressure, determines whether an eruption will be effusive or explosive.

We identified two main categories of volcanoes: central vent volcanoes and large-scale volcanic terrains. Shield volcanoes, stratovolcanoes, and rhyolite caldera complexes are characterized by increasing amounts of silica and, therefore, greater potential of exploding, with rhyolite caldera complexes being the source of the largest explosions in Earth's history. Monogenetic fields, large igneous provinces, and mid-ocean ridges are large-scale volcanic terrains. Most volcanoes are associated with spreading center volcanism, arc volcanism, or intraplate volcanism.

STUDY GUIDE

6-1 A volcano is any landform from which lava, gas, or ashes escape from underground, or have done so in the past.

- **Volcanoes** occur throughout Earth's seas and continents. Although their explosions are magnificent, and awesome to witness, these natural phenomena exhibit distinctive behaviors for reasons that we can understand only through patient application of scientific thinking.

- It is important to improve our understanding of the behavior of volcanoes in order to protect people who live near them. In addition, volcanoes provide us with information about other important natural processes, such as plate tectonics, climate change due to volcanism, and earthquakes related to the movement of magma.

- Since 1700, volcanic activity has killed more than 260,000 people worldwide and destroyed entire cities. Even though geologists have greatly improved ways to identify hazardous areas and warn of impending eruptions, increasing numbers of people face certain danger from volcanic activity. Today, the number of people at risk from volcanoes is estimated at 500 million. That means scientists continue to face a formidable challenge in providing reliable and timely warnings of eruptions.

6-2 There are three common types of magma: basaltic, andesitic, and rhyolitic.

- A magma's silica content (SiO_2) exerts a fundamental control on the behavior of **lava**, through silicate tetrahedron bonding. Silica content is often related to gas content, and the presence of gases in a magma source can result in massive eruptions and produce pyroclastic debris.

- Three types of magma are commonly associated with volcanoes: **basaltic**, **andesitic**, and **rhyolitic**. Basaltic magma normally erupts without explosion. The lava is very fluid and flows out of the volcano with ease. Andesitic magma tends to contain more gas and is viscous, therefore resistant to flow. It is often expelled in an **eruption column** as part of an explosive eruption. Rhyolitic magma is so viscous and gas-rich that it usually sets off massively explosive eruptions that devastate the nearby countryside.

6-3 Explosive eruptions are fueled by violent releases of volcanic gas.

- The gases expelled in an explosive eruption include carbon dioxide, water vapor, sulfur dioxide, and others.

- Magma bodies rise in the crust until they reach a point of neutral buoyancy. As gas expands and magma moves closer to the surface, the lowering of pressure and expansion of gases drive eruptions. The interaction among the viscosity, temperature, and gas content of magma determines whether an eruption will be **explosive** or **effusive** (characterized by relatively fluid lava flow).

6-4 Pyroclastic debris is produced by explosive eruptions.

- Pyroclastic debris is another volcanic product usually associated with explosive eruptions or with fire fountains that develop in gas-rich magma that does not discharge explosively. It consists of ash, lapilli, blocks, and bombs, all of which are called **tephra** if they are expelled through the atmosphere. These may cool and solidify on the ground, forming a rock called **tuff**. A **pyroclastic flow** can occur when the base of an **eruption column** collapses then avalanches down the slopes of the volcano, burning and burying everything in its path under hot ash and gases.

6-5 Volcanoes can be classified into six major types based on their shape, size, and origin.

- Geologists distinguish between two general classes of volcanoes: **central vent volcanoes** and **large-scale volcanic terrains**. Within each class, they have identified three types of volcanoes. Each type has a distinctive shape and size (morphology), behavior, and chemistry.

6-6 Shield volcanoes are a type of central vent volcano.

- **Shield volcanoes** are fed by low-silica, low-gas magmas that originate in the mantle. These produce basaltic lava in the form of pahoehoe and aa flows that are fluid and rarely explosive. Because of their low viscosity, basaltic lavas cannot hold a steep slope, so the resulting volcano morphology is broad and gentle. Occasionally, explosive **phreatomagmatic eruptions** occur as a result of sudden contact with groundwater.

6-7 Stratovolcanoes and rhyolite caldera complexes are central vent volcanoes.

- **Stratovolcanoes**, or composite volcanoes, consist of alternating andesitic lava flows and layers of explosively ejected pyroclastic deposits. The chemistry of the magma source is intermediate, making the lava viscous and difficult to erupt. The result is explosive eruptions due to the buildup of gases within the magma.

- **Rhyolite caldera complexes** are the most explosive of Earth's volcanoes, typically leaving only gaping holes in the crust as a record of their behavior. High-silica, high-gas magmas lead to massive explosions, followed by collapse of the volcanic edifice, producing an inverse volcano.

6-8 Large-scale volcanic terrains lack a central vent.

- Poorly understood by geologists, **monogenetic volcanic fields** are collections of hundreds of separate **vents**, lava flows, and **cinder cones** that are thought to be fed by a single magma source.

- **Large igneous provinces (LIPs)** are composed of flood basalts fed by massive mantle plumes. These are characterized by especially fluid basaltic lavas discharged over time, leading to the creation of large plateaus of basalt.

- Mid-ocean ridges develop at spreading centers and are characterized by basalt flows and the creation of young crust.

6-9 Most volcanoes are associated with spreading center volcanism, arc volcanism, or intraplate volcanism.

- Plate tectonics can help explain many aspects of volcanoes. In general, volcanoes are found in three types of plate settings. Spreading center volcanism is characterized by fluid basaltic magma that occurs at divergent plate margins; arc volcanism is characterized by explosive rhyolitic, andesitic, and basaltic magma that occurs at convergent margins; and intraplate volcanism is characterized by shield volcanoes, rhyolite caldera complexes, and monogenetic fields.

- Arc volcanism occurs in places where two plates converge. One plate, containing oceanic crust, descends beneath the adjacent plate, which can be composed of either continental or oceanic crust. If the overriding plate is oceanic, an island arc will be created. If the overriding plate is continental, a volcanic arc will develop.

- Geologists hypothesize that some hotspots are fed by a mantle plume with a source near the core-mantle boundary.

- Both convergent settings and intraplate hotspots are characterized by partial melting, yielding ample opportunity to produce high-silica magma and explosive volcanism.

6-10 Volcanic hazards threaten human communities.

- Volcanoes are extremely hazardous. Pyroclastic flows, lahars, ash falls, volcanic bombs, and massive lethal violent explosions are all sources of danger to humans, animals, and property.

KEY TERMS

andesitic magma (p. 155)
basaltic magma (p. 155)
calderas (p. 167)
central vent volcanoes (p. 152)
cinder cones (p. 164)
craters (p. 167)
effusive eruption (p. 153)
eruption column (p. 159)
explosive eruption (p. 153)
flood basalt (p. 170)

large igneous provinces (LIPs) (p. 170)
large-scale volcanic terrains (p. 152)
lava (p. 155)
monogenetic volcanic fields (p. 170)
phreatomagmatic eruption (p. 164)
Plinian eruption (p. 167)
pyroclastic debris (p. 155)
pyroclastic flows (p. 160)
rhyolite caldera complexes (p. 167)
rhyolitic magma (p. 155)

shield volcanoes (p. 163)
stratovolcanoes (p. 156)
tephra (p. 158)
tuff (p. 160)
vent (p. 152)
volcano (p. 152)
volcanologists (p. 152)
welded tuff (p. 160)

ASSESSING YOUR KNOWLEDGE

Please complete this exercise before coming to class. Identify the best answer to each question.

1. What is the best definition of a volcano?
   a. Any landform from which lava, gas, or ashes escape from underground, or has done so in the past.
   b. A large mountain that spews lava, gas, or ashes.
   c. Any feature on Earth that emits gas and lava, or has done so in the past.
   d. A hole in the ground from which lava escapes.
   e. A layer of ash on sloping ground.

2. Why are hotspots characterized by both explosive and nonexplosive volcanism?
   a. Because they always occur near seawater.
   b. Hotspots are never explosive.
   c. Because of their high temperature.
   d. Hotspots may have different types of magma sources.
   e. Hotspots are only explosive if they have a basaltic magma source.

3. Which of these volcanoes (a. or b.) is being fed by magma with the highest viscosity? _____

(a)　　　Associated Press　(b)　　　Beboy_ltd_iStockphoto

4. Why does this flow stand so high above the surrounding land?
   a. The surrounding land was forced down by the weight of the flows.
   b. The flows are composed of low-viscosity lava.
   c. The surrounding land was low, and the flows filled in valleys.
   d. The flows are composed of high-viscosity lava.
   e. It is actually many flows on top of each other.

5. What are the three products of volcanism?
   a. Gas, water, and dust
   b. Rock, lava, and intrusions
   c. Volcanic gas, lava, and pyroclastic debris
   d. Pyroclastic debris, fluid lava, and hardened lava
   e. Volcanoes, craters, and vents.

6. The viscosity of magma increases when:
   a. High silica content builds tetrahedral chains.
   b. Low silica content dissolves tetrahedral chains.
   c. High gas content breaks down tetrahedral chains.
   d. Low gas content breaks down tetrahedral chains.
   e. The number of tetrahedra decreases.

7. A Plinian-style eruption is characterized by:
   a. Phreatomagmatic explosions.
   b. Fissure eruptions and high volumes of lava.
   c. A massive eruption column of gas and pyroclastic debris.
   d. A lack of pyroclastic debris.
   e. Lava effusion.

8. These two types of lava are known as (a) _____ (aa, pahoehoe) and (b) _____ (aa, pahoehoe).

(a)　　USGS, Hawaiian Volcano Observatory　(b)　© Martin Rietze/Westend61/Co

9. True or False:
   _____ Shield volcanoes always occur at convergent margins.
   _____ The Mid–Atlantic Ridge is actually a monogenetic field.
   _____ Few volcanoes actually erupt lava; most erupt granite.
   _____ A large igneous province may mark the point at which a mantle plume first discharged lava onto Earth's crust.
   _____ Rhyolite caldera complexes most commonly are formed in places where continental crust is partially melted, producing a high-silica magma source.

10. A lava dome:
    a. Is formed by a large pile of cinders in the shape of a cone.
    b. Acts like a plug in a vent and so is usually explosively expelled.
    c. Is found only in shield volcanoes over hotspots.
    d. Is typically composed of low-silica basalt.
    e. All of the above.

11. Label the figure with these terms:
    Rhyolite caldera complex
    Island arc　　　　　　　　　　Mid-ocean ridge
    Stratovolcano　　　　　　　　Hotspot
    Shield volcano　　　　　　　　Volcanic arc

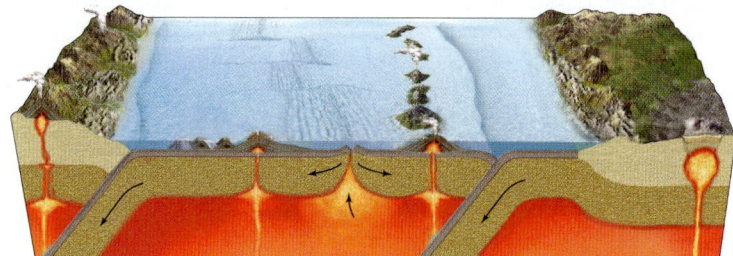

12. Yellowstone Volcano:
    a. Is associated with a hotspot under continental crust.
    b. Is an example of a rhyolite caldera complex.
    c. Is the site of the largest volcanic eruption in known geologic history.
    d. Still has an active magma chamber below ground.
    e. All of the above.

13. Fill in the blank to identify the most likely type of volcano to occur in each tectonic setting.

Tectonic Setting	Type of Volcano
Oceanic hotspot	_____ (shield volcano, strato-volcano, monogenetic field, cinder cone)
Volcanic arc	_____ (shield volcano, strato-volcano, monogenetic field, cinder cone)
Island arc	_____ (shield volcano, strato-volcano, monogenetic field, cinder cone)

14. Generally speaking, volcanic arc magma is more _____ (felsic, mafic) than island arc magma.

15. The African rift valley is an example of _____ (spreading center, arc, intraplate) volcanism.

16. Generally speaking, explosive eruptions develop from magma that has a high concentration of _____ (dissolved gas, dissolved iron, pahoehoe, ash content).

17. The photo shows an example of an _____ (explosive, effusive, phreatomagmatic) eruption.

18. Volcanic outgassing may contain the following gases (circle all that are correct). Water, carbon dioxide, oxygen, hydrogen, sulfur dioxide, nitrogen, dioxin

19. The shape of a volcano is largely due to the chemistry of the magma. True or False?

20. Someday, the Atlantic seafloor may begin to subduct beneath the eastern United States. This is most likely to produce:
    a. An island arc
    b. A hotspot
    c. A mid-ocean ridge
    d. A volcanic arc
    e. A large igneous province

FURTHER RESEARCH

Go to the Web and research one of these volcanoes. Then answer the questions below as part of a two-page report that you turn in to your instructor.

Vesuvius, Italy	Novarupta and Amak, Alaska
Kilauea, Hawaii	Unzen, Japan
Etna, Sicily	Rotorua, New Zealand
Mount St. Helens, Washington	Yellowstone, Wyoming
Krakatau, Indonesia	Juan de Fuca Ridge
Tambora, Indonesia	Eastern Pacific

1. What is the name of the volcano or eruption that you selected?

2. Describe the volcanic processes occurring in your volcano, and classify it using the classification system presented in this chapter.

3. Is the volcano currently active? If not, when did it last erupt? What is its eruption history?

4. What is its eruption style: explosive, effusive, or some other type? Please elaborate.

5. What type of magma is present? How does its chemistry influence the volcanic processes occurring in this volcano?

6. What is the tectonic setting?

7. Is the volcano near a human settlement? How is this significant?

8. Describe unique and important features of the activity or eruption.

9. Use the Web to discover additional information about your volcano.

10. Turn in your typed (not handwritten) report, complete with a title page, illustrations, reference sources, and other research information. Your instructor will detail the proper format for you.

ONLINE RESOURCES

Explore more about volcanoes on the following Web sites:

NASA Earth Observatory contains up-to-date information on the latest eruptions:
http://earthobservatory.nasa.gov/NaturalHazards

USGS Cascades Volcano Observatory is full of activity updates, event reports, and an excellent photo glossary:
http://vulcan.wr.usgs.gov

USGS Alaska Volcano Observatory is a storehouse of information about volcanoes:
www.avo.alaska.edu

USGS Hawaii Volcano Observatory contains excellent information about the Hawaiian hotspot, shield volcanoes, and the ongoing eruption of Kilauea Volcano:
http://hvo.wr.usgs.gov

Additional animations, videos, and other online resources are available at this book's companion Web site:
www.wiley.com/college/fletcher

This companion Web site also has more information about WileyPLUS *and other Wiley teaching and learning resources.*

WEATHERING

CHAPTER CONTENTS & LEARNING OBJECTIVES

GEOLOGY IN OUR LIVES

Weathering is the physical, chemical, and biological degradation of Earth's crust. Weathering produces soil, which yields much of our food and the ores of critical metals, such as aluminum and iron. Almost everything we humans need to live can be traced back to weathering: the air we breathe, food we eat, and clothes we wear, as well as paper, timber, medicines, shade, and much more. Unfortunately, erosion caused by human activities destroys soil, damages ecosystems, and devastates coastal environments faster than they can naturally rejuvenate; and deforestation depletes soil to such a degree that natural ecosystems can rarely recover on their own. By understanding weathering, we can learn to conserve geologic resources and protect natural environments.

Q Monument Valley Utah is the product of weathering. What weathering processes contributed to the development of these remarkable rock formations?

© John and Lisa Merrill/Corbis

7-1 Weathering Includes Physical, Chemical, and Biological Processes

LO 7-1 Compare and contrast the three types of weathering.

In this chapter we delve into the subject of **weathering**, a set of physical, chemical, and biological processes that break down rocks and minerals in Earth's crust to produce sediment, new minerals, soil, and dissolved ions and compounds.

Weathering

Weathering (**FIGURE 7.1**) produces sediment, and **erosion** transports it across the landscape. Sediment produced by weathering consists of broken pieces of rock, mineral grains released from the crust, and new minerals that crystallize during the weathering process, or later, from the dissolved compounds yielded by weathering.

Streams, rivers, waves, gravity, and wind are responsible for the erosion that carries sediments to their ultimate resting places, known as *environments of deposition* (which we will study in Chapter 8, "Sedimentary Rock"). It is in depositional environments, such as the ocean, that sedimentary rock is formed and enters the rock cycle.

An important result of weathering is the formation of **soil.** Soil is unconsolidated mineral and organic material constituting the uppermost layer on Earth's surface. It serves as a natural medium for the growth of land plants, making it is an essential resource. Almost everything we need to survive can be traced back to soil: the air we breathe, food we eat, and clothes we wear, as well as paper, timber, medicines, shade, and much more. But soil that is eroded by human activities also is a major pollutant of streams, wetlands, and coastal ecosystems.

It takes many centuries to form a few centimeters of soil but only minutes to carry it away. Fortunately, with careful management, sustainable farming practices, and an understanding of weathering and erosion processes, we can conserve soil for future generations.

Weathering can be categorized as physical, biological, and chemical:

Physical weathering occurs when rock is fragmented by physical processes that do not change its chemical composition. It involves the mechanical breakdown of minerals and rocks by a variety of processes. If, say, you wanted to "weather" a piece of paper using mechanical force, you would tear it into pieces.

Biological weathering occurs when rock disintegrates due to the chemical and/or physical activity of a living organism. The types of organisms that cause weathering range from bacteria to plants and animals. The biological weathering of a piece of paper might occur by a mouse nibbling at it or by mold growing on it and causing the paper to decay.

Chemical weathering is the chemical decomposition of minerals in rock. This process may result in the formation of: 1) new **sedimentary minerals** (minerals formed by weathering); 2) compounds that are dissolved in water; and 3) gases that escape to the atmosphere, are dissolved in water, or are trapped in cavities in soil and sedimentary deposits. To chemically weather a piece of paper, you could burn it, which would produce gases and a residue of ash.

Chemical, physical, and biological weathering work together to break down Earth's crust, although of the three chemical weathering is generally the most effective in that it attacks all surfaces that are exposed to gases and fluids.

Physical and biological weathering cause rock to fragment into particles, thereby increasing the surface area that is vulnerable to chemical weathering (**FIGURE 7.2**). Consequently, the effectiveness of chemical weathering is greatly enhanced by mechanical and biological processes.

In Figure 7.2, the effect of increasing the amount of exposed surface area is illustrated on the large block on the left, which has a

Darlene Cutshall/Shutterstock

FIGURE 7.1 Physical, chemical, and biological weathering cause rocks and minerals to decompose.

Describe the weathering evident in this photograph.

1 unit

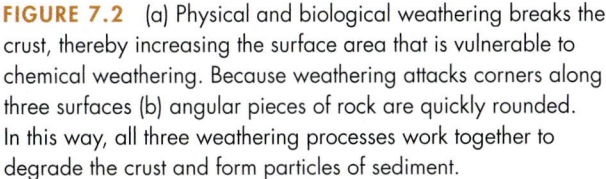

1/2 unit

1/4 unit

Original cube
(surface area 6 square units)

(a)

First subdivision into 8 cubes
(surface area 12 square units)

(b)

Second subdivision into 64 cubes
(surface area 24 square units)

Attack on
three sides

Attack on
two sides

Attack on
one side

Progressive reduction
of cube to sphere

FIGURE 7.2   (a) Physical and biological weathering breaks the crust, thereby increasing the surface area that is vulnerable to chemical weathering. Because weathering attacks corners along three surfaces (b) angular pieces of rock are quickly rounded. In this way, all three weathering processes work together to degrade the crust and form particles of sediment.

Why do the corners and edges of rock weather faster than a flat surface?

surface area of 6 square meters. When it is divided into 8 equal cubes, the surface area doubles, to 12 square meters; when it is divided into 64 equal cubes, the surface area doubles again, to 24 square meters. Continuing this process for 10 sequential halvings produces over 1 billion cubes, each 1 millimeter on a side (the size of a grain of sand).

The combined surface area of these cubes is 1000 times greater than that of the original cube. With over 1000 times more surface area to act on, chemical weathering becomes 1000 times more effective at destroying the rock. **FIGURE 7.3** contains common examples of everyday weathering.

K. Luzzi Paul/Shutterstock

Zyxx/Shutterstock

(a)

(b)

FIGURE 7.3 Weathering occurs all around us. (a) On this door, old paint is undergoing a combination of chemical and physical weathering. (b) This old car is rusting as a result of chemical weathering. (c) Marble statues are especially vulnerable to chemical weathering in humid climates. (d) The roots of this tree are breaking up the sidewalk by biological weathering.

agezinder/Age Fotostock America, Inc.

© EuroStyle Graphics/Alamy

(c)

(d)

EXPAND YOUR THINKING

Nearly everything becomes weathered when left exposed to the elements (including your skin, which you'll begin to notice as you grow older). **Describe a common example of weathering that you find outside your building.**

GEOLOGY IN ACTION

Space Weathering

On Earth's surface, weathering of exposed rocks is caused by the action of water, natural gases, and life. Where there is little atmosphere, such as on the Moon or asteroids, a different kind of weathering, called *space weathering*, occurs. Dr. Bruce Hapke, professor emeritus at the University of Pittsburgh, was the first scientist to correctly identify the process of space weathering.

The Moon's surface is bombarded by meteorites of all sizes, ranging from objects the size of mountains to tiny particles smaller than a grain of sand (called *micrometeorites*). These hit the Moon at speeds of over 16 kilometers per second. In addition to creating craters, the largest of which can be seen through telescopes, high-velocity impacts break up exposed rock into a fine powder referred to as *regolith*. (Regolith is the blanket of soil and loose rock fragments that overlays the bedrock; it is found on Earth as well as the Moon.) The entire lunar surface is covered with regolith, which can be seen in pictures and videos taken by astronauts who have landed on the Moon (**FIGURE 7.4**).

However, there is more to space weathering than the mere smashing of rocks. Material that is exposed to space also darkens over time. If a moon rock is ground into a fine powder similar to regolith, the powder will be considerably darker than the original rock. Thin layers of vaporized rock that coat the surfaces of grains of regolith cause this darkening. Freshly exposed rock surfaces on Earth also tend to darken with time, though they do so by chemical, not space, weathering.

On the Moon, rocks can be vaporized by two processes: meteorite impacts and radiation from the Sun. Enough heat is generated by high-speed impacts to melt and evaporate some particles of regolith. When the solar wind hits the lunar surface at nearly 500 kilometers per second, it knocks atoms off the rock surface in a process known as *sputtering*. Both evaporated and sputtered material coat surrounding grains of regolith, making them darker.

All lunar rocks contain iron, often combined with oxygen. That said, the processes of evaporation, sputtering, and redeposition of rock vapor result in the loss of some oxygen atoms, leaving the iron behind. The left-over iron forms tiny grains of metal less than one-millionth of a centimeter across. These tiny grains of metallic iron are very efficient absorbers of light. Hence, any object coated with these grains turns dark. If it were not for these fine iron particles, the Moon would be about three times brighter than it is.

FIGURE 7.4 Space weathering has produced the layer of weathered debris called regolith that covers the surface of the Moon.

Eugene Cernan/Corbis

Physical Weathering Causes Fragmentation of Rock

LO 7-2 Describe several types of physical weathering.

Physical weathering occurs in many different ways and plays the crucial role of increasing the surface area of rocks, which makes them more vulnerable to chemical decomposition. It also changes the appearance of Earth's surface as a result of several processes: pressure release, abrasion, freeze-thaw, hydraulic action, salt-crystal growth, and others.

Pressure Release

Rock is brittle, and it breaks when overlying pressure is released by tectonic forces or by erosion. This *pressure release* leads to the growth of fractures, known as **joints.** Joints are openings, or "partings," in rock where the two sides of the break are not offset, or laterally displaced.

If lateral displacement has occurred, the fracture is termed a *fault.* (See Chapter 10, where we discuss the process of faulting.) Sets of joints composed of repeating parallel fractures may occur, exposing large areas of crust to weathering (**FIGURE 7.5**).

Sheeted joints develop when rock is slowly uplifted by tectonic forces or by the removal of overlying layers by erosion. As the weight of overlying rock is released, the crust expands and fractures into flat horizontal slabs.

FIGURE 7.5 Sets of joints composed of repeating parallel fractures expose the crust to weathering.

How do joints affect the rate of rock weathering, and why?

Exfoliation occurs when these slabs shift and uncover the underlying rock. Half Dome in Yosemite National Park is part of a large exfoliated granite batholith that was exposed by glacial erosion. The removal of the overlying weight by glacial scouring resulted in sheeted joints and exfoliation (**FIGURE 7.6**).

Abrasion

Abrasion is an important agent of physical weathering. Abrasion occurs when sedimentary particles collide, leading to mechanical

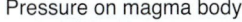
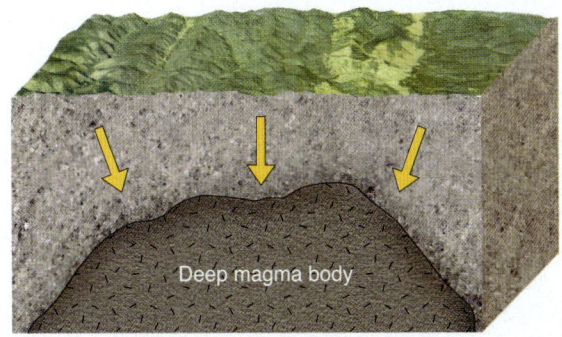
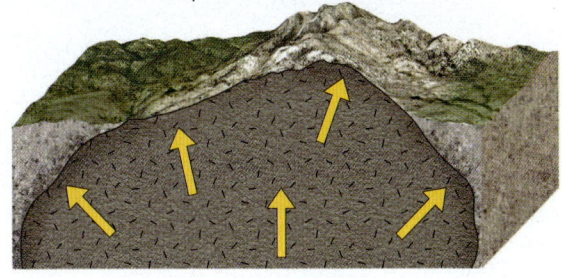

Gregory G. Dimijian/Photo Researchers

FIGURE 7.6 When overlying weight is removed, (a) rock expands and forms sheeted joints. Exfoliation (b) shapes the landscape in the Sierra Nevada in California.

The presence of exfoliation is evidence of pressure release. Say you observed exfoliation; use inductive reasoning to build a hypothesis that expresses a general principle.

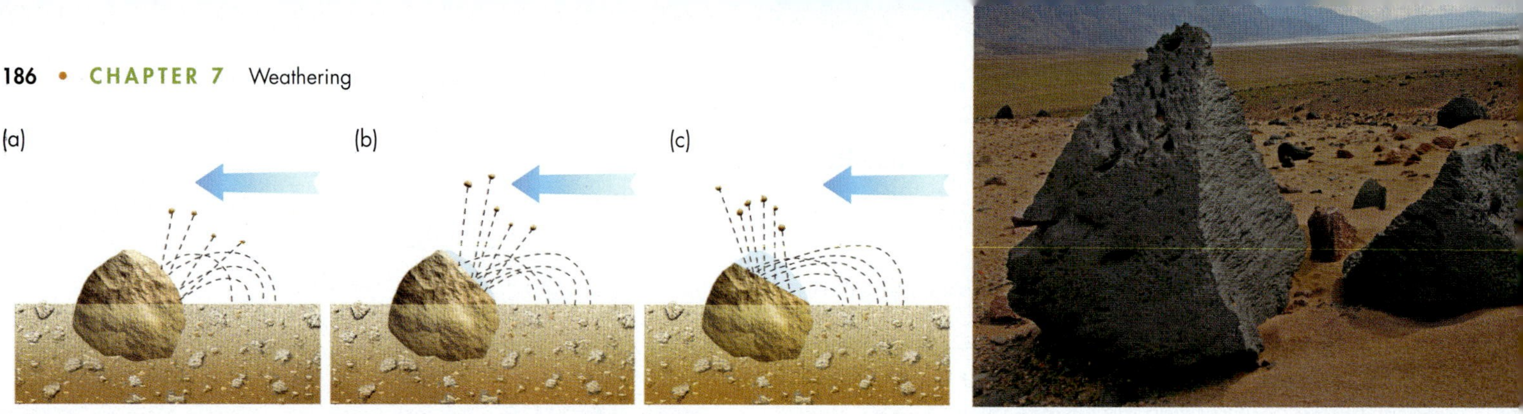

(a)   (b) 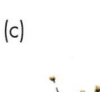 (c)

© Marli Bryant Miller

FIGURE 7.7 Ventifacts are formed when wind-blown particles abrade rock, (a) creating a flat surface (b) that faces in the direction of the prevailing wind (c).

🅠 **Draw an arrow on the photograph indicating the wind direction.**

wearing or grinding on their surfaces. This happens when small particles of rock carried by wind, water, or ice collide with larger rocks.

Blowing wind and running water are usually laden with suspended particles that abrade any surface they encounter, a process often referred to as "sandblasting." Abrasion creates sand and silt; and because it acts quickly, over geologic time it can cause significant and dramatic alteration of the crust. Rocks with unusual shapes and "fluted" or flat faces may have been abraded by wind-blown sediment (**FIGURE 7.7**). Such rocks are called **ventifacts.**

Freeze-Thaw

Ice wedging (**FIGURE 7.8**) occurs when water flows into a joint and freezes. Water increases in volume by 9% when it turns into ice, and

the growth of ice crystals forces the joint to split open. This process plays a major role in weathering the crust in temperate, arctic, and alpine regions.

Ice wedging is most effective at –5°C. It, along with gravity, is responsible for the formation of **talus,** slopes of fallen rock that collect at the base of cliffs and steep hillsides.

Hydraulic Action

On rocky shorelines, the powerful force of breaking waves drives water into cracks and fractures in the rock. Air trapped at the bottom of these openings is compressed against the rock and weakens it.

As the wave recedes, the pressurized air is released—explosively—cracking the rock and expelling fragments of the rock face. This pro-

© Marli Bryant Miller

FIGURE 7.8 Ice wedging on the steep side of a cliff causes blocks to fall and collect at the cliff's base. The resulting deposit of ice-wedged boulders is called a *talus slope.*

🅠 **What will this place look like eventually as frost wedging continues to weather the cliff face over time? What if crustal uplift occurs?**

FIGURE 7.9 When a wave strikes a rocky shoreline, the force of the water compresses air in cracks and openings of the rock. This exerts pressure, which can crack, break, splinter and detach rock particles. As the wave retreats, the air can decompress explosively, further weakening the rock. As cracks gradually widen, the amount of compressed air increases, and the explosive force of its release grows.

cess widens the crack so that more air is trapped by the next wave, leading to a greater explosive force. This hydraulic action damages cliffs and other types of coastal rock outcrops (**FIGURE 7.9**).

Growth of Salt Crystals

When saline water seeps into fractured crust and then evaporates, it may initiate the growth of salt crystals (**FIGURE 7.10**) that break the rock. Crystal growth usually accompanies evaporation, so for it to occur, heat must accumulate in the rock during the day. The salt crystals expand as they are heated and exert pressure on the surrounding rock.

The most effective salts are compounds of calcium chloride, magnesium sulfate, and sodium carbonate, all of which are found dissolved in seawater. Some of these salts can expand by three times or more. Salt-crystal growth is common in places where persistent salt-laden winds blow against sea cliffs and other types of rocky shorelines.

Questionable Factors in Physical Weathering

A number of other physical weathering phenomena are subjects of ongoing debate among geologists. In particular, weathering by expansion and contraction due to daily changes in rock temperature, known as *insolation weathering* (insolation refers to solar radiation), has been one of the most intensely disputed topics in research on rock weathering.

Previously, researchers thought that the continual heating and cooling of a rock—for example, on the floor of a desert—would cause the rock to first expand and then contract. This repeated swelling and shrinking would then lead to cracking and breakage of the outer surface. But repeated laboratory experiments have failed to confirm this hypothesis, leading scientists to question the role of insolation weathering as an important process.

Alternate wetting and drying of rocks, known as *slaking*, is also thought to play a role in weathering. Slaking occurs when successive layers of water molecules accumulate between the mineral grains of a rock. The increasing thickness of the water imposes tensional stress, which pulls the rock grains apart. Laboratory research has shown that slaking, in combination with naturally occur-

ring dissolved sodium sulfate (Na_2SO_4) found in soil moisture, can disintegrate samples of igneous and metamorphic crock in only 20 cycles of wetting and drying.

Weathering-Erosion Pathway

The various forms of physical and chemical weathering working together are effective at releasing fragments of rock from the crust and moving them into surface environments, from where they are later transported by erosion. This weathering-erosion pathway exists in many forms in many places; one example is shown in **FIGURE 7.11**.

Freeze-thaw weathering pries blocks and grains of rock loose from cliff walls, and they accumulate at the base as a talus slope. The talus moves slowly downhill under the force of gravity as it is undercut at its far edge by running water in a channel. As rock fragments fall into the channel, sediment in the moving water abrades the rock, reducing its size and producing particles of gravel, sand, and silt in the process.

FIGURE 7.10 Salt crystal growth occurs when seawater and other saline solutions seep into cracks and joints in rock and then evaporate, leaving behind salt crystals. Salt crystals expand as they are heated in the Sun, exerting pressure on the surrounding rock. This type of weathering may produce a honeycomb pattern of mechanical weathering known as *tafoni*.

These sedimentary particles are moved downstream during subsequent floods. Continued transport and abrasion in the channel reduce the average particle size of the sediments and they join the large volume of sediments that are moved from positions high in a watershed to positions downstream, perhaps as far as the ocean, whereupon they are deposited and become sedimentary rock.

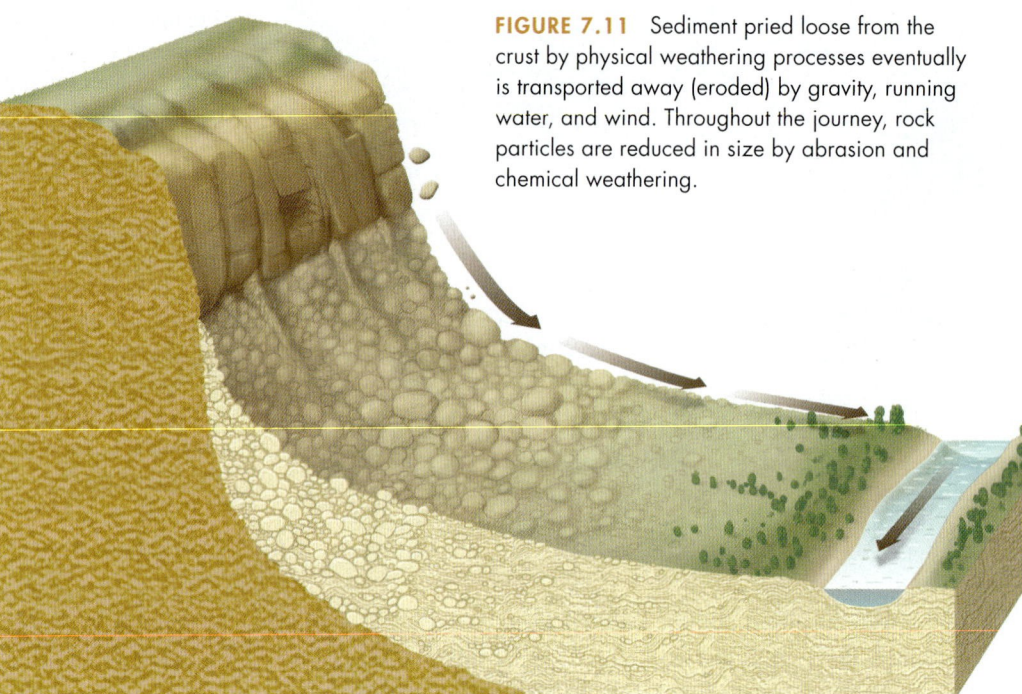

FIGURE 7.11 Sediment pried loose from the crust by physical weathering processes eventually is transported away (eroded) by gravity, running water, and wind. Throughout the journey, rock particles are reduced in size by abrasion and chemical weathering.

Q EXPAND YOUR THINKING

Washing clothes makes them look weathered. Is this physical, biological, or chemical weathering? Why?

7-3 Hydrolysis, Oxidation, and Dissolution Are Chemical Weathering Processes

LO 7-3 Describe the role of water in chemical weathering.

Most chemical weathering is the result of water interacting with minerals in a rock. Water can be a particularly effective agent of decomposition because of the nature of the water molecule.

The Water Molecule

A water molecule is formed when 2 atoms of hydrogen bond covalently with 1 atom of oxygen (**FIGURE 7.12**). The bonding involves the sharing of electrons by the hydrogen and oxygen atoms; however, in water, the sharing is not equal. The oxygen atom attracts the shared electrons more strongly than the hydrogen does, and the result is a slightly charged molecule (known as a *polarized molecule*). The 2 hydrogen atoms tend to be bonded strongly to one side of the oxygen atom (across an arc of ~104°), leaving the other side of the molecule open and able to form more bonds.

Close to the hydrogen atoms there is a partial positive charge; elsewhere, across the face of the oxygen atom, there is a partial negative charge. The negatively charged region tends to attract cations (positively charged ions), especially another hydrogen cation. This property also allows water to bond with ions in other molecules, such as metallic cations in most minerals.

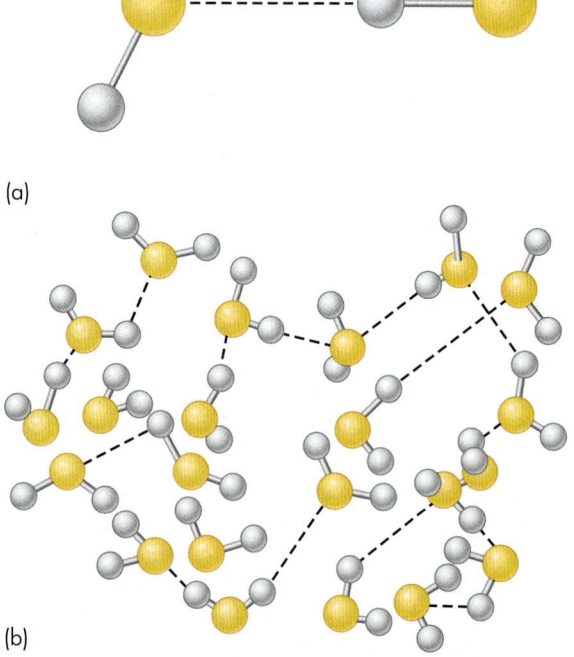

(a)

(b)

FIGURE 7.12 Water molecules (a) are polarized because hydrogen (gray) bonds to one side of the oxygen (gold) while the other side is exposed. (b) Water molecules bond to one another by sharing a hydrogen atom. A hydrogen atom of one molecule will bond to the open side of another molecule. When a number of hydrogen bonds act in unison, they create a strong combined effect, forming liquid water.

Q Does liquid water have a crystalline structure?

By bonding with, and removing cations from, a solid mineral surface, water is highly effective at dissolving many substances, earning its title as "the universal solvent." Water, therefore, plays a role in most chemical weathering reactions.

Hydrolysis

Hydrolysis is the most common chemical weathering process because it causes the decomposition of silicate minerals, the most widespread inorganic substances in Earth's crust. Many minerals are decomposed at least partially by hydrolysis (**FIGURE 7.13**).

During hydrolysis, ions in a mineral react with hydrogen (H^+) and hydroxyl (OH^-) ions in water. The hydrogen ions replace some of the cations in the mineral, thereby changing the mineral's composition. In addition, other ions and compounds may be dissolved from the mineral and carried away in the water.

Hydrolysis is not very effective in pure water, but when some acid (dissolved H^+) is added, it becomes an important agent of weathering. Acidic rainwater forms naturally when carbon dioxide in the atmosphere or the ground dissolves in water, to produce *carbonic acid*. Carbonic acid is formed by this reaction:

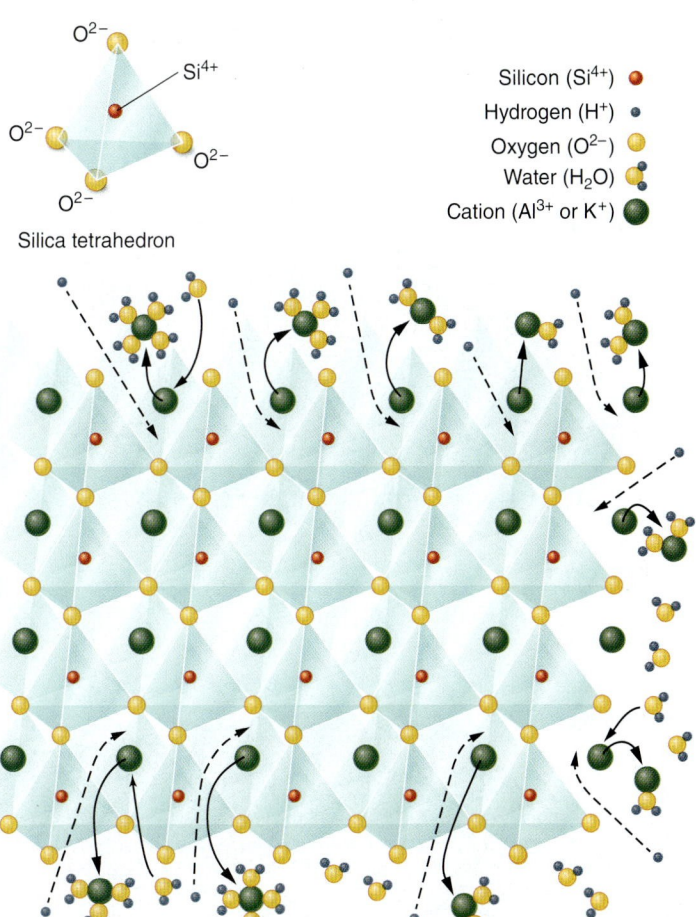

Silica tetrahedron

Silicon (Si^{4+}) ●
Hydrogen (H^+) ●
Oxygen (O^{2-}) ●
Water (H_2O) ●
Cation (Al^{3+} or K^+) ●

FIGURE 7.13 The polar nature of the water molecule enables it to easily bond with cations on a solid mineral surface. The solid mineral is dissolved when the ions are removed by joining with water molecules. In some cases, the cation is replaced with a hydrogen atom.

Why is water such an effective weathering agent?

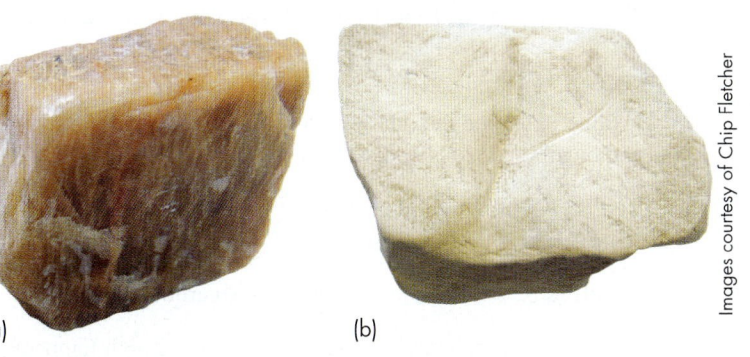

(a)   (b)

FIGURE 7.14 Hydrolysis of orthoclase feldspar (a) produces the sedimentary mineral kaolinite (china clay) (b).

CO_2 (gas in the atmosphere or ground) + H_2O (water vapor in the atmosphere or ground) = H_2CO_3 (carbonic acid; usually precipitates as moisture).

Hydrolysis breaks down feldspar, the most abundant mineral in Earth's crust, and creates **clay**, the most abundant sediment. Hydrolysis is the primary method by which some clay minerals, the primary minerals in the sedimentary rocks *shale* and *mudstone*, are formed.

Dissolved ions released during the hydrolysis of feldspar may combine to form other sedimentary minerals. The following equation is a hydrolysis reaction involving a common feldspar, orthoclase:

$$2KAlSi_3O_8 + 2H_2CO_3 + 9H_2O \rightleftharpoons Al_2Si_2O_5(OH)_4 + 4H_4SiO_4 + 2K^+ + 2HCO_3^-$$

orthoclase + carbonic acid + water (yield) clay (kaolinite) + silicic acid in solution + dissolved potassium and bicarbonate ions

In this reaction, the igneous mineral orthoclase decomposes to produce a clay mineral, *kaolinite* (also known as china clay). The physical properties of kaolinite are quite different from those of orthoclase (**FIGURE 7.14**). Unweathered orthoclase is a relatively hard mineral with two directions of cleavage, whereas crystals of kaolinite are soft, white, and microscopic.

Oxidation

Oxidation is another important chemical weathering process. Oxidation involves the loss of an electron from a cation in a crystal, and its use by free oxygen in the environment. In the process, the oxygen and the cation bond to become a new mineral that is a member of the *oxide mineral class*. Readily oxidized elements include iron, sulfur, and chromium.

Iron is the most commonly oxidized cation, illustrated by these two examples:

- Ferrous iron (Fe^{+2}) or ferric iron (Fe^{+3}), when combined with gaseous oxygen (O_2), yields the compound Fe_2O_3, *hematite*, a sedimentary mineral that is an important iron ore.

- Rusty brown oxides of iron develop on the surface of iron-containing rocks and minerals (**FIGURE 7.15**), as in the olivine variety *fayalite*, where iron released by hydrolysis undergoes oxidation to ferric oxide (hematite).

Images courtesy of Chip Fletcher

The next equations show hydrolysis followed by oxidation.

$$Fe_2SiO_4 + 4CO_2 + 4H_2O \rightleftharpoons 2Fe^{2+} + 4HCO_3^- + H_4SiO_4$$

fayalite + carbon dioxide + water (yield) dissolved iron and bicarbonate ions + dissolved silicic acid

$$2Fe^{2+} + 4HCO_3^- + \tfrac{1}{2}O_2 + 2H_2O \rightleftharpoons Fe_2O_3 + 4H_2CO_3$$

iron + bicarbonate in solution + gaseous oxygen + water (yield) ferric oxide mineral (hematite) + carbonic acid

Oxidation is accelerated by wet conditions and high temperatures. During the oxidation process, the volume of the mineral structure may increase, usually making the mineral softer and weaker and rendering it more vulnerable to other types of weathering.

Dissolution

Dissolution is a chemical weathering reaction in which carbonic acid dissolves the mineral *calcite*, usually found in limestone (a common sedimentary rock). Dissolution of calcite is similar to hydrolysis except that all the products are dissolved; there is no solid residue unless later precipitation (of the dissolved calcite) occurs at another location (e.g., forming *travertine* in a cave stalagmite).

Bicarbonate (HCO_3^-), a product of dissolution, is a major part of the dissolved chemical load of most streams. The next equation shows the dissolution of calcite.

$$CaCO_3 + H_2CO_3 \rightleftharpoons Ca^{2+} + 2HCO_3^-$$

calcite + carbonic acid (yield) dissolved calcium + dissolved bicarbonate

Dissolution of limestone on a large scale can yield a unique kind of landscape called **karst topography.** Karst is developed when crust composed of limestone bedrock experiences widespread dissolution. This occurs when carbonic acid, in the form of groundwater, percolates along joints and bedding planes.

As the rock dissolves, large underground caverns are created. This process is called *karstification*. Over time, the caverns grow, coalesce, and undermine an area, until finally the roof collapses, producing a depression called a **sinkhole** (FIGURE 7.16).

You can read more about the effects of dissolution as a result of excess carbon dioxide in the worlds oceans due to human use of fossil fuel as an energy source in the Earth Citizenship feature titled "Ocean Acidification."

FIGURE 7.15 Oxidation of iron often produces the brown and orange coloration in rocks and soil.

Francois Gohier/Photo Researchers

Q What is the common name for iron oxide?

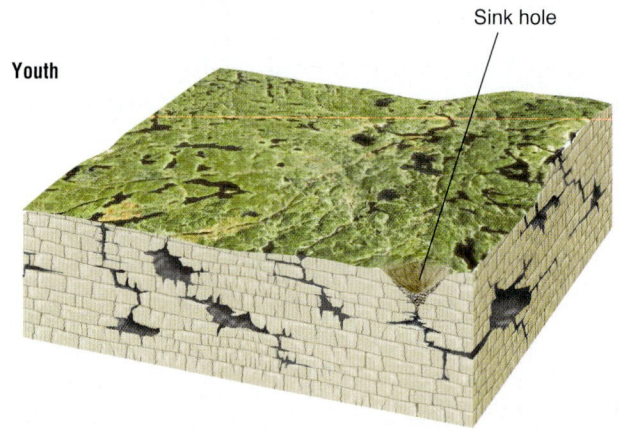

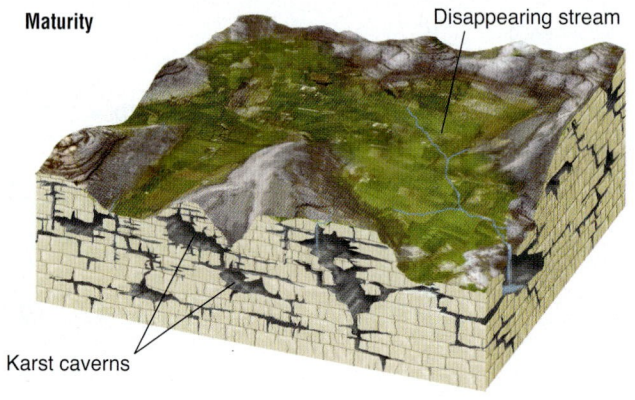

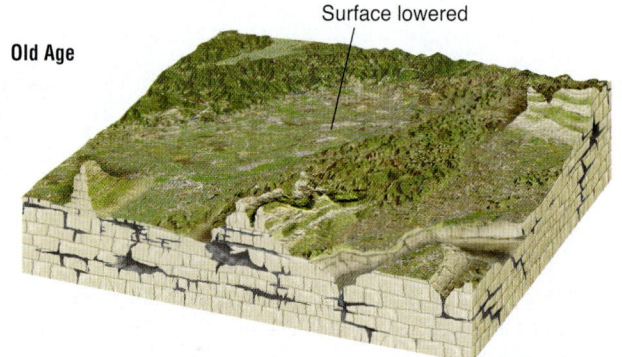

FIGURE 7.16 Limestone experiencing dissolution will develop karst topography. Karst topography progresses through three stages, labeled "youth," "maturity," and "old age." When acidic groundwater dissolves limestone, valleys, large caverns, sinkholes, and underground streams are created. As karst valleys widen and coalesce, the crust gradually weathers until eventually the ground surface is lowered and a new level is achieved.

Q What might happen to a building sitting on the roof of a sinkhole?

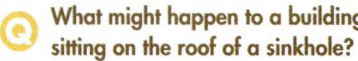

Q EXPAND YOUR THINKING

Why does rinsing your hands with water make them feel clean? Describe the chemical reactions taking place when you do this.

EARTH CITIZENSHIP

Ocean Acidification

Acid Bath

The same process involved in dissolution, namely the reaction of CO_2 and H_2O to produce H_2CO_3 (carbonic acid), is responsible for *ocean acidification*, a growing danger from global warming that is gaining worldwide attention.

Humans release CO_2 to the atmosphere—large quantities of it by burning coal and oil, making cement, and clearing forests to turn into farmland. This CO_2 traps heat and leads to global warming. (Chapter 14 has much more on the problems caused by global warming.) Furthermore, the ocean absorbs approximately one-third of our CO_2 emissions, leading to another set of problems related to lowering the *pH of water*.

The pH scale measures how *acidic* or *basic* a substance is, two extremes that describe chemicals, just as hot and cold are two extremes that describe temperature. Mixing acids and bases can cancel out their extreme effects, similar to the way mixing hot and cold water can moderate water temperature. The pH ranges from 0 to 14; a pH of 7 is *neutral*, a pH less than 7 is acidic, and a pH greater than 7 is basic.

Each whole number decrease of pH value is 10 times more acidic than the next higher value. For example, a pH of 4 is 10 times more acidic than a pH of 5, and 100 times (10 × 10) more acidic than a pH of 6.

The ocean has been absorbing excess CO_2 since the beginning of the Industrial Revolution, resulting in a decrease in the average pH of ocean surface waters of approximately 0.1 unit (from about 8.2 to 8.1), making them more acidic. Research indicates an additional drop of 0.2 to 0.3 is likely by the end of the century. This is the greatest change in millions of years.

Impact of Lowering Seawater pH

Lowering seawater pH with CO_2 has negative affects on biological processes such as photosynthesis, nutrient acquisition, growth, reproduction, and individual survival, depending on the degree of acidification. Studies show decreases in shell and skeletal growth in a range of marine organisms, including reef-building corals, commercially important mollusks such as oysters and mussels, and several types of plankton at the base of marine food webs. These members of the food chain are essentially being "weathered" by the very ocean water in which they have evolved to thrive.

Research data also suggest that there will be ecological winners and losers, leading to shifts in the composition and function of many marine ecosystems. Such changes could threaten coral reefs, fisheries, protected species, and other natural resources (**FIGURE 7.17**).

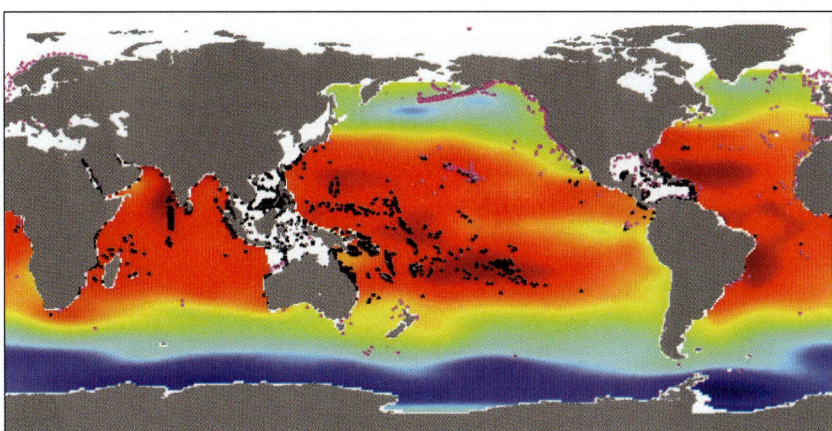

(a) Relative $CaCO_3$ saturation prior to industrialization.

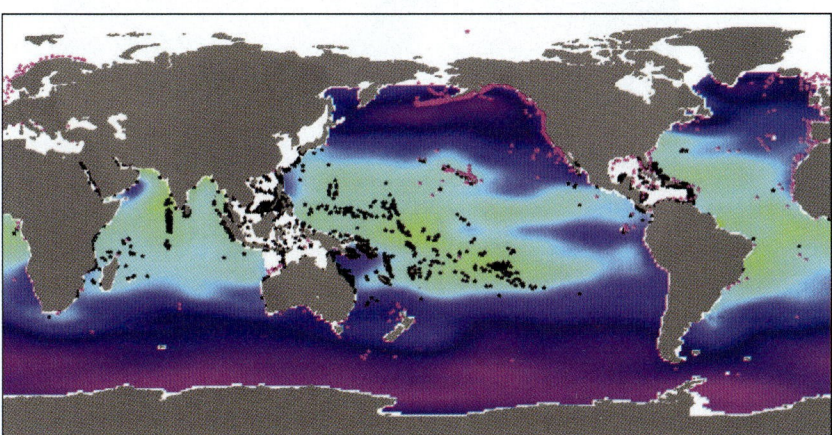

(b) Relative $CaCO_3$ saturation by the end of the century.

High $CaCO_3$ saturation (promotes shell growth)

Low $CaCO_3$ saturation (corrosive to shells)

FIGURE 7.17   Coral reefs (black and pink symbols) require the right levels of light and temperature and the presence of $CaCO_3$ in seawater. Dissolved CO_2 reduces the $CaCO_3$ saturation of seawater and affects the ability of all organisms to precipitate shells and other skeletal materials. The red/yellow equals high $CaCO_3$ saturation (480% to 330%), and the blue/purple equals low $CaCO_3$ saturation (180% to 50%; corrosive). As oceans acidify, the ability of organisms such as coral and certain types of plankton to precipitate shells and other skeletal materials decreases. By 2100, surface waters south of 60° S and portions of the North Pacific will become undersaturated (purple—corrosive) with respect to $CaCO_3$.

7-4 Biological Weathering Involves Both Chemical and Physical Processes, and Sedimentary Products Result from All Three Types of Weathering

LO 7-4 List four common weathering products, and describe how they are produced.

Biological weathering is the product of organisms causing weathering that is either chemical or physical in character, or a combination of the two.

Biological Weathering

The types of organisms that can precipitate biological weathering range from bacteria in soil and rock to plants to animals. Four of the more important biological weathering processes are:

1. *Movement and mixing of materials.* Burrowing organisms (**FIGURE 7.18**) cause soil particles to turn over, move to new locations, and change depth. This movement can introduce the materials to new weathering processes found at distinct depths under the ground and expose new particle surfaces to attack by chemical weathering.

2. *Simple breaking of particles.* Rocks can be fractured as a result of burrowing by animals or pressure from growing roots, known as **root wedging** (**FIGURE 7.19**).

3. *Production of carbon dioxide by animal respiration or organic decay.* Carbon dioxide raises the acidity of water, which then attacks and dissolves minerals and other compounds in rocks. When carbon dioxide combines with water, the resulting carbonic acid (H_2CO_3) is an effective chemical weathering agent of hydrolysis, dissolution, and oxidation.

4. *Changes in the moisture content of soils.* Organisms influence the moisture content of soils, and thus enhance weathering. Shade from leaves and stems, the presence of root masses, and high levels of organic material in soil all increase the amount of water in the soil. This higher moisture content, in turn, enhances physical and chemical weathering processes.

Products of Weathering

Weathering occurs because conditions at Earth's surface are different from the high-temperature and high-pressure conditions prevailing within the crust, where most igneous and metamorphic rocks and minerals are formed. The products of weathering are new minerals

FIGURE 7.18 Burrowing organisms produce carbon dioxide, overturn and mix soil particles, spread porosity, and increase soil moisture content.

Jacana/Photo Researchers

Why does overturning a soil particle make it more susceptible to weathering? What role does increasing soil moisture play?

FIGURE 7.19 Root wedging.

Geoffrey Bryant/Science Source

How would root wedging and rock joints work together to intensify weathering?

that are in equilibrium with surface conditions, and therefore are more likely to resist weathering themselves.

Chemical weathering yields *weathering products* that are significant components of the rock cycle and the sedimentary family of rocks. Some of these products are new sedimentary minerals that result from crystallization, such as **sedimentary quartz,** various types of *clay, calcite,* and *hematite.* Other weathering products include *dissolved compounds* and some *gases.*

Sedimentary quartz, hematite, and calcite are important natural cements that precipitate from groundwater and bind sedimentary particles to form solid sedimentary rock.

Clays are sheet silicates that trap water between layers. Clays have many economic uses in manufacturing, drilling, construction, and paper production; they are also important in crop production and natural ecosystems, as they are a major component of soils (**FIGURE 7.20**).

TABLE 7.1 Chemical Weathering Processes and Products

Example of weathering product	Dissolved material	Product	Process
Chert	Silica SiO$_2$	Sedimentary quartz (chert, agate, etc.)	Hydrolysis of quartz
Kaolinite	K$^+$, Silica SiO$_2$	Kaolinite clay	Hydrolysis of orthoclase feldspar
Hematite	4H$_2$CO$_3$	Hematite	Oxidation of olivine
Travertine	CaCO$_3$	Dissolved calcite, travertine (product of calcium carbonate precipitation)	Dissolution of calcite

Chemical weathering also produces dissolved ions and compounds, such as dissolved silica, cations, and bicarbonate, which are washed into the oceans and used by marine organisms to build shells (**TABLE 7.1**).

Q EXPAND YOUR THINKING

Biological weathering introduces organic material to regolith. When organic material decays, CO$_2$ is produced. Do you agree that this can greatly increase the rate of chemical weathering? Explain why or why not.

FIGURE 7.20 Many weathering products make up soil. Soils are a combination of mineral particles, clay, sand and rock fragments, organic debris, water, gas, and empty space (pores). Soil also contains nutrients released from bedrock by weathering, making them available to plants, which are eaten by animals. Thus, weathering is the essential first step in the global food chain.

7-5 Rocks and Minerals Can Be Ranked by Their Vulnerability to Weathering

LO 7-5 State the concept behind predicting mineral and rock stability on Earth's surface.

Rocks and minerals have complex origins. Those that crystallize at high temperatures and pressures, such as igneous and metamorphic minerals, are highly vulnerable to weathering on Earths surface. Sedimentary rocks and minerals typically are more stable in surface environments.

It has been estimated that 95% of the crust is composed of igneous and metamorphic rock. Even sedimentary rocks are often composed of igneous and metamorphic mineral grains. Having developed under conditions of high temperature and pressure, these materials are unstable in the low temperature and pressure of Earth's surface. Therefore, they tend to be decomposed by weathering (**FIGURE 7.21**).

Rock and Mineral Stability

Not all minerals decompose at the same rate. Various natural factors control how quickly any mineral will weather. *Climate* and *mineral chemistry* are the primary factors that regulate the rate at which various rocks and minerals are weathered.

Climate governs weathering and soil formation directly, through precipitation and temperature, and indirectly, through the kinds of plants, animals, and bacteria that cause biological weathering and influence the chemistry of groundwater. For instance, warm, wet climates promote rapid chemical weathering, whereas cool dry climates promote physical weathering.

A mineral's chemistry determines its vulnerability to specific weathering processes and the degree to which a rock is out of equilibrium with the conditions of the immediate environment. In fact, mineral chemistry is so important in controlling weathering that it is

FIGURE 7.22 Basalt is a mafic igneous rock and so is highly susceptible to weathering. The red coating on this basalt outcrop is hematite, an iron oxide formed by the process of oxidation.

possible to rank the vulnerability of minerals to weathering based on their chemistry.

TABLE 7.2 ranks a number of common minerals in order of their *stability,* or resistance to weathering. This ranking generally reflects the fact that many minerals formed at high temperatures and pressures (i.e., mafic and ultramafic minerals), which are the first to crystallize in a cooling magma chamber, are among the least stable on Earth's surface, where the environment differs most from their environment of crystallization (**FIGURE 7.22**).

The last minerals to crystallize (i.e., felsic minerals) are among the most stable on Earth's surface, where the environment is most similar to their environment of crystallization.

Sedimentary minerals are the most stable of all because they develop on Earth's surface as a result of weathering processes. These sedimentary minerals include oxides of silicon (SiO_2) quartz, which comes in many forms, one of which is *chert*; aluminum (Al_2O_3), known as alumina, the main component of the aluminum ore *bauxite*; and iron (Fe_2O_3), known as rust (iron oxide is the mineral hematite). Rust may not seem stable because it is easy to remove, but rust and other types of oxides typically will not weather; they simply grow with time.

The order of silicate mineral stability in Earth's surface environments is the reverse of that in Bowen's reaction series. Specifically, the most stable silicates from a weathering point of view are the last to crystallize in a cooling magma chamber, while those that are least stable silicates are the first to crystallize.

© Marli Bryant Miller

FIGURE 7.21 Granite, common in some mountain ranges, is not stable on Earth's surface and tends to weather quickly.

Why are these blocks of granite rounded? List the minerals that are common in granite and explain how each is likely to weather.

TABLE 7.2 Mineral and Rock Stability on Earth's Surface

Mineral Stability		Increasing Stability on Earth's Surface	Rock Stability	
Igneous and Metamorphic Minerals	Sedimentary Minerals		Igneous and Metamorphic Rocks	Sedimentary Rocks
Olivine	Halite		Basalt	Rock salt
Pyroxene	Calcite	↓	Granite	Limestone
Ca-Plagioclase feldspar	Hematite		Marble	Rock gypsum
Biotite	Kaolinite		Gneiss	Siltstone
Orthoclase feldspar	Bauxite		Schist	Shale
Quartz	Chert		Quartzite	Quartz sandstone

The ranking of mineral stability is referred to as the *Goldich stability series*, named for geologist *Samuel S. Goldich* (1909–2000).

Based on their mineral content, rocks also have predictable susceptibility to weathering, also shown in Table 7.2. Although mineralogy strongly influences the rate of a rock's decay, multiple weathering processes also affect it. Most rocks are vulnerable to some form of decomposition by physical, biological, and chemical weathering processes.

You can read more about the power of weathering in the Geology in Action feature, "Weathering Brings Down Mountains."

Clay Minerals

You are undoubtedly familiar with clay, and like most of us think of it as mud or small particles of soil. But did you know that *clay* is the name of an important group of minerals? Clay particles, which are indeed small, are actually sedimentary minerals that crystallize on Earth's surface. The most common clays are *kaolinite, montmorillonite,* and *illite.*

Clays are *phyllosilicates,* meaning that their crystalline structure consists of layers of silica tetrahedra organized in sheets. Typically, the silica sheets are sandwiched between layers of cations (**FIGURE 7.23**). The sheets are weakly bonded to these layers, so water molecules and other neutral atoms or molecules can be ensnared between them. Water

and other molecules trapped in the silicate structure cause clay to swell, which has the effect of making it soft to the touch. A well-known example is the clay mineral *talc,* used in talcum powder.

Although clays tend to be soft, they are remarkably resilient in surface environments. Members of this mineral group are the last to break down chemically at Earth's surface, and so are abundant in soils and sedimentary rocks. When mixed with water, clays become easily deformable and can be molded and shaped in ways that most people associate with so-called potter's clay.

Many everyday objects contain clay in some form. Clays are used in ceramics and as filler for paint, rubber, and plastic. But the largest quantities of clay are consumed by the paper industry, where kaolinite is used to produce the glossy paper on which most magazines are printed. Another clay, montmorillonite, is added to facial powder and to filler for paints and rubber. There even is clay in your toothpaste, to give it consistency.

Q EXPAND YOUR THINKING

Imagine that climate change causes a formerly arid region with limestone crust to experience greater rainfall over time. How will the weathering processes change? Should people living in the area be concerned? If so, why?

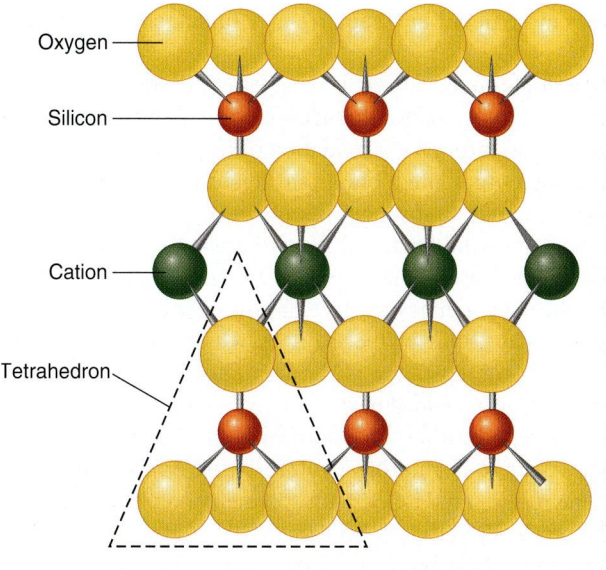

Oxygen
Silicon
Cation
Tetrahedron

(a)

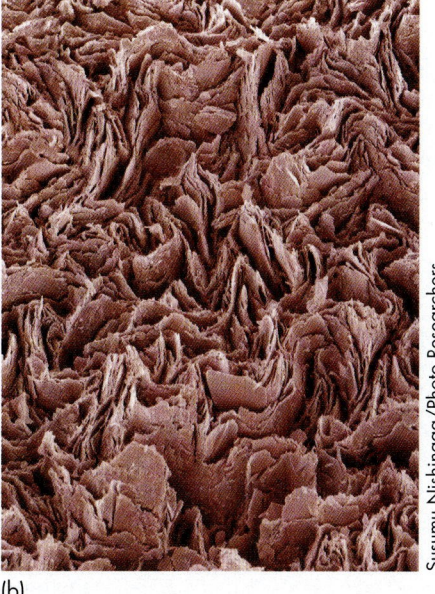

Susumu Nishinaga/Photo Researchers

(b)

FIGURE 7.23 (a) Clays comprise a group of minerals built of layers of silica tetrahedra. Different layers are connected by cations. (b) Clay consists of small particles shaped like platelets. Clay has properties of plasticity, compressibility, swelling, and shrinkage that make it useful for industrial purposes.

Q How is the silicate structure of clay related to its physical characteristics?

Bobkeenan Photography/Shutterstock

Weathering Brings Down Mountains

Earth's physical landscape is the product of both constructive and destructive physical and chemical processes that raise and lower the land. Plate tectonics processes are largely responsible for mountain building, while weathering and erosion level the land and, ultimately, take down mountains. **FIGURE 7.24** is an image of crumbling sedimentary rocks experiencing a combination of physical and chemical weathering.

FIGURE 7.24 Weathering is slowly destroying this sandstone ridge.

(7-6) The Effects of Weathering Can Produce Climate Change

LO 7-6 Describe the relationship between climate and weathering.

Weathering influences the chemistry of the atmosphere, oceans, and rocks. One reason Earth is habitable for humans is that weathering governs the amount of carbon dioxide (CO_2) in the atmosphere, and carbon dioxide influences the average temperature of the air.

Carbon Dioxide

To illustrate this concept, consider that although Venus is closer to the Sun than Earth, thick clouds of sulfuric acid reflect much of the incoming solar radiation so that, at its surface, Venus receives only half as much solar heat as Earth. Yet the average surface temperature on Venus is 460°C. Why? Because Venus's atmosphere is 96% carbon dioxide, a gas that traps heat in the same way as the glass panels of a greenhouse—which also explains why carbon dioxide is known as a **greenhouse gas.**

Earth's atmosphere, in contrast, is only 0.2% carbon dioxide, an ideal level for keeping the air warm, but not too warm. If the air contained no carbon dioxide, its average temperature would be 31°C cooler, making life as we know it impossible to sustain.

How does weathering control this important gas? Carbon dioxide is a major component of the **carbon cycle,** the production, storage, and movement of carbon on Earth. (We discuss the carbon cycle in more detail in Chapter 14, "Climate change," section 14–4.)

Over geologic time, carbon dioxide enters Earth's atmosphere through *volcanic outgassing*, the activity of hot springs and geothermal vents, and through the oxidation of organic carbon in sediments. Volcanic outgassing alone is sufficient to contribute all the carbon dioxide in our atmosphere in only 4000 years.

This input must be balanced by withdrawal of carbon dioxide from the atmosphere; otherwise, the amount of carbon dioxide present would have a runaway greenhouse effect, similar to that prevailing on Venus. One of the ways by which carbon dioxide is removed from the atmosphere is through chemical weathering of silicate minerals, as explained next.

As we discussed earlier, carbon dioxide in the atmosphere easily combines with water during hydrolysis to form carbonic acid (H_2CO_3). The reaction of H_2CO_3 with continental rocks (represented by $CaSiO_3$) produces $CaCO_3$ + SiO_2 and H_2O.

Streams carry these products to the ocean, where plankton use them to build shells, which fall to the seafloor when they die. These

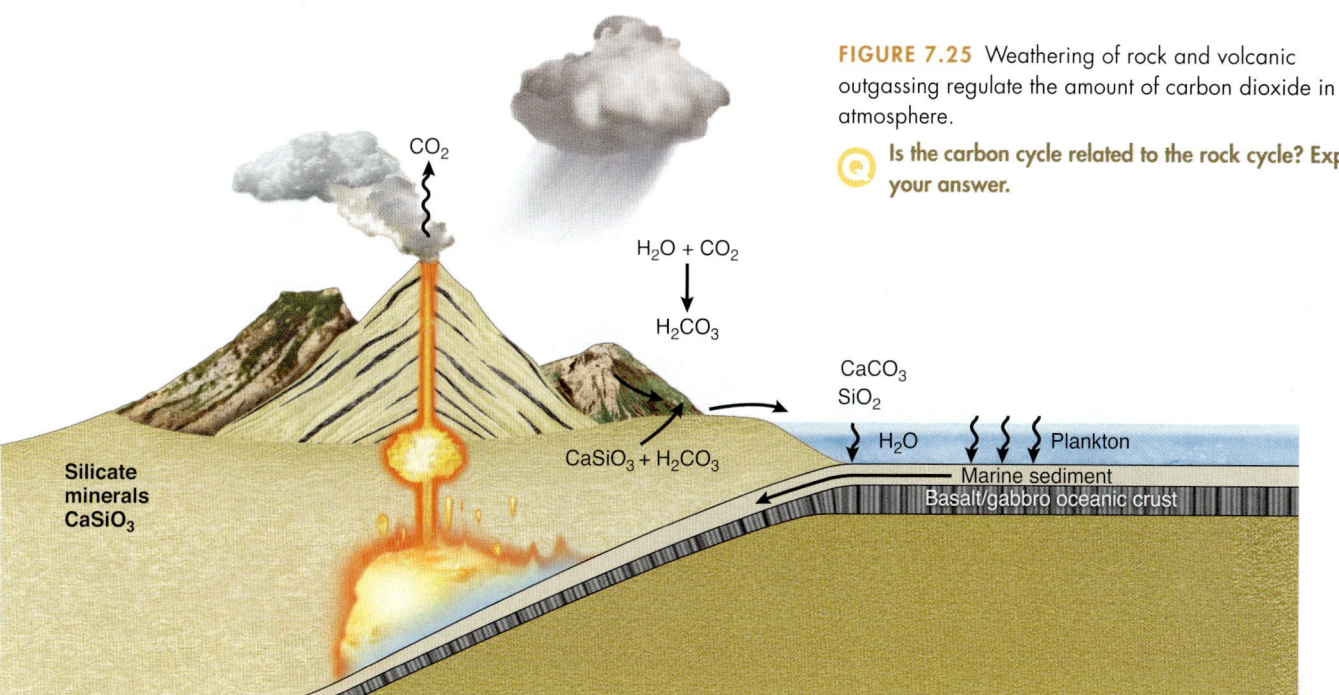

FIGURE 7.25 Weathering of rock and volcanic outgassing regulate the amount of carbon dioxide in the atmosphere.

Q Is the carbon cycle related to the rock cycle? Explain your answer.

biogenic sediments enter the rock cycle when the lithospheric plate they ride on is subducted and recycled.

Magma created by the subduction process migrates toward the crust and eventually can lead to volcanic outgassing of carbon dioxide through a volcanic arc (**FIGURE 7.25**).

Through this series of processes, carbon migrates through a complete turn of the carbon cycle.

Over geologic time, the rate of carbon dioxide removal by weathering must approximately equal the rate of carbon dioxide production by volcanoes. The balance maintained by chemical weathering and volcanic outgassing can be thought of as a thermostat governing Earth's climate.

However, this long-term equilibrium is punctuated by frequent and dramatic climate changes in the short term. Ice ages, warm periods, and extreme shifts from so-called hothouse climates to ice-house climates have been prevalent throughout Earth's history.

The Uplift Weathering Hypothesis

As noted, carbon dioxide is a greenhouse gas; as such, it traps heat and plays a significant role in regulating global climate. Volcanic outgassing is an important long-term source of carbon dioxide in the atmosphere, and weathering is one way to lower carbon dioxide levels.

The most important weathering process from the standpoint of lowering carbon dioxide levels is hydrolysis of continental silicate minerals. Hydrolysis draws carbon dioxide out of the atmosphere and stores it in the crust, in the form of carbon. Hence, the carbon dioxide content of the atmosphere can be decreased by exposing new crust to weathering.

Orogenesis, the building of mountain ranges by continental collision, exposes large areas of fresh silicate rock to hydrolysis (**FIGURE 7.26**), leading to withdrawal of carbon dioxide from the atmosphere. This idea is described by the *uplift weathering hypothesis*,

CORY RICHARDS/NationalGeographic

FIGURE 7.26 Weathering of the Himalayas, where crust has been newly exposed to carbon dioxide in the atmosphere, may be responsible for drawing down CO_2 to the lowest levels in Earth's history and causing a long, drawn-out period of global cooling several million years in duration.

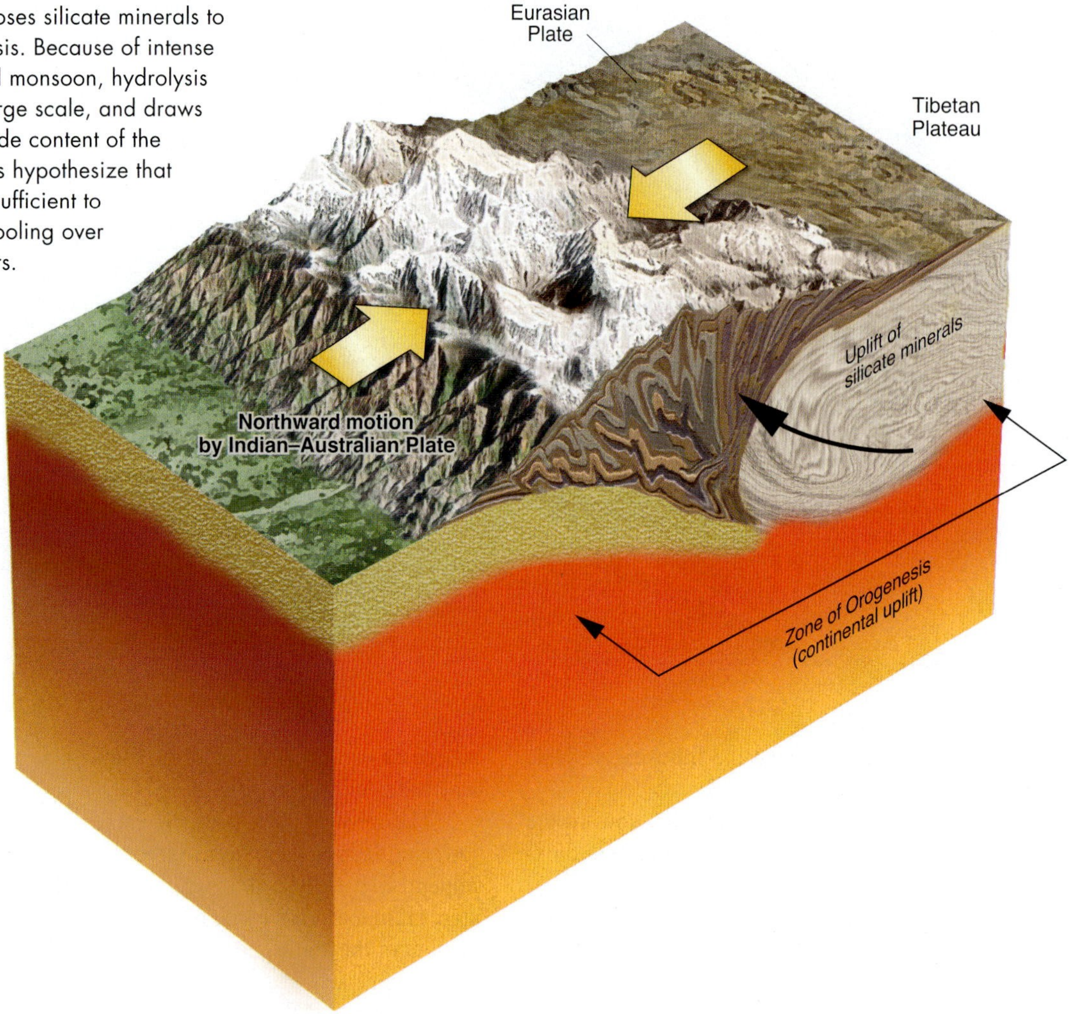

FIGURE 7.27 Orogenesis of the Tibetan–Himalayan region exposes silicate minerals to weathering by hydrolysis. Because of intense rains during the annual monsoon, hydrolysis occurs rapidly, on a large scale, and draws down the carbon dioxide content of the atmosphere. Geologists hypothesize that this process has been sufficient to produce slow global cooling over the past 55 million years.

which proposes that hydrolysis is an active agent in regulating global climate change.

The hypothesis asserts that the global rate of chemical weathering is dependent on the availability of fresh rock attacked by hydrolysis. In this case, tectonic uplift of mountains causes accelerated weathering, leading to drawdown of carbon dioxide and cooling of global climate.

Scientists have discovered that starting about 55 million years ago in the Paleogene Period, Earth's atmosphere began a long but steady cooling that continues today and is most evident at the poles and across the lower latitudes (tropics and subtropics) of both hemispheres. From this they infer that accelerated weathering resulting from uplift of the Himalayan range and Tibetan Plateau is causing a global decrease in atmospheric carbon dioxide, which is, in turn, cooling the atmosphere over millions of years (**FIGURE 7.27**).

Global Cooling?

You may find it confusing to learn that global *cooling* is occurring when all you hear about in the media these days is about global *warming*. The explanation lies in the time scales of the two trends.

Over millions of years, uplift of the Himalayas has cooled the atmosphere, playing a role in promoting several ice ages over the past million years. But more recently—over the past few centuries—pollution of the atmosphere with carbon dioxide and other greenhouse gases from the burning of *fossil fuels* has caused the atmosphere to warm. Thus, the uplift weathering hypothesis describes a process that operates over millions of years, while the modern global warming hypothesis describes a process that has been operating for only a few centuries. These two very different processes, both involving carbon dioxide, influence the global climate on vastly different time scales.

Q EXPAND YOUR THINKING

What evidence would you look for to support the hypothesis that global cooling has been taking place over the past 55 million years?

7-7 Weathering Produces Soil

LO **7-7** Compare and contrast the weathering processes in humid tropical and arctic climates.

Soil is a critical resource that provides food for the world's population. In fact, almost everything we need to live can be traced back to soil: the air we breathe, food we eat, and clothes we wear—plus paper, timber, medicines, shade, and more.

Components of Soil

Soil serves as a bridge between living things and the inanimate world. The dozen or so elements that are essential for sustaining animal and plant life come largely from soil: iron, magnesium, calcium, sodium, potassium, and others.

Plants and microorganisms take elements and compounds from soil and assimilate them into the food chain, thereby making them available to other plants and animals. Ultimately, the reason our food is capable of sustaining us day in and day out is that it is imbued with minerals and nutrients that come from soil and, ultimately, from weathering of rock. But soil is much more than a rock-weathering product. It is an ecosystem with a complex network of organisms that interact among themselves as well as with the gas, fluid, and mineral constituents.

Soil is a more multifaceted substance than most people realize. It is composed of more than just mineral particles or dirt (**FIGURE 7.28**).

A true soil is the product of a living environment. It consists of air, water, mineral particles, and organic material. The formation of soil is influenced by biological processes, the nature of the parent rock, climate, topography, and time.

Of these factors, the most important is *climate*. Climate is weather averaged over a long period, as expressed in the saying "Climate is what you expect, and weather is what you get."

Soil Formation Is Controlled by Climate

Climate is a critical factor not only in weathering but in soil formation as well (**FIGURE 7.29**). In regions where both temperature and rainfall are low, such as in cooler regions or at high elevations, mechanical weathering in the form of ice wedging and wind abrasion is the most influential weathering process. Chemical weathering (hydrolysis, dissolution, and oxidation) is most influential in regions where temperature and rainfall are both high, such as the humid tropics.

Hot, Arid Climates

Arid environments, such as deserts, allow little growth of vegetation and provide too little water to induce much chemical weathering. Consequently, these areas develop a special type of soil. Salts accumulate at the surface, due to evaporation, and erosion, frost, abrasion, and slaking break down the rocky surface into sand or gravel. Low moisture also means that fine particles are easily blown away while large particles remain behind to form a tightly packed layer known as *desert pavement*.

The negligible rainfall also means there is little biological activity and scarce organic matter. The result is a dry place full of sand, minerals formed by evaporation (such as halite and calcite), and desert pavement.

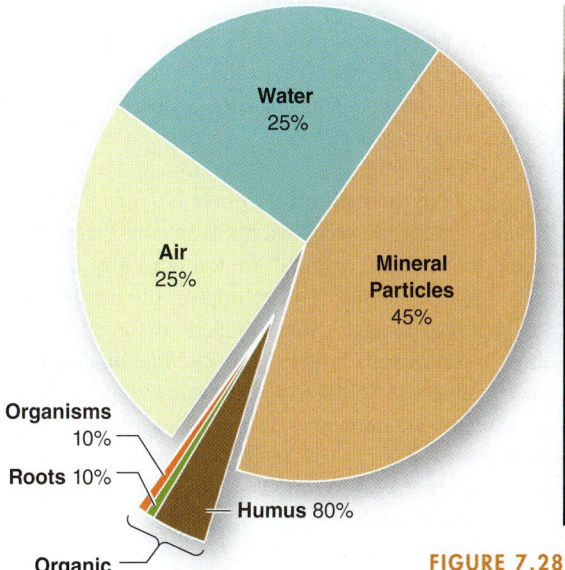

Water 25%

Air 25%

Mineral Particles 45%

Organisms 10%

Roots 10%

Humus 80%

Organic matter 5%

Michele Constantini/ZenShui/Corbis Images

FIGURE 7.28 Average soil contains four basic components: mineral particles, water, air, and organic matter. Organic matter can be subdivided further into humus, roots, and living organisms.

Describe the source of minerals that are found in soil.

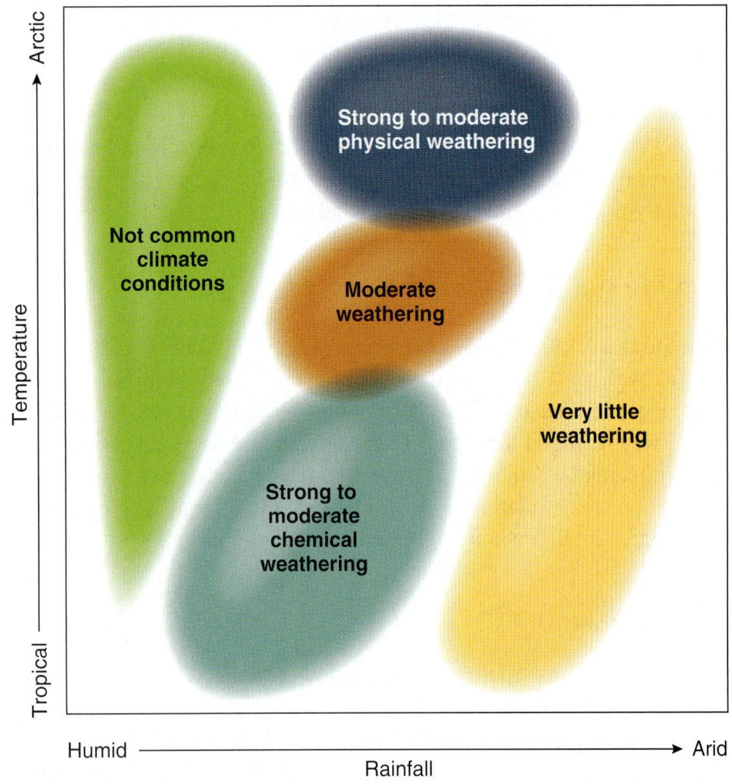

Temperature (vertical axis, from Tropical to Arctic)

Rainfall (horizontal axis, from Humid to Arid)

- Not common climate conditions
- Strong to moderate physical weathering
- Moderate weathering
- Strong to moderate chemical weathering
- Very little weathering

FIGURE 7.29 Temperature and rainfall (climate) control the stability of minerals at Earth's surface. Mineral stability, in turn, governs the rate of soil formation. The intensity of weathering differs across the world because of variations in climate and the geology of Earth's crust.

Would feldspar survive longer in the cold arctic or the humid tropics? In your answer, consider how weathering attacks it.

(a)

Siim Sepp//Shutterstock

(b)

© pierivb/iStockphoto

FIGURE 7.30 Bauxite (a) and laterite (b) are both reddish soils that are important sources of aluminum and iron ore, respectively.

Cold Climates

Arctic and alpine environments can also be dry, because in these regions water has turned to snow and ice and is unavailable for chemical weathering. Moreover, the biological activity of plants and microorganisms proceeds slowly in cold climates.

Organic material decays very slowly, giving it time to accrue and develop into wetlands known as *bogs* that contain thick accumulations of peat called *muskeg*. Mechanical breakdown by ice wedging is the major weathering process affecting the crust in cold climates. In the summer, however, weathering is accelerated by the presence of fluid water, organisms, and the long daylight of the high latitudes.

Hot, Humid Climates

In humid environments such as a tropical rain forest that receives extensive rainfall, the groundwater reaches almost to the surface for most of the year. Deep soils cannot develop, and most minerals and nutrients are stored in the living vegetation of the forest and in a rich, deep layer of leaf litter on the forest floor.

Vegetation can grow only when nutrients in this litter or the forest ecosystem become available for new growth—for example, when a tree sheds a leaf that then decomposes, or an animal dies and its body decomposes.

Tropical rain forest ecosystems are very rich, but they are also fragile. They depend on a tightly balanced and rapid exchange of organic material and nutrients from the forest floor into living organisms. Insects, tropical air temperature, and the heavy rainfall play important roles in accelerating the decomposition process.

Humid tropical climates lead to intense chemical weathering that produces soils largely composed of **insoluble residues,** or mineral products of weathering. These residues include thick soil crusts of iron oxide, known as **laterite,** and aluminum oxide, known as **bauxite** (**FIGURE 7.30**). In forming bauxite, an important ore of aluminum, the metallic cations usually are removed from a parent rock composed of orthoclase feldspar.

EXPAND YOUR THINKING

Some plants, such as spinach, are good sources of iron for our diet. Where does this iron come from, and why is it available?

7-8 The Soil Profile, Spheroidal Weathering, and Natural Arches Are Products of Weathering

LO 7-8 List and define the typical soil profile layers.

Soil typically develops a series of layers that reflect chemical, biological, and physical processes. These layers collectively are known as the **soil profile.**

The Soil Profile

Humid midlatitude climates that undergo seasonal freezing, as in much of the northern two-thirds of the United States, allow dead vegetation to accumulate in the soil. This occurs because microbes and insects have only the warm season during which to do their work of recycling the organic material, and this is not enough time to decompose it all. Consequently, the soil surface becomes rich in partially decayed organic plant debris, known as the **humus layer.**

As rain soaks through the humus layer, the water picks up dissolved ions and compounds and removes them by percolation, a process called *leaching*. Leached materials may recombine to form clay minerals, thus creating a set of layers, or *soil horizons* (**FIGURE 7.31**), that make up the soil profile.

The thickness of the soil profile depends on the climate and the nature of the parent rock, as well as the length of time over which it has developed. Note that not all the soil horizons shown in the figure and described here may be present at any given location, due to the variability of natural settings as well as disturbance by humans.

The uppermost horizon in the soil profile is the organic (O) horizon, or *humus layer*. It is composed of vegetation debris, leaf litter, and other organic material lying on the surface. This layer is not present in cultivated fields.

Below the humus layer is the A horizon, or *topsoil*. Usually, it is darker than lower layers, loose and crumbly, and contains varying amounts of organic matter. In cultivated fields, the plowed layer is topsoil. It is generally the most productive layer of the soil profile.

The E horizon is not found in all types of soil. This horizon experiences *eluviation*, the downward migration of dissolved compounds. Its main feature is loss of silicate clay, iron, aluminum, or some combination of these, leaving a nutrient-poor layer of sand and silt particles.

The B horizon, or *subsoil*, is usually lighter in color, dense, and low in organic content. Most of the material leached from the A horizon ends up in this zone.

In some arid climates, a K horizon may develop below the B horizon. The K horizon usually consists of *calcrete*, a hard, chalky mineral layer. Because calcrete is densely impregnated with calcium carbonate, it restricts water movement and tightly cements the grains. It is the result of heavy evaporation.

Still deeper is the C horizon, a transitional area between the soil and the parent rock of the crust. Partially disintegrated parent rock and mineral particles may be found in this horizon. The C horizon ends at *bedrock* (rock of the crust). Bedrock is solid sedimentary, igneous, or metamorphic rock beneath the soil.

Climate and the composition of the parent rock control the development of the soil profile. Forested regions with moderate rainfall tend to develop strong O and A horizons, along with well-developed B and C horizons. Grasslands have a restricted O horizon and wide topsoil or an A horizon. Arid desert regions develop a K horizon with a thin A horizon and no humus layer.

Spheroidal Weathering

Beneath the soil profile the bedrock often undergoes **spheroidal weathering.** This is a type of exfoliation that

O horizon (humus)
Decayed and loose organic material.

A horizon (topsoil)
Inorganic mineral particles mixed with some humus. Dark and organic.

E horizon (zone of eluviation)
Leaching of silicate clay, iron, aluminum; rich in sand and silt particles. Light colored.

B horizon (subsoil)
Dense, low in organic matter, clay accumulation.

C horizon
Transition zone between bedrock and soil. Partially altered bedrock.

Unweathered parent (bedrock)

FIGURE 7.31 A typical soil profile consists of a series of layers, or horizons. The O horizon is formed only when a well-developed source of organic matter is present. The topsoil, or A horizon, is most fully developed in grasslands, and is very thin in arid regions.

© Marli Bryant Miller

What are the primary factors that govern soil development?

FIGURE 7.32 Rectangular blocks outlined by joints are attacked by chemical weathering. Corners and edges are weathered most rapidly, and the block takes on a spheroid shape.

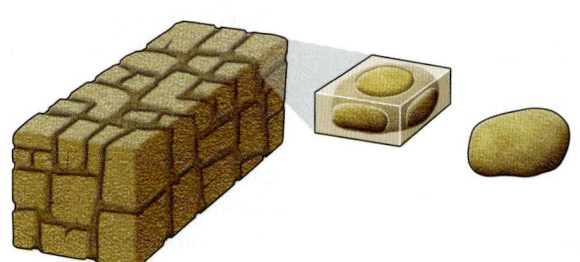

Q When weathering finally makes a sphere of this rock, will the bottom of the sphere weather as fast as the upright surfaces? Explain your answer.

© Marli Bryant Miller

occurs on rock characterized by intersecting sets of joints such as shown earlier in Figure 7.5. Spheroidal weathering occurs when rectangular blocks outlined by two or more perpendicular sets of joints are attacked by chemical weathering.

Chemical reactions can effectively weather the bedrock surface beneath the protection of the soil profile: Water percolating through the soil is chemically enhanced and so does not readily evaporate; additionally, geothermal heat is present to stimulate chemical reactions (such as hydrolysis). Spheroidal weathering also occurs on the surface (**FIGURE 7.32**).

Corners and edges are shaped by two, three, or more intersecting surfaces that are simultaneously attacked by weathering, and sharp angles quickly become smooth, rounded surfaces. Some researchers think this process may be enhanced by insolation weathering.

The rock disintegrates along *dilation* (expansion) *fractures* that conform to the surface topography. When a block has been weathered so that it is spherical, its entire surface is weathered evenly and, other than growing smaller with time, no further change in shape occurs.

Natural Arches

Natural arches (**FIGURE 7.33**) take shape when *sheeted* (sets of parallel) joints undergo weathering to produce narrow walls of rock called *fins*. Ice wedging causes the exposed rock on each side of a fin to gradually fall away. Eventually, the fin is cut through to produce a natural arch. Continued weathering causes the hole in the arch to grow until the arch becomes unstable and collapses.

Arches National Park in Utah has the greatest concentration of natural arches in the world. Its more than 200 arches range in size from 1 meter to over 89 meters wide. The reason that so many arches are present in this area is due to the particular conditions there: a well-developed parallel pattern of joints, low levels of precipitation, and a wide fluctuation in daily temperatures, with the result that water freezes and thaws regularly. Hence, ice wedging is the dominant weathering mechanism. Exposed joints are continually weathered into fins that, in turn, produce a landscape of arches.

To improve your understanding of weathering and soil production, work through the exercise in the Critical Thinking feature titled "Rock Weathering."

Q EXPAND YOUR THINKING

Why is it important to understand the soil profile?

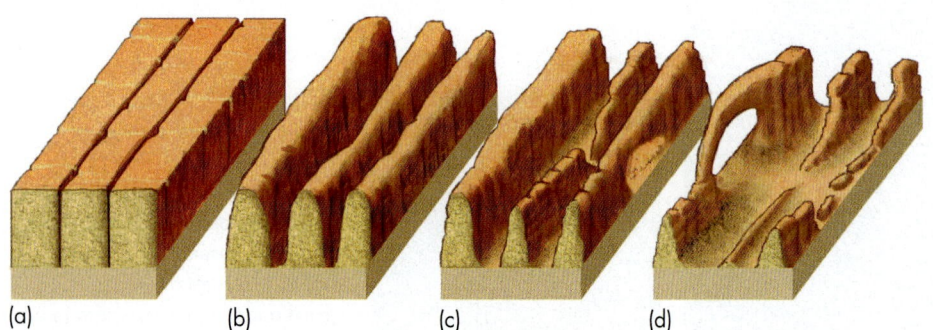

(a) (b) (c) (d)

FIGURE 7.33 Sheeted joints (a) weather into a set of fins (b). Exfoliation arches (arches that do not penetrate to the other side) develop in the sides of fins as a result of ice wedging (c). Eventually, exfoliation arches break through the fin, opening a window. The window grows into an arch that (d) eventually widens until it becomes unstable and collapses.

Q Why must rock have sheeted joints to eventually make a natural arch?

Bret Edge/Age Fotostock America, Inc.

7-9 Soil Erosion Is a Significant Problem

LO 7-9 Describe the various ways by which sediments are eroded.

Erosion is a process that removes sediments from their place of origin and transports them to a new location where they are deposited and eventually become sedimentary rock. Erosion is a natural process that is part of the rock cycle; but where human activities result in accelerated erosion, landscapes can be devastated and resources can be destroyed.

Active movement of particles by running water is called **fluvial erosion**. The term *fluvial* refers to water running in a *channel*, a subject covered in Chapter 17, "Surface Water." Other forms of erosion include *mass wasting* (the subject of Chapter 16), the movement of sediment by gravity—as occurs, for example, in a landslide or mudflow—and *eolian erosion* (covered in Chapter 19), in which sediment is moved by wind (*Aeolus* was the ancient Greek god of wind).

Erosion by Water

Fluvial erosion (**FIGURE 7.35**) occurs when particles of sediment are removed by flowing water in a channel. Water carries sediment in one of two ways: *suspended load* or *bed load*.

Suspension is a process in which sediments are kept in a perpetual state of movement in the water without touching the floor of the channel. Light (small) grains of sediment are held suspended in the water column by turbulence in streams and rivers.

FIGURE 7.36 Suspended sediment, usually composed of clay and silt particles, discolors water running in a channel. Eventually, suspended sediment in a river is carried to a delta or other depositional environment.

Bed load refers to sediments that roll or slide along the channel floor or bounce up temporarily into the water column. *Saltation* is a type of bed load in which sediments move by continually bouncing into the water column.

Together, suspended load and bed load move sediments to a final *environment of deposition*, where they come to rest and may eventually become sedimentary rock (discussed in Chapter 8).

Most streams transport the largest component of their sediment load in suspension (**FIGURE 7.36**). This is certainly true if the source of sediment yields large quantities of clay and silt. Small particles are easily kept in suspension by fast-flowing, turbulent waters. Suspended load appears as a cloud of sediment in the water.

Soil erosion takes place in four stages:

1. The impact of falling raindrops detaches clay, silt, and sand grains.

2. *Sheet erosion* removes grains from places where conspicuous water channels are lacking. Sheet erosion can be a serious cause of soil loss on steep slopes; it also operates effectively on very gentle slopes.

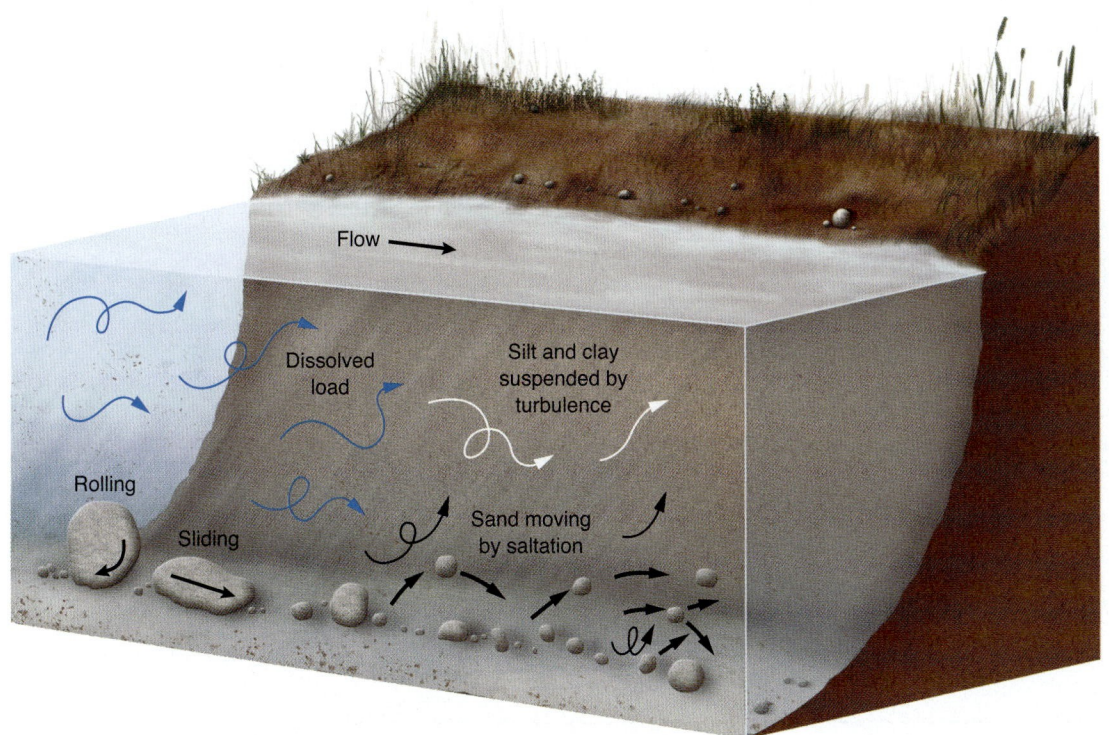

FIGURE 7.35 Sediments are moved downstream in bed load or suspended load. Saltation is a type of bed load in which sediments bounce into the water column for short periods.

Q Which type of sediment transport moves the greatest volume of sediment over the course of an average year: suspended load or bed load? Explain your reasoning.

CRITICAL THINKING

Rock Weathering

Study **FIGURE 7.34** and note the locations of the lettered sites before completing the exercise that follows.

Site A: Granite bedrock; high elevation; pronounced winter to summer temperature changes; steep slopes; little vegetation; freezing conditions in winter.

Site B: Basalt bedrock; hilly; forested; drained by streams; abundant rainfall; pronounced seasonal temperature changes.

Site C: Limestone bedrock; gentle slopes; grasslands; mild winters; hot summers; abundant rainfall; occasional drought lasting for several years.

Site D: Limestone bedrock; flat, parched land; little vegetation; occasional intense rainfall; normally arid; intense drought; hot, dry conditions all year.

Site E: Granite bedrock; abundant forest; hot, humid conditions; intense daily rainfall throughout the year; tropical climate.

1. Fill in the chart below describing the weathering conditions at each site.

2. Rank the bedrock at each site in order from most stable to least stable.

3. Describe the criteria you used for your ranking.

4. Predict the sediment composition produced at each site.

5. Rank the relative significance of physical, chemical, and biological weathering at each site, and explain your reasoning.

For each item, be prepared to argue your case in a group setting.

FIGURE 7.34 Typical weathering sites.

	General Climate Description	Most Important Climate Factors	Minerals Vulnerable to Weathering	Weathering Process(es)	Weathering Product(s)	Soil Profile Layers	Dominant Erosion Process(es)
Site A							
Site B							
Site C							
Site D							
Site E							

3. *Rill erosion* removes soil through many small, conspicuous channels, where runoff is concentrated.

4. *Gully erosion* (**FIGURE 7.37**) consists of downward cutting into the soil and the formation of deep channels. Gullies are a result of confined flow within the banks of a small channel. They develop in plow furrows, animal trails, vehicle ruts, and between rows of crop plants. In contrast to rills, gullies cannot be obliterated by ordinary tillage. Deep gullies grow rapidly and cause the loss of large quantities of soil.

Erosion by Wind

Eolian, or wind, erosion occurs when heavy winds (**FIGURE 7.38**) strip sediments and soil from the land. Wind erosion is a problem in many parts of the world, though it is worse in arid and semiarid regions where there is little vegetation to hold the sediment.

During the 1930s, a prolonged dry spell in North America culminated in dust storms and soil destruction of unprecedented proportions. The period, which came to be called the Dust Bowl, inflicted great hardship on the land and the people who lived on and worked it. Nearly 80 years later, wind erosion continues to threaten the sustainability of natural soil resources. As recently as the summer of 2012, wind erosion, coupled with prolonged drought, severely damaged agricultural land across the entire North American continent. Many scientists predict that with worsening global warming, dust bowl conditions will return with each dry season and grow in severity and geographical extent.

Soil Erosion

Soil erosion is a global problem. Each year about 170 million acres of cropland are eroded by wind and water at rates that exceed twice the tolerance level for sustainable agricultural production. On average, eolian erosion is responsible for about 40% of this loss. In the United States, wind erosion is a problem on about 73.6 million acres, damaging approximately 4.9 million acres annually.

Robert Harding Picture Library Ltd/Alamy

FIGURE 7.37 Gully erosion marks the start of true, confined (channelized) flow. Soil loss due to gully erosion is a global problem resulting from poor land conservation practices.

Ⓠ **Why is gully erosion a serious problem for farmers?**

Wind and water erosion remove the lighter, less dense constituents of soil, such as organic matter, clays, and silts—the most fertile part of the soil. In this way, erosion lowers soil productivity and leaves behind layers of sterile sand and gravel. In the United States, soil erosion is most widespread in the Great Plains states. But it is also a serious problem on sandy coastal areas, along river valleys, and in other environments. Soil erosion is often caused by poor land management. (To realize just how serious this problem is, read more about it in the Earth Citizenship feature, "Sediment Pollution.")

Ⓠ **EXPAND YOUR THINKING**

It has been said that soil erosion has brought down civilizations. What is the reasoning behind this?

FIGURE 7.38 Wind erosion, such as this gust front developing in Texas, causes massive damage to croplands.

© AP/Wide World Photos

Ⓠ **How might poor land management lead to wind erosion? How would you protect the soil on your land if you were a farmer in a windy area?**

EARTH CITIZENSHIP

Sediment Pollution

Soil is a valuable natural resource that is vital to the maintenance of the natural environment and many needs of society. Unfortunately, every day, hundreds of thousands of tons of unprotected soil are lost to erosion, causing a problem known as *sediment pollution* (**FIGURE 7.39**).

The situation in the state of Illinois provides a good illustration of sediment pollution. Based on a recent Illinois Water Quality Report, nearly 45% of the state's streams, and more than 75% of all lake areas, are adversely impacted by sediment pollution. Runoff from urban streets and construction sites are major sources of this pollution.

The following problems are related to sediment pollution:

- *Chemical pollutants*: Chemicals, such as pesticides, phosphorus, toxicants, and trace metals can be transported with sediment into lakes, streams, and the ocean, causing damage to aquatic ecosystems.

- *Construction runoff:* Construction sites are a major source of sediment and other polluted runoff. Soil erosion from a construction site without proper *sediment control practices* can average between 20 and 200 tons per acre per year. This is 10 to 20 times greater than typical soil losses on agricultural lands.

- *Harm to fish and aquatic plants*: Sediment in lakes and streams reduces sunlight penetration needed for aquatic plants; lowers survival rates for fish eggs; interferes with fish feeding habits; and clogs and damages fish gills, raising the rates of infection and disease. Sediment deposits destroy fish spawning areas; adversely impact aquatic insects that are at the base of the food chain; reduce channel capacity; and lower the overall quality of lakes, streams, and wetlands.

- *Damage to wetland mitigation sites*: Studies find that sediment deposition of less than 0.1 centimeter results in a 60% to 90%

reduction in wetland seed germination from new seedlings. Decreased species diversity is also a result of sediment deposition, often leading to less desirable species becoming prevalent.

- *Flooding*: Excess sediment reduces the capability of streams, wetlands, storm sewers, retention drainage ditches, and floodplains to accommodate storm waters. The result is flooding that proves hazardous to natural environments and developed regions.

© imagebroker/Alamy

FIGURE 7.39 The U.S. Environmental Protection Agency reports that the number-one pollutant in U.S. waterways is eroded soil.

7-10 There Are 12 Orders in the Soil Classification System

LO 7-10 Describe the dominant soil orders in polar soils, temperate soils, desert soils, and tropical soils.

It is critically important for governments to conserve and manage their soil resources if they are to meet the economic, agricultural, and environmental needs of their citizens. Finding ways to accomplish this goal has led to the science of **soil classification.**

Soil Orders

The U.S. Department of Agriculture classifies soils into 12 dominant *soil orders*, listed in **TABLE 7.3**. These categories are based on several factors, including mineral content, texture, organic composition, and moisture content.

In North America, six soil types account for approximately 75% of the soils (**FIGURE 7.40**). These include:

- *Mollisols* (grasslands), which serve as valuable agriculture lands in the Great Plains.

- *Alfisols* (fertile forests), which also support agriculture when they are cleared.

TABLE 7.3 The 12 Soil Orders

Order	Description	Percentage of U.S. Soil
Alfisols	Moderately leached, high-fertility soils with subsurface clay	13.9%
Andisols	Soils formed on volcanic ash	1.7%
Aridisols	Desert soils with clay, salt, gypsum, and calcium carbonate	8.3%
Entisols	Young soils with little or no development; lacking horizons	12.3%
Gelisols	Soils of very cold climates with permafrost	8.7%
Histosols	Highly organic soils with restricted drainage, such as wetlands	1.6%
Inceptisols	Soils on steep mountainous slopes with resistant bedrock	9.7%
Mollisols	Grassland soils with high organic content	21.5%
Oxisols	Intensely weathered tropical soils rich in iron and aluminum oxides	0.02%
Spodosols	Acid forest soils (coniferous forests) in cool, moist climates	3.5%
Ultisols	Intensely weathered soils with low fertility (humid climates)	9.2%
Vertisols	Clay-rich soils that shrink and swell with moisture changes	2.0%

- *Entisols* (young soils), which have not had sufficient time to develop horizons and tend to form on unconsolidated parent material such as sandy environments and loose rocky surfaces.

- *Inceptisols* (mountainous regions), which are thin soils on steep, rocky slopes.

- *Ultisols* (weathered forest soils), which are found in humid temperate and tropical environments and have experienced intense weathering.

- *Gelisols* (permafrost), which are found in arctic landscapes.

FIGURE 7.41 shows a map depicting the dominant global soil orders. By studying the map you will note several regional patterns in North America.

- New England and the Appalachian Mountain Belt are dominated by spodosols and inceptisols due to the cool, moist climate and abundance of coniferous forests.

- The humid temperate climate of the Southeast is dominated by ultisols that have been strongly leached.

- The eastern portion of the Midwest, including Ohio, Indiana, Illinois, and Missouri, are dominated by alfisols, which were formerly forested but have been cleared and now provide fertile farmlands.

- The western portion of the Midwest, including Kansas, Nebraska, and adjacent states, is dominated by mollisols, which formerly were prairies and have now been largely converted to farm and ranch lands.

- The western Rocky Mountain states and Southwest reflect the arid climate and the development of aridisols.

- The Pacific Northwest and California have highly variable topographies and widely varying climates. Consequently, the soil types also vary, to include andisols in volcanic regions, inceptisols in mountainous areas, alfisols in forested regions, and entisols in steep, rocky environments.

- Canada is dominated by gelisols in its northern region and mollisols, spodosols, and histosols in its southern region.

General Climate Conditions

This complex system of naming soils is easier to understand when described in terms of familiar climate zones. The 12 dominant soil orders can be classified into four general soil groups based on their affinity for certain climate conditions: polar soils, temperate soils, desert soils, and tropical soils.

1. *Polar soils* are dominated by mechanical weathering and the nature of water drainage. Because chemical weathering is much slower and weaker in a cold climate, clay production by feldspar hydrolysis is relatively unimportant. As a result, the clay-rich B

(a)

(b)

(c)

(d)

(e)

(f)

Large images (a)–(f): courtesy of Soil & Land Resources Division, University of Idaho. Inset images (a)-(f): USDA

FIGURE 7.40 The six most common soils types in North America: (a) mollisols, (b) alfisols, (c) entisols, (d) inceptisols, (e) ultisols, and (f) gelisols.

Q In which regions would each of these be common? What criteria will you use to make your predictions, and why?

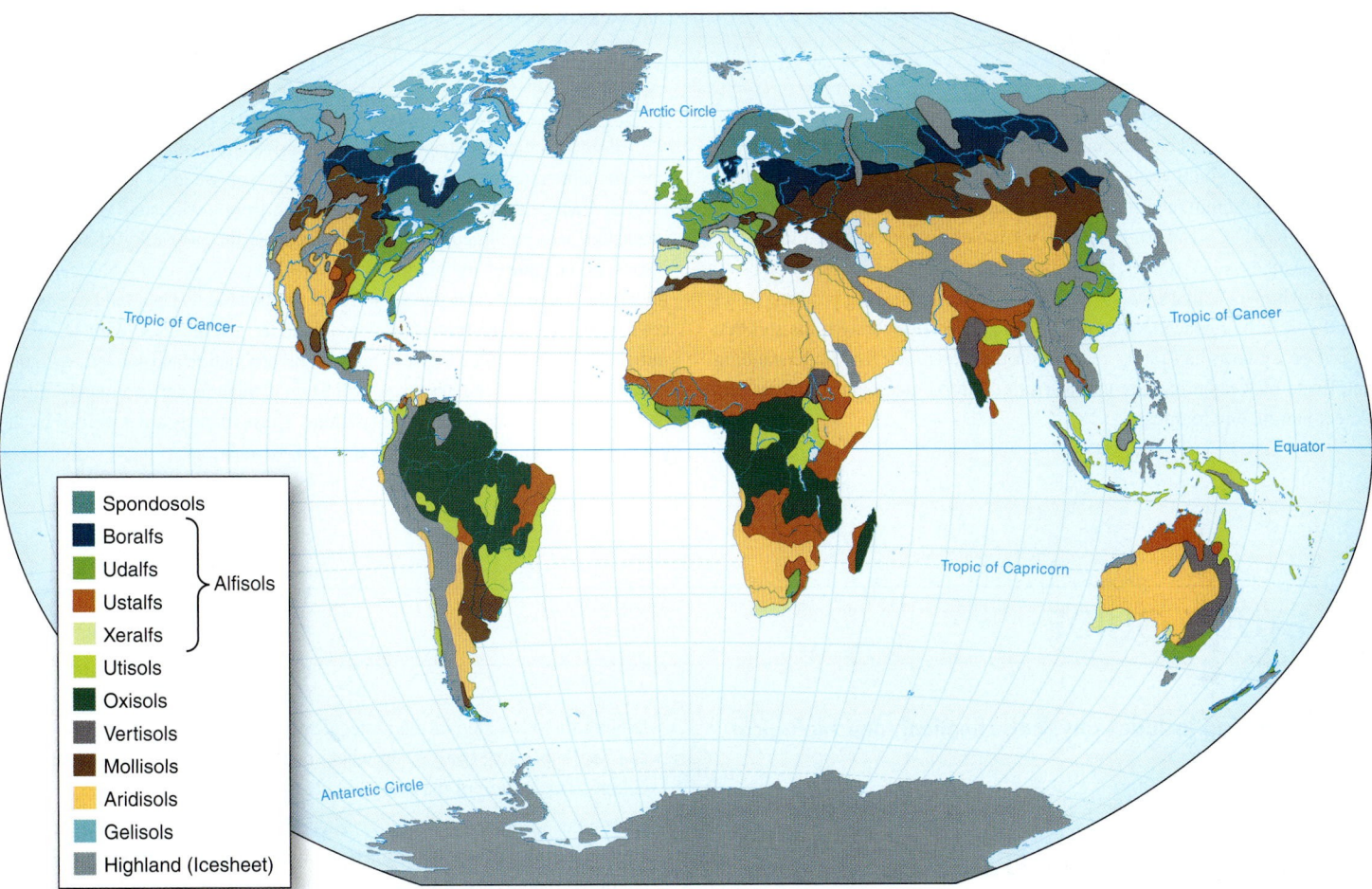

FIGURE 7.41 Maps of soil orders, like the global soil map here, assist governments in managing their precious soil resources. Maps also help scientists to understand the relationships between soil development and the influence of climate and local geology.

Why is it important to manage soil effectively?

horizon is not well developed in arid polar soils. These soils are usually classified as entisols because they lack well-developed horizons.

In humid polar settings, tundra vegetation may grow on permafrost that never thaws. These conditions prevent water drainage so that in the summer these soils become rich wetlands (histosols). Permafrost soils (which are perpetually frozen) are classified as gelisols. Where these soils are well drained (usually on slopes), A and B horizons lead to the formation of inceptisols.

2. *Temperate soils* are characterized by high organic content and well-developed soil horizons. Alfisols with a clay-rich B horizon develop in deciduous woodlands. Spodosols with an organic-rich A horizon and an iron-rich B horizon form in evergreen forests.

In cool mountainous regions and on steep slopes, slow rates of soil formation lead to entisols with poor profile development. Prairies and grasslands are characterized by mollisols with deep, rich A horizons. Subtropical regions with high rates of weathering produce heavily leached ultisols with strong B horizons.

3. *Desert soils* form in dry climates with high rates of evaporation. This reduces leaching, so calcium carbonate minerals (calcite) tend to collect. They accumulate below the B horizon and emerge as a distinctive K horizon composed of calcrete. These soils typically are classified as aridosols that are strongly alkaline compared to the more acidic soils in the temperate regions.

4. *Tropical soils* develop where there is high rainfall and warm temperatures that encourage rapid chemical reactions. Vertisols form with high clay content, causing the soil to alternately swell and shrink in wet and dry seasons. Elsewhere, where rainfall is more intense throughout the year (e.g., rain forests), oxisols (laterite and bauxite) develop that are infertile due to relentless leaching of essential nutrients out of the soil profile.

Q EXPAND YOUR THINKING

Predict which soil order is dominant where you live. Explain your reasoning.

LET'S REVIEW

Weathering is the physical, chemical, and biological degradation of the crust. It produces soil, which yields much of our food and the ores of critical metals, such as aluminum and iron. Almost everything humans need to live can be traced back to weathering: the air we breathe, food we eat, clothes we wear, as well as paper, timber, medicines, shade, and much more. But erosion caused by human activities destroys soil and other weathering products faster than they can be replaced; likewise, deforestation so severely ruins soil that natural ecosystems can rarely recover on their own. By first understanding the effects of weathering, we can learn to conserve geologic resources and protect natural environments.

To that end, in this chapter we have described how soil resources are lost at unsustainable rates to wind and water erosion when we are not attentive to the need to manage these processes. Turning land from a wild state with endemic vegetation into an agricultural state requires careful husbandry of fragile soil resources. The rate of soil production is scaled to geologic timespans: centuries, millennia, and eons. The rate of soil loss attributable to poor land-use practices is scaled to human timespans: seasons, years, and decades. Without careful mitigation, the mismatch between rates of soil production and soil loss will soon develop into a crisis. Soil resources will begin to dwindle and, eventually, disappear, to the detriment of future generations. These impacts are governed by the climate—the same climate that governs rock weathering.

STUDY GUIDE

7-1 Weathering includes physical, chemical, and biological processes.

- **Weathering** is a series of physical, biological, and chemical processes that modify rocks, minerals, and sediments in Earth's crust. Although weathering decays the crust, it also leads to the growth of new minerals through crystallization.
- **Chemical weathering** can attack only surfaces that are exposed to gases and fluids.
- **Physical** and **biological weathering** cause rock to fracture and fragment, thereby increasing the surface area that is vulnerable to chemical weathering.

7-2 Physical weathering causes fragmentation of rock.

- Physical weathering occurs in many different ways. **Joints** are openings or partings in rock where no lateral displacement has occurred. Sets of joints, composed of repeating parallel fractures, expose large areas of crust to weathering processes. As the weight of overlying rock is released, the crust expands and fractures into flat, horizontal slabs of sheeted joints. Exfoliation occurs when these slabs move and uncover the underlying rock. Abrasion occurs when some force causes two rocks to collide, leading to mechanical wearing or grinding of their surfaces. Growth of ice crystals, known as **ice wedging**, occurs when liquid water flows into joints and freezes, forcing the joints to split open further. Weathering by expansion and contraction due to daily temperature changes, known as insolation weathering, is the result of the physical inability of rocks to conduct heat well. Alternate wetting and drying of rocks, known as slaking, can be an important factor in weathering.

7-3 Hydrolysis, oxidation, and dissolution are chemical weathering processes.

- A water molecule is formed when 2 atoms of hydrogen bond with oxygen. Hydrogen atoms tend to bond strongly to one side of the oxygen atom,

leaving the other side of the molecule open. Close to the hydrogen atoms is a partial positive charge; elsewhere on the molecule is a partial negative charge. This region of the molecule attracts cations.

- Chemical weathering involves several common chemical reactions, among them **hydrolysis**, **oxidation**, and **dissolution**. Hydrolysis is the chemical reaction of a compound with acidic water, usually resulting in the formation of one or more new compounds. Oxidation involves the loss of an electron from a cation in a crystal, as well of the loss of its use by free oxygen in the environment. In the process, the oxygen and the cation bond to form a new mineral that is a member of the oxide mineral family. Dissolution is a chemical weathering process by which carbonic acid dissolves the mineral calcite, usually found in the sedimentary rock limestone.

- **Karst topography** emerges after limestone bedrock undergoes widespread dissolution as carbonic acid percolates along joints and bedding planes. As the rock dissolves, large underground caverns are created.

7-4 Biological weathering involves both chemical and physical processes, and sedimentary products result from all three types of weathering.

- Biological weathering is the product of organisms causing weathering that is either chemical or physical in nature, or a combination of the two. It includes breaking and movement of particles, exposure to carbon dioxide, and changes in the amount of moisture present.

- **Sedimentary quartz**, hematite, and calcite are important weathering products that act as natural cements. They precipitate from groundwater and bind sedimentary particles together to make solid sedimentary rock. Clay minerals, chert, hematite, and travertine also are valuable weathering products. Chemical weathering produces dissolved ions and compounds, such as dissolved silica and dissolved calcite, which are washed into the oceans and used by marine organisms to build shells.

7-5 Rocks and minerals can be ranked by their vulnerability to weathering.

- Weathering occurs because conditions at Earth's surface are different from those prevailing in the environments where most igneous and metamorphic rocks and minerals are formed.

- The order of stability for the silicate minerals in Earth's surface environments is the reverse of that in Bowen's reaction series. The most stable are the last to crystallize; the least stable are the first to crystallize.

7-6 The effects of weathering can produce climate change.

- Weathering is a geologic system that works with tectonic processes to change global climate, alter the chemistry of the atmosphere, and produce soil that is critical to sustaining life.

- The uplift weathering hypothesis asserts that the global rate of chemical weathering is dependent on the availability of fresh rock to be attacked by hydrolysis. The carbon dioxide content of the atmosphere is decreased when new crust is exposed to weathering by tectonic uplift during orogenesis. Orogenesis exposes large areas of fresh silicate-rich rock to hydrolysis, leading to withdrawal of carbon dioxide from the atmosphere.

7-7 Weathering produces soil.

- Soil consists of air, water, mineral particles, and organic material. The formation of soil is influenced by biological processes, the nature of the parent rock, climate, topography, and time.

- Climate is a critical factor in soil formation. In regions where temperature and rainfall are both low (such as in cooler regions or at high elevations), mechanical weathering is the most important weathering process. Chemical weathering, by contrast, is most important in regions where temperature and rainfall are both high, such as in the humid tropics.

- Humid tropical climates are marked by intense chemical weathering that produces soils composed largely of **insoluble residues**. These soils include thick soil crusts of iron oxides, known as **laterite**, and aluminum oxides, known as **bauxite**.

- Humid midlatitude climates that experience seasonal freezing allow vegetation debris to accumulate in the soil. The soil surface, known as the **humus layer**, becomes extremely rich in organic plant debris.

7-8 The soil profile, spheroidal weathering, and natural arches are products of weathering.

- As rain percolates through the humus layer, dissolved ions in the percolating water may not be removed; instead, they may recombine to form clay minerals that make up a set of layers, or soil horizons. These horizons are collectively known as the **soil profile**. The thickness of the soil profile varies, depending on the climate and bedrock.

- **Spheroidal weathering** occurs when rectangular blocks outlined by two perpendicular sets of joints are attacked by chemical weathering. Edges and corners are rapidly worn down by this process and become curved surfaces.

- Natural arches are shaped when joints undergo weathering that leaves behind narrow walls of rock, called fins. Ice wedging causes the exposed rock on each side of a fin to gradually fall away. Eventually, the fin is cut through and becomes a natural arch. Continued weathering causes the hole in the arch to grow larger, until, ultimately, the arch becomes unstable and collapses.

7-9 Soil erosion is a significant problem.

- Fluvial erosion occurs when particles of sediment are removed by flowing water in a channel. Water carries sediment in one of two ways: suspended load or bed load. Other forms of active erosion include mass wasting, in which sediment is moved by gravity, as in a landslide or mudflow, and eolian erosion, in which sediment is moved by wind.

- Soil erosion is a national and global problem. On croplands, about 170 million acres are eroded by wind and water at rates more than twice that of soil replenishment required to sustain agricultural production. On average, eolian erosion is responsible for about 40% of this loss. In the United States, wind erosion is a problem on about 73.6 million acres; it is responsible for moderate to severely damage of approximately 4.9 million acres annually.

7-10 There are 12 orders in the soil classification system.

- It is critically important for governments to conserve and manage the soil resources of their communities if they are to meet the economic, agricultural, and environmental needs of their citizens. Finding ways to accomplish this goal has led to the science of **soil classification**. Soils are classified by the U.S. Department of Agriculture into 12 dominant soil orders: alfisols, andisols, aridisols, entisols, gelisols, histosols, inceptisols, mollisols, oxisols, spodosols, ultisols, and vertisols.

KEY TERMS

bauxite (p. 200)
biological weathering (p. 182)
carbon cycle (p. 196)
chemical weathering (p. 182)
clay (p. 189)
dissolution (p. 190)
erosion (p. 182)
fluvial erosion (p. 203)
greenhouse gas (p. 196)
humus layer (p. 201)

hydrolysis (p. 189)
ice wedging (p. 186)
insoluble residues (p. 200)
joints (p. 185)
karst topography (p. 190)
laterite (p. 200)
oxidation (p. 189)
physical weathering (p. 182)
root wedging (p. 192)
sedimentary minerals (p. 182)

sedimentary quartz (p. 193)
sinkhole (p. 190)
soil (p. 182)
soil classification (p. 207)
soil profile (p. 201)
spheroidal weathering (p. 201)
talus (p. 186)
ventifacts (p. 186)
weathering (p. 182)

ASSESSING YOUR KNOWLEDGE

Please complete this exercise before coming to class. Identify the one best answer to each question.

1. Weathering consists of:
   a. Erosion, tectonics, and uplift.
   b. Chemical, biological, and physical degradation.
   c. Crust age, chemistry, and sedimentary minerals.
   d. Sedimentary quartz, hematite, and alumina.
   e. None of the above.

2. The pile of rocks at the base of this cliff is called _____ (talus, humus, vertisol); the rocks weathered off the cliff face by _____ (salt weathering, hydraulic action, abrasion, ice wedging).

©Marli Bryant Miller

3. Spheroidal weathering is caused by:
   a. Sand abrasion in running water.
   b. Crystal growth in cold climates.
   c. Chemical weathering of angular rocks.
   d. A combination of slaking and mass wasting.
   e. None of the above.

4. The chemical interaction of oxygen with other substances is known as:
   a. Dissolution.
   b. Hydrolysis.
   c. Saturation.
   d. Oxidation.
   e. None of the above.

5. The most important form of chemical weathering of silicate minerals is:
   a. Crystal growth.
   b. Slaking.
   c. Hydrolysis.
   d. Dissolution.
   e. Frost wedging.

6. _____ (Oxidation, Hydrolysis, Dissolution) causes feldspar to break down and produce clay.

Courtesy of Chip Fletcher

7. Insoluble residues are:
   a. Minerals produced by weathering.
   b. Dissolved compounds resulting from chemical weathering.
   c. Soils rich in organics.
   d. All of the above.
   e. Typically dissolved in hydraulic acid.

8. The tendency of silicates to weather on Earth's surface is predicted by:
   a. Mineral texture.
   b. Rock color and environment of deposition.
   c. Crystallization temperature.
   d. Tectonic setting.
   e. Their roundness.

9. In _____, rock breaks down into solid fragments by processes that do not change the rock's chemical composition.
   a. Chemical weathering
   b. Physical weathering
   c. Insolation weathering
   d. Space weathering

Gregory G. Dimijian/Photo Researchers

10. Which of the following statements about carbon dioxide is true?
    a. It is an important gas that regulates Earth's climate and influences groundwater chemistry.
    b. The amount present in the atmosphere is affected by the rate of crustal weathering.
    c. Its decrease has caused net global cooling over recent geologic history.
    d. It is a greenhouse gas.
    e. All of the above.

11. The variables that most affect the weathering process are rock composition and _____.
    a. Topography
    b. Surface area
    c. Living things
    d. Climate
    e. None of the above.

12. Which of the following statements about soil erosion is true?
    a. It is a form of pollution that affects biological communities.
    b. It is a major problem affecting million of acres of croplands.
    c. It threatens to impact food production.
    d. It takes centuries to make soil and only minutes to erode it.
    e. All of the above.

13. Soil profiles in temperate climates are characterized by:
   a. A rich humus layer.
   b. Calcrete horizons.
   c. A thin, stony A horizon.
   d. Aluminum and iron oxides in the upper layer.
   e. None of the above.

14. A dark-colored layer of mixed mineral and organic matter defines the _____ soil horizon.
   a. O
   b. A
   c. E
   d. B
   e. C

© Marli Bryant Miller

15. Karst topography is the result of:
   a. Soil erosion.
   b. Biological weathering of silicate rock
   c. Chemical weathering of carbonate rock.
   d. Spheroidal weathering.
   e. All of the above.

16. The major agricultural lands in the United States are based on which soil?
   a. Vertisols
   b. Gelisols
   c. Inceptisols
   d. Mollisols
   e. Oxisols

17. Aluminum ore comes from:
   a. Ice wedging.
   b. Hot, arid climates.
   c. Humid tropical settings.
   d. Physical weathering.
   e. All of the above.

18. Iron oxide and aluminum oxide soils are:
   a. Oxisols.
   b. Mollisols.
   c. Andisols.
   d. Typically in polar soils.
   e. None of the above.

19. The uplift weathering hypothesis:
   a. Explains global cooling.
   b. Explains global warming.
   c. Explains agricultural development.
   d. Causes orogenesis.
   e. None of these.

20. In soil, burrowing organisms turn over and break particles. They also produce organic material that decays and makes _____ (bauxite, hydrogen, carbon dioxide) gas. This combines with water to make _____ (kaolinite, carbonic acid, chert) which attacks silicate minerals by _____ (hydrolysis, dissolution, oxidation).

Jacana/Photo Researcher

FURTHER RESEARCH

1. Why is water an effective agent in chemical weathering?

2. How does the weathering of silicate minerals influence Earth's climate?

3. What are joints, and how are they formed? Why are they important in weathering?

4. What factors determine the stability of a mineral on Earth's surface?

5. Draw the rock cycle and indicate where weathering fits in to it. Design a special section that describes some of the details of weathering.

6. Draw an average soil profile for the area where you live.

7. Search the Web for an explanation of which U.S. government agency is responsible for soil classification. Study its mission.

ONLINE RESOURCES

Explore more about weathering on the following Web sites:

National Resources Conservation Service:
www.nrcs.usda.gov

National Soil Erosion Research Lab:
www.ars.usda.gov/main/site_main.htm?modecode=36-02-15-00

Additional animations, videos, and other online resources are available at this book's companion Web site:
www.wiley.com/college/fletcher

This companion Web site also has more information about WileyPLUS and other Wiley teaching and learning resources.

SEDIMENTARY ROCK

8–1 Sedimentary rock is formed from the weathered and eroded remains of Earth's crust.

> **LO 8-1** Describe the reasons that geologists study sedimentary rocks.

8–2 There are three common types of sediment: clastic, chemical, and biogenic.

> **LO 8-2** Compare and contrast the three types of sediment.

8–3 Sediments travel from source area to depositional environment.

> **LO 8-3** Describe the processes that act on sediments as they move from source areas to depositional environments.

8–4 Sediments change as they are transported across Earth's surface.

> **LO 8-4** List the ways that sediments change as they are transported across Earth's surface.

8–5 Clastic grains combine with chemical and biogenic sediments.

> **LO 8-5** Describe how the composition of sediments changes between the weathering site and the formation of sedimentary rock.

8–6 Sediment becomes rock during the sedimentary cycle.

> **LO 8-6** Describe the sedimentary cycle.

8–7 There are eight major types of clastic sedimentary rock.

> **LO 8-7** List the eight major types of clastic sedimentary rock, and give a brief description of each.

8–8 There are seven major types of chemical sedimentary rock and four major types of biogenic sedimentary rock.

> **LO 8-8** List the seven major types of chemical sedimentary rock and the four major types of biogenic sedimentary rock.

8–9 Sedimentary rocks preserve evidence of past depositional environments.

> **LO 8-9** Describe how a geologist uses inductive reasoning to interpret Earth's history using sedimentary rocks.

8–10 Primary sedimentary structures record environmental processes.

> **LO 8-10** List four common sedimentary structures, and explain their significance.

CHAPTER CONTENTS & LEARNING OBJECTIVES

GEOLOGY IN OUR LIVES

Sedimentary rock is formed from the weathered and eroded remains of the crust. It contains fossils and other evidence of the history of life; records changes in the environment, climate, and plate movement; and reveals many other details of how the planet has developed and transformed over time. Sedimentary rock also is the source of fossil fuels such as coal, petroleum, methane, and oil shale. We derive valuable metals from sedimentary accumulations, among them deposits of gold and platinum and the ores of iron and aluminum. Construction materials, including building stone, sand, and gravel to make road beds and cement for the building industry, all come from sedimentary rocks. Perhaps of greatest significance to us, however, is that most of our communities obtain freshwater for drinking, irrigation, and manufacturing from the groundwater often found in the pore spaces of sedimentary rocks.

Q Many of the rock layers you can see in this photograph are composed of sediments that accumulated on the seafloor in a shallow ocean. What evidence would reveal to a geologist that a rock formed in a shallow marine environment?

8-1 Sedimentary Rock Is Formed from the Weathered and Eroded Remains of Earth's Crust

LO 8-1 Describe the reasons that geologists study sedimentary rocks.

For billions of years, weathered particles of sediment have been eroded from the land—removed from mountains and hills through the action of gravity, wind, and running water—and eventually deposited in low-lying environments where they form sedimentary rock.

Sedimentary Rock Is a Critical Global Resource

Most of Earth's surface is covered with layers of loose **sediment,** particles of mineral, broken rock, and organic debris (**FIGURE 8.1a**); and over 75% of the land surface is covered with **sedimentary rock,** which is simply rock composed of sediment (**FIGURE 8.1b**).

Weathering destroys the crust and in the process produces little bits of rock and mineral grains that are washed downstream, where they eventually come to rest on the bottoms of rivers, lakes, and, ultimately, oceans. Over time, sediments accumulate as layer upon layer of particles that are deposited in low elevation environments across the planet's surface. As they are buried, the overlying pressure builds, and these layers *consolidate* and *compact*. Groundwater fills the pore spaces, and the grains are *cemented* by dissolved compounds in groundwater, to become, in due course, solid sedimentary rock.

Sedimentary rock and loose sediments can be "read" by geologists as records of earlier environmental conditions. Rock composed of gravel reveals the former presence of a stream or glacier; rock composed of **sand** tells of a past beach or sand dune; **mud** reveals ancient deltas, marshes, or deep lakes.

The greatest collection of sedimentary rock comes from the seafloor, where fossil plankton (**FIGURE 8.2**) collects to form *biogenic ooze,* and **clay** accumulates to form *abyssal clay.* In each of these cases, a geologist interprets the sediments and buried fossils to learn about former surface conditions on Earth. This chapter explains how sedimentary rock records the history of surface environments and life on Earth.

Sedimentary rock develops from the weathered and eroded remains of other igneous, metamorphic, and sedimentary rocks, as well as from organic sources. Many sedimentary rocks also contain both direct and indirect evidence of past organisms, such as *skeletal parts* or impressions of plants and animals (footprints, leaf imprints, etc.). Several sedimentary minerals crystallize from dissolved compounds in groundwater, or result from chemical reactions between the crust and gases in the atmosphere (such as oxygen and carbon dioxide).

The most common sedimentary minerals are formed by the weathering processes described in Chapter 7. These include **iron oxides** (such as *hematite*), **calcium carbonate** (*calcite*), and **sedimentary quartz.** These compounds become minerals that cement particles together to make sedimentary rock. Other sedimentary minerals are produced by living organisms that secrete a mineral coating, such as corals that produce calcite.

Sedimentary rocks composed of **silt** and clay harden when fine particles become consolidated and water is expelled by compaction. Although sedimentary rocks can be formed through several different processes, they generally take on characteristics that reflect specific conditions on Earth's surface. That is why geologists search them for clues to the history of Earth's surface environments.

Human society depends on many resources that come from sedimentary rock (**FIGURE 8.3**). Among the most important are the *fossil fuels* critical to industry and transportation: petroleum, coal, oil shale, and natural gas. Thirty-six percent of U.S. energy needs are met by petroleum, 20% by coal, and another 25% by natural gas (the rest of our energy comes from nuclear power—8%, and renewable

(a)

Hans Strand/Corbis Images

(b)

Phil Degginger/Alamy

FIGURE 8.1 (a) Loose sediment is found on Earth's surface in many types of environments. Here, sand forms a dune in the Namib Desert, Namibia. (b) Sediment collects in a sedimentary environment where it may consolidate and become hard sedimentary rock made up of individual particles cemented and/or compacted together. These sandstones once were sand dunes in Zion National Park, Utah.

Based on your knowledge of minerals that are resistant to weathering, what is the likely mineralogy of these sediments?

(a)

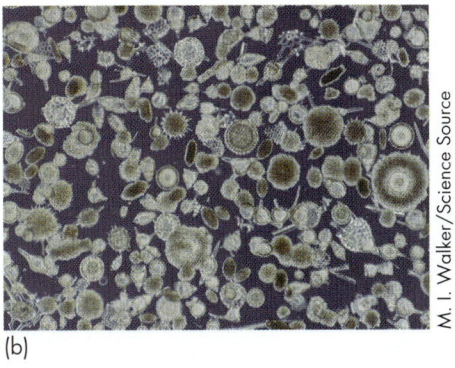

(b)

(c)

Biophoto Associates/Science Source

M. I. Walker/Science Source

Deepwater Canyons 2013 Expedition, NOAA-OER/BOEM/USGS

FIGURE 8.2 Biogenic ooze is composed of the remains of marine plankton. Accumulating on the seafloor, it comprises one of the greatest collections of sediment in the world, as shown here: (a) calcareous (composed of $CaCO_3$) nanoplankton—*coccolithophorids*; (b) siliceous (composed of SiO_2) nanoplankton—*diatoms*; and (c) samples of biogenic ooze taken at an over 2-kilometer depth.

sources—9%). Other sedimentary resources include accumulations of valuable metals, sand and gravel for construction, and the stored groundwater that becomes our drinking water.

Rocks and sediments can be thought of as pages in the book of Earth's history. Fossilized skeletal fragments of long-extinct plants and animals (many of them microscopic plankton and pollen) are evidence of the shifting climates and changing environments of our restless planet. Without sedimentary rocks, not only would we be unable to heat our homes or power our machines, but we would know

little about what Earth was like during the billions of years that preceded our short tenure here.

Q **EXPAND YOUR THINKING**

Use the texture and composition of sedimentary rock as described in section 8–1 to propose a classification system for sedimentary rock. What will you name your various rock types?

Lowell Georgia/Corbis Images

FIGURE 8.3 The fossil fuels that power modern society come entirely from sedimentary rock.

Q **List the various roles that fossil fuels play in your life.**

8-2 There Are Three Common Types of Sediment: Clastic, Chemical, and Biogenic

LO 8-2 Compare and contrast the three types of sediment.

Sedimentologists—geologists who study sediments—have identified three types of sediment based on their composition: **clastic sediments**, which are broken pieces of crust deposited by water, wind, ice, or some other physical process; **chemical sediments**, which are produced by inorganic (nonbiological) precipitation of dissolved compounds (e.g., through evaporation); and **biogenic sediments**, which are produced by organic (biological) precipitation of the remains of living organisms. We explore these three types of sediment in this section.

Clastic Sediments

Clastic sediments are solid pieces of preexisting rocks or minerals, or broken pieces of crust, that have been physically transported from one place to another (**FIGURE 8.4**). Since most of Earth's crust is composed of silicate minerals, the majority of clastic sediments are made up of these minerals. But this generalization may or may not be true from place to place.

The silicate minerals that are most likely to survive the rigors of weathering are those that are stable on Earth's surface and crystallize late, according to Bowen's reaction series. Unstable mafic minerals usually are not important constituents of most clastic sediments because they are vulnerable to chemical deterioration and do not readily survive the weathering process. These include ferromagnesian minerals such as olivine and amphibole and the calcium-rich members of the plagioclase feldspar series.

As we learned in Chapter 7, many igneous and metamorphic minerals formed at high temperatures or pressures are not stable under the low temperatures and pressures typical of sedimentary environments on Earth's surface: They break down rapidly and so are not commonly found in sedimentary rocks.

Stable minerals found in clastic sediments include quartz, orthoclase feldspar, muscovite, clay minerals, and some valuable *heavy minerals* (dense minerals such as garnet, zircon, tungsten, chromite, magnetite, tourmaline, and others). Also common in clastic sediments are eroded fragments of crustal rocks, known as *lithic fragments* (such as bits of shale, basalt, limestone, and others).

Chemical Sediments

Chemical sediments materialize as a result of chemical reactions unrelated to the actions of living organisms. Chemical sediments develop from water that is saturated with dissolved cations and anions. Chemical sediments form by crystallization that occurs when these ions develop covalent and ionic bonds and, thus, create chemical compounds, producing solid minerals such as calcite, quartz, and salt. The solid crystals then precipitate from the water to create sediment. This inorganic precipitation can occur in groundwater, in the ocean, in hot springs that flow onto Earth's surface, or even in freshwater lakes.

The most common chemical sediments are the result of *evaporation* in a lake or ocean (**FIGURE 8.5**). During the process of evaporation,

Courtesy of Chip Fletcher

FIGURE 8.4 Clastic sediment consists of the stable minerals quartz and orthoclase feldspar, as well as clay minerals, micas, rock fragments (called *lithic fragments*), and others.

Would you agree that sand composition might reflect climate? Explain.

Franck Guiziou/Hemis/Corbis Images

FIGURE 8.5 Evaporite deposits, such as these salts in the desiccated bed of a desert lake, consist of chemical sediments.

What is the mineral composition of evaporites?

(a)

(b)

(c)

FIGURE 8.6 Sandstones contain inorganically precipitated cements, a type of chemical sediment. These photos show the three most common forms of sandstone cement. Each type is shown in a rock sample and in a photo taken through a microscope: (a) calcium carbonate cemented sandstone; (b) silica cemented sandstone; and (c) iron oxide cemented sandstone.

What are the mineral names of these three types of cements?

water molecules change from liquid to gas; dissolved compounds containing calcium, silica, iron oxide, sodium, and chlorine (and others) are left behind. Consequently, any remaining water becomes enriched in these compounds, and they begin to crystallize.

Minerals known as **evaporites** are formed in the briny solution and then settle to the bottom as a layer of chemical sediments. The minerals halite, calcite, gypsum, and sodium and magnesium salts, among others, are all inorganically precipitated when water evaporates.

Inorganic precipitation also occurs in groundwater flowing through accumulations of porous sediments. A precipitated compound will grow around individual grains and fill empty spaces between grains, often cementing adjacent grains together. *Common mineral cements* (**FIGURE 8.6**) formed in this way are calcite, quartz, goethite, and hematite.

Most chemical sediments are deposited near the locations where they are formed, rather than being transported from afar. That is, they precipitate at the same location where they ultimately will be transformed into sedimentary rock. The dissolved compounds of which the sediments are composed have, however, typically traveled long distances from weathering sites located high in adjacent watersheds or discharged from hydrothermal vents at spreading centers.

Biogenic Sediments

Biogenic sediments are composed of the fossil remains of plants and animals (**FIGURE 8.7**). Biogenic sediments include skeletal fragments of reef-dwelling organisms such as corals and shells (usually made of calcite), highly organic plant remains that develop into coal (composed of carbon with various impurities, such as sulfur), and biogenic ooze found on the deep seafloor, which is composed of microscopic skeletal fragments of marine plankton (made of organically precipitated calcium carbonate or silica).

FIGURE 8.7 Biogenic sediments are composed of skeletal fragments, usually of marine organisms, such as plankton or reef-dwelling (as shown here) corals and algae.

What chemical compound(s) are most common in marine biogenic sediments?

EXPAND YOUR THINKING

What is similar about biogenic and chemical sediments that make them fundamentally different from clastic sediments?

8-3 Sediments Travel from Source Area to Depositional Environment

LO 8-3 Describe the processes that act on sediments as they move from source areas to depositional environments.

Sediments, especially clastic sediments, are produced in source areas and travel to depositional environments. Along the way they are changed and imprinted by various types of processes.

Sediments are produced in *source areas* where physical and biological weathering of the crust releases rock fragments, and chemical weathering leads to the development of new sedimentary minerals that are relatively stable in surface environments.

Sediments move from source areas to *depositional environments*; and as they travel through various surface environments, they experience several types of processes. These processes, such as abrasion, breakage, rounding, sorting, precipitation, mixing, and others, will imprint the sediment in ways that may be recognizable to geologists studying sedimentary rocks. **FIGURE 8.8** illustrates some of the surface environments where these processes operate.

FIGURE 8.8 Chemical and physical weathering produce sediments that pass through several types of surface environments along their journey to depositional environments. These environments include (a) fluvial environments, (b) glaciated regions, (c) coastal barrier islands and lagoons, (d) deltas, (e) the deep sea, and (f) eolian environments.

8-4 Sediments Change as They Are Transported Across Earth's Surface

LO 8-4 List the ways that sediments change as they are transported across Earth's surface.

From the moment they are freed by weathering and erosion to travel across Earth's surface, sediments become imprinted with the physical and chemical "signatures" of the various environments they encounter.

Particle Size Reflects Environmental Energy

Movement of air, water, and ice, along with gravity, is responsible for most of the work of transporting sediment from its origin in weathered crust toward sedimentary basins, also known as **environments of deposition,** where it finally comes to rest.

Strong currents move more and larger grains; weak currents move only smaller grains of clay and silt. Geologists describe the relative strength of natural processes as *environmental energy*. Gravity drives this transportation system, and its effect is relentless. Grain by grain, ion by ion, through the ages, even mountains are eventually worn down and delivered to depositional environments.

To denote the size of sedimentary particles (grains), geologists use common yet precise terms, including gravel, sand, silt, and clay, each of which denotes a particular range in the sizes of sedimentary particles (**FIGURE 8.9**). And mud indicates a mixture of silt and clay.

(a)

Andreas Molin/Nordic Photos//Getty Images, Inc.

(b)

© Freddy Eliasson/Matton Collection/Corbis

(c)

Andrew Dernie/Photodisc//Getty Images, Inc.

(d)

De Agostini / A. Dagli Orti/De Agostini Picture Library/Getty Images

FIGURE 8.9 (a) Gravel includes all grains with a large dimension greater than 2 millimeters. (b) Sand is between 2 and 0.06 millimeters. (c) Silt is smaller than 0.06 millimeters but larger than 0.004 millimeters. (d) Clay is 0.004 millimeters or finer.

In which environment might you find each of these grain sizes deposited? Give one example for each.

Gravel includes any particle that is more than 2 millimeters across in its largest dimension. In decreasing size order, gravel can consist of boulders, cobbles, and pebbles. Sand consists of particles between 2 and 0.06 millimeters across; further, geologists often subdivide sand into coarse, medium, and fine, based on specific size limits.

In practical terms, fine sand is about the smallest grain that can be easily seen with the naked eye. Silt is sediment that is finer than sand, and clay is the finest sediment of all. Mud, as just noted, is a combination of silt and clay.

As environmental energy builds up, the speed and turbulence of air and water currents increase. More and larger sedimentary particles roll and drag in the currents and are suspended above the bottom.

In high-energy environments, such as streams flowing at rates of over 50 centimeters per second and ocean beaches with strong waves, sedimentary deposits tend to collect gravel and sand. In environments with moderate energy levels, such as quiet streams (flowing at rates of 20 to 50 centimeters per second), lake beaches, many rivers, and lagoons, sedimentary deposits tend to contain more sand and mud.

In low-energy environments with currents flowing at rates of less than 20 centimeters per second, such as the deep seafloor, deep lake floors, wetlands, and parts of estuaries and deltas, sedimentary deposits contain no gravel and little sand and are dominated by mud.

Sedimentary deposits are governed not only by environmental energy but also by the abundance of sediment. That is, if most of the available sediment is gravel, the deposit will reflect this content. The same is true for other types of sediments, with the result that the final deposit reflects the combined influences of environmental energy and sediment availability.

As grains of sediment are transported, they undergo changes related to the energy of the environment. Two processes in particular occur in sediments: **sorting,** in which grains are separated by density and size, and **abrasion,** in which grains become smaller and more spherical the longer (in terms of time) and farther (in terms of distance) they are transported.

Sorting

Sorting is the tendency for environmental energy to separate grains according to size. A well-sorted deposit is one in which the grains are nearly all the same size (**FIGURE 8.10**). A poorly sorted deposit is

(a)

(b)

Ted Dayton Photography/Beateworks/Corbis Images

Dan Suzio/Photo Researchers

FIGURE 8.10 (a) If all grains in a sedimentary deposit are nearly the same size, it is considered well sorted. (b) If sediment contains pieces of gravel as well as sand and mud, such as in this stream channel, it is considered poorly sorted.

Why do windblown deposits tend to be well sorted?

TABLE 8.1 Features of Sedimentary Environments

Sedimentary Environment	Transportation Process	Sediment Size	Sorting	Abrasion (degree of rounding)
Wind-deposited dune	Wind	Fine to medium sand	Very strong	Strong
Ocean beach	Waves, currents, wind	Fine sand to gravel	Strong	Very strong
Wetland	Weakly circulated water, wind	Clay to silt	Strong	Weak
Stream channel	Running water	Silt to gravel	Moderate	Very strong
Glacier	Moving ice	Clay to large gravel (boulders)	Weak	Strong
Steep hillside	Mass wasting	Clay to large gravel (boulders)	Weak	Moderate

one in which the grains vary widely in size; such a deposit is often a mixture of mud, sand, and gravel.

Sandy dunes are an example of a well-sorted deposit. Sedimentary deposits resulting from a glacier are, in contrast, typically poorly sorted because the ice often carries clasts (fragments) of all sizes, from clay to boulders, simultaneously. **TABLE 8.1** itemizes the various features of sedimentary environments.

Water sorts particles according to their mass, determined by their size and composition. The size of particles with the same composition will dictate when moving currents have enough energy to erode them; when currents have too little energy to move them farther, the particles will be deposited.

Fine sand, in general, erodes sooner than large grains. Clay and silt, on the other hand, because they tend to form compact layers, are not easily eroded. In streams, lakes, and the ocean, sands usually collect with other sands, and gravels with other gravels. This sorting occurs because as the energy of moving water waxes and wanes, grains of equal mass (determined by density and size) tend to erode and be deposited together.

Wind is probably the most effective sorting mechanism for finer sediments. Windblown sand shapes fields of dunes, and windblown silt builds huge accumulations called *loess*.

Sorted gravel accumulations called *pavements* form when wind removes finer grains of sand and silt. Ice (in glaciers) is the poorest sorting mechanism; it transports and deposits sediments of all sizes with equal ease.

Abrasion

As sediment is transported, individual grains become subject to abrasion as they strike or rub against other particles. Abrasion produces rounding—smoothing the sharp edges or corners of grains. A well-rounded grain probably has traveled a great distance from its original source; an angular grain has not traveled as far. Boulders tend to be rounded more quickly than sand grains because they strike with greater force.

The greater the distance from their source area, the more likely sedimentary deposits are to become smaller grained and better sorted (**FIGURE 8.11**). For example, sediments in mountainous watersheds are poorly sorted, with angular clasts and an abundance of large-diameter particles. In contrast, the sediments far downstream are generally well rounded, finer grained, and better sorted.

This variation in sediment texture is due to the length and duration of transport and the sorting and abrasion that take place along the way. Different environments reflect various degrees of abrasion and sorting, depending on the energy and the type and abundance of sediment.

Boulders, gravel, and sand

Sand

Silt and clay

FIGURE 8.11 Sediment that is transported far from its source tends to be finer grained, more rounded, better sorted, and more highly abraded than sediment that is deposited close to its source.

Ⓠ Explain why grain size decreases in sediments far from their source.

Ⓠ **EXPAND YOUR THINKING**

Where do eroded grains eventually come to rest?

8-5 Clastic Grains Combine with Chemical and Biogenic Sediments

LO 8-5 Describe how the composition of sediments changes between the weathering site and the formation of sedimentary rock.

Sediments collecting in an environment of deposition may consist of clastic, biogenic, chemical and dissolved components. The character of the final deposit reflects environmental conditions along the route of sediment transport as well as during deposition.

Sediments Evolve as They Move Through Surface Environments

Sorting and abrasion change the character of clastic sediments as they travel from mountains to basins. Dissolved components also move from weathering sites, where they originate, into basins (such as the ocean), where they are deposited. Dissolved ions and compounds are deposited through either organic or inorganic precipitation. Both processes produce a solid that joins the sedimentary system as either chemical or biogenic sediment.

Among porous sediments in sedimentary basins, dissolved silica, calcium carbonate, and iron oxide may precipitate inorganically as cement that binds grains to one another (refer back to Figure 8.6). Inorganic precipitation is also important at locations where evaporation raises the concentration of dissolved ions to the point at which they spontaneously react to form crystalline solids.

One example of inorganic precipitation is the reaction of dissolved sodium with chlorine, which precipitates sodium chloride, or halite (salt). In the ocean, marine organisms combine dissolved calcium ions with dissolved bicarbonate ions to produce a wide variety of solid calcium carbonate shells. Some types of plankton, such as diatoms, also use dissolved silica in this way.

An important aspect of these organic and inorganic reactions is that they regulate the chemistry of ocean water. Oceans are fed by the discharge of streams and rivers around the world. Hydrothermal vents, the hot water springs that develop where there is submarine volcanism, also discharge into the world's oceans. In the course of geologic time, the input of these sources is balanced by evaporation, which removes water from the oceans and temporarily stores it in the atmosphere.

The same streams and hydrothermal vents are also sources of the dissolved chemicals that make seawater salty. Evaporation does not remove these chemicals, so without some mechanism to regulate them, they would perpetually build up in seawater, making the ocean saltier and saltier. That mechanism is precipitation by marine organisms that extract chemicals from seawater to make their shells and skeletons. These shells and skeletons, in turn, fall to the seafloor as biogenic sediment when the organisms die.

As chemicals are removed from the water by biogenic precipitation and stored in sedimentary rocks composed of fossil skeletal debris, the chemistry of seawater is regulated, through a combination of biologic, volcanic, and weathering processes, and remains fairly stable.

The Changing Composition of Sediments

As sediments accumulate in sedimentary basins, their composition changes, along with their texture. The composition of sedimentary deposits evolves over time as a result of two processes:

1. Grains are subjected to continuous chemical weathering as they are transported. This weathering serves to remove unstable grains (olivine, pyroxene, calcium-rich plagioclase, and amphibole) and raise the relative abundance level of more stable grains (quartz, clays, muscovite, and orthoclase).

2. New types of sediment are introduced, such as biogenic sediment (fossils and organic debris) produced within the sedimentary basin (**FIGURE 8.12**), local weathering products washed into the basin (soil, mineral assemblages with some unstable components, oxides, etc.), and locally precipitated chemical sediments.

Dissolved compounds mix with sediments in the basin and add cements or other types of chemical sediments that influence the overall composition of a deposit. The Critical Thinking feature titled "Sediment Deposition" provides a simple experiment to demonstrate how sediments of various sizes tend to accumulate in layers.

Organic sediments are most likely to originate *within* sedimentary basins. These sediments include vegetation preserved from decay in wetlands, which turns into peat (and, eventually, coal), and the remains of algae, bacteria, and other types of microscopic organisms, which may become petroleum. These materials mix with clastic sediments or accumulate in concentrations of organic sediment.

Q EXPAND YOUR THINKING

As described in this section, several factors control the salinity of seawater. Write a hypothesis that predicts the processes governing ocean salinity. Also, describe several ways to test your hypothesis.

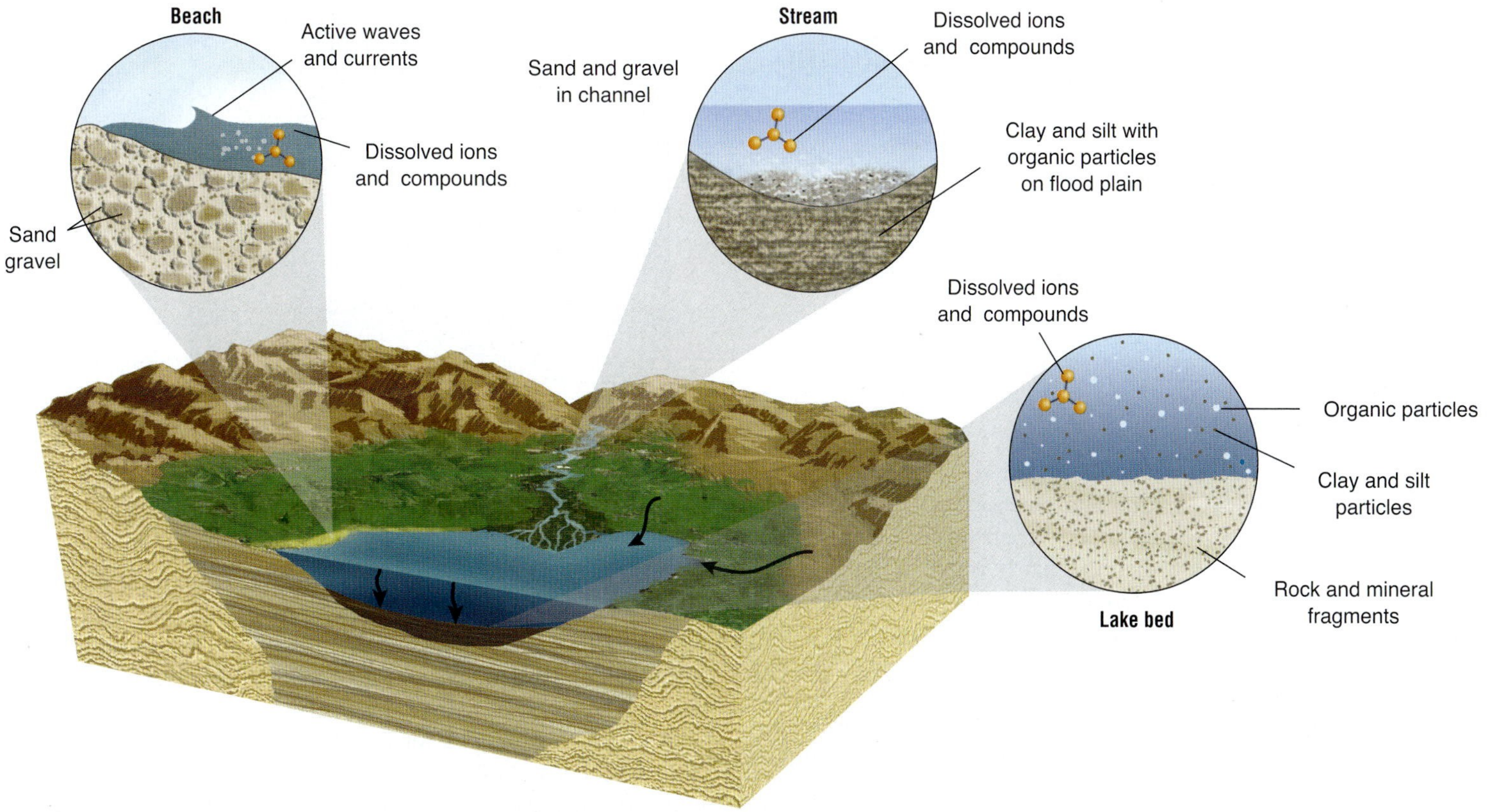

Beach
Active waves and currents
Dissolved ions and compounds
Sand gravel

Stream
Dissolved ions and compounds
Sand and gravel in channel
Clay and silt with organic particles on flood plain

Dissolved ions and compounds
Organic particles
Clay and silt particles
Rock and mineral fragments
Lake bed

FIGURE 8.12 The composition of sediment varies, depending on the mineralogy of grains transported into a basin and the chemical and biogenic sediments produced within the basin.

Q Describe how the composition of lake sediments would likely differ from that of stream sediments.

CRITICAL THINKING

Sediment Deposition

Conduct this simple experiment to see for yourself how sediments accumulate and why sedimentary rocks tend to form in layers.

1. Obtain approximately 1/4 liter (about 1 cup) of sediment consisting of a mixture of mud, sand, and gravel. Place it in a tall, narrow jar made of clear glass (**FIGURE 8.13**).

2. Add water to the jar until it is nearly full. Seal the jar with a lid or cover.

3. Shake the jar thoroughly so that the sediment and water are completely mixed.

4. Quickly turn the jar upright and set it on a flat surface.

5. Draw a diagram of the final accumulation of sediments and answer the following questions:

   a. How does the final accumulation differ from the sediment before it was mixed with water?

   b. What process caused this difference? How and where would the same process occur in nature?

   c. What feature of sedimentary rocks have you recreated? Find a picture of this feature in the book and interpret what it means.

FIGURE 8.13 Experimental setup to analyze sediment deposition.

8-6 Sediment Becomes Rock During the Sedimentary Cycle

LO 8-6 Describe the sedimentary cycle.

Because there are many types of sediment, there are many types of sedimentary rock.

Sedimentary Rock

Most sedimentary rock consists of some or all of four basic components: grains, cement, matrix, and pore spaces (**FIGURE 8.14**). Grains are particles of sediment (clasts, chemical, or biogenic); cement binds the grains together; the matrix is fine sediment (mud or other fine-grained material) that fills in some of the space around grains and cement; and pore spaces are empty spaces within the rock. **Porosity** is a measure of the amount of empty space contained in a rock.

When sediments accumulate and are buried, they eventually turn into solid rock. This process, called **lithification,** takes place in two ways:

1. *Cementation.* Larger particles, such as sand and gravel, are bound together by natural cements that are inorganically precipitated from dissolved compounds in groundwater.

2. *Compaction.* Smaller particles, such as silt and clay, are compacted and consolidated by pressure from the weight of overlying layers of sediment.

Most sedimentary rocks *lithify* (turn to stone) as a result of a combination of these two processes. As sediment is buried in the crust, it is subjected to heat and pressure, which produce physical and chemical changes. On average, temperature rises by 30°C for every kilometer of burial in the crust, and pressure increases by the equivalent of 1 atmosphere (the pressure of the atmosphere pressing down on Earth's surface at sea level) for every 4.4 meters of depth. At a depth of 4 kilometers, the temperature can be 120°C and the pressure can exceed 4,000 atmospheres. These high temperature and pressure levels cause cementation and compaction, which bind sediments and form sedimentary rock.

Compaction happens when sediment is compressed during burial (**FIGURE 8.15**). Clay and silt typically experience more compaction than do sand and gravel. Because mud is usually deposited in water, the pore spaces are filled with liquid. Compaction involves slow squeezing, with the result that this water is expelled and usually travels upward to locations where the compaction pressure is lower.

The resulting *dewatered* sediment is denser because it has lower porosity—less pore space between grains. Mud particles are shaped like flat dinner plates, which tend to align when compacted,

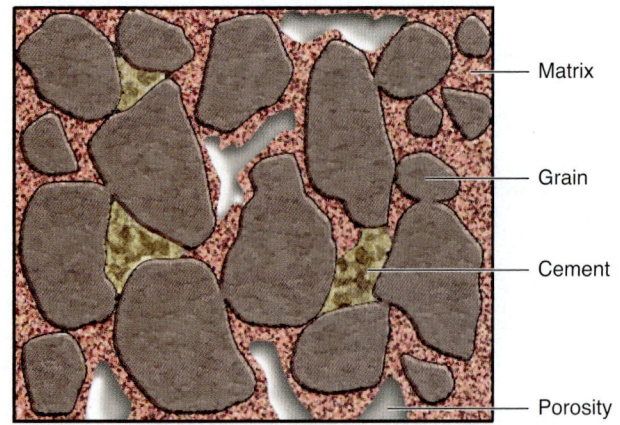

FIGURE 8.14 Sedimentary rock is composed of four constituents: grains, cement, matrix, and pore spaces.

Q **On what basis would you assign names to sedimentary rocks?**

forming thin layers of sediment called *laminations*. Particles adhere to one another with weak, grain-to-grain electrostatic forces.

As sands become compacted, they are reorganized into a closely packed stack of grains. Mud decreases in volume by as much as 50% to 80% during compaction, whereas sand may decrease in volume by 10% to 20%.

Cementation occurs when dissolved ions in groundwater inorganically precipitate to fill pore space (usually in sands), thus binding grains together. The most common cements come from dissolved compounds that originate within the surrounding sediments. Hence, silica (SiO_2) often cements quartz and feldspar grains; iron oxides (FeO or Fe_2O_3) often cement iron-rich grains; and calcium carbonate ($CaCO_3$) often cements calcite grains.

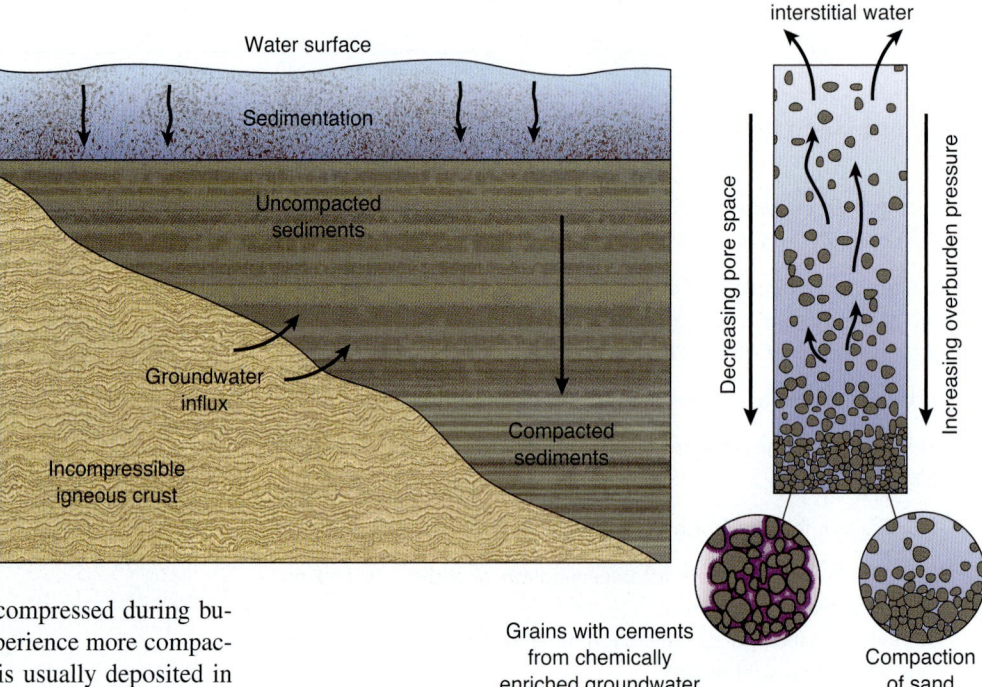

FIGURE 8.15 Compaction tends to drive water out of sediment, and cementation fills the empty space between grains with secondary minerals that precipitate from groundwater.

Q **How are sand and gravel grains lithified? How are mud particles lithified?**

The Sedimentary Cycle

The sedimentary cycle is a component of the rock cycle that perpetually supplies sediment from the weathering and erosion of tectonically uplifted crust (**FIGURE 8.16**). Each stage of the sedimentary cycle is characterized by an important set of processes that influence the resulting sedimentary rock. These processes include tectonic uplift, weathering, erosion and transportation, deposition, burial, and lithification followed by renewed uplift, which starts the cycle over again.

Most tectonic uplift is associated with plate boundary convergence. When crust is uplifted at a convergent plate margin, orogenesis or volcanism produces mountains. Mountainous slopes are immediately attacked by chemical, physical, and biological weathering, producing solid clasts and dissolved ions and compounds. Various physical processes of erosion (gravity, running water, wind, and others) remove these products of weathering.

Eroded sediments move across Earth's surface until they come to rest in an environment of deposition. As they move, the texture and composition of the sediments may be influenced by the chemical and physical characteristics of the surface environments they encounter.

Sediments eventually come to rest, accumulate, and form layers in response to varying rates of delivery. Sustained deposition through time requires a sedimentary basin where tectonics continuously lowers the crust so that it does not fill up. Buried sediment is subjected to compaction due to the pressure imposed by overlying sediments and to cementation by dissolved ions in groundwater. As a result of these processes, buried sediment becomes sedimentary rock.

The sedimentary cycle is renewed every time tectonic uplift occurs and sedimentary rocks are exposed to weathering. The cycle helps answer some important questions about the sedimentary system:

Why hasn't Earth's surface eroded down to sea level long ago? *Answer*: New mountains perpetually arise from orogenesis, arc volcanism occurring at convergent plate boundaries, and other types of mountain building. (Mountain building is discussed in Chapter 10.)

What is the relationship between plate tectonics and Earth surface environments? *Answer*: Mountains are a prolific source of sediment because they are exposed to intense weathering. The resulting sediment becomes an integral component of Earth's surface, in many cases influencing the character of environments and ecosystems.

Why is sedimentary rock such an effective recorder of Earth's history? *Answer*: The texture and composition of sedimentary rocks may be influenced by environmental conditions. They contain fossils and other constituents that are the result of geologic and biologic processes that were dominant at the time of deposition.

The sedimentary cycle is powered by: solar energy, which drives weathering; geothermal energy (and gravity), which drives plate motion; and gravity, which drives sediment erosion and transportation from mountains to sedimentary basins. It is estimated that some grains of highly resistant quartz and zircon have participated in as many as eight cycles of mountain building and erosion, repeatedly eroding from one sedimentary rock and eventually becoming part of another.

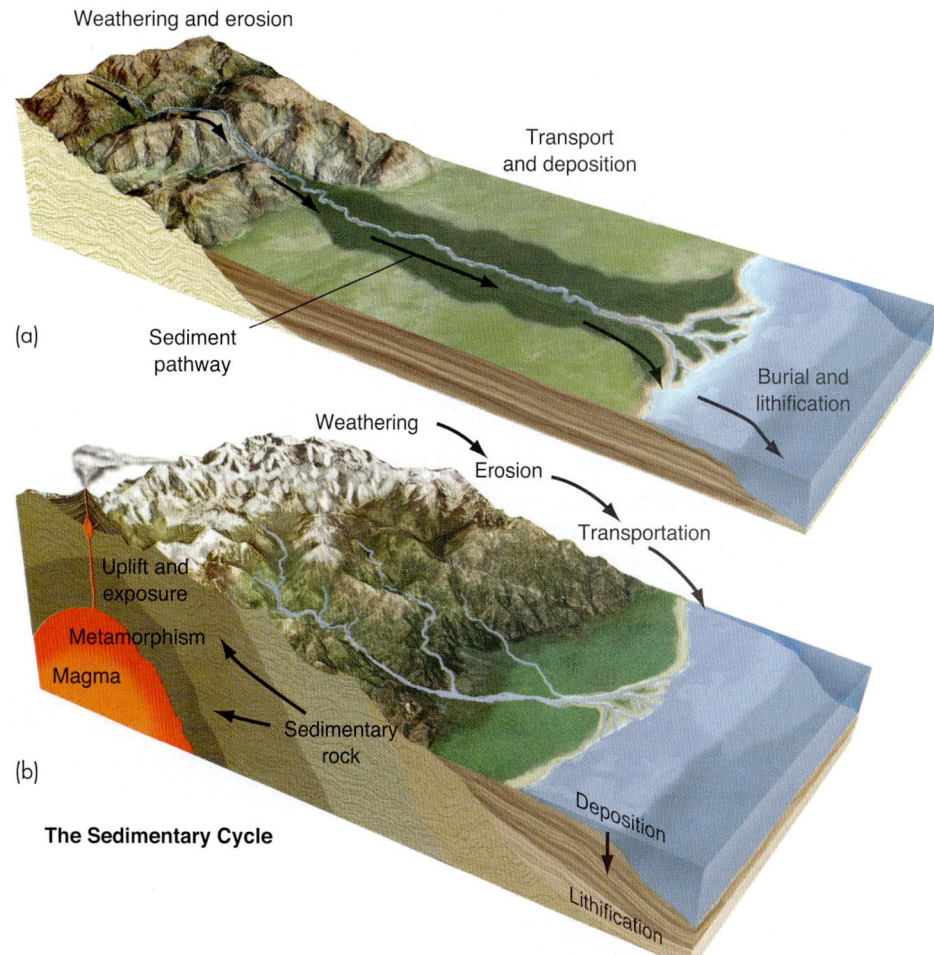

FIGURE 8.16 (a) Weathering forms sediment, which is moved by wind, water, and gravity into environments of deposition. (b) There sediment lithify and become solid sedimentary rock. Eventually they may be tectonically uplifted and subjected to a renewed cycle of weathering and erosion.

Q Why does uplift subject Earth's crust to renewed weathering and erosion?

Q EXPAND YOUR THINKING

Describe the relationship between the sedimentary cycle and plate tectonics.

(8-7) There Are Eight Major Types of Clastic Sedimentary Rock

LO 8-7 List the eight major types of clastic sedimentary rock, and give a brief description of each.

Sedimentary rocks are assigned names based on the lithology of their components—their grains and cements. A rock's *lithology* is a description of its composition (mineralogy) and texture (the size, shape, and arrangement of the sediments).

Clastic Sedimentary Rock

Lithified (literally, turned to stone) clastic sediments produce clastic sedimentary rock, listed in **TABLE 8.2** on the basis of grain size. Rocks in this category are distinguished by the size of their particles (**FIGURE 8.17**) as well as their mineral content, as shown in (**FIGURE 8.18, a–h**):

a. *Conglomerate* is composed of gravel particles that have been rounded by abrasion in stream or beach environments. Rounding of large, angular clasts indicates that they have been transported some distance from the weathered crust where they originated.

b. *Breccia* is similar to conglomerate, except that the gravel particles have very angular corners and edges, indicating that the clasts have not been transported far from their place of origin.

c.–e. Sandstone is commonly found in three varieties: *quartz sandstone, arkose,* and *lithic sandstone.* Quartz sandstone (c) is composed of quartz grains with minor amounts of other types of sediments, usually cemented by silica cement or iron oxide. Arkose (d) is rich in feldspar and quartz. It is typically coarse-grained, often with a matrix of clay and iron oxide, and pink or red in color. Arkose is usually formed by rapid weathering of granite. Lithic sandstone (e) is dark or gray, coarse-grained, and well-cemented, consisting of poorly sorted angular quartz and feldspar grains and a variety of dark rock and mineral fragments embedded in a clay matrix.

f.–g. Rocks consisting of clay and silt-size particles usually break into clumps or blocks. *Siltstone* (f) is composed of mostly silt particles, and *claystone* (g) is composed of clay particles. Many environments of deposition accumulate alternating layers of these fine-grained rocks as energy conditions wax and wane.

h. *Shale* is a sedimentary rock composed predominantly of clay particles; and usually silt is also present. Because shale is formed through compaction, the flat clay and silt particles are aligned along parallel planes.

Reading the Sands of Time

Sandstone can tell stories about its history. The sand found around active volcanoes consists of volcanic rock fragments, volcanic glass, and other igneous minerals (**FIGURE 8.19**). Beach sand from southern California consists of quartz (the most durable common mineral), some feldspar (also durable, but more easily weathered to clay), and other clastic minerals weathered from plutonic igneous rocks in nearby mountain ranges. In areas where there is no abundant source of land-based sediment, sand may be composed of shell fragments, coral, and marine skeletal fragments.

Recall from section 8–3 that the region where crust weathers and erodes to produce sediment is known as the *source area*. By analyzing sand grains in sediment (**TABLE 8.3**), geologists can infer the type of rock present at the source area, along with the

TABLE 8.2 Clastic Sedimentary Rocks

Rock	Texture	Composition
Conglomerate	Rounded gravel	Clastic, lithic fragments
Breccia	Angular gravel	Clastic, lithic fragments
Quartz sandstone	Sand	Quartz
Arkose sandstone	Sand	Quartz, feldspar
Lithic sandstone	Sand	Quartz, feldspar, lithic fragments
Siltstone	Silt, massive	Silica minerals
Claystone	Clay, massive	Clay minerals
Shale	Mud, laminated	Silica minerals and clay minerals

FIGURE 8.17 Clastic sedimentary rocks are named for the size of their clasts.

Sediment: Gravel | Sand | Silt | Clay and silt

Sedimentary rock: Conglomerate | Sandstone | Siltstone | Shale

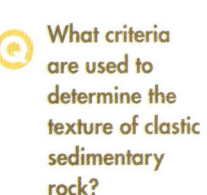
Q What criteria are used to determine the texture of clastic sedimentary rock?

All images courtesy of Chip Fletcher

FIGURE 8.18 The eight major clastic sedimentary rocks: (a) conglomerate, (b) breccia, (c) quartz sandstone, (d) arkose, (e) lithic sandstone, (f) siltstone, (g) claystone, and (h) shale.

TABLE 8.3 Inferring Geologic History from the Composition of Sand

Sand Type	Transport Energy and Distance	Weathering Intensity	Probable Source Rock
Quartz	High energy and distance	High	Granite or metamorphic rocks
Feldspar and quartz	High energy, moderate distance	Moderate to low	Granite or gneiss
Coarse sand	High energy, low distance	Moderate to low	Nearby, any rock type
Carbonate	Any energy, low distance	Low	Reef or marine organisms
Heavy minerals	High energy, any distance	Moderate to low	Igneous or metamorphic

climate, the relative rate of weathering, tectonic activity, processes acting on the sand within the environment of deposition, and the length of time the sand has resided in the depositional environment.

Quartz grains, for example, come from source rocks that are rich in quartz, such as granite, gneiss (a metamorphic rock), or other sandstones containing quartz. Feldspar sand in arkose comes from weathering granite or gneiss. Rock fragments in lithic sandstone come from fine-grained source rocks, such as shale, slate, basalt, rhyolite, andesite, chert, and various metamorphic rocks.

Using such information about the composition of sand, geologists can infer the climate in the source area. Weathering in humid climates will quickly produce clay, iron oxides, and aluminum oxides from unstable host rocks containing feldspar and minerals rich in iron and aluminum.

The presence of feldspar in sandstone indicates either that the climate was arid and that chemical weathering was not intense, or that erosion rates were rapid. (If tectonic uplift was fast, the resulting steep slopes would produce rapid erosion.) If quartz is dominant, the climate was probably humid, destroying all feldspars and other easily weathered minerals.

Courtesy of Chip Fletcher

FIGURE 8.19 Sand grains exhibit a wide range of mineral types.

Q EXPAND YOUR THINKING

What can the eight major types of clastic sedimentary rock tell you about their source areas? Interpret each.

8-8 There Are Seven Major Types of Chemical Sedimentary Rock and Four Major Types of Biogenic Sedimentary Rock

LO 8-8 List the seven major types of chemical sedimentary rock and the four major types of biogenic sedimentary rock.

For the most part, chemical sedimentary rocks are identified based on their mineral composition; in some cases, however, rock texture is the basis for identification (**TABLE 8.4**).

Chemical Sedimentary Rocks

The seven types of chemical sedimentary rocks are shown in **FIGURE 8.20 a–g**, and described as follows:

a. *Rock salt,* an evaporite, is composed of the mineral halite (a member of the halide mineral group, as explained in Chapter 4). It is typically formed by the evaporation of salty water or seawater that contains dissolved sodium and chlorine ions.

b. *Rock gypsum* is composed of the mineral gypsum ($CaSO_4 \cdot 2H_2O$), a member of the sulfate mineral group; like rock salt, it usually results from the evaporation of seawater.

c.–e. *Limestone* (c) is composed of the mineral calcite (from the carbonate mineral group). Although most limestone is bio-

TABLE 8.4 Chemical Sedimentary Rocks

Rock	Texture	Composition
Chert	Crystalline	Microcrystalline silica
Dolostone	Crystalline	Dolomite
Limestone	Can be crystalline or microcrystalline	Calcite
Micrite	Microcrystalline	Carbonate mud
Rock gypsum	Crystalline	Gypsum
Rock salt	Crystalline	Halite
Travertine	Microcrystalline	Calcite from saturated fluids

genic, there are two types of inorganic limestone: *travertine* (d) and *micrite* (e). Travertine is formed wherever water saturated with $CaCO_3$ evaporates. Under such conditions, layers of inorganically precipitated calcium carbonate accumulate, developing into terraces, towers, pools, and other odd shapes. Travertine is also formed in caves where groundwater enriched with $CaCO_3$ drips from the ceiling. In time, structures called stalactites grow and hang down from the ceiling. Water that drips to the floor may become *stalagmites*, which gradually extend upward toward the ceiling. When *stalagmites* and stalactites meet and merge, they become a *column*. Micrite is "lime mud," dense, dull-looking sediment composed of clay-size crystals of calcite.

f. *Dolostone* is quite similar to limestone but is composed mostly of the mineral *dolomite* [$CaMg(CO_3)_2$], a carbonate. It is formed when dissolved magnesium in groundwater is substituted for some of the calcium in the original limestone; it

All images courtesy of Chip Fletcher

Q What is the primary criterion for classifying chemical sedimentary rocks?

FIGURE 8.20 Chemical sedimentary rocks: (a) rock salt, (b) rock gypsum, (c) limestone, (d) travertine, (e) micrite, (f) dolostone, and (g) chert.

can also develop as a result of direct precipitation of dolomite. Most dolostone is a result of chemical alteration of limestone by magnesium-rich groundwater.

g. *Chert* is a sedimentary rock composed of microcrystalline quartz that comes from inorganic precipitation in groundwater. A biogenic form of chert originates in organic beds of marine silica-rich fossils of two types of plankton: radiolarians and diatoms. Black or gray chert is called *flint*; red chert is referred to as *jasper*.

TABLE 8.5 Biogenic Sedimentary Rocks

Rock	Texture	Composition
Chalk	Clay or mud	Skeletal coccolithophorids
Coal	Massive, blocky	Concentrated carbon
Coquina	Sand or gravel	Shell fragments
Limestone	Visible or microscopic skeletal fragments	Calcite

Biogenic Sedimentary Rock

The biogenic sedimentary rocks, listed in **TABLE 8.5** and shown in **FIGURE 8.21 a–d**, are also identified based on their mineral composition and texture, as follows:

a. *Coquina* is a form of limestone composed entirely of broken and worn shell fragments. The pieces usually are weakly cemented together, especially in recent deposits. Coquina consists of mechanically transported fossil shell debris with little or no matrix, loosely cemented so that the rock looks very porous. Typically the shells are those of clams and snails from a beach environment.

b. *Skeletal limestone* is composed of fossil fragments of carbonate organisms. Some of these are visible fragments, but many limestones are made up of microscopic skeletal fragments. Most limestone accumulates in marine environments from biogenic ooze, calcium carbonate made up of skeletal fragments of microscopic plankton. Skeletal limestone may also consist of large pieces of reef organisms such as coral or some other type of shelled fossil. Carbonate rock of all types accounts for about 15% of Earth's sedimentary crust.

c. *Chalk* is a biogenic limestone made up of the microscopic skeletons of coccolithophorids, single-celled plankton, of-

ten called "coccoliths." Coccolith skeletons are continuously settling toward the bottoms of oceans around the world. But deep water is so cold and acidic that the skeletons dissolve. Therefore, chalk accumulates only at shallower depths, which today means either along mid-ocean ridges or on shallow continental shelves.

d. *Coal* is an organic sedimentary rock that is the result of physical and chemical alteration of *peat* by processes involving bacterial decay, compaction, heat, and time. Peat deposits typically develop in freshwater wetlands where plant debris accumulates at a rate that exceeds that of bacterial decay. The rate of decay is low because the decay process uses up the oxygen in organic-rich water. For peat to become coal, it must be buried by sediment. Burial causes compaction, which squeezes out water. Continued burial plus heat and time cause complex hydrocarbon compounds in the deposit to break down and change in a variety of ways. Through this complex process, peat eventually is transformed into coal. (To deepen your understanding of coal, read the "Earth Citizenship" feature, "The Pros and Cons of Coal.")

(a)   (b)   (c)   (d)

All images courtesy of Chip Fletcher

FIGURE 8.21 Biogenic sedimentary rocks: (a) coquina, (b) skeletal limestone, (c) chalk, and (d) coal.

Each biogenic rock tells a story about the history of life on Earth. Imagine these four rocks were the same age (say, 10 million years old) and found within 100 kilometers of each other. Describe a logical geographic pattern for where they originally formed. What would they signify about past environments? Build a testable hypothesis.

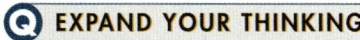

EXPAND YOUR THINKING

What is the dominant mineral in biogenic and chemical sedimentary rock? Where does it come from and why is it so abundant?

The Pros and Cons of Coal

On average, each person in the United States uses 9 kilograms of coal every day. Coal represents over 80% of America's recoverable fossil fuels, and 55% of the nation's electricity is generated by coal-burning utility companies. The nation today has 250 billion tons of accessible coal, enough to last at least 300 years at current rates of consumption. U.S. coal has a total energy value equal to that of about 2 trillion barrels of crude oil—more than triple the world's known oil reserves. This vast amount of coal makes the United States the world leader in known coal reserves.

Worldwide, coal is the most abundant and most widely distributed fossil fuel (**FIGURE 8.22**). The amount of recoverable global coal reserves is estimated at about 1081 billion tons. At current rates of production, the reserves could last for another two centuries. The United States, with 26%, and the former Soviet Union, with 23%, account for nearly half of global coal reserves. China (12%), Australia (8%), Ger-many (7%), South Africa (5%), and Poland (2%) also have significant amounts of the world's recoverable coal.

In the United States, coal is found in 38 states. Montana has the most, about 120 billion tons. Other top coal states, in descending order of known reserves, are Illinois, Wyoming, Kentucky, West Virginia, Pennsylvania, Ohio, Colorado, Texas, and Indiana.

What Is the Origin of Coal?

Coal is the product of ancient vegetation that has been altered over long periods of time. Dense layers of wet, matted vegetation growing in warm, humid wetlands are converted by microorganisms into a compressed and concentrated deposit known as *peat*, which eventually becomes coal.

Of course, much of the plant matter that accumulates on Earth's surface never turns into peat or coal because it is destroyed by fire or decomposed by organic decay. That is why geologists infer that the vast coal deposits found in ancient rocks must represent periods during which several favorable biological and physical processes occurred at the same time: broad tracts of low-lying, freshwater wetlands; prolific growth of vegetation in a moderate climate; and rapid burial and compaction of that vegetation by clastic sediments.

Coal is classified into three grades (**FIGURE 8.23**): *lignite* (brown coal), *bituminous* (soft coal), and *anthracite* (hard coal). These grades are formed in a continuous sequence from sedimentary (lignite, bituminous) to metamorphic rock (anthracite).

When peat is exposed to heat and pressure due to burial, it is compressed to about 20% of its original thickness and becomes lignite.

(a) Global coal reserves

(b)

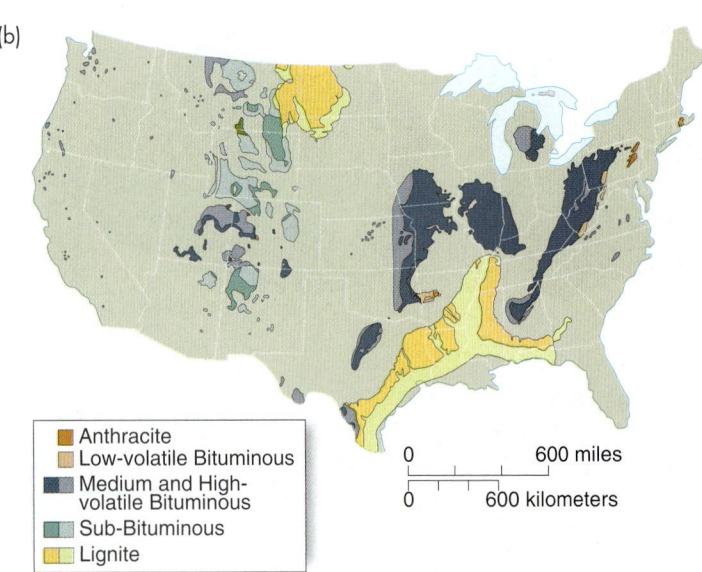

- Anthracite
- Low-volatile Bituminous
- Medium and High-volatile Bituminous
- Sub-Bituminous
- Lignite

0 ———— 600 miles
0 ———— 600 kilometers

FIGURE 8.22 (a) The United States and Russia account for nearly half of global coal reserves. (b) Total coal production in the United States has risen consistently from 602 million tons in 1970 to over 1 billion tons today.

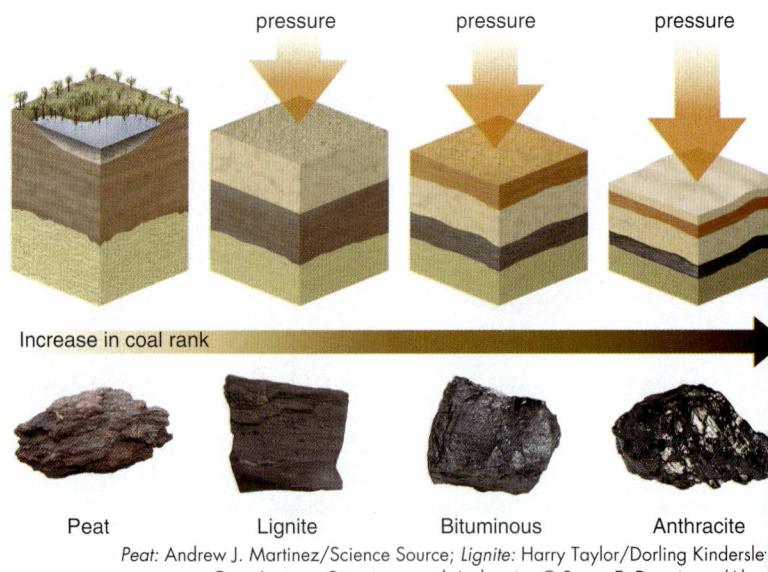

pressure   pressure   pressure

Increase in coal rank →

Peat   Lignite   Bituminous   Anthracite

Peat: Andrew J. Martinez/Science Source; *Lignite:* Harry Taylor/Dorling Kindersley/Getty Images; *Bituminous* and *Anthracite:* © Susan E. Degginger/Alam

FIGURE 8.23 Coal is made when organic debris is buried, compressed, and heated. This process drives out water and concentrates the carbon into bituminous coal. With further heating, anthracite coal is made.

FIGURE 8.24 Coal-fired power plants produce unhealthy emissions that threaten the health of neighboring communities and contribute to global warming.

FIGURE 8.25 Surface mining causes massive environmental damage. In West Virginia 1600 kilometers of streams have been buried under mine waste, and 300,000 acres of forests have been destroyed by surface coal mining.

As lignite is further compressed and heated, to 100°C to 200°C, it is transformed into bituminous coal. This process drives out water and volatile gases, concentrating the carbon. Longer exposure to high temperature further drives out gases and stimulates changes that produce anthracite. Anthracites are compressed to 5% to 10% of their original thickness and contain less than 10% water and volatile gases.

Problems Associated with Coal Use

Coal is abundant and so is a major energy source; but there are serious problems associated with burning it. One, coal produces many compounds when burned, some of which can cause cancer. Two, the residue from burning coal is a leading cause of smog, acid rain, and air toxicity. Annually, a single typical coal power plant generates these products (**FIGURE 8.24**):

- 3.7 million tons of carbon dioxide, the primary cause of global warming. This is equivalent to the amount of carbon dioxide contained in 161 million trees.

- 10,000 tons of sulfur dioxide, which causes acid rain that damages forests, lakes, and buildings, and forms small airborne particles that can cause respiratory ailments.

- 500 tons of airborne particles called fly ash, which causes bronchitis, asthma, and atmospheric haze.

- 10,200 tons of nitrogen oxide, equal to the amount emitted by half a million cars. Nitrogen oxide leads to the formation of ozone (smog) that inflames the lungs.

- 720 tons of carbon monoxide, a cause of headaches and stress for people with heart disease.

- 220 tons of hydrocarbons, volatile organic compounds that form ozone at ground level.

- 77 kilograms of mercury, of which just 1/70th of a teaspoon deposited on a 25-acre lake can make the fish unsafe to eat.

- 102 kilograms of arsenic, which causes cancer in 1 out of 100 people who drink water containing 50 parts per billion.

- 51 kilograms of lead, 2 kilograms of cadmium, other toxic metals, and traces of radioactive uranium.

Children are particularly susceptible to air pollution because they breathe more air per kilogram of body weight than adults do, and their lungs are still developing. Researchers at the Harvard School of Public Health found that people living within 48 kilometers of coal-powered utility smokestacks had a three to four times greater chance of dying prematurely than those living farther away.

In an effort to address these problems, industrial devices that "scrub" coal combustion have been designed to trap some of these pollutants. And the efficiency and environmental compatibility of coal-burning plants are improving as they increasingly turn to innovative combustion methods and coal liquefaction processes.

A third problem associated with coal is the environmental damage caused by mining it. About 60% of U.S. coal is stripped from surface mines; the rest comes from underground mines. Surface mining dramatically alters the landscape: Coal companies often remove entire mountaintops to expose the coal below, and then generally dump wastes into nearby valleys and streams (**FIGURE 8.25**). Acid mine drainage from waste rock damages watersheds and groundwater. In West Virginia, more than 300,000 acres of hardwood forests (half the size of Rhode Island) and 1600 kilometers of streams have been destroyed by strip mines and mountaintop mining.

In sum, although the United States has abundant coal reserves, there are sizable environmental and health costs associated with its extraction and use that are not easy to mitigate.

 Sedimentary Rocks Preserve Evidence of Past Depositional Environments

LO 8-9 Describe how a geologist uses inductive reasoning to interpret Earth's history using sedimentary rocks.

As noted earlier, every sedimentary rock bears the imprint of the natural forces that shaped it. By examining layers of sedimentary rock that accumulate over time, geologists can decipher Earth's history, as if reading from the pages of an encyclopedia.

Interpreting Sediments

Using *inductive reasoning* to interpret sediments, geologists can make a number of inferences. Where sandstones are found, they can infer there was once a river or beach; marine skeletal limestones indicate former seafloor. Organic sediments tell them where life has been active in the past; and poorly sorted but heavily abraded sediments might reveal the site of a former glacier. Consider further that seashells can be found in the high mountaintops of the Himalayas; that ancient reefs dot the Rocky Mountains; and that evidence of massive river deltas exists far from any modern coastline in the Appalachian Mountains.

Some of these sedimentary deposits contain valuable ores and fuel. Others encase fossils of long-extinct species (**FIGURE 8.26**). Throughout the world, sedimentary rocks present the most complete record of past Earth surface environments.

Continental Environments

FIGURE 8.27 a–f shows the various types of continental environments, described here:

a. *Glaciers* (studied in Chapter 15) are characterized by cold climate, moving ice that mechanically grinds the crust and plucks and abrades clasts, ranging in size from clay to boulders the size of houses, and voluminous discharges of sediment-laden water from the front of the ice (where it is perpetually melting). The sediment, which is usually poorly sorted, contains particles of many different sizes and an abundance of silt and clay.

b. *Deserts* and *dunes* (described more in Chapter 19) emerge in arid environments with little vegetation. In such environments, wind is especially effective in shaping the land surface. Windblown sand is abraded and well sorted, and erosion of sand by the action of wind concentrates gravel particles, forming a desert pavement.

c. *Alluvial fans* develop in arid environments where rainfall is infrequent but torrential. When it does rain, massive discharges of sediment emerge from steep valleys to form fan-shape deposits of mud, sand, and gravel.

d. *Streams* (see Chapter 17) are characterized by channels that carry sediment of all sizes. Sedimentary deposits within the channels typically consist of coarse, well-rounded gravel and sand. The adjoining floodplain is covered with fine-grained sand, silt, and clay.

e. *Wetlands* are characterized by standing water in which most of the sediment is organic, produced by abundant vegetation. Biogenic sediments such as peat tend to accumulate, as do clays and silts. Wetlands develop where the water table intersects the land surface.

f. *Lakes* are highly diverse settings that vary greatly in size and may be fresh or saline, shallow or deep. These environments accumulate clastic, chemical, and biogenic sediments. Some lakes

FIGURE 8.26 Sedimentary rocks contain fossils that provide important clues to past Earth environments. This rock is from a former lake in the Green River Formation in Wyoming. We know the formation (a distinct rock unit) was previously a lake because of the fine sediments (silt) and the types of fossils we find there. This fossil shows a larger fish eating a smaller fish that apparently died midmeal.

Jeff Foott/Discovery Channel Images/Getty Images

Q What might have killed this fish 48 million years ago? What evidence would you look for to support your hypothesis?

FIGURE 8.27 Continental environments of deposition: (a) glacier, (b) desert and dune, (c) alluvial fan, (d) stream, (e) wetland, and (f) lake.

Q What is the typical sediment type found in each of these environments: clastic, chemical, biogenic, or some combination?

dry up each year, producing chemical sediment through evaporation. Other lakes are deep and large and characterized by high-energy physical processes, such as frequent storms, that promote coarse clastic sediments.

Coastal Environments

Coastal environments (explored more in Chapter 20), shown in **FIGURE 8.28 a–e**, are described as follows:

a. *Deltas* contain deposits of clay, silt, and sand that take an irregular fan shape at locations where a stream flows into a standing body of water, such as a lake or the sea. Deltas develop where large amounts of sediment accumulate and are usually associated with major rivers, such as the Mississippi and the Nile.

b. *Beaches* are exposed to high-energy waves that deposit coarse sand. They tend to erode or accrete (grow seaward) according to the season and are frequently pounded by storms that cause substantial erosion.

c. *Barrier islands* are composed of coarse sand and are separated from the mainland by a barrier lagoon. Chains of barrier islands emerge from the accumulation of sand driven by waves and cur-

rents. They are separated by tidal inlets that control the flow of seawater into the lagoon.

d. *Lagoons* contain mud and fine sand with a large biogenic component. They are protected from the ocean by barrier islands, yet they experience strong tidal currents through inlets between them. Lagoons are also formed behind tropical barrier reefs.

e. *Tidal wetlands* contain silt and clay with large amounts of organic sediment and skeletal fragments of marine organisms, such as plankton. They support abundant growth of salt-tolerant plants that trap suspended mud that is carried onto the wetland surface by high tides during the course of the day.

Marine Environments

There are four types of marine environments (explored more in Chapter 21) shown in **FIGURE 8.29 a–d** and described here:

a. *Reefs* are massive biological constructions of calcium carbonate that extend along the edges of continental shelves or volcanic islands. They are built by corals and other invertebrates, and various types of algae in warm, sunlit seas in the tropics between latitudes 30° north and 30° south of the equator.

FIGURE 8.28 Coastal environments of deposition: (a) delta, (b) beach, (c) barrier island, (d) lagoon, and (e) tidal wetland.

What is the typical sediment type found in each of these environments: clastic chemical biogenic or a combination?

b. *Continental shelves* are the flooded edges of continents. They accumulate clastic, biogenic, and chemical sediments over long periods of geologic time. A shelf can be shallow (less than 200 meters deep), flat, and up to hundreds of kilometers wide. Al-though mapping the ocean floor is very difficult, by using the techniques described in the Geology in Our Lives feature, "Mapping the Seafloor Using Satellite Radar," geologists are able to decipher the topography of continental shelves and the deep sea.

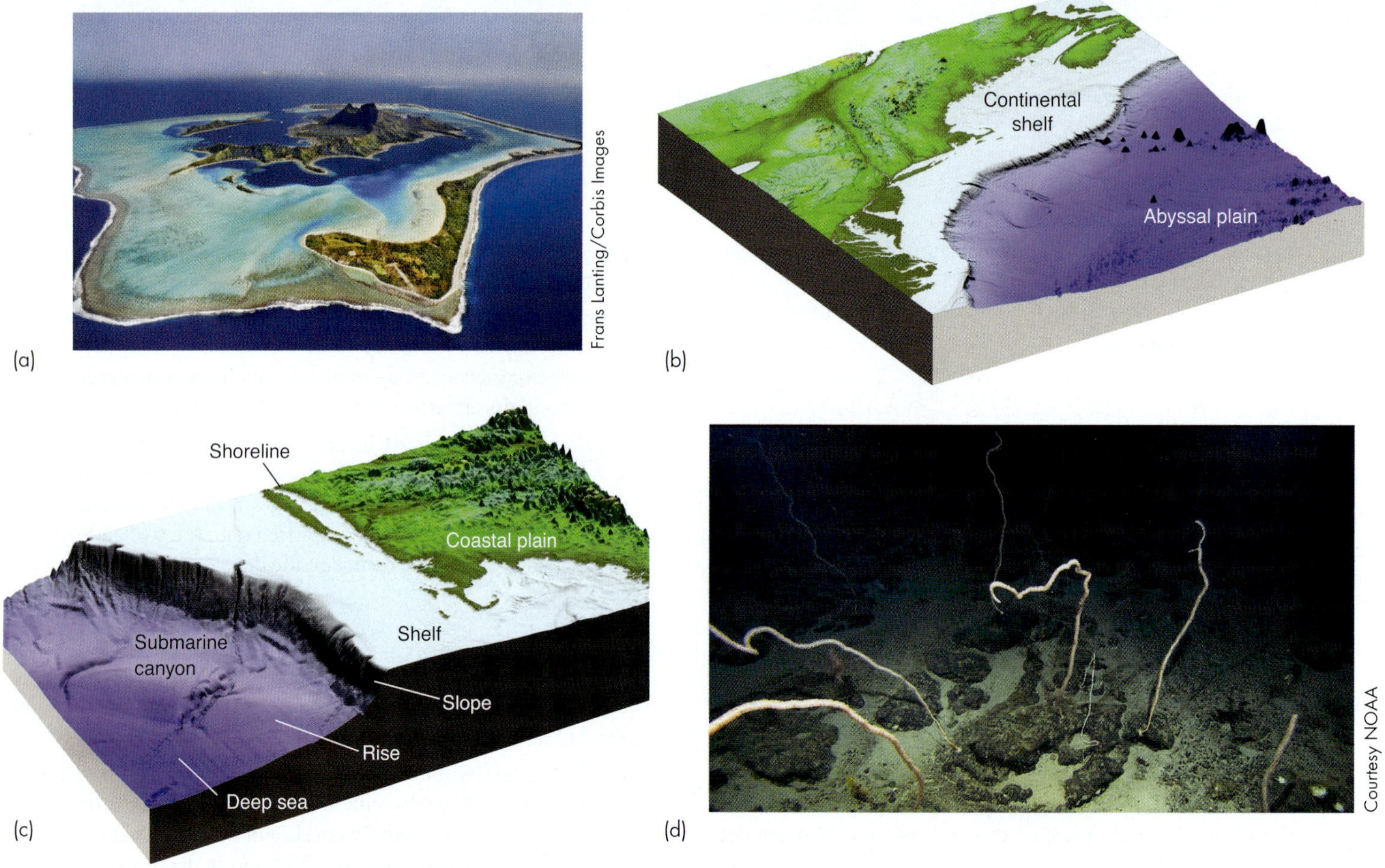

FIGURE 8.29 Marine environments of deposition: (a) reef, (b) continental shelves, (c) continental margin, and (d) deep-sea environments.

What is the typical sediment type found in each of these environments: clastic chemical biogenic or a combination?

c. *Continental margins* begin at the seaward edge of continental shelves. Margins are composed of the continental slope, which descends steeply, and the continental rise, which has a gentle slope. The rise is composed of biogenic marine sediments and clastic sediments delivered to the area by turbidity currents, undersea avalanches of sediment that originate on the slope.

d. *Deep-sea environments* include the deep floors of all the oceans (including oceanic trenches), where clastic and biogenous sediments collect. Biogenous sediments in the deep sea are the remains of plankton, which turn into organic mud on the seafloor. Abyssal plains, the deep flat seafloor between trenches and

mid-ocean ridges, typically collect abyssal clays and biogenous ooze. These ridges, where new oceanic crust is developed, have only thin sedimentary coatings because of their young age.

To expand your understanding of these environments, please complete the Critical Thinking exercise titled "Environments of Deposition."

EXPAND YOUR THINKING

Why do geologists study the record of past environments in sedimentary rocks? Identify some important reasons why you would study a sedimentary formation in detail.

 GEOLOGY IN *OUR LIVES*

Mapping the Seafloor Using Satellite Radar

We do not feel it, but as we move from place to place, we experience changes in gravity generated by variations in the density of the rock beneath our feet. Regions with denser rock exert a stronger gravitational pull. Our bodies are not sensitive enough to sense these changes, but water instantly "records" variations in gravity by accumulating in places where gravitational force is high. On the deep seafloor, high and low topographic features cause variations in the gravity field based on their greater or lesser mass. The surface of the ocean displays bumps, ridges, and depressions that mimic these features, even though they may be over 5 kilometers below the surface. These bumps and dips in the ocean surface are evidence of even minute variations in Earth's gravitational field resulting from fluctuations in seafloor topography.

A typical undersea volcano is 2000 meters tall and has a radius of some 20 kilometers. In contrast, the surface bump it creates cannot be seen with the naked eye because its sides slope very gently, but it can be measured using satellite technology. Mid-ocean ridges, deep-sea trenches, undersea volcanoes, oceanic fracture zones, and many other types of seafloor and tectonic features are reflected in the topography of the ocean surface (**FIGURE 8.30**).

How do scientists measure these tiny bumps and dips on the ocean surface? By using highly accurate radar mounted on a satellite. In 1985, the U.S. Navy launched the Geosat satellite to map the ocean surface at a horizontal resolution of 10 to 15 kilometers and a vertical resolution of 0.03 meters. Geosat orbits on a path that takes it nearly over the north and south poles 14.3 times per day. While in this orbit the satellite maps the sea surface at a speed of approximately 7 kilometers per second; that means it takes about 1.5 years to map the entire surface of Earth.

Two very precise distance measurements must be taken in order to map the topography of the ocean surface to such a high degree of accuracy. First, a global network of stations constantly tracks the location of the satellite; second, the satellite measures its height above the ocean surface using the two-way travel time of pulses of microwave radar. A sharp radar pulse emitted from the satellite measures a spot on the ocean surface 1 to 5 kilometers in diameter. The "footprint" of the pulse must be large enough to average out the local irregularities in the surface due to ocean waves. The satellite emits 1000 pulses per second to improve the accuracy of the measurement. The radar data are corrected for variations in the composition of the atmosphere (e.g., moisture content) that may cause changes in the radar pulse. The sea surface height is also corrected for tidal fluctuations.

This precise application of physics confirms definitively the theory of plate tectonics. The formation and movement of the plates explain almost every facet of seafloor topography revealed in radar maps.

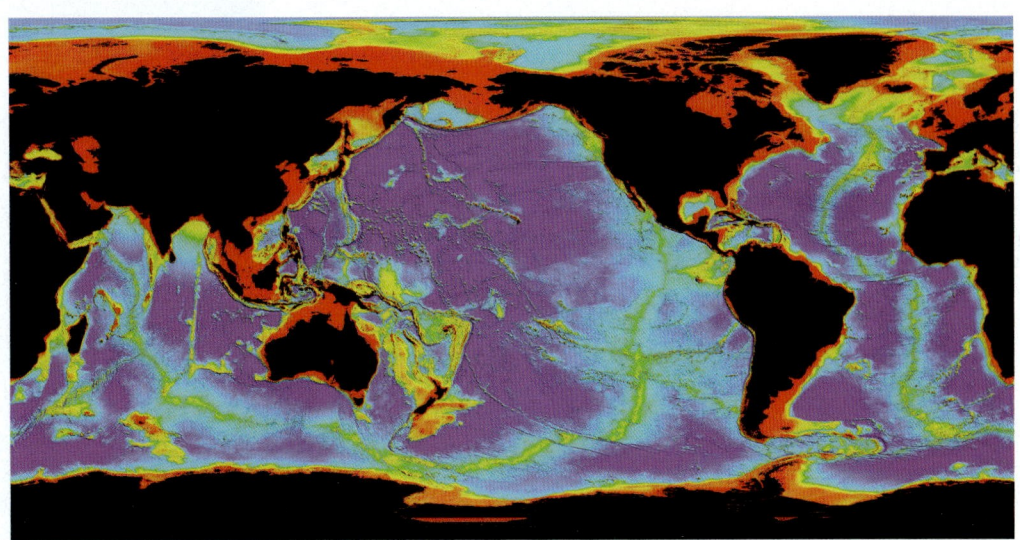

FIGURE 8.30 Most seafloor topography is caused by the creation or movement of lithospheric plates.

CRITICAL THINKING

Environments of Deposition

Clastic sediments are exposed to a wide range of natural conditions as they travel from their source to the depositional environments where they come to rest and are joined by chemical and biogenic sediments. Along the way, they are abraded, sorted, rounded, temporarily buried, eroded, and subjected to chemical weathering.

FIGURE 8.31 illustrates some of the many environments in the sedimentary cycle from the mountains to the seafloor.

With a partner, please work through the following exercise:

1. What specific characteristics could you use to distinguish between marine and nonmarine sedimentary deposits?

2. Imagine that you were asked to sample the sediments in a stream channel at three locations: the upper, middle, and lower portions of the stream. Assuming that granite and lithic sandstone are the sources of weathering for your sediments, describe the composition and texture at each location.

3. Choose four environments of deposition. For each, describe the texture and composition of the sediments you would likely find there.

4. Name the sedimentary rock that would be formed by the sediments in each of your four environments.

5. Drawing on your knowledge of natural resources, name three environments of deposition, the rock that will be formed in them, and the natural resource that could be extracted from that rock.

6. List three high-energy environments and three low-energy environments in Figure 8.31.

7. Consider the continental shelf, the streams, the desert, and the glacier. If the sediment source for each was granitic mountains, predict the texture and composition of sediments at each environment.

FIGURE 8.31 Environments of deposition where sediments collect and lithify.

8-10 Primary Sedimentary Structures Record Environmental Processes

LO **8-10** List four common sedimentary structures, and explain their significance.

Depending on the depositional environment, sedimentary rocks record certain features that reflect the physical processes that act on the sediments. These features are called **primary sedimentary structures**.

Primary Sedimentary Structures

Primary sedimentary structures include features such as *ripple marks* and *mud cracks*, which are formed in a sedimentary environment and are preserved in the solid rock after it lithifies. Some sedimentary structures are created by water or wind as it moves grains of sediment. Others are formed after sediments have been deposited.

Primary sedimentary structures provide information about the environmental conditions under which sediment is deposited—for example, whether they are formed in quiet water under low-energy conditions or in moving water under high-energy conditions. Many primary sedimentary structures emerge originally as bedforms, features that develop in the active environment as a result of wind or water shaping loose sediments. These bedforms include dunes, ripples, and sand waves.

The most basic sedimentary structure is *stratification*, the layering that occurs as sediments are deposited over time (**FIGURE 8.32**). You experimented with stratification earlier in this chapter, in the Critical Thinking feature titled "Sediment Deposition." Stratification is the most obvious feature of sedimentary rocks because the strata, or layers, are visible due to differences in the color or texture of the various layers.

Because grains fall through water or air under the influence of gravity, most sedimentary environments produce horizontal strata, with older sediment on the bottom and younger sediment on the top. However, when sediments accumulate on the steep slope of a sand dune, the front of a delta, or in certain kinds of stream channels, the beds may appear at an angle to the horizon. Such beds are called *cross beds* (**FIGURE 8.33**), and they give us a clue to the physical processes operating at the time when the sediments accumulated. Usually, cross beds indicate that blowing wind or running water led to the deposition of these sediments.

Graded beds develop when rapid deposition occurs in water containing sediment consisting of grains of many sizes. Larger, heavier grains fall through water more rapidly than smaller grains do, so the resulting bed has large particles at the bottom and smaller particles

FIGURE 8.32 Sedimentary rocks usually display layers of deposited sediments called strata. Stratification occurs because of changes in the grain size, sorting, color, or matrix from one layer to the next.

FIGURE 8.33 Cross bedding is apparent in this sandstone composed of fossil sand dunes at Zion National Park in Utah. The orientation of the beds indicates the former face of a dune migrating from left to right.

 Do strata reflect changes in environment?

 Which way did the wind blow in this ancient desert?

Tim Kiusalaas/Photographer's Choice/Getty Images

Bernhard Edmaie/Photo Researchers

at the top (**FIGURE 8.34**). You may have observed the formation of graded bedding if you used sandy soil in the "Sediment Deposition" experiment.

A ripple mark is an important sedimentary structure that shows up when running water or blowing wind shapes loose sediment into ripples. These ripples are commonly preserved during lithification (**FIGURE 8.35**) and provide clear evidence that grains were actively transported from their former environment.

No doubt you have noticed the cracks that develop in mud that lies drying in the sun. As the mud dries, it contracts in all directions and forms polygonal features known as mud cracks. These, too, can be preserved in sedimentary rock, showing that the same process occurred in past environments.

Q EXPAND YOUR THINKING

Ripple marks in streams and rivers tend to be asymmetrical: One slope is short and steep (facing downstream) and the other is long and gentle (facing upstream). On the shallow seafloor, however, ripples are symmetrical: both slopes are the same. Why is this?

© Marli Bryant Miller

FIGURE 8.34 This accumulation of stream sediment displays multiple graded beds, each with larger grains at the base and smaller grains at the top.

Q Why do graded beds form?

(a)

moodboard/Corbis Images

(b)

Tim Graham/Getty Images

(c)

Theo Allofs/Corbis Images

(d)

© Marli Bryant Miller

FIGURE 8.35 Bedforms: (a) modern ripple marks; (b) ancient ripple marks in sandstone; (c) modern mud cracks formed by desiccation; (d) ancient mud cracks lithified in sedimentary rock.

Q What kind of sediment typically forms mud cracks?

LET'S REVIEW

Sediments can yield valuable information about the environment where you live. If, say, you live in an earthquake-prone region, sediments near your home may reveal how often in the past damaging earthquakes have occurred in the area. If you live near the ocean, nearby sediments may record the pattern and frequency of hurricanes and tsunamis there. Sediments also can reveal how often a river floods, a mountain produces landslides, or a volcano erupts. By "reading" the sedimentary record in this way, geologists help to make our world safer, find critical natural resources, and learn about human origins. Keep this important idea in mind when we discuss geologic time in Chapter 12, and look more closely at Earth's history in Chapter 13.

STUDY GUIDE

8-1 Sedimentary rock is formed from the weathered and eroded remains of Earth's crust.

- Most of Earth's surface is covered with layers of loose **sediment**, and more than 75% of the land surface is **sedimentary rock**.

- From the moment they are freed by erosion to travel across Earth's surface, sediments are imprinted with the physical and chemical characteristics of the various environments they encounter. Geologists interpret these "signatures" and from them make inferences about geologic history.

- Sedimentary rock provides natural energy and mineral resources that are critically important to modern industry, technology, and society.

- Sedimentary rock forms from weathered and eroded remains of other igneous, metamorphic, and sedimentary rock. Many sedimentary rocks contain direct and indirect evidence of living organisms, such as skeletal parts or impressions of plants and animals (footprints, leaf imprints, etc.).

8-2 There are three common types of sediment: clastic, chemical, and biogenic.

The three types of sediment are defined by their composition:
- **Clastic sediments**, which are broken pieces of crust deposited by water, wind, ice, or other physical process. Minerals that are commonly found in clastic sediments include quartz, orthoclase feldspar, muscovite, clay minerals, and some heavy minerals. Also common in clastic sediments are eroded fragments of rocks, known as lithic fragments.
- **Chemical sediments**, which are produced by inorganic precipitation of dissolved compounds: calcite, gypsum, halite, microcrystalline quartz, iron oxide, and others.
- **Biogenic sediments**, which are produced by living organisms and are organically precipitated: calcite, chert.

8-3 Sediments travel from source area to depositional environment.

- Sediments are eroded from source areas and transported across Earth's surface to depositional environments.
- As they travel, sedimentary grains experience abrasion, breakage, rounding, sorting, precipitation, mixing, and other processes that imprint the sediment.
- These changes to the sediment are recognizable by geologists studying sedimentary rocks and are used by them to interpret the history of the sediments (and the past history of Earth's surface).

8-4 Sediments change as they are transported across Earth's surface.

- Clastic sediments are solid components of preexisting rock that have been transported from one location to another. Since most rock is composed of silicate minerals, most clastic sediment consists of these minerals.
- As sedimentary grains are transported, they undergo two kinds of changes related to the energy of the environment—the force of the wind or current carrying them:
  - They experience **sorting**, in which they are separated based on their density.
  - They experience **abrasion**, and the farther they are transported, the smaller and more spherical they become.

8-5 Clastic grains combine with chemical and biogenic sediments.

- Dissolved components move from weathering sites, where they are created, into sedimentary basins, where they are deposited. Dissolved ions and compounds may be deposited through either organic or inorganic precipitation. Both processes result in a crystalline solid that joins the sedimentary cycle as either chemical or biogenic sediment.
- Sediments that accumulate within sedimentary basins display changes in composition. Unstable grains (olivine, pyroxene, calcium-rich plagioclase, amphibole, and others) become less abundant, while stable grains (quartz, clays, muscovite, orthoclase) become more abundant. Biogenic sediments accumulate, and chemical sediments may become more abundant.

8-6 Sediment becomes rock during the sedimentary cycle.

Sedimentary rock is formed by **lithification** in two ways:
- Larger particles, such as sand and gravel, are bound together by natural cements that are inorganically precipitated from dissolved compounds in groundwater.
- Smaller particles, such as silt and clay, are compacted and consolidated by pressure exerted by the weight of overlying layers of sediment.

The sedimentary cycle is a component of the rock cycle that supplies sediment created by the weathering and erosion of tectonically uplifted crust. The cycle consists of tectonic uplift, weathering, erosion and transportation, deposition, burial, and lithification.

8-7 There are eight major types of clastic sedimentary rock.

- Sedimentary rocks are identified based on their lithology, the composition and texture of the particles and cements within them. Specific combinations of texture and composition are characteristic of particular types of sedimentary rock. The composition of clastic sedimentary rock can be interpreted in terms of the history of the sediments: transport energy and distance, weathering intensity, and probable source rock. These reveal vital information about Earth's history.

- Clastic rocks are distinguished on the basis of grain size as well as composition. They consist of conglomerate, breccia, quartz sandstone, arkose, lithic sandstone, siltstone, claystone, and shale.

8-8 There are seven major types of chemical sedimentary rock and four major types of biogenic sedimentary rock.

- Chemical sedimentary rocks are distinguished based primarily on their mineral composition; but in some cases, texture is the critical feature. The rocks in this category are rock salt, rock gypsum, travertine, micrite, dolostone, chert, and limestone.

- Biogenic sedimentary rocks are also distinguished based on their mineral composition and texture. The rocks in this category are limestone, coquina, chalk, and coal.

8-9 Sedimentary rocks preserve evidence of past depositional environments.

- Most sedimentary rocks reflect the environment in which they were formed, and in this way record past conditions on Earth's surface.

- A number of sedimentary environments are found on continents, along coasts, and in the oceans. Continental environments include glaciers, deserts and dunes, alluvial fans, streams, wetlands, and lakes. Coastal environments include deltas, barrier islands, lagoons, beaches, and tidal wetlands. Marine environments include reefs, continental shelves, continental margins, and the deep sea.

8-10 Primary sedimentary structures record environmental processes.

Primary sedimentary structures are physical features constructed in a sedimentary environment and preserved in the rock after it has lithified. They give geologists important clues for analyzing past surface conditions. Sedimentary structures include stratification, graded beds, ripple marks, and mud cracks.

KEY TERMS

abrasion (p. 222)
biogenic sediments (p. 218)
calcium carbonate (p. 216)
chemical sediments (p. 218)
clastic sediments (p. 218)
clay (p. 216)
environments of deposition (p. 221)

evaporites (p. 219)
iron oxides (p. 216)
lithification (p. 226)
mud (p. 216)
porosity (p. 226)
primary sedimentary structure (p. 240)
sand (p. 216)

sediment (p. 216)
sedimentary quartz (p. 216)
sedimentary rock (p. 216)
silt (p. 216)
sorting (p. 222)

ASSESSING YOUR KNOWLEDGE

Please complete this exercise before coming to class. Identify the best answer to each question.

1. Geologists study sedimentary rocks because:
   a. They provide a record of Earth's history.
   b. They are sources of fossil fuels.
   c. They may contain important mineral resources.
   d. They may contain fossils, revealing a history of life, including human evolution.
   e. All of the above.

2. _____ sediment forms from loose rock and mineral debris produced by weathering and erosion.
   a. Clastic
   b. Biogenic
   c. Chemical

Courtesy of Chip Fletcher

3. Sediments produced by the action of living organisms are called:
   a. Chemical sediments.
   b. Physical sediments.
   c. Clastic sediments.
   d. Biogenic sediments.
   e. None of the above.

4. Well-sorted and well-rounded sand grains indicate that sediment:
   a. Came from a nearby source area.
   b. Was deposited at the location where it was found.
   c. Traveled from a distant source area.
   d. Has not been influenced by weathering.
   e. None of the above.

5. "Lithification" refers to:
   a. The set of natural processes that turn sediment into rock.
   b. The processes of erosion and tectonic uplift.
   c. The effects of chemical weathering.
   d. All of the above.

6. Two processes can lead to the lithification of sediment. During _____, the weight of accumulating sediment reduces pores space and forces out water from sediment. _____ occurs when ions dissolved in solution precipitate out, forming minerals that hold the grains together.
   a. compaction; Recrystallization
   b. compaction; Cementation
   c. dehydration; Cementation
   d. cementation: Recrystallization

© 2010 Warren Rosenberg - Fundamental Photographs

Courtesy of Chip Fletcher

7. After being created by weathering, sediments:
   a. Experience more weathering.
   b. Combine with chemical sediments.
   c. Combine with biogenic sediments.
   d. Experience sorting and abrasion.
   e. All of the above.

8. The "sedimentary cycle" refers to:
   a. The interrupted erosion of sediments from mountainsides.
   b. The process of recycling sediments.
   c. The formation of rock through compaction of sediments.
   d. The formation of rock through chemical precipitation of sediments.
   e. None of the above.

9. In what type of tectonic setting would you expect to find large amounts of volcanic sediment?
   a. Glaciated plate interior
   b. Convergent margin–subduction zone

   c. Convergent margin–continental collision
   d. Passive continental margin

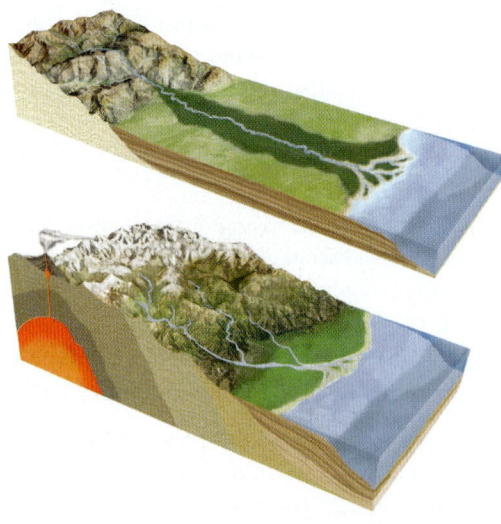

10. Accretionary prisms composed of clastic and biogenic sediment scraped off the oceanic crust are typical of what type of tectonic setting?
    a. Divergent margin–rift valley
    b. Convergent margin–subduction zone
    c. Convergent margin–continental collision
    d. Passive continental margin

Andreas Molin/Nordic Photos//Getty Images Inc.

© Freddy Eliasson/Matton Collection/Corbis

Andrew Dernie/Photodisc/Getty Images, Inc.

De Agostini/A. Dagli Orti/De Agostini Picture Library/Getty Images

11. Which of the following statements is correct?
    a. Clastic sedimentary rocks include arkose and lithic sandstone.
    b. Biogenic sedimentary rocks include coquina and coal.
    c. Chemical sedimentary rocks include travertine and micrite.
    d. Clastic sedimentary rock includes conglomerate and shale.
    e. All of the above.

12. Chemical sedimentary rocks include:
    a. Lithic sandstone.
    b. Coal.
    c. Breccia.
    d. Chert.
    e. None of the above.

13. Biogenic sedimentary rocks are formed by:
    a. Evaporation.
    b. Inorganic precipitation.
    c. Glaciers.
    d. Floods.
    e. None of the above.

14. Rock fragments are known as:
    a. Lithic fragments.
    b. Bioclastic sediments.
    c. Evaporites.
    d. Natural cements.
    e. None of the above.

15. Indicate the one most likely rock type for each of the sites shown.

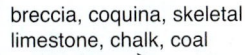

breccia, coquina, skeletal
limestone, chalk, coal

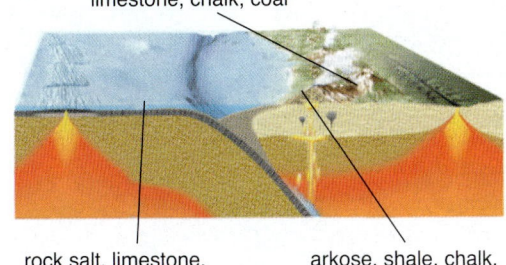

rock salt, limestone,
coquina, arkose

arkose, shale, chalk,
rock gypsum

16. Particle sizes are described using the following terms:
    a. Sand, gravel, lithic fragments, and natural cements.
    b. Gravel, sand, silt, and clay.
    c. Conglomerate, sandstone, arkose, and shale.

17. Continental environments include:
    a. Beaches, continental margins, streams, and lakes.
    b. Beaches, streams, lakes, and glaciers.
    c. Beaches, streams, deltas, and barrier islands.
    d. Streams, lakes, wetlands, and glaciers.
    e. None of the above.

18. Organisms play a significant role in the origin of _____ sedimentary rock.
    a. Clastic
    b. Biogenic
    c. Chemical
    d. Lithologic
    e. None of the above

19. Sedimentary rocks are classified by:
    a. Mineralogy and fossils.
    b. Cementation and compaction.
    c. Environments of precipitation and deposition.
    d. Composition and texture.
    e. None of the above.

20. Primary sedimentary structures are:
    a. Physical features of a rock related to the environment of deposition.
    b. Physical features of a rock related to the process of cementation.
    c. Chemical features of a rock produced by the motion of water and wind.
    d. Sediment forms produced by biogenic processes.
    e. None of the above.

FURTHER RESEARCH

1. What types of sedimentary environments are found in your area? What types of sediments accumulate there?

2. What role does calcium carbonate play in the rock cycle?

3. What types of sedimentary rock would you expect to find on the Moon? On Mars?

4. Choose a sedimentary environment and describe the primary sedimentary structures you would expect to find preserved in the resulting rocks.

5. Why do sediments tend to move from mountainous regions toward the sea?

6. As unconsolidated sediments are compacted, what happens to the water within the pore spaces?

7. How can quartz move through the sedimentary cycle several times in the course of geologic history?

ONLINE RESOURCES

Explore more about sedimentary rocks on the following Web sites:

U.S. Geological Survey Web page on sedimentary rocks:
http://wrgis.wr.usgs.gov/docs/parks/rxmin/rock2.html

National Space Society page on life in extreme environments:
www.astrobiology.com/extreme.html

NASA Earth Observatory page, displaying satellite images of modern environments:
http://earthobservatory.nasa.gov

Additional animations, videos, and other online resources are available at this book's companion Web site:
www.wiley.com/college/fletcher

This companion Web site also has more information about WileyPLUS and other Wiley teaching and learning resources.

METAMORPHIC ROCK

CHAPTER CONTENTS & LEARNING OBJECTIVES

GEOLOGY IN OUR LIVES

Wherever rocks are intruded by magma, buried deep in Earth's crust, or involved in processes at convergent margins, they may be exposed to rising temperature and pressure conditions. Increases in pressure and heat recrystallize igneous, sedimentary, and metamorphic minerals. This process makes metamorphic rock. Conditions that lead to metamorphism are linked to plate tectonics or deep burial within the crust, or both. Metamorphic rocks are the source of important building materials and mineral resources.

Q Metamorphic rock forms when the crust is exposed to rising pressure and temperature conditions. What processes can raise the pressure and temperature of the crust?

9-1 Metamorphic Rocks Are Composed of Sedimentary, Igneous, or Metamorphic Minerals That Have Recrystallized

LO 9-1 Describe the process of metamorphism.

Deep within the crust, pressures can reach a crushing 10,000 atmospheres (10,000 times the pressure of the atmosphere on Earth at sea level), with temperatures over 600°C. What happens when rocks are subjected to such high pressures and temperatures? How do they change, and are the changes predictable? What factors influence the evolution of rock under these extreme conditions? This chapter answers these and additional questions regarding rock **metamorphism.**

What Is Metamorphic Rock?

Geologists use the term *metamorphism* to describe changes that take place in the mineralogy and texture of rocks affected by temperature, pressure (also called **stress**), and/or *chemically active fluids*. In the world of rocks, metamorphism refers to changes that occur without melting, a phenomenon called *solid-state reaction* or solid-state change. Put another way, metamorphism requires that minerals remain solid as they evolve; if a rock melts, it becomes igneous.

Metamorphic rocks are sedimentary, igneous, or metamorphic rocks that take on a new texture and mineralogy in order to maintain equilibrium under changing conditions within the crust. Simply, minerals within a rock change in order to remain stable under new conditions. In most cases, this occurs through **recrystallization** of minerals in the original, or "parent," rock (**FIGURE 9.1**).

Recrystallization is the growth of new mineral grains in a rock that starts by recycling old minerals. Rarely does recrystallization involve only one mineral. Rather, a metamorphic rock develops from a new *assemblage* of minerals resulting from recrystallization.

Metamorphic rocks are a major source of building materials. Two of them, *slate* and *marble*, are commonly used as finishing stone

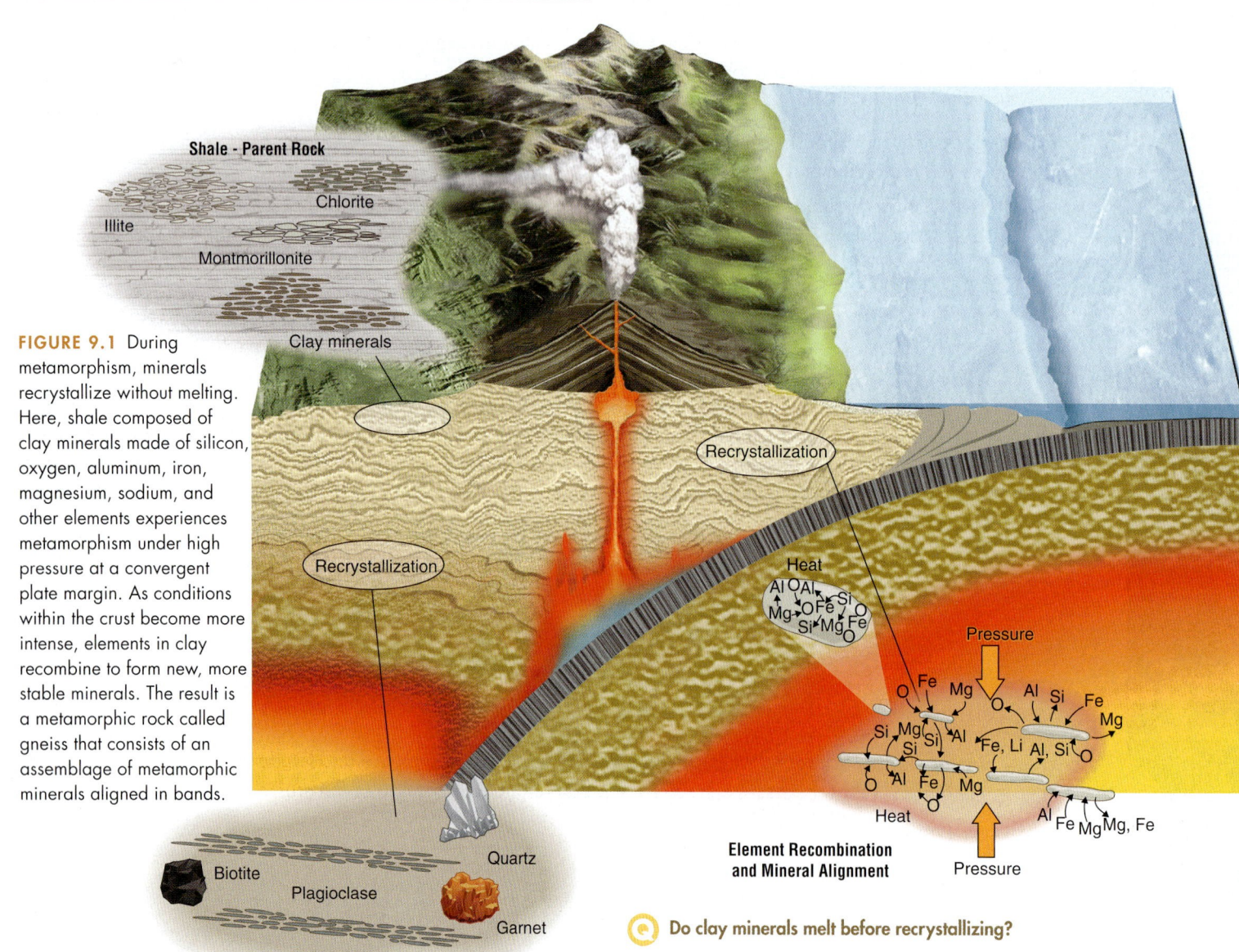

FIGURE 9.1 During metamorphism, minerals recrystallize without melting. Here, shale composed of clay minerals made of silicon, oxygen, aluminum, iron, magnesium, sodium, and other elements experiences metamorphism under high pressure at a convergent plate margin. As conditions within the crust become more intense, elements in clay recombine to form new, more stable minerals. The result is a metamorphic rock called gneiss that consists of an assemblage of metamorphic minerals aligned in bands.

Do clay minerals melt before recrystallizing?

in buildings and for statuary. (In the Geology in Our Lives feature, you'll read about a man whose artistry made him the "Master of Marble.") Metamorphism of ultramafic rock (magnesium- and iron-rich igneous rocks) produces the mineral chrysotile the principal source of *asbestos*, an important component of fire-retardant materials. Metamorphism of impure dolostone produces *talc*, a very soft silicate mineral that is widely used as filler in paints, rubber, paper, asphalt, and cosmetics. Metamorphism also has the effect of bringing together certain elements, such as metals, into dense accumulations that form valuable ores, such as the gold deposits located in the Abitibi region of Quebec, Canada.

Like the other types of rock we have studied, metamorphic rocks are named based on their texture and mineral composition. If a metamorphic rock is heavily squeezed by tectonic forces, newly crystalli-

zing minerals will align to create a texture known as **foliation.** Different types of foliation give rise to different types of rock (as we will see later in the chapter).

As a result of tectonic forces that pull, squeeze, and shear the crust, most rock that undergoes metamorphism acquires new texture as well as new mineralogy. The new minerals align as they are formed, layers develop, color bands of minerals appear, and crystals may stretch and even spin.

Features of the parent rock, such as vesicles, fossils, and primary sedimentary structures (cross-bedding, stratification, etc.), usually disappear. The metamorphic rock is, then, a child of the composition of its parent (also called *protolith*) and the conditions it has experienced. In this way, metamorphic rock opens a window into the environment of the deep crust.

Ⓠ EXPAND YOUR THINKING

How do metamorphosed rocks become exposed at the surface, where we can study them?

 GEOLOGY IN OUR LIVES

Master of Marble

The beauty and softness of *marble* (metamorphosed limestone) has made it the preferred medium of sculptors who created many of the world's greatest statuary. But perhaps the artist most revered for his ability to breathe warm life into cold stone was the Italian painter, poet, architect, and sculptor Michelangelo di Lodovico Buonarroti Simoni (1475–1564).

To find the marble for his masterpieces, Michelangelo visited the 1500-year-old quarry at Carrara in Tuscany, Italy (**FIGURE 9.2**), several times to search for ideal blocks of the stone. The marble at Carrara is renowned for its uniform and homogeneous crystal pattern, which can be predictably shaped. It is also soft and resistant to shattering and has a degree of transparency that lets light penetrate and scatter just below the stone surface, giving it the appearance of life. Finally, Carrara marble develops in a range of shades. Michelangelo took advantage of several important aspects of this stone to shape his stunning sculptures, including his two most famous: David and the Pietà.

Michelangelo's approach to sculpture was to liberate the human image he envisioned encased in the cold marble, just as he believed the human soul to be encased within the physical body. He would draw the figure on the face of a block of marble and, working from all sides, chisel away at the stone bit by bit to free that image from the depths of the rock.

The Pietà (Italian for "mercy") is considered by many to be Michelangelo's masterpiece. It is revered for the soft flowing lines of the Virgin's robes and the realistic depiction of the body of Jesus in her arms, after the crucifixion. The statue, carved from a solid block of Carrara marble that Michelangelo spent months looking for, is on display today in the Vatican in Rome. The figures in the Pietà are, however, out of proportion, owing to the difficulty of depicting a grown man cradled full length in a woman's lap. If Christ were carved to human scale, the

(a)

(b)

Mauro Fermariello/Photo Researchers   Araldo de Luca/Corbis Images

FIGURE 9.2 The marble quarry at Carrara, Italy (a) has been mined for building stone and sculpture since the time of ancient Rome. Michelangelo's most famous sculpture is the Pietà (b), which depicts the Virgin Mary holding the body of Christ after the crucifixion.

Virgin, standing, would be nearly 5 meters tall. But Michelangelo's great skill enabled him to conceal much of the Virgin's size within her drapery, so the figures look natural, and peaceful. Michelangelo's interpretation of this moment in Christian history was different from those of other artists. He chose to create a youthful, serene, and celestial Virgin Mary, rather than one appearing brokenhearted and somewhat older.

9-2 Metamorphism Is Caused by Heat and Pressure

LO 9-2 Compare and contrast regional and contact metamorphism.

Metamorphism is not an isolated event. Rather, it is the cumulative effect of all the continuous changes that a rock has undergone.

The Influence of Heat

As explained, metamorphic rock is the product of shifting pressure and temperature conditions that produce changes in the mineralogy and texture of a protolith. Hence, the temperature, pressure, and composition of the parent rock are primary factors in determining the mineral assemblage and texture of the resulting metamorphic rock.

Geothermal heat is conducted upward through Earth's interior, eventually radiating out to space. Heat rises with depth in the crust, along what is known as the *geothermal gradient*. The average geothermal gradient is an increase of about 30°C per kilometer of depth in the crust, although the actual level of increase varies in different tectonic settings. In volcanically active areas like island arcs, the temperature may rise by about 30°C to 50°C per kilometer. In areas such as ocean trenches, where cold oceanic crust is driven into the interior, the temperature may rise by as little as 5°C to 10°C per kilometer (**FIGURE 9.3**).

When sedimentary and igneous rocks are buried, they immediately begin to experience geothermal heat. As they remain buried, and temperatures continue to rise, solid-state reactions begin to take place. Temperatures between 150°C and 200°C (at depths of approximately 6 to 7 kilometers) cause clay minerals, the most common sediment, to become unstable. Clays recrystallize to become the metamorphic mineral *chlorite*, followed by *muscovite*. These minerals often mark the onset of metamorphism.

A second source of heat is plutonic intrusions. Magma that intrudes into the crust bakes the solid rock around it. Intruded crust is generically referred to as *country rock*. It experiences a type of metamorphism called **contact metamorphism,** or *thermal metamorphism*, under conditions that are related to high heat levels in the absence of greatly elevated pressure (**FIGURE 9.4**).

Contact metamorphism is a result of the high geothermal gradient surrounding a pluton. This type of metamorphism usually is restricted to relatively shallow depths (less than 10 kilometers), where there is a large contrast in temperature between the intruding magma and the cool surrounding country rock.

The area of rock affected by contact metamorphism is usually a baked region, known as an *aureole*. An aureole consists of zones of recrystallized minerals approximately concentric to the pluton. Multiple zones usually develop around an intrusion, each characterized by a distinct mineral assemblage reflecting the temperature gradient from the high-temperature magma to the unaltered country rock.

When a crystalline structure is heated, ions within it begin to vibrate. High heat leads to greater vibration, and when the bonds forming compounds are eventually broken, the rock melts. Before melting occurs, however, some atoms vibrate free and migrate to new locations within the mineral structure or in the rock. These atoms are prime candidates for recrystallization.

As this process continues, new atomic structures (new minerals) are formed that are stable at the higher temperature. Overall, the composition of the resulting metamorphic rock is very similar to that of the parent rock because the parent atoms have been recycled; however, the minerals within it are different. In addition, the composition of the rock may change when chemically active fluids are introduced. If the heating process ceases before melting can occur, a metamorphic rock is produced. If melting occurs, an igneous rock is produced.

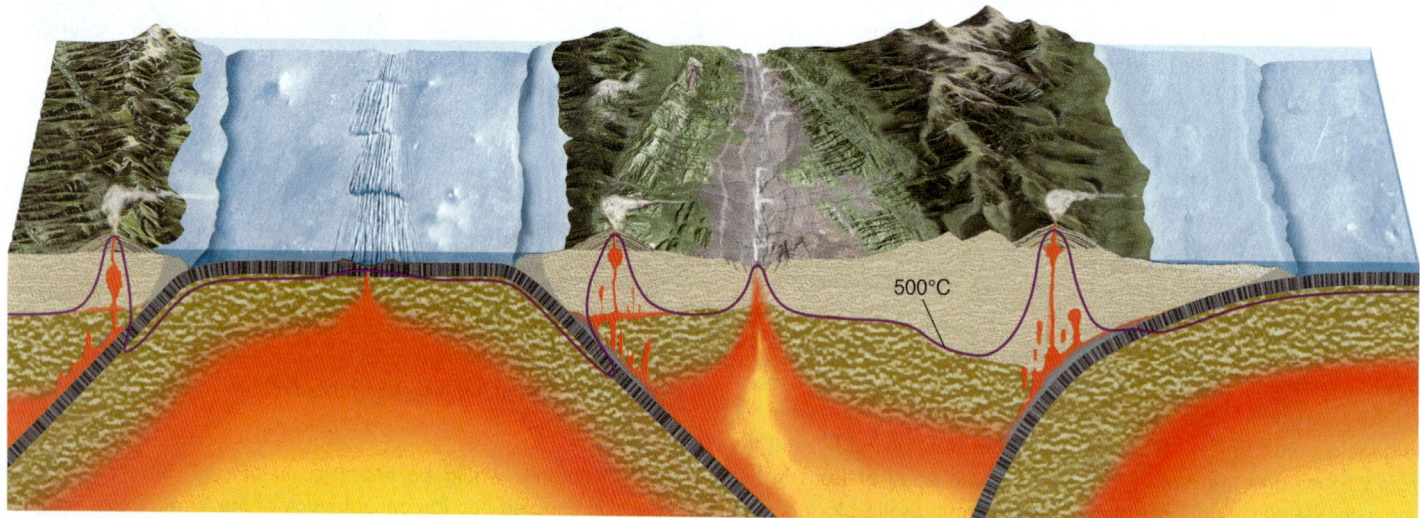

FIGURE 9.3 Heat within Earth's crust varies among tectonic settings. At spreading centers and volcanic arcs, the crust is warm near the surface. At subduction zones, cold seafloor descends into the mantle. The purple line represents the 500°C isotherm (a line connecting points with equal temperature).

Describe the processes that cause the 500°C isotherm to vary its location in the crust.

(label in figure: 500°C)

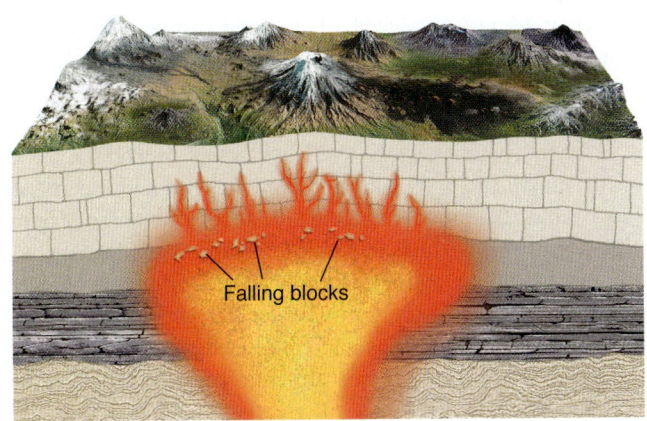

FIGURE 9.4 Contact metamorphism is characterized by high-temperature conditions surrounding plutonic intrusions.

(Q) **How might an intrusion affect groundwater in the vicinity?**

The Influence of Pressure

Within the crust, the pressure caused by the overlying layers of rock is called *lithostatic stress*; it acts on rock simultaneously and equally in all dimensions. Lithostatic stress increases with depth in the crust and plays an important role in metamorphism, closing the pore spaces in rocks, causing compaction and, at great depths, making minerals recrystallize to form more compact atomic structures.

Directed stress differs from lithostatic stress in that pressure acting on the crust is greatest in one direction. This leads to folding, bending, breaking, or even overturning of rock strata. Directed stress causes rocks to be shortened in the direction of principal stress, and conversely, elongated or lengthened in the direction perpendicular to the principal stress.

Directed stress at convergent margins is generated by the collision of two plates such that pressure is greatest in one direction. Metamorphic minerals growing under directed stress crystallize so that their long axis extends in the direction perpendicular to the principal stress. This creates foliation because minerals are lined up in an easily visible manner. (To picture foliation, think of the alignment of cards in a deck, or a bunch of pencils held together in a hand.)

Platy minerals (so named for their dinner-plate shape), such as members of the mica family, and elongate minerals (which are especially long in one direction, like a pencil), such as members of the amphibole family, are examples of foliation under directed stress (**FIGURE 9.5**). These minerals lengthen into the low-stress area and become elongated in a direction perpendicular to the direction of principal stress.

When pressure and temperature combine to alter rock over large areas, the result is **regional metamorphism.** This kind of metamorphism, which can extend over thousands of square kilometers, is not related to plutonism. Most regionally metamorphosed rocks occur in places where the crust has been deformed during orogenesis or by burial deep in the crust.

Two colliding plates (a convergent margin) create an extensive zone of high pressure (produced by compression, or "squeezing") characterized by directed stress in the crust, usually producing foliation. Thus, regionally metamorphosed rocks occur in the cores of orogenic mountain belts formed at continent-continent convergent margins and on the leading edges of continents and oceanic plates overriding subduction zones.

Regional metamorphism results from a general increase of temperature and pressure over a large area. Due to the great depths at which it occurs, high pressure seals the pore spaces in the rock, preventing water from circulating. That means groundwater may play only a limited role in transporting heat and dissolved ions involved in the recrystallization process. And in terms of volume, the process of regional metamorphism is far more significant than contact metamorphism; rocks produced in this way are distinguished by their foliated texture and mineral assemblage.

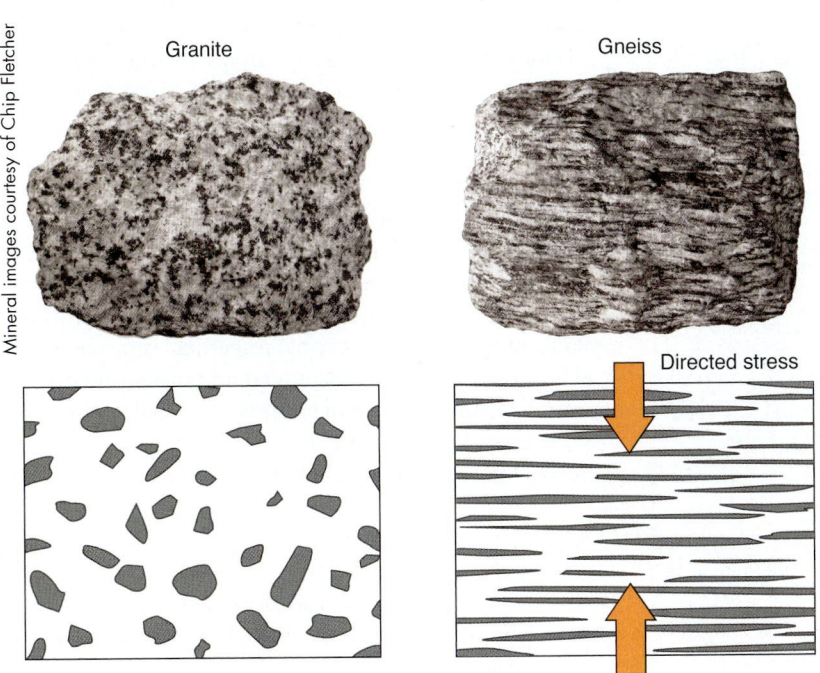

Mineral images courtesy of Chip Fletcher

Granite

Gneiss

Protolith

Directed stress

Dark and light banding

FIGURE 9.5 Minerals recrystallize toward the direction of lowest stress, which is perpendicular to the direction of principal stress. Directed stress is strongest in one direction and produces mineral layering that is perpendicular to the direction of highest stress. The gneiss here shows foliation; the granite does not.

(Q) **What was the direction of principal stress in Figure 9.5?**

(Q) **EXPAND YOUR THINKING**

Groundwater can transport heat and dissolved compounds. Explain how groundwater might influence contact metamorphism.

9-3 Chemically Active Fluids Transport Heat and Promote Recrystallization

LO 9-3 Describe the role of water in metamorphism.

Water is found throughout Earth's crust and it plays an important role in nearly all metamorphic reactions. This *groundwater* (the subject of Chapter 18) influences both the texture and composition of many types of sedimentary, igneous, and metamorphic rocks that experience metamorphism.

Chemically Active Fluids

Groundwater that comes into contact with intruding magma or that leaks out of marine sediment in subducting oceanic crust usually contains a rich medley of dissolved compounds and metallic cations. Such chemically enhanced waters are known as *chemically active fluids*. Chemically active fluids are especially important in recrystallizing the country rock surrounding an igneous intrusion. These fluids transport heat and influence the chemistry and mineralogy of newly forming metamorphic rocks.

Chemically active fluids come from groundwater when metamorphism is occurring in shallow crust (less than 2 kilometers in depth). They are also produced when water (usually steam) escapes from magma; when seawater gets buried in the crust with a subducting oceanic plate; or when water-rich minerals such as micas or amphiboles are dehydrated by heat in the crust and the escaping water circulates through the surrounding rock. The water found throughout the crust lowers the melting point of many minerals, thus promoting early partial melting of felsic compounds and the production of granitic magmas.

Chemically active fluids intensify mineral recrystallization by promoting the dissolution and migration of ions. During metamorphism, as temperature and pressure go up, water becomes more chemically reactive, and so dissolves ions and compounds more readily. For instance, when two mineral grains are squeezed together by directed stress, any water that is present will greatly accelerate the dissolution of ions from those grains. And because the water flows from areas of high stress to areas of low stress, those ions will be carried away. Dissolved ions may bond again in a nearby area of lower stress. As a result, as described earlier, minerals tend to lengthen into the low-stress area, growing in a direction perpendicular to the direction of principal stress.

Certain types of chemically active fluids, known as *hydrothermal fluids*, are produced by the interaction of groundwater and magma in the final stages of cooling. Hydrothermal fluids are hot and contain dissolved ions and compounds that react with country rock, causing crystallization of valuable *metals*, such as lead, silver, platinum, and gold. (Read more about these valuable metals in the Geology in Our Lives feature "Chemically Active Fluids Produce Important Metallic Ores.")

Hot water that is enriched in dissolved minerals will gain access to cracks and fractures in stressed rocks and line these openings with **hydrothermal vein fillings** of newly crystallized minerals (**FIGURE 9.6**). The dissolved compounds will greatly modify any solid-state reaction that is taking place. Chemical alteration in the presence of hydrothermal fluids is known as *metasomatism*.

Hydrothermal fluids causing metasomatism alter the baked zones of contact metamorphic rocks such that they contain many new types of metamorphic minerals that would not be formed if only dry country rock was metamorphosed. Hydrothermally precipitated minerals characterize the metamorphic rocks that develop around contact metamorphic zones. Hydrothermal fluids are particularly important in the metasomatism of limestone and dolomite.

Asbestos, an important but controversial industrial mineral, may be found as a hydrothermal vein deposit. Widely used as a fire retardant, certain types of asbestos have been identified as health hazards. The Geology in Our Lives feature "Asbestos: Real Danger or Overreaction?" addresses this issue in greater detail.

Water-rich magma zone

Marble

Granitic pluton

Limestone country rock

Mineral images courtesy of Chip Fletcher

FIGURE 9.6 Water plays an important role in nearly all metamorphic reactions. It lowers the melting point of many minerals, thus promoting early partial melting of felsic compounds and the production of granitic magmas. Water also transports dissolved compounds into cracks and fractures in overlying rocks, producing hydrothermal vein deposits that are rich in valuable minerals. Shown at the top of the diagram are veins of amethyst (purple), a type of quartz.

Q EXPAND YOUR THINKING

Where do fluids in Earth's crust come from?

Q How did minerals get into the fractures above the magma chamber?

GEOLOGY IN OUR LIVES

Asbestos: Real Danger or Overreaction?

To people around the world, the word "asbestos" elicits a reaction of fear. Once commonly used in everyday items such as toothpaste and as insulation in our homes, asbestos, which is actually various types of fibrous minerals from the amphibole and serpentine groups (FIGURE 9.7), is now viewed by many as only a cause of two principal types of cancer: malignancy of lung tissue, and mesothelioma, a cancer of the thin membrane that surrounds the lungs and other internal organs.

Asbestos fibers do not melt or ignite, making them useful in fireproof clothing and insulation, and mixed into plaster and paint to make fire-retardant building materials. There are six commercial varieties of asbestos: five made from a member of the igneous mineral amphibole (called "brown" and "blue" asbestos) and one made from the metamorphic mineral chrysotile (called "white" asbestos).

The asbestos panic was set in motion in 1986, when the U.S. Environmental Protection Agency (EPA) implemented the Asbestos Hazard Emergency Response Act, requiring all schools to be inspected for the presence of asbestos. The EPA implemented this act in response to studies that showed asbestos to be a carcinogen (cancer-causing substance). The studies had been conducted on mine workers in South Africa and western Australia who were exposed to high levels of blue and brown asbestos. They reported unusually high rates of mesothelioma, in some cases after less than a year of exposure. Later studies documented that when the thin, rodlike fibers of brown and blue asbestos are inhaled into the lungs, they are not then exhaled, nor do they break down; they can remain in the tissue indefinitely, potentially causing severe and even deadly lung disease.

What most people don't realize, however, is that the health risks of white asbestos (chrysotile), the most widely used industrial variety in the United States and elsewhere, are minimal to nonexistent, based on studies of workers in chrysotile mines in Canada and Italy. Mortality rates from mesothelioma and lung cancer among these workers vary insignificantly from those in the general population. Chrysotile fibers are curly and less rodlike than those of blue and brown asbestos; and they tend to break down if inhaled and are more easily expelled from the lungs.

Unfortunately, even though most of the asbestos used in the United States is relatively harmless chrysotile, it has been removed from schools and other public and private buildings at huge cost—$50 to $100 billion—and has depreciated the value of commercial buildings by $1 trillion. Worse, it has actually increased, not decreased, the risk of injury or death, from increased exposure to asbestos dusts during the dirty "ripouts" and from accidents experienced by workers removing the asbestos.

Does asbestos present a risk to students across the nation? Data indicate that the level of asbestos in school buildings is equivalent to that found outdoors, produced by natural sources in the crust. As noted, nearly all the asbestos found in schools is chrysotile, which is relatively harmless in low concentrations. More specifically, the risk of injury or illness from asbestos exposure is far lower than that of other common risks, such as school sports (100 times more dangerous), aircraft accidents (60 times more dangerous), and car accidents (320 times more deadly).

These data indicate that the asbestos panic was unwarranted, and has imposed an expensive burden on government agencies that feel compelled to remove asbestos from public buildings, even though everyday hazards are the source of far greater risks

FIGURE 9.7 (a) The mineral asbestos is found in at least six common forms. White asbestos (chrysotile) is the most frequent source of asbestos fibers used in schools and public buildings. Chrysotile fibers are not as hazardous as those from brown or blue forms of asbestos and have been proven to pose little to no health risk to exposed persons. (b) Nonetheless, worldwide, expensive asbestos removal programs continue to cost taxpayers billions of dollars each year.

Courtesy of Chip Fletcher

(a)

Ted Spiegel/Corbis Images

(b)

GEOLOGY IN OUR LIVES

Chemically Active Fluids Produce Important Metallic Ores

Economically valuable mineral deposits are formed in several ways. *Metallic ores* may be produced by chemically active fluids under conditions that lead to metamorphism of the nearby crust—such as magma intrusion. Two general categories of ore deposits occur in these conditions: *magmatic segregation deposits* and *hydrothermal deposits* (**FIGURE 9.8**).

Magmatic Segregation Mineral Deposits

A magmatic segregation deposit is formed during magma crystallization, usually in *cumulates* and *pegmatites*. Magmatic cumulates are formed as a result of *density segregation* and *immiscible melts*. A cumulate is a layer of dense minerals forming a concentrated deposit on the

© geoz/Alamy

(a) Chromite

DEA/A. RIZZI/De Agostini Picture Library/Getty Images

(b) Ilmenite

Vitaly Raduntsev/Shutterstock

(c) Magnetite

© PjrStudio/Alamy

(d) Chalcopyrite

© E.R. Degginger/Alamy

(e) Gold

© Nataliya Nikonova/Alamy

(f) Copper

John Cancalosi/ Photolibrary/Getty Images

(g) Lead

John W. Banagan/Photographer's Choice/Getty Images

FIGURE 9.8 Chemically active fluids associated with metamorphic conditions in the crust form important metallic ores. These ores are the source of critical metals used in the medical, auto, electronics, and computer industries, and others.

JOHN CANCALOSI/ National Geographic Stock

(i) Lithium

Harry Taylor/Dorling Kindersley/Getty Images

(h) Platinum

floor of a magma chamber. This happens when early-forming crystals either fall or are carried by currents to the floor of a magma chamber.

Magma begins life as a homogeneous liquid; then, magmatic segregation through crystallization produces cumulates of varying composition (**FIGURE 9.9**). One example, the Bushveld Igneous Complex of South Africa, consists of concentrated deposits of chromite (iron-magnesium-chromium oxide) and magnetite (iron oxide) encased in layers of feldspar.

A different kind of magmatic segregation involves immiscible melts that form cumulates. Cooling magma can precipitate droplets of a second magma with a different composition. Like oil and water, the two do not mix—they are "immiscible." When magma is saturated with the melt of a particular mineral, solid precipitation of that mineral occurs. If saturation is reached at a temperature higher than the melting point of a mineral, a drop of liquid precipitates instead of a solid mineral grain, and that liquid can develop rich ore deposits later in the cooling history.

Most immiscible magmas form iron sulfide, a source of copper, nickel, and platinum. Immiscible sulfide forms heavy liquid layers in magma in a manner similar to that by which a cumulate is formed; the result is a deposit of copper, nickel, and platinum metals.

Pegmatites develop in certain kinds of magma, especially granitic magma, that contain dissolved water. As this "wet" magma cools, the first minerals to crystallize (such as feldspar) tend to exclude water from their chemical makeup, leaving a water-rich residue. Some elements, such as lithium, beryllium, and niobium, are concentrated

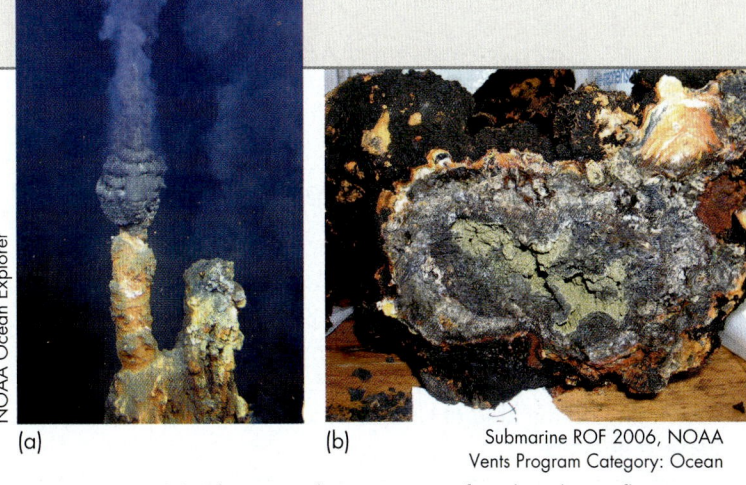

(a)   (b)

NOAA Ocean Explorer

Submarine ROF 2006, NOAA
Vents Program Category: Ocean

FIGURE 9.10 (a) Wherever volcanic action is found on the seafloor, seawater can penetrate the crust and come into contact with hot rocks and magma. (b) Deposits formed in this way are known as volcanogenic massive sulfide (VMS) deposits.

in this residue. If the residual magma migrates (say, into a fracture), it may form bodies of igneous rock enriched in rare elements, called *rare-metal pegmatites*. These rocks can be very coarse-grained, with individual crystals of mica, feldspar, and beryl over 1 meter long.

Hydrothermal Mineral Deposits

Hydrothermal deposits accumulate when hot, salty water dissolves and transports metallic ions that become ore minerals. The simplest hydrothermal deposit is a *vein*, which emerges when hydrothermal solutions precipitate compounds and native metals along the walls of a crack. Most veins occur where igneous plutons provide a heat source.

Another hydrothermal deposit is *porphyry copper*. Typically found in association with intrusive rock containing *disseminated minerals*, these deposits occur in fractured country rock containing a network of tiny, closely spaced veins of quartz. Each vein carries grains of the copper ore minerals chalcopyrite and bornite or the molybdenum ore molybdenite. The volume of mineralized rock is huge, potentially amounting to billions of tons of ore.

Volcanogenic massive sulfide (VMS) deposits (**FIGURE 9.10**) accumulate at volcanoes on the seafloor, in spreading center, hotspot, or subduction zone environments. Seawater penetrates the crust and becomes a hydrothermal fluid that dissolves and concentrates scarce metals. A hydrothermal vent develops where gushing, mineralized water precipitates an ore deposit onto the adjacent seafloor. VMS deposits are among the richest metal-yielding ores in the world, with as much as 20% of a deposit yielding copper, lead, zinc, gold, and silver.

FIGURE 9.9 Magmatic cumulates develop through two processes. Density segregation forms layers of dense minerals by settling of grains, in-place crystallization, and magmatic currents sweeping grains from the walls and roof within a pluton. Immiscible melts may develop if a compound reaches its saturation state before it passes its crystallization temperature. The result is a dense liquid that eventually becomes a concentrated mineral deposit.

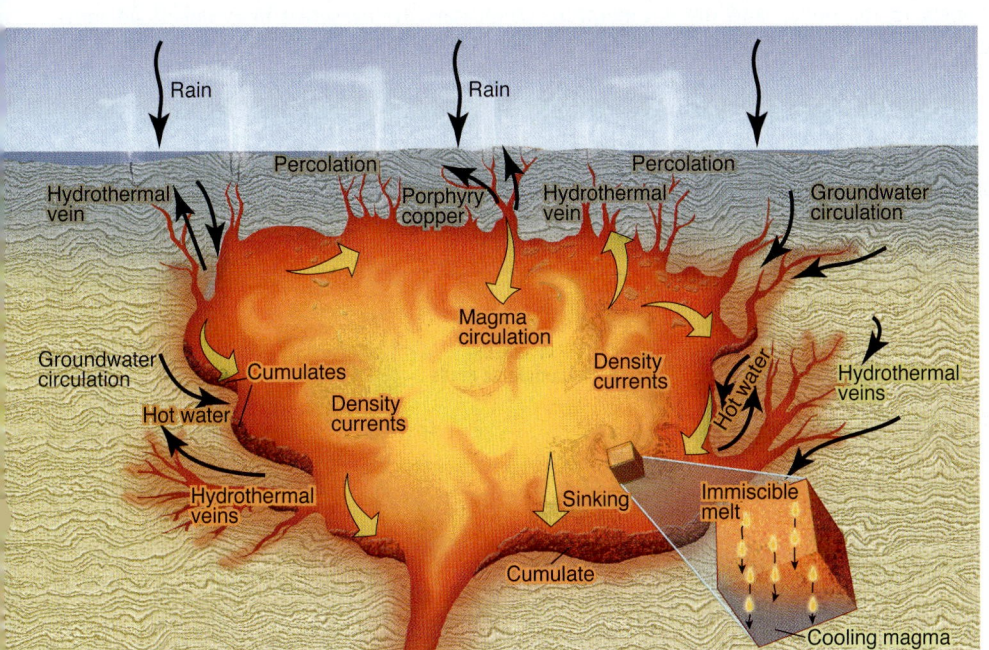

9-4 Rocks Evolve Through a Sequence of Metamorphic Grades

LO 9-4 Describe the metamorphic grades and how they are identified.

Different locations in Earth's crust experience different levels of heating and stress, some intense, others relatively mild. As a result, rock may experience different *grades* of metamorphism, depending on the kind of recrystallization that is stimulated by these conditions.

Gradational Change

The concept of gradational change, from **low-grade metamorphism** to **high-grade metamorphism,** reflects the sequence of mineral and texture changes that occur under different metamorphic conditions (**FIGURE 9.11**).

The changes that take place during metamorphism are recorded in the rock, in its textures and mineral assemblages. High-grade metamorphic rock is greatly altered from its original form, such that often it has a completely different mineralogy than that of its parent rock. That said, the mineralogy of the parent rock, the pressure and temperature conditions, and the presence of any chemically active fluids will govern the mineralogy of the metamorphic product. Low-grade metamorphic rock, while experiencing less severe conditions, nonetheless displays a unique texture and mineralogy.

Index Minerals

The change from a parent rock to a metamorphic product is characterized by evolving mineral assemblages. As explained earlier, the assemblage of minerals within a rock evolves in response to changing temperature and pressure conditions, leading geologists to reason that specific mineral assemblages serve as "geothermometers" and

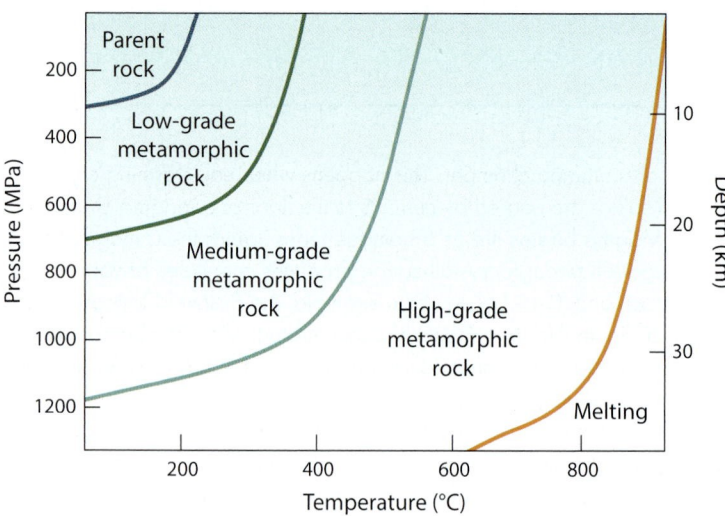

FIGURE 9.11 As rock is buried deeper in the crust, the temperature and pressure increase. Higher temperatures and pressures (MPa, metric unit of pressure) lead to metamorphism.

Q Why do texture and mineral assemblage change within metamorphic grade?

"geobarometers" of conditions within the crust. Moreover, certain minerals identify specific metamorphic conditions. This is one way geologists use critical thinking to interpret Earth's history as it is recorded in metamorphic rocks: They can formulate a simple hypothesis predicting conditions in the crust based on the presence of specific metamorphic minerals.

Minerals that serve as such "guideposts" are known as **index minerals.** Different types of parent rocks evolve different types of index minerals. Shale, for example, the most common sedimentary rock, develops these index minerals as it metamorphoses from lowest to highest metamorphic grade: *chlorite, muscovite, biotite, garnet, kyanite,* and *sillimanite* (**FIGURE 9.12**). These minerals serve as indicators of the degree of metamorphism experienced by the shale. For instance, chlorite tends to develop under low-pressure and low-temperature conditions. Its presence in a rock is an indicator of low-grade metamorphism. In contrast, sillimanite tends to

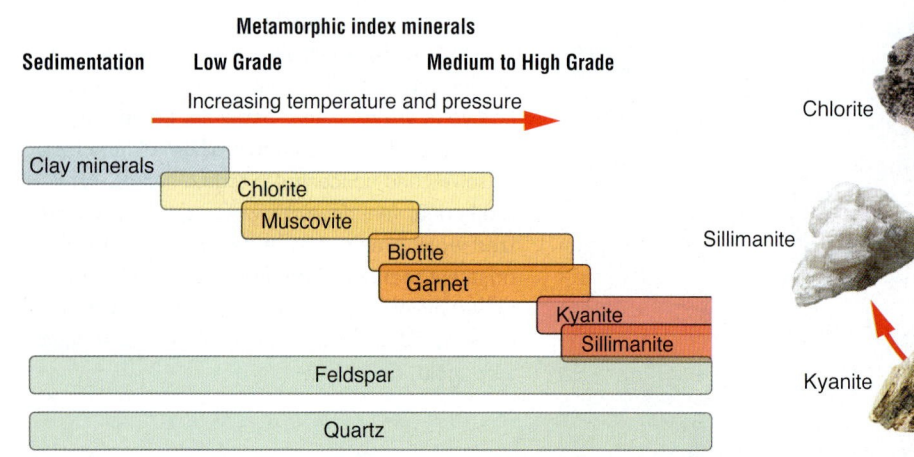

FIGURE 9.12 Metamorphic grade is a scale of metamorphic intensity that incorporates index minerals as geothermometers and geobarometers.

Q What characteristic makes these good index minerals?

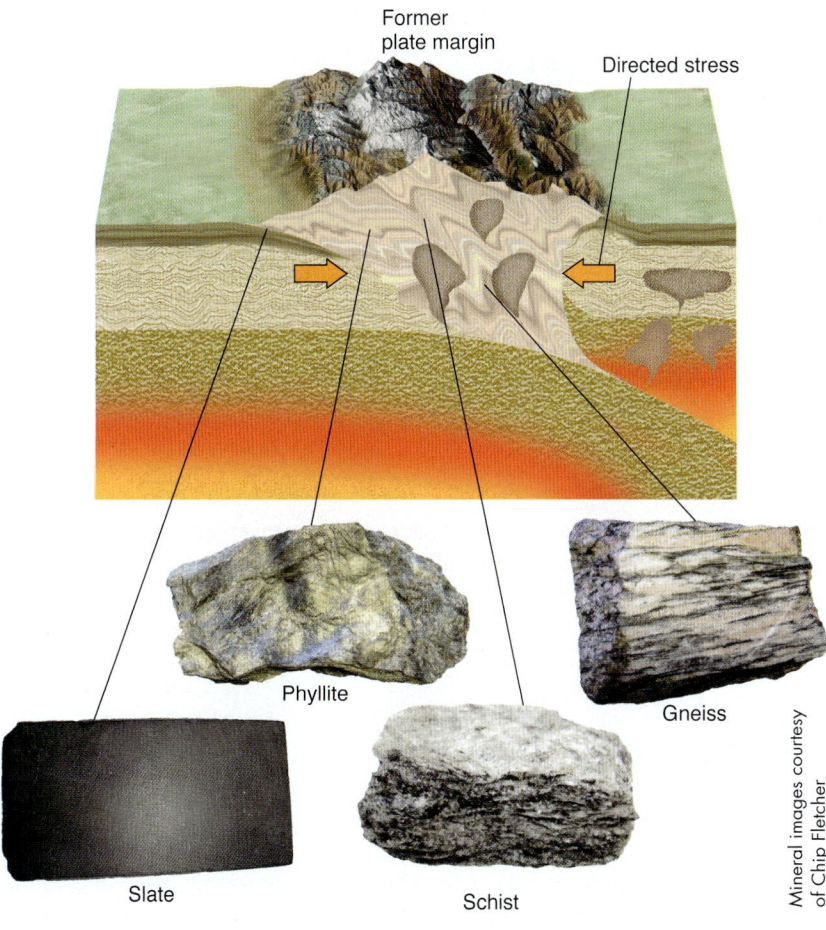

Mineral images courtesy of Chip Fletcher

FIGURE 9.13 Regional metamorphism of shale can produce a sequence of foliated rocks whose textures evolve with increasing pressure.

In general, what happens to the foliation of regionally metamorphosed rock as the grade increases?

form in high-pressure conditions, and its presence in a rock is an indicator of high-grade metamorphism.

Quartz and feldspar also appear in metamorphic products. However, because they tend to crystallize in low-grade as well as high-grade conditions, they are ubiquitous. They do not indicate a specific metamorphic condition and so are not considered index minerals.

Texture

Rocks exposed to directed stress (usually related to regional metamorphism) may develop foliation. Again, we will use a shale protolith as our example. Like the mineral assemblage, foliation in shale evolves with increasing metamorphic intensity. Foliated rocks have crystals oriented in near-parallel layers.

Shale evolves through four common foliated textures, in order of increasing metamorphic grade: *slate*, *phyllite*, *schist*, and *gneiss* (**FIGURE 9.13**). Those with large-grained textures (schist and gneiss) typify medium- to high-grade metamorphism; those with fine-grained textures (slate and phyllite) generally indicate low- to medium-grade metamorphism.

Contact metamorphism, and in some cases regional metamorphism, may produce nonfoliated products. Shale exposed to high temperature (such as during contact metamorphism) evolves into a fine-grained nonfoliated rock called *hornfels*. Another example of a nonfoliated rock is *anthracite* coal, the metamorphic product of bituminous coal. Large-grained nonfoliated rocks include *marble*, *quartzite*, and *metaconglomerate*. For the most part, they are identified on the basis of their mineral assemblage; we discuss them later in the chapter.

TABLE 9.1 provides a summary of the common parent rocks and their foliated and nonfoliated metamorphic products. These rocks and products are discussed in the next two sections.

TABLE 9.1 Protoliths and Metamorphic Products

Protolith	Low- to Medium-Grade	High-Grade
Foliated Product		
Fine-grained sedimentary (e.g., shale)	Slate, phyllite, schist	Gneiss
Igneous rock (e.g., basalt)	Greenschist, blueschist	Amphibolite, granulite, eclogite
Nonfoliated Product		
Conglomerate	Metaconglomerate	Becomes gneiss (foliated)
Sandstone	Quartzite	Quartzite
Shale, mudstone, siltstone	Hornfels	Hornfels
Bituminous coal	Anthracite	Anthracite
Igneous rock (e.g., basalt)	Hornfels	Hornfels
Limestone	Marble	Marble

EXPAND YOUR THINKING

Is a mineral that crystallizes in low-grade, medium-grade, and high-grade conditions a good index mineral? Why or why not?

9-5 Foliated Texture Is Produced by Directed Stress Related to Regional Metamorphism

LO 9-5 Compare and contrast the four principal types of foliated texture.

Just as igneous and sedimentary rocks are classified by their texture and composition, so are metamorphic rocks.

TABLE 9.2 Foliated Metamorphic Rocks

Types of Foliated Texture			
Slate	**Phyllite**	**Schist**	**Gneiss**
Fine grained; minerals not visible	Fine grained; minerals usually not visible	Medium to coarse grained; minerals visible	Medium to coarse grained; light and dark bands
Clay minerals, chlorite, muscovite	Chlorite, Muscovite	Muscovite, biotite, garnet, kyanite, and others	Feldspars, quartz, muscovite, biotite, ferromagnesian minerals
Dense	Satiny luster	Shiny luster	Banded

Foliated Texture

As mentioned earlier, metamorphic textures fall into two categories: foliated and nonfoliated. Foliated texture is produced by directed stress related to regional metamorphism (see **TABLE 9.2** and **FIGURE 9.14**), whereas **nonfoliated texture** is generally found in metamorphic rocks that are undergoing contact metamorphism, or in rocks that, when viewed with the naked eye, show no evidence of foliated texture.

Foliated, or *slaty*, texture is caused by the parallel orientation of microscopic grains. This texture is characteristic of slate, which has a tendency to separate along parallel planes. Slaty texture results in a property known as **rock cleavage,** also called *slaty cleavage* (**FIGURE 9.15**). Rock cleavage should not be confused with cleavage in a mineral, which is related to internal atomic structure. Rock cleavage develops because new minerals, usually chlorite, grow during metamorphism and align in a direction perpendicular to the dominant direction of stress. Hence, rock cleavage describes a relationship between minerals, whereas mineral cleavage describes the arrangement of atoms within a mineral.

Geologists use a compass to measure the orientation and direction of rock cleavage, which is usually caused by directed stress generated by plate convergence. This information is used to infer the past location of plate collision zones. Slaty cleavage, for example, has been used to reestablish the former position and movement of continents, and to reconstruct the supercontinent Pangaea.

Phyllite

Phyllitic texture refers to the parallel arrangement of platy minerals, usually micas (such as chlorite and muscovite) that are barely visible to the naked eye. The mineral arrangement produces a luster that is satiny, silky, and/or wavy. Abundant mica grains give rock specimens a silky sheen with phyllitic texture, representing low- to medium-grade metamorphism.

Schist

Schist is metamorphosed phyllite. *Schistose texture* refers to foliation resulting from near-parallel to parallel orientation of platy minerals like chlorite or muscovite. Quartz and amphibole are also common in schist. The average size of minerals is generally smaller in schist than in gneiss. The main difference, however, is that foliation is more distinct in schist, due to recrystallization and enlargement of chlorite and muscovite crystals.

Samples of schist often contain large crystals of minerals such as garnet and tourmaline. These develop as increasingly intense metamorphic conditions transform several different minerals into a new and distinct metamorphic mineral. There are a variety of schists, named for their dominant minerals. Examples include *garnet-mica schist*, *greenschist* (dominated by greenish chlorite), *blueschist* (dominated by the

(a) (b) (c) (d) (e)

Mineral images courtesy of Chip Fletcher

FIGURE 9.14 In foliated rocks, as directed stress increases, the texture evolves from a parent rock (a) shale to (b) slate to (c) phyllite to (d) schist and, finally, to (e) gneiss.

Q Why does directed stress tend to produce foliated texture?

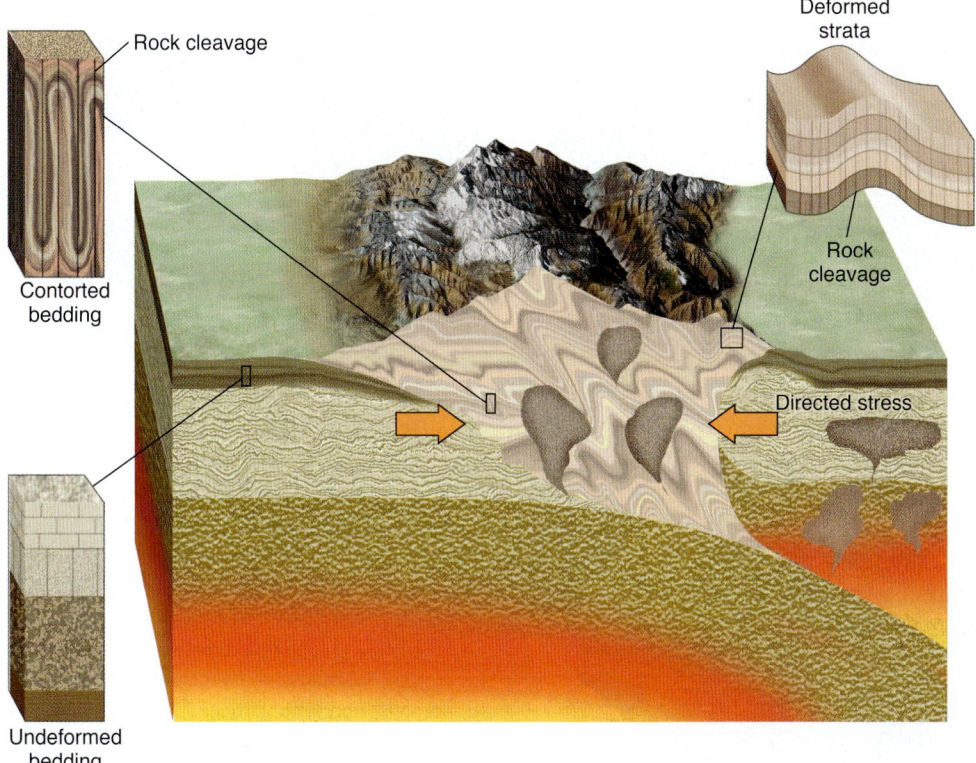

FIGURE 9.15 The growth of chlorite during low-grade metamorphism leads to cleavage. The study of rock cleavage allows geologists to reconstruct past tectonic conditions and former plate boundaries.

Q Would contact metamorphism produce rock cleavage? Explain your answer.

two minerals and is characterized by distinctive dark or light coloring, giving the rock a striped appearance. Light bands commonly contain quartz and feldspar; dark bands commonly contain hornblende and biotite.

Gneiss is a high-grade metamorphic rock that can evolve from a number of different protoliths, the most common being granite, diorite, and schist. Intense conditions cause solid-state migration of atoms, leading to the formation of mineral bands oriented perpendicular to the direction of principal stress. As gneiss approaches its melting point, the mineral bands bend and become contorted. Of course, if the temperature climbs too high, the rock will melt and return to the igneous phase of the rock cycle.

mineral glaucophane), and *biotite schist* (dominated by biotite). Each is formed under different pressure and temperature conditions.

Gneiss

Gneissic texture refers to coarse foliation in which minerals are segregated in discontinuous bands ranging from 1 millimeter to several centimeters in thickness. Each band is dominated by one or

Q **EXPAND YOUR THINKING**

Explain the relationship among recrystallization, rock cleavage, directed stress, and plate motion. State your answer in the form of a testable hypothesis.

9-6 Nonfoliated Rocks May Develop During Regional or Contact Metamorphism

LO 9-6 List the common nonfoliated rocks, and describe the products of contact metamorphism.

Metamorphic rocks in which there is no visible orientation of mineral crystals have a nonfoliated texture (TABLE 9.3).

Nonfoliated Rocks

Nonfoliated rocks (FIGURE 9.16) may develop during regional or contact metamorphism; therefore, nonfoliated texture does not always signal an absence of directed stress. This is because some minerals growing under directed stress do not develop foliation. For instance, when limestone or dolomite metamorphoses to become marble, calcite crystals enlarge but may not become foliated, even under directed stress. When quartz sandstone metamorphoses to become quartzite, the foliated quartz grains may not be visible unless viewed through a microscope.

Marble and quartzite usually contain uniformly sized grains of a single mineral (calcite, dolomite, quartz). Metamorphosed conglomerates may retain the original texture of the parent rock, including the outlines and colors of larger grains, such as pebbles.

TABLE 9.3 Nonfoliated Metamorphic Rocks

Marble	Quartzite	Metaconglomerate	Hornfels	Anthracite Coal
Medium to coarse grained; minerals visible	Medium to coarse grained minerals	Large clasts visible	Fine-grained; minerals not visible	Lack of minerals
Calcite ($CaCO_3$)	Quartz (SiO_2)	Anything conglomerate	Clay minerals, muscovite	Carbon-rich material
Mohs hardness of 3; bubbles rapidly with dilute hydrochloric acid	Mohs hardness of 7; breaks across grains	Breaks across and around grains	Dense, dark colored	Black, shiny; conchoidal fracture

(a)

(b)

(c)

(d)

(e)

Mineral images courtesy of Chip Fletcher

FIGURE 9.16 Nonfoliated metamorphic rocks: (a) marble, (b) quartzite, (c) metaconglomerate, (d) hornfels, and (e) anthracite.

Q Explain what factors lead to foliated texture and what factors lead to nonfoliated texture.

But because metamorphism has caused recrystallization of the matrix, the metamorphosed conglomerate is known as a *metaconglomerate*. In cases where metamorphism has deformed the shape of the large grains, the rock is referred to as a *stretched pebble conglomerate*.

Quartzite and metaconglomerate are distinguished from their sedimentary parent rocks by the fact that they break *across* the quartz grains, not around them. Marble has a crystalline appearance, and its mineral grains are larger than those of its limestone or dolomite parent rock. In many cases, calcite crystals become larger during metamorphism. Marble and quartzite are characterized by the interlocking nature of their crystals.

Hornfels, a nonfoliated rock formed by metamorphism of shale and other types of rocks, has a nondescript appearance. Usually a medium to dark shade of gray, it lacks foliation, and contains few recognizable minerals.

Anthracite, often overlooked in discussions of metamorphism, is a nonfoliated rock produced by metamorphism of bituminous coal.

Zoned Mineral Recrystallization

When a pluton, or magma body, intrudes into the crust, it can produce zones of minerals that reflect the chemical interaction of magma with country rock and chemically active fluids. The composition of the country rock is a big help in determining the mineralogy of the final metamorphic product, which may be economically important (**FIGURE 9.17**).

In dry limestone country rock, calcite and dolomite are stable at temperatures above 700°C. Therefore, contact metamorphism of a pure limestone or dolostone may yield no more than a coarsening (enlarging) of grain size, producing marble. However, most such country rock contains other materials besides calcite or dolomite. Often it contains silica in the form of chert, or feldspar sand

grains; calcite-dolomite-quartz mixtures are also common. When they undergo metamorphism, clay minerals in the country rock will add aluminum to the mix. And when water is present, mineral reactions are enhanced, and dissolved ionic compounds are transported throughout the contact zone.

In any of these cases, if the protolith is chemically complex, the resulting metamorphic product can be quite different from the parent. As one example, contact between a silicate-rich pluton and impure limestone country rock results in the formation of the mineral *wollastonite* and the production of carbon dioxide gas:

$$CaCO_3 + SiO_2 \longrightarrow CaSiO_3 + CO_2$$
$$\text{Calcite} + \text{Silica} \longrightarrow \text{Wollastonite} + \text{Carbon Dioxide}$$

This reaction occurs at temperatures of about 500°C. Thus, when wollastonite is present in a rock, it can be interpreted as a record of the interaction between a silica intrusion and limestone crust (**FIGURE 9.18**). Wollastonite, along with garnet and diopside, typically comprise the innermost zone of the aureole, and the minerals calcite, serpentinite, chlorite, dolomite, and olivine form assemblages that occur as parallel outer zones. Marble is often found at the outermost edges of intruded limestone.

Intrusions into silicate sedimentary rocks, such as shale and sandstone, also produce zoned aureoles. The contact metamorphic rock hornfels, containing pyroxene and mica, is found at locations where intruding magma meets shale.

Farther away, the quartz, clay, and carbonate components of shale and sandstone may be converted into biotite, chlorite, and calcite, as well as other minerals. In the cases we have described involving limestone, shale, and sandstone, the contact aureole zone surrounding a large pluton may be hundreds of meters wide.

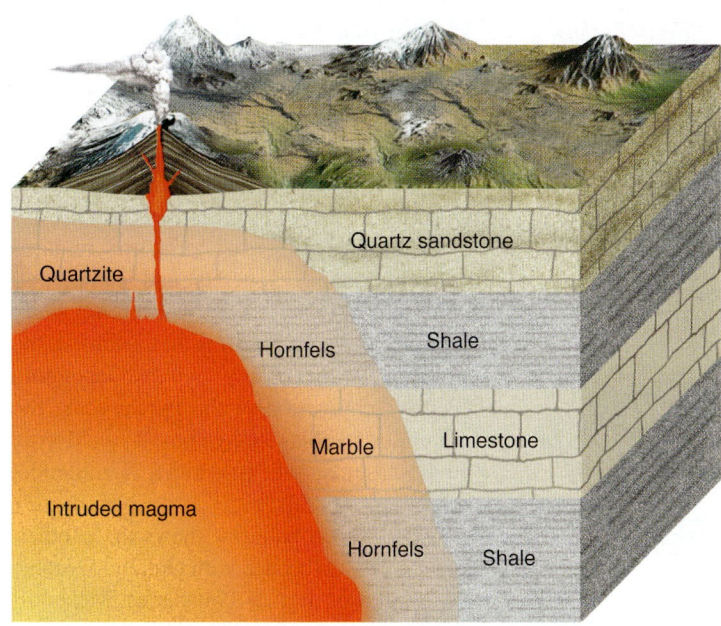

FIGURE 9.17 In general, contact metamorphism produces nonfoliated rocks.

What factors contribute to the growth of chemically diverse minerals in contact metamorphism?

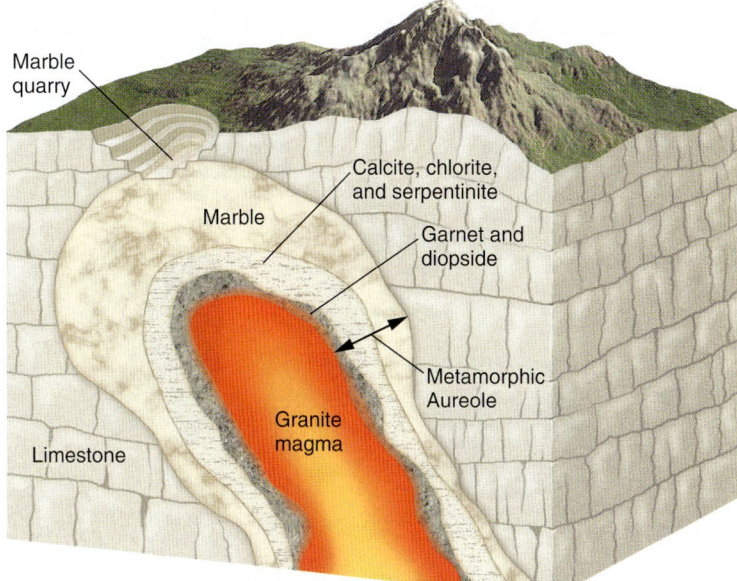

FIGURE 9.18 Impure limestone is altered by contact metamorphism. For example, the minerals garnet and diopside may characterize the baked zone closest to the magma. Calcite, chlorite, and serpentinite minerals may develop in the next zone. An outer zone consists of marble, which is metamorphosed calcite.

Marble is an economically important rock. How is it used?

9-7 The Relationship Between Mineral Assemblage and Metamorphic Grade Is Expressed by Metamorphic Facies

LO 9-7 Define the term metamorphic facies.

Depending on the composition of a protolith, the mineral assemblage of a metamorphic product reflects the temperature and pressure conditions prevailing during metamorphism. The relationship between mineral assemblage and metamorphic grade is expressed by the concept of **metamorphic facies**, a set of metamorphic mineral assemblages repeatedly found together and designating certain pressure and temperature conditions.

Metamorphic Facies

A metamorphic *facies* (Latin for "face") is a mineral assemblage that reflects specific pressure and temperature conditions in Earth's crust. A portion of crust that is exposed to metamorphic conditions, such as a convergent plate margin, will develop a range of facies, each of which represents a stage of low- to high-grade conditions. Hence, a *sequence* of metamorphic facies might be observed in a metamorphosed region reflecting the range of low-grade to high-grade conditions that have occurred.

The nature of those facies will depend on the *geothermal gradient*, the rate at which pressure and temperature increase with depth in the crust. A geothermal gradient characterized by a very rapid rise in temperature but relatively little change in pressure might be present around an igneous intrusion and would result in a specific texture and assemblage of minerals. Another geothermal gradient—for instance, one reflecting a rapid increase in pressure and relatively little change in temperature—would result in a very different texture and mineral assemblage and, therefore, different metamorphic facies (**FIGURE 9.19**).

The concept of metamorphic facies was developed by a Finnish geologist, *Pentti Eskola* (1883–1964), in 1915. In the course of a study of basalt rocks, Eskola identified the various pressure (P) and temperature (T) conditions that basalt may experience in the course of metamorphism on its way to melting. Figure 9.19 illustrates Eskola's concepts as they have been modified to account for the effects of plate tectonics. Although Eskola's work was largely with basalts, today, facies names are applied to many types of protoliths.

Note in Figure 9.19, in the upper left corner of the diagram, that both the lowest pressure and temperature are represented within the upper few kilometers of the crust. This region of P/T conditions is named the *zeolite facies*, after a mineral group found in lightly metamorphosed oceanic crust. As seafloor basalt undergoes early metamorphism associated with shallow burial in the crust, it produces zeolites, which are characterized by interlocking tetrahedra of silicate and aluminum oxide. Zeolites are distinctive for their capability to lose and absorb water without damage to their crystal structures. Remember that the zeolite facies, like all facies, comprises a conceptual zone and may consist of other types of minerals, depending on the composition of the parent rock.

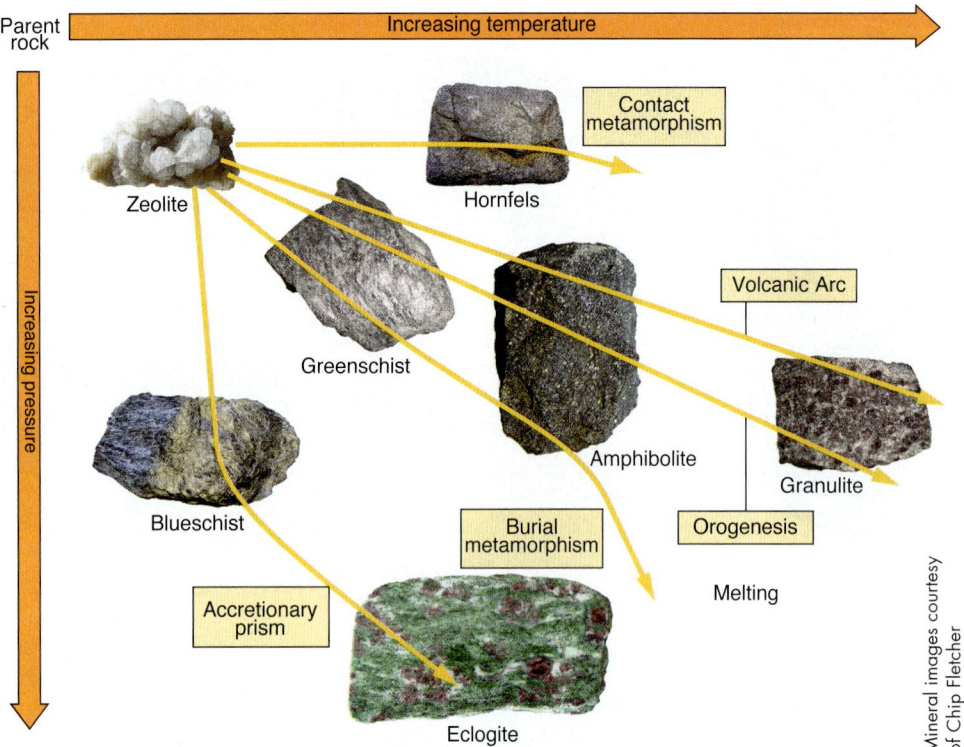

Parent rock

Increasing temperature →

Increasing pressure ↓

Zeolite

Contact metamorphism

Hornfels

Greenschist

Volcanic Arc

Amphibolite

Granulite

Blueschist

Burial metamorphism

Orogenesis

Melting

Accretionary prism

Eclogite

Mineral images courtesy of Chip Fletcher

FIGURE 9.19 Five basic pathways of metamorphism are: 1) contact metamorphism, 2) metamorphism typical of volcanic arcs, 3) metamorphism typical of collisional mountain belts (orogenesis), 4) burial metamorphism (metamorphism caused by deep burial in the continental crust), and 5) metamorphism typical of accretionary prisms. A rock that is undergoing metamorphic change along any of these pathways will pass through phases, or facies, characterized by crystallization of certain sets of minerals.

ⓠ **What is the origin of zeolite?**

Metamorphic Pathways

From this point of origin, five pathways of metamorphosis are diagramed. The uppermost pathway shows a typical geothermal gradient associated with contact metamorphism. Temperature rises rapidly toward the plutonic body, but pressure remains relatively stable. Igneous and sedimentary rocks in contact with a plutonic body metamorphose to hornfels and establish the *hornfels facies* (**FIGURE 9.20**). Depending on the composition of the parent rock, the mineral assemblage may consist of chlorite, plagioclase, quartz, pyroxene, hornblende, and other minerals.

The second pathway shows the geothermal gradient associated with plutonic-volcanic arc complexes at plate boundaries. Metamorphism progresses from the zeolite facies into the *greenschist facies*, consisting of chlorite, serpentine, quartz, biotite, plagioclase, and other minerals, depending on the parent rock. As metamorphism continues, the rock passes through the *amphibolite facies* and into the *granulite facies*. A zone of partial melting is encountered in the amphibolite facies, where minerals with low melting points will begin to form magma. This partial melting occurs when enough water is present to lower the melting point of minerals in the evolving rock.

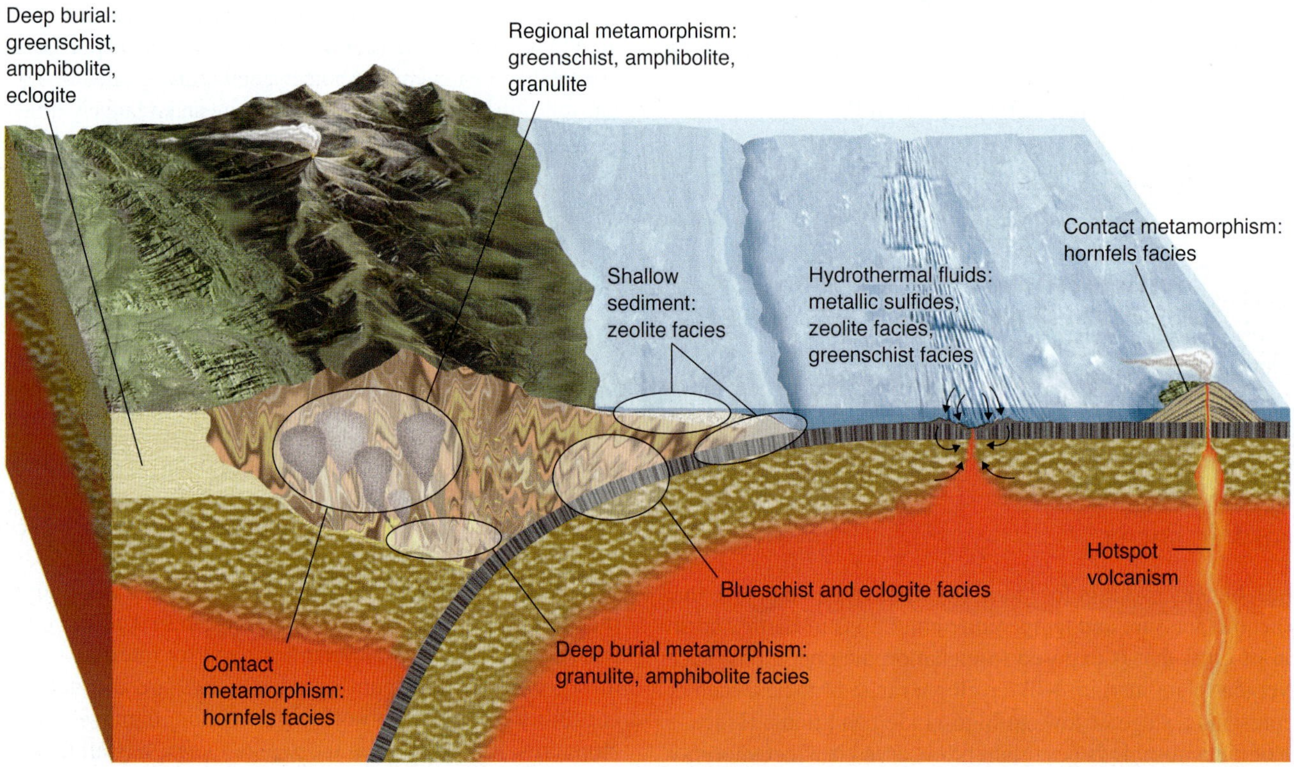

Deep burial: greenschist, amphibolite, eclogite

Regional metamorphism: greenschist, amphibolite, granulite

Shallow sediment: zeolite facies

Hydrothermal fluids: metallic sulfides, zeolite facies, greenschist facies

Contact metamorphism: hornfels facies

Contact metamorphism: hornfels facies

Deep burial metamorphism: granulite, amphibolite facies

Blueschist and eclogite facies

Hotspot volcanism

FIGURE 9.20 Metamorphic facies are typically related to tectonic environments or deep burial in continental crust.

ⓠ **What environmental conditions are found with deep burial, and how are they similar or different relative to a convergent margin?**

This type of partial melting is important in releasing the first magmas associated with igneous activity in subduction zones.

The third metamorphic pathway is associated with collisional mountain building during the process of orogenesis. The orogenesis pathway passes through the same low- and medium-grade facies that volcanic arc metamorphism does, but enters each facies at a higher pressure.

The fourth pathway is characteristic of burial in the crust under the influence of the geothermal gradient. Burial metamorphism typically occurs at plate interiors and results from pressures and temperatures in rock buried to depths of up to 30 to 40 kilometers. In this case, temperatures do not rise as rapidly as in the first three pathways, but the rock eventually encounters the same sequence of metamorphic facies until it enters the *eclogite facies*. The mineralogy of eclogite facies is dependent on that of the parent rock but may include olivine, pyroxene, garnet, quartz, and kyanite.

A fifth metamorphic pathway is encountered at plate convergent zones that build accretionary prisms. This pathway is characteristic of regional metamorphism at locations where oceanic sediments, on sub-ducting crust, mix with sediments from continental crust to build accretionary prisms. The resulting accumulation of sediment experiences a rapid rise in pressure and a relatively slow rise in temperature: Rocks progress from the zeolite facies to the *blueschist facies* and, finally, to the eclogite facies. Blueschists are encountered in the *forearc area* of a subduction zone, the region between the deep-sea trench and the volcanic arc. Depending on the exact composition of the parent rock, blueschists consist of the minerals glaucophane, sodium-rich amphibole (which has a bluish hue), chlorite, epidote, quartz, and kyanite.

The exercise in the Critical Thinking feature titled "Metamorphism and Tectonics" will demonstrate how you can apply your understanding of plate tectonics to the concept of metamorphic facies.

Ⓠ EXPAND YOUR THINKING

Metamorphic facies can be a difficult concept to understand. Explain it in the simplest terms possible.

9-8 Metamorphism Is Linked to Plate Tectonics

LO 9-8 Describe how metamorphic facies are related to plate tectonics.

Throughout our discussion of metamorphic rocks, we have connected specific tectonic processes to the pressure and temperature conditions that lead to metamorphism. Clearly, metamorphism and plate tectonics are intimately linked (FIGURE 9.21).

Divergent Margins

At spreading centers, hydrothermal fluids interact with mafic and ultramafic oceanic crust. Hydrothermal fluids lead to the deposition of valuable *metallic sulfide deposits*, which accumulate in fractures and fissures among the young basalts in the mid-ocean ridge. These deposits contain minerals from the native elements and sulfide classes. Where they are found on continents (at former convergent margins where ancient seafloor is preserved) they have been mined for economically valuable metals.

In most cases, these sites are characterized by low pressure and moderately elevated temperatures. The most significant metamorphic process occurring at mid-ocean ridges is the heat-driven circulation of hydrothermal fluids, leading to mineral reaction and recrystallization and the development of zeolite and greenschist facies.

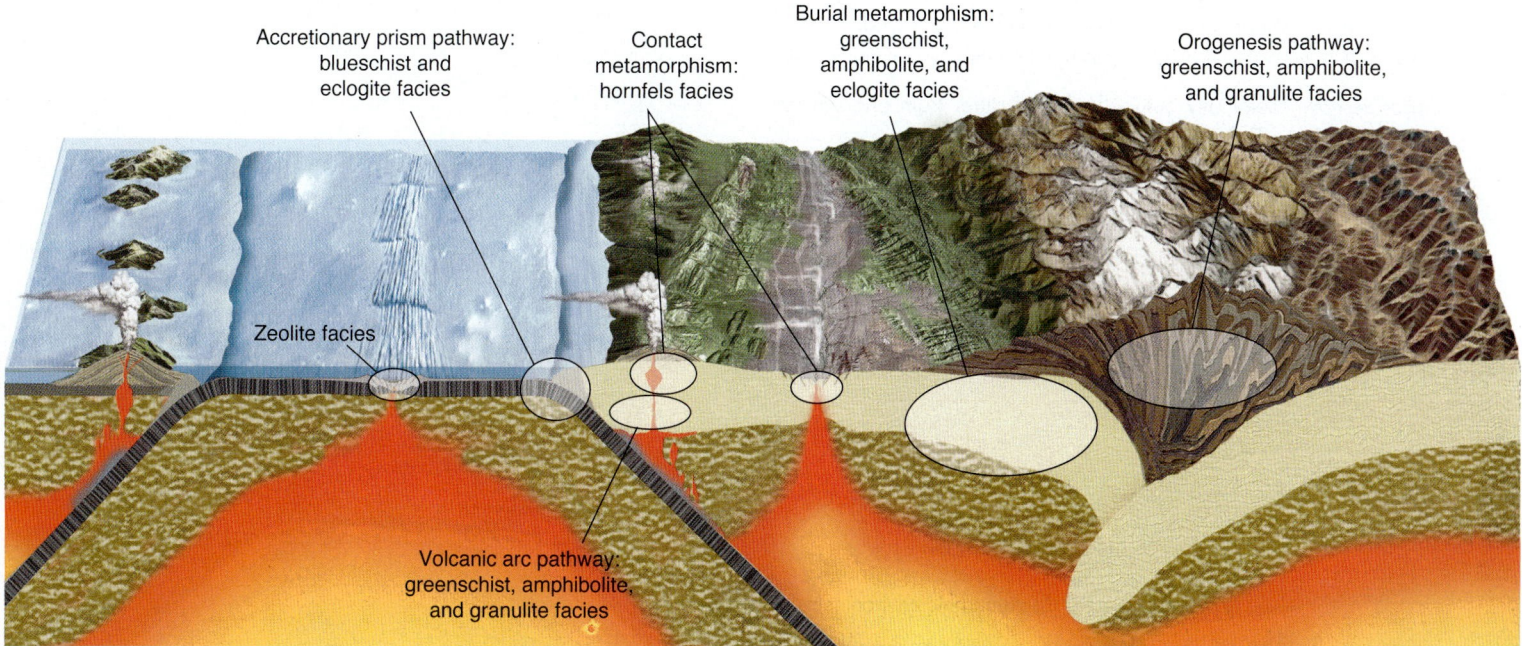

Accretionary prism pathway: blueschist and eclogite facies

Contact metamorphism: hornfels facies

Burial metamorphism: greenschist, amphibolite, and eclogite facies

Orogenesis pathway: greenschist, amphibolite, and granulite facies

Zeolite facies

Volcanic arc pathway: greenschist, amphibolite, and granulite facies

FIGURE 9.21 Metamorphism develops at divergent and convergent boundaries, as well as midplate settings.

Ⓠ **In which tectonic environment does pressure play the greatest role? What about heat?**

CRITICAL THINKING

Metamorphism and Tectonics

Metamorphic environments are closely tied to plate tectonics (FIGURE 9.22). Consider that nearly all types of metamorphism can occur at subduction zones. Regional and contact metamorphism is found in the temperature and pressure conditions that develop at convergent plate boundaries. Divergent plate boundaries, at spreading centers, are locations where zeolite and greenschist facies develop in the presence of hydrothermal circulation. The pressure and temperature conditions at convergent and divergent boundaries produce the majority of the metamorphic rocks in the crust.

Working with a partner, begin by examining Figure 9.22 and then work through the following exercise:

1. Identify where contact metamorphism is likely to occur.

2. Identify where regional metamorphism is likely to occur.

3. Identify where you would find each of the following: marble, hornfels, phyllite, metaconglomerate, gneiss. Explain why you picked these sites.

4. What pressure and temperature conditions and protoliths determine the location of the major metamorphic facies (zeolite, hornfels, blueschist, greenschist, amphibolite, eclogite, and granulite)? Indicate the location of each.

5. Describe the pressure and temperature conditions that prevail where burial metamorphism occurs.

6. What role do chemically active fluids play in the growth of metamorphic rocks at convergent and divergent boundaries? Where do these fluids come from?

7. Is the accretionary wedge a site of high temperature or high pressure, or both? What is the origin of the accretionary wedge, and what is its composition? Is it likely to have chemically active fluids?

8. Describe how chemically active fluids will differ from one site to another in Figure 9.22. Explain why they differ.

9. For each facies you have identified, describe its protolith, the P/T conditions in which it evolved, the role of chemically active fluids, and its present mineralogy.

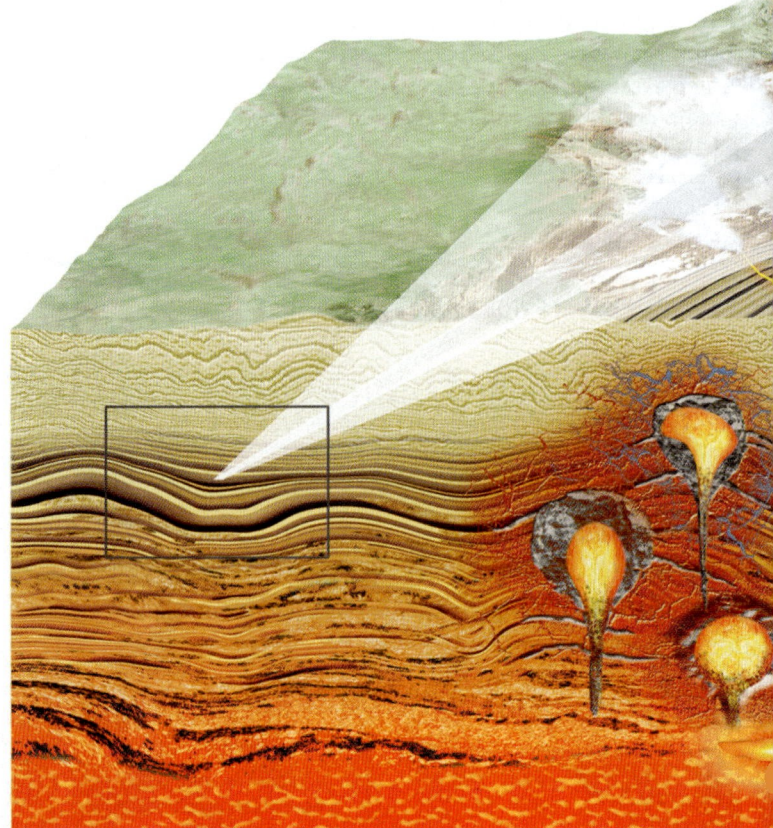

FIGURE 9.22 Metamorphic environments.

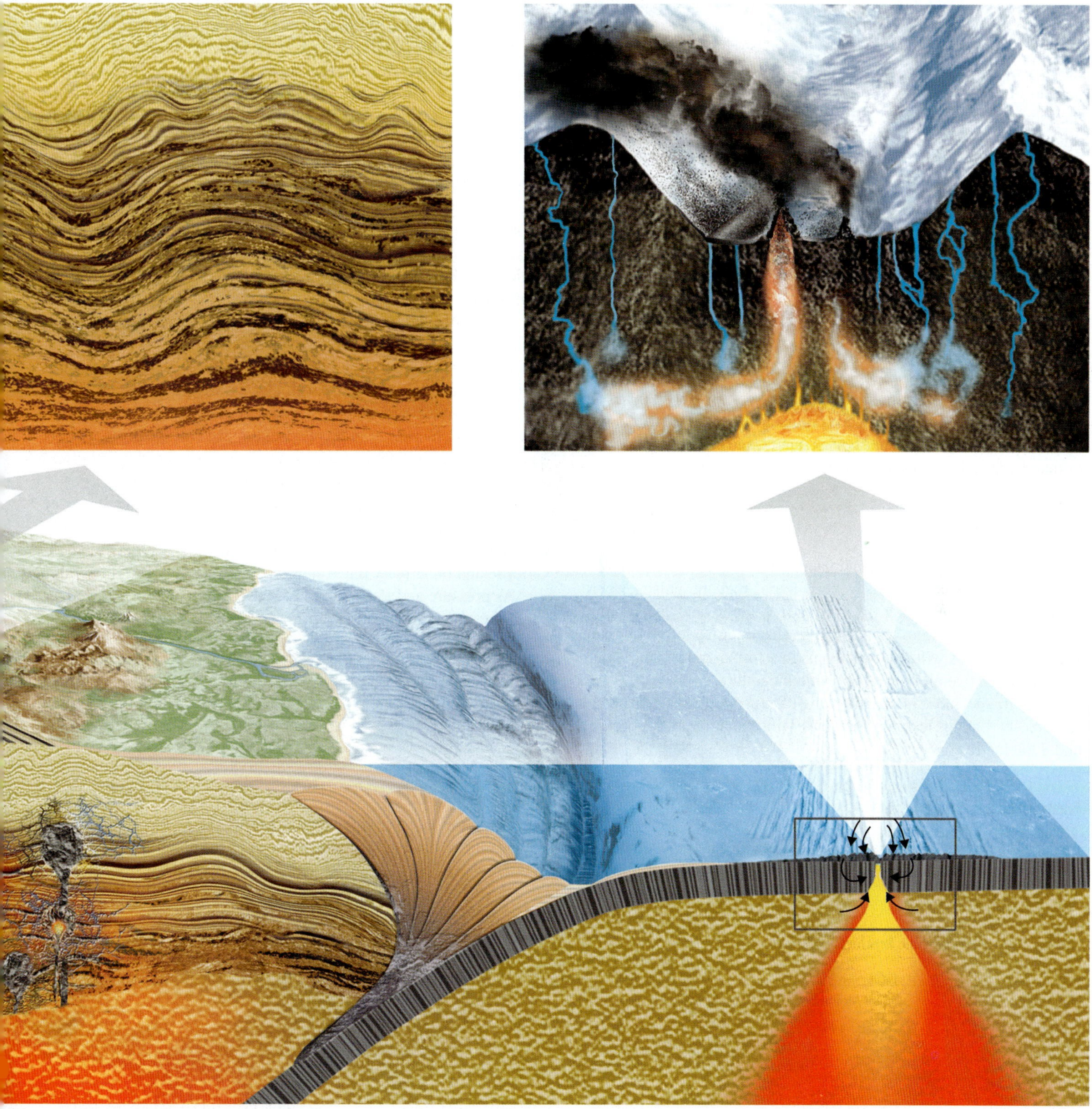

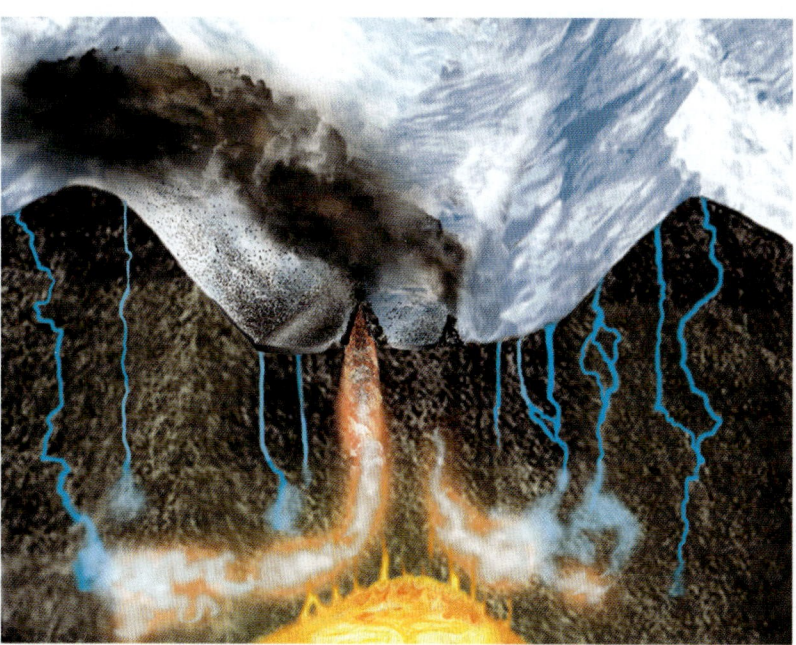

Convergent Margins

As oceanic crust moves away from a divergent margin, it is buried by marine sediment. Burial continues as both basaltic and sedimentary parent rocks move directly toward active subduction. During subduction, the crust is exposed to intensified directed stress, with an accompanying rise in temperature. Slate, phyllite, and schist textures develop during plate convergence as blueschist and eclogite facies evolve in the region of greatest directed stress.

Large-scale regional metamorphism takes place in the core of the overriding plate. Greenschist, amphibolite, and granulite metamorphic facies develop across the broad geographic band between the subduction zone and the interior region of the overriding plate. In this region, increases in pressure due to plate convergence are accompanied by temperature increases associated with partial melting of deeply buried continental lithosphere and descending oceanic crust.

Hotspots

Igneous intrusions are sites of contact metamorphism that are found at convergent boundaries and hotspots. The composition of intruded crust, the presence of chemically active fluids, and the level of thermal metamorphism occurring in contact aureoles determine the mineral assemblage that will develop at such sites.

Contact metamorphism in oceanic crust typically forms pyroxene and hornblende-rich hornfels, while intrusion into sedimentary country rock results in plagioclase, muscovite, quartz, and calcite-rich hornfels or marble. Intruded limestone results in the formation of marble.

Deep Burial

The stable interior of the continental lithosphere, far from plate boundaries, is characterized by deep burial metamorphism, where temperatures exceeding 600°C produce eclogite and amphibolite facies; if enough water is present, partial melting of high-grade schists and gneiss will occur. This magma forms granitic intrusions that add to the overall rise in regional temperature.

Basalt, the most abundant form of crust, offers a real-world example of linking tectonic locations with metamorphic processes under the facies concept.

Metamorphism of Basalt Crust

Basalt is the most common igneous rock, with many metamorphic products (TABLE 9.4). It is rich in the ferromagnesian silicates olivine, pyroxene, and calcium-rich plagioclase feldspar. These minerals are *anhydrous*—they lack water. This means that when water is present, they will readily use it to form new minerals. For instance, when chemically active fluids enter the metamorphic process, new assemblages of minerals develop from the ferromagnesium silicates present in basalt. We saw already that this led to zeolite and greenschist facies at the spreading center.

Water is present at convergent boundaries because it is trapped in the pores and bonded to the minerals in basalt and marine sedimentary rocks. In some cases, water is trapped in the atomic structure of minerals within the crust. Riding within an oceanic plate as it moves toward its destiny at a subduction zone, these minerals readily release their trapped water in the earliest stages of metamorphism during subduction.

Hot water is released from the subducting plate. This water lowers the melting point of rocks that it encounters in the upper mantle and crust in the overriding plate. Magma formation in the wedge of mantle above the subducting plate plays an important role in raising the regional temperature of the crust, as well as producing plutonic bodies that intrude into overlying rock, resulting in contact metamorphism.

In seafloor basalt that is undergoing regional metamorphism at a subduction zone, water is usually abundant, making possible the crystallization of new assemblages of *hydrous* (water-rich) zeolites. As basalt undergoes metamorphism in the presence of water, it evolves into low-grade greenschist, which contains a mineral assemblage consisting of chlorite, epidote, plagioclase, and calcite. Greenschist is equivalent in metamorphic grade to slate; it has distinctive foliation and a greenish hue produced by the green minerals chlorite and epidote.

As pressure and temperature increase during the later stages of subduction, chlorite is replaced by amphibole. The resulting rock, amphibolite, is dark and coarse-grained, lacking distinct foliation. Chlorite and other platy micas have been replaced by dark minerals.

As pressure and temperature increase, amphibole is replaced by pyroxene, and granulite, the high-grade member of the basalt sequence, is formed. Granulite consists largely of pyroxene, plagioclase, and garnet. Like amphibolite, granulite is weakly foliated. Continued increase in heat and pressure eventually will lead to partial melting and, ultimately, to the end of metamorphism.

ⓠ EXPAND YOUR THINKING

Are plate boundaries the only places where metamorphism occurs? If not, describe where else metamorphism occurs, and the source and role of chemically active fluids there.

TABLE 9.4 Metamorphic Products of Basalt

Parent Rock Basalt	Not Metamorphosed	Increasing Temperature and Pressure ⟶		
		Low Grade	Medium Grade	High Grade
Facies	Basalt + H₂O	Zeolite, greenschist	Amphibolite	Granulite
Foliation	None	Distinct schistosity	Indistinct; occasionally foliation is visible due to parallel amphibole crystals	Indistinct due to absence of micas
Size of grains	Visible with magnifier	Visible with magnifier	Visible with naked eye	Large and obvious plagioclase, garnet
Minerals	Olivine, pyroxene, plagioclase	Chlorite, epidote, calcite, plagioclase	Amphibole, plagioclase, quartz	

LET'S REVIEW

Metamorphic rocks contain a variety of different mineral assemblages. Exactly which minerals make up a metamorphic rock depends primarily on two factors: 1) the mineralogy and chemical composition of the parent rock, and any fluids present; and 2) the temperature and pressure prevailing during metamorphism (metamorphic grade). Metamorphism is a good example of how natural forces interact with natural materials to make predictable products. The same assemblage of protolith minerals will result in different metamorphic products, depending on the specific heat and temperature pathway they follow. Add fluids to the mix, and the variety of resulting products becomes even greater.

Yes, pressure and heat metamorphose rocks in the crust. But the memorable lesson is that geologic processes leave behind a trail of information that makes it possible, by using critical thinking, to discover how Earth works.

With the completion of Chapter 9, we come to the end of our survey of the three large families of rocks that make up the solid shell of Earth: igneous, sedimentary, and metamorphic. In the following chapters we investigate two important processes that shape the crust, mountain building and earthquakes.

STUDY GUIDE

9-1 Metamorphic rocks are composed of sedimentary, igneous, or metamorphic minerals that have recrystallized.

- **Metamorphic rocks** are sedimentary, igneous, or metamorphic rocks that have taken on a new texture and mineralogy in order to maintain equilibrium under changing conditions within Earth's crust. This occurs as a result of solid-state **recrystallization** of minerals in the original ("parent") rock. Recrystallization produces a new assemblage of minerals by recycling existing minerals.

- Metamorphic rocks represent a distinct phase in the life of the crust, and a key element of the rock cycle.

9-2 Metamorphism is caused by heat and pressure.

- Rocks undergoing **metamorphism** alter their form and mineralogy in order to achieve a more stable configuration when experiencing increased temperature and pressure and chemically active fluids.

- Heat rises with depth in the crust, along what is known as the geothermal gradient. The average geothermal gradient is an increase of about 30°C to 50°C per kilometer of depth in the crust; but the actual increase varies in different tectonic settings. At temperatures between 150°C and 200°C (at a depth of ~8 kilometers), clay minerals (the most ubiquitous sediment) become unstable and recrystallize to become the metamorphic minerals chlorite and muscovite, marking the onset of early metamorphism.

- Lithostatic **stress** intensifies with depth in the crust and plays an important role in metamorphism, closing the pore spaces in rocks, causing compaction, and at great depths, causing minerals to recrystallize into more compact atomic structures. Lithostatic stress deforms the rock equally in all directions. **Directed stress** differs from lithostatic stress in that pressure acting on the crust is greatest in one direction. This causes folding, bending, breaking, or even overturning of rock strata.

9-3 Chemically active fluids transport heat and promote recrystallization.

- Chemically active fluids too have an important role in metamorphism. They transport heat and have a profound influence on the chemistry and mineralogy of metamorphic rocks. Important fluids may come from

groundwater, if metamorphism is occurring in shallow crust (less than ~2 kilometers deep). They also may be produced by dewatering of magma or by dehydration of water-rich minerals, such as micas or amphiboles; or they may come from seawater.

- Hydrothermal fluids emerge from the interaction of groundwater and magma in the final stages of cooling. These fluids are hot and contain dissolved ions and compounds that react with country rock, causing crystallization of valuable metals, such as lead, silver, platinum, and gold.

- Chemical alteration in the presence of hydrothermal fluids is known as metasomatism.

9-4 Rocks evolve through a sequence of metamorphic grades.

- Rock may undergo different grades of metamorphism, depending on the conditions in which metamorphism takes place. The concept of gradational change, from **low-grade** to **high-grade metamorphism**, reflects the nature of the mineral and textural changes that occur.

- Shifting mineral assemblages characterize the change from parent rock to metamorphic product. *Index minerals* that track these changes include (from lowest to highest metamorphic grade) chlorite, muscovite, biotite, garnet, kyanite, and sillimanite. When identified in rock samples, these minerals serve as indicators of the degree of metamorphism experienced by the rock. For instance, the presence of chlorite indicates low-grade conditions while the presence of sillimanite indicates high-grade conditions.

- Shale evolves through four common foliated textures, in order of increasing metamorphic grade: slate, phyllite, schist, and gneiss. Those with large-grained textures (schist and gneiss) typify medium- to high-grade metamorphism; those with fine-grained textures (slate and phyllite) generally indicate low- to medium-grade metamorphism.

9-5 Foliated texture is produced by directed stress related to regional metamorphism.

- Within the crust, directed stress is generated by the collision of two plates or by simple burial in such a way that pressure is greatest in one direction. Rocks formed under directed stress usually exhibit a texture known as *foliation*, created by aligned minerals.

- *Regional metamorphism* occurs when high pressures and temperatures are generated at convergent tectonic margins. Two colliding plates create an extensive zone of high pressure characterized by directed stress in the deep crust. Directed stress typically causes the formation of foliation in the products of regional metamorphism.

9-6 Nonfoliated rocks may develop during regional or contact metamorphism.

- **Contact metamorphism** takes place when a rock is subjected to high temperature because of intrusion of an igneous plutonic body. In general, contact metamorphism is not accompanied by significant changes in pressure.

- **Nonfoliated** rocks that are fine grained include hornfels and anthracite coal. The large-grained nonfoliated rocks are marble, quartzite, amphibolite, and metaconglomerate. These rocks are, for the most part, identified based on their mineral assemblage.

- The result of contact metamorphism of shale is hornfels. When limestone is changed by contact metamorphism, the resulting rock is marble.

- Contact between a silicate-rich pluton and impure limestone country rock results in the formation of the mineral wollastonite and the production of carbon dioxide gas.

9-7 The relationship between mineral assemblage and metamorphic grade is expressed by metamorphic facies.

- Relationships between mineral assemblage and metamorphic grade are expressed by the concept of **metamorphic facies**, which states that, depending on the composition of the original rock, the mineral assemblage in the metamorphic rock reflects the temperature and pressure conditions prevailing during metamorphism.

- The major facies are: zeolite, hornfels, greenschist, amphibolite, granulite, blueschist, and eclogite.

9-8 Metamorphism is linked to plate tectonics.

- The pressure and temperature conditions that lead to metamorphism are linked to specific tectonic processes occurring at plate boundaries. Metamorphism and plate tectonics thus are intimately linked.

KEY TERMS

contact metamorphism (p. 250)
directed stress (p. 251)
foliation (p. 249)
high-grade metamorphism (p. 256)
hydrothermal vein fillings (p. 252)

index minerals (p. 256)
low-grade metamorphism (p. 256)
metamorphic facies (p. 261)
metamorphic rocks (p. 248)
metamorphism (p. 248)

nonfoliated texture (p. 257)
recrystallization (p. 248)
regional metamorphism (p. 251)
rock cleavage (p. 257)
stress (p. 248)

ASSESSING YOUR KNOWLEDGE

Please complete this exercise before coming to class. Identify one best answer to each question.

1. Metamorphic rocks are formed by increased:
   a. Pressure and cementation.
   b. Heat and melting.
   c. Pressure and heat.
   d. Cooling and solidification.
   e. None of the above.

2. Metamorphism occurs when:
   a. Minerals partially melt and quickly recrystallize.
   b. Recrystallization occurs in the solid state.
   c. Loose sediments grow new crystals that cement grains together.
   d. Igneous minerals have solidified.
   e. None of the above.

3. How does metamorphism differ from lithification?
   a. Lithification occurs at higher temperatures and pressures.
   b. Lithification occurs at lower temperatures and pressures.
   c. Metamorphism requires melting of preexisting rock.
   d. Lithification does not cause any significant changes in the rock or sediment.

4. Under which of the following conditions does metamorphism occur?
   a. High temperature and high pressure
   b. High temperature and low pressure
   c. Recrystallization of minerals
   d. All of the above answers are correct.

5. The metamorphic rock shown here is a _____ (schist, gneiss, zeolite). It was formed by _____ (contact, regional) metamorphism. The primary factor driving metamorphism of this rock was _____ (heat, pressure).

Courtesy of Chip Fletcher

6. What type of metamorphism is local in extent and results from the rise in temperature in country rock surrounding an igneous intrusion?
   a. Regional
   b. Contact
   c. Burial
   d. Metasomatism
   e. Plutonism

7. Chemically active fluids promote:
   a. Regional stress.
   b. Formation of oceanic intrusions.
   c. Stable conditions deep in the crust.
   d. Dissolution and migration of ions.
   e. None of the above.

8. What is the texture of the sequence of rocks pictured here (slaty, phyllitic, schistose, gneissic, no particular texture)?

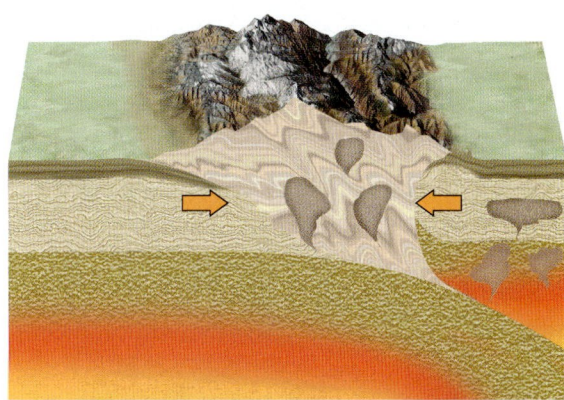

Courtesy of Chip Fletcher

9. The metamorphic index minerals are:
   a. Kaolinite, garnet, quartz, chlorite, biotite, and schist.
   b. Chlorite, garnet, sillimanite, hornfels, schist, and muscovite.
   c. Slate, phyllite, schist, chlorite, greenschist, and gneiss.
   d. Chlorite, muscovite, biotite, garnet, kyanite, and sillimanite.
   e. Gneiss, slate, chlorite, and quartz.

10. Foliated metamorphic rocks, in order of increasing metamorphic grade, are:
    a. Clay, chlorite, muscovite, biotite, garnet, and sillimanite.
    b. Marble, quartzite, mylonite, and gneiss.
    c. Slate, phyllite, schist, and gneiss.
    d. Shale, slate, quartzite, marble, and schist.
    e. Gneiss, slate, schist, chlorite, and phyllite.

11. Identify each of the metamorphic index minerals shown here (chlorite, muscovite, sillimanite, garnet, biotite, or kyanite).

Courtesy of Chip Fletcher

12. Marble is related to limestone in the same way that:
    a. Basalt is related to andesite.
    b. Hornfels is related to shale.
    c. Gravel is related to breccia.
    d. Gneiss is related to marble.
    e. Sandstone is related to basalt.

13. Which of the following statements about foliated rocks is correct?
    a. They reflect the influence of directed stress in the crust.
    b. They are usually formed within intruded country rock.
    c. They are the product of metasomatism.
    d. They rarely develop at convergent margins.
    e. None of the above.

14. The eclogite facies:
    a. Reflects low temperature and low pressure conditions.
    b. Develops only from carbonate sedimentary protoliths.
    c. Is formed in shallow conditions along the orogenesis pathway.
    d. Reflects conditions on the accretionary prism and burial metamorphism pathways.
    e. None of the above.

15. Which of the following usually develops when magma intrudes into shale?
    a. Hornfels
    b. Zeolite
    c. Schist
    d. Marble
    e. Subduction

16. Which of the following tectonic processes is/are most important to metamorphism?
    a. Plate rotation
    b. Sediment accumulation and erosion
    c. Subduction and plate convergence
    d. Paleomagnetic wandering
    e. Plate tectonics is not related to metamorphism.

17. Indicate on this illustration where the following rocks would be found: zeolite, eclogite, metaconglomerate, marble, gneiss.

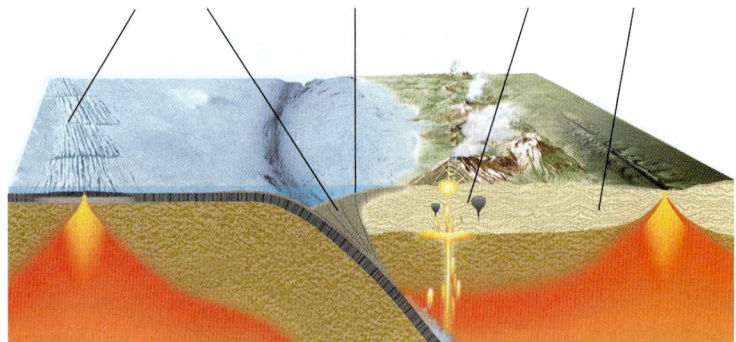

18. Common contact metamorphic rocks include:
    a. Zeolite, hornfels, and shale.
    b. Slate, gneiss, and marble.
    c. Quartzite, marble, and hornfels.
    d. Basalt, granulite, and blueschist.
    e. None of the above.

19. Regional metamorphosis of shale occurs in the following sequence:
    a. Zeolite – gneiss – slate
    b. Slate – phyllite – schist – gneiss
    c. Gneiss – marble – schist – hornfels
    d. Greenschist – slate – hornfels – basalt
    e. None of the above

20. List the metamorphic rocks produced by contact metamorphism of the following protoliths:
    a. Basalt _____
    b. Shale _____
    c. Limestone _____
    d. Quartz sandstone _____

FURTHER RESEARCH

1. Is metamorphism possible without plate tectonics?

2. Why does most marble not display visible foliation?

3. A convergent plate margin can produce both regional and contact metamorphism. Explain how this is possible.

4. Why does rock cleavage develop?

5. How do conditions produced by regional metamorphism differ from conditions produced by contact metamorphism?

6. What are the index minerals, and how do geologists interpret them?

7. Describe the concept of metamorphic facies. What are the five pathways of metamorphism?

8. How does metamorphism in midcontinent settings differ from metamorphism at convergent margins?

ONLINE RESOURCES

Explore more about metamorphic rocks on the following Web sites:

The U. S. Geological Survey site Geology School has links to myriad geological topics, including metamorphic rocks and metamorphism:

www.earth.ox.ac.uk/~davewa/metpet.html

The University of Oxford site compiles links on metamorphism:

www.earth.ox.ac.uk/~davewa/metpet.html

Additional animations, videos, and other online resources are available at this book's companion Web site:

www.wiley.com/college/fletcher

This companion Web site also has more information about WileyPLUS and other Wiley teaching and learning resources.

MOUNTAIN BUILDING

10-1 Rocks in Earth's crust are bent, stretched, and broken.

> **LO 10-1** Define the types of stress that are present in the crust.

10-2 Strain takes place in three stages: elastic deformation, ductile deformation, and fracture.

> **LO 10-2** Define the three stages of strain.

10-3 Strain in the crust produces joints, faults, and folds.

> **LO 10-3** Compare and contrast joints, faults, and folds.

10-4 Dip-slip and strike-slip faults are the most common types of faults.

> **LO 10-4** Compare and contrast dip-slip faults and strike-slip faults.

10-5 Rock folds are the result of ductile deformation.

> **LO 10-5** Describe simple folds and complex folds.

10-6 Outcrop patterns reveal the structure of the crust.

> **LO 10-6** Describe the use of strike and dip symbols in making a geologic map.

10-7 The San Andreas Fault is a plate boundary.

> **LO 10-7** Describe the boundary of the North American and the Pacific Plates.

10-8 Mountain building may be caused by volcanism, faulting, and folding.

> **LO 10-8** Compare and contrast volcanic mountains, fault-block mountains, and fold-and-thrust mountains.

10-9 Volcanic mountains are formed by volcanic products, not by deformation.

> **LO 10-9** Describe the geology of the Cascade Range.

10-10 Crustal extension formed the Basin and Range Province.

> **LO 10-10** Describe the origin of the Basin and Range Province.

10-11 Fold-and-thrust belts are the highest and most structurally complex mountain belts.

> **LO 10-11** Describe the origin of fold-and-thrust belts.

CHAPTER CONTENTS & LEARNING OBJECTIVES

GEOLOGY IN OUR LIVES

Stress (pressure) in Earth's crust causes rock to fold and break. This activates earthquakes and landslides, pushes up mountains, and challenges our ability to find and extract geologic resources. Past and present plate boundaries are regions where stress tends to be the greatest and where the crust has the most complicated structure. By understanding how stress affects the crust, we can improve our access to oil and coal, groundwater, metallic ores, and other geologic resources on which we depend. And by studying the structure of the crust, we can advance not only our knowledge of geologic hazards but of Earth's history, as well.

Q **What kind of plate boundary produces mountains like these?**

10-1 Rocks in Earth's Crust Are Bent, Stretched, and Broken

LO 10-1 Define the types of stress that are present in the crust.

You might be surprised to learn that the solid rock of Earth's crust may experience bending, pulling, and fracturing. Geologists refer to this process as **deformation** (**FIGURE 10.1**), and the forces that cause deformation, **stresses**.

Rocks Are Deformed by Stress

We first explored stress in Chapter 9, where we defined *lithostatic stress* (section 9–2) as uniform pressure acting on a rock simultaneously and equally from all directions. In many cases, rock may experience unequal stress due to tectonic forces. This *directed stress* is force acting on a rock to deform it in a particular direction. There are three basic kinds: **tensional stress** (stretching), **compressional stress** (squeezing), and **shear stress** (side-to-side shearing).

These stresses generally originate at plate boundaries and lead to faulting and folding of the crust—hence, to orogeny (mountain building). This chapter explores these subjects in more detail.

Stress is defined as "force applied over an area." All of the crust, oceanic and continental, is exposed to a complex set of stresses often related to plate tectonics; that means the crust at plate boundaries is usually highly deformed because plate interactions generate extreme stresses.

Some of the force at the boundaries can radiate throughout a plate, and may even be present at midplate settings. Former plate boundaries, such as a convergent, continent-continent boundary that is now far from the edge of a plate, may still experience stress related to their tectonic history.

In addition to stresses related to plate interaction, lithostatic stress, associated with the confining weight of rock (*confining stress*), is present throughout the crust. Confining stress increases with depth (**FIGURE 10.2A**).

Types of Differential Stress

Tensional stress (also called *extension* or *pull-apart stress*) is a force that tends to pull rock apart. Put another way, tension will lengthen or stretch the crust (**FIGURE 10.2B**). Crustal lengthening occurs when rock layers are bent—for example, when they are pushed up from below and have to stretch across the greater length. The region at the top of a fold (or the top of a dome) undergoes stretching as a layer changes shape to accommodate the greater length needed to extend around the circumference of a bend.

Crustal extension is found at divergent boundaries where two plates move in opposite directions or pull apart. The plate boundary is pushed upward by magma intrusion and by buoyancy associated with high heat flow through the crust. As new crust moves outward from the rift valley, it is pulled by gravity down the slopes of an oceanic ridge. The combination of all these processes generates tensional stress within the crust and results in fractures and thinning.

Compressional stress is a force that squeezes or shortens rock. Crustal shortening (**FIGURE 10.2C**) is common at convergent plate boundaries where folding, crushing, and fracturing of rock layers can occur. Compression thickens crust by forcing rock into a smaller space. Remember, however, that even though compression causes rock to fold, layers in the region of greatest curvature of a fold undergo tension and thinning as they extend around the bend. That means both compressional and tensional stresses are at work on the crust as it is shortened.

Shear stress (**FIGURE 10.2D**) is force that causes slippage, or "translation" (migration), of rock layers past one another. Crustal slippage produces rock faulting, a displacement or offset across a fracture zone. Shear stress is characteristic of *transform plate*

©Marli Bryant Miller

FIGURE 10.1 Stress in the crust leads to rock deformation. These sedimentary rocks have been folded (bent) and faulted (broken, and the two sides displaced relative to one another).

Q Describe the directions of stress that caused the folding and faulting in this photo.

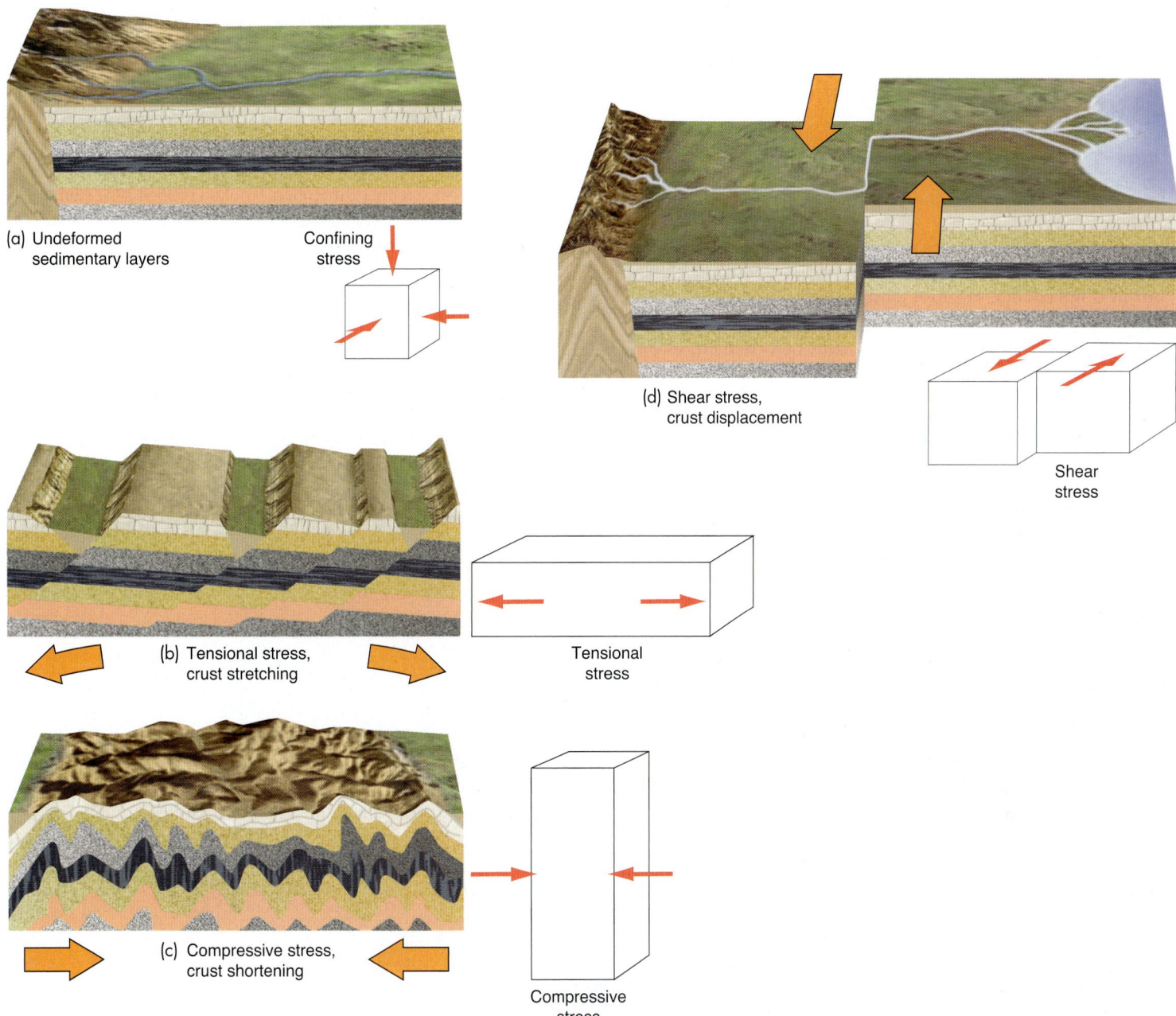

(a) Undeformed sedimentary layers

Confining stress

(d) Shear stress, crust displacement

Shear stress

(b) Tensional stress, crust stretching

Tensional stress

(c) Compressive stress, crust shortening

Compressive stress

FIGURE 10.2 (a) Confining stress increases with depth. (b) Tensional stress leads to lengthening or stretching (thinning) of the crust. (c) Compressional stress leads to shortening or squeezing (thickening) of the crust. (d) Shear stress leads to horizontal displacement.

Q *What kind of stress makes rock fold? What kind can make it fault?*

boundaries that move in opposite directions. As two plates grind past each other, shear stresses at the boundary fracture and crush the crust, creating broad fault zones like the San Andreas Fault Zone in California.

Q EXPAND YOUR THINKING

Stress deforms rock. How might rock react to stress that is applied very quickly compared to stress that is applied very slowly?

10-2 Strain Takes Place in Three Stages: Elastic Deformation, Ductile Deformation, and Fracture

LO 10-2 Define the three stages of strain.

When rocks are deformed, they are said to undergo **strain**. Strain is the change in shape and/or volume of a rock caused by stress. When rocks undergo strain, they pass through three successive stages of deformation: *elastic deformation*, *ductile deformation*, and *fracture*.

Elastic Deformation

Elastic deformation is fully reversible, meaning that the deformation is not permanent. Most rocks are elastically deformed when they first experience stress, which causes the atomic bonds within and between minerals to stretch, flex, and bend. When the stress is removed, the bonds snap back to their original form and the rock regains its original shape.

For instance, as a glacier grows on bedrock at the beginning of an ice age, the rock slowly yields to the growing weight of the ice over thousands of years and elastically deforms downward. When climate changes and the ice age ends, the crust rebounds elastically as the glacier retreats and the ice melts. This is called *glacioisostatic rebound.*

In most types of material, elastic behavior occurs under small amounts of stress, usually when the stress is first applied and before it has a chance to build up to significant levels. But as stress increases and the atomic bonds are further deformed, rocks (and other materials) approach their *elastic limit,* the point beyond which deformation becomes permanent; it is not reversible.

A spring bearing a weight is a good example of elastic behavior. As stress is applied to the spring by adding more and more weight, the increase in strain is displayed by the stretching of the spring. The spring continues to display elastic strain only as long as it fully recovers its shape when the weight is removed. As more weight is added, however, eventually the elastic limit of the spring will be exceeded, at which point, when the weight is removed, the spring will not fully recover its original shape.

The British scientist Sir Robert Hooke (1635–1703) studied various materials under states of stress. He showed that a graph, or plot, of stress (units of pressure) versus strain (units of deformation) yields a straight line as long as the elastic limit is not exceeded. Although this relationship is more easily demonstrated using a spring, Hooke's law holds true for rocks as well (**FIGURE 10.3**).

Ductile Deformation

Ductile deformation, also known as *plastic deformation*, occurs when a material accumulates strain that is not reversible. Plastic deformation takes the form of a permanent change in the size and/or shape of a rock that has passed its elastic limit. As further stress is applied, the deformation (strain) becomes increasingly pronounced. Most materials will stop exhibiting ductile deformation at some point and begin to fracture. Rocks exhibit ductility by stretching somewhat as they are folded (**FIGURE 10.4**).

Fracture

Fracture occurs when a material that is accumulating strain, or deforming, finally breaks. This stage is termed *brittle deformation.* Like ductile strain, brittle deformation is permanent. Once fracture has occurred, further stress widens the fracture but does not cause any additional change in the volume and/or shape of the rock. So, at the point of fracture, strain no longer grows as stress is applied; the fracture simply experiences greater displacement.

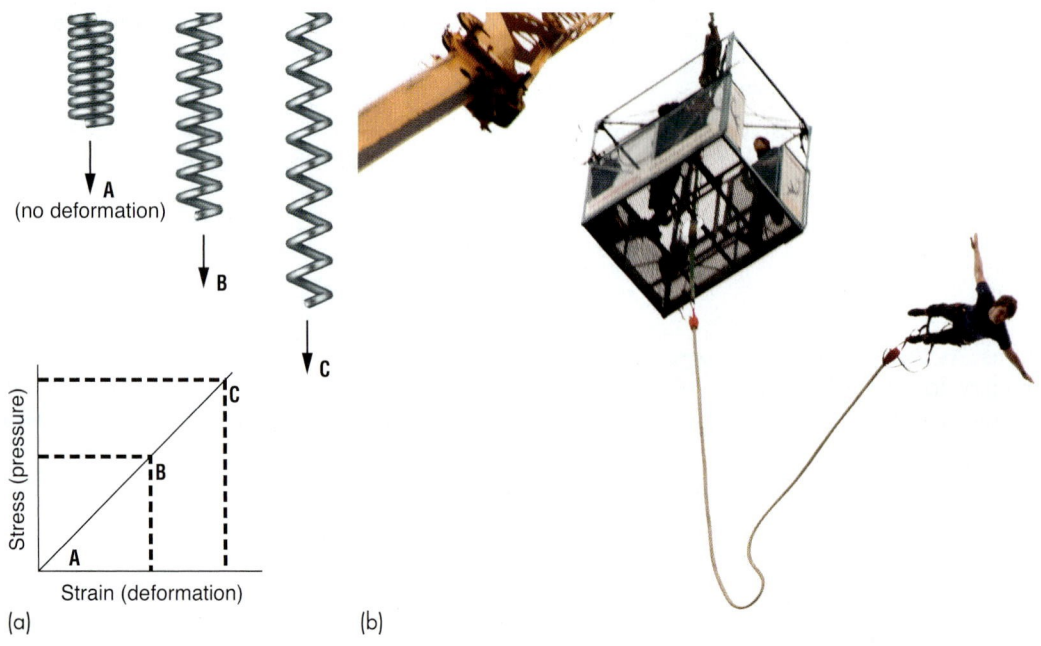

A (no deformation)

B

C

Stress (pressure)

Strain (deformation)

(a)

(b)

FIGURE 10.3 Hooke's law demonstrates that in elastic behavior, strain is proportional to stress. (a) The law predicts that as weights are added to a spring, the stress (weight) and strain (change in length) can be plotted as a straight line. (b) Hooke's law also makes it possible to calculate the progress made by a bungee jumper on an elastic line, as shown here.

Q Can you describe another example of elastic deformation?

FIGURE 10.4 Ductile strain in this metamorphosed gneiss in Norway is displayed by the many small folds developed in response to compressive stress.

Describe an example of ductile deformation in some common materials.

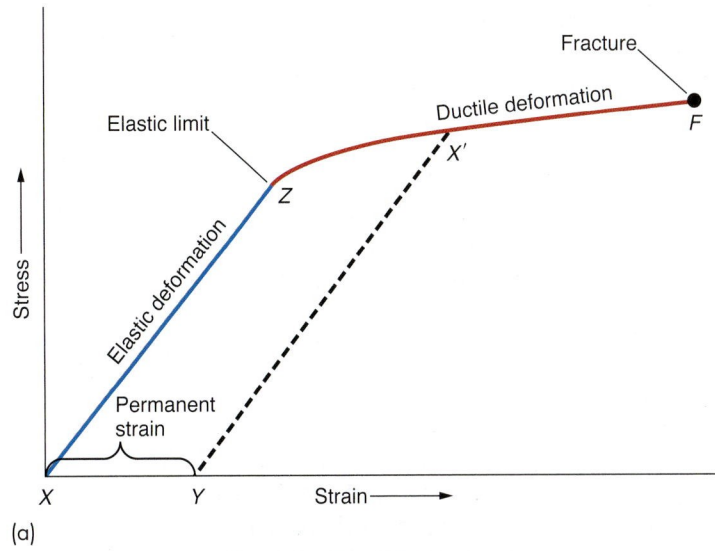

(a)

FIGURE 10.5 plots the rate of strain accumulation as stress is applied to a rock sample in a laboratory. Note that, at first, the rock responds elastically, showing linear strain accumulation (consistent with Hooke's law); but when the rock passes its elastic limit, it undergoes ductile deformation. Finally, as stress builds, the rock fractures. In the crust, many such fractures develop into **faults**—zones or planes of breakage across which layers of rock are displaced relative to one another.

Factors That Govern Strain

The type of strain experienced by a rock depends on several factors: temperature, confining pressure, strain rate, rock composition, and water. At high temperatures, rock exhibits greater *ductility* because molecules, and their bonds, stretch and move more readily at high temperatures than at lower temperatures. Cold rock tends to be brittle, and fractures more easily. Confining pressure has the same effect. At higher confining pressure—that is, under greater lithostatic stress—rock tends to be more ductile because the pressure of the surrounding environment hinders the formation of fractures.

The rate at which strain accumulates can also influence the behavior of rocks. When stress is applied rapidly, as with a sudden hammer blow, rocks and other solids tend to fracture more readily. When stress is applied slowly and strain accumulates slowly, atomic and molecular bonds have a chance to accommodate the stress by deforming.

The composition of a rock influences the type of strain it undergoes. Some minerals, such as quartz, olivine, and feldspars, are very brittle. Others, such as micas, clays, calcite, and gypsum, are more ductile. This difference is due largely to the types of atomic bonds holding the minerals together. The mineralogy of a rock is, thus, an important factor influencing the type of strain it undergoes.

(b)

FIGURE 10.5 (a) A typical sample of rock tested in a laboratory will exhibit a straight-line relationship between stress and strain (X to Z) when deforming elastically, as predicted by Hooke's law. At the elastic limit (Z) the rock experiences ductile deformation until the stress is removed at X'. At this point, the rock returns to an unstressed state along line X'Y but retains some permanent deformation (strain amount XY). If the rock continues to experience stress past point X', it will eventually fracture (point F). (b) A fault is formed when rock fracture leads to displacement across the fracture surface.

In Figure 10.5b does the fault reveal crustal shortening or crustal extension? What type of stress caused this?

Last, the presence or absence of water can influence how a rock is deformed. It is thought that water weakens bonds and may form a film around mineral grains. These properties allow slippage to take place, encouraging wet rocks to behave in a ductile manner, while dry rocks tend to be brittle.

EXPAND YOUR THINKING

With your finger push down on your forearm and then remove your finger. What kind of strain did your arm exhibit?

10-3 Strain in the Crust Produces Joints, Faults, and Folds

LO 10-3 Compare and contrast joints, faults, and folds.

Stress applied to the crust will cause rocks to deform. In most cases, rocks undergo ductile strain (folds) or fracture (joints and faults).

Joints and Faults

Joints are fractures in Earth's crust that display no net displacement between the two sides of the crack. Many joints are formed when confining stress is removed and rock expands. Other joints are a result of stress spread across a broad region within the crust, as when brittle rock attempts to conform to the curvature of Earth's surface.

Joints are common in the crust and can develop in all types of rock. Since most rock at Earth's surface is relatively cold, it is brittle and tends to fracture when stressed or, conversely, when stress is removed and the rock expands. As with any brittle material, rock joints will develop at any point of weakness, such as around a fossil, between minerals, or along bedding planes.

Joints can appear under a number of conditions. Igneous rock shrinks as it cools and develops sets of intersecting joints called *columnar jointing* that reflect the decrease in the rock's volume. When a large weight is removed from the crust, such as when a glacier melts, the confining pressure is reduced. The crust expands upward, and fractures that follow flaws in the rock take shape.

Joints emerge along the edge of cliffs, or at **outcrops** (places where layers of rock are exposed), because at such locations the crust is not confined and tends to expand outward. Joints can also develop as a result of tectonic stresses that travel far from plate boundaries and provoke fracturing at locations hundreds of kilometers away.

Joints enhance other geologic processes in important ways. Jointed rocks serve as easy routes for groundwater, enabling it to move along fracture planes in otherwise impermeable crust. Chemically active fluids that emanate from cooling magma often travel along fractures in the country rock and deposit hydrothermal veins of ore minerals.

Joints are attacked by acidic fluids and subjected to enhanced dissolution, hydrolysis, and other types of chemical weathering, as described in Chapter 7. Freeze-thaw processes lever joints open and

(a) *Martin Bond/Science Source*

(b) *David Parker/Photo Researchers*

(c) *©Marli Bryant Miller*

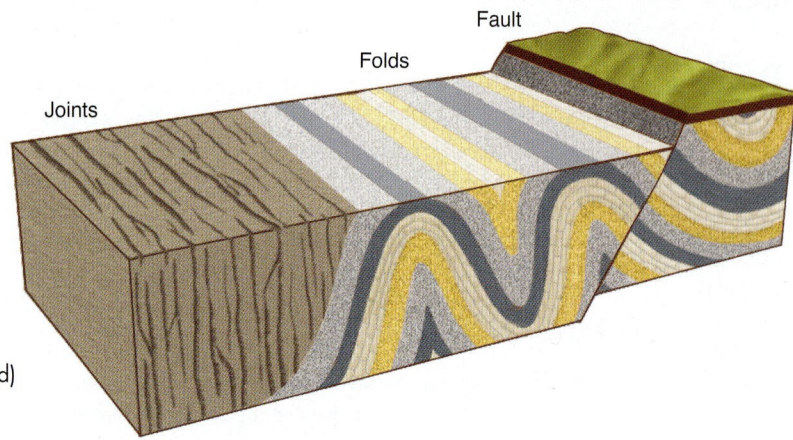
(d)

FIGURE 10.6 (a) Joints, (b) folds, and (c) faults are common geologic structures. (d) At times, joints, folds, and faults may occur in the same vicinity.

Q How does a geologist determine whether a fracture is a joint or a fault?

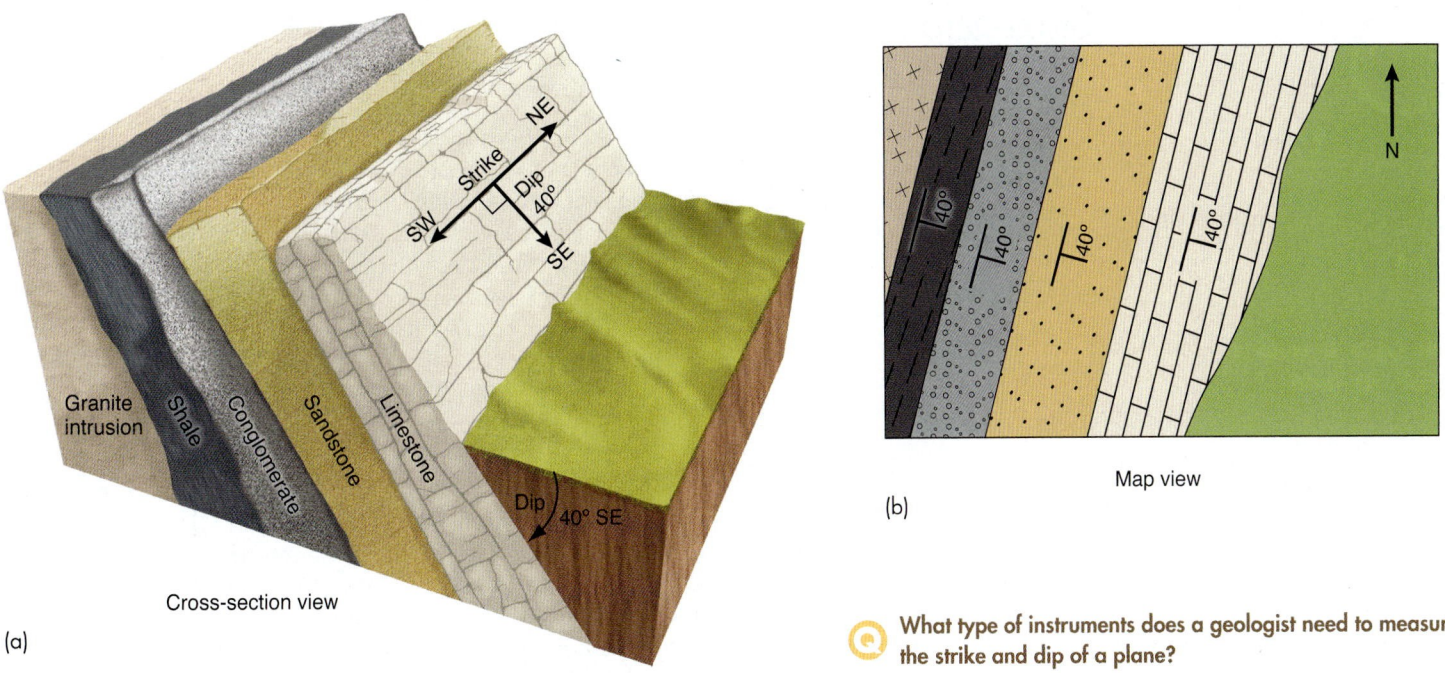

(a)

Cross-section view

(b)

Map view

FIGURE 10.7 (a) The line of strike (denoted by a compass bearing) marks the orientation of the trend of a geologic plane, such as a fault surface or (as shown) a rock layer. The line of dip (always oriented 90° to the strike) is also marked in proper compass orientation, along with the angle of inclination of the geologic plane. (b) Strike and dip symbols are used on a map to represent the spatial orientation of geologic features such as rock strata, faults, joints, and others. Because the top of a map always indicates north, the orientation of the strike and dip symbols represent the real-life orientation of the geologic feature.

take shape as natural arches; these processes also cause cliffs to collapse and talus to form.

Faults emerge when brittle rocks fracture and there is a distinct offset along the plane of breakage. In some cases, the offset is minor and the displacement is easily measured. In other cases, the displacement is large, making it difficult to obtain an accurate measure of the degree of displacement. When strain accumulates slowly and prevailing conditions, such as high temperature and pressure or the presence of water, allow rocks to experience ductile deformation, they may develop into one or many **folds** (**FIGURE 10.6**).

Strike and Dip

Geologists seek to understand the development of stresses within the crust in order to gain a better understanding of tectonic processes and their history. As a first step, they draw maps of geologic structures such as joints and faults. Maps are two-dimensional representations of three-dimensional surfaces oriented to Earth's coordinate system (north, south, east, and west). On a proper map, north is always at the top. To show crustal structure on a map, geologists have devised a simple symbol to display the spatial orientation of rock layers that make up folds, faults, and joints. This symbol has two parts: the **strike** and the **dip**.

Imagine the surface of a tilted rock layer, or a fault fracture, as a plane. A plane is two-dimensional and so can be defined in space using an x-axis and a y-axis oriented in terms of compass heading

(x-axis) and angle to the horizon (y-axis). Geologists call the x-axis the strike of the plane and the y-axis the dip of the plane, referring to the orientation of a geologic feature (**FIGURE 10.7**).

The strike of a planar body of rock, or stratum, is a line representing the intersection of that feature with the plane of the horizon. (Remember from geometry class that a line is formed at the intersection of two planes.) On a map, this is represented with a short straight line oriented to the compass direction of the strike. Strike is given as a *compass bearing* (e.g., N 30° E) or as a single three-digit number representing the *azimuth* (the compass bearing in degrees; e.g., 030°).

The dip gives the angle below the horizontal of a geologic feature. It can be thought of as the direction in which water would flow if rain fell on the fault surface or rock layer. On a map, the dip is a shorter perpendicular line off one side of the strike line. Often, the angle of inclination is written next to the dip line (e.g., 40°). The strike is always measured perpendicular to the dip. (We discuss the strike and dip symbols again a little later in the chapter, and explain how they are used to interpret various types of folds and faults.)

Describe the stress and strain history of Figure 10.6d. Note that it is not simple and involves at least two stages.

10-4 Dip-Slip and Strike-Slip Faults Are the Most Common Types of Faults

LO 10-4 Compare and contrast dip-slip faults and strike-slip faults.

Two types of faults are the most common: dip-slip faults and strike-slip faults.

Dip-Slip Faults

Dip-slip faults appear at locations where a fault plane is inclined, or at an angle to the horizon, and one block of crust is displaced up or down along the plane relative to another. In dip-slip faults, the displacement occurs up and down along the fault-dip surface—hence, the term *dip-slip*.

The main types of dip-slip faults are **normal faults** and **reverse faults;** in addition, there is a special type of reverse fault, known as a **thrust fault**. For any inclined fault, the block that hangs above the fault plane is the *hanging wall block,* and the block below it is the *footwall block* (**FIGURE 10.8**).

The terms used to name faults originated with miners, who found that faulted rock was easy to excavate and that fault surfaces often were filled with valuable minerals precipitated from fluids that migrate along the fault plane.

Miners named the ceiling of their tunnels the "hanging wall block," and the floor of a mineshaft the "footwall block." Normal faults and reverse faults are defined by the relative displacement of the hanging wall block to the footwall block:

1. In normal faults, the hanging wall block moves down relative to the footwall block;
2. In a reverse fault, the hanging wall block moves up.

FIGURE 10.9 displays the major types of faults. Normal faults develop in brittle rocks that are exposed to tensional stress. In the case of a normal fault, an inclined fracture emerges as the crust is pulled apart and the hanging wall block slips downward along the inclined fault plane relative to the footwall block. Such faults appear at locations where crustal stretching or lengthening is occurring. Rift valleys, for example, are places where crustal extension occurs and normal faults are common.

Reverse faults (**FIGURE 10.9B**) are caused by compressive stress; the hanging wall block is displaced upward along the fault plane relative to the footwall block. (Read about the force of a reverse fault in the Geology in Action feature, "The Sichuan Quake of 2008," where in China, 9 meters of reverse faulting on the Longmenshan Fault resulted in catastrophic damage.)

A thrust fault is a special case of a reverse fault in which the dip of the inclined fault plane is less than 15°. Thrust faults develop when stresses are distributed over a wide geographic region. Consequently, thrust faults can have considerable displacement, measuring hundreds of kilometers, and may result in the formation of major topographic features composed of rock layers in which older strata overlie younger ones.

Strike-Slip Faults

Strike-slip faults occur at locations where the motion on a fault plane is horizontal (**FIGURE 10.9D, 10.9E**). Two blocks of crust travel (slip) along the strike of a fault rather than along the dip. Shear stress acting on the crust causes strike-slip faults.

Strike-slip faults take two forms, depending on the relative displacement across the fault. To an observer standing on one side of the fault and looking across to the far side, where a block is relatively displaced and the block on the other side has moved to the left, the fault is a *left lateral strike-slip fault.* If the block has moved to the right, the fault is a *right lateral strike-slip fault.*

Among strike-slip faults are some of the world's most dangerous fault systems, such as the North Anatolian Fault in Turkey, the San Andreas Fault in California, and the Enriquillo–Plantain Garden Fault in Haiti. All three are renowned for setting off devastating earthquakes suddenly and with little warning when there is movement along the strike of these faults (**FIGURE 10.10**).

- In Turkey, right lateral motion along the North Anatolian Fault System generated a 45-second earthquake at 3:01 A.M. on August 17, 1999. The epicenter was southeast of Izmit, an industrial city near Istanbul. The earthquake was felt over a large area, as far east as Ankara, about 320 kilometers away. Unofficial estimates placed the death toll at between 30,000 and 40,000. Most of the deaths and injuries were caused by the collapse of commercial and residential buildings, typically four to eight stories high.

- At 5:00 P.M. on October 17, 1989, the San Andreas Fault in northern California suddenly slipped. The resulting quake was responsible for 62 deaths, 3757 injuries, and over $6 billion in damage. More than 18,000 homes and 2600 businesses were damaged, and about 3000 people were left homeless.

- On January 12, 2010, the Enriquillo–Plantain Garden Fault in Haiti broke open after 250 years of quietly building stress. The rupture was roughly 65 kilometers long with an average slip of

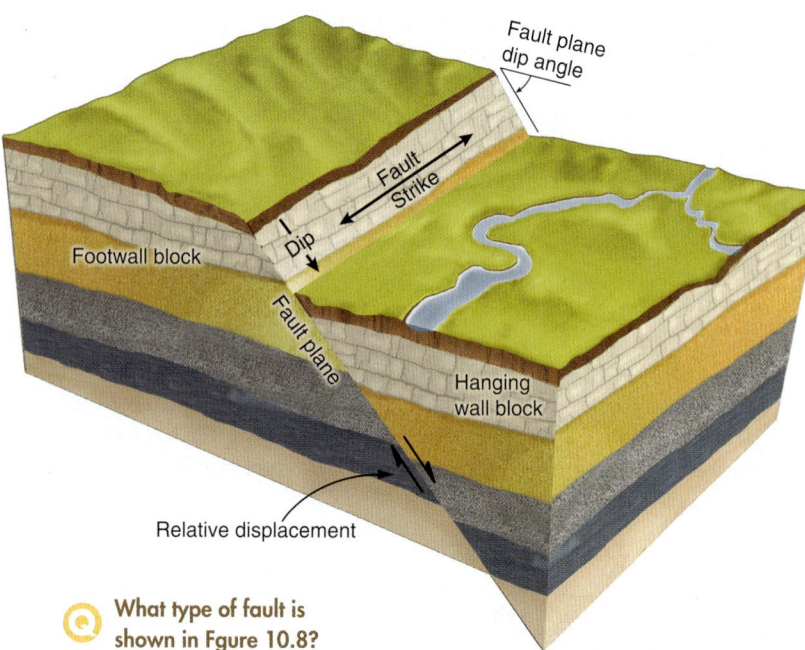

Q What type of fault is shown in Fgure 10.8?

FIGURE 10.8 The footwall block lies below the fault plane; the hanging wall block lies above it. The strike of a fault plane is perpendicular to the dip of the fault plane.

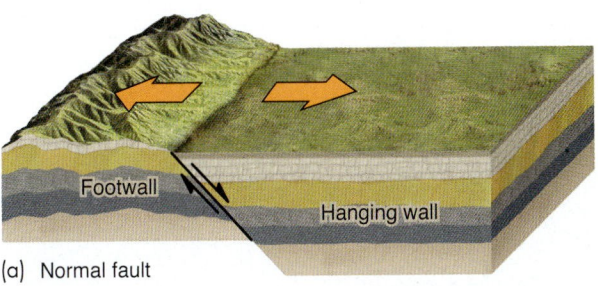

(a) Normal fault

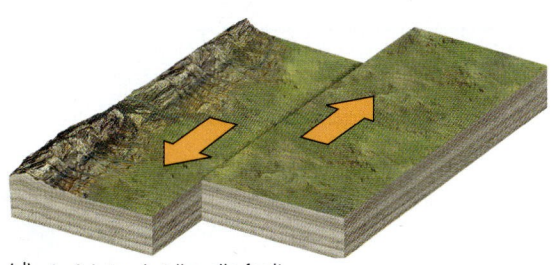

(d) Left lateral strike-slip fault

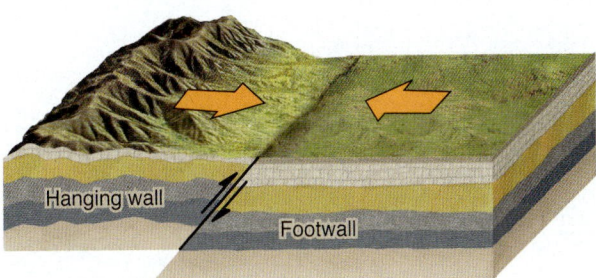

(b) Reverse fault

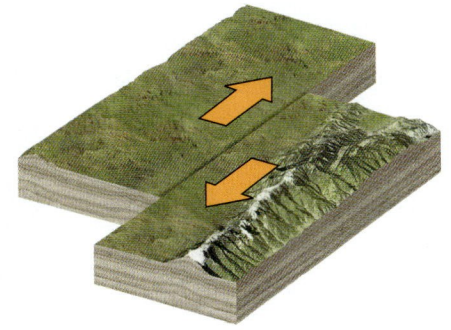

(e) Right lateral strike-slip fault

In Chapter 3 we described several types of plate boundaries. What type of faulting would be found at each?

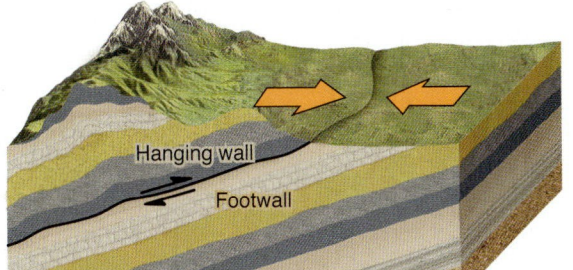

(c) Thrust fault

FIGURE 10.9 (a) Tensional stress leads to crustal lengthening and causes brittle rock to fracture in normal faults. (b) A reverse fault is formed under compressive stress, when the hanging wall moves up the inclined fault plane relative to the footwall. (c) A thrust fault develops when the angle of a reverse fault plane is 15° or less. Hence, a thrust fault is a low-angle reverse fault and may result in older rock layers lying above younger layers. (d) Shear stress in the crust will cause displacement that is horizontal rather than vertical. If the block on the far side of a fault is displaced to the left, it is a left lateral strike-slip fault. (e) If the block is displaced to the right, it is a right lateral strike-slip fault.

1.8 meters. By January 24, the devastated nation had endured 52 aftershocks, each of sufficient strength to cause extensive damage.

After all three of these quakes, teams of geologists rushed to the scenes to assist in rescue operations and study the evidence of fault movement, in hopes of reducing the damage from future quakes.

Evidence of Faulting

Finding a fault can be difficult for geologists. Evidence of faulting can be obvious or subtle, depending on the extent of fault movement, the amount of time that has passed since the last movement, and the nature of the ground surface where fault movement last occurred.

(a) Yannis Kontos/Sygma/Corbis

(b) © Roger Ressmeyer/Corbis

(c) ©AP/Wide World Photos

FIGURE 10.10 (a) A collapsed seven-story building in Turkey in 1999. (b) Buildings destroyed by the San Andreas Fault quake of 1989. (c) Devastation in Port-au-Prince, Haiti, where more than 200,000 people died in January, 2010.

Building collapse is a major hazard during an earthquake. Propose a hypothesis that relates ground motion to how a building fails. How would you test your hypothesis?

CRITICAL THINKING

Inferring the History of Broken Rock

Moab, Utah is an arid region where the local geology is well exposed. FIGURE 10.11 shows one of many outcrops, known as the "Moab Fault", that reveal the structure of the crust. Study the photo carefully before answering the following questions:

1. What type of stress caused this set of faults?

2. What is your evidence for identifying the movement of the two sides?

3. Describe the geological history of this place—what geologic event happened first, second, and so on?

4. Infer from your analysis a plausible tectonic history for the crust here. Describe this history.

5. What additional observations should you collect to test your hypothesis?

FIGURE 10.11 Moab Fault, Utah.

© SweetMango/iStockphoto

Weathering and sediment deposition at the surface also tend to obscure the features that develop when rocks rupture and slide past each other. Nevertheless, geologists have learned to look for three types of features that offer clear evidence of faulting:

- *Slickensides:* These are parallel grooves in rock that result from frictional sliding of one block past another. They typically appear on a flat, polished-looking surface that represents the fault plane; the orientation of the grooves indicates the direction of the most recent movement.

- *Fault breccia:* This is made up of loose, coarse fragments of rock that are formed in the fault zone by grinding and crushing due to fault movement. Fault breccia occurs on either or both sides of the fault plane, or may form a zone that obscures the fault plane entirely.

- *Fault gouge:* This is very fine-grained, pulverized rock that emerges from the grinding and crushing along the fault plane.

Evidence of faulting may also be apparent where two geologic rock units are joined, units that would otherwise not be in contact, such as two sedimentary rock formations of different ages, or intrusive rock in sharp contact with country rock lacking a hornfels or baked zone in between. Geologists also examine topographic maps and aerial photographs for linear features on the surface, such as *fault scarps*, cliffs or escarpments formed by a fault (FIGURE 10.12).

FIGURE 10.12 This fault scarp was formed by compressive stress in northeastern Tibet.

Q What type of faulting caused this scarp?

Q EXPAND YOUR THINKING

Imagine that the wall directly in front of you is lined up perfectly with the compass so that the left side points due west and the right side points due east. Now, in your mind, tilt the wall away from you 45°. Imagine the room filling halfway up with water, and that where the water level hits the tilted wall is a line of strike. Draw the strike and dip of the wall on a sheet of paper oriented to be a proper map. (Recall that a proper map depicts North to the top of the paper.)

GEOLOGY IN ACTION

The Sichuan Quake of 2008

On Monday, May 12, 2008, the most damaging earthquake to hit China since 1975 shook the ground violently for about two or three minutes, according to witnesses (**FIGURE 10.13**). The earthquake was generated by reverse faulting along the Longmenshan Fault, located in south-central China, in the province of Sichuan. Measurements indicate that the hanging wall shifted approximately 9 meters along the fault plane during the quake.

According to Chinese officials, the quake killed 87,000 people and injured another 400,000. Sadly, these figures include 158 earthquake relief workers who were killed in landslides as they tried to repair roads. The earthquake left at least 4.8 million people without housing, and the number could be as high as 11 million.

The seismicity (the relative frequency and distribution of earthquakes) of central and eastern Asia is caused by the northward movement of the Indian–Australian Plate at a rate of 5 centimeters per year as it collides with Eurasia. This results in the uplift of the Himalayan Mountain Belt and Tibetan Plateau and associated earthquake activity. The convergence taking place between these two plates is the reason China frequently suffers large and deadly earthquakes.

FIGURE 10.13 Many people were trapped in collapsed buildings following the Sichuan earthquake of 2008.

Rock Folds Are the Result of Ductile Deformation

LO 10-5 Describe simple folds and complex folds.

Rocks that have exceeded their elastic limit and behave in a ductile manner are likely to bend or contort; the resulting geologic structures are called *folds*.

Folds

Folds are the result of compressional stresses acting over long periods. But the *rate of strain accumulation* is low, so rocks that might normally experience brittle or elastic deformation can instead undergo ductile (or *plastic*) deformation and produce a number of different types of folds. Folds are most visible in rocks composed of layers.

For ductile deformation of rock to occur, a number of conditions must be met:

- The rock must have the capability to deform under pressure and heat.

- The temperature of the rock must be high; the higher the temperature, the more plastic it becomes.

- The level of pressure must not exceed the internal strength of the rock. If it does, fracturing will occur.

- Deformation must take place slowly.

Folds are described by their geometry and orientation (**FIGURE 10.14**). The sides of a fold are called *limbs*. The limbs intersect at the tightest part of the fold, called the *hinge*, through which runs the *axis* of the fold. If the axis is not horizontal, the fold is called a *plunging fold*. An imaginary plane that includes the axis and divides the fold as symmetrically as possible is called the *axial plane* of the fold.

Simple Folds

The simplest types of folds are symmetrical and upright. That is, their limbs dip away from their axes at the same angle, and their axial plane is vertical. Two types of simple folds are the **anticline**, defined as a convex upward arch with an axial plane that is vertical and has the oldest rocks in the center of the fold, and the **syncline**, defined as a concave upward arch with an axial plane that is vertical and has the youngest rocks in the center of the fold (**FIGURE 10.15**).

An upright, symmetrical anticline is a fold in which originally horizontal layers have been bent to shape an arch. The axial plane is vertical, and the two limbs of the anticline dip away from the center

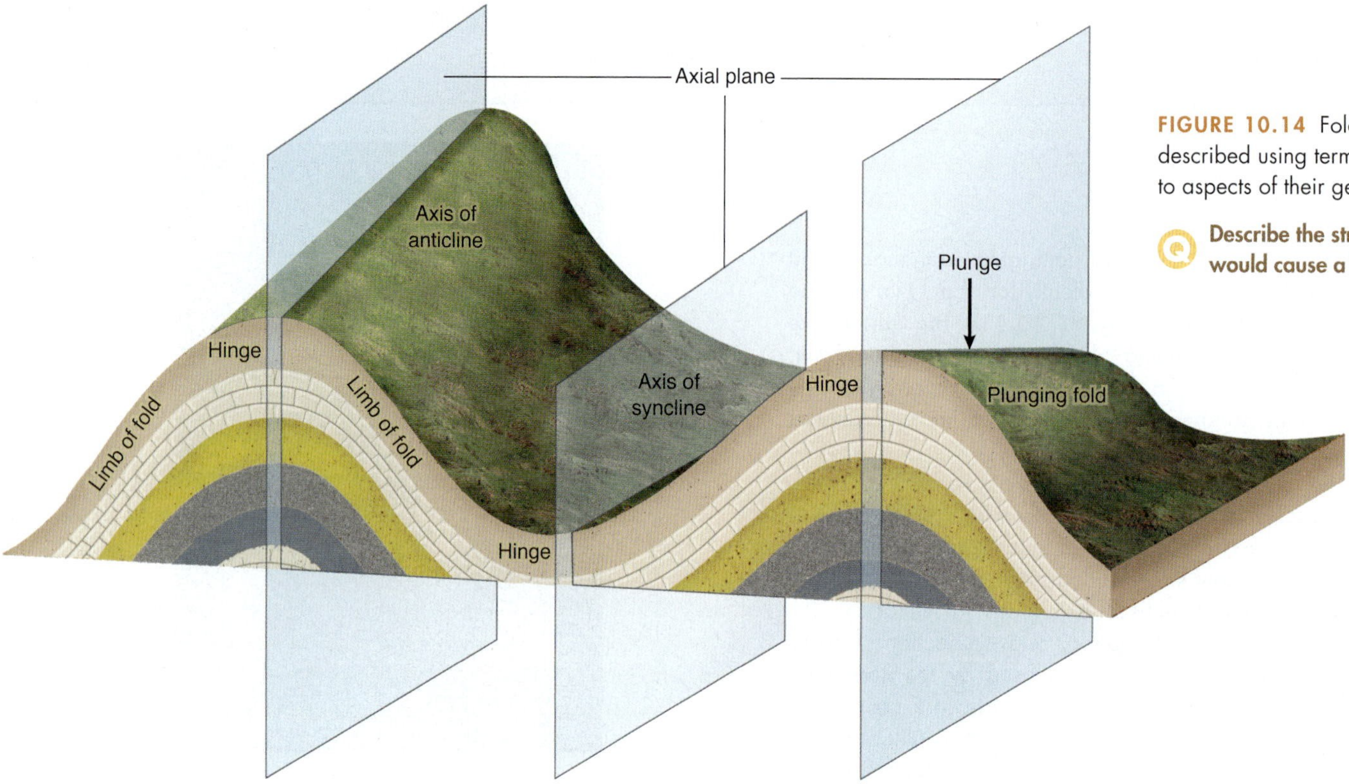

Axial plane

Axis of anticline

Hinge

Limb of fold

Limb of fold

Hinge

Axis of syncline

Hinge

Plunge

Hinge

Plunging fold

FIGURE 10.14 Folds are described using terms that refer to aspects of their geometry.

Q Describe the stresses that would cause a fold to plunge.

of the arch at the same angle. The inverse of an anticline is an upright, symmetrical syncline. A syncline is a concave upward bend in originally parallel layers of rock.

Another type of simple fold is the **monocline**, marked by a slight bend in parallel layers of rock. Monoclines develop at locations where horizontal strata are bent upward by a buried fault that does not break the surface, leaving the two limbs horizontal. The axial plane of the fold is tilted at an angle to the vertical. Note that the occurrence of a fault is a sign of brittle rock behavior, whereas the occurrence of folded layers is a sign of ductile behavior.

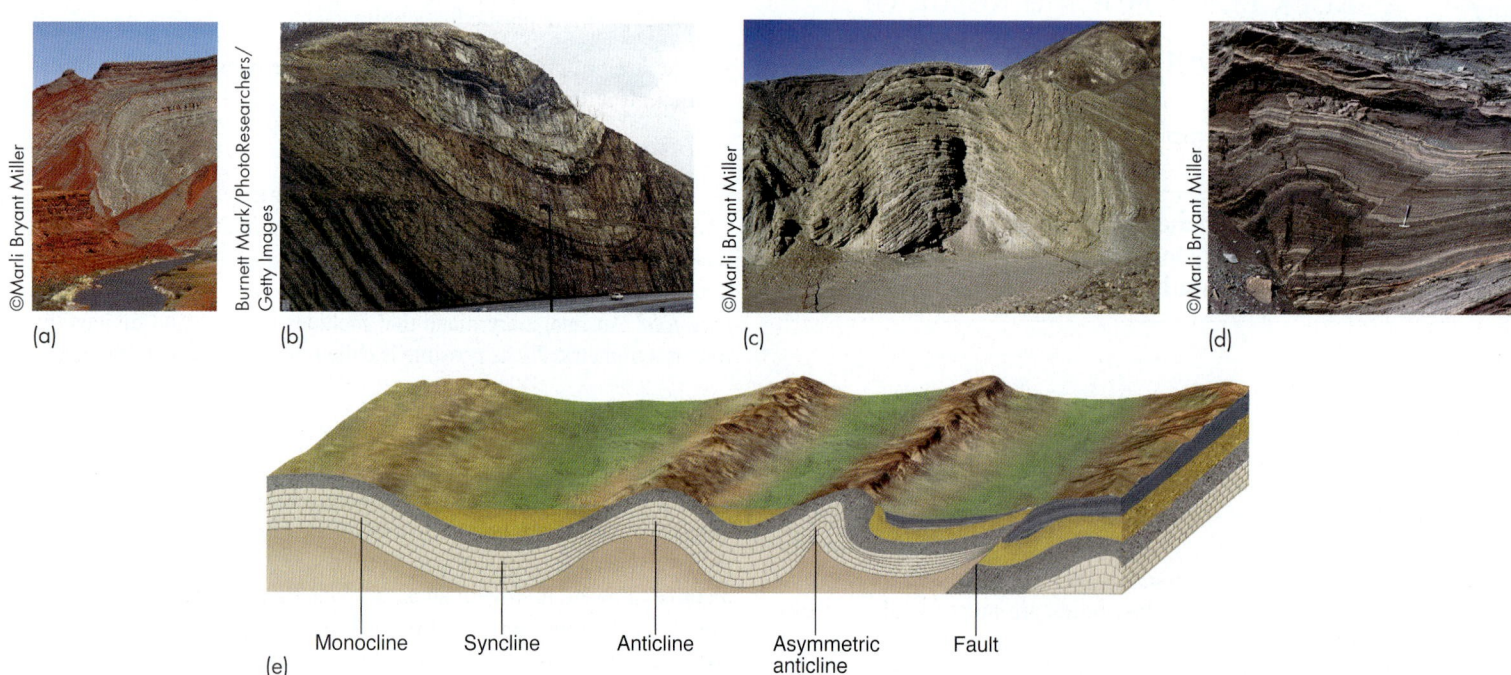

©Marli Bryant Miller

(a)

Burnett Mark/PhotoResearchers/ Getty Images

(b)

©Marli Bryant Miller

(c)

©Marli Bryant Miller

(d)

Monocline   Syncline   Anticline   Asymmetric anticline   Fault

(e)

FIGURE 10.15 (a) A monocline is a simple fold in which both limbs remain parallel to the horizon. (b) A syncline is a concave upward arch with an axial plane that is vertical. (c) An anticline is a convex upward arch with an axial plane that is vertical. An asymmetrical fold (see diagram) develops when one limb is steeper than the other. An overturned fold is formed when one limb is tilted beyond the vertical. (d) Some folds may be compressed so severely that the rock fractures, resulting in a faulted fold.

Q What types of stress are responsible for the structures in Figure 10.15? Did formation of these structures occur all at once or in stages?

Complex Folds

As stresses in the crust increase in magnitude and complexity, the resulting folds become more complicated (**FIGURE 10.16**). Asymmetrical anticlines and synclines can develop, as well as overturned structures in which one limb is tilted beyond the vertical.

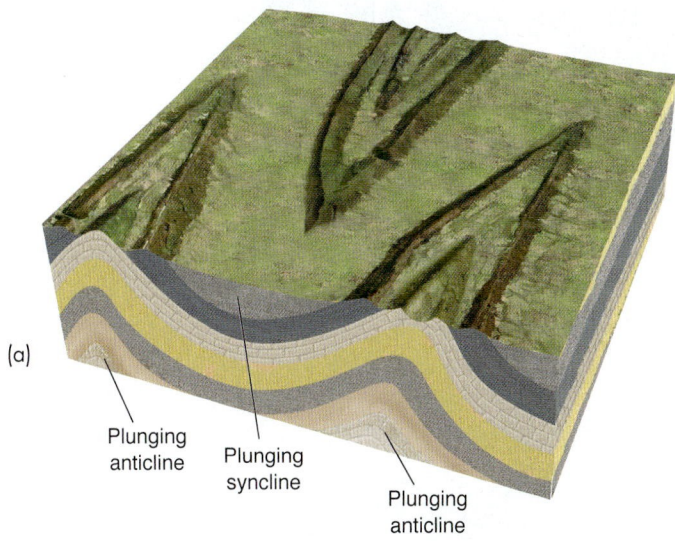

(a)

Plunging anticline

Plunging syncline

Plunging anticline

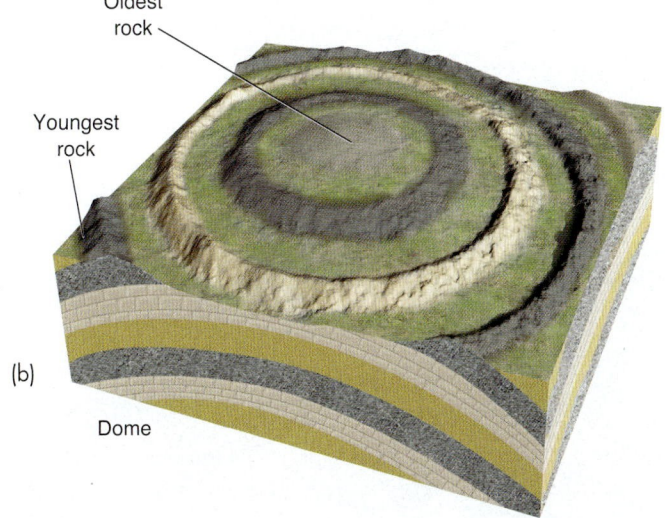

Oldest rock

Youngest rock

(b)

Dome

Plunging anticlines and *plunging synclines* have axes that are not horizontal and that intersect the surface. Groups of anticlines and synclines, known as *S folds*, can appear in ductile strata, with some folds being overturned and others remaining upright. Other types of complex folds include *domes*, circular or slightly elongated upward displacements of rock layers, and *basins*, circular or slightly elongated downward displacements of rock.

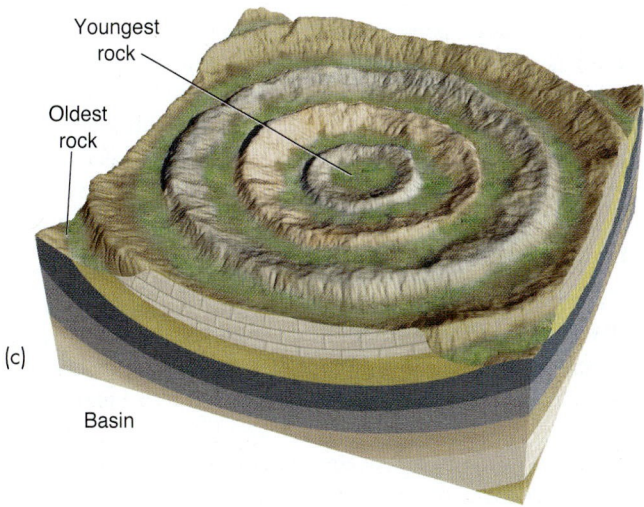

Youngest rock

Oldest rock

(c)

Basin

FIGURE 10.16 (a) Plunging folds have axes that are not horizontal. (b) A dome is a circular upwardly displaced fold. (c) A basin is a circular downwardly displaced fold.

Q **What geologic processes would make a dome or a basin?**

Q **EXPAND YOUR THINKING**

You have been asked to fold a granite kitchen countertop into the shape of an anticline. You have the technology you need at your disposal. How will you complete the task?

10-6 Outcrop Patterns Reveal the Structure of the Crust

LO 10-6 Describe the use of strike and dip symbols in making a geologic map.

It is one thing to discuss folds while observing their full three-dimensional geometry on a sheet of paper. It is quite another to piece together the nature of folded crust "in the field," without knowing what lies hidden underground. This is the challenge faced by *structural geologists*.

Structural Geology

Structural geology is concerned with determining the geometry of the crust, and from that inferring the nature of the stresses it has undergone. To do this, geologists must identify the nature of folds and faults in rocks that usually are belowground, and therefore not in full view.

Structural geologists analyze outcrops to establish the spatial orientation of rock layers and the organization (or "structure") of the crust belowground. From this information they infer the relative age of the layers and the stresses they have experienced.

Gravity acts uniformly on sedimentary particles settling through water, so we know that sedimentary rocks are originally formed as horizontal layers. (This is referred to as the *principle of original horizontality*, a concept we discuss more fully in Chapter 12.) When these previously flat-lying strata are tilted, they leave behind a record of their deformation history. Geologists measure the angle and orientation of tilting in order to determine the direction and source of the deforming stresses. More specifically, they determine the strike (trend) and dip (inclination) of the rock.

As we discussed briefly in section 10–3, geologists measure strike and dip by representing a layer of tilted rock as a geometric plane. The orientation of that plane in space can be defined using strike and dip.

- The dip is described by a compass azimuth (0° to 360°) and an angle to the horizon that varies from 0° (horizontal) to 90° (vertical). As mentioned earlier, the dip may be imagined as the direction and angle in which water would flow if rain were to fall on the rock layer.

- The strike is measured perpendicular to the dip. It may be envisioned as the line formed by the intersection of the plane of the rock layer with a horizontal plane. One way to determine a horizontal plane is to imagine flooding the outcrop with water so that an imaginary water level (a horizontal plane) intersects the surface of the tilted rock. The water's edge, formed where the two planes intersect, is the strike, and it is measured using a compass that indicates the orientation of the line relative to north.

Strike and dip are represented by symbols that can be placed on a **geological map**. Geological maps show the topography of an area, the lithology and age of rock units exposed at the surface or buried shallowly under loose sediment and soil, and their strike and dip (**FIGURE 10.17**).

Geological maps display the location of color-coded rock units and plot the measurements of strike and dip at outcrops. In **FIGURE 10.17A**, a geology student interprets the dip of rock layers to determine the geologic structure of the crust. This information is used to make a map (**FIGURE 10.17C**) where strike and dip symbols are plotted on outcrop locations.

In Figure 10.17c, adjacent dip indicators point away from one another to depict an anticline. Where adjacent dip symbols point toward each other, a syncline is depicted. Notice also how the rock units crop out in repeating patterns. Such patterns develop when erosion lowers Earth's surface by removing overlying rock and sediment so that portions of the structures are gone, and other portions, such as the limbs, are exposed to view.

Certain other aspects of outcroppings are useful for identifying the structure of the crust. The pattern of outcropping may reveal the orientation of the underlying structure. Parallel strips of outcrops indicate folds with horizontal axes—that is, the folds do not plunge. An outcrop pattern of Vs (**FIGURE 10.18**) indicates plunging folds. When a V opens in the direction of plunge, it is a syncline; a V that closes in the direction of plunge is an anticline.

If the relative age of rock layers is known, geologists can use this information to interpret the history of Earth's crust. If an outcrop pattern reveals younger rock layers (deposited first) between older rocks (deposited later), a syncline is present. If the pattern consists of older rock layers between younger layers, an anticline is present. Likewise, in the case of a V-outcrop pattern (a plunging fold), older rocks will be found in the middle of the V if an anticline is present, and younger rocks in the middle of the V if a syncline is present.

(a)

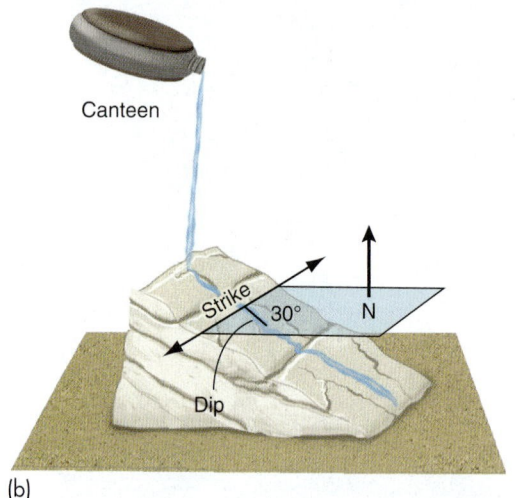
(b)

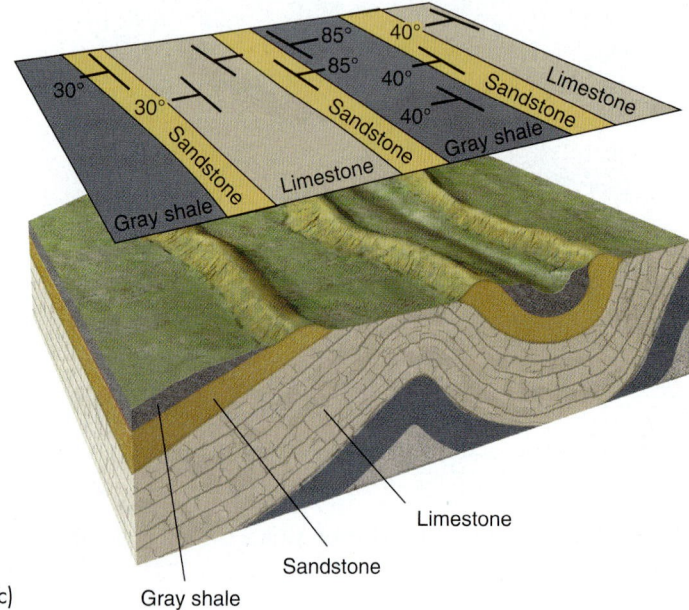
(c)

FIGURE 10.17 (a) These geology students are measuring the orientation of rock layers at an outcrop. (b) Strike is the compass bearing of a plane representing a rock layer at an outcrop. In this example, if north is toward the back of the outcrop (note north arrow), the strike is approximately northeast-southwest and the dip is 30° to the southeast. Dip is the direction in which water will flow across the outcrop. (c) A geological map shows the lithology, age, and strike and dip of rock units.

Q What types of folds are shown in Figure 10.17c?

Q **EXPAND YOUR THINKING**

What characteristics of maps make them such important and effective tools in structural geology?

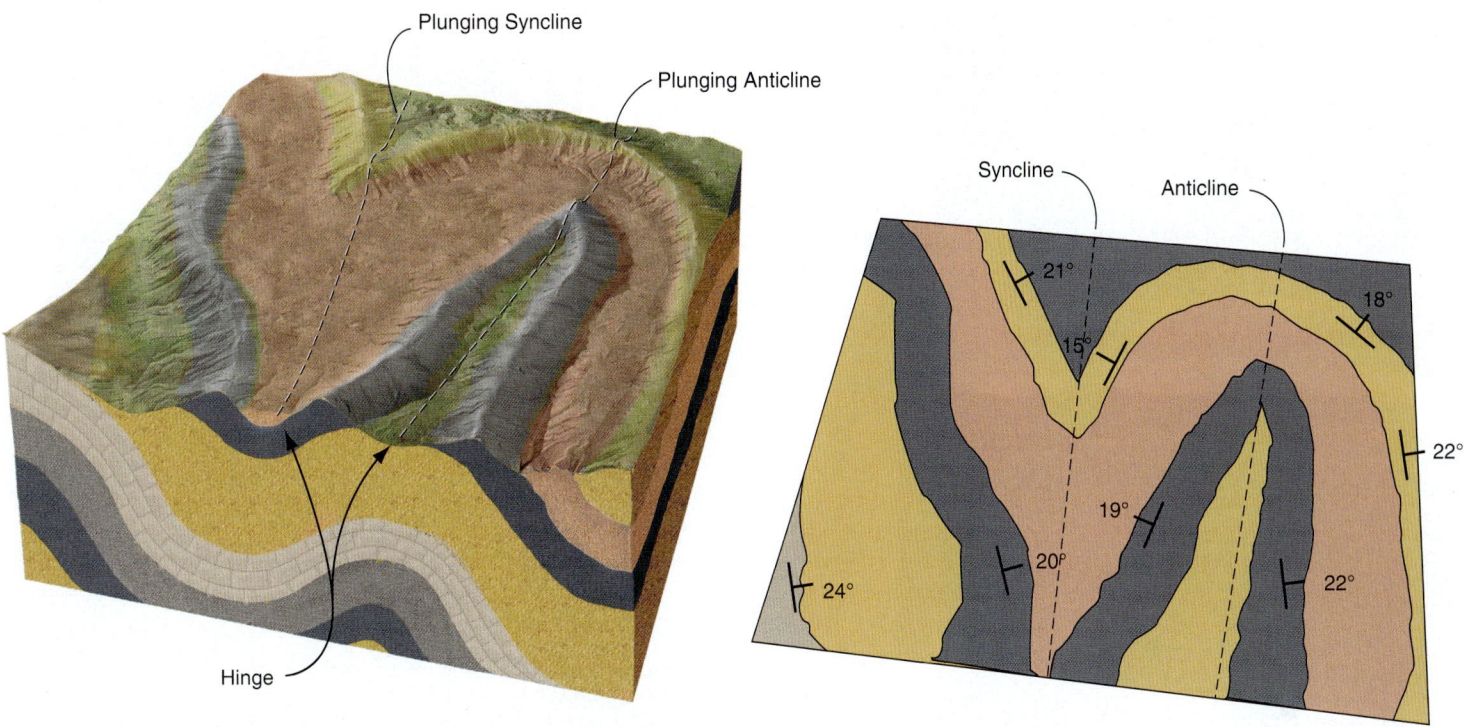

FIGURE 10.18 Erosion exposes internal rock layers forming outcrop patterns that reveal the orientation of underlying structures. These patterns can be mapped, and the strike and dip of the rocks can be used to reveal underlying folds.

Imagine you are mapping a dome: What would the strike and dip pattern be?

10-7 The San Andreas Fault Is a Plate Boundary

LO 10-7 Describe the boundary of the North American and the Pacific Plates.

The San Andreas Fault is a right lateral strike-slip fault that forms a transform boundary marking the contact between the North American and Pacific lithospheric plates.

San Andreas

The San Andreas Fault is a series of related faults across a broad region of western California that have formed as a result of the stress of plate motion; thus, the San Andreas Fault is more properly referred to as a *fault zone*. In general, geologists have identified south, central, and north segments of the fault, each with its own history and structural elements, though all three segments are closely related.

Land to the west of the fault lies on the Pacific Plate and moves toward the northwest. Land to the east of the fault lies on the North American Plate and moves to the southwest. Consequently, there is a component of compressive stress between the two plates, although most of the relative motion is the result of shear stress.

Displacement along the fault trace averages 33 to 37 millimeters a year. The westward component of motion on the North American Plate generates a compressive force that has raised the mountainous Coast Range. Uplift is also created by northward movement of the Pacific Plate, where it collides with the North American Plate, raising the Transverse Range in southern California and the Santa Cruz Mountains in northern California.

The San Andreas Fault exits land and moves onto the Pacific seafloor through Tomales Bay, creating the Point Reyes National Seashore north of San Francisco (the fault touches back onshore at points north). Movement along the fault leads to the emergence of numerous landforms, some of which are identified in **FIGURE 10.19**.

Movement along the fault also produces damaging earthquakes. The three most famous examples are the San Francisco earthquake in 1906 (causing the death of at least 3000 people and destroying over 80% of the city), the Loma Prieta earthquake in 1989 (causing 63 deaths and moderate damage), and the Northridge earthquake in 1994 (killing 57, injuring 8700, and costing $20 billion in damage).

© Roger Ressmeyer/CORBIS

(a)

Courtesy of Thule Scientific

(c)

Dr. Matthew A. d'Alessio & UC Berkeley Active Tectonics Team

(d)

(e)

Kathy Sloane/PhotoResearchers/Getty Images

(b)

Peter Menzel/Science Source

James Balog/The Image Bank/Getty Images

(f)

Garry Hayes

(g)

Photo copyrighted by A.L. Murat/Library of Congress
Prints and Photographs Division Washington

U.S. Geological Survey Open-File

(h)

(i)

FEMA News Photo

FIGURE 10.19 The San Andreas Fault is a right lateral strike-slip fault that marks the boundary between the Pacific Plate and the North American Plate. The fault runs 1300 kilometers through the state of California and exits offshore north of San Francisco. Compressive stresses along the fault (a) deform sedimentary strata and (b) produce pressure ridges in the crust along the fault line. In places, the contact of the two plates is exposed. Here (c), rocks of the Pacific Plate (left) lie adjacent to rocks of the North American Plate (right). Measurements by global positioning system sensors reveal (d) the motion of the two plates and associated faults in the San Francisco Bay region. Wherever the fault runs through developed neighborhoods there are indicators of displacement due to slippage along the fault, including: (e) a road that constantly requires repair, and (f) a sidewalk that shows the offset. Earthquakes along the fault have caused severe damage in the past: (g) the 1906 San Francisco earthquake, (h) the 1989 Loma Prieta earthquake, and (i) the 1994 Northridge earthquake.

10-8 Mountain Building May Be Caused by Volcanism, Faulting, and Folding

LO 10-8 Compare and contrast volcanic mountains, fault-block mountains, and fold-and-thrust mountains.

A *mountain* is an area of land that rises abruptly from the surrounding region.

What Is a Mountain?

Generally, a mountain is considered to be a landmass that projects at least 300 meters above the surrounding land. A *mountain range* consists of several closely spaced mountains with a common origin. **Mountain belts** are made up of several mountain ranges, usually with related histories. Familiar mountain belts include the *North American Cordillera* (including the Rockies and Sierra Nevada ranges), the *Himalayas*, the *Alps*, and the *Appalachian Mountains*.

Mountain-building processes fall into three general categories and produce three types of mountain belts: 1) **volcanic mountains**, which usually form at volcanic arcs above subduction zones, at divergent boundaries, or at hotspots; 2) **fault-block mountains**, which are characterized by tensional stress, crustal thinning, and normal faulting; and (3) **fold-and-thrust mountains**, which emerge at continental collision sites along convergent margins.

Volcanic Mountains

Volcanic mountains are the result of rising magma that breaks through Earth's surface and builds a volcano (the topic of Chapter 6). Volcanic mountains are usually related to hotspots, convergent tectonic boundaries, or divergent boundaries.

The volcanic mountains of Hawaii contain both the tallest mountain on Earth, when measured from base to summit: the Mauna Kea shield volcano, rising 10,203 meters from the Pacific seafloor to the highest point; and the most massive mountain on the planet: the Mauna Loa shield volcano, with a total volume of 40,000 square kilometers (**FIGURE 10.20**).

Fault-Block Mountains

Fault-block mountains are the product of tensional stresses that pull apart the crust in a horizontal direction, breaking it into a number of separate blocks. As each block moves vertically, it compensates for these tensional forces by undergoing normal faulting that yields sequences of upthrown ranges and downthrown valleys (**FIGURE 10.21**).

Fold-and-Thrust Mountains

Although volcanic and fault-block mountains are common, the largest mountain belts are shaped by tectonic forces that raise and deform the crust. These tectonic belts are known as fold-and-thrust mountains. They may occur as a single range (the Ural Mountain Range in Eurasia) or as a belt of several mountain ranges (the Rocky Mountains).

In general, fold-and-thrust mountains require tectonic convergence (plate collision), the only force great enough to generate the enormous stresses necessary to produce this scale of rock deformation and uplift.

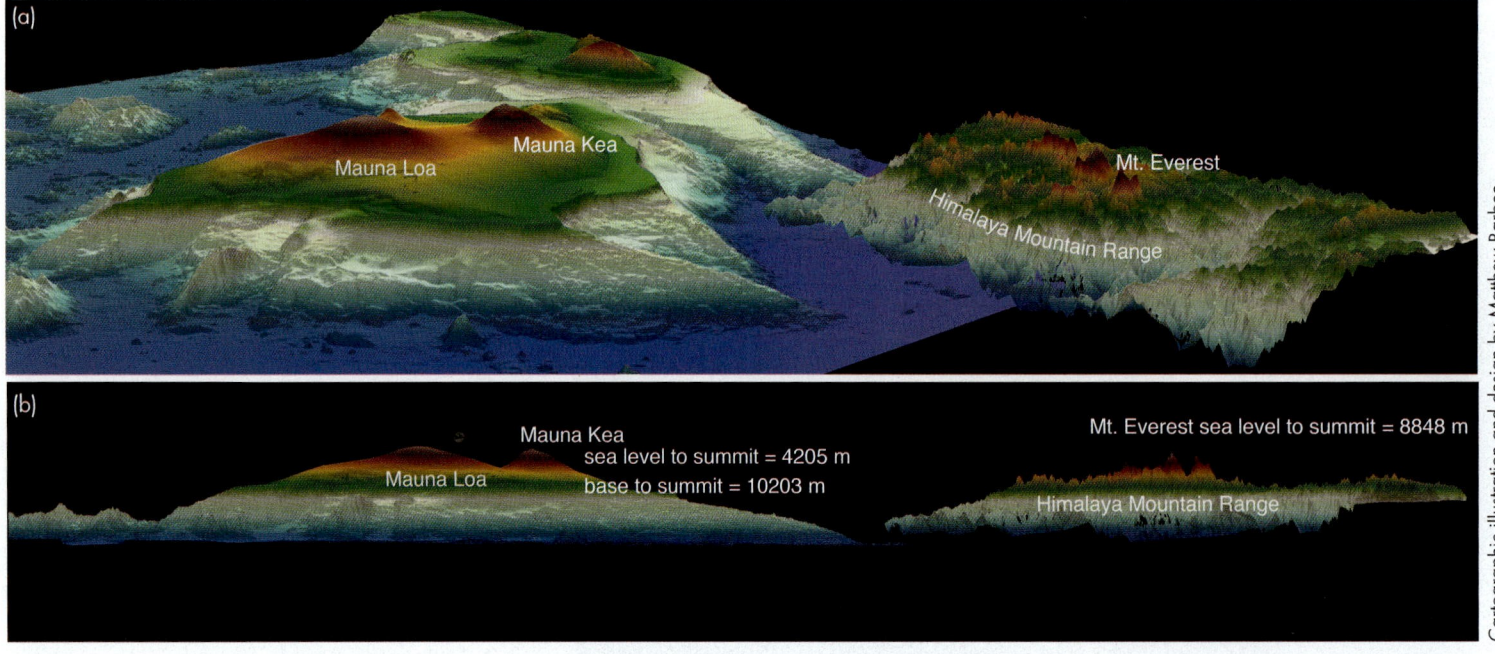

FIGURE 10.20 (a) The Mauna Loa shield volcano is the most massive mountain on the planet, with a total volume of 40,000 square kilometers. (b) Its neighbor, Mauna Kea, is the tallest mountain, with a total height from the seafloor to its summit of 10,203 meters, higher overall than Mount Everest.

What is the source of magma in Hawaiian volcanoes?

Cartographic illustration and design by Matthew Barbee

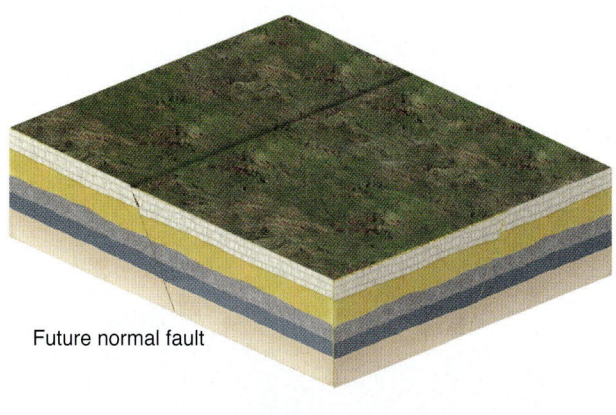

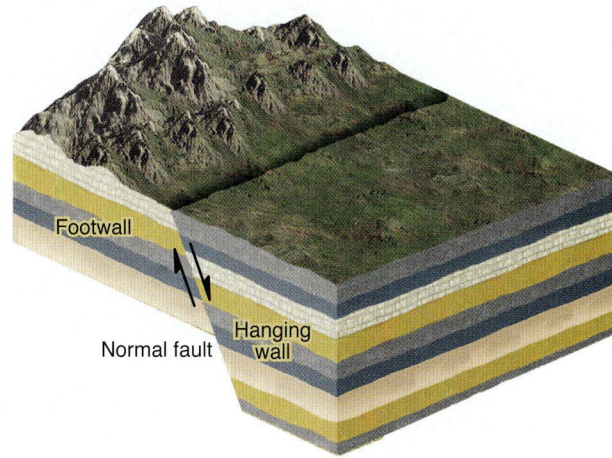

FIGURE 10.21 Fault-block mountains emerge in areas where the crust is stretched (or extended). Stretching leads to normal faulting, which produces blocks of crust that are displaced downward, making valleys. Blocks that stand higher, or are displaced upward, create mountain ranges.

> **What type of stress produces fault-block mountains?**

Mountain range

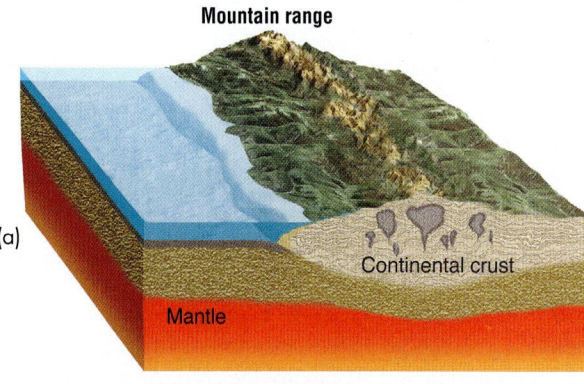

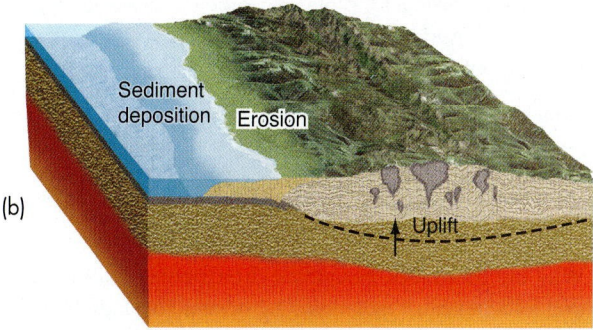

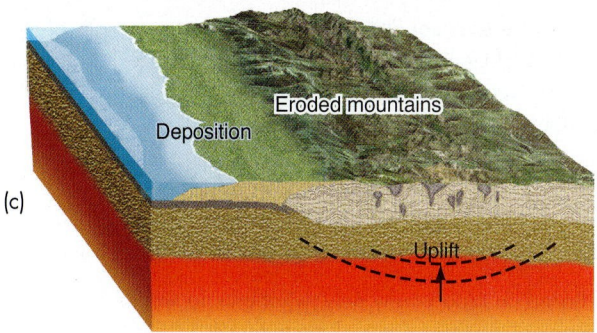

FIGURE 10.22 As erosion wears away the high elevation of a mountain belt, the loss of mass is compensated by vertical adjustment of the crust and shortening of the mountains' roots.

> **Which is more dense, felsic crust or ultramafic mantle? How is this fact relevant to the process of isostasy?**

Newly formed fold-and-thrust mountains are subject to weathering and erosion, which ultimately reduce their mass, as sediments are transported to other locations by combinations of wind, gravity, and running water. The crust compensates for this loss of mass by rising higher in the mantle (geologists use the term *rebound*) and in elevation, a process known as **isostasy** (from the Greek roots *iso,* "equal," and *stasis,* "standing"). Because of isostasy, the crust continually rebounds, offering up new rock to the processes of weathering and erosion by raising the crust until most of the rock in a mountain belt has been washed into the sea as sediment.

To test your understanding of mountain belts, complete the exercise in the Critical Thinking feature titled "Mountain Building."

Isostasy

As an iceberg that must have 90% of its mass below the ocean surface in order to support 10% of its mass above, mountains require deep roots, which add to the overall buoyancy of continental crust. These deep roots are necessary to support the mass of lofty mountain belts.

According to the principle of isostasy, mountain belts stand high above the surface because they have thick, buoyant roots of felsic crust

that extend into the mantle. Isostasy also means that the vertical position of the crust must perpetually change in order to adjust to changes in mass, a process termed *isostatic adjustment*. For instance, constant erosion of rock from mountains into the sea constitutes a shift of mass, and a mountain belt will rise to compensate for it (**FIGURE 10.22**).

As erosion wears mountains away, they rise in response to the loss of mass, and their roots shorten, until equilibrium is achieved, approaching the average elevation of continents (about 840 meters). Eventually, all that remains of the mountain belt is the exposed core of highly deformed and metamorphosed rock.

> **EXPAND YOUR THINKING**
>
> Describe a common example of isostasy that you encounter in your life.

CRITICAL THINKING

Mountain Building

There are three common types of mountain-building processes (FIGURE 10.23): Volcanic mountains are formed at locations where magma intrudes into the crust—at midplate hotspots, divergent plate boundaries, or convergent plate boundaries. Fault-block mountains are formed by extension and thinning of the crust. And fold-and-thrust belts develop where continental plate convergence leads to large-scale reverse faulting caused by compressive stress.

Working with a partner, complete the following exercise:

1. Which of these three mountain-building environments is most likely to develop folding? Why?

2. How do basic tectonic processes differ between volcanic and fold-and-thrust mountain belts?

3. Why are fault-block mountains characterized by normal faulting, and fold-and-thrust mountains by reverse faulting? How is this difference related to the type of stress at each setting?

4. Describe physical differences among all three types of mountains. Which of the three mountain-building processes produces the highest mountains? The most extensive mountain belt? Describe other differences.

5. Refer to a map of the world. Name two examples of each of the three types of mountains.

6. Label the tectonic- and mountain-building processes in Figure 10.23.

7. Based on topography only, what criteria would you use to identify each of the three types of mountains?

8. Do you expect that there should be a sequence of volcanic rocks in the fold-and-thrust belt? Explain your reasoning.

9. Explain why it is unlikely that a fault-block system and a fold-and-thrust system would develop simultaneously in the same place.

(a)

(c)

FIGURE 10.23 Three types of mountains: (a) fault-block mountains, (b) volcanic arc mountains, and (c) fold-and-thrust belt mountains.

(b)

10-9 Volcanic Mountains Are Formed by Volcanic Products, Not by Deformation

LO 10-9 Describe the geology of the Cascade Range.

Volcanic mountains are formed not by deformational processes but by volcanic products on Earth's surface.

Cascadia Subduction Zone

We reviewed volcanic processes thoroughly in Chapter 6, and here we will apply that knowledge to a discussion of the history of the most active volcanic system in the conterminous United States: the *Cascade Range*.

The Cascade Range of British Columbia, Washington, Oregon, and northern California consists of several active volcanoes as well as many apparently dormant volcanic centers that have been active during the late Neogene and Quaternary periods. This volcanic arc emerged above a subduction zone reaching from northern California to southern British Columbia. (You can read more about this development in the Geology in Action feature, "Strain Buildup at the Cascadia Subduction Zone.") It was formed by the subduction of the Explorer, Juan de Fuca, and Gorda Plates (fragments of the former Farallon Plate) beneath the western edge of the North American Plate (**FIGURE 10.24**).

Unlike the situation in typical subduction zones, no trench is present along the coast of the Cascade Range. Instead, *accreted terranes* (former continental slivers similar to Japan, and New Zealand and former island arcs similar to Indonesia that have converged onto the North American Plate) and an accretionary wedge have been uplifted and take shape as a series of coastal ranges and mountain peaks.

Accreted terranes are blocks of crust with relatively low density (andesite or granite). Thus, as the plate of which they are a part subducts, they cannot and instead, they accrete onto the leading edge of the overriding plate. Each terrane makes a ridge or range along the Pacific Northwest coast; over 20 such terranes have been identified. Inland from these coastal mountains are a series of valleys: Shasta, Willamette, and Puget. Each valley is a former forearc basin that has been uplifted (as part of the accreted terrane process) during the growth of the western margin of the continent as it moved to the west. Major cities are located in these valleys, including Portland, Seattle, and Vancouver.

The Cascade Range include more than a dozen large volcanoes. Although they share some general characteristics, each has unique geologic traits. There are two regions of current volcanic activity. Volcanoes are most active in Washington and northern Oregon, with a second region of eruptions occurring in northern California. In contrast, central and southern Oregon remain calm, as does southern British Columbia. The locations of the calm zones correspond to the positions of fracture zones that offset the Explorer, Juan de Fuca, and Gorda oceanic ridges.

The magma of the Cascades is generated by partial melting of the North American Plate. The typical lava of Cascade eruptions is andesite, which is characteristic of subduction zones around the world. Other types of lava that are present range in composition from basalt to rhyolite; they are products of various combinations of partial

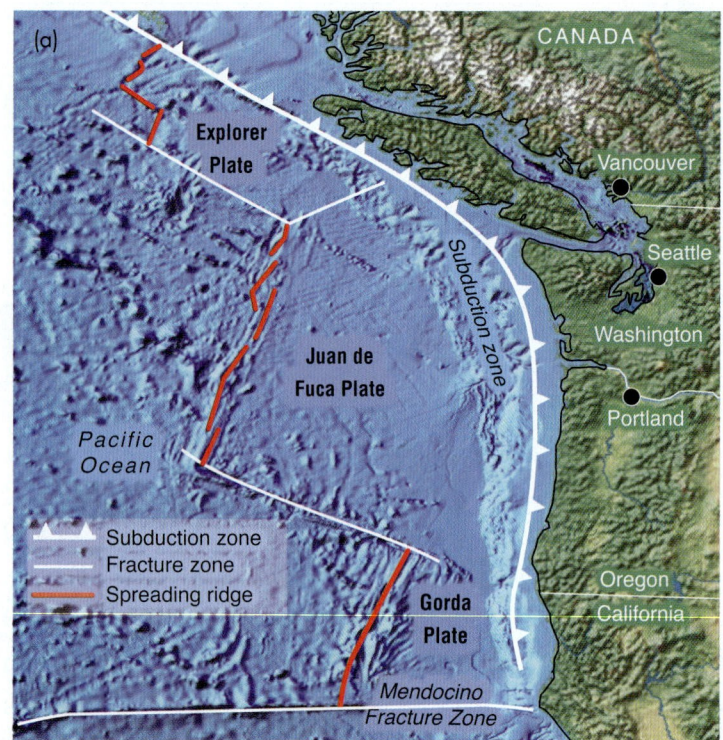

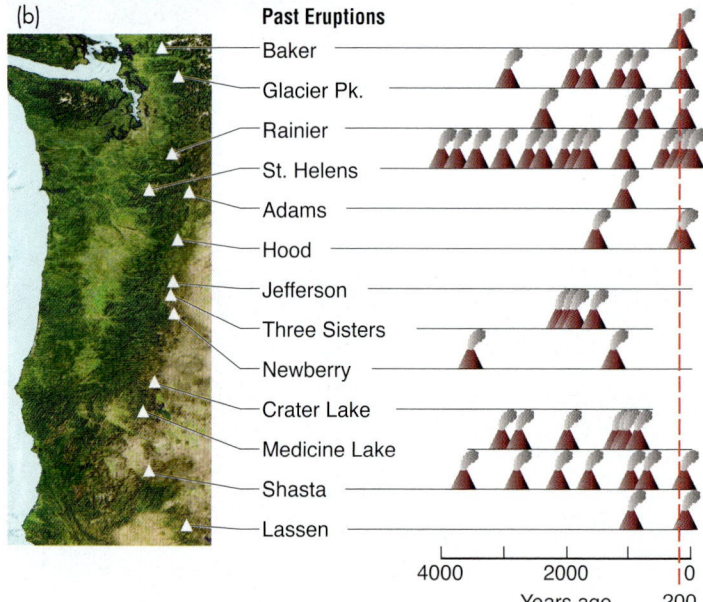

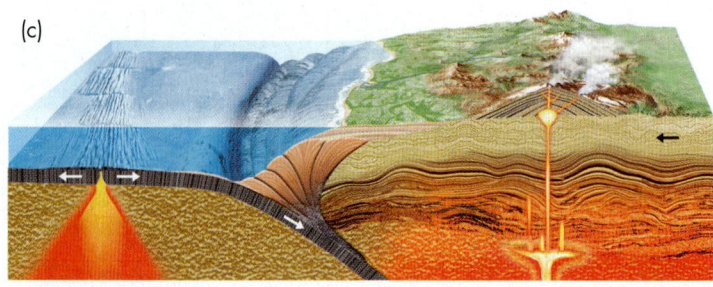

FIGURE 10.24 (a) The volcanic arc of the Pacific Northwest is built of magma generated by partial melting where the Explorer, Juan de Fuca, and Gorda Plates are subducted. (b) Thirteen principal volcanic centers characterize the Cascade region. (c) The Cascade Range comprises a volcanic arc.

What will happen in the Cascade Range when the Explorer, Juan de Fuca, and Gorda Plates are completely subducted?

melting and igneous evolution. Each major Cascade volcano has a distinct signature in terms of its lava composition.

Eruptions of basalt tend to be relatively calm events dominated by lava flows. Exceptions occur where such magmas encounter substantial groundwater (under snow cover or glaciers), which may create steam-driven *phreatomagmatic explosions.*

Andesite and rhyolite magmas are more viscous, and so tend to build up higher gas pressures. Such magmas may explode violently in catastrophic eruptions that release immense volumes of tephra, ash, and pumice. Glacier Peak, Mazama (Crater Lake), and Mount St. Helens (**FIGURE 10.24B**) have experienced this type of eruption.

Recent eruptions, notably that of Mount St. Helens in 1980, have provided geologists with important clues for better understanding an-

cient volcanism in the Cascade system. One result of the 1980 eruption was recognition of the importance of landslides in the development of volcanic landscapes. A huge section on the north side of Mount St. Helens slid away, ending up as chaotic topography many miles from the volcano itself. Accompanying the eruption were pyroclastic flows and mudflows that swept across the landscape. Similar events occurred at Mount Shasta and other Cascade volcanoes in prehistoric times.

Q EXPAND YOUR THINKING

Explain why different Cascade volcanoes would have different magma types.

 **GEOLOGY IN ACTION**

Strain Buildup at the Cascadia Subduction Zone

Subduction zones are dominated by three types of stress. Compressive stress develops at locations where two plates collide, with one (the denser of the two) sliding below the other. This sliding is called *interplate slip.* Tensional stress also builds up. The broad region of bending where one plate curves below another is characterized by extension; the front of the overriding plate can extend forward as well. Shear stress is high at the surface of the subducting plate, where it slides below the overriding plate.

The subduction process may not always be smooth, in which case two plates may become "locked." That is, interplate slip ceases as the top of the subducting plate and the bottom or front of the overriding plate won't move. In these cases, strain accumulates as pressure builds around the locked portion of the subduction zone until it unlocks.

Geologists hypothesize that a portion of the Cascadia Subduction Zone is locked, where the Juan de Fuca Plate subducts beneath the North American Plate along the coastline of southern British Columbia, Washington, and Oregon. Compressive stress at the subduction zone is active, and interplate slip has ceased, so strain is accumulating in the form of crustal shortening and vertical uplift of the ground surface (**FIGURE 10.25**). The land above the locked region is deforming elastically upward to such an extent that it can be measured using a network of global positioning sensors.

Based on the fact that the deformation is still in an elastic phase, Hooke's law predicts that it is fully reversible. This means that when the locked region once again experiences interplate slip, the land surface will recover elastically, probably rapidly, leading to the creation of a large and dangerous earthquake. Scientists have been able to determine, from a combination of geologic evidence and Japanese diaries, that the last time the locked crust was released, in 1700 AD, it rebounded so rapidly it set off a massive earthquake and tsunami (a seismic sea wave) that flooded not only the Cascadia coast but the coast of Japan as well (**FIGURE 10.26**). The same type of event is expected to occur again; exactly when remains a mystery.

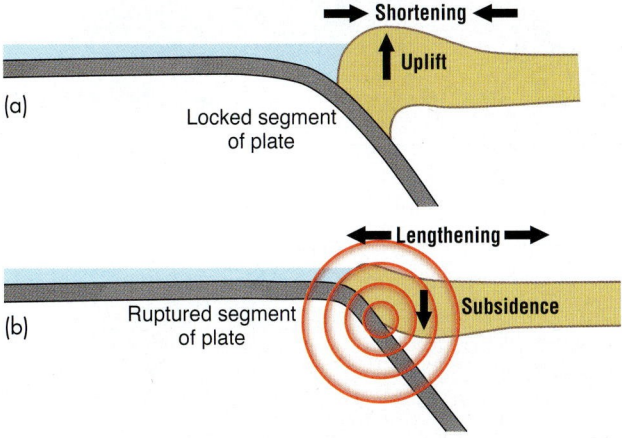

FIGURE 10.25 (a) Buildup of elastic strain along locked portions of the Cascadia Subduction Zone causes the crust to shorten and the land surface to rise. (b) When interplate slip occurs, the lock ruptures, the crust extends, and the surface subsides. This event may happen rapidly, setting off a damaging earthquake and, then, a tsunami. *Source:* Natural Resources Canada, http://earthquakescanada.nrcan.gc.ca/zones/cascadia/menace/fig-eng.php, September 7, 2008. Reproduced with permission of the Minister of Public Works and Government Services, 2010.

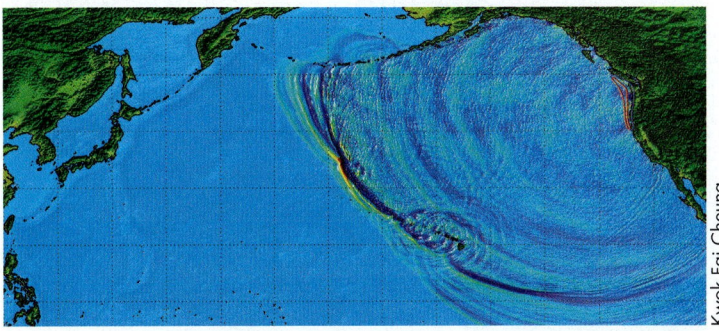

Kwok Fai Cheung

FIGURE 10.26 Researchers have determined that giant earthquakes have occurred along the Cascadia Subduction Zone. The most recent, on January 26, 1700, caused a tsunami that flooded portions of the North American coast and crossed the Pacific Ocean, damaging coastal Japan. This is a computer model of the 1700 tsunami as it crosses the Pacific.

10-10 Crustal Extension Formed the Basin and Range Province

LO **10-10** Describe the origin of the Basin and Range Province.

The Basin and Range Province of the western United States and Mexico experiences crustal extension related to plate movement. It is characterized by unique topography that reflects normal faulting of the crust.

The Basin and Range

In the Basin and Range Province (**FIGURE 10.27**), blocks of faulted crust displaced downward form linear valleys, known as *grabens*. These valleys are separated by parallel mountain ranges of high-standing blocks, known as *horsts*. In the U.S. portion of the Basin and Range Province, horsts take shape as approximately 400 short, straight mountain ranges separated by arid valleys. This topography is typical of crust that is being pushed upward from below, causing regional extension.

Sedimentology

Within many of the graben valleys are seasonal lakes that fill with water in the winter and spring and evaporate in summer and fall. Broad sedimentary deposits of precipitated minerals (evaporites) result from the evaporation, and in the wet season, clays, silts, and layers of algae that bloom in the wet conditions bury these deposits. Continued

faulting causes the valleys to drop down over time, and the layered deposits to become very thick. This is the sort of environment that produces *oil shale*, a fuel source that is potentially valuable but environmentally destructive to exploit (**FIGURE 10.28**).

The entire Basin and Range Province has an arid climate because of the *rain shadow* cast by the high Sierra Nevada range, which blocks and collects moisture coming off the Pacific Ocean to the west. However, in winter and spring, short, intense bursts of rainfall sweep tons of mud, sand, and gravel onto the valley floors. Over hundreds of thousands of years, as the ranges are worn down, the valleys fill with sediment. Slowly, the valley floors bury the slopes of the mountains in their own sediment until a new round of faulting rejuvenates the landscape and resurrects the fault-block mountains once again.

Tectonics

The *Sierra Nevada* of California mark the western border of the Basin and Range Province, and another province, called the *Colorado Plateau*, marks the eastern border. Nearly the entire southwestern portion of the United States, including all three of these regions, has been uplifted, and the crust extended (or stretched), over the last 20 million years due to plate tectonics. This tectonic activity has led to a widespread pattern of normal faulting and the formation of fault-block mountain ranges.

Regional uplift of the western United States is the result of the western movement of the North American Plate and its interaction with plates to the west. Approximately 30 million years ago, the central California coastline was an active subduction zone where the now-extinct *Farallon Plate* subducted beneath the North American Plate (**FIGURE 10.29**). Compressive stress associated with plate convergence caused crustal thickening east of the subduction site (western North America). This led to uplift of the Colorado Plateau and downward erosion of the Colorado River and other

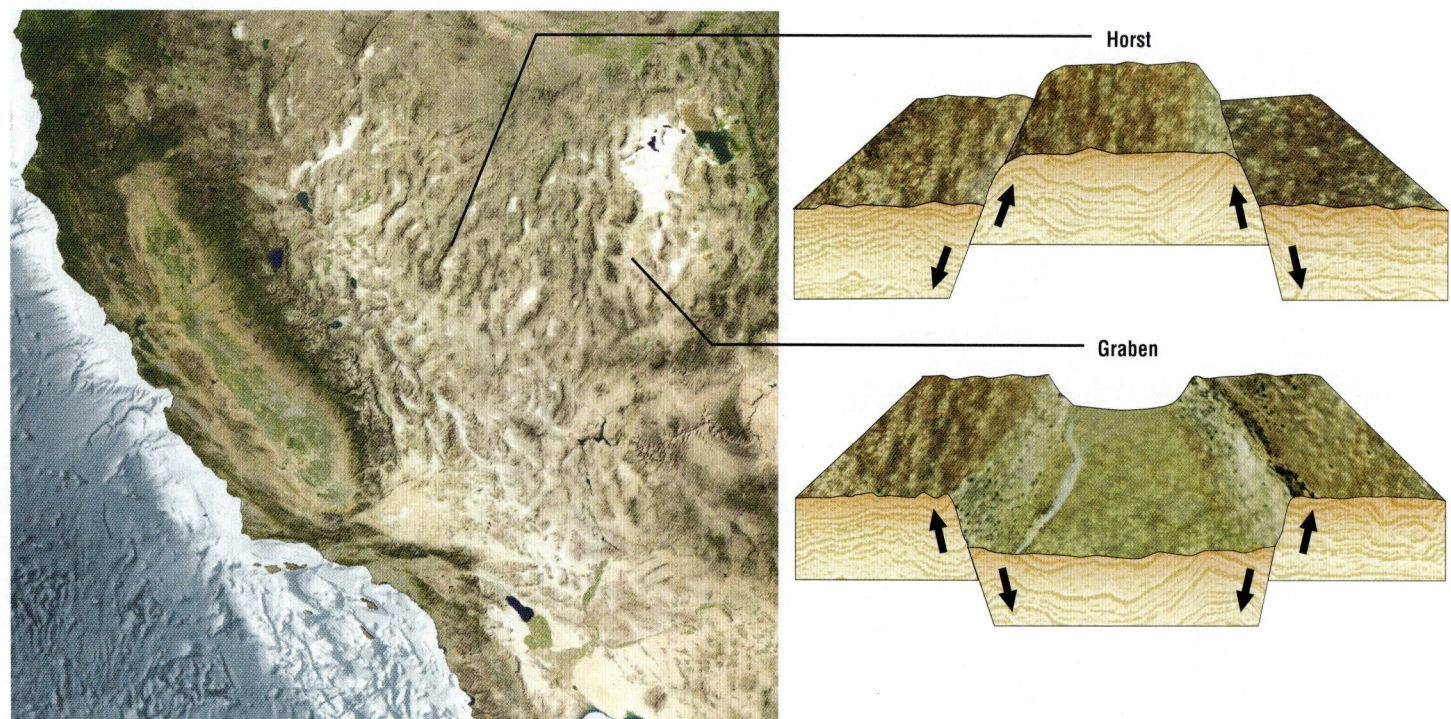

FIGURE 10.27 Crustal extension under Idaho, Nevada, and Utah has formed the Basin and Range Province. The province is characterized by normal faulting that creates numerous horsts and grabens.

 What is the cause of crustal extension in the Basin and Range Province?

FIGURE 10.28 The United States has the world's largest oil shale deposits, a potentially important fuel source. However, surface mining of oil shale is environmentally destructive.

> Why are oil shale deposits so thick in the Basin and Range Province?

stream systems in the region, which remained at their original elevation while the land around them rose. In this way the Grand Canyon took shape.

About 20 million years ago, subduction ended in the Los Angeles area as remaining fragments of the Farallon Plate became separate plates: the Explorer, Juan de Fuca, and Gorda Plates to the north, and the Cocos Plate to the south. (These plates still exist and are subducting under different portions of the North American Plate.) The former subduction zone was replaced by the present-day strike-slip faulting of the San Andreas transform boundary.

Over the next 20 million years, the transform boundary extended to the north and south, slowly replacing the subduction process along most of the California coast and adjoining north-south areas. Although some aspects of this transition are poorly understood, it is thought that the former Farallon Plate remained intact after subducting, perhaps at a very shallow angle, beneath the North American Plate. When subduction ceased, the last portion of

FIGURE 10.29 (a) Approximately 30 million years ago, the western boundary of the North American Plate experienced crustal thickening and compression related to subduction of the Farallon Plate. (b) Subduction of the rift zone formerly marking the boundary of the Farallon and Pacific Plates led to the development of a region of decompression melting and high heat flow under the North American continent. This lifted the continental crust, generating tension and producing the Basin and Range Province, as well as some volcanic action.

> What is the source of heat that extends the crust in the Basin and Range area?

the Farallon Plate under the continent became detached and interacted with the underside of the North American Plate, which continued to migrate to the west. This process may be continuing today.

When the subducted Farallon Plate became detached, an opening behind it was created in the upper mantle under the southwestern United States. This was the rift zone marking the former boundary of the Pacific and Farallon Plates. This region was characterized by decompression melting, high heat flow, and plutonism, which combined to raise the crust and produce a tensional stress environment. As upwelling magma pushed into the base of the crust, it arched upward and stretched and thinned, creating a broad rift that became the Basin and Range Province of Nevada, Utah, and parts of Idaho, New Mexico, Arizona, and Mexico. This tensional environment shaped the hundreds of fault-bounded horsts and grabens that characterize the rugged modern topography.

ⓠ EXPAND YOUR THINKING

A subducted plate under Nevada makes the Basin and Range Province; under Oregon, a different subducted plate makes the Cascade Range. Why do different geologic products result from the same subduction process?

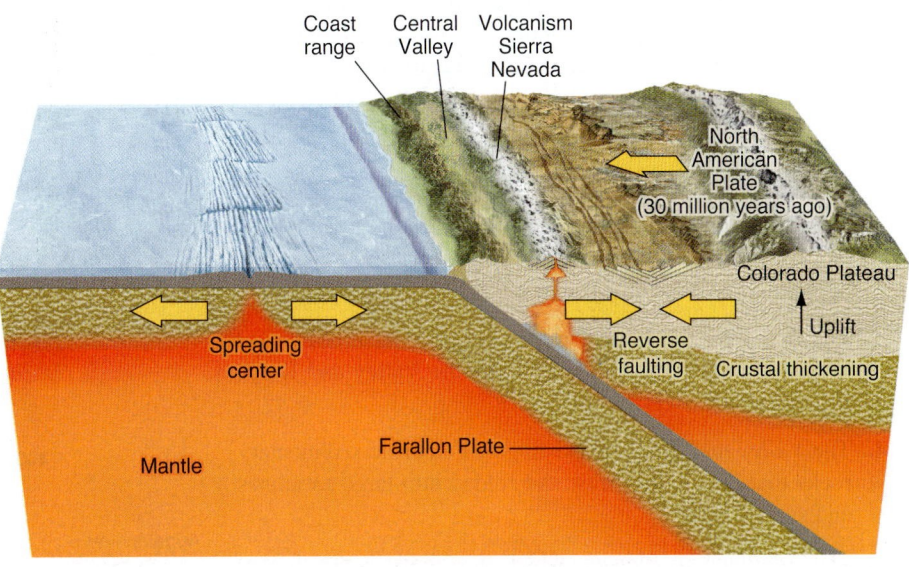

(a)

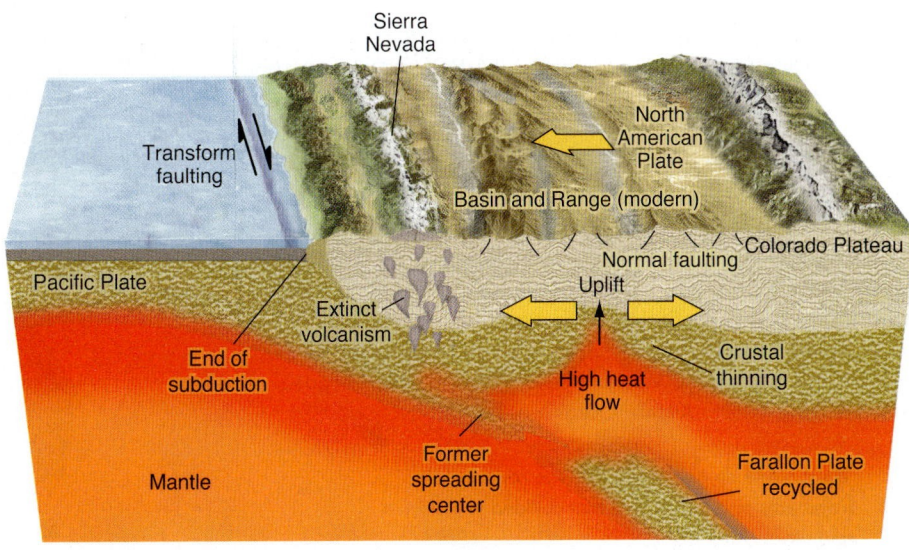

(b)

10-11 Fold-and-Thrust Belts Are the Highest and Most Structurally Complex Mountain Belts

LO 10-11 Describe the origin of fold-and-thrust belts.

Large compressional stresses are generated in the crust at convergent margins, especially where continents collide. The Himalayas (the highest mountains on Earth) were formed when the Indian–Australian Plate collided with the Eurasian Plate—a process that is still underway, as explained in the Geology in Action feature, "Raising Everest."

The Appalachian Belt

The Appalachian Mountain Belt extends from Alabama through Maine in the United States and continues across the southeastern provinces of Canada to Newfoundland. Geologists divide the Appalachians into five main provinces (**FIGURE 10.30**):

1. The *Coastal Plain*, extending from the seaward edge of the continental shelf to the foothills of the Appalachians;

2. The *Piedmont* (French for "foothills"), reaching from Alabama to New York and making up the easternmost portion of the mountainous topography;

3. The *Blue Ridge*, a narrow ridge composed of Precambrian rocks separating the Piedmont from the Valley and Ridge Province to the west;

4. The *Valley and Ridge*, consisting of Paleozoic sedimentary rocks that have been thrust and folded into large anticlines and synclines; and

5. The *Appalachian Plateau*, which lies above gently folded and tilted Paleozoic sedimentary strata.

The story of how the Appalachians came into being began nearly 1,100 million years ago during the Proterozoic eon when convergent tectonics created an early supercontinent named *Rodinia* (similar in origin to Pangaea). The Adirondack Mountains of New York State date from this time. By the start of the Phanerozoic eon, Rodinia began a process of rifting as the continents slowly separated.

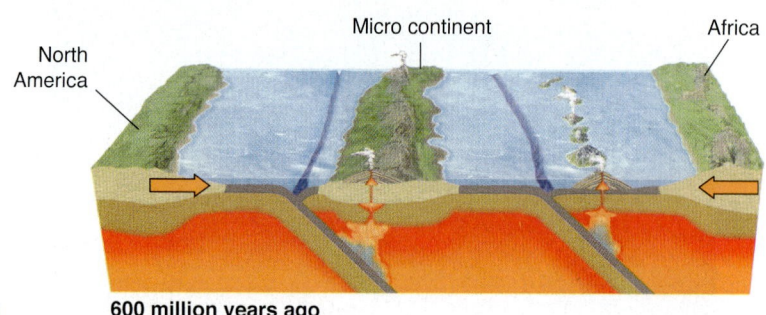

(a) **600 million years ago**

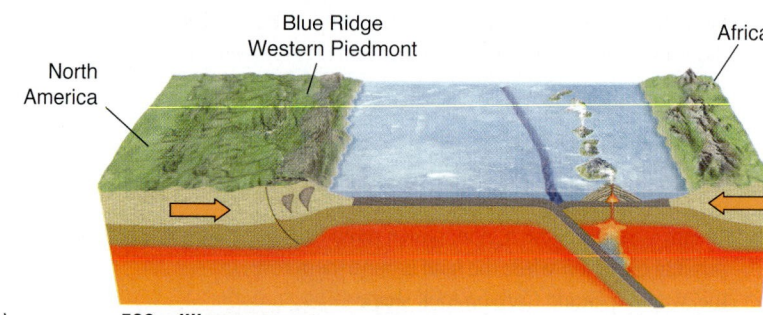

(b) **500 million years ago**

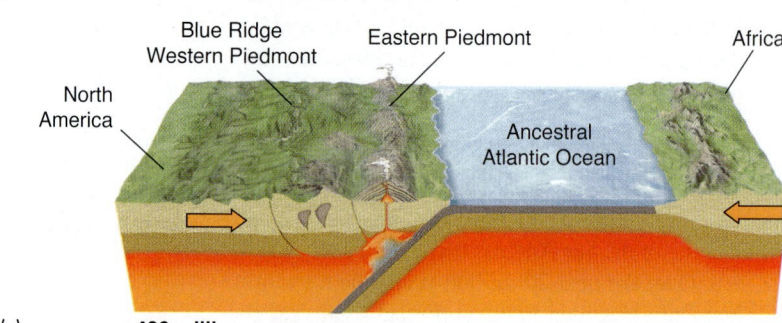

(c) **400 million years ago**

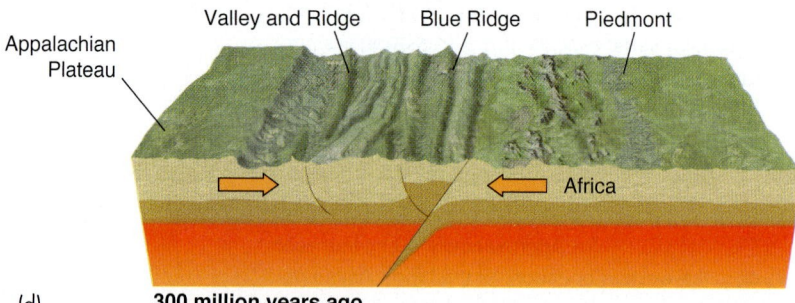

(d) **300 million years ago**

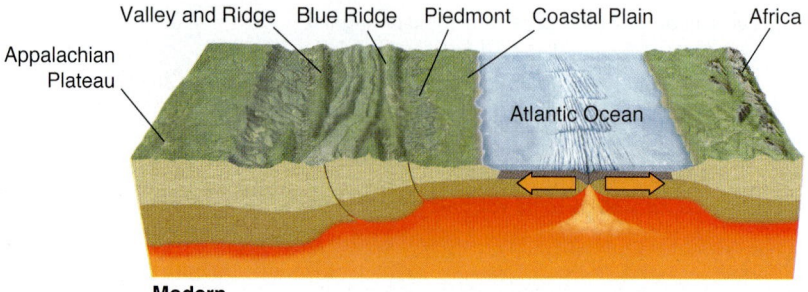

Modern

FIGURE 10.30 In the Proterozoic eon, the Iapetus Ocean opened during rifting of Rodinia (an early supercontinent). During the Paleozoic era, the Iapetus began to close, (a) squeezing a series of continental fragments and island arcs between the North American and African continents. Each collision was a mountain-building event: (b) the Taconic orogeny, (c) the Acadian orogeny, and (d) the Alleghenian orogeny. (e) Mountain building ended when Pangaea rifted and the modern Atlantic Ocean was born.

What name do geologists give the continental fragments and island arcs that collided with Africa and North America?

Over time, the early Appalachians eroded, depositing sediments in a widening rift basin—the predecessor of the Atlantic, known as the *Iapetus Ocean*. By the Ordovician period, the Iapetus basin, which had been widening for hundreds of millions of years, began to close as Africa, North America, and Europe moved together. This gave rise to the first of several phases of mountain building, or orogeny, that would become the Appalachian fold-and-thrust belt.

The tectonic history of the modern Appalachians (Figure 10.30) is divided into three main orogenies. Each orogeny begins with accumulation of thick marine sediments and volcanic deposits, continues with deformation and uplift of mountains, and culminates with the calming of tectonic conditions. A particular result of the crustal uplift associated with each orogeny was the production of sediment in great volumes due to rapid erosion of the steep highlands. Thus, each phase is marked by the buildup of a so-called *clastic wedge*, a massive accumulation of sediments that filled shallow seas on the continental (west) side of the Appalachians.

- The *Taconic Orogeny*, the first important tectonic activity in the Appalachians, took place during the Ordovician Period. Thrusting and folding occurred mainly in the northern portion of the mountain belt. Uplifted mountains shed sediment to the west, forming the Queenston clastic wedge near Albany, New York.
- The *Acadian Orogeny* was the major orogeny of the northern Appalachians. It occurred in the Devonian Period and was centered in New England. Sediments accumulated in the Catskill clastic wedge across southern New York and northern Pennsylvania.
- The *Alleghenian Orogeny* was the major mountain-building phase responsible for forming the southern Appalachians. It ended during the Pennsylvanian Period. A clastic sedimentary wedge spread over western Pennsylvania, West Virginia, Kentucky, and Tennessee.

All three of these orogenies are interpreted as collisions that occurred during the closing of the Iapetus Ocean between North America, Europe, and Africa as part of the development of Pangaea. Seafloor spreading, subduction-zone volcanism, and clastic wedges were all part of a long, drawn-out sequence of tectonic events.

Q EXPAND YOUR THINKING

What does the term *fold-and-thrust belt* describe?

GEOLOGY IN ACTION

Raising Everest

Long known as the highest peak on the planet, Mount Everest (**FIGURE 10.31**) has other names: to the Tibetans, it is Chomolungma, "goddess mother of the world"; in Nepal, it is known as Sagarmatha, "goddess of the sky."

The Indian mathematician and surveyor Radhanath Sikdar (1813–1870) was the first to identify Mount Everest as the world's tallest peak. In 1852, using trigonometric calculations, he measured the elevation as 8,840 meters (29,002 feet) above sea level. (His actual measurement was 29,000 feet exactly, but it was thought that this number would be viewed as a rounded adjustment, so an arbitrary 2 feet was added to it). For this accomplishment, some believe that the peak should be named after Sikdar, not Sir George Everest, who was the surveyor general of India in the early nineteenth century and is not known to have even laid eyes on the great peak. For more than a century, this measurement remained the mountain's officially accepted height. Then in 1954, the Indian surveyor B. L. Gulatee calculated a new height by averaging 12 measurements from around the mountain. His calculations adjusted the height upward by 8 meters to 8,848 meters.

On May 5, 1999, a team of climbers armed with a state-of-the-art global positioning system (GPS) reached the summit of Everest. Climbers Pete Athans and Bill Crouse were supported by five Sherpas (Nepalese guides), who carried the GPS equipment and an extra supply of oxygen. The team climbed through the night so that they would

Galen Rowel/Corbis

FIGURE 10.31 Mount Everest.

reach the top and make their measurements during the warmest part of the day. Based on their calculations, Athans and Crouse revised the height upward by 2 meters, to 8,850 meters. Since then, no significant change has been made to this measure. But future change seems inevitable, as the inexorable Indian Plate seems to be shifting Everest's horizontal position steadily and slightly to the northeast, at a rate of about 6 centimeters a year.

LET'S REVIEW

Stress generates strain in rock. Strain is exhibited by deformation, which may initially be elastic (i.e., fully reversible). However, if the elastic limit is exceeded, permanent deformation results, and rocks show ductile strain. Eventually, fracture may occur—the usual outcome—setting off sudden, damaging earthquakes. The greatest strain develops at plate boundaries, and these are regions of mountain building.

In the next chapter, "Earthquakes," we look into the sources and impacts of earthquakes, and their history of causing severe damage and loss of life over the centuries. As you prepare to read Chapter 11, keep in mind that every third or fourth day, on average, an earthquake occurs somewhere in the world, damaging buildings, roads, and human communities. The energy released by earthquakes is a direct result of strain.

STUDY GUIDE

10-1 Rocks in Earth's crust are bent, stretched, and broken.

- Rocks may be bent, pulled, and fractured. This process is called **deformation**; the forces that cause deformation are called **stresses**. Three types of directed stress act on rock and cause it to deform: **tensional stress, compressional stress**, and **shear stress**. In general, these stresses originate at plate boundaries.

10-2 Strain takes place in three stages: elastic deformation, ductile deformation, and fracture.

- Strain is the change in shape and/or volume of a rock caused by stress. When rocks undergo strain, they pass through three successive stages of deformation: elastic deformation, ductile deformation, and **fracture.**

- The type of strain experienced by a rock depends on temperature, confining pressure, rate of strain, presence of water, and composition of the rock. When the rock passes its elastic limit, it experiences ductile deformation. As stress builds, the rock fractures. In the crust, many such fractures develop into **faults**, zones or planes of breakage across which layers of rock are displaced relative to one another.

10-3 Strain in the crust produces joints, faults, and folds.

- Stress causes deformation in the form of ductile strain (folds) or fracture (joints and faults). **Joints** are fractures that display no net displacement between the two sides of the crack. Faults develop when brittle rocks fracture, and there is a distinct offset along the plane of breakage. Rocks are most likely to **fold** when: strain accumulates slowly; there is high temperature and pressure; the rock is composed of minerals that tend to be ductile; and fluids are present.

- Researchers create maps of geologic structures in order to better understand their stress history. To show crustal structure on a **geological map**, geologists use a simple two-part symbol, called the **strike** and the **dip**, to display the spatial orientation of the layers of rock that make up folds, faults, and joints. The strike of a body of rock is a line representing the intersection of that feature with the plane of the horizon. The dip gives the angle below the horizontal of a geologic feature. It can be thought of as the direction in which water would flow if rain fell on the fault surface or rock layer.

10-4 Dip-slip and strike-slip faults are the most common types of faults.

- **Dip-slip faults** may be either **normal faults** or **reverse faults;** a special type of reverse fault is known as a **thrust fault**. Normal faults de-velop in brittle rocks that are exposed to tensional stress. Reverse faults are caused by compressive stress; the hanging wall, which sits above the fault plane, is displaced upward along the fault plane relative to the footwall, which sits below the fault plane. A thrust fault is a special case of a reverse fault in which the dip of the inclined fault plane is less than 15°.

- **Strike-slip faults** occur where the relative motion on a fault plane is horizontal. There are left lateral strike-slip faults and right lateral strike-slip faults.

10-5 Rock folds are the result of ductile deformation.

- Folds are the result of compressional stresses acting over long periods. The simplest folds are the **anticline**, the **syncline**, and the **monocline**. Complex folds include asymmetrical anticlines and synclines, plunging anticlines and synclines, groups of anticlines and synclines, domes, basins, and contorted folds.

10-6 Outcrop patterns reveal the structure of the crust.

- To unravel the geometry and history of the crust, a structural geologist analyzes **outcrops**. Their goal is to determine the spatial orientation of rock layers and, from this, the relative age of the layers and the history of deforming stresses experienced by the rock.

- A layer of tilted rock can be represented with a plane. The orientation of that plane in space can be defined using strike and dip notations. The dip is described by a compass direction (0° to 360°) and an angle to the horizon, which varies from 0° (horizontal) to 90° (vertical).

- **A** geological map shows the topography of an area, the lithology and age of rock units exposed at the surface or buried shallowly under loose sediment and soil, and their strike and dip.

- Patterns of rock reveal the orientation of crustal structure. Parallel strips of outcrops indicate folds with horizontal axes. An outcrop pattern of Vs indicates plunging folds. When a V opens in the direction of the plunge, it is a syncline; one that closes in the direction of the plunge is an anticline.

10-7 The San Andreas Fault is a plate boundary.

- The San Andreas Fault is a right lateral strike-slip fault that forms a transform boundary marking the contact between the North American and Pacific lithospheric plates.

10-8 Mountain building may be caused by volcanism, faulting, and folding.

- Three types of mountains are commonly formed: **volcanic mountains, fault-block mountains**, and **fold-and-thrust mountains**.

- When land is raised during mountain building, it is immediately attacked by weathering and erosion. As the mass of the mountain is reduced, it compensates by rising higher in the mantle. This process is called **isostasy**. Due to isostasy, mountain belts stand high above the surface upon thick roots of felsic rock that extend deep into the underlying mantle. The vertical position of the crust must shift continuously in order to adjust to the changes in mass, a process referred to as isostatic adjustment. For instance, constant erosion of rock from mountains into the sea constitutes a shift of mass that is compensated for by a slight rise in the mountain belt.

10-9 Volcanic mountains are formed by volcanic products, not by deformation.

- Volcanic mountains are formed not by deformational processes but by volcanic products on Earth's surface.

- The Cascade Mountains of British Columbia, Washington, Oregon, and California consist of several active volcanoes, as well as many volcanic centers that were active during the Neogene and Quaternary periods.

This volcanic arc developed above a subduction zone that stretches from northern California to southern British Columbia and was formed by the recycling of the Juan de Fuca and Gorda Plates beneath the western edge of the North American Plate.

10-10 Crustal extension formed the Basin and Range Province.

- Fault-block mountain building results from tensional stress that stretches areas of continental crust. Mountains shaped in this way are found throughout the southwestern United States in the Basin and Range Province. As crust is stretched, it develops normal faults that may build horst and graben structures.

10-11 Fold-and-thrust belts are the highest and most structurally complex mountain belts.

- Large compressional stresses are generated in the crust at convergent margins, where continental crustal areas collide. Rocks located in the collision zone between two continental blocks are folded, faulted, and thrust-faulted as part of the process of crustal thickening. This process pushes peaks upward and secures deep roots to form fold-and-thrust mountains. The Appalachian Belt was built by this same fold-and-thrust process.

KEY TERMS

anticline (p. 283)
compressional stress (p. 274)
deformation (p. 274)
dip (p. 278)
dip-slip faults (p. 279)
fault-block mountains (p. 290)
faults (p. 276)
fold-and-thrust mountains (p. 290)
folds (p. 278)
fracture (p. 276)

geological map (p. 286)
isostasy (p. 291)
joints (p. 277)
monocline (p. 284)
mountain belts (p. 290)
normal faults (p. 279)
outcrops (p. 278)
reverse faults (p. 279)
shear stress (p. 274)
strain (p. 275)

stresses (p. 274)
strike (p. 278)
strike-slip faults (p. 281)
structural geology (p. 285)
syncline (p. 283)
tensional stress (p. 274)
thrust fault (p. 279)
volcanic mountains (p. 290)

ASSESSING YOUR KNOWLEDGE

Please complete this exercise before coming to class. Identify the one best answer to each question.

1. The three types of stresses are:
   a. Anticline, syncline, and monocline.
   b. Folds, faults, and fractures.
   c. Compressional, tensional, and shearing.
   d. Elastic, ductile, and brittle.
   e. Volcanic, fold and thrust, fault-block.

2. Label the tectonic environments on this diagram with the dominant stress: compressive, tensional, or shear.

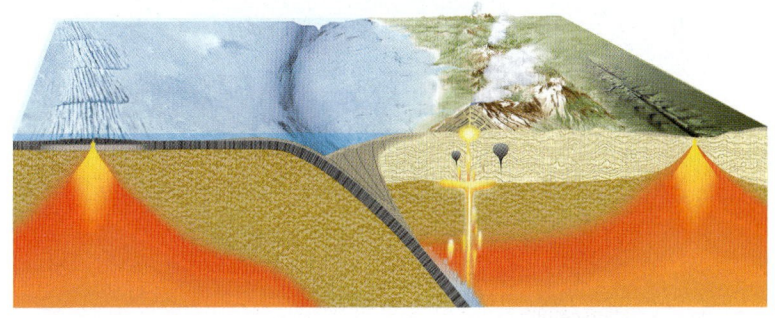

3. Strain is:
   a. A hotspot.
   b. Only exhibited where rock has been partially melted, usually in a subduction zone.
   c. A condition of being sheared.
   d. A change in shape and/or volume of a rock caused by stress.
   e. None of the above.

4. The three types of strain are:
   a. Elastic, ductile, and fracture.
   b. Fracture, folding, and faulting.
   c. Bending, breaking, and stretching.
   d. Compressional, tensional, and folding.
   e. Convergent, divergent, transform.

5. A reverse fault is formed when:
   a. The footwall moves upward relative to the hanging wall block.
   b. The fault plane shifts along the line of strike.
   c. The hanging wall is displaced upward along the fault plane.
   d. Tensional stress causes brittle strain.
   e. All of the above.

6. An anticline is the product of:
   a. Tensional stress.
   b. Compressive stress.
   c. Shear stress.
   d. Elastic stress.
   e. Anticlines are not made by stress.

7. How do dip-slip and strike-slip faults differ from one another?
   a. They describe different directions of movement along a fault plane.
   b. Each is a different type of fold.
   c. Only one of them involves ductile behavior.
   d. They are not different; they are the same thing.
   e. Both are the result of ductile strain.

8. A V-outcrop pattern indicates:
   a. An upright fold.
   b. A dip-slip fault.
   c. Shear stress.
   d. A plunging fold.
   e. That no folding has occurred.

9. Study the diagram here then answer the following questions. Note that north is to the top, or back, of this structure.
   a. Name the structures that characterize the crust or back.

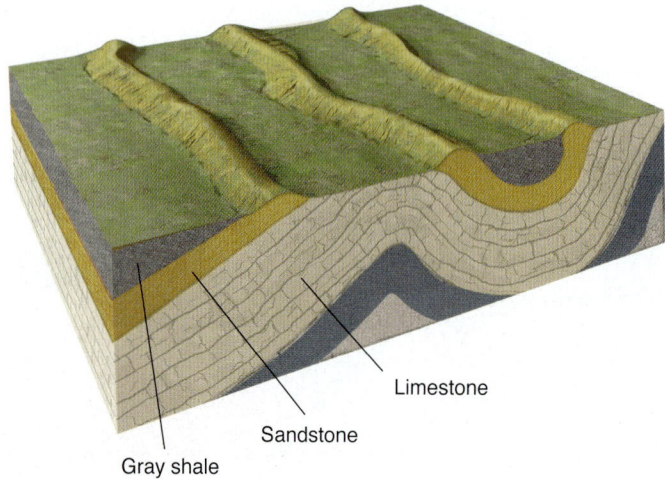

Limestone

Sandstone

Gray shale

   b. Which way are the rocks dipping in the left-hand portion of this diagram? (east, west, north, south).
   c. Which way are the rocks dipping in the right-hand portion of this diagram? (east, west, north, south).
   d. _____ stress made these geologic structures. (compressive, tensional, shear)

10. Why does the formation of a fold require ductile behavior?
    a. Because a rock must deform plastically around a bend.
    b. Because it involves a reverse fault.
    c. It does not require ductile behavior.
    d. A fold only forms under shear stress.
    e. A fold is a type of dip-slip fault.

11. What are strike and dip?
    a. Descriptions of partial melting.
    b. The result of volcanic mountain building.
    c. Symbols on a map indicating the orientation of a feature.
    d. The angle at which a fault is folded after elastic behavior.
    e. None of the above.

12. Label the fault type, and the stress in each of the following photographs.

Marli Bryant Miller

   a.

_____(normal, strike-slip, reverse);
_____(tension, shear, compression)

©Marli Bryant Miller

   b.

_____(normal, strike-slip, reverse);
_____(tension, shear, compression)

c.

_____(normal, strike-slip, reverse);

_____(tension, shear, compression)

d.

_____(normal, strike-slip, reverse);

_____(tension, shear, compression)

13. What types of faults are formed by tensional, compressive, and shear stresses, respectively?
   a. Reverse fault, strike-slip fault, normal fault
   b. Normal fault, strike-slip fault, reverse fault
   c. Strike-slip fault, normal fault, normal fault
   d. Normal fault, reverse fault, strike-slip fault
   e. All of the above

14. Referring to the illustration of a geologic setting:
   a. Identify the tectonic processes at this setting. (subduction, convergent boundary, volcanic arc, divergent boundary)
   b. What is the nature of stress and strain relationships? (compression, folding; tension, faulting; shear, faulting; compression, faulting)
   c. What is the type of faulting exhibited here? (normal, strike-slip, reverse, thrust)
   d. What is the source of stress? (upwelling magma, plate convergence, transform boundary, mid-plate hotspot)

15. Stress produces strain due to the fact that _____ applied over a geologic area results in _____ of that geologic area.
   a. Force; deformation
   b. Uplift; erosion
   c. Breaking; bending
   d. Action; dissolving
   e. Pushing; pulling

16. Volcanic mountains:
   a. Are found only at plate boundaries.
   b. Are found at plate boundaries and hotspots.
   c. Are formed by dip-slip faulting.
   d. Are found only at continent-continent convergent boundaries.
   e. None of the above.

17. The Cascade Range:
   a. Formed in the Precambrian eon.
   b. Formed at a divergent plate boundary.
   c. Is caused by fault-block movement.
   d. Comprises a volcanic arc.
   e. All of the above.

18. Fault-block mountain systems:
   a. Are caused by volcanic activity.
   b. Are related to transform plate boundaries.
   c. Are created by compressive stress.
   d. Are the product of tensional stress.
   e. There are no fault-block mountains.

19. What is a fold-and-thrust belt?
   a. Another name for a V-outcrop pattern
   b. A mountain belt at a spreading center
   c. A mountain belt at a continent-continent convergent margin
   d. The result of crustal thinning by tensional stress
   e. A mountain belt resulting from transform faulting

20. This is a photo of Mount Everest in Nepal. What type of mountain-building process produced this peak? (volcanism, fold and thrust, fault-block)

   What is the type of stress that operates here? (compression, tension, shear)

FURTHER RESEARCH

Test your skill at interpreting a geologic map in the following exercise:

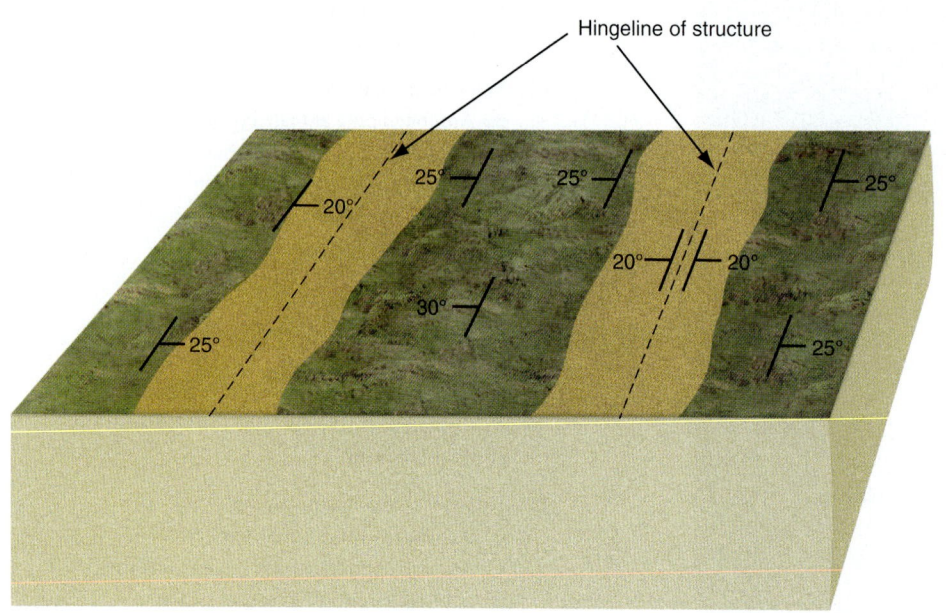

FIGURE 10.31 Identify (draw in) the structures that underlie this geologic map.

Layla Johnson is a sophomore majoring in education. She is taking a course in physical geology to fulfill her core science requirement. She knows that after she graduates and finds a job as an elementary schoolteacher, one of the subjects she will be teaching is earth science. So she is working hard now to learn as much geology as she can.

Johnson's geology teacher has asked her to interpret a geologic map and draw the types of structures found within the crust. After examining **FIGURE 10.31**, help her do the following:

1. Fill in the rock layers in the empty sides.

2. Label the structures in the crust.

3. Determine the approximate angle of dip of the rock layers on the left side of the cross section.

4. Draw arrows indicating the stresses acting on the crust. What type of stress is this?

5. Describe a tectonic setting that might produce this type of deformation in the crust.

ONLINE RESOURCES

Explore more about mountain building on the following Web sites:

The "Structural Geology" entry in the online free Web encyclopedia NASA's Earth Observatory, for images and discussion of Earth surface topography: http://earthobservatory.nasa.gov

Additional animations, videos, and other online resources are available at this book's companion Web site: www.wiley.com/college/fletcher

This companion Web site also has more information about WileyPLUS *and other Wiley teaching and learning resources.*

EARTHQUAKES

CHAPTER CONTENTS & LEARNING OBJECTIVES

GEOLOGY IN OUR LIVES

An earthquake is a sudden shaking of the crust set in motion by the rupture of a fault. Earthquakes produce seismic waves that scientists can measure to improve their understanding of Earth's interior. But they are also dangerous geologic hazards that are responsible for devastating many communities around the world. More than 380 major cities lie on or near unstable regions of Earth's crust, and, because of swelling urban populations, geologists fear that a major earthquake, capable of killing 1 million people, could occur this century. But crowded inner cities are not the only concern; global growth in the human population in the past 100 years means that people now live in areas that were previously remote and thought to be too dangerous for human habitation. By knowing the seismic risk to our communities, designing and constructing buildings that can withstand shaking, and establishing effective earthquake response and disaster assistance networks, we can better manage these hazards in the future.

Q **Where are earthquakes most likely to occur?**

An Earthquake Is a Sudden Shaking of Earth's Crust

LO 11-1 Explain why the risk from earthquakes has increased in recent decades.

An **earthquake** is a sudden shaking of Earth's crust typically caused by the abrupt release of strain that has slowly accumulated at a fault.

What Causes an Earthquake?

A trembling sensation forces you out of your sleep. Groggy, you realize that a mild but persistent vibration seems to be running through your bed. Creaks and groans emanate from the walls, and you hear the windows rattling. A low rumbling fills the air, sounding like a truck driving back and forth past your house. The vibrating intensifies. Two thoughts enter your mind, one right after the other: *The shaking isn't stopping*, and then, *It's getting stronger*. Now fully awake, you suddenly grasp the situation, punctuated by the unmistakable sound of dishes falling to the floor and breaking in the kitchen. Panic sets in as you rush from the room, yelling to other occupants of the house, "It's an earthquake!"

Recall from Chapter 10 that *strain* is the change in shape and/or volume of a rock caused by *stress* (pressure). When stress is applied slowly and continuously over time, most rocks will deform elastically. That is, they will rebound to their original shape if the stress is

removed or when the rock fractures. The same is true along a fault that may be "stuck": the elastic deformation accumulates until the fault becomes "unstuck" (**FIGURE 11.1**). When rock breaks or a fault abruptly ruptures, it releases strain in a form of energy called **seismic waves**, which radiate through Earth's interior and along its surface. Seismic waves cause the shaking felt during an earthquake.

Most earthquakes are associated with the rupture of a fault. But they may also result from volcanic action, landslides, or any other rapid release of energy that shakes the ground. Strain accumulation (elastic deformation) is persistent at the edges of lithospheric plates. That is why plate boundaries have a high frequency of earthquakes, and so are among the most dangerous places on Earth. *Intraplate earthquakes* also occur, and are capable of causing serious damage and loss of life.

Today, earthquakes endanger more people than ever before, for the simple reason that there are more people than ever before. The world's population has more than doubled in the past half century; and, already, half of the world's people live in densely developed urban centers. More than 380 major cities with swelling urban populations lie on or near unstable regions of Earth's crust, leading seismologists to fear that a devastating earthquake capable of killing 1 million people could occur in this century (**FIGURE 11.2**). In any given year, more than 3 million earthquakes occur throughout the world. Fortunately, most of these cause little or no damage because they are relatively small. California alone experiences about 60 minor tremors every day.

As the human population continues to grow, and urban centers to grow with it, the infrastructure on which modern life depends (e.g., high-rises, bridges, pipelines, communication towers, energy grids, emergency services, etc.) will expand into previously undeveloped regions. In concert with that expansion, the risks posed by earthquakes will rise, to unprecedented levels. A century ago, even a large

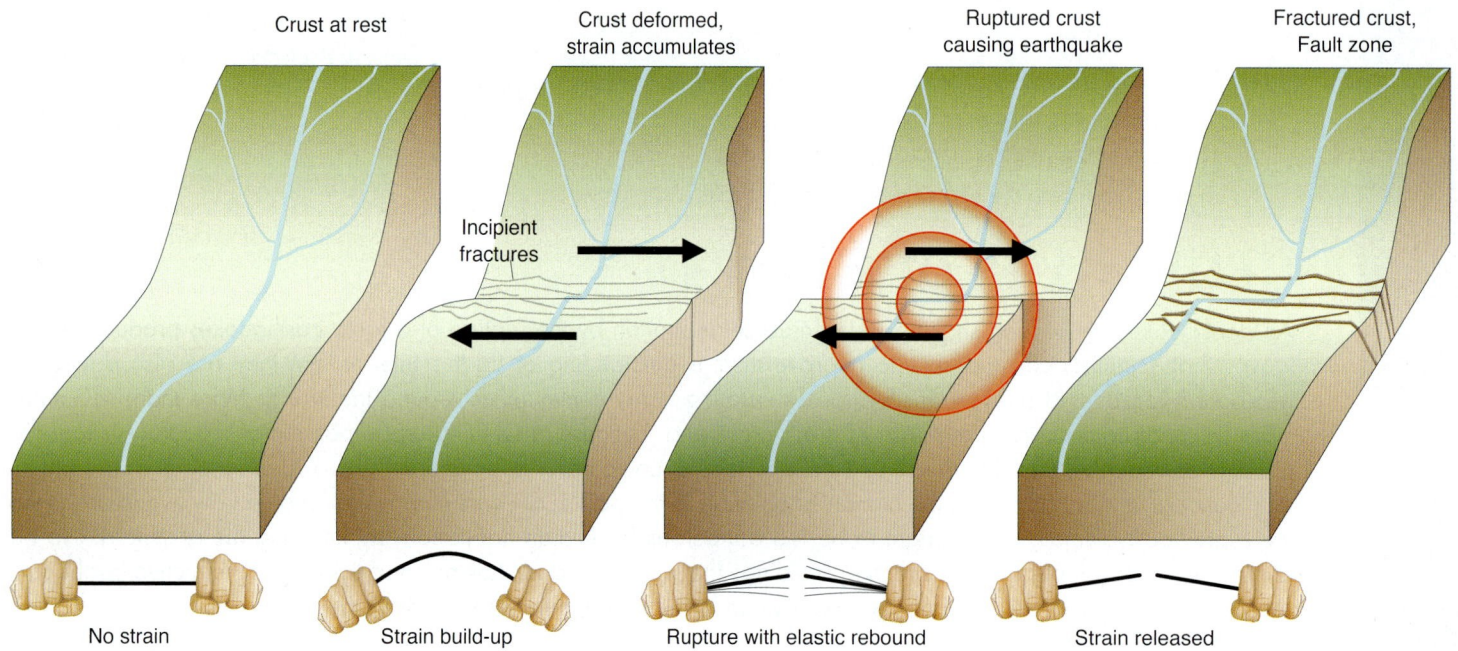

Crust at rest

Crust deformed, strain accumulates

Ruptured crust causing earthquake

Fractured crust, Fault zone

Incipient fractures

No strain

Strain build-up

Rupture with elastic rebound

Strain released

FIGURE 11.1 When Earth's crust is stressed (subjected to pressure), it will accumulate strain (deformation). When strain accumulates slowly, rocks can behave elastically—they spring back to their original shape when the stress is removed. This can happen when they break or when a fault ruptures. Bending and breaking a stick is the same process. The strain is released in the form of energy called seismic waves, which pass through Earth's interior and along its surface.

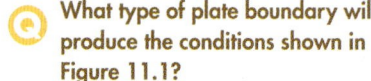
What type of plate boundary will produce the conditions shown in Figure 11.1?

FIGURE 11.2 Seismic waves instigate shaking, which can damage buildings.

Roger Ressmeyer/Corbis

How do buildings react when they are shaken?

earthquake could go unnoticed when it occurred in a remote area. But today, with few remote areas left on the planet, even a small earthquake may be felt by thousands of people; a large one, by millions.

Case in point is the 2008 *Sichuan earthquake*, which left over 80,000 dead and destroyed more than 4 million homes in a region that just two decades ago was largely open countryside. China's rapid economic expansion had turned the area into a heavily populated one; and unfortunately, buildings there were not constructed to withstand such violent shaking; worse, emergency services were unprepared for a seismic event of that magnitude.

By studying the nature of the destruction caused by past quakes (**TABLE 11.1**), we have learned how to build strong yet flexible buildings that can withstand strong shaking. Nevertheless, earthquakes occur all too frequently, and often with deadly consequences, as evidenced by a number of recent events: the 2003 Bam quake in Iran, which killed over 26,000 people; the 2004 Sumatra quake in Indonesia, which spawned a tsunami and killed 230,000; the 2005 Kashmir earthquake, which killed over 70,000; the 2008 Sichuan earthquake, which killed 87,000; the 2010 Haiti earthquake, which killed more than 230,000; and the 2010 Chile earthquake, which killed hundreds. These events leave no doubt that the risks associated with earthquakes remain high in our modern world.

How frequently do major quakes occur? Since 1900, 114 quakes have led to the deaths of more than 1,000 people each. That is a global average of more than one quake per year. Clearly, research on the distribution and causes of earthquakes is critically important to improving the safety of communities worldwide.

TABLE 11.1 Quakes Throughout History Causing at Least 50,000 Deaths

Year (Date)	Deaths	Location (Country, Area)	Comments
0856 (December 22)	200,000	Iran, Damghan	Few details available
0893 (March 23)	150,000	Iran, Ardabil	Few details available
1138 (August 9)	230,000	Syria, Aleppo	Few details available
1268 (no specific date)	60,000	Asia Minor, Silicia	Few details available
1290 (September 27)	100,000	China, Chihli	Few details available
1556 (January 23)	830,000	China, Shansi	Total building collapse in some towns; large ground fractures
1667 (November)	80,000	Caucasia, Shemakha	Few details available
1693 (January 11)	60,000	Italy, Sicily	Few details available
1727 (November 18)	77,000	Iran, Tabriz	Few details available
1755 (November 1)	70,000	Portugal, Lisbon	Set off deadly tsunami
1783 (February 4)	50,000	Italy, Calabria	Few details available
1908 (December 28)	72,000	Italy, Messina	Set off deadly tsunami
1920 (December 16)	200,000	China, Haiyuan	Major landslides
1923 (September 1)	143,000	Japan, Kwanto	Great Tokyo fire
1948 (October 5)	110,000	Soviet Union (Turkmenistan)	Widespread building collapse
1970 (May 31)	70,000	Chimbote, Peru	Great rockslide; floods
1976 (July 27)	255,000 reported; 655,000 probable	China, Tangshan	Highest death toll in the 20th century
1990 (June 20)	50,000	Western Iran	Landslides
2004 (December 26)	230,000	Sumatra	Set off deadly tsunami
2005 (October 8)	86,000	Kashmir, India, and Pakistan	3.3 million left homeless
2008 (May 12)	87,000	Sichuan, China	4.8 million homes destroyed
2010 (January 13)	>230,000	Haiti, Port-au-Prince	Primarily building collapse

EXPAND YOUR THINKING

What causes the shaking felt during an earthquake?

11-2 There Are Several Types of Earthquake Hazards

LO 11-2 List and describe the different types of earthquake hazards.

Scientists cannot predict *when* an earthquake will happen, but it is possible for them to define areas that are especially vulnerable to *earthquake hazards*, defined as dangerous events caused by earthquakes. The principal earthquake hazards are: 1) ground shaking, leading to building collapse; 2) landslides; 3) fire; and 4) tsunamis. We discuss each hazard in turn in this section.

Ground Shaking

Ground shaking caused by earthquakes actually poses little direct danger to people; an earthquake cannot shake a person to death. The danger lies in the effects shaking produces, which *can* lead to damage and loss of life.

One of these dangers is *sediment liquefaction*. Liquefaction occurs when shaking causes groundwater to rise to the surface and mix with sandy or muddy soil. The sediment loses its strength, and any buildings or roads on it will be lost as they sink into the wet, soupy quicksand (or quickclay). The sediment may firm up again after the earthquake, as the groundwater drains to its natural location deeper in the crust.

Shaking also sets vibration in motion at the base of a building, a dynamic process that many buildings are not designed to withstand. If you place strain onto the structure and then let it snap

CRITICAL THINKING

Investigating Earthquake Activity

Can you connect earthquake activity with tectonic processes? To answer this question, go to the U.S. Geological Services (USGS) Earthquake Hazards Program home page at http://earthquake.usgs.gov. There you will find the Real-time Earthquake Map:

1. Click on the map to see a display of earthquakes that have occurred around the world in the past week.

2. Use the slider bar on the left side to zoom in on the United States.

3. Move the map by "grabbing" it with your cursor. You can click on the map and enlarge any area to study the location, magnitude, and details about each quake that is shown; and click on a specific quake to get information such as its depth and time.

4. Pick one quake on the world map and one on the U.S. map and then answer the following questions for each quake.

   a. Where is this quake located?

   b. What is the tectonic environment at this location?

   c. Which tectonic processes most likely produced this quake?

5. Study the seismic hazard map in **FIGURE 11.3** (or go to http://earthquake.usgs.gov/hazards/products/graphic2pct50.jpg). Is the U.S. quake in a known hazard area? What is its risk level? Describe the probability of seismic shaking in this area.

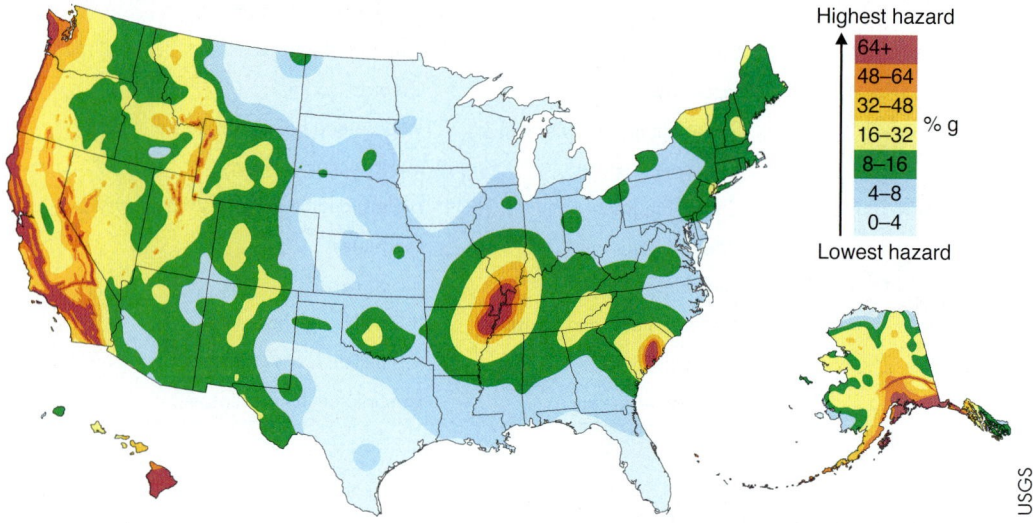

FIGURE 11.3 The U.S. Geological Survey uses data on past earthquakes to determine the probability of ground shaking within a certain period. Colors on this map show the level of horizontal shaking that have a 1-in-10 (10%) chance of being exceeded in a 50-year period. Shaking is expressed as a percentage of *g*, the acceleration of a falling object due to gravity. Areas in "hot" colors are at greatest risk for seismic shaking that can cause damage. This information is useful for purposes such as developing safer building codes.

Q Why does the West Coast have a high seismic shaking rating?

back, it will sway back and forth, at an amplitude that decreases with time. The swaying will happen at a particular frequency called the *natural frequency*. If the ground shakes with the same frequency as a building's natural frequency, the amplitude of sway will grow larger such that the shaking is in *resonance* with the building's natural frequency. Seismic shaking that matches a building's natural frequency imposes the most strain on the building, and so can lead to its collapse.

The U.S. Geological Survey (USGS) maintains an earthquake research program that defines probable levels of building shaking throughout the country. This agency produces maps of **seismic shaking hazards** that are used by engineers and city planners in designing and constructing buildings, bridges, highways, and utilities that are better able to withstand earthquakes (Figure 11.3). To learn more, go to http://eqhazmaps.usgs.gov. And to try your hand at using such information, work the exercise in the Critical Thinking feature, "Investigating Earthquake Activity."

Landslides

In addition to bringing down buildings, ground shaking can cause steep hillsides to collapse. This process, known as *mass wasting*, may, however, occur for many reasons other than earthquakes, such as: *undercutting* the base of a steep hill by, for example, lateral river erosion or highway construction; heavy rainfall that saturates and lubricates sediments or rock layers; even lightning strikes have been known to trigger a mass wasting event. Still, the mass wasting triggered by ground shaking during an earthquake is extremely common (**FIGURE 11.4**). (Mass wasting is discussed in greater detail in Chapter 16.) As noted previously, many communities today—tens of thousands, in fact—around the world are located at the base of steep hills and mountains in earthquake-prone areas. This is, to say the least, an extremely dangerous situation.

Fire

Fires that are ignited in the aftermath of an earthquake, by broken gas and power lines or tipped-over wood or coal stoves, also are very serious quake-related problems. Often, water lines and emergency equipment are damaged during an earthquake, and roads are rendered impassable, therefore making it impossible for firefighters and other emergency personnel to get to affected sites, much less do their work.

One horrible example of the damage fire can do in an earthquake-disrupted area occurred in Kobe, Japan, in 1995. According to official statistics, 5,472 people were killed and more than 400,000 were injured. Most of the destruction and deaths were not, however, a direct result of the earthquake but of the hundreds of fires that were set off in its wake. Newer buildings that had been built to withstand the seismic stress did so, but many of the city's older residences were traditional wooden structures. They gave way easily under their heavy tile roofs (designed only to withstand hurricanes), in many cases trapping their occupants. Even if rescue crews had been able to reach them, though, the water supply was not functioning, leaving fires to rage throughout neighborhoods for days after the quake, with authorities unable to control them (**FIGURE 11.5**).

Tsunamis

If an earthquake also generates a **tsunami**—a series of waves in the ocean triggered by rapid movement of the seafloor—coastal damage can be widespread, and deadly. Although most people call a tsunami a tidal wave, in fact it has nothing to do with the ocean tides. Any movement of the seafloor, such as an undersea landslide or a submarine volcanic explosion, can produce a tsunami. That said, the most common cause is faulting associated with an earthquake (**FIGURE 11.6**) that violently shifts the seafloor and displaces the overlying water column, launching a wave.

Mian Khursheed/Reuters/Corbis

FIGURE 11.4 These landslides were triggered by the 2005 Kashmir earthquake.

Although seismic shaking set off these landslides, undercutting had already made the cliff unstable. What caused the undercutting?

AFP/Getty Images, Inc.

FIGURE 11.5 Ruptured power and gas lines ignited hundreds of fires in Kobe, Japan, following the 1995 earthquake there.

Why was it difficult for emergency crews to suppress the fires in Kobe?

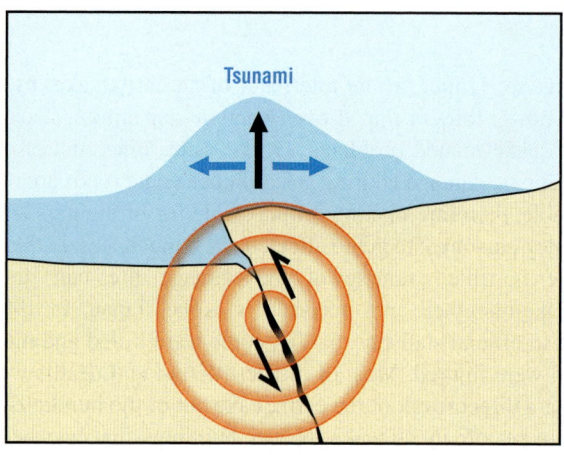

FIGURE 11.6 A tsunami is set off by sudden movement of the seafloor. Faulting associated with an earthquake is the most common cause of tsunamis.

> **What is the major difference between a 2-meter wind wave and a 2-meter tsunami? Why is one more dangerous?**

A tsunami is not to be confused with the more common *wind wave* that we are used to seeing on the sea surface. Tsunamis are imperceptible in the open ocean—except to instruments—but when they run aground, they have the force to flood across the shoreline for many minutes, inundating everything in their path.

The period of time between successive tsunami wave crests may be anywhere from five minutes to an hour or more, depending on how far away the waves originated; and the time from the front of the wave crest to the back may be many minutes. Hence, when one wave hits the shore, it floods the land for several minutes until the back of the wave arrives. The famous Sumatra earthquake in 2004 was accompanied by a deadly tsunami that flooded lands over 30 meters above sea level, destroying thousands of coastal villages.

EXPAND YOUR THINKING

How would you design a city so that it could recover quickly from an earthquake?

(11-3) The Elastic Rebound Theory Explains the Origin of Earthquakes

LO 11-3 Define the elastic rebound theory.

Rocks accumulate strain elastically. That is, after they rupture, they return to their original shape, and while doing so, they generate vibrations that become seismic waves.

Elastic Rebound Theory

Earthquakes occur when rocks in the crust break at a fault. The point where slippage first occurs is called the **earthquake focus**, and the **epicenter** is the geographic spot on Earth's surface that is located directly above the focus (**FIGURE 11.7**). When an earthquake occurs, energy is released from the focus and travels in all directions in the form of **surface waves** and **body waves**. Surface waves travel along Earth's surface like ripples travel across a pond. Body waves travel into and through Earth's interior, and when they emerge on the other side of the planet, they can be detected by instruments known as **seismometers**.

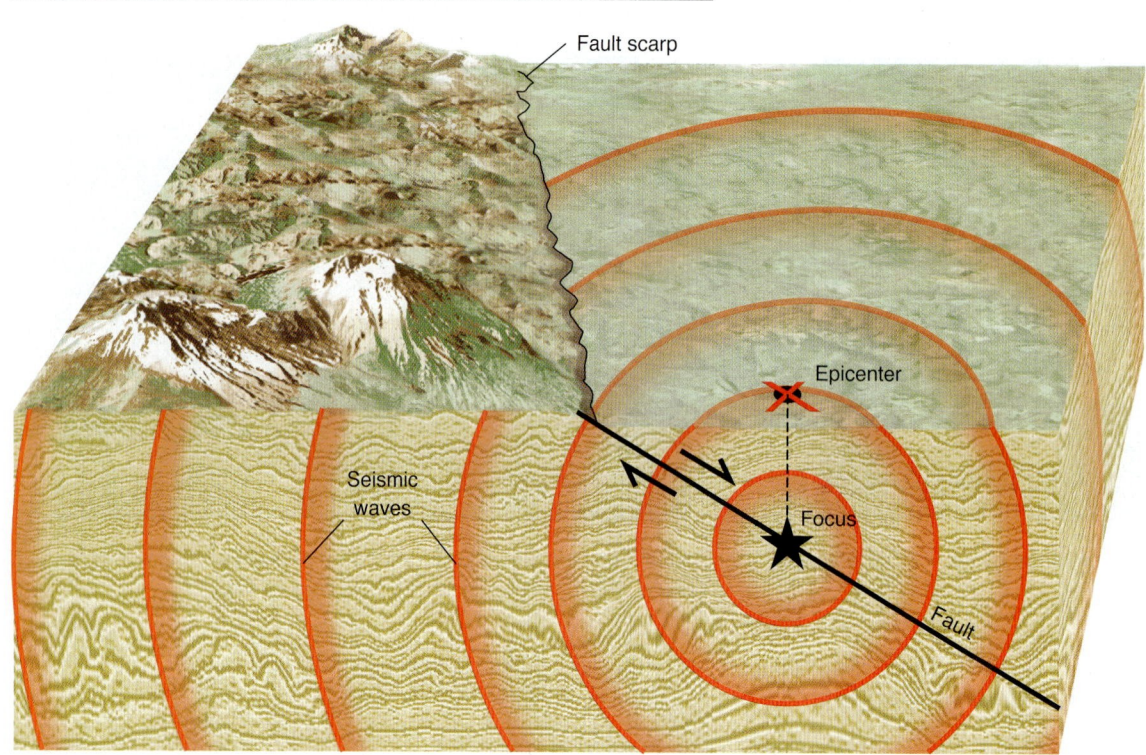

Fault scarp

Seismic waves

Epicenter

Focus

Fault

FIGURE 11.7 Earthquakes usually are associated with slippage along a fault.

> **What type of fault is shown in this figure?**

© Marli Bryant Miller

FIGURE 11.8 Strain retained in the crust is suddenly released when rocks fracture or slip along a fault. As the rock returns elastically to its original shape, it generates vibrations that turn into seismic waves that travel through Earth's interior and along Earth's surface.

If you were to study this fault plane closely, what evidence would you find that movement had occurred between the two sides of the fault?

Although we generally hear about earthquakes only when they are large and dramatically damaging, they are, in fact, an everyday occurrence on our planet. According to the U.S. Geological Survey, more than 3 million earthquakes occur every year. That translates to about 8,000 a day, or 1 every 11 seconds. The vast majority of them are extremely weak, and so go unnoticed, except by the sensitive seismometers designed to measure them. It is the big quakes, such as the 2010 Haiti earthquake, that occur in highly populated areas, and usually cause horrifying devastation, that grab our attention, and the attention of the media.

The **elastic rebound theory** describes the mechanism by which earthquakes are generated. The theory was formulated in the aftermath of the devastating 1906 San Francisco earthquake. After the quake, geologists discovered a fault that could be followed along the ground in a more or less straight line for 435 kilometers. The crust on one side of the fault had slipped laterally relative to the crust on the other side by up to 7 meters; the average amount of slippage was 4.7 meters.

The fault and slippage raised the curiosity of scientists, especially since prior to that time no one had been able to explain what happens within Earth's crust to set off earthquakes. Previously, it had been assumed that stresses causing earthquakes must be located close to the earthquakes themselves; the concept of plate tectonics was as yet unknown.

After studying the fault line created by the 1906 quake, Harry Fielding Reid (1859–1944) of Johns Hopkins University postulated that earthquakes are caused by distant forces. He suggested that these forces cause a gradual buildup of *elastic energy* in the crust over tens to thousands of years, slowly accumulating strain (deformation) and distorting the crust in much the same way that a stick stores elastic energy by bending before it breaks (refer back to Figure 11.1.)

Eventually, a fault (**FIGURE 11.8**), or preexisting weakness in the rock, is no longer able to accommodate any more strain, and it ruptures. Slippage of rock at the focus produces stress that travels along the fault, leading to displacement of both sides of the fault plane until the strain has been largely released. At the time of the earthquake, strained rock snaps back elastically into position, although it experiences net displacement (breakage) in one direction. Vibrations of the rock as it returns to its original shape trigger seismic waves. Reid described this process as "elastic rebound." It can be compared to gradually pulling a rubber band until it is stretched as far as possible; the strain is released when the band breaks and the two pieces snap back to their original shape.

Foreshocks and Aftershocks

A necessary requirement of residing in a region that is prone to earthquakes is learning to live with **foreshocks** and **aftershocks**. As strain accumulating in the crust approaches the elastic limit of the rock, earthquakes tend to occur in *clusters*. The largest quake in a cluster is the **mainshock**; those occurring before it are foreshocks, and those after it are, logically, aftershocks. Foreshocks generally are felt in the immediate vicinity of the focus of the mainshock, whereas aftershocks may be spread out along the rupture surface (the fault plane), as well as at other nearby faults (**FIGURE 11.9**).

You can learn about the largest fault rupture ever recorded by reading the Geology in Action feature titled "The Sumatra–Andaman Earthquake: The Quake That Moved the Entire Planet." Also read the Geology in Action feature "The Tohoku Earthquake: the fifth most powerful in recorded history."

EXPAND YOUR THINKING

What sequence of events characterizes the buildup of elastic energy in the crust?

(a)

Prior to earthquake

Foreshock foci

Focus of future large earthquake

Strain buildup and deformation

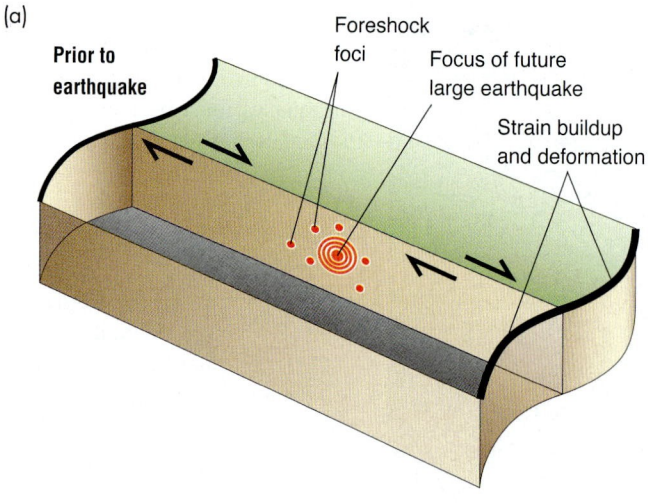

(b) **After earthquake**

Aftershock foci

Focus of recent large earthquake

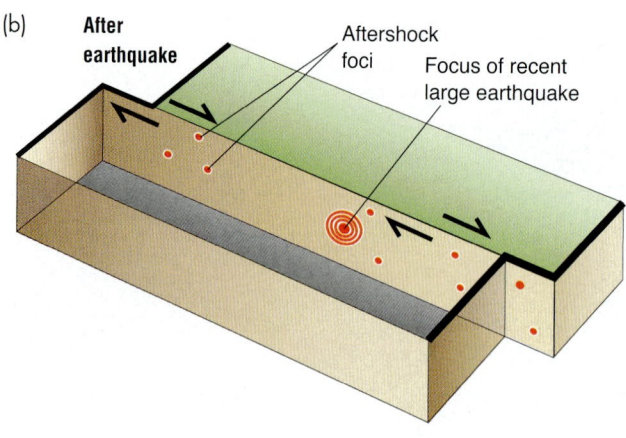

FIGURE 11.9 (a) Foreshocks tend to cluster around the future focus of a mainshock. (b) Aftershocks may be spread out along the rupture surface of the mainshock, as well as at nearby faults. (c) The mainshock is the largest among a cluster of earthquakes.

(c)

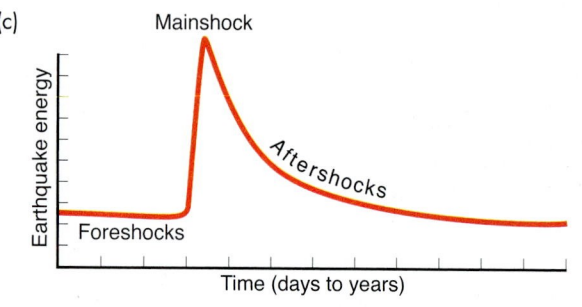

Mainshock

Earthquake energy

Aftershocks

Foreshocks

Time (days to years)

Q How can you tell whether an earthquake is a foreshock, mainshock, or aftershock?

GEOLOGY IN ACTION

The Sumatra–Andaman Earthquake: The Quake That Moved the Entire Planet

The Sumatra–Andaman earthquake of 2004 set many records: It was the most powerful quake in more than 40 years; it triggered a devastating tsunami that killed more people than any other natural catastrophe in modern history; it lasted an unbelievable 500 to 600 seconds (most quakes last only a few seconds); and it shook the ground everywhere on Earth's surface. Weeks later, the planet was still trembling. The quake was produced by the longest fault rupture ever observed: 1,255 kilometers—the distance from Los Angeles, California, to Portland, Oregon. The fault shifted continuously for 10 minutes, also a record.

The earthquake and the resulting tsunami (**FIGURE 11.10**), which swept across the Indian Ocean and was recorded on instruments around the world, killed approximately 230,000 people in 11 countries and left over 1 million people homeless. Ground movement of as much as 1 centimeter occurred everywhere on Earth's surface, although in most areas it was too small to be felt.

ASSOCIATED PRESS

FIGURE 11.10 The tsunami (shown is Banda Aceh, Indonesia); triggered by the 2004 Sumatra–Andaman earthquake devastated the shores of Indonesia, Sri Lanka, India, Thailand, and other countries, with waves up to 30 meters high. It caused serious damage and deaths as far away as Port Elizabeth, South Africa, 8,000 kilometers from the source.

The Tohoku Earthquake: The Fifth Most Powerful in Recorded History

At 2:46 on the afternoon of March 11, 2011, a powerful undersea *megathrust earthquake* occurred off the coast of Japan. The Tohoku earthquake and tsunami, as it's now known, centered on the Japan Trench, where the Pacific Plate subducts beneath the Okhotsk Plate. The event measured 9.03 on the Richter scale (described in section 11–8), making it the fifth most powerful earthquake in the world and the largest in Japan's history.

Megathrust earthquakes occur at ocean-ocean or ocean-continent convergent margins where the shallow dip of a subducting plate causes large sections to get "stuck" (as explained in Chapter 3, section 3–9). As strain slowly builds, the overlying plate is subjected to shortening and crustal thickening. Once the stuck portion of the plate releases, the overlying plate undergoes rapid elastic extension, often moving the seafloor and setting off a tsunami. Megathrust earthquakes are the most destructive type of quake, responsible for all six earthquakes since 1900 at magnitudes of 9.0 or greater.

At the epicenter of the Tohoku earthquake, the Pacific seafloor rose vertically about 7 meters. The quake generated a massive tsunami, with waves reaching 40.5 meters in height and traveling up to 10 kilometers inland. Elastic release of strain in the overriding plate led to crustal extension and as much as 1.2 meters of subsidence on the main Japanese island of Honshu, consequently exposing the human community to widespread flooding.

More dramatically, the strain release shifted portions of northeast Japan up to 2.4 meters closer to North America; the Pacific Plate shifted to the west by 20 to 40 meters. One published report concluded that the seafloor near the epicenter moved 50 meters to the east-southeast as a result of the earthquake.

Shaking and flooding precipitated by the combined effects of the earthquake and the tsunami resulted in massive, widespread damage (**FIGURE 11.11**):15,882 confirmed deaths, 6,142 injuries, and 2,668 missing persons; 129,225 buildings collapsed, 254,204 "half collapsed," and another 691,766 buildings were partially damaged. Over 4 million households were left without electricity, and 1.5 million without water; 7 meltdowns occurred at the nearby Fukushima Nuclear Power Plant, releasing radiation into the atmosphere and the ocean.

The World Bank estimated the total economic costs of the Tohoku earthquake and tsunami at $235 billion, making it the world's costliest natural disaster.

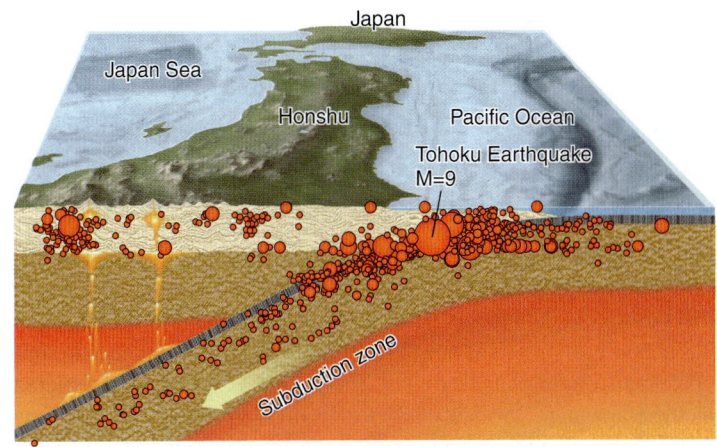

(a)

(b)

AFP/Getty Images

(c)

STR/EPA/Newscom

FIGURE 11.11 The March 11th Tohoku earthquake and its tectonic setting in detail. (a) Earthquake foci along the subducting Pacific Plate, are shown in red from the past 4 decades. The size of each circle is proportional to the earthquake magnitude. (b) Flooding by the tsunami reached an elevation of 40.5 meters, and (c) caused an estimated $235 billion in damage, making it the worlds most costly disaster.

11-4 Most Earthquakes Occur at Plate Boundaries, but Intraplate Seismicity Is Also Common

LO 11-4 Describe the relationship of earthquakes to plate boundaries.

Earthquakes usually are associated with slippage along a fault that is most often (but not always) related to a plate boundary. Plate movement, and the accumulation of strain at a plate boundary, is responsible for causing most earthquakes.

In 1969, researchers Muawia Barazangi of Cornell University, and James Dorman of Columbia University, published a map showing the locations of earthquakes occurring between 1961 and 1967. This map has become renowned as one of the primary pieces of evidence to support the theory of plate tectonics. The earthquake epicenters depicted in the Barazangi/Dorman map clearly lined up along long, narrow zones that define the boundaries of lithospheric plates. The interiors of the plates themselves are, for the most part, not prone to the high numbers of quakes that occur at plate boundaries (**FIGURE 11.12**), and parts of plate interiors are earthquake free, or *aseismic*. Still, plate interiors may experience some seismic activity.

Intraplate Earthquakes

Although the great majority of earthquakes occur at plate boundaries, some occur far from modern plate margins, as revealed in Figure 11.12. Many of these *intraplate earthquakes* probably are linked to stresses transmitted into plate interiors that originate at the plate

edge. Often these quakes have shallow foci and are related to the reactivation of old faults that are normally quiet. These faults may be left over from a period of tectonic action that has since ceased, or they may mark the establishment of a new region of deformation.

Intraplate volcanism, such as at a hotspot, is another source of intraplate *seismicity*, the relative distribution and frequency of earthquakes. The region around Yellowstone National Park, Wyoming, is famous for its frequent earthquakes, where visitors who have never before felt a quake are thrilled to experience their first while viewing the park's hot springs and geysers.

The Big Island of Hawaii, location of Kilauea Volcano at Volcano National Park, one of the world's most active volcanoes, is the most seismically active location in the United States despite the fact that it is thousands of kilometers from the nearest plate boundary. Hawaiian quakes are the result of magma moving through the crust and breaking rock as it intrudes, or when the crust ruptures under the great weight of a massive shield volcano (see Chapter 6).

Earlier in the chapter we mentioned one of the worst intraplate seismic events to occur in the modern era, the Sichuan earthquake that occurred on May 12, 2008, in that province in south-central China. As noted, it has been estimated that 87,000 lives were lost and that damage exceeded $20 billion. The event is thought to be related to the ongoing collision of the Indian–Australian Plate with the Eurasian Plate and the resulting fracturing of the Longmenshan Thrust Fault at a depth of 20 kilometers. This seismicity is the latest in a long history of damaging earthquakes extending through the Himalayan fold-and-thrust belt, across the Tibetan Plateau and northward into central and northern China.

The presence of modern plate boundaries cannot, however, satisfactorily explain the 1811 or 1812 earthquake at New Madrid, Missouri, or the 1886 earthquake at Charleston, South Carolina, in North America. Still, there may be a past tectonic explanation for these quakes.

The New Madrid earthquake occurred along a *failed rift* known as the *Reelfoot Rift*. A failed rift is a weakened zone of normal faults

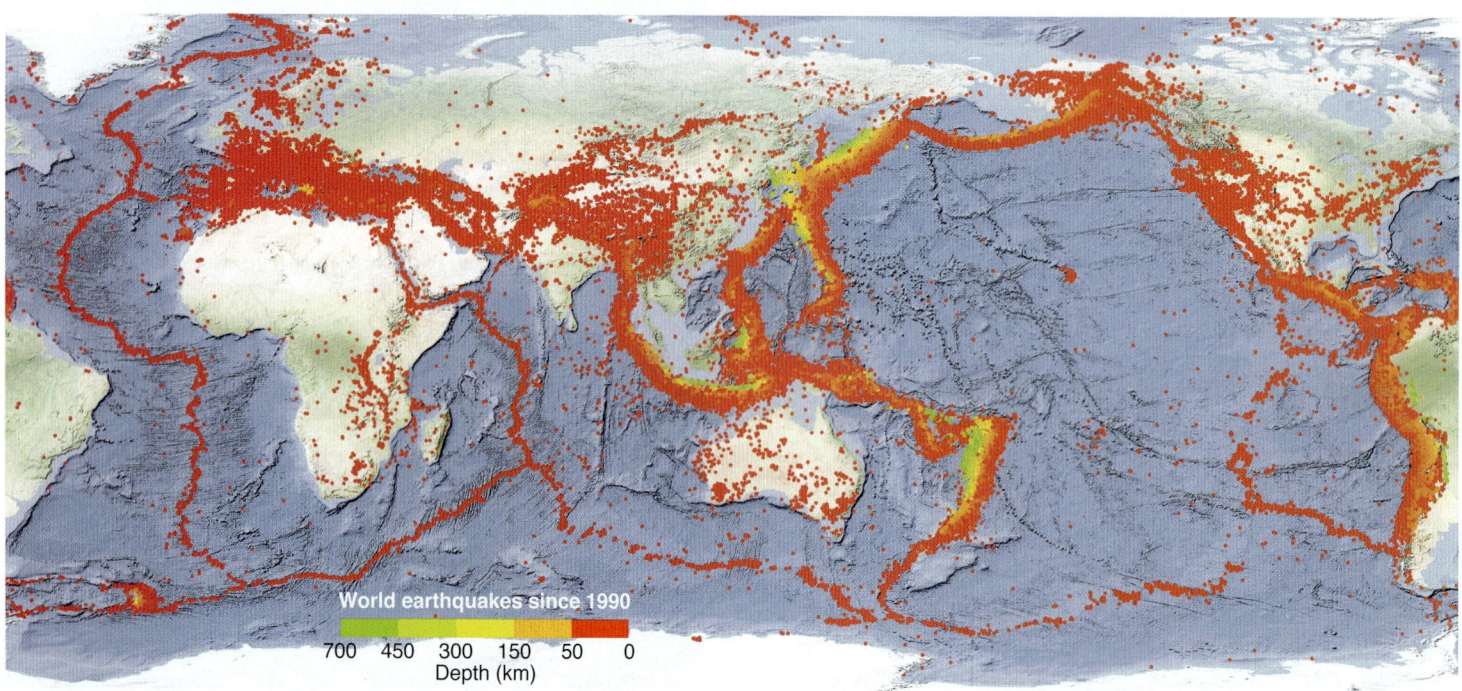

World earthquakes since 1990

700   450   300   150   50   0
Depth (km)

USGS

FIGURE 11.12 This map depicts the global distribution of foci—color coded by depth—for moderate to large earthquakes since 1990.

Deep earthquakes occur in areas where oceanic crust is being actively subducted. About 90% of all earthquakes occur at depths of up to 100 kilometers. Why do subduction zones have "deep" earthquakes?

(horsts and grabens) in the crust, in this case left behind from a partial separation of continents that took place 550 million years ago, at the end of the Proterozoic eon. This rift rests within the crust below Missouri, Arkansas, Tennessee, Kentucky, and Illinois.

The New Madrid fault area averages 2 earthquakes a week; but in the course of a typical year, only 8 to 10 quakes are felt because most seismic events in the region are rather minor. At times, though, the fault sets off larger earthquakes. Geologists who study earthquakes estimate that every 70 to 90 years there is a moderate to strong earthquake, and every 250 to 500 years a major earthquake, capable of causing severe damage.

On April 18, 2008, a moderate tremor was felt along the Wabash Valley Seismic Zone, an extension of the Reelfoot Rift located over 200 kilometers to the northeast. Strain released by this event may be related to stress along the Reelfoot Rift and other associated tectonic features in this intraplate setting.

The Charleston earthquake in 1886 is not so easily explained. Geologists have been unable to identify the fault zone along which this quake occurred. Yet quakes continue to occur in South Carolina: The state's seismic network has recorded approximately 10 per year over recent decades, 70% of which occurred in the high-intensity area of the 1886 Charleston earthquake. The strongest one happened on August 21, 1992. From 1989 to 1993, there was an increase in earthquake activity in the area, perhaps the result of stresses transmitted inward from the boundaries of the North American Plate.

Earthquake Prediction

How can plate tectonics help in earthquake prediction? To begin to answer that, consider first that earthquakes occur in four settings: 1) intraplate regions, 2) divergent settings, 3) convergent margins, and 4) transform settings. (Types 2 to 4 are discussed in the next section.) Thus, it is possible to identify regions of Earth's surface where large earthquakes can be expected to take place. We know that each year about 140 earthquakes of moderate strength or greater will occur in these settings, which together account for about 10% of Earth's surface.

Although it is possible to predict *where* quakes are likely to happen, on a worldwide basis it is not possible to determine precisely *when* they will occur. The cyclical processes that drive plate tectonics—strain buildup followed by an earthquake followed by more strain buildup—have been going on for hundreds of millions of years; as a result, plates have moved, on average, several centimeters per year. But at any moment in geologic time—say, the year 2015—we do not know exactly what point in the cycle of strain buildup and strain release has been reached by any given portion of crust.

Only by monitoring in great detail the stress and strain in small areas can scientists hope to predict when renewed seismic activity is likely to take place; even then, such predictions will amount to a statement of statistical probability, which lacks specific information about timing, location, or the amount of energy released.

In summary, plate tectonics is a powerful but broad tool for predicting earthquakes. It tells us where 90% of Earth's major earthquakes are likely to occur, but it cannot tell us much about when they will occur.

The Geology in Our Lives feature, "Dates of Major U.S. Earthquakes in the Past 50 Years," summarizes the details of several quakes in this country.

ⓠ EXPAND YOUR THINKING

Come up with a testable hypothesis explaining intraplate earthquakes.

 **GEOLOGY IN OUR LIVES**

Dates of Major U.S. Earthquakes in the Past 50 Years

- **October 15, 2006:** In Hawaii, numerous people suffered injuries; at least 1,173 buildings were damaged; and landslides blocked roads and isolated communities. Day-long power outages were reported throughout the state, and damage was estimated at $73 million.

- **November 3, 2002:** One of the largest earthquakes on U.S. soil occurred in central Alaska, leaving massive cracks in the ground but causing minimal damage to property and resulting in few injuries and no deaths.

- **February 28, 2001:** A strong earthquake southwest of Seattle damaged the Washington State Capitol, closed the Seattle–Tacoma International Airport, and injured about 400 people.

- **October 16, 1999:** A major earthquake in California's Mojave Desert derailed an Amtrak train and knocked out power to thousands. No serious damage or injuries were reported.

- **January 17, 1994:** A strong earthquake in the Northridge section of Los Angeles killed 60 people, injured 9,000, and caused $25 billion in damage.

- **June 28, 1992:** A major earthquake struck in Landers, California, and a second hit near Big Bear Lake in the San Bernardino Mountains; three people died.

- **October 18, 1989:** The so-called World Series earthquake shook the San Francisco Bay area, killed 63 people, and caused $7 billion in damage.

- **October 1, 1987:** An earthquake in Whittier, California, killed eight people.

- **February 9, 1971:** In California's San Fernando Valley, a strong quake left 65 people dead.

- **March 28, 1964:** Known as the Good Friday earthquake, the second largest quake in history struck near Prince William Sound, Alaska, killing 128 people.

11-5 Divergent, Convergent, and Transform Margins Are the Sites of Frequent Earthquake Activity

LO 11-5 Compare and contrast seismicity at divergent, convergent, and transverse plate boundaries.

Plate boundaries, as noted at the beginning of the chapter, are characterized by frequent earthquakes. In this section, we discuss divergent boundaries, convergent margins, and transform settings in more detail.

Divergent Seismicity

The narrow earthquake belt extending along the world's mid-ocean ridges provides direct support for the concept of seafloor spreading. The lithosphere is thin and weak at divergent boundaries, preventing large amounts of strain from accumulating; furthermore, seismic activity tends to be of low to moderate energy levels and to occur at shallow depths. Major earthquakes are, therefore, rare in this region, though moderate quakes are more common and can be extremely damaging. Volcanic activity along the axis of the ridges is associated with this type of seismicity.

Earthquakes at divergent boundaries are related to normal faulting, whereby blocks of crust along the rift valley walls slide inward in the overall tensional environment (**FIGURE 11.13**). Strike-slip faulting is also a source of seismicity at spreading centers, when crustal slippage occurs along the transform boundaries found on most mid-ocean ridge systems.

Convergent Seismicity

Earth's largest and most active seismicity occurs at convergent boundaries, where one plate subducts beneath another in a zone of accumulating compressive stress. These regions of plate interaction are characterized by progressively deeper earthquakes along a line proceeding from the trench across the overriding plate (**FIGURE 11.14**). In most cases, earthquake foci are located within or along the surface of the subducting plate. These foci define a steeply dipping plane called the **Wadati–Benioff Zone**, or WBZ, that corresponds to faulting associated with the subducting oceanic slab. The WBZ is named for the two seismologists who independently discovered the zones: Hugo Benioff (1899-1968) and Kiyoo Wadati (1902-1995).

The epicenters of deep-focus quakes are located far from the surface expression of the plate boundary, an illustration of how far below the overriding plate a subducted slab extends as it is recycled. These epicenters appear to be intraplate earthquakes but are in fact tied to the subduction of oceanic lithosphere deep below the surface, and are part of the process of tectonic convergence.

Transform Seismicity

Tectonic boundaries may be characterized by strike-slip faulting, whereby two plates slip past each other at a transform margin. At transforms, earthquakes are shallow, running only as deep as 25 kilometers. Two excellent examples of dangerous transform seismicity are found at the *San Andreas Fault* in California and the *North Anatolian Fault* in Turkey. Both are right lateral strike-slip faults (**FIGURE 11.15**).

At the San Andreas Fault, the Pacific Plate moves northwestward relative to the North American Plate. This movement is the impetus of earthquakes along the fault. The San Andreas is actually the "master" fault of a complex fracture network that cuts through western California. The entire San Andreas system is more than 1,200 kilometers long and extends to depths of about 16 kilometers. The fault zone is a complex corridor of crushed rock, ranging from a few hundred meters to 2 kilometers wide. Many smaller faults, the product of regional stresses emanating from the master fault, branch out from and rejoin the zone. During the 1906 San Francisco earthquake, roads, fences, and rows of trees that crossed the fault were all offset by several meters. In each case, the ground west of the fault moved northward relative to the ground east of the fault.

Turkey has a long history of large seismic events that occur as a series of progressive adjacent earthquakes. Starting in 1939, the North Anatolian Fault set off a sequence of major earthquakes. More recently, the eleventh, in 1999, killed over 17,000 and left approximately half a million people homeless. Researchers had estimated that there was a 12% chance of this earthquake occurring sometime in the 30 years between 1996 and 2026. On March 8, 2010, the North Anatolian Fault shook again, killing at least 51 people and

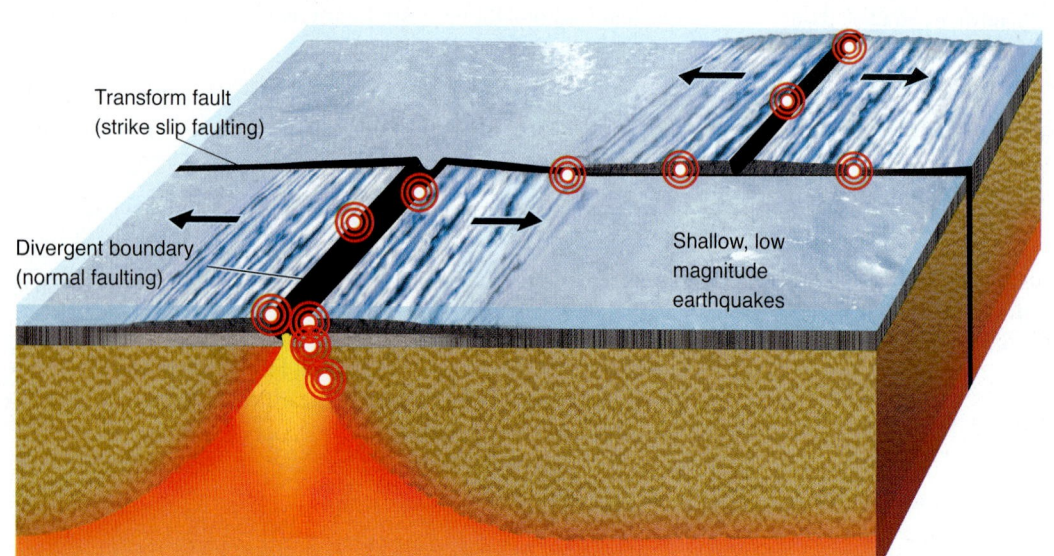

FIGURE 11.13 Normal faulting in the tensional stress environment of spreading centers causes shallow, low-energy earthquakes. Strike-slip faulting at transform faults also produces shallow earthquakes.

Explain the absence of deep-focus earthquakes at divergent and transform boundaries.

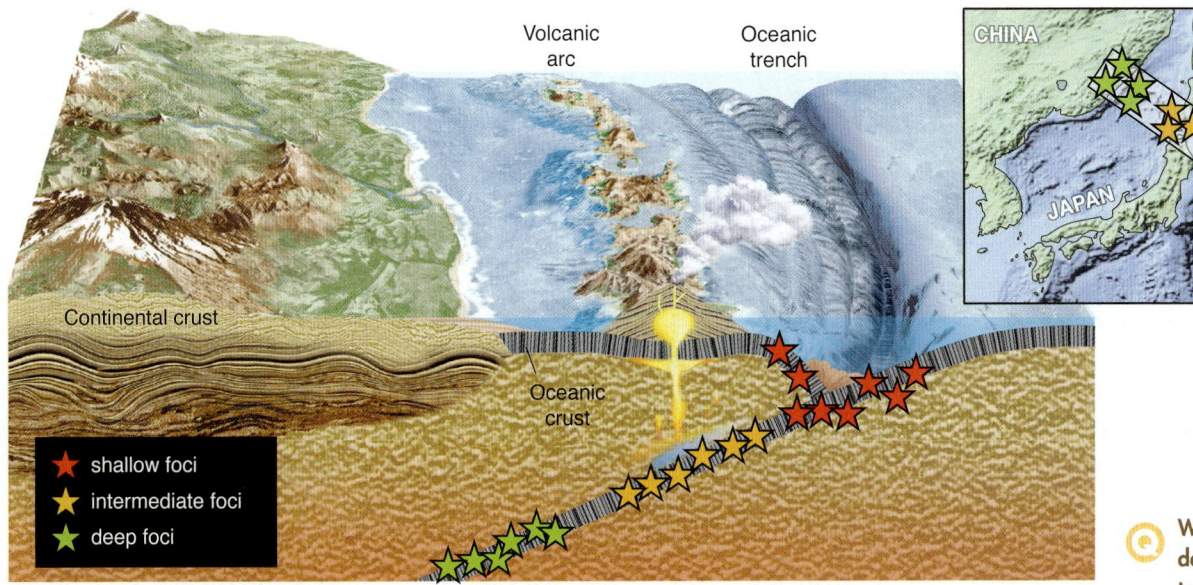

FIGURE 11.14 The subducting oceanic slab at convergent boundaries is the site of numerous earthquakes that extend from shallow to deep regions under the overriding plate.

shallow foci
intermediate foci
deep foci

Why are the epicenters of deep-subduction earthquakes typically located far from the surface expression of the plate boundary?

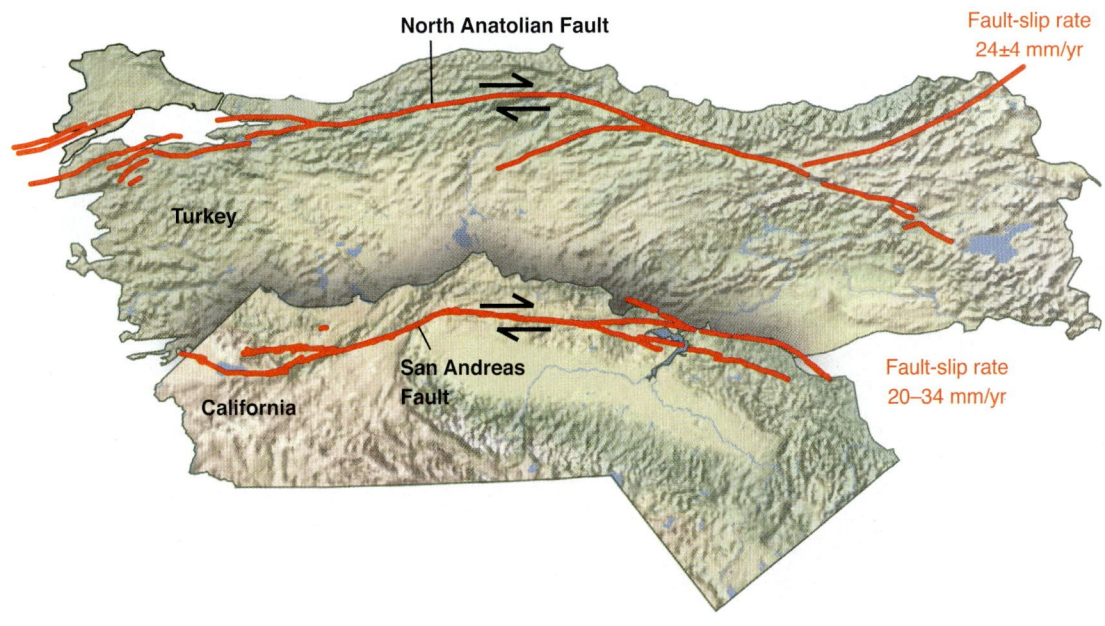

FIGURE 11.15 At the San Andreas and North Anatolian Faults, great earthquakes occur on only one section of the fault at a time. Offset along the fault plane accumulates in an uneven fashion, primarily by movement on first one, then another, section of the fault.

What type of faulting typically occurs at the San Andreas and Anatolian Faults?

destroying traditional stone and mud-brick homes, as well as minarets in at least 6 villages. The next day, over 100 aftershocks were felt.

In both the San Andreas and North Anatolian Faults, the sudden faulting that initiates a great earthquake occurs on only one section of the fault at a time. Total offset along the length of the transform accumulates in an uneven fashion, primarily through movement on first one, then another, section of the fault. The sections that are the sites of great earthquakes remain "locked" and quiet over a period of years as strain builds; then, in great lurches, the strain is released in the force of a devastating earthquake.

Along the San Andreas Fault, some portions of the fault accommodate strain buildup by constant creeping, rather than by sudden offsets that generate great earthquakes. In historical times, these creeping sections did not generate earthquakes of the magnitude seen on the locked sections. Rather, they inched forward in a series of *microseisms*, minor earthquakes that did not raise alarm.

Tackle the exercise in the next Critical Thinking feature, "Tectonic Settings for Earthquakes," to extend your understanding of the relationship between seismicity and plate tectonics.

Q EXPAND YOUR THINKING

Describe the stress and strain environment at divergent, convergent, and transform boundaries.

CRITICAL THINKING

Tectonic Settings for Earthquakes

Keeping in mind that earthquakes tend to occur at plate boundaries and some intraplate settings (**FIGURE 11.16**), work with a partner on the following problems:

1. Construct a table listing the characteristics of earthquake foci at divergent, convergent, transform, and intraplate settings. Describe each category in terms of depth of foci (shallow, intermediate, deep), frequency of occurrence (often, infrequent), tendency to produce major quakes (more likely, less likely), and likelihood of being heavily developed by humans (more likely, less likely).

2. Describe the nature of the prevailing stress at each of the four settings.

3. At each setting, determine whether strain typically accumulates over long periods. Explain.

4. What is the nature of strain release: localized or regional?

5. Describe the nature of rock breakage at each setting. That is, what type of faulting is likely to occur in an earthquake at that location?

6. The concepts of stress and strain establish an overarching framework for synthesizing a general hypothesis of earthquake occurrence. Write a general hypothesis describing earthquake occurrence, and design appropriate tests for it.

7. For each illustration in Figure 11.16, label the types of stresses, strains, faults, and other seismic, tectonic, and structural features.

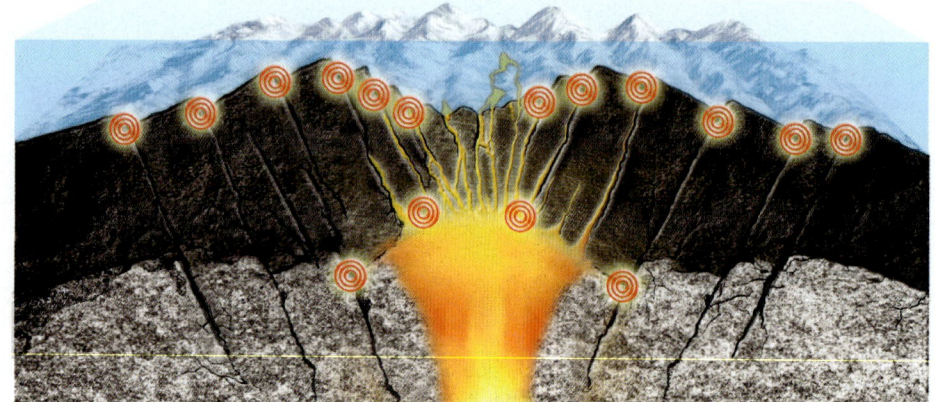

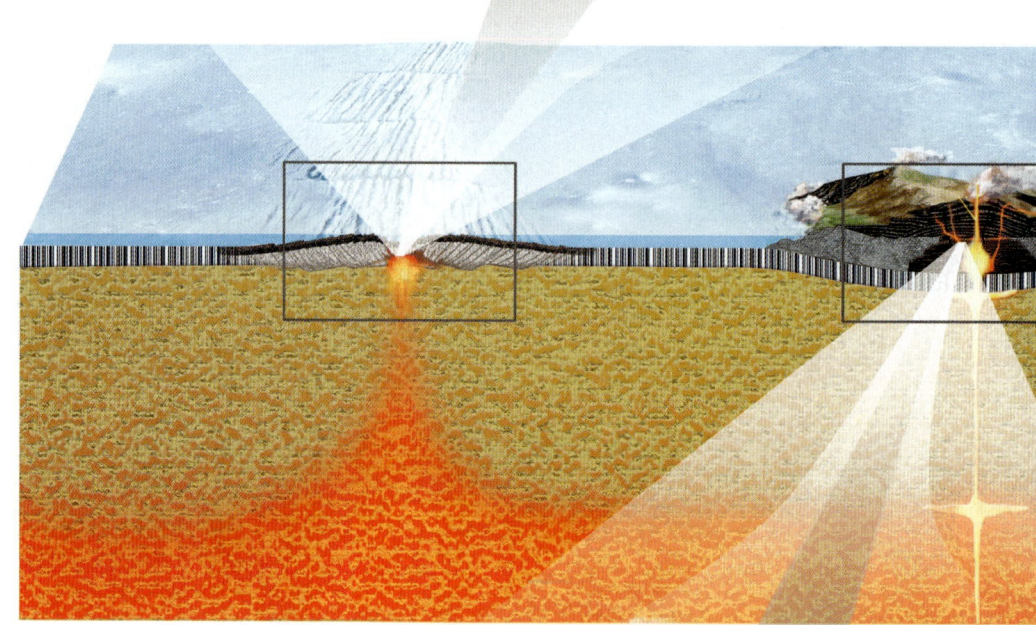

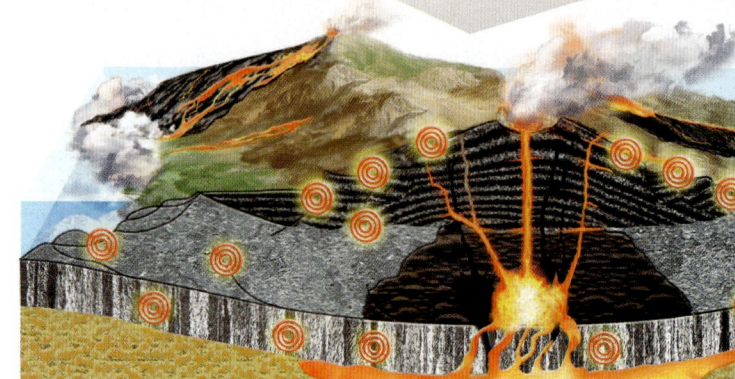

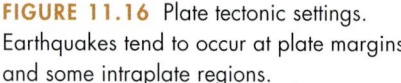

FIGURE 11.16 Plate tectonic settings. Earthquakes tend to occur at plate margins and some intraplate regions.

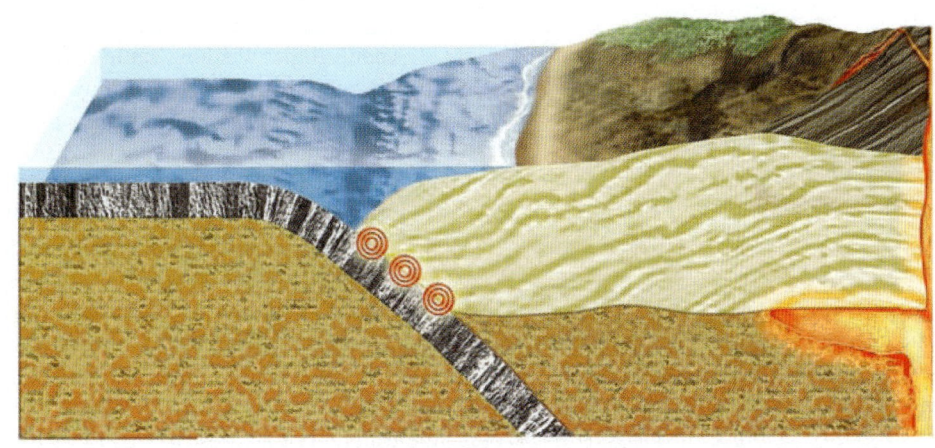

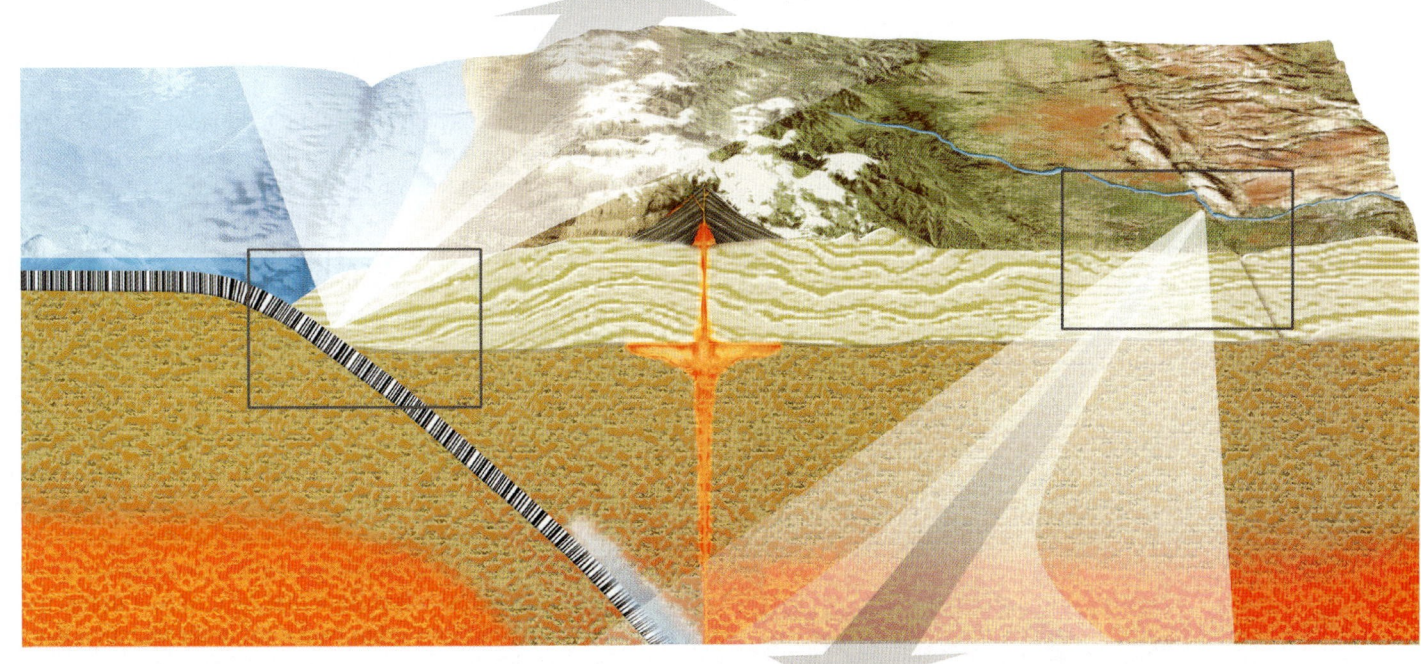

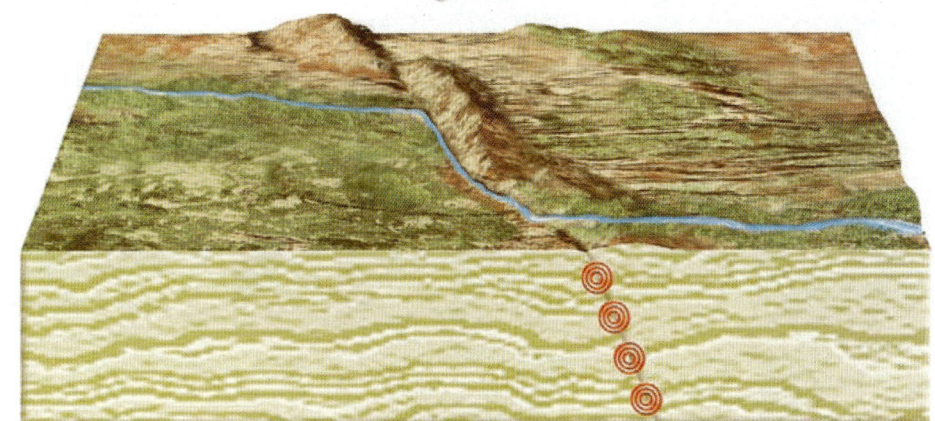

11-6 Earthquakes Produce Four Kinds of Seismic Waves

LO 11-6 List and describe the four kinds of seismic waves.

Earthquakes release energy in the form of seismic waves that travel outward from their foci through the crust and into Earth's interior (FIGURE 11.17). There are four kinds of seismic waves, and they all move in different ways.

Seismic Waves

Earthquakes radiate seismic energy (TABLE 11.2) in the form of two types of *surface waves*—**Love waves** and **Rayleigh waves**—and two types of *body waves*—primary, or **P waves**, and secondary, or **S waves**. Body waves travel through Earth's inner layers, whereas most (not all) of the energy in surface waves moves along the planet's surface.

TABLE 11.2 Properties of Seismic Waves

Type	Relative Speed	Passes Through/Along	Type of Motion
Body waves			
P wave	Fastest	Solid and liquid	Pushes and pulls
S wave	Second fastest	Solid only	Side to side
Surface waves			
Rayleigh wave	Third fastest	Earth's surface	Up and down, forward and back
Love wave	Slowest	Earth's surface	Side to side

The seismic P wave is the fastest kind of wave. It moves through solid rock as well as fluids, such as water or the liquid layer of Earth (the outer core). A P wave compresses and *dilates* (stretches) the material through which it moves. Hence, any material that resists changes in volume (such as solids and liquids) will transmit a P wave rather than absorb it.

A P wave pushes and pulls on the rock or liquid it moves through, just as sound waves push and pull on air (FIGURE 11.18). Have you ever noticed that windows rattle when a big clap of thunder booms? This happens because the sound waves push and pull on the window glass in much the same way that P waves push and pull on rock. Animals sometimes hear the P waves of an earthquake, whereas humans usually only feel the bump and rattle of these waves.

The S wave travels slower than a P wave and can move only through solid materials, such as rock. The S wave moves up and down or side to side. Hence, any material that is resistant to changes in shape will transmit an S wave. Liquids do not resist changes in shape, so they will absorb S waves rather than transmit them. For this reason, S waves travel through the solid layers of Earth (crust and mantle) but not the liquid layer (outer core). As we will discover later, the different properties of S and P waves (and large-energy surface waves) enable scientists to use seismic waves to explore the structure of Earth's interior.

Seismic surface waves are the slowest, and therefore arrive at a distant seismometer station after the P and S waves generated by a quake. They may cause a rolling motion, like that of swells at sea, or a very strong side-to-side shaking similar to that caused by S waves. Generally, surface waves cause the strongest vibrations and worst damage.

The fastest surface wave is the Love wave, named for *A. E. H. Love* (1863–1940), a British mathematician who, in 1911, worked out the mathematical model for this kind of wave (FIGURE 11.18). The Love wave is usually the first surface wave to be recorded by a seismograph. This type of wave moves the ground from side to side, leading to widespread damage to buildings, roads, and other solid objects.

The second kind of surface wave, the Rayleigh wave, was named for *John William Strutt, Lord Rayleigh* (1842–1919). In 1885, he made the mathematical calculations that predicted the existence of this kind of wave. A Rayleigh wave rolls along the ground just as a wave rolls across a lake or ocean. As it rolls, it moves the ground up and down and forward to backward in the direction that the wave is moving. Most of the shaking felt from an earthquake is due to the Rayleigh wave, which can be much larger than the other waves.

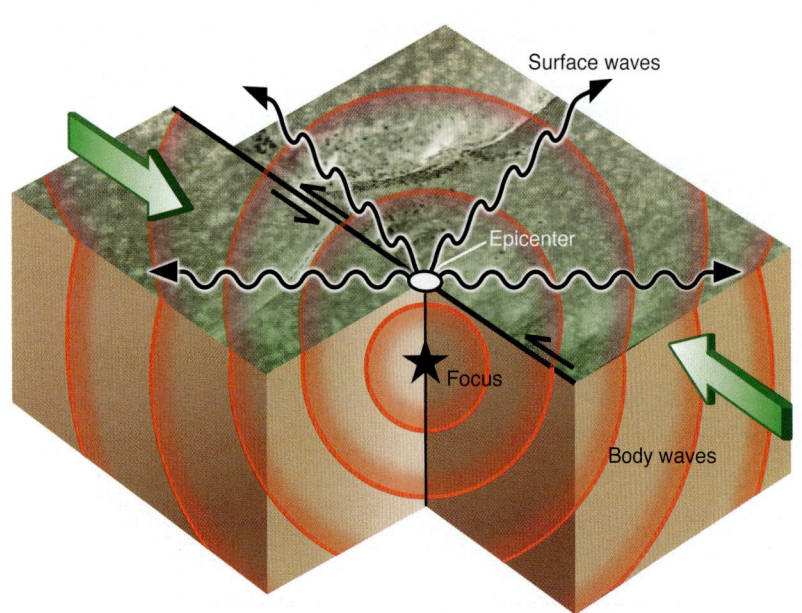

FIGURE 11.17 Earthquakes radiate seismic energy in the form of surface waves and body waves. There are two types of body waves: P (primary) waves and S (secondary) waves; and two types of surface waves: Love waves and Rayleigh waves.

Q How do these four types of waves differ from one another?

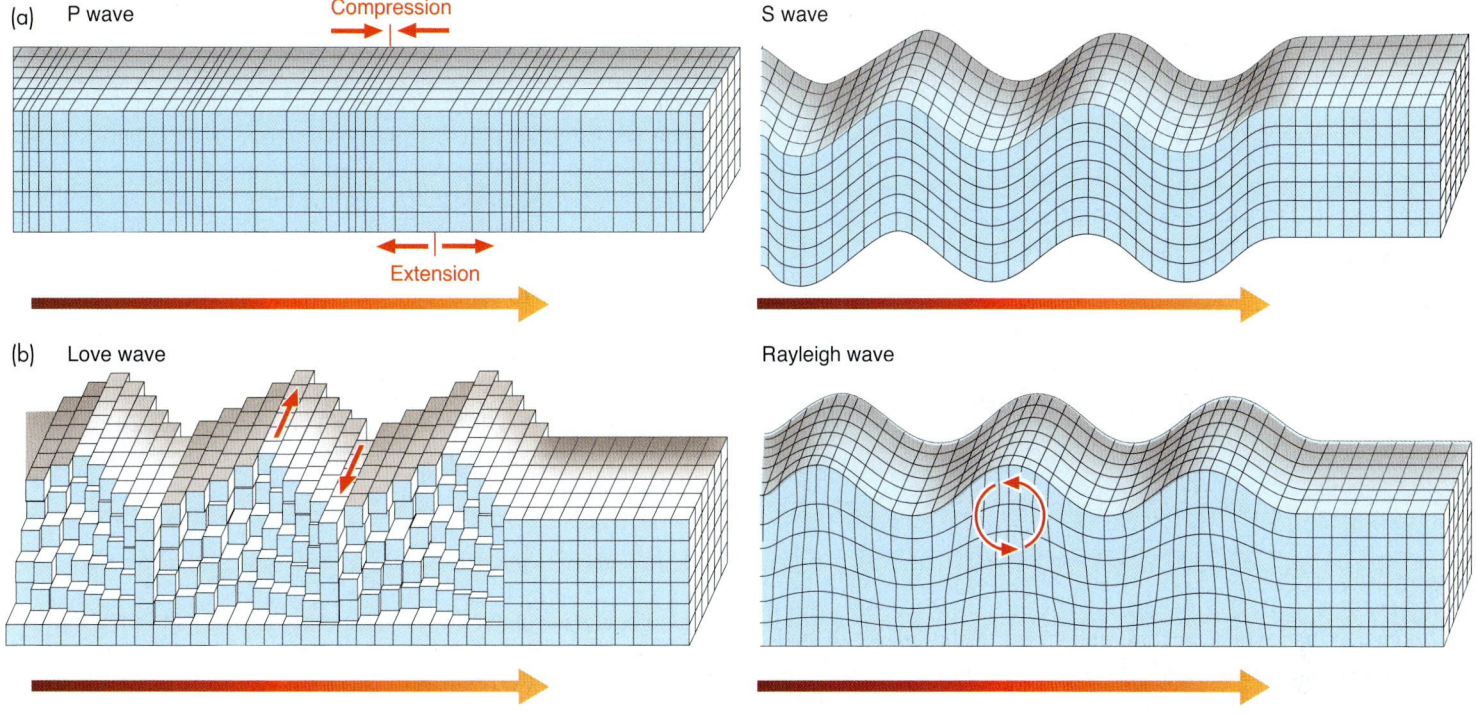

FIGURE 11.18 (a) The fastest wave is the P wave, which is able to travel through solids and liquids by alternating compression and extension. The S wave, which travels much slower than a P wave, is absorbed by liquids, but not by solids, because it causes material to change shape but not volume. (b) Love waves cause the ground to shift laterally, from side to side. Rayleigh waves cause the ground to roll up and down and shift forward and backward.

Which types of waves are responsible for the greatest damage to buildings?

Measuring Waves

A *seismogram* is a record of the arrival and size of body waves and surface waves; it is made by a seismometer. When an earthquake occurs, the first wave recorded on a seismogram is a P wave. The next wave is the S wave, followed by the surface waves (**FIGURE 11.19**). These waves arrive at different times because of their varying speeds. For instance, the velocity of P waves traveling through granite is about 6 kilometers per second, while S waves travel at about 3.5 kilometers per second. In most conditions encountered in the solid crust, P waves travel about 1.7 times faster than S waves. Surface waves travel at about 90% the speed of S waves.

EXPAND YOUR THINKING

Explain why **resisting** wave energy is the key to transmitting it.

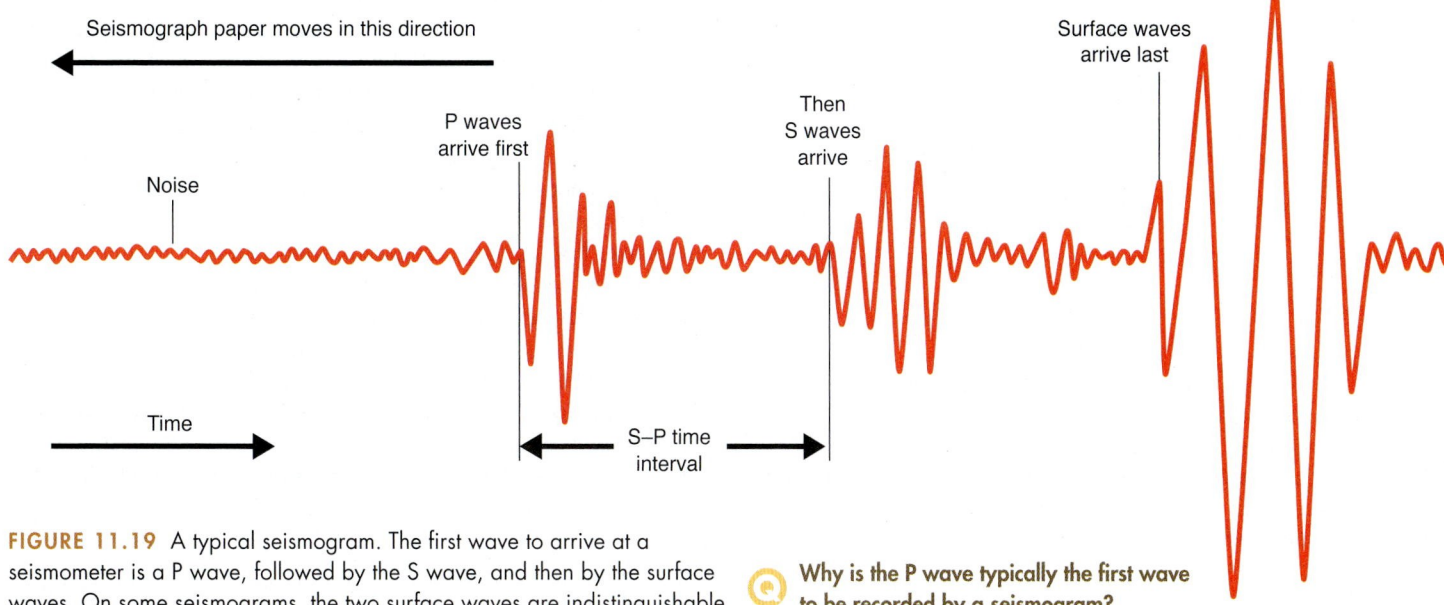

FIGURE 11.19 A typical seismogram. The first wave to arrive at a seismometer is a P wave, followed by the S wave, and then by the surface waves. On some seismograms, the two surface waves are indistinguishable.

Why is the P wave typically the first wave to be recorded by a seismogram?

11-7 Seismometers Are Instruments That Locate and Measure Earthquakes

LO 11-7 Describe seismometers and how they are used to locate the epicenter of an earthquake.

Because earthquakes can cause such terrible destruction, attempts have been made for centuries to measure them and understand how they work. Today, geologists have instruments they can use to measure and record the vibrations produced by earthquakes.

Seismometers

Seismometers can detect, record, and measure the arrival of both surface waves (Love and Rayleigh waves) and body waves (P and S waves). Using these measurements, scientists can determine the location of an earthquake's epicenter, the depth of its focus, time of occurrence, and type of faulting, and estimate how much energy the earthquake released.

Modern seismometers use highly precise electronic signals to measure seismic energy. (They are so precise they can even record the vibrations caused by breaking waves and changing tide levels on shorelines many kilometers away.) Taking advantage of inertia (the tendency of a body to maintain a state of rest or uniform motion unless acted upon by an external force), an electric current is used to keep a free-hanging weight moving along with the rest of the seismometer (which is attached to the ground) as shaking occurs. Variations in the level of current needed to keep the weight synchronized to the motions of the ground provide a sensitive record of seismic energy.

Since most seismic stations measure three components of vibration (east-west, north-south, and up-down), different directions of motion registered by a seismometer yield records of different kinds of seismic waves. To better understand this, imagine that a seismometer in Seattle records the energy emitted by an earthquake happening in Denver (**FIGURE 11.20**). The seismic waves approaching from the east will register as P waves on the east-west and up-down sensors, as Rayleigh waves on the east-west and up-down sensors, as S waves on all three sensors, and as Love waves on the east-west and north-south sensors.

Finding an Epicenter

Seismologists use networks of seismometer stations to pinpoint the location of an earthquake and estimate focus depth, exact time, type of faulting, and level of energy. These calculations are based on the arrival times of P and S waves. These waves travel at different velocities, so the spacing between their arrival times at a seismometer station is proportional to the distance from the station to an epicenter (**FIGURE 11.21**). Hence, it is possible for a single station to calculate the distance to an earthquake, but *not* the direction in which the waves are traveling.

To locate an epicenter, at least three seismic stations are needed. Each station calculates the distance to the epicenter. On a map, the location of each station is surrounded by a circle, whose radius is equal to the distance to the epicenter. The point at which the three circles intersect marks the location of the epicenter. This process is called *triangulation*.

Earthquake Intensity

There are many ways to measure how much energy was released during an earthquake. Scientists now have seismometers capable of helping them quantify the amount of energy released—a topic we discuss further in the next section. Before the widespread use of seismometers, it was possible only to say that an earthquake was big or small, based on the amount of shaking and damage that occurred.

Today, earthquake *intensity* is used as a qualitative measure of earthquake energy. More specifically, intensity is a measure of the physical effects of shaking and the amount of damage wreaked by an earthquake. For a single earthquake, surrounding areas may experience different levels of damage; thus, the intensity of an earthquake can vary at different locations. Intensity in a given area depends not only on the energy of the earthquake but also on the area's geology and its distance from the epicenter of the quake.

The intensity scale currently in use is the *Modified Mercalli (MM) Intensity Scale*, named after *Giuseppe Mercalli* (1850–1914), an Italian scientist who first described the scale in 1902. The MM scale was further developed in 1931 by the American seismologists *Harry Wood* (1879–1958) and Frank Neumann. Their ranking system is composed of 12 levels of intensity, ranging from imperceptible shaking to catastrophic destruction.

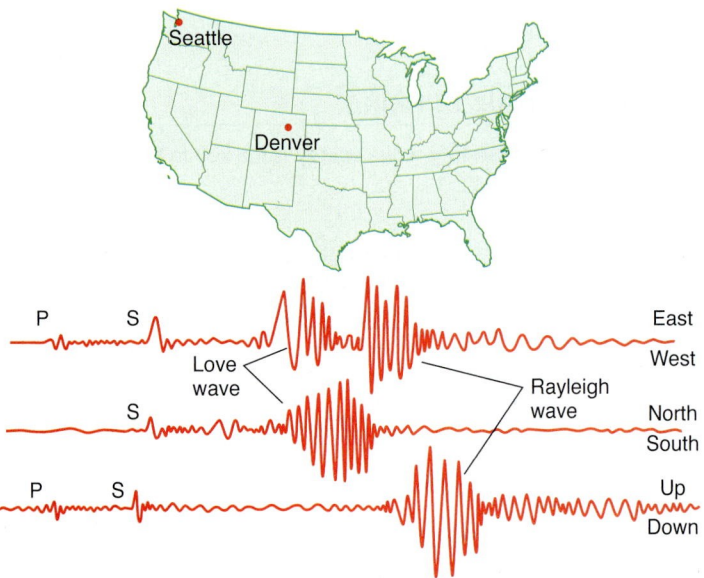

FIGURE 11.20 Seismometer stations measure vibrations in three directions. If an earthquake in Denver is measured by an instrument in Seattle, seismic waves approaching from the east will register as P waves on the east-west and up-down sensors, as Rayleigh waves on the east-west and up-down sensors, as S waves on all three sensors, and as Love waves on the east-west and north-south sensors.

Q Why do seismometers measure vibrations in three directions?

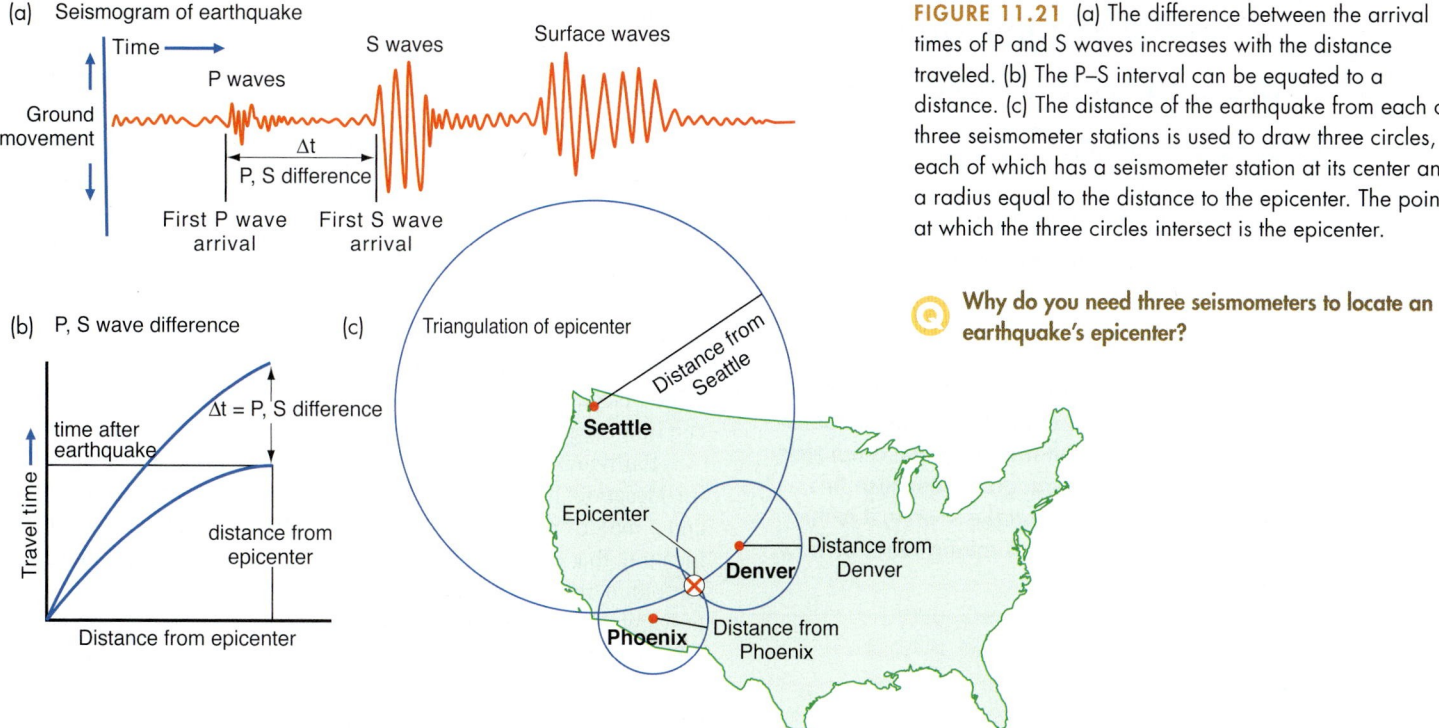

(a) Seismogram of earthquake

(b) P, S wave difference

(c) Triangulation of epicenter

FIGURE 11.21 (a) The difference between the arrival times of P and S waves increases with the distance traveled. (b) The P–S interval can be equated to a distance. (c) The distance of the earthquake from each of three seismometer stations is used to draw three circles, each of which has a seismometer station at its center and a radius equal to the distance to the epicenter. The point at which the three circles intersect is the epicenter.

Why do you need three seismometers to locate an earthquake's epicenter?

Each level is designated by the Roman numerals I to XII. The lower numerals of the intensity scale refer to how people feel an earthquake. The higher numerals of the scale are based on observed structural damage (**TABLE 11.3**). Note that the MM scale does not have a mathematical basis; rather, it is an arbitrary ranking based on observed effects, such as people being awakened by the quake, movement of furniture, damage to chimneys, and total destruction of buildings.

The MM scale is most meaningful to nonscientists in that it refers to effects actually experienced in a particular location. To develop maps of earthquake intensity, the U.S. Geological Survey (USGS) mails questionnaires, called *felt reports*, to postmasters in a region that has recently experienced an earthquake. Postmasters then distribute them throughout the disturbed area so that intensity values can be assigned based on firsthand reports. In general, the maximum intensity occurs near the quake's epicenter.

TABLE 11.3 Modified Mercalli Intensity Scale

   I. Not felt except by a very few persons under especially favorable conditions.

  II. Felt only by a few persons at rest, especially on upper floors of buildings.

 III. Felt quite noticeably by persons indoors, especially on upper floors of buildings. Many people do not recognize it as an earthquake. Standing motor cars may rock slightly; vibrations are similar to those from the passing of a truck. Duration estimated.

 IV. Felt indoors by many; outdoors by few during the day. At night, some people awakened. Dishes, windows, doors disturbed; walls make cracking sound. Sensation like heavy truck striking building. Standing motor cars rocked noticeably.

  V. Felt by nearly everyone; many awakened. Some dishes and windows broken. Unstable objects overturned. Pendulum clocks may stop.

 VI. Felt by all; many frightened. Some heavy furniture moved; a few reports of fallen plaster. Damage slight.

VII. Damage negligible in buildings of good design and construction; slight to moderate in well-built ordinary structures; considerable damage in poorly built or badly designed structures; some chimneys broken.

VIII. Damage slight in specially designed structures; considerable damage in ordinary, substantial buildings, with partial collapse. Damage great in poorly built structures. Collapse of chimneys, factory stacks, columns, monuments, walls. Heavy furniture overturned.

 IX. Damage considerable in specially designed structures; well-designed frame structures thrown out of plumb. Damage great in substantial buildings, with partial collapse. Buildings shifted off foundations.

  X. Some well-built wooden structures destroyed; most masonry and frame structures destroyed, with foundations. Rails bent.

 XI. Few, if any, (masonry) structures remain standing. Bridges destroyed. Rails bent greatly.

XII. Damage total. Lines of sight and level are distorted. Objects thrown into the air.

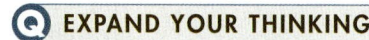

EXPAND YOUR THINKING

An earthquake raises public concern. Why is it important to quickly locate the epicenter?

11-8 Earthquake Magnitude Is Expressed as a Whole Number and a Decimal Fraction

LO 11-8 Compare and contrast the ways to characterize earthquakes: moment magnitude, Richter magnitude, and intensity.

Most seismologists measure and report the energy released by an earthquake by describing its *magnitude*. Magnitude is computed on the **Richter scale**, which measures the amplitude of the largest seismic wave on a seismogram. Because this number describes the size of an *earthquake source*, it is the same no matter where the earthquake is taking place or how powerful the shaking is.

Richter Magnitude

The *intensity* of an earthquake depends on local factors, such as types of buildings and the geologic setting. Naturally, these factors differ from one location to another, so the MM scale cannot be used to make objective comparisons between earthquakes in different places—say, India and New York. The Richter magnitude scale, which was devised in 1935 by Charles F. Richter (1900–1985) of the California Institute of Technology, solves this problem by assigning a single mathematical value for comparing the size of all earthquakes. The scale, now accepted for use around the world, calculates earthquake magnitude from the *logarithm of the maximum seismic wave amplitude* recorded by seismometers (**FIGURE 11.22**).

The starting point for Richter's design was the quantitative scale used by astronomers to describe the brightness of stars. He reasoned that a similar scale could be employed for studying earthquakes, for, like the brightness of stars, the magnitude of earthquakes varies over a wide range. In order to compress the scale to a more manageable range, Richter chose to use the logarithm of the seismic wave amplitude.

But to interpret seismic wave amplitude accurately, instruments everywhere must be calibrated to the same degree of sensitivity so that their records can be compared with one another. Adjustments are made to account for differences in the distance between various seismometers and the epicenter of an earthquake, as well as the type of displacement that takes place along a fault plane, both of which can influence the output of the seismogram.

Earthquake magnitude is expressed as a whole number and a decimal fraction; for example, a minor quake might rank 4.8 on the scale, and a major one 7.1. Keep in mind the Richter scale is logarithmic, meaning that each whole-number increase in magnitude represents a tenfold increase in the amplitude of a measured wave. That is, 1 unit of Richter magnitude translates to a difference factor of 10 in ground vibration. Thus, the vibration caused by an earthquake of magnitude 4 is 10 times larger than an earthquake of magnitude 3; and a magnitude 7 quake is 100 times greater than a magnitude 5 quake. In terms of energy released by an earthquake, one whole-number increase on the scale is equivalent to about 33 times more energy than the amount associated with the preceding whole-number value.

A *micro-earthquake* typically has a magnitude of about 2.0 or less. Such a quake is not commonly felt by people, and usually is recorded only on local seismometers. Sensitive seismometers all over the world are capable of recording events of about 4.5 or greater, and several thousand such shocks are recorded annually (**TABLE 11.4**). On average, one *great earthquake* (magnitude of 8.0 or greater) occurs somewhere in the world each year. A recent example of a great earthquake occurred on February 27, 2010, charting a Richter magnitude of 8.8, the seventh strongest in recorded history. It occurred in the South American nation of Chile, and was so strong it shifted Earth's axis of rotation by about 8 centimeters.

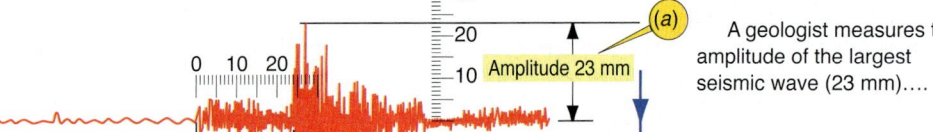

(a) A geologist measures the amplitude of the largest seismic wave (23 mm)....

(b) ......and the time interval between the P wave and S wave arrivals (24 s) to determine the distance from the epicenter to the station (210 km).

(c) By plotting the two measurements on these graphs and connecting the points, the geologist determines the Richter magnitude of the earthquake (5.0).

FIGURE 11.22 To assign a Richter magnitude to an earthquake, (a) a geologist measures the amplitude of the largest seismic wave on a seismogram. (b) The P–S wave interval is calculated to determine the distance to the epicenter. Then (c) a line is drawn to connect these two measurements on a graph, which identifies the Richter magnitude of the quake.

Why is the Richter magnitude preferable to the Modified Mercalli scale for comparing the size of two earthquakes in different locations?

TABLE 11.4 Earthquake Size and Number Since 1900

Descriptor	Richter Magnitude	Average Number Annually
Great	8–higher	1
Major	7–7.9	18
Strong	6–6.9	120
Moderate	5–5.9	800
Light	4–4.9	6,200 (estimated)
Minor	3–3.9	49,000 (estimated)
Very minor	< 3.0	1,000 per day

The Richter scale has no upper limit, nor is it used to express level of damage. A moderate quake in a densely populated area that is responsible for many deaths and considerable damage may be of the same magnitude as a quake in a remote area that does nothing more than shake the trees in a forest.

Earthquake Moment Magnitude

Another scale, called the *moment magnitude* scale, has been devised to enable more precise study of great earthquakes. Moment magnitude is a measure of the amount of energy released by an earthquake. The energy can be estimated based on the degree of rupture that a fault experiences. The moment magnitude tends to spread out the larger earthquakes along a more refined scale of measurement so that

they are not all grouped together, as they are at the top of the Richter scale (**FIGURE 11.23**). Scientists generally prefer moment magnitude over the Richter scale measure for its greater precision.

The moment magnitude describes a physical factor about an earthquake—specifically, the area of the fault that ruptured. This is calculated in part by multiplying the area of the fault's rupture surface by the distance over which the crust is displaced along the fault. That means the most accurate calculation of moment magnitude must be based on field measurements of the rupture surface *after* the event has taken place. But because many earthquakes originate from faults buried underground (known as *blind faults*), scientists have devised a way to estimate moment magnitude using the recordings on a seismogram.

As noted, large earthquakes occur far less frequently than small ones. As Figure 11.23 shows, each year approximately 1 million earthquakes occur with a moment magnitude greater than 2. The frequency of more powerful quakes decreases by an order of magnitude for every increase in the magnitude of the event. So, each year, approximately 100,000 earthquakes exceed magnitude 3, 10,000 quakes exceed magnitude 4, and 1,000 exceed magnitude 5.

These statistics suggest that events larger than magnitude 8 should occur, on average, once per year. In fact, they are even rarer than that, occurring only every three to five years. The largest quakes ever recorded—the 9.5 event in Chile (1960) and the 9.2 event in Alaska (1964)—are exceptionally uncommon events.

The exercise in the Critical Thinking feature titled "Earthquake Analysis" offers you an opportunity to use what you have learned to analyze a real earthquake that occurred in the western United States.

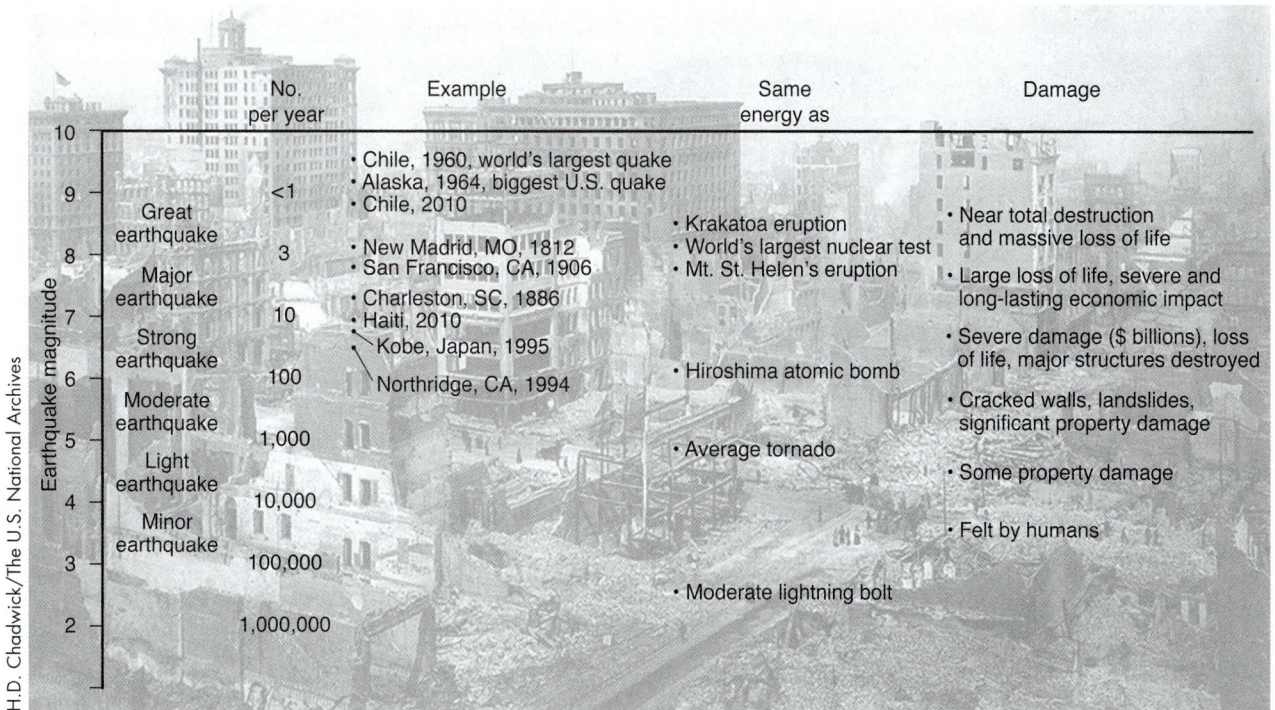

FIGURE 11.23 This chart shows a comparison of earthquakes in terms of magnitude and familiar events.

Some earthquakes last longer than others. What does this tell you in terms of what is happening on the rupturing fault?

EXPAND YOUR THINKING

What method is typically used by the media to characterize an earthquake? Why?

CRITICAL THINKING

Earthquake Analysis

You will need the following tools for this exercise: pencil, ruler, and a compass to draw circles.

An earthquake occurred in 1989 somewhere in the western United States. To conduct analysis of the quake, please follow the directions below and consult **FIGURE 11.24A**, which shows an example of how to measure the S–P interval on a seismogram, and **FIGURE 11.24B**, which is a chart for calculating the distance from the epicenter using the S–P interval.

1. Use the seismogram in **FIGURE 11.24C** from Eureka, California, **FIGURE 11.24D** from Elko, Nevada, and **FIGURE 11.24E** from Las Vegas, Nevada, to complete the following:

Station	S–P Interval (seconds)	Distance from Epicenter (kilometers)
Eureka, CA		
Elko, NV		
Las Vegas, NV		

2. On the map in **FIGURE 11.24F**, draw a circle around each seismograph station. The radius of each circle should be equal to the distance from the epicenter. Use the map scale to measure the radius of the circle. Find a map of the western United States and identify the location of the nearest town to the earthquake epicenter. Write it below:

    Epicenter: _____.

3. You need two measurements to calculate the magnitude of an earthquake: the S–P interval and the maximum amplitude of the seismic waves. Although only one amplitude measurement is necessary to estimate the magnitude of an earthquake, confidence in the result goes up if measurements from several seismograph stations are used. On each seismogram measure the maximum amplitude of the S wave (use the vertical axis and measure from 0 in one direction). Fill in the following:

Station	Distance from Epicenter (kilometers)	S-Wave Amplitude (millimeters)
Eureka, CA		
Elko, NV		
Las Vegas, NV		

4. Use **FIGURE 11.24G** to calculate the earthquake magnitude. Plot the station distance and S-wave amplitude from the above table. Draw a line between each pair of points to establish an estimated magnitude. Write the magnitude for each station below:

    Earthquake magnitude at:

    Eureka, CA   _____

    Elko, NV   _____

    Las Vegas, NV   _____

5. What is the average magnitude of the earthquake?

6. Write a one-paragraph press release describing your methodology and conclusions.

7. Using the location of the epicenter, look up the history of this actual event on the Web. What is the name of this earthquake, and when did it occur?

8. If assigned by your instructor, write a short paper recounting the history of this event.

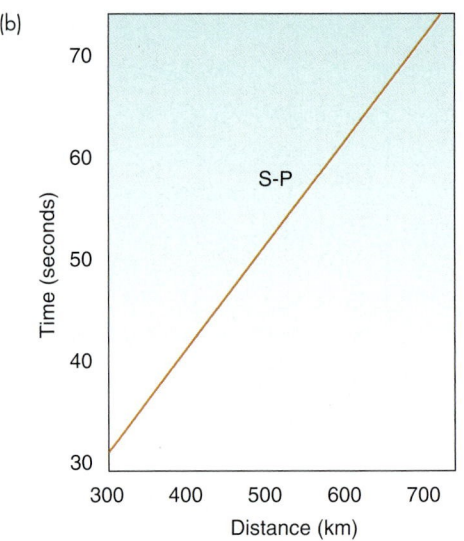

9. Based on the epicenter and what you have read, build a hypothesis that describes the cause of the earthquake. How would you test your hypothesis? Extrapolate the implications of this earthquake with regard to the seismic hazard in the surrounding region.

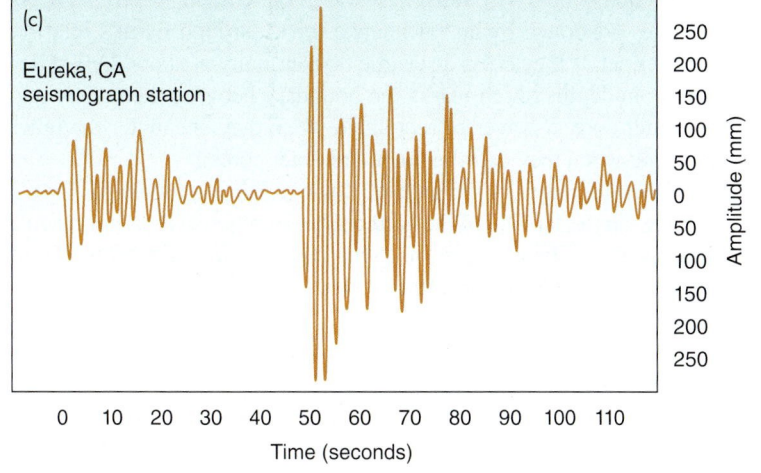

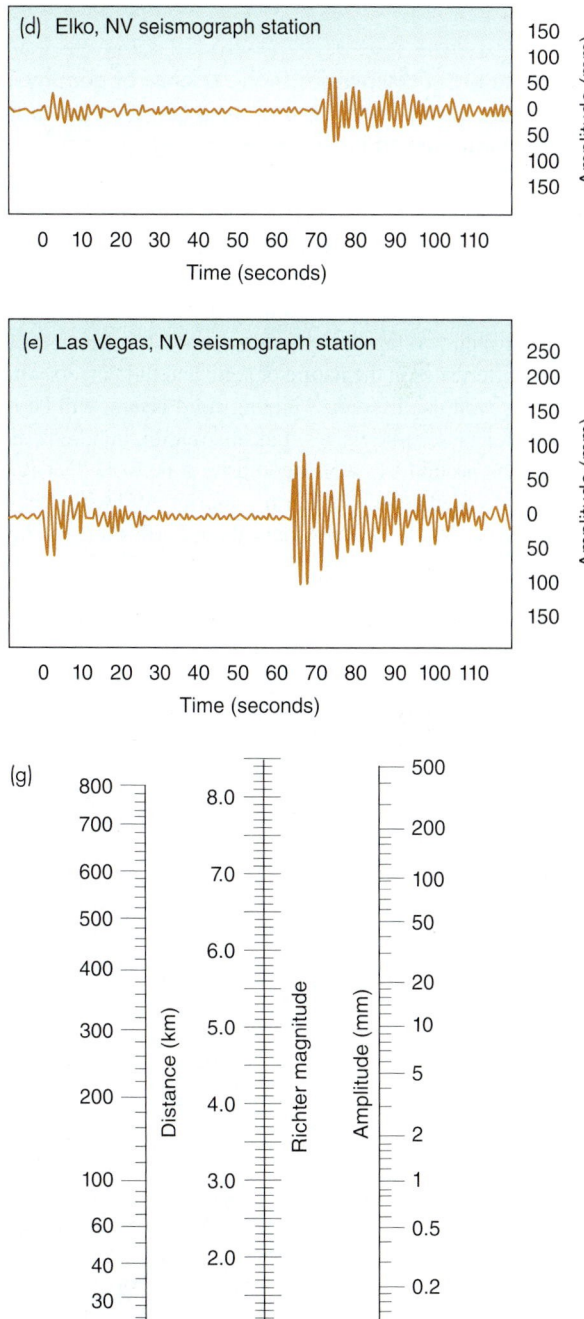

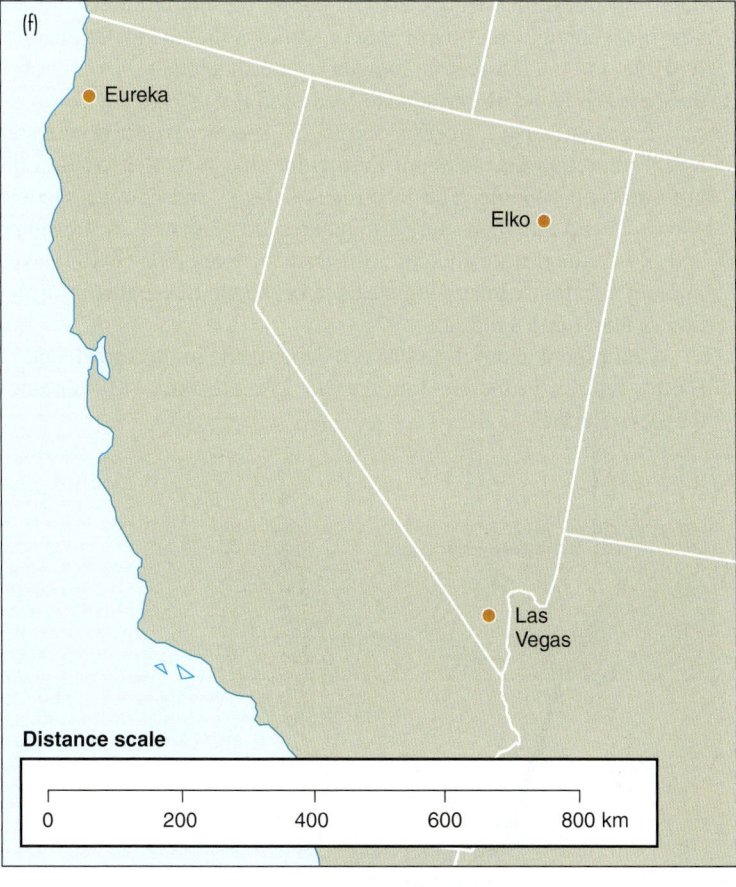

FIGURE 11.24 (a) Example of calculating the S–P interval. The S–P interval is 36 seconds as measured on the horizontal axis. (b) Relates S–P interval to distance. Notice 36 seconds is roughly equal to 350 kilometers. Seismograms from (c) Eureka, California, (d) Elko, Nevada, and (e) Las Vegas, Nevada. (f) Map of the western United States. (g) Nomogram (a graphical calculator) used to calculate the earthquake magnitude.

Seismology Is the Study of Seismic Waves Conducted to Improve Our Understanding of Earth's Interior

11-9

LO **11-9** Describe how seismic wave characteristics result in P-wave and S-wave shadow zones.

One of the great discoveries in the science of geophysics was that seismic waves from earthquakes can be used to improve our understanding of Earth's interior.

An important property of seismic waves is that their velocity changes when they encounter density differences in rock layers, due to changes in pressure, temperature, or composition. In addition, they *refract* (change direction) and *reflect* (bounce off) when passing across the interface between two layers in Earth's interior. It is because of this property that differences in the nature of solid and fluid layers within Earth's interior, as well as differences among solid layers, will be recorded in the behavior of seismic waves. For this reason, hundreds of seismometer stations around the world comprise a network that is capable of recording detailed aspects of Earth's interior. And with every new large earthquake, scientists gain new data about Earth's internal layers.

The P-Wave Shadow Zone

Wave refraction is caused by changes in wave velocity due to differences in rock density. Seismic waves have a higher velocity in denser rock and a lower velocity in less dense rock. If a seismic wave passes into rock in which its velocity will increase, it will be refracted upward relative to its original path. A wave passing into rock in which its velocity will decrease will be refracted downward. **FIGURE 11.25** illustrates this natural phenomenon.

Within Earth's interior, seismic wave velocities generally increase with depth because rock density generally increases with depth and waves travel faster in denser material. As a result, waves are continually refracted along curved paths (called *ray paths*) that arc gently back toward Earth's surface (**FIGURE 11.26A**). However, should there be a sudden change in rock density, the wave velocity will change suddenly in response; such an interface is called a **discontinuity**. Discontinuities found at well-mapped depths in the interior of the planet help us mark boundaries between various layers, such as core, mantle, and crust.

P waves change the *volume* of material that they encounter. Hence, any material that resists changing volume, including most liquids and solids, will transmit P waves rather than absorb them. As P waves generated by an earthquake travel through Earth's interior, they refract as they cross a seismic discontinuity at about 2,900 kilometers in depth, which marks the boundary between the mantle and the outer core. P waves refract again when they encounter the inner core and when they exit the inner and outer cores.

These changes in the path of P waves leave a region of Earth's surface, on the far side of the earthquake source, that does not receive any P waves. The result is a *P-wave shadow zone* on the far side of the core created by wave refraction (**FIGURE 11.26B**).

The S-Wave Shadow Zone

S waves change the shape of materials; therefore, liquids, which do not resist changes in shape, absorb S waves rather than transmit them. As a result, S waves that encounter the liquid outer core are absorbed, creating an *S-wave shadow zone* on the far side of the core (**FIGURE 11.26C**). Any seismometer located such that the outer core lies between it and an earthquake focus will not receive any S waves.

Together, wave refraction and the P- and S-wave shadow zones reveal important details about Earth's interior, facts that would otherwise be unavailable. The locations of the P- and S-wave shadow zones depend on where the earthquake occurs, as well as the shape and size of the mantle and core. Because there are many earthquakes in many locations, seismologists have been able to use this information to map Earth's interior.

Learn more about the advances being made in the knowledge of Earth's interior in the Geology in Our Lives feature, "Mysteries of the Deep Earth."

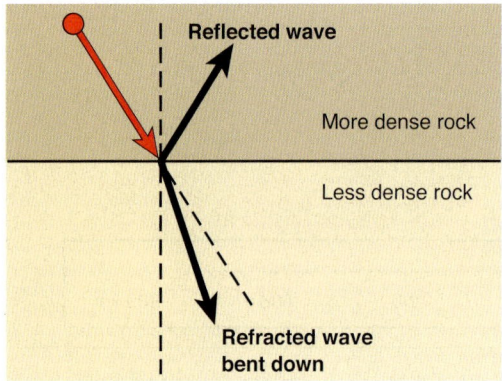

FIGURE 11.25 Seismic waves both reflect and refract when they encounter the boundary between two materials of differing density.

Sunlight refracts when it passes from the air into water. Which way does a light ray refract: upward or downward?

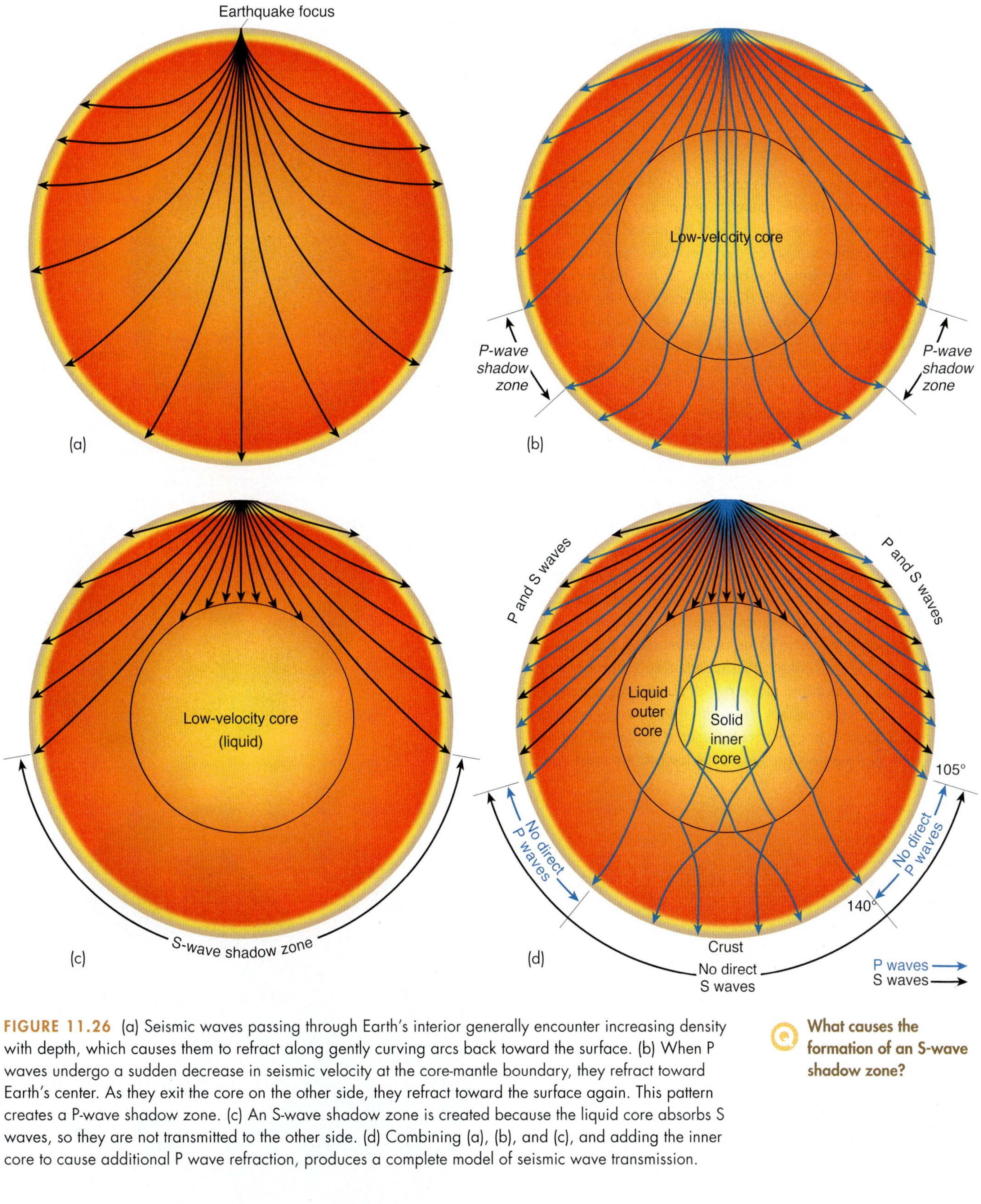

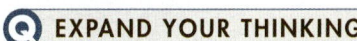

FIGURE 11.26 (a) Seismic waves passing through Earth's interior generally encounter increasing density with depth, which causes them to refract along gently curving arcs back toward the surface. (b) When P waves undergo a sudden decrease in seismic velocity at the core-mantle boundary, they refract toward Earth's center. As they exit the core on the other side, they refract toward the surface again. This pattern creates a P-wave shadow zone. (c) An S-wave shadow zone is created because the liquid core absorbs S waves, so they are not transmitted to the other side. (d) Combining (a), (b), and (c), and adding the inner core to cause additional P wave refraction, produces a complete model of seismic wave transmission.

What causes the formation of an S-wave shadow zone?

Q EXPAND YOUR THINKING

If Earth had no core, how would we know?

GEOLOGY IN OUR LIVES

Mysteries of the Deep Earth

With the use of mathematical models and images of the deep Earth produced based on seismic waves, scientists are at last converging on a uniform concept of what the lower mantle of the planet looks like. Long ago the scientific community agreed that the theory of plate tectonics does a good job of describing the behavior of the rocky coating of Earth's crust and the cycling of cold dense seafloor into the hot mantle. Scientists have, however, continued to argue over how heat and rock make its way back to the surface. Could immense plumes of hot rock be rising from the base of the mantle, the way waxy lumps in a lava lamp do?

Thanks to new studies that utilize the science of seismic tomography—the imaging of Earth's interior using energy released by large earthquakes—a new picture is emerging of how things work 2,900 kilometers below the surface. The idea that mantle plumes, upwellings of hot rock in the mantle, characterize the interior has been controversial among some researchers. Now, however, data such as that illustrated in **FIGURE 11.27A**, is helping to convince many that upward directed currents of rock are moving through the mantle.

What does the base of the mantle look like? Data suggest that there are two piles of rock, known as *Large Low Shear-wave Velocity Provinces* (LLSVPs) on either side of the planet marking the base of the mantle (**FIGURE 11.27B**). Further, it is hypothesized that this material has been pushed or guided into place by the bulldozing action of subducted slabs of oceanic crust that make their way to the base of the mantle. These LLSVPs, it is thought, are the source of plumes that rise off their surface and migrate through the mantle toward the crust.

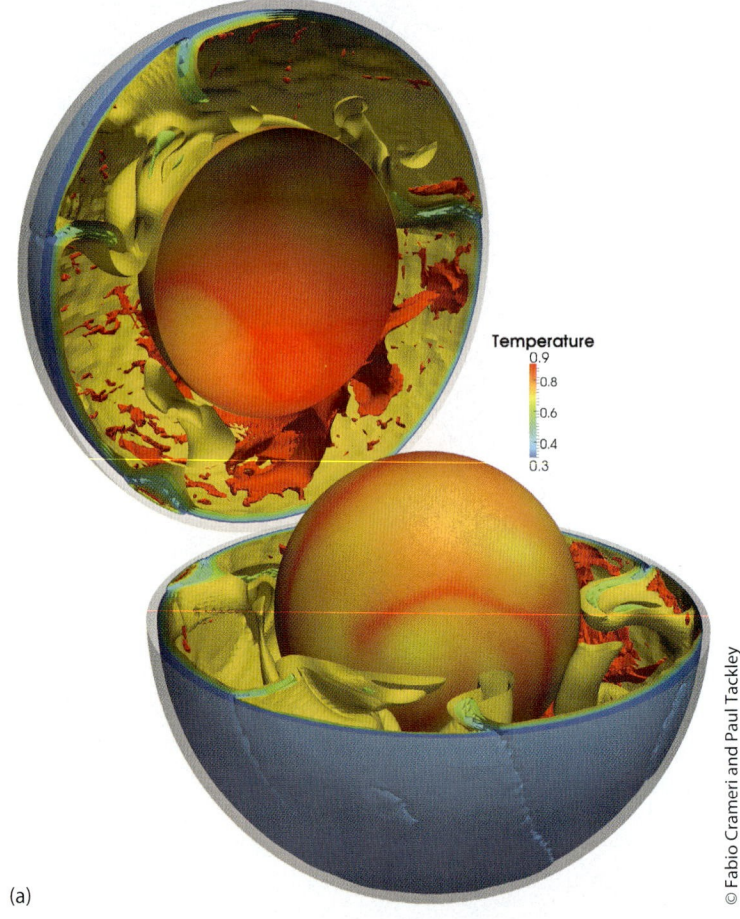

(a)

© Fabio Crameri and Paul Tackley

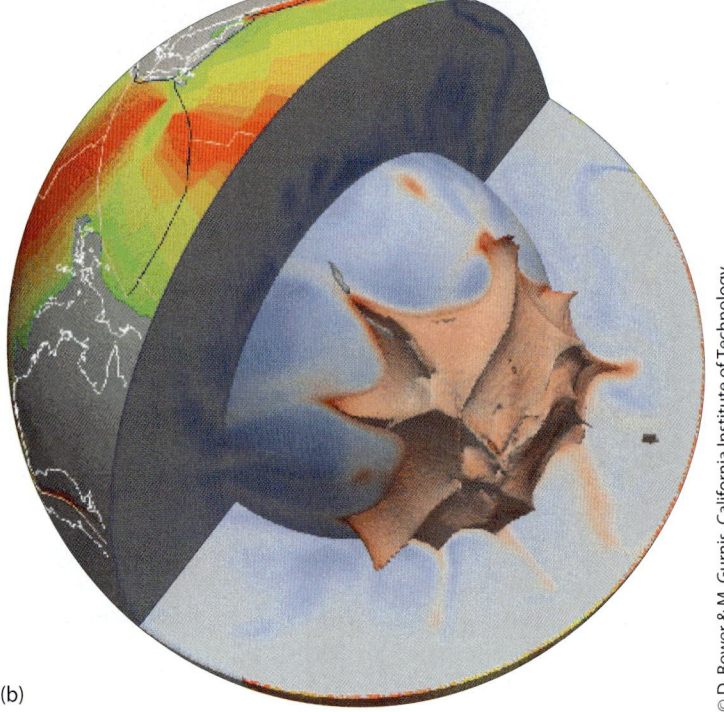

(b)

© D. Bower & M. Gurnis, California Institute of Technology

FIGURE 11.27 It has become widely accepted among researchers that Earth's interior is characterized by (a) regions of cold slabs (in yellow, cored with light blue) of lithospheric plate (dark blue when on the surface) descending into the mantle, and hot plumes (red) that rise from near the hot core. At the base of the mantle (b) cold descending slabs (in blue) have herded dense mantle rock into a pile, or LLSVP (brown), from which plumes of hot rock rise toward the crust.

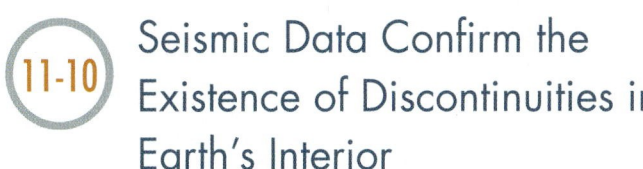

Seismic Data Confirm the Existence of Discontinuities in Earth's Interior

LO 11-10 List the major discontinuities in Earth's interior and their depths.

Seismologists have accumulated data that confirm the existence of several pervasive discontinuities marking distinct layers in Earth's interior. These layers are the result of variations in the chemistry and physical properties of rock lying beneath the crust.

The Moho

The shallowest major discontinuity marks the border between the crust and the mantle. This boundary was identified in the early twentieth century by a Croatian scientist named *Andrija Mohorovicic* (1857–1936).

Studying seismic records, Mohorovicic noticed the arrival of a pair of P and S waves, followed by another, slower, pair of P and S waves. The slower pair, he reasoned, had traveled more or less directly through the crust, and the faster set had probably refracted at a boundary in Earth's interior and traveled at some depth in a higher-velocity zone. Using critical thinking, he inferred that beneath the crust was a layer of denser rock (thus permitting higher seismic velocity) that must have refracted and sped along the first arriving set of waves. Based on this inference, Mohorovicic hypothesized the existence of a distinct boundary that separates the crust from some lower-lying portion of Earth's interior with a different rock composition.

The boundary between the crust and the mantle, its existence now confirmed many times since it was first described by Mohorovicic, is called the **Mohorovicic discontinuity**, or *Moho* for short. The depth of the Moho is known to vary from approximately 8 kilometers beneath ocean basins to 20 to 70 kilometers beneath the continents.

As we explained in Chapter 10, thicker portions of continental crust are found beneath mountain belts in the form of deep roots needed to sustain the mass of the crust. By comparing laboratory measurements of the velocity with which seismic waves travel through various kinds of rock, it is possible to infer the composition of the crust. Oceanic crust is generally composed of basalt, with the lower portion consisting of gabbro, while continental crust varies from granite near the top to gabbro near the Moho. These findings are consistent with other evidence, such as drilling, mapping of rock outcrops, and seismic studies based on artificial sound sources.

The Mantle

Beneath the Moho lies the mantle. Laboratory tests of rocks that are rich in dense minerals such as olivine, pyroxene, and garnet show seismic velocities of about 8 kilometers per second, while rocks of granite, gabbro, and basalt have velocities of 6 to 7 kilometers per second. These measurements are consistent with P-wave velocities that increase abruptly at the Moho from about 6 to 7 kilometers per second to 8 to 12 kilometers per second.

From this evidence, seismologists infer that the mantle is composed of dense, ultramafic rock such as *peridotite* (Chapter 5). The few actual samples of mantle rock that are available—mantle fragments found in magmas and diamond-rich *kimberlite pipes* of mantle rock that intrude the crust—confirm the composition of the mantle inferred from seismic studies.

Within the upper mantle, at a depth of about 100 kilometers, there is a significant, though subtle, decrease in both P- and S-wave velocities that marks the base of the lithosphere and the top of the *asthenosphere*. You will recall from Chapter 3 that the lithosphere is composed of both the crust and the upper part of the mantle. The lithosphere is brittle solid rock broken into plates that sit on a *low-velocity zone* (LVZ) marking the asthenosphere.

Lithospheric plates appear to float and move around on top of the more ductile asthenosphere. The seismic velocity of P waves increases below the Moho until the waves reach a depth of about 100 kilometers, beyond which velocity decreases to below 8 kilometers per second. This second discontinuity is thought to be the boundary between the base of the lithosphere and the top of the asthenosphere, marked by the LVZ (**FIGURE 11.28**).

Much has yet to be learned about the LVZ, but it is thought to extend to a depth of approximately 350 kilometers and to be composed of rock that is closer to its melting point than the rock above or below it. This makes rock in the LVZ less rigid and more ductile than the layers above and below.

One hypothesis for the origin of the LVZ is based on the fact that minerals of the mantle are susceptible to melting over a range of temperatures. Some amount of melting might have produced a liquid component that could act as a lubricant. If this were the case, it would increase the ductile behavior of the LVZ. However, because S waves, which do not travel through liquid, are successfully transmitted by the LVZ, we know that relatively little melting must occur.

At a depth of 400 kilometers, there is another abrupt, though small, increase in the velocities of both S and P waves that marks the lower boundary of the upper mantle and the top of the *mantle transition zone* (**FIGURE 11.29**). This boundary, known as the *400-kilometer discontinuity*, is thought to be the result of a denser reorganization of the atoms in the mineral olivine, inferred to be the most abundant mineral in the mantle. Laboratory experiments show that when samples of olivine are subjected to pressure and temperature conditions equivalent to those at a depth of 400 kilometers, the atoms recrystallize into a form known as the *spinel* mineral family, which is about 10% denser than olivine. The increase in P- and S-wave velocities at the 400-kilometer discontinuity is probably the result of the olivine-spinel transition and the resultant increase in density.

Another abrupt increase in seismic wave velocities occurs at a depth of 660 kilometers, marking the base of the mantle transition zone and the top of the lower mantle. It is uncertain whether this discontinuity, known as the *660-kilometer discontinuity*, is the result of a mineral change in the mantle or a change in chemistry, or both.

One hypothesis about this problem describes a switch from pyroxene in mantle rocks to a mineral known as *perovskite,* which crystallizes at high pressure. Pyroxene, a silicate mineral, has 4 oxygen atoms packed around 1 silicon atom (recall from Chapter 4). Perovskite has a denser structure, with 6 oxygen atoms packed on 1 silicon atom. Laboratory experiments suggest that minerals in the spinel family, supposedly formed at the 400-kilometer discontinuity,

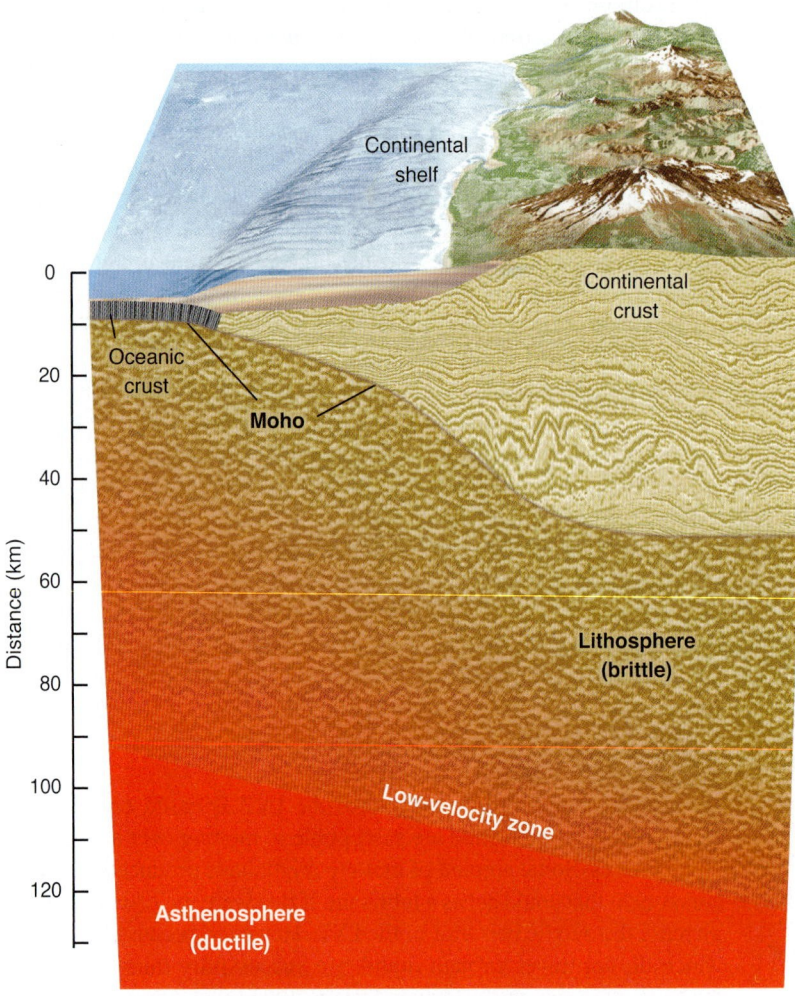

FIGURE 11.28 Below the Moho lies the lower lithosphere, which is a brittle solid, similar to the crust but denser, that composes the lower portion of plates. The lithosphere forms a boundary with the asthenosphere at the low-velocity zone (LVZ), which exhibits ductile deformation perhaps related to partial melting.

Ⓠ Why is the thickness of oceanic crust different from that of continental crust?

also reorganize themselves to form perovskite at pressures found at a depth of 660 kilometers.

The Core

At a depth of 2,900 kilometers, seismic P-wave velocities suddenly decrease, and S-wave velocities drop to zero. This is the top of the outer core. Seismologists infer that this layer is liquid, since S waves are not transmitted. A sudden increase in P-wave velocities at a depth of approximately 4,800 kilometers indicates that the inner core is solid, composed mostly of iron and small amounts of nickel and sulfur.

Ⓠ **EXPAND YOUR THINKING**

How do the seismic discontinuities correlate to the classic layers of Earth: crust, mantle, outer core, and inner core?

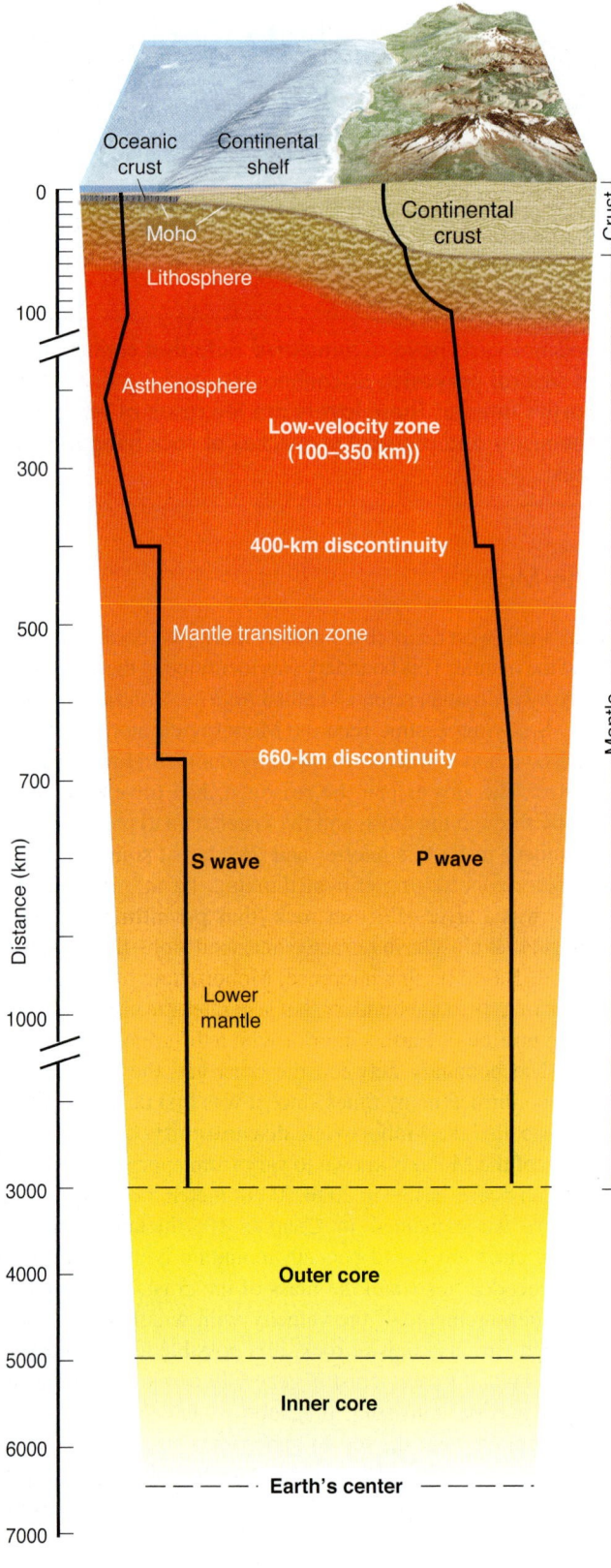

FIGURE 11.29 Changes in the seismic velocity of P and S waves mark discontinuities at 100 kilometers (the low-velocity zone), 400 kilometers (the base of the upper mantle), 660 kilometers (the top of the lower mantle), 2,900 kilometers (the top of the core), and 4,800 kilometers (the inner core boundary).

Ⓠ Why does the mineralogy of the mantle change from olivine to denser spinel and then to even denser perovskite?

Seismic Tomography Uses Seismic Data to Make Cross Sections of Earth's Interior

LO 11-11 Describe seismic tomography and what it reveals about Earth's interior.

In recent decades, the P and S waves from hundreds of earthquakes have been tracked at hundreds of seismometer stations, producing a global database of seismic velocities at all levels within Earth. These data offer an unprecedented view of the organization of Earth's interior.

Seismic Tomography

The science of **seismic tomography** utilizes seismic data to make cross-sectional slices through Earth in order to examine the nature of the planet's interior. *Tomography* means, literally, "drawing slices"—in this case, slices of Earth. The method is borrowed from the medical CAT-scan technique widely used in hospitals. CAT (computerized axial tomography) scans use small differences in X-rays that sweep the body to map variations in tissue and organs in order to construct cross sections and three-dimensional images of organs.

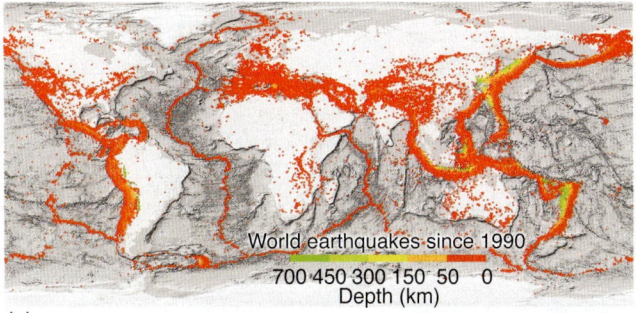

(a)   Global earthquake epicenters

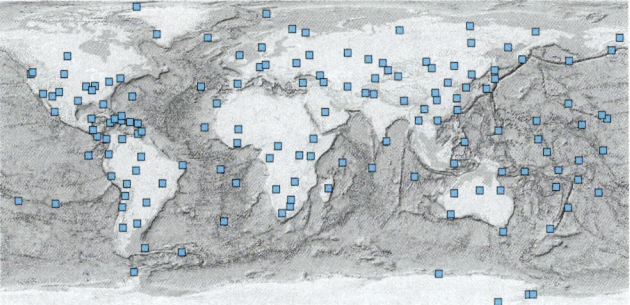

(b)   Global seismograph stations

FIGURE 11.30 The uneven distribution of earthquakes (a) largely along plate margins, and seismometer stations (b) largely in the northern hemisphere and continental settings, means that global tomographic data are rich in some areas and sparse in others.

 How would you solve the problem of the dearth of seismometers in oceanic areas?

Variations in the velocity of seismic waves depend on the temperature, pressure, and chemical composition of Earth's interior. But if the physical parameters that determine seismic velocities varied only with depth, our planet would be a very dull place, basically consisting of little more than parallel layers. No mountains would be built; there would be no active volcanism; and earthquakes would not occur. However, seismic tomography has shown us that Earth's interior is much more compelling (refer back to Figure 3.5). It is *lateral variations* in temperature, pressure, and composition (and, consequently, in seismic wave velocities) that create hot and cold regions in Earth's interior.

These regions reflect various phenomena such as mantle plumes, continental roots, subducted slabs, and other complexities inside the planet, all of which we've discussed in earlier chapters. Seismic tomography allows seismologists to map these features by determining the size, shape, and location of three-dimensional variations in wave speed.

But developing accurate three-dimensional images of Earth's interior is not without its problems, one of which stems from the uneven distribution of earthquakes and seismometers around the globe (**FIGURE 11.30**). Remember, the majority of earthquakes occur along plate margins; so, for large sections of crust, we have no source data for P- and S-wave velocities, simply because no large earthquakes have occurred in those areas in the modern era of instrumentation.

Another problem is that seismometer stations are located predominantly in continental settings in the northern hemisphere, meaning that data from the southern hemisphere and oceanic locations are underrepresented. Consequently, although seismic data are globally abundant, they are not necessarily always available in critical areas needed to define details of Earth's interior. Researchers are still experimenting with all types of seismic waves—Rayleigh waves, Love waves, P waves, and S waves—to determine which type offers the greatest possibility of deriving accurate images, so that we may continue to deepen our understanding of Earth's interior.

Tomographic mapping of Earth's interior has, fortunately, revealed new information about the organization of the mantle and core. Differences in the density of rock within the mantle indicate that the various discontinuities (discussed in section 11–10) possess rugged topography, some of which may be related to subducted slabs of oceanic crust that have fallen into the mantle.

FIGURE 11.31 shows the P-wave tomography of the mantle. Areas in blue depict cold, dense bodies of rock where P-wave velocities are high; areas in yellow and red reveal regions of hot, less dense rock, where P-wave velocities are low. This image displays a fallen slab (also called a *stagnant slab*) of the subducted Pacific Plate, lying on the 660-kilometer discontinuity under Japan and the Sea of Japan.

Stagnant slabs are thought by researchers to be former plates located in the mantle that are now largely inactive. Below this is an extensive high-velocity area in the lower mantle that is interpreted to be a "slab graveyard," where other cold slabs of subducted oceanic crust have fallen and collected on the core mantle boundary. Also shown are two mantle plumes: the Pacific plume and African plume.

As seen in Figure 11.31, tomographic mapping reveals features that are explained by plate tectonics theory. Cold regions below oceanic trenches appear to be descending slabs of oceanic lithosphere. Hot regions below mid-ocean ridges suggest the presence of thermally buoyant regions in the mantle supplying new rock to the lithosphere. Slab graveyards are identified where cold oceanic crust has not been fully recycled into the mantle.

FIGURE 11.30A uses variations in S-wave velocity to construct a tomographic cross section of the mantle below the North American

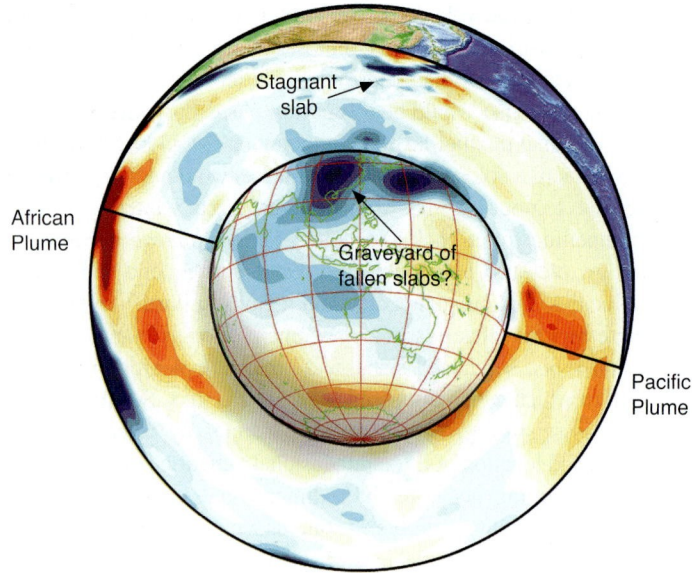

FIGURE 11.31 Earth's mantle mapped using the velocity of seismic P waves. Red areas consist of hot, low-density rock, and blue areas consist of cold, high-density rock. This cross section reveals two mantle plumes: the Pacific plume and African plume. Also displayed is a stagnant slab of the Pacific Plate where it has been subducted beneath Japan (the Honshu arc) and lies on the 660-kilometer discontinuity. Other stagnant slabs have fallen to the deep mantle and collected in a "slab graveyard" at the core-mantle boundary.

Ⓠ **Why does the mantle have low-density and high-density areas?**

Plate (NAM). The cross section reveals several features that are consistent with plate tectonics theory. At the surface, the cold, dense rock of the NAM is colored in dark blue. Beneath NAM, the figure shows remnants of the Farallon Plate (FAR) as a strong greenish-blue feature. In the western region, FAR is bluish and has stagnated on the 660-kilometer discontinuity. In the central and eastern regions, FAR is greenish-blue and has penetrated the mantle below the 660-kilometer discontinuity and extends to the central and lower mantle.

FIGURE 11.32B is an interpretation of how FAR, NAM, and the Pacific Plate (PAC) have interacted over the past 80 million years. Westward movement of NAM eventually overruns the spreading center at the divergent boundary of PAC and FAR. FAR subducts below North America between 80 and 40 million years ago; but by 20 million years ago, it has detached from the crust, and portions of it lie as a stagnant slab on the 660-kilometer discontinuity.

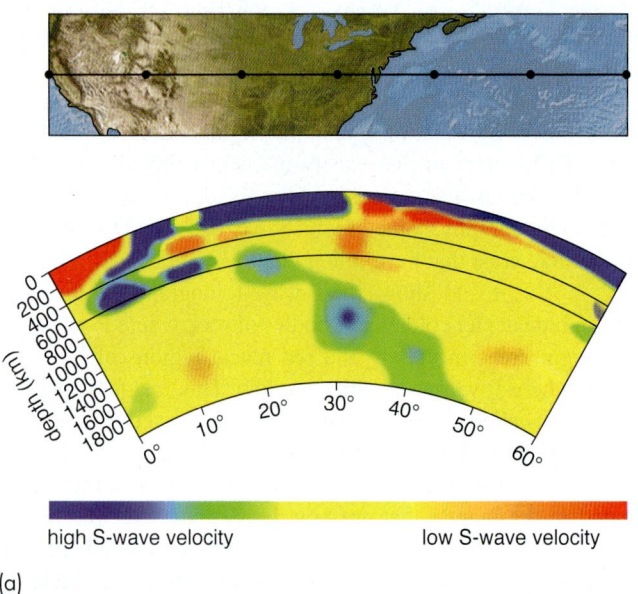

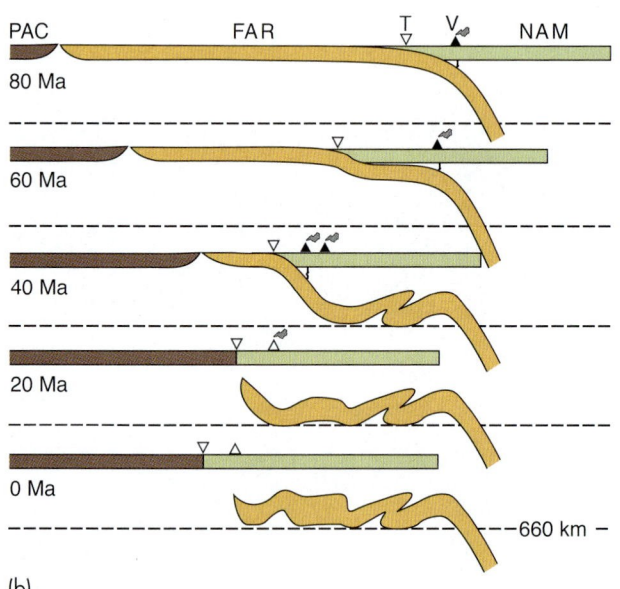

(a)

(b)

FIGURE 11.32 (a) Tomographic cross section of the mantle spanning the North American Plate. S-wave velocities are color-coded. Each dot on the map corresponds to a 10° shift in longitude shown under the cross section. (b) Interpretation of how the Farallon Plate (FAR), North American Plate (NAM), and the Pacific Plate (PAC) have interacted over the past 80 million years. T = trench and V = volcano. Source: Reprinted with permission from Christian Schmid, et al., "Fate of the Cenozoic Farallon Slab from a Comparison of Kinematic Thermal Modeling with Tomographic Images," in *Earth and Planetary Science Letters*, November 20, 2002. © 2002 Elsevier.

Ⓠ **What sorts of features do you see in Earth's interior in Figure 11.32a? Explain the processes responsible for these features.**

Ⓠ **EXPAND YOUR THINKING**

How do you reconcile the model of lava lamp convection that we discussed in Chapter 3 (see Figure 3.6) with major discontinuities consistent with a layered interior?

LET'S REVIEW

Earthquakes can cause widespread devastation and loss of life. Ground shaking, hillside failure, tsunamis, and fires all may result from seismic events. Over history, more than 20 earthquakes have each been responsible for the loss of more than 50,000 human lives.

An earthquake is a sudden shaking of Earth's crust, typically caused by the abrupt release of strain that has slowly accumulated in rock. Seismic energy, released by a quake, consists of surface waves and body waves that travel through Earth's interior and along the surface. Earthquakes occur at faults, many of which are related to the dynamic interaction of plate margins. But quakes also occur in plate interiors away from the margins. Seismic waves are measured by seismometers, hundreds of which, located throughout the world, monitor and record the passage of seismic waves set in motion by earthquakes, which occur every day at an average of 1 every 11 seconds.

STUDY GUIDE

11-1 An earthquake is a sudden shaking of Earth's crust.

- An **earthquake** is a sudden shaking of the crust. It typically is caused by the abrupt release of strain that has slowly accumulated in rock. Most rocks are elastic, so when stress is applied slowly and continuously over time, rocks will deform elastically. That is, they will rebound back to their original shape if the stress is removed or when the rock fractures.

- Most earthquakes are associated with the rupture of a fault. When rock breaks or a fault abruptly ruptures, it releases strain in a form of energy called **seismic waves**, which radiate through Earth's interior and along its surface. Seismic waves cause the shaking associated with an earthquake.

- More than 380 major cities lie on or near unstable regions of Earth's crust, leading seismologists to fear that a devastating earthquake capable of killing a million people could occur this century. With more infrastructure being built across greater expanses of Earth's surface to sustain human development, our communities become increasingly vulnerable to earthquake damage.

11-2 There are several types of earthquake hazards.

- *Ground shaking* caused by earthquakes can produce *sediment liquefaction*, *mass wasting*, and building collapse. Liquefaction occurs when groundwater rises to the surface and causes sediment to lose strength, leading to the collapse of buildings and roads. Any hillside that is overlain by weak or fractured bedrock, thick soil, or layers of unconsolidated sediment may experience mass wasting when exposed to ground shaking. Buildings that have a natural frequency that resonates with the shaking of an earthquake will likely collapse.

- The U.S. Geological Survey produces maps of **seismic shaking hazards** that are used by engineers and city planners to design and construct structures to withstand earthquake shaking.

- Fires ignited consequent to an earthquake can be a serious problem. Often, water lines and emergency equipment are damaged during an earthquake, making it impossible for firefighters and other emergency personnel to do their work.

- Earthquakes may generate a **tsunami**, a series of waves in the ocean set off when the seafloor is rapidly displaced. Tsunamis, which can be several meters high when they hit a shoreline, are capable of causing enormous damage.

11-3 The elastic rebound theory explains the origin of earthquakes.

- Geologists use the **elastic rebound theory** to explain why earthquakes occur. Distant forces cause a gradual buildup of elastic energy in the crust. Rocks accumulate strain, much as a stick stores elastic energy by bending. Eventually, a fault, or preexisting weakness in the rock, fails by fracturing.

- The **earthquake focus** is the location where rocks first fail. Slippage occurring at the focus is a source of stress that travels along the fault and causes displacement until the built-up strain is largely released. At the time of an earthquake, strained rock elastically snaps back into position. Vibrations of the rock returning elastically to its original shape set off seismic waves.

- As strain accumulating in the crust approaches the elastic limit of rocks, earthquakes tend to occur in clusters. The largest of a cluster is the **mainshock**; anything occurring before it is a **foreshock**, and anything occurring after it is an **aftershock**.

11-4 Most earthquakes occur at plate boundaries, but intraplate seismicity is also common.

- Earthquakes are not spread randomly across the globe. For the most part, they occur at locations where plate boundaries meet and brittle rocks of the crust fracture, as plate edges separate, collide, and shear across each other.

- Earthquakes occur in four kinds of plate settings: 1) intraplate regions, 2) divergent settings, 3) convergent margins, and 4) transform settings.

- Intraplate earthquakes often are related to stresses transmitted into plate interiors that originate at the plate edge. The impetus for many is the reactivation of old rift zones, or faults that are normally quiet. Such quakes tend to have shallow foci, and are left over from former tectonic action that has ceased; or they may mark the establishment of a new region of deformation. Intraplate seismicity also may be related to magma movement associated with hotspots in plate interiors, such as at Yellowstone or in Hawaii.

11-5 Divergent, convergent, and transform margins are the sites of frequent earthquake activity.

- At divergent boundaries, seismic activity tends to occur at shallow depths. The lithosphere at a divergent boundary is thin and weak, hence the large amounts of strain necessary to cause major earthquakes do not accumulate there. Nevertheless, quakes at divergent boundaries may cause significant damage.

- The largest and most active areas of seismicity are located at convergent boundaries, where one plate subducts beneath another in a zone of actively accumulating compressive stress. These regions of plate interaction are characterized by progressively deeper earthquakes along a line proceeding from the trench toward the overriding plate.

- Earthquake foci at convergent margins define a steeply dipping plane called the **Wadati–Benioff Zone**, or WBZ, that corresponds to faulting between the subducting oceanic slab and the upper mantle of the overriding plate.

- At transform boundaries, earthquakes are shallow, running only as deep as 25 kilometers. Two examples of dangerous transform seismicity are found at the San Andreas Fault in California and the North Anatolian Fault in Turkey. Both are right lateral strike-slip faults.

11-6 Earthquakes produce four kinds of seismic waves.

- Earthquakes radiate seismic energy in the form of two types of **surface waves** and two types of **body waves**. Surface waves consist of **Love waves** and **Rayleigh waves**. Body waves consist of primary waves, or **P waves**, and secondary waves, or **S waves**. Because liquids do not resist changes in shape, they absorb S waves rather than transmit them.

11-7 Seismometers are instruments that measure and locate earthquakes.

- A **seismometer** is an instrument that records seismic waves. A record of seismic waves is called a seismogram. One seismometer station will measure three types of ground motion: north-south, east-west, and up-down. Seismograms allow scientists to estimate the distance, direction, size, and type of faulting of an earthquake.

- Earthquake intensity is measured by the Modified Mercalli (MM) Intensity Scale, which ranks the level of damage caused by the quake on a scale from I to XII.

11-8 Earthquake magnitude is expressed as a whole number and a decimal fraction.

- The magnitude of a quake is determined using the **Richter scale**, which measures the largest wave amplitude on a seismogram. Earthquake magnitude is calculated from the logarithm of the maximum seismic wave amplitude recorded by seismometers.

- The moment magnitude is calculated in part by multiplying the area of the fault's rupture surface by the distance the crust has been displaced along

a fault. That means the most accurate calculation of moment magnitude is based on field measurements of the rupture surface.

11-9 Seismology is the study of seismic waves conducted to improve our understanding of Earth's interior.

- Seismic wave refraction is caused by changes in wave velocity due to differences in density. If a wave passes into rock of higher seismic velocity, the wave will be refracted upward relative to its original path. A wave passing into rock of lower seismic velocity will be refracted downward. **Wave refraction** is used to identify a **discontinuity.**

- P-wave and S-wave shadow zones on the far side of the core are created by wave refraction and absorption due to changes in the material through which they pass.

11-10 Seismic data confirm the existence of discontinuities in Earth's interior.

- Seismologists have confirmed the existence of several pervasive discontinuities that mark distinct layers in Earth's interior. The discontinuity between the crust and the mantle is named the **Mohorovicic discontinuity**, or Moho, at a depth of approximately 8 kilometers beneath ocean basins to 20 to 70 kilometers beneath continents. The base of the lithosphere and the top of the asthenosphere is marked by a discontinuity called the *low-velocity zone* (LVZ), at a depth of about 100 kilometers.

- At a depth of 400 kilometers, an abrupt, though small, increase in the velocities of both S and P waves marks the lower boundary of the upper mantle and the top of the *mantle transition zone*. Another abrupt increase in seismic wave velocities occurs at a depth of 660 kilometers, marking the base of the mantle transition zone and the top of the lower mantle. At a depth of 2,900 kilometers, seismic P-wave velocities suddenly decrease and S-wave velocities drop to zero. This is the top of the outer core. A sudden increase in P-wave velocities at a depth of approximately 4,800 kilometers indicates that the inner core is solid, composed mostly of iron and small amounts of nickel and sulfur.

11-11 Seismic tomography uses seismic data to make cross sections of Earth's interior.

- **Seismic tomography** uses changes in P- and S-wave velocity to make cross-sectional slices, which scientists use to examine the nature of rock in Earth's interior. Tomography reveals significant differences in rock density within the mantle, which researchers interpret as mantle plumes, stagnant slabs of subducted crust, and slab graveyards on the core-mantle boundary.

KEY TERMS

aftershocks (p. 313)
body waves (p. 312)
discontinuity (p. 330)
earthquake (p. 308)
earthquake focus (p. 312)
elastic rebound theory (p. 313)
epicenter (p. 312)
foreshocks (p. 313)

Love waves (p. 322)
mainshock (p. 313)
Mohorovicic discontinuity (p. 333)
P waves (p. 322)
Rayleigh waves (p. 322)
Richter scale (p. 326)
S waves (p. 322)
seismic shaking hazards (p. 311)

seismic tomography (p. 335)
seismic waves (p. 308)
seismometers (p. 312)
surface waves (p. 312)
tsunami (p. 311)
Wadati–Benioff Zone (p. 318)
wave refraction (p. 330)

ASSESSING YOUR KNOWLEDGE

Please complete this exercise before coming to class. Identify the best answer to each question.

1. Earthquake risk has increased in recent decades because:
   a. There are more earthquakes.
   b. Earthquakes have grown larger.
   c. The human population has expanded.
   d. Plate margins have become more dangerous.
   e. All of the above.

2. In regions lacking modern engineering standards, the greatest hazard during an earthquake is:
   a. Building collapse.
   b. Ground fractures.
   c. Violent shaking.
   d. Groundwater withdrawal.
   e. None of the above.

3. Earthquake hazards include:
   a. Shaking, fires, landslides, and tsunamis.
   b. Tsunamis, liquefaction.
   c. Lack of communication, lack of food and water due to damaged public services.
   d. Lack of emergency help, building collapse, mass wasting.
   e. All of the above.

4. What is an earthquake?
   a. A sonic boom
   b. Any rock fracture
   c. Sudden rapid shaking of Earth's crust
   d. Any breakage of rock
   e. None of the above

5. Why do subduction zones have "deep" earthquakes?
   a. Because divergent plate boundaries are fed by deep magma upwelling.
   b. Because subducted slabs of oceanic crust experience faulting at significant depths within the mantle.
   c. Because transform faults, like the San Andreas Fault, are very active.
   d. Because where plates converge they tend to become very thick and break in the asthenosphere.
   e. None of the above.

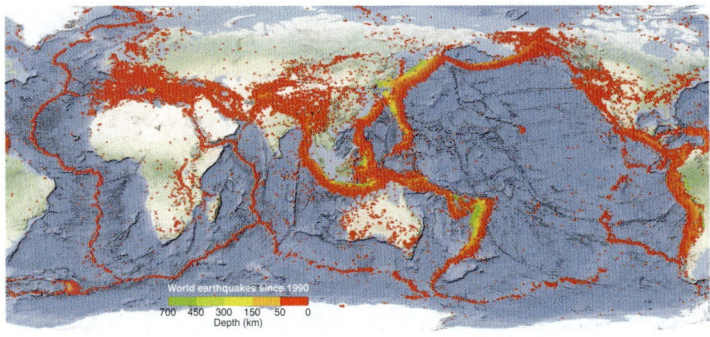

6. The elastic rebound theory:
   a. Describes how seismic waves pass through Earth's interior.
   b. Describes how folding in the crust reduces earthquake magnitude.
   c. Describes the behavior of faulting crust.
   d. Was proven wrong in the middle of the twentieth century.
   e. Was first defined by Archimedes.

7. Why does the west coast of the United States have a high seismic hazard rating?
   a. It is a region of a former plate boundary that is still active.
   b. There is active volcanism along the entire coastline.
   c. It is characterized by active plate boundaries of one type or another.
   d. For all of these reasons.

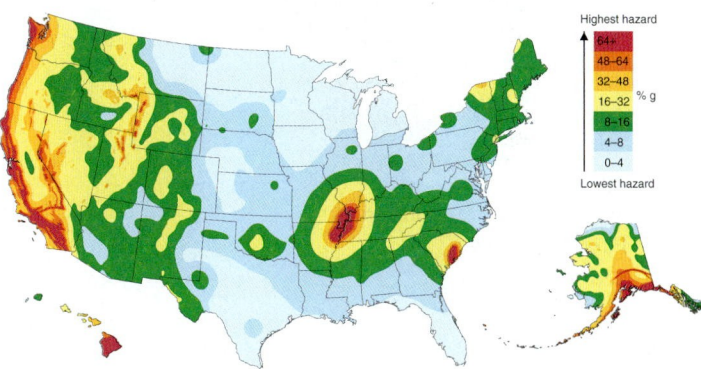

8. Earthquakes at divergent margins tend to be:
   a. Deep and strong.
   b. Produced by normal faulting.
   c. On the surface of a subducting slab.
   d. Related to compression.
   e. Caused by anticlines.

9. A U.S. city marked by intraplate seismicity is:
   a. San Francisco, California.
   b. Charleston, South Carolina.
   c. Los Angeles, California.
   d. Seattle, Washington.
   e. Portland, Oregon.

10. Seismicity at divergent, convergent, and transverse plate boundaries is caused by the following stresses (in correct order):
    a. Compressive, shear, and subductive
    b. Transverse, subductive, and decompression
    c. Tensional, compressional, and shearing
    d. Normal, reverse, and plunging
    e. None of the above

11. On this diagram, label Earth's internal structure using the following terms: mantle, asthenosphere, inner core, lithosphere, outer core, continental crust, oceanic crust, Moho, mantle transition zone.

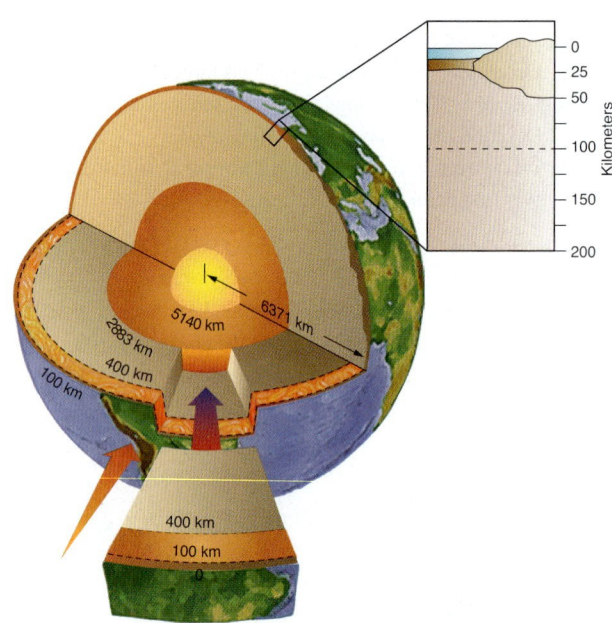

13. Finding an epicenter requires:
    a. Firsthand accounts.
    b. Seismic tomography.
    c. Wave refraction.
    d. Triangulation.
    e. Calculating earthquake intensity.

14. The first seismic wave recorded on a seismogram is the:
    a. Rayleigh wave.
    b. S wave.
    c. P wave.
    d. Love wave.
    e. All body waves arrive at the same time.

15. This diagram shows a seismogram of a hypothetical earthquake. On the seismogram, label the following:
    S–P interval
    First arrival of P wave
    First arrival of S wave
    Background noise
    First arrival of surface waves

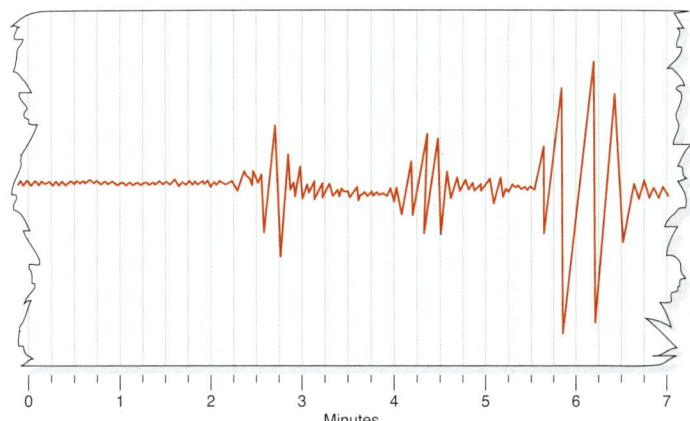

16. Earthquake magnitude is measured by:
    a. The Richter scale.
    b. The Modified Mercalli scale.
    c. The epicenter.
    d. Triangulation.
    e. Level of damage.

17. Seismic shadow zones are the result of:
    a. Wave refraction at discontinuities and absorption in the outer core.
    b. Wave reflection and interference in the mantle.
    c. Seismic amplitude causing discontinuities in the crust.
    d. Infrequent occurrence of large earthquakes.
    e. A lack of seismometers in the southern hemisphere.

18. Seismic discontinuities are found at the following depths, measured in kilometers (km):
    a. 8 km to 20–70 km, 100 km, 400 km, 660 km, 2,900 km, and 4,800 km
    b. 20 km to 50 km, 70 km, 3,500 km, and 2,900 km
    c. 20 km to 50 km, but only at subduction zones
    d. 8 km, 100 km, 1,000 km, 2,000 km, 3,000 km, and 4,000 km
    e. None of these

19. Seismic tomography reveals that Earth's interior is:
    a. Like a layer cake with four uniform layers.
    b. A well-mixed layer of liquid mantle with solid rock above it and below it.
    c. Highly complex, with hot regions and cool regions.
    d. Too complex to understand.
    e. Made of olivine entirely to the inner core.

20. On this image identify where changes in the seismic velocity of P and S waves mark discontinuities in Earth's interior.

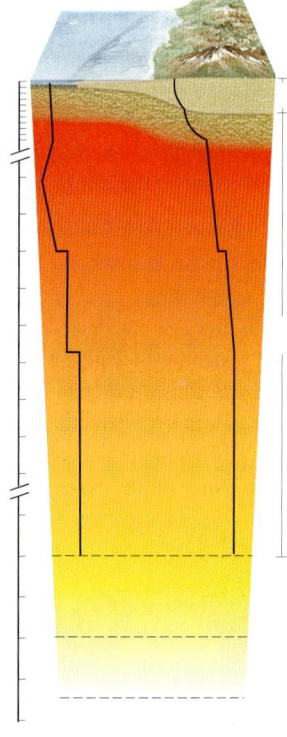

FURTHER RESEARCH

1. Are there likely to be earthquakes on other planets? Why or why not?

2. How do seismic waves destroy buildings?

3. What can we predict about future earthquakes?

4. How do earthquakes start?

5. What is meant by elastic rebound?

6. What factors contribute to the level of destruction caused by an earthquake?

ONLINE RESOURCES

Explore more about geology on the following Web sites:

Investigate research conducted by the U.S. Geological Survey at the USGS Earthquake Hazards Program:
http://earthquake.usgs.gov

Learn about quake hazards in Washington and Oregon at the Pacific Northwest Seismograph Network:
www.pnsn.org/INFO_GENERAL/eqhazards.html

Check out the Federal Emergency Management Agency's (FEMA) earthquake hazard reduction page:
www.fema.gov/plan/prevent/earthquake/index.shtm

Additional animations, videos, and other online resources are available at this book's companion Web site:
www.wiley.com/college/fletcher

This companion Web site also has more information about WileyPLUS *and other Wiley teaching and learning resources.*

GEOLOGIC TIME

CHAPTER CONTENTS & LEARNING OBJECTIVES

GEOLOGY IN OUR LIVES

Knowing the history of rock layers in the crust is critical to helping geologists predict the location of geologic resources and discern the events that make up Earth's past. Geologists analyze geologic time using two methods, both based on critical thinking: 1) relative dating, which determines the *order* of geologic events; and 2) radiometric dating, which calculates the *age* of a geologic event. These methods have revealed that Earth has a long and fascinating history, and that it is approximately 4.6 billion years old. But to fully unravel the history of Earth's crust it is necessary to integrate relative and radiometric dating, in order to find the resources upon which our communities depend on every day, such as groundwater, metals, coal, petroleum, and others.

Q What geologic events are represented in this scene?

12-1 Earth's History Is a Sequence of Geologic Events

LO 12-1 Describe early attempts to determine Earth's age.

Not so long ago, most people based their ideas about Earth's age on dogma, faith, and religious doctrine, rather than critical thinking. Through much of the eighteenth and nineteenth centuries, most Western cultures believed that the world had been brought into existence exactly 4004 years before the birth of Christ.

Early Notions About Earth's Age

How would you determine Earth's age? What information would you need? How would you apply scientific thinking to the potentially sparse data available to address this question?

Many early biblical scholars interpreted Earth's age by making detailed analyses of the Bible. Among the best known of these was *James Ussher* (1581–1656), the Archbishop Armagh, Ireland (**FIGURE 12.1**), who specified the date of Creation with great precision but, as it turned out, low accuracy. According to his analysis of the Bible, Earth came into existence at nightfall, Sunday evening, October 23, 4004 BC. Ussher arrived at this date by using a combination

The Granger Collection, New York

FIGURE 12.1 Based on a careful analysis of the Bible, Archbishop James Ussher concluded that God created Earth on Sunday, October 23, 4004 BC.

Q Ussher's dating of Earth's age is an example of precision. What is the difference between precision and accuracy?

of astronomical cycles, historical accounts, and the chronology reported in the Book of Genesis. He placed the beginning of Creation just 4000 years before the birth of Christ—the last "4" in 4004 was occasioned by the date of Christ's birth, believed to be 4 BC. Ussher's findings, which he wrote about between 1650 and 1658, became widely accepted because they were cited in the margin of the Great Edition of the English Bible (1701), where the citation remained without explanation for 200 years. It wasn't until 1900, upon the publication of a new edition, that Cambridge University Press removed the reference to Ussher's date.

Ussher's concepts went largely unchallenged at the time in part because they posed no threat to the social order. That is, they adhered to the logic of other theologians and did not question the authority of church and state. Still, not everyone agreed with Ussher's stated age for Earth. A number of scholars in various fields were beginning to consider the idea of *deep time*.

The discovery of deep time, the idea that the age of the universe can be measured in billions of years, goes back at least to the speculations of the German philosopher *Immanuel Kant* (1724–1804), the French mathematician and astronomer *Pierre Simon Laplace* (1749–1827), and the Scottish naturalist *James Hutton* (1726–1797). Kant and Laplace were the originators of the solar nebula hypothesis for the origin of the solar system (Chapter 2). Obviously, such an origin implies enormous amounts of time. Hutton proposed immensely long time spans to explain how observable rates of sediment erosion and deposition, and ongoing volcanic activity, might have created great valleys, thick sediment sequences, mountain ranges, and other features of Earth's surface.

Not all such speculators thought in terms of billions of years, however. *Leonardo da Vinci* (1452–1519), for one, calculated the rates of sedimentation in the Po River of Italy and concluded that it had taken some 200,000 years to form nearby rock deposits. More than two centuries later, in 1760, the Frenchman *Georges–Louis Leclerc de Buffon* (1707–1788) estimated Earth's age to be 75,000 years by calculating the time it would have taken for the planet to cool from the molten state. Next, in 1831, *Charles Lyell* (1797–1875) arrived at the age of 240 million years, based on changes recorded in fossils he found in rock in the English countryside. By 1901, *John Joly* (1857–1933) had calculated the rate of salt delivery from rivers to the ocean in order to estimate the time necessary to produce seawater. His conclusion: 90 to 120 million years.

Clearly, the age of Earth has long been a subject of debate among scientists, and between biblical creationists and scientists. Now, however, at least as far as scientists are concerned, that debate has been resolved: Calculations based on the phenomenon of radioactivity have shown conclusively that Earth was formed approximately 4.6 billion years ago (bya). Until the beginning of the twentieth century, scientists had only vague ideas of Earth's age. It was generally believed that Earth was essentially unchanged since its formation—that oceans were constant, that the continents remained unaltered, and that any given mountain or valley had stood, with little modification, since the beginning of time. Until advances in understanding radioactivity were made—thanks to the discoveries of Marie and Pierre Curie, along with Henri Becquerel, in 1898—there was no reliable method for determining the numerical age of any rock, fossil, geologic feature, or geologic event.

In the absence of any absolute age calculations, eighteenth- and nineteenth-century naturalists such as William Smith, Charles Lyell, James Hutton, and others developed the science of relative dating. Their investigations and methods convinced them that Earth was

(a) Calvin Larsen/Science Source

(b) David Davis/Photo Researchers

(c) Ken M. Johns/Photo Researchers

(d) G.R. Roberts/Photo Researchers

(e) Francois Gohier/Photo Researchers

Describe the geologic process at work in each photo.

FIGURE 12.2 Five fundamental geologic events can be ordered using relative dating. (a) Sediment deposition leads to the formation of sedimentary rock. (b) Erosion is slow but effective at removing portions of the rock record. (c) Igneous intrusion creates dikes and other types of intrusive bodies. (d) Faulting is revealed when two sides of a fracture are displaced relative to each other. (e) Deformation is produced by stress in the crust.

much older than most people imagined, and than earlier estimates had stated. Scientists today still apply the principles of relative dating developed by these pioneering geologists.

Five Fundamental Geologic Events

Relative dating is a system of *reasoning* that is used to determine the chronological sequence or order of a series of **geologic events,** notable occurrences of common geologic processes. Relative dating does not calculate a precise age; it only indicates that a particular geologic event occurred before or after another. It was not until a few decades ago, when numerical dating techniques based on radioactivity had significantly improved, that sequences of events could be dated in terms of "years ago."

Relative dating is based on the identification of five geologic events:

1. **Deposition** of sediments and the formation of sedimentary rock strata in environments of deposition (**FIGURE 12.2A**).

2. **Erosion** of the crust such that a gap, known as an **unconformity,** opens in the geologic record of past events (**FIGURE 12.2B**). The duration and extent of the gap depends on the duration and extent of the erosion. Any rock that is exposed to the weathering processes of the atmosphere and ocean is eroded.

3. **Intrusion** of plutonic and volcanic igneous rocks. Intrusion involves the movement of magma through the crust (**FIGURE 12.2C**). Intrusive rock typically cuts across layers of older country rock.

4. **Faulting** of the crust. A fault is a break or fracture where rock layers on one side of the break are displaced relative to rock layers on the other side. A fault may occur as an entire zone, as in the San Andreas Fault Zone of California, which may be over a kilometer wide and consists of several main faults and many small secondary faults. Or, a fault may be a single fracture that runs through the crust. The distinguishing feature of a fault is that it causes relative *movement* of the rocks on either side (**FIGURE 12.2D**).

5. **Deformation** and **uplift** of rock layers so that they become tilted, folded, or even turned upside down. Deformation occurs because of tectonic activity due to plate movement that can disrupt the crust at plate boundaries and within plate interiors (**FIGURE 12.2E**). This process *uplifts* the crust, subjecting it to weathering, erosion, and the formation of unconformities.

EXPAND YOUR THINKING

What is the fundamental difference between scientific thinking and religious thinking about a problem?

12-2 Geology Is the Science of Time

LO 12-2 Compare and contrast relative and radiometric dating.

Each field of science is defined by a single focus of study: Physics is the science of motion; chemistry is concerned with the composition of matter; and biology seeks to understand life. Geology is known as the science of time.

Time

The eighteenth-century naturalist James Hutton is often called the father of geologic time. He is credited with describing Earth as so old, and time so vast, that "We find no vestige of a beginning, no prospect of an end." Since then, however, through the integrated work of geologists, physicists, and chemists, a vestige of that beginning has been found; and furthermore, has established that Earth formed approximately 4.6 bya, an estimate based on several lines of evidence. The evidence itself is based on techniques used to determine the age of Moon rocks and meteorites, the abundance of lead (Pb) isotopes created by radioactive decay in Earth rocks, and the direct measurement of very old rocks and minerals in the continental crust (**FIGURE 2.3**). We examine each of these later in this chapter.

Geologists are trained to analyze and interpret evidence gleaned from the vast archive of Earth's rocks. Sedimentary rocks, in addition to recording environmental change, also contain fossils, recording a history of changing life on Earth. Igneous rocks document magma production, volcanism, oceanic crust, and important aspects of plate tectonics. Metamorphic rocks preserve evidence of past fold-and-thrust belts, continental collisions, and the pressure and temperature

conditions of deep burial within the crust. In a very real sense, then, geologists are Earth historians, for hidden among the minerals and strata of the crust is the writing of the ages (**FIGURE 12.4**). In this chapter we will examine the critical thinking skills geologists use to interpret this evidence and learn the history of our planet as it is written in Earth's *rock record*, the record of Earth's history preserved in the crust.

Dating

Imagine that your history professor gives you an assignment on the Revolutionary War. You recall reading a magazine article sometime last winter about the colonial army—probably, January. Pulling out a stack of magazines from the past year, you dig down through the pile, starting with the most recent issues and on to March and February; January *must* be next, you think.

Why did you think that? Because, with the more recent issues on the top, you assumed the older issues would be under them, with January at the bottom of the pile; so you worked backward—through time. Without realizing it, you were applying the geologic *principle of superposition*, perhaps the most powerful dating tool in our critical thinking toolbox.

This simple scenario illustrates the critical thinking that geologists employ to interpret the history of Earth's crust. Actually, they use two types of time analysis: **relative dating** and **radiometric dating.** Relative dating determines the order of formation (like the magazines—which appeared in sequence from most recent, to old, to oldest) among layers of rock based on their relationship to one another. Radiometric dating uses naturally occurring **radioactive decay** to estimate the age of geologic samples. (Radioactive decay is explained in section 12–6.) Relative and radiometric dating methods can be combined to build a potent understanding of the history of Earth through time.

Why Study Geologic Time?

The study of geologic time has direct implications in all our lives. The same tools that scientists use to understand the first organisms

FIGURE 12.3 The oldest Earth rocks ever found, located along the shore of Hudson Bay in northeastern Canada, formed approximately 4.28 billion years ago. They may be remnants of a portion of Earth's primordial crust—the first crust that formed on the surface of our planet.

Ⓠ This rock has been involved in several episodes of fold-and-thrust belt orogeny (mountain building). Which family of rock is it likely to be in: sedimentary, igneous, or metamorphic, and why?

FIGURE 12.4 The Grand Canyon reveals a thick sequence of igneous, metamorphic, and sedimentary rocks that preserve a history of changing geologic environments. The history of these rocks has been identified using a combination of relative and radiometric dating. The oldest among them formed over 1 billion years ago, recording events on Earth before the advent of multicellular life.

Ⓠ In which part of the Grand Canyon will you find the oldest rocks?

Courtesy Don Francis

Carol Polich/Lonely Planet Images

that populated early oceans and walked on ancient continents can be applied to find coal and oil, valuable metals, groundwater, and the other geologic resources humans depend on. Geologists have learned, for example, that petroleum is created by the exposure of highly organic layers of sediment to specific pressure and temperature conditions (the "oil window"). Sediments that are "cooked" in the crust too long may hold only methane, while those that are not left buried long enough will fail to be converted into petroleum. Thus, to find oil, you need to know the age of the rocks.

In some areas, sedimentary rock layers of a specific age are known to hold oil, so by being able to determine the age of potentially oil-bearing rocks, we simplify the process of oil exploration—resulting in a direct payoff at the pump. The same is true of coal. Earth's history features two great periods of coal formation: the Carboniferous and the Permian. Geologists searching for new coal seams will explore rocks from these periods to raise the probability of success. We can apply the same deductive reasoning in searching for economic minerals, construction materials, and other types of geologic resources.

Above and beyond the economic incentives for studying geologic time, human curiosity drives us to ask and answer questions about our origins, our history, and our place in Earth's sequence of natural events. It is humbling to realize that the entire span of human history represents less than 1% of the length of time that dinosaurs walked the land. Or, as Mark Twain once observed, "If the Eiffel Tower were now representing the world's age, the skin of paint on the pinnacle-knob at its summit would represent man's share of that age; and anybody would perceive that that skin was what the tower was built for." To make his point, Twain was being satirical in presuming that all of geologic history took place in preparation for the arrival of the human species. Clearly, though, he was a man who grasped the full magnitude of geologic time.

A Walk through Time

Imagine that Earth's entire history is contained in the last 12 hours (720 minutes) (**FIGURE 12.5**): Recorded human history began only 1.25 seconds ago. The first evidence of our species, *Homo sapiens*, appeared in the last 20 seconds. The dinosaurs disappeared from Earth 10 minutes ago, and the first primitive land plants appeared only 1 hour ago. The first evidence of multicelled life occurred in rocks 2 short hours ago. Oxygen accumulated in Earth's atmosphere about 4 hours ago, and the great continental landmasses had mostly reached their current size 6.5 hours ago. Life itself, the hallmark of our venerable planet, first appeared in its tenuous and fragile form as a tiny single-celled creature approximately 9 hours ago.

The remarkable thing about Earth's history is that the events many believe to be ancient—dinosaurs and primitive plants and huge marine reptiles, to name a few—are, in reality, the recent cousins of modern life-forms when compared to life's truly ancient ancestors. Just think: 10 hours of our "day" had already passed before the first multicelled life-forms evolved. When viewed from the colossal expanse of Earth's history, our species has been present for the mere blink of an eye.

EXPAND YOUR THINKING

Given the influence of plate tectonics, where would you search for Hadean and early Archean rocks? Why?

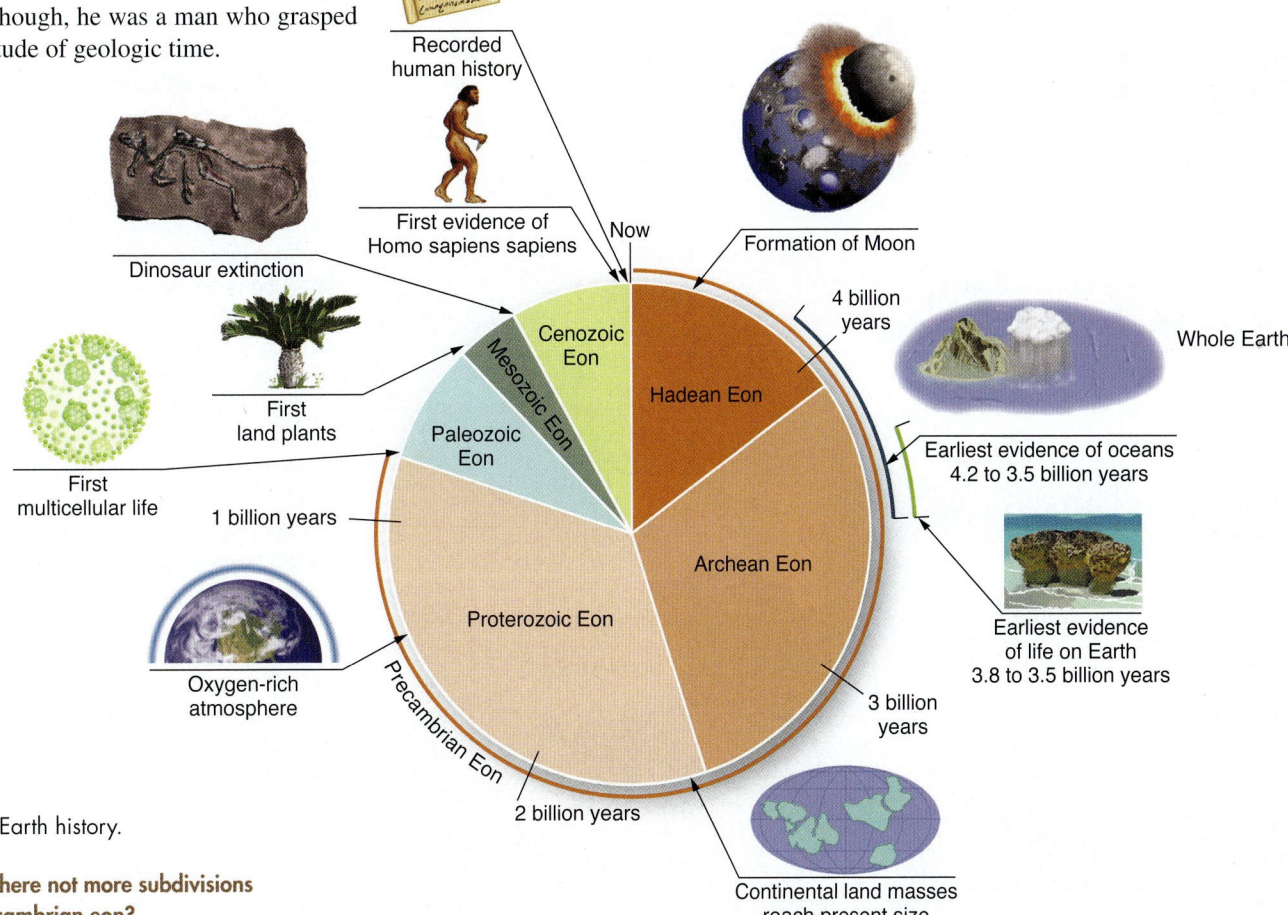

FIGURE 12.5 Earth history.

Why are there not more subdivisions in the Precambrian eon?

Seven Stratigraphic Principles Are Used in Relative Dating

12-3

LO 12-3 List and describe the seven stratigraphic principles.

Recognizing and identifying the five geologic events—deposition, erosion, intrusion, faulting, and deformation and uplift—in an outcropping of rock or a geological cross section of the crust is the first step in relative dating. But doing this is not always easy.

Principles of Relative Dating

The presence of sedimentary rock is an obvious indicator that deposition has taken place. Erosion, in contrast, can be more difficult to recognize. One reliable sign that erosion has occurred is where a layer of rock appears to be cut into and filled by the overlying layer. Occasionally, the presence of a layer of gravel is another indicator that erosion has occurred. Erosion also is recognizable when there is a significant age difference between two adjacent rock units. A marker of intrusion is the presence of an igneous dike, and faulting

and deformation are apparent from broken and displaced layers and folded layers, respectively.

To properly interpret a complex section of crust in which all or many of these events have played a role, geologists have established a set of rules to guide their thinking. These rules, called **stratigraphic principles,** are basic and relatively simplistic, but nonetheless important concepts that lay the foundation for relative dating. There are seven stratigraphic principles:

1. The *principle of superposition*. In a sequence of undisturbed sedimentary rocks, the oldest layer is at the bottom and the youngest layer is at the top (**FIGURE 12.6A**). This is sometimes referred to as "layer-cake geology": the bottom layers of a cake are placed down first, with subsequent layers placed on top. The layer on the bottom would be the oldest, with the layers overlying it sequentially younger (exactly like the layers of magazines above). An exception to this principle occurs when rock units have been disturbed and overturned by tectonic forces.

2. The *principle of original horizontality*. Sedimentary rocks are deposited in layers parallel to Earth's surface. There are some exceptions to this rule, but they are uncommon and can be recognized. When rocks are found in a nonhorizontal configuration, we can conclude that some geologic event has tilted them (**FIGURE 12.6B**). For instance, when two continents collide at a convergent margin, originally flat layers of sedimentary rock are bent and broken, so that they are tilted. This is the signal that the geologic event of deformation has taken place.

3. The *principle of original lateral continuity*. Sedimentary beds originally are laterally continuous within their environment of deposition. Hence, similar rock units at different locations may, in fact, be the same even though they are not now connected

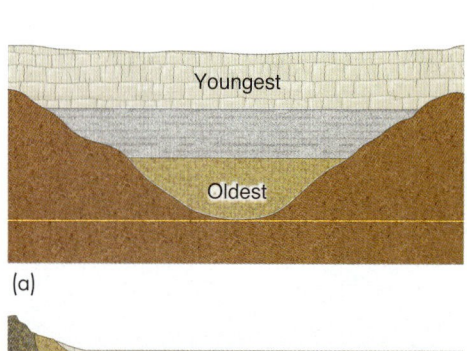

(a)

Continuous strata

(c)

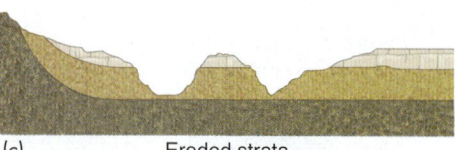

Eroded strata

Sandstone

Granite

(e)

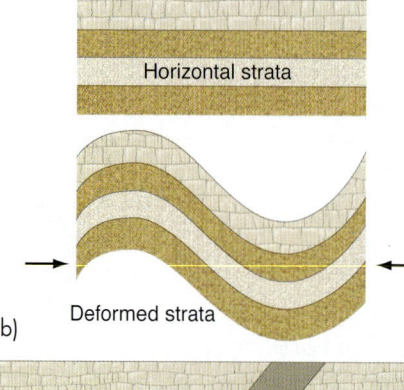

(b) Horizontal strata / Deformed strata

Dike

(d)

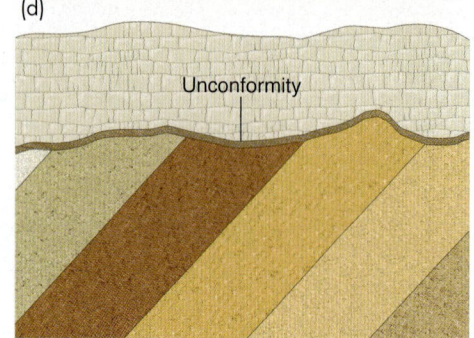

Unconformity

(f)

FIGURE 12.6 Illustration of six of the seven stratigraphic principles. (a) Principle of superposition: Among undisturbed layers of sedimentary rock, the lower layers are older than the upper layers. (b) Principle of original horizontality: Most sedimentary rocks are originally deposited as horizontal strata. Tilted strata are a sign of deformation. (c) Principle of original lateral continuity: Faulting, severe folding, and erosion may separate originally continuous strata. (d) Principle of cross-cutting relationships: Geologic events (intrusion, erosion, faulting, and deformation) that cut across preexisting rocks are younger than the rocks they cut across. (e) Principle of inclusions: Clasts of existing rock that are incorporated into a sedimentary layer or an intrusion come from rocks that are older than the sedimentary layer or intrusion into which they have been incorporated. (f) Principle of unconformities: Unconformities develop when erosion or nondeposition interrupts the continuity of the geologic record.

Q In each example can you identify the oldest geologic event?

(because erosion has removed some of the unit). Faulting, severe folding, and erosion may have separated the originally continuous beds. When a stream erodes the crust, for example, it may cut downward through layers of rock, to become a deep, wide canyon. Despite the distance between similar rock outcrops on either side of the canyon, the principle of lateral continuity tells us that they once formed a continuous layer of rock from the same environment (FIGURE 12.6C).

4. The *principle of cross-cutting relationships.* Rocks or other geologic features (e.g., erosion or intrusions) that cut across preexisting rocks are younger than the rocks they cut across. FIGURE 12.6D illustrates this concept well: A basalt dike intrudes layers of sedimentary rock. The principle of cross-cutting relationships tells us that the basalt is younger than the sedimentary beds because the dike cuts across them.

5. The *principle of inclusions.* Any part of an existing rock that is incorporated into a sedimentary layer or igneous intrusion is older than the sedimentary layer or intrusion into which it has been incorporated (FIGURE 12.6E). Consider that a granite batholith that intrudes sandstone crust may have pieces of sandstone incorporated within the cooled granite. This may be because of blocks of sandstone that fell off the walls of the magma chamber into the molten granite. A geologist would interpret this as a sign that the granite is younger than the sandstone inclusions.

6. The *principle of unconformities.* Unconformities develop in a geologic cross section when erosion or nondeposition (environments where no sediments have been deposited for extended periods of geologic time) interrupts the continuity of the geologic record (FIGURE 12.6F). The position of an unconformity in a relative dating sequence is determined by the principle of cross-cutting relationships. The presence of an unconformity indicates that a portion of the rock record has been removed, as if pages had been torn out of a book. Without numerical dating to assign an exact age to the rocks on either side of the unconformity, it is difficult to know how much of the rock record has been removed by erosion.

7. The *principle of fossil succession.* Plants and animals change through time; this change is termed **evolution.** (Evolution is discussed in greater detail in Chapter 13.) Rock layers record these changes because the fossils of plants and animals that lived at one time will not be found in rocks that were formed at a different time. This succession of fossils serves as a relative dating tool—the assemblage of fossils occurring in a rock can serve to identify and date that rock. This principle is incredibly useful and warrants more attention here, as well as its own illustration (FIGURE 12.7), which is why it was not included with the other six principles above.

The principle of fossil succession is an especially powerful geologic tool. If we begin at the present and examine increasingly older layers of rock (see Figure 12.7), we will come to a level where no fossils of humans are present. As we step further back in time, each step brings us to levels where no fossils of flowering plants are present; the next level has no birds; the next, no mammals; then no reptiles, no four-footed vertebrates, no land plants, no fish, and no shells. Finally, we reach ancient rocks, where there are no plants or animals whatsoever. Each of these steps takes us back in time, and the presence of each fossil group indicates a single, specific interval of geologic time.

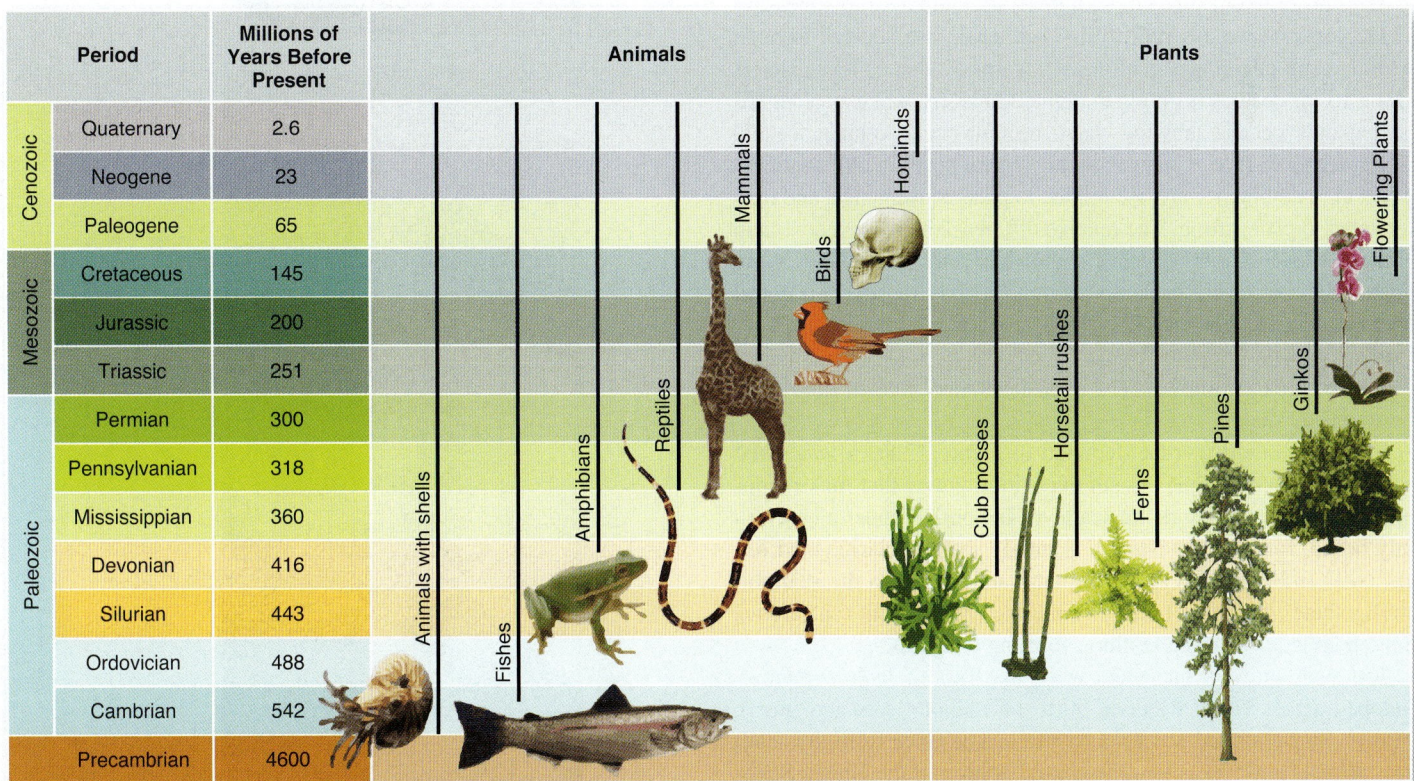

Period		Millions of Years Before Present	Animals	Plants
Cenozoic	Quaternary	2.6		
	Neogene	23		
	Paleogene	65		
Mesozoic	Cretaceous	145		
	Jurassic	200		
	Triassic	251		
Paleozoic	Permian	300		
	Pennsylvanian	318		
	Mississippian	360		
	Devonian	416		
	Silurian	443		
	Ordovician	488		
	Cambrian	542		
	Precambrian	4600		

FIGURE 12.7 Sedimentary strata may contain groups of fossils that are unique to that period of Earth's history.

Explain the relationship between evolution and the principle of fossil succession.

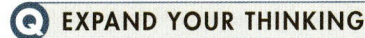 **EXPAND YOUR THINKING**

Explain how fossils help geologists establish the relative ages of different rock layers or outcrops.

12-4 Relative Dating Determines the Order of Geologic Events

LO 12-4 Describe the use of stratigraphic principles to construct a history of geologic events.

Geologists study the organization of the crust in order to unravel Earth history.

Imagine that you have been studying the geology of the land near your school for several years and, further, have constructed a geological cross section that depicts the layers and organization of rocks in the crust there. You find various types of sedimentary rocks, an unconformity, a basalt dike and sill, and evidence of uplift and deformation. By applying the stratigraphic principles, you can use this information to create a model of the history of the crust. Such a model is shown in **FIGURE 12.8**. It depicts the step-by-step sequence of geologic events that led to the formation of a section of the crust as it is today.

An Example of Relative Dating

FIGURE 12.8A begins the history of a section of crust by illustrating the deposition of sedimentary layers, A to D, during a period of rising sea level. The principle of superposition tells us that these layers were deposited in order, from A (first) to D (last).

Why do different types of sediments accumulate in distinct layers? Geologists have ascertained that sea levels rise and fall through time as a result of changes in climate: a warmer climate melts glaciers and warms the oceans, resulting in higher sea levels, while a colder climate expands icecaps, drawing water from the ocean to make ice and thus lowering sea level. Changes in the vertical position of a continent also occur; a continent may rise related to plate convergence, and may fall related to plate rifting. The relative difference between a continent and the sea level is referred to as *continental freeboard*. When warm climate (high sea level) and a low continent (low freeboard) occur at the same time, shallow seas can flood broad tracts of land. This occurred during the Ordovician and Cretaceous periods when shallow seas flooded the U.S. midwest and deposited thick layers of sandstone, limestone, and shale.

In the hypothetical case depicted in Figure 12.8a, a watershed environment (an energetic stream depositing gravel that makes conglomerate) is flooded by rising sea level and buried under a beach (a sandy beach, forming sandstone), a muddy shallow marine environment (silts and clays swept off the land and settled to the seafloor, forming shale), and, finally, a fully marine environment (calcareous plankton collecting on the seafloor, forming limestone).

Just such an event happened when the *Western Interior Seaway* developed across North America 90 to 65 million years ago (mya). A warm climate caused sea level to rise and advance from the south across what is now Texas into the central portion of the United States and all the way into Canada. This warm, shallow sea developed after the breakup of Pangaea, an event that involved widespread rifting and lowering of continental freeboard across North America (**FIGURE 12.9**).

In **FIGURE 12.8B**, an igneous sill, E, intrudes between layers C and D. The principle of cross-cutting relationships tells us that sill

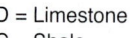

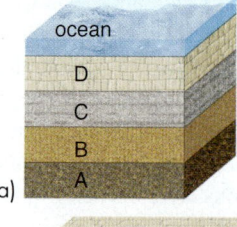
(a)

D = Limestone
C = Shale
B = Sandstone
A = Conglomerate

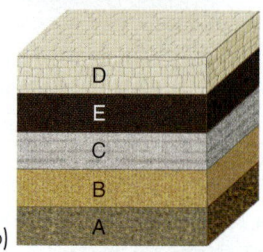
(b)

E = Sill, younger than the rocks it intrudes

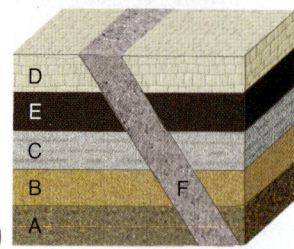
(c)

F = Intrusion of dike F

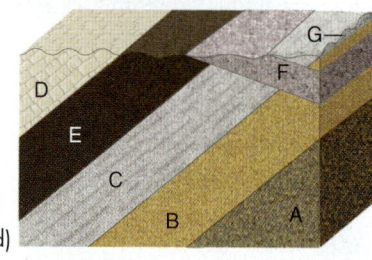
(d)

G = Unconformity

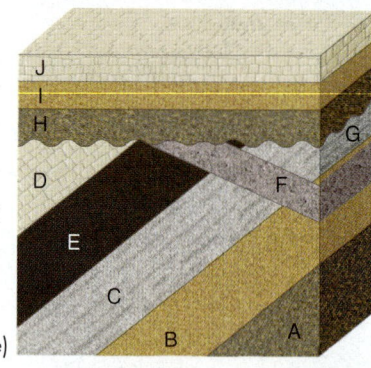
(e)

H = Conglomerate
I = Sandstone
J = Limestone

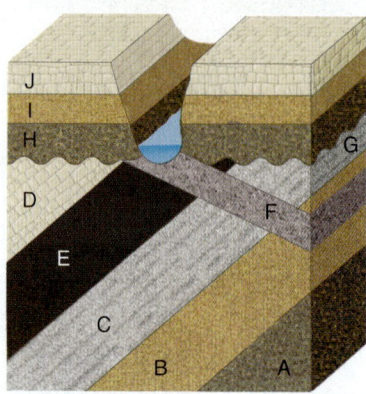
(f)

FIGURE 12.8 An example of relative dating, detailed in text.

Q Which of the five fundamental geologic events is not represented in this figure? Draw Figure 12.8f as if that fifth event *had* taken place.

FIGURE 12.9 The Western Interior Seaway extended from the Gulf of Mexico to the northern reaches of Canada about 90 to 65 million years ago. There is no modern analog to this shallow marine basin, which was about 1000 kilometers wide and 800 to 900 meters deep.

What types of sediment would preserve a record of the Western Interior Seaway?

E is younger than the strata into which it intrudes. **FIGURE 12.8C** reveals the intrusion of dike F. Again, the principle of cross-cutting relationships tells us that dike F must be younger than the rocks (layers A to E) into which it intrudes.

The transition from Figure 12.8c to **FIGURE 12.8D** involves a dramatic sequence of geologic events. From the principle of original horizontality we can infer that tectonic forces (perhaps plate convergence) caused the deformation (tilting) of all the layers that have been formed so far (layers A to F). These strata have been tilted at a high angle such that weathering and erosion have leveled the upturned strata and formed a flat surface environment. This becomes an unconformity (labeled G) when sedimentary strata H, I, and J are deposited by another episode of rising sea level in **FIGURE 12.8E**. In **FIGURE 12.8F**, layers H to J are eroded at the modern surface by a stream channel.

In constructing this geologic history, we identified four types of geologic events (deposition, erosion, deformation, and intrusion). In addition, we employed the stratigraphic principles of superposition, cross-cutting relationships, original horizontality, and unconformities to unravel the details of crust formation.

In summary, layers A, B, C, and D were laid down horizontally, with A deposited first, and D last. The intrusion of an igneous sill (labeled E) occurred next, followed by intrusion of a dike (F). A tectonic event occurred, probably an orogeny resulting from continental collision at a convergent margin, tilting and lifting the strata. This led to erosion of the land surface, forming an unconformity (G). Layers H to J were then deposited during a rise in sea level, H first and J last. These layers currently are being eroded by a modern stream cutting down through the rock layers and forming a canyon.

The First Geologic Map

William "Strata" Smith (1769–1839) made the world's first geologic map by applying these very same principles. He was born in Oxfordshire, England, on March 23, 1769. As a young man trained in surveying, he was employed to supervise the digging of canals in southern England. At the time, canals were dug by hand, making it crucial to know the local geology when planning the route of a new canal, to give diggers access to the softest rock. Smith observed that the fossils found in a section of sedimentary rock always appeared in a certain order, from the bottom to the top. This order of appearance could also be seen in other rock sections, even those on the other side of England; from this observation, he was able to predict the organization of the crust based on fossil evidence.

Why is it significant that fossils appear in a predictable sequence? Three factors are important in answering that question: 1) Fossils represent the remains of once-living organisms; 2) many fossils are the remains of extinct organisms—they belong to species that are no longer living anywhere on Earth; and 3) extinct fossilized organisms serve as unique timepieces for identifying and dating the rock layers in which they are found. Different fossils are found in rocks of different ages because life on Earth has changed throughout time. So when we find the same kinds of fossils in rocks from different places, we know that the rocks are the same age. By the early 1800s, Smith had proposed the principle of fossil succession and developed England's first geologic map (**FIGURE 12.10**).

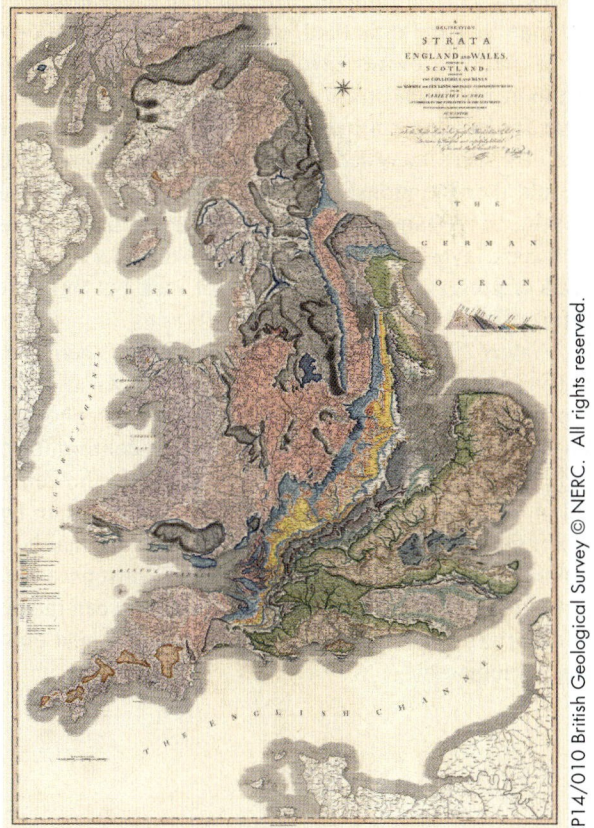

FIGURE 12.10 The first geologic map, developed by William Smith. Each color represents strata of a different age.

EXPAND YOUR THINKING

Describe a testable scientific hypothesis for why the Western Interior Seaway came into existence.

James Hutton Recognized the Meaning of Unconformities in the Geologic Record

12-5

LO 12-5 Explain the significance of Siccar Point, Scotland.

Scientists can put the geologic events and stratigraphic principles we have just reviewed to good use in unraveling the geologic history of complicated sections of Earth's crust.

The first person to apply the stratigraphic principles as a relative dating tool was one of the founders of the science of geology, James Hutton. Hutton was born to a Scottish merchant in Edinburgh, in June 1726. **FIGURE 12.11** shows the rocks at Siccar Point in Scotland where, in 1788, Hutton—now Sir James Hall—together with Scottish scientist and mathematician *John Playfair* (1748–1819), found what is now one of the most famous sites in the history of geology. Known today as *Hutton's Unconformity* and designated a Scottish National Heritage Site, this outcrop displays tilted strata that reveal missing units of geologic time.

Applying critical thinking, Hutton realized that the sequence of rocks at Siccar Point (and elsewhere in Scotland) recorded cycles of geologic events involving deposition of sediments on the seafloor, uplift and tilting to push up mountains, erosion by geologic forces, and further deposition and uplift. He saw that these cycles were recorded in the presence of rocks, unconformities, and their relationships to each other. Realizing that these must be due to the same geological forces operating at present, the exposed rock layers implied to him vast stretches of time, much of which is preserved not in the form of rock but in the form of unconformities.

At Siccar Point, shale from the Silurian period (turned vertically) is overlain by younger sandstone (known as the "old red sandstone") from the Devonian period. It was at this location that Hutton first understood the significance of unconformities in the geologic record. Note the vertical shale layers in the center foreground of the photo (a). These vertical beds are overlain by sandstones (c) that have also been tilted from their original horizontal orientation. Both the shale and sandstone beds must have been horizontal when they were originally deposited. Between them lies an unconformity (b) that Hutton realized recorded a great episode of missing geologic time.

Peter L. Kresan

FIGURE 12.11 The Hutton Unconformity at Siccar Point, Scotland.

Old red sandstone
345 million years

C

B

A

Silurian-aged
sandstone
425 million years

12-6 Radiometric Dating Uses Radioactive Decay to Estimate the Age of Geologic Samples

LO 12-6 Define the concept of half-life.

Numerical geologic time identifies the age of natural materials in "years before present." The most common tool used in determining numerical geologic time employs the natural phenomenon of radioactive decay; therefore, this type of age determination is called *radiometric dating.*

Radioactivity

Radioactive decay is a process wherein the unstable nucleus of an atom spontaneously emits radiation and a *subatomic particle*. This results in an atom of one type, called the *parent isotope*, transforming into an atom of another type, called the *daughter isotope*. During the first half of the twentieth century much of modern physics was devoted to exploring why this change occurs.

Let us review what we know about atoms. You may recall from Chapter 4 that the nucleus of each chemical element is composed of a specific number of protons, known as the *atomic number*, which is how the element is defined. Carbon (C) has 6 protons, nitrogen (N) has 7, oxygen (O) has 8, and so on. The nucleus of an element also contains a variable number of neutrons.

The number of protons plus the number of neutrons equals the *mass number*. Atoms of the same element that have different mass numbers (because they have different numbers of neutrons) are known as *isotopes* of that element. For example, carbon with 6 protons may have 6, 7, or 8 neutrons, yielding the isotopes ^{12}C, ^{13}C, and ^{14}C. However, ^{14}C is not stable, and its nucleus will decay radioactively by releasing radiation and a beta particle (an electron). When a neutron loses a beta particle it becomes a proton, and in the process the atomic number changes from 6 (carbon) to 7 (nitrogen), producing ^{14}N.

Within most igneous, sedimentary, and metamorphic mineral crystals are radioactive atoms called **radioisotopes.** Examples of radioisotopes include uranium (U) isotopes such as ^{238}U, and ^{235}U; thorium (Th) isotopes such as ^{232}Th; the potassium (K) isotope ^{40}K; and others. Each type of radioisotope decays at a unique, fixed rate. Translated that means that as time passes, the amount of parent isotope decreases and the amount of daughter isotope increases at a fixed rate within a mineral. If you know the rate of decay, the changing ratio of parent to daughter is a direct reflection of how much time has elapsed since the isotopes became trapped in the mineral.

Researchers, utilizing sensitive laboratory instruments to measure the abundance of parent and daughter isotopes in a mineral, use this ratio and the rate of decay to estimate the time since the mineral crystallized, and thereby the age of the geologic event it represents (such as intrusion or metamorphism, etc.). Using radioisotopes to date Earth materials is a powerful technique, one that has been employed to reveal many important events in Earth's history.

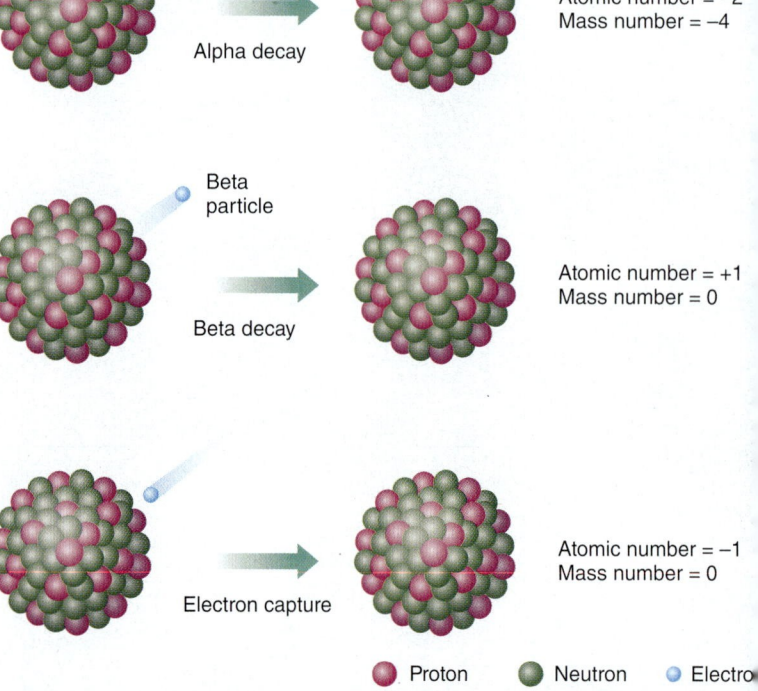

FIGURE 12.12 In alpha decay, the atomic number of the parent isotope decreases by 2 and the mass number decreases by 4. In beta decay, the atomic number increases by 1; the mass number remains unchanged. In electron capture decay, the atomic number decreases by 1; the mass number remains unchanged.

Q What aspect of radioactive decay makes it useful for dating geological materials? Explain your answer.

How does decay occur? There are three answers to that question because there are three types of decay: *alpha decay, beta decay,* and *electron capture* (**FIGURE 12.12**).

- In alpha decay, the nucleus of an atom emits a single alpha particle consisting of 2 protons and 2 neutrons. Alpha decay results in the formation of a daughter isotope with 2 fewer protons than the parent isotope (atomic number decreases by 2) and a decrease in the mass number by 4. The isotope ^{234}U (useful for dating fossil coral), for example, experiences alpha decay and produces the daughter isotope ^{230}Th. The atomic number of uranium is 92; that of thorium is 90 (2 fewer protons). (Read the Geology in Our Lives feature titled "Alpha Decay, Silent Killer" to learn about the environmental hazard associated with alpha decay.)

- During beta decay, the nucleus emits a negative beta particle and in so doing one of the neutrons changes into a proton. In this case, the parent isotope gains 1 proton and loses 1 neutron. Hence, the mass number does not change but the atomic number increases. For instance, ^{234}Th experiences beta decay and becomes the daughter isotope ^{234}Pa. Notice that protactinium–234 has the same mass as thorium–234, but the element Pa (atomic no. 91) is 1 proton heavier than Th (atomic no. 90).

- During electron capture, an atom loses 1 proton (it becomes a neutron) and is transformed into a lighter element. However, the mass number does not change because the total number of protons and neutrons remains the same. For example, the radioactive isotope ^{40}K decays through electron capture into ^{40}Ar. The potassium isotope loses 1 proton and becomes argon, but the mass number stays the same.

Half-Life

It is impossible to determine exactly when a single radioisotope will decay, so geologists express the probability of decay with the statistical **half-life.** The half-life is the probable time required for half of the radioisotopes in a sample to decay. Half-life is a value that has been measured for each type of radioisotope. As time passes, the amount of parent isotope in a sample decreases by half during each half-life. Therefore, after one half-life, the sample will contain half the number of atoms that were originally present. After two half-lives, one-fourth of the original number of atoms will be present. After three half-lives, one-eighth of the original amount will remain; after four half-lives, one-sixteenth will remain; and so forth (**FIGURE 12.13**).

Half-life is a statistical description that says: "There is a very high probability that half of the atoms of a radioisotope in a sample will decay within a defined period of time." Each radioactive isotope has been studied carefully in order to make a reasonably accurate estimate of its half-life.

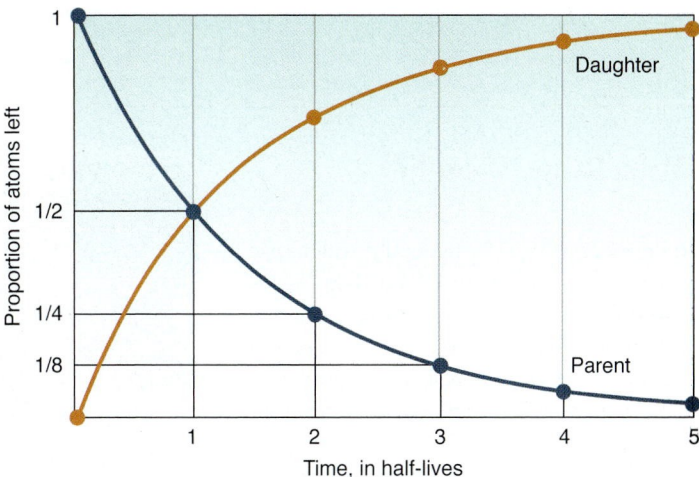

FIGURE 12.13 With each half-life, the amount of parent radioisotope in a sample decreases by half while the amount of daughter radioisotope increases by half.

Ⓠ If the amount of an isotope is 1/32 of its original amount, how many half-lives have passed?

Ⓠ **EXPAND YOUR THINKING**

If radioactive isotopes had variable rates of decay, would radiometric dating still be possible? Why or why not?

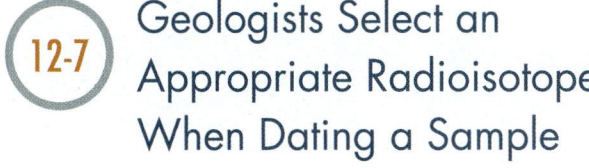

Geologists Select an Appropriate Radioisotope When Dating a Sample

LO 12-7 Compare and contrast primary and cosmogenic radioisotopes.

It is appropriate to date very old samples with radioisotopes that have long half-lives and to date young samples with radioisotopes that have short half-lives. It is also appropriate to use radioisotopes that are abundant (not rare or absent) in the sample being dated.

Dating Considerations

Not every type of radioisotope (**TABLE 12.1**) is appropriate to use for dating a geologic sample. One does not want to look for, say, ^{40}K in a rock that is known to contain little potassium. Another important consideration is that only certain radioisotopes will accurately characterize the time frame of the sample being dated. Put another way, young samples should be dated with short-lived radioisotopes, and old samples with long-lived radioisotopes. If a very young event is being dated, such as a lava flow approximately 500 to 1000 years old, it is not appropriate to pick a radioisotope with a half-life of a few billion years: there simply will not be enough daughter isotope to be measured accurately, because not enough time will have passed. Likewise, if a sample to be dated is several hundred million years old, such as an ancient granite pluton, it is not appropriate to try to date it using a radioisotope with a half-life of only a few

TABLE 12.1 Common Radioisotopes and Half-Lives Used in Dating Samples

Radioactive (Parent)	Product (Daughter)	Half-Life (Years)	Radioactive (Parent)	Product (Daughter)	Half-Life (Years)
Samarium–147	Neodymium–143	106 billion	Uranium–235	Lead–207	0.7 billion
Rubidium–87	Strontium–87	48.8 billion	Beryllium–10	Boron–10	1.52 million
Thorium–232	Lead–208	14 billion	Chlorine–36	Argon–36	300,000
Uranium–238	Lead–206	4.5 billion	Uranium–234	Thorium–230	248,000
Potassium–40	Argon–40	1.25 billion	Thorium–230	Radium–226	75,400
			Carbon–14	Nitrogen–14	5,730

GEOLOGY IN OUR LIVES

Alpha Decay, Silent Killer

Radon–222 (^{222}Rn) is a deadly radioactive gas that fills 1 of every 15 buildings in the United States. It undergoes alpha decay and has a half-life of only 3.8 days; that means if you live or work in a building infiltrated by ^{222}Rn, it is impossible to avoid inhaling this radioisotope. Alpha particles cause damage to sensitive lung tissue; in fact, breathing radon-laden air over a lifetime is thought to be the second leading cause of lung cancer, responsible for 15,000 to 22,000 deaths each year.

Produced by the decay of ^{238}U in the rock, soil, and groundwater beneath our communities (**FIGURE 12.14**), ^{222}Rn gas cannot be seen, smelled, or tasted, and as it filters up through the soil or outgases from our drinking water, it becomes trapped in the buildings where we live and work, and can build up to dangerous levels.

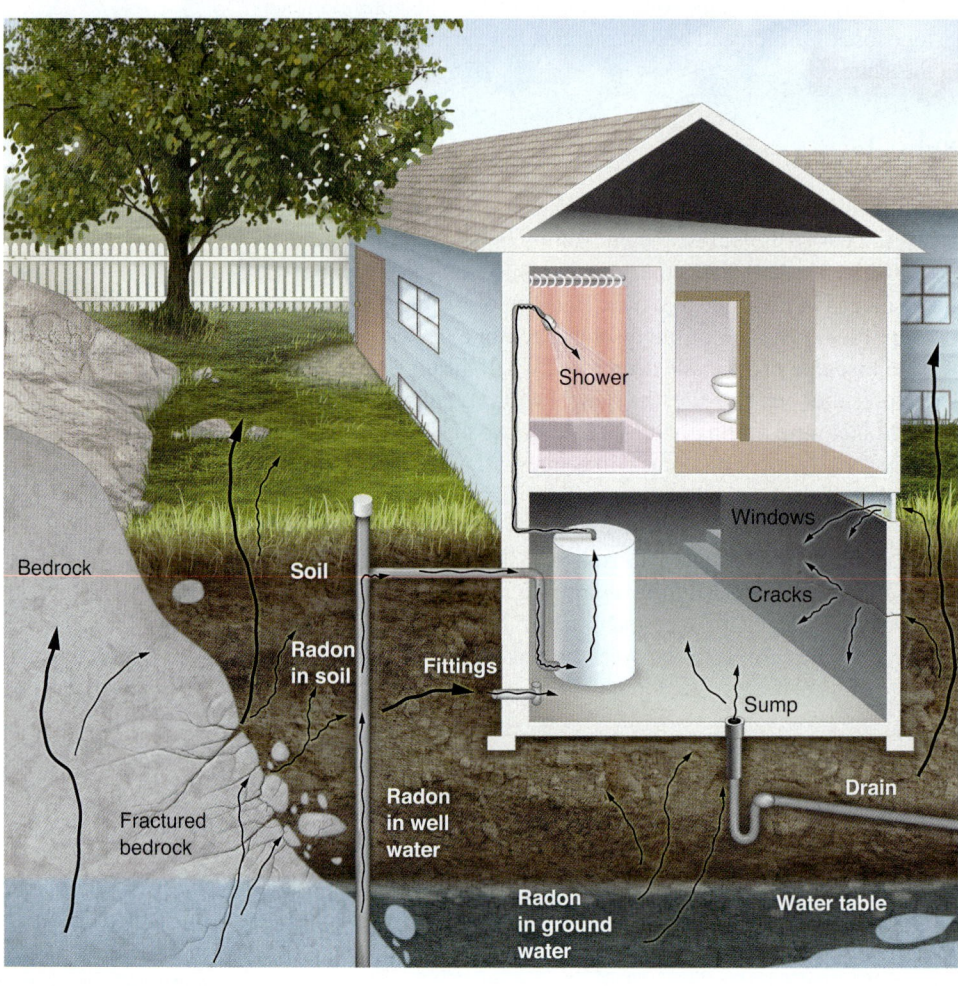

FIGURE 12.14 Radon may enter a building through several pathways.

---

thousand years. There will not be enough parent radioisotope left to be measured accurately, because too much time will have passed and most of the parent radioisotope will have decayed.

Isotopic Clocks

Scientists use an atom-measuring instrument called a *mass spectrometer* to assess the abundance of isotopes in a sample. Information about the ratio of parent radioisotopes to daughter isotopes in a sample is used to calculate an age based on the amount of time necessary for a given ratio to develop at the specified decay rate. Well over 40 radiometric dating techniques are in use around the world, each based on a different radioisotope. In general, useful isotopes fall into two categories: *primary radioisotopes* and *cosmogenic radioisotopes*.

Earth inherited the primary radioisotopes when it accreted during nebular condensation (see Chapter 2). These radioisotopes have been held in rocks for the entire 4.6 billion years of Earth's history. They are released as a rock melts, only to be incorporated into new rock whenever mineral crystallization occurs. The primary radioisotopes typically are used to date rocks from continental and oceanic crust.

Cosmogenic radioisotopes are formed by the interaction of subatomic particles (such as neutrons) produced by the Sun with nuclei of atoms in the atmosphere or on the surface of mineral grains in the crust that are exposed to the atmosphere. These isotopes have relatively short half-lives, so if they had not been continuously created by cosmic rays, they would not be present on Earth today. The cosmogenic radioisotopes typically are used to date young samples involved in physical, chemical, and biological processes occurring in surface environments.

The next two Critical Thinking exercises, "The Age of Geologic Events," and "Relative and Radiometric Dating," give you the opportunity to apply the principles of radiometric and relative dating. See if you can calculate the age of various rock types using the ratio of parents to daughters and the half-life.

Q EXPAND YOUR THINKING

Identify the type of geologic event or geologic material that could be dated using the half-life of each radioisotope pair in Table 12.1.

CRITICAL THINKING

The Age of Geologic Events

FIGURE 12.15 is a cross section of the crust consisting of sedimentary, igneous, and metamorphic rocks. You can determine the age of basalt, granite, and limestone mineral samples using the radioisotope ^{235}U. **TABLE 12.2** provides the result of laboratory isotopic analysis of the samples.

The percentage of remaining parent atoms is proportional to the age of each sample. This proportion, applied to the half-life, enables you to estimate the age of a sample.

Begin by using the formula

$$P_i\% = P_t \div (P_t + D_t)$$

to calculate $P_i\%$, the percentage of initial parent atoms remaining in the system. P_t is the number of parent atoms today; D_t is the number of daughter atoms today. Multiply the percentage of parent remaining times the half-life for ^{235}U (700 million years). Then subtract this number from the half-life. The answer, in years, gives you an estimate of the amount of time that has passed since the sample became a closed system—in other words, its age.

Now answer the following:

1. What are the ages (in years) of the basalt, granite, and limestone samples?

2. What can you infer about the ages of the sandstone and shale units?

3. What are the geologic events, and their proper sequence, that built this section of crust? Apply a combination of relative dating and radiometric dating to derive the geologic history.

4. In your geologic history, are there any events whose sequence is not clear? What additional information would help to clarify the correct sequence of events?

5. What is the very last event?

6. What are the igneous processes that could produce basalt and granite in the same place?

7. What are the sedimentary environments that might have produced this sequence of rocks?

8. Which stratigraphic principles did you use in constructing your geologic history?

TABLE 12.2 Results of Isotopic Analysis

Rock Unit	Number of Parent Atoms	Number of Daughter Atoms
Basalt	7,079,400	372,600
Granite	850,000	150,000
Limestone	2,100,000	900,000

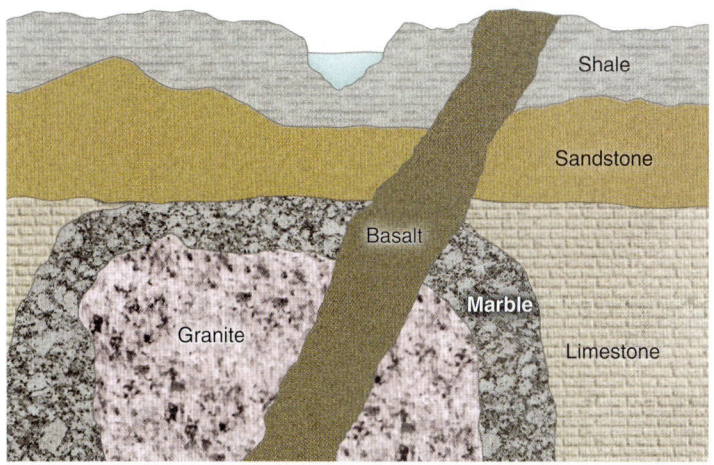

FIGURE 12.15 Cross section of the crust.

12-8 Accurate Dating Requires Understanding Sources of Uncertainty

LO 12-8 Identify potential sources of uncertainty in radiometric dating.

Igneous rocks are good candidates for radiometric dating because the ages of minerals within an igneous rock document the time when the mineral crystallized from a magma source. However, accurate dating also requires carefully factoring in sources of uncertainty.

Sources of Uncertainty

When molten rock cools and crystallizes, atoms are no longer free to move about. Radioisotopes of the right size and charge will be incorporated into the crystalline structure. Often, daughter isotopes will not be included because they are the wrong size, but this is not always the case. Hence, a newly formed mineral may be free of daughter isotopes, which simplifies the process of determining its age. Any daughter atoms resulting from radioactive decay *after* crystallization will be locked in place within the crystalline structure and so can be counted.

When a geologist takes a sample of rock and brings it to a laboratory for analysis and dating, he or she must begin by answering several questions: What is the relationship between the sample and the event it represents?

CRITICAL THINKING

Relative and Radiometric Dating

FIGURE 12.16 shows a section of crust with a complex geologic history. Sedimentary, metamorphic, and igneous rocks are all present, as are faulting, intrusions, and unconformities. Your job is to construct a complete geologic history of this region, using a combination of relative and radiometric dating. The rock types are defined using symbols that are explained in the legend.

TABLE 12.3 provides the half-life and the measured ratio of daughter to parent isotopes in a number of samples. The numerical ages you calculate can be assigned to geologic events, and they provide a framework for estimating the timing of other events that are recognized using relative dating. We will assume that the following relationship holds true for all the samples in this problem:

$$\text{Age} = \text{measure ratio} \times 1/\lambda$$
$$\lambda = 0.69/\text{half-life}$$

Once you have calculated the ages of samples A to H, construct a history of geologic events and assign them to their proper geologic periods in historical order (youngest at the top, oldest at the bottom of your list). You will have to estimate the age of some events, but that is the nature of critical thinking. Be sure you have a reason for each of your estimates. There may be some geologic events that are not obvious but must have occurred based on the available evidence.

Write a complete geologic history of this crust.

TABLE 12.3 Results of Isotopic Analysis

Sample	Isotope	Half-Life	Measured Ratio
A	Uranium–238–Lead–206	4.5×10^9	0.090
B	no values obtained	—	–
C	Potassium-Argon	1.3×10^9	0.006
D	Rubidium-Strontium	4.7×10^{10}	0.050
E	Potassium-Argon	1.3×10^9	0.097
F	Uranium–235–Lead–207	7.1×10^8	0.046
G	Carbon–14	5.7×10^3	1.870
H	Lead–210	2×10	2.070

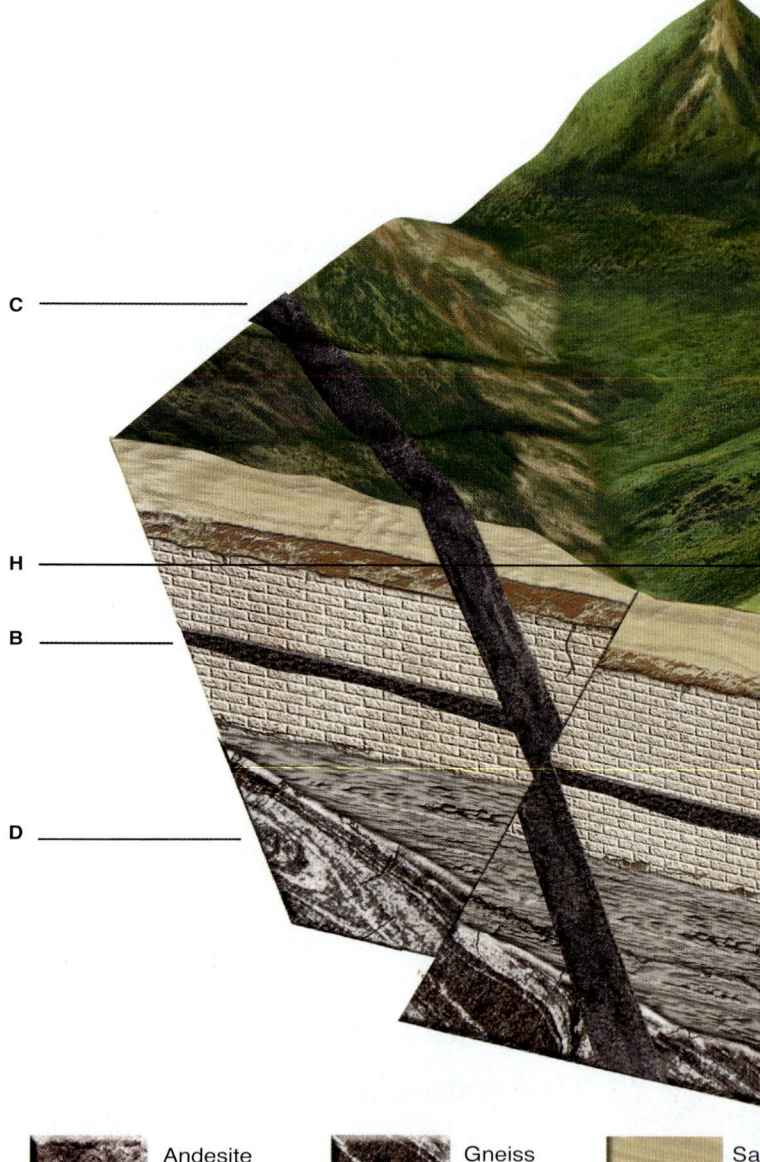

Andesite　　Gneiss　　Sar

Basalt　　Granite　　Sha

Conglomerate　　Limestone　　Silt

FIGURE 12.16 A section of crust with a complicated geologic history.

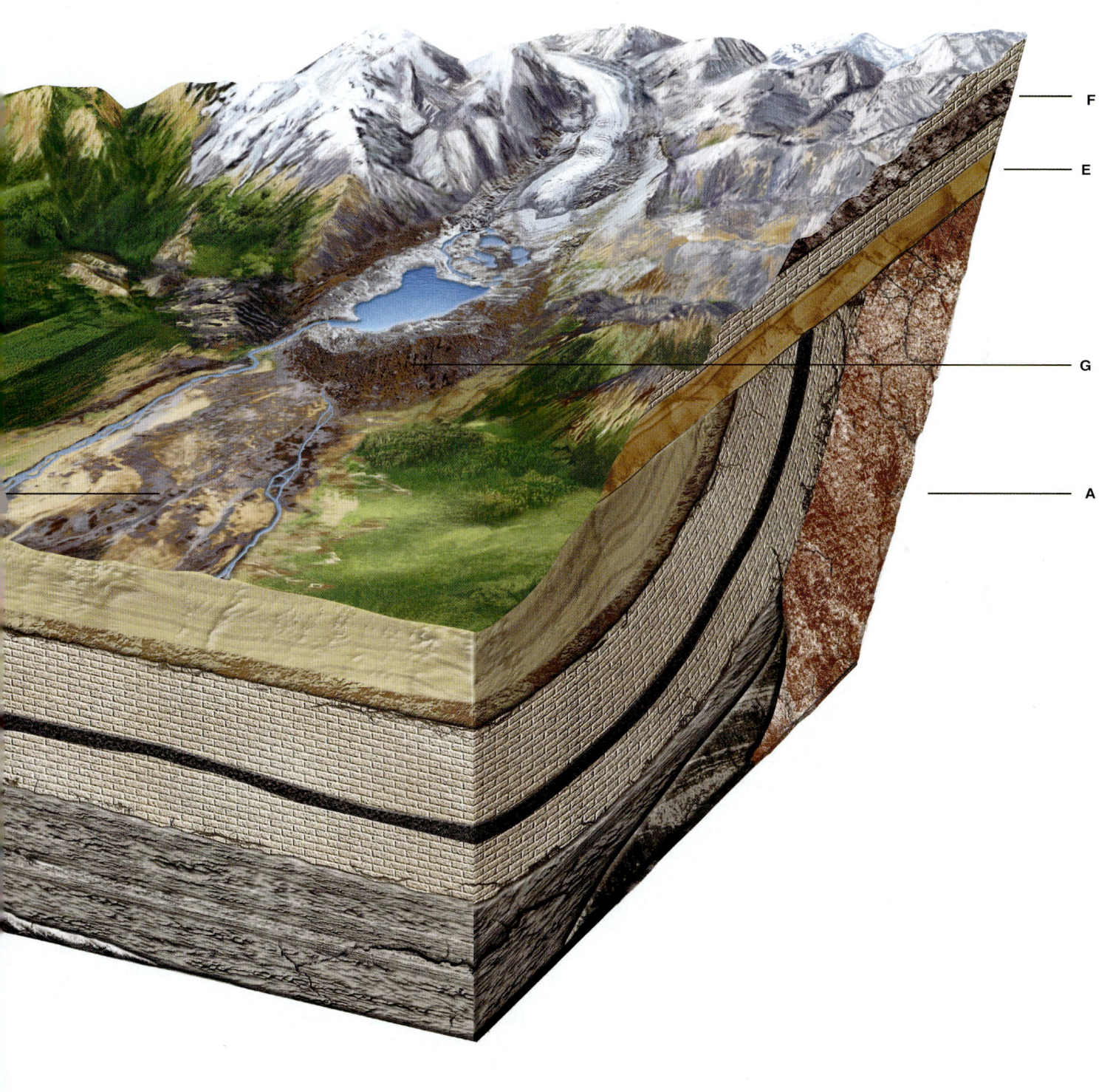

F

E

G

A

- What is the approximate age of the sample?
- Which radioisotope is appropriate for both the type of material being dated and the approximate age of the sample?
- What are the potential sources of uncertainty in calculating the sample's age, and how can the age be tested?

To determine an accurate age, the geologist assumes that the sample being dated has remained a *closed system* throughout its life. This means that neither parent nor daughter isotopes have left or entered the sample since its formation. If the sample has not remained a closed system, metamorphism or alteration by groundwater probably has occurred. If parent isotopes have left the sample, the calculated age will be too old; if daughter isotopes have left, the calculated age will be too young.

Metamorphic conditions of high heat and/or pressure, or the presence of groundwater (especially chemically active fluids) containing parent or daughter isotopes, can lead to *open-system behavior,* in which parent and daughter isotopes migrate in and out of a sample (**FIGURE 12.17**). And if daughter isotopes are completely driven from a sample during metamorphism, any calculated age would be useful for determining how much time has passed since metamorphism occurred—a potentially valid scientific goal.

Determining the age of a sample is a two-step process. First the ratio of parent atoms to daughter atoms of a selected element in a sample is calculated. Then the half-life is used to calculate the time it took to produce that ratio of parent atoms to daughter atoms. In actual practice, in some cases, it is reasonable to assume that no daughter isotope was present at the time of crystallization. But in other cases, the initial amount of the daughter product must be determined. Usually, the geologist will analyze several samples from one rock body. The differing amounts of parent and daughter isotopes found in these analyses can be used to estimate how much daughter isotope was present originally.

Now we will explore some specific radioisotopes that geologists use to determine the absolute age of a rock or fossil.

The Uranium Decay Chain

All isotopes of the element uranium are at least mildly radioactive. The isotope ^{238}U, which occurs naturally in most types of granite and soil in varying degrees, decays with a half-life of 4.5 billion years (**FIGURE 12.18**) and produces the stable daughter isotope ^{206}Pb. There are 14 separate steps involved in producing ^{206}Pb, each leading to the production and subsequent decay of a radioactive daughter isotope. These steps are referred to as a **decay chain.**

Figure 12.18 shows the decay chain of ^{238}U and the type of decay and half-life of each daughter isotope. The chain of decay products is formed as a result of the production of radioactive daughter isotopes. One daughter radioisotope decays to form another radioisotope, which in turn decays further until the decay process finally ends with an isotope that is stable. In this case, that stable isotope is ^{206}Pb. Some of the intermediate isotopes have very short half-lives; examples include ^{234}Pa, with a half-life of slightly more than one minute, and ^{214}Po, which decays in a few ten-thousandths of a second. The uranium decay chain is an example of how the process of radioactive decay can lead to the production of many daughter isotopes, each with its own distinctive half-life.

The use of uranium as an isotopic clock was the first method developed for dating very old rocks. As mentioned, the half-life of

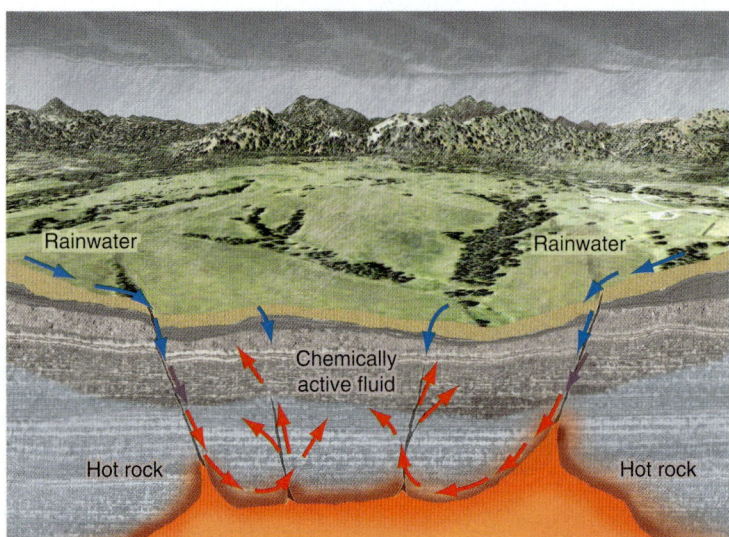

FIGURE 12.17 Open-system behavior can be caused by metamorphism, interaction with groundwater, and alteration by chemically active fluids. A sample that has experienced open-system behavior will not provide an accurate radiometric age of its original crystallization. It will instead reflect some combination of that age and the degree to which open-system behavior altered the parent-to-daughter ratio.

How could chemically active fluids influence the radiometric age of a rock sample?

^{238}U is 4.5 billion years; ^{235}U has a half-life of 700 million years, and ^{232}Th has a half-life of 14 billion years. Each decays, in steps, to form other elements, including polonium and, finally, lead. The different parent isotopes yield different daughter isotopes of lead: ^{238}U yields ^{206}Pb; ^{235}U yields ^{207}Pb; and ^{232}Th yields ^{208}Pb. To arrive at a sample's age, the amounts of these lead isotopes in the sample are measured and compared with the amounts of the parent isotopes remaining in the sample.

U–Pb dating is carried out on only a limited set of minerals: namely, zircon, quartz, and apatite. Zircon is a particularly valuable mineral for this purpose because when it crystallizes in magma it incorporates uranium into its crystal structure but does not incorporate significant amounts of lead. Therefore, the proportion of lead it contains is mostly produced by the radioactive decay of uranium, which thus serves as a useful radiometric clock. Zircon is also extremely resistant to chemical and physical weathering; and once it has crystallized, it is very difficult to add or remove lead or uranium. For these reasons, zircon crystals survive for a long time in geologic materials, and they act as a closed system that does not easily metamorphose.

The resistance of zircon to metamorphism, coupled with its longevity, has made it an important mineral for dating ancient rocks. Zircon grains, formed over 4 billion years ago, have been found in sandstones in Australia (see Figure 12.22). Zircon has been used to date other extremely old rocks (4.28 billion years in northeast Canada; refer back to Figure 12.1), as well as to determine the age of granite batholiths that form huge intrusive complexes on several continents.

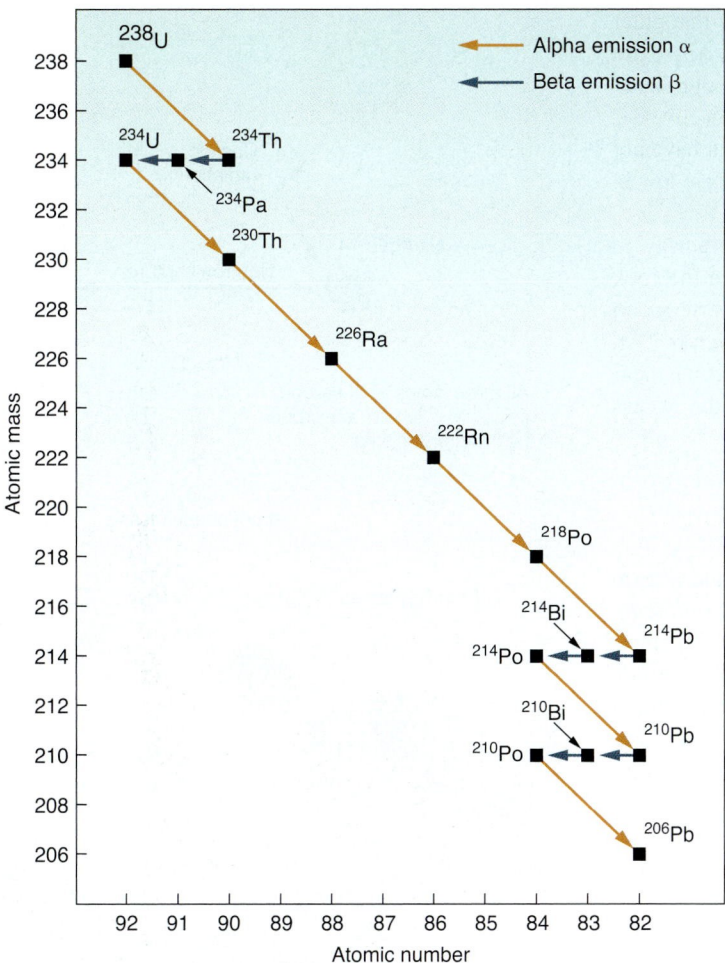

Nuclide	Half-Life
uranium—238	4.5×10^9 years
thorium—234	24.5 days
protactinium—234	1.14 minutes
uranium—234	2.33×10^5 years
thorium—230	8.3×10^4 years
radium—226	1590 years
radon—222	3.825 days
polonium—218	3.05 minutes
lead—214	26.8 minutes
bismuth—214	19.7 minutes
polonium—214	1.5×10^{-4} seconds
lead—210	22 years
bismuth—210	5 days
polonium—210	140 days
lead—206	stable

FIGURE 12.18 Fourteen decay steps are involved in the production of stable lead–206.

Ⓠ You want to know the age of sandstone composed of quartz and zircon grains that are cemented with silica. You determine the age of a zircon grain, a quartz grain, and the silica cement. What does each of these ages tell you? What are the implications of each date to your original question?

Ⓠ **EXPAND YOUR THINKING**

What is a closed system and why is it important for accurate dating?

Potassium-Argon and Carbon Serve as Important Isotopic Clocks

12-9

LO 12-9 Describe why radiocarbon dating is useful for dating relatively recent organic material.

Certain radioisotopes have been particularly useful in deciphering Earth's history: potassium (K)-argon (Ar) and radiocarbon are two examples.

Potassium-Argon

Potassium, which is abundant in the crust, has been used to date igneous events in Earth's history in thousands of locations. One isotope, ^{40}K, is radioactive and decays to form two daughter products, ^{40}Ca and ^{40}Ar, through two different methods of decay. The ratio of these two daughter products to each other is constant: 11.2% of the parent becomes ^{40}Ar, and 88.8% becomes ^{40}Ca. Argon is an inert, noble gas. When rock melts, argon tends to escape. Once the rock solidifies, it will trap new argon produced by radioactive decay. This process creates a potassium-argon "clock." The clock is reset to zero by melting followed by crystallization.

You might think that a geologist simply needs to measure the amounts of ^{40}K and ^{36}Ar in a rock to establish a date. However, in many cases, a small amount of argon remains in a rock when it hardens. This argon usually is trapped in tiny air bubbles that enter from the atmosphere (1% of the air we breathe is argon). Any extra argon from air bubbles must be taken into account if it is significant relative to the amount of daughter argon produced by radioactive decay. This would most likely be the case in young rocks that have not had time to produce much daughter argon, or in rocks that are low in parent potassium.

To be able to date the rock, geologists must have a way to determine how much *original* atmospheric argon was in the rock when it crystallized. They can do this because atmospheric argon has a couple of other isotopes, the most abundant of which is ^{36}Ar. We know that the ratio of ^{40}Ar to ^{36}Ar in air is 295. Thus, if we measure ^{36}Ar and ^{40}Ar, we can calculate and then subtract the excess atmospheric ^{40}Ar to obtain an accurate age.

Radiocarbon

It was in the 1940s that *Willard F. Libby* (1908–1980) and a team of scientists at the University of Chicago developed the *radiocarbon dating method*. It has since become the most powerful method of dating relatively young geologic and archaeological materials and, as a result, has greatly advanced our understanding of human chronology, climate change, and environmental processes that have occurred over the last 50,000 years. For his efforts, Libby won the 1960 Nobel Prize in chemistry. (Read the Geology in Action feature titled "Radiocarbon Dating the First Map of North America" to deepen your understanding of this crucial dating method.)

The radioisotope ^{14}C is formed in the atmosphere when high-energy particles from the Sun (mostly protons and neutrons) interact with nitrogen in the air (**FIGURE 12.19**). When a neutron hits the nucleus of nitrogen, it drives out a proton. This changes the nitrogen into an isotope of carbon with a mass number of 14. The carbon isotope is unstable and will undergo beta decay to become ^{14}N. The half-life for this reaction is 5,730 years.

Newly formed radiocarbon combines with oxygen to form radioactive carbon dioxide ($^{14}CO_2$). In today's atmosphere, one atom of ^{14}C exists for every trillion atoms of the most abundant isotope of the element carbon, ^{12}C. Plants absorb radiocarbon as they take up carbon dioxide during photosynthesis. The radiocarbon is then passed from plants to animals, and eventually into soil, through the food chain. Although the amount of radiocarbon in plant and animal cells is perpetually decreasing through radioactive decay, eating and breathing renew it as long as the organism is alive.

When an organism dies, the amount of radiocarbon in its tissue decreases consistently according to the half-life clock. Therefore, by measuring the residual ^{14}C in organic samples (bones, plant fragments, soil samples, wood, coral, shells, organic-rich mud, etc.)—provided they have not been contaminated by younger carbon, from bacteria or carbon-rich acid (carbonic acid) in the soil or groundwater, or older carbon (such as calcium carbonate from rocks)—it is possible to calculate the time elapsed since the plant or animal tissue was formed.

The amount of radiocarbon produced in Earth's atmosphere fluctuates through time, because solar radiation varies in intensity, the geomagnetic field fluctuates, and the amount of carbon dioxide (CO_2) in the atmosphere changes from year to year. This means that

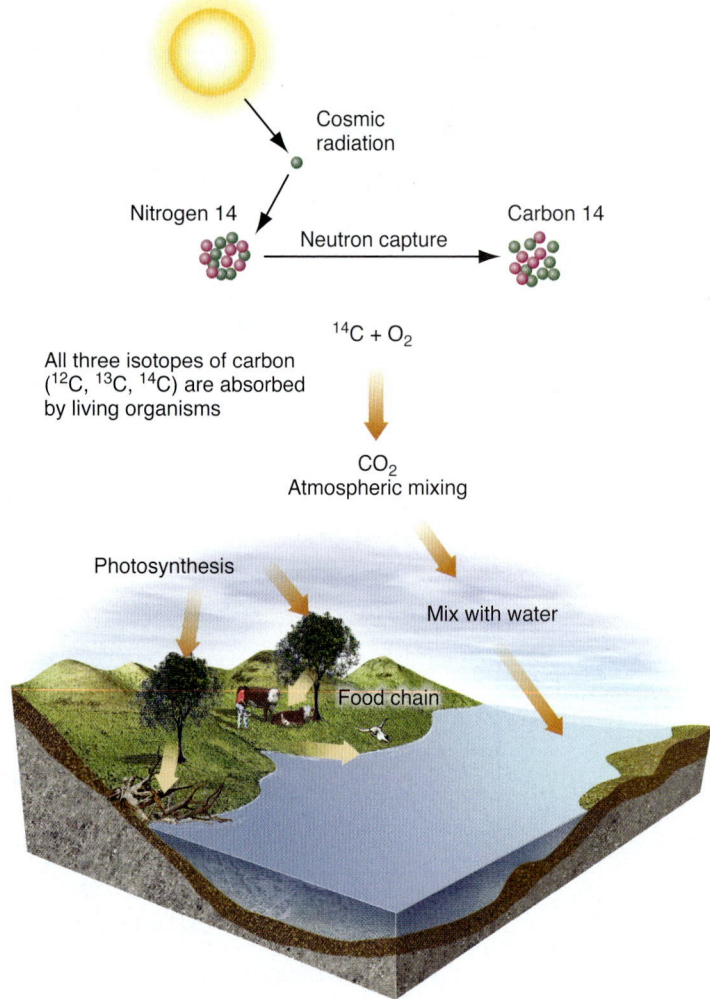

FIGURE 12.19 Radioisotope ^{14}C is formed in Earth's atmosphere, where it combines with oxygen to form radioactive CO_2. Radiocarbon will eventually enter the food chain and provide a means of dating fossil organic materials.

How does carbon dioxide with radioactive carbon enter the food chain?

the amount of ^{14}C entering the food chain is not consistent from one year to another.

To take this variability into account, geologists must calibrate a calculated radiocarbon age to an actual calendar year. Counting annual tree rings and dating the wood from each ring results in a calibration curve that can be used for this purpose. Since certain long-lived trees grow a new ring each year (principally, U.S. bristlecone pine and German and Irish oak), the amount of ^{14}C in each tree ring gives us an exact record of radiocarbon changes through time. In addition to tree rings, ^{14}C measurements of annually banded corals and annually layered organic-rich mud in lakes and bogs have extended the calibration to about 45,000 years—approximately the entire time span of the radiocarbon clock (**FIGURE 12.20**).

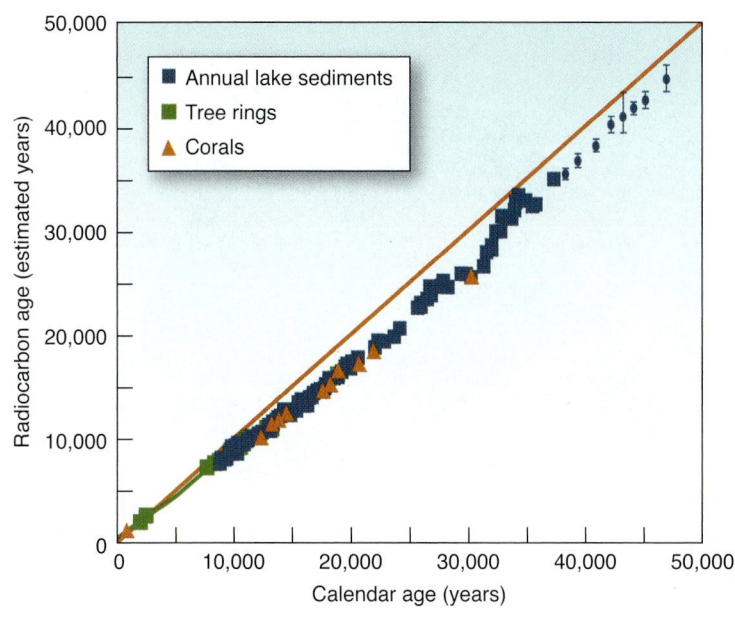

FIGURE 12.20 A radiocarbon age (y-axis) must be converted into actual calendar years (x-axis) to be accurate. The blue symbols are from annual layers of organic lake mud; the green data comes from annual tree rings; and the red symbols are from annually banded corals. If all the samples could be plotted on the straight line, ^{14}C years would exactly equal calendar years. But they do not, hence the need for a calibration system to correct radiocarbon years to actual calendar years.

Q Why is it necessary to calibrate a radiocarbon date with calendar years obtained by other means?

Q EXPAND YOUR THINKING

Why must the calibration record have annual resolution?

GEOLOGY IN ACTION

Radiocarbon Dating the First Map of North America

Scholars believe that the Vinland Map (**FIGURE 12.21**) is a fifteenth-century map depicting Viking exploration of North America half a century before the arrival of Columbus. If genuine, the Vinland Map is one of the great documents of Western civilization; if fake, it is an amazing forgery.

According to one team of scientists, the kind of ink used on the map was made only in the twentieth century. According to another, the parchment dates from the mid-1400s. Skeptics argue that the parchment's age is irrelevant, while others, who believe that the analysis of the ink is flawed, have introduced evidence that many medieval documents used similar ink.

The Vinland Map shows Europe, the Mediterranean, North Africa, and Greenland, all of which were known to travelers of the time. In the northwest Atlantic Ocean, however, it also shows the "Island of Vinland," which represents a part of present-day North America. Text on the map reads, in part:

By God's will, after a long voyage from the island of Greenland to the south toward the most distant remaining parts of the western ocean sea, sailing southward amidst the ice, the companions Bjarni and Leif Eiriksson discovered a new land, extremely fertile and even having vines, ... which island they named Vinland.

In 1995, scientists from the U.S. Department of Energy, the University of Arizona, and the Smithsonian Center for Materials Research and Education radiocarbon-dated the parchment on which the map was drawn. The scientists were allowed to trim a three-inch-long sliver off the bottom edge of the map for analysis. Using the National Science Foundation–University of Arizona's accelerator mass spectrometer, the scientists determined that the parchment dates to 1434 AD, plus or minus 11 years. The unusually high precision of the date was possible because it fell within a very favorable region of the ^{14}C calibration curve.

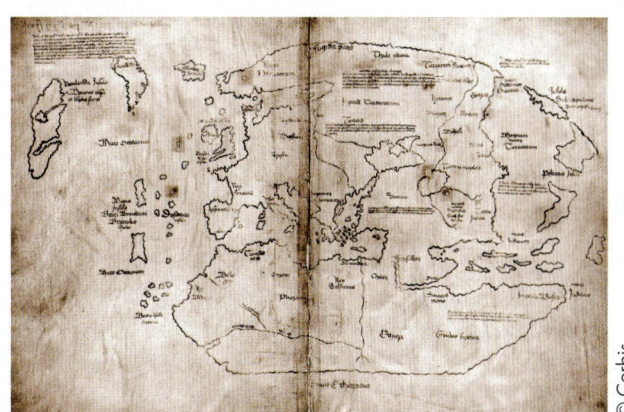

© Corbis

FIGURE 12.21 The oldest map showing North America. It depicts Europe (including Scandinavia), northern Africa, Asia, and the Far East. A previously unmapped portion of North America—the region around Labrador, Newfoundland, and Baffin Island—is displayed in the upper left corner of the map.

12-10 Scientists Arrived at Earth's Age Via Several Independent Observations

LO 12-10 Describe the evidence that Earth's age is about 4.6 billion years.

As stated throughout this chapter, the commonly accepted age of Earth is 4.5 to 4.6 billion years. We cannot adequately comprehend the value of this piece of information without acknowledging how it came to be—the intellectual process required to conceive where to find geologic records of Earth's age, the technological accomplishment of analyzing those records, and the efforts made to find independent data to test and refine initial findings. All these are potent examples of the power of critical thinking and the scientific process.

Our understanding of Earth's age comes from multiple lines of scientific evidence, including the: 1) age of primordial crust, 2) age of Moon rocks, 3) age of meteorites, 4) abundance of lead isotopes, and 5) consistency of these independent observations.

Primordial Crust

Despite widespread attempts, efforts to identify crust from the time of Earth's origin have failed. There are two reasons for this: 1) The oldest rocks were probably destroyed in the molten phase of Earth's early history (see Chapter 3); and 2) processes of the rock cycle—namely, erosion and plate tectonics—have recycled the crust and destroyed direct evidence of Earth's very early history.

The oldest whole rock of terrestrial origin (located in northeast Canada) developed 4.28 billion years ago (see Figure 12.1), but an older particle from Australia, a zircon grain sampled in ancient sedimentary layers, has been dated at 4.4 billion years old (**FIGURE 12.22**). Rocks this old are rare, however, and most old crust, found in the ancient cores of continents, dates from about 3.5 billion years ago.

Moon Rocks

Although scientists still have not established the Moon's exact origin, studying lunar samples helps us understand Earth's geology, because the two bodies probably are closely related. Tectonics never operated on the Moon, and wind and water are absent, meaning that lunar rocks have not been subjected to nearly as much recycling as those on Earth. Furthermore, lunar geology is complicated by the effects of meteoric bombardment and by the small number of samples available for geologists to work with—the Apollo space missions brought back only a few samples.

The oldest Moon rocks are from *lunar highlands* composed of old crust. The rocks were formed when the

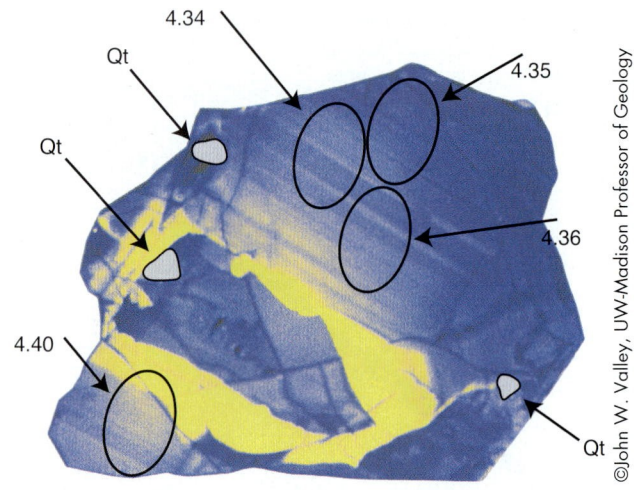

FIGURE 12.22 A tiny grain of zircon from Australia is 4.4 billion years old, the oldest date yet assigned to a rock from Earth. Circles identify other dated samples in billions of years before present, and Qt indicates quartz grains.

©John W. Valley, UW-Madison Professor of Geology

Why is it unlikely to find an Earth rock that dates from the time of Earth's origin?

early crust was partially or entirely molten, and they are mostly basalt in composition. **TABLE 12.4** shows that these rocks formed between 4.3 and 4.6 billion years ago.

If the solar system truly has a single origin, such that the Moon, Earth, and the other planetary bodies have a common origin and have since evolved, as proposed by the solar nebula hypothesis, the age of the oldest Moon rocks should reflect the age of Earth. The data in Table 12.4 are consistent with the oldest ages of Earth rocks that have been influenced by rock cycle processes, which probably destroyed the earliest components of the crust.

Meteorites

The majority of the 70 or so well-dated meteorites have been assigned ages of 4.4 to 4.7 billion years (**TABLE 12.5**). These meteorites, which are fragments of asteroids and represent some of the most primitive material in the solar system (**FIGURE 12.23**), have been dated by several independent radiometric dating methods using primary radioisotopes. Meteorites are, in a sense, ideal for age

TABLE 12.4 Oldest Moon Rocks

Mission	Dating Technique	Half-Life	Age (billions of years)
Apollo 17	Rubidium (Rb) Strontium (Sr)	48.8 billion	4.55 ± 0.1
Apollo 17	Rb–Sr	48.8 billion	4.60 ± 0.1
Apollo 17	Samarium (Sm) Neodymium (Nd)	106 billion	4.34 ± 0.05
Apollo 16	^{40}Ar-^{39}Ar	1.25 billion	4.47 ± 0.1

TABLE 12.5 Ages of Some Meteorites

Meteorite Name	Dating Technique	Age (billions of years)
Allende fragment	Ar–Ar	4.52 ± 0.02
Juvinas fragment	Sm–Nd	4.56 ± 0.08
Angra dos Reis fragment	Sm–Nd	4.55 ± 0.04
Mundrabrilla fragment	Ar–Ar	4.57 ± 0.06
Various ordinary chondrites	Various methods	4.6 to 4.69 ± 0.14

studies since there is very little chance that their composition was altered after their formation. Although meteorite ages have no direct bearing on Earth's age, they tell us something about the age of the solar system and, therefore, the age of the planets.

Geologists assume that meteorites and Moon rocks were not subjected to the extensive alteration that Earth rocks have experienced. Instead, they believe, their ages indicate when they were formed. Major objects within the solar system are thought to have been formed at the same time, and have since evolved, meaning that Earth must be the same age as meteorites and a bit older than the Moon—about 4.55 to 4.6 billion years old. Recall (Chapter 2) that the impact hypothesis for the origin of the Moon suggests that it is somewhat younger than Earth and was formed soon after the origin of the solar system. This history is also consistent with the absolute ages of these materials.

Lead Isotopes

A fourth line of evidence for Earth's age comes from the abundance of lead isotopes in Earth rocks (**FIGURE 12.24**). This method yields

FIGURE 12.23 Meteorites are among the most primitive materials in the solar system.

Q What is the primary source of meteorites?

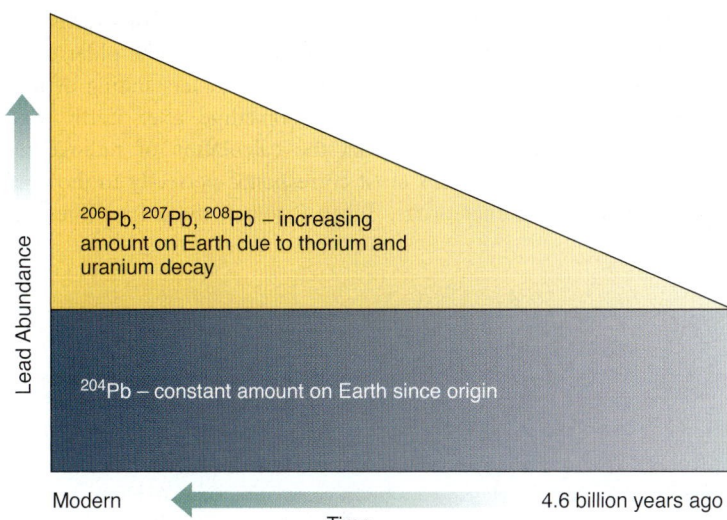

^{206}Pb, ^{207}Pb, ^{208}Pb – increasing amount on Earth due to thorium and uranium decay

^{204}Pb – constant amount on Earth since origin

Lead Abundance

Modern   Time   4.6 billion years ago

FIGURE 12.24 The accumulation of ^{206}Pb, ^{207}Pb, ^{208}Pb, and ^{204}Pb provides a measure of elapsed time since Earth was formed.

Q How would you test the lead hypothesis for Earth's age?

results that are consistent with both meteorite and lunar ages. Natural lead is a mixture of four stable isotopes; three of these isotopes (^{206}Pb, ^{207}Pb, and ^{208}Pb) result from the decay of radioisotopes of thorium and uranium. The fourth, ^{204}Pb, is not the result of radioactive decay; it is the product of *nucleosynthesis* in the core of an exploding star. This means that all the ^{204}Pb on Earth has been around since the planet's formation.

Based on extensive sampling of Earth's crust, geologists have determined the present-day abundances of the four lead isotopes relative to one another and to the parent isotopes (^{238}U, ^{235}U, and ^{232}Th) that produced three of them. Scientists cannot measure the original amount of lead on the planet, so instead they use meteorites to determine Earth's original lead content. Some primitive meteorites contain the four lead isotopes but low amounts of uranium and thorium. Scientists infer that the majority of lead in these meteorites is not the product of radioactive decay, and therefore has not changed since they were formed. Scientists believe that this lead content provides a reasonable approximation of Earth's original lead content, the so-called primordial lead.

By comparing the amounts of the four lead isotopes in primordial lead to their current amounts on Earth, scientists can determine how much lead has been added by radioactive decay since Earth was formed. Knowing the half-life of each parent, it is possible to calculate how long it took to create modern differences between the amounts of present-day and primordial lead for each of the three isotopes produced by radioactive decay. These calculations yield an age of about 4.6 billion years for Earth, which is consistent with the ages independently determined from meteorites and lunar rocks.

Consistency

It is noteworthy that lunar and meteorite ages are remarkably consistent—especially since they have been produced by different researchers using different radiometric techniques from different samples. In general, the oldest meteorite ages tend to be older than the oldest lunar ages, a fact that fits with our understanding of the formation of the Moon as occurring sometime after Earth was formed. It is also noteworthy that the calculation of radiogenic lead accumulation in Earth's crust corresponds perfectly to the age of Moon rocks and meteorites. Taken together, these four lines of evidence provide strong justification for the conclusion that Earth formed 4.55 to 4.6 billion years ago.

⊙ EXPAND YOUR THINKING

What geologic environment would you search to look for the oldest rock on Earth?

LET'S REVIEW

Geology is the science of time. We have learned that time can be measured in several ways. The study of relative time is based on the application of basic principles of stratigraphy that identify the step-by-step sequence of geologic events leading to the structure of the crust. Early naturalists applied these principles to interpreting the origin of rock outcrops and recognized that geologic events recorded immense periods of time. They were the first to propose that Earth was hundreds of millions of years old.

With improved understanding of the physics of radioactivity came the realization that minerals hold "isotopic clocks" that are useful for radiometric dating of geologic events. In a breathtaking example of critical thinking, radiometric dating and relative dating combined to address the fundamental question of Earth's age, which is now estimated to be 4.5 to 4.6 billion years. On the basis of these technical advances, geologists have built and refined the geologic time scale, which tells the story of changes that characterize the planet's past and the diverse life that populates its surface. That is the story told in the next chapter.

STUDY GUIDE

12-1 Earth's history is a sequence of geologic events.

- Early estimates of Earth's age were based on religious dogma. Scientific analysis of the problem of calculating Earth's age led to estimates of hundreds of millions of years.

- Major geologic events fall into five main classes: **intrusion**, **erosion**, **deposition**, **faulting**, and rock **deformation**.

12-2 Geology is the science of time.

- Every field of science is is defined by a single focus of study; geology is known as the science of time. Geologists are trained to interpret vast time spans in the details of Earth's rocks.

- Geologists apply two types of time analysis: **relative dating** and **radiometric dating**. Relative dating determines the order of formation (which came first, second, and so on) among layers of rock, based on their interrelationships. Radiometric dating uses radioisotopes within minerals to calculate their chronological age.

- The study of geologic time is conducted to find fossil fuels, precious metals, and other geologic resources. In addition, human curiosity demands answers to questions about our origin, our history, and our place in the sequence of natural events.

12-3 Seven stratigraphic principles are used in relative dating.

- **Stratigraphic principles** are used to guide scientific thinking when unraveling the history of crust formation.

- There are seven stratigraphic principles: the principle of superposition, the principle of original horizontality, the principle of cross-cutting relationships,

the principle of original lateral continuity, the principle of fossil succession, the principle of inclusions, and the principle of unconformities.

12-4 Relative dating determines the order of geologic events.

- A block of crust contains a complicated system of rock layers. Geologists can interpret these by applying stratigraphic principles to determine the sequence of geologic events that produced the crust.

12-5 James Hutton recognized the meaning of unconformities in the geologic record.

- James Hutton studied sequences of rocks on the Scottish coast and realized that they recorded cycles of geologic events involving deposition of sediments on the seafloor, uplift with tilting to form mountains, erosion by natural forces, and further deposition and uplift. He saw that these events were recorded in the presence of rocks and unconformities.

- Realizing that these events must be due to the same geological forces still operating, he inferred from the exposed rock layers vast stretches of time, some of which are recorded as unconformities.

12-6 Radiometric dating uses radioactive decay to estimate the age of geologic samples.

- **Radioactive decay** is a process wherein the unstable nucleus of an atom spontaneously emits radiation and a subatomic particle. This results in an atom of one type, called the parent isotope, transforming into an atom of another type, called the daughter isotope. The rate of radioactive decay is fixed for a given **radioisotope**. It produces daughter isotopes that can be counted, and makes it possible to estimate the original amounts of parent isotopes.

- There are three main types of radioactive decay: alpha decay, beta decay, and electron capture. Alpha decay emits a single alpha particle composed of 2 protons and 2 neutrons and produces a daughter isotope with 2 fewer protons. Beta decay occurs when the nucleus emits a negative beta particle that changes one of the neutrons into a proton. The parent isotope gains 1 proton at the expense of 1 neutron. In electron capture, an electron is drawn into the nucleus, where it combines with a proton and becomes a neutron.

- Because the rate of decay is fixed, the passage of time can be tracked by the reduction in parent isotopes and the increase in daughter isotopes. It is impossible, however, to determine exactly when a single atom will decay, so geologists have defined the statistical **half-life** of decay. The half-life is the time it takes for half the radioactive atoms in any sample to undergo radioactive decay.

12-7 Geologists select an appropriate radioisotope when dating a sample.

- Not every type of radioisotope can be used to date every type of rock. A geologist must choose among the many dating techniques available and select a radioisotope that is appropriate and useful for the material to be dated. Specifically, young materials should be dated with short-lived radioisotopes, and old materials with long-lived ones. In general, useful isotopes fall into two categories: primary radioisotopes and cosmogenic radioisotopes.

12-8 Accurate dating requires understanding various sources of uncertainty.

- To date a rock accurately, the geologist must answer several questions: What is the relationship between the mineral sample and the event it represents? What is the approximate age of the sample? Which radioisotope is appropriate for both the type of material being dated and the approximate age of the sample? What are the sources of potential error in a calculated age, and how can the age be tested?

- To arrive at an accurate age, the geologist assumes that the mineral being dated has remained a closed system throughout its life. This means that neither parent nor daughter isotopes have left or entered the mineral since its formation. When this is not the case, it is usually a sign that metamorphism has occurred.

- All isotopes of the element uranium are at least mildly radioactive. ^{238}U, which occurs naturally in most types of granite and soil in varying degrees, decays with a half-life of 4.5 billion years and produces the stable daughter isotope ^{206}Pb. This process consists of 14 steps, each leading to the production and subsequent decay of a radioactive daughter isotope. These steps are referred to as a **decay chain**.

12-9 Potassium-argon and carbon serve as important isotopic clocks.

- The potassium-argon decay pair is an effective tool for dating igneous rocks, and radiocarbon is an effective tool for dating recent organic materials in the time frame of the last 40,000 years.

12-10 Scientists arrived at Earth's age via several independent observations.

- The scientifically measured age of Earth is 4.55 to 4.6 billion years.

- Estimates of Earth's age are derived from several lines of evidence: 1) the age of primordial crust, 2) the age of Moon rocks, 3) the age of meteorites, 4) the abundance of lead isotopes, and 5) the consistency of these separate lines of evidence.

KEY TERMS

decay chain (p. 360)
deformation (p. 345)
deposition (p. 345)
erosion (p. 345)
evolution (p. 349)
faulting (p. 345)

geologic events (p. 345)
half-life (p. 354)
intrusion (p. 345)
radioactive decay (p. 346)
radioisotopes (p. 354)
radiometric dating (p. 346)

relative dating (p. 345)
stratigraphic principles (p. 348)
unconformity (p. 345)
uplift (p. 345)

ASSESSING YOUR KNOWLEDGE

Please complete this exercise before coming to class. Identify the best answer to each question.

1. Relative dating is the process of:
   a. Calculating the age of a rock sample.
   b. Determining how old a mineral is.
   c. Enumerating the sequence of events in the crust.
   d. Calculating when a mineral was renewed by metamorphism.
   e. Identifying the sources of uncertainty in a date.

2. Radiometric dating is the process of:
   a. Estimating the age of a sample using radioisotopes.
   b. Documenting the unique fossil assemblage in a rock.
   c. Determining the geologic events that formed a rock.
   d. Enumerating the sequence of geologic events in the crust.
   e. Assessing the rate of sediment accumulation in the ocean.

3. Early critical thinkers estimated Earth's age using:
   a. Calculations of the rate of delivery of salt to the sea.
   b. Estimates of sediment accumulation over time.
   c. Calculations of the time needed for Earth to cool.
   d. Assessments of time needed to allow for evolution, as recorded in rocks.
   e. All of these.

4. The five fundamental geologic events described by stratigraphic principles are:
   a. Dating, eroding, depositing, removing, and folding.
   b. Deformation, unconformities, superposition, deposition, and faulting.
   c. Deposition, erosion, deformation, faulting, and intrusion.
   d. Intrusion, deposition, erosion, tectonism, and unconformities.
   e. None of these.

5. Label the two decay sequences depicted in this diagram as either alpha emission or beta decay. For each decay sequence, also label the following:
   Parent nucleus
   Daughter nucleus
   Alpha particle (or) beta particle

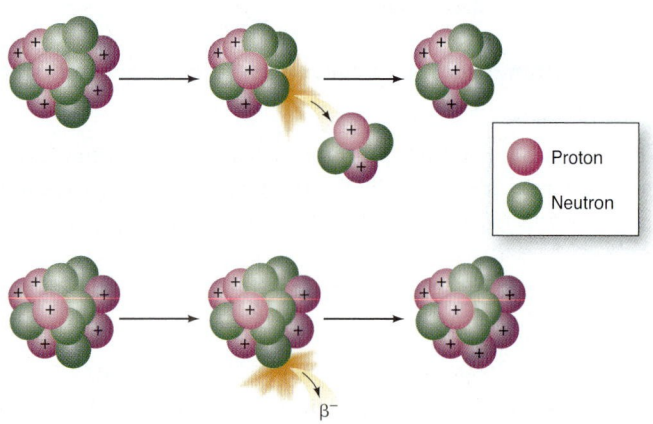

6. The principle of superposition states that:
   a. The lowest layers in an undeformed sequence are the oldest.
   b. The highest layers in an undeformed sequence are the youngest.
   c. A rock layer lying above another must be the younger of the two.
   d. Older rock units typically are found at the base of a sequence of rocks.
   e. All of the above.

7. When an intrusion invades the crust, its relative position in a sequence of geologic events is determined using the principle of:
   a. Unconformities.
   b. Superposition.
   c. Cross-cutting relationships.
   d. Original lateral continuity.
   e. None of these.

8. The principle of cross-cutting relationships says that_____
   a. Water-borne sediments are deposited in nearly horizontal layers.
   b. A sediment or sedimentary rock layer is younger than the layers below it and older than the layers above it.

Francois Gohier/Photo Researchers

c. A rock unit is older than a feature that disrupts it, such as a fault or igneous intrusion.
   d. A sediment or sedimentary rock layer is older than the layers below it and younger than the layers above it.

9. William "Strata" Smith is known for:
   a. His use of the principle of fossil succession.
   b. Using radiometric dating to determine the geology of England.
   c. Developing modern map-making techniques.
   d. Identifying England's earliest life-forms.
   e. None of these.

10. Explain how the principle of cross-cutting relationships is related to the principle of unconformities.
    a. Unconformities are made by intrusions that cut across strata.
    b. Unconformities cut across preexisting strata.
    c. These two principles are not related.
    d. Unconformities are younger than strata that lie above them.
    e. Unconformities are not used in geologic time.

11. Erosion, such as shown in the photo, can influence the rock record. Which of the following stratigraphic principles are relevant to understanding the role of erosion?
    a. Principle of cross-cutting relationships
    b. Principle of superposition
    c. Principle of original lateral continuity
    d. Principle of unconformities
    e. All of these

David Davis/Photo Researchers

12. To date a very old rock, a geologist should use an isotope that:
    a. Is very old.
    b. Is very young.
    c. Has a long half-life.
    d. Has a short half-life.
    e. Has been contaminated by groundwater.

13. A radioisotope with a short half-life should be used to date:
    a. A very old geologic event.
    b. A very young geologic event.
    c. A sample that has not had a closed system.
    d. The age of Moon rocks.
    e. Meteorites.

14. Typically, primary radioisotopes:
    a. Have shorter half-lives than cosmogenic radioisotopes.
    b. Have longer half-lives than cosmogenic radioisotopes.

c. Have the same half-lives as cosmogenic radioisotopes.

d. Come from the decay of cosmogenic radioisotopes.

e. Cannot be compared to cosmogenic radioisotopes.

15. One problem with radiometric dating is:

a. Open-system behavior.

b. Dangerous radioactivity.

c. Minerals contaminated by sand.

d. Dating a mineral with many types of radioactivity.

e. There are typically very few problems with radiometric dating.

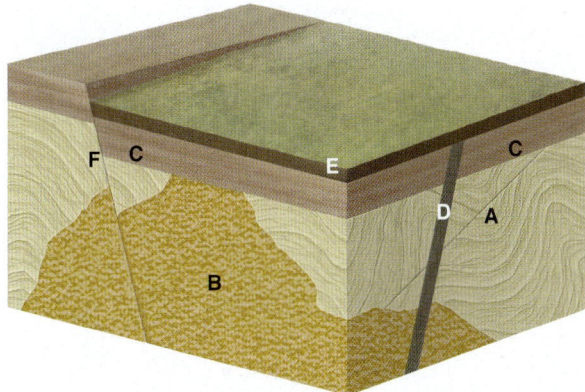

16. Indicate the geologic events represented in this section of crust. (deposition, erosion, faulting, deformation, intrusion)

17. Which method would a geologist use to date a fossil bone fragment of a mammoth?

a. Potassium-argon

b. Rubidium-strontium

c. There is no way to date bone.

d. ^{14}C

e. None of these.

18. Moon rocks are:

a. About the same age as the continents.

b. Evidence that Earth and Moon are unrelated.

c. Evidence that Earth's age is approximately 4.6 billion years.

d. Not datable because they are radioactive.

e. Not allowed to be dated by NASA.

19. Which of the following statements is true?

a. Moon rocks, meteorites, and continents are all the same age.

b. The oldest Earth rock is the age of Earth.

c. Earth, Moon, and meteorites all formed at the same time.

d. It is unlikely that any rocks are left on Earth from its origin.

e. Meteorites come from the Moon.

20. List the stratigraphic principles Hutton applied in developing his understanding of the geologic history at Siccar Point. _____

Peter L. Kresan

FURTHER RESEARCH

1. The half-life of ^{234}U decaying to ^{230}Th is 248,000 years. What is the age of a coral sample that has a $^{234}U/^{230}Th$ ratio of 1:3? What if the ratio were 3:1?

2. Why does radiometric dating rely on an unchanging rate of decay to be effective?

3. How can deformation interfere with use of the principle of superposition?

4. What are the major sources of uncertainty in radiometric dating?

ONLINE RESOURCES

Explore more about geologic time on the following Web sites:

Public Broadcast System (PBS) Evolution home page, "A Journey into Where We're From and Where We're Going": www.pbs.org/wgbh/evolution/index.html

The American Scientific Affiliation (ASA) article "Radiometric Dating—A Christian Perspective": www.asa3.org/ASA/resources/Wiens.html

U.S. Geologic Survey (USGS) on "Geologic Time": http://pubs.usgs.gov/gip/geotime/contents.html

Talk Origins document on "Radiometric Dating and the Geological Time Scale": www.talkorigins.org/faqs/dating.html

Additional animations, videos, and other online resources are available at this book's companion Web site: www.wiley.com/college/fletcher

This companion Web site also has more information about WileyPLUS and other Wiley teaching and learning resources.

EARTH'S HISTORY

CHAPTER CONTENTS & LEARNING OBJECTIVES

GEOLOGY IN OUR LIVES

By studying Earth's history we gain a sense of our place in, and a perspective about, the natural world. We learn that time is deep and that great change has occurred and continues to do so, both in geological environments and among the biological community. The fossil record provides strong evidence of the evolution of life—tens of thousands of fossils define animal and plant lineages that document evolutionary changes. These fossils prove that evolution is responsible for Earth's vast diversity of life-forms. There also are practical aspects to the study of Earth's history. Evolution guides our understanding and treatment of medical problems, for example. And geologists use fossils to identify the rocks that carry natural resources, such as oil, coal, and many others. Fossils tell us that significant environmental change leads to mass extinctions; that humans and other four-limbed animals share a common ancestor; that chimpanzees are more closely related to humans than to gorillas; and that, ultimately, all living organisms are related in the great web of life. In the end, this story tells us that if humans are to continue to live sustainably on this planet, it is absolutely crucial that we look far beyond the horizon defined by our short lives.

Dinosaurs represent more than just "terrible lizards." What do they mean to you?

Earth's History Has Been Unveiled by Scientists Applying the Tools of Critical Thinking

13-1

LO **13-1** Name the forebears of geology and describe their contributions.

The development of the dating tools we studied in Chapter 12, and their use by scientists, have led to fundamental advances in our knowledge of Earth's history—a field known as **historical geology**. Many of these advances came through the efforts of pioneering scientists who, a century and a half ago, walked the land, studied natural materials, and formulated many of the principles that define the science of geology today.

Two significant contributions of these forebears have become hallmarks of science: 1) the **theory of evolution**, changes in the inherited traits of a biological population from one generation to the next, and 2) the recognition that Earth is profoundly old, a characteristic that scientist and author *Stephen J. Gould* (1941-2002) called "deep time". This chapter takes you on a walk through deep time to examine the major tectonic and biologic events of the Precambrian and Phanerozoic eons (**FIGURE 13.1**).

The science of geology owes its origin to several visionary scholars (**FIGURE 13.2**) who viewed Earth as a complex system with a history characterized by perpetual change across great spans of time. They realized that to fully understand that history, scientists needed analytical tools based on critical thinking, and they set forth to create those tools.

Nicholas Steno

Nicholas Steno (1638–1686), considered the father of geology and stratigraphy (the study of rock layers and their geologic history), was a Danish priest and anatomist who became interested in "figured stones" (fossils), some of which looked familiar, others of which did not. He was the first to suggest that fossils that looked like living organisms had in fact once been alive. Steno also proposed the principles of original horizontality, superposition, and lateral continuity, which we studied in Chapter 12.

James Hutton

James Hutton (1726–1797), a Scottish physician, farmer, and geologist, was a staunch proponent of the concept that geologic processes alter Earth's surface. This was contrary to the common view in his time, called *Neptunism*, which held that all rocks and the shape of Earths surface had been formed by precipitation or sedimentation from a single great ocean (the biblical flood) and remained essentially unchanged from that time. Hutton proposed that geologic time was indefinitely long and that Earth was self-renewing: As mountains eroded away, new ones were uplifted; as the sea covered some lands, it receded from others. This is the basis of the *rock cycle* (Chapter 1). Hutton is remembered for his claim that Earth has "no vestige of a beginning—no prospect of an end."

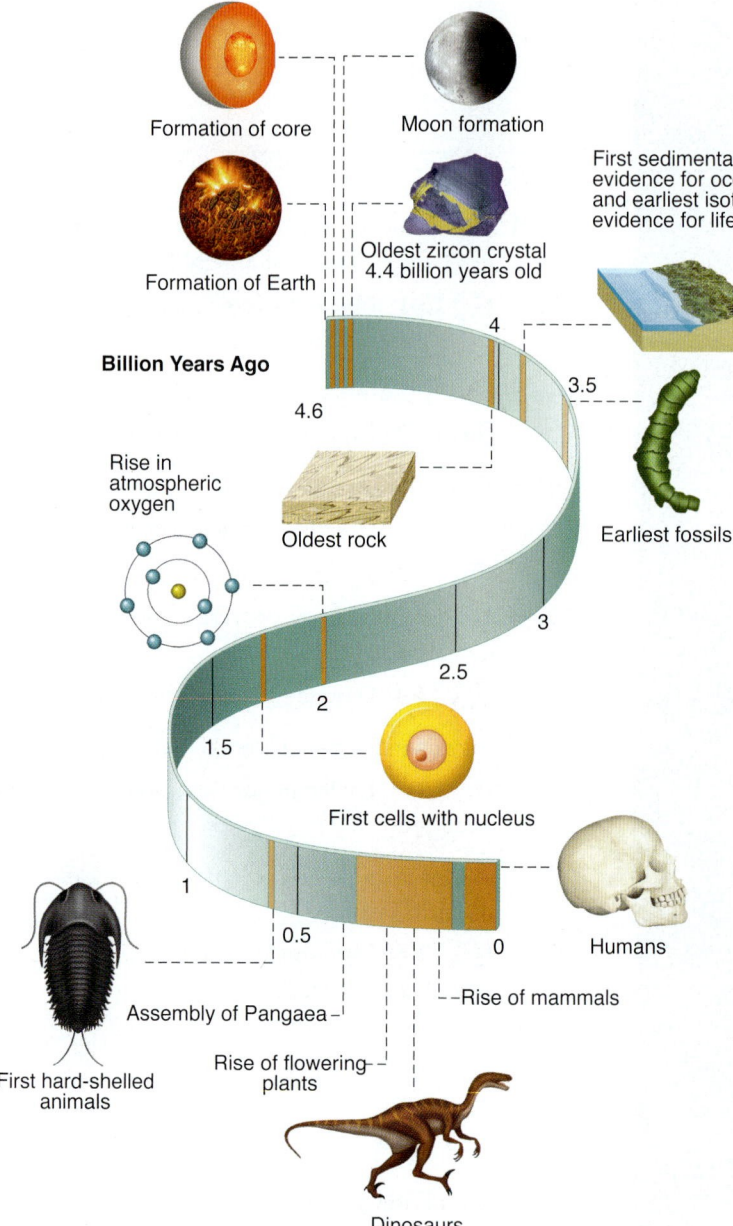

FIGURE 13.1 Earth's history is characterized by the evolution of life from simple early forms to highly diverse modern ecosystems. This history includes tectonic events that have built mountain systems, split and assembled continents, and formed oceans.

Formulate a hypothesis that describes long-term changes in living organisms over geologic time. How would you test this hypothesis?

Perhaps Hutton's most lasting contribution is the concept of **uniformitarianism,** which states that Earth's history is best explained by observations of modern processes. In other words, geologic principles have been uniform over time. *Sir Archibald Geikie* (1835–1924), another Scottish geologist and writer, summed up uniformitarianism in the oft-repeated statement: "The present is the key to the past."

(a)

(b)

(c)

(d)

The Granger Collection, New York

The Granger Collection, New York

George Bernard/Photo Researchers

Popperfoto/Getty Images, Inc.

FIGURE 13.2 (a) Nicholas Steno, (b) James Hutton, (c) Charles Lyell, (d) Charles Darwin.

Describe the fundamental advance in critical thinking that each of these early scientists contributed.

Charles Lyell

British lawyer and naturalist **Charles Lyell** (1797–1875) is often referred to as the father of modern geology. His five-volume *Principles of Geology* became an indispensable reference for every nineteenth-century geologist. Lyell explained the principle of cross-cutting relationships and the principle of inclusions, two of the seven stratigraphic principles introduced in Chapter 12. He was best known for his ability to explain the meanings of discoveries by other geologists. Notably, he expounded at great length on Hutton's concept of uniformitarianism, thereby bringing the idea into full flower among the scientists of his time. As one of his colleagues said, "We collect the data and Lyell teaches us the meaning of them."

Charles Darwin

British biologist **Charles Darwin** (1809–1882) is credited with developing the scientific theory of evolution, which accounts for changes seen in the fossil record. Today, Darwin's theory is a cornerstone of scientific thought, especially in the fields of geology, biology, and medicine.

In 1831, Darwin secured an unpaid position as naturalist aboard the HMS *Beagle*, a British ship that circumnavigated the globe. Subsequently, he published his observations in a book titled *A Naturalist's Voyage on the Beagle* (1839). Two decades later, in 1859, he published his seminal work on the theory of evolution, called *On the Origin of Species by Means of Natural Selection*, followed by *The Descent of Man*, in 1871, which explained that all living things develop from a very few simple forms over time through **natural selection.**

Natural selection is the process by which favorable traits that are *heritable* (passed on to offspring) become more common in successive generations of a population of reproducing organisms (**FIGURE 13.3**). Organisms that are best adapted to their environment (i.e., have variations that are most favorable for their survival) tend to survive and reproduce, transmitting their genetic characteristics to the next generation. Those organisms that are not as well suited to their environment die out. In this

way, populations change over time, a phenomenon that is predicted by the theory of evolution. Natural selection is, simply, the way by which evolution takes place. These changes are recorded in the geologic record of fossils, reflecting the principle of fossil succession.

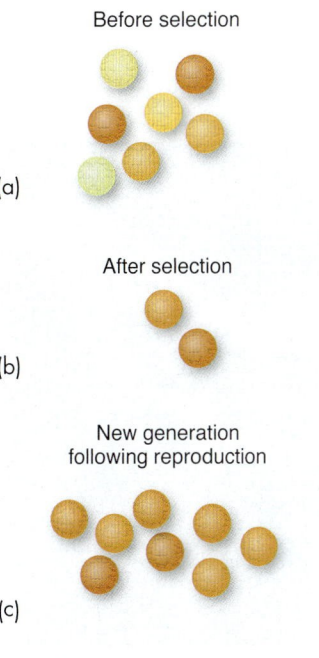

Before selection

(a)

After selection

(b)

New generation following reproduction

(c)

Adapted to environment

Low   High

FIGURE 13.3 Schematic representation of natural selection. (a) A population of reproducing organisms. (b) Population after natural selection takes place. (c) New generation of organisms following reproduction.

Disease is one mechanism of natural selection. Can you name some others?

EXPAND YOUR THINKING

Natural selection is at work all around us. Describe an example of natural selection that you encounter.

13-2 Fossils Preserve a Record of Past Life

LO 13-2 Explain the process of fossilization.

Fossils are the remains of animals and plants, or traces of their presence, that have been preserved in Earth's crust. *Fossilization* is the process that turns a once-living organism into a fossil. There are many fossils still to be found, yet only a tiny percentage of all the animals and plants that ever lived have been fossilized.

How Are Fossils Made?

Without fossils (**FIGURE 13.4**), our knowledge of Earth's history and the history of life on Earth would be very limited. Fossils give us details about how individual species come into existence, how they lived and interacted. In addition, because of the principle of fossil succession, they can tell us the age and history of sedimentary beds.

Fossils provide a record of how life changed through time, and offer strong evidence that supports and refines Darwin's theory of evolution. Evolution is the most significant natural process influencing Earth's biological communities throughout the long years of geologic history.

When an animal or plant dies, it is usually completely destroyed. Another animal may eat it, or it may decay. But when the remains of an animal are buried before they can be destroyed, and the conditions are just right, those remains will be preserved in rock as fossils. In some very rare cases, scientists have even found fossils of bird feathers and dinosaur skin (**FIGURE 13.5**). But usually only the *hard parts* of an animal, such as teeth, bones, and shells, become fossilized. Think about your own body. The parts that are most likely to become fossilized include your teeth, hard bones, and nails. The same is true of other animals.

Almost all fossils are preserved in sedimentary rocks because the melting associated with igneous rocks and the pressure of metamorphism tends to destroy fossils. Furthermore, fossilization does not always occur as simple burial. Imagine a cardboard coffee cup carelessly thrown into a mud puddle. There are four ways to fossilize the cup for later discovery: formation of a *cast*, formation of an *internal mold*, formation of an *external mold*, and *replacement* (**FIGURE 13.6**). Because the cup is not hard, it is unlikely that the cardboard will be preserved.

Replacement occurs when a hard mineral replaces soft parts, such as the cardboard of our cup. This happens when chemically active groundwater dissolves the original material of a body part and replaces it with a hard mineral of equal volume and shape. Replacement can be a very delicate and precise mode of fossilization, as it occurs atom by atom. The process may preserve the tiniest details of the fossil, much to the delight of the geologist who discovers it. Replacement also can preserve an exact replica of the fossil, such as a shell replaced by calcium carbonate or a tree limb replaced by silica.

Michael S. Yamashita/Corbis

FIGURE 13.4 The claws of *Allosaurus* do not have a sharp edge on the underside. This indicates that they were used principally for grasping (like those of hawks and owls) rather than cutting. *Allosaurus* was first discovered in 1869 in Grand County, Colorado. Large adults of the species attained lengths of over 10 meters and stood 5 to 6 meters high. They were the largest carnivores of the late Jurassic age.

If *Allosaurus* did not use its claws for cutting, what other feature of its anatomy must have been prominent on this meat eater?

epa/Corbis Images

FIGURE 13.5 In 1989, researchers uncovered a fossilized dinosaur nest in Patagonia, South America. Within the eggs was the preserved skin of embryonic dinosaurs from the late Cretaceous.

What special conditions are needed to preserve soft body parts as fossils?

FIGURE 13.6 A coffee cup can be preserved as a cast (a hardened sample of the mud that fills the cup), as an internal mold (the impression of the inside of the cup), as an external mold (an impression of the outside of the cup), or by replacement (a mineral is substituted for the cardboard by groundwater).

Ⓠ **Imagine you make a footprint in sand that is later filled in with mud. What kinds of fossil have you made?**

A mold is created when the coffee cup is pressed into the mud. If the cup dissolves, the impression of the outside of the cup will remain—an external mold (**FIGURE 13.7**). If the cup fills with mud, an imprint of the inside of the cup will create an internal mold. A mold is a reverse fossil: Ridges on the original fossil become grooves, knobs become depressions, and cavities become bumps.

L. K. Broman/Photo Researchers

FIGURE 13.7 Brachiopod (a marine invertebrate) fossils from the Paleozoic era are preserved in dolomite from Michigan.

Ⓠ **What type of fossilization preserves these brachiopods?**

Ⓠ **EXPAND YOUR THINKING**

The fact that (usually) only hard parts of an organism are preserved as a record of its existence tends to limit our understanding of past ecosystems. Why is this so?

13-3 There Are Several Lines of Evidence for Evolution

LO 13-3 Detail the geological evidence for evolution.

The early proponents of Darwin's theory had a difficult time convincing some of their colleagues that evolution was at work in the plant and animal kingdoms. One reason is that evolution can take generations to manifest its changes, making it difficult to observe evolutionary processes firsthand among communities of living organisms. Fossils offer a solution to this problem.

Phylogeny

A famous example of fossil evidence of evolution is the evolutionary lineage, or **phylogeny,** of the modern horse (**FIGURE 13.8**). Nearly 35 now-extinct ancestral species make up the family tree of the modern horse. Over time, a number of changes occurred among these ancestors, passed from one generation to the next, that produced traits characterizing the horse family as we know it today: the growth of a strong, grass-chewing set of teeth; an increase in size, strength, and speed; the development of a single toe (the hoof) from forebears originally having four toes; lengthening of the jaws to raise the eyes away from the mouth; and an increase in brain size.

The pattern of horse evolution is typical of many phylogenies. The horse phylogeny is a complex tree with numerous "side branches," some leading to extinct species and others leading to species closely related to modern *Equus*. This branched family tree is the result of random genomic variations and natural selection in a changing climate.

Evidence suggests that horses evolved at the same time that forests were giving way to grasslands across the continents. Consistent with Darwin's theory, changes in the environment exerted natural selection on early populations of horse ancestors so that individuals with characteristics favorable to survival in open grassy plains were more likely to endure and reproduce.

Homologous Structures

Another clue to evolution is seen in **FIGURE 13.9**. **Homologous structures,** similarity among characteristics of organisms resulting from their shared ancestry, are found in such diverse organisms as birds, swimming mammals, four-legged animals, humans, and insect-eating reptiles. This pattern suggests that these animals evolved from a common ancestor and that survival pressure (natural selection) has preferentially selected individuals possessing the specific traits and functions served by each limb.

In four-limbed vertebrates, for example, limb bones may vary in size and shape, but all contain the same number and position of specific bones. This phenomenon indicates that totally different families of organisms share a common ancestor. If this were not the case, it is difficult to imagine how specific and fundamentally unique selective pressures could have converged on a single basic blueprint for limb design.

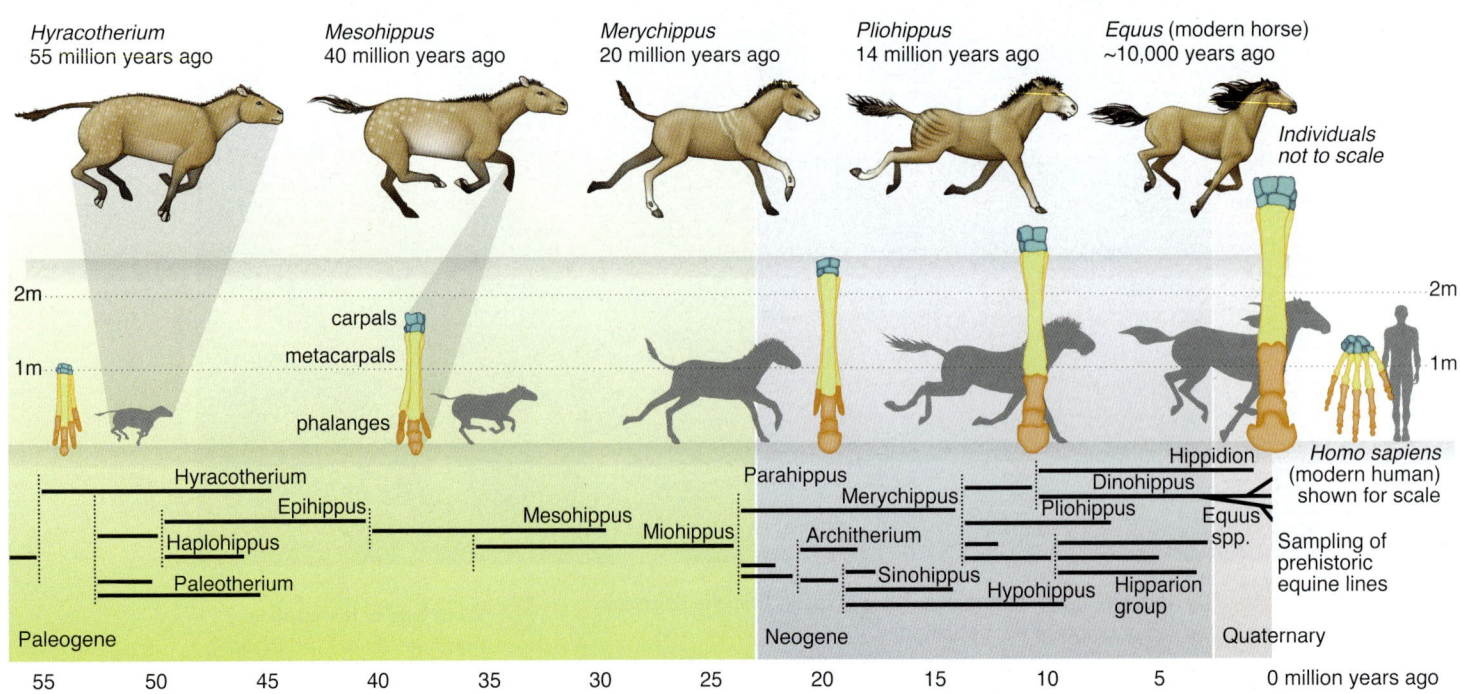

FIGURE 13.8 The phylogeny of the horse family shows consistent changes in the lower leg area, the development of molars designed for chewing tough grasses, and an increase in body size and strength.

Apply inductive reasoning—drawing general conclusions from specific observations—and use the changing characteristics of horse ancestors to describe the natural processes exerting selection pressure on the population.

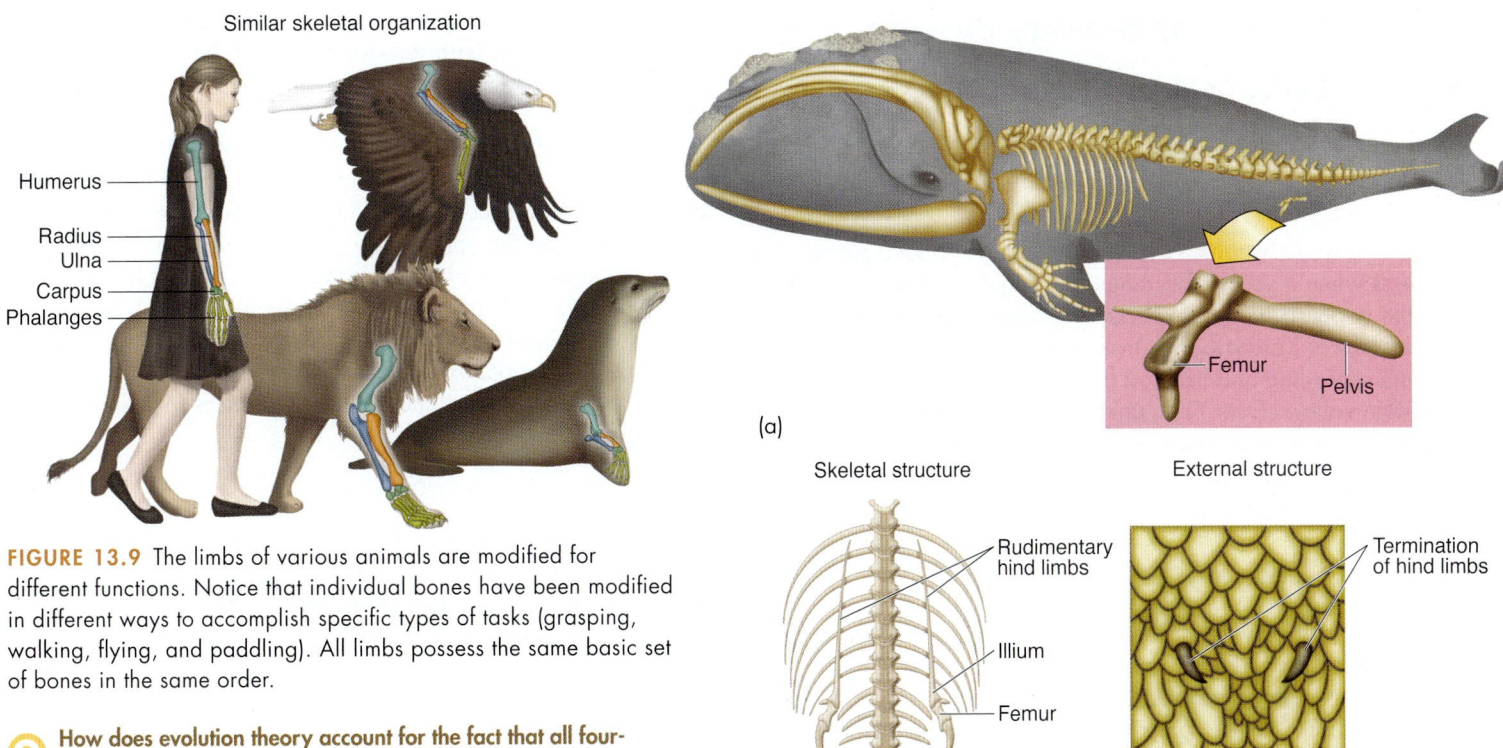

Similar skeletal organization

Humerus
Radius
Ulna
Carpus
Phalanges

FIGURE 13.9 The limbs of various animals are modified for different functions. Notice that individual bones have been modified in different ways to accomplish specific types of tasks (grasping, walking, flying, and paddling). All limbs possess the same basic set of bones in the same order.

How does evolution theory account for the fact that all four-limbed animals have the same basic set of bones?

(a)

Femur
Pelvis

Skeletal structure

Rudimentary hind limbs

Illium

Femur

(b)

External structure

Termination of hind limbs

FIGURE 13.10 (a) Whales possess a vestigial pelvis and femur originally designed for walking. Proof that modern whales have evolved from walking ancestors was found in 1994 in the form of a fossil whale that had flippers for front legs and long hind limbs with elongate toes for webbed feet. (b) Boa constrictors also have vestiges of legs.

The ridges on our upper lips are vestigial. What was their original form and function?

Vestigial Structures

In another important line of evidence for evolution, biologists identify **vestigial structures** in certain organisms as having resulted from evolution under selective pressure. Vestigial structures are the relics of body parts that were used by ancestral forms but are now nonessential (**FIGURE 13.10**). Natural selection does not exert enough pressure to completely eliminate vestigial parts once they have been replaced with newer functions; as a result, organisms tend to retain some aspect of the original structure. Humans possess more than 100 vestigial structures, including ear muscles, wisdom teeth, the appendix, the coccyx (tail vertebrae), hair, ridges on the upper lip, nipples on males, fingernails and toenails, and others.

Embryology

Evidence of evolution is also present in studies of *embryology*. In their early stages of development, embryos of various fish, birds, and mammals display strikingly similar characteristics. Human embryos have tails (which eventually grow into the base of the spine) and gill slits (which become the eustachian tube), as do other life-forms in their early stages. It is thought that all these animals have a distant common ancestor from which they inherited a set of genes that control early embryologic development (**FIGURE 13.11**). As embryo development progresses, other genes assume control and produce an individual of a particular species.

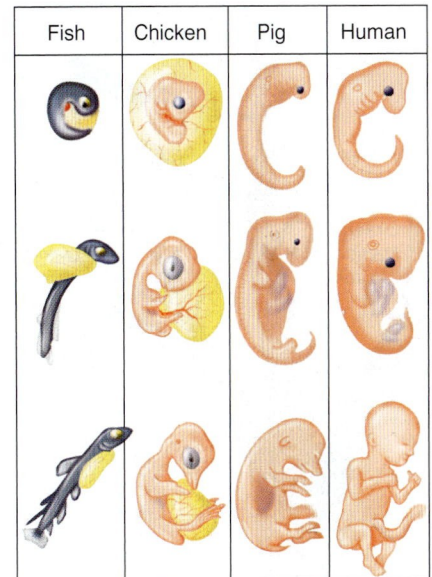

Fish	Chicken	Pig	Human

FIGURE 13.11 Many animals display remarkably similar features during their development as embryos. Humans, for instance, have a tail and primitive gill slits.

EXPAND YOUR THINKING

Do you find these lines of evidence for evolution personally compelling? What questions do you have?

13-4 Molecular Biology Provides Evidence of Evolution

One of the ways by which evolution is thought to take place happens when a species becomes separated into geographically isolated populations that experience different environmental conditions. For instance, two groups of the same species may become separated and isolated from one another in two mountain valleys. Over time, unique characteristics may arise among the separate populations, eventually causing them to become distinct and form new species.

Evolution Processes

Differences that arise in separate populations are the product of **genetic mutation,** random changes to genetic material—RNA or DNA—and **genetic variation,** differences in inherited traits among individuals in the population. Favorable variations are naturally selected through *environmental pressure* (they enhance the ability to survive in the environment) or *community pressure* (they enhance the ability to survive in a community of individuals). Variations that burden the organism with disadvantages will be selected *against*, making individuals with these variations less likely to survive and reproduce.

Controlled breeding is a modern form of isolation practiced by horse, dog, and cat breeders (and others). In controlled breeding, humans allow specific individual animals with certain traits to breed, increasing the opportunity for these traits to collect in the next generation. Controlled breeding is responsible for the more than 400 types of dogs that have been bred into existence by humans, breeds that range from the tiny Chihuahua to the massive Great Dane. Remarkably, the entire family of dogs evolved from a few wolves that were domesticated by humans in Asia less than 15,000 years ago.

Molecular Biology

In the field of molecular biology we find evidence of evolution employing the molecules of DNA, RNA, and protein that are present in every organism. If one organism has evolved from another, its DNA sequence will have similarities; and if two organisms are distantly related, their DNA will be different.

For example, children with the same parents are closely related and therefore have very similar DNA sequences. A comparison of the DNA of chimpanzees with that of gorillas and humans reveals that chimpanzees share as much as 96% of their genes with humans (more than with gorillas). This finding indicates that humans and chimpanzees are more closely related to each other than to gorillas.

Molecular comparisons such as this have allowed biologists to detail the evolution of life on a *relationship tree* (**FIGURE 13.12**).

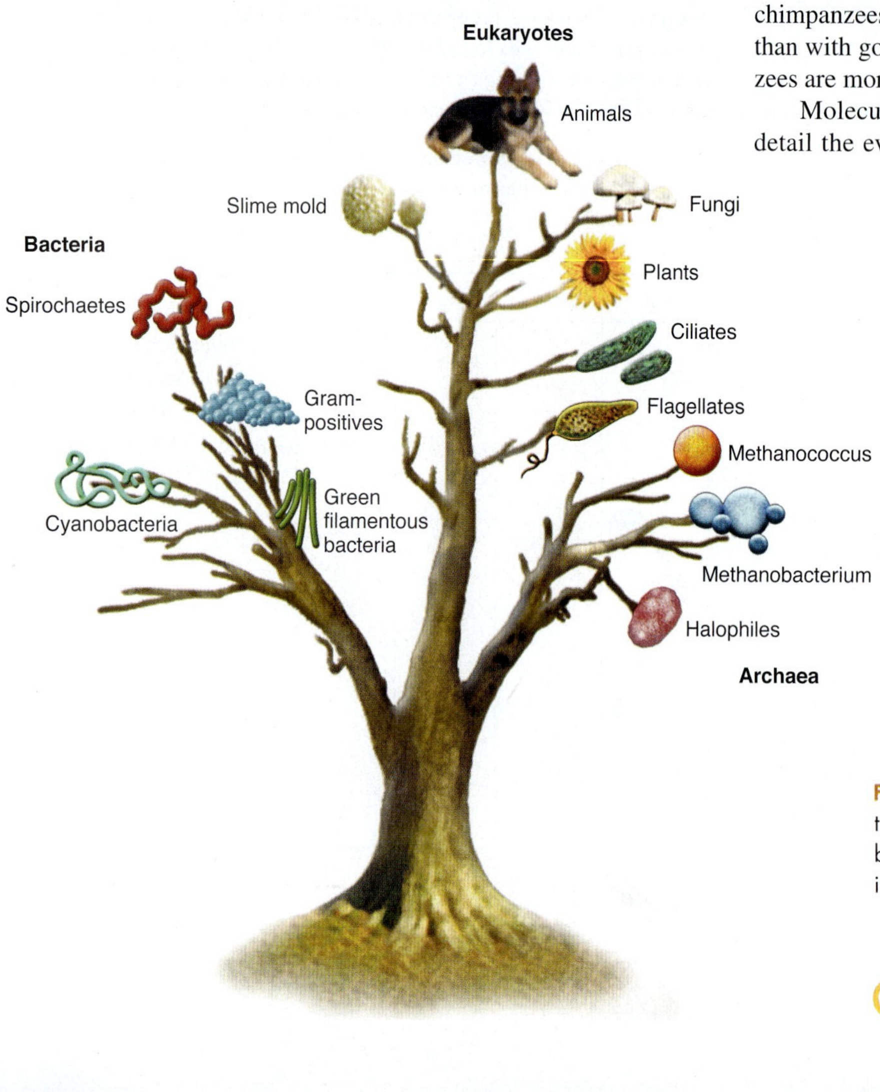

Eukaryotes

Animals

Slime mold

Fungi

Bacteria

Plants

Spirochaetes

Ciliates

Gram-positives

Flagellates

Methanococcus

Cyanobacteria

Green filamentous bacteria

Methanobacterium

Halophiles

Archaea

FIGURE 13.12 This simple portrayal of the relationship tree of life depicts the three major lineages of organisms—bacteria, archaea, and eukaryotes—that have been identified by molecular biologists.

 How do genetic mutation and genetic variation lead to evolution and, ultimately, the relationship tree of life?

Molecular comparisons permit researchers to draw conclusions about organisms whose common ancestors lived such a long time ago that little obvious similarity is apparent in their appearance today.

In the past three decades, biologists have identified three primary lineages of living organisms: 1) the *eukaryote* branch, which developed into animals, plants, fungi, and protists (including protozoa and most algae); 2) the relatively little-known *archaea* branch, organisms that had previously been found only in extreme environments, such as hot springs, but are now thought to occupy most ecological niches on Earth; and 3) the *bacteria* branch, a large group of single-celled microorganisms, some of which serve beneficial functions in animals and humans and others of which cause infections and disease.

AIDS Evolves

Why has it not been possible for scientists to find a cure for the common flu or AIDS? These diseases (and others) are caused by *viruses* that infect human victims. To fight viruses, we employ *vaccines*, specific chemicals that trigger the production of antibodies. *Antibodies* are "soldier" cells that kill a specific type of virus before it can do much damage. But viruses are living organisms that evolve, like all organisms, and so resist our efforts to control them. A well-documented current example is that of AIDS, caused by the human immunodeficiency virus, or HIV (**FIGURE 13.13**)—because HIV evolves, it escapes medical treatment.

Because viruses evolve, antibodies designed to attack one form of a virus will be useless against a new form of that virus. This is the reason vaccines have failed to eliminate the common flu or AIDS, for example. To be effective, vaccines against particular strains of a flu virus must be constantly updated, a process that is carried out each year in preparation for "flu season."

In the case of AIDS, as the HIV virus multiplies within a patient, some of the virus offspring will have genetic mutations and variations that are drug resistant (by genetic mutation or genetic variation). These variations can change those parts of the virus that antibodies are designed to recognize. All it takes is a very small change—one that otherwise is completely irrelevant to the structure or function of the virus—for the antibody to become ineffective against the virus.

The statistics of mutation can become overwhelming due to the huge number of viral cells that inhabit an infected person, coupled with the rapid turnover of the viral population—which can occur in a few hours or days. Put another way, evolution can completely change the viral "strain" in a matter of months. Although a patient might be able to fight off the first strain, the virus can evolve faster than scientists can develop medicines to fight it. In scientific parlance, the virus "escapes" drug therapy by evolving.

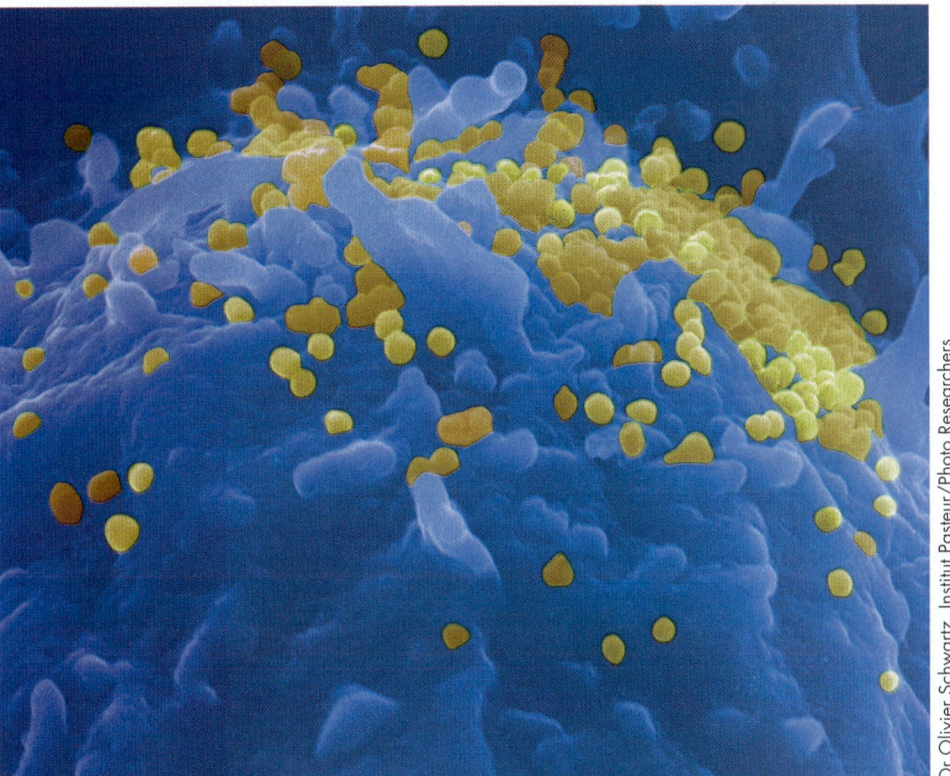

Dr Olivier Schwartz, Institut Pasteur/Photo Researchers

FIGURE 13.13 HIV, a living organism within the human body, evolves new strains that are resistant to drug treatment. It enters a cell and hijacks the cell's machinery to make more copies of the virus; the new virus particles (yellow) then burst from the membrane of the cell (blue), killing it.

Explain how evolution allows a virus to escape treatment by modern medicine.

Researchers are concerned with drug resistance not only in HIV but also in other viruses and bacteria. The problem has become more serious in recent decades, in part because physicians have been overprescribing antibiotics and antiviral agents, thereby inadvertently strengthening germs, which evolved new strains resistant to available treatments. Patients, too, have to shoulder some of the blame for the current situation. Too often, they demand antibiotics for colds or flu, which are not affected by the drugs. Also, patients frequently stop taking antibiotics before they have completed the full regimen, leaving behind a small amount of bacteria strong enough to resist the drugs in the patient's system—bacteria that may be passed on to others. Researchers note that the mechanisms of evolution need to be better understood before we can remedy this problem.

EXPAND YOUR THINKING

Imagine that the theory of evolution did not exist. How might medical researchers account for the capability of viruses and bacteria to escape treatment?

13-5 Mass Extinctions Influence the Evolution of Life

LO 13-5 List the major extinctions and when they occurred.

One of the more startling revelations of historical geology is that **mass extinctions**—events during which large numbers of species permanently die out within a very short period—have occurred several times during Earth's history.

Mass Extinctions

These events are more than curious anomalies. They exert a powerful influence on the direction of animal and plant evolution. Consider this: At the time of their extinction 65 million years ago (mya), dinosaurs had been the ruling class of animals on both land and sea for over 200 million years. (Modern humans, in contrast, are estimated to have been on Earth for only 200,000 years.) Had dinosaurs not gone extinct, and thus allowed mammals to spread and adapt to new habitats, humans and many of the animals we share the planet with would probably not exist.

Why is this so? Mass extinctions wipe out hundreds, even thousands, of dominant species (**FIGURE 13.14**). Once those species are absent, ecological niches open up to other species, in which they can live, grow, and reproduce. The dying-out of the dinosaurs eliminated the most important predators of mammals, mostly small shrew- and mouselike creatures. As a result, mammals flourished. Environments and resources not previously available to them were now ripe for the picking, and mammals experienced accelerated evolution and blossomed in diversity and distribution around the globe.

There have been at least 10 mass extinctions, 5 of which were particularly noteworthy. These 5 extinctions (**FIGURE 13.15**) occurred at the following times:

INTERFOTO/Alamy

FIGURE 13.14 *Dimetrodon* was a "synapsid" (a mammal-like reptile) that lived as a top predator in the food chain during the Permian period. It was not a dinosaur (even though it looks like one); unlike reptiles, which tend to gulp their food down without chewing, synapsids like *Dimetrodon* developed teeth capable of shearing meat into smaller pieces. These eventually gave rise to the various kinds of teeth present in modern mammals.

Q *Dimetrodon* went extinct during the Permian–Triassic extinction. What sort of events are capable of causing an extinction?

1. *Ordovician–Silurian extinction: 440–450 mya.* The Ordovician–Silurian extinction occurred at a time when life had just recently emerged out of the oceans and onto the continents. According to the fossil record, 25% of marine families and 60% of marine genera were lost.

2. *Late Devonian extinction: 360–375 mya.* The Late Devonian extinction is estimated to have killed 22% of marine families and 57% of marine genera.

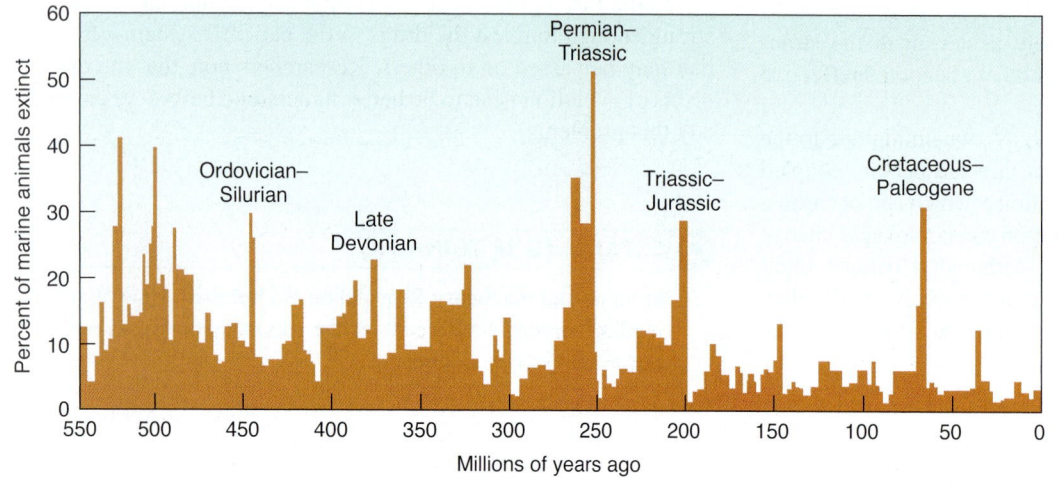

FIGURE 13.15 Five major mass extinctions have occurred during the last 500 million years of Earth's history.

Q Why are mass extinctions important in the course of evolution?

3. *Permian–Triassic extinction: 251 mya.* The Permian–Triassic extinction was the most catastrophic: 95% of all marine life on Earth died off, and 70% of land organisms became extinct. The Permian extinction killed 53% of marine families, 84% of marine genera, and an estimated 70% of land species, such as plants, insects, and vertebrate animals.

4. *Triassic–Jurassic extinction: 200 mya.* The Triassic–Jurassic extinction is thought to have been caused by massive floods of lava erupting from the central Atlantic seafloor, an event that triggered the opening of the Atlantic Ocean. The volcanism might have caused deadly global warming and the demise of 22% of marine families and 52% of marine genera.

5. *Cretaceous–Paleogene extinction: 65 mya.* The Cretaceous–Paleogene extinction—the one responsible for the demise of the dinosaurs—was likely caused by the impact of the huge asteroid (several kilometers wide) that created the *Chicxulub crater*, which is now buried below the surface of the Yucatan Peninsula and beneath the Gulf of Mexico (**FIGURE 13.16**). The collision would have kicked up a cloud of gas and dust that blocked the Sun for years and changed global environmental conditions. The extinction killed 16% of marine families, 47% of marine genera, and 18% of land vertebrate families, including the dinosaurs.

Richard Bizley/Photo Researchers

FIGURE 13.16 The impact of an extraterrestrial body striking Earth's surface can release billions of tons of ash and gas into the atmosphere, leading to global environmental changes that can cause mass extinctions.

On a global scale today, how would a large extraterrestrial impact disturb the human community?

Not all mass extinctions occur for the same reasons. Various hypotheses have been forwarded for each one, and are the subjects of intense scientific debate. In general, extinctions are thought to result from drastic global changes that follow catastrophic events, such as asteroid or comet impacts or massive volcanic eruptions.

For example, the Permian extinction has been linked to huge volcanic eruptions in what is now Siberia that spewed volcanic dust and droplets of sulfuric acid for up to 800,000 years. These eruptions would have blocked sunlight, thereby causing worldwide cooling and the advance of continental ice sheets. Water would have become locked up as ice in these massive glaciers, lowering the level of shallow seas and eliminating or changing marine habitats across the planet.

Other theories attempting to explain the Permian extinction include a massive meteor impact and a nearby supernova that bathed Earth in radioactivity and destroyed the ozone layer, which protects life-forms from ultraviolet radiation emitted by the Sun.

Whatever their cause, there can be no doubt that mass extinctions have occurred on Earth and have played a crucial role in shaping the character of the flora and fauna that are alive on the planet today.

EXPAND YOUR THINKING

It has been said that, today, humans are driving the greatest mass extinction in geologic history. How would you test this hypothesis, and how might you use the results of your study to help in ecosystem management?

13-6 The Geologic Time Scale Is the Calendar of Events in Earth's History

LO 13-6 Describe the geologic time scale and how it is organized.

The geologic time scale subdivides all time since the beginning of Earth's history—approximately 4.6 billion years ago (bya)—into named units of variable length.

Structure of the Geologic Time Scale

In Chapter 1 we learned about the **geologic time scale**—essentially, the calendar of events in Earth's history (**FIGURE 13.17**). Just as we divide time into days, weeks, months, and years, the time scale divides it into **epochs** (tens of thousands to millions of years), **periods** (millions to tens of millions of years), **eras** (tens of millions to hundreds of millions of years), and **eons** (hundreds of millions to billions of years). An important difference, however, is that geologic units are not fixed at a certain length; they vary in duration because they are determined by events, not by lengths of time.

The geologic time scale is based on *stratigraphy*, the correlation and classification of sedimentary strata. Fossils found in strata provide the chief means of establishing a time scale. As living things evolve over time, they change in appearance. The *principle of fossil succession* tells us that certain kinds of organisms existed only at specific times in Earth's history and, therefore, appear only in particular parts of the geologic record. In other words, evolution does not resurrect an organism once it has gone extinct. By correlating strata in which certain types of fossils are found, researchers can reconstruct the geologic history of various regions (and of Earth as a whole).

The geologic time scale consists of a succession of names that represent various intervals of time. The scale includes the ages, in years before present, marking the boundaries of the intervals. The names of many time units have been in use since the nineteenth century and come from locations where rocks of that age are common or were first described. The ages of all the time units were established during the twentieth century, and continue to be revised as more detailed information becomes available.

Time boundaries are based mostly on notable events in Earth's history as indicated by fossils: The appearance or disappearance of certain fossils may mark the end of one time unit and the beginning of another. Hence, as mentioned earlier, extinctions play an important role in defining geologic time units.

A remarkable characteristic of the time scale is that most of it is blank. The beginning of the Paleozoic era, placed where the first complex life is seen in the fossil record, occurred only 542 mya. The Paleozoic is preceded by 4 billion years of Earth's history, for which we largely lack detailed information.

Hadean Time

As we learned in Chapter 12, scientists arrive at Earth's age by applying the tools of radiometric dating to Moon rocks and meteorites, as well as to Earth rocks. Although few rocks on Earth remain from Hadean time (4.6 to 3.8 bya) because the processes of the rock cycle have largely destroyed them, it is nonetheless desirable to infer the nature of Earth's surface during that time.

Recall from Chapter 2 that we studied the origin of the solar system and the processes associated with planetesimal accretion. In brief, scientists deduce that Earth was initially a cool body of rock in space. But heat-producing processes, such as extraterrestrial bombardment, internal radioactive decay, and consolidation of the planet by gravity, led to the formation of a magma ocean as Earth reached a molten state.

When the planet cooled, its exterior solidified as a primitive crust characterized by widespread volcanism. The oldest rock yet identified, a 4.28 billion-year-old igneous rock from northeastern Canada, may be a piece of this primitive crust. That said, tiny crystals of zircon found in the Jack Hills conglomerate of western Australia challenge the accepted thinking. Zircon, remember, is a common mineral that is especially resistant to weathering. The Jack Hills sample has been dated to 4.4 bya—a time when Earth was supposedly molten. Contained within these minerals are isotopes of oxygen (^{16}O and ^{18}O) that allow scientists to estimate the temperatures of processes leading to the formation of magmas and rocks.

Geochemists measure the ratio of ^{18}O (a rare isotope representing about 0.2% of all oxygen on Earth) to ^{16}O (the common oxygen isotope, which accounts for about 99.8% of all oxygen). The proportions of ^{18}O and ^{16}O in a crystal depend on the temperature at which it was formed. The $^{18}O/^{16}O$ ratio in Earth's mantle—about 5.3—is well known, and geologists expected that the Jack Hills zircons would reflect the presumed molten nature of Earth's surface, with isotope ratios similar to those found in the mantle. But when scientists completed their analysis of the crystals, they discovered that their predictions had been wrong: the isotopic ratios ranged up to 7.4.

At first, researchers were confused by these results. What might these high oxygen isotope ratios mean? In younger rocks, the answer would be obvious, because such samples are common. Rocks at low temperatures on Earth's surface can acquire a high oxygen isotope ratio when they interact chemically with rain or ocean water. Those high ^{18}O rocks, if buried and melted, form magma that retains the high ^{18}O value, which is passed on to zircons during crystallization. Thus, liquid water and low temperatures are required for the formation of zircons and magmas with high ^{18}O content.

The presence of high oxygen isotope ratios in the Jack Hills zircons implies that liquid water must have existed on Earth's surface at least 400 million years earlier than the oldest known sedimentary rocks, those at Isua, Greenland. If this is correct, entire oceans may have existed, making Earth's early climate more like a sauna than a Hadean fireball. Research in this area continues, and the true nature of Earth's surface during Hadean time remains unresolved.

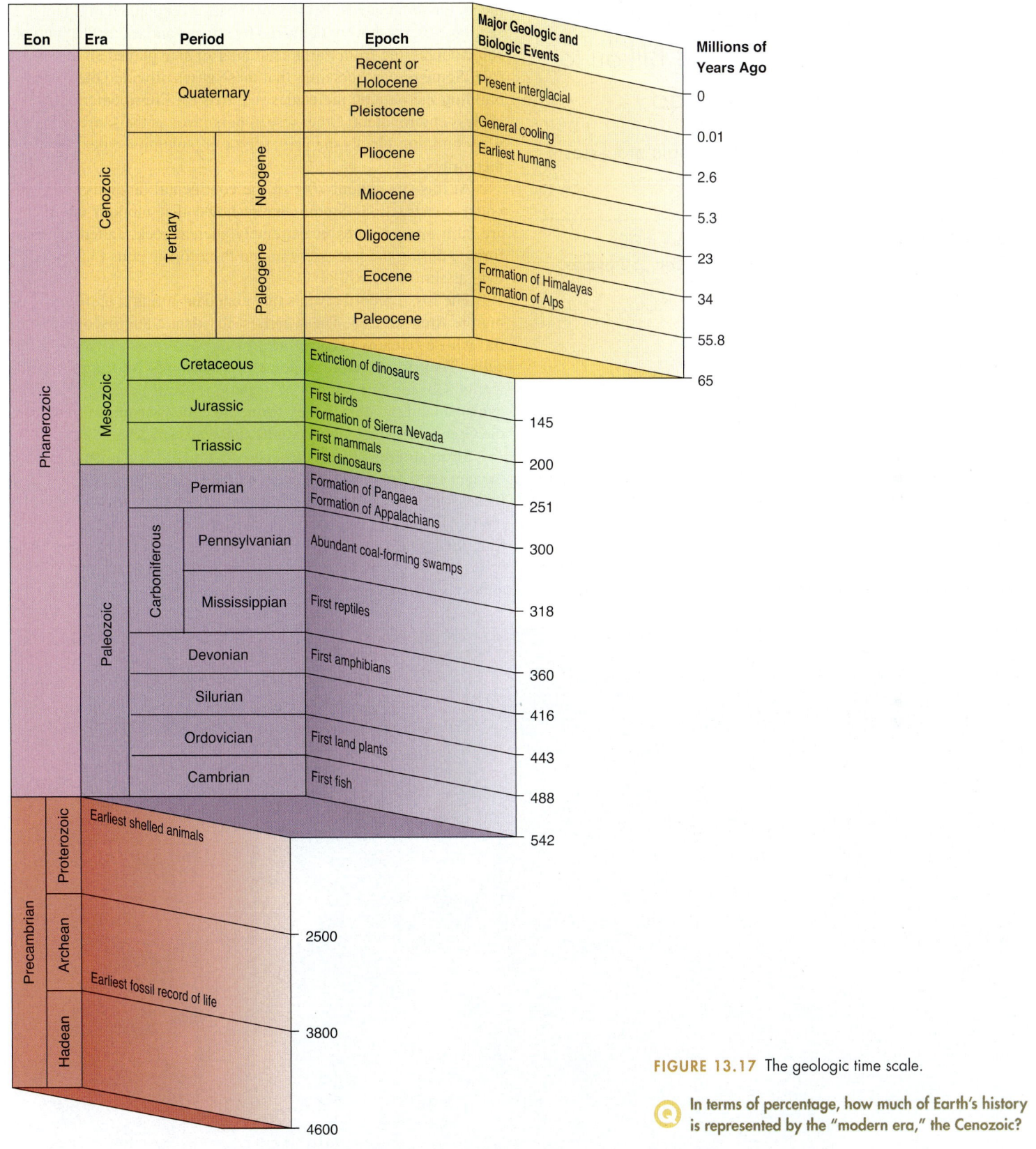

FIGURE 13.17 The geologic time scale.

In terms of percentage, how much of Earth's history is represented by the "modern era," the Cenozoic?

EXPAND YOUR THINKING

Which epoch, period, era, and eon do we live in?

The Archean and Proterozoic Eons Lasted from 3.8 Billion to 542 Million Years Ago

13-7

LO **13-7** Compare and contrast geologic processes occurring in the Archean and Proterozoic eons.

After the Hadean Eon, Earth entered the *Archean Eon* (3.8 to 2.5 bya) followed by the *Proterozoic Eon* (2.5 bya to 542 mya).

Archean Eon

Archean rocks are peculiar and interesting, because of their great age and the unique geologic processes that formed them. The surface of the Moon consists mainly of Archean rock. Interestingly, some ultramafic igneous Archean rocks present on Earth are similar to some of the Moon rocks collected by astronauts. This suggests that similar geologic processes occurred on the Moon and Earth during their earliest histories.

Earth's atmosphere was composed of noxious, unbreathable gases; methane, ammonia, carbon dioxide, and water vapor were the major components. For the most part, as explained in Chapter 2, section 2–6, these were delivered by comets and asteroids; but volcanic outgassing from Earth's hot interior also played an important role. Some researchers infer that these gases provided the basis for building the organic molecules required for life to emerge from a lifeless environment. Other researchers favor as the source of early life's building blocks the presence of organic molecules found on meteorites.

At this time about 70% of the continental landmass was still under construction. Studies have shown that modern continents are built around cores of extremely ancient rock, called **cratons,** formed during the late Archean and Proterozoic eons (3.2 to 0.542 bya) as shown in **FIGURE 13.18**.

The fossil record reveals that abundant life first appeared during the Archean Eon. These earliest fossils are microscopic bacteria, whose remains occur as strings of *blue-green cyano-bacteria* cells. They took shape as domelike structures by trapping sediments between fine strands of bacteria (**FIGURE 13.19**). These organisms, and the domes they form, are called **stromatolites,** from the Greek terms *stroma* (stratum) and *lithos* (rock).

Stromatolites, like other photosynthetic organisms, release oxygen as a by-product of metabolism. As oxygen released by these organisms accumulated in the atmosphere, it displaced the methane, ammonia, and carbon dioxide that were abundant at the time. Eventually, with the spread of other successful photosynthetic organisms throughout the shallow sunlit seas, our oxygen-rich atmosphere came into existence.

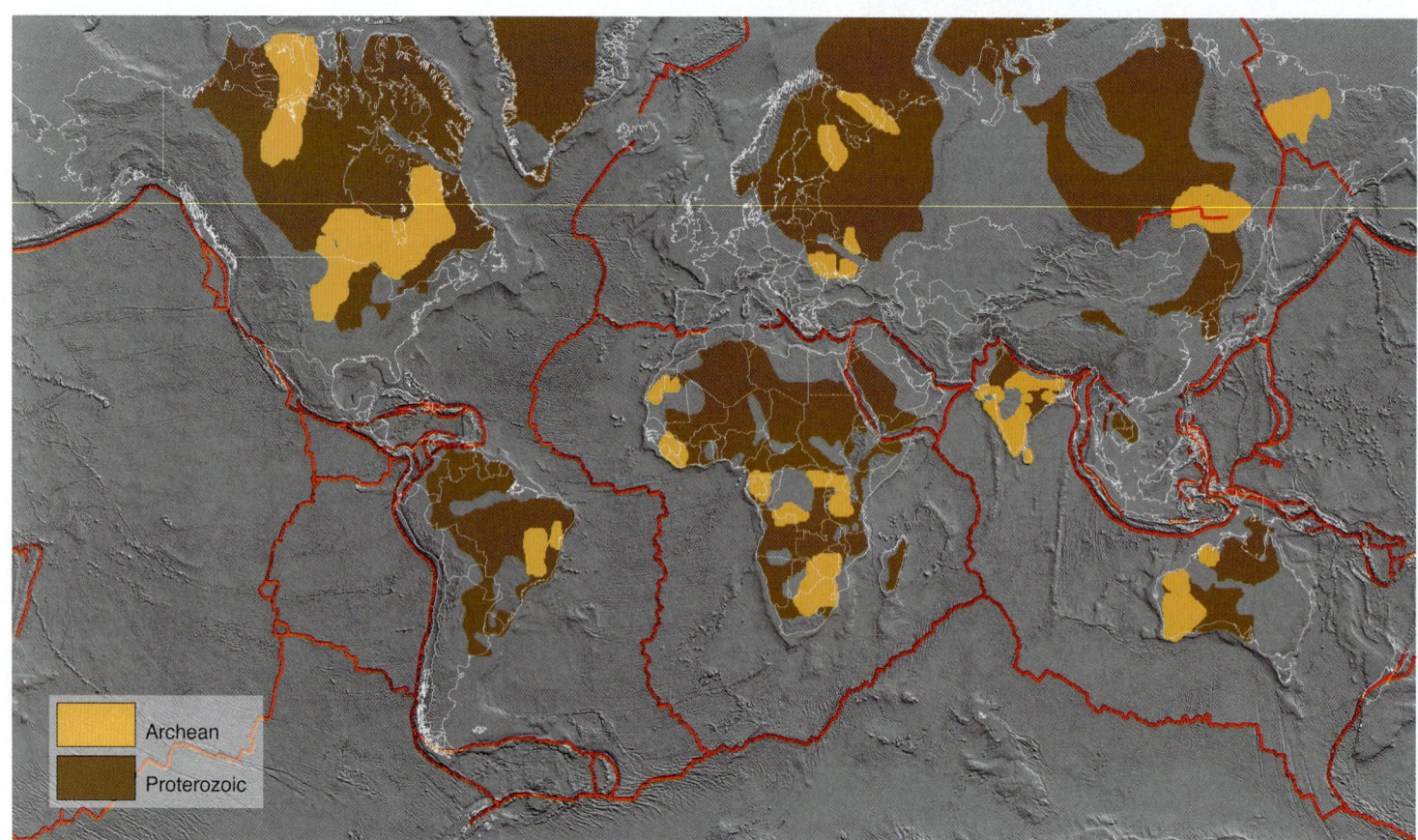

Archean

Proterozoic

FIGURE 13.18 The ancient cores of continents were formed in the late Archean and Proterozoic eons. Called cratons, today they are found in the interior of most of the major landmasses on Earth.

Describe tectonic processes that could build portions of the continents that are younger than the cratons.

(a)   Georgette Douwma/Photo Researchers

(b)   Sinclair Stammers/Photo Researchers

FIGURE 13.19 (a) Modern stromatolites, Shark Bay, Australia. (b) Fossil stromatolites. Stromatolites are formed by bacteria growing layers of cells that trap and bind sedimentary particles.

What role did stromatolites play in changing the chemistry of Proterozoic oceans and atmosphere?

The Proterozoic Eon

The Proterozoic Eon saw the continued development and growth of the continents (a process that took about 1 billion years), the emergence of a new atmosphere that was rich in oxygen, and the emergence of new mountain ranges. Although these early mountains were long ago worn down by erosion, the heavily metamorphosed rocks that composed their base can still be found.

The first known supercontinent, *Rodinia* (**FIGURE 13.20**), began forming approximately 1.3 bya; it began to break up due to plate divergence around 800 mya. Several preexisting continents converged to become Rodinia in an event known as the *Grenville Orogeny*. (Recall from Chapter 10 that an *orogeny* is a tectonic episode characterized by orogenesis, or mountain building.) Among them were the ancient cratons of Laurentia (North America), South America, Australia, Africa, Siberia, South China, and Antarctica.

Diversification of life was a hallmark of the Proterozoic Eon. Life progressed from single-celled organisms to the first multicellular, ocean-dwelling plants and animals. Photosynthetic plants flourishing in the sea produced oxygen gas, which changed the chemistry of air and seawater. As dissolved oxygen accumulated in the oceans, it combined with dissolved iron to form hematite (iron oxide), which settled onto the seafloor. Evidence for this change in seawater chemistry is the appearance of *banded iron formations* composed of layers of hematite (**FIGURE 13.21**) deposited in Proterozoic oceans.

FIGURE 13.20 During the Proterozoic Eon, between approximately 1.3 bya and 800 mya, continents were clustered as one large mass, called Rodinia, in the southern hemisphere.

Which tectonic process builds supercontinents, and which process breaks them up?

In the Paleozoic Era, Complex Life Emerged and the Continents Reorganized

13-8

LO 13-8 List the periods of the Paleozoic Era and a notable characteristic of each.

The *Phanerozoic Eon* (542 mya to present) consists of the Paleozoic, Mesozoic, and Cenozoic eras. Geologic history of the Phanerozoic is recorded in much greater detail by rocks and fossils than are earlier eons. Major mountain-building events are evident; there is widespread evidence of the past positions and movement of plates; and the detailed fossil record can be used to reconstruct the history of life (**FIGURE 13.22**).

Cambrian Period

Complex forms of life with shells and hard external body parts appeared in abundance about 542 mya, marking the beginning of the *Cambrian Period* (542 to 488 mya) of the **Paleozoic Era.** The Cambrian was a time of rapid and unprecedented evolutionary diversification. Of special note is a Cambrian rock unit in western Canada known as the *Burgess Shale*. Designated a World Heritage site because of its significance, the Burgess Shale records the "Cambrian explosion," the sudden appearance of a highly diverse animal assemblage (**FIGURE 13.23**) after billions of years of single-celled organisms.

The Ordovician Period

During the *Ordovician Period* (488 to 443 mya), the area north of the tropics was almost entirely ocean, and most of the world's land was collected into the southern supercontinent known as *Gondwana*. Throughout the Ordovician, Gondwana shifted toward the South Pole. Jawless fishes with thick bony, armored plates appeared, and corals and clams came into existence. Shallow seas covered much of North America and then receded, leaving behind thick limestone units.

The Silurian Period

During the *Silurian Period* (443 to 416 mya), global warming contributed to a substantial rise in sea levels; broad tracts of coral reefs made their first appearance; and with the first land animals (insects), sharks, and jawed fishes (both marine and freshwater) appeared. This is the first period for which there is strong evidence of life on land, including fossils of organisms related to modern-day spiders, centipedes, and leafless land plants.

Blue Gum Pictures/Alamy

FIGURE 13.21 Oxidized iron accumulated on the seafloor in the Proterozoic as microscopic marine plants produced Earth's first abundant atmospheric oxygen gas.

Q EXPAND YOUR THINKING

Why does the Moon's surface still consist of Archean rock?

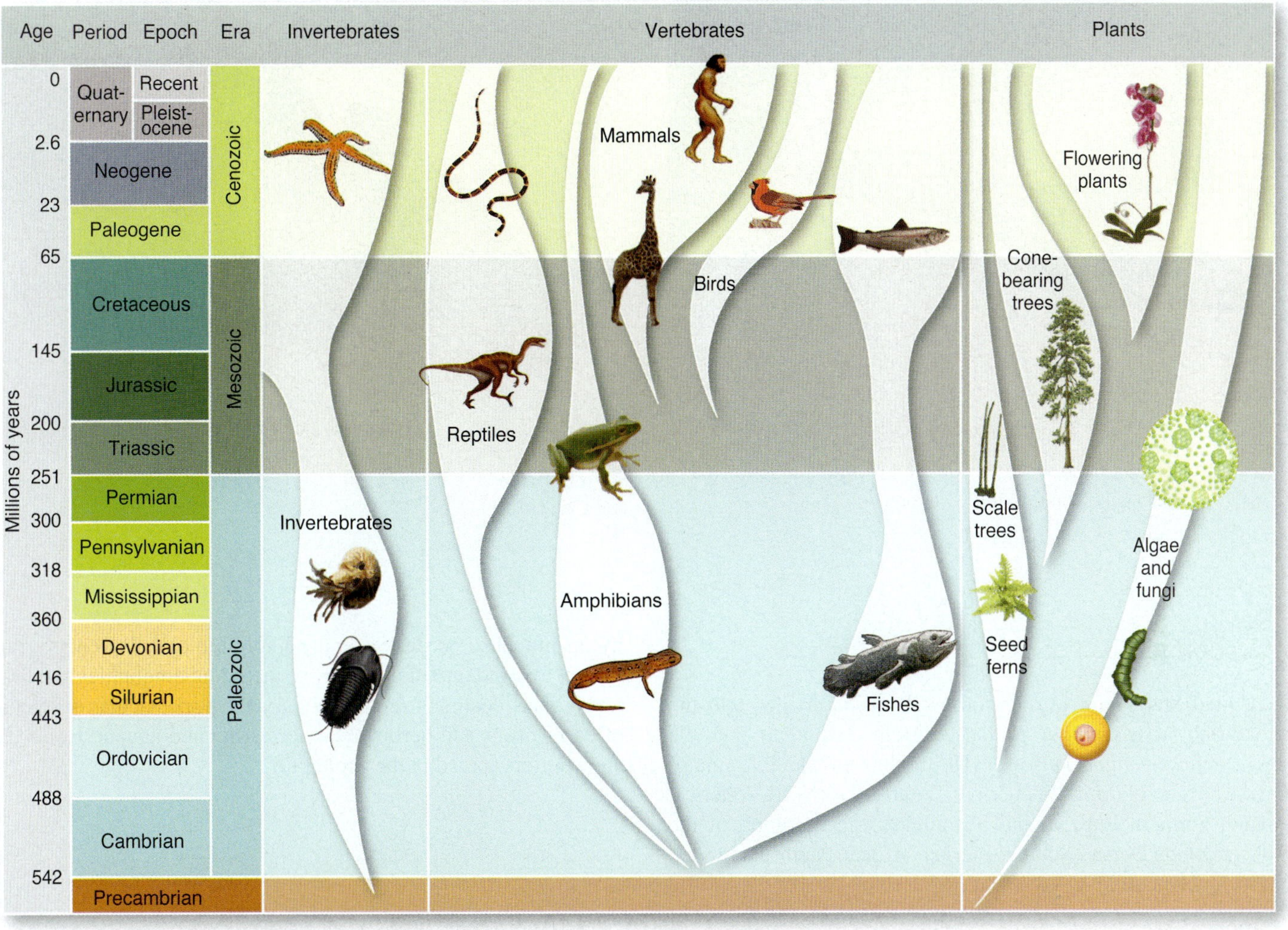

Age	Period	Epoch	Era	Invertebrates	Vertebrates	Plants

FIGURE 13.22 Following the Precambrian Eon, the Paleozoic Era witnessed an explosion of life-forms. Marine and continental life diversified during this era.

What important change in the fossil record marks the transition from the Precambrian Eon to the Phanerozoic Eon? Why is it notable?

The Devonian Period

The *Devonian Period* (416 to 360 mya) is known as the Age of Fish because abundant forms of armored fish, lungfish, and sharks are found preserved in rock layers from that time. The global ozone layer was formed (composed of the oxygen molecule O_3, which blocks harmful solar radiation), making possible the spread of air-breathing organisms (spiders, mites, and other insects) on the land. Fossils of the first transitional forms of life emerging from the water onto the land (ancestors of amphibians) are found in rocks dating from the end of the Devonian (**FIGURE 13.24**).

(a)

(b)

Publiphoto/Photo Researchers, Inc.

© Kevin Schafer/Alamy

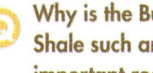

FIGURE 13.23
(a) The oldest community of multicelled organisms, found in the Burgess Shale, is estimated to be 542 to 535 mya. (b) The predatory *Anomalocaris* had a circular mouth ringed with fangs.

Why is the Burgess Shale such an important rock unit?

(a) Science Source/National Science Foundation/Photo Researchers   (b)   Ted Daeschler/National Science Foundation, Courtesy of VIREO

FIGURE 13.24 *Tiktaalik* roseae was a transitional organism between fish (with fins) and amphibians (with legs): (a) artist's depiction and (b) fossil form.

The Carboniferous Period

In the United States, the *Carboniferous Period* is divided into the limestone-rich *Mississippian Period* (360 to 318 mya) and the coal-rich *Pennsylvanian Period* (318 to 300 mya). Two major landmasses existed during this period: Euramerica (North America, Greenland, northern Europe, and Scandinavia) to the north of the equator, and Gondwana (South America, Africa, peninsular India, Australia, and Antarctica) to the south.

The Permian Period

The supercontinent Pangaea formed during the *Permian Period* (300 to 251 mya). It reached nearly from pole to pole and was surrounded by an immense ocean. Several notable biological events occurred: Insects evolved into modern forms, including dragonflies and beetles; amphibians declined; reptiles underwent a spectacular development of carnivorous and herbivorous, terrestrial and aquatic forms. Ferns and conifers spread in the cooler air.

Q EXPAND YOUR THINKING

Formulate a hypothesis that explains the development of shells and external hard parts among animals. How would you test your hypothesis?

13-9 In the Mesozoic Era, Biological Diversity Increased and Continents Reorganized

LO 13-9 Compare and contrast the three periods of the Mesozoic Era.

The **Mesozoic Era** (251 to 65 mya), composed of the *Triassic, Jurassic,* and *Cretaceous* periods, was characterized by the breakup of Pangaea and the formation of the Atlantic Ocean in the space created by the separation of North Africa, Europe, and North America. The Mid-Atlantic Ridge, characterized by plate divergence and seafloor spreading, was born at this time.

General Patterns

As North America moved westward, the leading (western) edge of the North American Plate collided with island arcs, continental fragments, and basalt crust, to build the Cordilleran thrust-and-fold belt and produce vigorous arc volcanism. This process continues today at the Cascadia Subduction Zone.

New groups of invertebrates appeared, including new forms of many established organisms such as marine corals, mollusks, and echinoderms. Land plants flourished, and modern types of seed-bearing trees and flowering plants emerged. Dinosaurs, mammals, and birds made their appearance, with the dinosaurs becoming dominant.

The end of the Mesozoic was marked by the Cretaceous–Paleogene extinction, a major episode of extinction that is widely thought to have been caused by a meteorite impact; this event defines the boundary between the Cretaceous and Paleogene periods.

The Triassic Period

The Triassic (251 to 200 mya), unlike previous periods, witnessed few significant tectonic events. Ammonites, now-extinct cephalopods (marine mollusks) with a chambered shell, evolved into diverse forms. Reptiles experienced an explosion of diversity that produced the dinosaurs, including marine reptiles such as ichthyosaurs and plesiosaurs.

The first mammals evolved from mammal-like reptiles. Pangaea covered nearly one-quarter of Earth's surface, then slowly began to break apart when continental rifting became widespread toward the end of the Triassic. The general climate was warm, becoming semi-arid to arid in continental areas.

The Jurassic Period

Named for the Jura Mountains on the border between France and Switzerland, where rocks of this age were first studied, the Jurassic Period (200 to 145 mya) became well known to the general public following the success of the movie *Jurassic Park*, in 1993. Its Hollywood celebrity aside, the Jurassic remains relevant to us today for two important reasons: its wealth of fossils and its economic importance. The oil fields of the North Sea, for instance, are Jurassic in age.

The Atlantic Ocean began to appear as North America separated from Africa and South America (**FIGURE 13.25**). Plate subduction along western North America caused the crust to fold, creating mountains in the western part of the continent. Reptiles adapted to the sea, air, and land. Dinosaurs were the dominant animal form on land. *Archaeopteryx*, the first bird, appeared, along with early amphibians, which were succeeded by the first frogs, toads, and salamanders. Mammals, small, shrewlike animals, were relegated to a minor role in the ecosystem. Forests of conifers and ginkgos became widespread. The seas and lands thrived with life-forms, most of which no longer exist.

The Cretaceous Period

During the Cretaceous Period (145 to 65 mya), the continents were shaped much as they are today. South America separated from Africa; the North Atlantic widened; and the *Tethys Sea* wrapped the globe at the equator. The westward motion of North America raised the ancestral Rocky Mountains and the Sierra Nevada. Sea levels rose, submerging 30% of Earth's current land surface (**FIGURE 13.26**).

Climates were generally warm, and the poles were free of ice. Dinosaurs became the dominant vertebrate life-form on Earth, extending their range across every continent. Gastropods (mollusks such as snails and slugs), corals, and sea urchins flourished. Early flowering plants evolved, including modern trees and modern insects. At the end of the Cretaceous, a mass extinction wiped out five major reptilian groups: dinosaurs, pterosaurs, ichthyosaurs, plesiosaurs (**FIGURE 13.27**), and mosasaurs.

Extinctions also occurred among ammonites, corals, and other invertebrates. Mammals, on the other hand, not only survived but flourished in the newly opened environments. The end of the Cretaceous brought the demise of many previously successful and diverse groups of organisms, such as nonavian dinosaurs and ammonites. This made it possible for groups that had previously played secondary roles to come to the forefront.

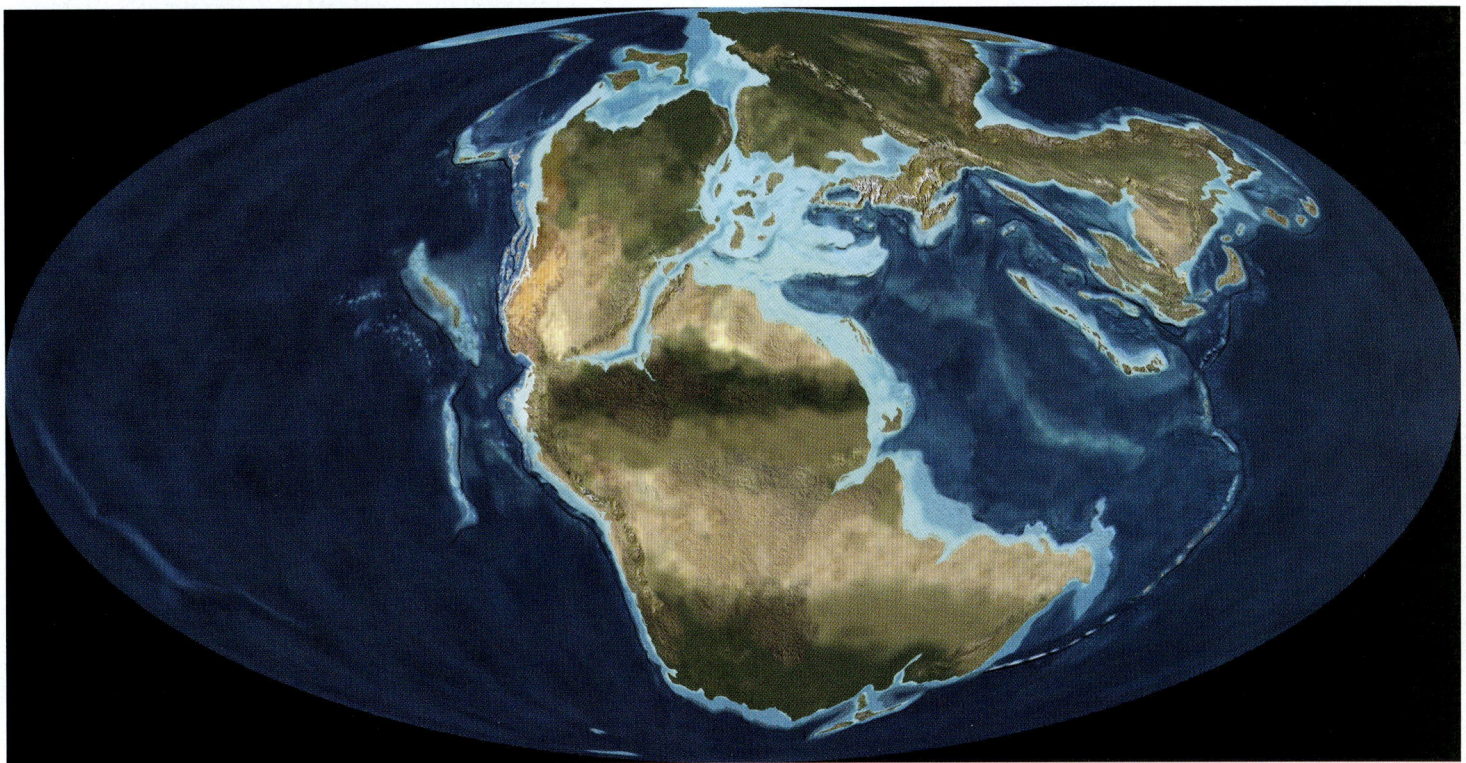

FIGURE 13.25 In the Early Jurassic, approximately 190 mya, the Atlantic Ocean was born as North America separated from Africa, Europe, and South America.

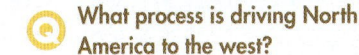 **What process is driving North America to the west?**

Ronald C. Blakey, Colorado Plateau Geosystems, Inc.

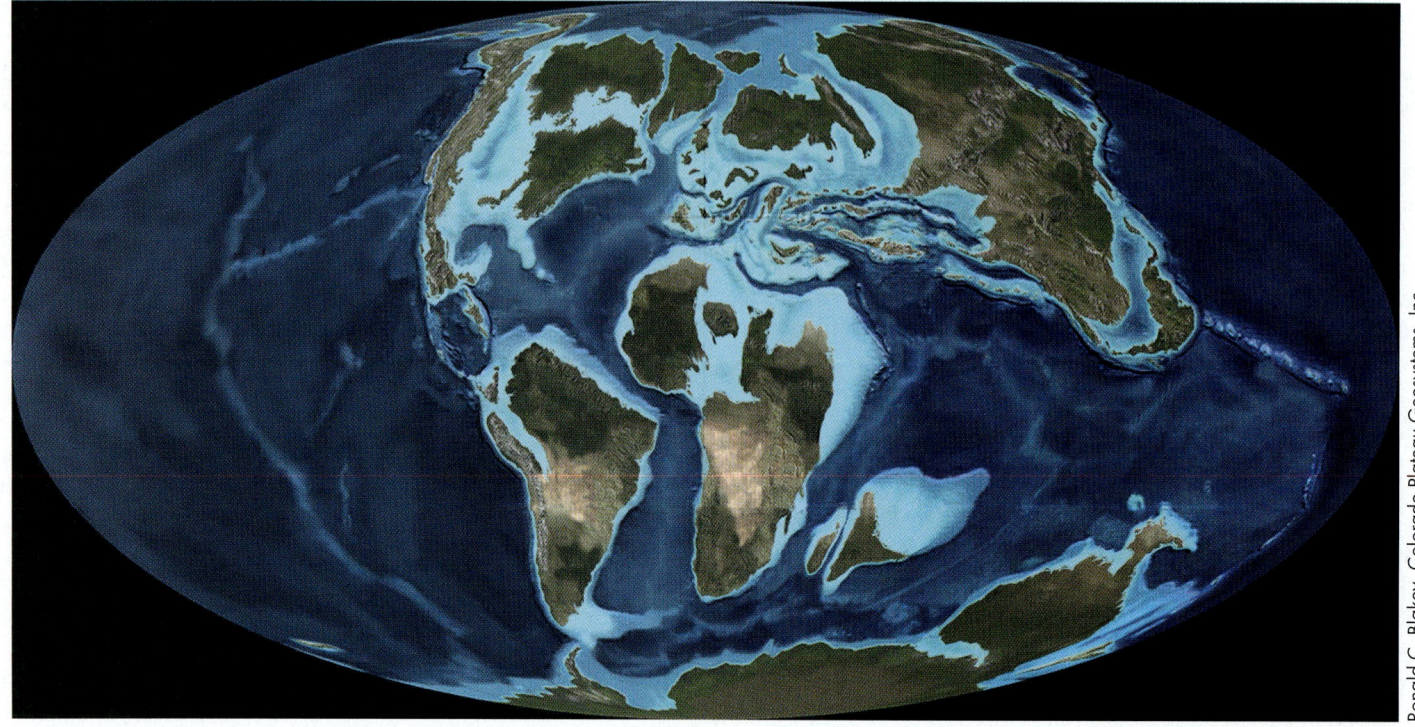

Ronald C. Blakey, Colorado Plateau Geosystems, Inc.

FIGURE 13.26 By the Cretaceous Period, approximately 90 mya, the continents had taken on their current configurations but were still moving to modern locations, largely due to continued widening of the Atlantic Ocean. Sea levels were higher then than they are today—approximately 30% of the present-day land surface was flooded. India, Australia, and much of the Middle East and southern Asia were still being formed.

 **Why was sea level much higher in the Cretaceous than it is today?**

Challenge yourself with the exercise in the Critical Thinking feature titled "Reconstructing Earth's History" to improve your understanding of Earth's history.

Christian Darkin/Photo Researchers

FIGURE 13.27 At the end of the Cretaceous, most marine reptiles, such as this plesiosaur, became extinct.

Q EXPAND YOUR THINKING

How might life today be different if the Cretaceous–Paleogene extinction had not happened?

13-10 Modern Humans Arose in the Cenozoic Era

LO 13-10 Recount the major events of the Cenozoic Era.

The **Cenozoic Era** (65 mya to the present) is relatively short compared to previous eras. During this time, the Atlantic Ocean continued to widen and push the North American Plate to the west as the mid-ocean ridge system laid new seafloor. This process activated additional processes, including subduction, mountain building, earthquakes, and igneous activity along the western coast of North America, all related to the continent's westward movement.

General Patterns

The end of the Cretaceous brought the extinction of many previously successful and diverse groups of organisms that had dominated marine and terrestrial environments. Consequently, niches in the food chain that previously had been held by Cretaceous life-forms opened up. New access to scarce resources made it possible for groups that previously had played secondary roles, such as birds, fish, and mammals, to move up the food chain and expand their diversity.

Modern groups of invertebrates appeared in the oceans, including the most common mollusks, echinoderms (e.g., starfishes, sea urchins), and crustaceans. Fish became abundant, flowering plants diversified, and the first grasses appeared. Birds and then mammals underwent major increases in diversity. Humans appeared approximately 4 mya—though it would be some time yet before "modern" humans appeared, about 200,000 years ago.

The Paleogene Period

The *Paleogene Period* (65 to 23 mya) consists of the Paleocene, Eocene, and Oligocene epochs. The period follows immediately after the Cretaceous extinction, which spelled the end of the dinosaurs and several other animal groups.

Lasting 42 million years, the Paleogene is notable for the evolution and diversification experienced by mammals, which evolved from relatively small, simple forms into a group of highly diverse animals. Mammals developed large, dominant forms specialized to marine, terrestrial, and airborne environments. The Paleogene was also a period in which birds underwent considerable change, to evolve into approximately their modern forms.

After the extinction of dinosaurs, mammals branched out into newly opened ecological niches. Cetaceans (baleen whales, toothed whales, dolphins) evolved from terrestrial meat-eating hoofed animals, as indicated by the presence of vestigial hips and femurs in modern whales and the discovery of intermediate fossil forms of whales. The first primates appeared in the early Paleogene, one of the first groups of mammals to have a placenta. Monkeys and the first true apes appeared more than 25 mya.

Early horses, goats, pigs, deer, primates, rodents, and carnivores also emerged. Plate tectonics and volcanic activity formed the Rocky Mountains in western North America, and collisions between the Indian and Asian Plates raised the Alpine–Himalayan mountain system. Antarctica and Australia separated and drifted apart, and Greenland split from North America.

Climates were subtropical and moist throughout North America and Europe. Continents converged to become the Middle East, and active volcanism characterized Central America. Spreading grasslands replaced forests over large areas on several continents, setting the stage for the evolution of modern species of horses.

The Neogene Period

The *Neogene Period* extends from 23 mya to the beginning of the *Quaternary Period* approximately 2.6 mya. The Neogene consists of the Miocene and Pliocene epochs. The climate cooled over the Neogene Period, although the exact causes are not well understood. One idea, the *uplift weathering hypothesis* (discussed in Chapter 7), suggests that a general decrease in atmospheric carbon dioxide occurred due to amplified weathering of newly exposed rock of the Himalayan fold-and-thrust belt.

The last opening between the Atlantic and Pacific oceans closed with the growth of new volcanic landforms in Central America. Closing the Pacific–Atlantic connection through Central America may have influenced global heat transport on oceanic currents, enhancing the cooling effect already underway.

The Quaternary Period

The Quaternary Period (2.6 mya to present) is notable for its series of **ice ages** and *interglacials* resulting from the advance and retreat of continental glaciers on every major landmass, except Australia.

During the last 500,000 years or so, there have been five or more separate glacial periods, occurring approximately every 100,000 years, a pattern that continues today (**FIGURE 13.28**). This pattern is related

Ronald C. Blakey, Colorado Plateau Geosystems, Inc.

FIGURE 13.28 Since late in the Quaternary Period, Earth has experienced climate shifts from ice ages to interglacials approximately every 100,000 years. The most recent ice age culminated 20,000 to 25,000 years ago with as much as 30% of Earth's land surface covered by glaciers.

How is Earth different during an ice age?

CRITICAL THINKING

Reconstructing Earth's History

Working with a partner and using the global reconstructions in **FIGURE 13.29** as a guide, complete the following exercise:

1. Describe the changes in continents that led to their modern configuration.

2. Speculate a bit and propose a hypothesis that describes a general trend followed by life during the course of evolution from the Archean to the late Quaternary (see next section).

3. How would the construction of a super-continent such as Pangaea contribute to the occurrence of a major extinction?

4. What general characteristics of life mark the Proterozoic and Phanerozoic eons?

5. How old is the Atlantic Ocean? How was it formed? What evidence would you look for to support your answers?

6. Summarize general trends in Earth's history in terms of the atmosphere, living organisms, and the crust.

7. What evidence would you look for to test the hypothesis that a supercontinent (Pangaea) existed circa 255 mya?

8. Construct an alternative hypothesis for the extinction of the dinosaurs. Identify the evidence you would need to support it.

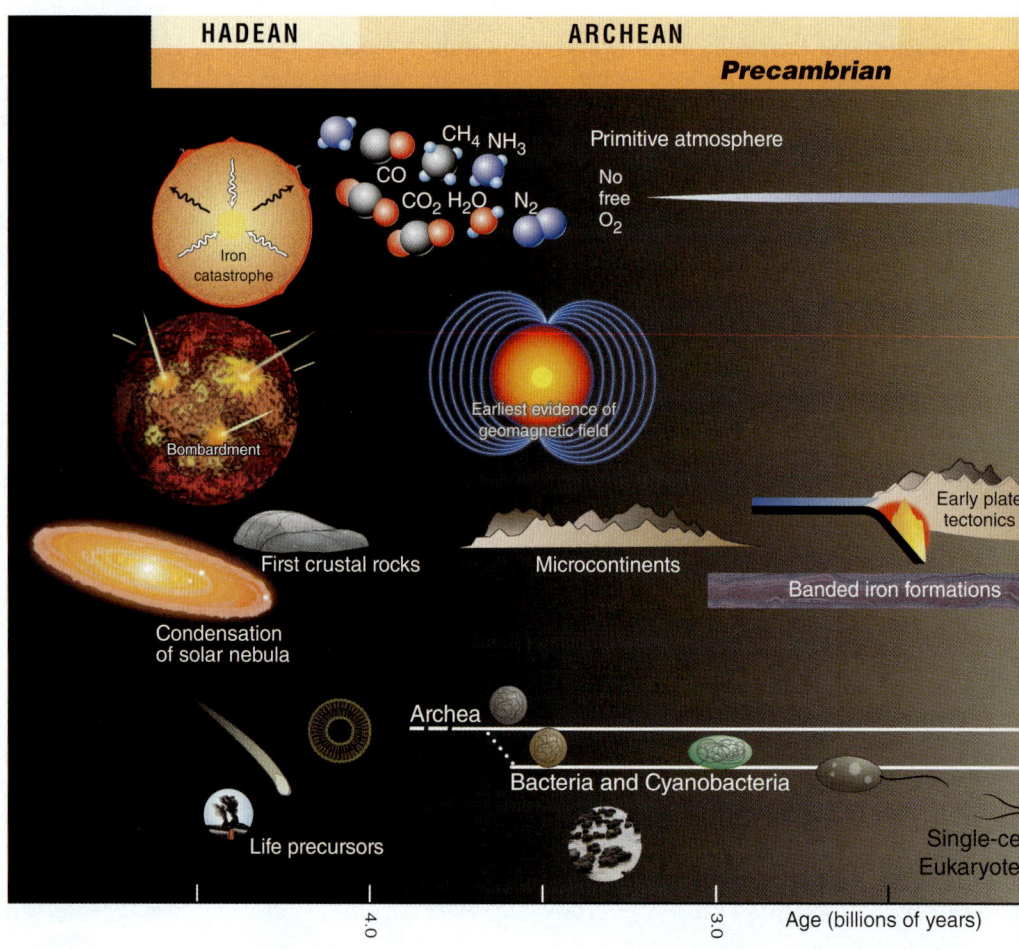

FIGURE 13.29 Major events in Earth history including: development of the atmosphere, mountain-building, plate tectonics, and evolution of living organisms.

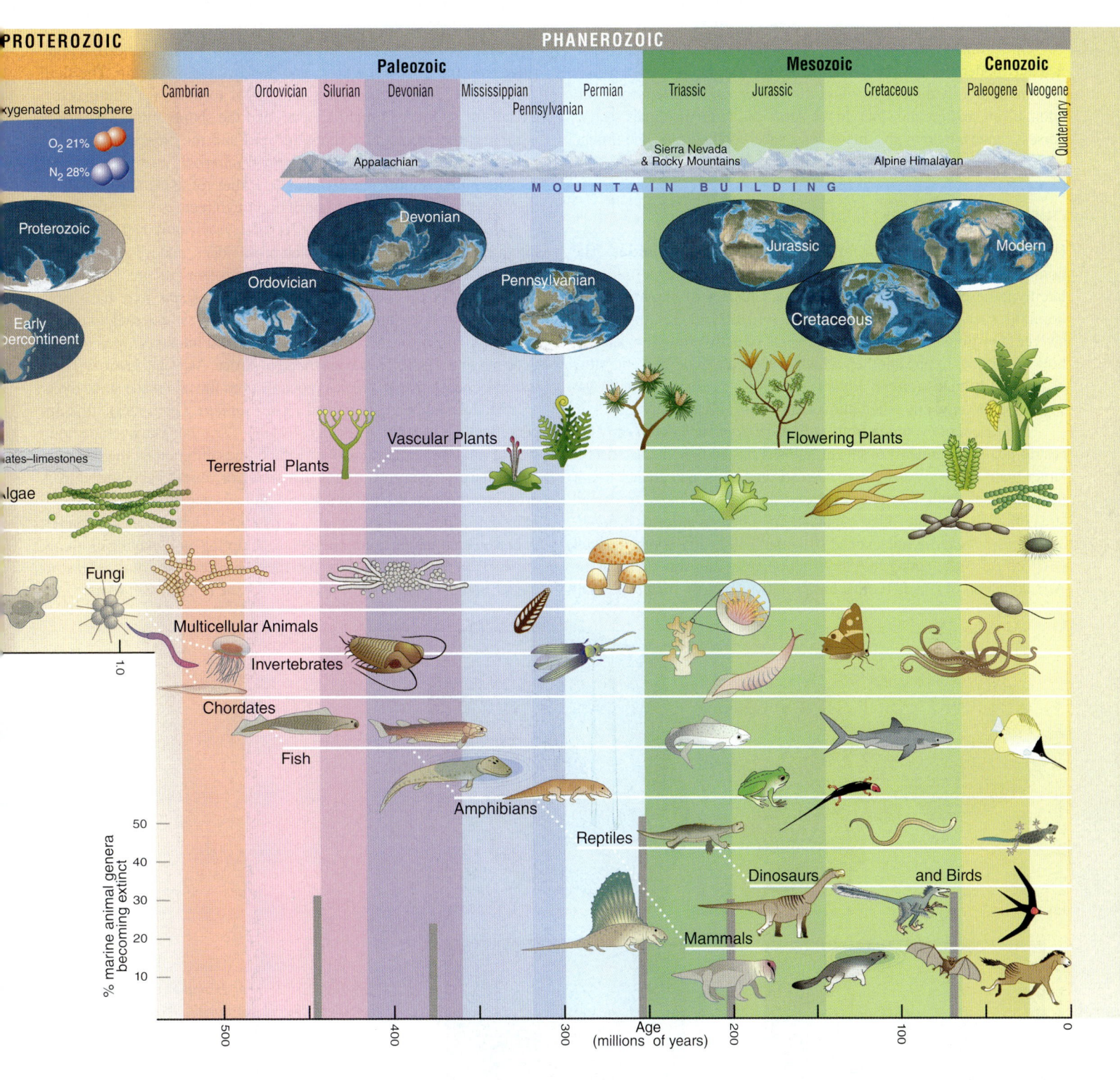

to variations in Earth's orbit around the Sun, which introduced changes in the amount and location of solar radiation reaching the planet's surface (discussed in Chapters 14 and 15). During ice ages, as much as 30% of Earth's surface is covered by glaciers, which distinctly alter the landscape. Lowered sea levels expose land bridges between Siberia and North America and among islands in southwestern Asia.

One probable human ancestor, *Homo habilis,* evolved and diversified into modern forms. Mammalian evolution featured species of woolly mammoth, woolly rhinoceros, musk ox, moose, reindeer, elephant, mastodon, bison, and ground sloth. In the Americas, large mammals such as horses, camels, mammoths, mastodons, saber-toothed cats, and ground sloths were extinct by the end of the final ice advance 20,000 years ago.

Human Origins

Humans are members of a family known as the **Hominidae.** The hominid fossil record is fragmentary, but enough is known to formulate hypotheses outlining the evolutionary history of humans. Many hominid species have been identified (**FIGURE 13.30**), but unanswered questions remain and are subjects of active research, such as how they relate to one another, which species are the direct ancestors of modern humans (*Homo sapiens sapiens*, a subspecies of *Homo sapiens*), and how and when human ancestors and apes diverged.

The record of evolution from early *Australopithecines* (an early ancestral genus to humans) to recent humans has certain characteristics: larger brain and body size; more sophisticated tools, and greater use of them; smaller tooth size; and decreased skeletal robustness, or overall strength. Currently, the oldest candidate for a human ancestor is *Sahelanthropus tchadensis* (nicknamed "Toumai"). A nearly complete skull (**FIGURE 13.30A**) dating to 6 to 7 million years ago was found in the African nation of Chad. Toumai has primitive features, such as a small brain (350 cubic centimeters) and pronounced brow ridges. These, plus its age, suggest that it is close to the common ancestor of humans and chimpanzees. However, until additional fossils from the same period are found, *paleontologists* (geologists who study past life) will not be able to determine whether Toumai represents an early ape or an early human.

Australopithecus afarensis lived between 3.9 and 3.0 mya (**FIGURE 13.30B**). It had an apelike face and a low forehead, a bony ridge over the eyes, a flat nose, and poorly developed chin. The skull (375 to 550 cubic centimeters) is similar to that of a chimpanzee, except for its more humanlike teeth and the shape of the jaw, which combines the rectangular shape of apes and the parabolic shape of humans. The pelvis and leg bones of *A. afarensis* resemble those of modern humans more closely than do earlier forms, leaving no doubt that these individuals walked upright on two feet.

Homo erectus (1.8 mya to 300,000 years ago) had a protruding jaw, large molars, no chin, thick brow ridges, and a long low skull (**FIGURE 13.30C**). Its brain size varied between 750 and 1,225 cubic centimeters. Fossil skeletons indicate that *H. erectus* may have been more efficient at walking than are modern humans, whose skeletons have adapted to allow for the birth of larger-brained infants. *H. erectus* was wide ranging; fossils have been found in Africa, Asia, and Europe. There is also evidence that *H. erectus* probably used fire and made stone tools.

Modern humans are named *Homo sapiens.* Our species first appeared about 200,000 years ago (**FIGURE 13.30D**). The average brain size of modern humans is about 1350 cubic centimeters; the forehead rises sharply; eyebrow ridges are small or absent; and the chin is prominent. About 40,000 years ago, humans began making more sophisticated tools, using a variety of materials, such as bone and antler, to produce implements for engraving, sculpting, and making clothing. Artwork, in the form of beads, carvings, clay figurines, musical instruments, and cave paintings, appeared over the next 20,000 years.

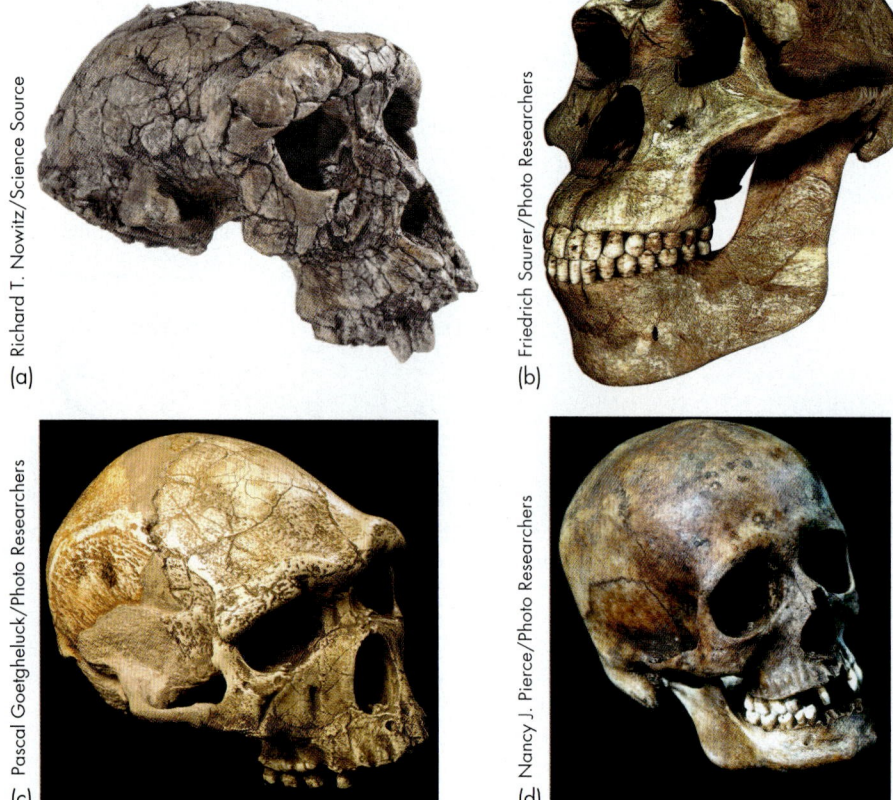

(a) Richard T. Nowitz/Science Source

(b) Friedrich Saurer/Photo Researchers

(c) Pascal Goetgheluck/Photo Researchers

(d) Nancy J. Pierce/Photo Researchers

FIGURE 13.30 Four examples of probable human ancestors: (a) *Sahelanthropus tchadensis,* 6 to 7 million years old; (b) *Australopithecus afarensis,* 3 to 4 million years old; (c) *Homo erectus,* 300,000 to 1.8 million years old; (d) *Homo sapiens sapiens,* modern human.

What looks different about the skull of *Homo sapiens sapiens* compared to the skulls of our ancestors?

LET'S REVIEW

As recently as 100 years ago, European cultures thought Earth was a young, essentially unchanging ball of rock. Then, as the ideas of early geologists such as Steno, Hutton, Lyell, Darwin, and others became more widely accepted, people came to understand that Earth is, in fact, a restless and evolving system with a long history characterized by perpetual change. The history of Earth can be described as the chemical and physical reorganization of the planet's interior, atmosphere, and oceans and the evolutionary change of life-forms.

The Precambrian, a wide expanse of time consisting of the first 88% of Earth's history, was a period of slow oxygenation of the atmosphere and ocean, the formation of an organized crust consisting of continents and ocean basins, and the early development of life. Since then, evolution has led to a perpetual increase in the diversity and complexity of life, punctuated by occasional, but significant, mass extinctions.

Geologic time is organized along the geologic time scale, the calendar of Earth's history, categorized in 4 eons, 3 eras, and 13 periods. Each of these witnessed special and distinctive episodes of evolving life, mountain building, and plate tectonics reorganization that shaped the character of our modern world.

STUDY GUIDE

13-1 Earth's history has been unveiled by scientists applying the tools of critical thinking.

- Earth's history is characterized by deep time and great change. The fossil record provides strong evidence of the evolution of life, the dynamic history of the crust, the interconnectedness of all living things, as well as the changing environmental conditions that led to mass extinctions on several occasions.

- The science of geology owes its origin to several visionary men who recognized that Earth was a complex system with a history marked by perpetual change. **Nicholas Steno** was the first to suggest that fossils that looked like living organisms had in fact once been living organisms. Steno also developed the principles of original horizontality, superposition, and lateral continuity. James Hutton proposed that geologic time was infinitely long and that Earth was like a self-renewing machine: As mountains eroded away, new ones were uplifted; as the sea covered some lands, it receded from others. **Charles Lyell**, often referred to as the father of modern geology, was able to craft sensible hypotheses from the observations of scientists of his day. Finally, **Charles Darwin** is credited with formulating the theory of **evolution**, which states that all living things develop from a very few simple forms over time through **natural selection**. Natural selection is the process by which favorable traits that are heritable (passed on to offspring) become more common in successive generations of a population of reproducing organisms.

13-2 Fossils preserve a record for past life.

- **Fossils** are the remains of animals and plants, or the record of their presence, preserved in the rocks of Earth. Fossilization is the process that turns a once-living creature into a fossil. There are many fossils to be found, even though only a tiny portion of all the animals and plants that ever lived have been fossilized.

- Usually only the hard parts of an animal, such as teeth, bones, and shells, become fossilized. These may be preserved in three ways: replacement, formation of a mold, or formation of a cast.

13-3 There are several lines of evidence for evolution.

- Fossils provide evidence of evolution. Consider that nearly 35 now-extinct ancestral species make up the family tree, or **phylogeny**, of the modern horse.

Consistent with Darwin's theory, natural changes in the environment through time exerted selective pressure on early communities of horse ancestors so that populations with certain characteristics survived at higher rates.

- The bone structure of such diverse organisms as birds, swimming mammals, four-legged animals, humans, and insect-eating reptiles also is evidence of evolution. This pattern of **homologous structures** suggests that these animals have evolved from a common ancestor and that natural selection preferentially selects individuals with the specific traits and functions served by each limb.

- Vestigial structures are body parts that were used by ancestral forms but are now nonessential. Selective pressure for completely eliminating vestigial parts is weak once they have been replaced with newer forms, so they tend to be retained in organisms for a long time.

- Evidence of evolution is also present in studies of embryology: in their early stages of development, embryos of various fish, birds, and mammals display strikingly similar characteristics.

13-4 Molecular biology provides evidence of evolution.

- Unique characteristics may arise among separate populations of the same species, causing them to become new species. These differences are the product of **genetic mutation** and **genetic variation** that are naturally selected as a consequence of environmental pressure or community pressure.

- Molecular biologists document the relationship of living organisms to their ancient ancestors. Three branches on the tree of living relationships describe all life-forms: eukaryotes, archaea, and bacteria.

- Modern medical treatments are based on an understanding that various parasitic microbes, viruses, and bacteria evolve, such that they can "escape"— treatment.

13-5 Mass extinctions influence the evolution of life.

- The history of life is characterized by five major **mass extinctions**: Ordovician–Silurian (440–450 mya), Late Devonian (360–375 mya), Permian–Triassic (251 mya), Triassic–Jurassic (200 mya), and Cretaceous–Paleogene (65 mya). These events, which led to the extinction of large percentages of all living species, exerted a powerful influence on the direction of animal and plant evolution for the species that survived.

- Mass extinctions do not all occur for the same reasons. Various hypotheses have been forwarded for each one, and they remain subjects of intense debate. In general, however, extinctions are thought to result from drastic global changes that follow catastrophic events, such as meteorite or comet impacts or massive volcanic eruptions.

13-6 The geologic time scale is the calendar of events in Earth's history.

- The **geologic time scale** subdivides all time since the beginning of Earth's history (4.6 bya) into named units of variable length: **epochs**, **periods**, **eras**, and **eons**.

- Fossils found in strata provide the chief means of establishing the geologic time scale. For instance, the appearance or disappearance of certain fossils may mark the end of one time unit and the beginning of another—which is why extinctions play an important role in defining geologic time units. The principle of fossil succession tells us that specific kinds of organisms are characteristic of certain times in Earth's history and particular parts of the geologic record. By correlating strata in which certain types of fossils are found, scientists can reconstruct the geologic history of various regions and of Earth as a whole.

- Hadean time corresponds to the period between Earth's formation, 4.6 bya, to the start of the Archean Eon, 3.8 bya. Few Hadean rocks survive, but based on theoretical arguments, the period was characterized by an initially cool Earth that gradually was heated past the melting point of iron and so developed a molten surface. This event was followed by cooling and the establishment of the first hard crust and very early oceans.

13-7 The Archean and Proterozoic eons lasted from 3.8 billion to 542 million years ago.

- The atmosphere during the Archean Eon was composed of unbreathable gases: methane, ammonia, carbon dioxide, and water vapor. These came from extraterrestrial impacts by comets and asteroids, as well as volcanic outgassing from the hot planet interior. The Proterozoic Eon saw the development and growth of the continents (a process that took about 1 billion years), the emergence of a new atmosphere rich in oxygen, and the building of new mountain ranges.

13-8 In the Paleozoic Era, complex life emerged and the continents reorganized.

- The Phanerozoic Eon is recorded in much greater detail by rocks and fossils than earlier eons. It contains information for reconstructing events in shorter time intervals consisting of the **Paleozoic**, **Mesozoic**, and **Cenozoic** eras, each of which comprise several periods of variable length.

- The Cambrian Period is the time when most of the major groups of animals first appear in the fossil record. Evolution was focused on marine invertebrates and fish. The Silurian Period witnessed the first coral reefs, the evolution of fishes, the first evidence of life on land, including relatives of spiders and centipedes, and the earliest fossils of vascular plants. The Devonian Period is known as the Age of Fish because various forms of armored fish, lungfish, and sharks are found preserved in rock layers from the time. During this period, the global ozone layer formed, allowing the spread of air-breathing spiders and mites on land. The Carboniferous Period witnessed the convergence of major continental areas and the flourishing of insects and reptiles. In the Permian Period, the supercontinent Pangaea formed, surrounded by an immense world ocean.

13-9 In the Mesozoic Era, biological diversity increased and continents reorganized.

- The **Mesozoic Era** is characterized by the breakup of Pangaea and the formation of the Atlantic Ocean.

- The Triassic Period had few significant geologic events. The reptiles experienced an explosion of diversity, producing the dinosaurs, including marine reptiles, such as ichthyosaurs and plesiosaurs. The first mammals evolved from mammal-like reptiles. In the Jurassic Period, reptiles adapted to the sea, the air, and the land. Dinosaurs were the dominant animal form. Mammals played a minor role in the ecosystem, in the form of small, shrewlike animals. Conifers and gingkoes became widespread. During the Cretaceous Period, the continents were shaped much as they are today. The Tethys Sea wrapped the globe at the equator, and the westward motion of North America raised ancestral Rocky Mountains and the Sierra Nevada.

13-10 Modern humans arose in the Cenozoic Era.

- The Cenozoic is relatively short compared to previous eras. During this time, the Atlantic Ocean continued to widen and push North America to the west as the mid-ocean ridge system laid new seafloor. After the extinction of dinosaurs, mammals expanded into newly opened ecological niches in the Paleogene Period. The climate cooled over the Neogene Period, although the exact causes are not well understood. In the Neogene, the last opening between the Atlantic and Pacific oceans closed, with volcanism in Central America uniting North and South America.

- During the Quaternary Period, Earth entered a series of **ice ages**. The fossil record of human evolution is characterized by larger brain and body size; more sophisticated tools, and greater use of them; smaller tooth size; and decreasing skeletal robustness, or overall strength.

KEY TERMS

ASSESSING YOUR KNOWLEDGE

Please complete the following exercise before coming to class. Identify the one best answer to each question.

1. Nicholas Steno was the originator of:
   a. Igneous geology.
   b. Stratigraphy.
   c. Metamorphism.
   d. Planetary geology.
   e. Radiometric dating.

2. The term *fossil* refers to:
   a. Any evidence of evolution.
   b. Any evidence of changes in species over time.
   c. Any trace of soft parts of organisms.
   d. Any evidence of past life on Earth.
   e. None of the above.

3. This fossilized bird and frog retain many of their soft parts, such as skin, and feathers. The fossils were found in shale beds near Darmstadt, Germany. Please answer the following questions.
   a. What aspect of shale depositional conditions promotes excellent preservation? (high energy, well oxygenated, coarse-grained sediments, low-energy conditions)
   b. The age of these rocks is probably _____. (Archean, Proterozoic, Paleozoic, Mesozoic, Cenozoic)
   c. The preservation of soft tissues indicates _____. (these animals were buried slowly, predators were abundant in the region, these animals were buried rapidly, there was little to no compaction)

Jonathan Blair/NG Image Collection

© Jonathan W. Blair/Corbis

4. Evidence for evolution includes:
   a. Vestigial structures, homologous structures, and phylogeny.
   b. Phylogeny, intrusion, and atmospheric oxidation.
   c. Vestigial structures, population growth, and global climate change.
   d. Phylogeny and fossil fragments remaining after mass extinctions.
   e. Written accounts.

5. Natural selection is:
   a. The tendency of populations with favorable variations to survive.
   b. The tendency of weaker individuals to produce more offspring.
   c. The tendency of a species to improve over time through random mating.

d. The tendency of weak species to dominate over stronger individuals.
e. The tendency of life to develop machines.

6. Evolution is marked by:
   a. Changes in the inherited traits of a population from one generation to the next.
   b. Changes in physical traits due to individual effort.
   c. Genetic variations that are not passed on to future generations.
   d. Emergence of stronger individuals as dominant in a community.
   e. Divine intervention.

7. Mass extinctions are important because they:
   a. Reduce competition and allow rapid evolution of surviving species.
   b. Generally cause the death of carnivorous species.
   c. Stop evolution because so many species remain.
   d. Lead to the formation of large landmasses.
   e. Close the Tethys Seaway.

8. The longest episode in Earth's history is the:
   a. Cenozoic.
   b. Paleozoic.
   c. Proterozoic.
   d. Quaternary.
   e. Holocene.

9. The Age of Fish was the:
   a. Precambrian.
   b. Silurian.
   c. Mesozoic.
   d. Devonian.
   e. Quaternary.

10. This 2-meter-long fossilized dinosaur (Oviraptor) was found curled around a nest containing at least 20 eggs. This is strong evidence that _____ (dinosaurs cared for their young, dinosaurs ate their young, dinosaurs did not build nests, dinosaurs were solitary individuals).

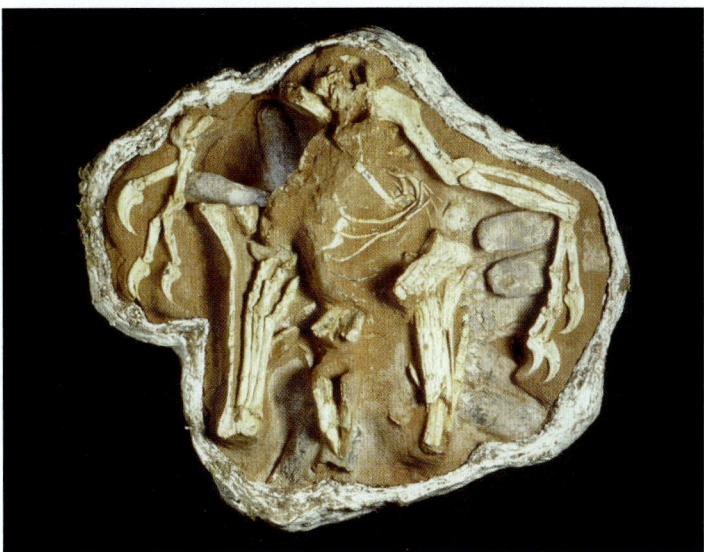

© Associated Press

11. What characteristics of life on Earth mark the Archean eon?
    a. Life was highly diverse.
    b. Life was confined to land alone.
    c. Life was characterized by increasing diversity among plants and insects.
    d. Life was characterized by simple forms, such as stromatolites.
    e. Mammals increased in diversity.

12. A major development at the end of the Precambrian was the:
    a. Evolution of birds on land.
    b. Emergence of the first complex life-forms.
    c. Cooling of the crust.
    d. Appearance of whales and other marine mammals.
    e. First water on Earth.

13. Cratons are:
    a. Regions of excessive volcanic activity.
    b. Precambrian rock at the core of most continents.
    c. Systems of mountain ranges formed by tectonic convergence.
    d. Heavily metamorphosed roots of young divergent zones.
    e. Island arcs and hotspots.

14. Following the break-up of Pangaea, much of the resulting continental reorganization was due to _____ (widening of the Pacific Ocean, break-up of Europe into smaller plates, widening of the Atlantic Ocean, closure of the Central American Isthmus).

Ronald C. Blakey, Colorado Plateau Geosystems, Inc.

15. The phylogeny of the horse shows:
    a. Decreasing size, less strength, and slower speed over time.
    b. Lengthening of the forelegs to adapt to life in forests.
    c. Increasing size, greater strength, and greater speed to adapt to life in grasslands.
    d. Random changes in response to the breakup of Pangaea.
    e. Smaller, and a stealthier body arrangement.

16. During the Cenozoic:
    a. Mammals went extinct.
    b. Dinosaurs were dominant.
    c. The first plants and animals developed in the ocean.
    d. We are not sure what happened in the Cenozoic since it is the oldest era.
    e. None of the above.

17. This geologist is holding a fossil of a trilobite, one of the first animals to develop a hard covering, presumably to defend against predators. Trilobites were an extremely long-lasting and numerous class of arthropods; they are ubiquitous among Cambrian fossils. This 34-centimeter-long specimen of *Olenellus getzi* was found in Lancaster County, Pennsylvania, and is Lower Cambrian in age. Most likely, this animal was preserved as a fossil because _____ (it had no soft tissue, its color blended into the rock, there were so many of them, it had a hard covering).

Sinclair Stammers/Science Source

18. Which of these defines uniformitarianism?
    a. The past is a key to the present.
    b. Understanding the future requires understanding the present.
    c. The study of fossils in order to predict evolution.
    d. Earth's history is best explained by observations of modern processes.
    e. A description of evolution.

19. As a mechanism for evolution, natural selection requires the passing on of specific traits from one generation to the next. These specific traits may reflect _____ passed on through an organism's _____.
    a. genetic variations; metabolism
    b. genetic mutations; DNA
    c. natural selection; lifestyle
    d. metabolism; DNA
    e. phylogeny; embryology

20. During which time did human ancestors emerge?
    a. The Paleozoic
    b. The Mesozoic
    c. The Neogene
    d. The Paleogene
    e. The Archean

FURTHER RESEARCH

1. What is the name of the rock unit that contains the earliest multicellular living community?

2. What was the Tethys Sea?

3. Which continents made up Pangaea?

4. How does natural selection cause evolution?

5. Which characteristics of the human immunodeficiency virus (HIV) allow it to evolve quickly?

6. What was the dominant class of animals during the Cenozoic era?

7. How can plate tectonics influence evolution?

8. Geologists recognize that "most of the geologic time scale is blank." What does this mean, and why would this be?

ONLINE RESOURCES

Learn more about Earth's history on the following Web sites:

"Geologic Time," a U.S. Geological Survey online publication, provides a thorough introductory discussion of many aspects of geologic time: http://pubs.usgs.gov/gip/geotime

Smithsonian National Museum of Natural History offers an online tour of Earth's history at this very educational and easy-to-navigate site: http://www.nmnh.si.edu/paleo/geotime/index.htm

Additional animations, videos, and other online resources are available at this book's companion Web site: www.wiley.com/college/fletcher

This companion Web site also has more information about WileyPLUS and other Wiley teaching and learning resources.

CLIMATE CHANGE

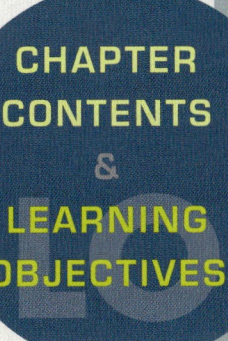

CHAPTER CONTENTS & LEARNING OBJECTIVES

GEOLOGY IN OUR LIVES

Global warming causes climate change and is leading to rising sea levels, extreme and dangerous weather patterns, and negative repercussions to ecosystems worldwide. Climate change poses a critical challenge to both our economy and our environment. The weight of scientific evidence indicates that global warming is caused by the release of greenhouse gases into the atmosphere by industrial activities. Only by limiting greenhouse gas production can we prevent the worst of the future impacts of global warming. If we humans want to preserve our planet in a state similar to that upon which civilization developed and to which life on Earth has adapted, we will have to substantially reduce greenhouse gas production. This is the pressing global challenge that the present generation of decision makers must rise to and, ultimately, meet.

Q Among scientists, there is no controversy about the cause or existence of climate change. But the public media continues to debate the subject. Why?

14-1 Climate Change Causes Shifts in Environmental Processes That Affect the Whole Earth

LO 14-1 Describe the cause of modern climate change.

There are many ways that human activities and behavior affect Earth's environment and its natural resources.

Global climate change is any large-scale shift in climate over time, whether natural or as a consequence of human activity. Changes in global climate that are occurring today are of particular importance because they are caused by human actions and they jeopardize the natural balance of Earth's oceans, land surface, and atmosphere; this then leads to large-scale transformations in ice and vegetation cover, sea level, and weather patterns— and in time, puts even human welfare at risk. Many experts believe that understanding global climate change, and taking action to manage it, is the most crucial challenge facing humanity in the twenty-first century.

Human Impacts

On a typical day we lose about 300 square kilometers of rainforest to logging (1 acre per second), 186 square kilometers of land to encroaching deserts, and numerous species to extinction. And on that same typical day, the world's human population increases by more than 200,000; we add 100 million tons of carbon dioxide to the atmosphere (**FIGURE 14.1**) and burn an average of 84.4 million barrels of oil (1000 barrels per second). By the end of that day, Earth's freshwater, soil, and ocean have become more acidic, its natural resources more depleted, and its temperature a little hotter.

These unrelenting negative effects on Earth's ecosystems and natural resources have led researchers to conclude that our planet is perched dangerously close to the edge of a tipping point, a planetary-scale critical transition set in motion by human consumption of natural resources. In a major article published in 2012 in the journal *Nature*, researchers stated, "Humans now dominate Earth, changing it in ways that threaten its ability to sustain us and other species." Scientists also warn that human population growth, widespread destruction of natural ecosystems, and climate change are pushing Earth's ecosystems and resources toward a situation that is irreversible.

The following startling facts about global environmental changes serve to underscore the seriousness of this situation:

- In the spring of 2013, the average atmospheric carbon dioxide (CO_2) concentration reached 400 parts per million (ppm), growing at an average annual rate of about 2.1 ppm.

- CO_2 levels for the past century have been at their highest Earthly concentration in approximately 15 million years, as measured in cores of glacier ice and seafloor sediments. Fifteen million years ago, sea level was 23 to 36 meters higher, and the global temperature was 2.8°C to 5.5°C warmer.

- Since 1880, the global mean annual air temperature has risen approximately 0.8°C.

- A 1°C change in atmospheric temperature caused by carbon dioxide (CO_2) will stimulate a water vapor increase, raising the temperature another 1°C. This is an example of a climate process called *positive feedback*.

- At an increase of 2°C global mean temperature, many climatologists believe that large portions of Earth's surface will become uninhabitable, due to drought, extreme and dangerous weather patterns, flooding due to several centuries of unstoppable sea level rise, lack of food, scarcity of freshwater, and heat waves.

- During the past three decades, Earth's surface temperature has trended upward about 0.2°C per decade. The average temperature in 2013 was about 14.6°C, warmer by 0.6°C than the mid-twentieth-century baseline.

- Because of climate change:

  - Glaciers are melting.

  - Air and ocean temperatures are rising.

  - The incidence of drought has increased about 10%.

  - Deadly heat waves are more frequent.

  - Snow cover is retreating.

  - Regions where water is frozen for at least one month each year are shrinking.

  - Permafrost is retreating poleward.

  - Plant and animal species are shifting poleward and to higher elevations.

  - Plants are leafing out and blooming earlier each year.

  - The tropics have expanded.

  - Atmospheric humidity is on the rise.

  - The global water cycle has accelerated.

  - Storm tracks are shifting poleward.

  - Extreme weather events are more frequent and widespread.

  - Daily record high temperatures occur twice as often as record lows.

Bruce Forster/Getty Images, Inc.

FIGURE 14.1 Natural environments sustain life on Earth. Humans, in turn, must employ critical thinking to minimize their activities and behaviors that have negative impacts on the environment.

What negative impact to the environment are you responsible for?

- Sea level is rising, and at an accelerated rate.

- Global wind speed has accelerated.

- Earth environments are changing on a global scale.

- The world's human population reached 7 billion in 2011, and is expected to reach 8 billion by 2020. At present, humanity withdraws natural resources from Earth at a rate that exceeds by about 25% the natural rate of renewal of those resources.

What do these facts have to do with physical geology? For starters, it is geologists who are experts in the natural processes that maintain the health of Earth's environment, and it is to these geologists that experts in other fields turn for factual guidance on how to manage natural resources to ensure they are not lost forever, and manage and avoid natural hazards that threaten human communities.

Global Warming

You have no doubt heard the term **global warming**, both in school and in the media. The term refers to a rise in the average temperature of Earth's surface, including the deep and shallow portions of the oceans. National Aeronautics and Space Administration (NASA) scientists report that 2013 was the seventh warmest of any year since 1880, extending a long-term trend of rising global temperatures. With the exception of 1988, the nine warmest years in the 132-year record all have occurred since 2000, with 2010 and 2005 ranking as the hottest on record.

According to published reports, researchers have concluded that Earth's average surface temperature today is most likely the highest of the past 11,000 years. They have also found that the rate of **sea level rise**, a reflection of heat stored in the ocean and ice melting on land, has accelerated, more than doubling in recent decades. Scientific evidence furthermore indicates that natural influences cannot explain the observed warming trend; rather, this evidence points to human activities as being responsible for the release of heat-trapping gases to the atmosphere that produce the warming.

Climate and Weather

It is important to distinguish between climate and weather, as many people use the terms interchangeably. *Weather* is the short-term state of the atmosphere at a given location. It affects the well-being of humans, plants, and animals, and the quality of our food and water supply. Weather is somewhat predictable thanks to our understanding of Earth's global climate patterns.

Climate is the long-term average weather pattern in a particular region, and is the product of interactions among land, ocean, atmosphere, ice, and the biosphere. Climate is described by many weather elements, among them temperature, precipitation, humidity, sunshine, and wind.

Both climate and weather result from processes that accumulate and move heat within and between the atmosphere and the oceans.

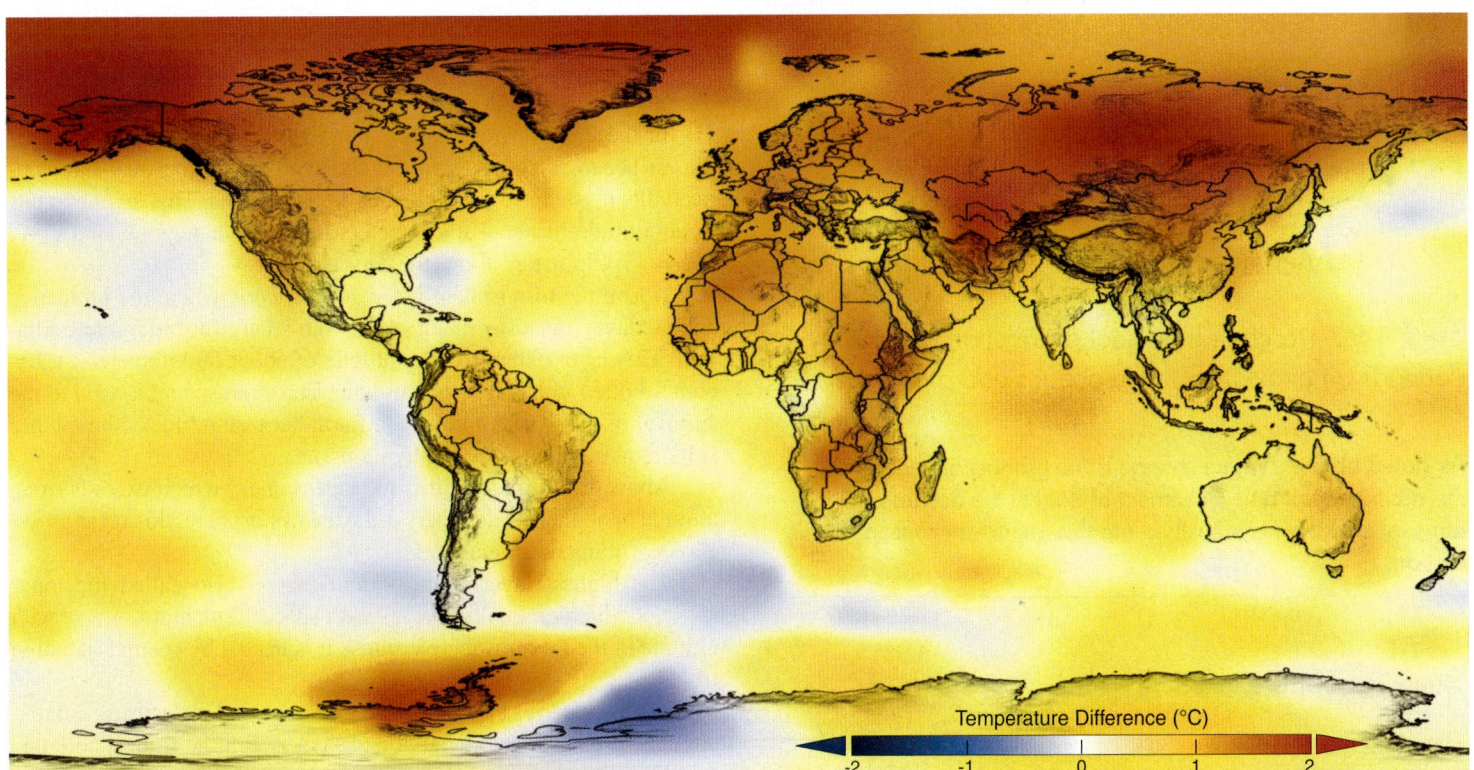

Temperature Difference (°C)

-2   -1   0   1   2

FIGURE 14.2 NASA scientists say that 2013 was the ninth warmest of any year since 1880, extending a long-term trend of rising global temperatures. With the exception of 1988, the nine warmest years in the 132-year record all have occurred since 2000, with 2010 and 2005 ranking as the hottest on record. This map represents global temperature anomalies averaged from 2008 through 2012. Warmer areas are in red, cooler are in blue. The greatest temperature increases occurred in the Arctic and a portion of Antarctica.

Q What are some consequences of the Arctic and Antarctic warming faster than the rest of the world?

Global Climate Change

The circulation of Earth's atmosphere and oceans links all the planet's living organisms and environments, from soil at the equator to ice at the poles. Despite its vastness—Earth is 40,075 kilometers in circumference and has a surface area of 509,600,000 square kilometers—the poles and tropics, deserts and forests, continents and oceans are all *connected* by global processes. The term *climate change* refers to alterations in these processes on a wholesale scale, encompassing the entire Earth.

Our planet is dynamic, meaning that it is constantly changing, as it has been throughout its 4.6-billion-year history. For most of Earth's history, those changes have been natural, and many of them have been stupendous (such as the movement of continents and the evolution of life).

The natural processes that bring about global climate change include plate tectonics, volcanic eruptions, solar output cycles, extraterrestrial impacts, and variations in Earth's orbit. But global climate change is also precipitated by human activities.

On modern Earth, human activities are responsible for major global changes in land use, air and water quality, and the quantity of natural resources, particularly in the last two centuries. There is scientific consensus that human activities are also altering Earth's climate, largely due to rising levels of carbon dioxide and other **greenhouse gases** (atmospheric gas that traps heat and therefore warms the atmosphere) released by the burning of *fossil fuels*.

Studies indicate that the climate change observed during the twentieth and early twenty-first centuries is attributable to a combination of changes in solar radiation, volcanic activity, land use, and increases in atmospheric greenhouse gases. Of these, greenhouse gases stand out as the dominant long-term influence.

Why Study Climate Change?

The distribution of heat on our planet is important in every region and every environment on Earth (**FIGURE 14.2**). The total amount of heat and its variation across the planet surface drives global winds that circulate the atmosphere and control regional weather patterns, growing seasons, and living conditions.

Earth is at the right distance from the Sun (about 148 million kilometers), with the right combination of gases in its atmosphere, and with water covering more than 70% of the planet's surface, to allow the origin and evolution of life and the resources necessary to sustain life on it. So far as we know, no other planet in our solar system has the thermal, physical, and chemical conditions that enable life to exist. These conditions are what make our so-called blue planet unique and habitable. By studying climate change, we gain the knowledge we need to sustain and enhance this livable condition, rather than counteract it.

Q EXPAND YOUR THINKING

Formulate a testable hypothesis that explains the origin of climate change.

14-2 Heat Circulation in the Atmosphere and Oceans Maintains Earth's Climate

LO 14-2 Describe how heat distribution on Earth depends on circulation in the oceans and atmosphere.

As stated above, climate change is the product of changes in the accumulation and movement of heat in the oceans and atmosphere. We delve further into this important concept in this section.

The Atmosphere

The atmosphere is the envelope of gases that surrounds Earth, extending from the surface to an altitude of about 145 kilometers (**FIGURE 14.3**). Around the world, the composition of the atmosphere is similar, but when looked at vertically, it does not appear as a uniform blanket of air. It can best be described as having several layers, each with distinct properties, such as temperature and chemical composition. The red line in Figure 14.3 shows how atmospheric temperature changes with altitude.

In the layer nearest Earth, the *troposphere*, or "weather zone," the air becomes colder with increasing altitude; you may have noticed this if you have ever hiked in the mountains. This layer extends to an altitude of about 8 kilometers in the polar regions, and up to nearly 17 kilometers above the equator.

Above the troposphere is the *stratosphere*, where the protective "ozone layer" absorbs much of the Sun's harmful ultraviolet radiation. This layer extends to an altitude of about 50 kilometers; it becomes hotter with increasing altitude. The stratosphere is vital to the survival of plants and animals on Earth because it blocks the intense solar radiation that damages living tissue.

Above the stratosphere is the *mesosphere*, which extends to an altitude of about 80 kilometers. Like the troposphere, this layer grows cooler with increasing altitude.

The highest layer is the *thermosphere* (also called the ionosphere), which gradually merges with space. Temperatures *rise* with altitude in the thermosphere because it is heated by cosmic radiation from space.

The boundaries between these layers are called "pauses." For example, the tropopause is the boundary between the troposphere and the stratosphere.

There is very little vertical mixing of gases in the atmosphere. This means that one layer can be warming while at the same time another is cooling. For example, global warming in the troposphere, the layer closest to Earth's surface, sets off cooling in the stratosphere because as more heat is trapped in the lower atmosphere, less heat reaches the upper atmosphere. In fact, some parts of the upper

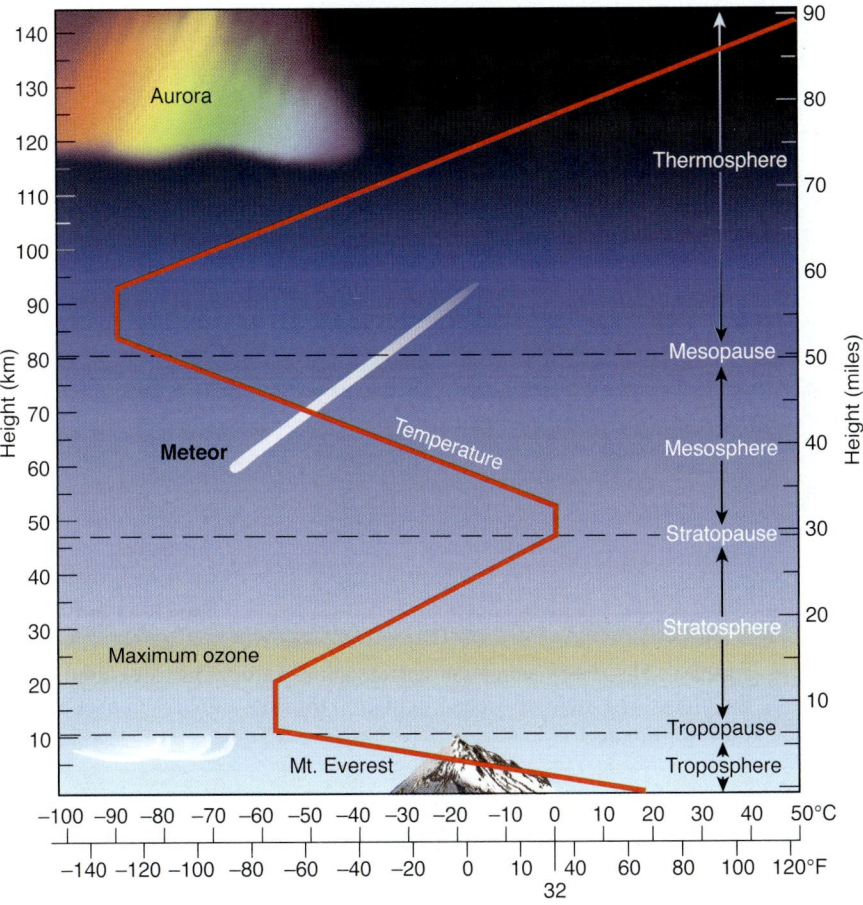

FIGURE 14.3 Vertical structure of Earth's atmosphere.

Ⓠ **Why would Earth appear to be cooling to an observer in space?**

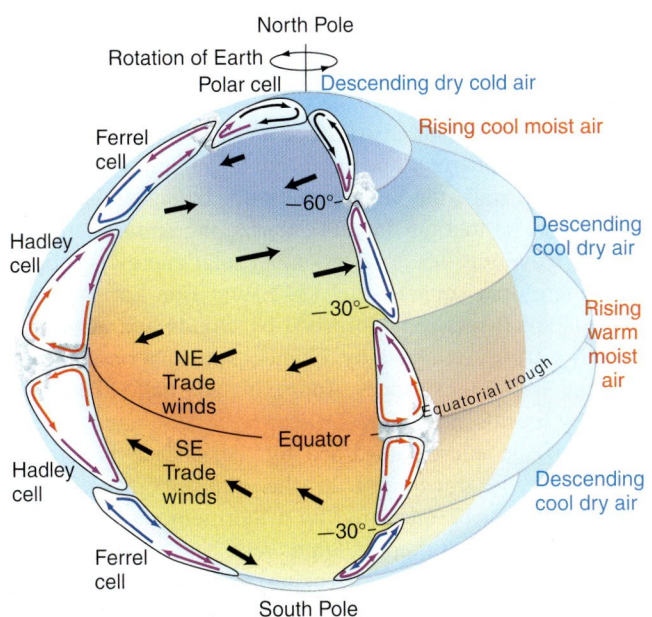

FIGURE 14.4 General circulation of the troposphere.

Ⓠ **How would climate be different if there were no general circulation of the atmosphere?**

atmosphere appear to be cooling at a rate of about 0.05°C per decade, even as the troposphere warms. To an observer in space, then, Earth would appear to be cooling, because greenhouse gases are trapping heat near the surface, causing the upper atmosphere to get colder.

Global Winds

In the troposphere, global winds circulate the atmosphere, mixing water vapor and heat and interacting with the ocean surface. Atmospheric circulation near Earth's surface is vigorous—air can travel around the world in less than a month (**FIGURE 14.4**). The atmosphere thus is an essential component of climate; moreover, it is the most rapidly changing and dynamic of Earth's physical systems, constantly interacting with Earth's other systems—the hydrosphere, biosphere, and lithosphere.

Global circulation is essentially driven by heat from the Sun and by the rotation of Earth. The worldwide system of winds that transport warm air from the equator, where solar heating is greatest, toward the cooler high latitudes is called the *general circulation of the atmosphere*. This pattern gives rise to Earth's climate zones.

The general circulation consists of a number of *cells* (distinct regions of air flow), such as the *Hadley cell* (shown in Figure 14.4). The impact of sunlight is strongest nearest the equator. Air that has been heated there rises and spreads out to the north and south. After cooling, the air sinks back to Earth's surface within the subtropical climate zone, between latitudes 25° and 40°.

This cool, dry descending air stabilizes the atmosphere at these sites and tends to prevent the formation of clouds or rainfall. That is why many of the world's desert climates are found in the subtropical climate zone under these dry descending air masses (a topic we discuss further in Chapter 19). Surface air from subtropical regions returns to the equator to replace the rising air, thus completing the cycle of air circulation within the Hadley cell.

This flow of air creates what are known as the **trade winds,** named several centuries ago by seamen on ships carrying goods around the world who found these winds reliable for planning their routes. North of the Hadley cell is the *Ferrel cell*; and north of this is the *Polar cell*, both of which operate in a fashion similar to the Hadley cell.

Global Water

The **hydrosphere** consists of all the liquid water on Earth—in the ocean, rivers, and lakes. Like the atmosphere, the ocean is characterized by a pattern of circulation that transports heat around the globe and affects Earth's climate (**FIGURE 14.5**). Global ocean **thermohaline circulation** is like a giant conveyor belt driven by the temperature ("thermo") and salinity ("haline") of ocean water.

Thermohaline circulation starts in the North Atlantic Ocean. When the warm surface water of the *Gulf Stream* (a current running north along the eastern coast of North America) reaches the cold polar North Atlantic, it cools and sinks. During its journey, Gulf Stream water has been evaporating (making it saltier) and cooling; therefore, it has become denser, and it sinks readily.

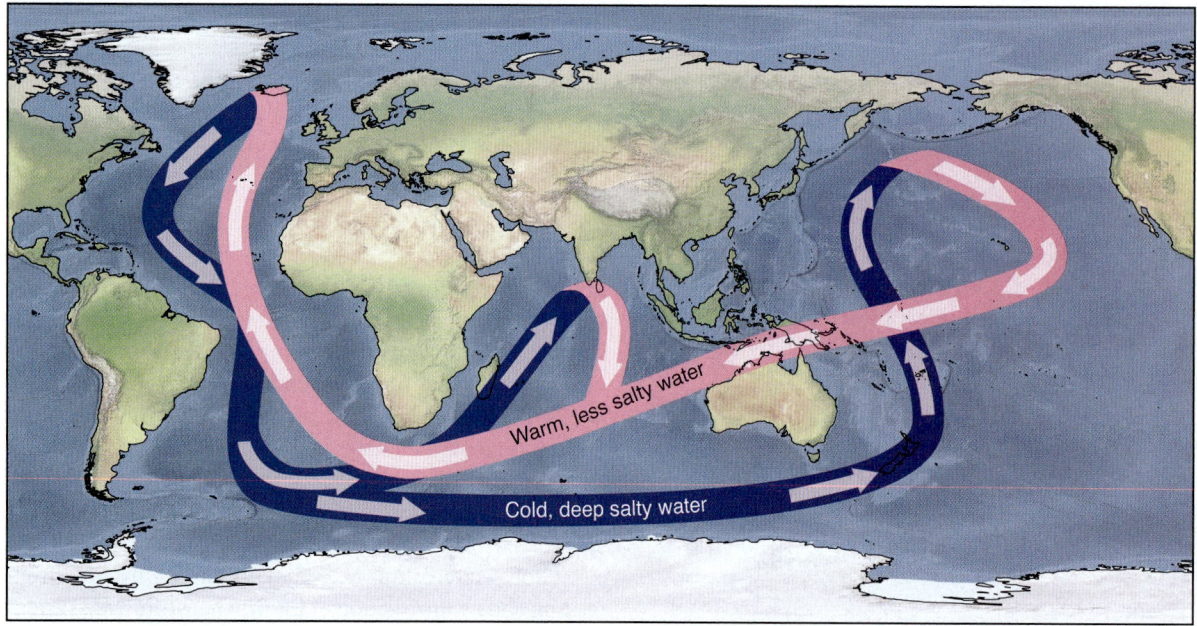

Warm, less salty water

Cold, deep salty water

FIGURE 14.5
Thermohaline circulation carries heat around the globe and affects climate.

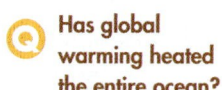

Has global warming heated the entire ocean?

When this water reaches the deep ocean, it travels southward as a cold, deep, salty current called the *North Atlantic Deep Water*. In the southern hemisphere, it is joined by cold, salty water from Antarctica and enters the Indian Ocean. The current flows eastward past Australia into the Pacific Ocean, eventually rising in the North Pacific. There it becomes a warm surface current and flows westward around Africa into the Atlantic.

It may take up to 1600 years for water to complete the entire global cycle. On its journey, the current transports heat, solid particles, and dissolved compounds (such as carbon dioxide) around the globe. The state of ocean circulation thus has a major impact on Earth's climate.

EXPAND YOUR THINKING

What is the impact of thermohaline circulation on the climate in northern Europe?

14-3   The Greenhouse Effect Is at the Heart of Earth's Climate System

LO **14-3** Explain what causes the amount of greenhouse gas in the atmosphere to change.

The **greenhouse effect** is the atmosphere's natural capability to store heat radiated from Earth.

The Heat Budget

The heat absorbed by certain gases (such as water vapor, carbon dioxide, and methane) maintains Earth's *surface temperature* at an average (and comfortable) 14°C. In contrast, as we learned in Chapter 2, a runaway greenhouse effect on Venus is responsible for raising the temperature there to a scalding 477°C, which is one reason why life is unlikely to exist on that planet.

Earth's greenhouse effect is more moderate and, therefore, conducive to life, which depends on the presence of liquid water. Without the greenhouse effect, the average temperature of Earth's atmosphere would be –18°C, below the freezing point of water.

Among all the known planets, conditions conducive to life exist only on Earth; and they exist because of Earth's *heat budget*, a complex balance of heat distribution and exchange among the atmosphere, hydrosphere, lithosphere, and biosphere (**FIGURE 14.6**).

In a natural state, the Sun's radiation is balanced at the top of the atmosphere so that the amount entering the atmosphere equals the amount leaving it. The total incoming solar energy is about 340 watts of energy per square meter (340 W/m²) on Earth's surface. Part of this solar energy is absorbed by clouds and atmospheric gases, and part is reflected by clouds, atmospheric gases, and Earth's land and water surfaces. Approximately half (170 W/m²) is absorbed by Earth's surface.

Some of the energy absorbed by the surface is reradiated upward; some is transferred to the atmosphere as *sensible heat*, heat that can be measured by a thermometer; and some is transferred to the atmosphere as *latent heat*, heat that is released by processes such as evaporation, freezing, melting, condensation, or sublimation. The atmosphere radiates this energy in all directions. When balance is achieved in the atmosphere, the total radiation leaving the top of the atmosphere equals the 340 W/m² received from the Sun.

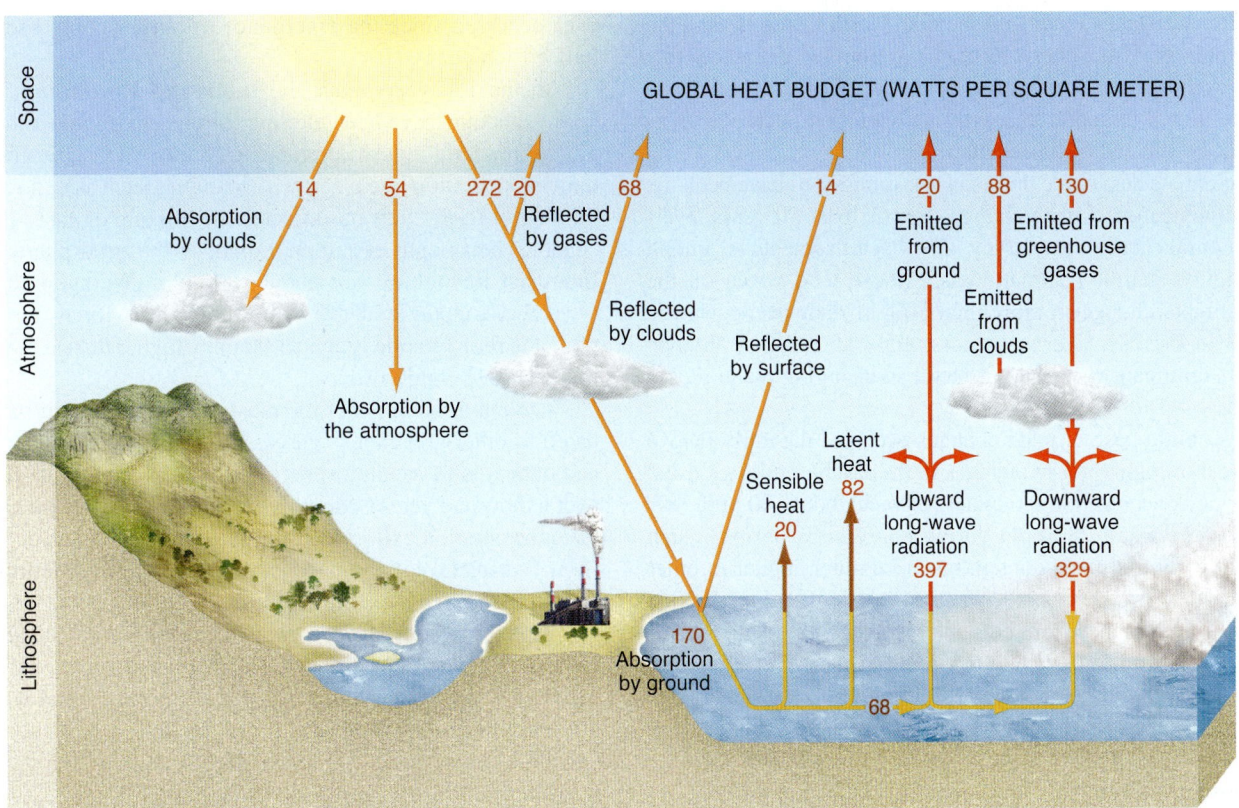

FIGURE 14.6 Earth's heat budget.

What type of radiation do the greenhouse gases trap?

Most of the light energy that penetrates Earth's atmosphere is *short-wave* ultraviolet (UV) radiation, which is mostly absorbed by the protective ozone (O_3) layer in the stratosphere. Of the radiation that reaches the troposphere, more than half is reflected back into space. About 45% reaches Earth's surface and is absorbed by the oceans and land, then is reradiated back into the atmosphere in the form of *long-wave infrared (IR) radiation*. This long-wave IR radiation is absorbed by the greenhouse gases in the atmosphere. The heat absorbed by these gases warms the surface and the lower atmosphere, maintaining a surface temperature that sustains life.

Earth's atmosphere is composed mostly of oxygen and nitrogen, but neither of these gases absorbs infrared energy, so neither plays a role in warming Earth. Six principal greenhouse gases in Earth's atmosphere absorb long-wave radiation and keep the planet warm:

1. Carbon dioxide (CO_2)
2. Methane (CH_4)
3. Nitrous oxide (N_2O)
4. Ozone (O_3)
5. Fluorocarbons
6. Water vapor (H_2O)

Combined, these gases make up about 1% of the atmosphere. Because greenhouse gases are efficient at trapping long-wave IR radiation from Earth's surface, even their small percentage is enough to keep temperatures in the ideal range for liquid water—and life—to exist on Earth.

In conjunction with these six gases have to be mentioned cooling aerosols, which though not a gas, nonetheless play an important role in Earth's heat budget.

Carbon Dioxide (CO_2)

The amount of carbon dioxide in the atmosphere has varied widely during Earth's history, and began doing so long before modern humans inhabited the planet. Natural sources of carbon dioxide include volcanic outgassing, animal respiration, and decay of organic matter. Scientists can measure the history of atmospheric carbon dioxide concentrations by analyzing air bubbles trapped in ice cores (**FIGURE 14.7**) and employing other techniques using fossils (such as marine sediments composed of fossil plankton). These methods have helped scientists decipher long-term trends in carbon dioxide variability and global climate change.

Additional carbon dioxide is released to the atmosphere by human activities—in particular, burning fossil fuels (oil, natural gas, and coal), solid waste, and wood (such as during **deforestation**, the destruction of Earth's forested areas). These *anthropogenic emissions* are at the center of research on global warming.

Although the sources and heat-trapping properties of greenhouse gases are undisputed, scientists are not sure how Earth's climate responds to increasing concentrations of the various gases. There is widespread consensus among scientists around the world (including leading climate researchers in the United States like those at the

National Climatic Data Center in Asheville, North Carolina, and the National Center for Atmospheric Research in Boulder, Colorado) that if levels of anthropogenic emissions of carbon dioxide continue to rise, average global temperatures will rise with them, perhaps by as much as 6°C, by the end of this century.

Most scientists also agree that industrial emissions have been the dominant influence on climate change over the past 50 years, overwhelming natural causes. In a survey of 3146 earth scientists, among the most highly qualified climatologists (those who wrote on the subject of climate change in more than 50% of their peer-reviewed publications in the past five years), over 95% agreed that "human activity is a significant contributing factor in changing mean global temperatures."

In the geologic past, climate changes occurred naturally (as we discuss more thoroughly in Chapter 15). In the past half million years, carbon dioxide concentration remained between about 180 ppm during *glacial periods* and 280 ppm during *interglacial periods*, such as today. But carbon dioxide content has been much higher in other periods of Earth's history. Consider that estimates of carbon dioxide content during the Phanerozoic eon, which are based on the chemistry of fossilized soils, fossil plants, and fossil shells of plankton, indicate that concentrations as high as 1000 to 4000 ppm may have occurred for sustained periods; twice this level may even have been reached.

The cause of such high levels is controversial: Episodes of extreme global volcanism, changes in land surface area due to plate tectonics, absence of polar ice, mountain building, and other mechanisms all have been suggested. What is clear is that the level of only 180 ppm during glaciations is not far from the lowest that has

ever occurred since the emergence of macroscopic life in the last half billion years.

In the 150 years since the **Industrial Revolution** in the mid-1800s, humans have been altering Earth's environment through their agricultural and industrial practices. Furthermore, the growth of the human population in concert with activities such as deforestation and burning of fossil fuels has affected the mixture of gases in the atmosphere. The amount of carbon dioxide in the atmosphere prior to the Industrial Revolution was about 280 ppm. (We know this from ice cores that contain evidence of past climates.) Today, the concentration of carbon dioxide is about 400 ppm, higher than at any other time in the past 15 million years.

Carbon dioxide is not the most effective absorber of heat, compared to other greenhouse gases, but it is one of the most abundant, and once it is in the atmosphere it stays for a long time (centuries to over a thousand years). Monthly records of atmospheric carbon dioxide concentration collected at the Mauna Loa Observatory in Hawaii show seasonal oscillations superimposed on a long-term increase in carbon dioxide in the atmosphere (**FIGURE 14.8**).

This increase is attributable to the activities of humans since the Industrial Revolution, primarily, as just noted above, the burning of fossil fuels and deforestation. Of all the greenhouse gases released by human activities, carbon dioxide is the largest individual contributor to the enhanced greenhouse effect, accounting for about 60%—a trend that shows no signs of slowing.

Alarmingly, the increase in global emissions of carbon dioxide from fossil fuels over the past five years was four times greater than the increase over the preceding 10 years. Simply put, rather than recognizing that the enhanced greenhouse effect is a potentially dangerous trend that can be curtailed by reducing our emissions of heat-trapping gases, humans are accelerating their carbon-burning activities, thereby escalating the production of carbon dioxide and spewing it into the atmosphere at annually record-setting amounts.

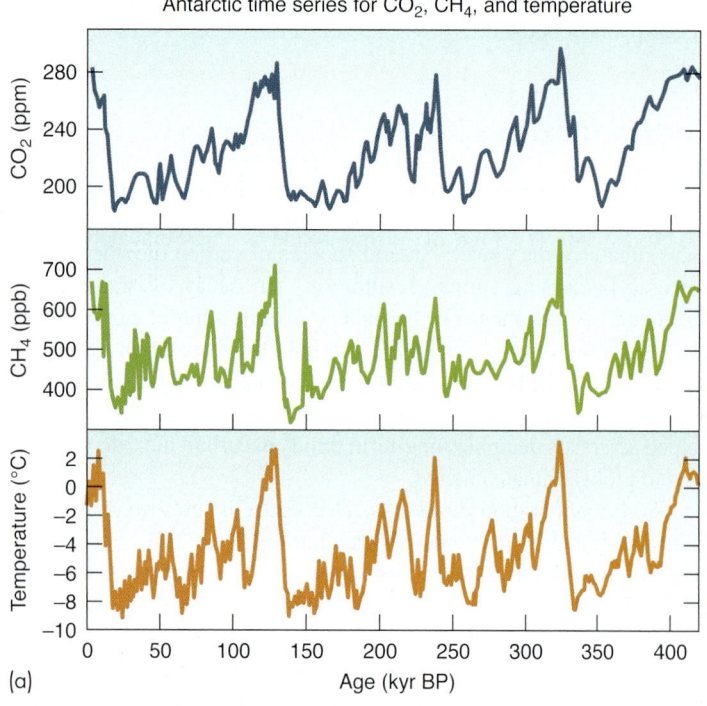

Antarctic time series for CO₂, CH₄, and temperature

Karim Agabi/Photo Researchers

FIGURE 14.7 (a) Global carbon dioxide content (CO₂ in parts per million [ppm]), methane content (CH₄ in parts per billion [ppb]), and temperature (in °C) over the past 400,000 years have been measured using air bubbles trapped in ice in Antarctica. (b) Scientists drill ice cores on mountain glaciers, as well as in Greenland and Antarctica, to obtain evidence of past atmospheric composition.

What climate processes are reflected in the peaks and valleys of the ice core record?

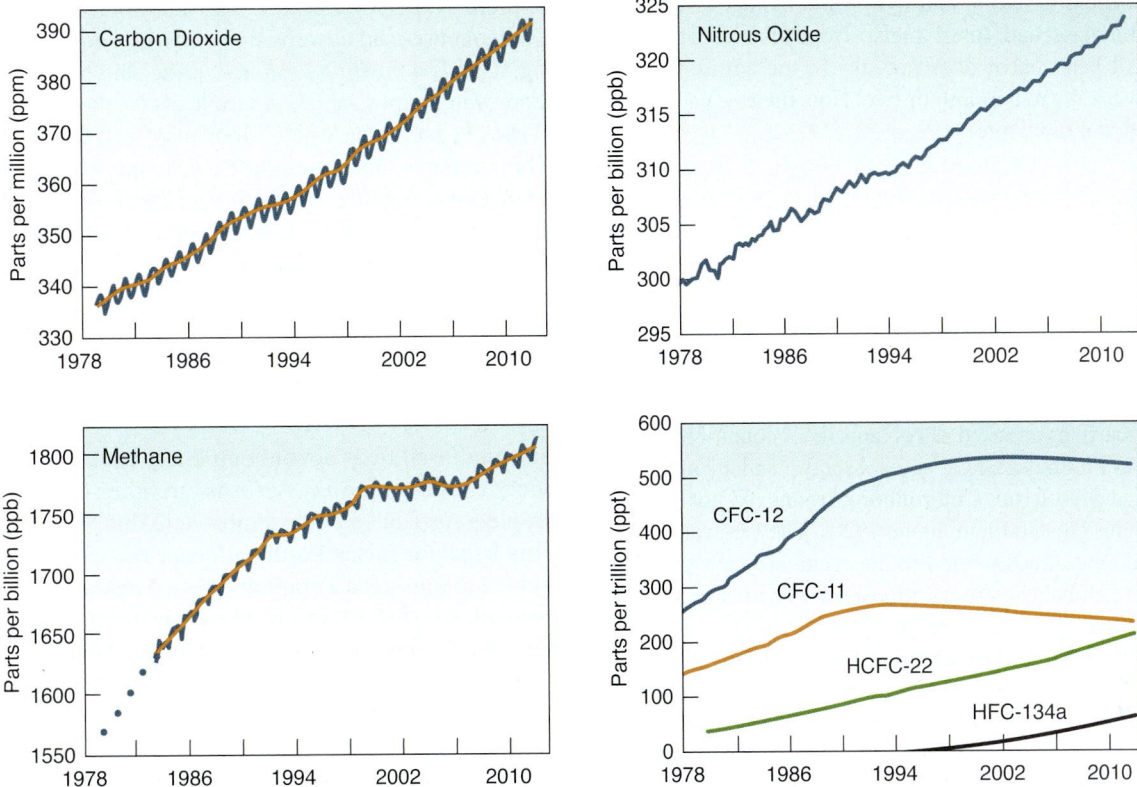

FIGURE 14.8 Concentration of the most important greenhouse gases in Earth's atmosphere. Clockwise from upper left: Levels of carbon dioxide and nitrous oxide continue to climb. Levels of chlorofluorocarbons (CFCs) have declined since the Montreal Protocol (discussed below) was implemented, in 1987. The concentration of methane slowed late in the twentieth century due to droughts and a temporary decline in industrial emissions but has since returned to its previous pattern of steady increases.

What is the predominant cause of the growing concentration of greenhouse gas?

Methane (CH$_4$)

Natural sources of methane include microbial and insect activity in wetlands and soils, wildfires, and the release of gases stored in ocean sediments. The current global atmospheric concentration of methane is 1.8 ppm, more than double what it was prior to the Industrial Revolution.

Methane levels rose steadily in the 1980s; then, in the 1990s, the rate of increase slowed, and came close to zero from 2000 to 2007. Researchers attribute this drop to a temporary reduction in emissions during the 1990s related to the decline of industry and farming after the former Soviet Union collapsed, along with a slowdown in wetland emissions during prolonged droughts. Since 2007 methane levels have been on the rise again, and scientists warn that a more typical rate of increase will have a marked impact on climate.

Methane is responsible for about 20% of global warming, second only to carbon dioxide. About 60% of annual methane emissions come from anthropogenic sources. Human activities that release methane into the atmosphere include deforestation, mining and burning fossil fuels, processing human wastes, raising livestock, and cultivating rice in paddies (industrial wetlands).

Methane is also trapped in ice, glaciers, seafloor sediment, and tundra; as melting occurs, the gas is released from these frozen sources to the atmosphere. Unlike carbon dioxide, methane is destroyed by reactions with other chemicals in the atmosphere and soil, so its atmospheric lifetime is only about 10 to 15 years.

Nitrous Oxide (N$_2$O)

The main natural source of nitrous oxide is microbial activity in swamps, soil, the ocean surface, and rain forests. Human sources of this greenhouse gas include fertilizers, industrial production of nitric acid and nylon, and burning fossil fuels and solid waste. The current concentration of N$_2$O is 323 parts per billion (ppb). It has gone up by 16% since the beginning of the Industrial Revolution and is responsible for 4% to 6% of global warming.

Ozone (O$_3$)

Ozone in the stratosphere absorbs and protects Earth from harmful UV radiation; but ozone in the troposphere is also a pollutant. Because it absorbs heat in the lower atmosphere, it is considered a greenhouse gas. Natural sources of ozone include chemical reactions that occur among carbon monoxide (CO), hydrocarbons, and nitrous oxides, as well as lightning and wildfires.

Human activities raise ozone concentrations indirectly by emitting pollutants that are precursors of ozone, including carbon monoxide, sulfur dioxide (SO_2), and hydrocarbons that result from the burning of biomass and fossil fuels. Tropospheric ozone is a strong absorber of heat, but it does not stay in the atmosphere for long—only a few weeks to a month or two. Nonetheless, its concentration is growing at a rapid rate.

Fluorocarbons

A number of very powerful heat-absorbing greenhouse gases in the atmosphere do *not* occur naturally. They include chlorofluorocarbons (CFCs), hydrofluorocarbons (HFCs), perfluorocarbons (PFCs), and sulfur hexafluoride (SF_6), all of which are generated by industrial processes.

Chlorofluorocarbons are used as coolants in air conditioning (the chemical known as Freon is a CFC), aerosol sprays, and the manufacture of plastics and Styrofoam. Chlorofluorocarbons did not exist on Earth before humans created them, in the 1920s. They are very stable compounds, have long atmospheric lifetimes, and are now abundant enough to instigate global changes in atmospheric chemistry, the biosphere, and climate.

Fluorocarbons contribute to warming by enhancing the greenhouse effect in the troposphere. Chlorofluorocarbons also react chemically with and destroy ozone in the stratosphere, creating the "ozone hole" over the southern hemisphere. Stratospheric ozone is important because it blocks 95% of the Sun's harmful UV rays from reaching Earth's surface. Loss of ozone allows more UV radiation to reach Earth's surface, where it is harmful to living organisms (e.g., by causing skin cancer), influences photosynthesis, and contributes to global warming.

The good news is that many of the effects of CFCs are reversible. Thanks to the *Montreal Protocol*, signed by 27 nations in 1987, CFCs were recognized as dangerous pollutants, and their production and use was substantially reduced. The United States, one of the signers of the protocol, banned the use of CFCs in aerosols and ceased producing them by 1995. Unfortunately, chlorofluorocarbons have lifetimes of 75 to 150 years, so those already in the atmosphere may contribute to ozone depletion for decades.

Water Vapor (H_2O)

Earth's climate is able to support life because of the greenhouse effect and the availability of water. Water vapor is a key component of both of these processes. It is, in fact, a powerful natural greenhouse gas—the most abundant one—and an important link between Earth's surface and its atmosphere.

The concentration of water in the atmosphere is constantly changing, controlled by the balance between evaporation and rainfall. The average water molecule spends only about nine days in the air before precipitating back to Earth's surface.

The amount of water in the atmosphere is not directly affected by human activities, but it does respond to *feedbacks* related to human activities. A **climate feedback** is a secondary change that occurs within the climate system in response to a primary change, which can either be *positive* or *negative*.

For example, as the temperature of Earth's surface and atmosphere rises, the atmosphere is able to hold more water vapor—that is, more evaporation occurs. The additional water vapor, acting as a greenhouse gas, absorbs more energy and causes further warming. This is a *positive feedback* in the sense that warming leads to more water vapor, which in turn leads to more warming, and so on. However, increased water vapor may lead to greater cloud cover, which reflects more sunlight back to space. This condition would favor global cooling, a *negative feedback*. Which of these two opposite effects dominate is still a subject of research, though there are strong indications that clouds promote warming as a positive feedback.

Cooling Aerosols

Burning fossil fuels not only produces heat-trapping gases, it produces *cooling aerosols*. Aerosols are fine solid particles or liquid droplets suspended in the atmosphere that "scatter" (reflect) light. This behavior raises Earth's *albedo*, the tendency to reflect sunlight, and thus have a cooling effect. Most anthropogenic aerosols are sulfates (SO_4) that are released with the pollution resulting from the burning of coal and petroleum. So much aerosol production accompanied industrial growth in the middle of the twentieth century that global cooling occurred between the 1950s and the 1970s.

Recall from Chapter 6 that volcanic eruptions can have the same effect. They blast huge clouds of particles and gases (including sulfur dioxide) into the atmosphere. In the stratosphere, sulfur dioxide converts to tiny persistent sulfuric acid (sulfate) particles that reflect sunlight. Particularly large eruptions can produce widespread cooling. Mount Pinatubo, remember, in the Philippines, which erupted in June 1991, cooled the planet nearly an entire degree, temporarily offsetting global warming.

Global Warming Potential of the Principal Greenhouse Gases

Molecule for molecule, some greenhouse gases are stronger than others. Each differs in its capability to absorb heat and the length of time it resides in the atmosphere. The capability to absorb heat and warm the atmosphere is expressed by a gas's *global warming potential* (GWP), usually compared to carbon dioxide over some given time horizon. For example, methane traps 21 times more heat per molecule than carbon dioxide. Nitrous oxide absorbs 270 times more heat per molecule than carbon dioxide. Fluorocarbons are the most heat-absorbent, with GWPs up to 30,000 times stronger than those of carbon dioxide.

Understanding the GWPs of various gases is a useful way to describe the impact of human emissions and identify which changes can have the most positive outcome.

> **Q EXPAND YOUR THINKING**
>
> Identify two natural geologic processes that can alter the greenhouse effect on Earth. Describe how they work.

The Global Carbon Cycle Describes How Carbon Moves Through Natural Systems

LO 14-4 List the components of the global carbon cycle.

Natural movement of nutrients, various dissolved compounds, and certain elements through living and nonliving systems are essential for life on Earth.

The Carbon Cycle

Most of the chemical compounds that occur on our planet move among the atmosphere, hydrosphere, lithosphere, and biosphere, and through *ecosystems* (communities of animals and plants and the environment they inhabit), which combine aspects of all these "spheres."

As a reflection of the many biologic, geologic, and chemical exchanges and reactions involved, this movement is known as *global biogeochemical cycling*. The key elements required for life move through biogeochemical cycles; they include oxygen, carbon, phosphorus, sulfur, and nitrogen. The rates at which elements and compounds move between places where they are temporarily stored (*reservoirs* or *sinks*) and where they are exchanged (*processes*) can be measured directly and modeled using computer programs.

One of the most important cycles that affects global climate is that of the element carbon (**FIGURE 14.9**). The **global carbon cycle** comprises the many forms that carbon takes in the reservoirs and processes for carbon found on Earth. These include:

- *Rocks* in the lithosphere, such as limestone and carbon-rich shale
- *Gases* in the atmosphere, such as carbon dioxide and methane
- *Carbon dioxide* dissolved in water (oceans and fresh water)
- *Organic matter* in the biosphere, such as the simple carbohydrate glucose ($C_6H_{12}O_6$), found in plants and animals

Most of the carbon on Earth is contained in the rocks of the lithosphere, where it has been deposited slowly over millions of years in

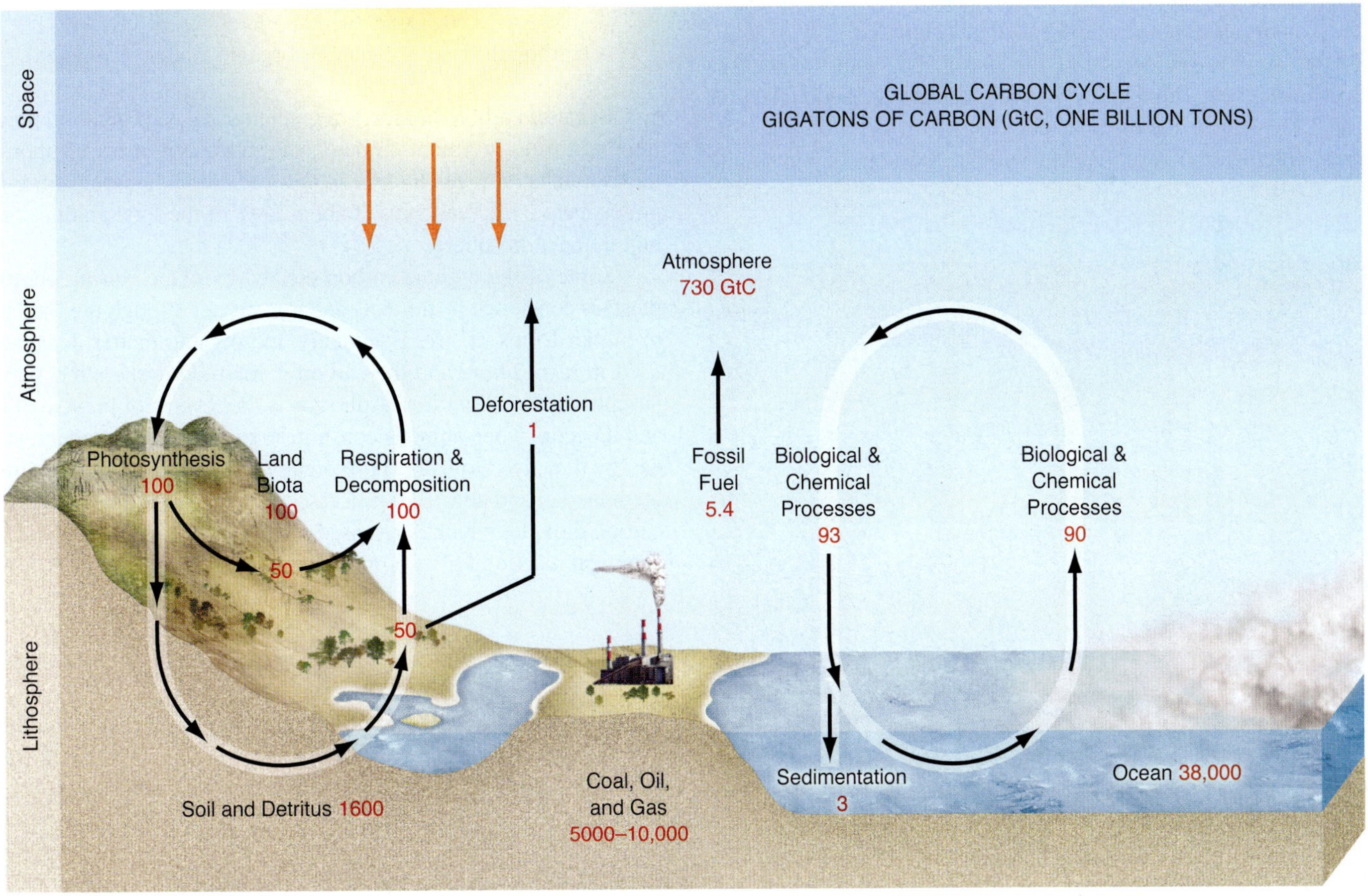

FIGURE 14.9 Carbon is cycled through Earth's atmosphere, oceans, lithosphere, and biosphere. The values given here reflect global carbon reservoirs in gigatons (1 Gt C = 1012 kg). Annual exchange and accumulation rates are given in gigatons of carbon per year (Gt C/year), calculated over a decade.

What is the largest reservoir of carbon?

the form of dead organisms (mostly plankton). Carbon is stored in the lithosphere in two forms: 1) *Oxidized carbon* is buried as carbonate, such as limestone, which is composed of calcium carbonate ($CaCO_3$); 2) *reduced carbon* is buried as organic matter, such as dead plant and animal tissue.

Most of Earth's carbon is contained in limestone, which provides effective long-term storage for carbon that has been taken from the atmosphere and transferred to the lithosphere. The process of storing carbon in this way occurs in several steps. Gaseous carbon dioxide in the atmosphere is constantly entering the surface waters of the hydrosphere, where it dissolves into the *bicarbonate ion* (HCO_3-) in this (simplified) chemical reaction:

$$2CO_2 + 2H_2O \rightarrow 2HCO_3^- + 2H^+$$

The bicarbonate ion combines with dissolved calcium (Ca^{2+}) in seawater to form calcium carbonate ($CaCO_3$, calcite), in a reaction called *calcification*:

$$2HCO_3^- + Ca^{2+} \rightarrow CaCO_3 + CO_2 + H_2O$$

In the first reaction, two molecules of CO_2 are taken from the atmosphere; in the second, only one molecule of CO_2 is released. This means that as limestone is formed, some atmospheric carbon dioxide is trapped and buried in the most stable of forms: rock. Coral and other marine organisms, such as mollusks, some types of algae, and the plankton *foraminifera*, are excellent calcifiers.

Most calcification occurs in the ocean, but some also takes place in freshwater. Have you ever seen stalagmites and stalactites in caves? They are made of limestone that was formed by the same chemical calcification reaction, but without the help of plants and animals (FIGURE 14.10).

FIGURE 14.10 An example of calcification is the formation of stalagmites and stalactites in caves. Cave deposits are composed of travertine, a type of calcium carbonate that is created by inorganic precipitation in freshwater.

J. Debru/Getty Images, Inc.

Q **What weathering process breaks down limestone?**

The movement of carbon does not end there. Limestone may eventually be broken down by weathering, as you learned in Chapter 7. This process consumes atmospheric carbon dioxide in a chemical reaction that is essentially the reverse of calcification:

$$CaCO_3 + CO_2 + H_2O \rightarrow Ca^{2+} + 2HCO_3^-$$

The weathering of silicate rocks (which account for most of the rocks in the lithosphere) also uses carbon dioxide:

$$CaSiO_3 + 2CO_2 + H_2O \rightarrow Ca^{2+} + 2HCO_3^- + SiO_2$$

Cycling of carbon also happens in the biosphere. Plants and some forms of bacteria can use inorganic carbon dioxide and convert it into organic carbon (such as carbohydrates and proteins), which is then consumed by all other forms of life, from zooplankton to humans, through the food chain.

During *photosynthesis*, plants remove carbon dioxide from the atmosphere and convert it into organic carbon in plant tissues. Photosynthesis occurs on land in trees, grasses, and aquatic (freshwater) plants, and in phytoplankton, algae, and kelp in ocean surface waters that are penetrated by sunlight. The reaction requires sunlight and chlorophyll, and in its simplest form can be represented in this way:

$$6CO_2 + 6H_2O \rightarrow C_6H_{12}O_6 + O_2$$
(carbon dioxide + water → organic matter + oxygen)

Because carbon dioxide is a greenhouse gas, vegetation plays an important role in global climate. Through the process of photosynthesis, plants remove *200 billion tons* of carbon dioxide from Earth's atmosphere each year. This is about 26% of the total amount of carbon in the atmosphere.

Some of the organic carbon created by plants during photosynthesis is consumed by animals and transferred through the food chain to higher forms of life. Eventually, the organic matter decays or is used in *respiration*, and the carbon is returned to the atmosphere as gaseous carbon dioxide. Respiration is the reverse of photosynthesis, and it occurs when animals consume organic material to generate the energy they need to live. These organisms (from bacteria to humans) breathe, die, and decay, all processes that convert organic carbon into carbon dioxide, which is released back in the atmosphere. The basic chemical reaction for respiration is:

$$C_6H_{12}O_6 + O_2 \rightarrow 6CO_2 + 6H_2O$$
(organic matter + oxygen → carbon dioxide + water)

The cycling of carbon through photosynthesis and respiration is so rapid and efficient that all of the carbon dioxide in the atmosphere passes through the biosphere every four to five years.

Q **EXPAND YOUR THINKING**

Explain why the carbon cycle is an important area of global warming research.

14-5 Modeling Improves Our Understanding of Climate Change

LO 14-5 Describe global circulation models.

You are no doubt starting to realize that Earth's climate system is very complex, with cycles and feedbacks and reservoirs (some natural, some related to human activities) that all interact with one another over different lengths of time. Global climate models are used to simplify these complexities and make predictions about Earths future climate.

Climate Models

Researchers who study global climate change often use climate models, called **global circulation models (GCMs),** computer-based mathematical programs that simulate the physics (behavior and interaction) of Earth's oceans, land, and atmosphere. These models consist of thousands of mathematical equations that are calculated by programs on supercomputers to simulate the processes that govern Earth's climate system.

The supercomputers receive input in the form of data on ocean currents and temperature; the concentration of carbon dioxide and other greenhouse gases in the atmosphere; the amount of sunlight; and the cover of vegetation, ice, and snow that affect the heating of Earth's surface; and many other parameters (**FIGURE 14.11**).

Climate models are designed to simulate climate on a range of scales, from global to regional (hundreds of kilometers). But few GCMs regularly tackle climate changes at the local level (tens of kilometers). Most break up the atmosphere into 10 or 20 vertical levels between Earth's surface and outer space, where those cycles, feedbacks, and reservoirs are represented by complex calculations.

Similarly, the ocean depths and land surface are described using grid cells (**FIGURE 14.12**). Many models also feature human population growth, economic behavior, human health, and resource use. The model's output might include predictions of long-term precipitation patterns, or a map of the sea level 100 years from now, or an estimate of future global temperatures at a certain concentration of greenhouse gases.

GCMs can simulate surface temperatures and compare the results to observed changes. In **FIGURE 14.13**, 100 years of observed temperature changes are plotted as a black line. Two different model results are plotted in red and blue. Blue simulations were produced using only natural factors: solar variation and volcanic activity. They do not match the observed temperature changes very well.

Red simulations were produced with a combination of natural *and* human factors, including anthropogenic emission of greenhouse gases and other products of industrial pollution. It is clear that the combination of human and natural factors results in the best match with measured temperatures, leading to the conclusion that human pollution by greenhouse gases is at least partially responsible for global warming.

The Sun's Role

As noted earlier, energy from the Sun drives Earth's heat budget, so it is not surprising that changes in **insolation,** the amount of energy Earth receives from the Sun, have an effect on global temperatures and climates. We observe this daily as temperatures fall after sunset, and seasonally as Earth's distance from the Sun and the planet's tilt are reflected in both temperatures and climate. Although solar output may vary, on average about the same amount of solar energy has reached our planet for millions of years. Scientists have measured this *solar constant* and found it to be approximately 1360 watts per square meter (W/m^2) of Earth's surface.

In the decades since satellite measurements began, in 1979, natural fluctuations of only 0.1% in the solar constant have been recorded. Greenhouse gases are about four times more effective at warming Earth's surface than

FIGURE 14.11 General circulation models are used for weather forecasting, simulating climate, and predicting climate change. Models must take many factors into account, such as how the atmosphere, the oceans, the land, ecosystems, ice, topography, and energy from the Sun all affect one another and Earth's climate.

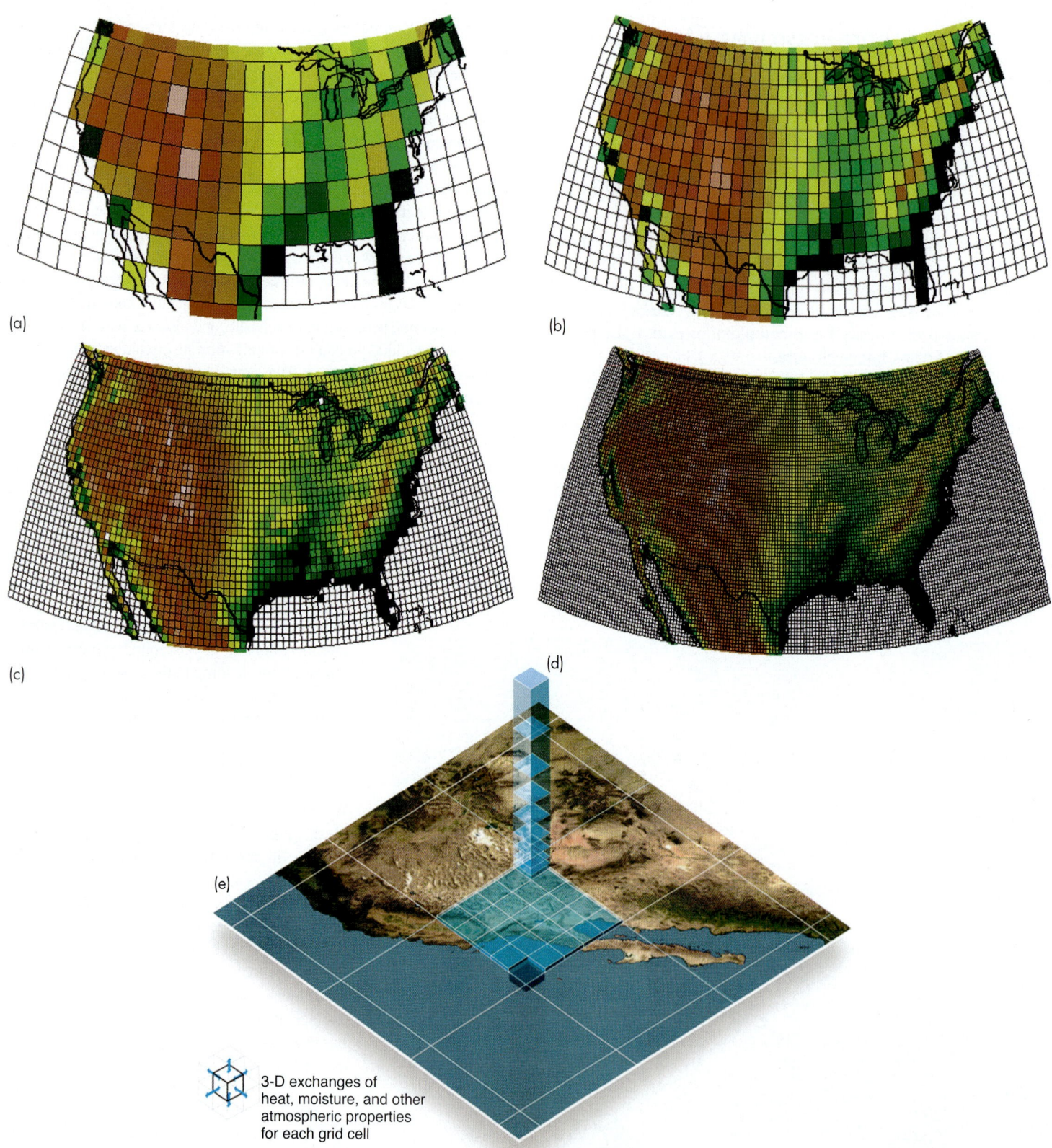

(a)

(b)

(c)

(d)

(e)

3-D exchanges of
heat, moisture, and other
atmospheric properties
for each grid cell

FIGURE 14.12 The resolution of climate models has increased over time. (a) In the 1990s, models used the T42 grid, where temperature, moisture, and other processes were simulated in grid boxes corresponding to about 200 by 300 kilometers. (b) For the important International Panel on Climate Change 2007 report, models like the Community Climate System Model (one of the world's leading GCMs) at the National Center for Atmospheric Research used the T85 resolution, with grid boxes corresponding to about 100 by 150 kilometers. As models improve, better resolution allows more realistic climate processes, which makes regional ((c) T170 and (d) T340) climate projections more accurate. (e) Computer models have the capability to reach high into a virtual atmosphere and deep into a virtual ocean. They simulate climate by dividing the world into three-dimensional grid boxes, measuring physical processes such as temperature at each grid point. Such models can be used to simulate changes in climate over years, decades, or even centuries. *Source*: University Corporation for Atmospheric Research (UCAR) http://www2.ucar.edu/climate/faq/aren-t-computer-models-used-predict-climate-really-simplistic#mediaterms, accessed July 10, 2012. Copyright © 2012 University Corporation for Atmospheric Research, NCAR/CGD. Figure by Gay Strand.

this variation in solar output, so most scientists agree that the effect of solar variability on global climate is small by comparison.

Although it does not appear that the Sun's output has varied enough to cause significant climate change, this does not mean that Earth's exposure to solar heating has been constant. For example, between 10,000 and 20,000 years ago, when Earth's climate shifted from an *ice age* to the *interglacial* of today, global temperatures have gone up by several degrees; and atmospheric carbon dioxide concentration rose from 180 to 280 ppm (**FIGURE 14.14**).

In fact, as illustrated in Figure 14.7, for the past several hundred thousand years, Earth's climate has shifted dramatically between glacial and interglacial states approximately every 100,000 years. This record is the result of cyclical shifts in the intensity and contrast of the seasons, which in turn are due to changes in the geometry of Earth and the Sun. For instance, mild winters and cool summers tend to promote ice buildup from one year to the next, while hot summers tend to discourage it; such patterns can lead to ice ages and interglacials.

What is the likely cause of the observed climate changes in the figure?

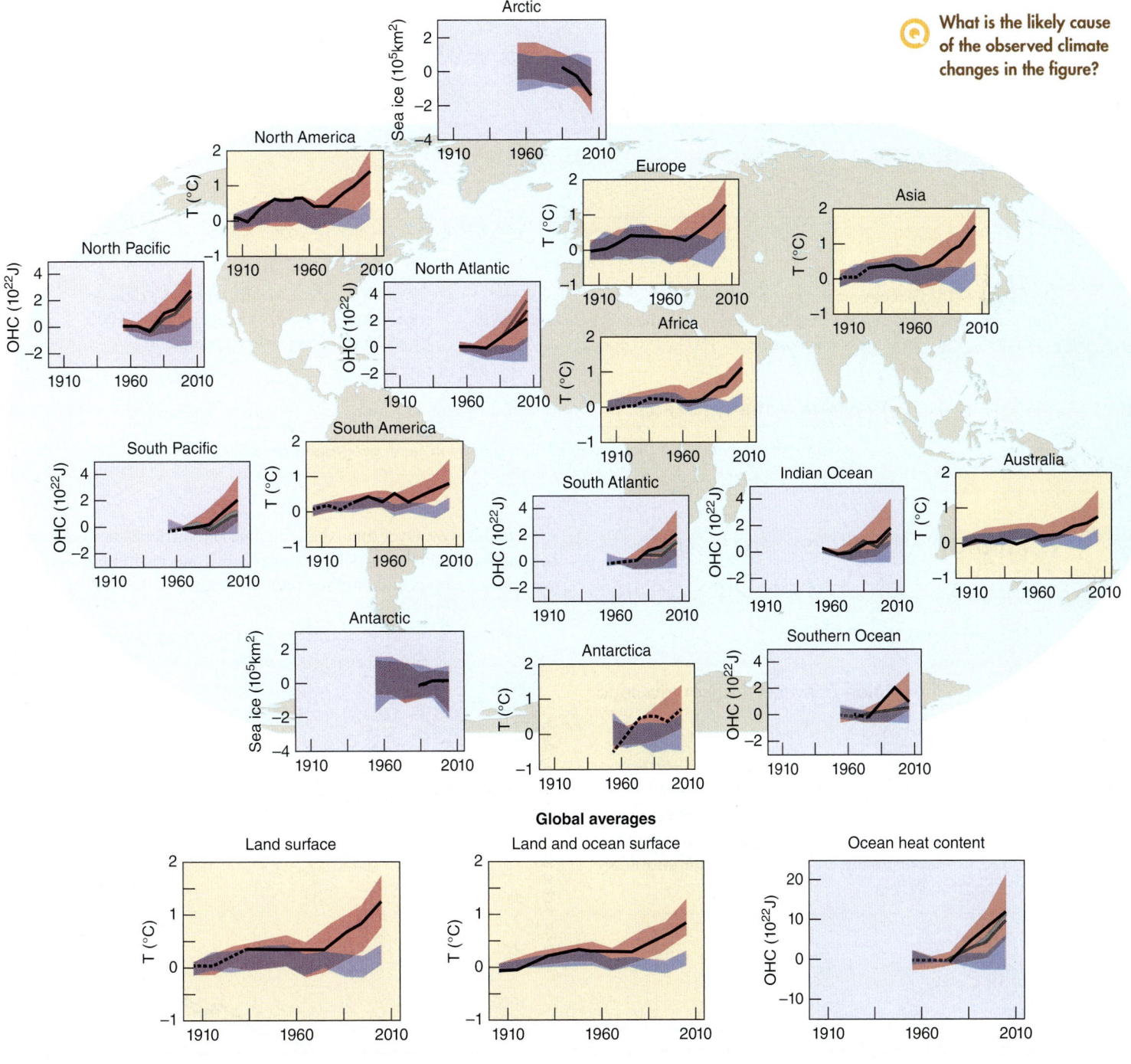

FIGURE 14.13 Simulating surface temperature using mathematical models (blue = natural factors only; red = natural and human factors combined) and comparing the results to measured changes (temperature = black line) can yield insight into causes of major temperature changes. The red band shows that human factors combined with natural factors account best for observed temperature changes. *Source*: Climate Change 2013: The Physical Science Basis. Working Group I Contribution to the Fifth Assessment Report of the Intergovernmental Panel on Climate Change, Figure SPM.4. Cambridge University Press.

Models using only natural forcings
Models using both natural and anthropogenic forcings
Observations

Earth does not orbit the Sun in a perfect circle. It wobbles, tilts, has eccentricities and other irregularities due to the varying gravitational influence of other planets. These cyclical changes in Earth–Sun orbital geometry do not affect the total amount of solar energy reaching Earth. Rather, they control the *location* and *timing* of solar energy around the planet, thus influencing whether seasons are particularly hot or cold (known as *seasonal intensity*) and contrasts between the seasons.

In moving toward a glacial climate, changes in the Earth–Sun geometry allow snowfall from one winter to survive mild summers and last until the following winter. This pattern leads to net accumulation of ice, a necessary condition for glacier formation. Other changes in orbital geometry lead to the end of ice ages and the onset of interglacials. We study this process in more detail in Chapter 15.

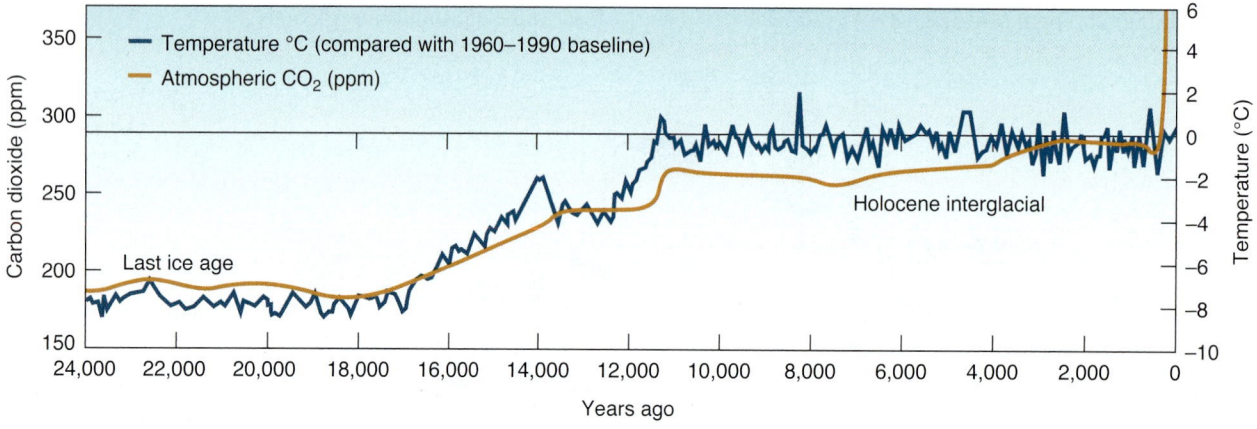

FIGURE 14.14 Global temperatures rose when Earth's climate shifted from the last ice age to the current interglacial. At the same time, the atmospheric carbon dioxide concentration increased from 180 to 280 ppm.

Q What is the cause of the temperature change that occurred between approximately 20,000 and 10,000 years before the present?

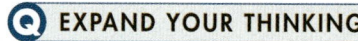

Q EXPAND YOUR THINKING

Are GCMs effective for predicting future climate change in small geographic areas?

Human Activities Have Raised the Level of Carbon Dioxide in the Atmosphere

LO 14-6 Identify the human activities that contribute to climate change.

The population growth of humans, coupled with our ever-greater consumption of resources and seemingly endless production of heat-trapping gases, is having a serious negative impact on our planet—specifically, the processes that enable it to function as a healthy system.

Burning Fossil Fuels

Many people still do not acknowledge that human activities can seriously damage the health of our planet. They still do not understand how everyday activities like driving cars, heating and cooling homes, and harvesting natural resources can affect our environment on a global scale. Here's how: Remember all the carbon that exists in the form of organic matter in the biosphere? Much of that carbon was buried in sediments or sedimentary rocks and became part of the lithosphere for millions of years. Four trillion tons of carbon was stored in Earth's

crust in this way. Over time, these organic carbon deposits were transformed into so-called fossil fuels—coal, oil, and natural gas—which we humans extract and burn to produce energy (**FIGURE 14.15**).

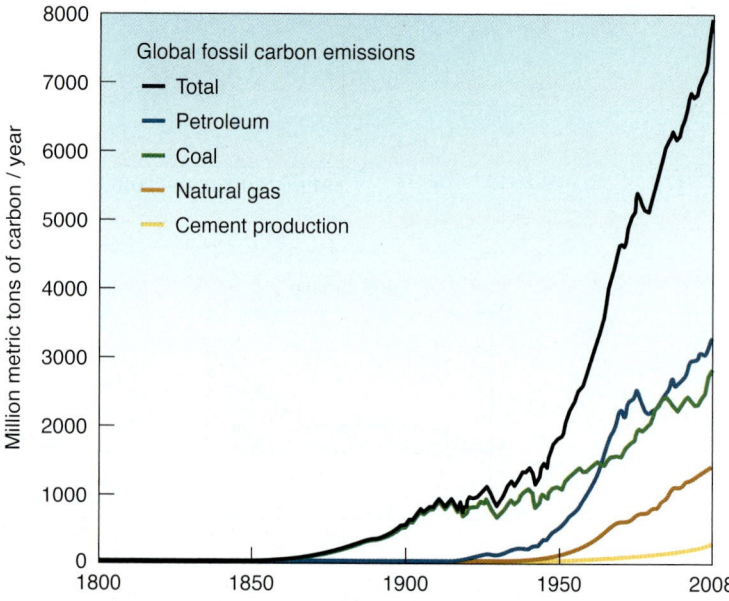

FIGURE 14.15 Global annual fossil carbon emissions, in million metric tons of carbon. The carbon dioxide released from cement production, for example, results from the thermal decomposition of limestone into lime.

Q What is the largest source of global fossil carbon emissions?

Burning fossil fuels and organic biomass is one of the primary activities humans perform to produce usable forms of energy to meet our power needs—for electricity, heat, industry, and motorized vehicles. Unfortunately, burning fossil fuels releases greenhouse gases to the troposphere and contributes to global warming and climate change.

Coal, too, like oil, has been buried for millions of years. Formed from plant material, it is the most abundant of the fossil fuels, supplying 30% of energy globally. At current levels of consumption, coal supplies could last for another 200 to 500 years. The problem is, burning coal releases large amounts of carbon dioxide, soot, sulfur, and other pollutants into the atmosphere. Coal, in fact, is considered one of the dirtiest forms of energy because of its detrimental effects on the environment and human health. (Coal power plants are the single largest source of mercury pollution in the United States.) In the future, our ability to use coal without contributing to global warming hinges on whether we can develop methods to prevent and/or trap its dirty emissions.

Most oil and methane had their origins in the ocean as the remains of *phytoplankton* (microscopic floating algae) that were deposited on the seafloor and deeply buried. At the elevated temperatures and pressures that prevail during long-term burial, the organic material is converted into oil and methane. These fossil fuels are a natural resource that we extract from the crust by means of drilling and mining.

Burning fossil fuels is similar to respiration but requires the addition of heat, denoted by the symbol Δ, as shown in this equation:

$$C_6H_{12}O_6 + O_2 + \Delta \rightarrow 6CO_2 + 6H_2O$$
(organic matter + oxygen + heat $\rightarrow$ carbon dioxide + water)

By burning these fuels, we speed up the natural cycling of carbon between the lithosphere and the atmosphere and add carbon dioxide, fine particles, and pollutants to the atmosphere at a much faster rate than would occur through normal tectonic and volcanic cycles. Furthermore, these energy resources are nonrenewable—that is, it takes millions of years to form fossil fuels, and they exist in limited amounts within the crust.

Respiration and the burning of fossil fuels are not the only sources of carbon dioxide. Recall that volcanic eruptions and metamorphic processes also introduce carbon dioxide into the atmosphere, and have done so throughout Earth's history. Clearly then, carbon dioxide is a natural component of Earth's atmosphere, so why does it matter that human activities add more? How do we know what carbon dioxide levels in the atmosphere were in the past? And why should we care?

Adding Carbon Dioxide

Activities such as burning fossil fuels, deforestation, and various other land-use practices account for about 12% of the total amount of carbon dioxide emitted to the atmosphere annually. About 83% of global human greenhouse gas emissions are produced by 25 countries, listed in TABLE 14.1.

Over the past 200 years, carbon dioxide levels in the atmosphere have risen by over 40% (from 280 ppm to more than 390 ppm), and they continue to rise by approximately 2.1% each year. Scientists now believe that human activities are responsible for these observed escalations in atmospheric carbon dioxide.

Scientists using GCMs predict that by the end of the twenty-first century, we could see carbon dioxide concentrations between 490 and 1260 ppm (75% to 350% higher than natural concentrations). The global warming associated with this rise in carbon dioxide levels will be greater and faster than the natural fluctuations seen in recent geologic history.

The global carbon cycle stabilizes carbon dioxide levels in the atmosphere and ocean fairly well over the long term, (thousands to millions of years) but there is no doubt that rapid, short-term (annual to decadal) additions of carbon dioxide to the atmosphere do have an effect on global climate. Even seemingly small effects, such as higher global average temperatures of just a few degrees, can have a marked effect on climate, winds, ocean circulation, and regional weather patterns.

In addition, because biogeochemical cycles are tightly interconnected, interacting with and affecting one another over very short periods (the time it takes to draw a breath) as well as very long ones (millions of years), changes in the carbon cycle affect the cycles of other elements and compounds. So, though most of the threat to

TABLE 14.1 Top 10 Greenhouse Emitters

Country	Percentage of Global Emissions	Tons per Person per Year	Percentage of World Gross Domestic Product	Percentage of World Population
China	21.5	1.1	10.7	20.9
United States	20.2	6.6	21.9	4.7
Other countries of the European Union	13.8	2.8	21.9	7.5
Russia	5.5	3.6	2.2	2.4
India	5.3	0.5	6.3	16.8
Japan	4.6	2.9	7.2	2.1
Germany	2.8	3.2	4.7	1.4
United Kingdom	2.0	6.3	1.9	<0.8
Canada	1.9	1.3	2.8	2.8
South Korea	1.7	3.1	3.1	1.0

climate comes from the carbon dioxide released by the burning of fossil fuels, human activities also leave their mark on the global carbon cycle, and thus global climate, in other ways.

Ongoing Population Growth

Between 1930 and 2000, the number of humans on Earth tripled, from 2 billion to 6.1 billion (**FIGURE 14.16**). More people leads to more industrial development and, thus, to much higher levels of consumption of natural resources—rising by an astonishing 1000% in the past 70 years. The human population is expected to continue to grow, estimated to reach 9 billion people worldwide by 2050, and 11 billion by 2100. Needless to say, this will put even more intense pressure on the natural resources upon which we humans depend. Of those, energy is one of the most important to us; it also happens to be the one whose consumption results in the worst environmental pollution and greatest depletion of resources.

The developed nations account for only 25% of the world's population but use more than 70% of the total energy produced. The United States consumes 20% of the total to sustain its population of 300 million, which amounts to less than 5% of the world's population. In contrast, India's population of more than 1 billion (17% of world population) uses only 2% of global energy.

Oil, coal, gas, hydropower, and nuclear power comprise what is called *commercial energy*, a commodity traded in the global marketplace. Oil supplies meet about 40% of the world's energy requirements, including 96% of the energy needed for transportation. There are approximately 700 million cars on the world's roads today, a number that is expected to jump to more than 1.2 billion by 2025. The global demand for, and consumption of, oil and gasoline will likely double in the same period. At the same time, gasoline has become much more expensive in recent years. Notably, in 2008, the chief executive of Royal Dutch Shell Oil Company announced that by 2015 the rate of oil consumption will exceed the rate of oil production, essentially spelling the end of the era of affordable, readily available gasoline.

Consider these other statistics: Coal supplies 30% of world energy needs and natural gas 20%, meaning that fossil-fuel burning accounts for 90% of the global commercial energy trade. Keep in mind, fossil fuels are nonrenewable sources of energy; there is a finite (limited) amount of these materials on Earth, and substantial cost and effort must be spent to find, extract, and process these materials. To date, the world's population has consumed over 875 billion barrels of oil, and experts estimate the global volume of recoverable oil at about 1,000 billion barrels. Doing the math, at the current rate of consumption, all the oil on Earth—which, remember, was formed over a period of millions of years—could be depleted in fewer than 40 years.

We must also factor in among Earth's resources its natural environments, ecosystems, and habitats, because they sustain the complex web of life that supports humanity. Global changes are taking places in these resources as well. The World Resources Institute estimates that 50% of the world's coastal ecosystems are severely at risk. The average global temperature has gone up by 0.8°C since the late nineteenth century, and by about 0.5°C over the past 40 years; and scientists expect it to rise by an additional 1°C to 6°C during the next 100 years.

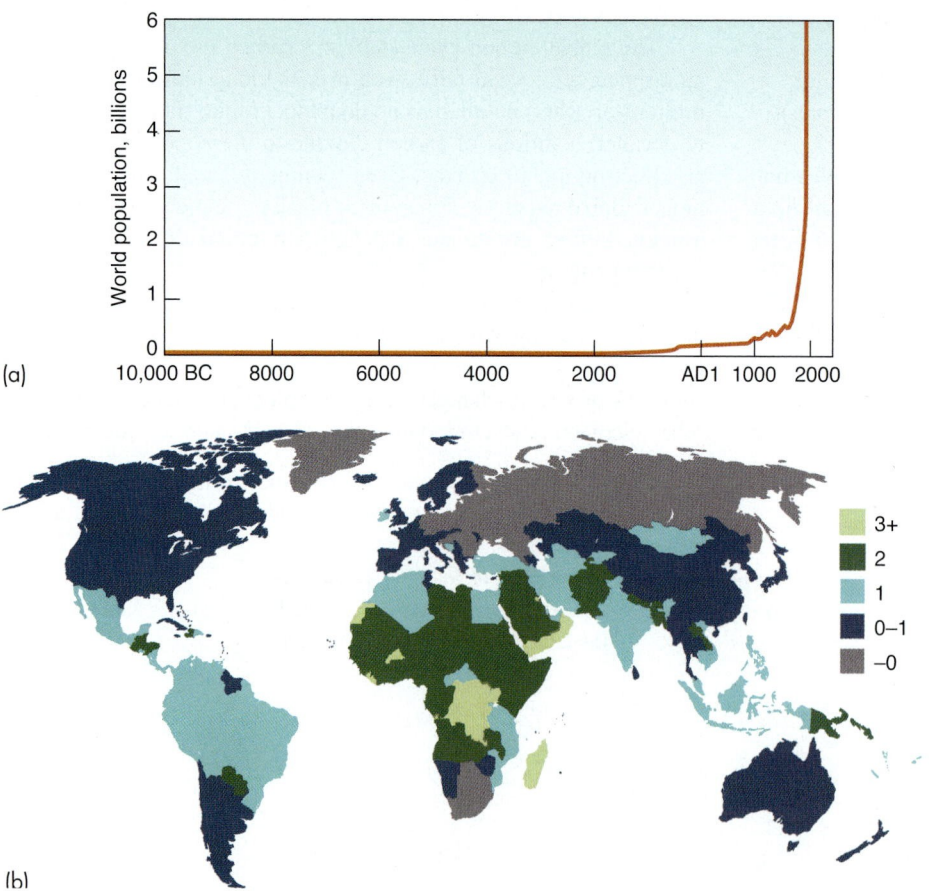

FIGURE 14.16 Human population growth. (a) The rate of human population growth rose rapidly in the twentieth century due to medical advances and greater agricultural productivity. (b) Percentage of annual population growth rates.

Why is population growth relevant to the consumption of fossil fuels?

To understand the significance of these statistics, consider that at the peak of the ice age about 21,000 years ago, the average global temperature was only 5°C to 6°C colder than it is today. Under those conditions, glaciers covered most of North America and Europe, and woolly mammoths flourished. The warming of 5°C to 6°C from the last ice age to the present interglacial took approximately 5,000 to 8,000 years. We humans are poised to accomplish the same thing in only a century.

Changing Land Uses

The term **land use** refers to how we humans, in a sense, consume Earth's surface—its urban, rural, agricultural, forest, and wilderness areas. On a global scale, carbon dioxide released by the dramatic alterations we make to the land surface (e.g., deforestation to harvest timber and clear land for crops and pastures) represents about 18% of total annual carbon dioxide emissions. This constitutes one-third of the total emissions from developing countries and more than 60% of emissions from the least developed countries.

Deforestation is one of the most damaging practices by which humans have contributed to global warming. It intensifies the greenhouse effect in two ways: 1) Burning trees en masse releases carbon dioxide; and 2) dead trees no longer photosynthesize (store) carbon dioxide from the atmosphere.

Already, one-fifth of our planet's forests have been cleared, including 35% of all temperate forests and 25% of all subtropical and deciduous forests. And be aware that these numbers represent *averages*; in some parts of the world, 50% to 100% of the natural forest has been cleared (**FIGURE 14.17**). Consider, for example, that in the 1700s, wooded forests occupied 1.7 million square kilometers between the Atlantic coast and the Mississippi River. Today, only 1.0 million square kilometers remain. (On a more hopeful note, in the U.S. Northeast, forests are returning). Removing forests also erodes sediment, speeds runoff during rainy seasons, and depletes the natural nutrients in soil.

The negative consequences of poor land use are not limited to forest environments. Polluted runoff and eroded sediment affect marine and freshwater ecosystems as well. Coral reefs below deforested watersheds around the world are stressed by sediment-laden water running off mountains that have been cleared of trees and vegetation. Salmon, trout, and other river fish depend on clean, clear streams in which to lay their eggs, and fish stocks have been in serious decline around the United States because of deteriorating stream habitats.

Rhett A. Butler/Mongabay.com

FIGURE 14.17 These hills in Madagascar have been completely stripped of their forests. Deforestation leads to soil erosion, higher concentrations of atmospheric carbon dioxide, and ecosystem destruction.

Humans need wood. How would you manage a forest to meet that need yet minimize the negative impacts of harvesting the timber in it?

Fortunately, growing global awareness of the problems caused by changes in land use has put forestry issues front and center on the political agendas of the United Nations and the world's major developed nations. Awareness and education are the first steps toward addressing the damage that's been done and ensuring that future actions will have more positive effects on the natural environment and all the forms of life it supports. To that end, in the next section we explore additional issues related to this crucial topic. But first, ask yourself, what is the source of the global warming problem? Then put yourself in the shoes of a decision maker by working through the exercise in the Critical Thinking feature titled "Where Do Greenhouse Gases Come From?"

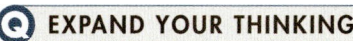

Q EXPAND YOUR THINKING

What human activities have added to the amount of greenhouse gases in the atmosphere?

CRITICAL THINKING

Where Do Greenhouse Gases Come From?

Working with a partner, use **FIGURE 14.18** to fill in **TABLE 14.2** with percentages, then answer the following questions:

1. What are the major sources of each type of gas?

2. Based on these data, what do you think are the easiest first steps to take to reduce greenhouse gas emissions?

3. How would you go about taking these steps?

4. What are the potential consequences of taking these steps?

5. Who would be affected by the first steps you recommend? How? What questions might they have, and how would you answer them?

6. Do you need additional information to answer questions 2 to 5?

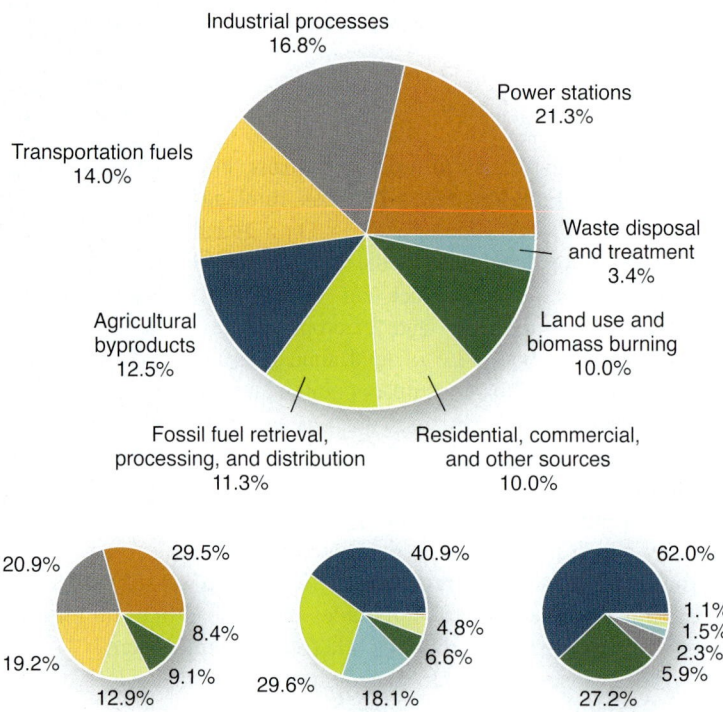

FIGURE 14.18 Relative percentages of anthropogenic greenhouse gases from eight categories of sources.

TABLE 14.2 Sources of Greenhouse Gases Annually, by Percentage

Gas Source	Type of Greenhouse Gas		
	Carbon Dioxide	Methane	Nitrous Oxide
Industrial processes			
Transportation fuels			
Power stations			
Waste disposal and treatment			
Land use and biomass burning			
Agricultural byproducts			
Fossil fuel retrieval, processing, and distribution			
Residential, commercial, and other sources			

14-7 Earth's Atmospheric Temperature Has Risen by About 0.9°C in the Past 100 Years

LO 14-7 Identify some patterns and consequences of climate change.

According to published research, global surface temperatures have risen about 0.9°C since the late nineteenth century, and the linear trend for the past 50 years of 0.2°C per decade has been nearly twice that for the past 100 years.

Global Warming

As pointed out early in the chapter, the years 2005 and 2010 were the warmest on record; moreover, the past 10 years have been the warmest decade since record-keeping began. Scientific analysis of modern and past climates has shown that Earth is warmer now than at any time in the past 1300 years, and probably in the past 11,000 years (**FIGURE 14.19**).

The Melting Arctic

Even though we've been talking about climate change throughout this chapter, it's important to point out that warming has not been uniform around the globe (**FIGURE 14.20**). High latitudes are heating up higher and faster than low latitudes, and the northern hemisphere faster than the southern hemisphere. But it is the Arctic that has warmed faster than any other region on Earth: by nearly 2.7°C in the last 30 years. The most rapidly warming location on the planet is Summit Camp, the highest point on the Greenland ice sheet. According to estimates, by the year 2025 there is a 50% probability that the entire surface of the Greenland ice sheet will enter a state of long-term meltdown.

The accelerated heating in the polar regions has occurred partly because as permafrost thaws, glaciers melt, snowfall declines, and sea ice disappears, the area of white reflective surfaces shrinks, so that the atmosphere absorbs (rather than reflects) more of the Sun's energy. This process is called *Arctic amplification*, which you can read more about in the Earth Citizenship feature, "The Cascading Effects of Arctic Amplification." This is the kind of feedback that makes predicting what will happen in the future so complicated.

Glaciers and sea ice all over the world are melting at the fastest rates ever measured. Alaska has 2,000 large glaciers, and it is estimated that 99% of them are "retreating"—that is, melting. Snow cover in the northern hemisphere has shrunk by about 10% since 1966.

In 2012, the melting rates of Arctic sea ice and Greenland's massive icecap reached record highs. In September 2012, the extent of sea ice dropped to 3.4 million square kilometers, from the long-term average of 7 million square kilometers. This, the lowest extent yet recorded in the satellite record, continues the trend of a 13% decrease per decade (**FIGURE 14.21**). Summer sea ice extent is vital because, among other things, it reflects sunlight, thereby keeping the Arctic region cool and moderating the rate of warming.

The outlook for the future climate of both the Arctic and the Antarctic is worrisome, to say the least. GCMs predict that the warming trend will not only continue, but intensify, with these regions heating

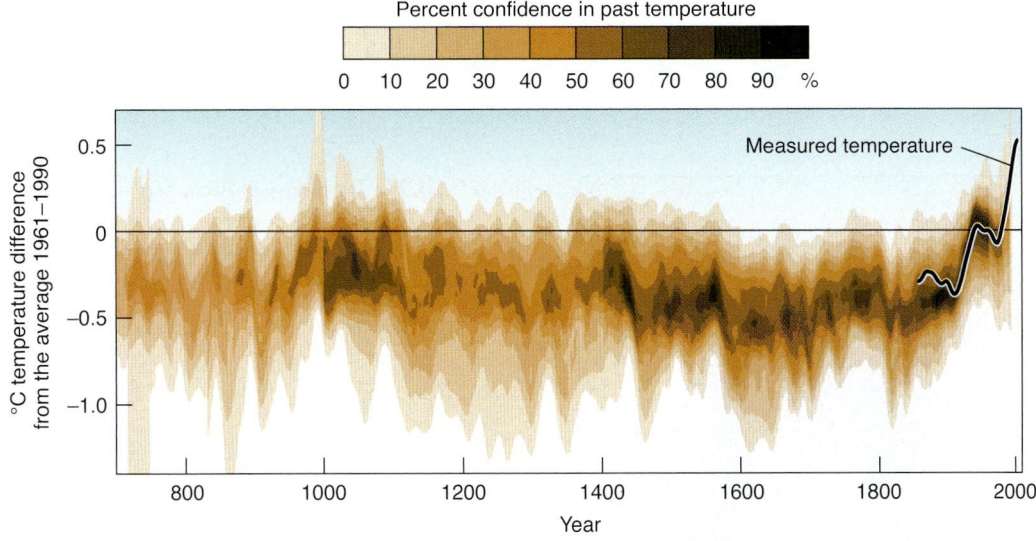

FIGURE 14.19 Reconstructions of global temperature (using geologic materials such as tree rings and corals that provide a record of temperature) indicate that Earth is warmer now than at any time in the past 1300 years. The black line is the record registered by modern reliable instruments; the level of brown shading represents probability of the temperature reconstruction. *Source*: Climate Change 2007: The Physical Science Basis. Working Group I Contribution to the Fourth Assessment Report of the Intergovernmental Panel on Climate Change, Figure 6.10. Cambridge University Press.

Q Describe the patterns you see in this global temperature record.

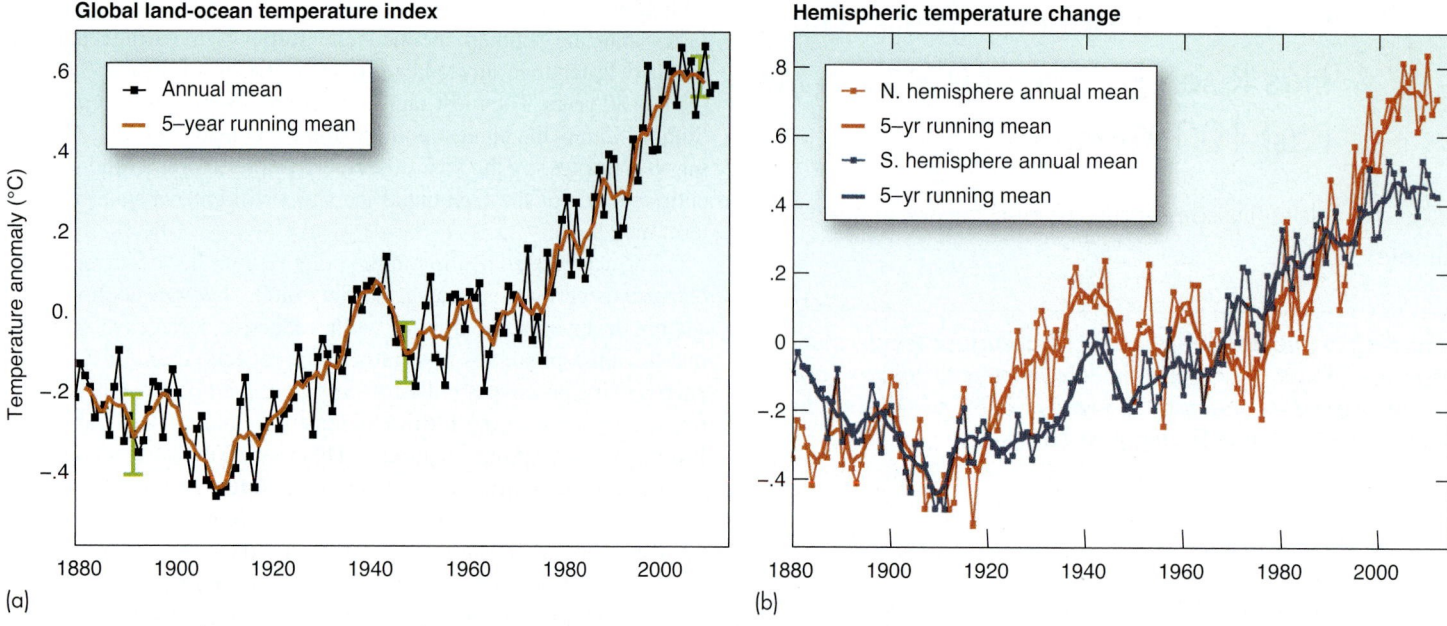

FIGURE 14.20 (a) Throughout the last five decades, global average temperatures have trended upward about 0.2°C per decade. Since 1880, the year that modern scientific instrumentation became available to monitor temperatures precisely, a clear warming trend has been evident, though a leveling off was noted between the 1940s and 1970s. In total, average global temperatures have gone up by about 0.9°C since 1880. (b) The northern hemisphere has warmed faster than the southern hemisphere. Researchers speculate that this is due to the unequal distribution of landmass: with less land, heat in the southern hemisphere is more readily absorbed by the ocean, thus moderating the air temperature.

Q **Where on Earth is the rate of warming the highest? What repercussions will it have?**

up by 4°C to 7°C in the next 100 years. One negative effect of this warming and melting will be on the distribution of ice-dwelling animals, such as polar bears and penguins, as well as the feeding and migration patterns of many species of birds and mammals.

Humans living in high latitudes will feel the effects in a more practical sense, in terms of food availability and problems in ice-dependent infrastructure, including oil rigs and ice roads. In addition, polar warming impacts the weather systems that operate

at lower latitudes—by increasing periods of drought and storminess, for instance—and the global circulation of heat throughout the atmosphere and oceans.

Rising Sea Levels

Sea level has been rising at the rate of 1 to 2 millimeters per year for the past 100 years. Translated, this means that during the last 100

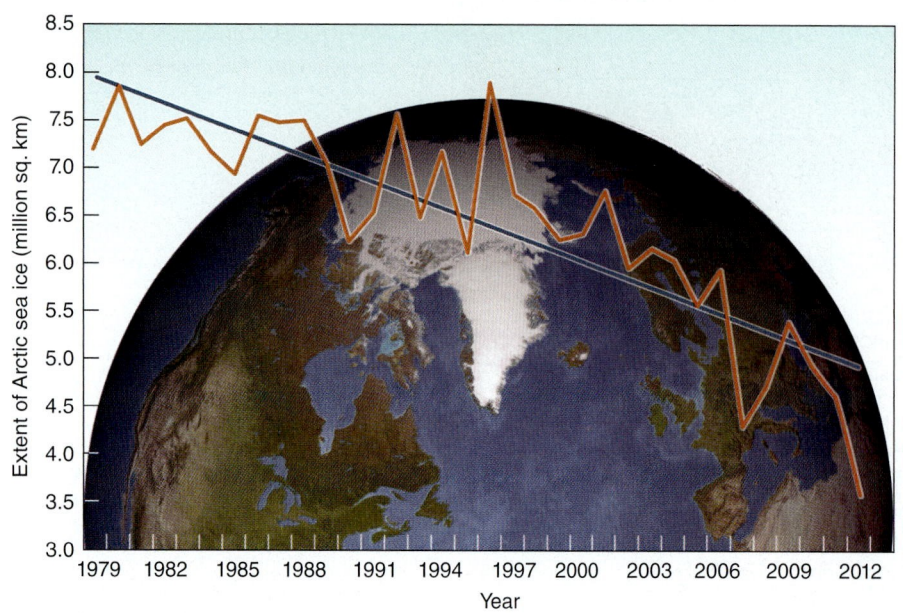

FIGURE 14.21 The history of Arctic sea ice.

Q **Describe the positive feedback to climate caused by melting sea ice.**

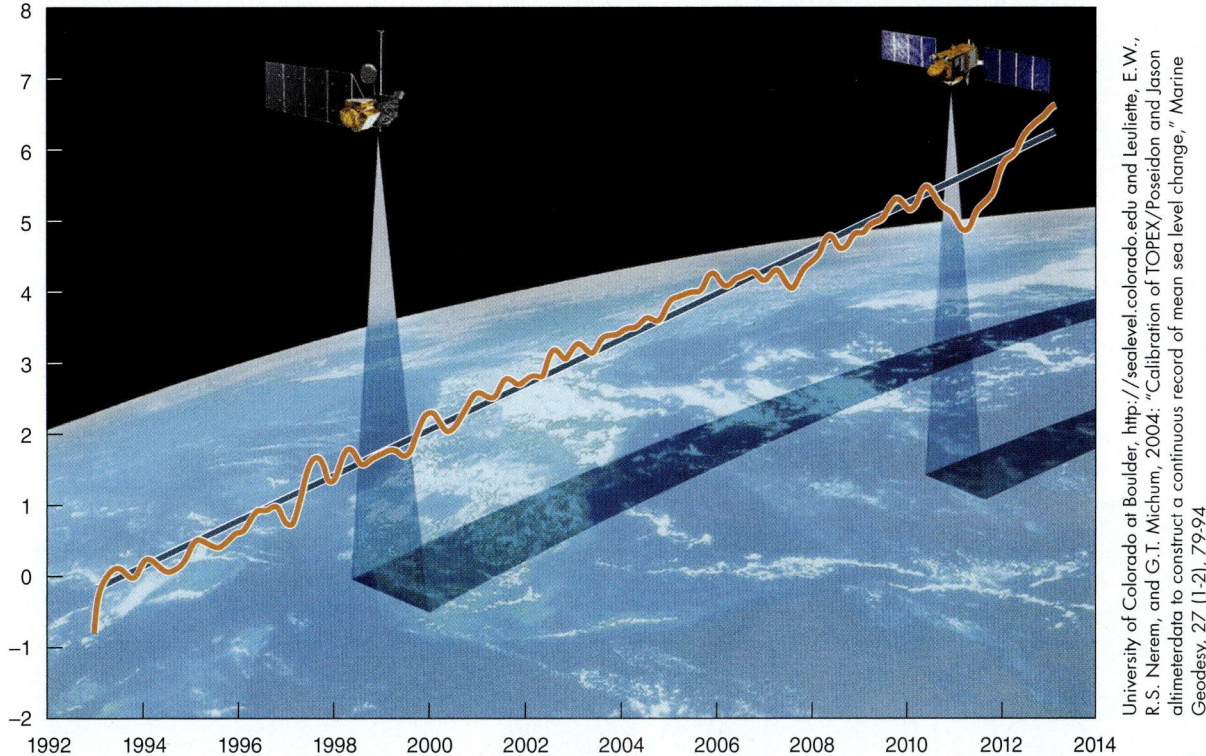

University of Colorado at Boulder, http://sealevel.colorado.edu and Leuliette, E.W., R.S. Nerem, and G.T. Michum, 2004: "Calibration of TOPEX/Poseidon and Jason altimeterdata to construct a continuous record of mean sea level change," Marine Geodesy, 27 (1-2), 79-94

FIGURE 14.22 Satellites have been mapping the ocean surface for over 15 years. The average rate of global sea level rise is over 3 millimeters per year.

Why is rising sea level a problem?

years, the level of the sea has risen by about 15 centimeters worldwide. Recent studies have found that the global rate of sea level rise is now over 3 millimeters per year, approximately twice the rate of the past century (**FIGURE 14.22**).

Global warming can lead to rising sea level in two ways: thermal expansion of ocean water, and melting of ice sheets and other types of glaciers. At present, scientists believe that these two causes have an approximately equal influence on global sea level rise. Further global warming of 1°C to 6°C, predicted to occur by 2100, probably will mean that the rate of global sea level rise will continue to accelerate. Estimates of how high differ but they generally range from 70 to 200 centimeters by the end of the century.

Rising sea levels threaten coastal development and, therefore, the human populations living and working in these areas—very large numbers: More than half of the world's population lives within 60 kilometers of the ocean. In the United States alone, 90% of the population lives within 160 kilometers of a coast.

The costs of rising sea level on coastal communities can be counted in financial as well as environmental and sociological factors. Millions of dollars are spent moving and protecting coastal infrastructure, such as roads and buildings. Home insurance in coastal communities is extremely expensive to account for the high risks of property damage and loss.

Think of New York City, for example, which has almost 1000 kilometers of coastline. Sea level rise there will cause flooding of roads, subways, airports, and buildings. Flood control policies in coastal regions have thus become a very important topic in such communities around the world. Superstorm Sandy, which occurred in late October of 2012, is a damaging example of this phenomenon. Sandy was the second costliest hurricane in U.S. history; at least 286 people were killed along its path and damage has been calculated at over $68 billion. The level of storm damage has been tied to global warming in four ways: 1) sea level rise over the past century caused coastal flooding to be worse than it otherwise would have been; 2) warmer air fueled the strength of the storm; 3) Arctic amplification caused high atmospheric pressure over Greenland and directed the storm to run the length of the U.S. east coast, causing damage the entire way; and (4) warmer sea surface temperatures further enhanced the energy of the storm core.

EXPAND YOUR THINKING

What is the evidence for global warming?

GEOLOGY IN OUR LIVES

The Cascading Effects of Arctic Amplification

The Arctic has experienced warming at a more rapid pace than the rest of the planet. Every year new records are set for reduced extent of sea ice, melting on the Greenland ice sheet, storminess in the Arctic Sea, and drought and associated high land temperatures in North America, Europe, and Asia.

Relative to lower latitudes, air temperatures in the Arctic deviate from historical averages by a factor of 2 or more. This phenomenon is known as *Arctic amplification* (**FIGURE 14.23**), a term that embodies these facts: atmospheric carbon dioxide concentration in the Arctic is higher than the global average (over 400 ppm); temperatures are rising faster than anywhere else on Earth; and warming seawater is releasing methane that bubbles through the sea ice into the air.

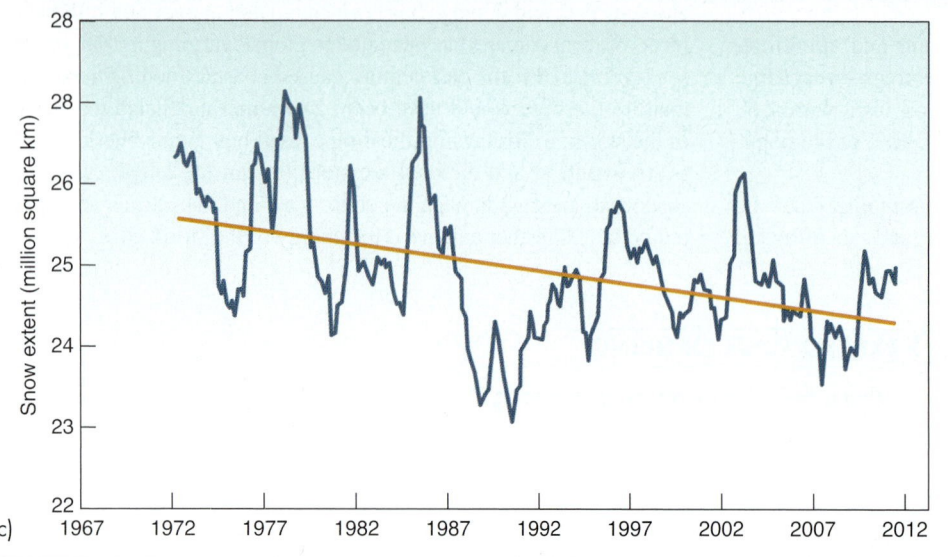

(a)

(c)

FIGURE 14.23 (a) In the Arctic, sea ice and snow cover have been declining for at least 30 years. (b) Arctic ice and snow reflects sunlight, helping to cool the planet. As this ice and snow melt, less sunlight gets reflected into space. It is instead absorbed into the oceans and land, raising the overall temperature and fueling further melting. This results in a positive feedback to warming, which causes the loss of the sea ice to be self-compounding—the more it disappears, the more likely it is to continue to disappear. (c) Annual average snow extent for the northern hemisphere. The trend indicates a decline of approximately 1.3 million square kilometers from 1972 to 2010, or 34,000 square kilometers decline per year. This snow cover decline is both a partial cause, and consequence, of Arctic amplification.

Science Source

NASA Goddard Space Flight Center

(d)

(d) Melting of the Greenland ice sheet. Areas in red are experiencing long-term melting; darker red indicates a longer melt season, from 0 extra days (white) to 75 extra days (dark red). The boundary between annual dry snow and melting snow is rising in elevation by over 40 meters per year.

(e) Because of arctic amplification, the jet stream has slowed and developed large meanders. Slow-moving meanders establish extreme weather patterns such as long heat waves, persistent drought, and intense snow and rain storms.

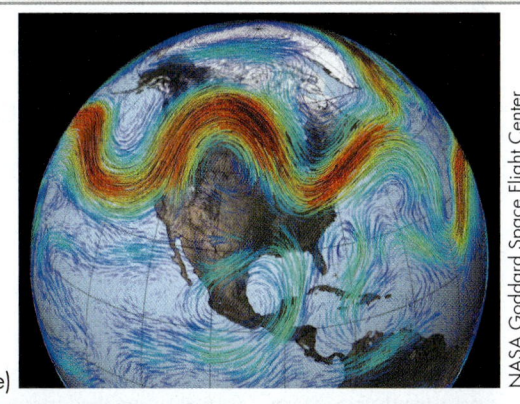

(e)

NASA Goddard Space Flight Center

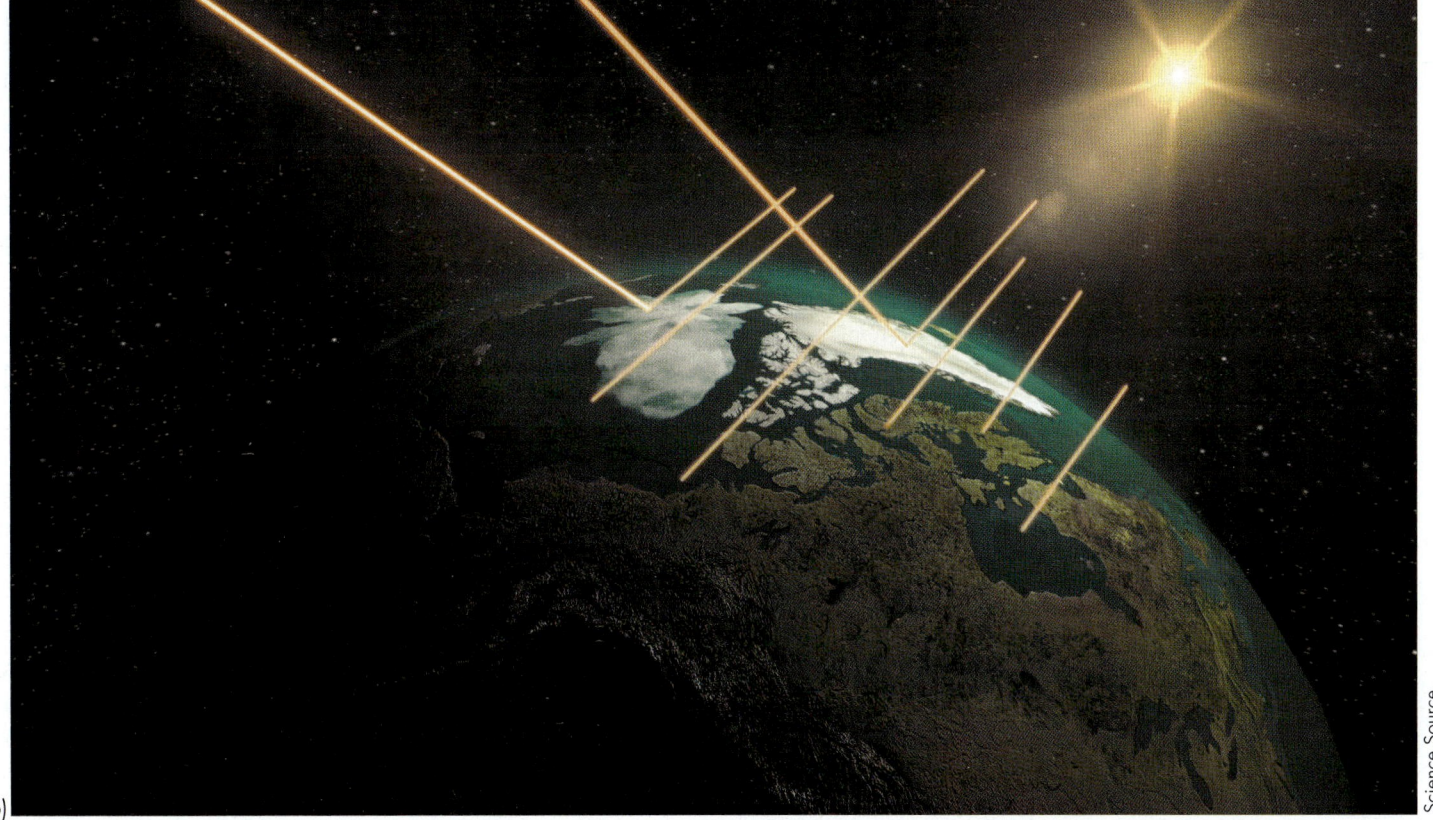

(b)

Science Source

(f) One of the extreme weather patterns established by the change in the jet stream is a punishing drought that extended across the United States. Its negative effects were felt in food production, farming, the economy, and water resources. This historic drought is but one example of the cascading effects of Arctic amplification.

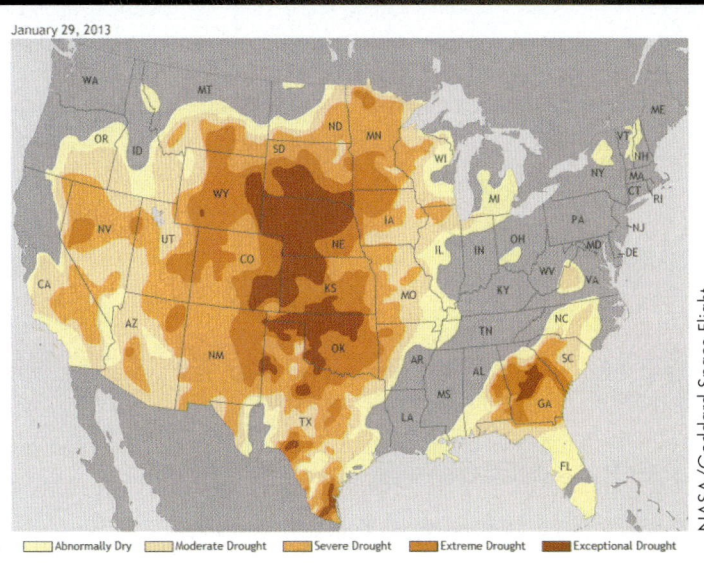

January 29, 2013

(f)

☐ Abnormally Dry ☐ Moderate Drought ☐ Severe Drought ☐ Extreme Drought ☐ Exceptional Drought

NASA/Goddard Space Flight Center Scientific Visualization Studio

14-8 Climate Change Leads to Ocean Acidification and Warming, Glacier Melting, Weather Changes, and Other Negative Impacts

LO 14-8 Describe how global warming is affecting the ocean, the weather, and the world's glaciers.

Global warming is wreaking havoc on the oceans, glaciers, and weather patterns, threatening the overall health of all our planet's ecosystems. We begin our discussion with the impacts to the ocean.

Ocean Impacts

In the oceans, sea surface temperature (SST) has risen by an average of 0.6°C in the past 100 years (**FIGURE 14.24**), and the acidity of the ocean surface has increased tenfold.

Simply put, corals cannot tolerate severely warming waters. The stress of the rising temperature leads to a phenomenon known as *bleaching*, whereby the corals expel the symbiotic algae that live in their tissues and provide them with a food source. In 1997 and 1998, coral bleaching was observed in almost all of the world's reefs in response to high SSTs during that record-setting year. An estimated 16% of the world's corals died during that strong bleaching event, an unprecedented occurrence.

The oceans have absorbed about 40% of the carbon dioxide emitted by humans over the past two centuries, and over 90% of the heat trapped by excess greenhouse gases. Higher levels of acidity, brought on by dissolved carbon dioxide that mixes with seawater to form *carbonic acid* (H_2CO_3), makes it difficult for calcifying organisms (corals, mollusks, many types of plankton) to secrete the calcium carbonate ($CaCO_3$) they need for their skeletal systems. This **ocean acidification** is one of the consequences of carbon dioxide buildup, which could, in time, have a powerful impact on the world's ocean ecology, an ecology that depends on the secretion of calcium carbonate by thousands of different species, not just those in reefs but in the open ocean as well.

In the past century, more than half of the reefs in the Caribbean and Red Sea have gone from pristine to near extinction. The loss of healthy coral reefs is felt by all the species that dwell there (such as turtles, seals, mollusks, crabs, and fish), as well as the animals that depend on reef habitats as a food source (including seabirds, mammals, and humans). One-quarter of all sea animals spend time in coral reef environments during their life cycle. There are economic consequences as well. Together, the income generated by tourism to coral reefs and by commercial fisheries adds up to billions of dollars in revenue annually. Marine biodiversity, food supplies, and economics thus may all reflect the changes to our global climate.

But reef loss is a complex issue. Reefs can suffer from coastal pollution, overfishing, and other types of human stresses. Researchers have yet to calculate exactly what the cost is of warming temperatures, ocean acidity, and other anthropogenic repercussions on global reef health; what is known is that the factors in the equation are all negative.

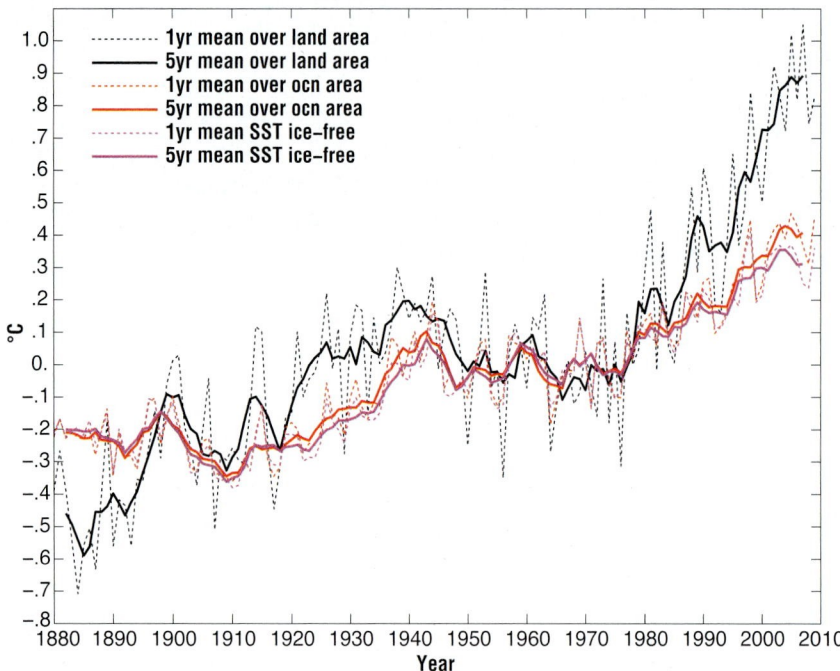

Surface Temperature Anomalies (base 1951–1980)

- 1yr mean over land area
- 5yr mean over land area
- 1yr mean over ocn area
- 5yr mean over ocn area
- 1yr mean SST ice-free
- 5yr mean SST ice-free

FIGURE 14.24 Annual and five-year average global temperature changes for the land (black), the global oceans (red), and ice-free oceans (pink).

Why is ocean warming slower than land warming?

Droughts, Storms, and Extreme Weather Patterns

Even small shifts in climate can result in major changes in much of what humans depend on to live, including a clean water supply and food crops. Most scientists agree that, due to global warming, we can expect more frequent heat waves, droughts, extreme storm events, water stress, wildfires, changes in agriculture and vegetation, sea level rise, coastal erosion and more. Climate and weather will become more unstable, hence less predictable, because the difference in temperature between the poles and the equator drives global winds and atmospheric circulation, which distribute heat and water (precipitation) around the world. Changes in the heat budget will lead to changes in wind and water circulation, leading to drought in some areas and more and stronger storms in others.

Between 1997 and 2007, 8 of the 10 hurricane seasons saw above-average storm activity. Was this intensified activity caused by global warming? Some climate experts say yes, believing that this trend was fueled by warming of the sea surface in the Atlantic. Elsewhere in the United States severe droughts punctuated the situation in the last two decades: Record dry spells were recorded in 30 states in the spring of 2004, and in 80% of the nation in 2012 and 2013. What

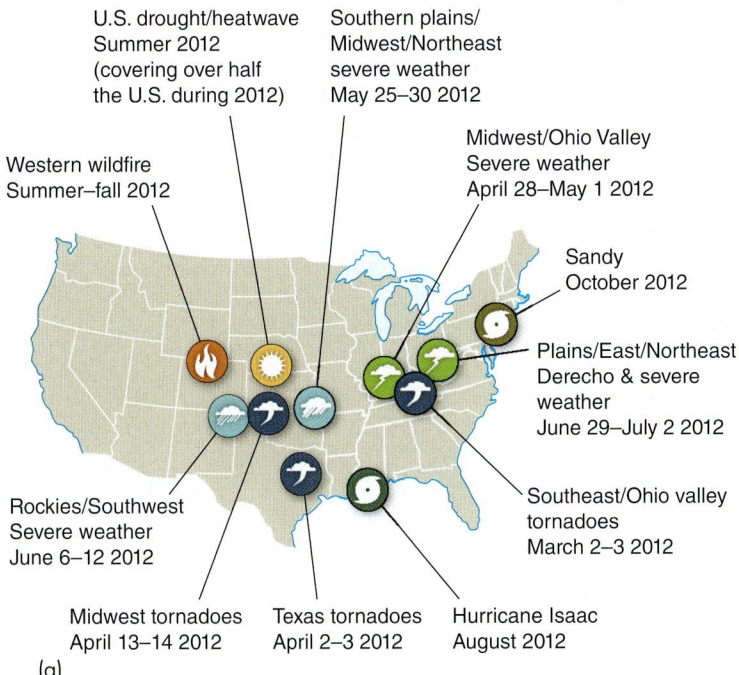

(a)

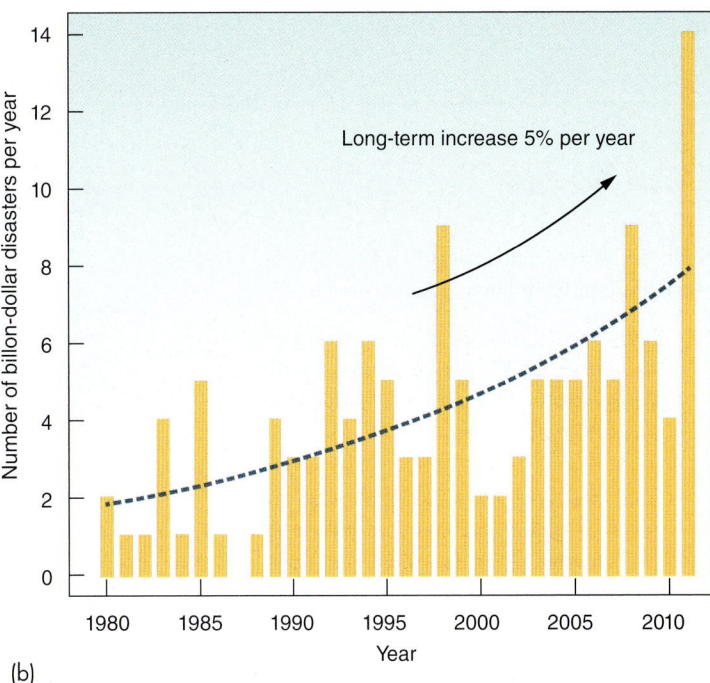

(b)

FIGURE 14.25 According to the U.S. National Climatic Data Center, the frequency of billion-dollar climate/weather disasters in the United States is climbing at a rate of 5% per year. (a) Map of billion-dollar U.S. weather and climate disasters, 2012; (b) increasing trend (5% per year) of billion-dollar climate/weather disasters in the United States.

other recent climate anomalies (uncommon events) have occurred? You can answer this for yourself by completing the exercise in the Critical Thinking feature, "Global Warming."

Based on computer models, scientific experts suggest that these types of extreme events and weather anomalies will become more common in the future as the global climate continues to change, leading to more widespread property damage and threatening the health and well-being of human communities around the world. It has been estimated that the economic costs of extreme weather events in the United States during the 1990s alone added up to close to $300 billion (**FIGURE 14.25**).

Melting Ice

Studies show that the net loss of about 148 cubic kilometers of water per year from the Antarctic ice sheet outpaces its capacity to accumulate snow. Put more simply, the Antarctic ice sheet is melting, not growing.

Three broad ice-covered regions characterize Antarctica: the Antarctic Peninsula, West Antarctica, and East Antarctica. In West Antarctica, the rate of ice loss has speeded up by 59% over the past decade, to about 132 billion metric tons a year, while the yearly loss along the peninsula has gone up by 140%, to 60 billion metric tons. Researchers state: "Without doubt, Antarctica as a whole is now losing ice yearly, and each year it's losing more." Studies also show that outlet glaciers feeding into the ocean from Antarctica and Greenland are accelerating, approximately doubling their discharge of ice between 2002 and 2008. Only in East Antarctica is the ice currently relatively stable, with interior snowfall approximately equaling melting by warm ocean currents along the coastal regions.

How does human population growth affect global warming?

The total volume of land-based ice in the Arctic has been estimated at about 3.1 million cubic kilometers. If this much ice were all to melt, the corresponding sea level rise would be about 8 meters. Alarmingly, research now shows that there is a 50% probability that the entire Greenland ice sheet surface will transition into a state of total melt by the year 2025, with no future accumulation of dry snow (the key type of ice needed to build a glacier).

Most arctic glaciers and icecaps have been in decline since the early 1960s, a trend that has been gaining momentum in the past two decades. The extent of seasonal surface melt on the Greenland ice sheet has been observed by satellite since 1979, showing the trend doubling over the past decade. In recent years record levels of summertime melting at higher altitudes have been noted on Greenland. More specifically, melting in areas above 2,000 meters has increased by 150% from the long-term average, and has occurred on 25 to 30 more days than the average for the previous 19 years. The highest point on the Greenland ice sheet, known as Summit Camp, is now the fastest-warming location on Earth's surface.

Impacts to the Biosphere

There are all too many clear signs of the impacts caused by climate change in the biosphere. They include degradation and shrinkage of wildlife habitats, changes in the distribution of food sources for animals and people, disruption of the timing of animal and bird migration patterns, and **desertification**, the process of becoming a desert.

Did you know that as much as one-third of the world is considered desert? And many deserts are expanding in concert with the warming of the global climate. The Sahara desert in North Africa, which is larger than the United States, has expanded by more than 650,000 square kilometers in the past 50 years. In the 1990s, the Gobi desert in China expanded by more than 50,000 square kilometers, an area larger than New Jersey and Massachusetts combined.

CRITICAL THINKING

Global Warming

With a partner, and using **FIGURE 14.26**, work through the following exercise:

1. Study the significant climate events of November 2013 shown in the figure. Make a list that summarizes the *types* of events that occurred.

2. Compare your list to the potential impacts of global warming described in this chapter. Discuss differences and similarities.

3. Discuss the longer-term impacts of sea level rise—both positive and negative—and how a higher sea level of, say, 0.5 meter, in the year 2065 will influence some of the events shown on this map.

4. Identify three events or anomalies and discuss how they may change as warming continues over the next couple of decades.

5. Describe events that appear to be consistent with a warming world and those that do not.

6. Identify four events or anomalies and describe the impact of each if they were to become regular events due to global warming.

7. Considering these events and your reading of this chapter, construct a testable alternative hypothesis that explains climate patterns of the past half-century.

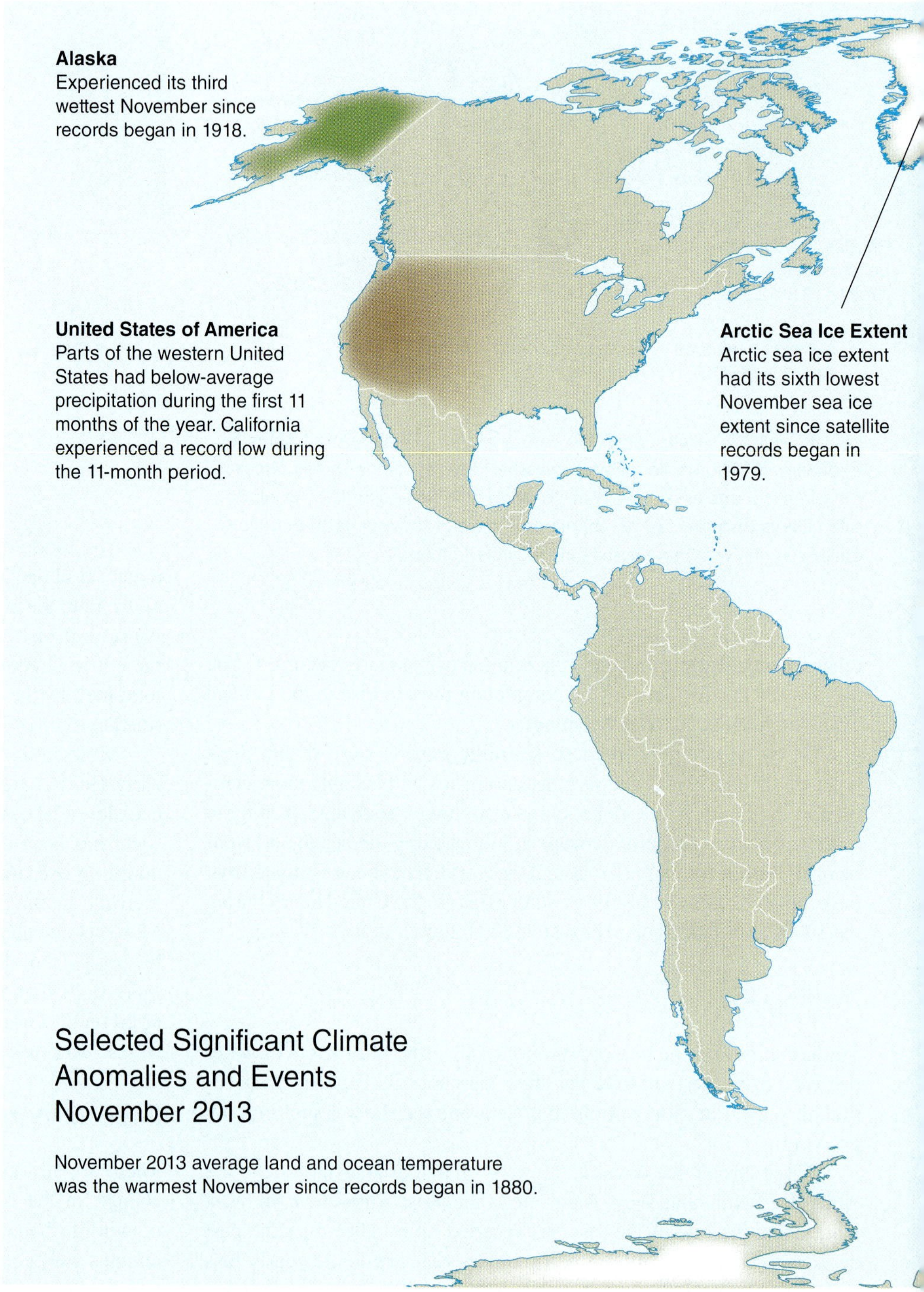

Alaska
Experienced its third wettest November since records began in 1918.

United States of America
Parts of the western United States had below-average precipitation during the first 11 months of the year. California experienced a record low during the 11-month period.

Arctic Sea Ice Extent
Arctic sea ice extent had its sixth lowest November sea ice extent since satellite records began in 1979.

Selected Significant Climate Anomalies and Events November 2013

November 2013 average land and ocean temperature was the warmest November since records began in 1880.

FIGURE 14.26 Significant weather and climate events in November, 2013.

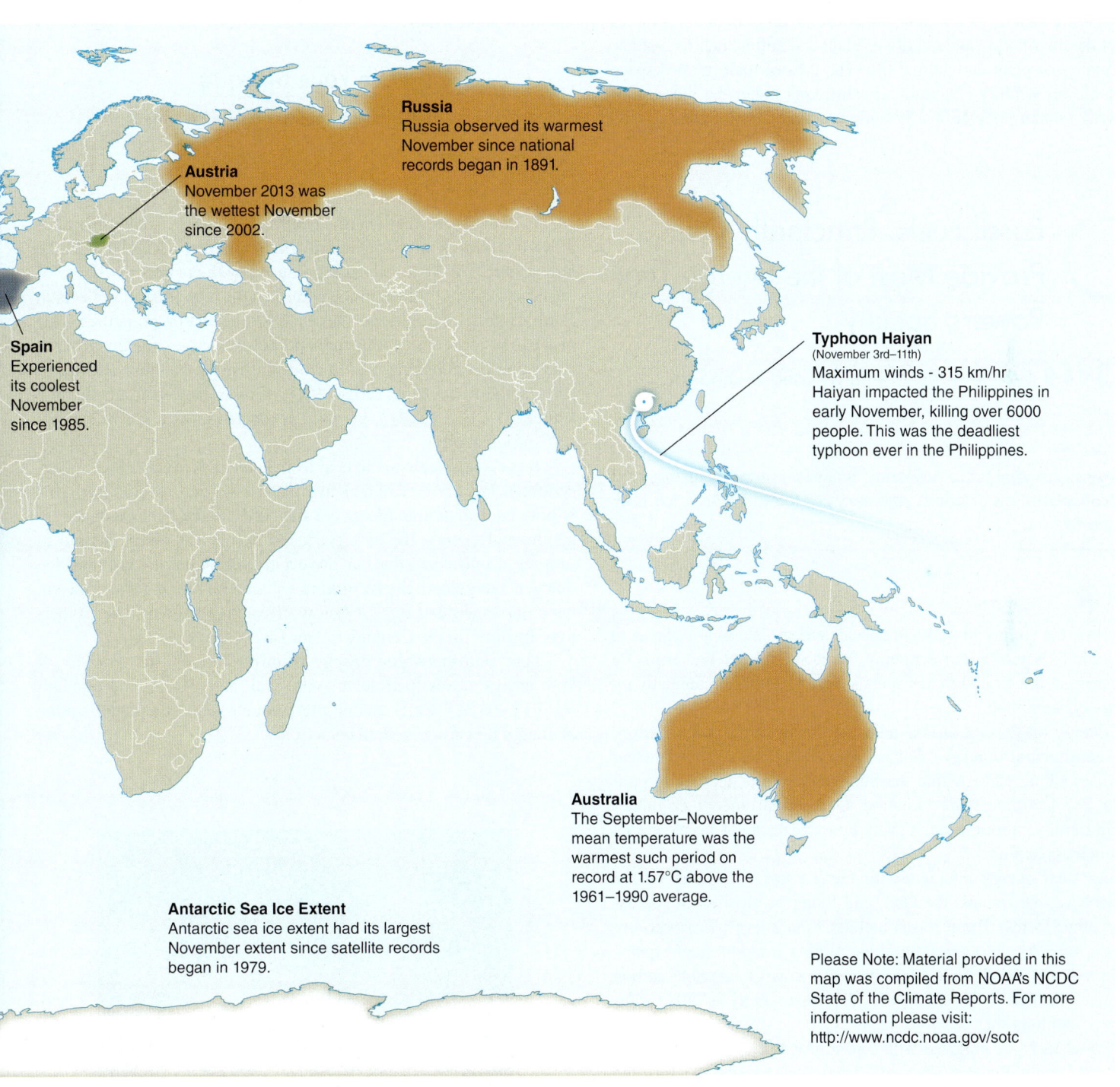

Russia
Russia observed its warmest November since national records began in 1891.

Austria
November 2013 was the wettest November since 2002.

Spain
Experienced its coolest November since 1985.

Typhoon Haiyan
(November 3rd–11th)
Maximum winds - 315 km/hr
Haiyan impacted the Philippines in early November, killing over 6000 people. This was the deadliest typhoon ever in the Philippines.

Australia
The September–November mean temperature was the warmest such period on record at 1.57°C above the 1961–1990 average.

Antarctic Sea Ice Extent
Antarctic sea ice extent had its largest November extent since satellite records began in 1979.

Please Note: Material provided in this map was compiled from NOAA's NCDC State of the Climate Reports. For more information please visit:
http://www.ncdc.noaa.gov/sotc

In many cases, global warming and poor land use by humans jointly set desertification in motion. Changes in the distribution of precipitation around the planet may be exacerbated by activities such as overfarming, failure to rotate crops, overgrazing, and deforestation. All these activities play roles in stripping soils of moisture and nutrients, leading to desertification. (Desertification is discussed further in Chapter 19.)

The ways humans use land and consume other natural resources affect the distribution and quality of plant and animal habitats—thus, the quality of plant and animal life. The area of undeveloped space available for wildlife is continually shrinking under the pressure of growing human populations. Pollution, coupled with damming and diverting water for human use, stress essential freshwater systems. The loss and degradation of habitat quality alters the distribution of food sources and severely impacts wildlife populations. For example, seabird populations around the world are in decline, due largely, researchers believe, to changes in global atmospheric and ocean temperatures that affect the availability of food and the timing of migration cues. Similar declines have been noted in fish and marine mammal populations.

Q EXPAND YOUR THINKING

Describe the impacts of global warming.

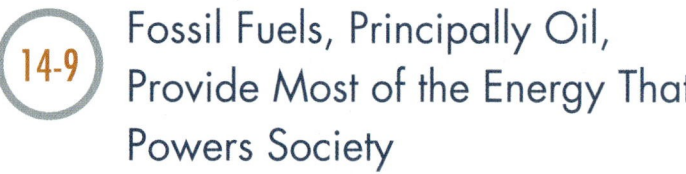

14-9 Fossil Fuels, Principally Oil, Provide Most of the Energy That Powers Society

LO 14-9 List the forms of energy used in modern society.

Energy is the vital force powering business, manufacturing, and the transportation of goods and services to North America and the world.

Energy

Together, the supply of and demand for energy is a major player in our nation's security and economy. So it should not be surprising to learn that the United States spends over $500 billion annually on energy consumption.

Energy on Earth's surface takes two main forms: **solar energy** and **geothermal energy**. Solar energy originates with nuclear reactions taking place within the Sun's core and is delivered to our planet as electromagnetic radiation. Geothermal energy originates within Earth's interior through the radioactive decay of atoms; it, too, is a nuclear reaction.

As solar energy interacts with Earth's surface—including the atmosphere, the oceans, the land, and living organisms—it can assume other forms. These forms include *wind energy, hydroelectric energy, biomass energy* (produced by burning wood or other organic products), and **fossil fuels,** concentrated forms of burnable carbon contained in rock. Solar energy can also exist simply as *direct solar energy*—sunlight that generates heat.

Fossil fuels are enriched with carbon from the remains of organisms that lived millions of years ago. These fuels include coal, oil, and natural gas. They provide 86% of all the energy consumed in the United States: nearly two-thirds of the electricity and virtually all transportation fuels (gasoline, jet fuel, etc.). Moreover, our nation's reliance on fossil fuels, especially oil, to power the economy is on track to grow in the next two decades, even in the face of aggressive development of other energy sources. That's a disturbing fact given that geologists, energy economists, and government officials are debating whether supplies of affordable oil will be exhausted in a few decades, or even a few years.

In contrast to solar energy, geothermal energy is heat that radiates from Earth's interior and through the surface, where it enters the atmosphere. Geothermal heat is produced by the radioactive decay of unstable nuclei of atoms in the core, mantle, and crust. It collects in Earth's interior because rock has excellent insulation properties. This heat can be tapped for commercial purposes by accessing hot groundwater that is circulating near igneous intrusions. Notably, Reykjavik (the capital of Iceland) derives 99% of its energy from geothermal sources (**FIGURE 14.27**), giving it a reputation as a clean, energy-self-sufficient city.

It is also possible to tap heat from nuclear decay by making concentrated *fuel rods* of radioactive materials, such as uranium. This is how *nuclear power plants* are fueled. Like the heat produced by geothermal energy, the heat produced by nuclear decay is used to turn water into steam that can drive a *turbine*. A turbine is a machine that generates electricity by turning a shaft with running water, steam, wind, or some other type of energy. (Read more about "The Ubiquitous Turbine" in the Geology in Our Lives feature.)

Four **nonrenewable energy resources** supply the majority of U.S. energy needs: petroleum (oil), coal, natural gas, and uranium (**FIGURE 14.28**). Chief among these is petroleum, the major source of energy powering modern society.

FIGURE 14.27 The Nesjavellir Geothermal Power Station in Iceland provides the space heating and hot water needs of Reykjavik, Iceland. Geothermal heating meets the needs of about 87% of all buildings in the country.

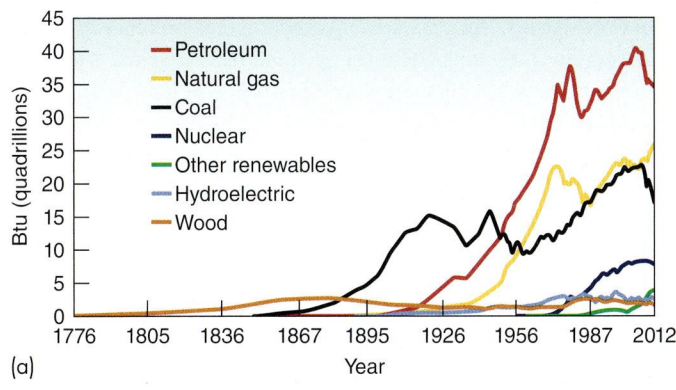

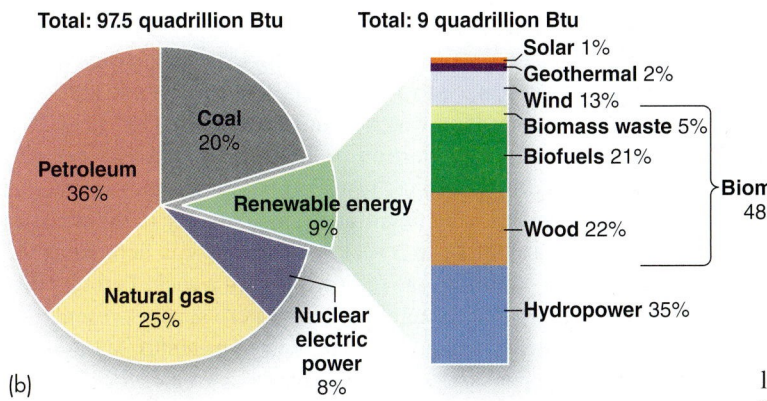

FIGURE 14.28 Energy consumption in the United States: (a) Energy consumption (British thermal units Btu) patterns have changed significantly over the history of the United States as new energy sources have been developed and as uses of energy have changed. Notice the recent decline in the use of petroleum and coal and the rise in other renewables and natural gas. (b) Today, the primary sources of energy used in the U.S. come from a mixture of fossil fuels (coal, petroleum, and natural gas) and nuclear power and renewable power.

Where does the energy come from where you live?

Petroleum

There is no other way to put it: The world is addicted to **petroleum,** more commonly known as oil. Altogether, the world now consumes a staggering *1000 barrels of oil per second.* That means that by the time you finish reading this sentence, nearly one-quarter million barrels of oil will have entered the air as a mixture of carbon dioxide, carbon monoxide, nitrous oxide, sulfur dioxide, unburned hydrocarbons, and particulate matter known as *black soot.*

Upon entering the atmosphere, these pollutants become major contributors to human-induced global warming, urban smog, and human respiratory disease. Ironically, after oil was first discovered in the middle of the nineteenth century, it was quickly adopted by industrialized nations because it was cheaper and cleaner than coal, the primary source of energy at the time. Now, with the price of oil skyrocketing, many predict a return in the future to reliance on coal power. Others want to see clean energy sources, such as wind and solar power, replace dirty fossil fuels.

The word "petroleum" translates to "rock oil," from the Latin *petra,* "stone," and *oleum,* "oil." Places where natural gas and oil escape from Earth's surface, called *seeps,* have been known from early times (**FIGURE 14.29**). The Sumerians, Assyrians, and Babylonians used crude oil collected from large seeps on the Euphrates River in what is now Iraq. The Egyptians employed oil as a wound dressing, liniment, and

Martin Shields/Science Source

FIGURE 14.29 Oil seeps, where petroleum naturally flows out of the ground, are found in many countries; they gave early oil hunters clues about where to drill.

Why did oil quickly replace coal as the primary source of energy after it was discovered?

laxative. Spanish explorers discovered oil pools in present-day Cuba, Mexico, Bolivia, and Peru. Native Americans in what is now New York and Pennsylvania tapped oil for medicinal purposes.

By the nineteenth century, the Industrial Revolution was underway. The need to lubricate moving machinery, as well as illuminate dark factories, spurred the search for new ways to obtain oil. Energy to drive machinery, formerly supplied by wood and coal, was supplanted by energy supplied by oil. Liquid petroleum was more concentrated and easier to transport than any fuel previously available.

These conditions set the stage for drilling the first oil well, which was done by *Edwin L. Drake* (1819–1880) and *William A. Smith* (1812–1890) in Titusville, Pennsylvania (**FIGURE 14.30**). Drake and Smith struck oil at a depth of 21 meters on August 27, 1859. With the spread of their drilling techniques, inexpensive oil flowed from the crust and was processed at already existing coal-oil refineries (where kerosene was produced from coal). By the end of the nineteenth century, oil wells had been drilled in 14 states and across Europe and East Asia.

FIGURE 14.30 Edwin L. Drake, manager of the Seneca Oil Well Company, drilled the first oil well in Titusville, PA, with William A. Smith, a local blacksmith, in summer 1859. In bringing forth liquid petroleum to the economy as a viable natural resource, this event ultimately changed the world.

Library of Congress Prints and Photographs Division Washington, D.C. 20540 US

Where Does Oil Come From? Petroleum comes from plankton. Oil and natural gas is the product of compressing and heating carbon-rich deposits of single-celled plants, such as *diatoms* and blue-green algae, and single-celled animals, such as *foraminifera*, over geologic time. This plankton once lived in environments of marine, brackish, or freshwater; after they died, they were buried rapidly, which prevented decay by oxidation and enabled later conversion into petroleum.

Anaerobic microorganisms in silt and clay convert buried organic matter into *kerogen* and *methane* (natural gas). Kerogen is a waxy, dark-colored product of chemically altered plant and animal debris. In time, as kerogen is buried deeper and subjected to higher temperatures and greater pressures, it experiences thermal degradation and *cracking* (a process in which heavy organic molecules are broken up into lighter ones).

Depending on the amount and type of organic matter, cracking occurs at depths between 760 and 4880 meters and temperatures between 65°C and 150°C, an environment termed the *oil window*. The resulting compounds, or kerogen products, share a basic molecular structure of carbon and hydrogen atoms linked in a ring in which atoms of sulfur, oxygen, and nitrogen are found. With continued cracking in the oil window, these products evolve into thousands of different organic compounds, all built of hydrogen and carbon atoms; these constitute a class of compounds known as *hydrocarbons*, of which the most important is oil.

Source Beds and Reservoir Beds Hydrocarbons in the form of oil are expelled from their *source bed*, the place where they are originally deposited and formed (usually, a deposit of organic shale) by pressure related to compaction of the sediments. Oil migrates through coarse-grained *carrier beds*, usually consisting of sandstone or carbonates (**FIGURE 14.31**), until it is trapped in a porous *reservoir bed* (also usually, sandstone or carbonate) and forms a *reservoir*. In some cases, however, it may migrate to the surface and become a seep.

Fore-arc basins, located at convergent plate boundaries (refer back to section 3–6), may be sites of abundant biogenic sediment deposition. Because of their geologic history, fore-arc basins form major oil provinces.

An idealized fore-arc basin consists of organic-rich source rocks, regionally extensive sandstone or limestone units that serve as reservoir beds, and a unit of impermeable sedimentary rock that seals the sequence (preventing seeps and placing the system under pressure). As shown in **FIGURE 14.31A–C**, during subduction the overriding plate actively deforms (a), producing uplifted crust, which undergoes weathering and erosion, (b) yielding sediments to the basin. These bury source rocks, creating reservoir beds and a seal, and (c) allowing for oil migration to the shallow flanks of the basin over distances of several hundred kilometers.

Most buried sediments are saturated with water, meaning that oil migration takes place in an environment of groundwater. And because oil is lighter than water, in time it floats to the top and accumulates in the highest portion of a reservoir bed. The porosity and permeability of carrier and reservoir beds are important factors in the migration and accumulation of oil. Porosity is the percentage of open (pore) space in a rock; permeability is defined as the degree to which a fluid is able to flow through a rock. Permeability depends on the number of interconnected pore spaces present.

A rock can have high porosity and low permeability (an example is shale, in which pore spaces are not connected), or moderate porosity and high permeability (as in some sandstones, in which pore spaces are connected). Most petroleum is found in *clastic reservoirs* (composed of sandstone and siltstone), where the oil resides in pores between sediment grains. Next in abundance are *carbonate reservoirs* (composed of limestone and dolomite), where the oil resides in karst cavities formed by dissolution of the rock during weathering, as well as between sediment grains. Reservoir rocks usually have a porosity ranging from 5%

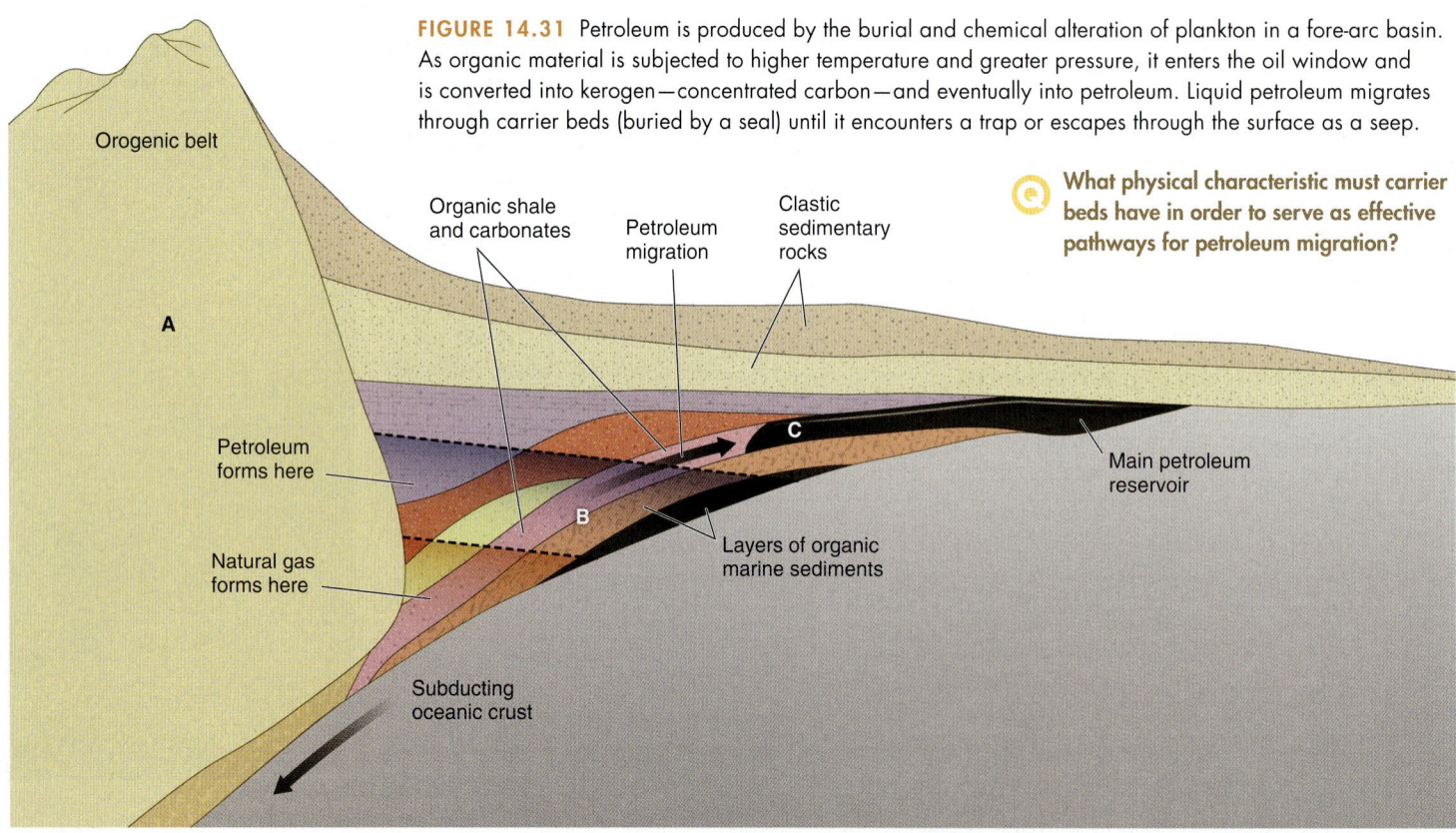

FIGURE 14.31 Petroleum is produced by the burial and chemical alteration of plankton in a fore-arc basin. As organic material is subjected to higher temperature and greater pressure, it enters the oil window and is converted into kerogen—concentrated carbon—and eventually into petroleum. Liquid petroleum migrates through carrier beds (buried by a seal) until it encounters a trap or escapes through the surface as a seep.

Orogenic belt

Organic shale and carbonates

Petroleum migration

Clastic sedimentary rocks

Q What physical characteristic must carrier beds have in order to serve as effective pathways for petroleum migration?

A

Petroleum forms here

Natural gas forms here

B

C

Layers of organic marine sediments

Main petroleum reservoir

Subducting oceanic crust

to 30%, but not all pore spaces are occupied by oil; a certain amount of groundwater cannot be displaced and so is always present.

The Status of the World Oil Supply   Two overriding principles apply to world oil supplies: 1) Most oil is found in a few large fields, and most fields are small; and 2) as exploration progresses, the average size of discoveries shrinks, as does the amount of oil found by exploratory drilling. In any region, the large fields usually are discovered first.

Although geological limitations ultimately govern the amount of oil available to the world, the price of oil is controlled by the nations that produce it. These nations have banded together to form alliances with the common goal of keeping oil prices stable in order to secure steady income for members. In other words, these nations control the availability of oil to the world and, therefore, the price, to ensure a profit.

The largest alliance is the *Organization of Petroleum Exporting Countries*, or OPEC. OPEC includes 13 nations in the Middle East, South America, Africa, and Southeast Asia. Together, this group controls approximately two-thirds of the world's oil reserves and over 40% of the oil production.

A total of about 50,000 oil fields have been discovered worldwide; however, the global impact of over 90% of them is insignificant. The largest oil fields—the "supergiants"—hold 5 billion or more barrels of recoverable oil; the "world-class giants" contain 500 million to 5 billion barrels. Approximately 280 world-class giant fields have been discovered. These, plus the supergiants, hold 80% of the world's known recoverable oil. Less than 5% of the known fields contain roughly 95% of the world's known oil.

Fewer than 40 supergiant fields are known worldwide, yet before being pumped these contained about half of all the oil that had been discovered up to that time. The Arabian–Iranian Sedimentary Basin in the Middle East contains two-thirds of these supergiant fields (**FIGURE 14.32**). The remaining supergiants are found in the United States (2), Russia (2), Mexico (2), Libya (1), Algeria (1), Venezuela (1), and China (1).

Oil Consumption   Although the United States consumes roughly 19.5 million barrels of oil a day, mostly to power its 200 million automobiles, it produces only about 5 million barrels a day, enough to supply a little more than 25% of its needs. The other roughly 75%, about 14.5 million barrels a day, has to be imported from other countries.

Canada provides the majority of our petroleum imports, with a little more than 19%. About 18% of our oil imports come from Persian Gulf countries (Saudi Arabia, Bahrain, Iraq, and others). Mexico (10%), Venezuela (9%), Nigeria (7.7%), and several other international sources meet the remainder of our oil needs. Meanwhile, the global demand for oil is growing,

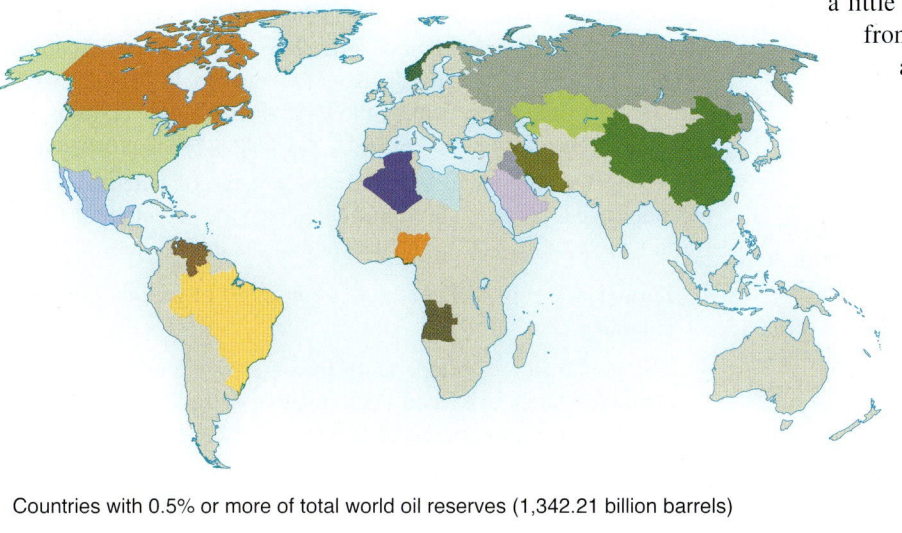

Countries with 0.5% or more of total world oil reserves (1,342.21 billion barrels)

(a)

Remaining world oil reserves

Saudia Arabia (19.87)
Canada (13.27)
Iran (10.14)
Iraq (8.57)
Kuwait (7.75)
Venezuela (7.4)
United Arab Emirates (7.29)
Russia (4.47)
Libya (3.25)
Nigeria (2.7)
Kazakhstan (2.24)
U.S. (1.59)
China (1.19)
Qatar (1.13)
Brazil (0.94)
Algeria (0.91)
Mexico (0.78)
Angola (0.67)
Azerbaijan (0.52)
Norway (0.5)

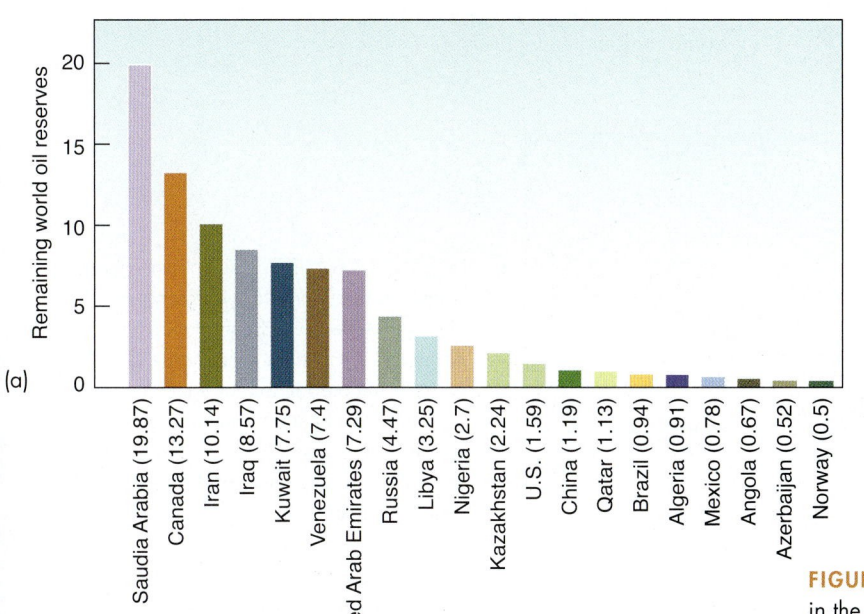

(b)

Thailand (1.1)
Taiwan (1.1)
Australia (1.1)
Netherlands (1.2)
Indonesia (1.4)
Spain (1.9)
Iran (2.0)
Italy (2.0)
United Kingdom (2.1)
France (2.3)
South Korea (2.6)
Mexico (2.5)
Saudi Arabia (2.7)
Canada (2.8)
Brazil (2.8)
Germany (2.9)
India (3.2)
Russia (3.4)
Japan (5.9)
China (0.9)
United States (24.3)
Rest of World (21.9)

% World oil consumption (total = 85,085,664 bbl/day)

FIGURE 14.32   (a) Of the world's oil reserves, the majority are located in the Middle East, Canada, South America, and Russia. (b) The United States is by far the world's most oil-hungry nation.

with India, China, and South Korea in particular fueling recent sharp increases in demand. It is inevitable, then, that competition will heat up among nations for a diminishing world oil supply.

The first 200 billion barrels of oil were produced in the 109 years from 1859 to 1968. Now that same amount is produced in less than a decade, at a rate of 22 billion barrels a year. **FIGURE 14.32A** shows the world oil supply. Estimates of total undiscovered oil resources, made by exploration geologists who have researched geologic deposits around the world, range from 275 billion to 1.469 trillion barrels.

The world's total oil supply (past, present, and future) amounts to about 2.39 trillion barrels. Of this, it is estimated that 77% has already been discovered and 30% has already been produced and consumed. If this estimate proves to be reasonably accurate, current rates of oil production could be sustained until about the middle of this century. At that point, a shortage of oil will force a production decline and major changes in the lifestyles of both oil-producing and oil-consuming nations. However, significant increases in oil usage will shorten the time to "peak oil". (Economists have given the name "peak oil" to the time when the maximum rate of petroleum production is reached, after which the rate of production is expected to enter terminal decline).

Over 50% of world oil supplies are located in the Middle East. North America's supplies add up to distant second, and almost half of that has already been produced. Eastern Europe, because of large deposits in Russia, is well endowed with oil, whereas Western Europe is not. Most of its oil is in the North Sea, where it is difficult and expensive to access. Africa, Asia, and South America have moderate amounts of oil. Large undiscovered oil resources are believed to exist in North America, Eastern Europe, and the Middle East.

Q EXPAND YOUR THINKING

How are you planning to manage the problem of increasingly expensive gasoline in your daily life?

14-10 Multiple Efforts Are Underway to Manage Climate Change

LO 14-10 Define carbon quotas, emission trading, and renewable energy.

The collective effort to reduce the release of greenhouse gases to the atmosphere is known as *mitigation*. Mitigation steps include formulating international treaties that commit nations to lowering greenhouse gas production, replacing carbon-based energy with noncarbon or *renewable energy* sources, and other requirements.

Intergovernmental Panel on Climate Change

In 1988, the United Nations and the World Meteorological Organization established the **Intergovernmental Panel on Climate Change (IPCC)** to monitor the complex issue of global warming. The IPCC's first major report, issued in 1990, indicated that there was broad international agreement that human activities were having a marked effect on global climate. Almost a decade later, in 2001, the IPCC projected future warming under various carbon dioxide *emissions scenarios*, economic models that project future greenhouse gas production. In 2014, the IPCC published its fifth assessment of global climate change, updating the emissions scenarios of 2001 and 2007, and making future projections.

Using GCM's, the IPCC analyzed a range of potential future greenhouse gas concentrations and calculated the potential for future warming (**FIGURE 14.33**). The results indicate that:

- Further warming will continue if emissions of greenhouse gases continue.

- The global surface temperature increase by the end of the 21st century is likely to exceed 1.5°C relative to the 1850 to 1900 period for most scenarios, and is likely to exceed 2.0°C for many scenarios

- The global water cycle will change, with increases in disparity between wet and dry regions, as well as wet and dry seasons, with some regional exceptions.

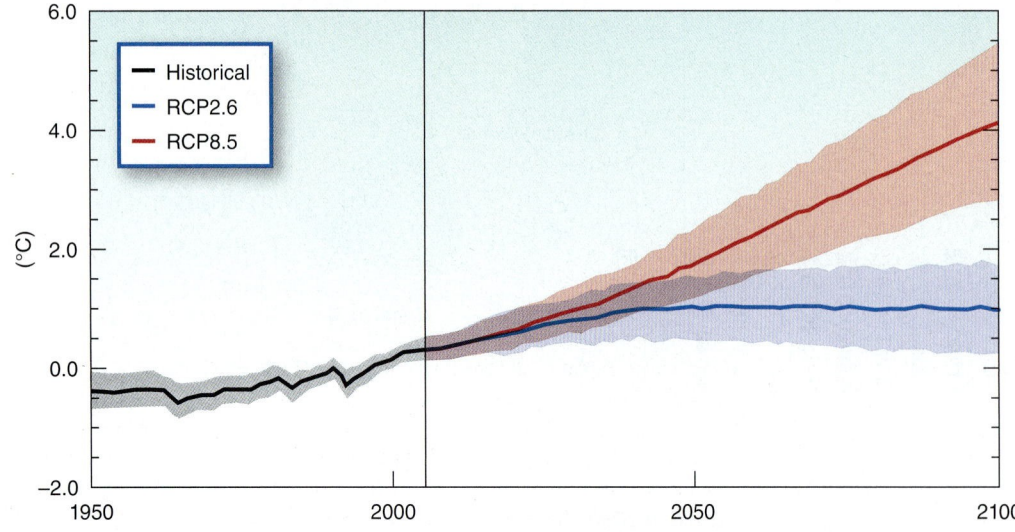

FIGURE 14.33 IPCC projections of global temperature change for two scenarios: RCP (representative concentration pathway) 2.6, where greenhouse gases cause a net global heat budget increase of 2.6 watts per square meter of Earth's surface—equivalent to a carbon dioxide concentration maximum of 421 ppm; and RCP8.5, equivalent to 936 ppm by 2100.

Q Why does this projection of future temperature grow less certain with time?

- The oceans will continue to warm, with heat extending to the deep ocean, affecting circulation patterns.

- Decreases are very likely in Arctic sea ice cover, Northern Hemisphere spring snow cover, and global glacier volume.

- Global mean sea level will continue to rise at a rate very likely to exceed the rate of the past four decades.

- Changes in climate will cause an increase in the rate of CO_2 production. Increased uptake by the oceans will increase the acidification of the oceans.

Future surface temperatures will be largely determined by cumulative CO_2, which means climate change will continue even if CO_2 emissions are stopped.

The Kyoto Protocol

In 1997, representatives of the United States and 83 other nations met in Kyoto, Japan, to discuss global climate change. The result of their deliberations was the *Kyoto Protocol*, an international agreement that carbon dioxide emissions should be regulated. The protocol proposed establishing *carbon quotas* for each country based on its population and level of industrialization. A carbon quota is a fixed permissible amount of CO_2 that a country is allowed to release to the atmosphere each year.

For example, the protocol required that by 2012 industrialized countries like the United States reduce their carbon dioxide emissions to 7% below the levels measured in 1990, and by 20% by 2020. As the leading developed nation, the United States has historically produced the bulk of the world's carbon dioxide emissions. Then, in 2007, China's output surpassed that of the United States; and powerful economic growth in India is rapidly escalating emission production there as well.

The United States never signed the Kyoto Protocol (because the U.S. Congress at the time refused to endorse the treaty), yet it reached the 2012 goal set by it, due in part to an economic recession and improved automobile fuel efficiency.

Renewable Energy

Before 1973, most industrial nations enjoyed almost unlimited access to fossil fuel supplies. Coal and oil, which together met over 90% of U.S. energy needs, were thought to be plentiful, despite rising demand. Then, during the following decade, the United States endured two severe oil shortages arising from political tensions in the Middle East, where, as noted, a large share of the world's oil supply is located. This drove up gasoline prices, put the economy into a tailspin, and sent U.S. lawmakers and government agencies scrabbling to develop new energy policies.

From this turmoil, and with crucial help from the government, the modern **renewable energy** industry was born. Renewable energy from inexhaustible sources such as the Sun and wind offered a tempting solution to the nation's energy problems (**FIGURE 14.34**). Renewable energy sources include:

- *Solar power*, which relies on the Sun's energy to provide heat, light, hot water, electricity, and even cooling for homes, businesses, and industry. Solar technologies include absorptive materials that convert sunlight into electricity (photovoltaic solar cells), mirrors that concentrate solar power, sunlight used to heat and cool buildings, and solar heating of hot water and living space.

- *Wind power*, which is used to generate electricity, charge batteries, pump water, and grind grain. Large wind turbines in *wind farms* generate electricity that can be sold to customers. The United States has over 6,300 megawatts of wind-generating capacity.

- *Bioenergy* technologies, which burn crops grown specifically as energy sources to make electricity, fuel, heat, chemicals, and other materials. In the United States, bioenergy accounts for 3% of energy production and ranks with hydropower as a crucial renewable energy source.

- *Geothermal energy*, which is harnessed from underground heat to produce electricity that can be sold. It also powers geothermal heat pumps that employ the near-constant temperature of soil or surface water to regulate the indoor climate in buildings. In the United States, most geothermal resources are concentrated in the West, but heat pumps can be used almost anywhere.

(a) ssuaphotos/Shutterstock

(b) © Topic Photo Agency/Corbis

FIGURE 14.34 Fields of (a) wind farms and (b) solar collectors incorporate advanced technologies, but the cost of these energy sources remains higher than that of fossil fuels. Until the cost of renewable energy drops, or the price of fossil fuel goes up even higher, renewable energy sources will probably not be effective competitors in the energy market.

🔘 **Why are solar and wind energy considered renewable?**

- *Hydrogen fuel cells*, which work like batteries but do not run down or need to be recharged. As long as fuel (hydrogen) is supplied, they will produce electricity and heat. Fuel cells can be used to power vehicles or to provide electricity and heat to buildings.

- *Hydroelectric power*, which is generated by flowing water, was for many years the largest source of renewable energy in the United States. However, bioenergy has begun to replace hydroelectric power as a nationally prominent source of renewable energy. U.S. hydroelectric facilities can supply 28 million households with electricity, the equivalent of nearly 500 million barrels of oil. Total U.S. hydropower capacity amounts to about 95,000 megawatts. Turbine design research continues to improve this energy source.

- *Ocean energy*, which draws on the energy of ocean waves and tides or on the heat stored in the ocean.

Despite decades of research and development, renewable energy to this date plays a minor role in the energy market. It accounts for only 1.6 % of global energy production (**FIGURE 14.35**). These energy sources are not widely used because they cost more than fossil fuel energy, sometimes two to four times as much.

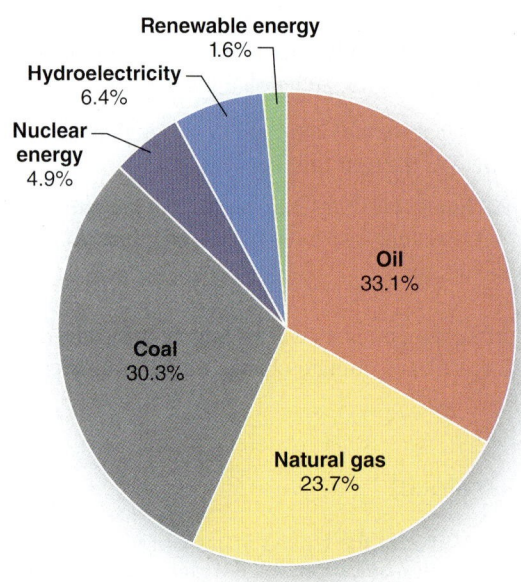

FIGURE 14.35 Globally, over 87% of new energy still comes from fossil fuels despite recent increases in renewable energy production.

 **EXPAND YOUR THINKING**

Describe four ways to significantly reduce greenhouse gas emissions. What are the costs and benefits of each? You will have to do some independent research (on the Web) to answer this adequately.

GEOLOGY IN OUR LIVES

The Ubiquitous Turbine

If you take a physics course, you will learn that by spinning a coil of copper wire between the two poles of a magnet you can make an electric current. This principle was discovered in the 1800s by physicist *Michael Faraday* (1791–1867). The machine that performs this work is known as a *turbine*, and it is employed by the power industry to make the electricity that is such an integral part of our lives. Approximately 80% of all electricity used by the world today is generated by the steam turbine (**FIGURE 14.36**), wherein the coil is turned by steam pushing fan blades inside the turbine. Making steam requires boiling water, and the heat for this can come from burning coal, oil, natural gas, vegetation, geothermal heat, nuclear fission—in other words, nearly every source of power we have discussed in this chapter. Even wind, waves, tide currents, and falling water can be used to turn a coil inside a magnet to create an electrical current.

FIGURE 14.36 Steam turbines consist of fan blades that are turned by steam to generate electricity. Steam can be created by nuclear fission, fossil fuels, biomass, and geothermal energy. Turbines can also be turned by wind, ocean currents and waves, and falling water.

LET'S REVIEW

The phenomenon of global warming, attributable to heat-trapping greenhouse gas production by humans, is dangerously altering Earth's climate, leading to rising sea levels, changes in weather patterns, and serious negative impacts on ecosystems. Climate change poses an extraordinarily difficult challenge to our economy and our environment. Only by limiting greenhouse gas production can we prevent the worst future repercussions of climate change. If our goal is to preserve a planet similar to that upon which civilization developed, and to which life on Earth is adapted, greenhouse gas production will have to be dramatically reduced.

With the help of computer models, scientists are predicting that temperatures will continue to rise during the twenty-first century and beyond. This trend is described in the fifth report of the Intergovernmental Panel on Climate Change (IPCC), which is based on the findings of an international team of climate scientists.

Meeting the challenge of climate change will require sustained worldwide effort over decades: on the part of governments, which must guide their cultures and societies as the effects of climate change unfold and as technological solutions are developed; on the part of industry, which must innovate, manufacture, and operate under new conditions in which climate change will drive many decisions; and on the part of the public, which must begin to follow a more environmentally conscious path, in deciding what they purchase and how and where they live.

In the next chapter we study the environment that is most sensitive to climate change: glaciers.

STUDY GUIDE

14-1 Climate change causes shifts in environmental processes that affect the whole Earth.

- **Global warming** refers to the rise in the average temperature of Earth's surface, including the oceans. The decade 2000 to 2009 was the warmest recorded since 1880. The year 2005 was the hottest ever measured, breaking the record set in 1998; 2009 was the second warmest (tying with several others).

- Climate is the long-term average pattern of weather of a particular region (say, over 30 years) and is the product of interactions among land, ocean, atmosphere, ice, and biosphere. More specifically, it is the product of processes that accumulate and move heat within and between the atmosphere and ocean.

- Human activities are responsible for significant global changes in land use, air and water quality, and the quantity of natural resources, particularly in the past two centuries. There is scientific consensus that human activities are also altering Earth's climate, largely due to rising levels of carbon dioxide and other **greenhouse gases** released into the atmosphere by the burning of fossil fuels.

14-2 Heat circulation in the atmosphere and oceans maintains Earth's climate.

- Climate change is set in motion by shifts in the accumulation and movement of heat in the ocean and atmosphere. To understand both natural and human influences on global climate, we must explore the physical processes governing the atmosphere and the ocean.

- There is very little vertical mixing of gases in the atmosphere. This means that one layer can be warming while another is cooling at the same time. The general circulation of air at the surface is broken up into a number of cells, of which the most common is the Hadley cell.

- **Thermohaline circulation** starts in the North Atlantic Ocean. When the warm surface water of the Gulf Stream reaches the cold polar North Atlantic, it cools and sinks. It travels southward as a cold, deep, salty current called the North Atlantic Deep Water. In the southern hemisphere, the current turns and flows northward past Australia into the Pacific Ocean. Eventually, it rises in the North Pacific, becoming a warm surface current that flows westward around Africa. It may take about 1600 years for water to complete the entire global cycle.

14-3 The greenhouse effect is at the heart of Earth's climate system.

- The **greenhouse effect** is the atmosphere's capability to store heat radiated from Earth. The heat absorbed by certain gases—such as water vapor (H_2O), carbon dioxide (CO_2), and methane (CH_4)—maintains Earth's temperature at an average (and comfortable) 14°C. Without it, Earth's average temperature would be −18°C.

- There are six principal greenhouse gases in Earth's atmosphere that absorb long-wave radiation and keep Earth warm: water vapor, carbon dioxide, ozone (O_3), methane , fluorocarbons, and nitrous oxide (N_2O). Combined, these gases make up about 1% of the atmosphere.

14-4 The global carbon cycle describes how carbon moves through natural systems.

- The **global carbon cycle** comprises the forms (and exchange rates) that carbon takes in the reservoirs (storage points) and sources (points of origin) for carbon on Earth.

- Most of the carbon on Earth is contained in the rocks of the lithosphere; it was deposited slowly over millions of years. Carbon is stored in the lithosphere in two forms: Oxidized carbon is buried as carbonate, such as limestone, which is composed of calcium carbonate ($CaCO_3$); reduced carbon is buried in the form of organic matter, such as dead plant and animal tissue, and may be incorporated in shale and other organic-rich forms, such as fossil fuels.

- Carbon is also found in gases in the atmosphere, such as carbon dioxide and methane; carbon dioxide dissolved in the hydrosphere (oceans and freshwater); and organic matter in the biosphere, such as the simple carbohydrate glucose ($C_6H_{12}O_6$), in plants and animals.

14-5 Modeling improves our understanding of climate change.

- Earth's climate system is very complex, with cycles and feedbacks and reservoirs that all interact with one another over different lengths of time. Scientists who study climate change use climate models, computer-based mathematical programs that simulate the behavior and interaction of Earth's oceans and atmosphere.

- The concentration of carbon dioxide in the atmosphere has varied naturally by a factor of 10 in the last 600 million years; today it is close to its minimum level, suggesting that we are in a relatively cool period in Phanerozoic history. Because of its role in warming Earth, carbon dioxide has affected global climate on long-time scales. Significant global changes in climate, sea level, and atmospheric composition have been part of Earth's history for millions of years.

14-6 Human activities have raised the level of carbon dioxide in the atmosphere.

- Four trillion tons of carbon are stored in the rocks of Earth's crust. The burning of fossil fuels (such as coal, oil, and gas) and organic biomass (such as trees during **deforestation**) are among the activities that humans undertake to produce usable forms of energy for electricity, industry, and motorized vehicles. These activities release the stored carbon into the atmosphere.

- In the last 200 years, carbon dioxide levels in the atmosphere have risen by 40% (from 280 to over 370 ppm), and they continue to rise by approximately 2.1% each year.

- As a result of the still-growing human population and accompanying industrial development, the consumption of natural resources has increased by 1000% in the last 70 years. Human population growth worldwide is expected to continue, reaching 9 billion people by 2050 and 11 billion by 2100. Such numbers will put even greater pressure on Earth's natural resources.

- Carbon dioxide released by changes in **land use** represents about 18% of total annual emissions. This source constitutes one-third of the total emissions from developing countries and more than 60% of emissions from the least developed countries.

14-7 Earth's atmospheric temperature has risen by about 0.9°C in the past 100 years.

- According to published research, global surface temperatures have gone up about 0.9°C since the late nineteenth century; furthermore, the linear trend for the past 50 years of 0.2°C per decade is nearly twice that for the past 100 years. The year 2009 was only a fraction of a degree cooler than 2005, the warmest year on record, and tied with a cluster of other years—1998, 2002, 2003, 2006, and 2007—as the second warmest year since record-keeping began. January 2000 to December 2009 was the warmest decade on record. Scientific analysis of modern and past climates has shown that Earth is warmer now than at any time in the past 1,300 years.

- High latitudes are warming faster than low latitudes. The Arctic has warmed faster than any other region on Earth—by nearly 2.7°C in the last 30 years alone.

- Glaciers and sea ice all over the world are melting at the fastest rates ever measured. Alaska has 2000 large glaciers, and it is estimated that 99% of them are "retreating," or melting. The extent of snow cover in the northern hemisphere has declined by about 10% since 1966. In 2007, record high rates were recorded for the melting of Arctic sea ice and Greenland's massive icecap.

- Sea level has been rising at the rate of 1 to 3 millimeters per year for the past 100 years, up to 20 centimeters worldwide.

14-8 Climate change leads to ocean acidification and warming, glacier melting, weather changes, and other negative impacts.

- Sea surface temperature (SST) has gone up by an average of 0.6°C in the last 100 years. **Ocean acidification** threatens marine organisms that rely on calcium carbonate.

- The clear signs and consequences of climate change in the biosphere include degradation and shrinkage of wildlife habitats, alterations in the distribution of food sources for animals and people, disruption of the timing of migration patterns, and **desertification.**

14-9 Fossil fuels, principally oil, provide most of the energy that powers society.

- Beginning in the 1970s, U.S. consumption of energy began to rapidly outpace domestic production, leading to heavier reliance on imported energy sources. Currently, 36% of U.S. energy needs are met by petroleum, 20% by coal, 25% by natural gas, 8% by nuclear power, and 9% by various types of renewable energy.

- Over the past half-century, most of the growth in energy sources has occurred in petroleum, natural gas, coal, and nuclear power.

- Petroleum comes from plankton. Oil and natural gas is the product of compressing and heating carbon-rich deposits of single-celled plants, such as *diatoms* and blue-green algae, and single-celled animals, such as *foraminifera*, over geologic time.

- Hydrocarbons in the form of oil are expelled from their *source bed*, where they are originally deposited and formed (usually, a deposit of organic shale), by pressure related to compaction of the sediments. Oil migrates through coarse-grained *carrier beds*, usually consisting of sandstone or carbonates, until it is trapped in a porous *reservoir bed* (also usually sandstone or carbonate) and forms a *reservoir*. In some cases, however, it may migrate to the surface and become a seep.

- Less than 5% of the known fields contain roughly 95% of the world's known oil.

- Two overriding principles apply to world oil supplies: 1) Most oil is found in a few large fields, and most fields are small; and 2) as exploration progresses, the average size of discoveries shrinks, as does the amount of oil found by exploratory drilling. In any region, the large fields usually are discovered first.

14-10 Multiple efforts are underway to manage climate change.

- The **Intergovernmental Panel on Climate Change (IPCC)** was established in 1988 by the United Nations and the World Meteorological Organization to address the complex issue of global climate change, particularly global warming.

- The Kyoto Protocol is an international agreement stating that carbon dioxide emissions resulting from human activities are seriously affecting Earth's climate and should be regulated over the long term.

- Renewable energy sources include: solar power, wind power, bioenergy, geothermal energy, hydrogen fuel cells, hydroelectric power, and ocean energy.

- Despite decades of research and development, renewable energy plays a small role in the energy market. It accounts for only 1.6% of global energy production.

KEY TERMS

climate feedback (p. 410)
deforestation (p. 406)
desertification (p. 427)
fossil fuels (p. 430)
geothermal energy (p. 430)
global carbon cycle (p. 411)
global circulation models (GCMs) (p. 413)
global climate change (p. 402)
global warming (p. 403)

greenhouse effect (p. 406)
greenhouse gases (p. 404)
hydrosphere (p. 405)
Industrial Revolution (p. 408)
insolation (p. 413)
Intergovernmental Panel on Climate Change (IPCC) (p. 434)
land use (p. 419)
nonrenewable energy sources (p. 430)

ocean acidification (p. 426)
petroleum (p. 431)
renewable energy (p. 435)
sea level rise (p. 403)
solar energy (p. 430)
thermohaline circulation (p. 405)
trade winds (p. 405)

ASSESSING YOUR KNOWLEDGE

Please complete this exercise before coming to class. Identify the best answer to each question.

1. Earth's atmosphere has warmed by about:
   a. 9°C in the last 100 years.
   b. 0.09°C in the last 1,000 years.
   c. 90°C in the last 100,000 years.
   d. 0.9°C in the last 100 years.
   e. Earth's atmosphere has not warmed.

2. Identify the greenhouse gas that is considered the most responsible for climate change_____.

Bruce Forster/Getty Images, Inc.

3. The carbon dioxide content of the atmosphere has risen in recent decades. This is likely due to:
   a. Increased volcanic outgassing.
   b. Decreased eccentricity and obliquity.
   c. Increased burning of fossil fuels.
   d. The breakup of Pangaea.
   e. None of the above.

4. The country that emits the most greenhouse gases is:
   a. China.
   b. Brazil.
   c. Russia.
   d. United States.
   e. India.

5. Heat-trapping gases in the _____ (stratosphere, troposphere, mesosphere) are causing cooling in the _____ (stratosphere, troposphere, mesosphere).

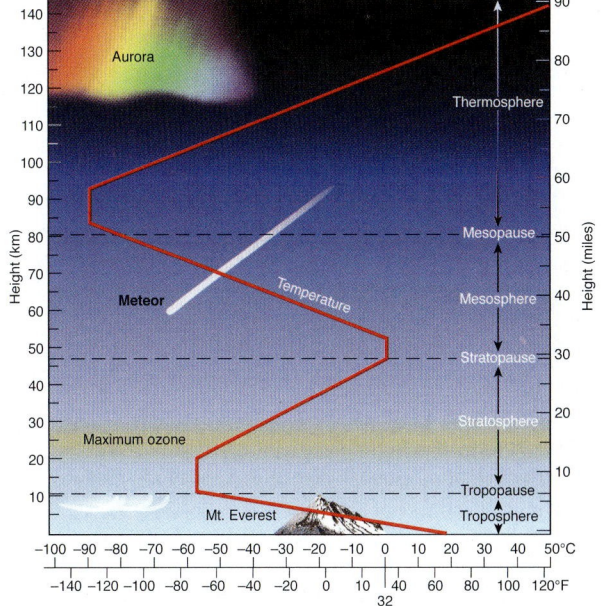

6. Heat circulation in the atmosphere governs climate. It comprises:
   a. Global circulation in the form of a number of "cells."
   b. Thermohaline circulation, in which air takes tens of thousands of years to circle the globe.
   c. Trade winds, which "trade" heat vertically from the troposphere into the ionosphere.
   d. Oceanic upwelling, which releases heat into the atmosphere from warm deep waters.
   e. There is very little heat circulation in the atmosphere.

7. The greenhouse effect:
   a. Is the atmosphere's natural capability to store heat radiated from Earth.
   b. Is governed in large part by the capability of ice to absorb heat.
   c. Includes the process of short-wave radiation from the biosphere.
   d. Does not include heat production by the Sun.
   e. All of the above.

8. The most important greenhouse gases are:
   a. CO_2, CH_4, CFC's, O_3, N_2O, and H_2O.
   b. CO_4, FH_4, H_2O, CO_3, N_2O, and SO_2.
   c. CO_2, CO, NH_3, O_2, N_3O, and HO.
   d. HSO, CO_3, CHO_2, N_3O, and NH_4.
   e. Only CO_2.

9. It is only since 1950 that greenhouses gases have caused climate change. _____ True or False.

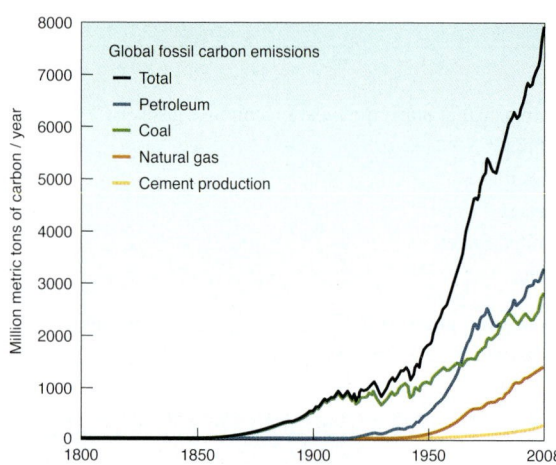

10. The atmosphere has several layers. These layers are the:
    a. Lithosphere, biosphere, ionosphere, and thermosphere.
    b. Troposphere, stratosphere, mesosphere, and thermosphere.
    c. Troposphere, ionosphere, hydrosphere, and thermosphere.
    d. Biosphere, lithosphere, atmosphere, and thermosphere.
    e. All of the above.

11. Fluctuations in solar output in the twentieth century are:
    a. Entirely responsible for global warming.
    b. A relatively small component of modern climate change.
    c. Not presently monitored.
    d. Responsible for the rise in greenhouse gas production.
    e. Solar heating does not fluctuate.

12. Climate is:
    a. The long-term average pattern of weather.
    b. The result of interactions among land, ocean, atmosphere, ice, and biosphere.
    c. The product of processes that accumulate and move heat within and between the atmosphere and ocean.
    d. Changing due to global warming.
    e. All of the above.

13. Climate change:
    a. Is changing the carbon cycle.
    b. Is happening faster in the Arctic.
    c. Is the result of global warming.
    d. Has changed the timing of the seasons.
    e. All of the above.

14. The northern hemisphere is warming faster than the southern hemisphere because _____ (it has more land, it has more people, it has more ocean).

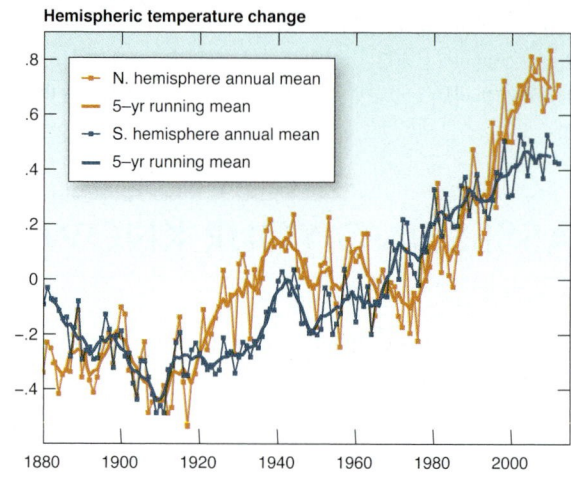

15. Which of the following statements about sea level rise is true?
    a. It is not yet occurring but is likely to occur in the future.
    b. It is occurring now and is one of the major hazards associated with global warming.
    c. It is occurring now at the highest rate in the last half-million years.
    d. It is a major cause of global warming.
    e. It is not a major worry.

16. Ways to mitigate climate change include:
    a. Reducing use of fossil fuels.
    b. Increasing use of renewable energy forms.
    c. Improving energy efficiency.
    d. Agreeing to global carbon-reduction treaties to make changes on a worldwide basis.
    e. All of the above.

17. Which of these reactions describes the calcification process?
    a. $2CO_2 + 2H_2O \rightarrow 2HCO_3^- + 2H^+$
    b. $CaCO_3 + CO_2 + H_2O \rightarrow Ca^{2+} + 2HCO_3^-$
    c. $2HCO_3^- + Ca^{2+} \rightarrow CaCO_3 + CO_2 + H_2O$
    d. $H_2O + 2HCO_3^- \rightarrow 2HCO_3^- + 2H^+$
    e. None of the above

18. A carbon quota is:
    a. The amount of carbon a single person is allowed to produce in one year.
    b. A limit of greenhouse gas emissions for each city in the world.
    c. A target greenhouse gas emission level for each country.
    d. The amount of greenhouse gas that will cause a certain level of damage.
    e. None of the above.

19. Some consequences of climate change include:
    a. Sea level rise and coral bleaching.
    b. Ocean acidification and glacier retreat.
    c. Shifts in species ranges and competition.
    d. Drought, storms, and rainfall changes.
    e. All of the above.

20. Arctic amplification is caused by:
    a. Increased solar wind.
    b. Rapid warming in the northern hemisphere.

    c. Decreased snow and ice cover.
    d. Accelerated sea level rise.
    e. None of these.

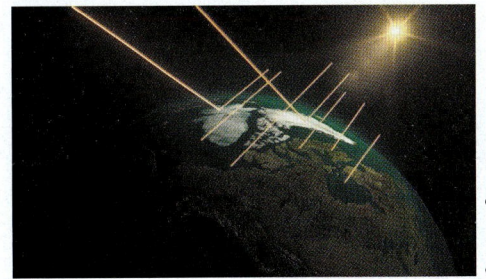
Science Source

FURTHER RESEARCH

1. Illustrate how Earth's climate has changed over the following time scales by filling in the table.

Time Scale	Pattern of Temperature Change	Primary Causes of Change	Scientific Evidence of the Change	Resulting Environmental Changes
50 years				
1,500 years				
21,000 years				
125,000 years				
400,000 years				

2. How could climate change affect us in the future?

3. How could greenhouse gas emissions be reduced?

4. What human activities contribute to climate change?

5. How is climate change related to anomalous weather and climate events?

ONLINE RESOURCES

Learn more about global warming on the following Web sites.

The U.S. Global Change Research Program, National Climate Assessment provides a review of the latest research and impacts to U.S. sectors:
www.globalchange.gov

The Pew Center on Global Climate Change advocates an aggressive approach to mitigating climate change:
www.pewclimate.org

The George C. Marshall Institute advocates a moderate approach to climate change:
www.marshall.org/subcategory.php?id=9

The Intergovernmental Panel on Climate Change provides the latest research on climate change:
www.ipcc.ch

The debate over climate change is summarized at:
www.opendemocracy.net/climate_change

Additional animations, videos, and other online resources are available at this book's companion Web site:
www.wiley.com/college/fletcher

This companion Web site also has more information about WileyPLUS and other Wiley teaching and learning resources.

GLACIERS AND PALEOCLIMATOLOGY

CHAPTER CONTENTS & LEARNING OBJECTIVES

GEOLOGY IN OUR LIVES

A glacier is a large, long-lasting river of ice; it is formed on land, undergoes internal deformation, erodes Earth's crust and deposits sediment, and, ultimately, creates glacial landforms. In this way, glaciers have shaped the mountains and nearby lands where they expanded during past periods of climate change. Earth's recent history has been characterized by cool climate periods, called ice ages, during which glaciers around the world expanded, and warm periods, called interglacials, during which glaciers retreated. By studying the history of climate changes, scientists learn how natural systems respond to shifts in temperature and ice volume. Today, most of the world's ice sheets, alpine glaciers, ice shelves, and sea ice are shrinking in reaction to global warming.

Q Glaciers are rivers of ice that sculpt the land surface by eroding the crust and depositing sediments as they move. Almost every mountain system in the world has evidence of past glaciation. What would evidence of past glaciation look like?

Mira/Alamy

15-1 A Glacier Is a River of Ice

LO 15-1 Describe glaciers and where they occur today.

Glaciers are large bodies of ice that slide and deform plastically under the force of gravity. As they move they carve the land and deposit sediments, creating **glacial landforms** that provide evidence of their past movements. We came to understand this phenomenon thanks to the work of a Swiss zoologist named Louis Agassiz.

Louis Agassiz

In 1837, a young **Louis Agassiz** (1807–1873) used critical thinking to develop a radical new hypothesis, one that would explain many aspects of northern European (and later North American) sedimentary geology and *geomorphology* (the study of landforms), which until then had no unifying scientific explanation. He suggested that northern Europe had once been covered by thick layers of ice, similar to those covering Greenland, during a period that he called the *ice age*.

Having spent a great deal of time walking among the many **glaciers** that filled the valleys of the Alps in his native Switzerland, Agassiz observed that these "rivers of ice" were in motion. He saw that their fronts advanced and retreated at various times, and that trapped within their icy grip were large amounts of bedrock, gravel, sand, and mud.

Agassiz also observed glacial sediments deposited many kilometers from their source. He compared the *glacial striations* (gouges in bedrock made by rocks suspended in glacial ice) exposed by retreating glaciers to identical gouges he found in bedrock far from any ice. From these astute observations, he was able to prove that the deep valleys and sharp peaks and ridges of the high Alps were the result of glacial erosion.

Today, Agassiz's *theory of the ice ages* is well established among scientists studying the **Quaternary period**, the most recent period of geologic time (2.6 million years ago to present). But he had to endure two decades of criticism from his peers before his original ideas would come to form the basis of scientific thinking about the origins of modern surface environments, as well as global **paleoclimatology**, the study of past climates.

How Glaciers Develop

As stated in the Geology in Our Lives introduction, a glacier is a large, long-lasting river of ice that is formed on land, undergoes internal deformation, and creates glacial landforms (**FIGURE 15.1**). Glaciers are made of compressed, recrystallized snow, and usually transport a large load of sediments; they range in length and width from several hundred meters to hundreds of kilometers. Most glaciers are several thousand years old, though the glaciers covering Greenland and Antarctica have been in existence in one form or another for hundreds of thousands of years.

Glaciers develop on land at high elevations (mountains) and high latitudes (the Arctic and Antarctic) above the *snowline*, the elevation above which snow tends to accumulate from one year to the next rather than completely melting during the summer. Even mountains in the tropics, such as the Andes of Peru, have a snowline; likewise, mountainous regions in equatorial New Guinea and Africa have glaciers.

zbindere/iStockphoto

FIGURE 15.1 A glacier is a large, long-lasting river of ice made of compressed, recrystallized snow.

 What are the dark stripes in the middle of the glacier? Develop a hypothesis about how they form.

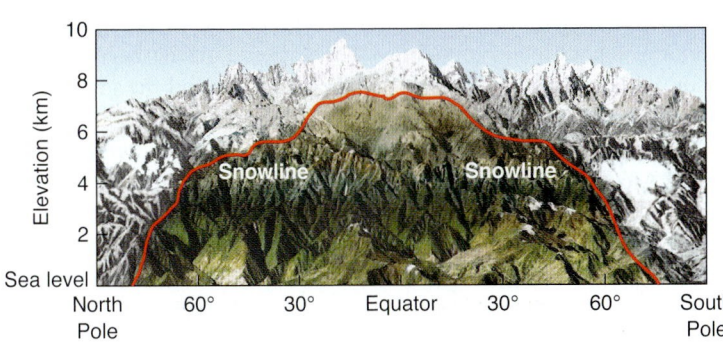

FIGURE 15.2 (a) In today's climate, glaciers form at elevations above 5,000 meters between 0° and 30° latitude. At higher latitudes, they may form near sea level. (b) Permafrost is permanently frozen soil with a seasonal active layer that can support plant life.

Why does permafrost develop? What climate conditions are necessary?

Snowlines occur at lower elevations in temperate and Arctic regions than in the tropics. In fact, the farther one travels from the equator, the lower the snowline, until reaching the High Arctic, where the ground at sea level is permanently frozen in what is known as **permafrost** (**FIGURE 15.2**).

Permafrost is soil whose temperature remains at or below the freezing point of water for more than two years. Understandably, then, it is found at high latitudes and on mountain slopes at high elevations in *periglacial environments* (environments dominated by ground ice and freeze-thaw processes, as in the tundra). The upper portion of permafrost is called the *active layer*. This layer is typically 0.6 to 4 meters thick, seasonally thaws and refreezes, and can support plant life for part of the year. In northern Siberia, the permafrost is over 1.4 kilometers thick. Today, approximately 20% of Earth's land surface is covered by permafrost or glacial ice.

Above the snowline, snow accumulates from one year to the next in relatively large quantities. This is a fundamental requirement for a glacier to develop, enabling the snow cover to become so thick that the lower layers recrystallize and become *glacier ice*. This process may involve repeated freezing and thawing, constituting a type of granular ice called *névé*, which, with continued compression, fuses to become *firn*. The weight of the overlying mass eventually causes firn to recrystallize as glacier ice (**FIGURE 15.3**).

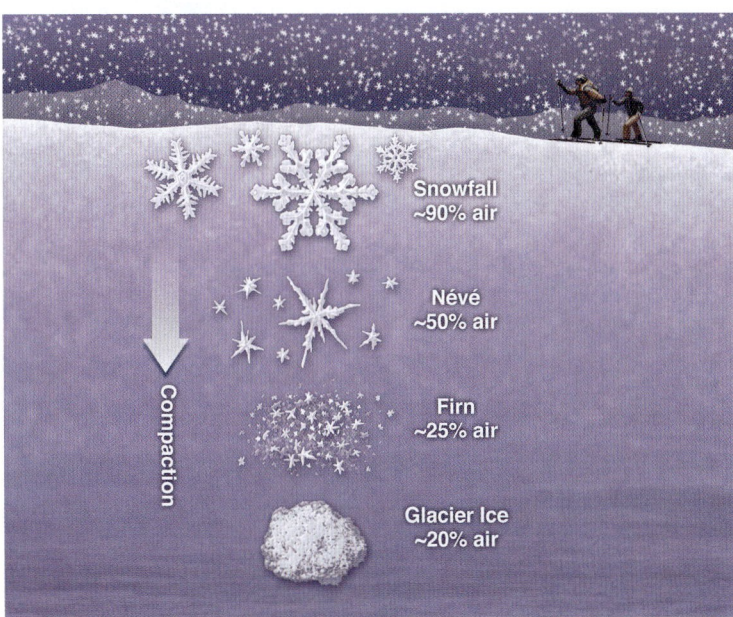

FIGURE 15.3 Snow will fuse under pressure to form névé and firn, which eventually recrystallizes as glacier ice.

Why does snow turn to ice as it is buried?

Q EXPAND YOUR THINKING

Describe the theory of the ice ages. What is the evidence supporting the theory? Can you provide an alternative hypothesis? How would you test your hypothesis?

As Ice Moves, It Erodes the Underlying Crust

15-2

As ice moves, it erodes the underlying bedrock and accumulates a large load of sediment that is carried toward the glacier *terminus*, or front.

Ice Movement

Ice behaves as a brittle solid (i.e., it fractures) until it reaches thicknesses of about 40 to 60 meters or more. At these thicknesses the lower layers of ice in a glacier experience *plastic flow* as weakly bonded layers of stacked water (H_2O) molecules begin to slide past each other (**FIGURE 15.4**).

Another way by which a glacier moves is *basal sliding*. Basal sliding is dominant in glaciers whose temperature hovers around the melting point. It sets in motion when portions of the glacier move across the crust, which is lubricated by a layer of **meltwater.**

Meltwater develops when increased pressure toward the base of a glacier lowers the melting point of ice; hence, the ice at the base melts more readily than the ice above it. Geothermal heat from Earth's interior also contributes to melting. The thicker a glacier becomes, the more efficient it is at trapping geothermal heat, and the greater the role of this heat in creating meltwater.

Keep in mind that as ice moves, it erodes the underlying bedrock. Most of this erosion occurs through two processes: plucking and abrasion (**FIGURE 15.5**). *Plucking* occurs when meltwater beneath the glacier penetrates fractures and crevices in bedrock and the resulting *ice wedging* (expansion of water when it freezes) breaks off pieces of rock. When the freezing water expands, it pries loose blocks of rock, which the moving ice then picks up.

Blocks of rock carried in the ice cause *abrasion* as they grind and pulverize the bedrock beneath a glacier, creating scars termed *glacial striations*. Pulverized rock, known as *rock flour*, is so abundant that the meltwater flowing away from a glacier is usually bluish-gray in color, a reflection of the suspended sediments. Through this erosion, sediments of all sizes become part of the glacier's load, so that the glacial deposits are poorly sorted.

Types of Glaciers

Glaciers are found in a wide diversity of forms and environments. However, there are two principal types: **alpine glaciers**, which are found in mountainous areas, and **continental glaciers**, which cover larger areas and are not confined to the high elevations of mountains. Most of the concepts we discuss in this chapter apply equally to both types (**FIGURE 15.6**).

Within these two principal categories of alpine and continental glaciers are many types. *Valley glaciers* are the smallest type of alpine glacier (refer back to Figure 15.1). The ice in *temperate glaciers* is near its melting point throughout the year, from its surface to its base; therefore, small variations in temperature—for example, from one season to the next—produce abundant water flowing from the front of the glacier, and basal slip is the dominant mode of ice movement.

(a)

Crevasse field

50 m

Brittle deformation
Plastic deformation

Flow direction

Ice below melting point

Ice near melting point

Bedrock obstruction

Water/mud film causing basal sliding

(b)

Canadian Space Agency/NASA/Ohio State University, Jet Propulsion Laboratory, Alaska SAR facility

FIGURE 15.4 The temperature near the base of a glacier is an important factor in how it moves. (a) If the ice is near the melting point, it is more likely to move by basal sliding. If it is well below the melting point, the glacier is more likely to move by plastic flow. (b) Movement of the world's largest glacier, Lambert Glacier in East Antarctica, has been mapped by satellite. The smaller tributary glaciers flow at relatively low velocities (green, 100 to 300 meters/year), whereas the downstream portion of the main Lambert Glacier flows at higher rates (blue to red, 400 to 800 meters/year).

Describe in the form of a testable hypothesis how glacier movement occurs.

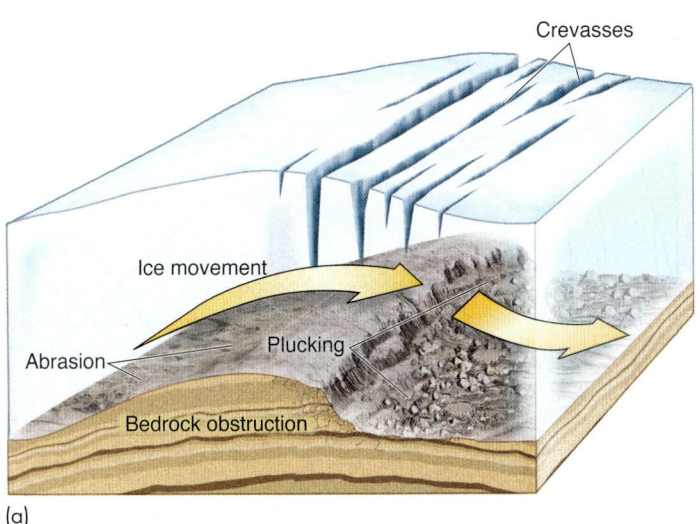

Crevasses

Ice movement

Abrasion

Plucking

Bedrock obstruction

(a)

Alan Majchrowicz/Age fotostock

(b)

Lee H. Rentz/Photoshot

(c)

FIGURE 15.5 (a) Glaciers erode bedrock through plucking and abrasion; meltwater at the base of the glacier enters fractures in the bedrock. When it freezes, ice wedging forces pieces of bedrock to break off and become incorporated into the flowing ice. Suspended blocks of rock and sediment abrade the bedrock, producing rock flour (b) and creating glacial striations (c).

Q **What glacial erosion process produces rock flour?**

The ice in *polar glaciers*, in contrast, is always below freezing; consequently, melting is rare, and ice loss occurs by *sublimation,* by which a solid (ice) turns into a gas (water vapor) without melting first. *Subpolar* glaciers have characteristics of both temperate and polar glaciers. They usually experience a seasonal zone of melting near the surface, which generates some meltwater but little to no basal melt.

Larger *icecaps* can cover an entire mountain, a range of mountains, or even a volcano—provided it is not active. The edges of icecaps typically are characterized by tongues of ice called *outlet glaciers* that flow outward from the main ice mass, usually through a mountain pass. Outlet glaciers are often very large, like the Lambert Glacier shown in Figure 15.4, an outlet glacier of the East Antarctic ice sheet.

The largest glaciers are called *continental ice sheets*; they completely cover the landscape and are so massive that the topography of the land may not visibly influence them except near their edges. They are referred to as ice sheets because they have the geometry of a bed sheet: They are thin relative to their great surface area.

The only continental ice sheets in existence today are found on Greenland and Antarctica. These glaciers are so large that if all the ice on Greenland were to melt, it would raise global sea level by about 7 meters; the ice on Antarctica, if it melted, would raise sea level by over 60 meters. The Geology in Action feature, "The Antarctic ice Sheet," has more on this.

Robert Harding Picture Library/SuperStock

(a)

Marvin Dembinsky Photo Associates/Alamy

(b)

FIGURE 15.6 There are two principal types of glaciers: (a) alpine glaciers, which form in mountain systems, and (b) continental glaciers, such as the Greenland ice sheet, which are remnants of the last ice age, when they were much larger.

Q **How do alpine glaciers differ from continental glaciers?**

Q **EXPAND YOUR THINKING**

How would you prove that glaciers cause erosion?

The Antarctic Ice Sheet

The Antarctic ice sheet consists of three geographic regions: *West Antarctica*, the *Antarctic Peninsula*, and *East Antarctica*. Recent data from satellite monitoring has allowed researchers to map the rate of ice drainage across the entire continent. This is critical information to have in order to determine the rate at which Antarctic ice is contributing to global sea level rise.

The map in **FIGURE 15.7** reveals areas where ice moves as quickly as a few kilometers per year (shown in bright purple and red) and other areas where movement is limited to a few centimeters per year (in pink). The fastest-moving glaciers are along the edge of the ice, and of these, the Pine Island and Thwaites glaciers are moving most rapidly.

Channels of fast-flowing ice extend far inland. These act as tributaries feeding large glacial rivers. The rapid flow of these inland tributaries surprised researchers, as they had previously thought this ice moved slowly and would ultimately limit the amount of ice contributing to global sea level rise. Thanks to this map, they now realize that Antarctic glaciers are capable of flowing at very high rates toward the ocean.

In West Antarctica the effects of climate change are dramatic (**FIGURE 15.8**), and warming of the atmosphere has been rapid. Ice loss in the region has increased by 59% over the past decade, to about 132 billion metric tons per year. One study identified Antarctica's Byrd Station as one of the fastest-warming regions on Earth.

Along the Antarctic Peninsula, yearly ice loss has increased by 140%, to 60 billion metric tons, and 87% of the glaciers there are retreating. In East Antarctica, melting is largely confined to the seaward edge of the massive ice sheet (where melting is caused by warm ocean currents), but every year the melting encroaches farther inland and onto higher elevations.

Despite expectations that snowfall in Antarctica would be more plentiful in the warmer atmosphere, and thereby offset melting, there has been no statistically significant increase in snowfall; to the contrary, recent data suggest that it actually has decreased slightly. Overall, Antarctica is losing a total of 100 cubic kilometers of ice each year, and the rate of loss is accelerating.

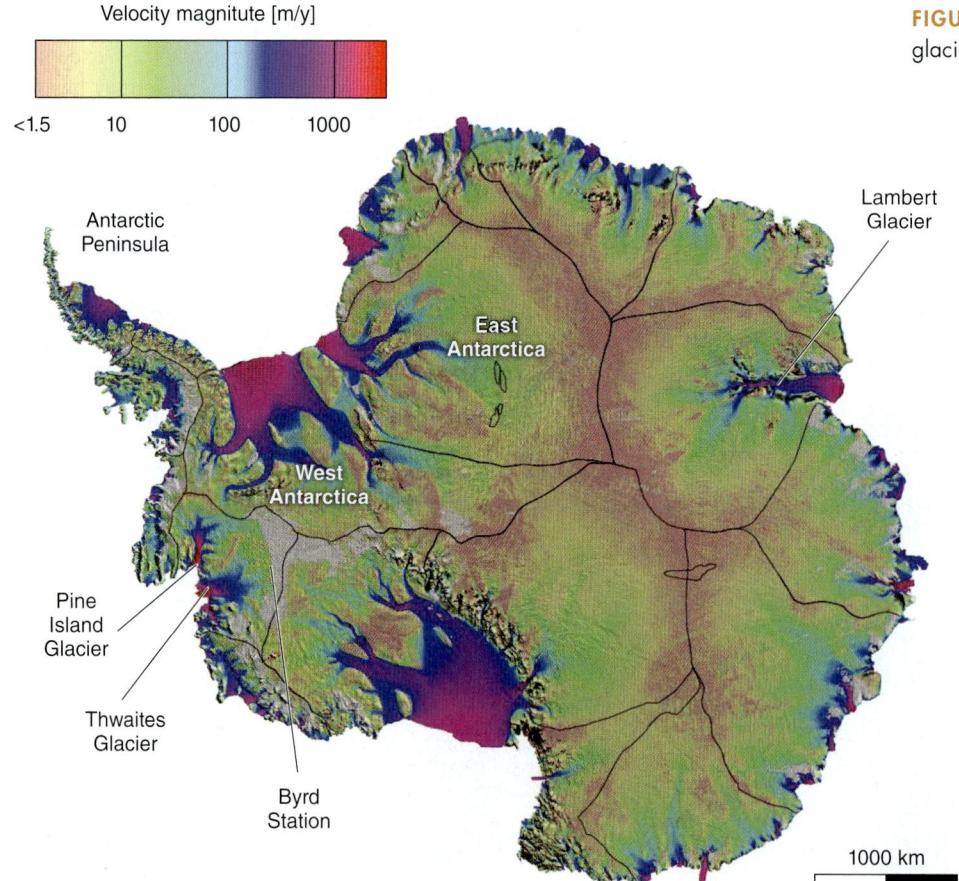

Velocity magnitude [m/y]

<1.5   10   100   1000

Antarctic Peninsula

Lambert Glacier

East Antarctica

West Antarctica

Pine Island Glacier

Thwaites Glacier

Byrd Station

NASA/JPL-Caltech/UCI

1000 km

FIGURE 15.7 Map of Antarctic glacier speed (meters/year).

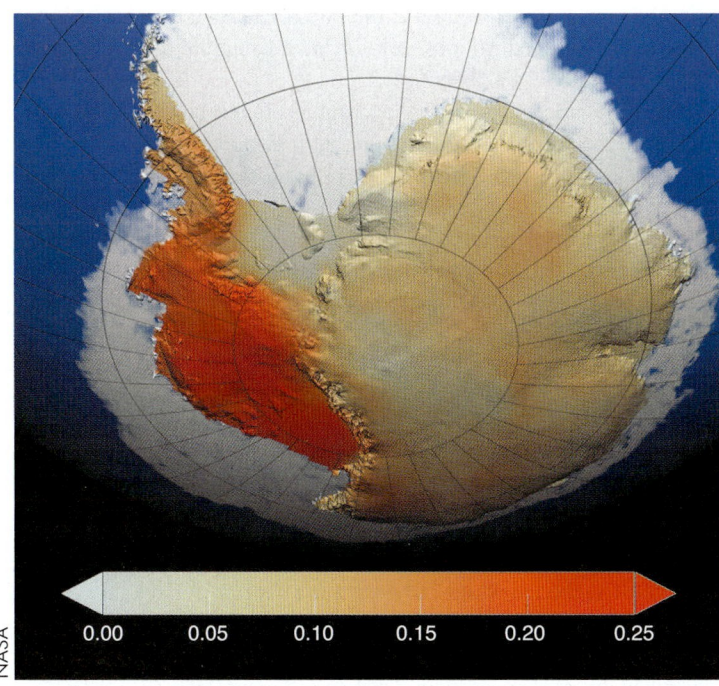

NASA

NASA/DMS

FIGURE 15.10 This crack (recording the collapse of an ice shelf) is spreading across the ice shelf of Pine Island Glacier, one of the most rapidly retreating glaciers in Antarctica. The photo, taken by a NASA mission to map ice shelf collapse, shows boulderlike blocks of ice that fell into the rift when the shelf split. NASA reports that the crack is approximately 30 kilometers long, 73 meters wide, and 57 meters deep. Scientists expect the crack to propagate and the Pine Island Glacier ice shelf to calve an iceberg of more than 777 square kilometers.

FIGURE 15.8 Antarctica has warmed at a rate of about 0.12°C per decade since 1957, for a total average temperature change of 0.5°C.

NASA/Goddard CGI Lab

FIGURE 15.9 Antarctic ice shelves are melting under the heat of warm air from above and warm ocean currents from below. The shelves are indicated by the rainbow color; red is thicker (greater than 550 meters), while blue is thinner (less than 200 meters). When ice shelves melt, the inland glaciers that feed them drain ice into the sea more rapidly, contributing to sea level rise.

Ice loss occurs by several mechanisms, including surface melting by warm air temperatures, melting on the underside of ice shelves by warm ocean currents (**FIGURE 15.9**), and increased flow into the sea by rapidly moving glaciers that empty into coastal ocean waters. An *ice shelf* is a thick floating platform of ice that forms where a glacier flows into the ocean. These glaciers eventually crack off large tabular pieces of ice as their ice shelves collapse (**FIGURE 15.10**).

As the continent of Antarctica continues to experience both warmer air temperatures and ocean currents, the amount of ice will continue to shrink. Ice, like all other forms of mass, can be weighed. Thus, it makes sense to try to weigh the Antarctic ice sheet as it loses mass in order to improve our understanding of how fast it is disappearing. Of course, trying to weigh the Antarctic ice sheet is not an easy task. Nevertheless, the European Space Agency and NASA have jointly launched a satellite mission to do just that.

Called the Gravity Recovery and Climate Experiment (GRACE), the mission consists of two satellites that were launched in 2002 to take detailed measurements of Earth's gravity field. GRACE satellites fly over the Antarctic and Greenland ice sheets every 10 days and collect detailed measurements of its total change in mass over time. Changes in mass on Antarctica are attributable to changes in the amount of ice. Gravity data show that Antarctica has been losing more than 100 cubic kilometers of ice each year since 2002, and at an accelerated rate.

Ice Moves Through the Interior of a Glacier as If on a One-Way Conveyor Belt

LO **15-3** Detail how ice moves in a glacier and explain how this movement affects the front of the glacier.

The interior of an active glacier can be thought of as a one-way conveyor belt, continuously delivering ice and sediments to its front, called the *terminus*. (More on the conveyor belt concept in section 15–10.)

Ice Movement

In a glacier, new ice originates in the **zone of accumulation**, which usually accounts for 60% to 70% of the glacier surface area, and moves internally to the **zone of wastage**, where ice is lost not only by melting but also by sublimation and *calving,* by which large blocks of ice break off the front of the glacier. Between the zone of accumulation and the zone of wastage is the *equilibrium line,* marked by the elevation at which accumulation and wasting are approximately equal (**FIGURE 15.11**).

Although ice within a glacier usually moves forward or is stagnant, the terminus will retreat if the rate of wastage exceeds the rate of ice delivery. The terminus will advance when the rate of delivery exceeds the rate of wastage.

The rate of ice delivery to the terminus often is controlled by snow buildup in the zone of accumulation, but it also may respond to short-term accelerations and decelerations in basal sliding and plastic flow. That means that in an active glacier, *ice is always moving forward* through the interior, and the terminus advances or retreats, depending on the relative rates of ice delivery and wastage; this process is characterized as the **glacier mass balance**. Ice flow may slow and stop, but a glacier will not "flow" uphill or somehow retreat other than by wasting.

As ice moves forward within the body of a glacier, it delivers sediment to the glacier's surface along fractures in the zone of wastage. Sediment is transported from the base of the glacier to the surface along a curved internal trajectory, aided by the constant wastage at the surface and a series of upward-curving fractures within the ice, along which shearing action causes sediment to migrate upward. In this way, sediment is delivered from the interior of the glacier and accumulates on the surface as the ice wastes, to form a *lag deposit* of boulders, gravel, sand, and mud. This deposit can become so thick and pervasive that it completely blankets the surface. For this reason, the surface of many glaciers in the zone of wastage is rich in sediment (**FIGURE 15.12**).

Vestiges of the Last Ice Age

Approximately 160,000 glaciers exist today, and they can be found in approximately 47 countries and on every continent except Australia. But more than 94% of all

FIGURE 15.11 An active glacier continuously feeds ice from the zone of accumulation to the zone of wastage. The glacier's terminus advances or retreats based on the balance between wasting and the internal delivery of ice from the zone of accumulation. The interior (cross section) of a glacier has two important regions. The region of plastic deformation lies below a depth of about 50 to 60 meters. (Even a glacier that moves by basal sliding has a region of plastic deformation.) Above this is the region of fracture. The rate of ice movement is slowest at the base of the glacier due to friction on the bedrock.

Describe ice movement inside a glacier that is retreating.

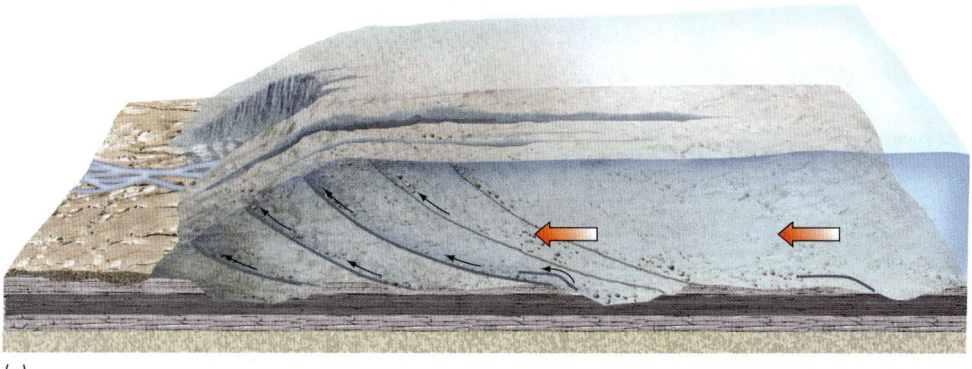

(b) © FLPA/Terry Whittaker/Age Fotostock America, Inc.

FIGURE 15.12 (a) Sediment is delivered to the zone of wastage along internal shear planes. (b) Poorly sorted sediment accumulates on the surface of glaciers.

Describe the sources, sinks, pathways, and fate of sediments in an alpine glacier.

the ice on Earth is locked up in the continental glaciers on Greenland and Antarctica.

Greenland and Antarctica are vestiges of massive ice sheets during the last ice age. The weight of these glaciers has so depressed the underlying bedrock that interior land area of Greenland lies more than 300 meters below sea level, and portions of Antarctica lie 3 kilometers below sea level. If you were to remove all the ice from these two places, the ocean would flow in and cover the majority of the land. Eventually, the land would *glacioisostatically rebound* (a term based on the concept of isostasy, which we studied in Chapter 10, referring to recovery of the crust once ice has retreated); but this process would take several thousand years.

The ice on Greenland and Antarctica has two components: 1) thick inland ice sheets that rest on the crust; and 2) thinner floating ice shelves, ice streams, outlet glaciers, and glacier tongues that take shape at the edges of the sheets (**FIGURE 15.13**). As successive layers of snow build up in the interior regions where the main ice sheets develop, the layers beneath are compressed gradually into solid ice. Snow input is balanced by glacial outflow at the edges, so unless the climate changes, the height of these ice sheets stays approximately constant through time.

The ice flowing away from the main accumulation area in a continental glacier follows a pressure gradient toward the outer edges of the glacier. This process differs from that which occurs in alpine glaciers, where ice flow is generally downhill.

In the case of continental glaciers, flow is away from the area of highest compressive stress and toward areas of lower stress along the coastline. There, ice either melts or is carried away as *icebergs,* which also eventually melt, thus returning the water to the ocean from whence it came.

Outflow from accumulation sites is organized into drainage basins separated by *ice divides* (usually, mountain ranges or stagnant ice) that concentrate the flow of ice into either narrow mountain-bounded outlet glaciers or fast-moving *ice streams* surrounded by slow-moving ice, rather than by rock walls.

Coastal settings where outlet glaciers are located are often deeply eroded into steep valleys known as *fjords* that are flooded by the ocean. In Antarctica, however, much of the outward-flowing ice has reached the coast and spread over the surface of the ocean to form floating ice shelves that are attached to ice on land. There are ice shelves along more than half of Antarctica's coast but very few in Greenland.

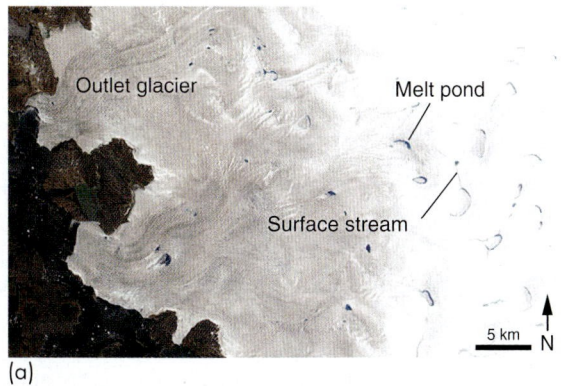

FIGURE 15.13 (a) Where the Greenland ice sheet flows into the ocean it is organized into numerous outlet glaciers that break up into icebergs. (b) Outlet glaciers flow into the sea from the Antarctic Peninsula and feed floating ice shelves that are attached to their glacier sources. When these shelves break off, they may form massive icebergs.

What drives ice flow in a continental glacier? In an alpine glacier? Does this difference affect sediment transport?

EXPAND YOUR THINKING

How would the equilibrium line respond to a warming climate? To a cooling climate? Why?

15-4 Glacial Landforms Are Widespread and Attest to Past Episodes of Glaciation

LO 15-4 Compare and contrast depositional and erosional features formed by glaciers.

Over the past 500,000 years or so, Earth's history has been characterized by great swings in global climate, from extreme states of cold (ice ages) to warm periods called **interglacials.**

Glacial History

Ice ages are characterized by the growth of massive continental ice sheets reaching across North America and northern Europe. At their maximum, these glaciers were over 4 kilometers thick in places. Accompanying the spread of ice sheets was dramatic expansion of alpine

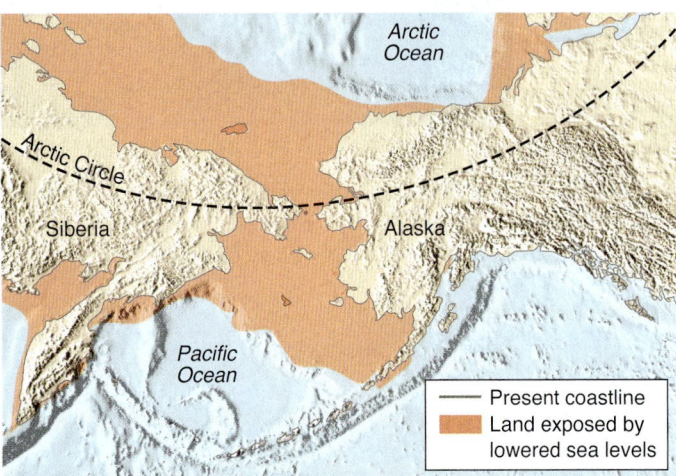

(a)

(b)

glaciers, many of which reached into icecaps that covered large areas of mountainous territory, down into valleys and onto lands surrounding the mountains.

The formation of all this ice required a vast reservoir of water, and the oceans were the obvious source of that water. Evaporation from the oceans fed snow precipitation in the cold climate, and glaciers expanded around the world. As a result, sea level fell by perhaps as much as 130 meters, exposing shallow seafloor around the continents.

In several places newly exposed seafloor connected adjacent lands that were previously separated by water. These "land bridges" allowed early communities of humans and animals to migrate to new lands. If you had been alive then, you could have walked on the newly exposed lands between Siberia and Alaska (a land bridge presumed to have aided the peopling of the Americas), from France into England and from there to Ireland, and from Malaysia across Indonesia and on to Borneo (**FIGURE 15.14**). In many places, today's shore is tens or even hundreds of kilometers from where it existed during the ice ages.

We currently live in an interglacial period called the *Holocene epoch*, which began about 10,000 years ago. The last ice age began approximately 75,000 years ago and peaked between 20,000 and 30,000 years ago. For 50,000 years the last ice age dominated Earth's climate. It was a time of great changes to the landscape, as glaciers expanded and contracted, leaving myriad *erosional* and *depositional* glacial landforms.

Erosional Glacial Features

During the last ice age, alpine glaciers carved the bedrock of mountainous regions into numerous glacial landforms. Now that the ice age has ended and most glaciers have retreated, these exposed lands are being carved again, by running water. *Glacial valleys* that have been eroded by alpine glaciers have a distinct U-shape in cross section (**FIGURE 15.15**). Valleys shaped by stream erosion, in contrast, have a distinct V-shape. In many cases, a mountain valley combines both shapes if it was once glaciated but is now being eroded by running water.

Mountainous regions contain many glacial landforms (**FIGURE 15.16**). A *cirque* is a bowl-shape, amphitheaterlike depression shaped by glacial plucking and abrasion at the head of a glacier. Cirques mark the birthplace of valley glaciers where conditions are favorable for perennial snow accumulation and the growth of glacial ice.

In the northern hemisphere, many cirques are on the north-facing sides of mountains, where there is less direct sunlight than on south-facing slopes. Cirques may contain a small lake, called a *tarn*. These high-elevation lakes, often set in idyllic alpine meadows, are known for their water clarity and beauty.

Where three or more cirques carve out opposite sides of a mountain, the resulting sharp peak is a *horn*. Two adjacent valleys that are

FIGURE 15.14 During the last ice age, sea level fell by approximately 130 meters. Immense tracts of land were uncovered around (a) the island nation of Indonesia and (b) the Bering Strait, exposing important land bridges, across which early humans migrated to new lands.

Describe changes in the water cycle that occur during an ice age.

James Steinberg/Photo Researchers

FIGURE 15.15 The broad U-shape of this valley is a result of glacial erosion during the last ice age. During the current Holocene interglacial, running water is resculpting the lower portion of the valley. The final shape of the valley will be a combination of both types of erosional processes.

What does the shape of a valley reveal about its history?

filled with glacial ice may carve a sharp ridgeline called an *arête*. These landscapes often include *hanging valleys* formed during the last ice age, when a glacier in a smaller tributary valley joined a larger glacier in the main valley. The *tributary glacier* would not have an opportunity to erode its base to the floor of the main valley, so when glacial ice melted, the floor of the tributary valley was left "hanging." Spectacular waterfalls plunging to the floor of the main valley often mark hanging valleys.

Depositional Features

Glaciers transport sediment of all sizes (**FIGURE 15.17**), from huge boulders the size of buildings to tiny clay particles. They carry this material either on the surface of the ice or embedded within the body of the glacier. As a result, depositional features range from very poorly sorted to well sorted, and the nature of the sorting depends on whether the sediment was transported by water or ice (or a combination of the two), for how long, and what happened to it after it was released from the ice by wasting. All sediment deposited as a result of glacial erosion is called *glacial drift*.

Glacial drift includes **glacial till,** unstratified and unsorted sediment that has been deposited directly by melting ice. Till consists of a chaotic mixture of angular rocks in a matrix of mud and sand produced by abrasion and plucking at the base of the glacier.

Glacial erratics are blocks of bedrock with a foreign lithology that have been directly deposited by melting ice. Some erratics are found hundreds of kilometers from their source. By mapping the distribution of erratics and using their lithology to trace them back to their source, geologists can determine the movement of the now vanished ice that carried these erratics.

One of the principal glacial landforms is the **glacial moraine** (**FIGURE 15.18**). A moraine is a deposit of till that marks a former position of the ice. There are several types of moraines. *Ground*

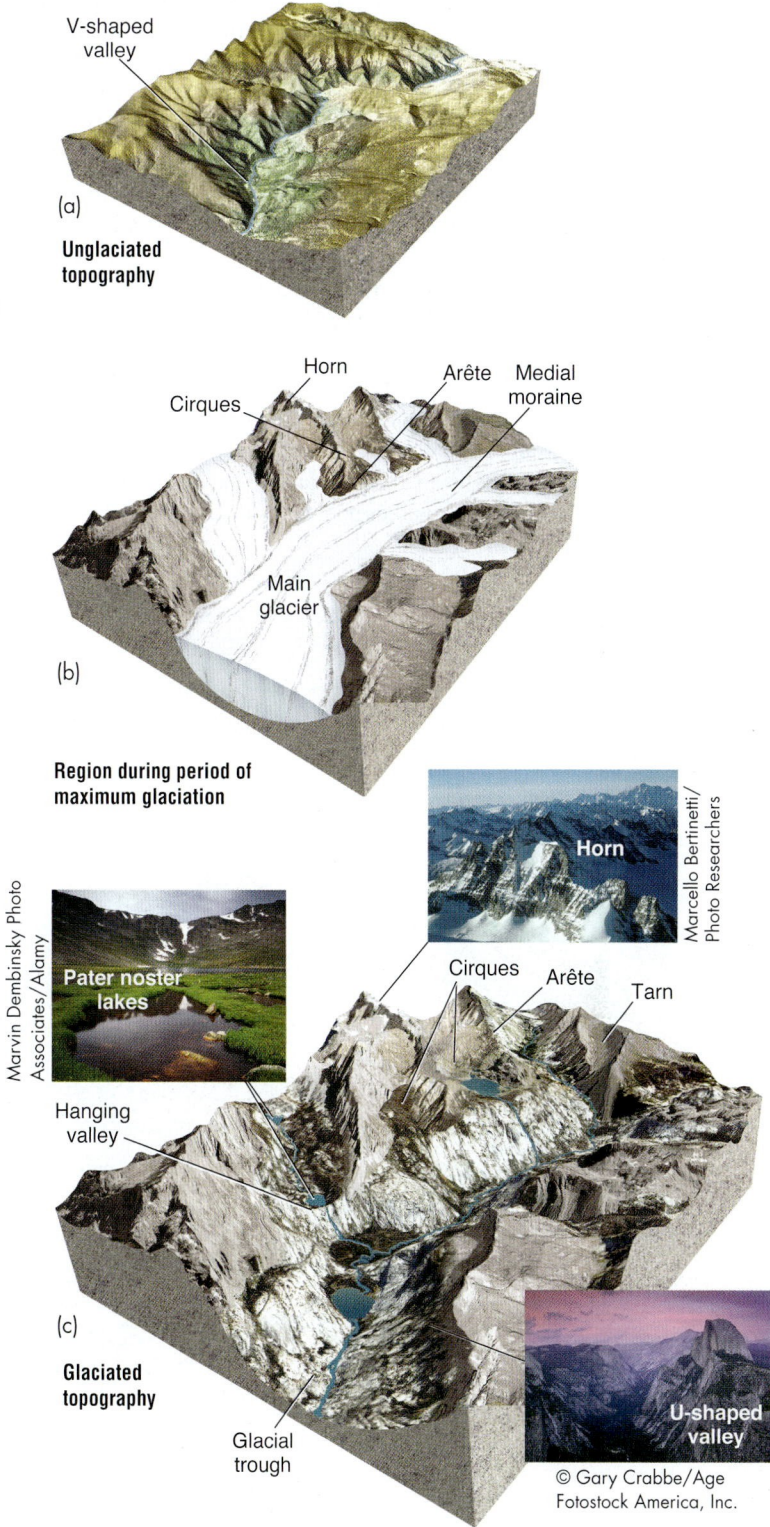

FIGURE 15.16 (a) Unglaciated topography is usually characterized by V-shape stream valleys and low, rounded mountains. (b) During an ice age, valley glaciers carve and excavate the rock. (c) When ice melts, the resulting topography is characterized by dramatically steeper terrain, high relief, and sheer walls.

Formulate a hypothesis that predicts mountainous landforms carved by erosional processes.

moraines tend to be flat deposits with a hummocky topography deposited beneath moving ice. An *end moraine* is a ridge of till marking the front edge of ice movement. *Terminal moraines* are ridges of till left behind at the front edge of a retreating glacier at the line of its farthest advance.

There may be several end moraines marking various stopping points during the history of a glacier, but there is only one terminal moraine that marks its farthest advance. *Lateral moraines* are deposits of till marking the edges of a glacier against a valley wall; a *medial moraine* is formed when two valley glaciers coalesce, and till from each mingles to create a boundary between the two.

Medial moraines are seen as black streaks in systems of valley glaciers composed of two or more tributary ice streams. All types of moraines, except the ground moraine, usually are found in a ridge composed of till. The shape and position of a moraine allow scientists to reconstruct the size, shape, and extent of the former glacier. Radiometric dating (primarily using ^{14}C) of material such as fossil wood or peat within a moraine also enables geologists to determine the timing of glacier advance and retreat.

Sediments within glacial drift can be picked up and moved by streams of meltwater that emanate from a glacier. The resulting deposit is known as *stratified drift*. One form of stratified drift is the *outwash plain* (**FIGURE 15.19**). Streams running from the end of a melting glacier usually are choked with sediment, and become *braided streams*, channels with a large sediment load forming numerous islands, as well as sand and gravel bars (discussed in Chapter 17). These streams deposit poorly sorted stratified sediment in an outwash plain—referred to simply as **glacial outwash**. If meltwater streams erode into outwash deposits, the banks form river terraces called *outwash terraces*.

Depressions called *kettle holes* may develop in outwash where a large block of ice has been buried by sediment and later melts. Kettle holes may later fill with water to become *kettle lakes*, in which fine-grained sediment usually collects.

Streams and lakes may even appear on top of a stagnant glacier and accumulate stratified sediment on the ice. When the glacier melts, these deposits are set down on the ground surface and become hills known as *kames*. Sediment in kames is stratified, because the ice melts uniformly. Former stream deposits on the ice become *kame terraces*. It is common to find glaciated landforms composing *kame and kettle topography* where kettle lakes and kame deposits are spread across the land.

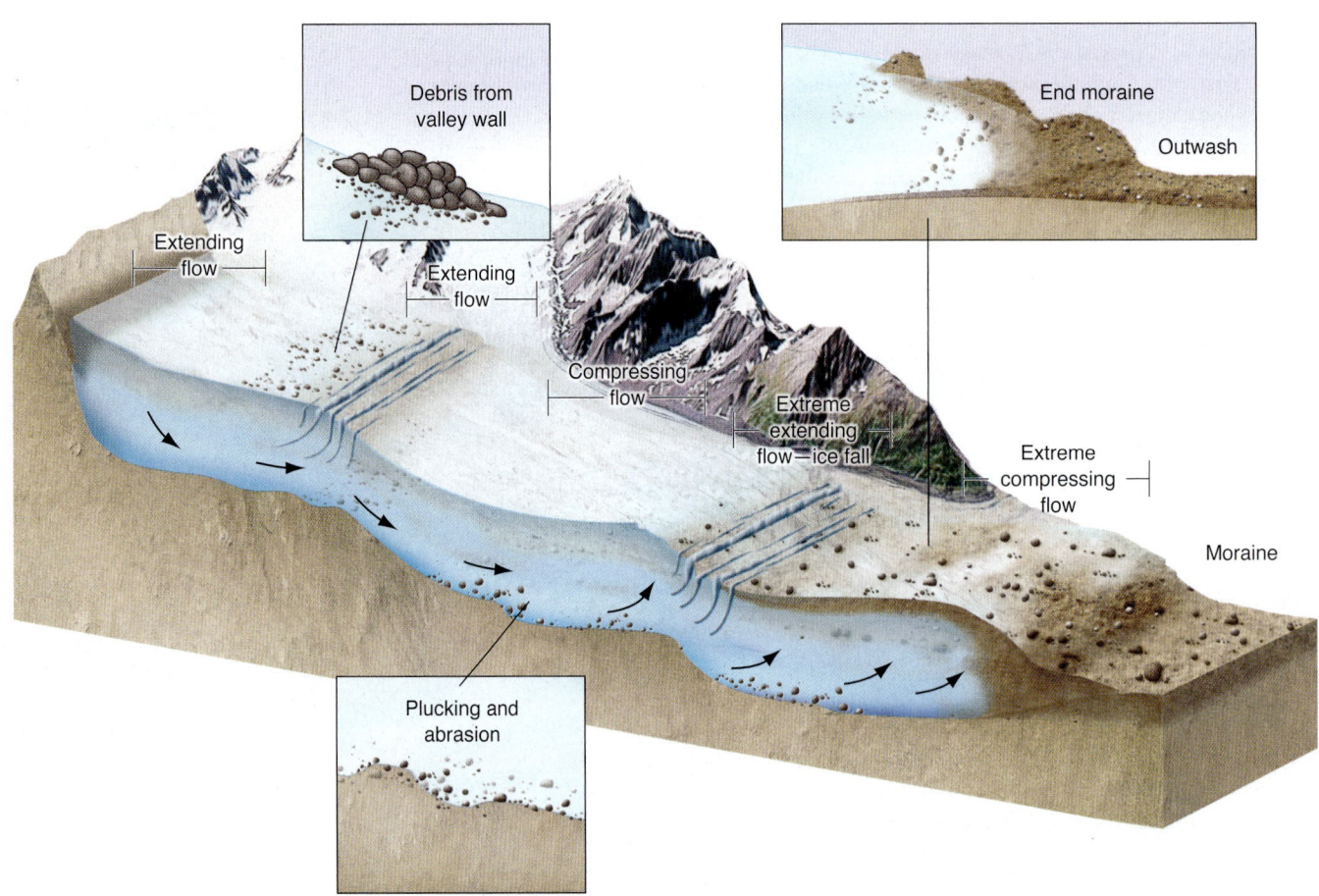

FIGURE 15.17 Glaciers transport sediment of all sizes. As the ice moves over uneven terrain, portions of the glacier compress and extend, introducing changes in the surface topography of the glacier, as well as differences in the internal movement of the ice. Glacial sediments are eroded from bedrock through plucking and abrasion; they may fall from surrounding valley walls onto the ice, and they may be carried to the front of the glacier by the internal movement of ice into the zone of wastage. When the ice retreats, the resulting deposit is known as glacial drift.

List and describe the ways that glaciers move and deposit sediment.

(a)

(b)

(c)

© Marli Bryant Miller

© Marli Bryant Miller

iStockphoto

FIGURE 15.18 (a) Glacial till is unstratified and unsorted sediment that has been deposited directly by wasting ice. (b) Moraines are deposits of till formerly in direct contact with ice. This photo shows a terminal moraine and lateral moraine. (c) When several valley glaciers coalesce, their medial moraines form strips of sediment (till) on the surface of the ice.

Q **What determines the composition and texture of sediment in all moraines?**

Eskers are long, sinuous ridges of sand and gravel deposited by streams that previously ran under or within a glacier. As the ice melts, the channel deposits are lowered to the ground, to become an esker. The land surface beneath a moving continental ice sheet can be molded into smooth, elongated shapes called *drumlins.*

Other types of glacial deposits include *glaciomarine drift,* which collects when glaciers reach the ocean or a lake. After entering the water, glaciers typically calve off large icebergs that float on the surface until they melt. Upon melting, the rock debris in the iceberg is deposited on the seafloor or lakebed as an unsorted layer of sediment. Sometimes, single large rock fragments fall out on the floor of the water body; these are called *drop stones.* Fjords are narrow inlets along the seacoast that were once occupied by a valley glacier, called a *fjord glacier.*

To test your comprehension of what you've learned in this section, complete the exercise in the Critical Thinking feature, "Interpreting Glacial Landscapes."

Q **EXPAND YOUR THINKING**

How would you determine the age of the past advance and retreat of an alpine glacier, and what would the significance of that history be?

FIGURE 15.19 Glacial landforms associated with sediment deposition.

Q **What is the difference between "till" and "outwash"?**

Wayne Lynch/Age Fotostock America, Inc.

Esker

Kevin Horan/The Image Bank/Getty Images

Drumlin field

Terminal moraine

Outwash plain

Retreating glacier

Ground moraine

Carlyn Iverson/Photo Researchers

Kettle lake

© Darrell Gulin/Corbis

Kame

CRITICAL THINKING

Interpreting Glacial Landscapes

Working with a partner, complete the following exercise, based on what you see in **FIGURE 15.20**:

1. Label the glacial landforms.

2. Describe the past maximum extent of glaciation. What is your evidence?

3. Several glaciers are depicted in the figure. Indicate which, if any, are advancing or receding.

4. What evidence or criteria did you use to arrive at your answer to number 3? Specify additional information you may need to arrive at a more accurate answer.

5. Describe sediment sources and sediment characteristics in Figure 15.20.

6. As a glaciologist you are asked to study this region and report on the impacts of global warming. Specify the data you will collect, the methods you will use, and the hypothesis you are testing. Consider questions such as: How will monitoring play a role in your project? How will you establish a history of glacier trends? How will you project/predict future changes?

7. Assuming that global warming is causing the glacial recession here, how will this impact human communities that rely on these glaciers for their water resources? What might be the short-term and long-term effects? How might such communities prepare today to address these issues?

FIGURE 15.20 Glacial landforms.

15-5 The Majority of Glaciers and Other Ice Features Are Retreating in Reaction to Global Warming

LO 15-5 Describe the response of various ice environments to global warming.

Data show that the majority of glaciers and other ice features on Earth are retreating due to global warming; and on Greenland and Antarctica, the rate of retreat has accelerated in the past decade. The retreat of alpine glaciers, the shrinking of sea ice, the breakage of ice shelves, and the melting of polar ice sheets all provide evidence of this phenomenon.

Retreat of Alpine Glaciers

Several research groups monitor the world's glaciers, and the most recent data reveal that alpine glaciers are disappearing (**FIGURE 15.21**). Alpine glaciers are recognized as sensitive indicators of climate, such that changes to their size over long periods point to a warming or cooling climate.

More specifically, researchers track a glacier's *mass balance* to discover glacier trends. The mass balance is the difference between accumulation and loss of snow on a glacier over a period of time (usually one year). When wastage exceeds snow accumulation, a glacier loses mass; it thins, and its leading edge (terminus) retreats. You can explore more about the concept of mass balance in the Critical Thinking Exercise, "Mass Balance of a Glacier."

Statistics attest that worldwide, glaciers are losing more mass than they are gaining (**FIGURE 15.22**), with 2013 marking the twenty-third consecutive year of negative mass balance for glaciers around the world. Over the past decade the average rate of melting and

(a)   NASA

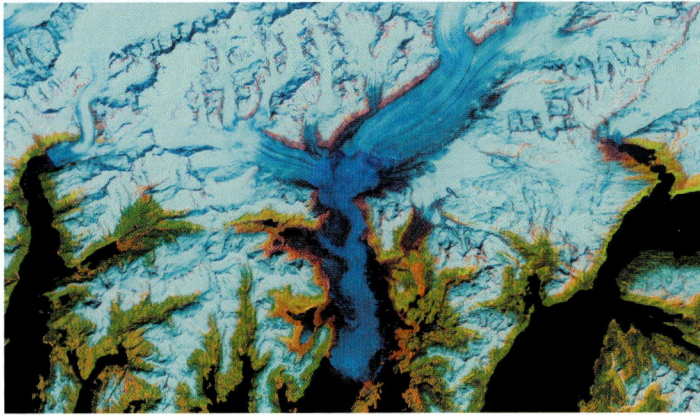

(b)   NASA

FIGURE 15.22 The Columbia Glacier empties into Prince William Sound in southeastern Alaska. It is one of the most rapidly changing glaciers in the world. Between 1986 (a) and 2011 (b), the glacier terminus retreated more than 20 kilometers. The Columbia Glacier has also thinned substantially, as shown by the expansion of brown bedrock areas in the Landsat images. Since the 1980s, the glacier has lost about half of its total thickness and volume.

What evidence would you expect to find in alpine valleys that glaciers are retreating?

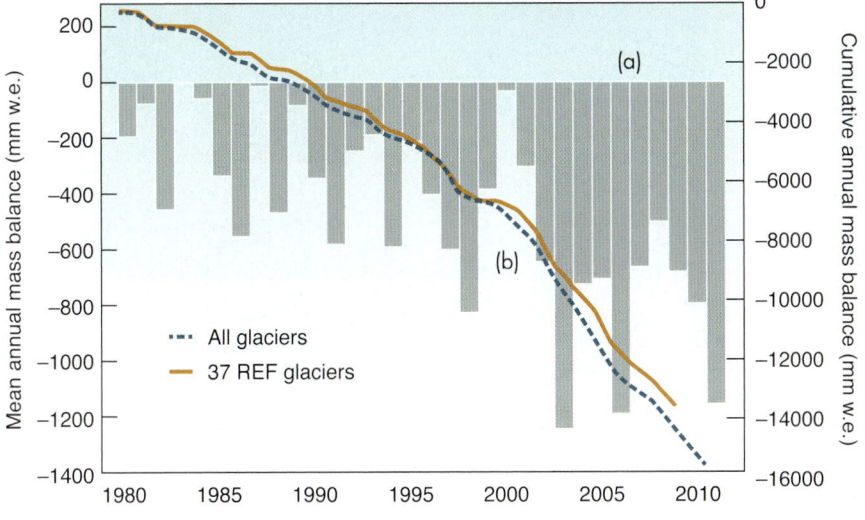

FIGURE 15.21 The World Glacier Monitoring Service collects data on global alpine glacier change over time. Its data show that (a) the annual global glacier mean mass balance (in millimeters water equivalent) has been decreasing since 1980. (b) The same data, plotted as cumulative mass balance, show a total reduction in thickness of over 12 meters for all glaciers globally (blue line) and for 37 intensely researched reference glaciers (REF, orange line).

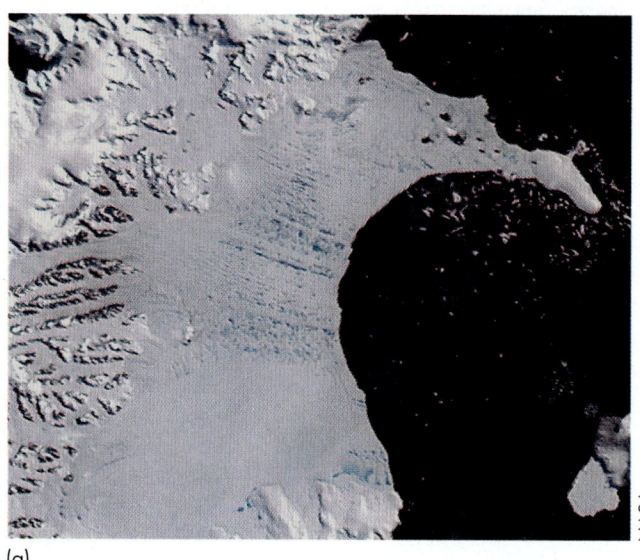

NASA/Goddard Scientific Visualization Studio

September 16, 2012

30 yr average position

FIGURE 15.23 Analysis of satellite data by NASA showed that the Arctic sea ice extent shrunk to 3.41 million square kilometers in 2012, the lowest in recorded history. Since 1979, more than 20% of the polar icecap has melted away.

thinning among alpine glaciers more than doubled; an average of 1 to 1.2 meters per year of ice thinning occurred, as compared to earlier losses averaging half a meter per year.

Shrinking of Arctic Sea Ice

In the summer of 2007, the area covered by Arctic sea ice shrank to a 29-year low, significantly below the minimum set in 2005. By the summer of 2012, the retreat of summer sea ice once again set an all-time low (**FIGURE 15.23**), establishing the smallest surface area of summer Arctic sea ice in recorded history.

Ironically, the accelerating retreat of sea ice may be attributable to changes in climate brought on by the lack of sea ice itself (refer back to the Geology in our Lives feature, "The Cascading Effects of Arctic Amplification," in Chapter 14). When there is less sea ice in the summer, the Arctic Ocean receives more heat from the Sun. The warmer water makes it harder for the ice to recover in the winter; hence, there is a greater likelihood that sea ice will retreat during the summer. This process repeats itself year after year.

Breakage of Ice Shelves

Ice shelves are thick slabs of ice that are attached to coastlines and extend out over the ocean. In the natural course of events, ice shelves often break off and become large icebergs. Beginning in the mid-1990s, however, some ice shelves began exhibiting a new behavior: rapid disintegration into small pieces, most likely as the result of warming temperatures (**FIGURE 15.24**).

In the Antarctic summer of 2002, the 220-meter-thick Larsen B ice shelf on the Antarctic Peninsula broke up. Within barely one month's time, it had disintegrated into about 3250 square kilometers of fragmented ice, an area significantly larger than the State of Rhode Island. An event of this magnitude had never been witnessed before. As a consequence of this breakage, glaciers in the adjacent mountains have surged forward, thinning at an accelerated rate because ice shelves no longer buttress where the glaciers enter the ocean. Since 2002, seven other ice shelves have collapsed around the Antarctic Peninsula.

Melting of Polar Ice Sheets

As mentioned earlier, massive continental glaciers are found in only two locations: Greenland and Antarctica. These vestiges of the last ice age are the intense focus of research because in the rapidly warm-

(a)   (b)

NASA   *NASA*

FIGURE 15.24 In early 2002, satellite imagery revealed that the Larsen B ice shelf, a large floating ice mass on the eastern side of the Antarctic Peninsula, had shattered and separated from the continent. A total of about 3250 square kilometers of shelf area disintegrated in a 35-day period beginning on January 31, 2002. Years after the event, the shelf continues to deteriorate; it has lost a total of 5700 square kilometers and is now about 40% of its previous size. (a) Larsen B, January 2002; (b) Larsen B, March 2002.

When an ice shelf collapses, what happens to the glaciers on its landward side?

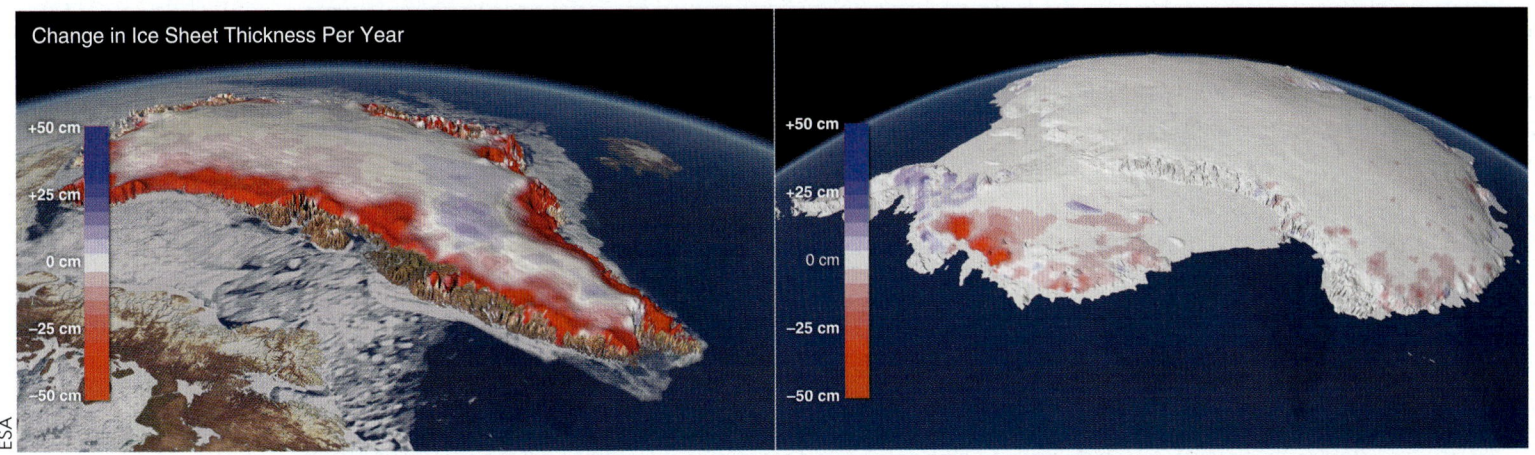

FIGURE 15.25 This image shows changes in ice sheet thickness per year for Greenland (left) and Antarctica (right).

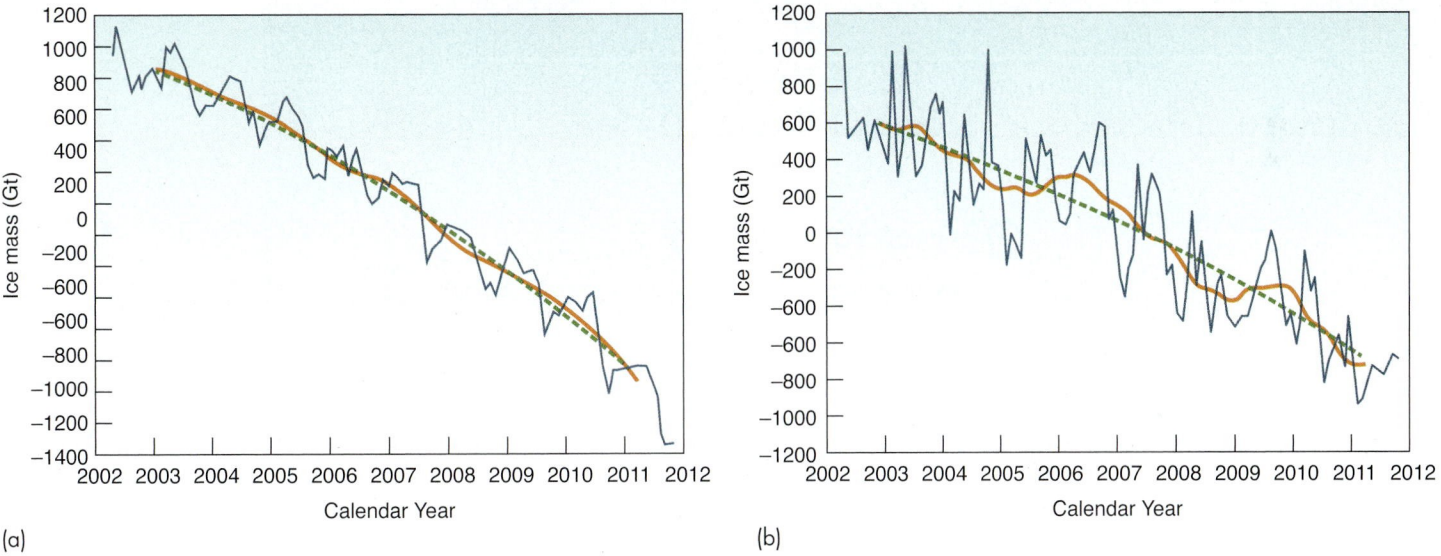

FIGURE 15.26 Changes in the mass of the (a) Greenland and (b) Antarctic ice sheets as detected by the GRACE satellite mission since 2002.

ing atmosphere, the faster these glaciers melt (**FIGURE 15.25**), the faster global sea level is going to rise and threaten billions of coastal residents worldwide. Unfortunately, in the harsh, forbidding environment on these ice sheets, it is difficult to make observations regarding the stability of the ice in a warming atmosphere. Additionally, it is difficult to extrapolate the findings of ground-based observations to the scale of the entire ice sheet. Still, satellite observations help formulate a solution to these problems, and provide detailed and accurate information on recent changes.

Greenland's mass balance has been negative since the early 1990s, with coastal regions thinning and losing glacial ice after what is believed to be a period of near-balance through much of the twentieth century. This is due to a combination of increased surface melting and accelerated ice discharge into the oceans. The interior of the ice sheet, above 2000 meters, also appears to be approaching a state of net decline. The line of equilibrium on the ice sheet is rising in elevation by over 40 meters per year, and accelerating through time. Given the current rate of warming, researchers have calculated that the surface of the entire Greenland ice sheet has a 50% probability of entering a state of irreversible decline by 2025.

An analysis of GRACE satellite gravity data (**FIGURE 15.26A**) indicates that Greenland is losing mass at 222 (±9) gigatons per year for the period 2002–2010. (You can read more about the Greenland ice sheet in the Geology in Action feature).

The Antarctic ice sheet is also vulnerable to ocean warming. Researchers estimate an escalation in the rate of melting along the undersides of floating ice shelves of 10 meters per year for every 1°C of ocean warming. This is seen in the Amundsen Sea region of West Antarctica over the past two decades, with thinning rates of tens of meters per year near the *grounding line* (the border between the land-based portion of a glacier and the ice shelf portion) of the Thwaites and Pine Island glaciers. This wastage generated a wave of thinning that traveled inland, accelerating ice flow and contributing to an overall loss of ice in West Antarctica, detected by the GRACE satellite (**FIGURE 15.26B**). Antarctic ice losses are estimated at 165 (±72) gigatons per year from 2002–2010.

As you may imagine, the loss of ice from Greenland and Antarctica is contributing to global sea level rise (**FIGURE 15.27**), approximately doubling that of the twentieth century.

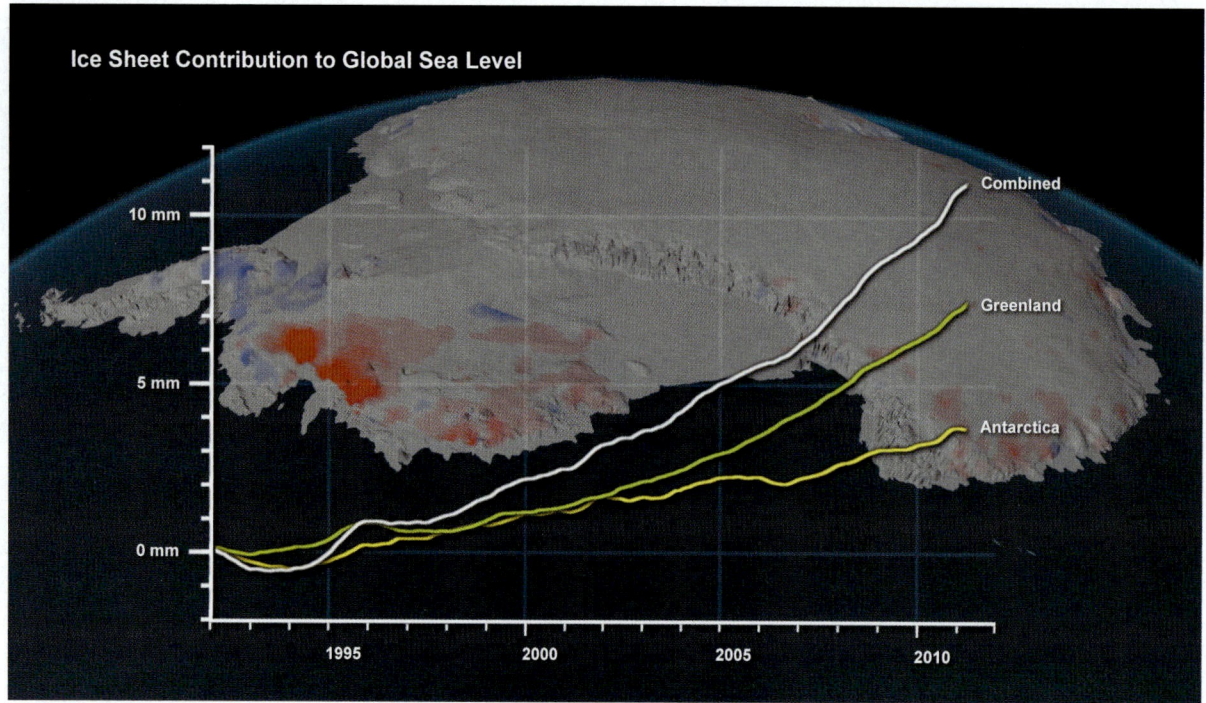

Ice Sheet Contribution to Global Sea Level

ESA/NASA/Planetary Visions

FIGURE 15.27
The loss of ice from Greenland and Antarctica is contributing to the worldwide acceleration in sea level rise since the twentieth century.

Q EXPAND YOUR THINKING

How does warming air contribute to glacier thinning?

15-6 The Ratio of Oxygen Isotopes in Glacial Ice and Deep-Sea Sediments Acts as a Proxy for Global Climate History

LO 15-6 Describe how oxygen isotopes serve as a proxy for global climate history.

Carbon dioxide levels today are the highest they have been since the Miocene epoch, 15 million years ago, and paleoclimatologists, scientists who study the climate of past ages, are using this information, and much more, to better understand Earth's history.

Clues from Paleoclimatology

Although carbon dioxide levels in the atmosphere have undergone dramatic shifts in the past several million years, not since 15 million years ago have they reached the high levels prevailing today. We know this because scientists can measure the past carbon dioxide content of the atmosphere in samples obtained by drilling in continental ice sheets and alpine glaciers, as well as in biogenic sediment composed of fossilized plankton on the seafloor (**FIGURE 15.28**).

In *ice cores*, scientists measure the past carbon dioxide content of the atmosphere directly from bubbles of air trapped during the formation of glacial ice. The longest ice cores (over 3 kilometers in length) come from Antarctica. In this sense, carbon dioxide is being used as a **climate proxy**—a representative of climate in the past—because it is directly related to the heat-trapping capability of the atmosphere.

Cores provide other measures of past climate as well. Fossil snow, for one, contains information about the temperature of the atmosphere and the amount of sunlight-blocking dust; and deep-sea cores can record changes in global ice volume and ocean chemistry.

Relevance of Oxygen Isotopes and Global Ice Volume

Deep-sea sediment is composed of the microscopic shells of fossil plankton. The chemistry of these shells—for instance, tiny plankton from the phylum *foraminifera*, shown in **FIGURE 15.28b**—leaves chemical clues to the climate prevailing when they were formed. Cores of these sediments offer a record of climate history extending hundreds of thousands to millions of years back through time.

FIGURE 15.28 Cores of ice and deep-sea sediments contain evidence of past climate. (a) Scientists from several nations have established collaborative drilling programs on the Greenland and Antarctic Ice Sheets as well as on high-elevation icecaps in mountains. (b) The Integrated Ocean Drilling Program is funded by a consortium of nations interested in using seafloor sediments to improve our understanding of Earth history.

Marc Steinmetz/Aurora Photos Inc.

Marc Steinmetz/Aurora Photos Inc.

Carlos Muñoz-Yagüe/Photo Researchers

Pasquale Sorrentino/PhotoResearchers

Marc Steinmetz/Aurora Photos Inc.

(a)

Q What is the purpose of extracting cores of old ice and marine sediments?

© JAMSTEC/IODP

© JAMSTEC/IODP

Thierry Berrod, Mona Lisa-Production/Science Source

(b)

Photo of IODP Nankai Trough Seismogenic Zone Experiment

Foraminifera use dissolved compounds and ions in seawater to precipitate microscopic shells of calcite ($CaCO_3$). Both the $CaCO_3$ of a foraminifer's skeleton and a molecule of water (H_2O) in seawater contain oxygen (O). In nature, oxygen occurs most commonly as the isotope ^{16}O, but it is also found as ^{17}O and ^{18}O. Water molecules composed of the heavier isotope ^{18}O do not evaporate as readily as those composed of the lighter isotope ^{16}O. Likewise, in atmospheric water vapor, heavier water molecules with ^{18}O tend to precipitate (as rain and snow) more readily than those composed of lighter ^{16}O (**FIGURE 15.29**).

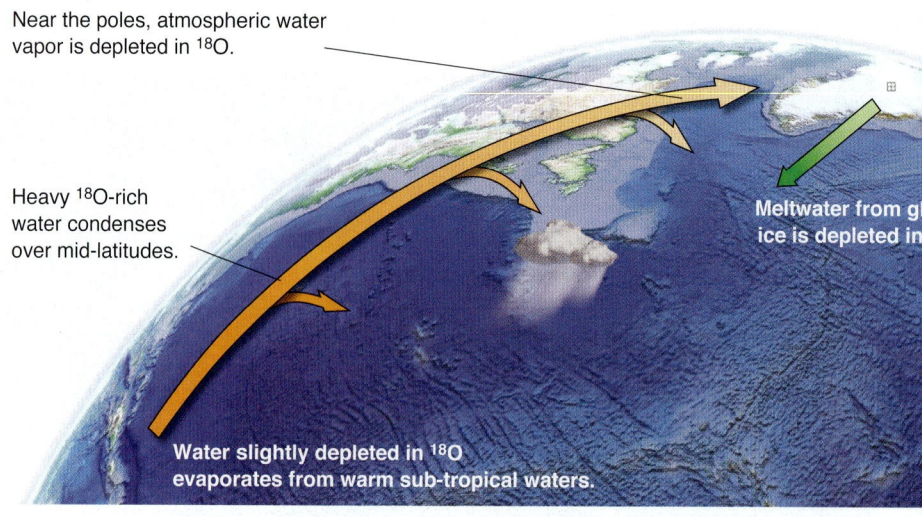

FIGURE 15.29 The $H_2^{18}O$ water molecule does not evaporate as readily as the $H_2^{16}O$ molecule. Once in the atmosphere, water vapor composed of $H_2^{18}O$ tends to condense and precipitate more readily in cooling air than does a molecule of $H_2^{16}O$. Because most water vapor originates from the tropical ocean, by the time it travels to high latitudes and high elevations where glaciers form, it is enriched in $H_2^{16}O$ relative to seawater. Hence, snow and ice are also enriched in the $H_2^{16}O$ molecule.

The molecules ^{16}O and ^{18}O behave differently in the water cycle of evaporation and precipitation. How does this reveal clues about climate?

Both evaporation and precipitation of oxygen isotopes occur in relation to temperature. $H_2^{18}O$ tends to be left behind when water vapor is formed during the evaporation of seawater, and it tends to be the first molecule to condense when rain and snow are forming. Since most water vapor in the atmosphere is formed by evaporation in the tropical ocean, that means that by the time it travels the long distance to the high latitudes and elevations where ice sheets and glaciers are located, it is relatively depleted in $H_2^{18}O$ and enriched in $H_2^{16}O$. It also means that during an ice age, vast amounts of $H_2^{16}O$ are locked up in global ice sheets for thousands of years. At the same time, the oceans are relatively enriched in $H_2^{18}O$. Since the ratio of ^{18}O to ^{16}O in the shells of foraminifera mimics the ratio of these isotopes in seawater, the oxygen isotope content of these shells provides a record of changes in global ice volume through time.

Role of Oxygen Isotopes and Atmospheric Temperature

Oxygen isotopes in fossil foraminifera leave us a record of global ice volume; and in ice cores oxygen isotopes do the same for changes in air temperature above the glacier. Because the atmosphere is so well mixed, the isotopic content of air above a glacier is indicative of the temperature of the atmosphere; therefore, the isotopic content of snow is useful as a proxy for global atmospheric temperature.

At the poles, as an air mass cools and water vapor condenses to snow, molecules of $H_2^{18}O$ condense more readily than do molecules of $H_2^{16}O$, depending on the temperature of the air. Typically, above a glacier, the condensation falls out of a cloud as snow. Thus, the oxygen isotopic content of snow (measured as the ratio of ^{18}O to ^{16}O) can serve as a proxy for air temperature; consequently, cores of glacial ice can be read as documents of the variations in air temperature through time.

Given that global ice volume and air temperature are related, the records of oxygen isotopes in foraminifera and glacial ice show similar patterns. These records provide researchers with two independent proxies for the history of global climate.

Many researchers have tested and verified the history of global ice volume preserved in deep-sea cores and the history of temperature preserved in glacial cores from every corner of the planet. Past episodes of cooler temperature reflected in ice cores strongly correlate to periods of increased global ice volume in marine sediments. Likewise, past episodes of warmer climate correlate well to periods of decreased ice volume.

An example of this record is shown in **FIGURE 15.30**. The variation in abundance of oxygen isotopes in ice (a proxy for atmospheric

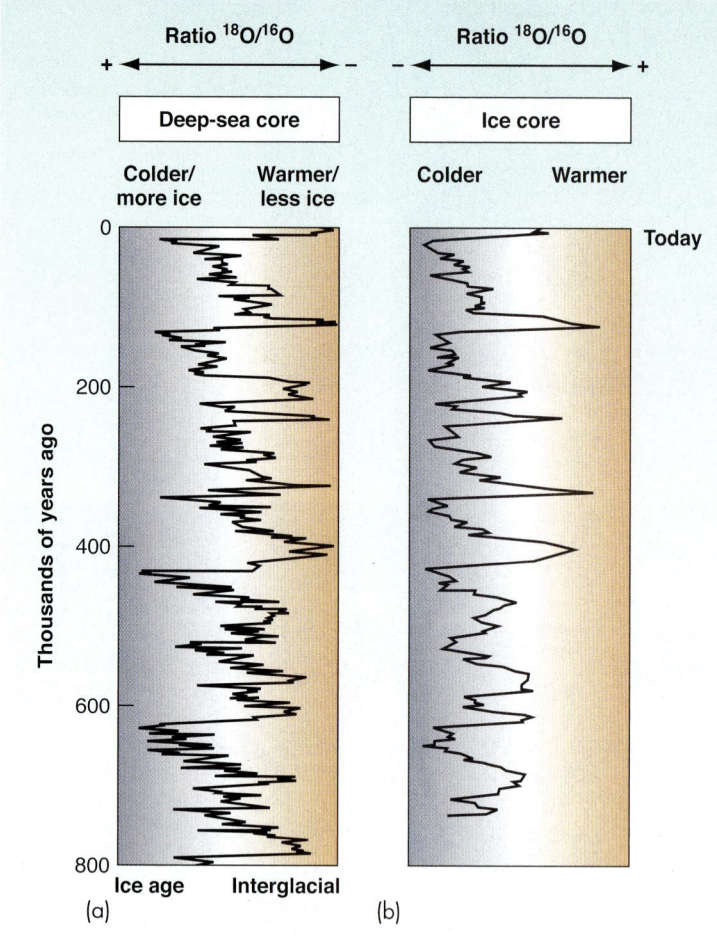

FIGURE 15.30 (a) The ratio of ^{18}O to ^{16}O in deep-sea cores of fossil foraminifera serves as a proxy for global ice volume. (b) The ratio of ^{18}O to ^{16}O in cores of glacial ice documents changes in atmospheric temperature, confirming that lower ice volume in deep-sea cores correlates to times of warmer atmosphere, while greater ice volume recorded in deep-sea cores correlates to times of cooler atmosphere.

temperature) and in marine foraminifera (a proxy for global ice volume) do indeed display strong agreement and so serve as a global guide to researchers as they work to interpret past climate patterns and events.

EXPAND YOUR THINKING

Identify other types of climate proxies you can think of.

GEOLOGY IN ACTION

The Greenland Ice Sheet

Since 1979 scientists have been tracking the extent of summer melting of the Greenland ice sheet (**FIGURE 15.31**). In 2007, the extent of melting broke the record set in 2005 by 10%, making it the season of greatest melting ever recorded. Melting on portions of the glacier rose 150% above the long-term average, with melting occurring on 25 to 30 more days in 2007 than the average for the previous 19 years.

In the past decade, the total mass deficit (the annual difference between snowfall and melting) tripled, and the amount of ice lost in 2008 was nearly three times the amount lost in 2007. In 2009, scientists announced that Greenland's ice was melting at a rate three times faster than it was only five years earlier. By 2010 a new record of melt-

ing was set on Greenland; and by the end of 2012 that record was surpassed again, by nearly twice as much. Clearly, the rate of melting on Greenland is accelerating.

In 2012, melting set a new record when 97% of the massive ice sheet surface temporarily experienced melting for approximately two weeks. Researchers analyzing this event concluded that surface of the Greenland ice sheet has a 50% probability of entering permanent decline by the year 2025. This state would be achieved when the equilibrium line, which is currently migrating upward in elevation at an annual rate of over 40 meters per year, reaches the highest point on the ice, thereby putting the entire glacier surface into the zone of wastage.

(a)

FIGURE 15.31 Melting on the Greenland ice sheet has tripled over the past decade. Greenland's surface melting in 2012 was intense, far in excess of any earlier year in the satellite record since 1979. (a) In July 2012, a very unusual weather event occurred. For a few days, 97% of the entire ice sheet experienced surface melting, a far more extensive meltdown than anything observed in the satellite record. (b) The standardized melting index is calculated by multiplying the number of days that melting occurs by the area where melting is detected. This approach compares a year's melt index value to a long-term average. The 2012 melt index was +2.4, compared to the 1979–2012 average, nearly twice the previous melt index record, set in 2010, of +1.3.

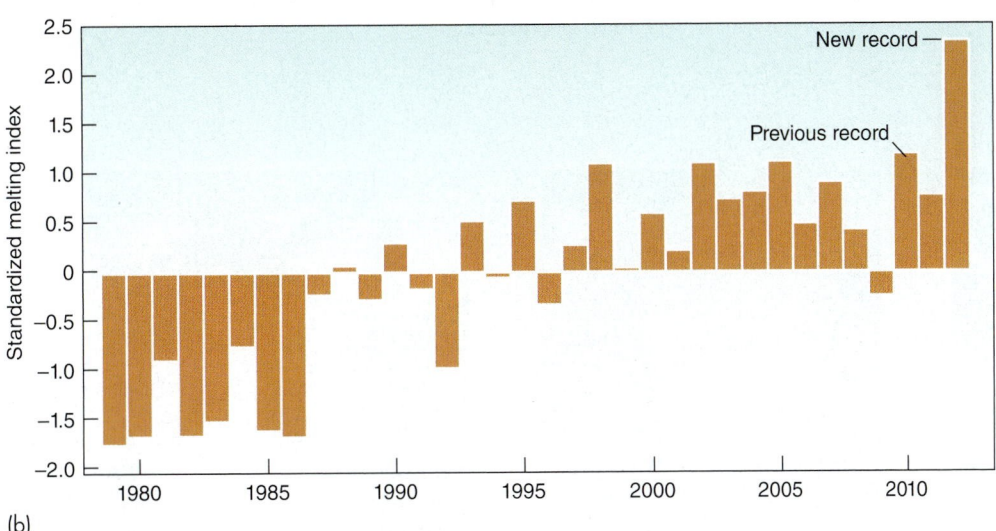

(b)

CRITICAL THINKING

Mass Balance of a Glacier

The U.S. Geological Survey (USGS) operates a long-term program to monitor glacier mass balance at certain key *benchmark glaciers* in the United States. *Glacier mass balance* is a term that refers to year-to-year changes in ice volume. Changes in ice volume occur when glacier wasting and accumulation increase or decrease from one year to the next. South Cascade Glacier in the State of Washington (FIGURE 15.32) is a long-term, high-quality mass balance monitoring site operated by the USGS.

Figure 15.32 shows five photos of South Cascade Glacier from 1928, 1958, 1979, 2003, and 2006. A comparison of these reveals much about the recent history of this glacier. FIGURE 15.33 shows monitoring data collected by glaciologists with the USGS, revealing trends in annual mass balance, cumulative mass balance, and air temperature over the past several decades.

Working with a partner, follow the directions below before answering the questions.

Directions

1. Using a marker, label these features on each photo: zone of wastage, zone of accumulation, lateral moraine, ground moraine, crevasse field, glacial drift, glacial outwash, recessional moraine.

2. Draw a dashed line on each photo marking the boundary between the zone of wastage and the zone of accumulation. Ignore patches of snow surrounded by bare ice. Label this line the "snowline." Along the edge of the glacier, draw a line where the glacier surface meets the rock wall on the left side of the valley.

Questions

1. What evidence did you use to delineate the zone of wastage and the zone of accumulation?

2. During the elapsed time, has the snowline generally remained stationary, moved to a lower elevation, or moved to a higher elevation? What does this indicate about the glacier?

3. Comparing the photographs from 1928 and 1979, what is the evidence that the glacier has thinned in the zone of wastage?

4. Comparing the photographs from 1979 and 2006, what is the evidence that the glacier has thinned in the zone of wastage?

5. How would you describe the behavior of the glacier terminus over the elapsed time? What about the glacial processes that caused this change?

6. Referring to Figure 15.33, answer the following:

   a. What is the relationship between Figures 15.33a and 15.33b?

   b. Why is it that in Figure 15.33a, the plots do not show steep slopes, but in Figure 15.33b, the plot does show a steep slope?

7. Compare the trends in Figure 15.33b with your interpretation of evidence in Figure 15.33. Do they agree or disagree? What do these two independent lines of evidence indicate about South Cascade Glacier? Is there any additional evidence that supports your conclusion?

8. State a hypothesis accounting for your observations of the glacier for 1928–2006.

9. How would you describe a research project in which a glaciologist tests your hypothesis? Write a proposal asking for funding to conduct this research.

(a)

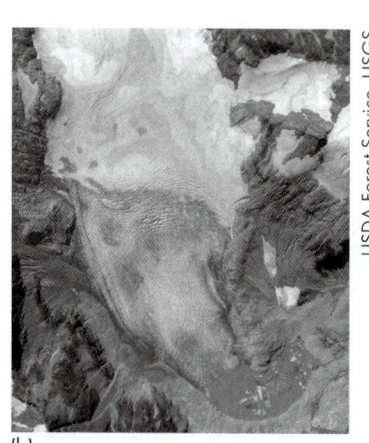

(b)

(c)

(d)   (e)

FIGURE 15.32 Photos of South Cascade Glacier from (a) 1928, (b) 1958, (c) 1979, (d) 2003, and (e) 2006.

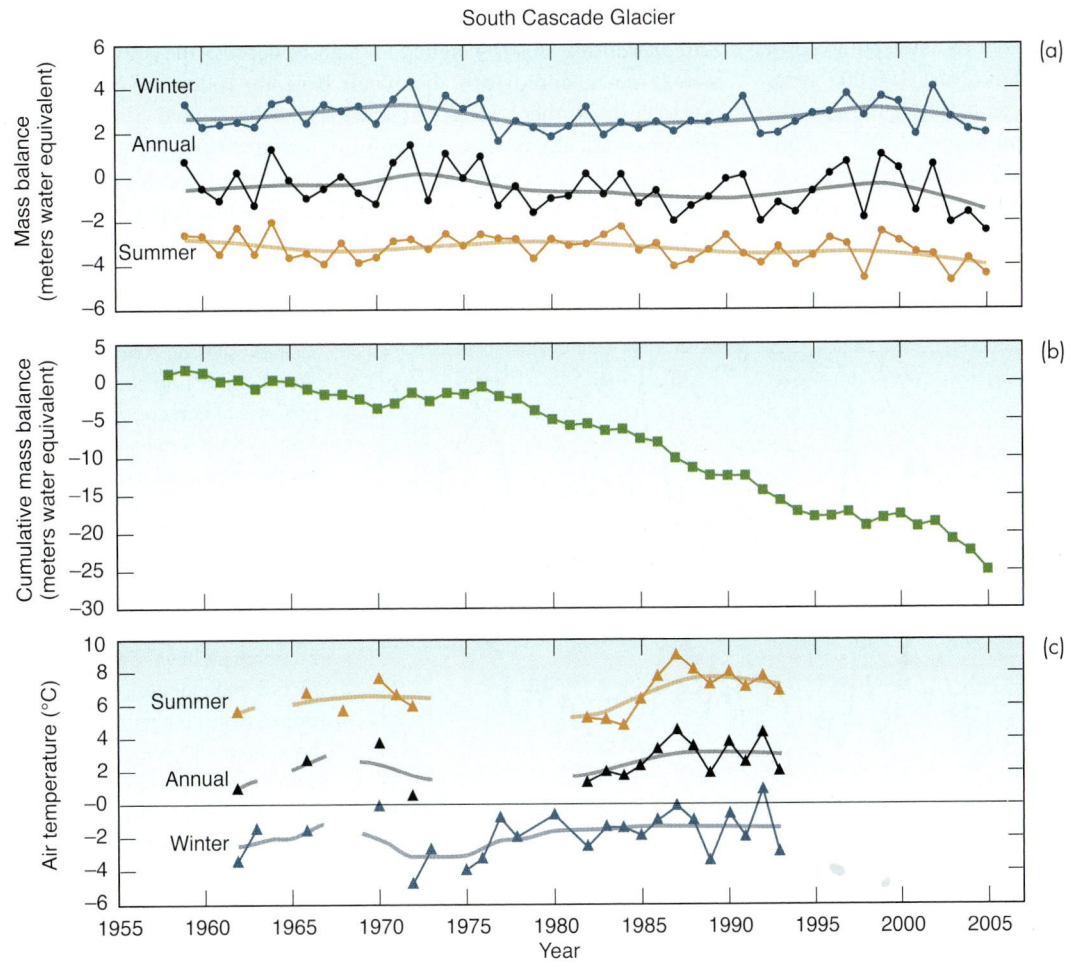

FIGURE 15.33 Observational measurements at South Cascade Glacier: (a) mass balance; (b) cumulative mass balance; and (c) air temperature.

15-7 Earth's Recent History Has Been Characterized by Alternating Cycles of Ice Ages and Interglacials

LO 15-7 Define marine isotopic stages and explain what they indicate.

The microscopic plankton called foraminifera, introduced in the preceding section, reveal an intriguing history of global ice volume that corresponds well with ice core records from both Greenland and Antarctica.

Climate Cycles

As explained in the previous section, scientists use *geologic proxies* for climate to improve their understanding of how past climate has changed through time. Geologic proxies, natural archives of climate information, include corals, marine plankton, ice cores, speleothems, sediments, and other types of geologic materials that record some aspect of past climate change.

The marine plankton foraminifera and ice cores from glaciers show that global climate change is characterized by alternating warm episodes and ice ages that occur approximately every 100,000 years (**FIGURE 15.34**). To identify these episodes, scientists have developed a numbering system divided into **marine isotopic stages,** MIS for short. Odd-numbered stages are warm; even-numbered stages are cool. We are living in the latest warm episode, known as MIS 1; the previous ice age is known as MIS 2. Figure 15.34 illustrates this identification system.

Earth's history of cooling and warming has several important features:

1. Major glacial and interglacial cycles are repeated approximately every 100,000 years.

2. Numerous minor episodes of cooling (called *stadials*) of which MIS 4 is an example, and warming (called *interstadials*) of which MIS 3 is an example, are spaced throughout the entire record.

3. Global ice volume during the peak of the last interglacial, approximately 125,000 years ago, was lower than at present, and global climate was warmer.

4. Following the last interglacial, global climate deteriorated in a long, drawn-out cooling phase culminating in MIS 2, approximately 20,000 to 30,000 years ago, with a major glaciation.

5. The current interglacial has lasted approximately 10,000 years.

Interglacial Cycles

During the last ice age (defined as MIS 2, 3, and 4), snow and ice accumulation built up massive continental glaciers in Europe, North America, Greenland, and Antarctica (**FIGURE 15.35**). In places, snow and ice accumulations were 3 to 4 kilometers thick and as high as some modern mountain ranges.

The ice sheet across eastern and central Canada, known as the *Laurentide ice sheet* (farther to the west it was called the *Cordilleran ice sheet*), and the ice sheet across northern Europe, known as the *Fennoscandian ice sheet*, weighed enough to depress the crust as much as 600 meters and deform the mantle beneath. Today, the crust under portions of northern Europe, Canada, and the United States is still glacioisostatically rebounding upward, now that the weight is gone.

Recall that the current interglacial, the Holocene Epoch, began about 10,000 years ago. But the Pleistocene Epoch (starting approximately 2.6 million years ago and ending at the onset of the Holocene Epoch) was a

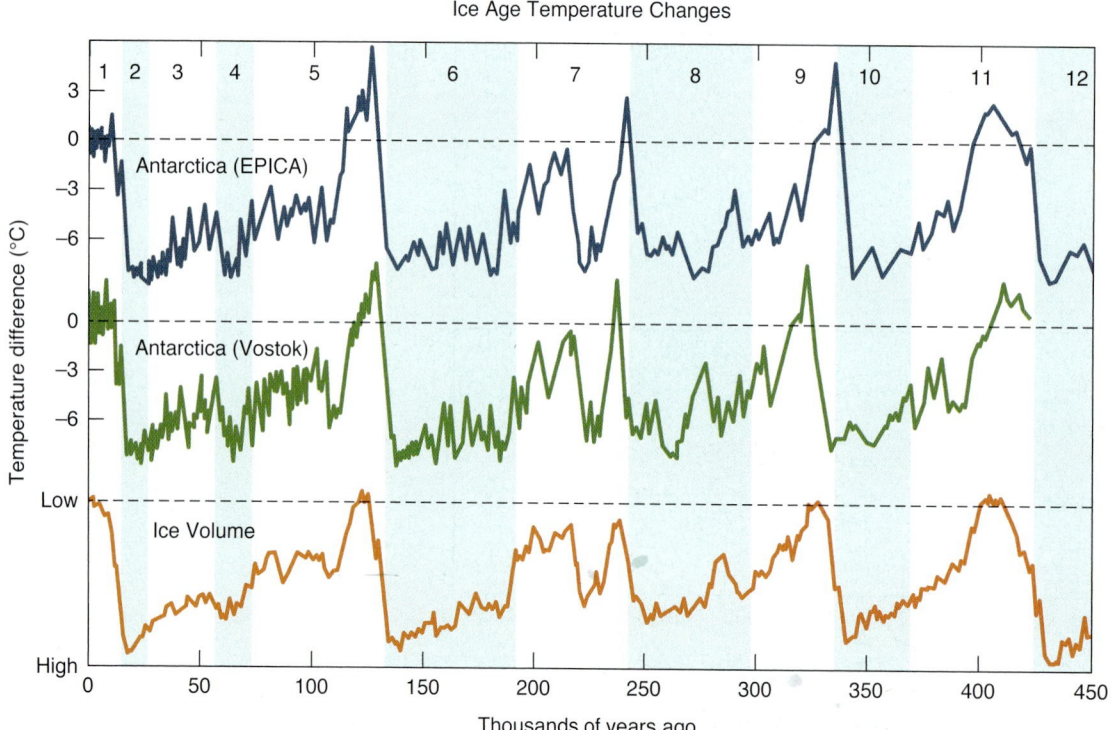

Ice Age Temperature Changes

Antarctica (EPICA)

Antarctica (Vostok)

Ice Volume

Temperature difference (°C)

Thousands of years ago

FIGURE 15.34 Ice cores from multiple sites on Antarctica (European coring site, EPICA; Russian coring site, Vostok) record atmospheric temperature, and marine cores record global ice volume. These sets of cores reveal similar histories of major warm-cold cycles occurring approximately every 100,000 years. Researchers have divided the record into marine isotopic stages that are numbered beginning with MIS 1, the current warm period. Odd-numbered stages are warm, and even-numbered stages are cool. MIS numbers appear along the top of the figure.

Explain the term marine isotopic stage.

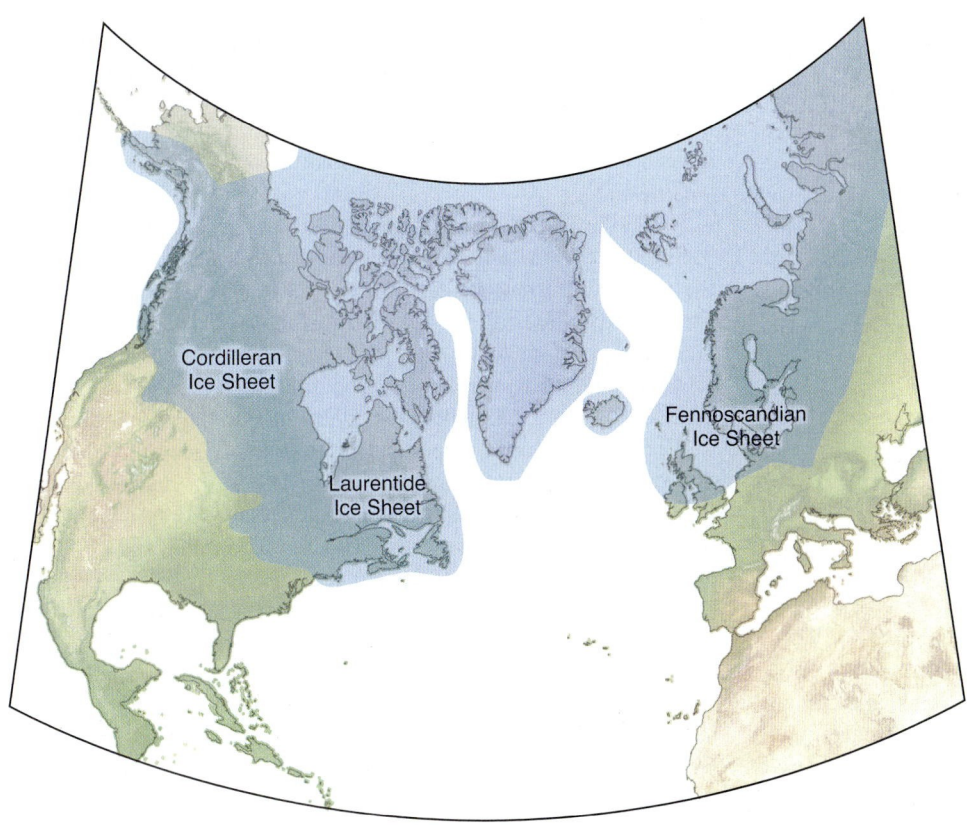

FIGURE 15.35 During the last ice age, continental ice sheets developed over much of North America and northern Europe.

What evidence would geologists have likely used to reconstruct the position of past ice sheets?

have been approximately five glaciations in this period.

During the length of a typical 100,000-year cycle, climate gradually cools and ice slowly expands until it reaches peak cooling. At the peak, glaciers cover most of Canada southward to the Great Lakes. Iceland, Scandinavia, and the British Isles are also convered. In the southern hemisphere, part of Chile is covered, and the ice of Antarctica covers part of what is now the Southern Ocean. In mountainous regions, the snowline lowers by 1,000 meters in altitude from the warmest to the coldest periods of a cycle.

Once ice cover reaches a maximum during a glacial episode, within a couple of thousand years, global temperature rises again and the glaciers retreat to their minimum extent and volume (**FIGURE 15.36**). The last ice age culminated about 20,000 to 30,000 years ago, and by approximately 5000 years ago, most of the ice had melted (except for remnants on Greenland and Antarctica). Since then glaciers generally have retreated to their smallest extent, with the exception of short-term climate fluctuations such as the so-called Little Ice Age, a cool period that lasted from about the sixteenth to the nineteenth century and was focused largely in the Northern Atlantic region.

time of many glacial and interglacial episodes. Sedimentary deposits are fragmentary because each new ice age tends to erode and bury the evidence of those that came before. Nevertheless, based on evidence from glacial deposits on land in North America and Europe, geologists have been able to document only four glaciations during the Pleistocene.

The more detailed and well-preserved marine oxygen isotope record suggests that over the past 500,000 years, each glacial-interglacial cycle has lasted about 100,000 years; that means there

FIGURE 15.36 The history of retreating ice in North America after the end of the last ice age.

Where in the United States would you expect to find glacial landforms? What would they look like?

Q EXPAND YOUR THINKING

Describe the nature of global climate change over the past half million years. How would you test the accuracy your description?

15-8 During the Last Interglacial, Climate Was Warmer and Sea Level Was Higher Than at Present

LO 15-8 List characteristics of the last interglacial that are relevant to understanding the consequences of modern global warming.

The last interglacial period occurred between approximately 130,000 and 75,000 years ago.

Last Interglacial

The last interglacial lasted approximately 55,000 years; and climate during this period was not always warm. Rather, researchers have identified five major phases consisting of three interstadials and two stadials. These show up clearly in the ice core records, as well as the deep-sea record. **FIGURE 15.37** shows these phases, using the identification system for marine isotopic stages. The last interglacial is called MIS 5, and the stadials and interstadials are labeled MIS 5a to 5e. Of these, MIS 5e was the warmest, and most like the Holocene Epoch.

Marine isotopic stage 5e is a good example of a warm period with characteristics similar to those of our current interglacial. And because it is also a relatively recent event, it is possible to recover rocks and sediments that record climate conditions from that time have not been subjected to extensive erosion, metamorphism, or weathering.

MIS 5e lasted approximately 12,000 years, from 130,000 to 118,000 years ago, and the average age of fossil corals from around the world that grew at that time is 123,000 years. **FIGURE 15.38** shows a fossil reef on the Hawaiian island of Oahu that illustrates another important feature of MIS 5e: Sea level was higher than present, leading researchers to conclude that climate was somewhat warmer and that melted ice contributed to the higher sea level.

This conclusion is supported by deep cores of ice from Antarctica that preserve temperature records from MIS 5e. If you look back at Figure 15.34, you will notice that ice core records from Antarctica and marine sediment both indicate that MIS 5e was warmer, and with lower ice volume, than the present-day climate.

The more far-reaching relevance of MIS 5e is that it has been cited as a possible analog for a future climate under continued global warming. Studies have shown that carbon dioxide concentrations in the atmosphere were relatively high (though not as high as they are today due to the contribution of industrial greenhouse gases), temperatures were higher than at present, and sea level was as much as 5.5 to 9 meters higher.

Scientists study MIS 5e to develop an accurate estimate of the duration of the last interglacial period and to deepen our understanding of its primary causes. Both of these goals are intended to establish a basis for improving *global circulation models* (Chapter 14) that are used to project climate change.

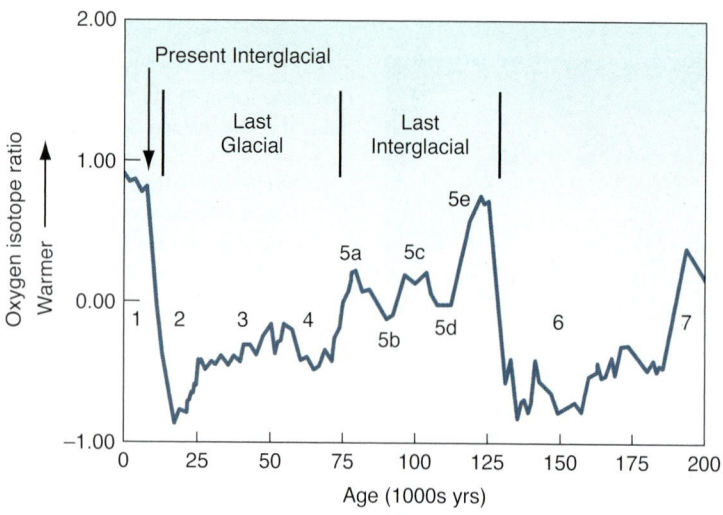

FIGURE 15.37 The last interglacial consisted of five stadials and interstadials, named MIS 5a to 5e. The last glacial consisted of two stadials, MIS 4 and MIS 2, as well as one interstadial, MIS 3. The current interglacial is MIS 1.

FIGURE 15.38 This rocky shoreline in Hawaii is composed of limestone formed by a fossil reef that grew under higher-than-present sea level during MIS 5e. The yellow line circles a fossil coral head from MIS 5e.

 Describe the general history of global climate over the last interglacial cycle.

Explain how this fossil reef can be used to reconstruct the past position of sea level.

MIS 5e as Analog for This Century

Computer models of climate change during MIS 5e indicate that sea level rise started with melting of the Greenland ice sheet, not the Antarctic ice sheet. Research also suggests that ice sheets across both the Arctic and Antarctic could melt more quickly than expected this century because temperatures are likely to rise higher than they did during MIS 5e, especially in the polar regions.

If these predictions are correct, by 2100, the Arctic could warm by 3°C to 5°C in summer. During MIS 5e, meltwater from Greenland and other Arctic sources raised sea level by as much as 4 meters. However, since global sea level actually rose 5.5 to 9 meters, researchers have concluded that Antarctic melting and thermal expansion of warm seawater must have been responsible for the remainder of the rise in sea level.

This conclusion is supported by the discovery of marine plankton fossils and isotopes formed by solar radiation (possible only if ice retreat exposed Antarctic bedrock to sunlight) beneath the West Antarctic ice sheet, indicating that parts of the ice disappeared at some point over the last several hundred thousand years.

The rise in sea levels produced by Arctic warming and melting could have floated, and thus destabilized, ice shelves sitting on the shallow continental shelf of Antarctica. Ice shelves that float do not buttress their glaciers, which may accelerate the rate at which they flow into the sea, a condition scientists refer to as "glacial collapse." If such a process occurred today, it could lead to rapid sea level rise, and more glacial collapse—a positive feedback response.

In the past few years, sea level has begun rising more rapidly; now it is climbing at a rate of over 3 centimeters per decade. Recent studies also have found accelerated rates of glacial retreat along the margins of both the Greenland and Antarctica ice sheets.

During MIS 5e, the level of global warming needed to initiate this melting was less than 3.5°C above modern summer temperatures, similar to the level predicted to be reached by midcentury if carbon dioxide levels continue to rise unchecked. The extent of Greenland ice sheet melting that led to higher sea levels is shown in **FIGURE 15.39**. According to this reconstruction, sea levels rose at more than 1.6 meters per century—a rate that would be catastrophic for coastal communities worldwide if it were to happen today.

Computer modeling reveals several other features of MIS 5e that may portend global conditions by the end of the twenty-first century: Global carbon dioxide rose by 1% per year (similar to the current rate of rise); 2100 will be significantly warmer than MIS 5e, so Greenland is already headed toward a state similar to that depicted in Figure 15.39; and the West Antarctic ice sheet will also contribute significantly to global sea level rise by 2100.

The later part of MIS 5e was characterized by precipitous changes in global climate. Although the period has been studied intensively, global climate during MIS 5a–5d still is poorly understood. Researchers can only speculate that temperatures during MIS 5d and 5b were markedly cooler than current temperatures, that global ice volume expanded, and that global sea level dropped, perhaps by as much as 25 meters below the current level. These were, in effect, mini ice ages that lasted a few thousand to 10,000 years each. MIS 5b was likely cooler than present, and ice volume was greater than currently. MIS 5a may have had sea level that was 1 meter higher than present, and is thus also studied as a potential analog for our warming world.

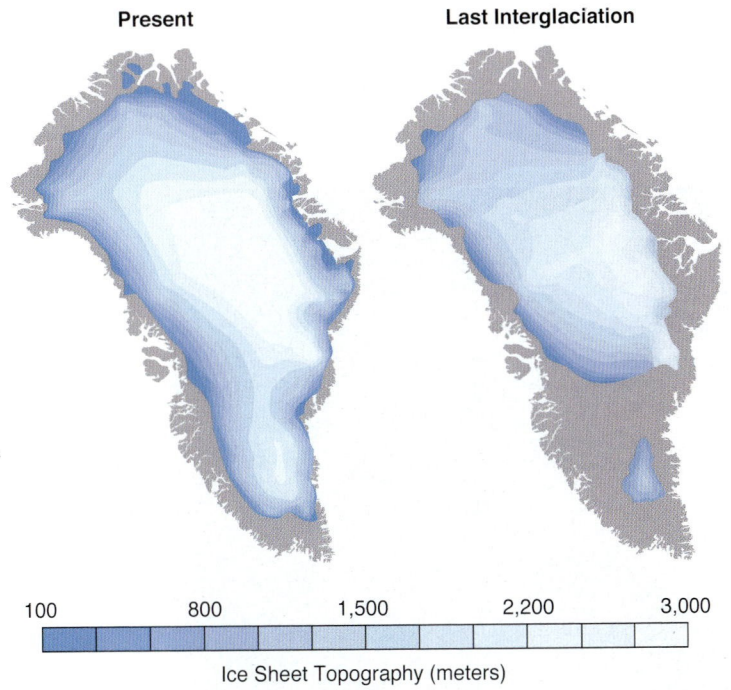

Present   **Last Interglaciation**

100   800   1,500   2,200   3,000

Ice Sheet Topography (meters)

FIGURE 15.39 Computer models that simulate climate during MIS 5e indicate that melting of the Greenland ice sheet was responsible for a global sea level rise of approximately 4 meters.

If you were asked to measure a rate of glacier retreat on the ground in Greenland, how would you do it?

EXPAND YOUR THINKING

What do climate researchers hope to learn through the study of MIS 5e?

15-9 Paleoclimate Is Influenced by the Timing and Location of Sunlight Reaching Earth

LO 15-9 Define the Earth–Sun orbital parameters.

Regular and predictable differences in Earth's exposure to solar radiation over the past half million years have played an important role in the past climate.

Influence of Insolation and Axial Tilt

Scientists are still uncertain about all the factors that drive variations in *paleoclimate*. But they do agree that the past 500,000 years or so have been characterized by particularly intense glacial-interglacial cycles occurring approximately every 100,000 years. This time frame has also been characterized by stadial-interstadial episodes that have occurred more often.

Fluctuations in the timing and location of *insolation* (solar radiation received at Earth's surface) initiated by regular shifts in the geometry of Earth's orbit around the Sun are the most likely source of these climate instabilities. This solar variable was neatly described by the Serbian mathematician *Milutin Milankovitch* (1879–1958) in 1930.

To understand these **orbital parameters,** we must appreciate the effect of Earth's tilted axis on the level of insolation received through the year. Earth's axis is tilted an average of 23.5° from the vertical (**FIGURE 15.40**). As Earth orbits the Sun, this tilt means that during one part of the year the northern hemisphere is tilted toward the Sun and receives greater insolation (summer), whereas six months later, it is tilted away from the Sun and receives less insolation (winter); the reverse applies to the southern hemisphere. These annual differences in insolation create the seasons.

Orbital Parameters

Three aspects of Earth's orbit geometry change in regular cycles under the influence of the combined gravity of Earth, the Moon, the Sun, and the other planets. These orbital parameters dictate the timing and location of insolation reaching Earth's surface, a process known as *solar forcing*, and thereby regulate climate (**FIGURE 15.41**).

Earth's orbit shifts from a nearly perfect circle to a more elliptical shape and back again in a 100,000-year cycle (and a 400,000-year cycle) known as *eccentricity*. Eccentricity affects the level of insolation received at *aphelion* (the point in its orbit at which Earth is farthest from the Sun) and at *perihelion* (the point in Earth's orbit at which it is closest to the Sun).

The effect of eccentricity is to shift the seasonal contrast in the northern and southern hemispheres. For example, when Earth's orbit is more elliptical, one hemisphere will have hot summers and cold winters; the other will have moderate summers and moderate winters. When the orbit is more circular, both hemispheres will have similar contrasts in seasonal temperature.

The second orbital parameter is called *obliquity*. The angle of Earth's axis of spin varies its tilt between 21.5° and 24.5° every 41,000 years. Changes in obliquity prompt major changes in the seasonal distribution of sunlight at high latitudes and in the length of the winter dark period at the poles. They have little effect on low latitudes.

Finally, Earth's axis of spin slowly wobbles. Like a spinning top running out of energy, the axis wobbles toward and away from the Sun over the span of 19,000 to 23,000 years. This is known as *precession*. Precession affects the timing of aphelion and perihelion, and has important implications for climate because it affects the seasonal balance of insolation. For example, when perihelion falls in January, winters in the northern hemisphere and summers in the southern hemisphere

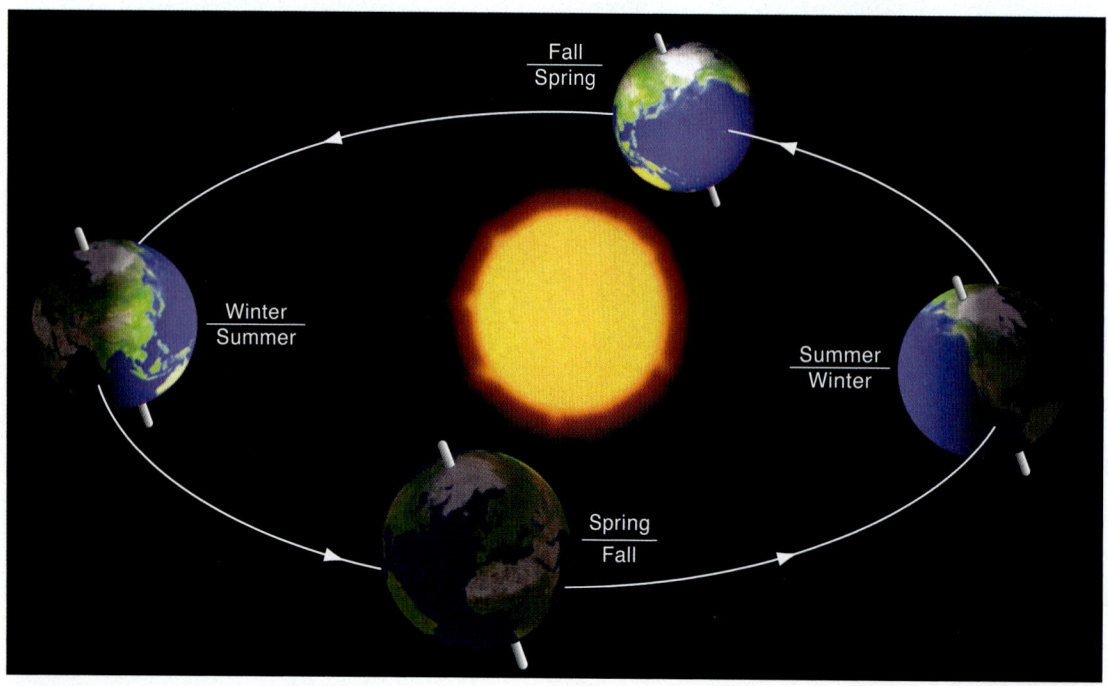

FIGURE 15.40 Earth's seasons are the result of a tilt in the planet's axis.

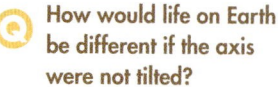 How would life on Earth be different if the axis were not tilted?

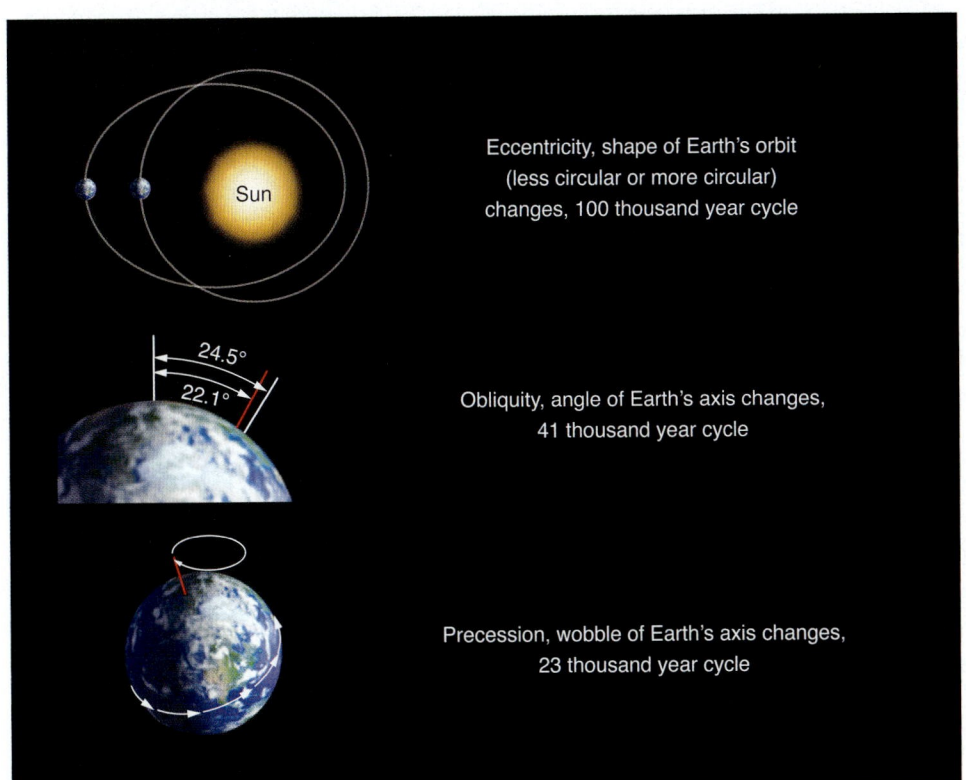

FIGURE 15.41 The primary orbital parameters driving climate changes over the past half million years are eccentricity, obliquity, and precession.

What is the principal effect of shifts in Earth's orbit?

are slightly warmer than the corresponding seasons in the opposite hemispheres.

The effects of precession on the amount of radiation reaching Earth are closely linked to the effects of obliquity (variations in tilt). The combined variation in these two factors leads to radiation changes of up to 15% at high latitudes, greatly influencing the growth and melting of ice sheets.

Milankovitch theorized that ice ages occur when orbital variations cause lands in the region of north latitude 65° (the approximate latitude of central Canada and northern Europe) to receive less sunshine in the summer. This allows snow to build up from year to year, to become glacial ice. Why is the northern hemisphere the crucial location for glacier formation? Because most of the continents are located in the northern hemisphere, and glaciers develop on land, not water.

Based on this reasoning and his calculations, Milankovitch predicted that the ice ages would peak every 100,000 and 41,000 years, with additional significant variations every 19,000 to 23,000 years. And, as shown in **FIGURE 15.42**, which plots variations in all three orbital parameters across the past 1 million years, Milankovitch's predictions have been verified by the ocean sediment record of global ice volume.

Ice ages and interglacials occur roughly every 100,000 years, and the timing of stadials and interstadials varies on the higher frequencies approximated by his predictions of 41,000, 19,000, and 23,00 years—although that exact timing is not preserved in the sediment or ice core records.

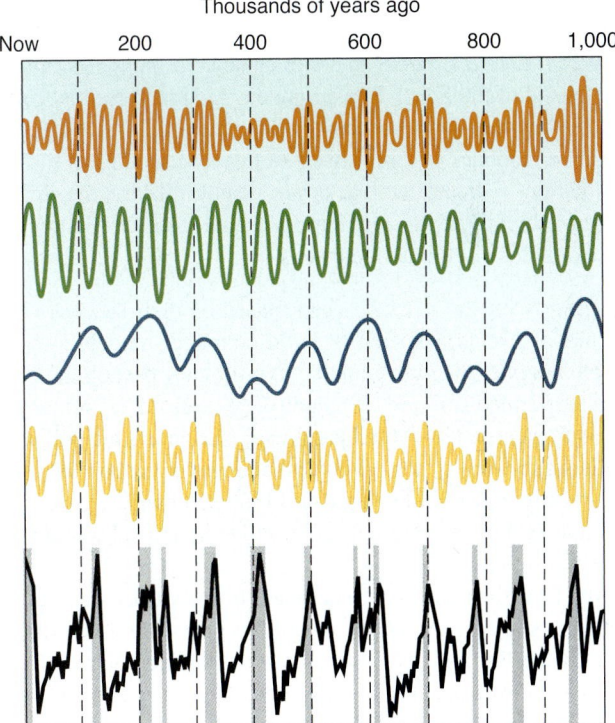

FIGURE 15.42 This graph shows the relative solar forcing attributable to precession (red), obliquity (green), and eccentricity (blue). The cumulative solar forcing (insolation) at 65° north latitude caused by the combined influence of the three orbital parameters is plotted in yellow. Black shows a history of global ice volume with interglacials (vertical gray bars) showing regular variations in climate approximately every 100,000 years, and more often, just as Milankovitch predicted.

Examine the figure: Does solar forcing account for all changes in ice volume?

EXPAND YOUR THINKING

Describe how plate tectonics and orbital parameters may combine to govern paleoclimate.

15-10 Together, Orbital Forcing and Climate Feedbacks Produced Paleoclimate

LO 15-10 Describe climate feedbacks and give examples.

Earth's climate is not solely a result of insolation; positive and negative climate feedbacks also play an important role.

Influence of Climate Feedbacks

If you compare the total solar forcing shown earlier in Figure 15.42 (yellow line) to the paleoclimate record in marine sediments (black line), you will notice that the timing and magnitude of the two do not exactly match. For instance, the solar forcing that led to MIS 5e is considerably greater than the forcing that is producing the modern interglacial. Also, the drop in insolation at MIS 5d is greater than the low at the peak of the last ice age 20,000 to 30,000 years ago, yet the last ice age was much colder. What is at the root of these disparities? The answer is that Earth's climate is not driven solely by insolation.

Recall that Earth's environmental system exercises positive and negative *climate feedbacks* that can enhance or suppress the timing and intensity of the Earth–Sun geometry. A climate feedback (as discussed in Chapter 14) is a process taking place on Earth that amplifies (a positive feedback) or minimizes (a negative feedback) the effects of insolation. Climate feedbacks are responsible for the difference between orbital forcing and Earth's actual climate. Following are some examples of climate feedbacks.

Emerging from the Last Ice Age Based on their success with climate models, scientists tried to build computer models to simulate paleoclimate, but they were unable to reproduce past climate change unless they added shifts in carbon dioxide levels to accompany those in insolation caused by orbital parameters. Scientists are still working to understand what is behind natural changes in carbon dioxide levels; currently, most believe that past climate changes were initiated by orbital forcing and then enhanced and maintained by a natural rise of greenhouse gases.

One idea for relating paleoclimate to feedbacks was developed by scientists analyzing a core of marine sediments from the ocean floor near the Philippines. That area of the Pacific contains foraminifera that live in tropical surface water. When they die and settle to the bottom, their remains serve as a record of changing tropical air temperatures. But the record is complicated by the fact that different types of foraminifera living on the deep seafloor are bathed in *bottom waters* (water that travels along the seafloor, not at the surface) fed from the Southern Ocean near Antarctica. These foraminifera record the temperature of the cold southern waters, yet the fossils of both types of foraminifera are deposited together on the seafloor.

Upon radiocarbon dating both types of foraminifera, scientists found that water from the Antarctic region warmed before waters in the tropics—by as much as 1000 to 1300 years earlier. The explanation for this difference, they believe, is a positive climate feedback.

More specifically, predictable variations in Earth's eccentricity and obliquity increased the insolation at high southern latitudes during spring in the southern hemisphere. That then warmed the Southern Ocean. As a result, sea ice shrank back toward Antarctica (**FIGURE 15.43**), uncovering ocean waters that had been isolated from the atmosphere for millennia. As the Southern Ocean warmed, it released great quantities of dissolved carbon dioxide into the atmosphere. (Remember, cold water can hold more dissolved gas than warm water.) The released gas proceeded to warm the whole world. This process was responsible for driving climate out of its glacial state and into an interglacial state. It explains how small temperature changes caused by orbital parameters led to a positive feedback in global carbon dioxide that warmed the world.

Rapid Climate Change When scientists first analyzed paleoclimate evidence in marine and glacial oxygen isotope records, they discovered that the Milankovitch theory predicted the occurrence of ice ages and interglacials with remarkable accuracy. But they also found something that required additional explanation: Some climate changes appeared to have occurred very rapidly. Based on the fact that the Milankovitch theory tied climate change to slow and regular variations in Earth's orbit, it was assumed that climate variations would likewise be slow and regular. The discovery of rapid changes was, therefore, a surprise. Here again, the answer lay in the climate feedback system.

Cores show that while it took thousands of years for Earth to emerge totally from the last ice age and warm to today's balmy climate, fully one-third to one-half of the warming—about 10°C at Greenland—occurred within mere decades, at least according to ice records in Greenland (**FIGURE 15.44**).

At approximately 12,800 years ago, following the last ice age, temperatures in much of the northern hemisphere rapidly returned to near-glacial conditions and stayed there during a climate event called the *Younger Dryas* (named after the alpine flower *Dryas octopetala*).

FIGURE 15.43 Increased insolation at high latitudes caused sea ice to retreat and the seawater to release carbon dioxide to the atmosphere. This positive feedback was responsible for Earth's transition from an ice age to an interglacial.

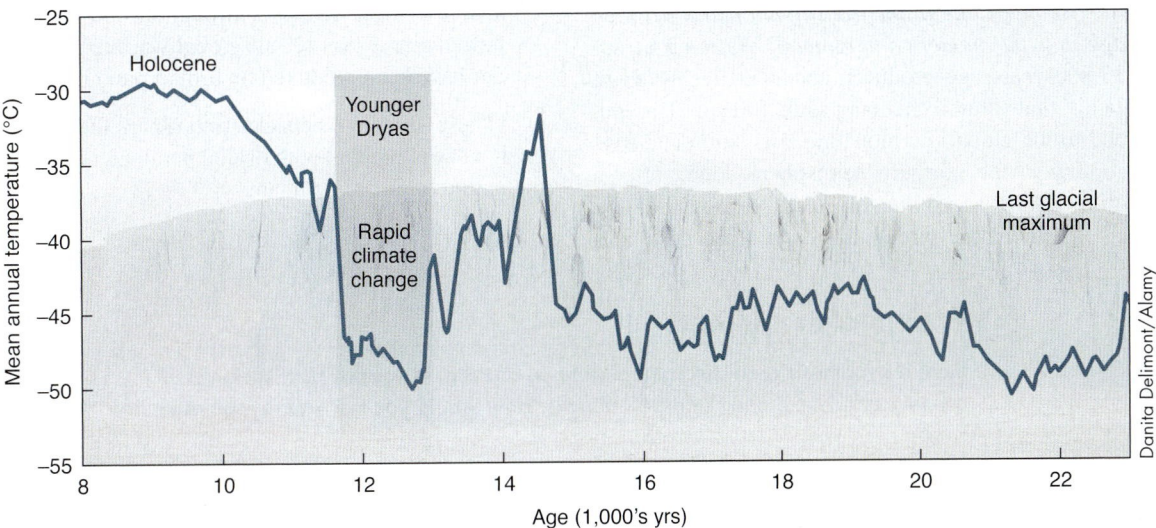

FIGURE 15.44 The Younger Dryas climate event was a dramatic cooling that lasted approximately 1300 years during the transition between glacial and interglacial states. The return to warm conditions was equally rapid, occurring in less than a human generation.

Why is understanding rapid climate change important?

The cool episode lasted about 1300 years, and by 11,500 years ago temperatures had warmed again.

Ice core records show that the recovery to warm conditions occurred with startling speed—less than a human generation. Should changes of this magnitude occur today, they would have a huge impact on modern human societies, such that there is urgent need today to understand such abrupt climate events so that scientists might be able to predict them.

Global thermohaline circulation is thought to be a major driver of rapid climate change. Today, warm water near the equator in the Atlantic Ocean is carried to the north on the *Gulf Stream,* which flows on the surface from southwest to northeast in the western Atlantic. The Gulf Stream releases heat into the atmosphere through evaporation, and the heat, in turn, moves across northern Europe and moderates the climate. As a result, the Gulf Stream becomes cool and salty as it approaches Iceland and Greenland. This makes it very dense; consequently, it sinks deep into the North Atlantic (making a current called the *North Atlantic Deep Water*) before it can freeze. From there it flows southward toward the equator. The Gulf Stream continuously replaces the sinking water, warming Europe and setting up the global oceanic conveyor belt we call the thermohaline circulation.

As we learned in Chapter 14, thermohaline circulation transports heat around the planet and so plays an important role in global climatology. Acting as a conveyor belt carrying heat from the equator into the North Atlantic, it raises Arctic temperatures, limiting the growth of ice sheets. At the same time, influxes of freshwater from melting ice on the lands that surround the North Atlantic (such as Greenland) can slow or shut down the circulation by preventing the formation of deep water. Freshwater in the North Atlantic decreases the flow of the Gulf Stream, and weakens its role as a source of heat, leading to cooling in the northern hemisphere, thereby regulating snowfall in the crucial region where ice sheets shrink and grow (N 65°). That means a shutdown of the thermohaline circulation could produce a negative climate feedback pattern, beginning with ice melting (warming), which, ironically, leads to glaciation (cooling).

The key to keeping the belt moving is the saltiness of the water. Saltier water becomes denser and, so, sinks. Many scientists believe that if too much freshwater entered the North Atlantic—for example, from melting Arctic glaciers and sea ice—the surface water would freeze before it could become dense enough to sink toward the bottom. If the water in the north did not sink, the Gulf Stream eventually would stop moving warm water northward, leaving Northern Europe cold and dry within a single decade.

This hypothesis of rapid climate change is called the *conveyor belt hypothesis*, and the paleoclimate record found in ocean sediment cores appears to support it. Paleoclimate studies have shown that, in the past, when heat circulation in the North Atlantic Ocean slowed, the climate of northern Europe changed. Although the last ice age peaked about 20,000 to 30,000 years ago, the warming trend that followed it was interrupted by cold spells at 17,500 years ago and again at 12,800 years ago (the Younger Dryas). These cold spells happened just after melting ice had diluted the salty North Atlantic water, slowing the ocean conveyor belt. It is this idea that inspired the 2004 movie *The Day After Tomorrow*, which depicts freshwater from melting ice, caused by global warming, stopping the thermohaline circulation, which in turn, produces deadly cooling (an unlikely scenario) in the North Atlantic.

In summary, we have explored two types of climate feedback:

1. *Positive climate feedback in the Antarctic that ended the last ice age.* Predictable variations in Earth's tilt and orbit caused warming, which triggered the withdrawal of sea ice in the Southern Ocean. This led to additional warming of ocean water, reducing its capability to hold dissolved carbon dioxide. The carbon dioxide escaped into the atmosphere and warmed the planet beyond the temperatures that would have been achieved by orbital parameters alone.

2. *Negative climate feedback late in the transitional phase between the last ice age and the modern interglacial.* Warming at approximately 12,800 years ago produced abundant freshwater in the North Atlantic that diluted the salty Gulf Stream. This put a temporary end to the global thermohaline circulation and triggered rapid cooling in the northern hemisphere. Later, after a period of cooling lasting approximately 1300 years, called the Younger Dryas, thermohaline circulation once again became a source of heat transport throughout the world's oceans. The renewed circulation set off global warming so rapidly that it could not have been predicted by orbital parameters alone.

These climate feedbacks, and others that are still being discovered, work in parallel with orbital parameters to ultimately determine the nature of Earthly climate.

EXPAND YOUR THINKING

State a hypothesis predicting global climate, one that integrates plate tectonics, orbital parameters, and climate feedbacks. How would you test your hypothesis?

LET'S REVIEW

A glacier is a large, long-lasting river of ice that is formed on land; it undergoes internal deformation, erodes the crust and deposits sediment, and creates glacial landforms. In this way, glaciers have shaped the mountains and nearby lands, where they expanded during past climate changes. Earth's recent history has been characterized alternately by cool climate periods, called ice ages, during which glaciers around the world expanded, and warm periods, called interglacials, during which glaciers retreated. By studying the history of climate changes, scientists learn how natural systems respond to shifts in temperature and ice volume. Today, most of the world's ice sheets, alpine glaciers, ice shelves, and sea ice are dwindling in reaction to global warming.

Past climate, called *paleoclimate*, led to the repeated expansion and contraction of ice sheets on the continents. These expansions and contractions sculpted the mountains, deposited sediments across the landscape, and produced much of the topography of northern lands that we see today. Changes in climate that produced past glaciations were caused by differences in the amount of sunlight reaching Earth (insolation), a result of shifting orbital parameters. But orbital parameters alone are not sufficient to set in motion the rapid and dramatic changes in climate revealed by ice cores and marine sediments. These changes also are influenced by climate feedbacks that enhance or diminish the effects of changing insolation. Indeed, today most of the world's ice sheets, alpine glaciers, and ice shelves are melting in response to a dramatic climate feedback process: human pollution of the atmosphere with greenhouse gases.

STUDY GUIDE

15-1 A glacier is a river of ice.

- **Louis Agassiz** proposed the "theory of the ice ages" and pointed out that the deep valleys and sharp peaks and ridges of the mountains of Europe and North America were once buried in ice, just as Greenland is today.

- A *glacier* is a large, long-lasting river of ice that is formed on land, undergoes internal deformation, and creates **glacial landforms**. Glaciers range in length and width from several hundred meters to hundreds of kilometers, depending on their type. Most glaciers are several thousand years old, but those covering Greenland and Antarctica have been in existence in one form or another for hundreds of thousands of years.

15-2 As ice moves, it erodes the underlying crust.

- Ice behaves as a brittle solid (i.e., it fractures) until it reaches thicknesses of 40 to 60 meters or more. At these thicknesses, the lower layers of ice experience plastic flow. Basal sliding occurs when portions of the glacier move across the crust on a layer of **meltwater**.

- As ice moves, it causes erosion to the underlying bedrock through plucking and abrasion. Pulverized rock, known as rock flour, is composed of silt-size grains of ground-up rock. Sediments of all sizes become part of the glacier's load; as a result, many glacial deposits are poorly sorted.

- Glaciers are found in a wide variety of types and environments. However, there are two principal types of glaciers: **alpine glaciers**, which are found in mountainous areas, and **continental glaciers**, which cover larger areas and are not confined to the high elevations of mountains.

15-3 Ice moves through the interior of a glacier as if on a one-way conveyor belt.

- The interior of an active glacier can be thought of as a one-way conveyor belt, continuously delivering ice, rock, and sediments to the front of the glacier. This system feeds new ice from the **zone of accumulation**, usually accounting for 60% to 70% of the glacier's surface area, to the **zone of wastage**, where ice is lost not only by melting but also by sublimation and calving (when large blocks of ice break off the front). Between the zone of accumulation and the zone of wastage is the equilibrium line (also known as the snowline), marked by the elevation at which accumulation and wasting are approximately equal.

- Approximately 160,000 glaciers exist today; they are present on every continent except Australia and in approximately 47 of the world's countries. Yet more than 94% of all the ice on Earth is locked up in the continental glaciers covering Greenland and Antarctica.

15-4 Glacial landforms are widespread and attest to past episodes of glaciation.

- Earth's history has been characterized by wide swings in global climate, from extreme states of cold (ice ages) to warm periods called **interglacials**. Ice ages are characterized by the growth of massive continental ice sheets reaching across North America and northern Europe.

- During ice ages, sea level is lowered by as much as 130 meters, exposing lands between Siberia and Alaska, from Ireland to England and France, and from Malaysia to Indonesia to Borneo. In many places, the shoreline was tens to hundreds of kilometers seaward of its present location.

- Glacial valleys that have been eroded by alpine glaciers are U-shaped in cross section. Valleys that have been shaped by stream erosion have a V-shape. A mountain valley will have a combination of both shapes if it was once glaciated but is now being eroded by a stream channel.

- All sediment that has been deposited by glacial erosion is called glacial drift. Glacial drift includes **glacial till**, unstratified and unsorted sediment that is deposited directly by melting ice. Glacial erratics are blocks of bedrock deposited by melting ice and that have a foreign lithology. A **glacial moraine** is a deposit of till that marks a former position of the ice. Ground moraines tend to be flat deposits beneath moving ice; an end moraine, or terminal moraine, is a ridge of till that marks the downhill extent of ice movement. Lateral moraines are deposits of till that mark the edges of the glacier against a valley wall; a medial moraine is formed when two valley glaciers coalesce, and till from each of them combines to draw a boundary between the two. Other depositional features include **glacial outwash** plains, kames, eskers, kettle lakes, and drumlins.

15-5 The majority of glaciers and other ice features are retreating in reaction to global warming.

- Global warming threatens the stability of glaciers around the world. Scientists are monitoring alpine glaciers, ice sheets, Arctic sea ice, and ice shelves. Data show that the majority of glaciers and other ice features are retreating due to global warming.

- In 2006–2007, the average rate of melting and thinning of alpine glaciers more than doubled. In 2007, the extent of melting on Greenland broke the record set in 2005 by 10%, making it the season of most extensive melting ever recorded. In West Antarctica, ice loss has increased by 59% over the past decade, to about 132 billion metric tons a year, while the yearly loss along the Antarctic Peninsula has gone up by 140%, to 60 billion metric tons, and 87% of glaciers there are retreating. In East Antarctica, melting is largely confined to the seaward edge of the ice (where it is caused by warm ocean currents), but every year melting encroaches inland and onto higher elevations.

15-6 The ratio of oxygen isotopes in glacial ice and deep-sea sediments acts as a proxy for global climate history.

- Foraminifera in deep-sea sediment use dissolved compounds and ions to precipitate microscopic shells made of calcite ($CaCO_3$). Later, scientists can use the oxygen isotope content of these shells to track changes in global ice volume. Oxygen isotopes in ice cores leave a record of changing air temperatures; and because global ice volume and air temperature are related, the records of oxygen isotopes in foraminifera and glacial ice show similar patterns.

15-7 Earth's recent history has been characterized by alternating cycles of ice ages and interglacials.

- Oxygen isotope records from marine sediments and ice cores show Earth's climate alternating between warm periods and ice ages. These are identified using a numbering system called **marine isotopic stages**, or MIS. Odd-numbered stages are warm, even-numbered stages are cool.

- Major glacial and interglacial cycles are repeated approximately every 100,000 years. Minor episodes of cooling (stadials) and warming (interstadials) occur as well. Global ice volume during the last interglacial was lower than at present, and climate was warmer. Following the last interglacial, climate entered a cooling phase, which culminated 20,000 to 30,000 years ago with a major glaciation. The current interglacial has lasted approximately 10,000 years.

15-8 During the last interglacial, climate was warmer and sea level was higher than at present.

- The last interglacial occurred 130,000 to 75,000 years ago and consisted of three interstadials and two stadials. The last interglacial (MIS 5) consisted of a sequence of stadials and interstadials labeled MIS 5a–5e. MIS 5e is the best example of a warm period with characteristics similar to those of the current interglacial.

- Computer models simulating climate change during the last interglacial show that sea-level rise started with melting of the Greenland ice sheet, not the Antarctic ice sheet. Results from the simulations suggest that ice sheets across the Arctic and Antarctic could melt more quickly than expected during this century because temperatures are likely to rise higher than they did during the last interglacial. By 2100, Arctic summers may be as warm as they were 130,000 years ago, when sea level eventually was 5.5 to 9 meters higher than it is today.

15-9 Paleoclimate is influenced by the timing and location of sunlight reaching Earth.

- Differences in Earth's exposure to solar radiation match the timing of climate swings, and thus play an important role in paleoclimate.

- The past 500,000 years was marked by particularly intense glacial-interglacial cycles spaced approximately every 100,000 years, with stadial and interstadial episodes occurring at greater frequency. Fluctuations in insolation introduced by changes in the geometry of Earth's solar orbit are the most likely cause of instabilities in Earth's climate.

- Solar variables were described by the Serbian mathematician Milutin Milankovitch. He calculated the timing of three components of Earth's solar orbit that contribute to changes in climate: a cycle known as eccentricity; a change in the angle of the axis of spin called obliquity; and a wobble in Earth's axis known as precession.

- Small variations in insolation related to eccentricity, obliquity, and precession influence the amount of sunlight each hemisphere receives. Ice ages occur when lands at N 65° (the latitude of central Canada and northern Europe) receive less sunshine in summer than usual.

- Milankovitch predicted that ice ages peak every 100,000 and 41,000 years, with additional significant variations every 19,000 to 23,000 years.

15-10 Together, orbital forcing and climate feedbacks produced paleoclimate.

- Paleoclimatic events do not exactly match insolation because Earth's environmental system exercises positive and negative climate feedbacks that can enhance or suppress the timing and intensity of the Earth–Sun geometry. A climate feedback is a process that takes place on Earth that amplifies (a positive feedback) or minimizes (a negative feedback) the effect of the orbital parameters. Climate feedbacks are responsible for the difference between orbital forcing and Earth's actual climate.

- A positive climate feedback ended the ice age. Predictable variations in Earth's orbit and tilt increased the amount of sunlight hitting southern latitudes, causing Antarctic sea ice to retreat. The reduced ice cover resulted in warming of ocean water, which allowed dissolved carbon dioxide to escape into the atmosphere. The released gas proceeded to warm the whole world.

- Milankovitch's theory ties climate change to slow and regular variations in Earth's orbit. Thermohaline circulation may contribute to rapid climate change. Acting as a conveyor belt, it carries heat from the equator toward the poles; there it raises Arctic temperatures, discouraging the growth of ice sheets. Influxes of freshwater slow or shut down the circulation by preventing bottom water formation. The cessation of the conveyor belt cools the northern hemisphere and regulates snowfall at N 65°.

KEY TERMS

alpine glaciers (p. 446)
climate proxy (p. 460)
continental glaciers (p. 446)
glacial landforms (p. 444)
glacial moraine (p. 453)
glacial outwash (p. 454)
glacial till (p. 453)

glacier mass balance (p. 450)
glaciers (p. 444)
interglacials (p. 452)
Louis Agassiz (p. 444)
marine isotopic stages (p. 466)
meltwater (p. 446)
orbital parameters (p. 470)

paleoclimatology (p. 444)
permafrost (p. 445)
Quaternary period (p. 444)
zone of accumulation (p. 450)
zone of wastage (p. 450)

ASSESSING YOUR KNOWLEDGE

Please complete this exercise before coming to class. Identify the best answer to each question.

1. What is a glacier?
   a. A frozen river
   b. Snow on a steep slope that is slowly sliding downhill
   c. An ice avalanche composed of compressed recrystallized snow
   d. A large, long-lasting river of ice
   e. A type of ice in mountains that moves up and down hills

2. The smallest type of alpine glacier is a(n):
   a. Temperate glacier.
   b. Icecap.
   c. Outlet glacier.
   d. Polar glacier.
   e. Valley glacier.

3. The boundary between the zone of wastage and the zone of accumulation is called the _____ (ablation, level, equilibrium, terminus) line.

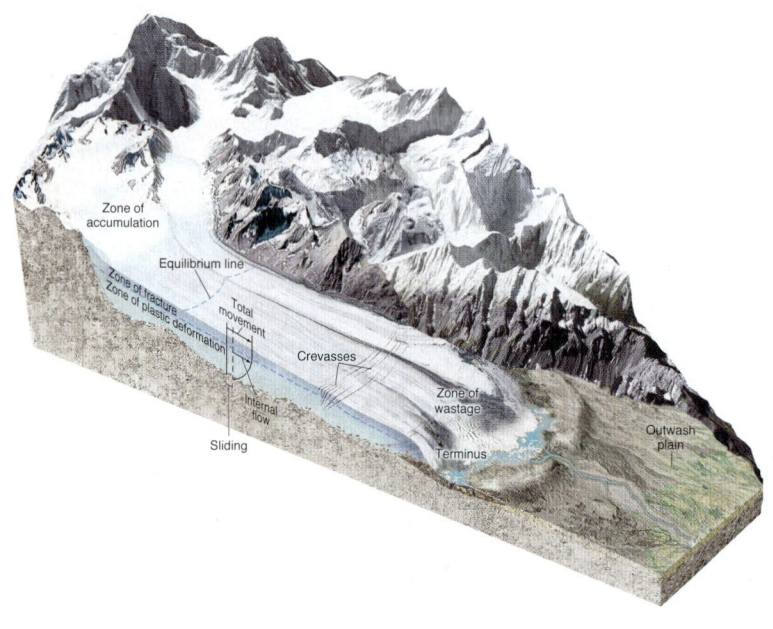

4. True or False: Ice in a glacier may flow both downhill and up-hill depending on the balance between accumulation and wastage.

   _____

5. Ice within a glacier is:
   a. Perpetually moving forward toward the terminus or is stagnant.
   b. Moving backward and forward, depending on the rate of supply and melting.
   c. Mostly immobile.
   d. Perpetually frozen to the bedrock beneath.
   e. Ice is not the primary component of a glacier.

6. A moraine is:
   a. A deposit of till that was in contact with glacial ice.
   b. A deposit of shale that forms under glacial ice.
   c. Stratified but unsorted conglomerate.
   d. A deposit of sand formed only when a glacier advances.
   e. A hump of old ice.

7. Which glacial landforms will be left by a retreating continental ice sheet? (There may be more than one answer.)
   a. Drumlins, eskers, and terminal moraines
   b. Ground moraine, kettle lakes, end moraines
   c. Deposits of till, outwash plain, glacial erratic
   d. Horns, arêtes, and tarns
   e. Hanging valleys, U-shaped valley

8. In response to global warming, alpine glaciers around the world are:
   a. Generally retreating.
   b. Relatively stable.
   c. Generally advancing.
   d. Moving in random ways.
   e. None of the above.

9. Retreat of Arctic sea ice:
   a. Has no effect on global warming.
   b. Causes sea level to fall due to melting sea ice.
   c. May accelerate snowfall on the Antarctic ice sheet.
   d. May accelerate growth of the Greenland ice sheet.
   e. Will change albedo and be a positive feedback to warming.

10. Glacial till is generally:
    a. Well sorted.
    b. Poorly sorted but well stratified.
    c. Well stratified and well sorted.
    d. Poorly sorted and poorly stratified.
    e. None of the above.

11. Oxygen isotopes serve as a climate proxy because:
    a. When they freeze in sea ice, they can be detected by satellites.
    b. Oxygen isotopes are sensitive to temperature, evaporation, and precipitation.
    c. Some oxygen isotopes are never detected.
    d. Oxygen isotopes preserve a natural record of plankton abundance.
    e. None of the above.

12. Marine isotopic stages:
    a. Are not useful in studying paleoclimate.
    b. Are expressed in double digits for warm periods and in single digits for cool periods.
    c. Are used to identify glacials and interglacials and stadials and interstadials.
    d. Occur principally during ice ages, when glaciers expand.
    e. Are used to name glacier advances and retreats.

13. During the interstadial 5e:
    a. Sea level was lower and temperature was warmer.
    b. Sea level was about the same and temperature was higher.
    c. Sea level was higher and temperature was warmer.
    d. Greenland was larger and sea level was lower.
    e. None of the above.

14. Which of the following statements about orbital parameters are true? (There may be more than one answer.)
    a. They govern the weather due to their relationship to the Moon.
    b. They include precession, obliquity, and axial tilt.
    c. They produce climate cycles with a 100,000-year periodicity.
    d. They correlate with observations of paleoclimate.
    e. They do not influence global climate.

15. The Younger Dryas:
    a. Was a short-term climatic warming at the end of the last ice age.
    b. Occurred during the last interglacial.
    c. Was a cool period that followed the end of the last ice age.
    d. Is the scientific name for modern global warming.
    e. Occurred during the last ice age.

16. During the transition from the last ice age to the present interglacial, the retreat of Antarctic sea ice acted as a _____ (positive, negative, unimportant) feedback.

17. Climate feedbacks:
    a. May magnify or suppress the influence of orbital parameters.
    b. Explain the timing of major climate cycles.
    c. Are the results of the gravitational influence of the nearby planets.
    d. Rarely influenced the nature of the paleoclimate.
    e. Comprise the process that controls Earth's exposure to insolation.

18. For a glacier to advance:
    a. Snowfall must slow and then stop.
    b. Retreat must equal advance.
    c. Wastage must exceed accumulation.
    d. Accumulation must exceed wastage.
    e. The internal conveyor belt must reverse direction.

19. Rapid climate changes:
    a. Are well predicted by orbital parameters.
    b. Have not been shown to be real phenomena.
    c. Are caused by climate feedbacks, not by orbital parameters.
    d. Are important only in the tropics.
    e. Are typical of changes in insolation.

20. True or False: Global warming is causing Greenland to experience melting at lower elevation, but enough snow continues to fall each year at higher elevations to offset this trend. _____

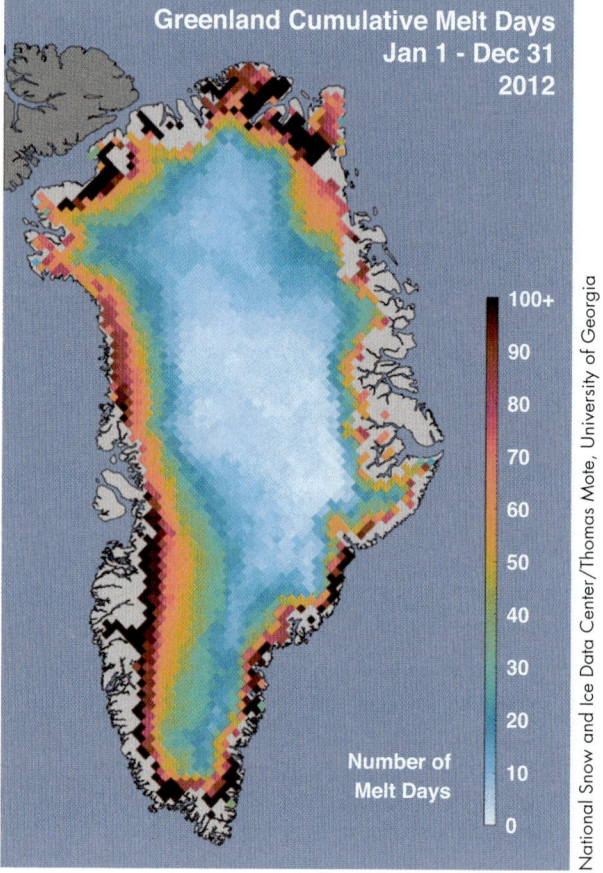

FURTHER RESEARCH

1. Use the Internet to find the glacier closest to your location. Is it advancing or retreating?

2. Do you live in a coastal community? Has your local newspaper or TV news station presented any stories about sea level rise?

3. Research the Internet, interview local geology professors, or ask the state geological survey in your locality to learn what your region was like during the last ice age. Is there any evidence of processes that were active in your area during the last ice age?

4. How does the landscape of Canada reflect the influence of the Laurentide or Cordilleran ice sheet?

ONLINE RESOURCES

Explore more about glaciers and paleoclimatology on the following Web sites:

National Snow and Ice Data Center, "All About Glaciers":
http://nsidc.org/glaciers

U.S. Geological Survey "Benchmark Glaciers" Web site:
http://ak.water.usgs.gov/glaciology

U.S. Geological Survey, "Satellite Image Atlas of Glaciers of the World":
http://pubs.usgs.gov/fs/2005/3056

NASA Earth Observatory, "Glaciers, Climate Change, and Sea Level Rise":
http://earthobservatory.nasa.gov/IOTD/view.php?id=5668

National Climatic Data Center, "Paleoclimatology":
www.ncdc.noaa.gov/paleo/paleo.html

NASA Earth Observatory, "Paleoclimatology: Introduction":
http://earthobservatory.nasa.gov/Features/Paleoclimatology

Additional animations, videos, and other online resources are available at this book's companion Web site:
www.wiley.com/college/fletcher

This companion Web site also has more additional information about WileyPLUS and other Wiley teaching and learning resources.

MASS WASTING

16-1 Mass wasting is the movement of rock and soil down a slope under the force of gravity.

LO 16-1 Define mass wasting.

16-2 Creep, solifluction, and slumping are common types of mass wasting.

LO 16-2 Define *creep*, *solifluction*, and *slumping*.

16-3 Fast-moving mass wasting events tend to be the most dangerous.

LO 16-3 Distinguish among mass wasting processes that are flows, slides, and falls.

16-4 Avalanches, lahars, and submarine landslides are special types of mass wasting processes.

LO 16-4 Compare and contrast avalanches, lahars, and submarine landslides.

16-5 Several factors contribute to unstable slopes.

LO 16-5 Itemize the factors that contribute to mass wasting.

16-6 Mass wasting processes vary in speed and moisture content.

LO 16-6 Describe how various types of mass wasting processes differ in terms of their speed and moisture content.

16-7 Human activities are often the cause of mass wasting.

LO 16-7 List some ways humans cause mass wasting.

16-8 Research improves knowledge of mass wasting and contributes to the development of mitigation practices.

LO 16-8 Describe methods of mitigating mass wasting hazards.

CHAPTER CONTENTS & LEARNING OBJECTIVES

GEOLOGY IN OUR LIVES

Mass wasting is the set of processes that move weathered rock, sediment, and soil down a slope under the force of gravity. Mass wasting can be hazardous to humans if they or their buildings and roads are in the way of these processes. Over the past century, the world's population has grown enormously, and many people now live in neighborhoods and other types of residential communities that formerly were considered too dangerous for human habitation. As buildable land becomes more scarce, people are increasingly moving into environments that are more exposed to mass wasting and other geologic hazards, putting them at even greater risk of losing property—and life. The best way to manage mass wasting hazards is to know the characteristics of a hazardous slope and avoid building there. Taking such a precaution is called "hazard avoidance"; it is the single most effective—and inexpensive—approach to managing geologic hazards.

Q What geologic factors might contribute to causing a landslide such as this?

Genilson Araujo/Globo/Getty Images, Inc.

16-1 Mass Wasting Is the Movement of Rock and Soil Down a Slope Under the Force of Gravity

LO 16-1 Define mass wasting.

The process by which gravity pulls soil, debris, sediment, and broken rock (collectively known as *regolith*) down a hillside or cliff is called mass wasting. And the best way to convey the force of this phenomenon is with a true story.

The Danger of Mass Wasting

It was a typical sunny day along the California coast in January 2010. Taking a break from hiking the beach, Jenny relaxed against the base of the sandy cliffs to enjoy a snack. While she ate, and without warning, a thick, heavy mass of muddy earth and debris slumped down the cliff face, burying her up to her neck. Hearing her screams, people from a nearby trailer park rushed down the slope and, using their bare hands, worked to free her from the crushing weight that entombed her. Once freed, and still dazed, Jenny was rushed to the hospital, where she was treated for broken ribs, a dislocated hip, skin abrasions, and a concussion.

Jenny was one of the lucky ones, for in a typical year, slope failure is responsible for 25 to 50 deaths and over $1.5 billion in damages in the United States alone. In many other countries, losses tend to be even higher due to irresponsible land-use practices and building codes that ignore geologic hazards. The **mass wasting** process, which, driven by gravity, pulls **regolith** down a hillside or cliff, is shown in **FIGURE 16.1**.

Prevention Through Avoidance

Mountain building, weathering, and erosion are perpetually altering the slope of the land. And, inevitably, gravity drags loose and unstable accumulations of rock and sediment downhill. But mass wasting is not a single phenomenon; it occurs in several variations (discussed in the next section), from the instantaneous event that trapped Jenny on the beach to longer and slower processes that shift soil and regolith to the base of slopes over the course of years, decades, and centuries.

All mass wasting events can, however, be hazardous to humans when they, or their communities and infrastructure, are in the path of these events. Mass wasting is another example of a *geologic hazard* that is best dealt with simply by avoiding it. We can counteract the threat posed by mass wasting by, first, learning to recognize the characteristics of an unstable slope and, then, electing not to build or engage in other activities there. This approach is termed **avoidance**.

Impact of Population Growth

As we have pointed out in previous chapters, over the past century, the world's population has grown astronomically, and many people now live in places that were formerly considered too dangerous or risky for human habitation. As buildable land becomes scarce, communities are expanding into hazardous environments, turning them into neighborhoods and residential centers. These environments include stream valleys that are prone to flooding, the slopes of active volcanoes, earthquake fault zones, shorelines where a storm-driven ocean is a constant threat, and hillsides vulnerable to mass wasting (**FIGURE 16.2**).

By moving our communities onto risky hillsides, we alter the landscape in ways that make slopes unstable and more prone to mass wasting. Put another way, we not only choose to live in dangerous places, by reshaping them, we make them even more dangerous. Read on to learn more about this. **TABLE 16.1** enumerates just how dangerous this practice can be. And note, this list contains just a few of the many landslide events that have resulted in the loss of human life.

The Cost of Neglect: Vaiont Dam

Although it happened over 50 years ago, the Vaiont Dam disaster still serves as a startling reminder of the tragic consequences of neglecting to consider **slope stability** in engineering projects. One of the world's highest dams, at 266 meters, Vaiont Dam spans the Vaiont River Valley in the Italian Alps. The purpose of the dam was to trap the river in order to create a reservoir and generate hydroelectric power for neighbouring communities.

FIGURE 16.1 Mass wasting is defined as the movement of rock and soil down a slope under the force of gravity.

Human activity caused this landslide. List three human activities that might be responsible for mass wasting of this magnitude.

USGS

Reuters/Landov LLC

FIGURE 16.2 As human populations expand into previously undeveloped areas, activities such as building this wilderness road across an unstable slope may put people directly in the path of geologic hazards. In early 2009, 33 coffee pickers walking home from work were killed by this landslide in northern Guatemala.

Q How could building this road have contributed to the landslide?

TABLE 16.1 Recent Landslide Events

Year	Location	Types	Deaths
1993	Papua New Guinea	Earthquake-triggered landslide	37
1996	Washington State	Sea cliff collapse	9
1997	Australia	Landslide	18
2001	El Salvador	Earthquake-triggered landslide	1,200
2002	Russia	Landslide	111
2005	California	Landslide triggered by heavy rains	16
2006	Philippines	Huge debris slide triggered by heavy rains and an earthquake	1,126
2010	Uganda	Landslide triggered by heavy rains	100 to 300

Shortly after it was constructed, and the narrow valley behind the dam began to fill with water, large cracks were seen opening in the soil near the top of Monte Toc, the mountainside rising 1,000 meters above the reservoir. Engineers grew concerned that landslides into the reservoir might reduce its capacity. As it turned out, that was the least of their worries.

When Vaiont Dam was completed, and the reservoir filled with water, groundwater in the surrounding valley and its slopes began to rise. This had the effect of weakening the strength of rock on the valley walls as they became saturated and as water pressure separated internal fracture surfaces. As summer turned to autumn, heavy rains added to the weight of rock and soil perched on steep slopes above the reservoir valley.

By early October, engineers monitoring the hillside measured shifts in the rock at the startling rate of 40 centimeters per day. Sensing danger, wildlife abandoned the area, and the hills fell silent. Now more alarmed, engineers opened the dam's floodgates in an effort to lower the lake level, but the water continued to rise. The accelerating downward movement of the mountainside was swelling the reservoir.

On October 9, 1963, at about 10:15 PM, a massive block of soil and rock 2 kilometers long, 1.5 kilometers wide, and several hundred meters thick accelerated down Mount Toc into the reservoir. There

it launched a titanic wave 100 meters high that overtopped the dam and raced through the valley below. Without warning it slammed into the town of Longarone, killing 2600 people. The entire event lasted only six minutes.

With fatal consequences, the engineers who designed the reservoir neglected to account for the geology of the valley's slopes. A steeply dipping layer of highly fractured limestone rested on a bed of clay that became slick and unsteady when wet. Over the centuries, river erosion had removed rock at the base of the hillside that previously buttressed the limestone, a process known as **undercutting**. Rainfall and the rising water table reduced the friction holding the limestone, causing the clay layer to lose strength. Although the landslide might have eventually occurred naturally, the reservoir accelerated the process. In short, this was the wrong place to build a dam.

Amazingly, the dam survived the giant wave and still stands; what did not survive was the town of Longarone down the valley and many of the people who lived there. Subsequently, the reservoir was closed and abandoned, but the Vaiont catastrophe has not been forgotten. As word of the disaster spread worldwide, its causes were closely analyzed; thereafter, greater attention was focused on the science of **geological engineering**, the study of geological processes and the role they play in engineering projects and safety. The *natural processes* and *geological products* associated with mass wasting have since become the subject of classroom lessons in universities around the world, and engineers have learned to highlight the importance of geological analysis in designing major construction projects, such as dams, highways, and bridges.

Gain deeper insight into this event by taking the Critical Thinking exercise called "Vaiont Dam."

Q EXPAND YOUR THINKING

The debris that hurt Jenny came down with no warning. What could have triggered it?

CRITICAL THINKING

Vaiont Dam

The Google Earth image in **FIGURE 16.3** shows the Vaiont River Valley in Italy and the landslide that occurred there. Please study the image before completing the following exercise:

1. With a marker, outline the scar left by the landslide.

2. Draw an outline around the regolith that slid into the Vaiont River Valley.

3. Label the remaining portion of the reservoir; now outline the edges of the total reservoir if the landslide had not happened.

4. The landslide launched a giant wave that swept over the dam and killed 2600 people in the town of Longarone. Show with arrows and labels the sequence of events, the path of the landslide, and the direction of the wave.

5. Imagine you are a geological engineer: What would you study, and why, to ensure that a reservoir you are building is safe? Be detailed and specific in your answer.

FIGURE 16.3 Vaiont Dam was the highest in Europe. Shortly after it was constructed, a massive landslide set off a 100 meter-high wave, killing 2600 people and blocking the center portion of the Vaiont reservoir.

© CNES 2010, distribution Astrium Services/Spot Image S.A.

16-2 Creep, Solifluction, and Slumping Are Common Types of Mass Wasting

LO 16-2 Define creep, solifluction, and slumping.

Most regolith is the product of weathering; that means mass wasting is the first type of erosion to move these sediments from the place where they were formed. We discuss creep, solifluction, and slumping in this section, three common types of mass wasting. But we begin by examining how the movement of regolith shapes valleys.

The Effect of Mass Wasting on Valleys

Given that mass wasting is the movement of regolith down a slope under the force of gravity, and that most regolith is the product of weathering, mass wasting is the first type of erosion to move these sediments from their point of origin.

The effect of mass wasting on valleys, in particular, is profound. Without it, valleys cut by streams would be narrow, steep-sided gorges with vertical walls. But the tendency for rock and regolith to become unstable and move to the base of slopes (where they are carried away by running water) leads to the widening and expansion that occurs in stream valleys (**FIGURE 16.4**). As we learned in Chapter 15, glacial erosion is another important process that sculpts valleys in mountain systems around the world. Consequently, most canyons and valleys have been shaped by a combination of glacial and stream erosion and mass wasting.

FIGURE 16.4 Mass wasting creates broad valleys from steep gorges that have been cut by streams. The width of the Grand Canyon is a result of mass wasting.

🔵 **Erosion by the Colorado River has made the Grand Canyon deep. What type of mass wasting has made the canyon wide?**

Several variations of mass wasting occur. Although the word "landslide" is commonly used to describe any kind of slope movement, geologists distinguish among several types of mass wasting processes. Knowing the characteristics, and the causes and outcomes of each of these processes, is the first step toward successfully avoiding them.

Creep

Soil **creep** is the slow, downslope migration of soil under the influence of gravity. Creep occurs over periods of time ranging from months to centuries. Curved tree trunks, cracks in slopes (**FIGURE 16.5**), tilted power poles, and bent fences or walls are familiar indicators that creep is taking place.

The mechanics of soil creep depend on the role of gravity and the rise and fall of the slope surface. Because water increases in volume by over 9% when it freezes, the surface of a slope will expand upward in winter—a process known as *frost heave*—and contract again in summer. Alternate wetting and drying of clay in the soil can have the same effect, although clay may expand when wet. These processes work in essentially the same way: Individual soil or rock particles are raised at right angles to the slope by swelling or expansion of the soil, and settle vertically downward during compaction or contraction. The net result is slow downslope creep.

Solifluction

Solifluction is a type of soil creep that occurs in water-saturated regolith in cold climates. Most solifluction is seen in *permafrost* zones. Recall from Chapter 15 that permafrost is ground that is permanently frozen but develops a thawed layer in the summer.

Solifluction produces distinctive wave forms of slowly sliding soil on hill slopes where the top soil remains saturated with water for long periods but may slide on an underlying layer of weakness (usually ice) (**FIGURE 16.6**).

During warm weather, the ground in permafrost areas will begin to thaw from the surface downward. Much of the freshly melted water cannot be absorbed by or move through the permafrost layer, so it saturates the upper layer of vegetation and regolith. The saturated layer flows down even the slightest of slopes as it slips over the frozen layer beneath it. The result is seen as *lobes*, wave forms, of slowly flowing soil, the distinctive sign of solifluction.

Slumps

A **slump** occurs when regolith suddenly drops a short distance down a slope, usually as a cohesive block of earth that simultaneously slides and rotates along a *failure surface* shaped like the bowl of a spoon (**FIGURE 16.7**). As material shifts downward, one or more crescent-shape *headwall scarps* develop at the upslope end of the slide. The base of the slide is characterized by the *toe* of the slump.

In a typical slump, the upper surface of the rotating block remains relatively undisturbed. Slumps leave a curved scar or depression on the slope. They can be isolated or occur in large *slump complexes* covering thousands of square meters.

Slumps are especially common in places where clay-rich sediments are exposed along a steep bank that has been undercut. They often occur as a result of human activities (such as deforestation) and

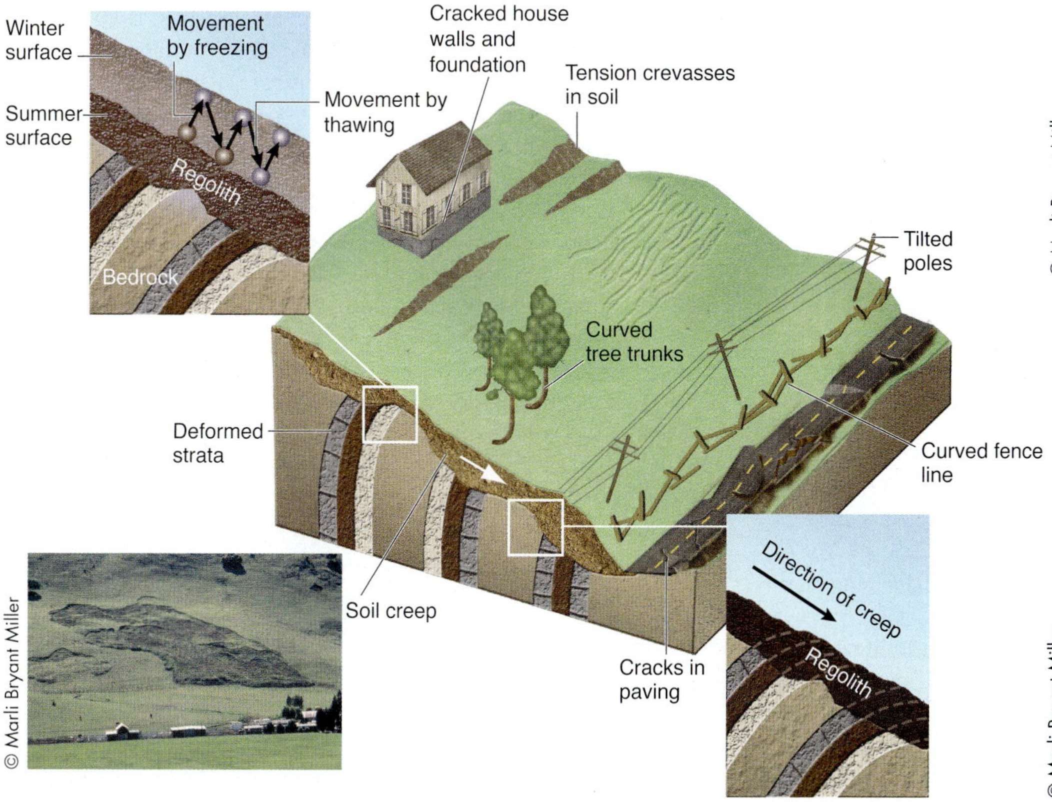

Winter surface

Movement by freezing

Cracked house walls and foundation

Tension crevasses in soil

Summer surface

Movement by thawing

Regolith

Bedrock

Tilted poles

Curved tree trunks

Deformed strata

Curved fence line

© Marli Bryant Miller

Soil creep

Direction of creep

Regolith

Cracks in paving

FIGURE 16.5 Soil creep often is marked by warped fence lines, soil ripples and cracks, broken road surfaces and building walls, tilted poles and trees, and curved tree trunks.

© Marli Bryant Miller

What environmental conditions promote soil creep?

are common along roads where hillsides have been overly steepened by undercutting at the base during construction. They are also common along riverbanks and coastlines, where erosion has undercut the slopes. Heavy rains and earthquakes can trigger slumps.

Ⓠ EXPAND YOUR THINKING

Describe a geologic environment where creep, slump, and solifluction are all taking place near each other. What triggers might be at work?

© Marli Bryant Miller

FIGURE 16.6 Prevalent in polar regions and some high mountains, solifluction is a very slow type of slope failure that occurs in areas of permafrost where saturated soil flows over a frozen layer of earth beneath it.

Describe a home with design features that are resistant to damage by solifluction.

(a)

Slump

(b)

© Marli Bryant Miller

FIGURE 16.7 (a) A slump is defined by a curved failure surface and a rotating block (or series of blocks). (b) The tops of slump blocks remain undisturbed even though they have been displaced downhill by several meters. Note the headwall scarp.

 What are the distinguishing characteristics of a slump?

16-3 Fast-Moving Mass Wasting Events Tend to Be the Most Dangerous

LO 16-3 Distinguish among mass wasting processes that are flows, slides, and falls.

Mud flows and debris flows are rivers of rock, earth, and other detritus saturated with water, and they can be extremely dangerous to anyone or anything in their paths.

Mud Flows and Debris Flows

On February 17, 2006, after 10 days of heavy rains and a minor earthquake, a **debris flow** set loose on the Philippine island of Leyte, with deadly consequences—many lives were lost and a great deal of property was damaged (**FIGURE 16.8**). An entire elementary school filled with students and their teachers (1126, total) was buried under a massive debris flow that surged out of the nearby hills and fanned out across the fields where the school was located.

Both **mud flows** and debris flows are rivers swollen with rock, earth, and other detritus saturated with water. The difference between them is that mud flows tend to have lower viscosity (i.e., they are more watery) than debris flows because they are composed principally of mud; debris flows, in contrast, carry clasts, ranging in size from clay particles to boulders and, often, a large amount of woody debris.

The two types of flows develop when water accumulates rapidly in the ground during heavy rainfall or rapid snowmelt, changing stable regolith into a surging river of sediment, or slurry. They can move at high speeds, and strike with little or no warning.

The distinguishing feature of either a debris or mud flow is that the material contained within tends to exhibit *turbulent flow characteristics*. Although other types of mass wasting processes are chaotic, they may not display true flow characteristics of a fluid, whereas a debris or mud flow does exhibit these characteristics. Both types of flows typically consist of a well-mixed, turbulent mass of saturated regolith that literally *flows*, rather than *slides* (**FIGURE 16.9**).

Mud flows, due to their low viscosity, can travel several tens of kilometers from their source, growing in size as they pick up trees, boulders, cars, and other materials. Debris flows typically do not travel so far. Both mud and debris flows result from heavy rains in areas with an abundance of unconsolidated sediment, such as steep slopes that have been deforested. In these areas, it is common after a heavy rain for streams to turn into mud or debris flows as they gather up loose sediment and eventually break free of their banks.

As we learned in Chapter 6, flows of muddy ash called *lahars* also can result from volcanic eruptions that cause melting of snow or ice on the volcano's slopes or the draining of a crater lake.

Rock Slides and Debris Slides

Debris slides and **rock slides** result when a slope fails along a plane of weakness. Collectively, debris slides and rock slides are commonly known as *landslides*, but they have differences, described as follows:

- A debris slide is characterized by unconsolidated rock, debris, and regolith that have moved downslope along a relatively shallow failure plane. Debris slides carve steep, unvegetated scars in the upper region of the slide and irregular, hummocky (bumpy) deposits in the toe region. Debris slide scars are likely to "ravel"—continue to pull apart into separate pieces—and remain unvegetated for many years. The mass wasting event at Vaiont Dam, reviewed earlier, was a debris slide.

- A rock slide typically lacks much debris; usually, it consists of blocks of rock in a chaotic mass. Such slides are most likely to occur on slopes with an incline of over 65% where unconsolidated regolith overlies shallow soil or bedrock. The shallow slide surface is usually less than 5 meters deep. The probability of sliding is low in places where bedrock is exposed, unless weak bedding planes and extensive bedrock joints and fractures parallel the slope.

Both types of slides can develop on a bedding plane, a clay layer, a buried erosional surface, or a *joint* surface (**FIGURE 16.10**). Recall from earlier chapters that joints are fractures in the crust that result from the cooling or expansion of a rock layer.

Rockfalls and Debris Falls

Rockfalls (**FIGURE 16.11**) occur when an accumulation of consolidated rock is dislodged and falls through the air—free-falls—under the force of gravity. **Debris falls** are similar, except that they usually involve a mixture of soil, regolith, vegetation, and rocks. A rockfall may consist of a single rock or a mass of rocks, and the falling rocks may dislodge others as they collide with the cliff face. Because this process involves free-falling through the air, rockfalls and debris falls are associated with steep cliffs.

FIGURE 16.8 This debris flow on the Philippine island of Leyte buried a school, full of children and their teachers, under 40 meters of water-saturated dirt, mud, trees, regolith, and other debris. They could not be rescued.

Based on field evidence, how would you tell the difference between a debris flow and a mud flow?

(a)

(b)

FIGURE 16.9 (a) This Indonesian village was destroyed by a mud flow. (b) The debris flow shown here damaged the town of Glenwood Springs, Colorado.

What geologic clues would you look for to assess whether your home was vulnerable to a debris flow or a debris slide?

Rock and debris slides

Bedding or joint planes

(a)

(b)

FIGURE 16.10

(a) Rock or debris slides tend to occur along a plane of weakness, such as a bedding surface, an unconformity, a clay layer, or a joint.
(b) This debris/rock slide occurred in the spring of 2002 along McAuley Creek, British Columbia, Canada.

(a)

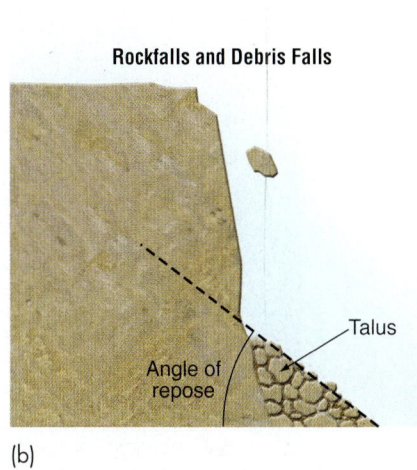

Rockfalls and Debris Falls

Talus

Angle of repose

(b)

(c)

FIGURE 16.11 (a) Rockfall. (b) When a single rock or a mass of debris free-falls through the air, it may dislodge others as it collides with the cliff face. (c) Debris fall.

Rockfalls and debris falls are identified by one primary feature: What is it?

EXPAND YOUR THINKING

The people of Leyte in the Philippines were buried under a debris flow. What conditions would lead to the development of a debris flow? What would trigger such an event, and how would you avoid it?

16-4 Avalanches, Lahars, and Submarine Landslides Are Special Types of Mass Wasting Processes

LO 16-4 Compare and contrast avalanches, lahars, and submarine landslides.

Avalanches, lahars, and submarine landslides occur under particular geologic conditions and may cause catastrophic damage. We address the specifics of these three types of mass wasting in turn.

Avalanches

An **avalanche** is a type of debris flow; it is a fast-flowing, fluidized (by air, not water) mass of snow, ice, air, and occasionally some regolith, that cascades down a mountainside following the collapse of a snowfield (**FIGURE 16.12**). Large avalanches, moving over 10 million tons of snow, can travel at speeds exceeding 300 kilometers per hour.

Snow accumulates in layers, some of which are stronger than others due to differences in the weather conditions that generated the snowfall. As snow builds up on a steep slope, a weak layer may fail under the increasing weight of the snow, in reaction to vibration, or due to some other cause. The overlying mass detaches from the slope, slides over the weak layer, and gains strength and size as it gathers more material on its downhill journey.

Loose snow avalanches are the most common type. They consist of newly fallen snow that has not had time to compact. These avalanches usually develop in steep terrain on high slopes and originate at a single detachment point. The avalanche gradually widens on its way down the slope as it accumulates more snow, forming a teardrop shape.

While loose snow avalanches may be the most common, it is *slab avalanches*, consisting of a strong, stiff layer of snow known as a *slab*, that account for about 90% of avalanche-related fatalities. Slabs develop when the wind deposits snow on a *lee slope* (the mountain side not facing the wind) in compact masses. The avalanche starts with a crack that quickly grows and spreads across the slab, in some cases releasing a solid mass of snow hundreds of meters long and several meters thick.

Still another type of avalanche, a *powder snow avalanche*, develops as air is mixed with the snow at the front of an avalanche, becoming a powder cloud, a turbulent suspension of snow particles. The largest avalanches are usually of this type.

Lahars

We first discussed **lahars** in Chapter 6, but they bear revisiting here since they cause such catastrophic damage. A lahar (**FIGURE 16.13**) is a type of mud flow composed of pyroclastic ash, lapilii (gravel-size volcanic particles), and water that flows down the slopes of a volcano.

Lahars have the behavior of concrete; they are fluid when moving but become solid upon coming to rest. Lahars can be created by snow and ice that is melted by a pyroclastic flow during an eruption, or by a flood set off by a melting glacier, a flash flood in a stream, or a heavy rainfall. The key feature of a lahar is that it consists of mud originating in an existing volcanic ash deposit.

Lahars often start in a stream channel, but as they gather mass and speed, they break out of the channel when their momentum prevents them from turning. This is when they become most dangerous to a local population. Several volcanoes near population centers are considered particularly dangerous as potential sources of lahars, among them Mount Rainier in the United States, Mount Ruapehu in New Zealand, and Mount Galunggung in Indonesia.

Galen Rowell/Mountain Light/Alamy

FIGURE 16.12 Powder snow avalanches, such as this one, tend to be the largest type of avalanche.

What can trigger an avalanche?

© Roger Ressmeyer/CORBIS

FIGURE 16.13 A lahar is a mud flow composed of pyroclastic material and water that flows down the slopes of a volcano. This lahar originated on the slopes of Mount Pinatubo in the Philippines in 1991.

How do geologists determine whether an area is vulnerable to a lahar?

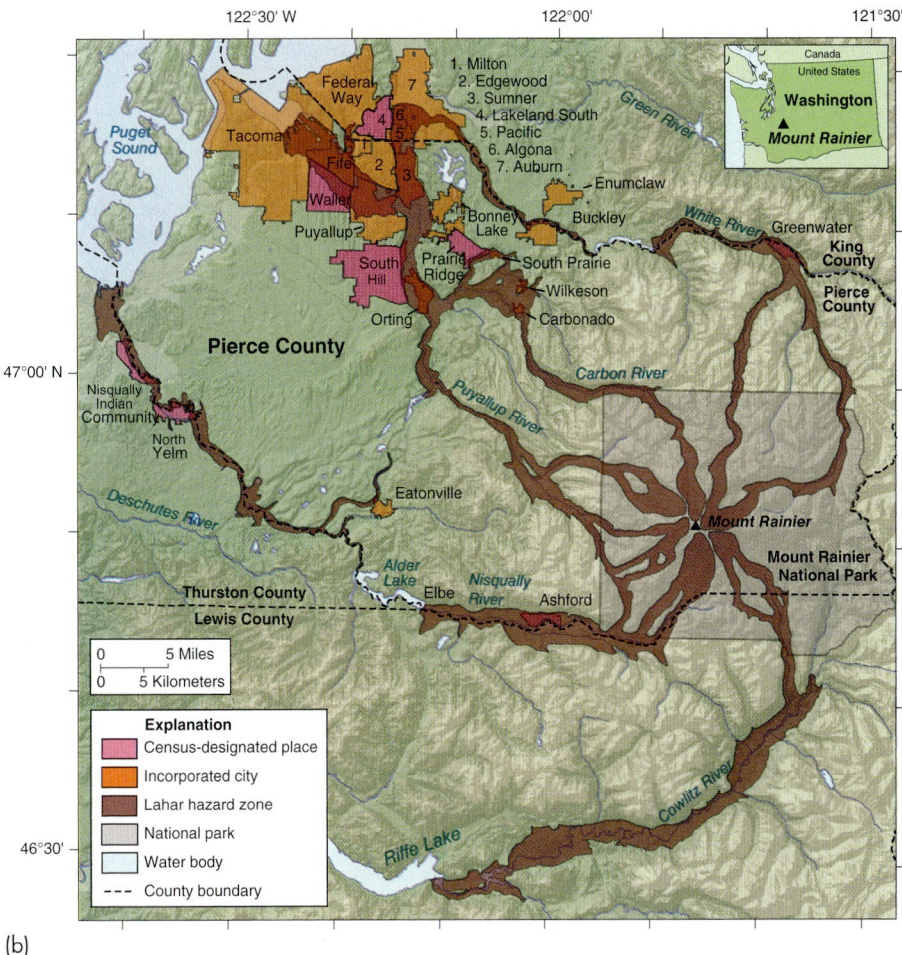

David Wieprecht/USGS

(a)

(b)

FIGURE 16.14 (a) Several hundred thousand people live on 500-year-old lahar deposits in the Puyallup Valley, Washington, in the shadow of Mount Rainier. (b) Are these communities vulnerable to future lahar events?

At the base of Mount Rainier, several hundred thousand people live in the Puyallup Valley, near Seattle (**FIGURE 16.14**), where towns have been built on lahar deposits that are only about 500 years old. It is believed that lahars flow through the valley every 500 to 1000 years, meaning that these communities face considerable risk. Local authorities have set up a warning system and instituted evacuation plans to be put into effect at the first sign of a developing lahar.

Submarine Landslides

Among the awe-inspiring natural features of the Hawaiian Islands are the dramatically beautiful *palis*, Hawaiian for "cliffs." These sheer, vertical rock faces of ancient basalt lava flows are hundreds of meters high. Many palis drop directly into the sea (**FIGURE 16.15**). They are made by *submarine landslides*, a type of debris flow.

(a)

Koshtra Tolle/Getty Images, Inc.

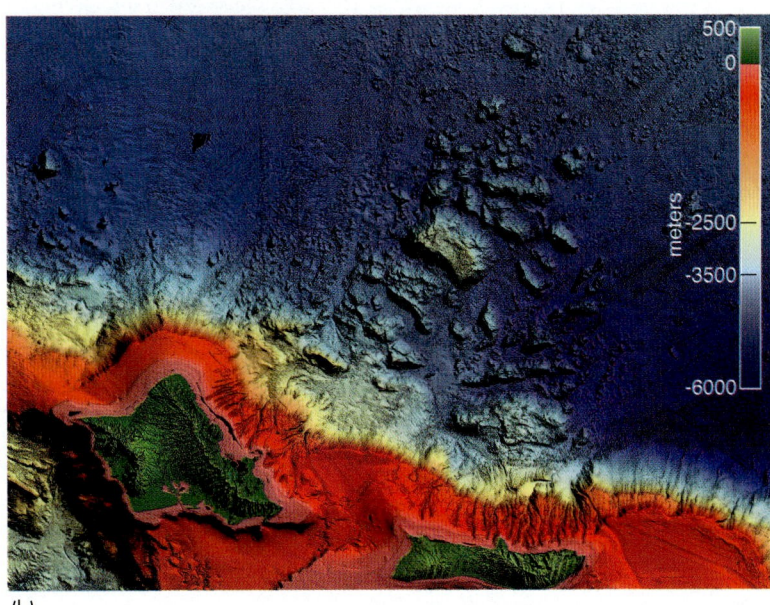

(b)

FIGURE 16.15 The Hawaiian palis (a) are the highest sea cliffs in the world. A large debris field (b) off the windward side of the islands of Oahu and Molokai is connected to each island at the location of two large sea cliffs.

What is the evidence for submarine landslides in Hawaii?

The geologic history of these immense cliffs is a tale of mass wasting on a prodigious scale. The cliffs are formed by vast debris flows that take away whole pieces of an island. To understand this process, recall the discussion of mineralogy in Chapter 4 and igneous rocks in Chapter 5. The Hawaiian Islands are composed of two types of mafic igneous rock: extrusive and intrusive. The extrusive rock (basalt) is lava that solidifies within a few minutes of reaching the atmosphere. With so little time in which to crystallize, the lava, though it consists of basalt, has the texture of glass and is very brittle because it lacks a crystalline structure.

Although no one was present when these palis were formed—they are thousands to millions of years old depending on which Hawaiian island they are found—it is hypothesized that swelling of a volcano and deep-seated earthquakes associated with magma intrusion prior to an eruption occasionally cause a flank of the volcano literally to break off and slide into the sea. Such an occurrence is possible because the islands are essentially piles of glassy layers. Today's palis are the eroded remnants of those fractures, which have since retreated and been reshaped by weathering due to waves, wind, and runoff.

Several observations provide evidence in support of this hypothesis. Maps of the seafloor show topography that is consistent with a landslide source. Fields of chaotic debris can be traced back to the islands where some of the highest palis are located. In addition, the main shield volcanoes composing the islands have incomplete outlines. Rather than having the broad circular or oval shape typically found at the base of a shield volcano, each of the two shield volcanoes forming the island of Oahu shows only half its base, and each forms a crescent rather than an oval. The same is true of the island of Molokai. Finally, by mapping the seafloor, marine scientists have been able to identify the remnants of more than 25 giant landslides surrounding the Hawaiian Islands. These slides were some of the largest on Earth, and most took place within the past 4 million years.

Ⓠ EXPAND YOUR THINKING

What features would you design into a community to make it less vulnerable to a lahar?

16-5 Several Factors Contribute to Unstable Slopes

LO 16-5 Itemize the factors that contribute to mass wasting.

Slope failure is attributable to a number of factors, including gravity, water content, and oversteepening; the last two require triggers to set off mass wasting, which include vibration, ice wedging, and others.

Gravity

Mass wasting, as explained previously, is governed by *gravity*. A fundamental force of nature, gravity attracts objects toward one another and, on Earth's surface, draws materials downhill. Gravity can be divided into two components for objects resting on sloping surfaces (**FIGURE 16.16**): one *parallel* to the slope (g_s) and the other *perpendicular* to the slope (g_p). On steep slopes (>45°), the force parallel to the slope will be greatest and will tend to pull objects downhill. On gentler slopes (<45°) the component perpendicular to the slope will be greatest and will tend to hold the object in place.

In concert with Earth's gravity, which determines the stability of materials on a slope, *friction* and *density* play important roles in determining when and how slope movement may occur. Friction, generated by resistance to movement at the point of contact between grains of material, tends to be a stabilizing factor in slope deposits. But when frictional forces are temporarily or permanently disrupted, mass wasting may occur. If the amount of friction is low, an object is more likely to move. In contrast, mass wasting is less likely if the amount of friction is high.

Water

Friction can be disrupted in a number of ways. Heavy rainfall can saturate rock and sediment on a slope and increase its weight so that it exceeds the force holding it in place. The same rainfall may lubricate the surface on which a deposit rests, also disrupting the friction that holds it in place. Slight shaking from a nearby earthquake, and desiccation in hot sunlight, causing sediment to shrink and crack, are other ways friction on a slope may be disrupted.

The density of sedimentary deposits is lowered when they become saturated with water. That is, their overall density approaches that of water, and they may achieve a state of temporary buoyancy. It is at this moment when gravity may control the deposit and lead to slope failure. *Water content* is, thus, an important controlling factor that influences the stability of Earth materials on slopes. Water destroys the cohesion that exists when one grain of sediment is in contact with another.

FIGURE 16.16 On steep slopes, the gravity component parallel to the slope (g_s) moves materials downhill. On gentler slopes, the gravity component perpendicular to the slope (g_p) tends to hold materials in place.

Ⓠ **What factors can reduce friction on a slope?**

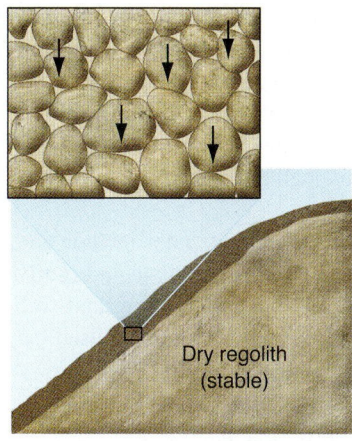
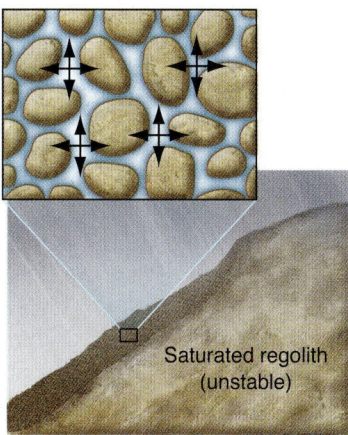

Dry regolith
(stable)

Saturated regolith
(unstable)

FIGURE 16.17 As the pore space between grains fills with water, water pressure lifts one grain off another in a process called dilation. This can cause a slope to fail.

Describe the forces at work in saturated regolith with building pore pressure.

FIGURE 16.18 Landslides along coastal cliffs occur when high waves batter the shoreline. Coastal slopes are oversteepened when wave erosion removes material from its base. This is known as undercutting.

Describe several processes, both human and natural, that produce oversteepening.

Aerial Archives/Alamy

Triggers of Slope Failure

As noted, water content, gravity, and oversteepening control slope stability; they may create conditions in which regolith or rock is poised on the threshold of failing. But to initiate a mass wasting event, some kind of *trigger* is needed.

The most common trigger of slope failure is *vibration*. Vibration, or ground shaking, can be set off by passing trucks or trains, an earthquake, a breaking wave, the rumbling of a volcano, or even thunder. Another common trigger is melting ice. Ice will hold unstable materials together, but with seasonal or even daily warming, ice can melt and loosen its grip on failure-prone slope deposits.

Another ice-related trigger of mass wasting is *ice wedging*, the physical weathering process by which ice expansion breaks off pieces of rock (see Chapter 7). It may occur when direct morning sunlight warms a mountain slope and melts ice that was formed during the previous evening or overnight. This process of daily ice wedging followed by slope failure is responsible for much of the fallen rock that collects at the base of steep cliffs. The resulting apron of loose rock is known as **talus** or scree (**FIGURE 16.19**).

As the pore space between grains fills with water, water pressure increases. Pressure can build so much that pore water literally lifts one grain off another, in a process called *dilation* (expansion). Once grains become dilated, the entire assemblage of rock and regolith gains buoyancy, and gravity causes the slope to fail (**FIGURE 16.17**).

Water plays other roles in the slope failure scenario. As rock and sediment on a slope become saturated, they gain weight, and once in motion, that weight contributes to the *momentum* of the mass. Momentum is the tendency of a moving mass to continue moving. A large mass of saturated rock and sediment weighing millions of tons gains a powerful momentum that can be hard to stop. Heavy, water-saturated lahars and mud flows have been known to travel dozens of kilometers at speeds of over 100 kilometers per hour, crossing ridges, filling valleys, and burying entire landscapes as the momentum of the mass carries it far beyond the slope on which it originated.

Water also is involved in altering the physical properties of regolith: What was formerly stable becomes unstable. Dry clay is among the strongest sedimentary deposits because of its grain-to-grain cohesion. But once it becomes wet, clay turns into a slick, cohesionless mass that not only possesses little internal strength but actually acts as a lubricant that enhances downslope movement of other materials.

Oversteepening

A third factor that controls slope stability is **oversteepening.** Loose sediment tends to be stable on slopes less than the *angle of repose*, the slope beyond which an unconsolidated mass will spontaneously fail. Coarse-grained sediment tends to have a steeper angle of repose, while fine-grained sediment tends to have a lower angle of repose.

In general, angles of repose vary between 25° and 40°, depending on the content of the sediment. Adding material to the top of a sedimentary accumulation or *undercutting* the slope at its base has the effect of sharpening the angle of the slope to the point at which its stability is exceeded. Most often, this leads to slope failure and, frequently, to catastrophic mass wasting. Examples of undercutting a slope at the base are common where roadways are carved into valley walls or wave erosion on coastlines undercuts the base of a cliff (**FIGURE 16.18**).

EXPAND YOUR THINKING

Describe the roles of water in mass wasting.

Gregory G. Dimijian/Photo Researchers

FIGURE 16.19 Talus lines the base of this cliff.

Use evidence in this photo and your own reasoning to write a hypothesis explaining the origin of this pile of debris. How would you test your hypothesis?

16-6 Mass Wasting Processes Vary in Speed and Moisture Content

LO 16-6 Describe how various types of mass wasting processes differ in terms of their speed and moisture content.

Mass wasting processes are categorized according to their relative speed (fast to slow) and moisture content (wet to dry)

Types of Mass Wasting Processes

The fastest mass wasting processes are mud flows, debris flows, and debris falls and rockfalls (**FIGURE 16.20**). These move at rates rang- ing from 1 meter per second to several tens of meters per second. Mud flows require that muddy sediments be saturated with ground- water, whereas rockfalls can be a completely dry process. The speed of these types of failures makes them the most hazardous, as they can overwhelm individuals and even entire villages in a matter of minutes, burying victims as well as causing catastrophic property damage.

Mass wasting processes that move at moderate rates include de- bris slides, rock slides, and slumps. The slowest forms include soli- fluction, which requires a high degree of water saturation, and soil creep, which is a largely dry process, although it can be caused by alternating wetting and drying or freezing and thawing.

Protection from these hazards requires awareness of their pres- ence, mapping and monitoring of controlling factors and triggers, and, finally, engineering practices focused on altering the environ- mental processes that lead to slope failure. You can read about the efforts of one community to prevent such a disaster in the Earth Citi- zenship feature, "El Salvador's Deadly Landslide."

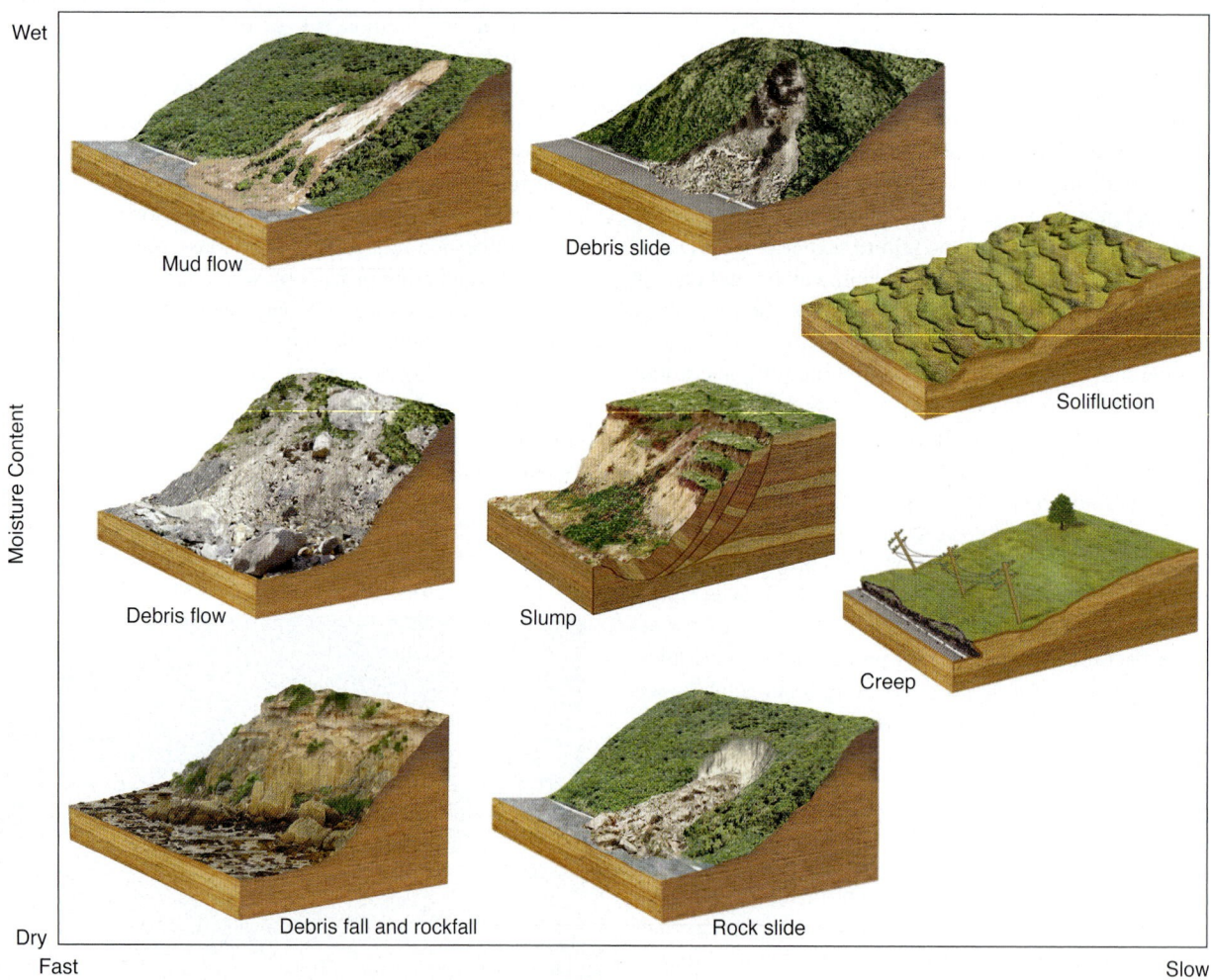

Wet

Moisture Content

Dry

Fast   Relative Speed   Slow

Mud flow

Debris slide

Solifluction

Debris flow

Slump

Creep

Debris fall and rockfall

Rock slide

FIGURE 16.20 Typical slope failure processes can be categorized based on moisture content and relative speed.

All these mass wasting events require a trigger. Give some examples of a mass wasting trigger, along with a discussion of factors (including human) that promote mass wasting.

EARTH CITIZENSHIP

El Salvador's Deadly Landslide

On January 13, 2001, a large earthquake off the coast of El Salvador triggered a debris slide of rock and soil that roared down on the city of Santa Tecla. In an instant, hundreds of homes and more than 1200 people were buried. International search teams worked for weeks to rescue the living, provide care for the injured, and retrieve bodies (**FIGURE 16.21**).

Local residents had warned authorities for years that *deforestation*, hillside *undercutting* to make room for new homes, *dirt fill* placed at the hilltop, and lack of proper geological engineering analysis would lead to a disaster if the hillside should fail in the next earthquake. "The developers kept digging further and further into the base of the mountainside despite everything we said, and the government let them do it," lamented Miguel Cordero, a survivor who lost his family.

Local groups of citizens filed a lawsuit to halt developers, and although the court ruled that development must cease, the ruling was only temporary. Soon it was overturned, and developers went on clearing trees and cutting away the base of the slope, eventually resulting in the deadly debris slide. It is hoped that a careful analysis of this event will teach lessons to help prevent future such catastrophes.

©AP/Wide World Photos

© AP/Wide World Photos

FIGURE 16.21 A large earthquake triggered the collapse of this oversteepened hillside. More than 1200 people were buried under the debris slide at Santa Tecla, El Salvador, in 2001.

EXPAND YOUR THINKING

As a geological engineer responsible for analyzing potential mass wasting hazards on a nearby hillside, why would you pay special attention to the speed and moisture content of potential slope processes? Describe the methods you would use to assess the hazard potential of a slope.

16-7 Human Activities Are Often the Cause of Mass Wasting

LO 16-7 List some ways humans cause mass wasting.

Understanding the details of the relationship between local geology and mass wasting processes can lead to better *land-use planning.*

Land-Use Planning

Many geologists are involved in *land-use planning.* Their job is to assess the presence of geologic hazards and resources, plan appropriate modifications to land so as to mitigate hazards and conserve resources, and recommend sustainable practices to ensure a safe and productive future for land use.

One goal of effective land-use planning is to minimize human vulnerability to geologic hazards, including slope failure. Examples of high quality planning include analyzing the geology of hillsides before building on them; electing not to build on or below unstable slopes; and avoiding actions that may cause slope failure, such as undercutting the base of a hillside or removing vegetation that holds regolith in place. Clearly, then, it is crucial when developing the land to be familiar with the various types of mass wasting processes, their underlying causes, factors that affect slope stability, and what humans can do to reduce their vulnerability and risk from slope failure.

Human Activities

Simply put, the human tendency to alter hillsides often leads to mass wasting. Even global climate change has an impact on slope stability (see the Earth Citizenship feature, "Global Warming and Mass Wasting"). **FIGURE 16.22** shows several examples of how development on steep hillsides may create an unsafe situation. In the photo you can see, for example, a large water tank built high on a hillside, to take advantage of the gravity force necessary to supply a community with water pressure. This is a common practice; most water delivery to homes and businesses relies on natural water pressure from an elevated storage tank. Either a tank is built on a nearby hillside or, where the land around a community has low relief, a high-standing tank is constructed on steel legs, often 30 meters high or more. Water is mechanically pumped into the tank and allowed to naturally flow downhill to be used by homes and businesses.

On a hillside, however, the weight of a water tank may destabilize the slope and thus cause oversteepening. In addition, any leakage from the tank or delivery lines may saturate soil and rock (regolith) on the slope, or lubricate underlying planes of weakness. The results may not be apparent for years, though the mass wasting that eventually occurs may be sudden and catastrophic; or it may take the form of slow creep that is barely perceptible yet nonetheless causes damage.

When houses or other buildings are constructed on slopes, it is common practice to cut into the slope to create a level *bench.* Dirt and gravel are then added to lay a foundation for features such as garages, pools, and even the main structure of the buildings. This practice can oversteepen the hillside because it cuts into the toe of the upslope regolith and rock. The addition of dirt fill oversteepens the slope and adds weight to unstable portions of the hillside.

Another problem that develops is the result of in-ground waste disposal, which involves moving fluid waste from a building into a temporary holding tank that is buried nearby. The tank is perforated so that waste fluid seeps into the ground. The soil acts as a natural filter that both cleans the wastewater and serves as a growth environment for microbes and various bacteria that consume the organic waste. These systems are commonly referred to as *septic tanks*, *cesspools*, and *leach fields.*

In-ground waste disposal frequently is used in communities on hill slopes because of the difficulty of extending waste delivery pipes up steep slopes to carry waste materials to a centralized wastewater processing plant. As wastewater from multiple homes on a hillside soaks into the ground, it can saturate regolith, increase the weight of the unconsolidated material, and further weaken and lubricate underlying zones of weak rock.

EARTH CITIZENSHIP

Global Warming and Mass Wasting

As global warming heats the air at Earth's surface, rates of evaporation and precipitation rise. Greater rates of evaporation put more water in motion through the water cycle. In many places, this has increased the intensity of rainfall. Because water is a significant factor in mass wasting, especially acting as a trigger in many cases, heavier rain storms may heighten the tendency for mass wasting in certain areas.

While it may sound counterintuitive, a warmer world produces both wetter and drier conditions. Even though total global precipitation increases, the regional and seasonal distribution of precipitation changes, and more precipitation falls in heavier rains (which can cause flooding and mass wasting), rather than light events. In the past century, averaged over the United States, total precipitation has increased by about 7%, while the heaviest 1% of rain events increased by nearly 20%. This has been especially noteworthy in the Northeast, where the annual number of days with very heavy precipitation has gone up by 58% in the past 50 years.

Flooding often occurs when heavy precipitation persists for weeks to months in large river basins. Such extended periods of heavy precipitation have also been growing in frequency over the past century, most notably in the past two to three decades in the United States. Areas where these conditions combine with other types of triggers, such as earthquakes, may be at greater risk to mass wasting due to global warming.

FIGURE 16.22 Human activities may undermine the stability of regolith and rock, leading to mass wasting.

Ⓠ **Identify the various ways human activities undermine slope stability.**

Driveways, streets, and other roadways carved into hillsides oversteepen the slope as well, and undermine the base of regolith deposits and rock strata. Overall, construction on hillsides heightens the vulnerability of buildings, both on the slope and below it, to hazardous slope failure events.

To test your comprehension of this important topic, complete the exercise in the Critical Thinking feature, "Mass Wasting."

Ⓠ **EXPAND YOUR THINKING**

Describe some mitigation steps that a community can take to minimize the threat of damage and injury caused by mass wasting.

CRITICAL THINKING

Mass Wasting

Working with a partner, and referring to **FIGURE 16.23**, complete the following exercise:

1. Describe the difference between a mud flow and a debris flow. What conditions would lead to the formation of one rather than the other?

2. What natural conditions influence mass wasting? How are these conditions related to the *shape* of a mass movement?

3. How do weathering, rock types, relief, climate, soil type, and various trigger mechanisms combine to influence mass wasting where you live?

4. Construct a table listing the mass wasting processes shown in Figure 16.23 and the conditions that govern them. Use a marker to number the processes in the figure.

5. List the hazards associated with mass wasting. How are human activities and structures vulnerable to mass wasting?

6. You are the mass wasting expert employed by a city in the Appalachian Mountains. Describe the research program you have set up to improve understanding of mass wasting in nearby hills and valleys.

7. As the expert in question 6, you have to convince the town council each year to fund your program. Write one compelling paragraph explaining why they should continue to spend money on your program.

FIGURE 16.23

Research Improves Knowledge of Mass Wasting and Contributes to the Development of Mitigation Practices

LO 16-8 Describe methods of mitigating mass wasting hazards.

A number of commonsense practices can reduce the risk of mass wasting, and the National Landslide Hazards Program, developed by the U.S. Geological Survey, is a good place to start to learn about them.

National Landslide Hazards Program

The first step in mitigating any geological hazard is to identify locations where the risk level is high. To that end, the U.S. Geological Survey (USGS) and partner agencies have created the *National Landslide Hazards Program.* The program's goals are to provide information, encourage research, and promote public education about the hazards related to mass wasting.

The USGS established an effective approach to understanding all types of mass wasting. It begins by identifying areas that are generally prone to unstable slopes:

- On existing landslides
- On or at the base of slopes
- In or at the base of minor drainage hollows
- At the base or top of an old fill slope
- At the base or top of a steep-cut slope
- On developed hillsides where leach-field septic systems are used

The USGS also maintains a mass wasting mapping program that identifies areas of landslide incidence and susceptibility (**FIGURE 16.24**).

On the map, two color scales are labeled: Landslide Incidence and Landslide Susceptibility/Incidence. Under the first, Landslide Incidence, past occurrences of mass wasting are shaded as follows:

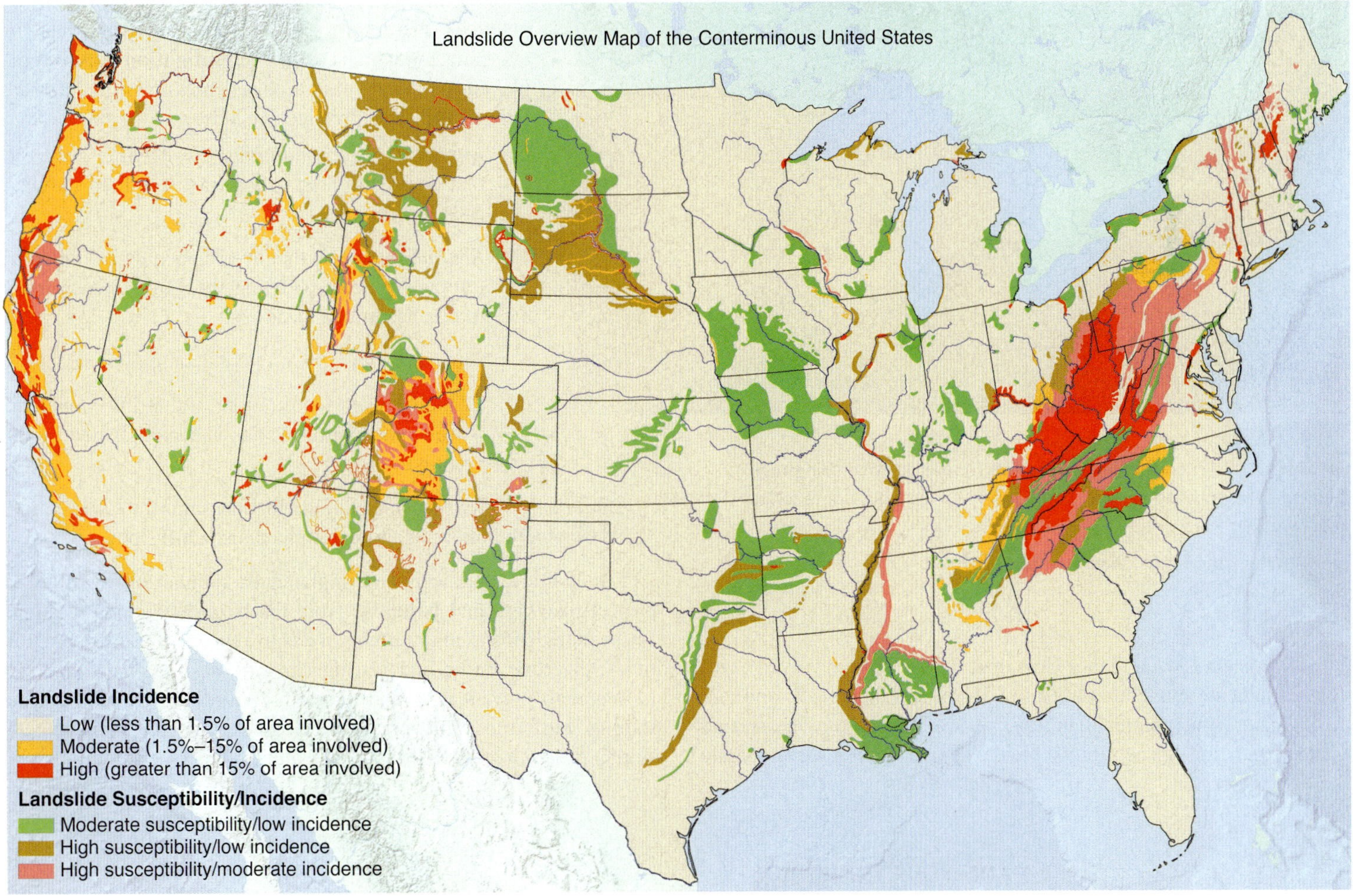

Landslide Overview Map of the Conterminous United States

Landslide Incidence
- Low (less than 1.5% of area involved)
- Moderate (1.5%–15% of area involved)
- High (greater than 15% of area involved)

Landslide Susceptibility/Incidence
- Moderate susceptibility/low incidence
- High susceptibility/low incidence
- High susceptibility/moderate incidence

FIGURE 16.24 National map of landslide susceptibility.

Which regions of the nation are most susceptible to mass wasting? In general, what are the characteristics of these areas that promote mass wasting?

Justin Kase/Alamy

(a)

© Marli Bryant Miller

(b)

FIGURE 16.25 Rock netting (a) protects roadways from falling rock. Shotcrete (b), sprayed cement, can be used on unstable slopes to limit mass wasting.

Q Why would vegetation on a slope help to stabilize it?

tan, to indicate low incidence (less than 1.5% of area involved); yellow, to indicate moderate (1.5% to 15% of area involved); and red, to indicate high (greater than 15% of area involved). The second scale, Landslide Susceptibility/Incidence, uses green for moderate susceptibility/low incidence; brown for high susceptibility/low incidence; and pink for high susceptibility/moderate incidence.

The term *susceptibility* refers to the risk of exposure to mass wasting hazards for a location. The term *incidence* refers to frequency rate of mass wasting at that location. Say your home is located at the base of a steep cliff, but a mass wasting event has never been recorded there: It would be ranked as a location of high susceptibility/low incidence, and so would be shaded in brown on the map of landslide susceptibility. With maps like this one, it is possible to decrease vulnerability to mass wasting events by avoiding construction in regions that are mapped as susceptible to landslide hazards.

Once a region has been identified as prone to mass wasting, local governments, communities, and homeowners can take several steps to lower vulnerability to the hazard. The first and most obvious step, as we've said throughout this chapter, is avoidance. Simply put, the single best way to protect against mass wasting hazards is to avoid them in the first place. We can accomplish this by recognizing, and acknowledging, the signs of potentially hazardous slopes and electing not to build or develop in these areas.

Unfortunately, because of decades of past building practices, there are many sites where it is too late for avoidance, in which case **direct mitigation** must be employed to reduce vulnerability to mass wasting hazards. Direct mitigation practices include:

- *Introducing and maintaining vegetation cover* so that dense and deep root growth can help stabilize a slope.

- *Regrading* land so that oversteepening is limited or removed by redistributing regolith and rock mass to less hazardous locations.

- *Preventing undercutting* at the base of a hill.

- *Reducing water* infiltration so that hill slopes do not become saturated; this is accomplished by diverting drainage.

- *Stabilizing hillsides* using engineering measures (**FIGURE 16.25**) designed to hold rock and regolith against a hill slope.

In addition, engineers can implement a number of measures to protect roadways and buildings from the hazards of mass wasting. These include installing *retaining walls* to keep fallen rock debris off roads; diverting traffic under avalanche and *rockfall sheds;* erecting *rock nets* to prevent rock and debris falls; drilling in *rock bolts* to stabilize cliff faces; and spraying rock faces with *shotcrete* (cement slurry) to stabilize weathering rock walls.

Q EXPAND YOUR THINKING

What is the best way to reduce vulnerability to mass wasting? Why is it the "best"?
What problems prevent this method from being used in *all* cases?

LET'S REVIEW

When the gravitational force acting on a slope exceeds the slope's capability to resist it, slope failure occurs. Mass wasting may take the form of rapid downslope movement of rock and/or regolith or slower types of movement that may continue for decades. It is a major geologic hazard that is responsible for over 25 fatalities and $1.5 billion in damage every year in the United States alone. Landslides and other forms of mass wasting cause damage to bridges, roadways, buildings, and public infrastructure. By recognizing the signs of a potentially hazardous hill slope and electing not to build or develop in the area, we can protect ourselves and our communities from the dangerous consequences of unstable slopes.

STUDY GUIDE

16-1 Mass wasting is the movement of rock and soil down a slope under the force of gravity.

- In a typical year, landslides cause over 25 to 50 deaths in and cost $1.5 billion in damage in the United States. In many other countries, the toll is higher due to poor land-use practices and building codes that ignore geologic hazards.

- **Regolith** consists of soil, debris, sediment, and broken rock. **Mass wasting** is the movement of regolith down a slope under the force of gravity. Mass wasting is an example of a geologic hazard that is best dealt with **avoidance**—learning to recognize the signs of an unstable slope and not building or engaging in any sort of unsafe activity there.

- The world's population has grown tremendously, and people now live in places that formerly were considered too dangerous or risky for human habitation. In addition to living in dangerous places, we also alter them in ways that make them even more dangerous.

16-2 Creep, solifluction, and slumping are common types of mass wasting.

- Most canyons and valleys have been shaped by a combination of glacial and stream erosion and mass wasting.

- Soil **creep** is the slow downslope migration of soil under the influence of gravity. Creep occurs over periods ranging from months to centuries.

- **Solifluction** is a type of soil creep that occurs in cold climates when regolith is saturated with water. Most solifluction is seen in permafrost zones. Solifluction produces distinctive lobes on hill slopes where the soil remains saturated with water for long periods but is capable of sliding on an underlying layer of ice.

- A **slump** occurs when regolith suddenly drops a short distance down a slope, usually as a cohesive block of earth that simultaneously slides and rotates along a failure surface shaped like the bowl of a spoon. As material shifts downward, one or more crescent-shape headwall scarps develop at the upslope end of the slide. The base of the slide is characterized by the toe of the slump.

16-3 Fast-moving mass wasting events tend to be the most dangerous.

- A **mud flow** is a river of rock, earth, and other debris saturated with water. Mud flows tend to have low viscosity because they are very watery. **Debris flows** are also saturated with water and exhibit flow characteristics, but they carry clasts ranging in size from clay particles to boulders, and often a large amount of woody detritus as well. Both types of flows develop when water rapidly accumulates in the ground during heavy rainfall or rapid snowmelt, changing stable regolith into a flowing river of sediment, or slurry.

- **Rock slides** and **debris slides** result when a slope fails along a plane of weakness. A debris slide is characterized by unconsolidated rock, debris, and regolith that have moved downslope along a relatively shallow failure plane. A rock slide typically lacks much debris, usually consisting of blocks of rock in a chaotic mass.

- **Rockfalls** occur when an accumulation of consolidated rock is dislodged and falls through the air, or free-falls, under the force of gravity. **Debris falls** are similar, except that they usually involve a mixture of soil, regolith, vegetation, and rocks.

16-4 Avalanches, lahars, and submarine landslides are special types of mass wasting processes.

- An **avalanche** is a fast-flowing, fluidized (by air, not water) mass of snow, ice, air, and occasionally some regolith, that cascades down a mountainside following the collapse of a snowfield. As snow builds up on a steep slope, a weak layer may fail due to the increasing weight, or from a vibration or some other cause. The overlying mass detaches itself from the slope, slides along the weak layer, and gains strength and size as it gathers additional material on its downhill journey.

- A **lahar** is a mud flow composed of pyroclastic material and water that moves down the slopes of a volcano. Lahars have the consistency of concrete: they are fluid when moving but solidify after they have stopped flowing.

- Submarine landslides are vast debris flows that take away whole pieces of an island in a single event. It is hypothesized that swelling of a volcano and deep-seated earthquakes associated with magma intrusion may cause a flank of a volcano to literally break off and slide into the sea. Such an event is possible because a volcanic island is essentially a pile of glassy talus.

16-5 Several factors contribute to unstable slopes.

- In all cases, mass wasting is governed by gravity. Gravity can be divided into two components for objects resting on sloping surfaces: one component parallel to the slope (g_s) and one is perpendicular to the slope (g_p). On steep slopes, the force parallel to the slope will be greatest and will tend to pull objects downhill. On gentler slopes, the component perpendicular to the slope will be greatest.

- Friction tends to hold regolith in place, but it can be disrupted in a number of ways. Heavy rainfall can saturate rock and sediment on a slope and increase its weight so that it exceeds the force holding it in place. The same rainfall may lubricate the surface on which a deposit rests, also disrupting the friction that holds it in place. Slight shaking from a nearby earthquake, or desiccation in hot sunlight, causing sediment to shrink and crack, may also disrupt friction on a slope.

- Water destroys the cohesion that exists when one grain of sediment is in contact with another. As the pore space between grains fills with water, water pressure increases. As a result, pore water literally lifts one grain off another, in a process called dilation.

- Water content and oversteepening affect slope stability. A trigger is needed to initiate a mass wasting event. The most common trigger of slope failure is vibration.

16-6 Mass wasting processes vary in speed and moisture content.

- We can categorize mass wasting processes based on their relative speed (fast to slow) and moisture content (wet to dry). The fastest processes are mud flows, debris flows, and debris falls and rockfalls. Mass wasting processes that move at moderate rates include debris slides, rock slides, and slumps.

The slowest forms include solifluction, which requires a high degree of water saturation, and soil creep, a largely dry process, although it can be caused by alternating wetting and drying as well as freezing and thawing.

16-7 Human activities are often the cause of mass wasting.

- One goal of land-use planning is to reduce human vulnerability to slope failure. Examples include analyzing the geology of hillsides before building on them, not building on or below unstable slopes, and avoiding actions that may cause slope failure, such as undercutting the base of a hillside or removing vegetation that holds soil in place.

16-8 Research improves knowledge of mass wasting and contributes to the development of mitigation practices.

- Areas generally prone to landslides are found on existing landslides, on or at the base of slopes, in or at the base of minor drainage hollows, at the base or top of an old fill slope, at the base or top of a steep cut slope, and on developed hillsides where leach field septic systems are used.

- The best way to protect against mass wasting hazards is to avoid them in the first place. This is accomplished by recognizing, and acknowledging, the signs of potentially hazardous hill slopes and electing not to build or develop in these areas.

KEY TERMS

avalanche (p. 490)
avoidance (p. 482)
creep (p. 486)
debris falls (p. 488)
debris flow (p. 488)
debris slides (p. 488)
direct mitigation (p. 500)

geological engineering (p. 483)
lahars (p. 490)
mass wasting (p. 482)
mud flows (p. 488)
oversteepening (p. 493)
regolith (p. 482)
rockfalls (p. 488)

rock slides (p. 488)
slope stability (p. 482)
slump (p. 486)
solifluction (p. 486)
talus (p. 493)
undercutting (p. 483)

ASSESSING YOUR KNOWLEDGE

Please complete the following exercise before coming to class. Identify the best answer to each question.

1. What is mass wasting?
   a. Chemical and physical weathering of dirt and rock
   b. Movement of regolith down a slope under the force of gravity
   c. Weathering of regolith by rainfall
   d. Oversaturation of loose rock and soil by water
   e. Flash flooding by a stream

2. Very slow movement of rock and soil down a hillside without the formation of a scarp is termed a(n):
   a. Slump.
   b. Debris slide.
   c. Creep.
   d. Earthflow.
   e. Avalanche.

3. What type of mass wasting event occurred in this photo?
   a. Slump.
   b. Debris fall.
   c. Rock fall.
   d. Earthflow.
   e. Avalanche.

Bill Hatcher/NG Image Collection

4. The primary requirement for solifluction is:
   a. A tropical environment.
   b. A steep hillside above a river channel.
   c. Glaciation.
   d. Permafrost.
   e. None of the above.

5. Flows, slides, and falls are characterized by (respectively):
   a. High water content, high shaking, and glaciation.
   b. High water content, lower water content, permafrost.
   c. Free-fall through the air and fluidization.
   d. Turbulent flow, sliding on a plane of weakness, and free-falling.
   e. Fluidization, turbulent flow, and sliding on a plane of weakness.

6. What is a dangerous type of mass wasting on active volcanoes?
   a. Creep
   b. Lahars
   c. Rock slide
   d. Solifluction
   e. Rockfalls

7. Which of the following is most prone to slope failure?
   a. Conglomerate
   b. Dipping fractured limestone on a clay bed
   c. Horizontally bedded quartz sandstone
   d. Basalt dike
   e. Marine sediment

8. Which of the following contributes to mass wasting?
   a. High water content
   b. Vibration
   c. Strata inclined downhill
   d. Oversteepened slope
   e. All of the above.

9. This is a photo of _____.
   a. Solifluction
   b. Debris flow
   c. Rock slide
   d. Creep
   e. None of the above.

© Marli Bryant Miller

10. What is the difference between a slump and a debris slide?
    a. The presence or absence of water determines whether it is a slump or debris slide.
    b. Slumps occur only on permafrost.
    c. A debris slide is chaotic, whereas a slump consists of rotating but coherent blocks.
    d. Slumps consist of solid rock; debris slides do not.
    e. All of the above.

11. A mass wasting process that is fast and has high water content is:
    a. Solifluction.
    b. Slumping.
    c. Creep.
    d. Mud flow.
    e. Debris fall.

12. Talus is a result of:
    a. Mud flow.
    b. Rockfall.
    c. Earth flow.
    d. Soil creep.
    e. None of these.

13. What triggers mass wasting?
    a. Desiccation
    b. Vibration
    c. Saturation
    d. Freeze-thaw
    e. All of the above

14. When developing an area, it is best to _____ any mass wasting hazard that may be present.
    a. Mitigate
    b. Avoid
    c. Fix
    d. Engineer
    e. Oversteepen

15. **True or False**: Mudflows are an example of a relatively slow type of mass wasting with low moisture content. _____

David McNew/Getty Images, Inc.

16. Human causes of mass wasting include:
    a. Cutting the base of a hillside.
    b. Overweighting the top of a slope.
    c. Saturating a slope.
    d. Removing vegetation.
    e. All of the above.

17. The potential for mass wasting can be reduced by:
    a. Limiting water infiltration.
    b. Vegetating hillsides.
    c. Regrading land.
    d. Installing rock netting.
    e. All of the above.

18. Submarine landslides are the result of:
    a. Coastal erosion of island shorelines.
    b. Explosive volcanism.
    c. Solifluction.
    d. Lahars.
    e. Glassy texture of basalt.

Koshtra Tolle/Getty Images, Inc.

19. Imagine a steep stream valley whose long axis is oriented north-south and whose sedimentary rock layers are dipping to the east. On which side of the valley is a rock slide more likely to occur?
    a. Southwest
    b. Northwest
    c. West
    d. East
    e. None of the above.

20. The talus at the base of this cliff is most likely a form of:
    a. Debris slide.
    b. Rock fall.
    c. Creep.
    d. Debris flow.
    e. None of the above.

Gregory G. Dimjian/Photo Researchers

FURTHER RESEARCH

1. What tectonic settings lead to frequent mass wasting?

2. Would mass wasting occur on other planets? Why or why not? What factors would govern the types of mass wasting that take place on other planets?

3. In your area, what factors trigger mass wasting, and what types of mass wasting events are most common?

4. What steps can people take to reduce vulnerability to mass wasting?

5. Is a snow avalanche a type of mass wasting? Why or why not?

ONLINE RESOURCES

Explore more about mass wasting on the following Web sites:

U.S. Geological Survey Landslide Hazards Program:
http://landslides.usgs.gov

National Landslide Map:
http://landslides.usgs.gov/learning/nationalmap

Additional animations, videos, and other online resources are available at this book's companion Web site:
www.wiley.com/college/fletcher

This companion Web site also has more information about WileyPLUS and other Wiley teaching and learning resources.

17

SURFACE WATER

CHAPTER CONTENTS & LEARNING OBJECTIVES

GEOLOGY IN OUR LIVES

Water is naturally cleansed and renewed as it moves through the hydrologic cycle—evapotranspiration, condensation, precipitation, infiltration, and runoff—among the atmosphere, the ocean, and the crust. Unfortunately, the rate of water use by humans may outpace the natural rate of water renewal. As a result, water is becoming scarcer in dozens of nations; worse is that polluted runoff is transporting pathogens and contaminants into natural ecosystems and human drinking water supplies around the world. Another widespread problem is stream flooding, which threatens communities built on floodplains, the natural extensions of healthy channels. Although periodic flooding is a normal process on floodplains, it is best to avoid building on them because attempts to stop flooding usually result in environmental damage, as well as aggravate flooding problems in the next community downstream.

Q How does water move through the natural world?

Adam Jones/Getty Images, Inc.

17-1 The Hydrologic Cycle Moves Water Between the Atmosphere, the Ocean, and the Crust

LO 17-1 List and describe the five processes in the hydrologic cycle.

Under the influence of gravity and the Sun's energy, water moves restlessly between the atmosphere, the ocean, and Earth's crust in what is termed the hydrologic cycle.

Hydrologic Cycle

Would you believe that a dinosaur once may have ingested the very same molecules that were in your last drink of water? This is possible because from the time water vapor condensed on the cooling Earth, in the late Hadean Eon, the water environment has been in motion. This global process, called the **hydrologic cycle** (**FIGURE 17.1**), consists of five major processes—condensation, precipitation, infiltration, runoff, and evapotranspiration—that keep water continuously moving through Earth's environments.

Water covers 71% of Earth's surface; as such it is the dominant agent governing environmental processes. Woven into water's geologic journey are the many ways we humans rely on clean water: We use it for manufacturing, to irrigate crops, as a source of energy, for recreation, and, of course, for drinking. Although nature continually refreshes water resources, the rate at which we use it generally outpaces the natural rate at which it is renewed. For this reason, it is important that we learn to manage water carefully as a valuable resource, not only for our own use today but for that of future generations. To do this, we must understand the hydrologic cycle and the nature of surface water, the topic of this chapter, and groundwater, the topic of the next.

The Five Hydrologic Cycle Processes

Each year the hydrologic cycle circulates nearly 577,000 cubic kilometers of water (about 152 quadrillion gallons). A crucial part of the cycle is the storage of water in *natural reservoirs* such as the ground, the ocean, and various forms of ice. After leaving the atmosphere as rain, snow, or some other type of condensation, water may run quickly to the sea in channels called **streams**; or it may be held in a lake for a hundred years, in a glacier for thousands of years, or in the ground for 10,000 years or more; or it may evaporate immediately. Regardless of how long the water may be detained, it is eventually released to reenter the hydrologic cycle.

Five natural processes keep water moving through the hydrologic cycle: condensation, precipitation, infiltration, runoff, and evapotranspiration. *Condensation* occurs when water changes from vapor (gas) to liquid. Water vapor moves mostly by convection; that is, warm, humid air will rise and cool air will descend. As warm water vapor rises into the atmosphere, it loses heat and cools. This causes the rate of condensation to exceed the rate of evaporation; thus, the vapor changes into liquid or ice.

You can see condensation in action as water drops form on the outside of a cold glass or can. This water obviously does not come from inside the glass or can; it comes from warm air touching the cool surface of the container: Water vapor in warm air condenses into liquid water on the outside of the cool container. The same process forms clouds. When warm water vapor in the air cools and condenses, it creates clouds of microscopic droplets held aloft by air currents. As condensation continues, the droplets grow in size until they are too large and heavy to stay in the air; the result is rain.

- *Precipitation* is water that falls from the atmosphere as rain or snow. Sometimes it evaporates before reaching the ground. More often, it reaches Earth's surface, adding to streams and

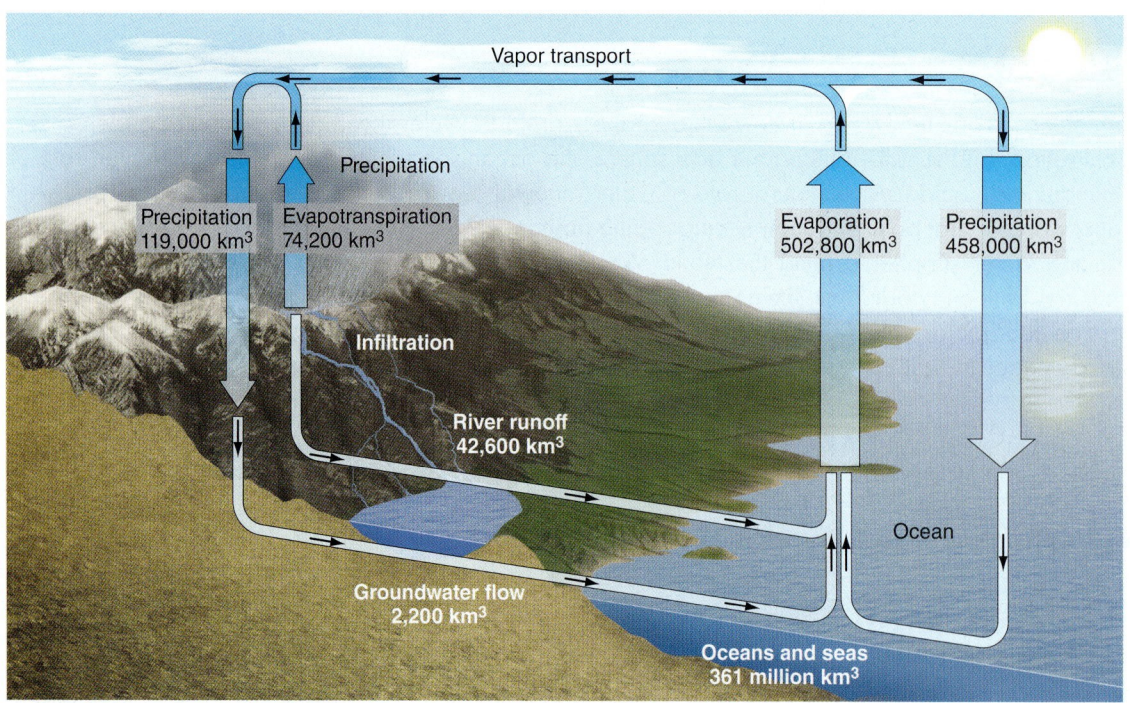

FIGURE 17.1 Nearly 577,000 cubic kilometers (km³) of water circulates through the hydrologic cycle every year. Researchers estimate that each year about 502,800 cubic kilometers of water evaporates over the oceans and seas, 90% of which (458,000 cubic kilometers) returns directly to the oceans through precipitation, while the remainder (44,800 cubic kilometers) falls over land. Over land, plant transpiration (65,200 cubic kilometers) and evaporation (9000 cubic kilometers) together total about 74,200 cubic kilometers. Around 35% of this, or 44,800 cubic kilometers, is eventually returned to the oceans as runoff from rivers, groundwater, and glaciers.

Labels in figure:
Vapor transport
Precipitation
Precipitation 119,000 km³
Evapotranspiration 74,200 km³
Infiltration
River runoff 42,600 km³
Groundwater flow 2,200 km³
Evaporation 502,800 km³
Precipitation 458,000 km³
Ocean
Oceans and seas 361 million km³

What are the major processes in the hydrologic cycle?

lakes, or soaks into the soil to become *groundwater* (water temporarily stored in the crust).

- *Infiltration* occurs when water seeps into the ground. The amount of water that infiltrates the soil varies with land slope, vegetation, soil type, and rock type, and is dependent on whether the soil is already saturated with water. The amount of open space within the soil and rock of the crust determines the amount of water that can be stored in the ground, as water is stored in the small spaces between the grains.

- *Runoff* is precipitation that reaches the surface but does not infiltrate the soil. Runoff also comes from melted snow and ice. During heavy precipitation, soils become saturated with water so that additional water cannot soak in and must flow on the surface. Due to gravity, surface water travels downhill. Hence, runoff drains downward into streams, lakes, and, eventually, the ocean. Normally, less than 20% of rainfall runs off the surface; the remaining 80% soaks into the ground or evaporates. During times of abnormally high rainfall, however, runoff can approach 100% of rainfall.

- Water vapor reenters the atmosphere by *evapotranspiration*. Water moves into the atmosphere by evaporating from the ground and transpiring from plants. Transpiration is part of plant metabolism. It occurs when plants take in water through their roots and release it through their leaves, a process that can clean water by removing contaminants and pollution. Evaporation occurs when energy from the Sun heats water, activating water molecules to such a degree that some of them rise into the atmosphere as water vapor.

Freshwater for Human Use

Freshwater resources are, literally, of vital importance to humans—we need it to live. Yet the supply is very limited: The total volume of water on Earth is about 1.4 billion cubic kilometers; of this amount, only 2.5% is freshwater (35 million cubic kilometers). According to the United Nations, the total usable freshwater supply for ecosystems and humans is only 200,000 cubic kilometers, or less than 1% of all freshwater resources and only 0.01% of all the water on Earth. Fully 68.9% of all freshwater is locked up in the form of ice and permanent snow cover in mountainous areas, the Antarctic, and Arctic regions. Another 30.8% is stored underground in the form of groundwater of various types, including *soil moisture*, shallow and deep aquifers, *wetlands*, and permafrost. Freshwater lakes and rivers contain only 0.3% of the world's freshwater.

You can learn more about water use and its abundance—or lack thereof—in the Earth Citizenship feature, "Water Facts."

Q EXPAND YOUR THINKING

What keeps water moving through the hydrologic cycle?

EARTH CITIZENSHIP

Water Facts

Water is a major factor in our daily lives. In wealthy countries, it is so familiar and so readily available, in water fountains and sinks everywhere, that most of us rarely give it a second thought. But perhaps we should, given that the total usable freshwater supply for ecosystems and humans constitutes only 0.01% of all the water on Earth.

To expand your knowledge of this most common, and most valuable, of Earth's natural resources, familiarize yourself with the following list of water-related facts:

- About 80% of all evaporation is from oceans; 20% comes from water and plants on the land.

- Water vapor is the third most abundant gas in the atmosphere (after nitrogen and oxygen).

- If the world's total supply of water were poured on the 50 United States, the land surface would be submerged to a depth of 144 kilometers.

- If all the water vapor in the atmosphere were to fall all at once, Earth would be covered with only about 2.54 centimeters of water.

- About 75% of the human body consists of water.

- Each day, people in the United States drink about 416 million liters of water.

- Showering/bathing is one of the largest (27%) domestic uses of water.

- The average American uses 530 to 640 liters of water per day (including industrial and agricultural needs).

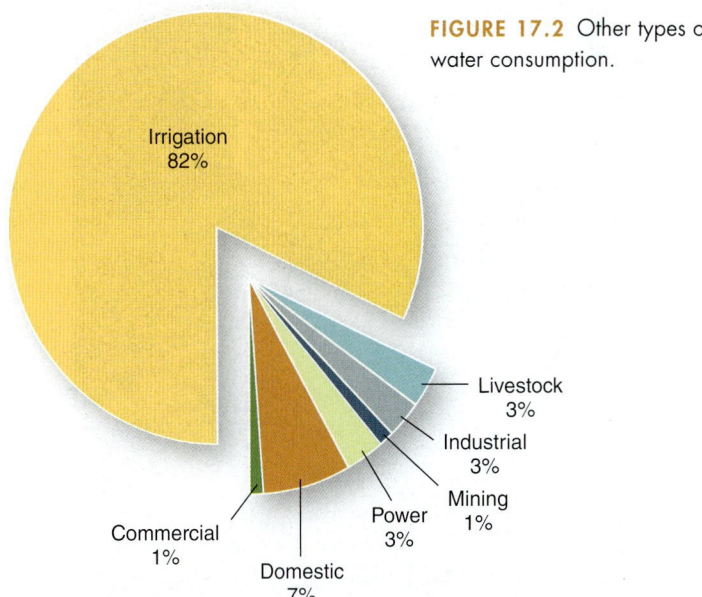

FIGURE 17.2 Other types of water consumption.

Irrigation 82%
Livestock 3%
Industrial 3%
Mining 1%
Power 3%
Domestic 7%
Commercial 1%

- A leaky faucet can waste 380 liters a day.

- An average family of four uses 3335 liters of water per week just to flush the toilet.

- The average shower uses about 75 liters of water; 150 liters are used in 10 minutes.

- You use about 19 liters of water if you leave it running while brushing your teeth.

17-2 Runoff Enters Channels That Join Other Channels to Form a Drainage System

LO 17-2 Explain the concept of drainage systems.

Water is a hard-working compound, eroding and transporting sediments and dissolved molecules and ions toward the ocean in the form of runoff.

Runoff

Every year, 36,000 cubic kilometers of water runs off the land worldwide (**FIGURE 17.3**). It has been calculated that the energy of all this water is equal to 9 billion kilowatts. Much of this energy is spent eroding the land. Think of it in these terms: It is the equivalent of one small tractor with a single-horsepower engine hauling dirt off each 12,000-square-meter parcel of land (3 acres) worldwide every day and night for an entire year. It is easy to conclude from this that water is capable of removing enormous quantities of soil and sediment from Earth's surface. Combined, weathering and runoff-induced erosion are the most important force at work sculpting the land surface.

As we learned in Chapters 7 and 8, when water erodes land, it picks up sedimentary particles and carries them away in the flow of streams. A stream is any flowing body of water that follows a **channel,** the physical confines of flowing water consisting of a bed and banks; a *river* is a major branch of a stream system.

Worldwide, streams carry about 16 billion tons of sedimentary particles and 2 to 4 billion tons of dissolved ions every year. Weathered rock and regolith supply this material, and when streams carry it away grain by grain, the land is worn down. Stream erosion, working over geologic time, has chiseled many mountain ranges down to the level of flat plains. As is evident to anyone gazing out of an airplane window, much of Earth's landscape is the product of stream erosion.

Drainage Systems

Steams do not flow in isolation; they are part of a large network of channels of various sizes known as a **drainage system.** Drainage systems are fed by surface runoff that quickly becomes organized into channels.

The total area feeding water to a stream is called the **watershed,** or *drainage basin* (**FIGURE 17.4**). Within a watershed, all runoff drains into the same stream. Every point on Earth's surface is part of a watershed, since all the rain falling on land must drain somewhere. Perhaps the best example is the watershed for the Mississippi River, which drains two-thirds of the continental United States as it flows 3781 kilometers to the Gulf of Mexico. But nested within the Mississippi watershed are many smaller watersheds that drain every channel from every slope of every hill and ridge in the Midwestern states.

Watersheds are shaped roughly like bowls separated by *drainage divides*; the divide outlines the rim of the bowl. Water flows down the inside of the bowl and collects at the low point—the stream system—and is then channeled out of the bowl. Drainage divides are the topographic highs, such as ridges, that force water to drain in separate directions into different watersheds. The Continental Divide, for instance, separates the drainage basins of streams that flow to the Atlantic and the Gulf of Mexico from those that flow to the Pacific.

Drainage systems can be described using a simple scheme, called *stream ordering*, that is widely used by *hydrologists*, scientists who study water in nature (**FIGURE 17.5**). The lowest-order stream is defined as one with no **tributaries;** this is a *first-order stream,* also called *headwaters*. A second-order stream is formed by the *confluence* (or juncture) of two first-order streams; and a third-order stream is formed by the confluence of two second-order streams. A drainage basin can be described using similar criteria.

A basin is ranked by the highest-order stream it contains. Hence, a *second-order basin* contains first- and second-order streams but no third-order stream. Scientists have discovered that the relationship between the number of stream segments in one order and the next, called the *bifurcation ratio*, is consistently around 3:1. This has been called the *law of stream numbers*.

Getty Images, Inc.

FIGURE 17.3 Channels carry runoff across the land to the sea. In the process, the water erodes and transports sediment and dissolved chemicals.

Describe the work of water as it runs across the land.

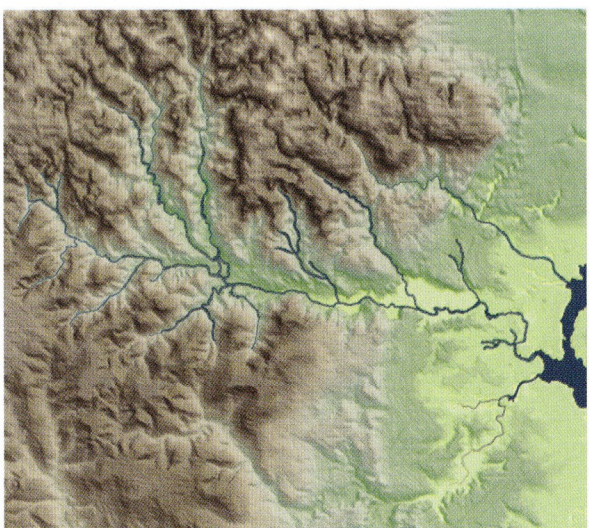

FIGURE 17.4 Smaller watersheds are nested within larger watersheds to create a drainage system.

Within a watershed the volume of runoff tends to increase from top to bottom. Why is this?

Drainage Patterns

Hydrologists have noticed that drainage patterns reflect the local topography and nature of the bedrock. The *dendritic*—having a branching structure like a tree—drainage pattern is the most common (**FIGURE 17.6**). It develops where the underlying rock has a uniform resistance to erosion and lacks features (such as joints or fractures) that could significantly influence drainage. Essentially, the dendritic pattern is the drainage pattern that water establishes for itself in the absence of other factors.

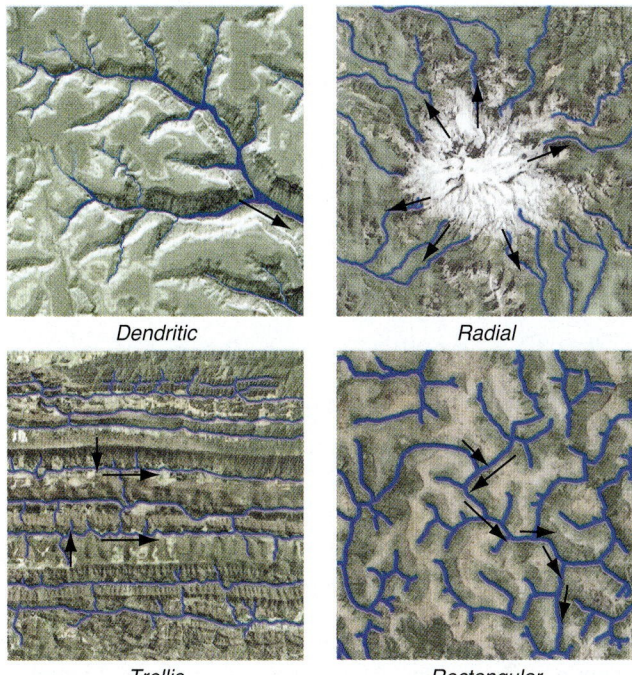

Dendritic   Radial

Trellis   Rectangular

FIGURE 17.6 The character of the underlying geology and topography controls the geometry of a drainage pattern. Geologists learn to recognize drainage patterns as clues to the nature of the underlying crust.

Are bedrock fractures important in drainage geometry? Why or why not?

A *trellis* pattern develops where bedrock exerts strong control over streams because of the structure of the underlying geology; channels align themselves parallel to fractures in the bedrock, with minor tributaries approaching at right angles. *Radial* drainage occurs where runoff flows away from a dome or hilltop in all directions.

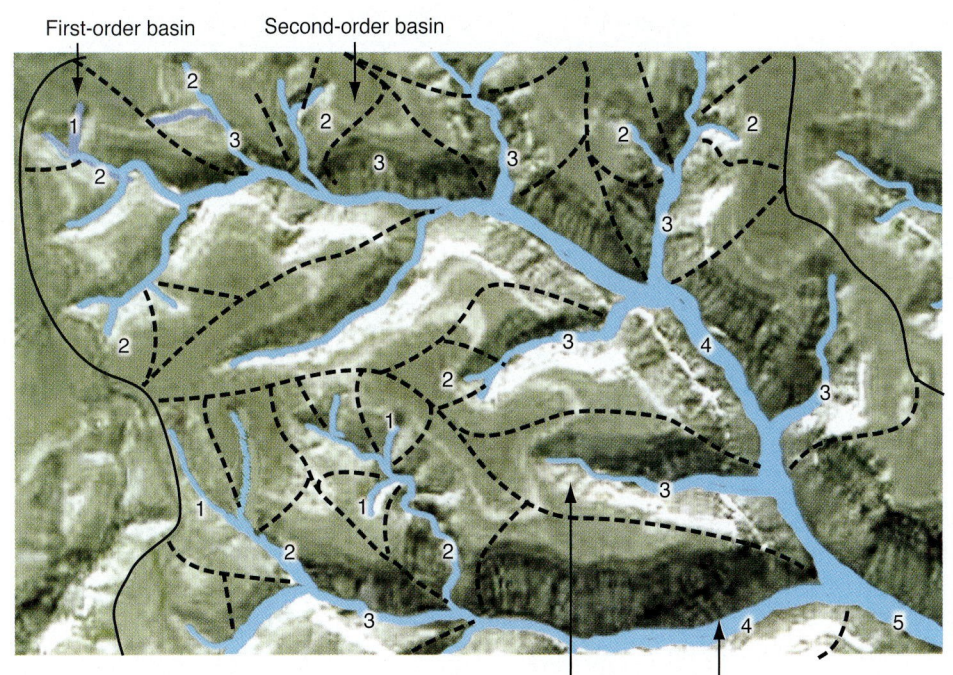

First-order basin   Second-order basin

Third-order basin   Fourth-order basin

State a hypothesis about how the topography of the land controls the drainage pattern in a watershed.

FIGURE 17.5 The system of stream ordering identifies the rank of a channel within a drainage system.

Areas where the crust has parallel faults or repeated sets of joints cause streams to take on a gridlike or *rectangular* pattern. There are several other drainage patterns as well, all dependent on the nature of the underlying geology.

Q EXPAND YOUR THINKING

How would you use the law of stream numbers to help manage streams?

17-3 Discharge Is the Amount of Water Passing a Given Point in a Measured Period of Time

LO 17-3 Identify how discharge, flow, and channel characteristics vary between headwaters and base level.

A stream continually adjusts its shape and path as the amount of water passing through the channel changes.

Base Level

A stream channel is a conduit that allows water and sediment to move from a *source area* to a base level. The **base level** is the theoretical lowest level to which the land will erode globally; it is sea level, the average level of the ocean surface.

So-called local base level is the lowest point a particular channel will reach—usually, the point where it joins with another channel. The local base level also may be a lake, marsh, the sea, or a reservoir. The ultimate base level is the ocean. If all streams were allowed to flow continuously through time, with no new uplift of the land, eventually they would carve their channels down to sea level.

Discharge

The volume of water (per unit of time) passing any point on a stream is called the **discharge**. Discharge is measured in units of volume divided by time (cubic meters per second) and is defined as the amount of water passing a given point over a measured period of time. The discharge, Q, is equal to the cross-sectional area, A (square meters), of a channel (width [meters] × average depth [meters]) times the average velocity, V (meters per second), of the flow.

$$Q = A \times V$$

By measuring stream discharge, it is possible to improve our understanding of flow characteristics that are useful in managing a stream as a resource. During certain seasons of the year, streams tend to flood. In other seasons, they can become dry. By collecting data on stream discharge over a period of years, authorities responsible for managing stream resources learn how much water a healthy stream typically carries at various times of the year. This information improves their ability to make sustainable decisions about activities that

may withdraw water from the system (such as irrigation) or contribute water to it (such as runoff from paved areas).

Flow

Water flowing within a channel may be either *laminar*, in which all water molecules travel along parallel and uniform flow paths, or *turbulent*, in which individual water molecules follow irregular paths (**FIGURE 17.7**).

Turbulent flow can keep sediment suspended in the water column longer than laminar flow can, and it therefore escalates erosion of the stream bottom and channel walls. The average velocity of laminar flow is generally greater than that of turbulent flow because turbulent flow is characterized by some water movement that is directed upstream in the form of eddies, against the mean flow going downstream. *Turbulent eddies* also direct flow vertically from the bed toward the surface (**FIGURE 17.8**).

Because eddies contain flow components that are not directed downstream, they tend to lower the efficiency of the flow and produce locally chaotic water movement. However, the same turbulent action enhances the water's sediment-carrying capacity. Turbulent eddies generally scour the bottom and suspend sediment, and the chaotic water movement keeps particles suspended in upward-directed currents.

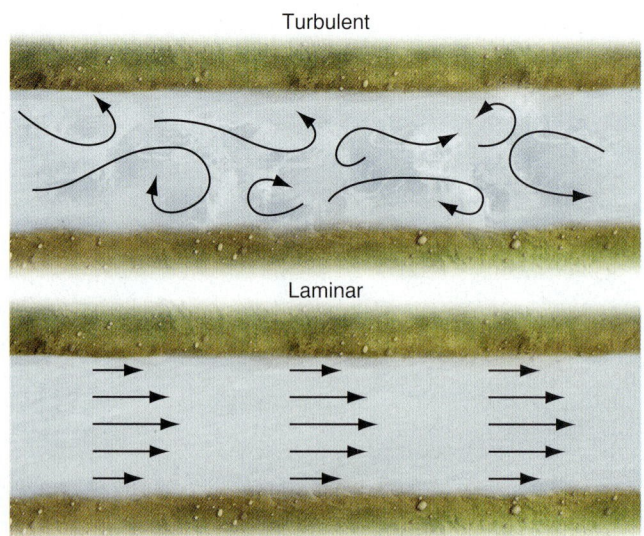

Turbulent

Laminar

FIGURE 17.7 Turbulent flow is characterized by water motion that goes against the mean flow direction or is directed vertically from the channel bed toward the water's surface. Laminar flow occurs when water molecules travel along uniform flow paths. This is a map view of two types of flow in channels.

Q Why does turbulent flow cause more erosion?

FIGURE 17.8 This turbulent flow is created by the high velocity of the water flowing over a ledge of bedrock in the stream channel.

FIGURE 17.9 The amount of water flowing in the Mississippi watershed increases as the water moves downstream because of the addition of tributaries.

550 m³/s
1400 m³/s
2800 m³/s
4250 m³/s
8550 m³/s

NASA

How do stream gradient and discharge change with distance downstream?

Q Which pathways do water particles follow in turbulent flow?

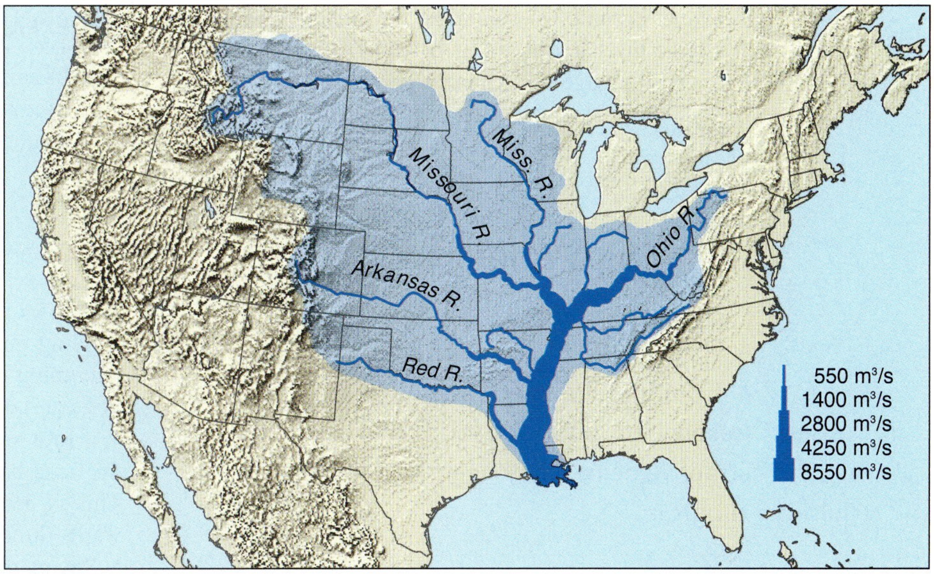

Channels Characteristics

In a channel, both width and depth expand as the flow moves downstream. This occurs because the amount of water flowing in a channel increases downstream as additional tributaries join the system (**FIGURE 17.9**).

Stream channels have different characteristics, depending on three primary variables: *gradient*, *discharge*, and *sediment load*. Channel gradient, or slope, is the angle between the channel bed or floor and a horizontal plane. The slope determines the flow velocity of the water in the channel. A steeper gradient produces an increase in flow velocity, which in turn increases the amount of sediment moved by the stream.

The cross-sectional shape of a channel changes as the stream becomes deeper and wider due to the influx of tributaries. The deepest parts of a channel develop where velocity is highest because of the capability of faster water to erode sediment and deepen the channel. The cross-sectional shape of a channel varies with its location in the stream and with changes in the discharge (**FIGURE 17.10**). As discharge increases, the width and

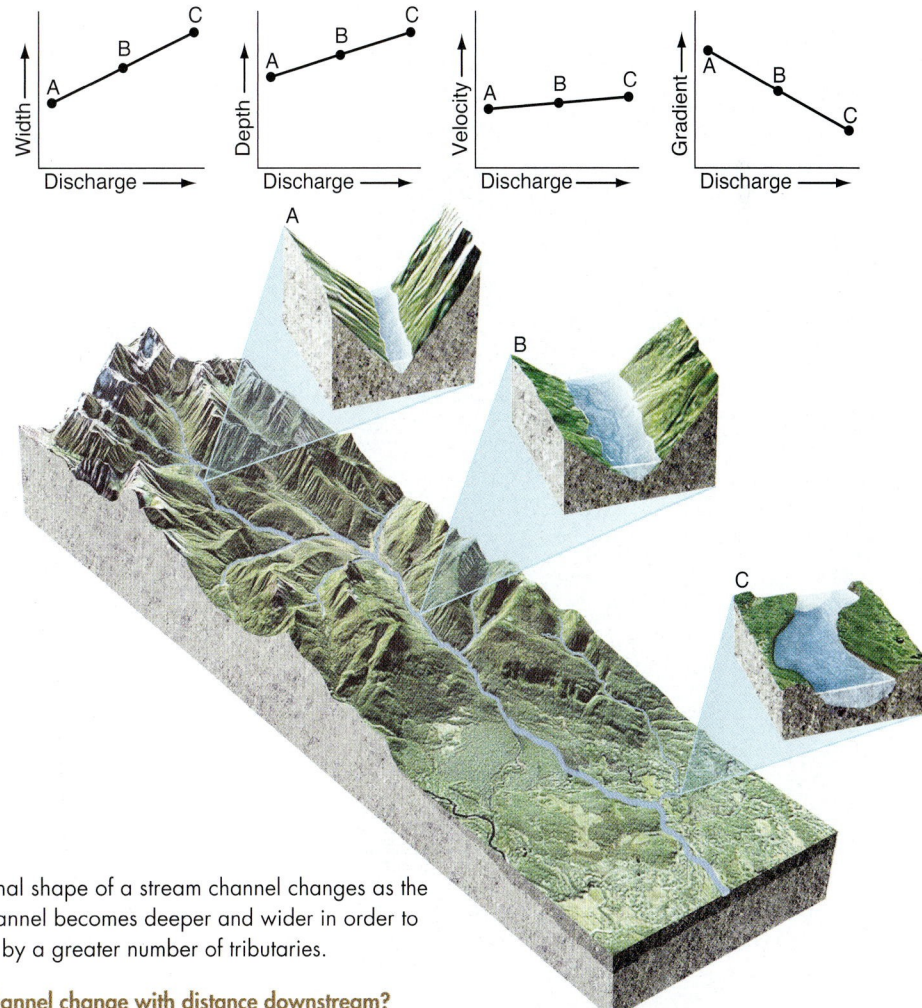

FIGURE 17.10 The cross-sectional shape of a stream channel changes as the water flows downstream. The channel becomes deeper and wider in order to carry additional water delivered by a greater number of tributaries.

Q How does the shape of a channel change with distance downstream?

depth of the channel expand faster than the flow velocity. A typical channel with a large discharge flows across relatively low-gradient slopes. That means discharge and channel gradient tend to change inversely: When one increases, the other decreases.

Q EXPAND YOUR THINKING

Describe the field methods you would use to obtain the data necessary for calculating the discharge in a stream in your neighborhood.

17-4 Running Water Erodes Sediment

LO 17-4 Compare the various types of sediment transport (suspended load, bed load, and dissolved load) with the current velocities they need as described in the Hjulstrom diagram.

Erosion is of great importance to humans, as it plays a major role in the management of natural resource problems such as soil loss, stabilization of hillsides, and unwanted sediment buildup in areas such as lakes, coastlines, and coral reefs.

Sediment Load

Within a channel, the point of maximum velocity shifts position depending on the characteristics of the channel. In straight segments, maximum flow is near the surface in the center, away from the friction caused by the bed and walls.

In a sinuous segment, the maximum velocity migrates toward the outer bank and lies somewhat below the surface (**FIGURE 17.11**). This is because *momentum* (the tendency to continue moving forward) directs the water toward the outside of a curve, causing it to collide with the far bank.

A stream carries a sediment load that may be transported as either bed load or suspended load. A stream also transports a large volume of dissolved load (**FIGURE 17.12**). *Suspended load* consists of particles that are carried along hanging in the water. The size of these particles depends on their density and the velocity of the stream. Turbulent eddies in higher-velocity currents keep particles suspended and support larger and denser particles. The suspended load is what gives most streams their muddy appearance and brown or red color.

Bed load consists of large particles that, most of the time, remain on the stream bed even as they are moved by the water. They move by sliding or rolling, called *traction*, or jumping, known as *saltation*, as a result of collisions between particles. As stream velocity increases, sediment that was carried as bed load tends to become suspended, and sediment that was not transported at lower velocities begins to move as bed load.

In general, as stream velocity increases, more sediment is transported, and grains move through the stages of traction, then saltation, and then suspension. Bed load moves at only a small fraction of the average flow velocity of the stream. Suspended load, in contrast, moves at a large fraction of the stream velocity; therefore, bed load (5% to 10%) is generally less important than suspended load (90% to 95%) in terms of total sediment load of a stream.

Dissolved load consists of ions that have entered the water as a result of chemical weathering of rocks. This load is invisible

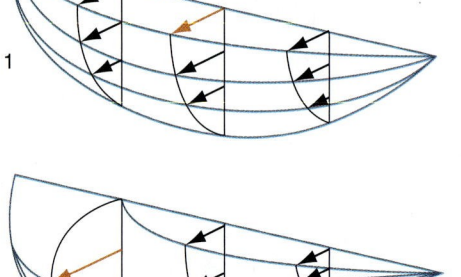
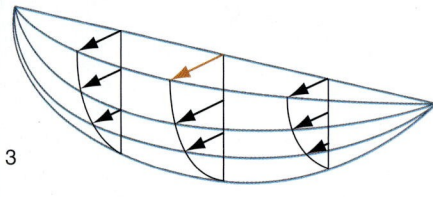
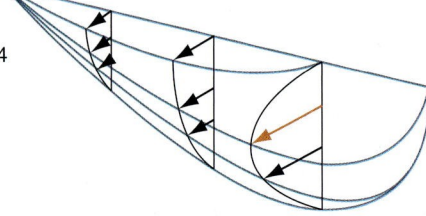

Thalweg
Water velocity

FIGURE 17.11 In curved segments of channel, maximum flow velocity (red arrow) migrates toward the outer bank and lies below the surface.

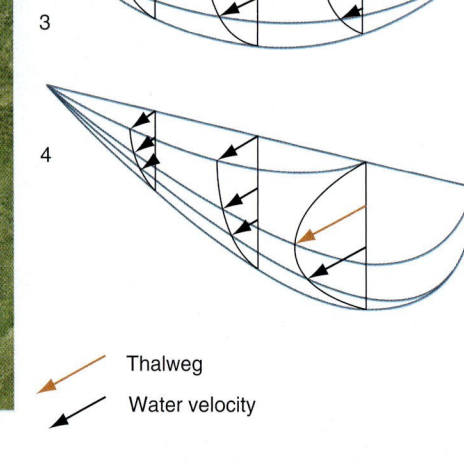

Q How does the outer bank of a channel in a curved segment change with time?

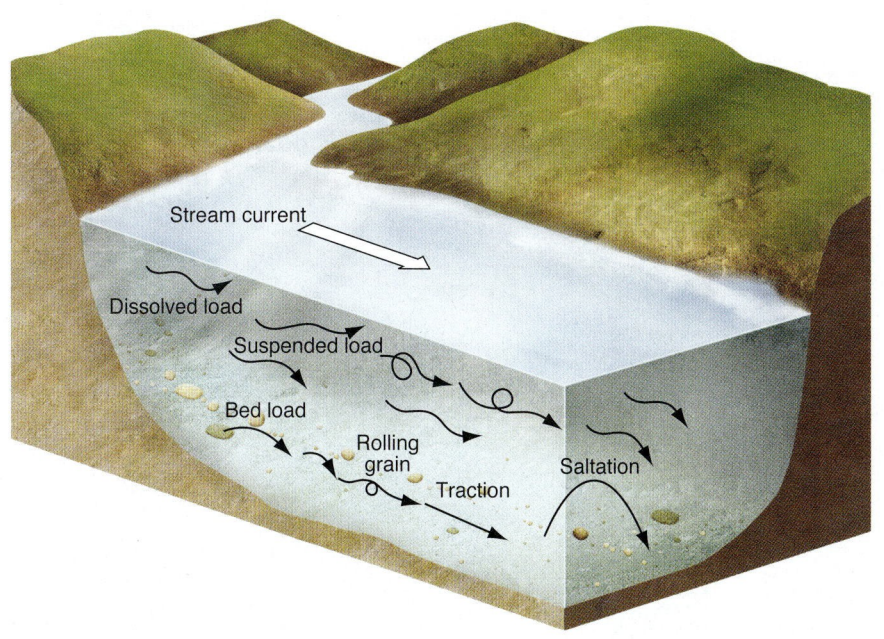

FIGURE 17.12 Sediment load consists of suspended load and bed load. Bed load includes saltation, particles jumping into the water column, and traction, grains moving by sliding and rolling along the bed. Dissolved load consists of dissolved compounds.

Does bed load or suspended load carry the largest volume of sediment in a stream?

because the ions are dissolved in the water. It consists mainly of HCO^{-3} (bicarbonate ions), calcium (Ca^{+2}), sulfate (SO_4^{-2}), chloride (Cl^-), sodium (Na^+), magnesium (Mg^{+2}), and potassium (K^+). Eventually, these ions are carried to the oceans, and give the water its salty taste. Streams fed by groundwater that has flowed through the crust generally carry a higher dissolved load than do those whose only source is runoff.

Sediment Erosion

In answer to the need to control landscape erosion, geologists have developed a simple model of the nature of erosion. This model is known as the *Hjulstrom diagram* (**FIGURE 17.13**), named for Swedish geographer *Filip Hjulström* (1902–1982).

The Hjulstrom diagram plots the relationship between water velocity and sediment size, and shows two curves representing: 1) the approximate stream velocity needed to erode sediments of varying sizes from the streambed, and 2) the approximate velocity required to continue to transport sediments of varying sizes once they are moving in the channel. As you can see, Figure 17.13 is divided into three regions. The tan region (erosion) shows the water velocity needed to erode sediment of various sizes from the bed. The blue region (transportation) shows water velocities needed to keep sediments moving once they have been eroded. Notice that it takes less velocity to keep sediments moving than it does to erode them in the first place. The green region (deposition) depicts the velocities at which sediment typically comes to rest on the bed—that is, when they are deposited.

The Hjulstrom diagram reveals some interesting facts about erosion and deposition in moving water. For one, note that once mud and silt have been eroded, they will stay in motion even if the velocity of the water decreases drastically. It is not as easy to transport coarser sediments (sand and gravel), as they require nearly as much water velocity as is required to erode the grains in the first place. Also note that it takes as much water energy to erode large pieces of gravel as it does to erode clay. Clay particles have a platy "habit," or shape, and thus tend to lay flat on the stream bed and have laminar flow directly above them. That makes clay particles relatively hard to erode.

One of the surprising facts revealed by the Hjulstrom diagram is that sand is the easiest sediment to erode, even though sand grains are heavier than silt or clay. The explanation for this is, as it turns out, simple: Sand grains, being larger than grains of silt or clay, stick up from the bed into the moving water. As a result, they travel along the bed before silt and clay do. Although sand grains are usually rounded, both clay and silt consist of flat particles that do not stick up into the moving water. It is also thought that turbulent eddies containing strong currents that dig into the bed are needed to erode clay because clay particles have such a low profile.

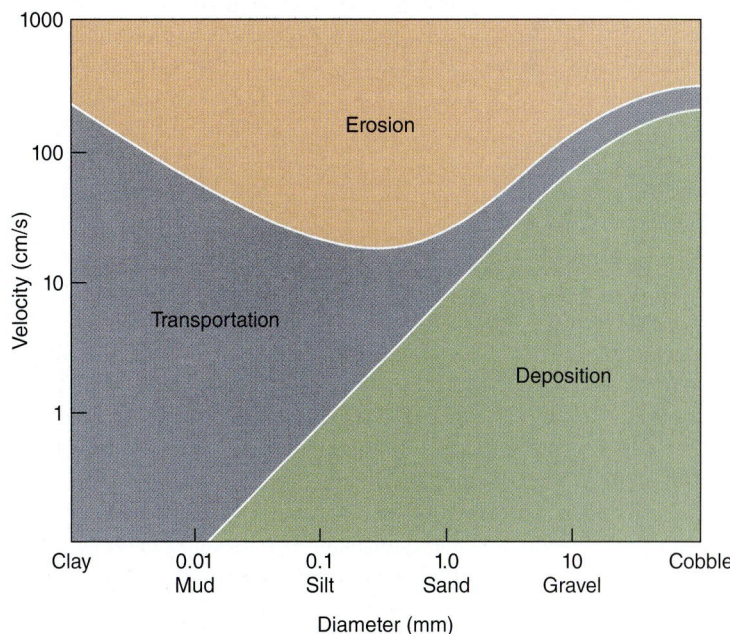

FIGURE 17.13 Hjulstrom diagram.

Which is harder to erode: gravel or clay?

EXPAND YOUR THINKING

Describe the research program that must have been conducted to arrive at the Hjulstrom diagram.

17-5 There Are Three Types of Stream Channels: Straight, Meandering, and Braided

LO 17-5 Describe how meandering channels change with time, and explain what causes the development of a braided channel.

The three general types of stream channels are: straight, meandering, and braided.

Channel Descriptions

Straight channels are rare, found usually only as short segments of otherwise meandering channels, or where the underlying topography and structure of the bedrock (such as a fault or fracture pattern) force the channel to be straight.

The reason that straight channels are rare is that water naturally flows in a sinuous fashion due to minute differences in flow characteristics. You can observe this phenomenon when a raindrop runs down the windshield of your car. It follows a snaky, curving path. Even in straight channel segments, water meanders or migrates from side to side down the length of the channel (**FIGURE 17.14**).

Flow velocity is highest in the zone overlying the deepest part of the stream. For that reason, sediment there is readily eroded, leaving behind a *pool* or *scour depression* in the channel. If one were to draw a line running along the channel and connecting the deepest parts of the stream, it would mark the natural direction (the profile) of the watercourse. Such a line is called the **thalweg**, German for "valley line". The thalweg is almost always the line of fastest flow in a river.

Where stream velocity is low, sediment will be deposited to form a *bar*. Alternating erosion and deposition along the channel creates a specific type of channel shape, with regularly spaced pools and bars. The path of highest-velocity flow is a line that connects the pools and goes around the bars.

Meandering channels are the most common characteristic of a stream that is free to roam across a valley floor (**FIGURE 17.15**). Meandering channels develop most readily on low-gradient slopes composed of easily eroded sediment. Like all objects with mass, water in a channel possesses momentum. This means that as water enters a bend in a channel, momentum will keep it moving in a straight line so that it collides with the far bank.

This intersection of water and erodible bank leads to a constant, persistent behavior for the outer bank of a meander bend to erode, and thus *migrate outward* even more. From this behavior derives the term *cutbank* for the outer bank of a meander

Simultaneously, sediment deposition takes place along the inside bank of a meander bend, where the flow velocity is lowest. The resulting sediment deposit produces an exposed bar, called a *point bar*. Point bars tend to migrate inward at nearly the same pace that the cutbank migrates outward; that is why the width of the channel stays relatively stable even as it migrates across the valley floor.

If you consider the consequences of a perpetually outwardly migrating channel, you will realize that it cannot go on forever.

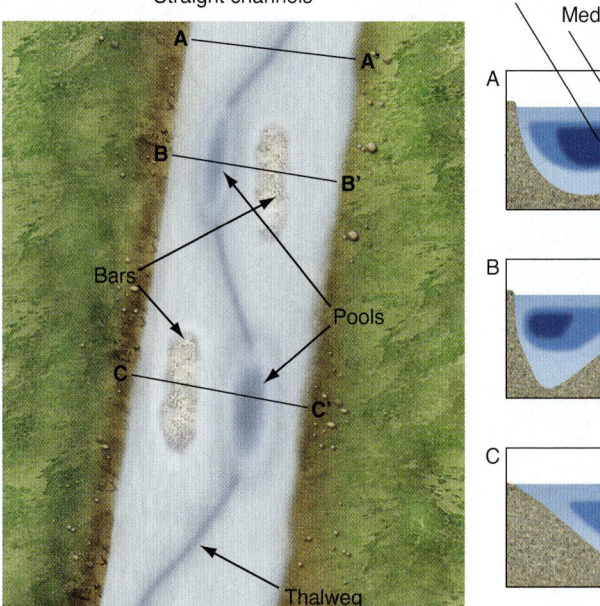

FIGURE 17.14 Water tends to flow in a sinuous fashion even in straight channels. This means that the highest-velocity flow migrates from side to side. Consequently, the channel is characterized by alternating pools (erosion) and bars (deposition).

Why are straight channels rare in nature?

Eventually, the stream will become so sinuous and tortuous that the movement of water through the valley will become inefficient. At its extreme, the channel will loop from one side of a valley to the other, with greatly reduced efficiency in the movement of water through the drainage system. Nature, ever an effective solver of problems, resolves this potential crisis by cutting off a highly sinuous meander at the neck.

FIGURE 17.16 shows how erosion on the outside of a highly sinuous meander bend eventually will cut across the neck between two meanders and shape a new, straighter, and more efficient channel. When this occurs, the cutoff meander bend forms an **oxbow lake**. In time, the oxbow will become a *meander scar* filled with a wetland as it accumulates organic plant material and sediment delivered by runoff and wind.

The third type of channel is a **braided channel**. Braided channels are shaped when a stream contains more sediment than it can readily transport. The excess sediment accumulates in the channel

Meandering channels

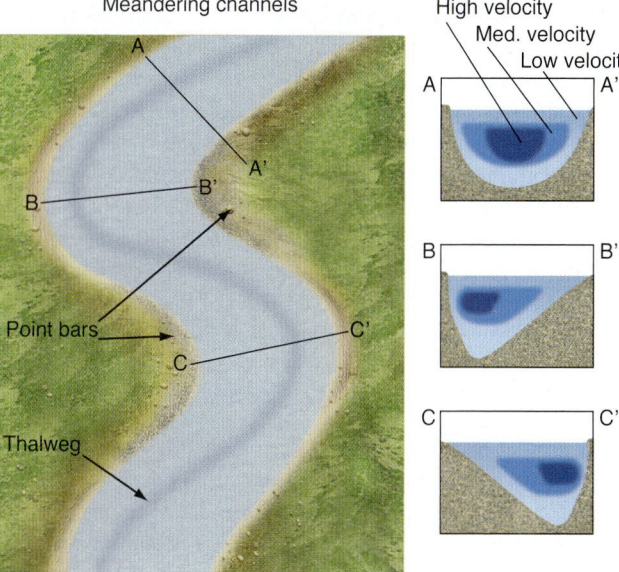

High velocity
Med. velocity
Low velocity

FIGURE 17.15 When a stream channel is able to roam freely, it tends to develop a meandering channel. Meandering channels readily develop on low-gradient slopes composed of easily eroded sediment.

What is the effect of momentum on water flow?

and constructs bars and islands that split the water into multiple channels. This gives the channel a "braided" appearance (**FIGURE 17.17**).

Braided streams are characterized by highly variable discharge, easily eroded banks, and excessive sediment. The sediment load is carried primarily during periods of high discharge. When the rate of discharge returns to normal, sediment is deposited in the form of bars and islands. The water is forced to flow in a braided pattern around the emergent deposits, dividing and reuniting as it moves downstream.

During periods of high discharge, the entire channel may contain water, and the islands become covered. As they erode and are reshaped under the high waters, sediment is redistributed throughout the channel. Usually, when the water level lowers and deposits re-emerge, the location and shape of islands and bars are entirely different from what they were previously.

EXPAND YOUR THINKING

State a hypothesis describing (summarizing what you know about) water flow in a channel. Describe how you would test your hypothesis.

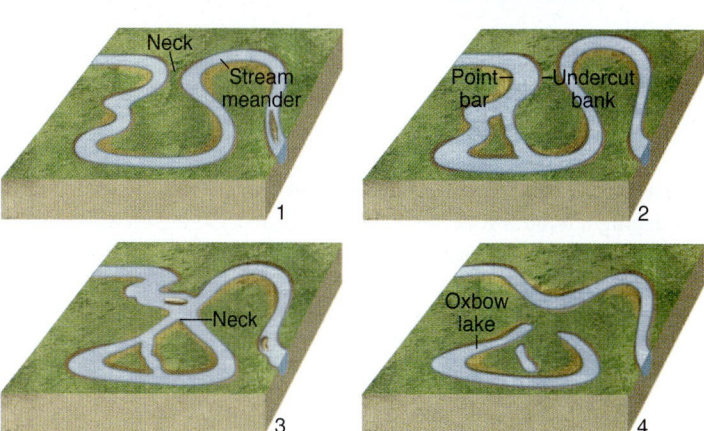

FIGURE 17.16 Formation of an oxbow lake by meander cutoff.

How does a meander cutoff change with time?

Braided channel

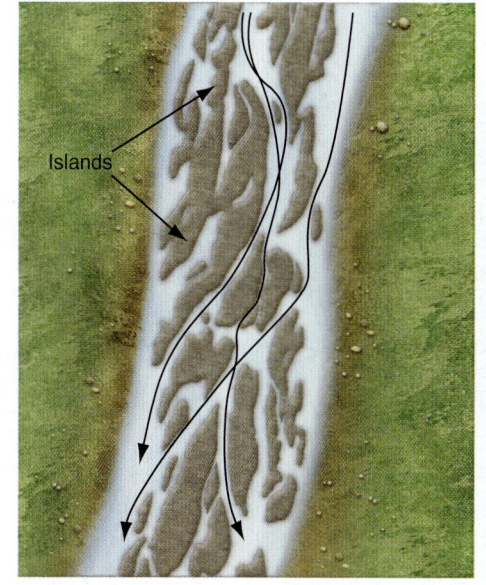

Islands

iStockphoto

FIGURE 17.17 A braided channel is characterized by bars, islands, and multiple channels. Braided streams develop because excess sediment that cannot be transported by the stream flow builds bars and islands that split the water into many interweaving channels.

Describe the effect of high discharge on a braided channel.

17-6 Flooding Is a Natural Process in Healthy Streams

LO 17-6 Identify the causes of flooding and the hazards associated with building on a floodplain.

Flooding is a natural process for healthy streams; but when humans build communities along riverbanks, they make flooding worse, and thereby turn a natural process into a geologic hazard.

Flooding

The rate of discharge in a channel can change rapidly, potentially producing a **flood**. Flooding occurs when discharge increases at such a rate that there is too much water for a channel to carry within its *banks*. Excess water then flows outside of the channel onto the adjoining land, known as the **floodplain** (**FIGURE 17.18**), which is described more fully in the next subsection.

Flooding happens for many reasons, but in most cases, it is the result of high discharge in several tributary streams feeding a single river channel that is incapable of handling the excess water without overflowing its banks.

There are three common causes of flooding: 1) an intense but short rainstorm in the headwaters of a drainage basin, leading to downstream flooding because multiple tributaries carrying high amounts of discharge feed into a single channel that floods; 2) prolonged rainfall in a drainage basin that saturates the ground and forces all additional rainfall to flow in channels, producing high discharge in many tributaries that cause flooding in a river they feed; and 3) a winter with heavy snowfall ending in a series of very warm weeks, during which accumulated snow melts rapidly and many tributaries carry the high discharge into a single channel, which then floods.

Each of these three situations results in a "wave" of high discharge that moves down a drainage system into a single, high-order river channel. The wave of high discharge builds over several hours to a day or more and may peak within the space of only an hour or two (**FIGURE 17.19**).

A flooding channel is observed first as slowly rising water of greater velocity that contains a great deal of suspended sediment. Within a few hours, or in some cases a day or more, water rises to the limit of the stream banks (a condition called *bankful discharge*) and then overruns the banks as a flood. In the following hours or days, water returns to its normal discharge condition as the flood wave moves downstream. The flood may grow downstream if more tributaries continue to add water. On particularly large rivers, the migrating flood may last over a few weeks.

As a stream overtops its banks during a flood, the velocity of the flow is high at first but then suddenly drops as the water flows out over the gentle gradient of the floodplain. Consequent to the sudden decrease in velocity, coarse-grained suspended sediment such as sand is deposited along the riverbank, eventually building up a *natural levee* (**FIGURE 17.20**). Natural levees provide some protection from flooding because they raise the relative height of the channel banks.

Floodplains

Floodplains are flat areas adjacent to stream channels that are composed primarily of sediments deposited during floods. A number of features are found on a floodplain. Some are depositional, such as natural levees, former point bars that have been abandoned by meander cutoff, and wetlands located in meander scars. Other features, such as oxbow lakes, have been created by past meander processes.

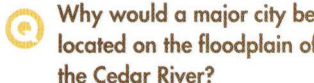
David Greedy/Getty Images, Inc.

FIGURE 17.18 Stream flooding occurs when discharge increases so much that water overflows the channel. In 2008, the Cedar River flooded downtown Cedar Rapids, Iowa.

Why would a major city be located on the floodplain of the Cedar River?

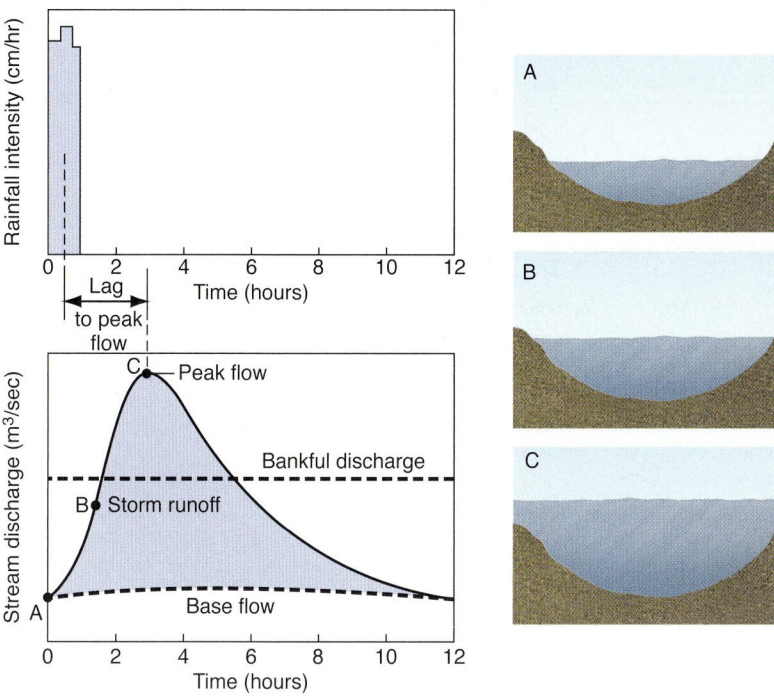

FIGURE 17.19 Intense rainfall in the headwaters of a drainage basin increases the discharge in tributaries feeding a high-order channel. There may be a lag of several hours to a day or more between when the rain falls and when flooding occurs further downstream. Normal discharge (a) will increase (b) until bankful discharge is exceeded. The peak of the flood (c) occurs several hours after the initial increase in discharge and then slowly tails off.

Why is flooding described as a "wave"?

Floodplains develop when streams overtop their levees and spread sediment-laden floodwater over the land surface. When the floodwaters retreat, stream velocities are reduced and a layer of **alluvium**, sediment that originates from a stream, is deposited across the floodplain.

Over time, repeated flood cycles result in the deposition of successive layers of alluvium, so that the floodplain's elevation rises. Floodplains, then, are composed of a combination of point bar sands and gravels and alluvium deposits of silt and clay. A floodplain is as much a part of a stream system as bark is part of a tree, because the floodplain is where a stream stores excess water and sediment during a flood. More, it is the environment where a stream is free to roam and form meanders, oxbows, point bars, and other natural channel features.

Efforts to curtail the hazards of stream flooding by confining a channel with *artificial levees,* built by humans to keep flood water from causing damage, instead actually damage fertile floodplains by preventing the continued deposition of nutrient-rich sediment and the annual cycle of natural wetting. Floodwater that is prevented from flowing onto a floodplain is passed downstream to the next community, where it causes more severe flooding. Floodplains that are deprived of sediment eventually lower in elevation and, ironically, experience more frequent flooding, thereby heightening the hazard risk to local communities.

As explained in Chapter 16 in regard to mass wasting, the best way to solve flooding problems is to *avoid* them altogether, by electing not to build communities or other developments in flood-prone regions such as floodplains. In the long run, taking the avoidance approach is cheaper and less damaging to the environment than attempting to control the natural flooding process.

To challenge what you've learned about flooding, complete the Critical Thinking feature titled "The 100-Year Floodplain."

EXPAND YOUR THINKING

Explain why building on a floodplain is detrimental to both humans and the environment.

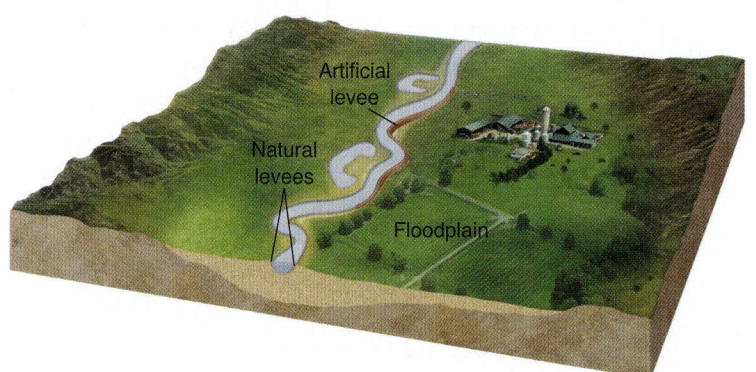

FIGURE 17.20 A floodplain consists of natural levees, oxbow lakes, wetlands in meander scars, point bars, and flood sediments.

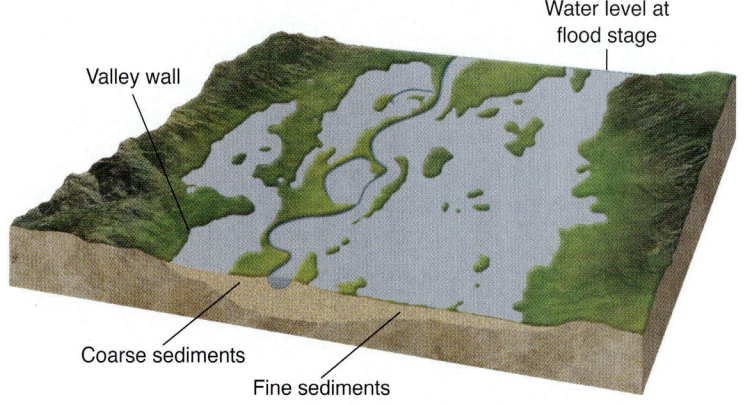

What is the geologic composition of a floodplain?

CRITICAL THINKING

The 100-Year Floodplain

You have been hired to analyze Deer Creek, where developers propose to build new housing. But no new development is allowed in the stream's *100-year floodplain*. That means that any area with a greater than 1% chance of flooding in a single year is off-limits.

To determine the legal location for construction of new homes, you must calculate the height, or elevation, of floods as measured by a *stream gauge*, which has recorded the maximum water level every year since 1935. Here are the steps:

1. Calculate the percent probability (*P*) of annual recurrence for each of the flood events listed in **TABLE 17.1.**

$$P = 100 \times M/(n + 1)$$

where *M* is the flood magnitude shown in column 3 and *n* is the total number of years, 69.

2. Use the graph paper provided in **FIGURE 17.21** to plot the data in Table 17.1. Plot the stage or elevation of the flood and the value of *P* for the same year. Be sure to use the scale at the top of the graph, labeled *Percent probability of recurrence (P)*, to locate the *P* value.

3. Draw a straight line through the data points that best represents the trend of the data.

A flood level recurrence interval (*RI*), or time interval between floods that are similar in size, is calculated using the equation

$$RI = (n + 1)/M.$$

The recurrence interval is also equal to 100/*P*. Values for the *RI* are given on the lower axis of Figure 17.21.

TABLE 17.1 Deer Creek Flood History

Year	Maximum Stage meters	Magnitude M	P
1936	6.8	30	
1946	5.5	48	
1959	2.5	67	
1964	4.9	54	
1976	3.4	65	
1984	8.4	8	
1985	7.6	17	
1993	10.2	2	

Source: U.S. Dept. of the Interior, USGS, 2000.

4. The elevation of the stream gauge on Deer Creek is 210.31 meters. Use your line to determine the elevation (or stage) of the 100-year floodplain.

5. Add the elevation of the 100-year floodplain to the elevation of the stream gauge and convert the answer into units of feet (multiply by 3.28). Your answer is the elevation below which is illegal to develop the land. On the map (**FIGURE 17.22**), land elevation is contoured in feet using a 10-foot contour interval. In pencil, outline and shade the land area that lies at or below the elevation of the 100-year floodplain.

Based on your calculations, answer the following:

1. Why is there a law making it illegal to build in the 100-year floodplain?

2. Why should a person who does not live in the floodplain care whether someone else builds there?

3. Why does the government have an obligation to identify geologic hazards for its citizens and to limit their ability to engage in certain activities, such as building in hazardous locations?

4. You have mapped lands vulnerable to the 100-year flood. What additional hazards might threaten lands in the map area? Be sure to include in your answer considerations of population growth, global warming, stream meandering, potential upstream changes, slope stability, and other possibilities. To illustrate your analysis, mark on the map areas where these threats may be found.

5. Based on your analysis, indicate where you recommend future development of new housing be located.

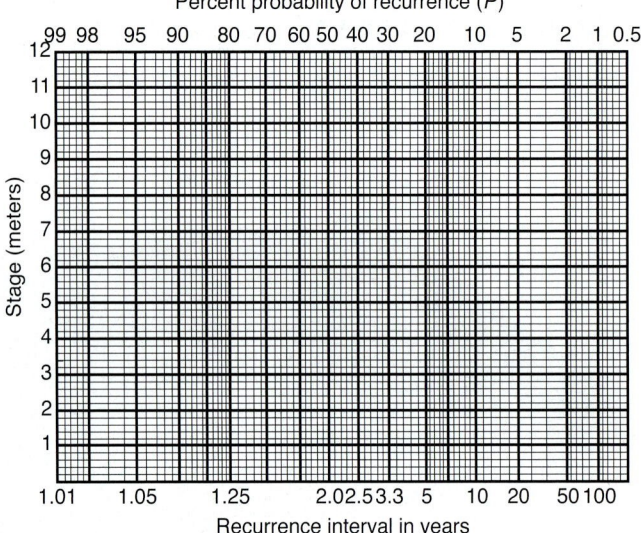

FIGURE 17.21 Probability graph paper.

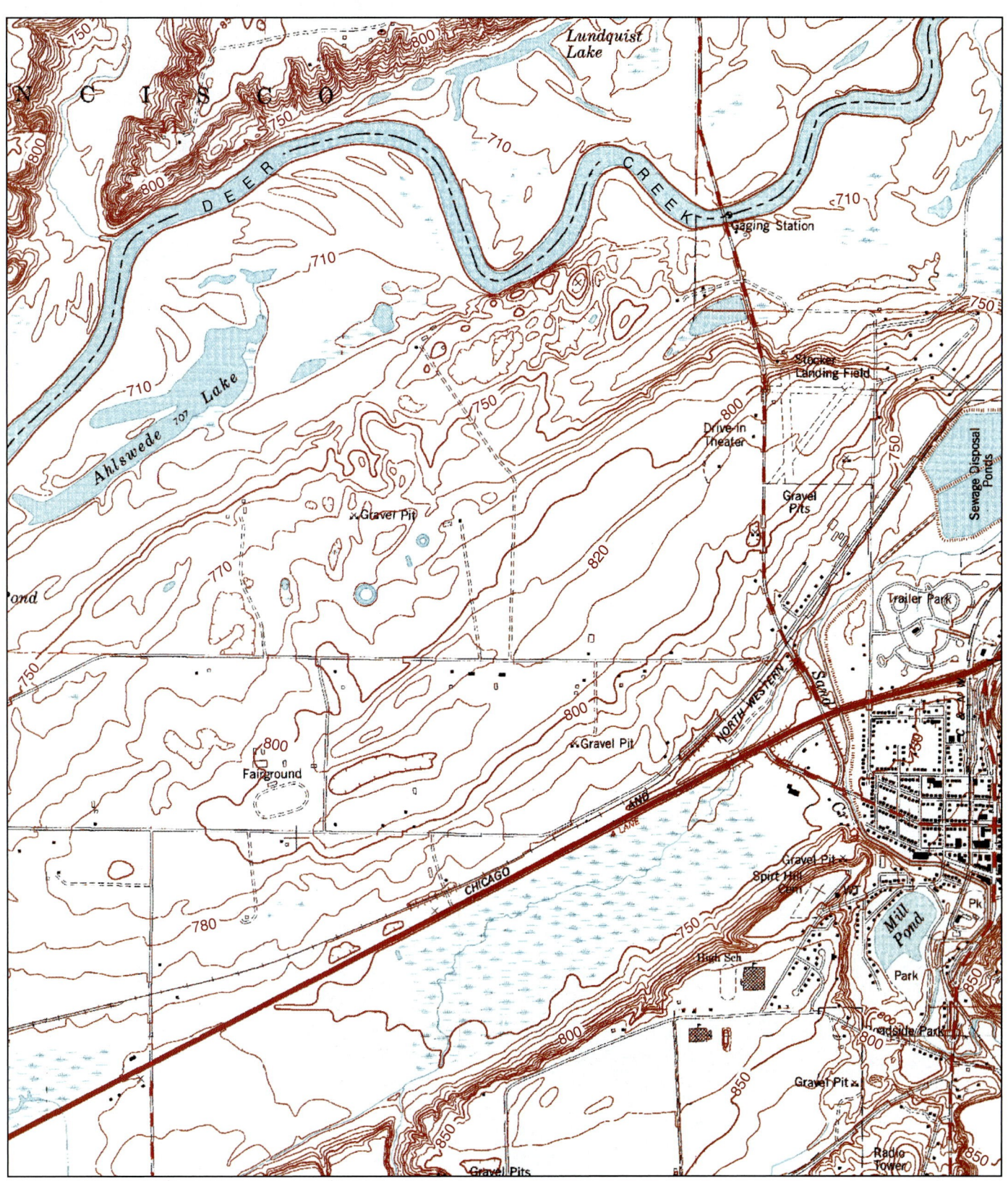

FIGURE 17.22 Deer Creek map. Contour interval 10 feet. *Source:* U.S. Dept. of the Interior, USGS, 2000.

17-7 Streams May Develop a Graded Profile

LO **17-7** Define the concept of a graded stream.

Streams are generally steeper in their upper reaches and gentler in their lower reaches This defines a characteristic profile, one that is gently concave with a flattened lower end.

Stream Profile

From its headwaters to where a channel connects with the sea, a stream system has a distinctive cross-sectional shape (**FIGURE 17.23**). Generally, in their upper reaches, stream valleys are narrow and steep, while in their lower reaches, they are wider and have gentler slopes.

Because of the steep gradient, water in the upper reaches of a stream profile flows at high velocity. This rapid flow promotes *downward cutting* (channel formation) and *headward erosion* (rather than meandering). The stream valley, therefore, is typically V-shaped in cross section. Headward erosion is the tendency of a stream channel to lengthen upslope in reaction to greater erosion at valley headwaters.

The exact shape of the valley (i.e., wide or narrow) is also determined by mass wasting, which as you'll recall from Chapter 16, depends on hill slope processes and the stability of regolith. As a channel cuts down and back, the valley it has formed tends to widen. This widening favors meandering, so, with time, meandering takes precedence over headward erosion in sculpting the landscape.

An interesting consequence of headward erosion is **stream piracy** (or *stream capture*), so called because it occurs when headward erosion breaches a drainage divide, intersects another channel, and "captures" its flow. Piracy expands drainage area by diverting runoff from neighboring basins. With increased runoff, headward erosion may accelerate and capture additional neighboring streams (**FIGURE 17.24**).

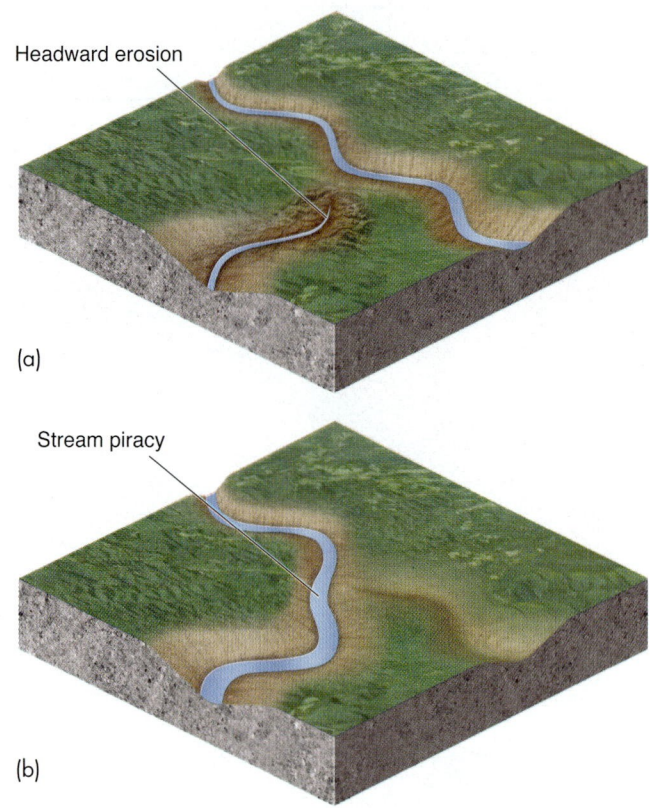

(a)

(b)

FIGURE 17.24 (a) As a stream erodes in a headward direction, it may (b) intersect a neighboring drainage system and capture its discharge.

Describe how headward erosion changes drainage basins.

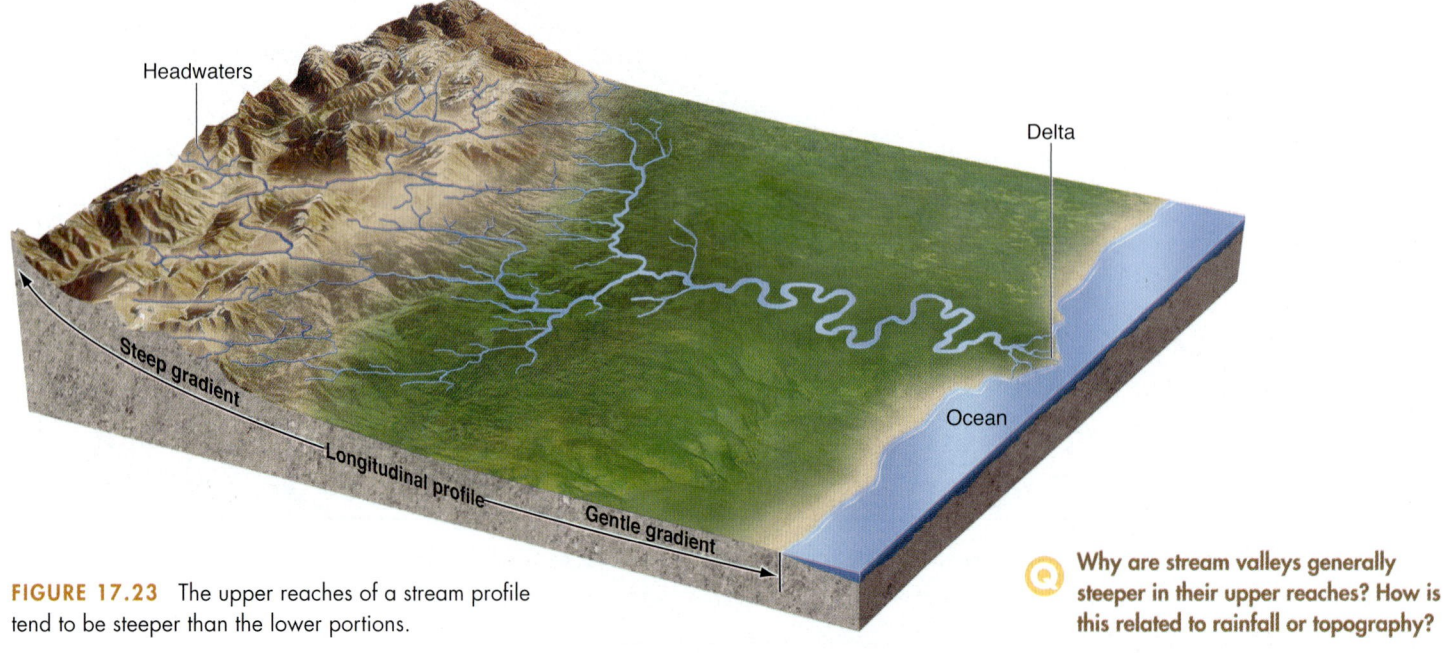

FIGURE 17.23 The upper reaches of a stream profile tend to be steeper than the lower portions.

Why are stream valleys generally steeper in their upper reaches? How is this related to rainfall or topography?

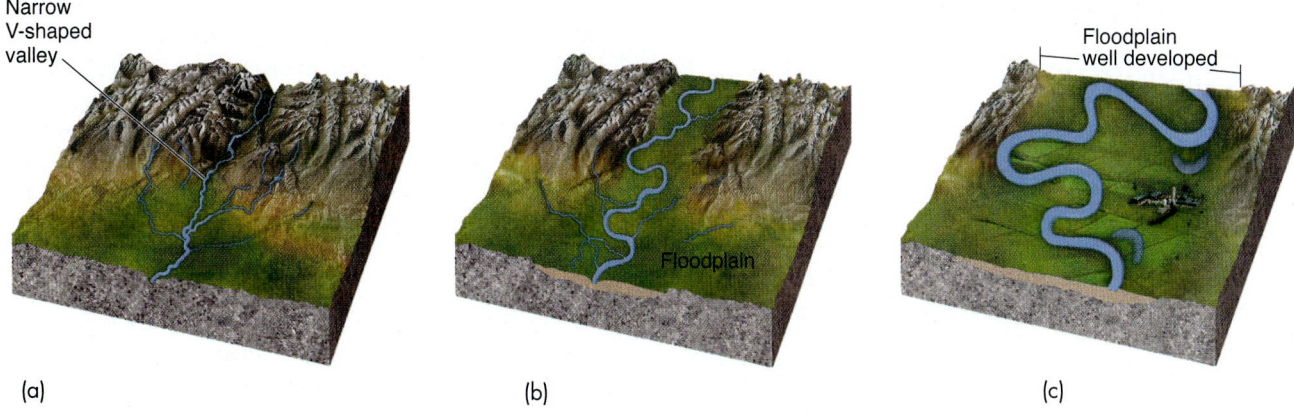

FIGURE 17.25 (a) A tectonically uplifting region will develop a fluvial system characterized by a steep gradient, migrating waterfalls and rapids, steep V-shaped valleys, and gorges. (b) Mass wasting and undercutting of slopes widen valleys, and meandering becomes more pronounced. (c) Eventually, the gradient decreases as a wide floodplain develops, and the stream flows through a floodplain composed of its own alluvium.

Q Describe how crustal uplift can affect the evolution of a fluvial landscape.

In the middle reaches of a stream valley, two important changes usually occur: The stream develops an **alluvial channel**, meaning that it flows through its own alluvium (stream sediment), and the channel has a greater tendency to erode laterally (meander), creating a wide, flat-bottomed valley (**FIGURE 17.25**). In the lower reaches, the floodplain grows wider, the meanders more pronounced, and the gradient becomes very gentle.

The floodplain is an important site of sediment storage within a fluvial system. Continual deposition of sediment in a floodplain is known as *aggradation*, the thickening accumulation of floodplain sediment through deposition. Yet it is possible for a stream to attain a state of *dynamic equilibrium*: If the amount of deposition is balanced by the amount of erosion, such that there is no net gain or loss, then sediment will be transported through the system without a net buildup of the floodplain.

Graded Streams

Scientists who study streams have defined the concept of the **graded stream** (or *graded profile*). They apply this term to streams that have apparently achieved, throughout long segments, a state of dynamic equilibrium between the rate of sediment transport (erosion) and the rate of sediment supply (deposition).

Stream grading always is defined over a specific segment of channel. Defining a graded stream over an entire drainage system would have no meaning because, generally speaking, the lower parts of a drainage system always deposit, while the uppermost parts always erode. In contrast, over a particular segment, a graded stream has no net erosion or deposition; the input of sediments into the segment is the same as the output, and erosion equals deposition over that segment.

A graded stream maintains an equilibrium between the processes of erosion and deposition and, therefore, between aggradation and degradation (downward erosion of a streambed and floodplain). If excess sediment enters a stream, the stream will store the excess in the form of islands and bars, which are propelled downstream during times of high discharge (a braided channel).

A stream that has a deficit of sediment will cut downward into its bed. A high gradient generally leads to a stream that erodes its bed. A lower gradient may produce a graded stream that successfully transports all the available sediment and so does not erode the bed.

In a very simple way, this concept describes the evolution of the **fluvial** (from the Latin for "river" or "flowing water") landscape depicted in Figure 17.25. In their early stages, as tectonic activity uplifts an area and streams have steep gradients, channels are characterized by waterfalls and rapids in high-relief topography. These features promote abrasion and sediment production. Gorges and canyons are carved, and active headward erosion is prevalent.

Waterfalls migrate upstream as resistant rock ledges are eroded. Mass wasting widens the valley as slopes are oversteepened by stream undercutting, and collapse into the channel. Over time, a valley widens and the landscape is altered so that the stream gradient is lessened and a floodplain develops. The stream reaches equilibrium when its sediment load matches the streams capability to carry sediment. The channel does not cut into its bed, nor does the floodplain accumulate sediment deposits. The stream is then said to have a graded profile.

Q **EXPAND YOUR THINKING**

How would you determine whether a stream has achieved a graded profile? Describe the research you would conduct and the data you would collect.

17-8 Fluvial Processes Adjust to Changes in Base Level

LO 17-8 Explain the influence of base level changes, such as dams and waterfalls, on fluvial processes.

The stream profile is sensitive to changes in the land surface. Dams, waterfalls, and tectonic uplift all transform fluvial processes.

Approaching Base Level

Earlier, we introduced the concept of *base level*, the elevation below which a stream can no longer erode the land. As a stream approaches its base level (the ocean, a lake, or another stream), it develops a low gradient.

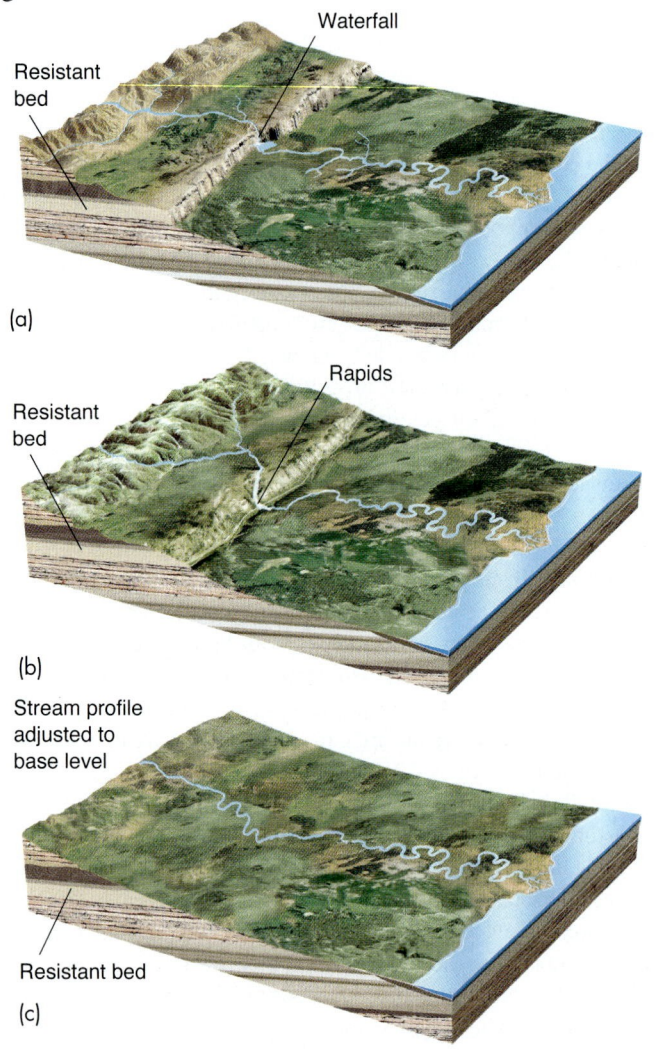

(a)

(b)

(c)

FIGURE 17.26 A waterfall (a) represents a local base level. The waterfall will erode and migrate upstream (b) until it lies below the gradient of the river (c).

As a field geologist in the area of panel (c), what evidence would convince you that the area once had a waterfall?

A waterfall formed by a resistant layer of rock can be considered a local base level. Until the rock has been eroded, the stream can erode no lower and will develop a low gradient above the waterfall. Eventually, it will erode through the rock, and the waterfall will migrate upstream until it reaches an elevation equal to that of the graded profile of the stream. Once the bed of the graded channel matches the surface of the resistant layer, the waterfall will no longer exist (**FIGURE 17.26**).

One benefit of running water is its capability to turn *turbines*, machines that produce electric power (recall the description of turbines in Chapter 14). *Hydroelectric dams* furnish a vertical drop in order to accelerate the velocity of water and turn turbines to generate electricity. In addition, such dams create *reservoirs*, stored water in the form of a lake, that yield recreational opportunities and drinking water sources for urban areas. But dams also cause problems.

Reservoirs inundate river valleys and swamp what once was valuable farmland (**FIGURE 17.27**). Dams also lead to the destruction of ecosystems adapted to a river environment. Flooded archeological sites may be lost forever. And, since reservoirs represent a local base level, the velocity of streams immediately drops to zero when they enter a reservoir, depositing sediments in the process. With time, as streams continue to deliver sediment, a reservoir will fill with sand and mud, and *dredging* will be required in order to extend its life.

Downstream of a dam, a channel will be deprived of its normal load of sediment because of its deposition in the reservoir. The stream below a dam will experience an increase in velocity and a decrease in sediment load, both of which cause it to erode (or *incise*) into its channel. What previously had been a graded profile now becomes a highly erosive stream that cuts into its bed.

Upstream of the reservoir, the stream undergoes a rise in base level, and the profile of the stream valley accumulates sediment. This means that the stream will build a higher floodplain and lower its gradient as it deposits sediment.

Incised Channels and Terraces

Once a stream has established a graded profile, it is vulnerable to variations in sediment availability, shifts in base level, and tectonic alterations in land level. Each of these changes can disrupt the equilibrium of the graded profile.

Uplift of the land related to tectonic movement of Earth's crust, such as at a convergent plate boundary, alters a stream's gradient. Even if the land does not tilt as it is uplifting, the gradient must increase because of the stream's higher elevation relative to its base level. An increased gradient results in greater flow velocity within the channel, causing channel erosion. As a channel incises its bed, it will cut into the underlying bedrock and develop steep banks (**FIGURE 17.28**).

Alluvial terraces develop when a graded stream incises its floodplain. Incision can be the result of decreased sediment load, lowered base level, or increased gradient due to uplift. Terraces are former floodplain deposits that are exposed when a stream incises its channel. A channel is said to become *rejuvenated* when the gradient is increased or the sediment load is decreased and the stream cuts through its floodplain.

Q EXPAND YOUR THINKING

What are the upstream and downstream problems that may arise due to the building of a dam?

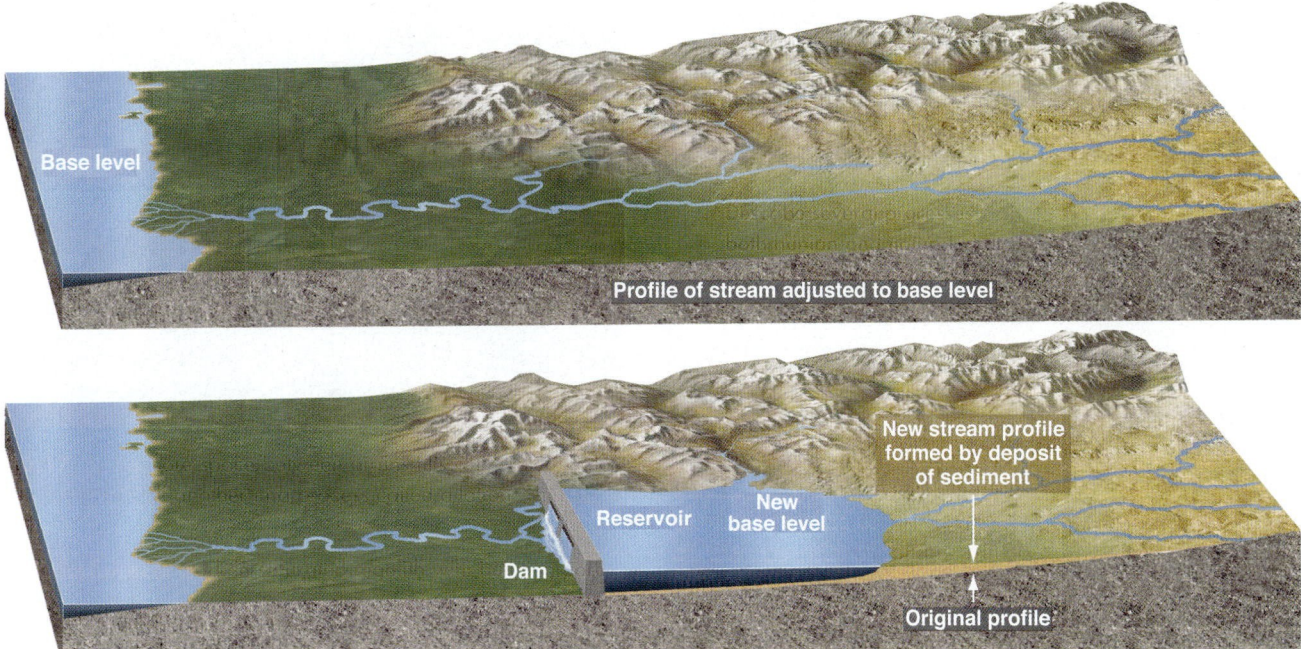

FIGURE 17.27 Below a dam, the stream is deprived of sediment and erodes into its bed. Above the dam, the stream experiences a rise in base level and the stream channel accumulates sediment.

Why do reservoirs require periodic dredging?

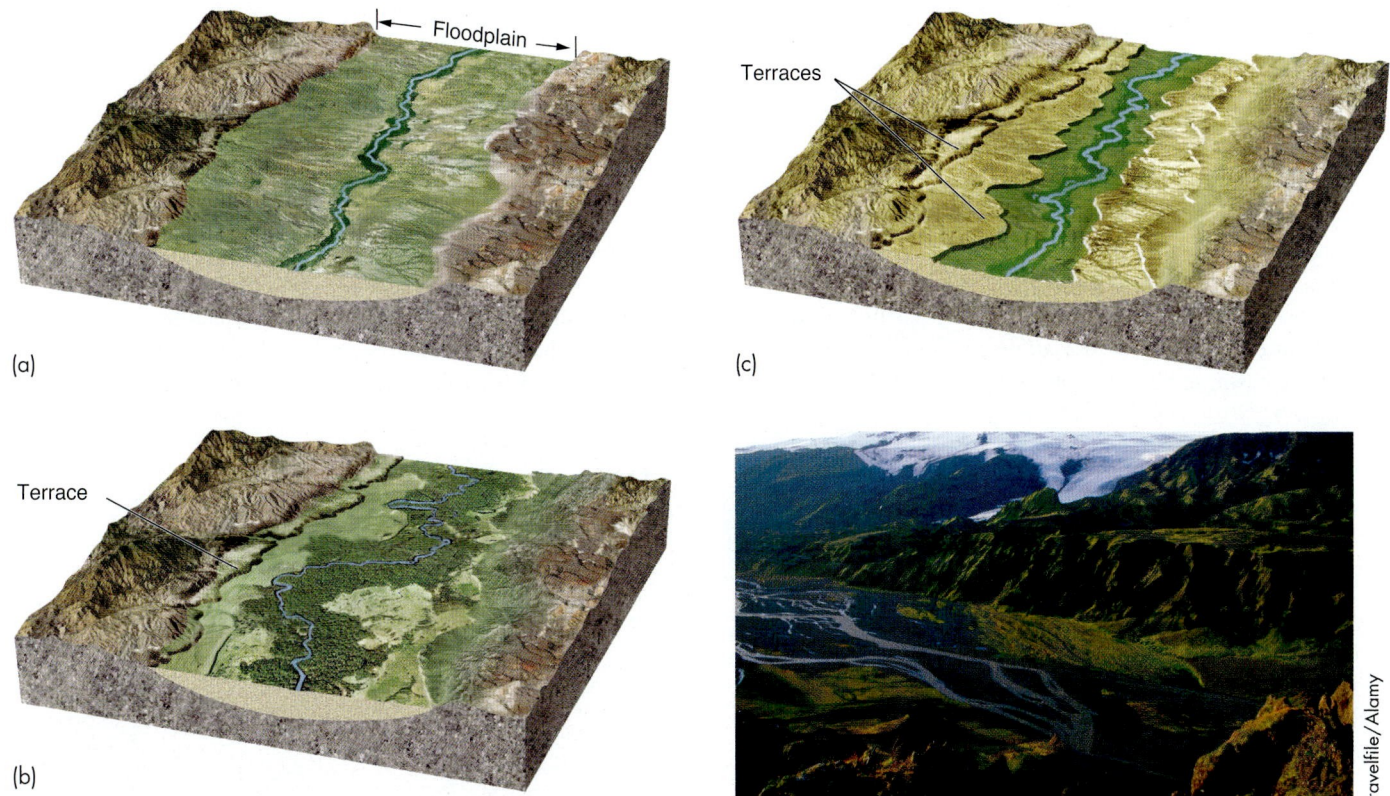

FIGURE 17.28 (a) Graded profile in equilibrium with the sediment load. (b) Decrease in sediment load or land uplift or lowering base level initiates stream incision. (c) Alluvial terraces develop after prolonged incision.

What causes the formation of an alluvial terrace?

17-9 Fluvial Sediment Builds Alluvial Fans and Deltas

LO 17-9 Describe an alluvial fan and the avulsion process in delta building.

When a stream experiences a sudden reduction in velocity, it will deposit its sediment load and build a delta or alluvial fan.

Alluvial Fan

An **alluvial fan** is a fan- or cone-shaped deposit of stream sediment. It is a semicircular, gently sloping cone of fluvial sediment deposited when a confined stream leaves a narrow canyon or gulley and enters a flat plain or valley floor. The change in channel confinement and greater infiltration of water reduce stream power and lead to deposition of the sediment load. Sediment of all sizes, from boulders to mud, is deposited at the base of the valley, forming a low, conical apron of clastic debris (**FIGURE 17.29**).

Alluvial fans are large features, as wide as 10 kilometers across. The sediment composing the fan is subtly sorted so that coarse material, such as boulders and cobbles, tend to be deposited near the valley mouth, where the stream first experiences a decline in velocity. Finer particles, such as sand and mud, are deposited across the fan surface, and may even be carried to the distant edge of the fan and out onto the floor of the adjacent valley.

Typically, braided channels drain the sides and center of the fan surface. Alluvial fans are most likely to develop in areas where intermittent but powerful rainstorms suddenly load stream channels with unsorted alluvium containing intermixed boulders, cobbles, gravel, sand, and mud.

In arid regions, rainfall generally is related to the buildup of large thunderheads that contain abundant precipitation, and there is no groundwater contribution to stream discharge. Consequently, streams flow only when there is direct runoff from a rainstorm; and the intense nature of rainfall means that there is little time for runoff to soak into the ground. Nearly all precipitation becomes runoff into a drainage basin, setting off a phenomenon known as *flash flooding.*

© Marli Bryant Miller

FIGURE 17.29 An alluvial fan.

Q **What conditions lead to the formation of an alluvial fan?**

Major floods do not happen every year. Often, several years, or even decades, may pass between flood events. In between these floods, physical and chemical weathering litters mountain slopes with loose sediment. When a storm does occur, it washes this sediment into dry stream gullies. There, high-gradient, high-velocity flow turns into a raging flood that sweeps sediment of all sizes down the narrow valley. Such flash floods are very hazardous. Hikers report hearing a sound like a high-speed locomotive approaching as a flash flood roars down narrow canyons.

Deltas

A **delta** is a sedimentary deposit that takes shape when a stream flows into standing water (i.e., reaches its base level) and deposits its load of sediment. It is so named because it fans outward from its point of origin at a channel mouth, such that it resembles the Greek letter delta (Δ). With continued deposition over time, a delta can extend out from the coast and become a large geologic feature.

Deltas accumulate three types of sedimentary beds: *bottomset beds, foreset beds,* and *topset beds* (**FIGURE 17.30**). As a river

FIGURE 17.30 A simple delta is composed of three types of beds: bottomset beds, foreset beds, and topset beds.

Q **What is the difference between a delta and an alluvial fan?**

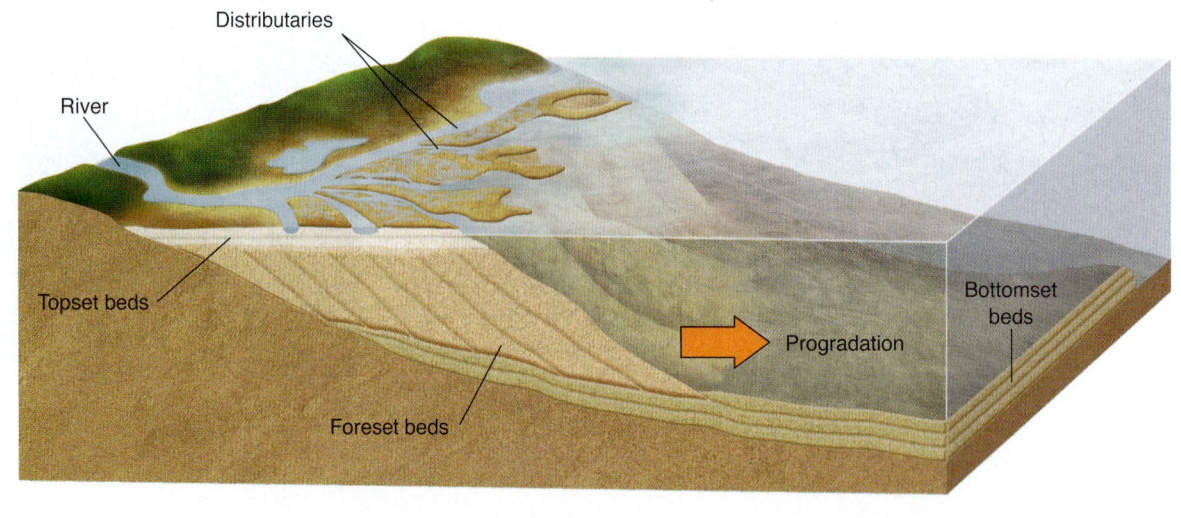

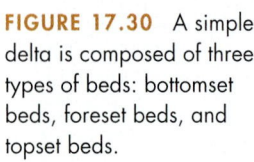
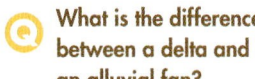

encounters its base level, silt and clay carried in the channel settle onto the basin floor to form bottomset beds. Coarser grains of sand accumulate closer to the river's mouth and build out a series of angled beds into the basin and bury bottomset beds. These are foreset beds, and they represent the forward-building front of the delta; above these are topset beds, which consist of flat-lying sand and mud layers.

Topset beds constitute the *delta plain* and, if the delta is formed in the ocean, are host to freshwater and saltwater **wetlands.** The most famous delta in the United States is the *Mississippi Delta* on the Gulf of Mexico coastline (**FIGURE 17.31**).

The Mississippi River drains over 3 million square kilometers of the U.S. interior. Over time, the river has delivered immense amounts of sediment to the ocean through a major channel reaching into the Gulf of Mexico. Approximately seven similar channels have released sediment over the past 8000 years, each with an average life span of 1000 years (**FIGURE 17.32**).

In time, each channel is abandoned and a new one started through the process of *avulsion*, which is similar to the meander cutoff process. Specifically, a flood that has swept down the river arrives at the flat delta plain, overflows its banks, and establishes a new route to the sea, leading to *avulsion*, or abandonment, of the old channel. The new pathway represents a more efficient route, and so the old channel is forsaken. Eventually, the old channel fills with sediment and becomes part of the vast wetland on the delta plain. The Mississippi River has experienced seven major avulsion events, each of which has produced a new channel in the delta system.

Apply your knowledge of surface water to the Critical Thinking exercise "Channel Systems."

Image courtesy Liam Gumley, Space Science and Engineering Center, University of Wisconsin-Madison and the MODIS science team.

FIGURE 17.31 The Mississippi River drains the interior of the United States. Sediments carried by the river accumulate at the point where the river enters the ocean, its base level, and build the Mississippi River Delta.

How is the size of the Mississippi Delta related to the size of its watershed?

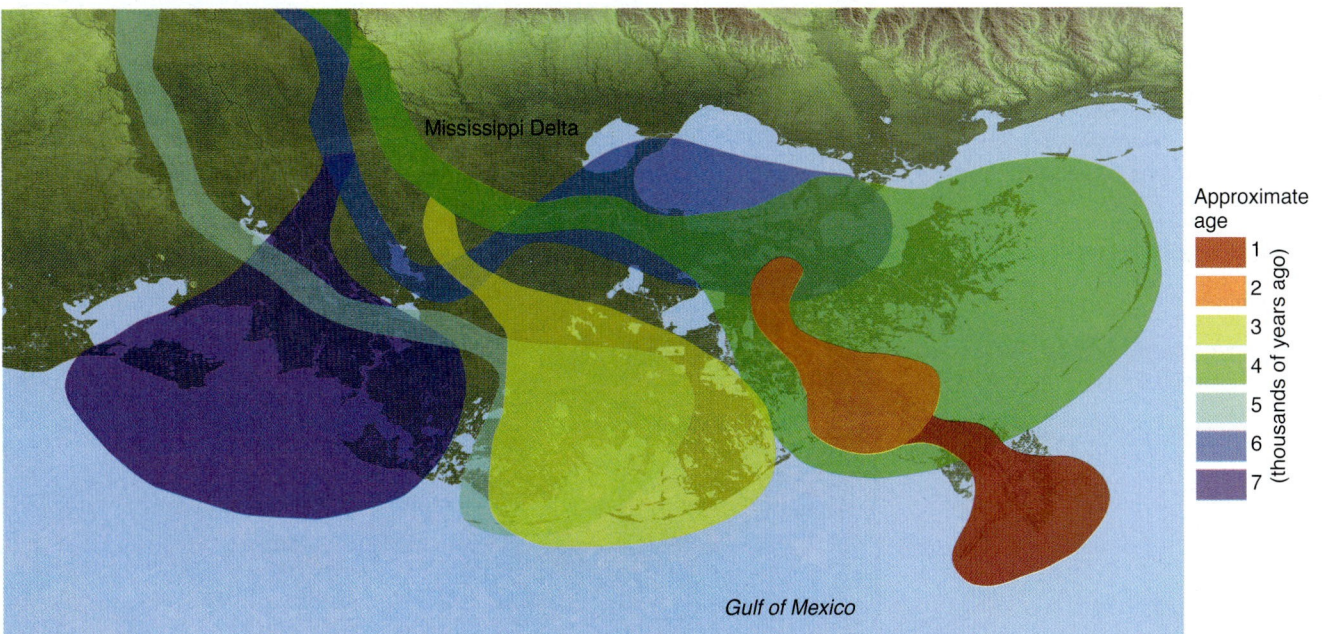

Mississippi Delta

Gulf of Mexico

Approximate age

1
2
3
4
5
6
7

(thousands of years ago)

FIGURE 17.32 Avulsion events have resulted in the abandonment of major distributary channels (lobes) and the establishment of new ones in the Mississippi Delta. Seven former delta lobes, 7 (oldest) through 1 (youngest), are outlined here, each of which lasted an average of 1000 years before being abandoned as a result of avulsion.

What causes avulsion?

EXPAND YOUR THINKING

Develop a hypothesis that predicts the impact of building artificial levees along the Mississippi River and delta. How would this affect downstream communities, delta region ecosystems, and coastal waters?

CRITICAL THINKING

Channel Systems

Together with a partner, work through the following exercise:

1. Describe how the size of eroded sediment grains varies with changes in stream velocity.

2. What is a graded profile, and what would you measure to determine whether a stream is graded?

3. Describe the changing nature of sediment deposition in a stream from its headwaters to its base level.

4. What controls the development of straight, meandering, and braided channels?

5. On **FIGURE 17.33**, label components of the hydrological cycle, fluvial landforms, fluvial processes, and geologic hazards.

6. Pick one watershed in the scene in the figure. Combine your knowledge of channel systems with your understanding of weathering and sedimentary environments to formulate a hypothesis describing the evolution of your watershed. In the watershed history:

   a. How might tectonics play a role?

   b. How might climate play a role?

   c. Where and how are humans exposed to natural hazards?

   d. How might mass wasting play a role?

FIGURE 17.33 Typical components of a channel system.

17-10 Water Problems Exist on a Global Scale

LO **17-10** Describe the freshwater problems that are growing worldwide.

Freshwater is a basic human need, fundamental to life itself. But Earth has only a finite supply of freshwater, and it is diminishing due to overuse, pollution, and drought.

Freshwater

Freshwater is stored in the ground, in various surface environments such as lakes, glaciers, and streams, and in the atmosphere (more on this in Chapter 18). The growing shortage of freshwater is such that, according to the United Nations, more than 2.8 billion people in 48 countries will face scarcity conditions or *water stress*—defined as a condition in which the population consumes more than 10% of its total water supply per year—by 2025. Of these countries, 40 are in West Asia, North Africa, or sub-Saharan Africa (**FIGURE 17.34**).

Over the next two decades, population growth and, with it, greater demand for freshwater are projected to push all west Asian countries into water scarcity conditions. By 2050, the number of countries facing water stress or scarcity could rise to 54, with a combined population of 4 billion—about 40% of the projected global population of 9.4 billion.

Restricted access to freshwater inevitably will incite political disputes, even warfare. The projected level of water stress by the middle of this century is a very serious issue, one with the potential to disrupt the peaceful coexistence of neighboring nations around the world. These problems will not go away by themselves. They can be solved, however, through early planning, new technologies for water management, and population control.

It is important to point out here that some people contend that ocean saltwater also is available for human use; the reality is that the high amount of energy needed to convert saltwater into freshwater (a process known as *desalinization*) makes it prohibitively expensive for all but the wealthiest communities.

Polluted Runoff

Storm water runoff is water that flows overland during a rainstorm. Overland flow during storms is a natural part of the water cycle, one

Freshwater Stress

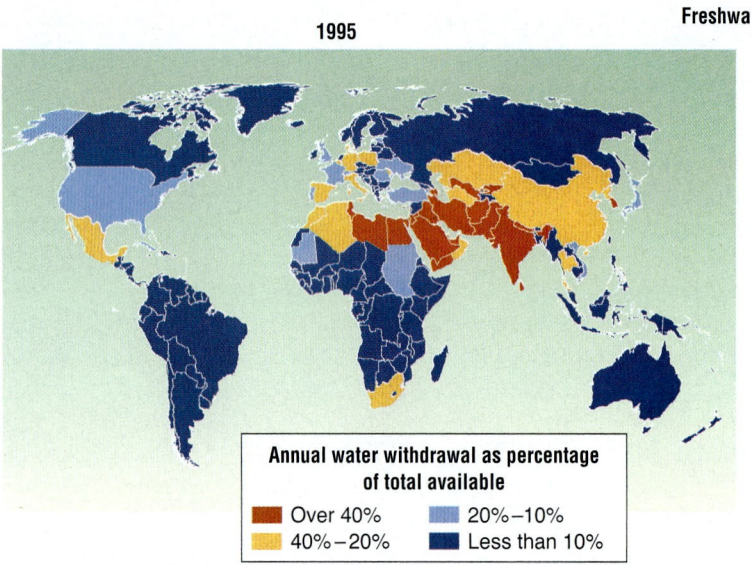

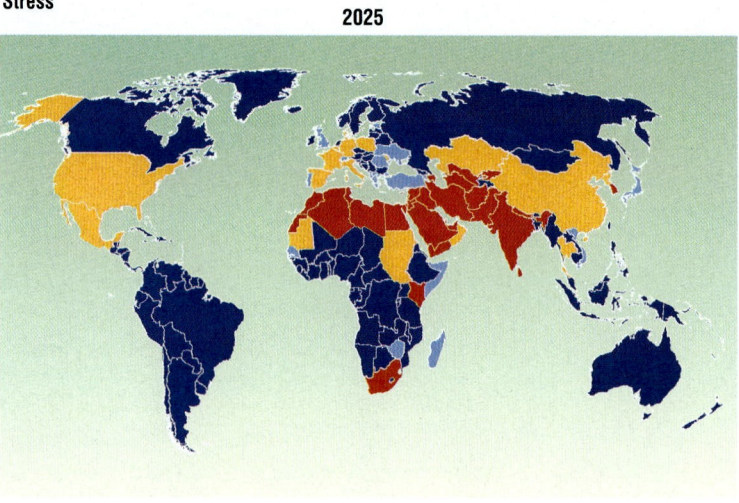

1995

2025

Annual water withdrawal as percentage of total available
- Over 40%
- 40%–20%
- 20%–10%
- Less than 10%

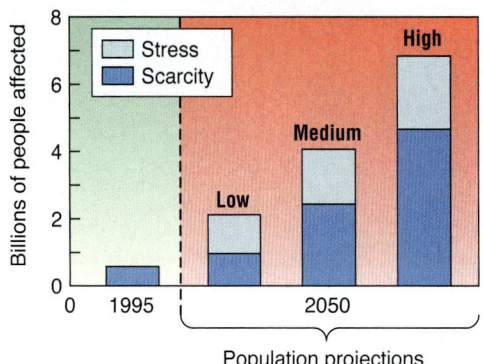

People Suffering from Water Stress and Scarcity

FIGURE 17.34 Freshwater stress is projected to intensify by 2025 to the point at which two out of every three people on Earth will face water scarcity.

Q **Why is water scarcity growing as a global problem?**

that serves many beneficial purposes, among them clearing stream channels of blockages and delivering sediment to floodplains during floods. Unfortunately, human activities that alter the land surface also can change the natural character of storm water runoff in ways that can be damaging both to the environment and to human communities (FIGURE 17.35).

On a heavily vegetated property that is in its natural state, for example, storm water runoff moves across the land slowly, allowing for some infiltration into the ground and absorption by vegetation. The vegetation partially buffers downstream portions of the drainage system from the full impact of a rainstorm. Of course, as explained earlier, in arid regions that lack vegetation, storm runoff may lead to flash floods.

If this same property is then developed—buildings are constructed and the land surface is paved—the hydrologic character of the runoff will change. Storm water will no longer be buffered by the friction of vegetation and soil, and no longer be absorbed by porous regolith. Instead, smooth, paved surfaces will prevent infiltration and accelerate flow across the ground.

Impervious surfaces, such as storm sewers, channels, sidewalks, curbs, and gutters, will force runoff downhill into fewer and fewer conduits. Without infiltration, runoff discharge will flow at higher velocity across urbanized surfaces. Eventually, urbanized flow will enter—with devastating consequences—a natural environment such as a stream channel, a wetland, or a seashore estuary. Many of these natural settings lack the capacity to withstand the high discharge and high velocity runoff that occurs under these conditions, and their ecosystems are damaged or destroyed entirely.

Studies have shown that storm water runoff is a major source of pollution, causing declines in fisheries, prompting restrictions on swimming and other recreational activities, and limiting our ability to enjoy many of the other benefits that water provides. Storm water runoff, furthermore, contains contaminants that can introduce problems in any natural environment receiving the water.

Contaminated runoff is referred to as nonpoint source pollution (NPS), or **polluted runoff**. Polluted runoff may contain bacteria, oil, detergent, pesticides, heavy metals, animal and human feces, and dirt. These pollutants come from a variety of sources, including roads, gutters, lawns, construction sites, and even the atmosphere. Unlike wastewater entering treatment plants, polluted storm water runoff does not receive any treatment before it is washed into creeks, rivers, and lakes. NPS pollution can have serious negative impacts on drinking water supplies, recreation, fish and wildlife. It is, in fact, the leading cause of most of today's water quality problems.

For many years, efforts to control the discharge of storm water focused on quantity (flood control) more than quality (pollutant and sediment control). More recently, awareness of the need to improve *water quality* has grown. As a result, federal, state, and local governments have established programs and instituted laws designed to reduce the types and amounts of pollutants contained in storm water discharges into waterways; these programs promote the concept and the practice of *preventing pollution at the source*, before it can cause environmental or health problems.

Anoxic Dead Zones

Off the mouths of several major river systems around the world, "dead zones" are forming, places where water in contact with the seafloor ("bottom water") is *anoxic* (has no dissolved oxygen); simply put, areas with such low concentrations of dissolved oxygen cannot support most marine life (FIGURE 17.36).

The cause of anoxic bottom waters has been traced to nutrients entering the ocean from agriculture products (fertilizer) and human sewage in stream water. These nutrients feed a community of *phytoplankton* (microscopic algae) that float on the surface of the ocean. These plankton thrive in the abundant pollutants delivered by streams, and the rapid turnover of their populations means that the amount of dead algae falling to the bottom is very large.

© Aleksandar Jaksic/iStockphoto

FIGURE 17.35 Polluted runoff collects in streams, lakes, and the ocean.

iStockphoto

FIGURE 17.36 Anoxic waters cause the death of thousands of marine organisms, especially fish.

 What are some ways to control polluted runoff?

What is the cause of "dead zones"?

Bacteria on the seafloor use dissolved oxygen to decay this organic matter. The decay is so prolific that the water is robbed of all oxygen, suffocating other forms of life—crabs, clams, shrimp, zooplankton, and fish. Entire ecosystems can die off in the space of one afternoon.

Anoxic zones (**FIGURE 17.37**) are spreading worldwide; their number has doubled between 2000 and 2010. One major culprit behind these high rates of anoxia is the widespread use of fertile floodplain soils for agriculture. Almost all the floodplains along America's

major river systems are used for agriculture. Given the inevitability of regular flooding, clearly this is not a wise use of the land. Nutrients, fertilizers, insecticides, herbicides, and other chemicals used in farming produce excess chemical residues that enter watersheds and are swept into the ocean. Freshwater aquatic and coastal marine ecosystems are exposed to these residues in ways that are not yet well understood, but evidence suggests that negative consequences are widespread.

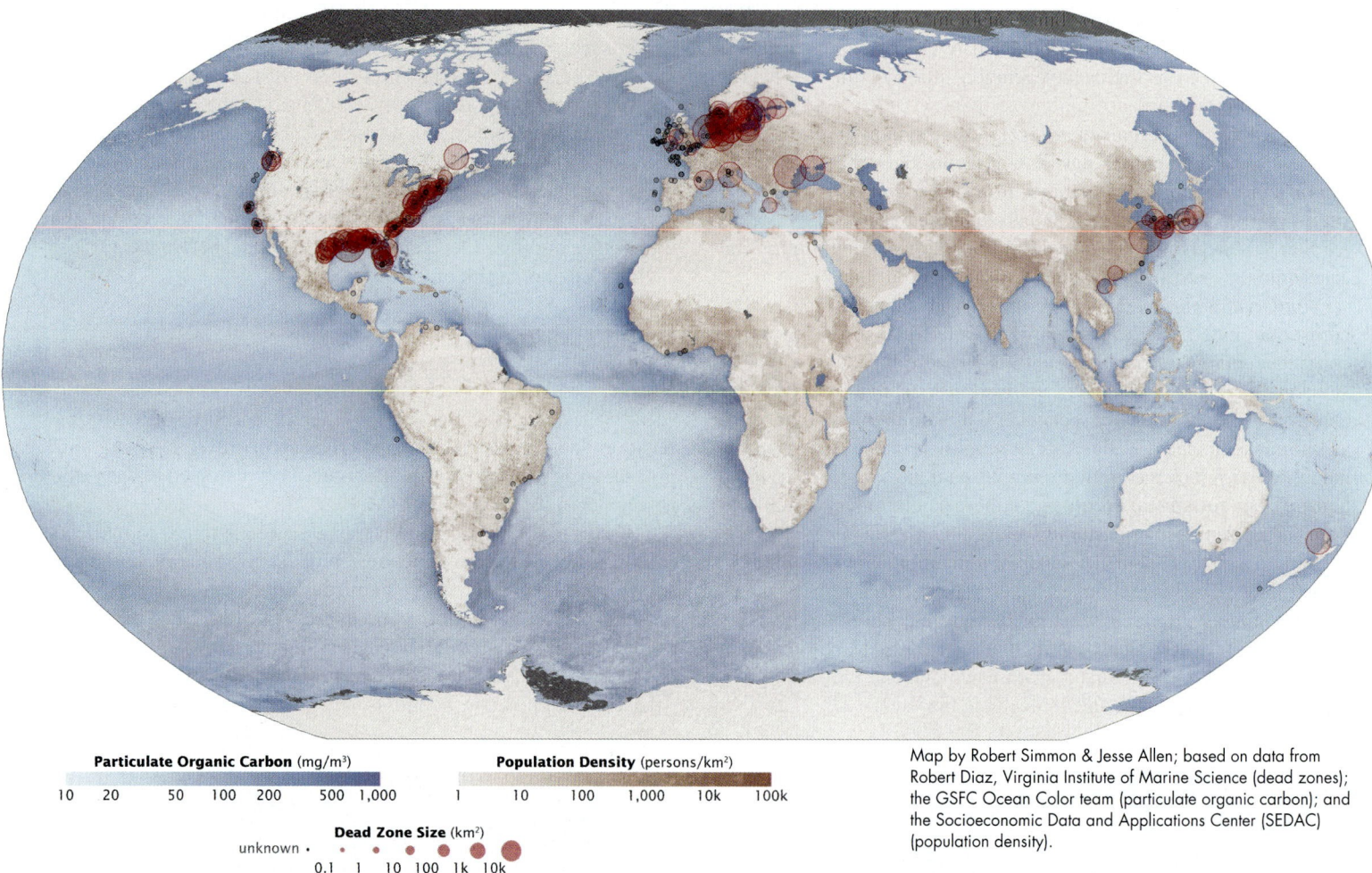

Map by Robert Simmon & Jesse Allen; based on data from Robert Diaz, Virginia Institute of Marine Science (dead zones); the GSFC Ocean Color team (particulate organic carbon); and the Socioeconomic Data and Applications Center (SEDAC) (population density).

FIGURE 17.37 The size and number of marine dead zones—areas where marine waters are so polluted by streams that sea creatures can't survive—have exploded in the past half-century. Researchers have found more than 400 hypoxic zones in the world. The hypoxic zone in the Gulf of Mexico is the largest in the United States and one of the largest globally. Red circles on this map show the location and size of many of our planet's dead zones. Black dots indicate where dead zones have been observed, though their size is unknown.

Q EXPAND YOUR THINKING

What are the water problems of concern where you live?

LET'S REVIEW

Flowing water, or runoff, is responsible for shaping more of Earth's land surface than any other process. Runoff is part of the hydrologic cycle, a geologic concept that describes the movement of water through Earth's environments. Runoff becomes channelized flow on the land surface. Water flowing in a channel is known as a stream, and streams are part of a larger drainage system that transports a load of sediment and dissolved compounds to its base level.

Freshwater is a highly valuable resource, but one that is subject to pollution and overuse—a theme we explore further in Chapter 18. Although nature continually refreshes water resources, humans frequently overuse them, making it critically important that we carefully manage water so that it remains available for our use, as well as for that of future generations.

STUDY GUIDE

17-1 The hydrologic cycle moves water between the atmosphere, the ocean, and the crust.

- Water covers 71% of Earth's surface and is the dominant agent governing environmental processes. Although nature continually refreshes water resources, rates of human usage outpace natural rates of renewal. For this reason, it is important to carefully manage water as a valuable resource. To that end, it is necessary to understand the **hydrologic cycle** and the nature of channelized water.

- Five natural processes keep water moving through the hydrologic cycle: condensation, precipitation, infiltration, runoff, and evapotranspiration.

17-2 Runoff enters channels that join other channels to form a drainage system.

- Every year, 36,000 cubic kilometers of water runs off the land. This water does an enormous amount of work by eroding and transporting sediments and dissolved ions toward the ocean.

- A **stream** is any flowing body of water following a **channel**, and a river is a major branch of a stream system. Worldwide, streams carry about 16 billion tons of sediment particles and 2 to 4 billion tons of dissolved ions every year. Streams do not flow in isolation; they are part of a large network of channels of various sizes known as a **drainage system**.

17-3 Discharge is the amount of water passing a given point in a measured period of time.

- **Base level** is the theoretical lowest level toward which erosion of Earth's surface constantly progresses but seldom, if ever, reaches. Local base level is the lowest point a particular channel will reach. The ultimate base level is the ocean. If all streams were allowed to flow continuously through time, with no new uplift of the land, they eventually would carve their channels down to sea level.

- The **discharge** (Q) is equal to the cross-sectional area (A) of a channel (width times average depth [m^2]) times the average velocity (V) of the flow [m/sec]. Discharge is measured in cubic meters of water per second [m^3/sec].

17-4 Running water erodes sediment.

- Stream channels have varied characteristics, depending on three primary variables: gradient, discharge, and sediment load. Bed load consists of large particles that remain on the stream bed most of the time. They move by sliding or rolling (also called traction) or by jumping (also known as saltation) as a result of collisions between particles. Sediment moves between bed load and suspended load as the velocity of the stream changes, depending on discharge. Dissolved load consists of ions introduced into the water through the chemical weathering of rocks.

- The Hjulstrom diagram reveals that once mud and silt have been eroded, they stay in motion even if the flow velocity drops drastically. It is not as easy to transport coarser sediments (sand and gravel), for they require nearly as much water velocity as is needed to erode the grains in the first place. It takes as much water energy to erode large pieces of gravel as it does to erode clay.

17-5 There are three types of stream channels: straight, meandering, and braided.

- Three general types of stream channels can be described: **straight**, **meandering**, and **braided**.

- Meandering channels are the most common characteristic of a stream that is free to roam across a valley floor. Meandering channels develop most readily on low-gradient slopes with an easily eroded bed and banks. As water enters a bend in a channel, momentum will keep it moving in a straight line so that it collides with the far bank. This intersection of water and erodible bank leads to a tendency of the outer bank of a meander bend to erode and thus migrate outward.

- Braided channels are formed when a stream is carrying excess sediment. Because the stream is unable to transport the excess sediment, the sediment accumulates in the channel and forms bars and islands that split the water into multiple channels.

17-6 Flooding is a natural process in healthy streams.

- Discharge in a channel can change rapidly, potentially producing a flood. Flooding occurs when discharge increases to the point at which there is too much water for a channel to carry within its banks. Excess water flows outside of the channel onto the adjoining land, known as a **floodplain**.

- A number of features are found on a floodplain. Some of these are depositional, such as natural levees, former point bars that have been abandoned as a result of meander cutoff, and wetlands located in meander scars. Other features, such as oxbow lakes, are created by past erosion. When floodwaters retreat, stream velocities are lowered and a layer of **alluvium**, sediment that originates from a stream, is deposited across the floodplain. Repeated flood cycles over time result in the deposition of many successive layers of alluvium, so the elevation of the floodplain rises. Floodplains, then, are composed of point bar sands and gravels and flood deposits of silt and clay.

17-7 Streams may develop a graded profile.

- From its headwaters to the point where a channel connects with the sea, a stream system will have a characteristic profile that is gently concave with a flattened lower end. Generally, in their upper reaches, stream valleys tend to be narrow and steep, while in lower reaches, they are wider and have gentler slopes.

- A **graded stream** is one that has achieved, throughout long reaches, a state of dynamic equilibrium between the rate of sediment transport (erosion) and the rate of sediment supply (deposition). Stream grading is always defined over a specific segment of channel. Defining a graded stream over an entire drainage system would have no meaning because the lower parts always deposit while the uppermost parts always erode.

17-8 Fluvial processes adjust to changes in base level.

- A waterfall formed by a resistant layer of rock can be considered a local base level. Until the rock has been eroded, the stream can erode no lower and will develop a low gradient above the waterfall. Eventually, the stream will erode through the rock, and the waterfall will migrate upstream until it reaches an elevation equal to the elevation of the stream's graded profile.

Once the bed of the graded channel matches the surface of the resistant layer, the waterfall will no longer exist.

- Uplift of the land related to tectonic movement of Earth's crust, such as during continental collision, changes a stream's gradient. An increased gradient results in greater flow velocity within the channel, which in turn causes channel erosion. As a channel erodes its bed, it becomes incised.

- Alluvial terraces develop when a graded stream incises its floodplain. Incision can be the result of decreased sediment load, lowered base level, or increased gradient due to tectonic uplift. Terraces are former floodplain deposits that are exposed when a stream incises its channel. A channel is said to become rejuvenated when the gradient is increased or sediment load is decreased and it cuts through its floodplain.

17-9 Fluvial sediment builds alluvial fans and deltas.

- An **alluvial fan** is a semicircular, gently sloping cone of channel sediment deposited when a high gradient stream leaves a narrow valley and enters a flat plain or valley floor. The significant change in gradient causes a drop in flow velocity. As a result, the stream loses its capability to carry its former sediment load. Sediment of all sizes, from boulders to mud, is deposited at the base of the valley, taking shape as a low conical apron of clastic debris.

- A **delta** is a sedimentary deposit that is formed when a stream flows into standing water (i.e., reaches its base level) and deposits its sediment load. With continued deposition over time, the deposit can extend into the basin and become a large geologic feature with major environmental impacts.

17-10 Water problems exist on a global scale.

- **Storm water runoff** is water that flows overland during a rainstorm. Overland flow during storms is a natural part of the water cycle that serves many beneficial purposes, such as clearing stream channels of entanglements and blockages and delivering sediment to floodplains during floods. However, alterations to the land surface wrought by human activities can change the natural character of storm water runoff in ways that can be damaging to both the environment and human habitation.

- According to the United Nations, more than 2.8 billion people in 48 countries will face water stress or scarcity conditions by 2025. Of these countries, 40 are in West Asia, North Africa, or sub-Saharan Africa.

- Contaminated runoff is referred to as nonpoint source pollution, or **polluted runoff**. Nonpoint source pollution occurs when water runs over the land or through the ground, picking up contaminants such as bacteria, oil, detergent, pesticides, animal feces, and dirt. Pollution can have serious negative effects on drinking water supplies, recreation facilities, fish and wildlife. It is the leading cause of most of today's water quality problems.

KEY TERMS

alluvial channel (p. 523)
alluvial fan (p. 526)
alluvium (p. 519)
base level (p. 512)
braided channel (p. 516)
channel (p. 510)
delta (p. 526)
discharge (p. 512)
drainage system (p. 510)

flood (p. 518)
floodplain (p. 518)
fluvial (p. 523)
graded stream (p. 523)
hydrologic cycle (p. 508)
meandering channels (p. 516)
oxbow lake (p. 516)
polluted runoff (p. 531)
storm water runoff (p. 530)

straight channels (p. 516)
stream piracy (p. 522)
streams (p. 508)
thalweg (p. 516)
tributaries (p. 511)
watershed (p. 510)
wetlands (p. 527)

ASSESSING YOUR KNOWLEDGE

Please complete the following exercise before coming to class. Identify the best answer to each question.

1. The hydrologic cycle describes:
   a. The movement of water through the environment.
   b. Condensation, runoff, evapotranspiration, precipitation, and infiltration.
   c. The water in various natural reservoirs.
   d. The water being exchanged in natural processes.
   e. All of the above.

2. On this diagram, label the hydrologic cycle (1 through 6) using the following terms: precipitation, surface runoff, infiltration, cloud formation, land evaporation and transpiration, ocean evaporation.

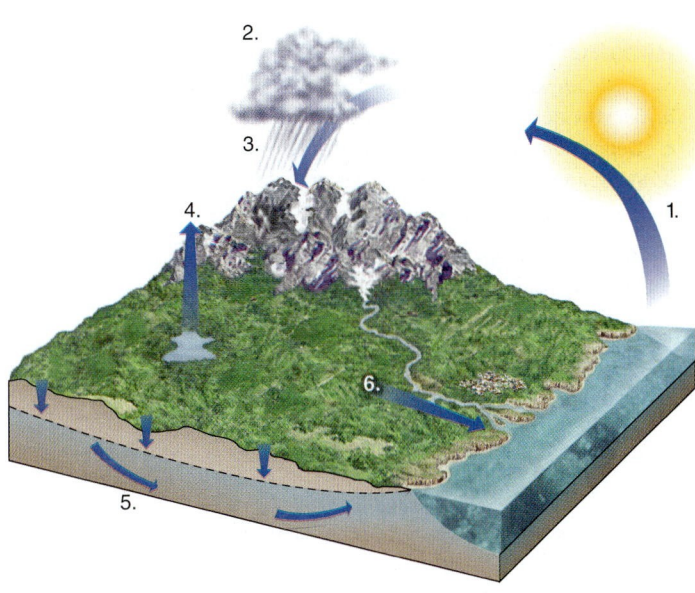

3. Within a drainage system:
   a. All water flows away from the largest channel.
   b. All runoff flows in only one channel.
   c. Water comes only from the same storm.
   d. All runoff drains into the same stream.
   e. All runoff infiltrates.

4. When a stream experiences an increase in gradient, it will:
   a. Flow faster and erode its channel.
   b. Develop a graded profile.
   c. Produce more flash floods.
   d. Erode all its alluvial fans.
   e. Form a delta.

5. A graded profile is one in which:
   a. Headward erosion is prevalent.
   b. Rejuvenation has occurred.
   c. Stream piracy is taking place.
   d. Erosion is stronger than deposition.
   e. Sediment deposition is equal to erosion.

6. On this diagram, label each of the landforms associated with a meandering stream system: alluvium, oxbow lake, floodplain, cutoff, meander neck, natural levees.

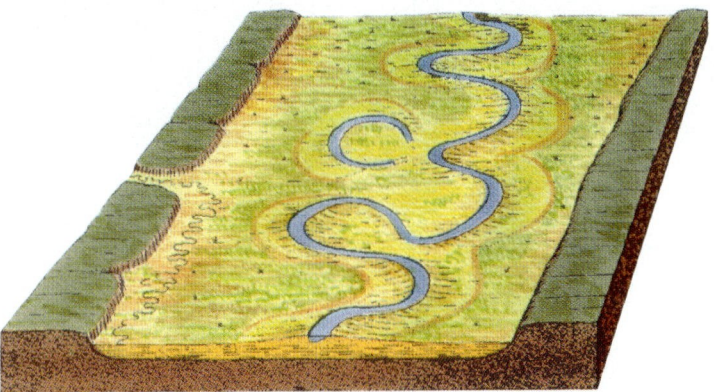

7. Between headwaters and base level, a channel will:
   a. Widen and deepen.
   b. Develop a flatter slope.
   c. Tend to accumulate finer-grained sediment.
   d. Collect discharge from more tributaries.
   e. All of the above.

8. The Hjulstrom diagram tells us that:
   a. Clay and gravel require about the same water velocity to erode.
   b. Clay is more easily eroded than gravel.
   c. Gravel is more easily eroded than clay.
   d. Sand is harder to erode than clay.
   e. All sediments erode at the same water velocity.

9. Why is clay difficult to erode but easy to transport?
   a. Because the particles are so heavy.
   b. Because the particles stick to the channel bed and then exhibit saltation.
   c. Actually, clay is easy to erode.
   d. Because clay sticks to the channel bed but stays in suspension once it has been eroded.
   e. None of the above.

10. The lowest level to which a stream can erode is known as the:
    a. Hjulstrom diagram.
    b. Topset bed.
    c. Graded profile.
    d. Base level.
    e. Floodplain.

11. This photo of the 2011 Mississippi River flood shows sediment being deposited on the floodplain. The pattern of sedimentation is typically as follows:
    a. Fine-grained sediments are deposited near the main channel, and coarse-grained sediments far from it.
    b. Sand is deposited in the meanders, and gravel is deposited in the straight channel segments.
    c. Sand deposition forms natural levees, and mud is deposited farther from the channel.
    d. Floods are erosional not depositional.
    e. None of the above.

12. Channels generally take one of three forms:
    a. Oxbow, meander scar, and alluvial fan.
    b. Meandering, braided, and straight.
    c. Turbulent, laminar, and graded.
    d. Alluvial, aggraded, and erosional.
    e. None of the above.

13. Braided channels form when the:
    a. Sediment load is small.
    b. Capability of the stream to transport sediment is larger than the amount of sediment.
    c. Capability of the stream to resist erosion is great.
    d. Capability of the stream to transport sediment is exceeded by the amount of sediment.
    e. Stream discharge is very high.

14. The discharge of a channel that is 75 meters wide, 3.2 meters deep, with a flow velocity of 0.34 meters per second is:
    a. 81.6 cubic meters per second.
    b. 8.16 cubic meters per second.
    c. 24.0 cubic meters per second.
    d. 240 cubic meters per second.
    e. None of the above.

15. The lower reaches of a watershed are characterized by:
    a. Steeper gradient and reduced sediment supply.
    b. Gentler gradient and shallower water.
    c. Gentler gradient and greater width.
    d. Greater sediment supply and lower discharge.
    e. None of the above.

16. Flooding is a result of:
    a. Prolonged rainfall saturating the ground.
    b. Rapid melting of winter snow and ice.
    c. Rapid and heavy rainfall.
    d. Presence of upstream artificial levees that pass the flood wave downstream.
    e. All of the above.

17. If base level lowers:
    a. A stream will rejuvenate.
    b. The floodplain will collect excess sediment.
    c. Deposition will become greater than erosion.
    d. A stream will decrease its discharge.
    e. A stream will decrease its gradient.

18. Avulsion occurs when:
    a. Flooding causes a stream to decrease discharge.
    b. Flooding causes a stream to build an alluvial fan.
    c. Flooding causes a stream to rejuvenate.
    d. Flooding causes a stream to establish a graded profile.
    e. Flooding causes a stream to establish a new channel to reach base level.

19. A graded stream is one in which:
    a. The entire channel is accumulating sediment.
    b. There are few meanders.
    c. The channel has no net erosion or deposition.
    d. The floodplain has reached the end of its growth.
    e. None of the above.

20. Freshwater stress is:
    a. A condition in which a stream flows only occasionally because of poor groundwater flow.
    b. A lack of sufficient freshwater to support the natural ecosystem.
    c. A situation in which humans use more water than can be supplied in the next season.
    d. A situation in which a population consumes more than 10% of its total water supply per year.
    e. A situation where water pollution has caused a "dead zone."

© Shaun Lowe/iStockphoto

FURTHER RESEARCH

1. How do deposits in meandering streams compare with the deposits in braided streams?

2. How does the gradient of a stream change as it flows from steep headwaters into the ocean?

3. Why does a stream tend to develop a graded profile?

4. Each year streams and other erosional processes remove about 10.75 cubic kilometers of sediment from the continents. The total volume of continents above sea level is approximately 93,000,000 cubic kilometers. Thus, continents should erode down to sea level in about 8.6 million years. Of course, if this were true, then the world's continents would now all lie at sea level. What is wrong with this reasoning?

5. List at least three factors that contribute to flooding.

6. Explain the logic behind the statement, "Floods today are more frequent than they were 100 years ago."

ONLINE RESOURCES

Explore more about surface water on the following Web sites:

U.S. Geological Survey (USGS), "Significant floods in the U.S. during the 20th century":
http://ks.water.usgs.gov/Kansas/pubs/fact-sheets/fs.024-00.html

USGS, "Natural Hazards—Floods":
www.usgs.gov/hazards/floods

USGS, "Flood Hazards—A National Threat":
http://pubs.usgs.gov/fs/2006/3026

Additional animations, videos, and other online resources are available at this book's companion Web site:
www.wiley.com/college/fletcher

This companion Web site also has more information about WileyPLUS and other Wiley teaching and learning resources.

GROUNDWATER

CHAPTER CONTENTS & LEARNING OBJECTIVES

GEOLOGY IN OUR LIVES

Groundwater is one of our most precious natural resources, and it is our most important source of drinking water. In addition, it supplies water for irrigation, industry, manufacturing, and domestic uses, as well as to meet most other human water needs. Less than 3% of the water on Earth is freshwater; about 75% is frozen in glaciers; approximately 25% is groundwater; and only 0.005% is surface water (lakes and streams). Groundwater is highly vulnerable to overuse and pollution, for three reasons: Most of it moves slowly, usually, less than 0.3 meters per day; it has been in the ground for decades to centuries; and it is not quickly renewed by the hydrologic cycle. Furthermore, the rate at which we use it generally outpaces the natural rate of groundwater replenishment, thus requiring the implementation of conservation measures in nearly every community. Properly conserving groundwater requires careful analysis of where it is located, how it moves, and how it is recharged. The central question leading groundwater science is: "How shall we use groundwater today so that it is remains available for generations tomorrow?"

Q **What are the physical characteristics of groundwater? How does it move, and how can we conserve it for use by future generations?**

18-1   Groundwater Is Our Most Important Source of Freshwater

LO 18-1 Itemize the general features of groundwater.

Groundwater is our most important source of freshwater and, therefore, requires that we value it as such, beginning by understanding its vulnerability to human-generated contaminants, and then working to ensure its sustainability.

The Importance of Groundwater

There is water beneath our feet—a lot of it. In fact, so much water lies within the tiny pore spaces and fractures of Earth's crust that about 95% of the total supply of freshwater of the United States is taken out of the ground.

Rainfall is the main source of fresh **groundwater**. About 25% of the rainfall in the United States becomes groundwater. That is equal to about 1.1 quadrillion liters annually. Of this, about 102 trillion liters are withdrawn for use in each year. Three-quarters of U.S. cities count on groundwater as part of their water supply. *Water wells* drilled into the crust extract groundwater using pumps. More than 800,000 new water wells are drilled in the United States each year.

Most groundwater is found in unconsolidated sands and gravels, which account for nearly 90% of all groundwater sources. Other sources include porous sandstone, limestone, and highly fractured crystalline and volcanic rock.

The United States Geological Survey estimates that groundwater in the upper 800 meters of the continental crust is 3000 times more plentiful than the water found in all the world's rivers at any given time. This water is not located in vast subterranean lakes and pools; rather, it is found as microscopic coatings of moisture clinging to individual mineral grains and filling the tiniest cracks and fractures within rocks and sediments of the crust. Because groundwater is critically important to our lives, geologists around the world study it in order to develop policies for its sustainable use.

As we saw in Chapter 17, when rain falls to the ground, the water does not stop moving: Some of it flows along the surface as runoff and enters streams or lakes; some of it is used by plants; some evaporates and returns to the atmosphere; and some sinks into the ground. Imagine pouring a glass of water onto a pile of sand. Where does the water go? It travels into the spaces among the particles of sand and becomes groundwater.

The Water Table

Groundwater is sited underground in cracks, pores, and spaces in soil, sediments, and rocks. The area where water fills these spaces is called the *saturated zone*. The top of this zone is called the **water table**; the top of the water table is the *capillary fringe*, a narrow zone where water seeps upward, pulled by surface tension, to fill empty pore spaces.

The water table may be only a meter below the ground's surface, or it may be hundreds of meters down (**FIGURE 18.1**). The portion of crust above the water table where pore spaces and other openings are filled with both air and water is called the *unsaturated zone*.

Groundwater can be found almost everywhere. As just noted, the water table may be deep or shallow; it may also rise or fall, depending on many factors; near the coast, for example, the water table rises and falls with the tide and with groups of waves. In humid areas, the water table is a subtle reflection of the ground's topography (**FIGURE 18.2**)—that is, where there are hills, the water table is high,

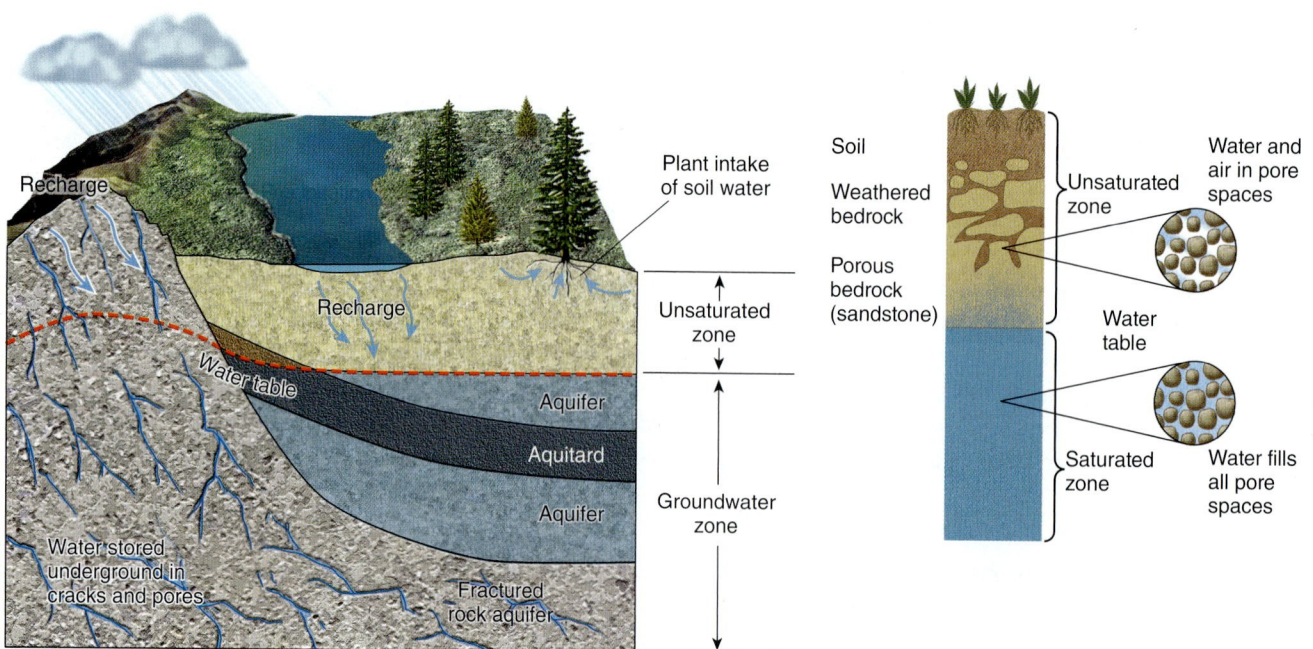

FIGURE 18.1 Water in the ground is located in the saturated zone beneath the water table and in the unsaturated zone above the water table.

What processes are capable of causing the water table to rise and fall?

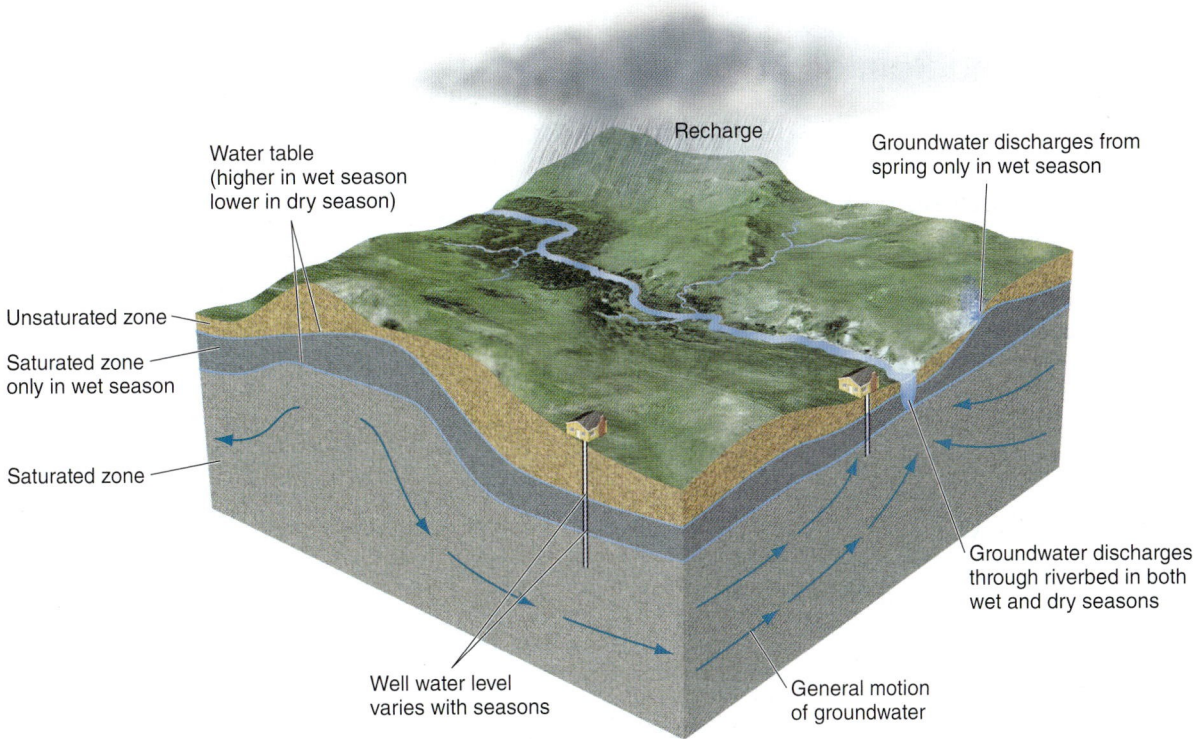

Recharge

Water table
(higher in wet season
lower in dry season)

Groundwater discharges from
spring only in wet season

Unsaturated zone

Saturated zone
only in wet season

Saturated zone

Groundwater discharges
through riverbed in both
wet and dry seasons

Well water level
varies with seasons

General motion
of groundwater

FIGURE 18.2 The water table rises during wet seasons and falls during dry seasons.

🔵 **What force drives the flow of water in the ground?**

and where there are valleys, the water table is low. Heavy rains or melting snow may cause the water table to rise, or an extended period of dry weather may prompt it to fall. When the water table falls, streams and wetlands dry up, lake levels lower, and the ground becomes parched and dry.

Water Wells

If you live in or near an urban area, you probably never give much thought to where your water comes from, how it is delivered, or whether it is clean. One of the services provided by local government, paid for by taxes, is the delivery of clean water; and most city governments are so successful at this that there is rarely a reason for the average city dweller to worry about water. All you need to know is how to open the tap at the sink and fill a glass with water. Still, it's important not to lose sight of the fact that water arrives at your tap as the result of a lot of hard work by many people, in concert with the application of modern technology.

In most cases, your water is pumped from a water well drilled into the ground. Water wells are drilled by *truck-mounted drill rigs* (**FIGURE 18.3**), which use various types of cutting heads on the end of a spinning shaft to bore through soil, sediment, sedimentary rock, and even crystalline igneous and metamorphic rock to access the water table. The *borehole* must penetrate beyond the water table to

account for variations in water table height related to seasonal shifts in elevation. Once a well has been drilled, it is lined with a pipe that is perforated to allow the inflow of water while keeping sand and other sediments out. A pump is installed to withdraw the water; a gravel layer is added along the walls of the borehole to block fine sediment; and the top of the well is sealed to prevent shallow, contaminated water from seeping into the well.

FIGURE 18.3 Water wells supply over 95% of U.S. freshwater needs.

🔵 **Enumerate the necessary steps in developing a useful water well.**

© PhotoSpin, Inc/Alamy

🔵 **EXPAND YOUR THINKING**

Geologists search for groundwater. Describe the data you would need to develop groundwater resources for a small town.

18-2 Groundwater Is Fed by Snowmelt and Rainfall in Areas of Recharge

LO 18-2 Describe the processes by which the water table gains and loses water.

Rainfall and snowmelt add new water to the groundwater system through the process of recharge.

Recharge

Recharge may occur in several ways. In most cases, runoff soaks into the ground through *percolation* (also known as *infiltration*) and adds to the water table below. Areas of high rainfall, where percolation enters the saturated zone, can recharge a groundwater system that extends for miles into the surrounding countryside.

High-rainfall areas that recharge the water table often are located at higher elevations, where warm moist air is forced to flow into cooler (higher) levels of the atmosphere. There it condenses as rain or snow, falls to the land surface, and creates a recharge area. Recharge also can occur in places where streams, lakes, and wetlands act as a source of water for the groundwater system (**FIGURE 18.4**).

Eventually, groundwater may reappear above the ground, through a process called *groundwater discharge*. Groundwater may flow into streams, wetlands, lakes, and oceans, or it may be dis-

charged in the form of springs and flowing wells. In this way, groundwater discharge can contribute to surface water and runoff. Notably, in dry periods, groundwater may supply the entire flow of streams. It has been estimated that about 50% of the nation's streams and rivers acquire a portion of their flow from groundwater. At all times of the year, in fact, the nature of groundwater flow through the crust and the position of the water table have a profound effect on the volume of surface runoff.

Although the rate of discharge determines the volume of water moving from the saturated zone into streams, it is the rate of *recharge* that determines the volume of water running across the surface. When it rains, for instance, the volume of water running into streams and rivers depends on how much rainfall the underground rock and sediment layers can absorb. When there is more water on the surface than can be absorbed into the ground, it runs off into streams and lakes.

Streams that recharge the groundwater system are known as *losing streams* or *influent streams*. In losing streams, the height of the water table is lower than the stream's water surface. Water seeps into the ground through the stream channel bed. In locations where the water table is so low that it is located below the base of the channel, it often bulges upward beneath the recharge area.

In *gaining streams*, or *effluent streams*, discharge from the water table contributes partly or entirely to their flow. Here, the water table must be at an elevation above the stream's surface. Water that enters a stream system from the water table through the channel bed is known as **base flow**.

A third situation features both water gain and loss along different portions of the same channel. Often, a stream gains water during wet seasons of the year and loses it during dry seasons (**FIGURE 18.5**).

Wetlands

Once thought of as useless swamps and wastelands, today **wetlands** are recognized as critically important components of natural ecosystems and so are protected under federal law. Wetlands develop in places where the water table intersects the land surface. Generally, they are lands where saturation with water is the dominant factor in determining the nature of soil development and the types of plant and animal communities living on and in the soil.

As with streams, groundwater interactions with wetlands can be grouped into three categories that focus on gains and losses of water (**FIGURE 18.6**):

1. Wetlands may serve as a source of recharge in places where surface water seeps into the ground and feeds the groundwater system.

2. Alternatively, the groundwater system may be a source of water for the wetland.

3. It is also possible for groundwater to feed a wetland from the upslope side, flow through the wetland, and seep back into the ground from the downslope side of the wetland.

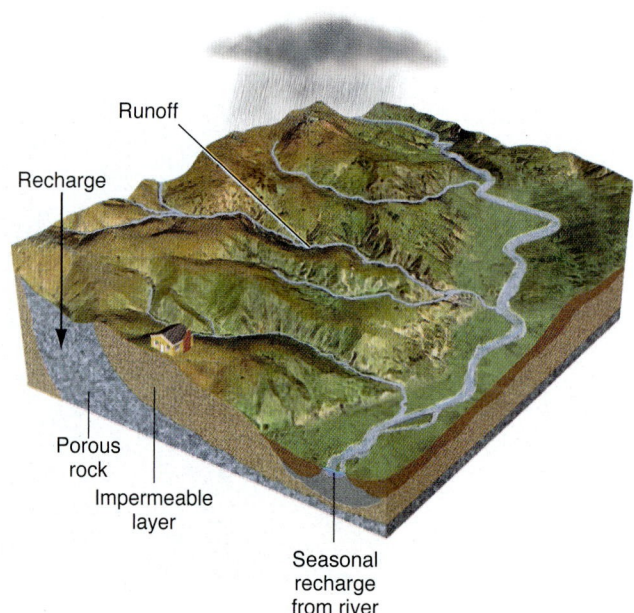

FIGURE 18.4 The groundwater system gains water from recharge areas that may be located in regions of high humidity or near lakes, streams, and wetlands.

Labels in figure: Runoff · Recharge · Porous rock · Impermeable layer · Seasonal recharge from river

What are the most common ways to recharge groundwater?

Q EXPAND YOUR THINKING

If you were the water manager for the city you live in, how would you determine the recharge area for the groundwater that your citizens drink?

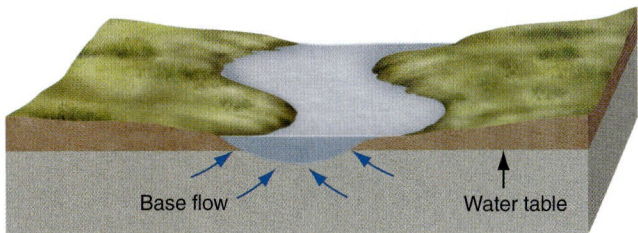

(a) Gaining stream

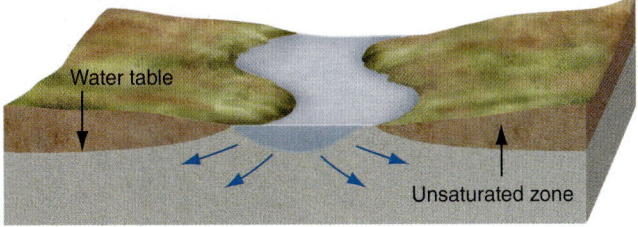

(b) Losing stream (connected)

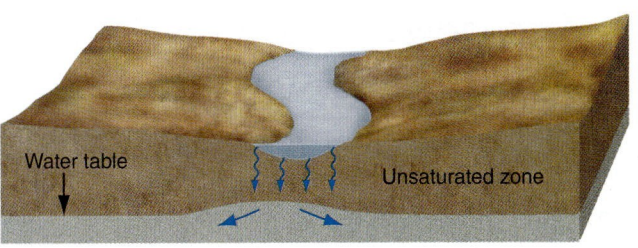

(c) Losing stream (disconnected)

FIGURE 18.5 (a) Gaining streams experience base flow where groundwater flows into the stream. (b) Losing streams contribute a portion of their discharge to the groundwater table by seeping through the channel bed. (c) The water table often bulges upward beneath a losing stream where the water table is not connected to the channel bed.

Under what conditions would a stream receive base flow during part of the year but provide recharge at other times?

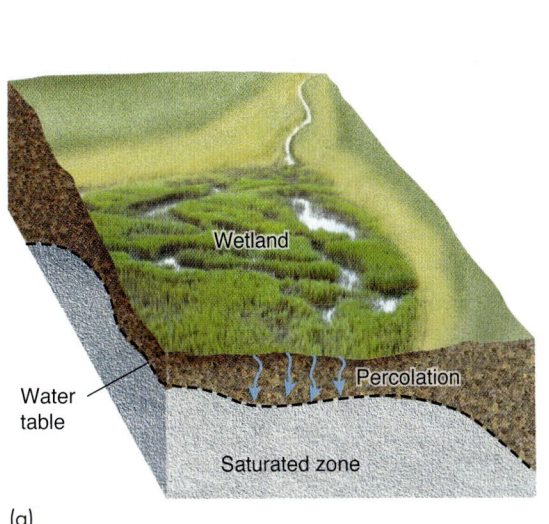

(a)

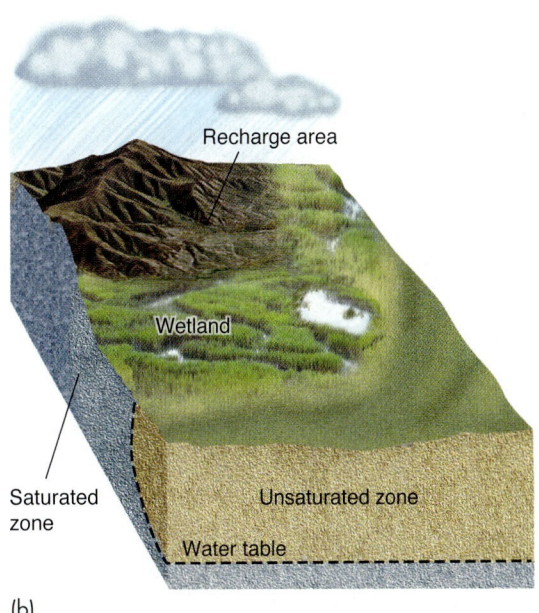

(b)

(c)

FIGURE 18.6 (a) A wetland may recharge groundwater. (b) Groundwater may contribute water to a wetland. In this case, the wetland is essentially the surface outcropping of the water table. (c) Some wetlands are fed by groundwater discharge on their upslope side; that water then recharges the groundwater table on the downhill side.

How does the water table contribute to wetland development?

18-3 Groundwater Moves in Response to Gravity and Hydraulic Pressure

LO 18-3 Identify the factors that influence groundwater movement.

Groundwater flows in response to hydraulic pressure, moving from regions of high pressure to regions of low pressure.

Groundwater Movement

Groundwater may move in response to gravity as it seeps downward toward the water table; in most cases, however, groundwater flows in response to **hydraulic pressure.** That is, it migrates from regions of high pressure to regions of low pressure. This enables it to move upward (against gravity) within the crust (**FIGURE 18.7**); but because it has to travel along tortuous pathways within rock and sediment, it does not progress as quickly as it would on the surface.

Although water exists everywhere beneath the water table, some parts of the saturated zone contain more than others. An **aquifer** is saturated crust that can produce useful quantities of water when tapped by a well. Aquifers may be small, extending under only a few hectares, or very large, underlying thousands of square kilometers of Earth's surface.

The capability of rock or sediment to transmit groundwater is called **permeability**, and it depends on the size of the spaces in an aquifer and how well the spaces are connected. Aquifers typically consist of highly permeable materials, such as gravel, sand, sandstone, or fractured rock such as limestone. These are permeable because they contain large connected spaces that allow water to flow with relative ease. In some permeable materials, groundwater may move several meters in a day; in others, it may move only a few centimeters in a century (**FIGURE 18.8**).

The amount of empty space within an aquifer is known as **porosity**. In sedimentary rocks, porosity depends on cementation, sorting, and the shape of the grains. Large, well-rounded grains have high porosity and so can store abundant water. Poorly sorted sediments have lower porosity. Cements, which crystallize in the pore spaces between grains, tend to result in lower porosity (**FIGURE 18.9**).

Porosity is an important characteristic that may control permeability. Highly porous rock or sediment with connected pores is permeable. But when the pores are not connected, permeability is low.

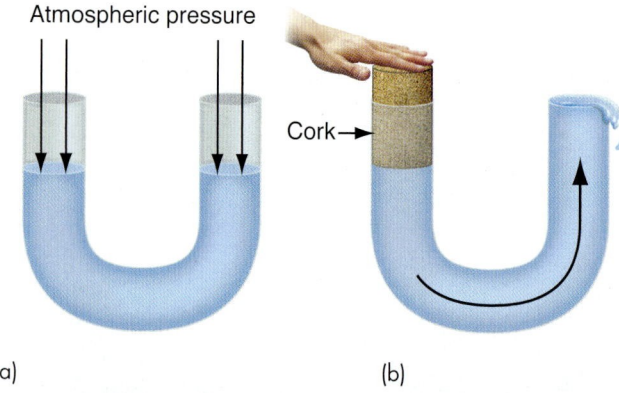
(a)　　　　　　　　(b)

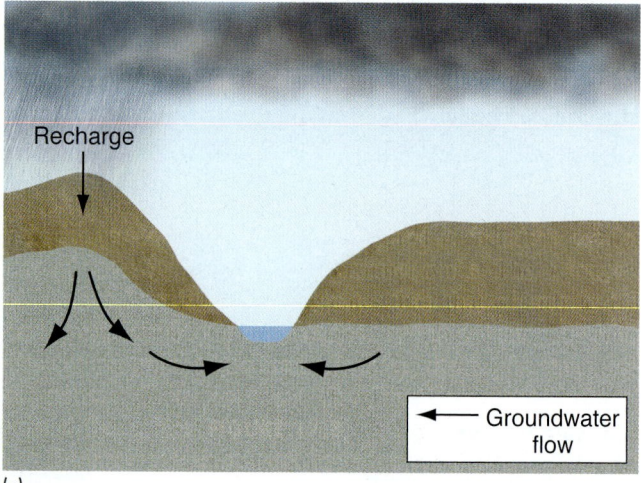
(c)

FIGURE 18.7 Groundwater moves from areas of high pressure to areas of low pressure. (a) Water will not flow if there is an even distribution of pressure acting on the surface. (b) When pressure is applied to one part of the surface, the water will flow toward an area of lower pressure. (c) Recharge creates an area of high pressure.

Q **What processes may control the amount of time water spends in the ground?**

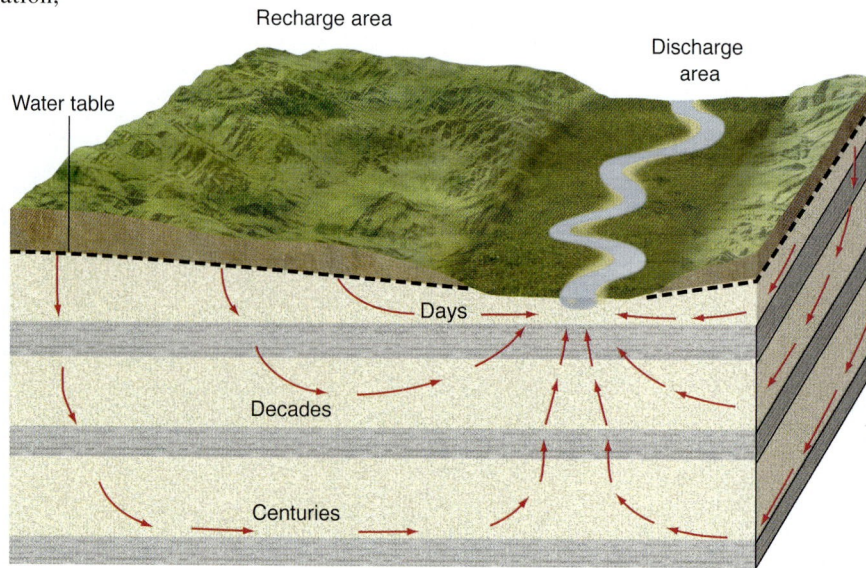

FIGURE 18.8 Groundwater deep in the crust tends to be older (has been there longer) than water that is closer to the water table.

 What factors influence permeability?

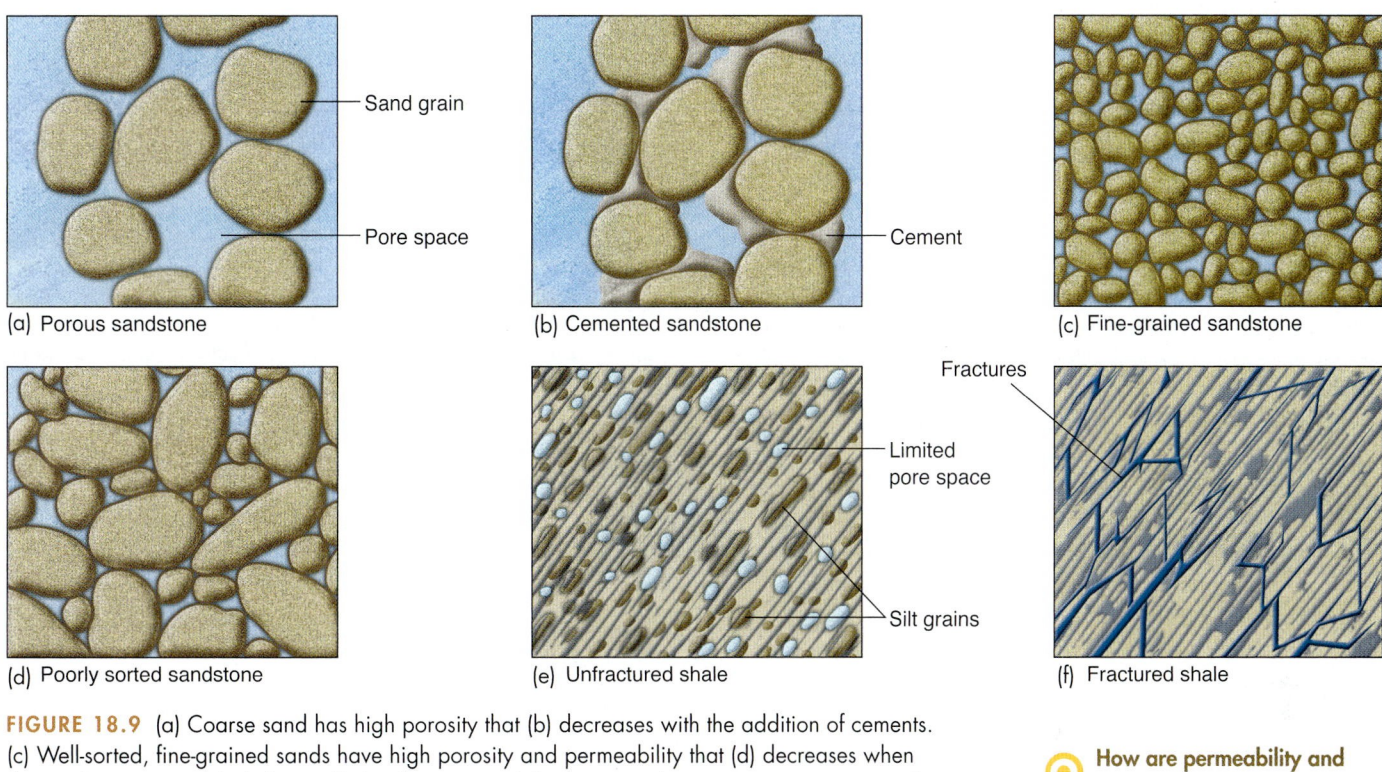

(a) Porous sandstone — Sand grain, Pore space

(b) Cemented sandstone — Cement

(c) Fine-grained sandstone

(d) Poorly sorted sandstone

(e) Unfractured shale — Limited pore space, Silt grains

(f) Fractured shale — Fractures

FIGURE 18.9 (a) Coarse sand has high porosity that (b) decreases with the addition of cements. (c) Well-sorted, fine-grained sands have high porosity and permeability that (d) decreases when there is less sorting. (e) Shale usually has low permeability because the pores are not connected. Permeability can increase (f) if the shale is fractured, with the fractures opening new pathways for the movement of groundwater.

How are permeability and porosity related to one another?

Take time now to test your growing knowledge of the elements of groundwater by taking the Critical Thinking exercise, "Groundwater Dynamics."

Hydraulic Gradient

Clay, shale, and other fine-grained rocks and sediments may have high porosity and low permeability because the many pore spaces in clay are not connected. Consequently, clay tends to inhibit groundwater flow. In igneous and metamorphic rock, porosity and permeability are typically very low due to the intergrown nature of the minerals. However, if the rock is fractured, the lines of breakage can connect, enabling high permeability to develop.

The rate at which groundwater moves depends on both the permeability of the rock and the **hydraulic gradient**, the difference in elevation between two points in an aquifer divided by their distance (**FIGURE 18.10**). The velocity (V) of groundwater flow is defined as:

$$V = K(h_2 - h_1)/L$$

The value K is the *coefficient of permeability,* and it describes the permeability of the aquifer. The hydraulic gradient, $(h_2 - h_1)/L$, is essentially the slope of the water table between two points, h_2 and h_1. L is the distance between the two points. If V is multiplied by an area, (A), through which water flows [$Q = AK(h_2 - h_1)/L$], the groundwater discharge (Q) is defined. This relationship is known as **Darcy's law**, named for French engineer *Henry Darcy* (1803–1858), and is usually measured in cubic meters per day.

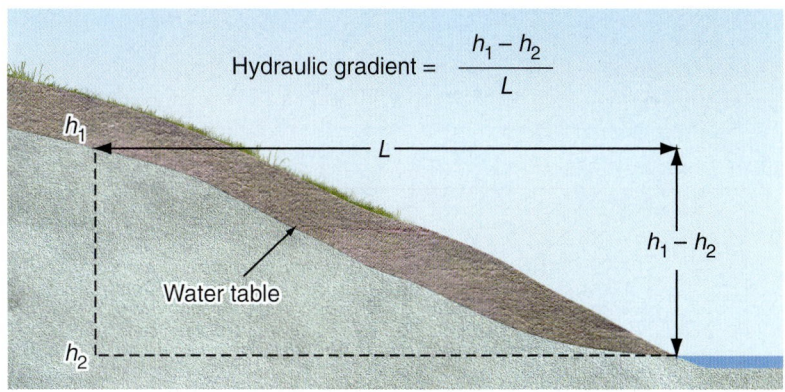

Hydraulic gradient = $\dfrac{h_1 - h_2}{L}$

FIGURE 18.10 Calculated according to Darcy's law, the hydraulic gradient is the difference in elevation between two points on the water table divided by their distance from each other.

How might Darcy's law be useful in planning a program of groundwater conservation?

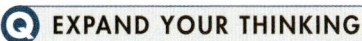

EXPAND YOUR THINKING

Can groundwater flow "uphill" underground? Why or why not?

CRITICAL THINKING

Groundwater Dynamics

The fact that groundwater flows from areas of high pressure to areas of low pressure makes it possible to draw maps showing the direction of groundwater flow using the surface contours of the water table. More specifically, the elevation of the water table can be mapped using wells, springs, lakes, and other evidence of its height. When the surface of the water table is contoured with lines that connect points at the same elevation, a line drawn perpendicular to the contours is assumed to indicate the direction of flow. This assumption is based on the hypothesis that water will flow from higher elevations toward lower elevations (**FIGURE 18.11**); and on the water table, different elevations represent differences in pressure—that is, like runoff, groundwater flows across contours rather than along them.

FIGURE 18.12 contains a map of an area in Florida where *limestone crust* has been heavily *karstified:* chemical weathering has dissolved the rock and created many caverns, sinkholes, and depressions. The climate in the area is humid, so there is high annual rainfall. Groundwater has filled the depressions and caverns with a high water table, such that numerous lakes and ponds follow the shape of the weathered crust. These various bodies of water are connected by the water table.

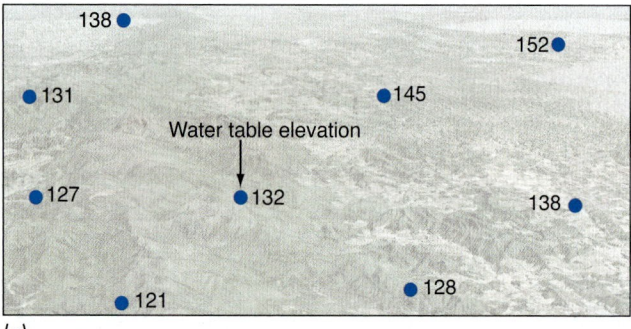

(a)

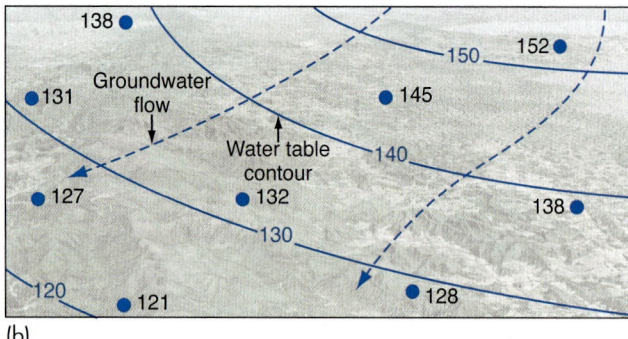

(b)

FIGURE 18.11 Using data from wells, springs, and other sources, it is possible to make a contour map of the water table. In most cases, groundwater will flow perpendicular to the contour lines.

The map shows several features:

- Two roads: CCC Road and Route 52
- A small town located at the intersection of Route 52 and CCC Road
- Elevations of the water surface in lakes and ponds of the region
- A wetland, shown in dark green
- Cedar Lake and Long Lake, popular locations for recreation and swimming
- Clear Lake, which supplies the town with drinking water
- One contour line of the water table at a 20-meter elevation

Based on the preceding information, and referring to Figures 18.11 and 18.12, do the following:

1. Contour the water table: With a pencil, add additional contours to this map using a 2-meter contour interval. The 20-meter contour line has been added to help you get started.

2. Draw dashed arrows predicting groundwater flow. Keep in mind that groundwater flow follows the hydraulic gradient: it flows along the shortest path from a higher elevation to a lower elevation of the water table (see Figure 18.11).

3. Identify major drainage directions in the water table. Place two or three long, bold arrows on the map showing them, and assign them a compass direction at the tip of the arrow (e.g., NW, SSE, etc.).

4. Calculate the average gradient along each of your drainage directions in meters of elevation change per kilometer of distance. Write these on the map. Which direction has the steepest gradient? What are the implications of this with regard to groundwater movement?

5. Imagine that a southbound tanker truck carrying pesticide is rounding the bend on CCC Road north of the town. Traveling too fast, it tips over, spilling its contents into the wetland (shown in dark green). Once the poison enters the groundwater system, in which direction will the toxic plume move? Note in particular that the town at the intersection of CCC Road and Route 52 draws its water from Clear Lake. Is its water supply in danger? As a consultant, propose a method for cleaning up the groundwater and removing the contamination.

6. The town is planning to build a new sewage processing plant and needs to identify a safe location. The plant will treat liquid waste from the town to the "secondary level." This means that after treatment, the effluent will not be drinkable but neither will it be highly toxic. Natural cleansing and dilution in the groundwater system ultimately will clean it, so the plant will inject its treated sewage into the ground. Nevertheless, it is important that the sewage be injected where it will not end up in the drinking supply or in recreational waters. Yet the plant must be near to the town so that building a

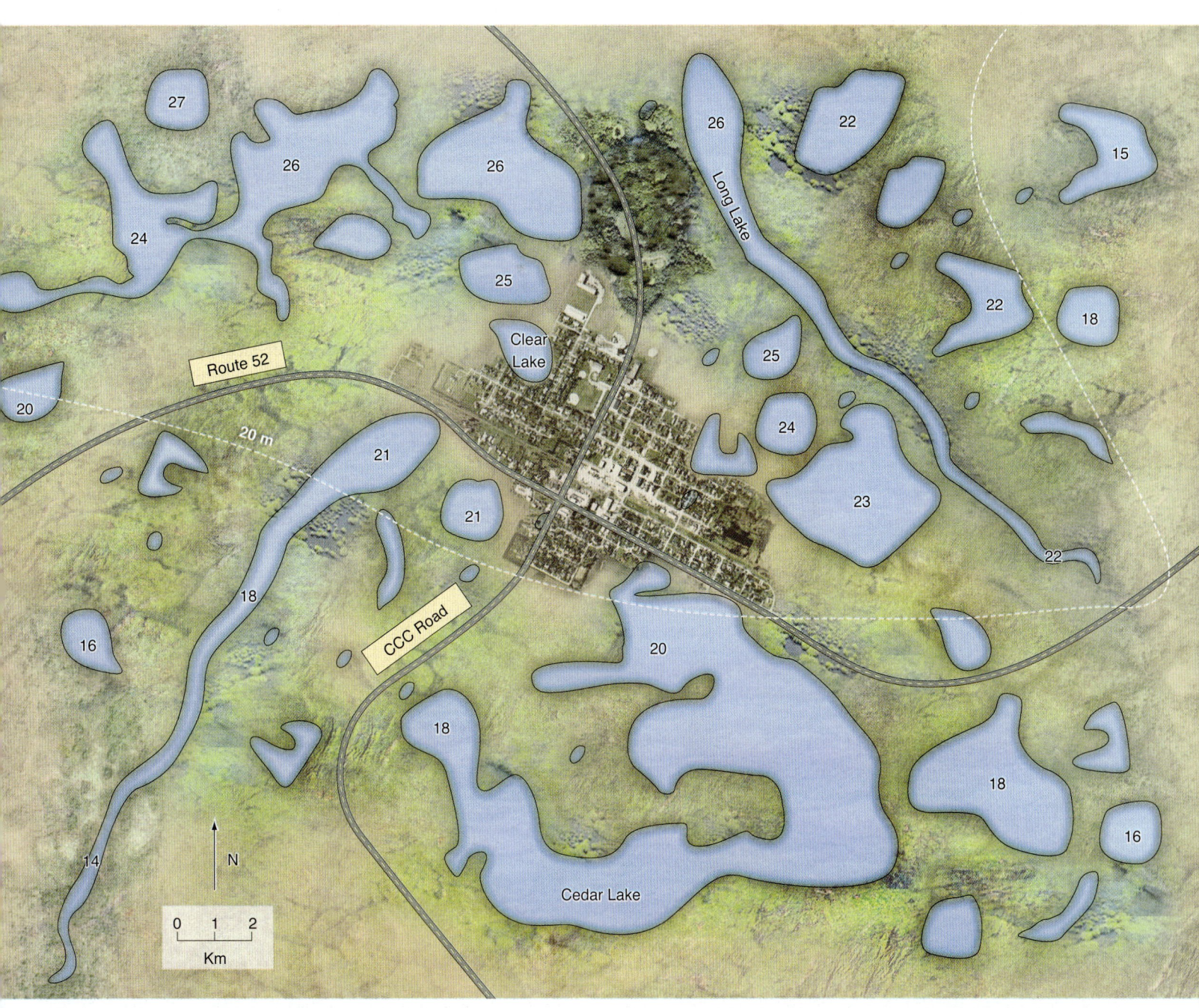

FIGURE 18.12 Water table map.

sewage delivery pipeline to the plant is not too expensive. Where do you recommend that the plant and the injection well be located? What issues should you consider?

7. The town is planning to construct a landfill for solid waste disposal. It will be lined with clay and an impervious liner. It is important that the landfill be located so that leaks cannot enter the drinking water or interfere with the recreational use of Long Lake and Cedar Lake, both of which are popular places for swimming. Where do you recommend the town build the landfill? Why should it be lined with clay?

(18-4) Porous Media and Fractured Aquifers Hold Groundwater

LO 18-4 Compare and contrast different types of aquifers.

An aquifer—from two Latin words, *aqua* (water) and *ferre* (to bear or carry)—is an underground layer of water-bearing permeable rock or unconsolidated materials (gravel or sand) from which groundwater can be extracted using a water well.

Aquifers

Aquifers, which literally carry water underground, exist in several forms: a layer of gravel or sand, a layer of sandstone or fractured limestone, or a fractured layer of crystalline (igneous or metamorphic) rock. Geologists who study groundwater, known as **hydrogeologists**, generally define two types of aquifers in terms of their physical characteristics: *porous media* and *fractured aquifers*.

Porous media (**FIGURE 18.13**) are aquifers that consist of aggregates of particles, such as layers or beds of sand and gravel. These make excellent, high-yielding aquifers because the water is stored within, and moves through, the openings between individual grains. Porous media in which the grains are not connected to each other are considered *unconsolidated*, whereas those with cemented or compacted grains are called *consolidated*. Sandstones are examples of consolidated porous media; loose sands are examples of unconsolidated porous media.

Fractured aquifers consist of strata, typically of limestone, cemented sandstone, or crystalline rock, in which groundwater moves through cracks, joints, or fractures in otherwise solid rock. Limestone often forms fractured aquifers where the cracks are the result of chemical weathering due to groundwater dissolution; over time, these cracks may expand into large channels or even caverns.

Porous media such as sandstone may become so highly cemented or recrystallized that all of the original pore space is filled. In such cases, the rock is no longer a porous medium. However, if it contains cracks, it still can act as a fractured aquifer. Igneous and metamorphic rock that is highly fractured or jointed also can act as a fractured aquifer.

Layers of impermeable rock or sediment form an **aquitard,** an important factor in determining the movement and location of usable groundwater. Aquitards usually are made up of layers of shale or clay that prevent the free flow of groundwater in an aquifer (**FIGURE 18.14**). A *confined aquifer* is one that lies beneath an aquitard.

A special kind of confined aquifer, an *artesian aquifer*, is one in which water escapes to the surface, driven by the hydraulic gradient of the water table. An artesian aquifer has internal pressure because of the confining bed that prevents water flow. This pressure is due to the recharge area being at a higher level than the rest of the aquifer. The force of gravity pulls the higher water downward, creating a steep hydraulic gradient inside the aquifer. This is why **artesian springs** (a spring from which water flows under natural pressure without pumping) flow by themselves; the pressure forces the water out of the ground, and it will rise to a level known as the *potentiometric surface*.

An important environmental issue concerning groundwater is contamination, which you can learn more about by reading the Earth Citizenship feature titled "The Superfund Program."

(a)

(b)

Fletcher & Baylis,/Photo Researchers

Marli Miller

FIGURE 18.13 (a) Unconsolidated porous media such as these sand and gravel beds may form aquifers that are rich sources of groundwater. (b) Fractured aquifers are formed when cracks, joints, and fractures open up zones of permeability in otherwise impermeable rock.

What types of rock-forming environments would lead to the development of porous media and fractured aquifers?

EXPAND YOUR THINKING

Toxic chemicals have spilled onto the ground in your area. How will you determine whether your drinking water is threatened?

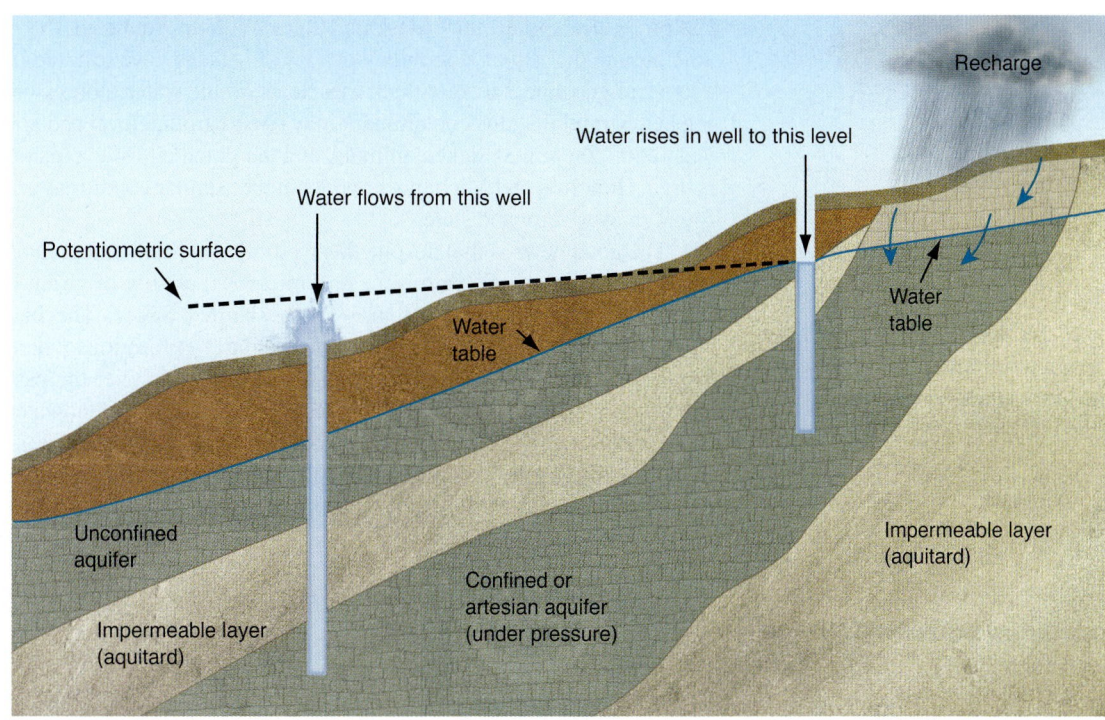

FIGURE 18.14 Water in a confined aquifer is restricted by an aquitard. An unconfined aquifer is one in which the water table can freely move up or down. An artesian aquifer develops when hydraulic pressure within an aquifer is high because the recharge area lies at a higher elevation than the aquifer. The high pressure causes the free flow of water through a spring if the confining bed is breached.

Q How does an artesian aquifer differ from a normal aquifer?

EARTH CITIZENSHIP

The Superfund Program

Throughout history, people have dumped chemicals and waste in the ground and disposed of them in shallow pits without regard to the possibility that it could lead to groundwater contamination. On tens of thousands of properties across the nation, these practices were common, continuous, and intensive. Now we recognize that disposing of toxic waste in the ground threatens human health and the environment, as shown and described in **FIGURE 18.15**.

Hazardous waste sites resulting from chemical dumping include abandoned warehouses, landfills, and undocumented disposal locations. Citizen concern over the extent of this problem led Congress to establish the **Superfund program** in 1980. The goal of the program is to locate, investigate, and clean up the worst toxic waste sites nationwide. The Environmental Protection Agency (EPA) administers the Superfund Program in cooperation with individual states and tribal governments. Within the EPA, the program is managed by the Office of Superfund Remediation Technology Innovation (OSRTI).

After more than a quarter century of effort, the Superfund Program has assessed 44,148 hazardous waste sites across the country and completed the total cleanup of 757 sites, with another 1450 in the final phases of cleanup. The Superfund Program also has taken more than 6400 actions to immediately reduce threats to public health and the environment. Over 70% of the sites it has placed on its National Priority List have cleanup work underway or completed by responsible parties.

A report to Congress lists the cost of carrying out the Superfund hazardous waste cleanup program at $14 to $16.4 billion from 2000

FIGURE 18.15 A bulldozer pushes soil away from one of the tanks used to hold toxic waste in Love Canal, New York. Throughout the 1950s and 1960s, the Hooker Chemical Company illegally dumped 21,000 tons of poisonous chemicals in an old canal. Later, the company sold the site to the Niagara Falls School Board; subsequently, the canal was filled with dirt, and an elementary school and a new suburb were built on it. After families in the area began to experience severe medical problems, it was found that the drinking water, soil, and even puddles of rainwater were filled with toxic waste. Now the town has been abandoned and the area is a Superfund site.

through 2009. Researchers suggest that direct remediation will account for at least 40% of the total estimated cost, but administration, staff, management, and support expenses will add up to another 18% to 25%. The annual Superfund budget is approximately $1.5 billion.

18-5 Groundwater Is Vulnerable to Several Sources of Pollution

LO 18-5 Identify ways by which pollutants enter the groundwater system.

About one-quarter of Earth's population drinks contaminated water. Recall that groundwater may move slowly and reside in the crust for long periods of time, two factors that make it vulnerable to contamination. Water is very effective at dissolving compounds and, therefore, is highly susceptible to carrying pollutants.

Pollution

Depending on the relative densities of water and a pollutant, groundwater is vulnerable to multiple sources of pollution that may float on the water table, dissolve within, or sink to the base of, an aquifer (**FIGURE 18.16**). It does not take much to pollute groundwater. Just 1 liter of oil can contaminate up to 1 million liters of drinking water or cause an oil slick extending over more than 8000 square meters. Among the many pollutant sources contributing to water contamination, one of the most common is underground gasoline storage tanks at gas stations, which may leak contaminants at every location where people drive cars. Alarmingly, it is thought that about 70% of all underground storage tanks leak gasoline and other toxic chemicals into the ground.

In Michigan alone, it is estimated that backyard mechanics dump more used automobile oil into the environment each year than the Exxon *Valdez* spilled into Alaska's Prince William Sound in 1989. And around the city of Honolulu, decades of leakage have resulted in a layer of gasoline 2 meters thick that floats on the water table. During heavy rainfalls, this contaminant may rise to ground level and kill vegetation; the fumes sicken animals, and the potential for explosion is high. Honolulu and Michigan are not unique; similar conditions are found in every city and state.

The good news is that despite these problems, the chemical (pollutants) and biological (pathogens and microbes) quality of groundwater is acceptable for most uses in the United States. The bad news is that in most urbanized areas, particularly in locations where groundwater is shallow, water quality has deteriorated over the past century due to human activities. The best way to care for groundwater is to understand the natural and human processes that influence it. Only by continuing to learn about this most precious of resources can we recover, sustain, and conserve it for the future.

Saltwater Intrusion

Another water contaminant issue is that, at present, 6 out of 10 people live within 60 kilometers of a seacoast, and more than two-thirds of the population of developing countries lives in the vicinity of the sea. The growing concentration of human settlements in coastal areas imposes heavy pressure to use coastal groundwater, leading to the *overpumping* of aquifers.

When water is pumped from an aquifer at a rate that exceeds the permeability of rock in the area, the water table around the well will lower. This process creates a **cone of depression** that steepens the hydraulic gradient and rapidly draws groundwater toward the well (**FIGURE 18.17**). The cone of depression will enlarge as long as the aquifer is unable to restore pumped water at a rate that matches the rate of withdrawal.

Continued pumping at an excessive rate in coastal areas will cause *saltwater intrusion*, which damages the quality of the aquifer. Saltwater intrusion occurs when freshwater is withdrawn faster than it can be recharged near a coastline. Saltwater generally intrudes upward and landward into an aquifer and around a well, although it may occur "passively" whenever there is a general lowering of the water table near a coastline.

The *transition zone* at the base of the aquifer (the interface where freshwater naturally mixes with seawater as it is discharged to the sea) naturally descends landward, causing the aquifer to be shaped like a wedge at the coastline. Saltwater intrusion is usually first noticed when water wells tap into this zone.

Many communities in coastal areas have been, or potentially could be, affected by saltwater intrusion precipitated by heavy pumping of groundwater. Saltwater intrusion threatens major sources of freshwater that coastal communities depend on. Sea level rise occurring as a result of global warming accelerates this process by raising the transition zone under the aquifer.

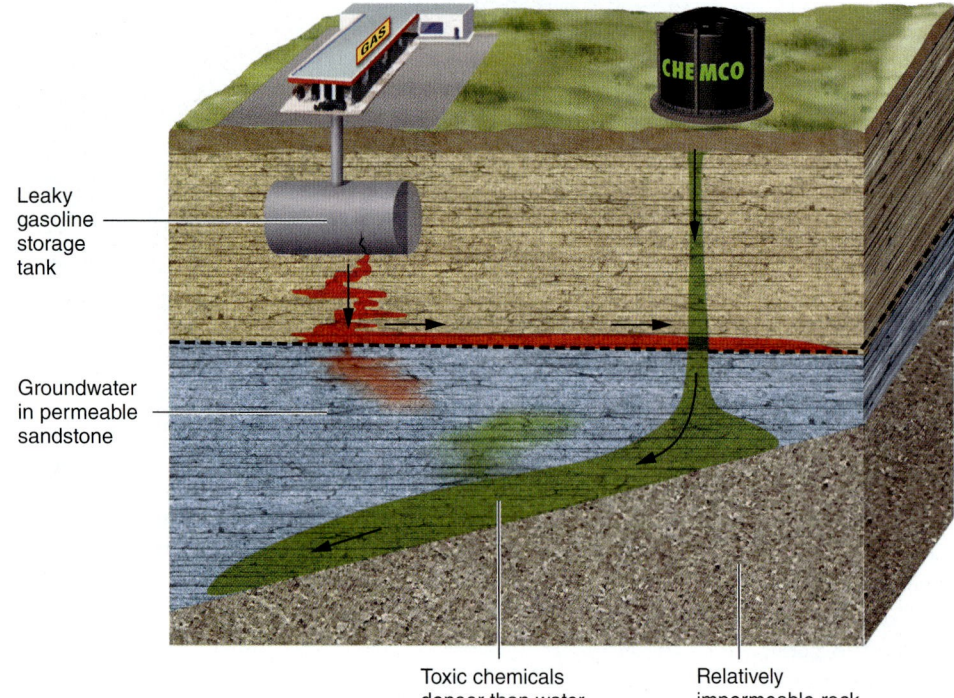

Leaky gasoline storage tank

Groundwater in permeable sandstone

Toxic chemicals denser than water

Relatively impermeable rock

FIGURE 18.16 Differences in density between freshwater and chemical contaminants determine where a pollutant is concentrated. Pollutants also easily dissolve into the main water body of an aquifer.

🔎 **Describe how the density of a pollutant governs its role in contaminating groundwater.**

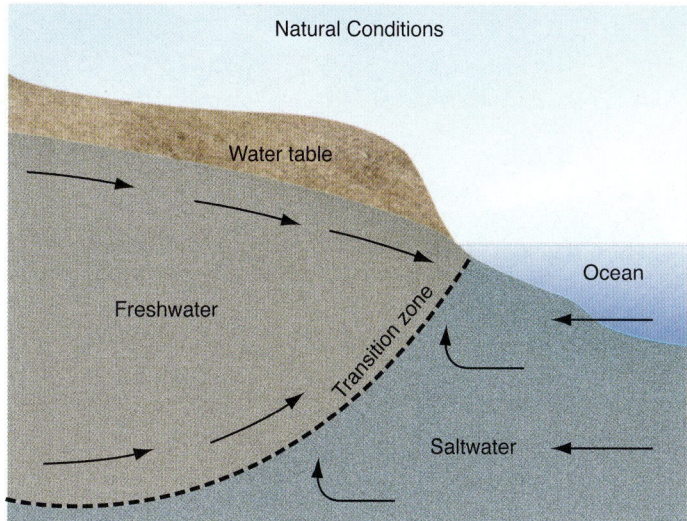
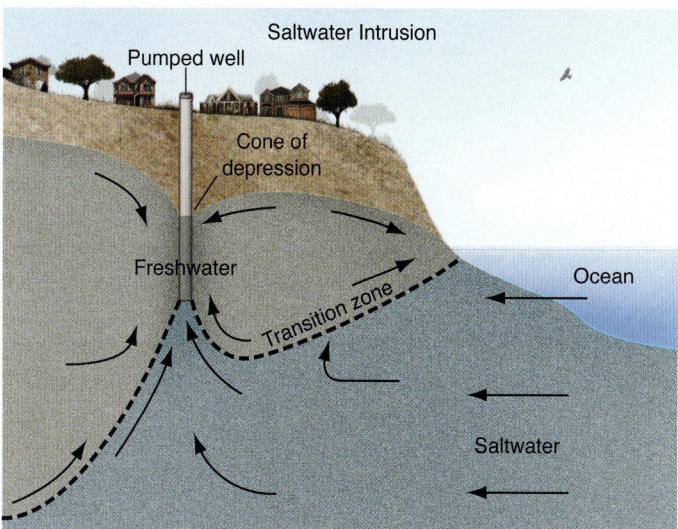

FIGURE 18.17 Population growth in coastal areas puts a high demand on groundwater, leading to saltwater intrusion and deterioration of water quality.

Describe how sea level rise and population growth combine to threaten coastal groundwater resources.

Dissolved Minerals

Another characteristic of water is that it is an effective solvent, giving it the capability to dissolve minerals from the rocks and sediments it encounters. For this reason, most groundwater contains dissolved minerals and gases that give it the slightly tangy taste many people enjoy. Without these dissolved substances, the water would taste "flat."

The most commonly dissolved constituents include the elements sodium (Na), iron (Fe), calcium (Ca), magnesium (Mg), potassium (K), and chloride (Cl), and the compounds bicarbonate (HCO_3) and sulfate (SO_4). Most water tastes bad if the level of dissolved minerals exceeds 1,000 milligrams per liter. When the level of dissolved constituents reaches a few thousand milligrams per liter, the water tastes salty, but it still may be used in areas where less mineralized water is not available.

Groundwater that contains high levels of dissolved minerals, known as *hard water*, often leaves coatings of mineral precipitates (minerals that were formerly dissolved in the water and have formed a solid layer) on plumbing fixtures. Some wells and springs contain very high concentrations of dissolved minerals and cannot be tolerated by humans and other animals or plants. Another source of hard water is deep aquifers that contain very salty water.

Sewage Waste

Fortunately, in general, groundwater is less vulnerable to bacterial pollution than surface water because the soil, sediments, and rocks through which groundwater percolates in the unsaturated and saturated zones serve to screen out many types of bacteria. But bacteria do occasionally invade groundwater, sometimes in dangerously high concentrations.

A common cause of bacterial contamination is direct disposal of liquid sewage from a home or business into the ground. Most liquid sewage generated by human waste enters a community system of buried pipes under roads through which it is pumped to a local sewage treatment plant. However, many rural and suburban homes are not connected to community sewage disposal lines. Instead, these buildings dispose of their liquid waste via *septic systems* and *cesspools*, which rely on the properties of soils and rocks in the unsaturated zone to filter out harmful bacteria (**FIGURE 18.18**). In-ground sewage disposal sys-

tems like these allow liquid sewage to cleanse naturally as it percolates through the unsaturated zone. But if sewage enters an aquifer before it is completely purified, it may contaminate a source of drinking water.

The Earth Citizenship feature, "The Water We Use," offers much more information about water issues of worldwide concern.

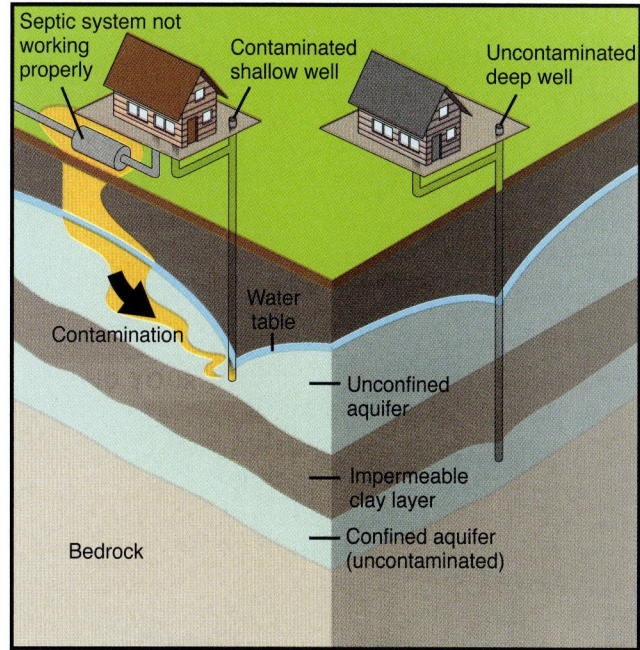

FIGURE 18.18 Domestic sewage in cesspools and septic systems may contaminate drinking water.

Why does a cone of depression tend to draw pollutants toward a well?

ⓠ EXPAND YOUR THINKING

As a consultant you are concerned about future freshwater availability for coastal towns in Florida. What data do you need to assess groundwater resources for the future? State a hypothesis you want to test.

EARTH CITIZENSHIP

The Water We Use

The United States consumes more water than any other nation. After a morning of showering, making coffee, using the toilet, and brushing our teeth, on average, we each use more than 30 gallons of freshwater. By the end of the day, after doing laundry, watering the lawn, and cleaning the dishes, we each use, on average, 150 gallons (**FIGURE 18.19**). Compare this to daily water use per capita elsewhere: England, 40 gallons; China, 22 gallons; and Kenya, 13 gallons.

Such statistics have led U.S. experts to agree that the demand for water is greater than the supply, and that groundwater levels in this country are falling faster than at any time in the past century. Managing water is, therefore, a growing concern in the nation, and communities all across it are starting to face challenges regarding water supplies and the need to update aging water treatment and delivery systems. Many of the states with projected population growth (**FIGURE 18.20**) also have higher per-capita water use and can expect greater demand for water resources.

Strains on water supplies and aging water treatment systems can lead to a variety of negative consequences for communities:

- Higher water prices, to ensure continued access to a reliable and safe supply

- Tighter water restrictions, to manage shortages

- Seasonal loss of recreational areas like lakes and rivers when the human demand for water conflicts with recreational needs

- Expensive water treatment projects, to obtain, transport, and store freshwater when demand exceeds available capacity

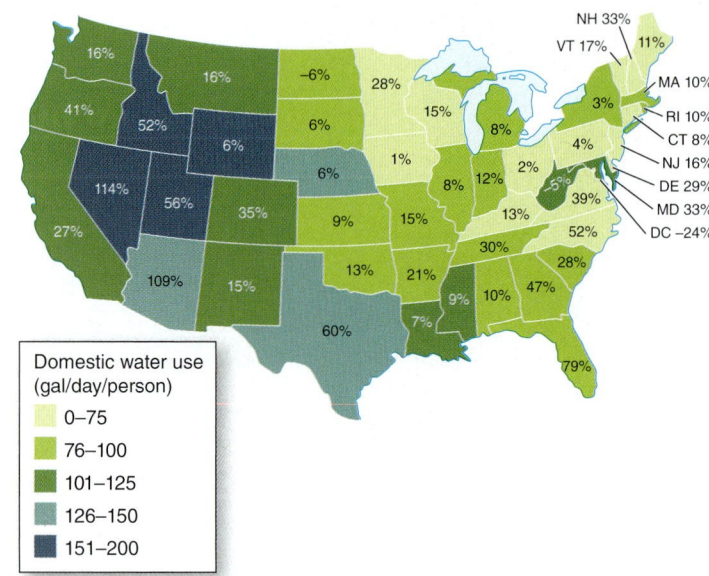

Domestic water use (gal/day/person)
- 0–75
- 76–100
- 101–125
- 126–150
- 151–200

FIGURE 18.20 U.S. water use in gallons per person per day and projected percent population change by 2030.

Another serious water-related problem is that climate change is leading to reduced rainfall in semiarid and arid regions of the United States and elsewhere in the world. Incidences of drought are spreading, and a freshwater crisis looming, especially in areas where water resources have been scarce in the past. Over the last few years, 36 states have declared water shortages, and much of the nation has been suffering an unrelenting drought, which has been compared to the 1930's Dust Bowl period (**FIGURE 18.21**).

In this country, 23% of freshwater is supplied by groundwater (79 billion gallons per day), and 77% is supplied by surface water (270 billion gallons per day). This water was put to use in many ways, including: generating electricity, agriculture irrigation, public water supply, industrial purposes, aquaculture, home consumption, and for livestock. But it is the thermoelectric power industry that "uses"—as opposed to "consumes"— the most freshwater in the United States (41%), to cool its electricity-generating equipment. It is important to make the distinction between "use" and "consume" in this case, for the water used by the thermoelectric power industry is usually returned to the water bodies it was taken from; thus, most of it is not actually consumed. *Consumption* is defined as the amount of water withdrawal that is not returned to the source, and generally lost to evaporation.

Irrigation, in contrast, is the overall largest consumer of freshwater, at about 81% of all fresh surface water (**FIGURE 18.22**). Put another way, nearly two-thirds of the fresh groundwater withdrawals in the U.S. are for the purpose of irrigation. More than one-half of the groundwater for irrigation is withdrawn in just four states: California, Nebraska, Arkansas, and Texas. In 25 states, irrigation is the largest consumer of fresh groundwater. To understand why, consider that it takes approximately 6 gallons of water to grow a single serving of lettuce, 49 gallons to produce just one 8-ounce glass of milk, and more than 2600 gallons of water to produce a single serving of steak. Nationwide, groundwater withdrawals for irrigation are about 3.5 times greater than for public supply.

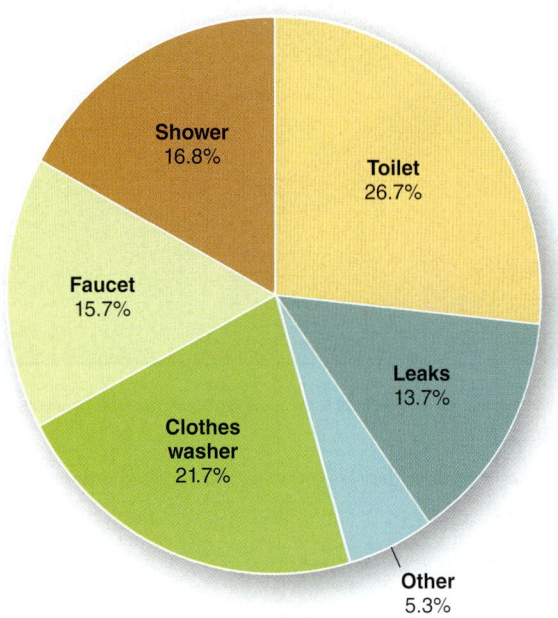

FIGURE 18.19 Americans use large quantities of water both inside and outside their homes. The average family of 4 can consume 400 gallons of water every day, approximately 70% of which, on average, is used indoors.

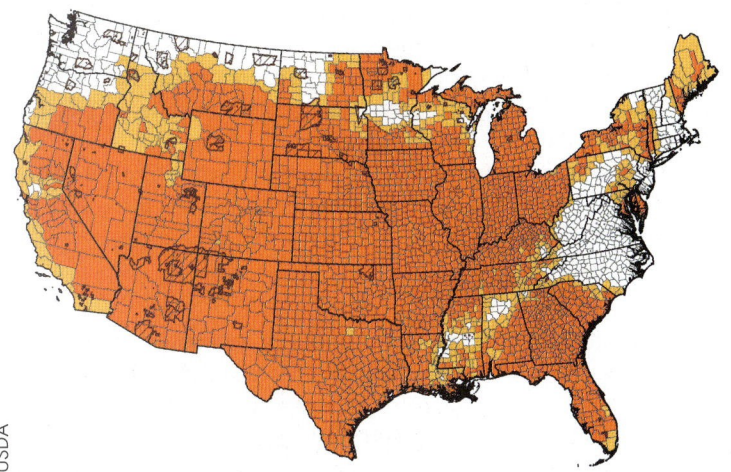

FIGURE 18.21 U.S. counties (red; yellow, neighboring counties) officially designated by the U.S. Department of Agriculture as in drought conditions during 2012.

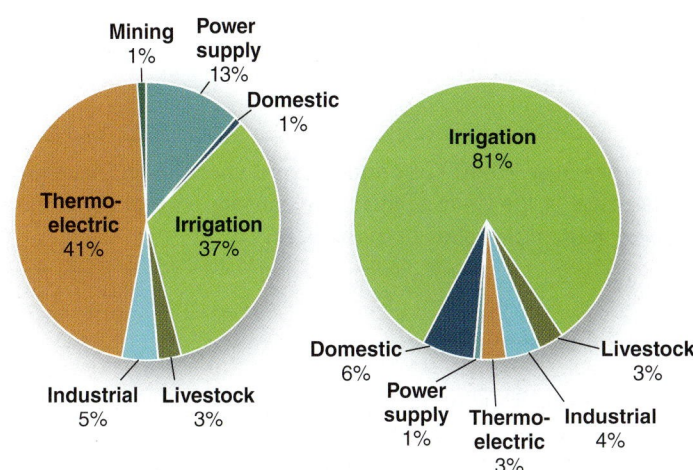

FIGURE 18.22 (a) Total annual freshwater withdrawals in the United States (b) freshwater consumption.

Internationally water withdrawals, too, are predicted to increase, by 50% by 2025 in developing countries and 18% in developed countries. Some of the current pressure on water resources comes from growing demands for animal feed to support meat production, which requires 8 to 10 times more water than grain production. But irrigation is not the only source of pressure on water resources. Over 1.4 billion people currently live in river basins where their water use exceeds minimum recharge levels, leading to the desiccation of rivers and depletion of groundwater. In 60% of European cities with more than 100,000 people, groundwater is being used at a faster rate than it can be replenished.

The preceding are just some of the water use—or, more accurately, overuse—issues tied to a global freshwater crisis, one that is being exacerbated by climate change (**FIGURE 18.23**). It is projected that by the year 2050, on all continents and in both developed and developing nations, the decline in water availability will reach 20% and more.

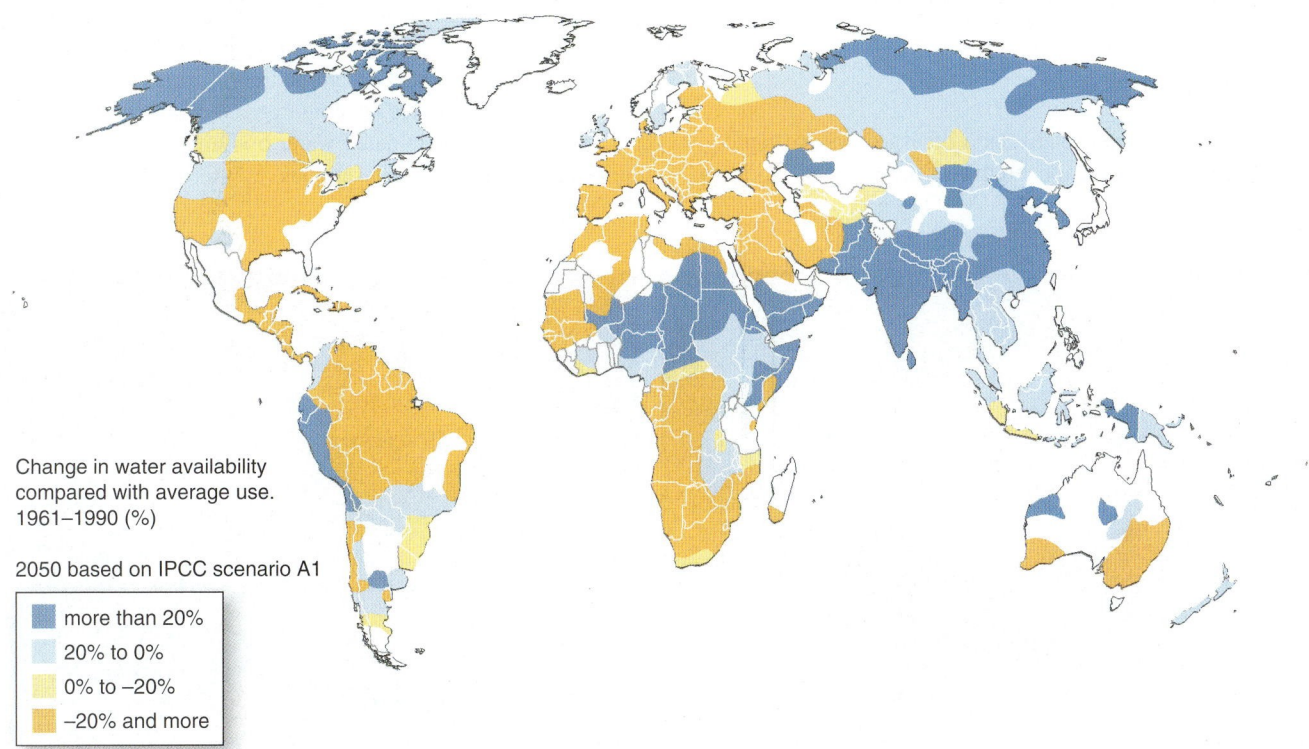

Change in water availability compared with average use. 1961–1990 (%)

2050 based on IPCC scenario A1

- more than 20%
- 20% to 0%
- 0% to −20%
- −20% and more

FIGURE 18.23 The contribution of climate change to declining water availability in the year 2050 compared to the average of 1961–1990, in percentage, based on projections by the Intergovernmental Panel on Climate Change (IPCC).

18-6 Common Human Activities Contaminate Groundwater

LO 18-6 List human activities that contaminate groundwater.

There are over 20,000 abandoned and uncontrolled hazardous waste sites in the United States, the consequence of contamination by human activities and practices.

Liquid Contamination

Before the enactment of modern laws controlling waste disposal and monitoring water quality, it was common practice to dispose of liquid and solid wastes in the ground. Even today illegal dumping of chemicals and contaminants continues on hidden and unused lands. Illegal disposal practices range from disposal of hundreds of drums of highly toxic chemicals to dumping used engine oil in the backyard (**FIGURE 18.24**).

Often, a simple hole in the ground is used to hide disposed chemicals. But containers eventually decay and leak their toxic contents into the ground. Common contaminants include old engine oil, pesticides, paint thinners, cleaning compounds, lubricants, inks, and manufactured toxic substances. All liquids dumped in the ground will percolate through the unsaturated zone, often accelerated by rainfall and melting snow, and form a **contaminant plume**. A contaminant plume will follow either the same hydraulic gradient that drives groundwater flow in aquifer systems, or the hydraulic gradient leading to a well if there is a cone of depression in the area.

Salt Contamination

Salt, because it absorbs water readily and breaks down ice rapidly, is commonly spread on roads and highways all over the world to combat icy conditions and make driving on them safer. Unfortunately, the resulting salty water turns into polluted runoff that percolates into the unsaturated zone and eventually enters the groundwater system, causing contamination.

High chloride levels are found in well water due to this problem, and drinking water may develop a salty, unpleasant taste. Fortunately, there is a solution: employ alternative deicing materials, such as sand; use salt more sparingly; and better manage salt stockpiles so that they do not pose a threat to drinking water.

To expand your knowledge of this issue, take the Critical Thinking exercise, "Your Drinking Water," at the end of this discussion.

Chemical Contamination

Since World War II, the growth of industry, technology, the human population, and with them, water use, has intensified the stress on both land and water resources. In many communities the quality of groundwater has been degraded. The chemical contamination of groundwater resources can be traced to multiple sources, among them: municipal and industrial wastes disposed in **landfills;** chemical fertilizers, herbicides, and pesticides used in agriculture; and underground storage tanks at gas stations (**FIGURE 18.25**).

Landfills are excavations in the ground meant to hold waste generated by a community. Older versions had few safeguards against contamination seeping out of landfills and into the groundwater system; modern construction techniques use clay barriers and waterproof linings to prevent this from happening, but leaks still may develop, and any of a wide variety of toxic liquids may leach out of a landfill. The list is long: bacteria, viruses, organic chemicals that can disrupt the human *endocrine system* (the system of hormone production and use), nitrates, dissolved metals, various cleansing and lubricating agents, and many other potentially dangerous chemicals.

The threat posed by pesticides, herbicides, and other poisons employed to kill pests, weeds, and insects is widespread. These products may be applied in excessive amounts or during inappropriate times in the crop-growing cycle, leading to groundwater contamination and consumption of contaminated water by humans, as well as to pollution of nearby streams and wetlands.

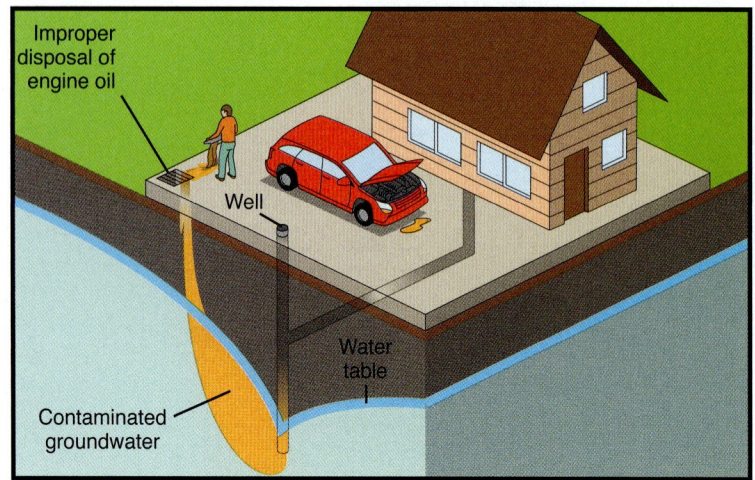

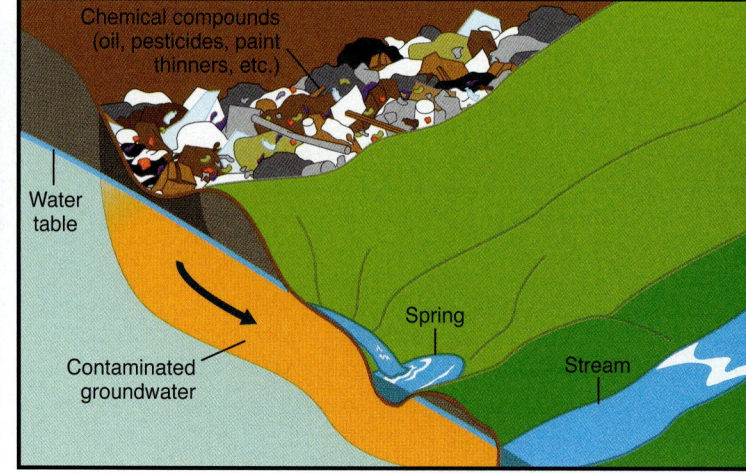

FIGURE 18.24 Dumping chemicals and other unwanted liquids in backyards and shallow pits creates a contaminant plume that may lead to groundwater contamination.

 What is a contaminant plume and how does it move?

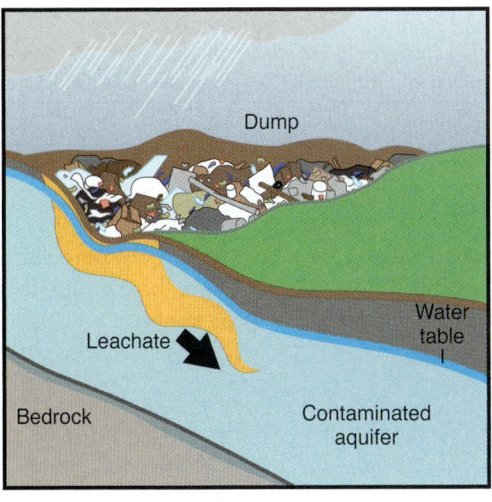 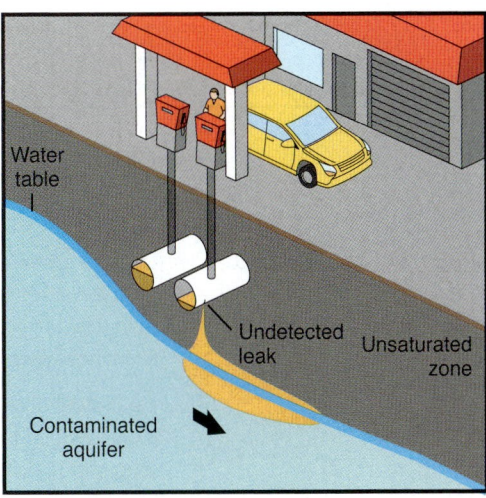 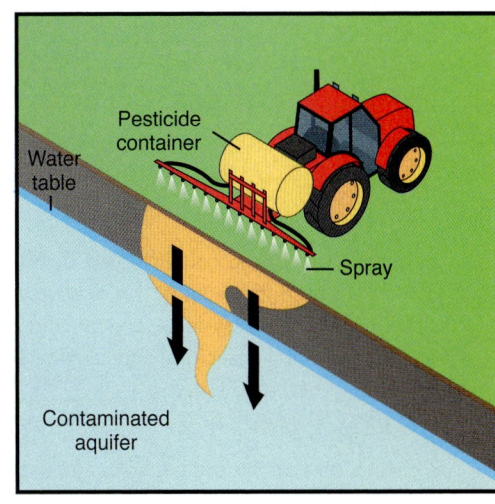

FIGURE 18.25 Chemical substances from landfills, underground storage tanks, and agricultural practices can pollute the groundwater system.

Describe a method of controlling contamination plumes by pumping groundwater. What aspect of the water table would you be manipulating?

Pesticide use in the United States can be broken down as follows: government, 7%; domestic, 8%; industrial and commercial, 17%, and agricultural, 68%. Pollution by pesticides can be reduced by employing best practices, which include: careful application of pesticides, restricted use near water wells, and implementation of alternative pest control methods.

The problem of underground storage tanks emerged as a result of the need of filling stations to store large volumes of gasoline. Since the boom in the automobile industry after World War II, many of these tanks have aged and developed leaks. Gasoline, of course, is highly toxic to plants and animals, making this a widespread problem. It is estimated that there are over 10 million underground storage tanks in the United States, and that 70% of them leak (**FIGURE 18.26**). Fortunately, today, programs exist to replace old tanks with newer ones made of durable and resistant materials. Bit by bit, this problem is being brought under control.

EXPAND YOUR THINKING

A contaminant plume probably has formed under a gas station in your community. As city water manager, how will you assess the threat to your city's drinking water?

Earth Gallery Environment/Alamy

FIGURE 18.26 To reduce groundwater contamination, leaky underground storage tanks at gasoline stations around the nation are being replaced.

What is the density difference between gasoline and water? Where is gasoline likely to collect in an aquifer?

CRITICAL THINKING

Your Drinking Water

Groundwater can be contaminated by chemical products, animal and human waste, and bacteria. Major sources for these pollutants include storage tanks, septic systems, animal feed lots, hazardous waste disposal sites, landfills, pesticides and fertilizers, road salt, industrial chemicals, and urban runoff.

Referring to **FIGURE 18.27**, work with a partner to complete the following:

1. Make a list of the sources of contamination in Figure 18.27.

2. How many of these sources are found where you live?

3. Can you list additional sources of groundwater contamination that are not shown in the figure?

4. The contamination plumes shown have city managers concerned. Formulate a method of controlling the plumes and eventually stopping their movement.

5. Do you agree that this area has historically lacked an overarching plan for managing water? Why or why not? Describe the necessary elements for such a plan. Write a list of steps to ensure future safe drinking water in this region.

6. As mayor of the city, you know that just controlling the spread of groundwater contamination (question 4) is a short-term fix. What longer-term actions do you propose for sustainably managing the water resource? Write your plan in the form of a press release.

7. Participate in a class forum on managing water.

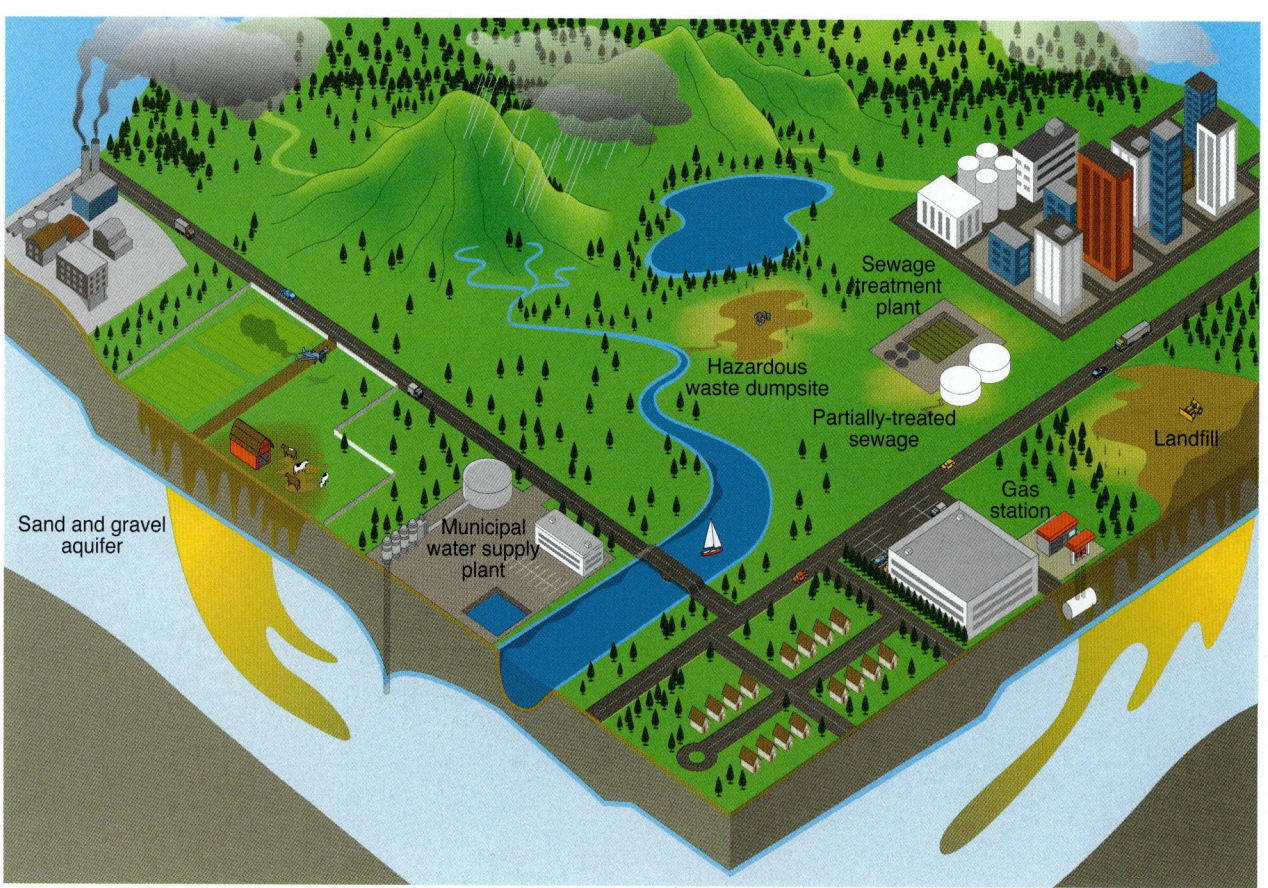

FIGURE 18.27 Potential sources of groundwater contamination.

18-7 Groundwater Remediation Incorporates Several Types of Treatment

LO 18-7 Explain methods of cleaning groundwater.

Widespread contamination of the groundwater we depend on for drinking, irrigation, and manufacturing has led to the development of a modern industry dedicated to groundwater remediation.

Remeditation

With over half of the U.S. population depending on groundwater for its drinking water, and as an important source of water for irrigation and manufacturing, it should be no surprise that the intense and often uncontrolled contamination of portions of the nation's groundwater supplies has led to the development of an entire industry dedicated to **groundwater remediation** (**FIGURE 18.28**).

Threats to groundwater are becoming more prevalent as the population grows, cities expand, and the consumer economy demands more goods and services. Estimates of remediation costs at U.S. government sites alone (such as military bases) range into the hundreds of billions of dollars. Thus, protecting the quality of groundwater supplies is a problem of broad societal importance.

Remediation methods are extremely expensive, and often it is difficult to predict how effective they will be. These methods rely, variably, on chemical, physical, and biological processes. Among the more commonly employed methods are:

- *Air sparging.* This method involves injecting gas (usually air or oxygen) into the saturated zone in order to mobilize contaminants that are volatile or easily stirred into a gaseous phase.

- *Air stripping.* With this method, groundwater is sprayed into the air so that it can vent off gases from pollutants such as gasoline and other petroleum-based pollutants. The water is then allowed to infiltrate back into the ground.

- *Directional wells.* Directional wells are drilled at near-horizontal angles in order to gain access to contaminant plumes and either pump them out or apply other methods of remediation.

- *Recirculation wells.* Employing this method requires circulating contaminated groundwater up to the surface and then back down into the ground. It allows gases to be vented off, pollutants to be filtered, or contaminated water to be filled with bubbles, thereby converting dissolved compounds into gases that can then be vented off.

- *Aquifer fracturing.* Techniques for enhancing permeability by fracturing aquifer rock yield improved access to pollutants so that they can be removed.

- *Injection wells.* By treating a contaminated aquifer with chemical additives, a specific pollutant can be cleansed or stripped. Injection wells of clean water or a chemical additive also are used to create an artificial hydraulic gradient in order to drive a contamination plume in a certain direction so that it can be vented, sparged, vacuumed, isolated, or treated in some other way.

- *Permeable reactive barriers.* This method involves using trenches or other barriers treated with a reactive substance in order to neutralize a pollutant. It often requires that a shallow aquitard be present to provide access to the entire aquifer.

- *Bioremediation.* Microbes and plant tissues can be used to neutralize various types of pollutants. Phytoremediation utilizes plants to absorb metals onto root systems and take up contaminants into leaves and stems. Other plants and microbes neutralize and decompose organic compounds into simple forms that are not hazardous.

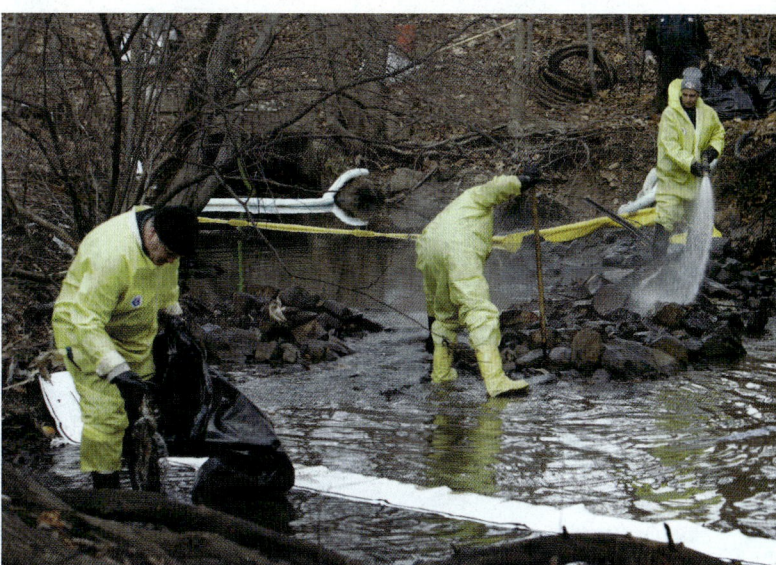

(a) USGS (b) © AP/Wide World Photos

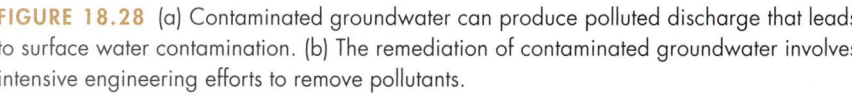

FIGURE 18.28 (a) Contaminated groundwater can produce polluted discharge that leads to surface water contamination. (b) The remediation of contaminated groundwater involves intensive engineering efforts to remove pollutants.

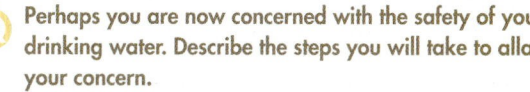

Perhaps you are now concerned with the safety of your drinking water. Describe the steps you will take to allay your concern.

- *Thermal or electrical treatments*. Injection of hot steam, heated water, radio frequency, or electrical resistance can cause the breakdown of various pollutants. Steam can mobilize volatile (gas-forming) compounds into a gaseous phase. Electrical probes in the ground can cause ions to migrate toward the probes, thereby breaking down chemical compounds.

Another technique, **natural attenuation,** is frequently employed to circumvent the high cost and logistical problems associated with engineered remediation. Basically, this approach involves shutting down any source of ongoing pollution, isolating a contamination plume, and allowing natural processes to take over and lower the toxic potential of contaminants. Although this method may be perceived by some as "walking away from the problem," often it is a rational approach based on cost and the decision that the aquifer can be abandoned.

A comprehensive source of information about major aquifers in the United States is available from the U.S. Geological Survey. Read more about it in the Geology in Our Lives feature, "Groundwater Atlas of the United States."

Q EXPAND YOUR THINKING

Describe the factors that should be considered in making a decision to allow natural attenuation of a contaminated aquifer.

GEOLOGY IN OUR LIVES

Groundwater Atlas of the United States

The *Groundwater Atlas of the United States*, published by the U.S. Geological Survey, provides a summary of the most important information available for each principal aquifer, or rock unit that will yield usable quantities of water to wells, throughout the 50 states, Puerto Rico, and the U.S. Virgin Islands.

The atlas is a comprehensive summary of the nation's groundwater resources and a basic reference on its major aquifers. It is useful to federal, state, and local officials with responsibilities for water allocation, waste disposal, and *wellhead protection* (safeguarding wells from contamination).

The atlas consists of an introductory chapter and 13 descriptive chapters, each covering a multistate region of the country. The principal aquifers, mapped as shown in **FIGURE 18.29**, occur in six types of permeable geologic materials: 1) unconsolidated deposits of sand and gravel, 2) semiconsolidated sand, 3) sandstone, 4) carbonate rocks, 5) interbedded sandstone and carbonate rocks, and 6) basalt and other types of volcanic rocks. Rocks and deposits with minimal permeability, which are not considered to be aquifers, consist of intrusive igneous rocks, metamorphic rocks, shale, siltstone, evaporite deposits, silt, and clay.

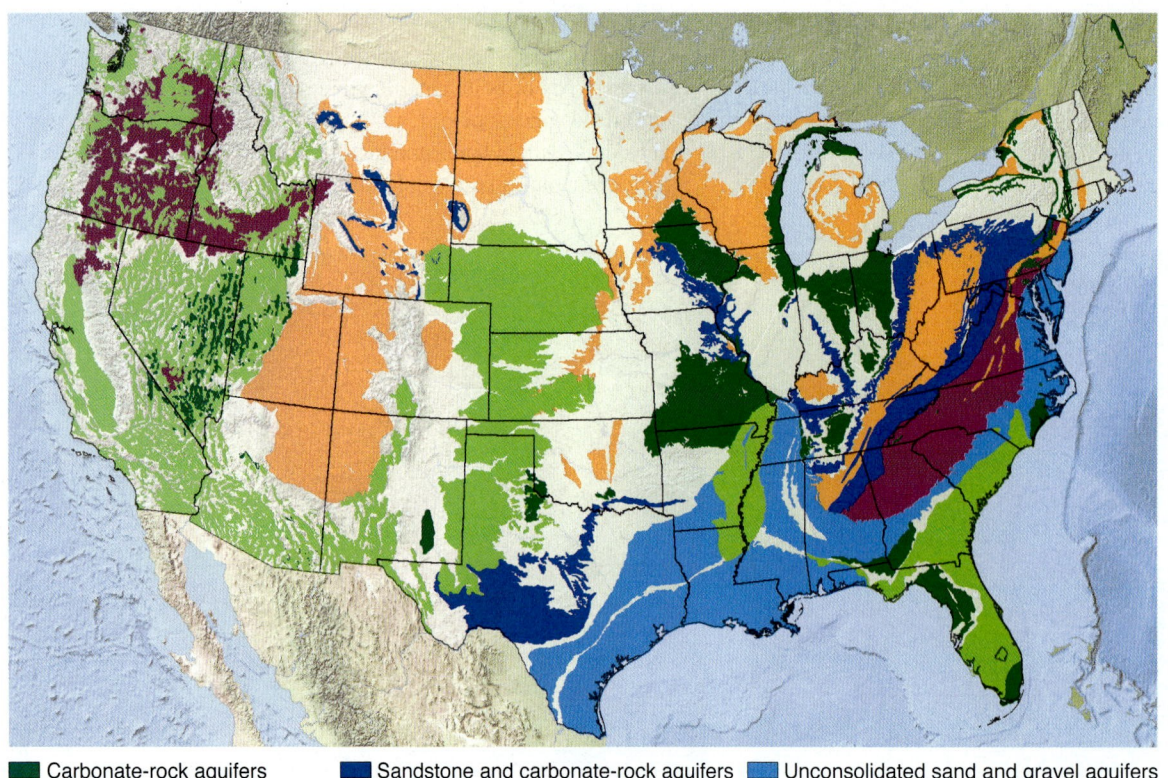

■ Carbonate-rock aquifers
■ Basalts and other volcanic-rock aquifers
☐ Other rocks
■ Sandstone and carbonate-rock aquifers
■ Sandstone aquifers
■ Unconsolidated sand and gravel aquifers
■ Semiconsolidated sand aquifers

Adapted from: Principal Aquifers of the 48 Conterminous United States, Hawaii, Puerto Rico, and the U.S. Virgin Islands.
Online linkage: http://nationalatlas.gov/atlasftp.html

FIGURE 18.29 Major aquifers of the United States as mapped in the *Groundwater Atlas of the United States*. *Source*: U.S. Geological Survey, 2009.

18-8 Groundwater Is Responsible for Producing Karst Topography and Springs

LO 18-8 Identify the major features of karst topography.

Groundwater is so effective at dissolving limestone that it can completely alter the topography of the land and the character of Earth's crust, adding to the landscape karst regions and natural springs.

Karst Geology

Groundwater has significant influence on the development of geologic features in the crust. The chemistry of water in the ground is capable of causing hydrolysis, dissolution, and the oxidation of many minerals. The dissolution mechanism is so strong that it can completely dissolve vast tracts of carbonate rock and modify the topography of the land and the character of the crust.

We first explored **karst geology** in Chapter 7 in our discussion of dissolution as a type of chemical weathering. Recall that karst is a distinctive geologic form that develops when groundwater dissolves *carbonate rock* (i.e., limestone, dolomite, marble). The dissolution process, occurring over many thousands of years, results in unusual surface and subsurface features, including sinkholes, vertical shafts, disappearing streams, subterranean drainage systems, caves, springs, and unusually steep hillsides and slopes.

Over time, as dissolution removes carbonate rock, and groundwater flows through dissolved rock with high permeability, aquifers are transformed from *diffuse-flow aquifers,* with water moving as laminar flow through small openings, into *conduit-flow aquifers,* with water moving primarily as turbulent flow through well-developed permeability channels called *conduit systems*.

Groundwater flow eventually advances toward discharge points at springs, submarine openings, and surface runoff sites. With time, the process of dissolution lowers the water table as broad cavities open in the aquifer and fill with water. When the water table sinks below the level of surface streams, runoff moves into cave systems below ground level, creating disappearing streams.

As more time passes, more of the surface drainage is diverted underground, and stream valleys virtually disappear, to be replaced by closed basins called **sinkholes**. Sinkholes vary from small cylindrical pits to large conical basins that collect and funnel runoff into karst aquifers. They mark the location of underground flow routes, aligning on the surface and signaling the presence of a subsurface cavern. As sinkholes coalesce, they eventually form a steep-walled valley, leaving the surrounding topography as a high-relief remnant of the former land surface (**FIGURE 18.30**).

Sinkholes may appear without warning. When the roof of a sinkhole collapses houses, roads, and businesses can disappear into the hole. This has caused the death of several people and is an expensive and common hazard in Florida and other locations with carbonate bedrock.

As explained in Chapter 7, the process of karst formation involves dissolution of carbon dioxide (CO_2) into the groundwater, creating a

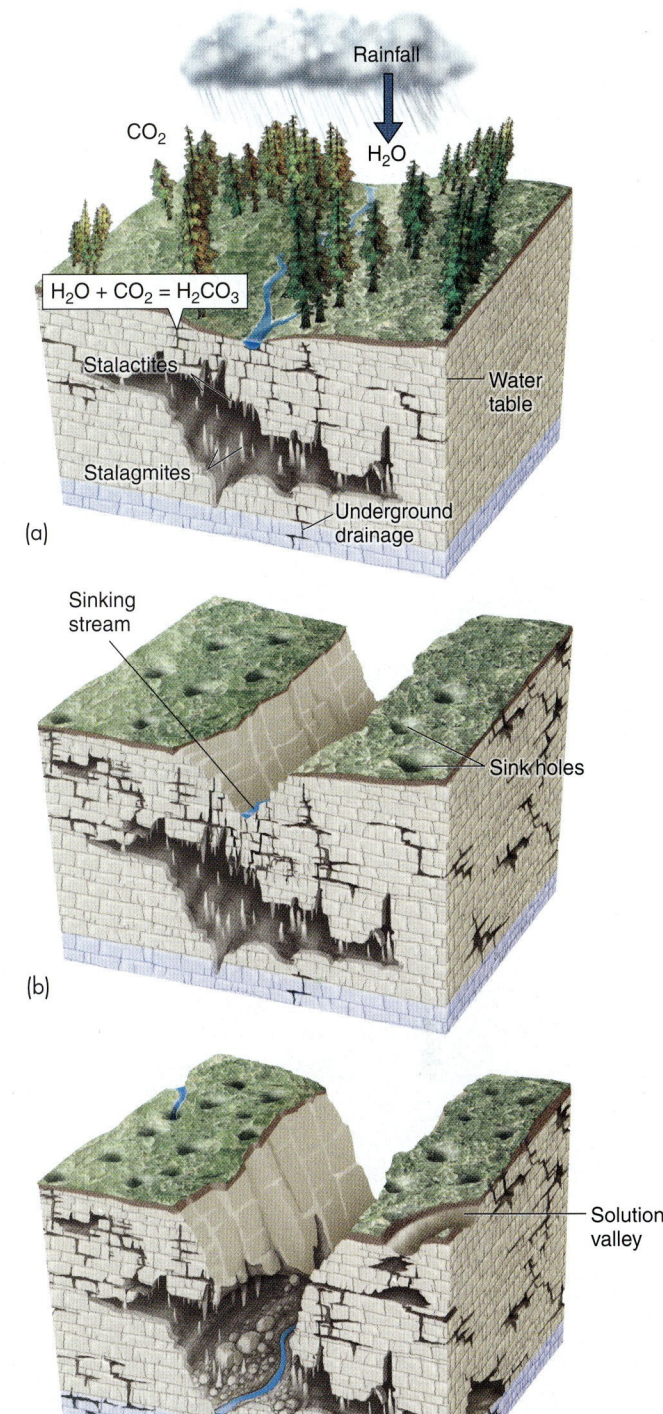

FIGURE 18.30 (a) Rainfall absorbs carbon dioxide in the atmosphere and soil and becomes carbonic acid. (b) Carbonic acid dissolves carbonate rock and thus contributes to the formation of karst topography (c).

mild *carbonic acid*. Carbonic acid starts out as rain falling through the atmosphere, picking up CO_2 that dissolves in the droplets. After it hits the ground and percolates through the soil, it accumulates more CO_2 from the decay of organic tissue produced by vegetation and animals. The CO_2-enriched groundwater (H_2O) in the unsaturated zone develops a weak solution of carbonic acid: $H_2O + CO_2 = H_2CO_3$.

Infiltrating water naturally exploits any cracks or crevices in the carbonate bedrock. Over time, with a continuous supply of CO_2-enriched water, carbonate bedrock begins to dissolve. Openings in the bedrock increase in size, and an underground drainage system develops that is a precursor to a type of cavern growth called *epikarst*. This drainage system allows more water to pass, further accelerating the dissolution process and leading to the formation of caverns and conduit flow systems. Eventually, the full development of subsurface caves, sinkholes, and vertical shafts leads to karst topography.

Springs

Natural **springs** are places where the water table intersects Earth's surface, producing water that flows on the land. Springs were the main source of freshwater for early civilizations and pioneers, which explains why many towns and villages have grown up around them. Likewise, knowledge that a groundwater resource existed nearby led communities to develop water wells to support continued population growth. Today, in less developed parts of the world, springs are still the principal—and sometimes the only—source of water for groups of families and small villages.

The discharge from springs may fluctuate according to seasonal and long-term periods of aridity and humidity, which cause variations in the elevation of the water table. It is important to understand the geology of springs so that sustainable use of these resources can be developed. Most springs emerge at locations where the water table is directed toward the surface by the structure of rock layers that hold groundwater (**FIGURE 18.31**).

EXPAND YOUR THINKING

Explain why groundwater in karst is highly vulnerable to contamination.

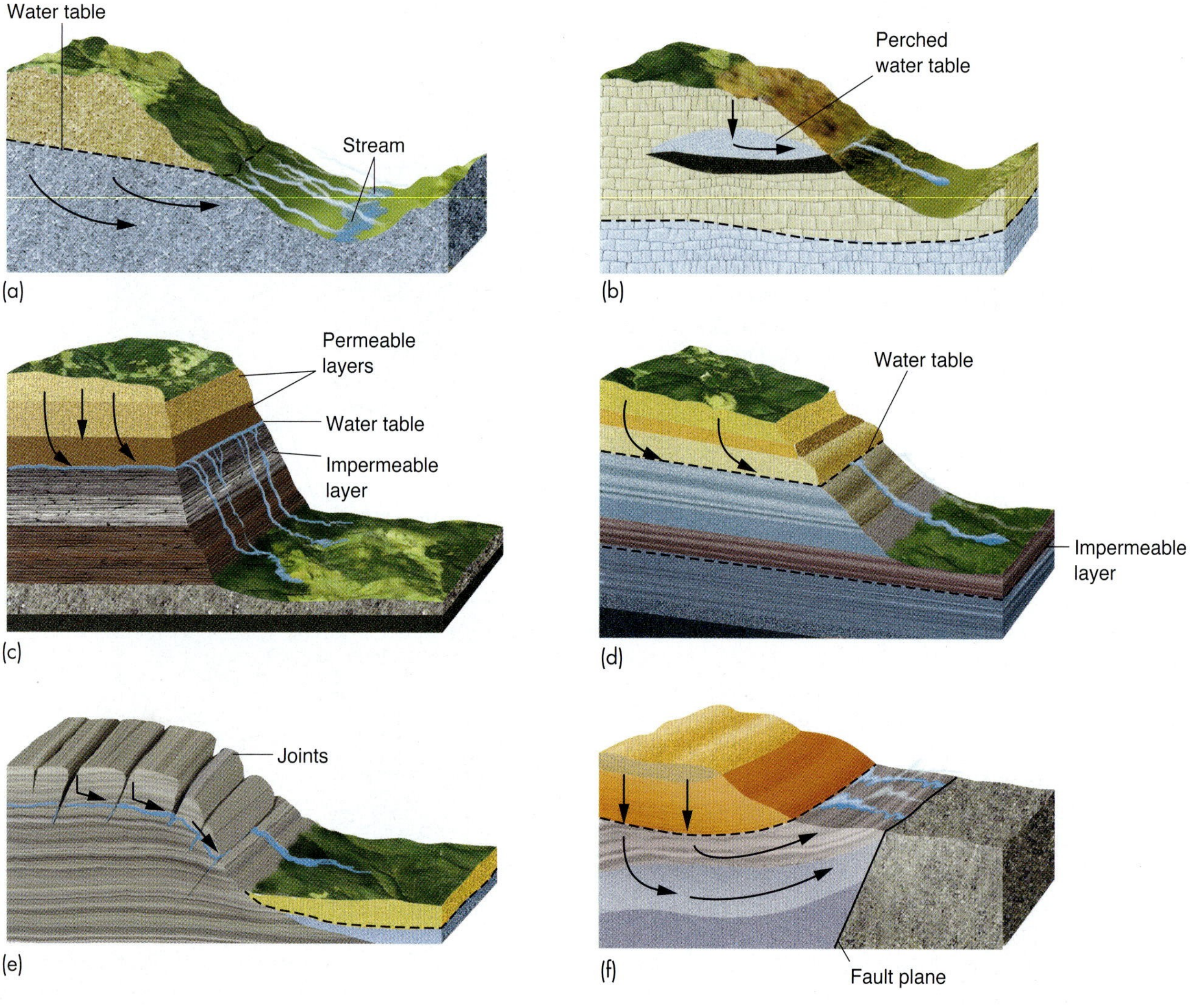

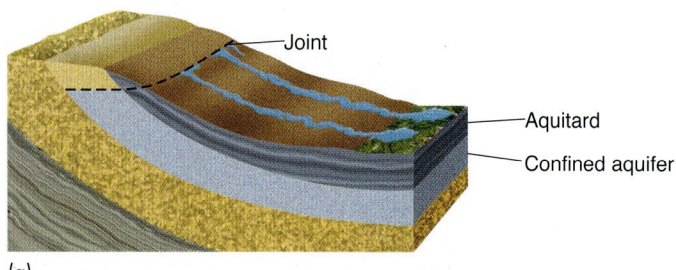

FIGURE 18.31 There are several types of springs: (a) A groundwater discharge area will form a spring. (b) A spring will develop where a water table perched on an aquitard intersects a slope. (c) As groundwater moves along the surface of an aquitard, it may emerge as a spring. (d) Springs appear where an aquifer crops out on a hillside. (e) If a fracture pattern directs groundwater flow toward a hillside, a spring may develop. (f) If aquifer flow is directed toward the surface along a fault plane, a spring may form. (g) When water in a confined aquifer migrates along a joint, a spring may emerge.

Hydrothermal Activity and Cave Formation Are Groundwater Processes

18-9

LO **18-9** Describe the major features of geysers and cave deposits.

Geysers, fumaroles (also called solfataras), and hot springs are features created by hydrothermal processes, the interaction of groundwater with young (hot) volcanic rocks.

Hydrothermal Processes

In a volcanically active area, runoff percolates through the unsaturated zone and may come into contact with high-temperature rock surrounding magma, creating hydrothermal processes. The magma may be active, or it may be recently solidified but still hot. As the water gains heat, it either loses density and rises toward the surface along cracks and fractures, becoming a *hot spring,* or turns into steam and erupts more forcefully as a **geyser** (from the Icelandic word *geysa,* "to gush") (**FIGURE 18.32**).

Geysers are relatively rare. There are only about 50 known geyser fields on Earth, and of those approximately two-thirds contain five or fewer active geysers. Yellowstone National Park in Wyoming has, by nearly an order of magnitude, more geysers than any other known field.

At least three essential conditions must be met for a geyser to develop, but many additional factors influence the character of geyser eruptions. The basic elements of a geyser are: 1) a supply of hot and cold water, 2) a heat source, and 3) a reservoir and the plumbing system associated with it.

Geyser eruptions are of two types: those that arise from *pool geysers* and those that arise from *columnar geysers.* Pool geysers require two separate water sources: one bringing large amounts of shallow, cool water, and another bringing a small quantity of boiling water from deeper in the crust. The two waters mix in the reservoir until the temperature of the entire system gradually rises and the water fills the reservoir. Heating continues, and when the system reaches a critical temperature, the hot water rises through the crust and erupts as steam in reaction to the reduction in pressure at the surface.

Columnar geysers (**FIGURE 18.33**) originate when hot water in the geyser's plumbing system mixes with cooler water from the surface. When the water reaches the boiling point, steam is caught in a constriction in the geyser's plumbing system. Constrictions in the channel are essential for columnar geysers to erupt. The pressure builds until it lifts overlying water up and out of the channel so that steam eventually escapes in an explosive eruption. The eruption continues until either the reservoir runs out of water or the temperature of the system drops below the boiling point.

Fumaroles (also called *solfataras*) are vents that allow volcanic gases, such as water vapor and hydrogen sulfide (H_2S), to escape into the atmosphere. Hydrogen sulfide readily oxidizes to become sulfuric acid (H_2SO_4) and native sulfur (S) in the atmosphere, leaving a brilliant yellow coating on rocks around a vent. If the water and heat sources are persistent, fumaroles may last for decades to centuries. If, on the other hand, they occur atop a quickly cooling volcanic deposit, they may disappear within weeks.

Hot springs appear at locations where groundwater comes into contact with a heat source in the crust. The temperature and discharge of hot springs depend on several factors, including the permeability of the aquifer at the spring, the rate of heating, and the nature and quantity of mixing with cool groundwater near the surface.

Caves and Cave Deposits

A **cave** is a natural opening in the crust that extends beyond the range of light and is large enough to permit entry by humans. The scientific study of caves is known as *spelology* (from the Greek words *spelaion,* "cave," and *logos,* "study"), and includes analysis using biological, geological, and archeological techniques.

(a)

(b)

(c)

FIGURE 18.32 (a) A geyser is an eruptive explosion of hot groundwater created by the release of steam. (b) Fumaroles emerge at locations where volcanic gases issue from the ground. (c) Hot springs are found in places where groundwater comes into contact with a heat source, and the heating causes the water to rise to the surface as it cools.

What is the difference between a hot spring and a geyser?

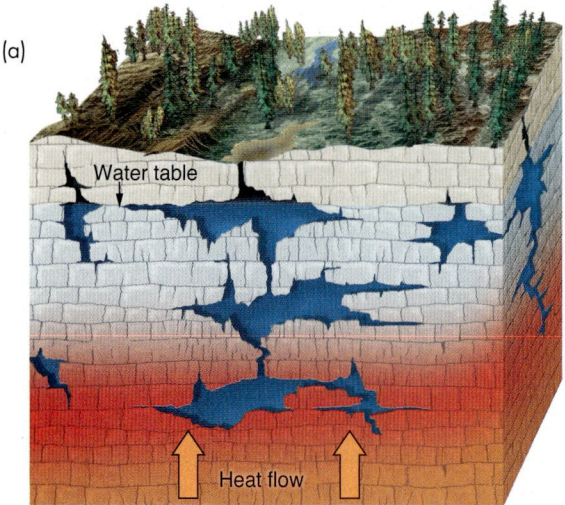

(a)

Water table

Heat flow

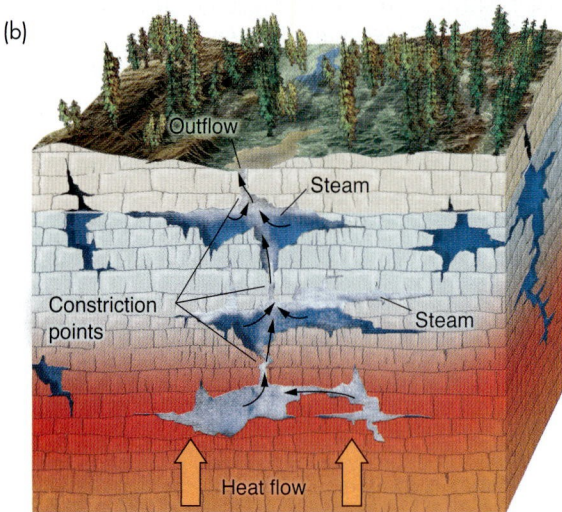

(b)

Outflow

Steam

Constriction
points

Steam

Heat flow

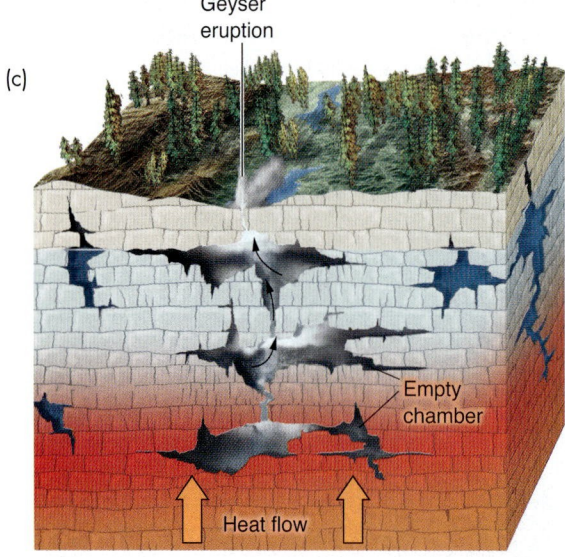

Geyser
eruption

(c)

Empty
chamber

Heat flow

FIGURE 18.33 (a) Columnar geysers erupt when groundwater turns into stream above a heat source but is prevented from exiting by a constriction (b) in the fracture system leading to the surface. When the steam eventually is freed, the rapid decrease in pressure lowers the boiling point of the remaining groundwater, causing it to erupt explosively (c) as a columnar geyser.

Caves occur in a wide range of rock types and for a variety of reasons. Among their variations are ice caves, limestone caverns, lava tubes, sea caves, eolian caves made by windblown sand, and caves formed by joints and fracture patterns. But by far the most common type of cave develops in relation to the karst process, and is called *limestone caverns*. Perhaps the two most famous caves in the United States are *Mammoth Cave* in Kentucky and *Carlsbad Caverns* in New Mexico, which together are visited by over 2 million people per year. Both are limestone caverns.

Groundwater that dissolves limestone strata creates limestone caverns. These cave systems are a form of karst that can extend for hundreds of kilometers, be divided into rooms of immense size, and contain fascinating and beautiful features. Mammoth Cave, the most extensive in the world, runs for 540 kilometers, while Carlsbad Caverns comprises a single room that covers an area equivalent to about 14 football fields, with a ceiling high enough to contain the U.S. Capitol.

In caves, dissolved calcium carbonate in groundwater precipitates on the floor, the roof, or the walls, creating complex "dripstone" features called *speleothems*. The most familiar of these are **stalactites** and **stalagmites.**

Stalactites hang downward from a ceiling and take shape as drop after drop of water slowly trickles through cracks in the cave roof. As each drop of water hangs from the ceiling, it evaporates and gives off carbon dioxide, causing the precipitation of a thin film of calcite.

Stalagmites grow upward from the floor of a cave, generally as a result of water dripping from overhanging stalactites. A *column* develops when a stalactite and a stalagmite grow until they join.

A *curtain* or *drapery* begins to form on an inclined ceiling when the drops of water trickle along a sloping roof. Gradually, a thin sheet of calcite grows downward from the ceiling and hangs in decorative folds, like a drape (**FIGURE 18.34**).

© Rocky Reston/Alamy

FIGURE 18.34 A complex network of speleothems—stalactites, stalagmites, and columns—make up cave deposits.

(Q) **What is the chemical composition of these features? How do they form?**

(Q) **EXPAND YOUR THINKING**

Are speleothems igneous, sedimentary, or metamorphic? Explain your reasoning.

LET'S REVIEW

Groundwater is one of our most precious natural resources. It is the most important source of drinking water, and supplies water for irrigation, industry, manufacturing, domestic uses, and to meet most other water needs in human society. Of all the freshwater that exists, about 75% is stored in polar ice and glaciers; about 25% is stored as groundwater. Groundwater is highly vulnerable to overuse and pollution, for three reasons:1) most of it moves slowly (usually less than 0.3 meters per day); 2) it has been in the ground for decades to centuries; 3) it is not quickly renewed by the hydrologic cycle. Rates of human usage generally outpace natural rates of groundwater replenishment, thus making it necessary to put conservation measures in place in nearly every community. To properly conserve groundwater requires careful analysis of where it is located, how it moves, and how it is recharged. The central question in groundwater science is: How shall we use groundwater today so that it remains available for future generations?

As noted, groundwater moves slowly and stays in the ground for long periods, making it vulnerable to contamination. Sources of contamination are widespread and usually related to chemical wastes and sewage finding their way to the water table. When a well pumps the water table at a rate that exceeds the permeability of the aquifer, a cone of depression is formed, which then creates a localized hydraulic gradient pulling water and any contamination toward the well. Groundwater contamination is so prevalent in communities around the world that major efforts are underway to monitor and clean up groundwater pollutants, usually at great expense. For these reasons, it is crucial that you do not take your water for granted. Remember that groundwater was used to grow your food, and probably is the source of the water you drink, not to mention to manufacture all the things you need to live.

STUDY GUIDE

18-1 Groundwater is our most important source of freshwater.

- Groundwater is found underground in cracks, pores, and spaces in soil, sediments, and rocks. The area in which water fills these spaces is called the saturated zone. The top of this zone is called the **water table**.

- About 25% of rainfall in the United States becomes groundwater. Three-quarters of the cities in the country use groundwater as part of their water supply. About 95% of the total supply of freshwater of the United States is taken out of the ground.

- In most cases, water is pumped from the ground by a water well. Water wells are drilled by truck-mounted drill rigs, which use various types of cutting heads on the end of a spinning shaft to bore through soil, sediment, sedimentary rock, and even crystalline igneous and metamorphic rock to gain access to the water table.

18-2 Groundwater is fed by snowmelt and rainfall in areas of recharge.

- Rainfall and snowmelt add water to the groundwater system through the process of **recharge**. Recharge may occur in several ways, but in most cases, runoff soaks into the ground through percolation and contributes to the water table below.

- Eventually, groundwater reappears above the ground in a process called groundwater discharge. Groundwater may flow into streams, wetlands, lakes, and oceans, or it may be discharged in the form of springs and flowing wells.

- Groundwater supplies many streams, lakes, and wetlands. In fact, about 30% of stream flow in the United States is fed by groundwater discharge.

- Wetlands are critically important components of natural ecosystems and are protected under federal law. Generally, wetlands are sites where saturation with water is the dominant factor determining the nature of soil development and the types of plant and animal communities that live in the soil and on its surface.

18-3 Groundwater moves in response to gravity and hydraulic pressure.

- In most locations where water moves within the saturated zone, groundwater flows in response to **hydraulic pressure**—it moves from an area of high pressure into areas of lower pressure.

- An **aquifer** is an underground formation of saturated crust that can produce useful quantities of water when tapped by a well.

- A measure of the capability of rock or sediment to transmit groundwater is called **permeability**, and it depends on the size of the spaces in the soil or rock of an aquifer and how well the spaces are connected. Aquifers typically consist of highly permeable materials such as gravel, sand, sandstone, or fractured rock like limestone.

- The amount of empty space in an aquifer is its **porosity**. In sediments or sedimentary rocks, porosity depends on the degree of cementation, the sorting, and the shape of the grains.

- The rate at which groundwater moves through an aquifer depends on both the permeability of the rock and the **hydraulic gradient**, defined as the difference in elevation between two points in an aquifer divided by their distance from each other.

18-4 Porous media and fractured aquifers hold groundwater.

- **Hydrogeologists** define two types of aquifers in terms of their physical characteristics: porous media and fractured aquifers.

- Important in determining the movement and location of usable groundwater are layers of impermeable rock or sediment called **aquitards**. Aquitards usually are formed by layers of shale or clay that prevent the free flow of groundwater in an aquifer. A confined aquifer is one that lies below an impermeable layer.

- A special kind of confined aquifer, an artesian aquifer, is found at locations where the escape of water to the surface is driven by the hydraulic gradient of the water table.

18-5 Groundwater is vulnerable to several sources of pollution.

- About one-quarter of Earth's population drinks contaminated water.

- Water is highly effective at dissolving compounds, so it often carries toxic compounds that are unhealthy for plants and animals. For instance, leaking underground storage tanks have contributed to contamination in every state. It is thought that about 70% of all underground storage tanks leak gasoline and other toxic chemicals into the ground.

- When water is pumped from an aquifer at a rate that exceeds the permeability of the surrounding soil and rock, the water table around the well will lower. This practice creates a **cone of depression** that steepens the hydraulic gradient and draws water toward the well.

- Continued pumping at an excessive rate in coastal areas will cause saltwater intrusion that damages the quality of the aquifer. Saltwater intrusion occurs when freshwater is withdrawn faster than it can be recharged near a coastline.

- One of the most common forms of bacterial contamination occurs through the disposal of liquid sewage from a home or business. Many rural and suburban homes are not connected to community sewage disposal lines. Instead, these buildings dispose of their liquid waste using septic and cesspool systems that rely on the properties of soils and rocks in the unsaturated zone to filter out harmful wastes.

18-6 Common human activities contaminate groundwater.

- There are more than 20,000 known abandoned and uncontrolled hazardous waste sites in the United States. Illegal disposal practices range from disposing hundreds of drums of highly toxic chemicals to dumping used engine oil in the backyard.

- **Contaminant plumes** follow the same hydraulic gradient that drives groundwater flow in aquifer systems. These contaminants may include used engine oil and other chemicals, pesticides, paint thinners, cleaning compounds, lubricants, inks, and manufactured toxic substances. A contaminant plume will follow the hydraulic gradient leading to a well if a cone of depression is present.

18-7 Groundwater remediation includes several types of treatment.

- The intense and often uncontrolled contamination of portions of the nation's groundwater supplies has led to the development of an entire industry dedicated to **groundwater remediation**.

- Estimates of remediation costs at U.S. government sites alone can reach hundreds of billions of dollars. Protecting the quality of groundwater supplies is a problem of broad societal importance.

18-8 Groundwater is responsible for producing karst topography and springs.

- The chemistry of water in the ground is capable of causing hydrolysis, dissolution, and even oxidation of many minerals in crustal rocks. The dissolution mechanism is so strong that it can completely dissolve vast tracts of carbonate rock and change the topography of the land and the character of the crust.

- **Springs** and **karst geology** are produced by groundwater.

18-9 Hydrothermal activity and cave formation are groundwater processes.

- **Geysers**, fumaroles, and hot springs are features created by the interaction of groundwater with young volcanic rocks. Typically, runoff percolates through the unsaturated zone and comes into contact with high-temperature rock surrounding a magma body.

- The most common types of **caves** are related to the karst process. These are called limestone caverns. The scientific study of caves is termed spelology and includes the investigation of the biology, geology, and archeology of caves. Perhaps the two most famous caves in the United States are Mammoth Cave in Kentucky and Carlsbad Caverns in New Mexico.

KEY TERMS

aquifer (p. 544)
aquitard (p. 548)
artesian springs (p. 548)
base flow (p. 542)
cave (p. 561)
cone of depression (p. 550)
contaminant plume (p. 555)
Darcy's law (p. 545)
geysers (p. 561)

groundwater (p. 540)
groundwater remediation (p. 557)
hydraulic gradient (p. 545)
hydraulic pressure (p. 544)
hydrogeologists (p. 548)
karst geology (p. 559)
landfills (p. 554)
natural attenuation (p. 558)
permeability (p. 544)

porosity (p. 544)
recharge (p. 542)
sinkholes (p. 559)
springs (p. 560)
stalactites (p. 562)
stalagmites (p. 562)
Superfund Program (p. 549)
water table (p. 540)
wetlands (p. 542)

ASSESSING YOUR KNOWLEDGE

Please complete the following exercise before coming to class. Identify the best answer to each question.

1. Groundwater typically is found as:
   a. Large underground lakes.
   b. Large underground rivers.
   c. Small pockets under rain forests.
   d. Rims of water around individual grains and in fractures.
   e. Vapor in pore spaces in the crust.

2. Which of the following three rocks is most likely to make a good aquifer?

(a)   (b)   (c)

Courtesy of Chip Fletcher

   a. Granite
   b. Sandstone
   c. Gneiss

3. Groundwater is lost by:
   a. Base flow and groundwater discharge.
   b. Evaporation.
   c. Capillary flow.
   d. Darcy's law.
   e. All types of streams.

4. Which of the following is a characteristic of wetlands?
   a. Wetlands may recharge groundwater.
   b. Wetlands may be fed by groundwater discharge.
   c. Both a and b.
   d. Neither a nor b.
   e. None of the above.

5. Permeability is:
   a. The rate of groundwater flow above the water table.
   b. A measure of the capability of rock or sediment to transmit groundwater.
   c. The size of pore spaces.
   d. The cementation of grains in the capillary fringe.
   e. None of the above.

6. Which of these sediments has the highest porosity? _____ Which has the highest permeability? _____

 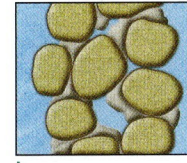 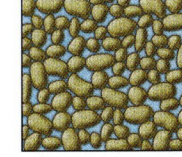

a.   b.   c.

7. Groundwater flows in response to:
   a. Gravity and density.
   b. Gravity and porosity.
   c. Permeability and topography.
   d. Gaseous pressure and slope.
   e. Gravity and hydraulic pressure.

8. An aquifer is:
   a. Any water in the ground.
   b. Any water at the water table.
   c. A useful source of groundwater.
   d. Water that is confined by aquitards.
   e. Groundwater that can be pumped from the ground.

9. Groundwater contamination may result from:
   a. Salt intrusion, illegal chemical disposal.
   b. Agricultural waste, pesticides.
   c. Landfill seepage, polluted runoff.
   d. Human sewage, leaky underground storage tanks.
   e. All of the above.

10. Once human-generated contamination enters the groundwater system, it may:
    a. Form a plume.
    b. Last for centuries.
    c. Be stripped off if it forms a gas.
    d. Be allowed to naturally attenuate.
    e. All of the above.

11. Groundwater remediation is:
    a. The process of finding new aquifers.
    b. Recharge by polluted runoff.
    c. Cleanup of sewage before releasing it to a treatment plant.
    d. Flow in the direction of hydraulic conductivity.
    e. Cleanup of polluted groundwater.

12. Which of the following factors do you need to calculate the hydraulic gradient between two wells?

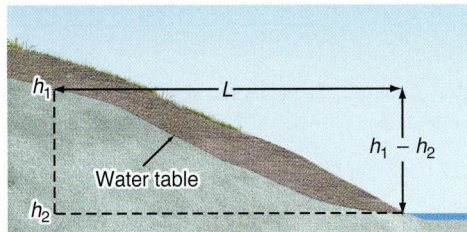

    a. Porosity and permeability
    b. Water table elevation
    c. Recharge and discharge
    d. Distance between wells
    e. Flow velocity and hydraulic pressure

13. Geysers:
    a. Are the reason caves have speleothems.
    b. May occur as nonviolent pools of warm water.
    c. Are the main reason groundwater migrates.
    d. Are violent eruptions of hot groundwater.
    e. All of the above.

14. Most liquid freshwater is found:
    a. In rivers.
    b. In lakes and streams.
    c. In the ocean.
    d. As groundwater.
    e. During percolation.

15. Porosity is:
    a. The percentage of empty space in the crust.
    b. The percentage of connected space of groundwater movement in the crust.
    c. The percentage of crustal space filled with groundwater.
    d. The percentage of pore space in the crust located at the water table.
    e. The water in the capillary fringe.

16. The primary reason a contaminant plume moves toward a well is because of the:

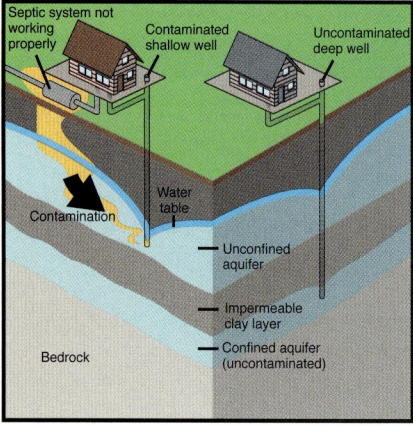

    a. Porosity.
    b. Density difference between the groundwater and the pollutant.
    c. Hydraulic gradient.

    d. Potentiometric surface of the confined aquifer.
    e. Proximity of the recharge area.

17. The method for mapping the direction of groundwater flow at the water table is to:
    a. Draw arrows parallel to the contours of the water table.
    b. Draw arrows at 45° to the contours of the water table.
    c. Draw arrows at 90° to the contours of the water table.
    d. Draw arrows that point downhill on the surface topography above the water table.
    e. None of the above.

18. An aquitard may lead to formation of an artesian well because:
    a. It allows atmospheric pressure to drain an unconfined aquifer.
    b. It confines an aquifer and closes off groundwater recharge.
    c. It confines an aquifer so that hydraulic pressure increases, leading to artesian flow.
    d. It causes karstification of the crust.
    e. All of the above.

19. Stalagmites and stalactites form primarily because of:
    a. Sediment deposition.
    b. Evaporation.
    c. Increased recharge.
    d. Fall of the water table.
    e. Decreased recharge.

20. Rainfall absorbs _____ to become acidic and dissolve limestone.
    a. Sulfur dioxide
    b. Oxygen
    c. Nitrous oxide
    d. Carbon dioxide
    e. Nitrogen

FURTHER RESEARCH

1. Describe the evolution of a landscape in which groundwater is the dominant agent of change.

2. Where does the water that you drink come from? You will have to contact the local city or state water supply office and ask the source of the water supplied to your address.

3. What is the closest Superfund site to you? How is it being remediated?

4. Under what geologic/strata-forming conditions would a confined aquifer be created? (Hint: Describe environments of deposition.)

5. What is the difference between porosity and permeability? How are they related?

ONLINE RESOURCES

Explore more about groundwater on the following Web sites:

The Groundwater Foundation:
www.groundwater.org

National Groundwater Association:
www.ngwa.org

U.S. Geological Survey (USGS), The Groundwater Atlas of the United States:
http://pubs.usgs.gov/ha/ha730/index.html

USGS groundwater information pages:
http://water.usgs.gov/ogw

Additional animations, videos, and other online resources are available at this book's companion Web site:
www.wiley.com/college/fletcher

This companion Web site also has more information about WileyPLUS and other Wiley teaching and learning resources.

19

DESERTS AND WIND

CHAPTER CONTENTS & LEARNING OBJECTIVES

GEOLOGY IN OUR LIVES

A desert is a landscape that receives little rainfall, has sparse vegetation, and is unable to support significant populations of animals. Deserts are more widespread than many people realize. In fact, they take up one-third of Earth's land surface, and their range is expanding. As more people move into marginally habitable areas bordering deserts, poor land management (such as overgrazing), and changes in the water cycle due to global warming, can permanently damage native vegetation and soil. These factors promote desertification, a problem expected to worsen in the future as the human population expands worldwide and global warming continues to alter patterns of rainfall distribution.

Q What factors contribute to the formation of deserts?

19-1 Deserts May Be Hot or Cold, but Low Precipitation Is a Common Trait

LO 19-1 Define the term desert and describe a desert environment.

Deserts comprise a global-scale environment that covers approximately one-third of Earth's land surface, and their range is expanding.

Desert Characteristics

A **desert** or arid region (**FIGURE 19.1**) is defined as one that receives less than 25 centimeters of precipitation per year, or as an area where the rate of evapotranspiration exceeds the rate of precipitation. Another defining feature of deserts is that vegetation in them is so sparse that they are unable to support significant populations of animals. An important distinction is that slightly wetter regions, receiving on average 25 to 51 centimeters of precipitation per year, are considered *semiarid*.

The 10 largest deserts—listed in **TABLE 19.1** and mapped in **FIGURE 19.2**—are spread across the planet. Given that the most common image of a desert that comes to mind are places like the Sahara (number 3 on the list) and the Great Basin Desert of North America (number 10), it may come as a surprise to note that the two at the top of the list, Antarctica and the Arctic, are covered with water (in the form of ice and snow) even though they receive little annual precipitation. (You can read about the history of the Sahara in the "Geology in Action" feature titled "The Great Sahara.")

FIGURE 19.1 Deserts are characterized by a lack of precipitation and sparse vegetation. The Atacama Desert, in Chile, is the driest place on Earth.

Mark Chivers/Age Fotostock America, Inc.

State a hypothesis explaining the location of the largest deserts on Earth

TABLE 19.1 World's 10 Largest Deserts

Rank	Desert	Area (square kilometers)
1	Antarctic Desert (Antarctica)	13,829,430
2	Arctic Desert (regions of permafrost above the Arctic Circle)	13,700,000+
3	Great Sahara (Africa)	9,100,000+
4	Arabian Desert (Middle East)	2,330,000
5	Gobi (Asia)	1,300,000
6	Kalahari Desert (Africa)	900,000
7	Patagonian Desert (South America)	670,000
8	Great Victorian Desert (Australia)	647,000
9	Syrian Desert (Middle East)	520,000
10	Great Basin Desert (North America)	492,000

Deserts have a wide daily and seasonal temperature range. High daytime temperatures (up to 45°C) followed by low nighttime temperatures (down to 0°C) are attributable to the low humidity. In the absence of water, the dry desert air has few clouds to block sunlight during the day or to trap much heat radiating from the ground at night. That is why deserts heat intensely during the day, when most of the sunlight reaches the ground; but as soon as the Sun sets, heat rapidly escaping from the surface cools deserts quickly.

Why do we study deserts? First, deserts are a major global environmental feature—as noted, covering approximately one-third of Earth's land surface. Second, they have major impacts, positive and negative, on human populations living both near and far; and, third, shifts in desert boundaries are a cause of great concern to neighboring populations.

One of those impacts is the phenomenon of expanding deserts, known as **desertification**, which affects the distribution of available natural resources (especially water and food). It can lead to political unrest and outright conflict as one group of people, in order to maintain its own supply level, appropriates resources from another group, or migrates into the resource-rich lands of another country. On the other hand, groups of people may also come together over a landscape of shifting resources. A case in point is the Middle East, where otherwise unfriendly neighbors have been, out of necessity, working together to manage water resources in the region. Expand your understanding of this issue in the Earth Citizenship feature, "The Global Water Crisis."

Another problematic impact is the environmental damage that can result from too great a demand for scarce resources in a desert. For instance, the Colorado River has nearly dried up, and no longer reaches the Gulf of California, because so many people have been tapping its water for drinking and irrigation for so long.

Lack of water has also contributed to the collapse of agriculture in many areas of the world, resulting in famine, political unrest, and civil strife. Primary examples of this negative impact are nations bordering the Sahara; Niger, Chad, Sudan, and Ethiopia, among others, seem to be particularly susceptible to the changing environment.

An old saying from the American West, "Whiskey's for drinkin', water's for fightin'," still rings true today as people all over the world increasingly must cope with shrinking water resources, the defining characteristic of a desert.

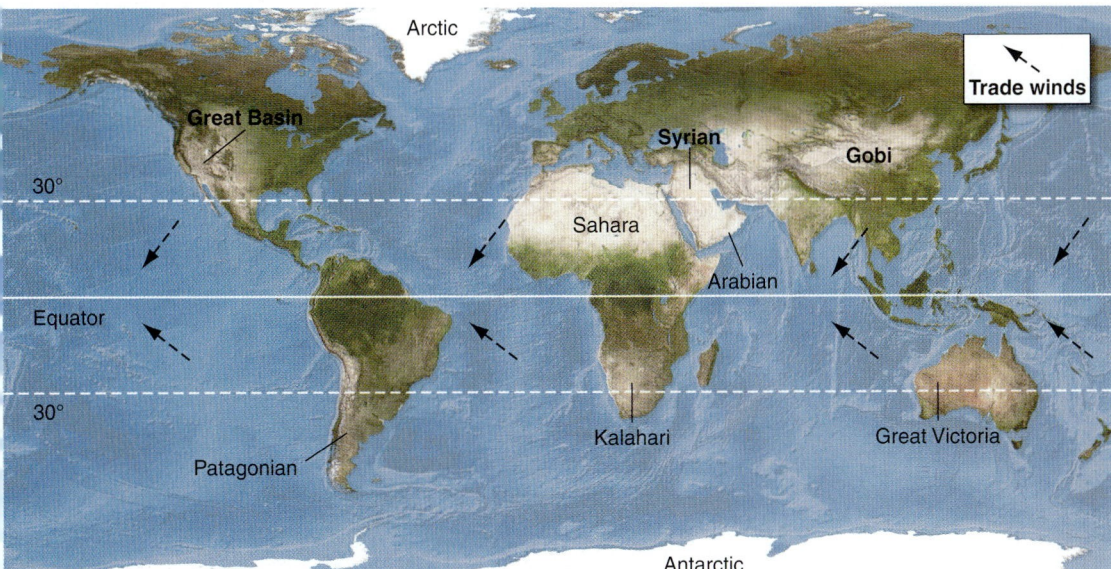

FIGURE 19.2 Deserts are found on every continent.

Q What are the physical characteristics of a desert?

Q EXPAND YOUR THINKING

Describe how desertification would impact or change specific aspects of the community where you live.

 GEOLOGY IN ACTION

The Great Sahara

It may be hard to imagine now, but the Sahara was a lush and humid place, between 5,000 and 10,000 years ago. The region then had plenty of rainfall and was teeming with large animals, such as giraffes, gazelles, lions, elephants, and even hippopotamus and crocodiles. Populations of early farmers and herders lived throughout the area (**FIGURE 19.3**).

We know about the large animals in the Sahara because their fossilized bones have been found at various sites across the desert, often associated with sediment deposited in former lakes and floodplains. From this discovery scientists have concluded that the Sahara must have been wetter in order to support populations of such large animals. But we did not truly appreciate the amount of water that must have been involved until the Space Shuttle was equipped with downward-looking radar capable of penetrating the dry desert sands and revealing ancient riverbeds. These ancient river channels are not visible on the ground, but their networks (**FIGURE 19.4**) are readily apparent when imaged by the radar of the high-flying Space Shuttle.

To explain the transformation of the Sahara, it lies under the influence of the yearly *monsoons* (seasons of high rainfall in the tropics). The monsoons in Africa and elsewhere are directly attributable to heating during the summer: Air over land becomes warmer and rises, pulling in cool wet air from the ocean, which brings rain.

What was different thousands of years ago? Recall the orbital parameters described in Chapter 15; subtle variations in Earth's orbit altered the path of the monsoon and led to changes in rainfall. Approximately 10,000 years ago, a change in *insolation* (sunlight received on Earth's surface) caused the African monsoon to shift roughly 600 kilometers to the north. This brought summer rains to what is now the Sahara, creating an environment similar to the grasslands of lush modern East Africa and filling large rivers and lakes. About 5000 years ago, the monsoons shifted back to the south and, by 4000 years ago, the Sahara was as dry as it is today.

FIGURE 19.3 In the now-uninhabitable central Sahara, 5,000-year-old rock art preserves images depicting ancient cultures that once lived in the region.

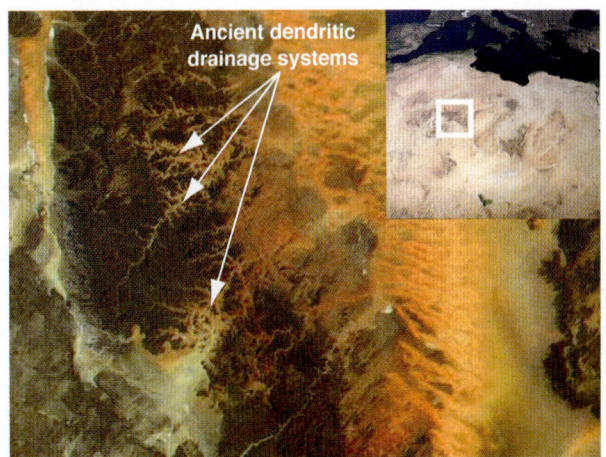

FIGURE 19.4 Ancient dendritic drainage patterns in the Acacus–Amsak region of the central Sahara are evidence of a wetter past.

EARTH CITIZENSHIP

The Global Water Crisis

Fifty years ago the global population was half what it is today. At the same time, water was widely viewed as a limitless resource, one perpetually cleansed by the hydrologic cycle. Humans in general, then, also ate less meat (a water-intensive product); fewer of them were personally wealthy and consequently they consumed fewer calories. Therefore, they needed less water to produce their food, and meet the freshwater requirements of irrigation, industry, and domestic uses. Only one-third as much water, for example, was drawn from rivers as we take today.

In stark contrast now, there are over 7 billion people on the planet; megacities (those with over 10 million people) dot every continent; and more people are wealthy. All these statistics add up to much greater food consumption and increasing competition for water to support industry, urbanization, and the production of biofuel crops. For example, for many around the world today, meat is on their dinner plates on a regular basis (**FIGURE 19.5**).

The High Cost of the Water Crisis

Many problems are directly related to the global water crisis:

- Aquifers are experiencing dropping water tables (**FIGURE 19.6**) for a number of reasons: the widespread use of mechanical pumps, the lack of managed water withdrawal at sustainable rates, growing demand for irrigation, and drops in recharge rates related to drought and global warming.

- Rivers are running dry in places (**FIGURE 19.7**) where the demand for freshwater exceeds availability.

- Global warming has caused the loss of glaciers (a major source of freshwater in Asia, India, Alaska, and South America) and the shrinkage of lakes and other freshwater bodies.

For the time being, there is still an abundant total freshwater supply. But the number of areas where it has become polluted, saline, and otherwise unsuitable or unavailable for drinking, industry, or agriculture is climbing—at an alarming rate.

Already, nearly a billion people do not have access to clean and safe water (**FIGURE 19.8**); 37% of those people live in Sub-Saharan Africa. Half of the world's hospital beds are occupied by people suffering from a water-related disease. In developing countries, as much

Gary Crabbe / Enlightened Images

FIGURE 19.5 Concentrated animal feeding operations, major consumers of fresh water, are growing worldwide in response to the demand for meat as a food source.

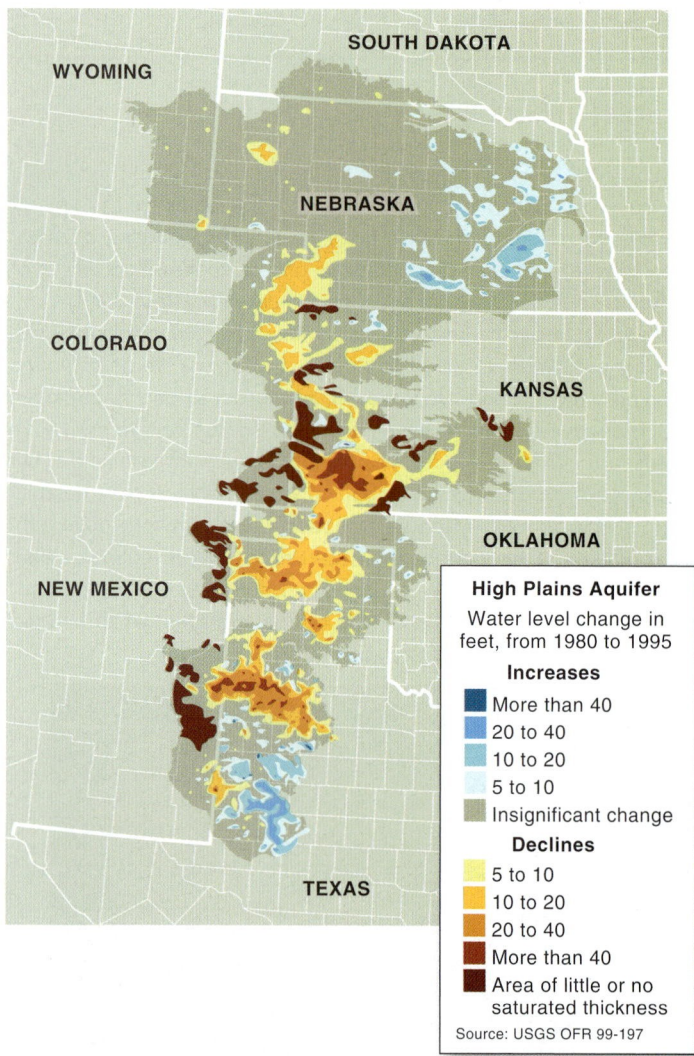

High Plains Aquifer
Water level change in feet, from 1980 to 1995

Increases
- More than 40
- 20 to 40
- 10 to 20
- 5 to 10
- Insignificant change

Declines
- 5 to 10
- 10 to 20
- 20 to 40
- More than 40
- Area of little or no saturated thickness

Source: USGS OFR 99-197

FIGURE 19.6 The Ogallala aquifer underlies about 27% of the irrigated land in the United States, and yields about 30% of all groundwater used for irrigation by the nation. The aquifer system supplies drinking water to 82% of the 2.3 million people who live in the High Plains region. However, water levels at the aquifer are declining, with some estimating that the remaining volume is sufficient for as little as 25 years.

FIGURE 19.7 The Murray River provides water to Australia's most productive agricultural zone, widely known as the nation's "food bowl." Here, water withdrawal for irrigation has resulted in rising salinity, and disruptions and diversions have so seriously reduced the flow that silt deposition sometimes closes the river's mouth.

FIGURE 19.8 In south China's Guangdong Province, villagers form a line to fill containers with fresh drinking water. This scene is being played out across the globe, where the only reliable water supplies are trucked in by governments or private entrepreneurs.

as 80% of illnesses are linked to poor water and sanitation conditions. Nearly 1 out of every 5 deaths under the age of 5 worldwide is due to a water-related disease, such as giardia, schistosomiasis, botulism, E. coli infection, dysentery, typhoid fever, and others (**FIGURE 19.9**). The problem of access to safe drinking water is getting worse, not better; over the next two decades, the average supply of water per person will drop by a third, possibly condemning millions of people to an avoidable premature death.

The Effects of Evapotranspiration

Global warming enhances evapotranspiration—the movement of water into the atmosphere under the influence of evaporation and transpiration. What does this mean in terms of the water crisis? It is expected to lead to more frequent incidences of drought in dry areas,

and expansion of dry areas themselves. Precipitation has been declining in the tropics and subtropics since 1970. Southern Africa, the Sahel region of Africa (more on this region later in the chapter), southern Asia, the Mediterranean, and the U.S. Southwest, to name five, are getting drier. Even areas that remain relatively wet now may experience long, dry conditions between extreme precipitation events.

Scientists expect the amount of land affected by drought to grow by midcentury—and water resources in affected areas to decline as much as 30% (**FIGURE 9.10**). These changes are anticipated to occur partly because of expansion of the Hadley cell (discussed in the next section) in which warm air in the tropics rises, loses moisture to tropical thunderstorms, and descends in the subtropics as dry air. As jet streams continue to shift to higher latitudes, and storm patterns along with them, semiarid and desert areas are expected to expand.

FIGURE 19.9 More than 3.4 million people die each year from water, sanitation, and hygiene-related causes. Nearly all deaths, 99%, occur in the developing world. There are 780 million people who lack access to an improved water source, approximately 1 in 9 people.

FIGURE 19.10 Climate models project that for every 0.56°C of warming, the length of periods with no rain will extend globally by 2.6%. The incidence of drought has increased globally by 4% in the past half century.

19-2 Atmospheric Moisture Circulation Determines the Location of Most Deserts

LO 19-2 Explain global atmospheric circulation.

Global atmospheric circulation is a primary factor in variations in rainfall and, hence, the location of deserts.

Atmospheric Circulation

We discussed the atmospheric circulation system in Chapter 14, but it is worth reviewing this system here, to help explain the distribution of deserts on Earth. Recall that the three basic components of the system are the *Hadley cell*, the *Ferrel cell*, and the *Polar cell* (**FIGURE 19.11**). There is one of each cell type in the northern hemisphere and one of each in the southern hemisphere.

Atmospheric circulation starts with the basic principle that air heated by the Sun rises at the equator, where solar heating is greatest.

As the air moves toward the poles, it cools and eventually sinks. Rising air causes low air pressure (at the equator), whereas sinking air causes high air pressure (at the poles).

If Earth were perfectly still and smooth, we might have a single cell in each hemisphere where hot air rises at the equator, moves north or south toward the poles, and then sinks to ground level as it cools at the poles. This air would then flow back to the equator along the ground surface. We would see this pattern expressed in the northern hemisphere as a constant north wind, and in the southern hemisphere as a constant south wind. Fortunately, however, Earth is neither still nor smooth. Earth spins on its axis, causing the changes of day and night, and large mountain ranges deflect the direction of surface winds. Life on Earth is much more interesting this way.

By the time an air mass that has risen at the equator has traveled to about 30° latitude, it has cooled sufficiently to sink back to Earth's surface, forming an area of high pressure. When this air reaches the surface, it must flow away, and it moves back either toward the equator or toward the pole. The air that flows back to the equator will be reheated and rise again to repeat the process. This completes the Hadley cell.

At the poles, cold, dense air descends, producing a high-pressure area. Air will flow away from the high pressure and toward the equator. By the time this air nears 60° latitude, it begins to meet

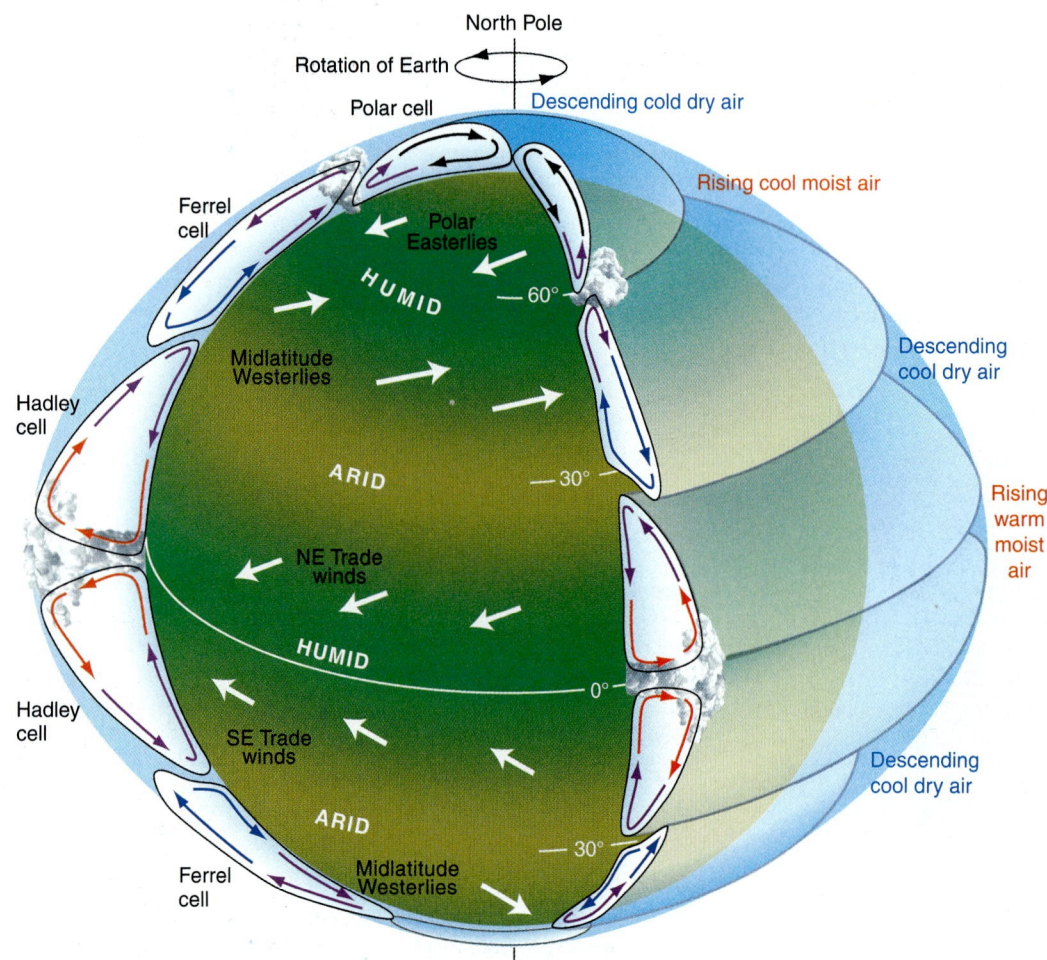

FIGURE 19.11 The Hadley, Ferrel, and Polar cells are the three major atmospheric convection cells. Deserts are concentrated where these cells produce sinking, dry air masses.

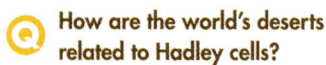 **How are the world's deserts related to Hadley cells?**

FIGURE 19.12 Many of Earth's deserts are clustered around 30° latitude due to the sinking air produced where the Hadley and the Ferrel cells meet.

Q How do surface climates compare or contrast with the atmospheric circulation system?

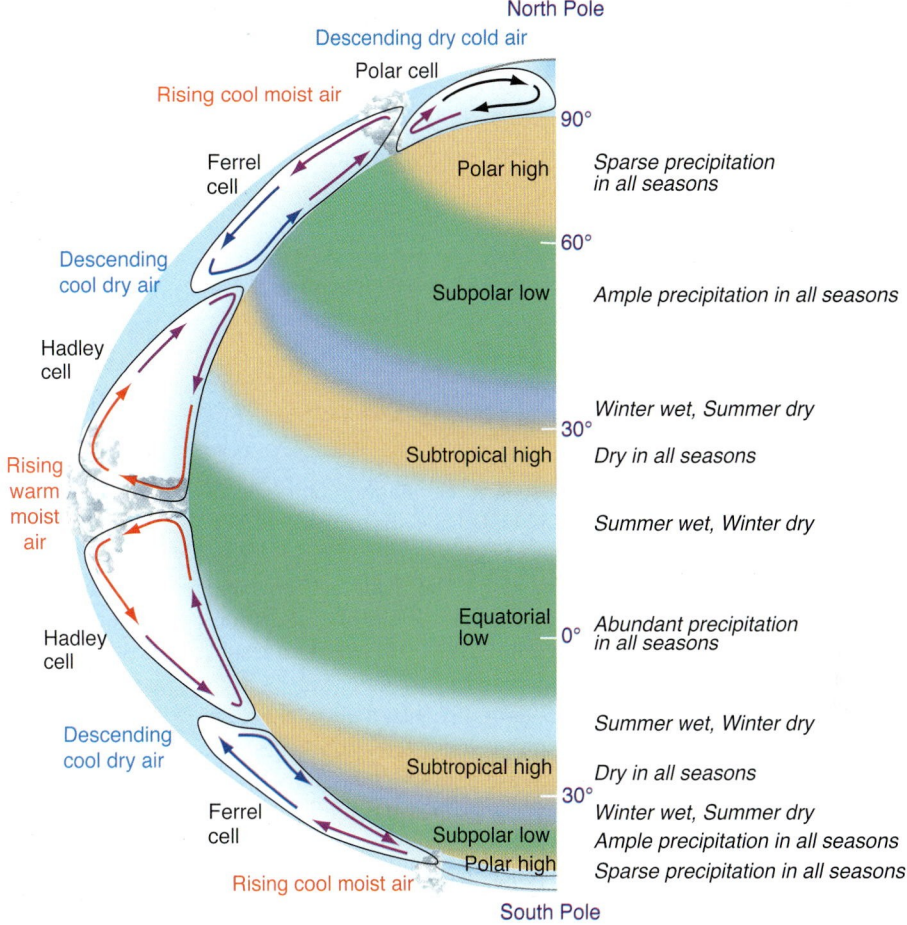

the air flowing poleward from the Hadley cell. When these two air masses meet, they have nowhere to go but up. As they rise, they cool and lose moisture, causing high precipitation. Once high in the atmosphere, they must head poleward, where they will cool and sink again, or toward the equator, where they will meet the flow heading poleward from the equator and sink. The circulatory cell sinking at the poles and rising at 60° latitude is called the Polar cell. The final cell, sinking at 30° latitude and rising at 60° latitude, is named the Ferrel cell, after American meteorologist William Ferrel (1817–1891).

In 1856, Ferrel demonstrated that, due to the rotation of Earth, air and water currents moving distances of tens to hundreds of miles tend to be deflected to the right in the northern hemisphere and to the left in the southern hemisphere. This phenomenon is known as the **Coriolis effect**. When surface winds in a Hadley cell are moving south in the northern hemisphere, and are deflected to the right, they turn westward and are called the *northeast trade winds*. In the southern hemisphere, they turn left, to become the *southeast trade winds*. The surface winds in the northern hemisphere's Ferrel cell are moving north, and when deflected right they become the midlatitude *westerlies*. The surface winds in the northern Polar cell are heading south, and when deflected right become the *polar easterlies*. Examine

a globe to convince yourself of these patterns and to figure out which part of the global atmospheric circulation system you live in.

The Influence of Patterns on Deserts

What do these patterns have to do with deserts? As air rises, it cools and expands. This is attributable to the increased distance from the warming effects of Earth's surface and the lower air pressure found at higher altitudes. As a rising air mass cools and expands, so does the water vapor contained in it. As the water vapor cools and expands, more water condenses than evaporates, causing water droplets and then clouds to form. Continued condensation produces precipitation, and it rains or snows. Therefore, in areas where relatively warm moist air is rising, such as near the equator and around 60° latitude, there will be a lot of precipitation.

The opposite is also true: Air warms and contracts as it sinks closer to Earth's surface. This causes evaporation to exceed condensation. No clouds will form in locations with lots of sinking air. These areas, such as at the poles and around 30° latitude, will have few clouds and little precipitation, thus forming a great belt of arid climate (and deserts) that girdles the globe. Many of the world's deserts are clustered around 30° N and 30° S latitudes for this reason (**FIGURE 19.12**).

Q EXPAND YOUR THINKING

What are the parts or main features of the northern hemisphere atmospheric circulation system?

19-3 Several Factors Contribute to Desert Formation

LO 19-3 Itemize the factors that contribute to desert formation.

The orographic effect, moisture transport, air temperature, proximity to cold seawater, and poor human management all contribute to the formation of deserts.

Contributing Factors

Several factors in addition to atmospheric circulation contribute to the location of deserts. The first is called the **orographic effect**, which causes a *rain shadow*. It occurs when a mountain range forces the prevailing winds to rise up over them. The rising air cools and expands, and the rate of condensation exceeds the rate of evaporation, causing clouds to form and rain to fall (**FIGURE 19.13**). As a result, the windward side of these mountains can be extremely wet.

On the leeward side of the mountains, in contrast, the sinking air warms and compresses—behaving just as it did when sinking as part of the Hadley cell. The resulting increase in evaporation limits cloud formation and rainfall, leading to arid and semiarid conditions.

The dry area downwind from the mountains can be very well defined; it is called a rain shadow, reflecting the lack of rain. In the western United States, much of the desert in Nevada lies in the rain shadow of the Sierra Nevada mountains of eastern California. This range traps moisture coming off the Pacific Ocean in winds that blow from west to east, the midlatitude westerlies.

The second factor contributing to desert formation has to do with the distance moisture is transported in the atmosphere. Since moisture gets into the atmosphere principally by evaporation from the ocean, distance from the ocean is another cause of low rainfall. The vast mountainous central Asian deserts of Kazakhstan, Afghanistan, Mongolia, and the Tarim Basin and Gobi Desert of northern China are far downwind from any oceans (**FIGURE 19.14**). Their extreme interior location ensures that by the time air masses reach them, most of the moisture has been extracted from the air.

Polar deserts form as a result of a third important factor: that is, the very cold air descending over polar regions contains little water vapor. This dry air, and the minimal water vapor it does contain, warms and expands as it descends, further inhibiting cloud formation.

A fourth factor contributing to desert formation is a cold ocean current next to a tropical coast. Cold ocean currents flow toward the equator along the west coast of most continents. Cold air above these ocean regions will move onshore over hot land and quickly and dramatically heat up and expand. This process causes high rates of evaporation that produces few clouds and little rain. The deserts along the west coasts of South America (Peru and Chile) and Africa (Namibia and Angola) are particularly dry, in part due to the presence of a cold current just offshore.

Another influential factor in desert formation is poor management of farmland by humans. Read more about this issue in the Earth Citizenship feature titled "The Dust Bowl."

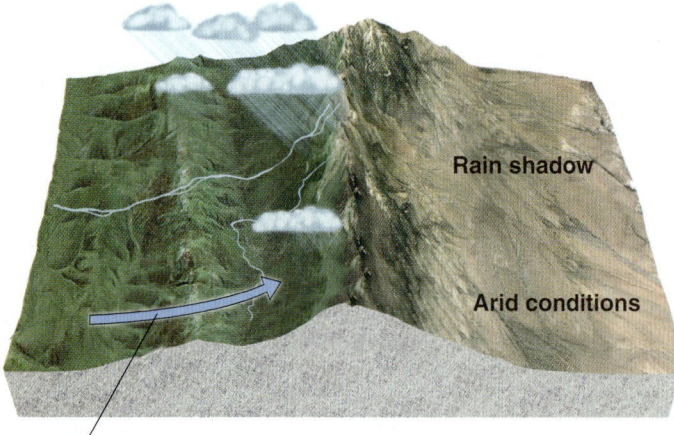

Orographic uplift

FIGURE 19.13 Moist air is forced upward by orographic uplift in the presence of mountain ranges. A rain shadow is produced on the downwind side of the mountains.

Why does a rain shadow develop on the downwind side of large topographic features?

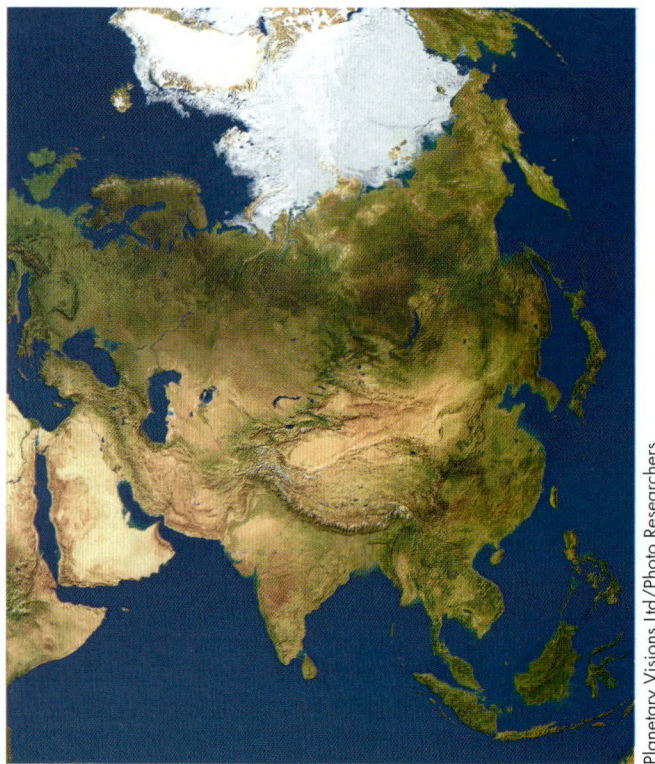

Planetary Visions Ltd/Photo Researchers

FIGURE 19.14 The great deserts of central Asia are far from any moisture source.

What factors contribute to the great deserts of central Asia?

Q EXPAND YOUR THINKING

State a hypothesis explaining (accounting for) the climate in your area. How would you test your hypothesis?

EARTH CITIZENSHIP

The Dust Bowl

Known to early European explorers as the "Great Desert," the plains of central North America frequently experience prolonged drought. Often a decade or more in length, these droughts greatly reduce the native vegetation, allowing the regions strong winds to produce large dust storms. During the nineteenth and early twentieth centuries, in the aftermath of repeated periods of drought in the plains, the region was swept over and over by immense dust clouds. And as it turned out, these were only a prelude for what was to come.

By the late 1920s, advances in agricultural technology were reshaping the plains into the "bread basket of the world." New-design tractors pulling disk plows expanded the acreage of land cultivated for wheat. In time, however, increased wheat production depressed world wheat prices; the reaction of farmers was to grow as much wheat as they could to avoid losing money.

In good years, rain was sufficient to support these wheat crops on the Great Plains—barely. Then, in 1931, disaster struck; no rain fell. When spring returned in 1932, the combination of thoroughly tilled, bone-dry soil and strong winds severely eroded farmland. So much soil was picked up and blown around that the term "black blizzard" was coined to describe the storms.

The relentless, choking dust got into everything. Farmers hung wet sheets over windows in an effort—largely futile—to keep dust out; children wore dust masks to and from school; and crops blew away with soil that was too dry to stay anchored (FIGURE 19.15). The Dust Bowl of the 1930s had begun.

The crisis in the farming economy brought about by the Dust Bowl was so severe that the federal government was forced to take steps to try to limit the damage. Farmers were paid to plant and cultivate native trees that could hold soil and act as wind barriers. Mortgages were refinanced to help farmers facing foreclosure. The Federal Surplus Relief Corporation was created to pay for and distribute surplus agricultural goods, as well as to feed and clothe those in need.

Experts estimated that 850 million tons of topsoil were blown off the surface of the Great Plains in 1935 alone. Over 1 million square kilometers of agricultural land was significantly damaged by topsoil loss. The dust-storm damage peaked on April 14, 1935, with so-called Black Sunday, when the worst "black blizzard" swept through. Less than two weeks later, Congress instituted Soil Conservation Districts, whereby farmers were paid to use techniques that conserve the soil.

By the time the series of droughts ended, in the fall of 1939, the American landscape had been dramatically transformed. And a mass migration out of the drought-ravaged Midwest set California on its path to becoming an agricultural, cultural, and political giant. Farming techniques also were revised, to preserve what everyone now knew was valuable, irreplaceable topsoil.

The Dust Bowl is an extreme example of desertification in America, attributable to a combination of drought and faulty agricultural practices. It illustrates what has happened, and is happening, in many areas around the world where agricultural practices that ignore the potential for soil erosion clash with the environment.

Science Source/Photo Researchers

FIGURE 19.15 The Dust Bowl was an agricultural, ecological, and economic disaster in the Great Plains region of North America.

19-4 Each Desert Has Unique Characteristics

LO 19-4 Identify the principal desert types.

There are many types of deserts, including trade wind deserts, midlatitude deserts, rain shadow deserts, coastal deserts, monsoon deserts, and polar deserts.

Desert Type Descriptions

Every researcher has his or her favorite classification system for the several types of deserts; the one we will adopt for the purpose of this discussion is the system proposed by the U.S. Geological Survey (USGS) that identifies deserts by their geographical location and the dominant weather pattern governing the availability of water.

Recall from the previous section that the factors that lead to desert formation are global atmospheric circulation, which produces areas of dry descending air; regions that fall in the shadow of an orographic barrier; dry polar regions (also related to high atmospheric pressure); remote continental areas located far from the ocean; and tropical coastal areas adjacent to cold ocean currents. Each of these factors is related to the types of deserts defined by the USGS system.

To reiterate from above, the desert types we cover here are trade wind, midlatitude, rain shadow, coastal, monsoon, and polar deserts.

Trade wind deserts are found in the northern and southern hemispheres in the belt of the trade winds. Trade winds originate at the subtropical high-pressure zones near 30° N and 30° S latitude. These air currents flow toward the equator out of the northeast in the northern hemisphere and the southeast in the southern hemisphere.

Trade winds are the product of dry descending air masses at the boundaries of the Hadley and Ferrel circulation cells. As the air masses descend, they heat up and their rate of evaporation rises; hence, they hold more moisture, and areas below them lose moisture rather than gain it. Trade winds are the extension of this dry air. As they sweep toward the equator, they dissipate much of the cloud cover they encounter, evaporating moisture and allowing more sunlight to heat the land. The Kalahari in southern Africa is an example of a trade wind desert (**FIGURE 19.16**).

Midlatitude deserts occur between 30° and 50° N and S poleward of the subtropical high-pressure zones. These deserts lie in remote continental regions far from oceans and have a wide range of annual temperatures; they often experience snowfall in the winter and scorching temperatures in the summer.

The Gobi of southern Mongolia is a typical midlatitude desert (**FIGURE 19.17**). These deserts suffer from a lack of precipitation because the descending air masses at the boundary of the Hadley and Ferrel cells are dry. Winds at midlatitude deserts tend to be westerlies, originating at the boundary of the two cells. At midlatitude deserts, the air flows poleward from the high-pressure zone rather than equatorward, as is the case for trade wind deserts.

Rain shadow deserts appear where topographic barriers drive winds up to cool high altitudes, where the rate of moisture condensation exceeds the rate of evaporation within the air mass. Since most available moisture goes into cloud formation, the air that descends on the lee side is dry, forming a rain shadow desert. The Great Basin Desert of North America (**FIGURE 19.18**) has emerged in the rain shadow of the Sierra Nevada mountains.

The fourth type of desert, called a *coastal desert*, is found on the western edges of continents in both the northern and southern hemispheres where cold ocean currents come close to shore (**FIGURE 19.19**). The cold air above these currents is drawn over the land by heating, and as the air mass rises, it leads to strong evaporation of surface water. The high rates of evaporation produce few clouds and little rain.

FIGURE 19.16 The Kalahari is a trade wind desert that lies between 30° S latitude and the equator in southern Africa.

Explain why trade winds are dry.

Reinhard Dirscherl/WaterFrame/Getty Images

Pixtal/SuperStock

FIGURE 19.17 Lying north of the subtropical high-pressure zones, midlatitude deserts have a wide range of annual temperatures. The Gobi is the fifth largest desert in the world.

What is the primary cause of the midlatitude deserts? Classify each of the deserts in Table 19.1 in terms of the principal desert types.

FIGURE 19.18 The Great Basin Desert of North America lies in the rain shadow of the Sierra Nevada in California.

Ⓠ **What causes the predominantly western air flow over the Sierras?**

The Indian *monsoon* is a seasonal wind that flows off the Indian Ocean and Arabian Sea and prevails for several months in the summer. It brings intense rainfall produced by condensation as the warm, moisture-laden air lifts to high altitudes. *Monsoon deserts* are found where the rain is blocked by mountain ranges that cast a rain shadow.

India, a nation whose climate is controlled by the annual monsoon, is home to the Rajasthan Desert, and neighboring Pakistan is home to the Thar Desert. Both of these monsoon deserts are located in the rain shadow of the Aravalli Range, of western India.

Polar deserts develop in the Arctic and Antarctic where the cold climate prevents air from holding even a small amount of moisture. Rain or snowfall freezes so quickly that the surface water, which would otherwise support plants and animals, is locked up in frozen landscapes of ice.

Antarctica is the world's largest polar desert. Precipitation is in the form of snow, but in terms of the equivalent amount of water, the average annual precipitation is only about 50 millimeters, less than the Sahara receives. Along the coast, the amount of precipitation increases, but still adds up to only about 200 millimeters. Unlike other deserts, there is little evaporation in Antarctica; and in the past, the relatively little snow that did fall did not melt; rather, it built up over hundreds and thousands of years into enormously thick ice sheets. Now, however, as revealed by data from satellites, the overall rate of melting exceeds the rate of ice accumulation on the Antarctic continent, due to global warming.

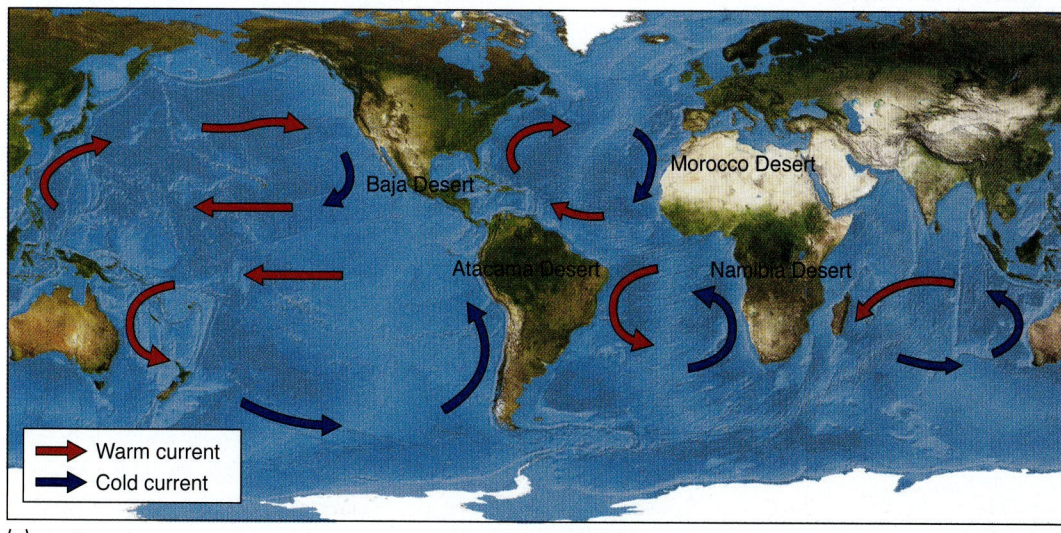

(a)

(b)

Janos Csernoch/Alamy

FIGURE 19.19 (a) Where cold ocean currents approach land, cold air will heat and generate intense evaporation. (b) The Atacama Desert, located in coastal Chile, is one of the driest places on Earth.

Ⓠ **Four coastal deserts are shown. What do you think is the likely influence of atmospheric circulation on each? (Refer back to section 19–2.)**

Ⓠ **EXPAND YOUR THINKING**

One effect of climate change is to warm the atmosphere and oceans. Consider how this will affect each of the processes responsible for the principal desert types, and hypothesize how each of the deserts listed in Table 19.1 is likely to be different in the future. How would you monitor the expected changes?

19-5 Wind Is an Important Geological Agent

LO 19-5 Describe sedimentary processes in deserts.

The lack of water and vegetation in deserts makes the wind a powerful agent of erosion and deposition in these landscapes. Consequently, many geologic features in the desert are characteristic of wind erosion, deposition, or moving sediment.

The Power of Wind

Unlike water, wind is not confined to a channel, so it can affect broad tracts of land. Wind is much less dense than water, however, and so generally it moves grains of sand and silt only. Larger grains are too heavy to budge, unless the wind speed reaches extreme levels.

There are two methods by which wind transports sediment. Which method prevails is determined by the speed of the wind and the size of the sediment. We have already encountered the first method in our study of water erosion in Chapters 7 and 17; it is called *saltation,* and refers to the process by which grains move short distances before falling back to the ground. Saltation occurs close to the ground with grains that are sand size (0.625 to 2 millimeters in diameter).

Under the force of wind erosion—just as with water erosion—when a grain lands, it may dislodge other grains and cause them to saltate downwind; and when they land, they in turn dislodge more grains. This sequence of motion can appear as if the grains are leaping downwind—like a beach ball bouncing across the sand.

If the grains are small (typically, the size of silt), they may not fall back to the ground but instead be carried *in suspension* by the wind. This process is enhanced when the wind is blowing at high speed and is turbulent. Suspended grains can reach great altitudes and travel thousands of kilometers. Dust suspended by the wind during storms in the Sahara and Gobi have crossed the Atlantic and Pacific oceans.

Once grains begin to accumulate—often around an obstacle of some sort, such as vegetation—**ripples** (centimeter-scale features) and **dunes** (meter-scale features) can develop (**FIGURE 19.20**). A simple ripple or dune has two angled slopes, one facing upwind (the *stoss* slope) and another facing downwind (the *lee* slope). Wind will continuously push grains up the stoss slope to the top of the ripple until it collapses under its own weight. This occurs by avalanching down the *slip face* on the lee side.

The collapsing sand comes to rest when it reaches a slope of about 30° to 34° (the *angle of repose*). Every pile of loose particles has a unique angle of repose, based on the properties of its material composition, such as grain size and roundness. Ripples will grow into dunes as the wind speed picks up and sand availability increases. The repeating cycle of sand inching up the windward side to the dune crest, then sliding down the slip face, allows the dune (or ripple) to move forward, migrating in the direction the wind blows.

Deposition of windblown sediments leaves highly sorted deposits where the wind slows down. Sediments of a particular size are left in a location appropriate to that size; sediments of other sizes are deposited likewise. As wind velocity slows, large grains are deposited first; small grains are transported farther.

Loess (pronounced "lus," as in luck) is a notable example of a windblown deposit. Loess deposits are composed of windblown silt exposed by repeated advances and retreats of continental glaciers. Loess deposits create particularly rich soil, explaining the substantial loess deposits in major agriculture areas in Asia (the Yellow River valley), northern Europe (the plains of Germany, Poland, Ukraine, and western Russia), and North America (the American Midwest) (**FIGURE 19.21**).

If all the fine material on a desert surface is transported away by the wind, what is left is a *lag deposit* composed of coarse sediment (often, gravel size) that the wind is unable to move. When a lag deposit is extensive, it is called a **desert pavement** (**FIGURE 19.22**).

Sand grains are the primary cause of *abrasion* in the desert. Their size and weight usually prevents them from being lifted more than 1 meter off the ground—the reason that abrasion is confined close to the ground. Rock outcrops may be sculpted by sand abrasion, shaping what is known as a *yardang* (**FIGURE 19.23**). In places with a predominant wind direction, rocks will be abraded on one side. Pebbles, cobbles, and boulders with such flat wind-abraded surfaces are called *ventifacts,* visible in Figure 19.22. Shifting winds or a rock that has been turned may cause a ventifact to have more than one flat surface.

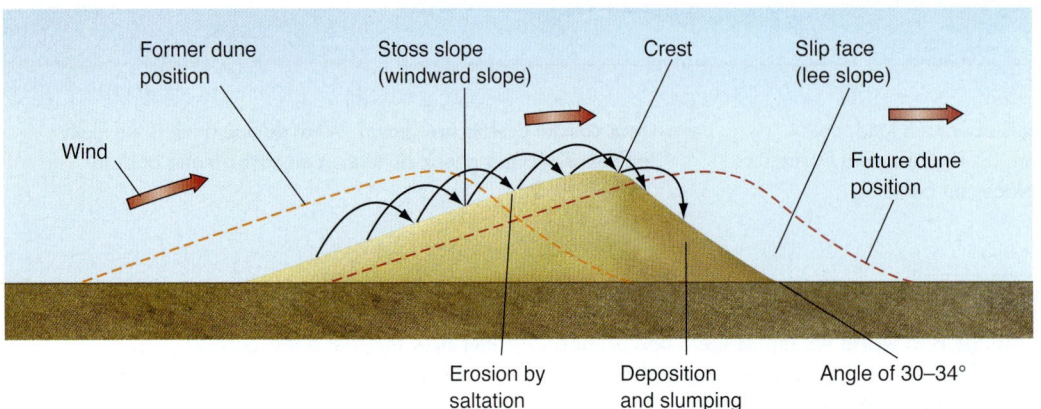

FIGURE 19.20 Ripples and dunes migrate downwind as sand travels up the windward slope, collects at the top of the pile until it becomes unstable, and then avalanches down the leeward slope.

As a ripple or dune migrates with the wind, what would you predict the internal structure (or organization of the grains) would look like?

(a)

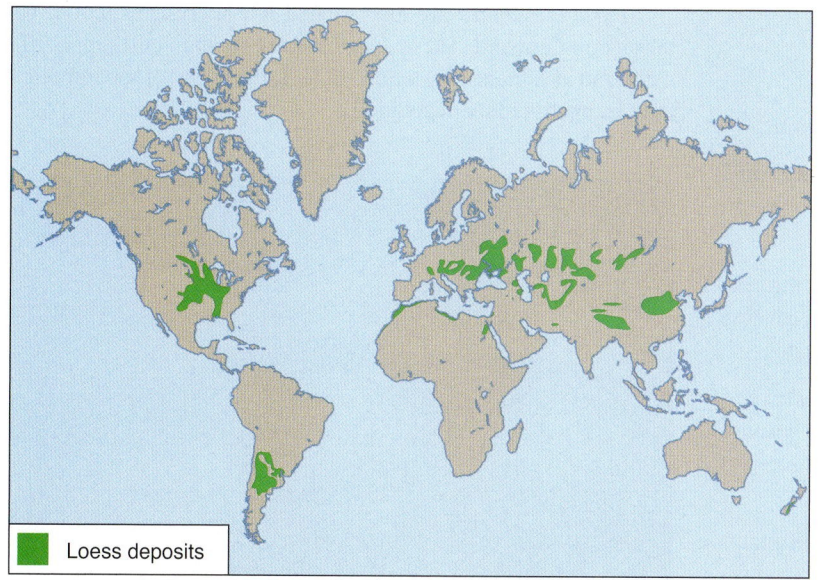

(b)

(c)

PIXTAL/Age Fotostock America, Inc.

FIGURE 19.21 Loess is silt deposited by the wind. (a) Wind will deposit heavier grains, such as sand, first, but may continue to transport silt for great distances, piling up loess deposits. (b) Many loess deposits are found downwind of formerly glaciated areas. (c) Loess landscapes can be massive in scale, dominating the geology of the surface environments.

Ⓠ **Why does loess develop in formerly glaciated regions?**

Reg Morrison/AUSCAPE

FIGURE 19.22 Desert pavement forms where the wind erodes fine sediments and leaves behind a lag deposit of coarse grains.

Ⓠ **What conditions promote the development of desert pavement?**

Mike P Shepherd/Alamy

FIGURE 19.23 A yardang is a wind-abraded ridge appearing in a desert environment.

Ⓠ **How does a wind-abraded ridge form?**

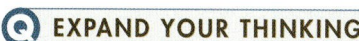

Ⓠ **EXPAND YOUR THINKING**

Desert pavement is also referred to as "armoring." Why? What is the net effect of armoring on sediment transport?

19-6 Sand Dunes Reflect Sediment Availability and Dominant Wind Direction

LO 19-6 Define the primary types of dunes.

Sand dunes, some of the most spectacular and dynamic landscapes found in arid regions, immediately come to mind for most people when they think of deserts.

Sand Dune Formation

Four conditions need to be met in order for **sand dunes** to form:

1. It is important to have abundant *loose sediment* (usually, sand size). Environments with abundant loose sediment include beaches, river channels, and deserts.

2. There must be *sufficient energy* to move the sediment. In the case of deserts, this would be wind energy, but dunes also can develop underwater in rivers from sand transported by running water.

3. There must be some *obstacle around which sand accumulates*. Areas with sparse vegetation commonly develop dunes because individual bushes can trap moving sand and start the process of building a dune. Dunes also take shape around rocks.

4. Desert dunes require a *dry climate*. Moisture sticks sand grains together, and larger grains, or clumps of grains, are more difficult to transport by wind. Moisture also facilitates plant growth, which stabilizes loose sediment and prevents it from moving (**FIGURE 19.24**).

Dune Types and Shapes

Dunes are classified on the basis of shape, but not all dunes can be categorized because many take irregular shapes. There are four prin-

cipal types and their accompanying shapes, listed and described in **TABLE 19.2**. The first three are asymmetrical with a gently sloping windward side and a steeply sloping leeward face. The fourth type, the *star dune*, is complex, displaying arms and ridges constructed by wind blowing from several different directions.

Crescentic dunes (**FIGURE 19.25**) are the most common. They are generally wider than they are long and come in two varieties. *Barchan dunes* are shaped like a crescent moon with the horns pointing downwind. These dunes, which emerge where sediment is limited, may be separated from one another by lengthy distances of bare rock. *Transverse dunes* are long, wavy linear dunes oriented perpendicular to the wind direction. Like barchan dunes, transverse dunes have their steep slip face on the concave side of the dune.

Parabolic dunes are also shaped like a crescent moon, but the horns point upwind. These dunes require an abundant supply of sediment and sufficient vegetation. In fact, the horns of a parabolic dune may be anchored by vegetation.

Jan Tove/Johner Images//Getty Images, Inc.

FIGURE 19.24 The formation of dunes requires loose sediment (usually, sand), energy to move the sediment (usually, wind), an obstacle to trap sand (often, a bush), and a dry climate.

 What criteria are used for classifying various types of dunes?

TABLE 19.2 Sand Dunes

Dune Type	Shape	Wind Characteristics	Environment of Deposition	Grain Characteristics
1. Crescentic				
a. Barchan	Crescent moon; horns point downwind	Constant wind from one direction	Limited vegetation and sediment, often appearing on flat bare rock	Well-sorted; very fine to medium sand
b. Transverse	Long, wavy, linear; oriented perpendicular to wind direction	Constant, moderate wind from one direction	Limited vegetation; more sediment than barchan	Well-sorted; very fine to medium sand
2. Parabolic	Crescent moon; horns point upwind	Variable strength, unidirectional wind	Abundant supply of sand and vegetation common	Well-sorted; very fine to medium sand
3. Longitudinal	Large, straight, or sinuous sand ridge; generally much longer than wide	Strong, steady winds that blow from two directions	Form in parallel sets of sand ridges	Well-sorted; very fine to medium sand
4. Star	Pyramid; three or more arms radiate from a peaked center	Wind blows from several different directions	Grow taller, rather than migrating	Well-sorted; very fine to medium sand

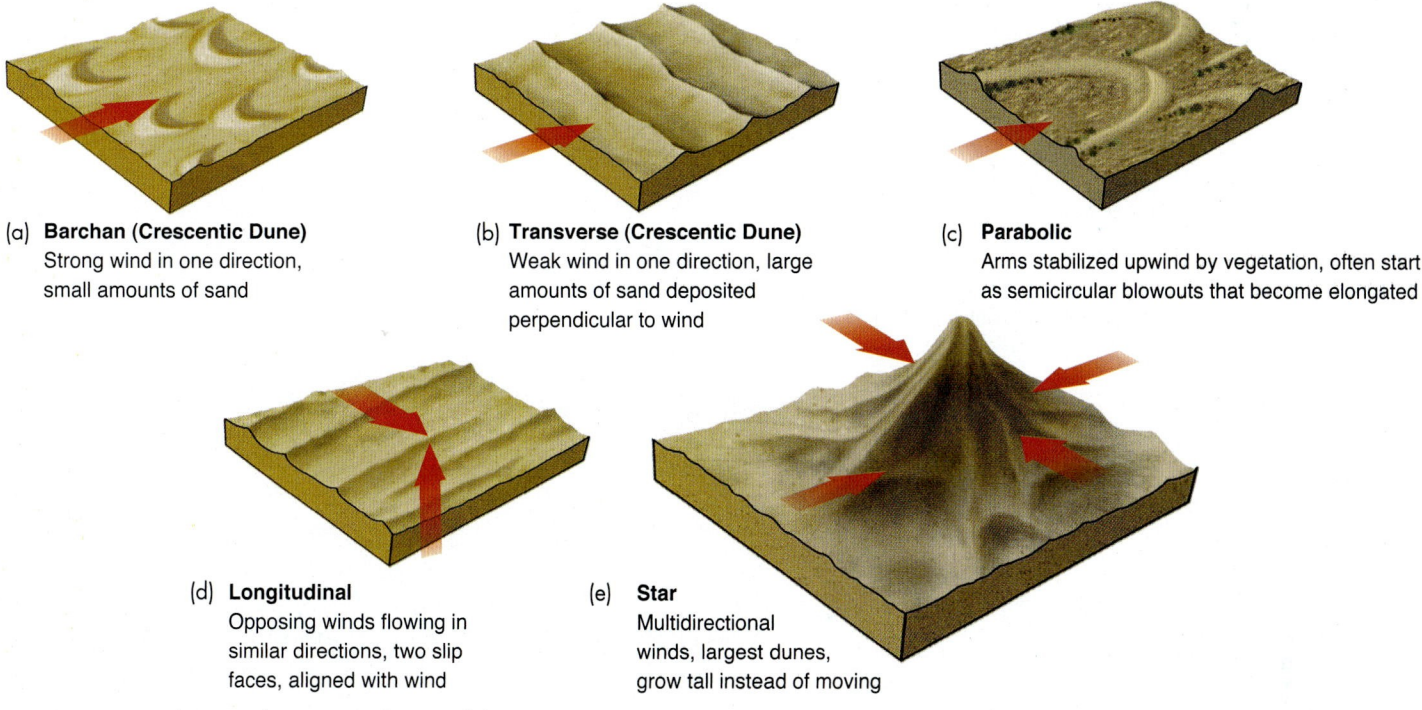

(a) **Barchan (Crescentic Dune)**
Strong wind in one direction, small amounts of sand

(b) **Transverse (Crescentic Dune)**
Weak wind in one direction, large amounts of sand deposited perpendicular to wind

(c) **Parabolic**
Arms stabilized upwind by vegetation, often start as semicircular blowouts that become elongated

(d) **Longitudinal**
Opposing winds flowing in similar directions, two slip faces, aligned with wind

(e) **Star**
Multidirectional winds, largest dunes, grow tall instead of moving

FIGURE 19.25 There are four principal types of dunes: crescentic, including (a) barchan and (b) transverse; parabolic (c), longitudinal (d); and star (e).

Q **Will a dune form if vegetation is present?**

Longitudinal dunes are long, straight, or slightly sinuous sand ridges generally much longer than wide. They usually develop in areas with two directions of wind blowing over the year, often associated with separate seasons. The long axis of these dunes is parallel to the direction of net sand movement.

Star dunes are pyramid-shape sand mounds with steep slip faces on three or more arms that radiate from the elevated center of the mound. They are located in regions where the wind blows from several different directions.

Sand dunes can form in large *sand seas,* regions (greater than 125 square kilometers) of windblown sand that contain numerous, very large dunes where sand covers more than 20% of the ground surface (**FIGURE 19.26**). Also known by the Arabic word *erg*, sand seas differ from *dune fields* mainly in size and complexity of dune shapes. Sand seas develop downwind from large volumes of loose, dry sand. Dry and abandoned riverbeds, floodplains, glacial outwash plains, dry lakes, and beaches are all excellent sources of sand. Smaller dune fields, in contrast, are considered local features and, as such, have smaller and simpler dunes than sand seas.

The Critical Thinking feature, "Interpreting Dunes," gives you an opportunity to analyze more desert features.

Image courtesy NASA/GSFC/MITI/ERSDAC/JAROS, and U.S./Japan ASTER Science Team

FIGURE 19.26 A sand sea, or erg, is probably what most people imagine when they picture a desert. Although not particularly common, a sand sea is the archetypical and most dramatic desert landscape. This is the Namib Erg in southern Africa.

Q **What conditions lead to the formation of an erg?**

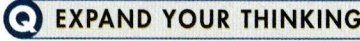

Q **EXPAND YOUR THINKING**

What would happen to a parabolic dune if the wind direction changed and precipitation decreased?

CRITICAL THINKING

Interpreting Dunes

FIGURE 19.27 depicts a landscape of dune types. Your assignment is to analyze the morphology of these dunes and interpret the environmental conditions. Work with a partner to complete the following exercise:

1. Name each dune type.

2. Describe the environmental conditions leading to each dune type.

   a. Draw arrows indicating the dominant wind for each dune type.

   b. Describe the apparent sediment abundance related to each dune type.

   c. Indicate sediment sources, the role of vegetation, and transitional dune forms. Describe each.

3. Considering all the above factors, and based on the several deserts we have studied:

   a. Where might this desert be located?

   b. Describe the desert-forming processes active there.

   c. List the features in this scene that are consistent with the location you have identified.

   d. List the features in this scene that are not consistent with the location you have identified.

4. Make a value decision about whether the scene here is realistic. Could this assemblage of dune morphologies be found in this configuration? What criteria will you consider in arriving at your decision? Support your decision with specific observations and formulate an answer based on the weight of evidence. Lay out the case both for and against this being a realistic scene. Incorporate actual processes operating at the location you identified in your answer to question 3.

5. Sketch a scene including all five dune types that you consider realistic, and assign it to an actual geographic location. Label all the features and use arrows and other indicators to show the relevant desert-forming processes.

FIGURE 19.27 Dune types that may be found in a sand sea.

19-7 Paradoxically, Arid Landforms Are Shaped by Water

LO 19-7 Discuss the role of water in desert landscapes.

Although desert landscapes may look very different from humid landscapes, they are shaped by the same geologic forces.

Arid Processes

Compared to humid landscapes, the aridity of the desert environment emphasizes different aspects of the same processes and highlights the degree to which the different climates affect those processes.

Chemical weathering is an important agent of change in humid climates. The lack of moisture and organic acids from decaying plants restricts the influence of chemical weathering in arid climates, but it is not entirely absent. Over time, clay minerals and soils do form, and iron-rich minerals oxidize to create the rich red colors so prevalent in some deserts.

Permanent streams are, of course, very rare in desert environments, due to low rainfall. In their place are **ephemeral streams**, in which water flows only after rainfall events (**FIGURE 19.28**). An ephemeral stream may have water in it for only a few hours or days each year, usually immediately after a rainstorm.

Normally, desert rains fall hard, and briefly, so the water does not soak into the hard-baked soil. Much of it becomes surface flow; and because vegetation is thin, runoff into dry streambeds is rapid. These dry streambeds fill quickly with fast-moving water, setting off *flash floods* in the wake of the heavy downpour. They are called, variously, *wash,* in English; *arroyo,* in Spanish; *wadi,* in Arabic; *oued,* in French; *vadi,* in Hebrew; and *nullah,* in Hindi.

© Marli Bryant Miller

FIGURE 19.28 Because of low rainfall in the desert, streams in them tend to be ephemeral, flowing only after heavy rainstorms.

○ **What are the characteristics of flow in an ephemeral stream?**

Unlike floods on perennial streams, which may take days to reach their crests and then subside, flash floods may rise and fall within a few hours. And the lack of vegetation helps the fast-moving water to erode unanchored soil and rocks at a much greater rate than for a similar-size flood in a humid climate.

Another consequence of low rainfall is that rivers in deserts tend to be small, and disappear before they reach the ocean. Low rainfall, coupled with high evaporation and infiltration into the streambed, means that more water is leaving the stream than entering it; such a stream, unless quite large to begin with, will quickly disappear.

The Nile and Colorado rivers are notable examples of large desert rivers that originate outside of a desert environment. In these cases, the rivers start in mountains with snow or rainfall supplying enough water to the rivers to overcome losses occurring as they traverse the desert regions. Today, of these two great waterways, only the Nile River reaches the sea; the Colorado River no longer does due to water removal by humans for irrigation and drinking.

Despite low rainfall, most erosion in deserts is nevertheless brought about by flowing water. Wind erosion is more of a force in deserts than in other environments, but flowing water is still the main erosive agent in deserts.

Landform Evolution

The Basin and Range Province of the southwestern United States (recall Chapter 10) features many particularly well-developed examples of desert landscapes and their evolution. This region is a result of tectonic forces spreading and thinning Earth's crust and precipitating the emergence of fault-block mountains.

As the name implies, the Basin and Range Province is characterized by hundreds of small north-south trending ranges separated by long and narrow basins. Much of this area is downwind of the Sierra Nevada mountains and thus lies within its rain shadow. Furthermore, the north-south orientation of the ranges produces multiple rain shadow effects against the dominant winds that blow west to east (westerlies). Consequently, much of this area is extremely dry and so serves as an excellent case study of desert landscapes.

Ridge crests in the Basin and Range Province have sufficiently high elevations to receive snow and rain. Therefore, they are chemically and physically weathered, and the resulting sediment is eroded during flash floods and deposited in the basins. The basins have flat floors, speeding the deposit of sediment at the mouth of discharging streams.

As this sediment builds up, it takes shape as an *alluvial fan.* Coarse sediment is deposited near the top of the fan where the stream emerges from the mountains; fine sediment is deposited at the base of the fan or on the floor of the basin. Through time, alluvial fans grow and spread along the base of the range.

When a series of alluvial fans from adjacent stream valleys grow large enough to join together, they form a *bajada* (**FIGURE 19.29**). Bajadas are broad, gently sloping depositional surfaces lining the entire front of a mountain range where they are crossed by meandering dry streambeds.

After a rain event, the floor of a basin may hold a shallow lake called a **playa lake** (Spanish for "beach"), a body of water controlled by rainfall and evaporation. Since these basins often have no drainage outlets, evaporation is the primary means by which water leaves playa lakes.

When a playa lake evaporates, it leaves a thin layer of organic-rich mud on the floor of the basin, which eventually becomes *oil shale.*

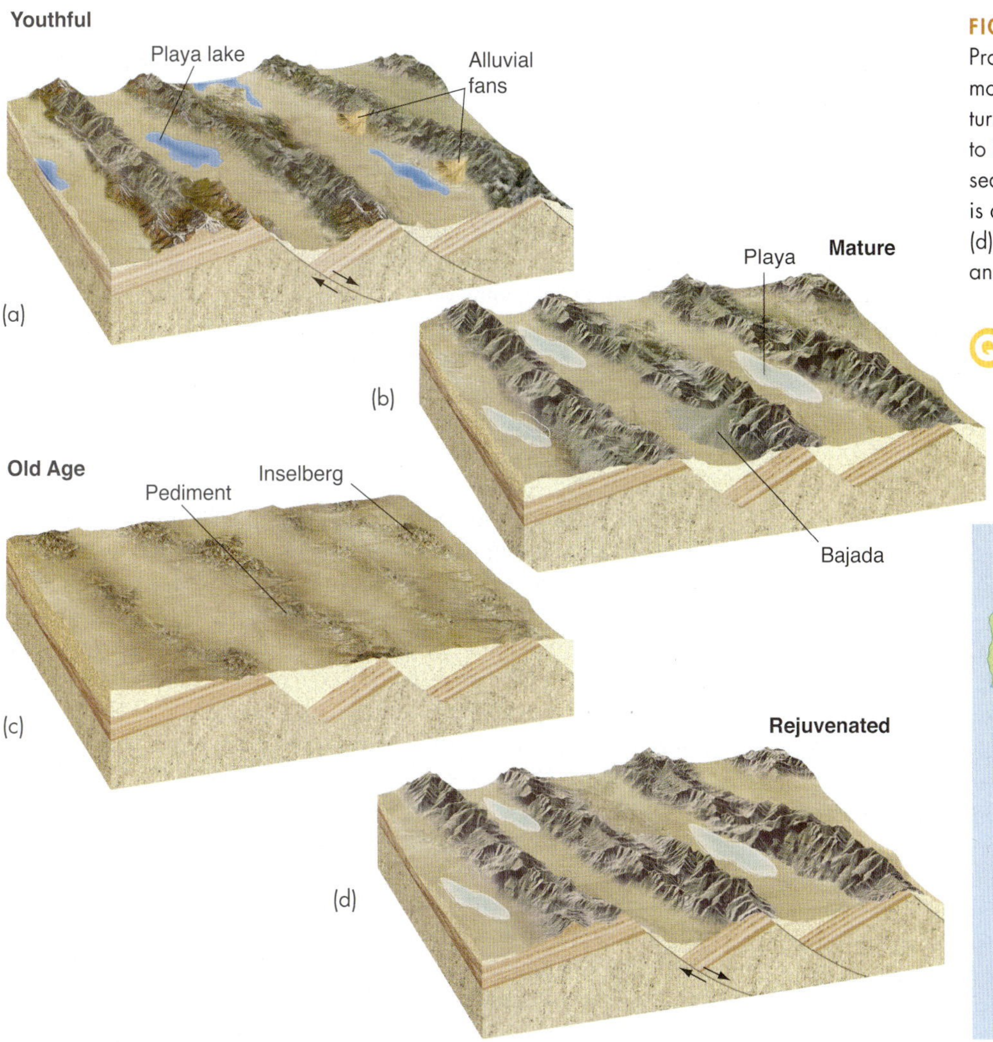

Youthful

Playa lake

Alluvial fans

(a)

Mature

Playa

(b)

Old Age

Inselberg

Pediment

(c)

Bajada

Rejuvenated

(d)

FIGURE 19.29 (a) The Basin and Range Province consists of hundreds of short fault-block mountains and narrow basins that fill with water, turning into playa lakes. (b) Weathering tends to lower the mountains and fill the basins with sediment. (c) Burial continues until all that is left is an erosional remnant called an inselberg. (d) Normal faulting raises the mountains again and rejuvenates the landscape.

Q **How would this landscape ultimately look if there were no future faulting?**

Basin and Range

Remember, the much coarser gravel and sand have been deposited on the alluvial fans, or bajada, and so they do not reach the playa lakes.

If the water flowing into the playa lake contains a significant amount of salt, a layer of white salt will be left on the floor of the basin when it evaporates. This basin floor with its very flat surface composed of hard-baked mud and salt is called a *playa*. Death Valley in California and the Bonneville Salt Flats of Utah are excellent examples of playas (**FIGURE 19.30**).

Over time, the ranges erode and the basins fill with their sediments. Eventually, all that may be visible above the flat plain of sediment is a small tip of rock called an *inselberg* (German for "island mountain"). Perhaps the best-known inselberg is Ayers Rock in Australia. Where the sediment over the eroded bedrock is thin or discontinuous, a *pediment* is formed. A pediment may look similar to a bajada, but it is a bedrock surface with a thin veneer of alluvial sediment, rather than a thick layer of loose sediment, the defining characteristic of a bajada.

The final step in this landscape evolution is when normal faulting once again lifts the mountains into high, sharp peaks, and the process begins anew.

FIGURE 19.30 A playa is a flat, dry lake bed of hard, mud-cracked clay and salt.

Q **Develop a hypothesis that predicts the stratigraphy of a playa. How can you test your hypothesis?**

Stephen Marks/Getty Images, Inc.

Q **EXPAND YOUR THINKING**

Why would the layers of organic mud in playa lakes lead to the formation of oil shale? What are the geologic steps involved from silt deposition to making oil shale?

19-8 Desertification Threatens All Six Inhabited Continents

LO 19-8 Describe the conditions leading to desertification.

Currently, over 250 million people experience the direct consequences of desertification. Many of them are the world's most destitute and vulnerable citizens.

Desertification

Recall from early in the chapter that desertification is the process by which land loses its vegetation and turns into a desert. (Remember that the second characteristic defining a desert is a lack of vegetation.) Characteristics of desertification include destruction of native vegetation (**FIGURE 19.31**), unusually high rates of soil erosion, declines in surface water supplies, rising levels of water and topsoil saltiness, and widespread lowering of groundwater tables.

Although desertification can be a natural process, progressing over thousands or millions of years, today it frequently has a combination of natural and human causes. Notably, plants native to a semiarid environment are adapted to occasional drought conditions, but global warming can lead to variations in the distribution of water abundance and hamper the capability of vegetation to survive in already marginal conditions. And when human activities such as farming and ranching weaken, damage, or clear much of the native vegetation in an area, a few drought years may kill off remaining natural plants (most crucially, those edible by humans and animals), until the area takes on many characteristics of a desert, even if it does not meet the overriding definition of a desert (i.e., less than 25 centimeters of rain per year).

Although desertification is a more serious problem in poor, heavily populated countries, more than 35% of the semiarid regions of

FIGURE 19.31 Desertification is the process by which land loses its vegetation and turns into a desert.

GSFC Collection/Alamy

ⓠ **What conditions lead to an increase in desertification?**

North America have experienced "severe" desertification, and all six inhabited continents have large areas threatened by this process (**FIGURE 19.32**).

Case in Point: The Sahel

The Sahel region of North Africa delineates the border between the shifting sands of the Sahara in the north and the moist tropical rain forest regions of central Africa. (Two Arabic words are the source of the terms for these regions: *Sahara* is Arabic for "desert," and **sahel** for "border" or "margin.") This swath stretches from the Atlantic to the Indian Ocean and includes parts or all of the countries of Senegal, Mauritania, Mali, Burkina Faso, Niger, Nigeria, Chad, Sudan, Ethiopia, Eritrea, Djibouti, and Somalia.

More than 300 million people live in this region of dry and unstable weather, many of them poor farmers or nomadic herders. They rely on local rains, or migrate to follow the rains. Unfortunately, the number of people here exceeds the carrying capacity of the land, and their huge livestock herds denude vegetation that is already stressed by low rainfall. And during the periodic multiyear droughts, the remaining native vegetation withers, crops fail, and millions of livestock animals die. Ultimately, in the wake of livestock losses, crop failures, and low water supplies, millions of people die too, from thirst and starvation, all brought on by the desertification of their lands (**FIGURE 19.33**).

Many of the worst famines and humanitarian crises of the last 50 years have occurred in the Sahel during droughts. These crises, instigated by desertification, frequently were compounded by political strife, started in no small part by individuals under stress stemming from desertification. In an effort to combat desertification, and thereby diffuse ensuing political unrest, the United Nations adopted the *Convention to Combat Desertification*, which over 110 countries have joined. Its purpose is to ensure the long-term productivity of inhabited dry areas.

Recognizing that human activities contribute to the process of desertification, the Convention promotes behaviors that reduce those contributions. It also recognizes that desertification adds to the lethal nature of political unrest and the likelihood of civil war. Clearly, starving people will find it more difficult to get along with their neighbors, and desertification is one of the basic causes of their starvation.

The United Nations also is implementing programs in individual countries based on the threats posed by desertification in specific ecosystems. Local communities are involved to help ensure that the solutions will be implemented in these areas. Each plan includes detailed instructions for how to manage natural resources such as soil, water, forests, livestock, crops, wildlife, mineral resources—even tourist resources. Human socioeconomic concerns are also addressed, among them: farm policies, social and economic infrastructure, natural resource access rights, environmental economics, population growth, and local community participation in natural resource management.

These programs highlight the complex web of processes that combine to create the conditions for desertification. They also acknowledge that local communities know their own regions better than any agency, and that even the best-laid plans will be doomed to failure if local communities are not asked to participate in plan development.

The Sahel is not the only region threatened by desertification. Every continent on Earth has large areas suffering due to this hazardous process. But the experiences in the Sahel can serve as a mirror of the strife that is emerging in other areas of the world where desertification is taking place.

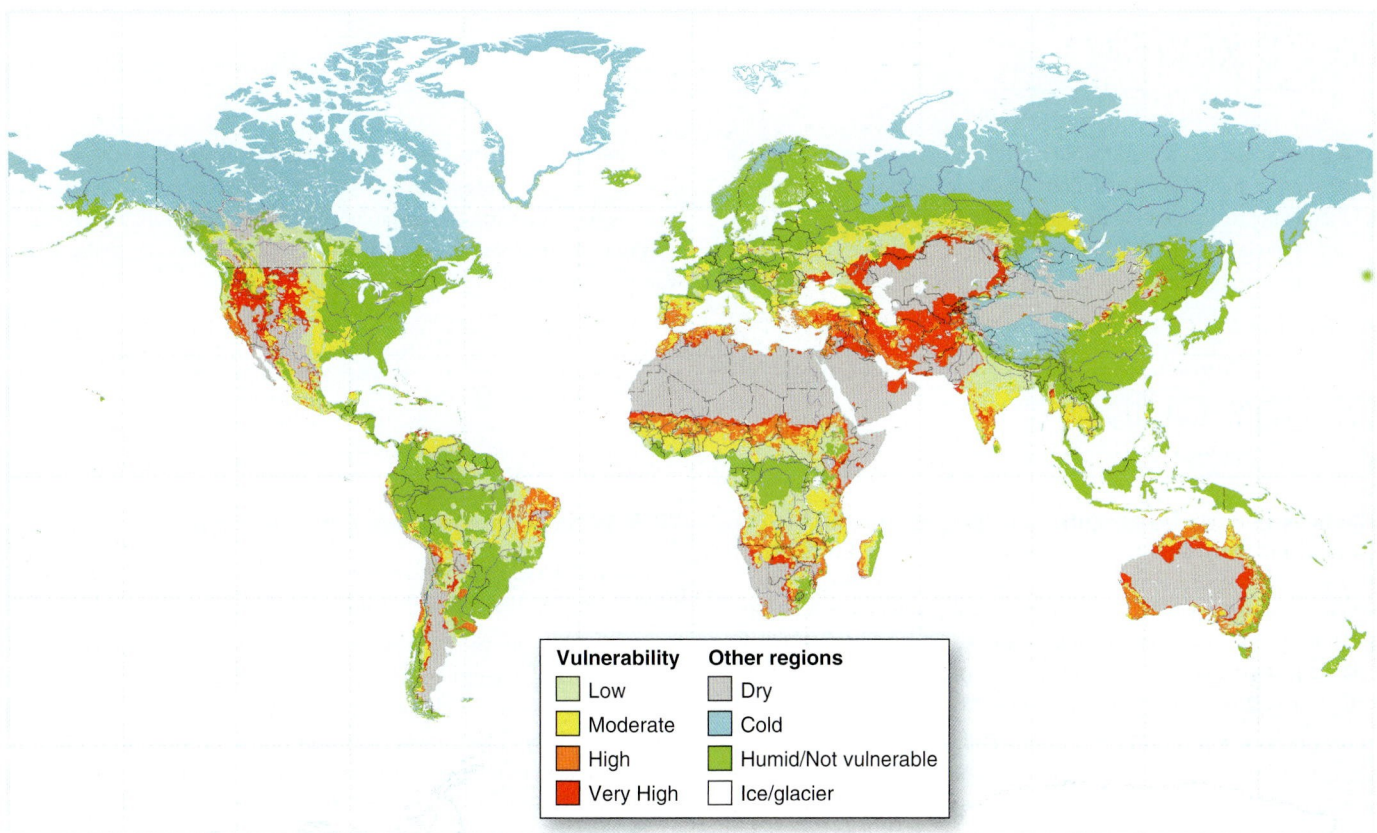

FIGURE 19.32 Desertification vulnerability: all six inhabited continents have large areas threatened by desertification.

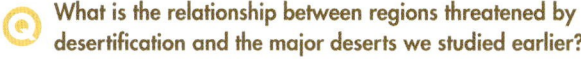 What is the relationship between regions threatened by desertification and the major deserts we studied earlier?

United Nations Environment Programme

FIGURE 19.33 The advance of desertification in the Sahel threatens food and water supplies, producing a population of "environmental refugees." Displaced populations settle on the outskirts of existing towns, as here in El Fasher, northern Sudan, where the distinguishing feature is white plastic sheeting. These new arrivals add to an already heavy environmental burden on the surrounding desert environment.

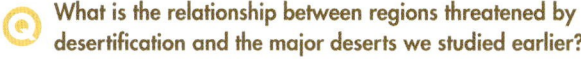 Identify some potential consequences of an influx of environmental refugees to a village already stressed by desertification.

EXPAND YOUR THINKING

How would desertification in American states that border one another potentially lead to political strife? Give specific examples.

LET'S REVIEW

A desert is a landscape that receives little rainfall, has sparse vegetation, and is unable to support significant populations of animals. Deserts are more widespread than many realize, taking up one-third of Earth's land surface. And their range is growing. As larger numbers of people swell into marginally habitable areas bordering deserts, poor land management (such as overgrazing) and changes in the water cycle due to global warming can permanently damage the native vegetation and soil. These factors lead to desertification. With human populations expanding worldwide, and global warming changing rainfall distribution, desertification is likely to be a problem that only worsens in the future.

STUDY GUIDE

19-1 Deserts may be hot or cold, but low precipitation is a common trait.

- A **desert** is an arid region that receives less than 25 centimeters of precipitation per year, and where vegetation is so sparse that the area is unable to support significant populations of animals. Slightly wetter regions, receiving between 25 and 51 centimeters of precipitation, are considered semiarid.

- Expanding deserts, a process known as **desertification**, alter the distribution of natural resources (especially water), which can lead to political unrest when one group of people appropriates the resources of another, or migrates into new, resource-rich lands.

- Deserts experience a wide daily and seasonal temperature range. High daytime temperatures (up to 45ºC), followed by low nighttime temperatures (down to 0ºC), are the result of the low humidity.

19-2 Atmospheric moisture circulation determines the location of most deserts.

- Global atmospheric circulation creates the general characteristics of weather everywhere. The basic components of the system are the Hadley cell, the Ferrel cell, and the Polar cell. There is one of each cell type in the northern hemisphere and one of each cell type in the southern hemisphere.

- Air warms and contracts as it sinks closer to Earth's surface. This causes evaporation to exceed condensation. No clouds will form in locations with a lot of sinking air. These areas, such as at the poles and 30° latitude, will have few clouds and little precipitation, thus forming a great belt of deserts that girdles the globe. Many of the world's deserts are clustered around 30° N and 30° S latitudes for this reason.

19-3 Several factors contribute to desert formation.

- In addition to the location of atmospheric circulation cells, a number of other factors contribute to desert formation: the **orographic effect**, which causes a rain shadow; the distance moisture is transported in the atmosphere; the very cold air descending over polar regions, which contains little water vapor; the presence of a cold ocean current next to a tropical coast; and poor human management of farmland.

19-4 Each desert has unique characteristics.

- There are several types of deserts—trade wind deserts, midlatitude deserts, rain shadow deserts, coastal deserts, monsoon deserts, and polar deserts—each with distinguishing attributes.

19-5 Wind is an important geological agent.

- A lack of water and vegetation makes wind a particularly powerful agent of erosion and deposition in deserts.

- Saltation, the process by which sand grains move short distances before falling back to the ground, occurs close to the ground. Under the force of wind erosion—just as happens during water erosion—when a grain lands, it may dislodge other grains and cause them to move downwind, and in turn dislodge more grains when they land. If the grains are small (typically, silt size), the grains may not fall back to the ground but be carried in suspension by the wind.

- If all the fine material on the surface is transported away by the wind, all that will be left is a lag deposit composed of coarse sediment, which the wind is unable to transport. An extensive lag deposit is called a **desert pavement**.

19-6 Sand dunes reflect sediment availability and dominant wind direction.

- In order for **sand dunes** to form, there must be: 1) abundant loose sediment, 2) sufficient energy to move the sediment, 3) some obstacle around which sand accumulates, and 4) a dry climate.

- There are four principal types of dunes: 1) crescentic, consisting of barchan and transverse dunes; 2) parabolic dunes; 3) longitudinal dunes; and 4) star dunes.

19-7 Paradoxically, arid landforms are shaped by water.

- Desert landscapes are formed by the same geologic forces that shape humid landscapes. The desert environment emphasizes different aspects of the same processes, highlighting the degree to which physical, chemical, and biological processes are variously affected by contrasting climates.

- Low rainfall means that permanent streams are very rare in desert environments. Rather, **ephemeral streams**, where flowing water is present only after rainfall events, are the norm in deserts. These streams quickly fill with fast-moving water, setting off flash floods as a result of heavy downpours that are common in deserts.

- The Basin and Range Province in the United States is characterized by many short, steep mountain ranges and long narrow basins. After a rain event, the floor of a basin may hold a shallow lake called a **playa lake**; but since these basins often have no drainage outlets, evaporation is the primary means by which water leaves such lakes.

- Over time, the ranges erode and fill the basins with their sediments. Eventually, the ranges erode down and the basins fill up, until all that may be visible above the flat plain of sediment is a small tip of rock called an inselberg. Finally, normal faulting once again uplifts the mountains into high, sharp peaks, and the process begins anew.

19-8 Desertification threatens all six inhabited continents.

- Desertification is the process by which land loses its vegetation and turns into a desert environment. Characteristics of desertification include de-

struction of native vegetation, unusually high rates of soil erosion, declines in surface water supplies, rising levels of water and topsoil saltiness, and widespread lowering of groundwater tables.

- More than 300 million people live in the **Sahel** region of dry and unstable weather, many of them poor farmers or nomadic herders. They rely on local rains, or migrate to follow the rains. The number of people, however, exceeds the carrying capacity of the land, and their large livestock herds overgraze the vegetation already stressed by low rainfall.

KEY TERMS

Coriolis effect (p. 575)
desert (p. 570)
desert pavement (p. 580)
desertification (p. 570)

dunes (p. 580)
ephemeral streams (p. 586)
loess (p. 580)
orographic effect (p. 576)

playa lake (p. 586)
ripples (p. 580)
sahel (p. 588)
sand dunes (p. 582)

ASSESSING YOUR KNOWLEDGE

Please complete this exercise before coming to class. Identify the best answer to each question.

1. A desert:
   a. Is hot.
   b. Is remote.
   c. Is covered by drifting sand.
   d. Receives less than 25 centimeters of rain a year.
   e. Is any place covered by sand dunes.

2. A typical sand dune:
   a. Has a gentle windward face.
   b. Has a steep leeward face.
   c. Migrates downwind.
   d. All of the above.

3. In the photo, the wind is blowing dominantly:
   a. Left to right.
   b. Right to left.
   c. Top to bottom.
   d. Wind direction here changes seasonally.
   e. None of the above.

Carsten Peter/NG Image Collection

4. Global atmospheric circulation tends to produce arid areas at:
   a. 30° latitude.
   b. 20° latitude.
   c. 40° latitude.
   d. The equator.
   e. None of these locations.

5. A Hadley cell:
   a. Is found only in the northern hemisphere.
   b. Develops between the equator and about 30° latitude.
   c. Develops between 30° and 70° latitude.
   d. Is a cell of moist falling air above the equator.
   e. Is found above every desert.

6. The orographic effect is:
   a. When winds turn to the right in the southern hemisphere.
   b. The evaporation that occurs above a cold coastal current near the tropics.
   c. When a mountain range forces moisture to rise and condense, creating a rain shadow.
   d. When a mountain range causes moisture to rise and evaporate, forming a polar desert.
   e. Where the rate of evaporation exceeds the rate of condensation.

7. As warm air rises into cooler air, the rate of (evaporation; condensation; rising; sinking) exceeds the rate of (evaporation; condensation; rising; sinking)_____.

8. The principal types of deserts include:
   a. Continental deserts and sand deserts.
   b. Coriolis deserts and oceanic deserts.
   c. Plateau deserts and valley deserts.
   d. High-elevation deserts and plateau deserts.
   e. Trade wind deserts and monsoon deserts.

9. A barchan dune is identifiable by:
   a. Its crescent shape, with horns pointing downwind.
   b. Its crescent shape, with horns pointing upwind.
   c. A long, linear ridge formed by winds blowing from two different directions.
   d. Its composition of loess.
   e. The absence of any vegetation.

10. An erg is:
    a. A type of dune.
    b. A region of North Africa near the Sahel.
    c. The name of a dry wind.
    d. A wind-abraded landform.
    e. The Arabic name for a sand sea.

11. Sediment transport:
    a. Has characteristics similar to transport in running water.
    b. Involves particle saltation and suspension.
    c. Typically entails turbulent air to move larger particles.
    d. Produces ripples and dunes.
    e. All of the above.

12. Why are there no important loess deposits in Africa?
    a. There were no continental glaciers in Africa.
    b. African continental glaciers did not produce loess.
    c. African glaciers deposited their loess in Europe.
    d. African loess turned into clay.
    e. None of the above.

13. Sand ripples migrate by:
    a. Sand moving around the outside edge of a pile and accumulating in front.
    b. Sand ripples do not migrate.
    c. Wind moving sand as a single pile.
    d. Sand moving up the stoss slope and avalanching down the lee slope.
    e. Sand saltating into a single form.

14. Star dunes:
    a. Are shaped like pyramids, with three or more arms that radiate from a peaked center.
    b. Are formed by wind that blows from several different directions.
    c. Grow taller rather than migrating.
    d. Require a rich source of sand.
    e. All of the above.

15. Water in a desert:
    a. Is not an important feature.
    b. Is mostly taken up by vegetation.
    c. Is involved in ripple and dune formation.
    d. Is second only to wind in shaping the desert surface.
    e. Is the primary agent shaping desert landforms.

16. Loess:
    a. Is windblown silt that accumulates in thick deposits.
    b. Is windblown sand that forms ergs.
    c. Collects only in arid regions below Hadley cells.
    d. Is common primarily in polar deserts.
    e. None of the above.

17. A _____ is formed when wind removes finer sediments, leaving behind a layer of coarse sediments.
    a. Yardang
    b. Barchan
    c. Dust bowl
    d. Pavement
    e. Erg

Reg Morrison/AUSCAPE

18. Desert pavement develops from:
    a. Past glacial deposits.
    b. Seasonal flooding.
    c. Erosion of sand by persistent wind.
    d. Freeze-thaw processes.
    e. Desert pavement is not well understood by scientists studying the feature.

19. Desertification is:
    a. The consequence of contraction of a desert and exposure of unvegetated regions.
    b. Primarily caused by warfare.
    c. Not related to global warming.
    d. A serious problem in the Arctic.
    e. Typically related to climate change and/or poor land management.

20. The formation of dunes usually requires:
   a. Loose sediment.
   b. Energy to move sediment.
   c. An obstacle to trap sediment.
   d. A dry climate.
   e. All the above.

Jan Tove/Johner Images//Getty Images, Inc.

FURTHER RESEARCH

1. How can you tell whether a dune is a barchan or a parabolic type?

2. How can you tell which direction the wind blew from when shaping a dune?

3. The Atacama Desert of northern Chile and the Mojave Desert of the western United States were both created, in part, by the orographic effect of nearby mountain ranges. Explain how this is possible given that the Atacama lies to the west of its mountains and the Mojave lies to the east of its mountains.

4. The Atacama Desert is usually drier than the Mojave. Give a couple of possible reasons why this is the case.

5. Explain the old American West saying, "Whiskey's for drinkin', water's for fightin'," within the context of today's water shortage.

ONLINE RESOURCES

Explore more about deserts on the following Web sites:

U.S. Geological Survey publication on the geology and resources of deserts:
http://pubs.usgs.gov/gip/deserts

The United Nations Convention to Combat Desertification:
www.un.org/ecosocdev/geninfo/sustdev/desert.htm

A description of desertification and its impact on human communities:
www.greenfacts.org/en/desertification

Additional animations, videos, and other online resources are available at this book's companion Web site:
www.wiley.com/college/fletcher

This companion Web site also has more information about WileyPLUS and other Wiley teaching and learning resources.

COASTAL GEOLOGY

CHAPTER CONTENTS & LEARNING OBJECTIVES

Cameron Davidson/The Image Bank/Getty Images

GEOLOGY IN OUR LIVES

America's ocean and Great Lakes coastlines are home to almost 153 million people, about 53% of the total U.S. population. Surging population growth in the coastal zone exposes more people to the dangers of geologic hazards—storms, hurricanes, tsunamis, and others—than in any other geological environment. The world's coasts are home to fragile ecosystems, beautiful vistas, and pristine waters, but also to major, growing cities, all coexisting in a narrow and constricted space. Expanding human communities compete for more space at the expense of extraordinary wild lands. Problems with coastal erosion, waste disposal, a dependency on imported food and water, and rising sea level are just some of the problems emerging in these areas. Management strategies to ensure the long-term viability of our coasts are largely science based, making scientists essential partners in building sustainable coastal communities.

Q What will happen in the future as human populations in the coastal zone continue to grow and sea level continues to rise?

20-1 Change Occurs Constantly Along the Global Shoreline

LO 20-1 Identify the many environmental influences on the world's coastlines.

Coastlines are unique and complex environmental systems that come under the influence of geologic, atmospheric, oceanographic, and biologic processes and materials, making them ever-changing environments. It has been said that "the only constant on the coastline is change."

Coastal Environments

Coastal lands, because of their temperate climate (air temperature is moderated by heat stored in the ocean) as well as their beautiful vistas and abundant resources, have become the most crowded and developed areas in the world. Fifty percent of the population of the industrialized world lives within 100 kilometers of a coast, a number that is expected to grow by 15% during the next two decades. Over 37 million people and 19 million homes were added to U.S. coastal areas

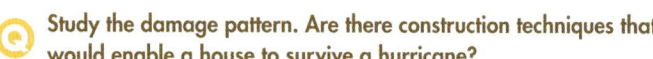

FIGURE 20.1 Hurricane Ike made landfall on September 13, 2008, across the barrier islands of Texas. The storm launched large, destructive waves and strong winds and raised the ocean level more than 3 meters above normal. Broad segments of the Texas coast were heavily damaged, and roads, homes, and businesses in many communities were destroyed.

Q Study the damage pattern. Are there construction techniques that would enable a house to survive a hurricane?

FIGURE 20.2 Trash like this washes up on beaches around the world because humans have been polluting the ocean for decades.

Q What are the potential consequences of dumping garbage in the ocean?

over the last three decades. This narrow fringe—comprising less than one-fifth of the contiguous land area of the United States—accounts for over half of the nation's population and housing supply.

Yet coasts are very vulnerable to destructive storms and tsunamis (**FIGURE 20.1**) that sweep inland from the sea, to coastal erosion that slowly chisels away at beaches and cliffs, and to accelerating sea level rise due to global warming. Furthermore, development pressure on the coast has accelerated pollution of coastal water bodies, depletion of coastal fisheries, and shifts in sedimentary processes that support vital ecosystems.

With so much of our culture and economy focused today on the shoreline, it is essential that geologists be active participants in advancing the scientific understanding of coastal environments and providing reliable data that can be applied to improving management plans and programs. To that end, in recent decades, a new field of study, **coastal geology**, has emerged. Coastal geology is concerned with understanding coastal processes (such as how waves and currents move sediment), the geologic history of coastal areas, and how humans live in coastal environments (**FIGURE 20.2**).

Coastal Processes

Sediments in the *coastal zone* may come from watersheds (typically, siliciclastic) as well as from the ocean (typically, carbonate). Winds, waves, and currents transport and deposit these sediments, thereby shaping various coastal environments—beaches, estuaries, deltas, wetlands, and reefs. Since the last ice age, a rising global sea level (over 120 meters) has flooded river mouths, drowned low-lying coastal lands, and shifted the coastline landward, in many cases over hundreds of kilometers.

Winds, waves, tides, sea level change, and currents act as agents of change, collectively called **coastal processes**. They sculpt the shoreline through erosion and deposition of sediment and by flooding low areas. Understanding how coastal processes and sedimentary materials interact is crucial to interpreting the geologic history of coastal systems and finding ways for people to live in the coastal zone in a sustainable manner.

Coastal Terminology

Scientists employ a number of specific terms when describing the coastal zone. Here, we focus on beaches because that is where we can most readily observe the processes associated with waves; but the concepts that apply to beaches apply to other types of coastal environments as well (**FIGURE 20.3**).

A **beach** is an unconsolidated accumulation of sand and gravel along the shoreline. In this discussion we emphasize the beaches most familiar to many, those that develop along the shores of oceans; but beaches also form in freshwater environments such as lakes and rivers.

Most beaches are shaped and maintained by waves. Many people do not realize that beaches consist of two portions: the submerged and *subaerial* (exposed to the atmosphere). The submerged portion extends seaward past the *breaker zone*, the region where waves from the deep ocean turn into breakers. After breaking, a wave enters the *surf zone* as a *bore* (a frothy wave front moving toward the shoreline). It then moves toward the beach and enters the *swash zone*. There the energy of the bore generates horizontal surging of the ocean surface up and down the sloping *foreshore* of the beach.

The landward extent of the foreshore is marked by a *berm*, a ridge of sand built by waves at high tide that deliver sand to the top of the foreshore. Landward of the berm is the *backshore,* which is characterized by a *coastal dune*, an accumulation of wind-blown sand into a low hill vegetated with salt-tolerant plants.

Q **EXPAND YOUR THINKING**

Sea level rise has accelerated in response to global warming. Describe the ways in which sea level rise will affect coastal cities.

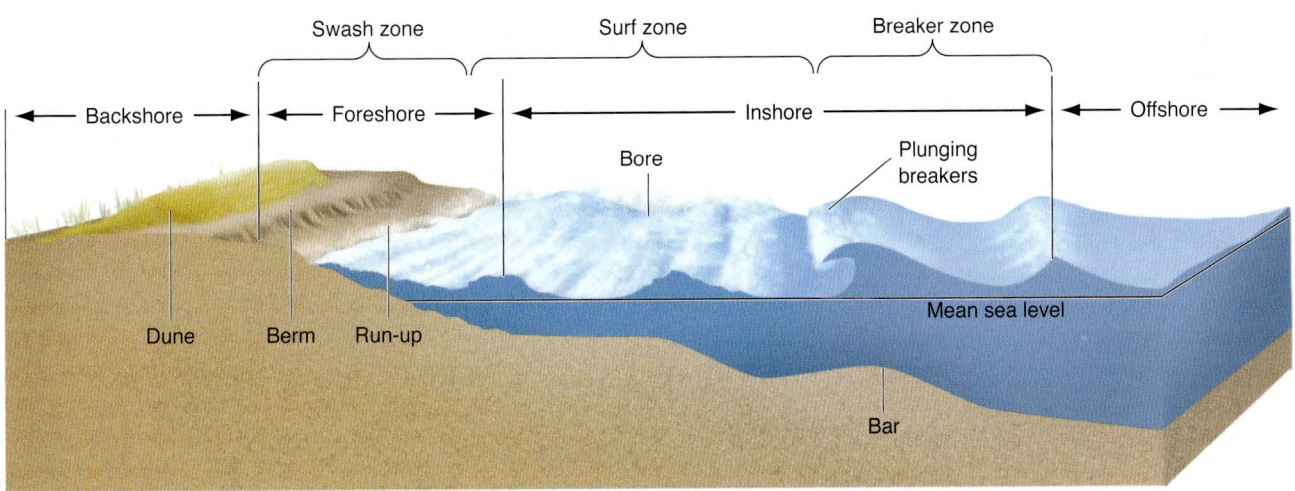

FIGURE 20.3 The coastal zone consists of several environments in which sediments, waves, and currents interact to produce coastal processes.

Q **What kind of sediments collect on beaches? Where do they come from?**

20-2 Wave Energy Is the Dominant Force Driving Natural Coastal Change

LO 20-2 Describe the transformation of waves as they travel from deeper to shallower water.

Waves on the ocean surface are generated by the friction of air blowing across the water—the reason they are called wind waves.

Wind Waves

Perhaps you have watched a breeze move across a quiet body of water, gently disturbing the surface and forming small ripples. As it does

so, it is transferring energy from the atmosphere into the water by displacing individual molecules of water. The molecules move in a circle, called an *orbital,* that is largest at the surface and gets smaller with depth (**FIGURE 20.4**).

A moving wave looks as if it is transporting water, but it is not. Just as music travels through the air without making wind, a wave travels across the ocean surface without making a current. (In the surf zone, however, waves do make currents.) Rather, energy (waves) radiates outward from the point at which the wind interacts with the ocean surface, but water does not. The size of a wave is governed by *wind velocity*, *fetch* (the distance over which the wind blows), and *duration* of time that the wind disturbs the water.

Waves are described by their *height*, the vertical distance between the peak of a crest and the bottom of a trough; their *wavelength* (L), the horizontal distance between two successive waves; and their *period*, the time it takes for two successive wave crests to pass a given point in space. In general, the circular water motion generated by a wave extends to a depth equal to half the wavelength (L/2).

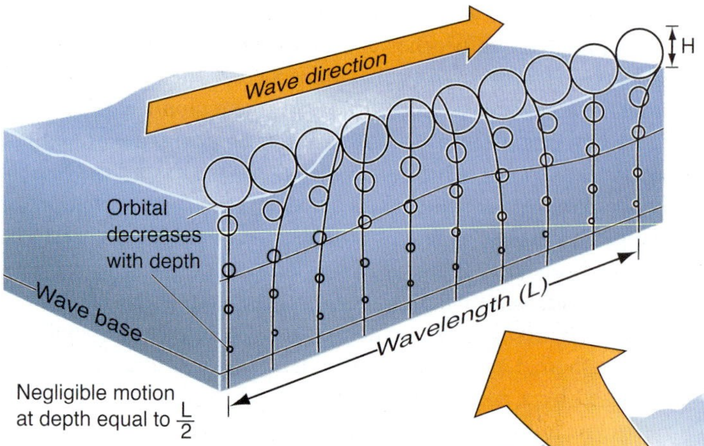

Orbital decreases with depth

Wave direction

H

Wave base

Negligible motion at depth equal to $\frac{L}{2}$

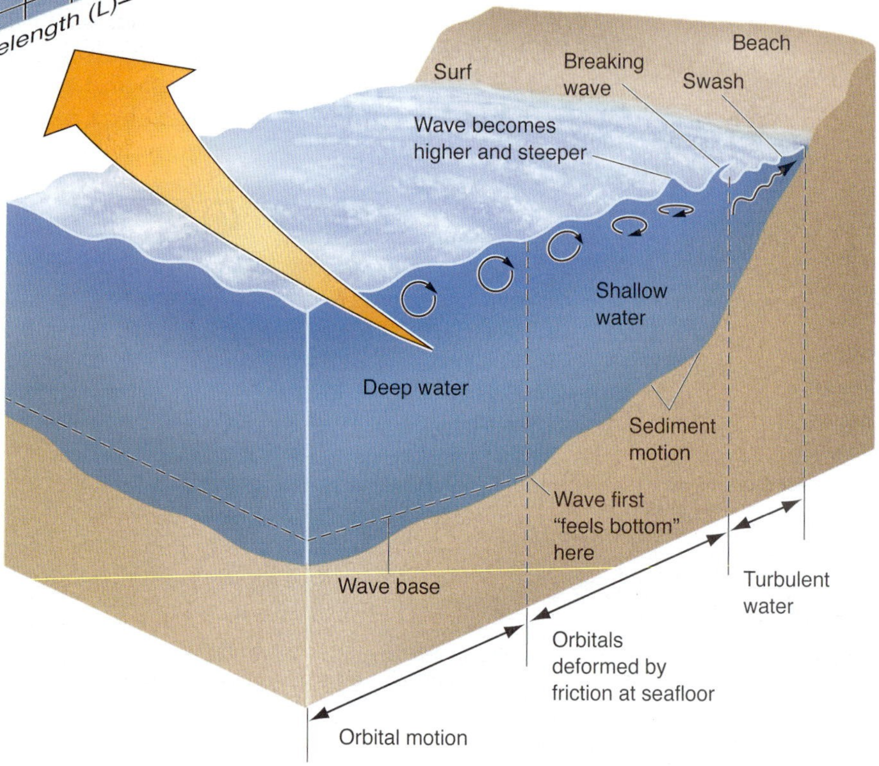

Wavelength (L)

Surf

Breaking wave

Beach

Swash

Wave becomes higher and steeper

Shallow water

Deep water

Sediment motion

Wave first "feels bottom" here

Wave base

Turbulent water

Orbitals deformed by friction at seafloor

Orbital motion

What factors govern the size of a wave?

FIGURE 20.4 A wave is described in terms of its height, length, and period (the time between successive wave crests). Particles of water disturbed by the wind travel in a circle called an orbital. A shoaling wave encounters the bottom at depths of less than half the wavelength. This slows the forward movement of the wave and causes it to become higher and to develop a steep face.

The ocean surface in a wave-generating area, called a *sea*, is chaotic and disturbed due to high wind velocities. A sea consists of steep, sharp-crested waves of many different heights, lengths, and periods. As waves move out of the area of generation in the absence of local wind, they become *swells*, long, regular, symmetrical waves with periods ranging from 5 to 20 seconds. Swells are noticeable as the smooth rolling action of the ocean surface. Each swell transfers energy horizontally across the ocean surface (**FIGURE 20.5**).

The shape and speed of a wave are governed by the displacement of water particles and controlled by two characteristics: wavelength and water depth. In deep water (where the depth is greater than the wavelength), a swell takes the form of a symmetrical sine wave (refer to Figure 20.4). The form of a wave changes dramatically when it enters shallow water (where the water depth is less than half the wavelength).

The nautical term for shallow water is *shoal*, and a wave entering shallow water is said to be a *shoaling wave*. Shoal water inter-

FIGURE 20.5 (a) Waves are generated by strong winds in a chaotic environment called a sea. (b) Beyond the sea, wave energy is organized into swells, which appear as a gently rolling motion of the ocean surface.

How does a wave change as it enters shallow water?

(a)

Mike Hill/Getty Images, Inc.

(b)

Jason Edwards/National Geographic/Getty Images, Inc.

feres with the movement of water particles at the base of the wave, slowing its forward motion. Interaction with the bottom changes a water particle orbital into an ellipse (a flattened circle); and at the seafloor, water particles experience back-and-forth movement that may transport sediment on the seafloor. As a wave enters shallow water, it slows, and the following wave, moving faster, will "catch up" with it. The wavelength decreases and the wave's height increases. Its crest becomes narrow and pointed with a steep face, while the trough becomes wide and flat. Because the wavelength decreases and the speed slows the wave period remains unchanged.

In very shallow water, the energy from sea and swell is released in the surf zone, where waves become *breakers*. When a wave's height is approximately equal to the depth of the water, the wave "breaks": the peak of the crest pitches forward and tumbles down the face of the wave in the direction in which the wave is moving (usually, landward) and breaks into a mass of aerated water. A wave breaks when it becomes overly steep. Wave breaking occurs because the velocities of water particles in the crest exceed the velocity of the wave and the crest surges ahead.

Global Wind and Wave Patterns

The biggest waves are generated by strong winds blowing across vast ocean distances. **FIGURE 20.6** shows a map of global wind speeds and wave heights as measured by satellite radar pulses. Wind speed is determined by the strength of a return pulse reflected from the ocean surface. Calm areas serve as good reflectors and return a strong pulse; a rough sea will scatter a radar signal and return a weak pulse. The radar instrument uses these differences to measure wind speed and wave height.

The Southern Ocean, with its intense belt of winds and vast fetch uninterrupted by continents, tends to generate the highest winds and waves on the planet—exceeding 15 meters per second and 6 meters in height. Strong winds and high waves are also associated with stormy regions in the North Atlantic and North Pacific oceans. These are indicated by the most intense red tones on the map. In general, there is a high degree of correlation between wind speed and wave height. The weakest winds and lowest waves (represented by magenta and dark blue) are found in the western tropical Pacific Ocean, the tropical Atlantic Ocean, and the tropical Indian Ocean. These areas are characterized by trade winds that tend to be persistent but relatively low in velocity.

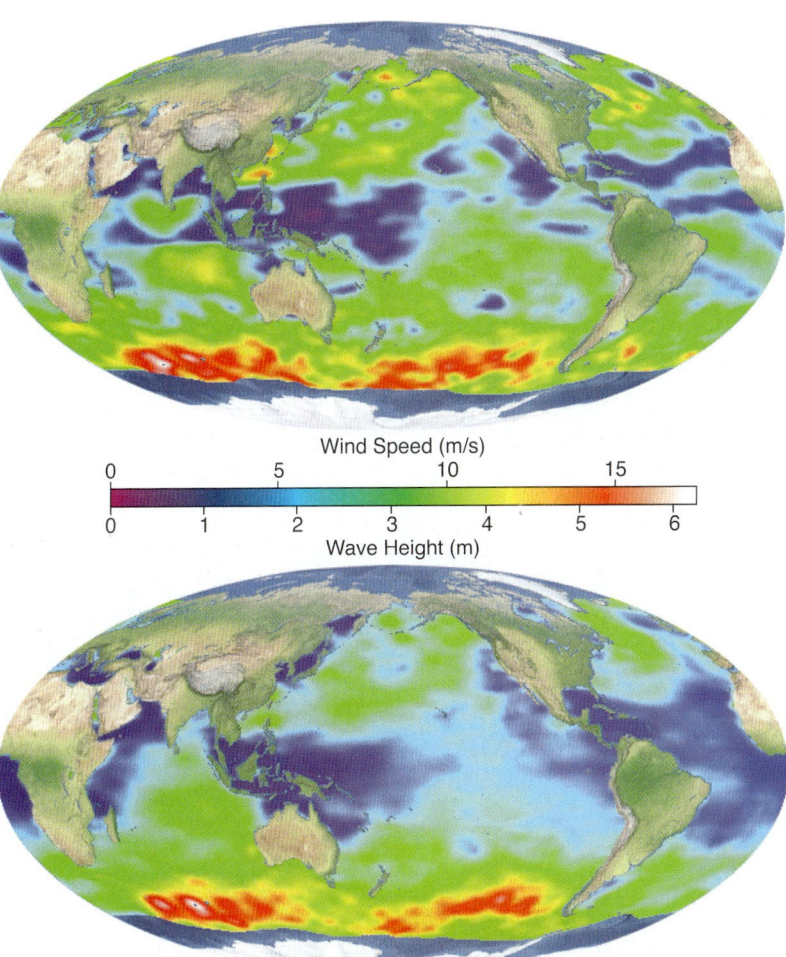

FIGURE 20.6 Radar mapping of the ocean surface identifies areas of greatest wind speed (top) and wave height (bottom).

Why do the world's largest waves occur in the Southern Ocean?

EXPAND YOUR THINKING

Referring to Figure 20.6, if you were to sail around the world and visit your favorite places, which route would you pick as safest?

20-3 Wave Refraction and Wave-Generated Currents Occur in Shallow Water

LO 20-3 Define wave refraction, and describe how wave-generated currents are made.

When a wave enters shallow water, it is subject to refraction and may produce wave-generated currents.

Wave Refraction

Wave refraction occurs when the wave crest bends until it is aligned parallel to the contours of the seafloor. For straight coasts with parallel contours on the shallow seafloor, refraction tends to orient waves so that the wave crest is parallel to the coastline.

The process of **refraction** is set in motion when a wave approaches a shoreline at an angle to the bottom contours. As the portion of the wave in shallow water experiences shoaling, it slows under frictional resistance from the bottom. The rest of the wave, still traveling unhindered in deeper water, tends to change its direction like a door slowly swinging on its hinge. The rapidly traveling portion of the wave swings landward and lines up with bottom contours so that the wave crest and bottom contours remain parallel (**FIGURE 20.7**).

When a wave approaches a *rocky headland* that sticks out into the ocean beyond the surrounding shoreline, the wave front will begin to refract when it first encounters the seafloor. This leads to *convergence* of wave energy on the headland. The wave crest takes on the orientation of the bottom contours and expends energy breaking against all sides of the headland, causing it to erode. Simultaneously, the portion of the wave crest moving into the adjacent bay also experiences refraction. In this

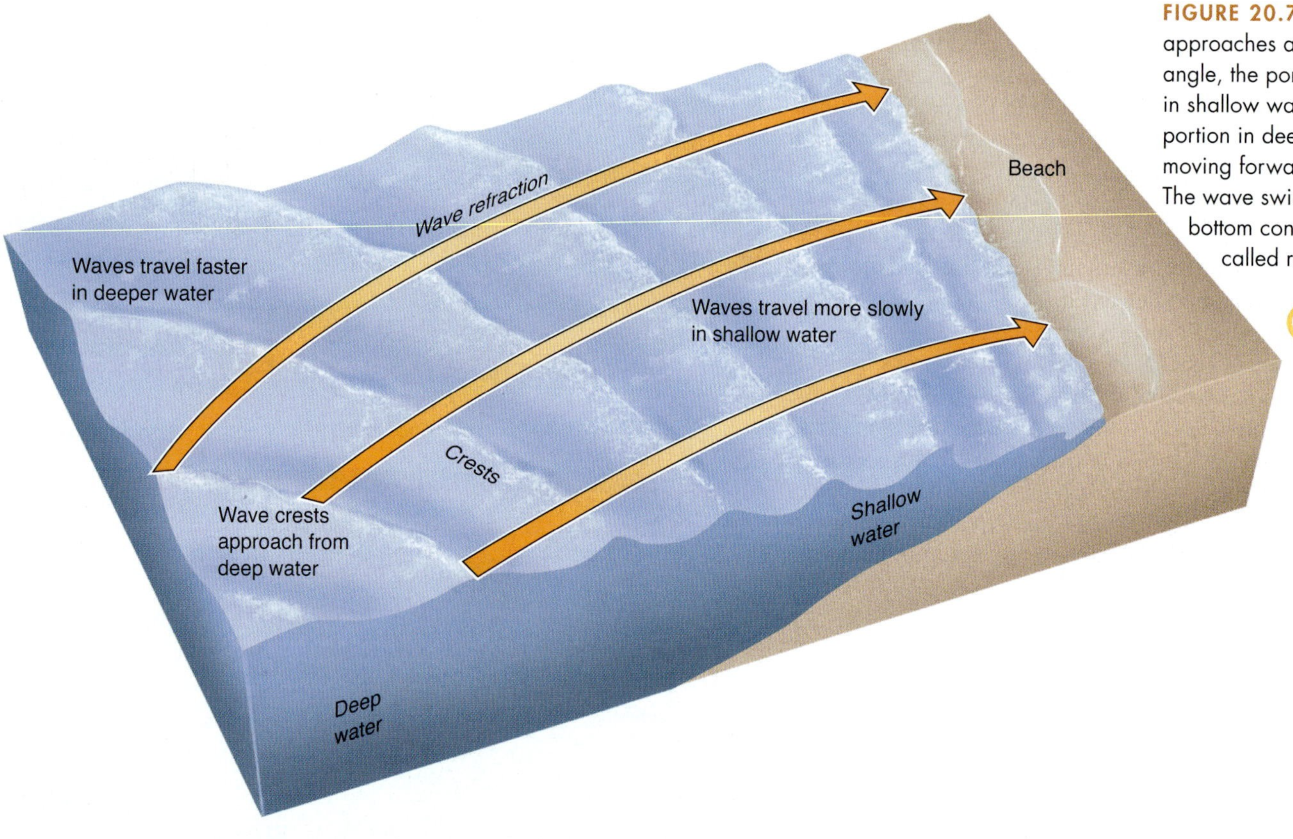

FIGURE 20.7 As a wave approaches a shoreline at an angle, the portion of the wave in shallow water slows while the portion in deeper water continues moving forward uninterrupted. The wave swings parallel to the bottom contours, a process called refraction.

Ⓠ **Explain why the wave crest lines up with bottom contours.**

Wave-Generated Currents

Water particles at the crest of a wave move forward with the wave, while particles in the trough move in the opposite direction (refer back to Figure 20.4). Together, these patterns define a not quite closed circle. The orbital motion of a water particle during the passage of a wave carves a slightly open curve rather than a perfectly closed circle. The momentum of a wave drives this forward movement of the water, called "Stokes drift," after Sir George Gabriel Stokes (1819–1903) an Irish mathematician and physicist, who formulated this effect. In deep water, this effect is minor, but in shallow water, it causes water to pile up along a coastline.

Water movement and sediment transport along the coast are highly complex. In the surf zone, there is net shoreward transport of water near the surface associated with the movement of bores. Most of this shoreward movement occurs under the crests of the bores because water particle motion there is directed onshore. This motion produces a gradual advance of the water mass toward a beach. Water piles up against the shore, raising the water level and creating *sea level setup*, which in turn leads to the formation of **wave-generated currents**.

Of course, water cannot pile up against the coast forever. A current that flows along the bottom in the seaward direction counterbalances the onshore movement of water in the surf zone. This is the *undertow*. **FIGURE 20.9** provides a simple two-dimensional model of circulation in the surf zone—it does not account for the delivery of sand onto a beach. Sand must move landward in the surf zone in order for beaches to be stable and to recover after erosion by large waves. Beach recovery following a storm has been observed along many shorelines. If the net movement of sand along

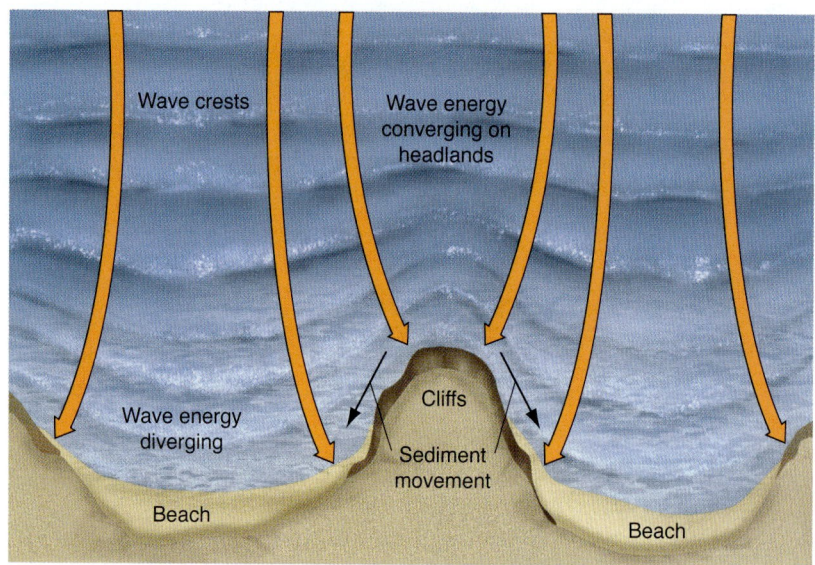

FIGURE 20.8 As a wave approaches a rocky headland, refraction causes the energy to converge on the headland and to diverge in the adjoining bay. Hence, the headland erodes and the bay accumulates the resulting sediment as a beach.

Ⓠ **How does sand transport in Figure 20.8 compare and contrast with transport in Figure 20.9? What are the implications of each for the beach?**

case, however, the wave energy *diverges* across a broad area of shoreline. The wave energy is effectively reduced in the embayment because the energy spreads out as the wave arcs into the bay (**FIGURE 20.8**).

the seabed were always directed offshore in the surf zone (as in the figure), beaches would not exist.

How does sand move toward a beach? Researchers continue to investigate the way beaches function, but the answer may be found in two areas: 1) Three-dimensional surf zone circulation may create some regions with onshore sand movement and others with offshore sand movement; and 2) the onshore water velocities under the bore crests may be very important in transporting sediment toward the beach. Undertow appears to operate most strongly during storms. This fact may account for the erosion that is often

observed during high waves. A particularly strong offshore flow during intense storm conditions might transport sand so far offshore that it is permanently lost from the beach, resulting in net recession of the shoreline.

EXPAND YOUR THINKING

If rocky headlands experience more wave energy than beaches do, what will eventually become of them? What is the long-term trend of shoreline evolution?

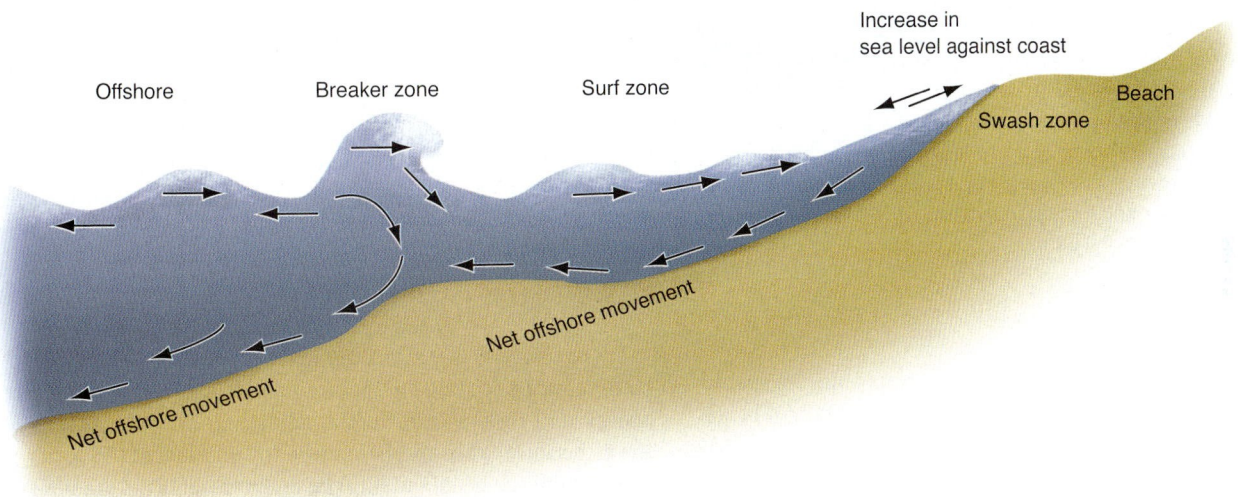

FIGURE 20.9 Breakers and bores cause net movement of water toward the shoreline, raising sea level along the coast and creating "setup." In the aftermath, the water surface gently slopes from the shoreline toward deeper water. Setup establishes an offshore-directed pressure gradient that drives a current along the seabed in the seaward direction called undertow.

How do offshore-directed currents affect sand transport in the coastal zone?

20-4 Longshore Currents and Rip Currents Transport Sediment in the Surf Zone

LO 20-4 Define and describe longshore currents and rip currents.

Wave-generated currents may include rip currents and longshore currents, both of which transport sediment and change the shape of the coastal system.

Longshore Currents and Rip Currents

When waves approach a shoreline from an angle, they generate a **longshore current** that moves along the shoreline in the direction of wave movement (**FIGURE 20.10**). This current develops in part following the periodic release of sea level setup as a volume of water

that moves along the shoreline, and in part as response to the momentum of a moving wave that is directed obliquely to the shoreline.

The longshore current may transport along the shoreline sand that has been suspended in the water column by breaking waves. This movement is important because it establishes a sand-sharing system along a beach, whereby one segment of a beach can contribute sand to another, helping to counteract the effects of erosion in which sand is lost offshore. At other times of the year, when waves approach from a different direction, the sand may return.

Grains of sand moving along the foreshore follow an asymmetrical or sawtooth path in a longshore current. They are moved up the beach by the uprush of a wave in the swash zone at an angle directed along the shoreline, but their return path is perpendicular to the shoreline, following a line directed by gravity. This process of sediment transport is called **longshore drift**.

Another important consequence of sea level setup along a shoreline is the development of **rip currents** (**FIGURE 20.11**). As water piles up along a beach, it is released periodically in the form of offshore-directed currents that surge through the surf zone. These currents carry suspended sand and floating debris, explaining why rip currents can be detected as dirty, rough, or dark water that extends

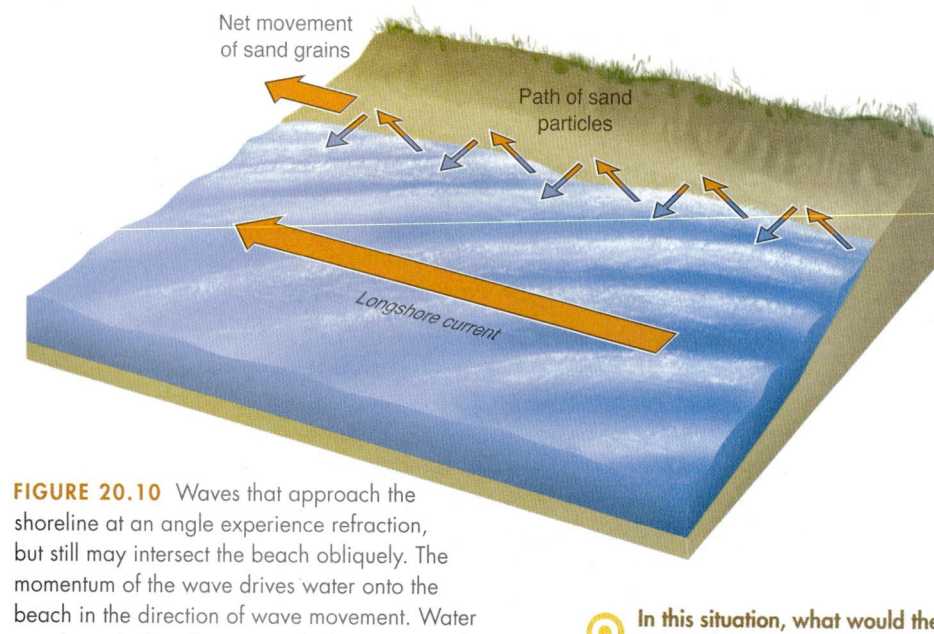

FIGURE 20.10 Waves that approach the shoreline at an angle experience refraction, but still may intersect the beach obliquely. The momentum of the wave drives water onto the beach in the direction of wave movement. Water travels up the beach at an angle but returns to the ocean, under the force of gravity, perpendicular to the shoreline.

In this situation, what would the effect be if a jetty (a solid wall) were built across the beach and out into the ocean?

through a surf zone beyond the breakers. Rip currents may be spaced along a beach in a recurring fashion such that a long stretch of beach may develop several rip currents.

As water moves offshore in a rip current, it excavates a channel in the sandy seafloor, like a stream channel, and will keep the channel open for several hours or days. The water depth is greater in the rip channel than in the surrounding surf zone, where there are shallow areas called **sand bars**, so the water appears darker and moves offshore at a relatively high velocity.

Between rip channels, water moves onshore under waves and builds an accumulation of sand in a sand bar. The difference in depth

between the sand bar and the rip channel may be greater than the average height of a person. For this reason, and because of the strong current, rip channels are hazardous to swimmers; people drown in them every year on beaches around the country. There is a way to escape from a rip current, however: avoid swimming against it and instead swim out of it sideways, along the shoreline to the nearby shallow sand bar.

Shorelines Straighten over Time

Another form of wave-generated current develops on a shoreline that has a rocky headland and an adjoining bay, as described earlier (see Figure 20.8). On rocky embayed coasts, two areas of refraction occur: one where energy *converges* on the headland and a second where energy *diverges* in an embayment. As a result of these two areas of refraction, sand is eroded from the headland and moves into the embayment, where it builds a beach. How does this sand move? It is propelled through the action of wave-generated currents related to setup at the headland.

The convergence of energy leads to sea level setup around headlands. That is, sea level at the headland is higher due to greater wave energy and momentum. In the adjoining embayment, sea level is lower because wave momentum is lower. Hence, the water surface slopes from the headland into the embayment. This gradient drives a longshore current that carries eroded sediment from the headland into the adjoining bay.

Over time, as sand accumulates in the bay, a beach is built and slowly fills the embayed area. When this process is carried to its conclusion, as the headland perpetually erodes back and the embayment slowly fills with sand, the *shoreline straightens over time* (**FIGURE 20.12**); this is an accepted law of coastal geology.

FIGURE 20.11 Sea level setup is released periodically by offshore-directed circulation. (a) Rip currents remove this water from the surf zone and carry suspended sediment and debris beyond the breaker zone. (b) Dark channels or the presence of turbid water traveling through the breaker zone are signs of rip currents. Sand bars may develop between rip currents.

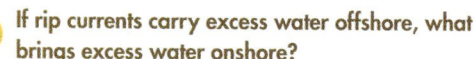

If rip currents carry excess water offshore, what brings excess water onshore?

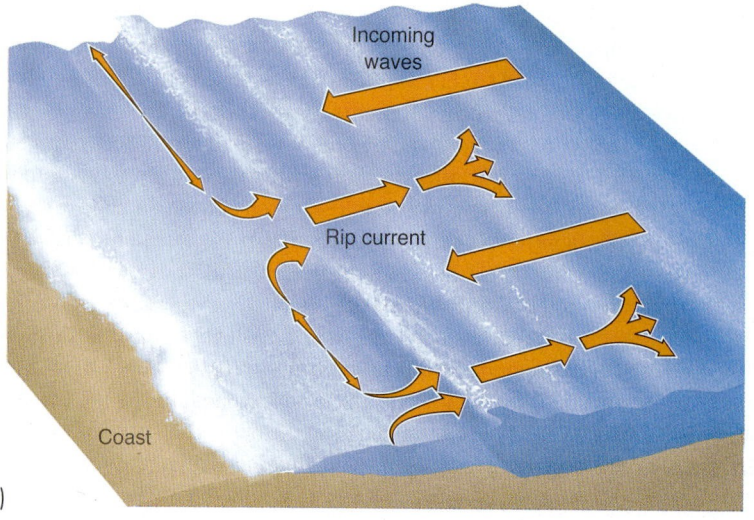

(a)

(b)

Lance Dettle

"Coasts Straighten with Time"

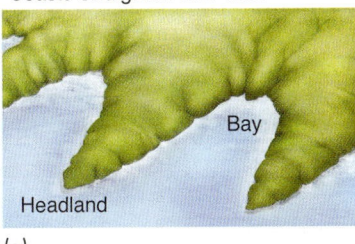

(a)

Bay

Headland

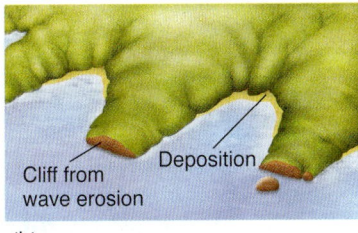

(b)

Cliff from wave erosion

Deposition

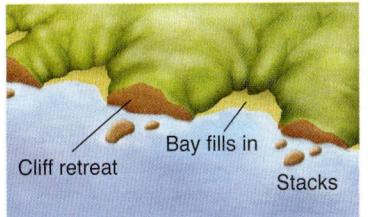

(c)

Cliff retreat

Bay fills in

Stacks

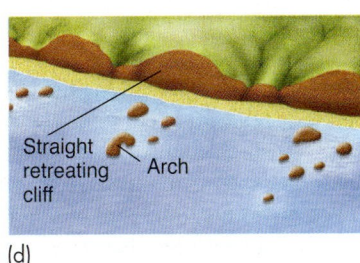

(d)

Straight retreating cliff

Arch

FIGURE 20.12 Wave-generated currents move sand eroded from a rocky headland into the adjoining embayment and build a beach. Over time, the headland is worn down and the bay fills with sand. Eventually, the shoreline becomes a long, straight beach. Portions of the headland that are resistant to erosion are stranded offshore, where they take shape as sea stacks and arches.

Q **What is the eventual fate of sea stacks and sea arches?**

Q EXPAND YOUR THINKING

Considering all that you have learned about coastal processes, draw an illustration that captures sediment transport on a beach. Label all components and write an accompanying description.

20-5 Gravity and Inertia Generate Two Tides Every Day

LO 20-5 Compare and contrast the forces that generate tides.

The oceans rise and fall twice each day in response to *tide-raising forces* related to the gravitational attraction among the Moon, the Sun, and Earth.

Tides

Tides are regular and predictable oscillations in the sea surface caused by gravitational forces exerted by the Moon and the Sun and Earth's rotation. Newton's law of universal gravitation states that the force of gravity existing between two bodies is directly proportional to their masses, and inversely proportional to the square of the distance between them. This means that the greater the mass of the objects and the closer they are to each other, the greater the gravitational attraction between them.

The Moon, though not large on a planetary scale, is near Earth and therefore exerts a significant gravitational attraction on it. The Sun, though not near Earth, is large and also exerts a significant attraction, but about half that of the Moon (**FIGURE 20.13**).

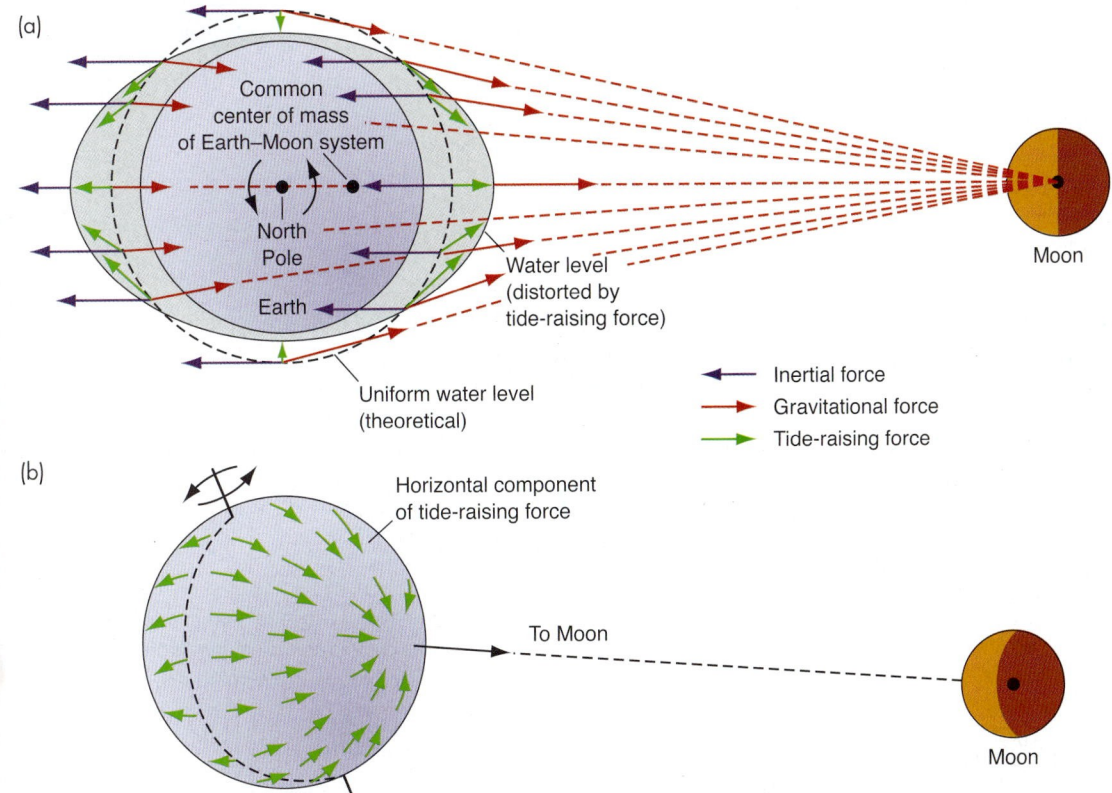

(a)

Common center of mass of Earth–Moon system

North Pole

Earth

Water level (distorted by tide-raising force)

Uniform water level (theoretical)

Moon

⟶ Inertial force
⟶ Gravitational force
⟶ Tide-raising force

(b)

Horizontal component of tide-raising force

To Moon

Moon

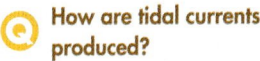

FIGURE 20.13 (a) Two tide-raising forces are responsible for the lunar ocean tide: the gravitational attractive force (red arrow) and the inertial force (blue arrow). The balance between these two opposing forces (green arrow) defines the tide-raising force. On the side of Earth facing the Moon, the gravitational attraction force is greater than the inertial force, and a tidal bulge is created in the ocean surface. On the side facing away from the Moon, the inertial force is greater and likewise creates a tidal bulge. Earth rotates beneath its watery envelope and encounters these two bulges (high tide) twice every day. (b) Green arrows display the theoretical movement of water toward the tidal bulge. As continents rotate into the tidal bulges, this water must flow within, throughout, and around coastlines in the form of a series of tidal currents.

Q **How are tidal currents produced?**

Lunar Tide

Gravity is only one of the major tide-raising forces responsible for creating tides. Another is *inertial force*, which counterbalances gravity in the Earth–Moon system. Inertia is the tendency of moving objects to continue moving in a straight line. Together, gravity and inertia are responsible for the creation of two major tidal bulges on Earth's surface.

The attraction between Earth and the Moon is strongest on the side facing the Moon. Water can immediately respond to the Moon's pull; hence, gravitational attraction causes water on Earth's nearest side to be pulled toward the Moon. As a result, the oceans "bulge out" on the side facing the Moon. However, at the same time that gravitational force draws the water closer to the Moon, inertial force works to keep the water in place. But the gravitational force is stronger so the water is pulled toward the Moon on Earth's near side.

On Earth's opposite side, the gravitational force is weaker because the Moon is farther away. Here, inertial force exceeds gravitational force, and the water tries to keep going in a straight line, moving away from Earth and also forming a bulge.

Many people believe that the tides travel around the planet, when actually it is Earth that rotates on its axis and slides under the watery bulges created by the Earth–Moon system. Because Earth is rotating while this is happening, two tides normally occur each day: one when Earth slides under the lunar bulge of the oceans and the second when Earth slides under the inertial bulge on the opposite side.

Solar Tide

The Sun also plays an important role in generating tide-raising forces. Two tidal bulges are created by the Earth–Sun tidal system: a gravitational bulge facing the Sun and an inertial bulge on Earth's opposite side. Although these tidal bulges produce tidal oscillations in their own right, their more important role is to modulate, or influence, the lunar tide-raising forces.

The interaction of the forces generated by the Moon and Sun is too complex to cover fully in this chapter, which serves only as an introduction to tidal theory; the important takeaway point from this brief overview is that when the Sun–Earth–Moon system is aligned, the tide range will be higher than usual. This is called *spring tide*. When the solar and lunar components are at 90° to each other, the tide range will be lower than usual. This is called *neap tide*. Both are illustrated in **FIGURE 20.14**.

Coastlines encounter the tidal bulges as Earth rotates. As a bulge approaches, coastal environments experience *tidal currents*, whereby water fills bays, inlets, lagoons, and other environments of the shoreline. Arriving currents are called the *flooding tides*. As a tidal bulge passes a coastal point, the water drains from these environments during *ebbing tides*.

ⓠ EXPAND YOUR THINKING

What role might tidal currents play in the coastal processes we studied earlier?

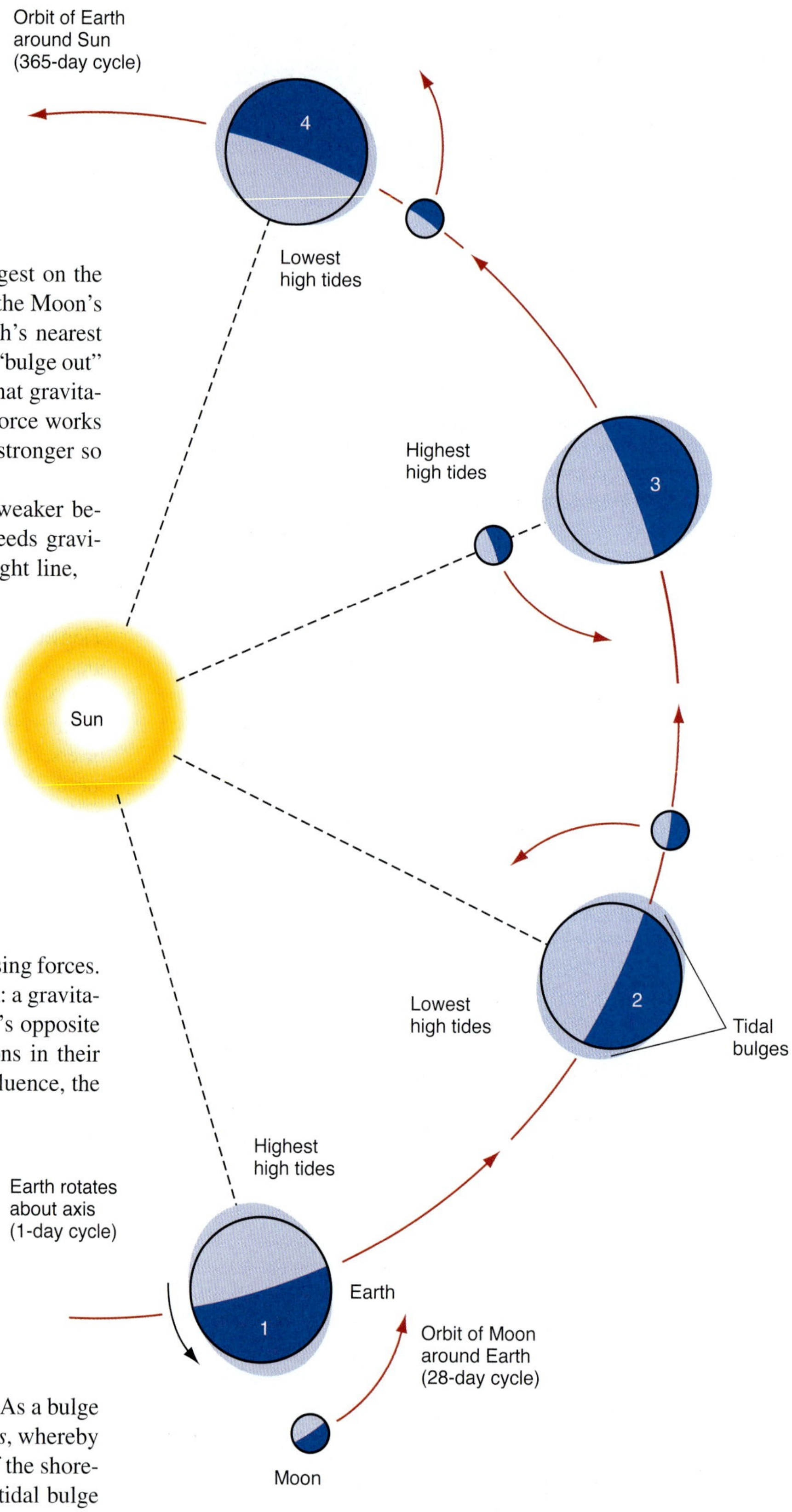

Orbit of Earth around Sun (365-day cycle)

Lowest high tides

Highest high tides

Lowest high tides

Tidal bulges

Highest high tides

Earth rotates about axis (1-day cycle)

Earth

Orbit of Moon around Earth (28-day cycle)

Moon

FIGURE 20.14 Tidal bulges on the ocean surface related to the gravitational and inertial forces of the Sun–Earth system modulate the lunar tide. When the Sun, Earth, and Moon are aligned, the spring tide is established. When the Moon is at 90° to the Sun–Earth alignment, neap tide is established.

 How many times each month does spring tide occur?

20-6   Hurricanes and Tropical Storms Cause Catastrophic Damage to Coastal Areas

LO 20-6 Describe the nature and impact of hurricanes.

Called "the greatest storm on Earth," a hurricane is capable of annihilating coastal areas with sustained winds of over 200 kilometers per hour, intense rainfall, and flooding ocean waters pushing huge waves. Remarkably, during its life cycle, a hurricane can expend as much energy as 10,000 nuclear bombs.

Hurricanes

Few events in nature compare to the destructive force of a **hurricane** (**FIGURE 20.15**). The term itself says it all: it is derived from the name of a god of evil, *Huracan*, recognized by the Tainos, an aboriginal tribe in Central America.

The Atlantic and Gulf coasts of the United States are well known for their vulnerability to hurricane and tropical storm damage, but, in fact, any coastline in a region where hurricanes develop and travel is also vulnerable. Notably, three regions in the northern hemisphere are highly likely to spawn hurricanes: the tropical Atlantic, the eastern tropical Pacific, and the western tropical Pacific. In the southern hemisphere, hurricanes develop in two primary areas: the western tropical Pacific and the Indian Ocean (**FIGURE 20.16**).

For a storm to be considered a hurricane, its winds must reach a maximum sustained speed of 119 kilometers per hour or higher. Hurricanes tend to come to life around a preexisting atmospheric disturbance in warm tropical oceans where there is high humidity in the atmosphere, light winds above the storm, and a high rate of condensation in the atmosphere. If the "right" conditions last long enough, a hurricane can produce violent winds that give rise to massive waves, and torrential rains that produce damaging floods.

There are, on average, six Atlantic hurricanes each year; over a three-year period, approximately five hurricanes strike the U.S. coastline between Texas and Maine. When hurricanes move onto land, the heavy rain, strong winds, and large waves can damage buildings, trees, and cars—and jeopardize life.

Once on land, however, a hurricane weakens because its source of heat—the warm ocean water—is gone. At this point, it becomes a *tropical storm*, defined as a storm with wind speeds between 63 and 117 kilometers per hour. These storms are also highly destructive, as they dump immense quantities of rain on the land and batter the coast with waves.

Accompanying the large waves is a high sea level phenomenon called **storm surge** (**FIGURE 20.17**). Storm surge is a combined effect that includes low atmospheric pressure above the ocean surface, causing a bulge of water to travel beneath a storm, plus the effect of wind pushing the water in the direction of the storm's forward movement.

Sea level setup due to wave momentum also contributes to storm surge. The combined processes of wind shear, low pressure, and setup can raise the water level several meters along a coastline. Add to this waves of 6 to 10 meters high or more launched by the winds of a storm, and it is easy to understand why hurricanes and tropical storms are so dangerous when they reach the shore, especially one with low topography. They cause massive damage and flooding of low-lying coastlines, which is compounded if they hit at high tide (**FIGURE 20.18**).

Though not the only type of storm that damages coastal areas, hurricanes are among the worst (**FIGURE 20.19**). Their high winds and storm surge are a deadly duo, a primary determinant of hurricane-inflicted loss of life and property damage. When a hurricane makes landfall along low-lying coastal lands, ocean waters sweep across beaches and roads and into adjacent communities. Heavy rains compound the impact by saturating the ground and setting off flooding of nearby streams. Coastal erosion is generated by high waves and storm surge that can strip a beach of its sand and undermine homes, roadways, and businesses. You can read about one of the deadliest and most destructive storms to date in the Earth Citizenship feature, "Superstorm Sandy—A Bad Storm Made Catastrophic by Global Warming."

How can such damage be mitigated? The simplest and most direct way is one we've discussed before in regard to mass wasting: *avoid the hazard* by not developing the shoreline. If we stopped building communities on the edge of the ocean, we would greatly minimize the suffering, loss of life, and expensive damage caused by hurricanes and other types of coastal hazards.

FIGURE 20.15 Hurricane Katrina hit the Gulf of Mexico in 2005.

Science Source/Photo Researchers

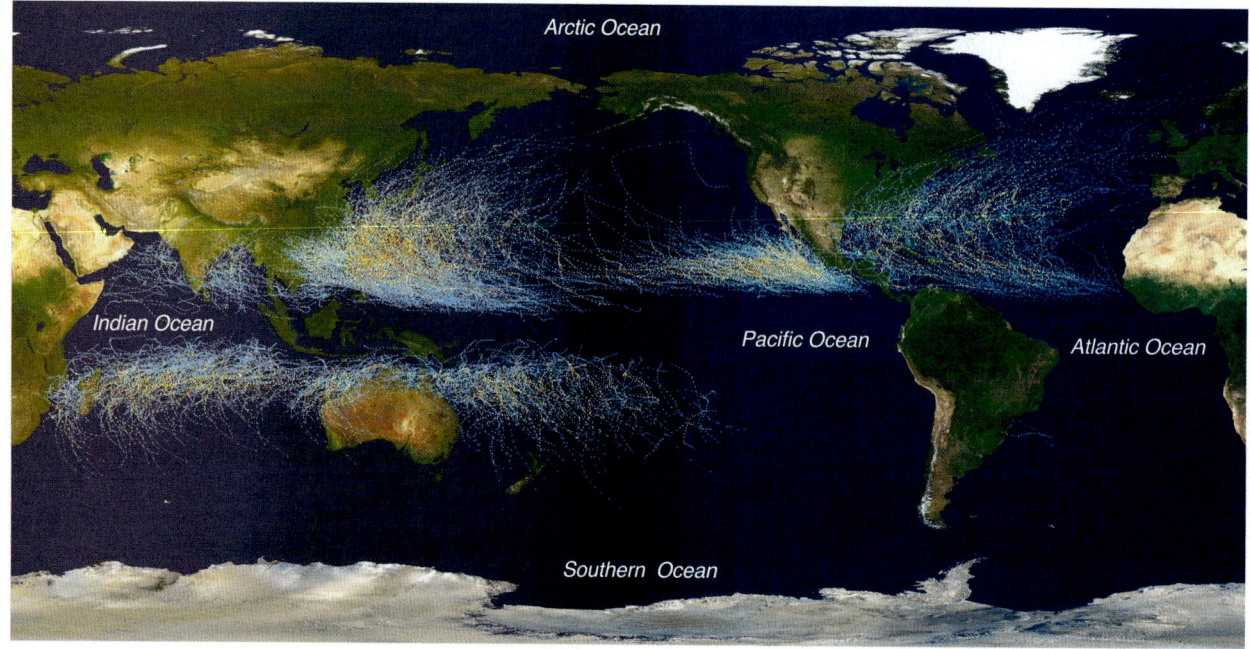

NASA

FIGURE 20.16 There are three major areas where hurricanes form in the northern hemisphere and two in the southern hemisphere.

What do hurricanes do to coastal environments when they strike?

© AP/Wide World Photos

FIGURE 20.17 Hurricane Katrina hit New Orleans, Louisiana, in 2005. It was one of the five deadliest and costliest hurricanes in the history of the United States. The combination of high waves, wind shear on the water, high tide, sea level setup by wave momentum, and low atmospheric pressure on the ocean surface propelled a storm surge that flooded 80% of the city.

How does flooding damage a community?

FIGURE 20.19 Hurricanes damage coastal communities with high winds and storm surge. Hurricane Ike hit this Texas community in 2008.

What might have made it possible for this single home left standing to survive the flooding of Hurricane Ike?

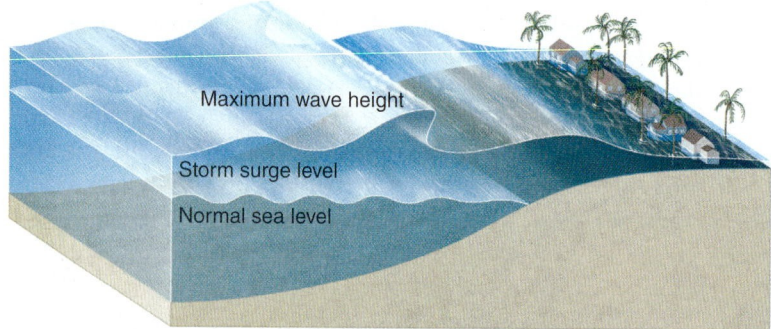

FIGURE 20.18 Storm surge is set off by the combined effects of high wind, causing water to pile up above the normal sea level, and lower atmospheric pressure in a storm center, which forces the ocean surface to bulge upward.

Smiley N. Pool-Pool/Getty Images, Inc.

EXPAND YOUR THINKING

Imagine that you have to build a beachfront home in a hurricane-prone area. What special features would you incorporate in your design? Explain their purpose. Find information on the Web to help you come up with an answer.

EARTH CITIZENSHIP

Superstorm Sandy—A Bad Storm Made Catastrophic by Global Warming

Hurricane Sandy (**FIGURE 20.20**), "born" on October 24, 2012, was the deadliest and most destructive storm of the 2012 hurricane season, and the second costliest in U.S. history.

Sandy developed in the south Caribbean and traveled north across the islands of Jamaica, Hispaniola, Cuba, and the Bahamas, and continued offshore of the East Coast of the United States. As it moved northward, storm surge damaged beach communities along the Delaware and New Jersey coasts; it peaked at 4 meters above sea level, accompanied by record-setting rainfall and flooding as it ran aground in New York City.

Scientists have attributed the high damage levels caused by Sandy to the influence of global warming, citing these factors:

1. At least 30 centimeters of storm surge flooding was the consequence of a century of sea level rise related to global warming.

2. High sea surface temperatures due to global warming infused the storm with extra energy.

3. Warm air carries high levels of water vapor, attributable to climate change. This fueled intense rains and flooding.

4. Hurricanes normally veer east into the Atlantic, but high pressure over Greenland, a result of arctic amplification, forced Sandy to hug the U.S. coastline.

5. Warmer air accelerated the wind speed, producing severe storm surge along the flooded coastline.

In its final days, Hurricane Sandy combined with a second storm that approached from the west. The combined force, dubbed a "superstorm," was a result of Sandy interacting with, and eventually overtaking, a trough, or continental low-pressure system, that had traveled across the United States. The two wrapped around each other, merging into a single, deep cyclone, 1800 kilometers across and extending from the ground to 1200 meters in the sky. This added to the size, damage area, and duration of the event, with the superstorm taking several days to wind down over Pennsylvania and New York.

(b)

(c)

FIGURE 20.20 (a) Hurricane Sandy formed on October 22, 2013, and (b) traveled north across the Caribbean region and offshore of several U.S. coastal states. The storm was responsible for over $68 billion in damage (b, c) and 285 deaths by the time it hit New York City on October 29, one week later.

(a)

Stocktrek Images/Getty Images

U S Air Force photo/Master Sgt Mark C Olsen/a/ SuperStock

Copyright © State of New Jersey, 1996–2014

20-7 Sea Level Rise Since the Last Ice Age Has Shaped Most Coastlines, and Continues to Do So

LO 20-7 Discuss how sea level rise affects coastlines.

One of the most influential geological processes shaping coastal environments around the world is the rise in sea level since the end of the last ice age.

History of Sea Level Rise

The last ice age reached its peak between 20,000 to 30,000 years ago, and rapid global warming began shortly thereafter. Sea level at that time was more than 120 meters below the present level, so much of the world's shallow seafloor and continental shelves were exposed (**FIGURE 20.21**). As sea level rose following the end of the last ice age, these lands were flooded and became the new seafloor.

The last ice age, which began about 80,000 years ago, was punctuated by short-lived warming spells, but overall it lasted nearly 60,000 years. Between about 20,000 and 6000 years ago, warm climate change caused glaciers around the world to shrink and seas to expand. The rising sea level flooded river mouths and continental shelves on every coastline of the world. This action shaped the look and character of modern coastal environments. Barrier islands emerged, estuaries widened, tidal wetlands accumulated sediment, and coral reefs flourished in warm seas.

Sea level is still rising today, and will continue to do so in the centuries ahead. We know this because greenhouse gas-induced global warming is causing the worlds glaciers to melt, which contributes water to the ocean, and it heats the ocean, which expands ocean water. These two processes are the main causes of global sea level rise.

Today, rising seas threaten coastal wetlands, estuaries, islands, beaches, and all types of coastal environments. Frequent flooding activated by storms and high tides, accelerated erosion, and saltwater intrusion into streams and aquifers endanger cities, ports, coastal communities, and other types of human development. Needless to say, these threats have dramatic economic and environmental consequences; as such, it is crucial that we study records of past sea level change to improve our understanding of how coastal environments evolve during periods of sea level rise.

Future of Sea Level Rise

Tide gauge data from around the world show that during the twentieth century, on average, sea level rose between 10 and 20 centimeters. Models of future climate change suggest that this increase was the precursor to an extended period (many centuries) of accelerated sea level rise beginning in the twenty-first century. Indeed, early in this century, using a combination of tide gauge and satellite data, researchers identified acceleration in global sea level rise. Some of the factors they cite in relation to this observation include the following:

- There is 30% more carbon dioxide (CO_2) in the atmosphere now than there was 100 years ago. As explained in Chapter 14, this greenhouse gas is effective at trapping heat in the lower atmosphere against Earth's surface. Researchers report that this past decade was the warmest in over 120 years; and 2010 tied with several other years as the warmest ever measured. Such a strong warming trend raises sea level because water expands as it warms, and when the ocean surface expands, the easiest direction to go is up.

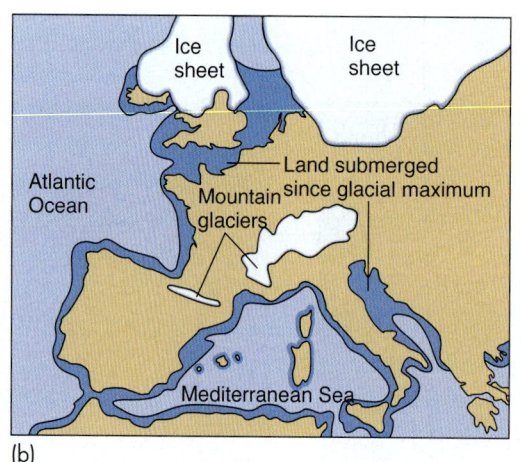
(a) (b)

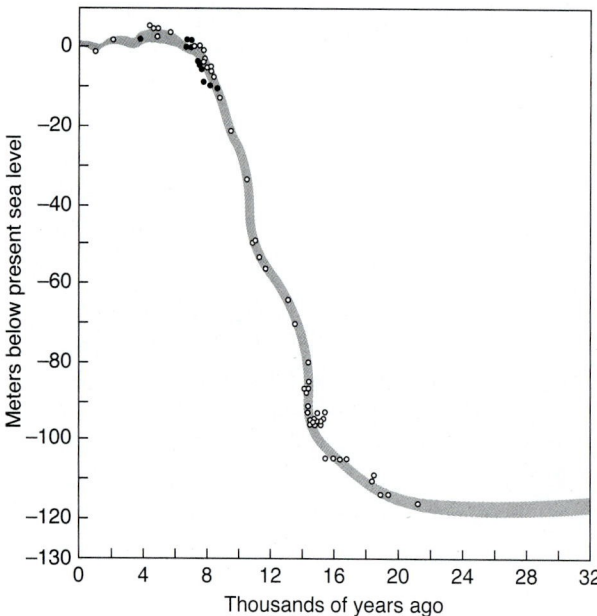
(c)

FIGURE 20.21 When the last ice age ended, melting of the great ice sheets added more than 120 meters of water to the world's oceans. As the water level climbed, coastlines retreated landward, giving birth to modern coastal environments. (a) The eastern coast of the United States at the peak of the last ice age approximately 20,000 to 30,000 years ago. (b) The coast of Europe and the Mediterranean Sea during the last ice age. (c) Dates of fossil corals (circles) that grow near sea level tell the history of sea level change since the last ice age.

 What effect does rising sea level have on a coast?

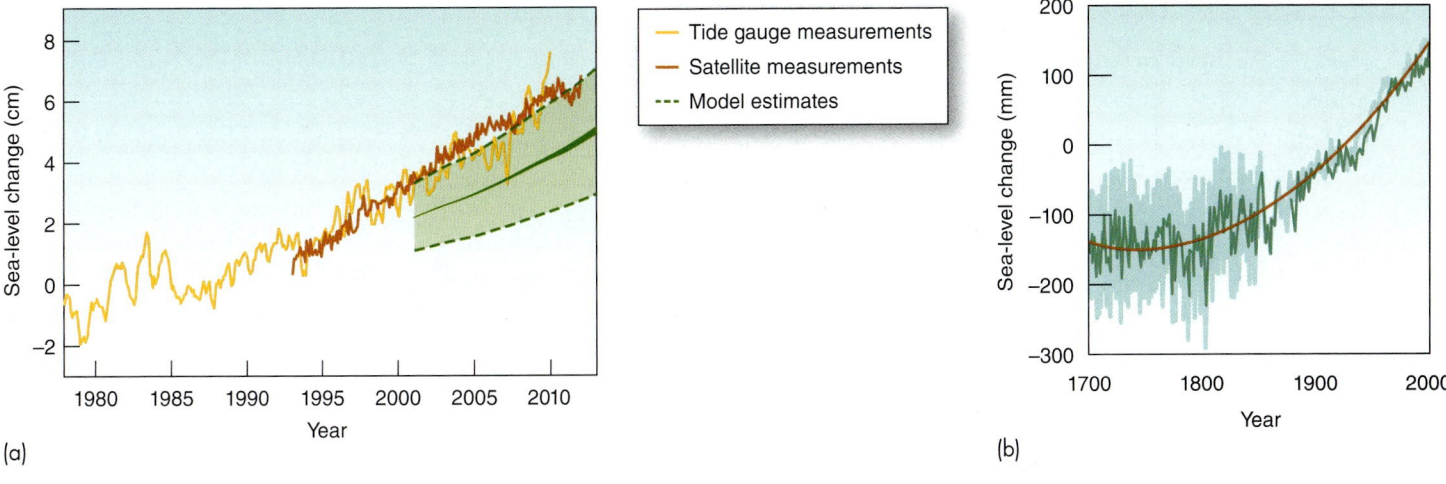

(a)

(b)

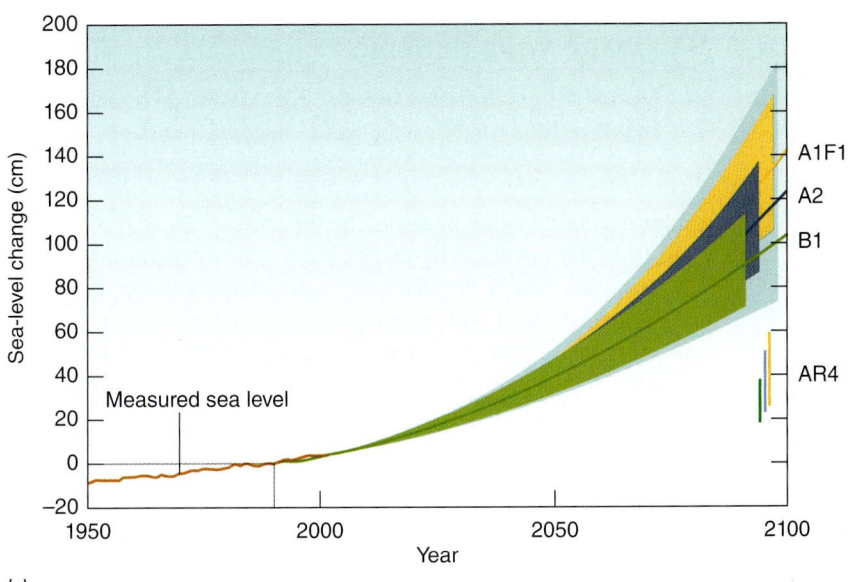

(c)

FIGURE 20.22 (a) The rate of modern sea level rise (gold and brown lines) is exceeding the worst-case scenario of computer models from the Intergovernmental Panel on Climate Change (green band). (b) Studies have found that the modern acceleration in the rate of sea level rise may have started over 200 years ago (uncertainty of the study is shown in light green). (c) Comparisons of global temperature and sea level change over the past 130 years have been used to project future sea level position. Research indicates that sea level may rise between 0.75 and 1.8 meters by 2100, depending on the production rate of greenhouse gases going forward. The future sea level projection of the 2007 IPCC Assessment Report 4 (AR4) is considered an underestimate of the threat of sea level rise. B1 is a best-case scenario of greenhouse gas production; A2 and A1F1 are moderate and worst-case scenarios, respectively (uncertainty shown in shades).

Q What options do coastal cities have when faced with sea level rise?

- Warming also melts ice. Studies have found that both the Greenland and Antarctic ice sheets are experiencing melting, and the meltwater is pouring into the oceans at an accelerating rate. Alarmingly, the rate of melting on Greenland alone has tripled in the last decade. Alpine glaciers are likewise dissolving around the world, and that meltwater, too, is flowing into the oceans.

- These three factors: 1) melting of Greenland and Antarctica, 2) melting of Alpine glaciers, and 3) expansion of warm ocean water, are each responsible for approximately 1/3 of the current global sea-level rise.

- The rate of sea level rise today, as measured by satellites mapping the ocean surface, is the highest ever observed: 32 centimeters per century, approximately twice that recorded in the twentieth century, and likely to accelerate even further.

Based on these statistics and observations, scientists have projected that by the end of this century sea level will rise between 0.75 and 1.8 meters, depending on the rate at which we humans continue to produce excess greenhouse gases (**FIGURE 20.22**).

Physical Impact of Sea Level Rise

The physical effects of sea level rise along the coast can be grouped into five categories: 1) inundation of low-lying areas, 2) erosion of beaches and bluffs, 3) salt intrusion into aquifers and surface waters, 4) higher water tables leading to groundwater inundation, and 5) worse flooding and storm damage. All of these are issues of grave concern, but the first two have had, and continue to have, the most dramatic impacts on coastal regions worldwide. Specifically, tidal wetlands, mapped by satellite, are shrinking due to flooding on the shores of estuaries around the world, and beach erosion has become a global problem, plaguing coastal communities and the tourism industry.

To date, the relationship between rising sea level and coastal erosion is still not well understood, but here's what we do know: A beach cross section, called a **beach profile**, has a characteristic shape that depends on the size of sand grains and the energy of waves. This profile can be described by the distance offshore at which waves first affect the profile (L) and the depth there (D). Typically, the ratio L/D can vary between 10 and 200. As sea level rises the profile shifts landward to

FIGURE 20.23 As sea level rises, the average beach profile will shift landward, eroding the land. The amount of erosion may be approximately 100 times the amount of sea level rise.

What will happen to houses or roads built next to an eroding beach?

regain the L/D ratio that is natural for that setting; it achieves this ratio by eroding the shoreline (**FIGURE 20.23**). If L/D is approximately 100, the change in D due to sea level rise (usually in millimeters) can translate into horizontal beach erosion that is two orders of magnitude greater than D. With global sea level rise currently at over 3 millimeters per year, this translates into approximately 30 centimeters per year of erosion, or 3 meters per decade. In many coastal communities this puts at risk homes, businesses, roads and highways, and natural ecosystems.

EXPAND YOUR THINKING

Most of the goods traded between countries travel by ship. How could sea level rise affect the global shipping industry?

20-8 Barrier Islands Migrate with Rising Sea Level

LO 20-8 Define barrier islands, and explain how they migrate.

Coastal processes related to waves, storms, tides, and sea level rise interact with the underlying geology of the coastline to produce various types of *coastal environments*. Low-elevation coastlines with an abundance of sand tend to form barrier islands characterized by sandy beaches, tidal inlets, lagoons, and tidal marshes.

Barrier Islands

Barrier islands (**FIGURE 20.24**) are long, low, narrow sandy islands with one shoreline that faces the open ocean and another that faces a saltwater *tidal lagoon* protected by the barrier. The barrier is separated from the mainland by one or more *tidal inlets* through which flows seawater driven by differences in tide level between the lagoon and the open ocean. Changes in tide generate the movement of strong tidal currents through the inlet, exchanging water and sediment between the lagoon and the ocean.

Sediments carried by flooding tide from the ocean into the lagoon will collect at the landward end of the inlet in a deposit known as a *flood-tide delta*. Those that collect on the seaward end of the inlet, carried by ebbing tide, form an *ebb-tide delta*. *Tidal deltas* are important sources of sediment for the adjacent lagoon and ocean shoreline and seafloor.

Barrier islands are found on coastlines around the world. In the United States, there are chains of barrier islands along the east and south coasts of North America, where they extend from New England down the Atlantic coast, around the Gulf of Mexico, and south to Mexico. They include the *Outer Banks* of North Carolina (**FIGURE 20.25**), a long line of sandy barrier islands extending over 280 kilometers, that protects the broad waters of *Pamlico Sound*, a tidal lagoon.

FIGURE 20.24 (a) Many barrier islands are heavily developed because they are popular destinations for vacationers. (b) Differences in water level created by the tides drive circulation between a lagoon and the open ocean. Sediments deposited at the ocean and lagoon ends of the tidal inlet are known as tidal deltas.

Describe the role of storms in a barrier-lagoon setting.

(a)

(b)

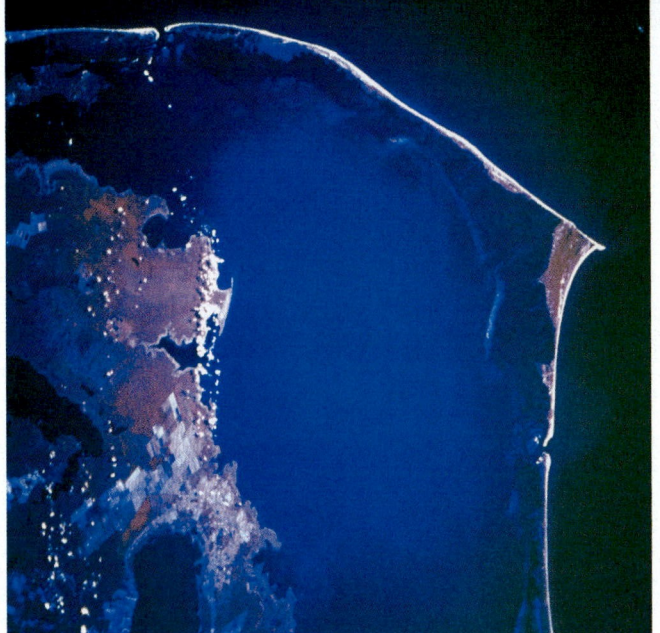

FIGURE 20.25 The long, thin line of barrier islands located off the Atlantic coast of North Carolina is called the Outer Banks. These islands protect a broad tidal lagoon, Pamlico Sound, which exchanges water with the ocean through tidal inlets.

Describe the role of tides in a barrier-lagoon setting.

Barrier islands are constantly changing environments that shift their positions when they are overrun by storm surge. When a storm such as a hurricane encounters a low barrier island, the floodwaters carry sand from the ocean side of the island and deposit it on the lagoon side. This process is called **washover**, and the resulting sandy deposit is called an *overwash fan*.

Sand also moves offshore during storms, as well as into the lagoon through tidal inlets under nonstorm conditions. Like a giant treadmill, washover forces barrier islands to roll landward by eroding on the ocean side and depositing on the lagoon side while at the same time losing some sand offshore. This process has been referred to as *barrier rollover*. Barrier rollover, driven by storm surge, washover, and sand transport through tidal inlets into lagoons, allows an island to migrate landward under conditions of sea level rise, and thereby avoid "drowning" in place (**FIGURE 20.26**).

Barrier islands are composed of several habitats, typically arranged as zones oriented parallel to the shore (**FIGURE 20.27**). Immediately landward of the ocean beach is a line of sand dunes that store sand delivered by winds blowing off the sea. Inland lies a wide sand-covered plain vegetated by drought-tolerant grasses or a forest thicket.

On the lagoon side, **tidal wetlands** appear on the surface of old overwash fans. Salt-tolerant plants grow in the muddy substrate of these marshes, which are flooded by high tide on a daily basis. As these plants die and their remains accumulate over time, they form *peat* between the levels of high and low tide. Eventually, barrier rollover buries the peat under overwash fans and migrating wind-blown dunes as the island shifts landward during periods of rising sea level. The emergence of these peat layers on the ocean side of the island after hundreds or thousands of years of burial provides undeniable proof of barrier rollover.

EXPAND YOUR THINKING

Explain how the process of barrier rollover would differ under slowly rising sea level versus rapidly rising sea level.

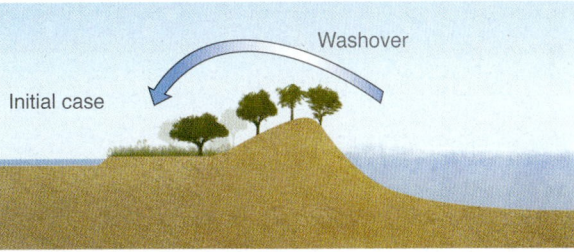

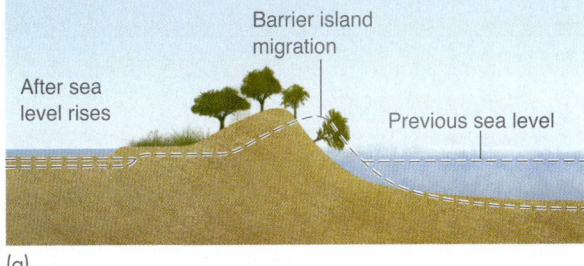

(a)

(b)

FIGURE 20.26 (a) Storms drive sand from the seaward side of a barrier island to the lagoon side. This process, called *washover*, creates *overwash fans*. (b) As sea level rises over several centuries, the process of washover during storms allows an island to migrate with rising sea level.

Given how barrier islands migrate with rising sea level, are they safe places to build communities? Explain your answer.

FIGURE 20.27 Barrier islands consist of several zones or habitats. On the seaward side of the island is a beach and surf zone. Landward of the beach is a system of sand dunes and a wide sand-covered plain vegetated by drought-tolerant grasses or a forest thicket; on the lagoon side are tidal wetlands growing on former overwash fans.

What happens to tidal wetlands when a barrier island rolls over them?

20-9 Rocky Shorelines, Estuaries, and Tidal Wetlands Are Important Coastal Environments

LO 20-9 Identify several types of coastal environments.

Rocky shorelines develop where steep hills or mountains descend directly into the sea. Estuaries are the mouths of former stream channels that have been flooded by sea level rise. Tidal wetlands are mud-dominated environments in quiet water settings. All three are crucial coastal environments in terms of coastal health and biological diversity.

Rocky Shorelines

Rocky shorelines (**FIGURE 20.28**), familiar around the world wherever steep hills or mountains descend directly into the sea, are characterized by types of rock that differ in their capability to withstand wave erosion. Rocks that succumb to erosion more easily will retreat, while the more resistant types will protrude into the surf. *Sea stacks*, composed of resistant rock that can withstand the pounding of waves and bioerosion by intertidal organisms, are shaped in this manner. Sea stacks mark the previous position of the shoreline, before erosion forced the coast to retreat landward.

As a rocky shoreline retreats under the force of erosion, it leaves behind a flat, wave-scoured platform marking the position between high and low tides. This *wave-cut platform* continues to widen as the shoreline retreats. Platforms expand as the sea erodes the base of a cliff by carving a wave-cut notch.

FIGURE 20.28 Made up of erosion-resistant rock, sea stacks mark the locations of former shorelines, which have shifted landward as a result of erosion.

Ⓠ **How might a sea stack influence coastal processes in its immediate vicinity?**

FIGURE 20.29 As wave erosion cuts away at the base of a cliff, carving a notch, it creates a broad rocky platform marking the retreat of the rocky shoreline.

Ⓠ **Explain the role of the tides in shaping a wave-cut platform.**

As the notch undercuts the cliff face, the weight of the unsupported rock above eventually collapses in a rock slide, a type of *mass wasting*. The sea removes the slide debris, and the notch-eroding process begins again. As the cliff retreats, the very base of the notch remains, leaving a wide bench—a wave-cut platform (**FIGURE 20.29**). Eventually, the platform becomes so wide that the sea expends too much energy reaching the cliff, and the shoreline retreat slows, and may end.

Estuaries

As ocean waters rose after the last ice age, they flooded coastlines and pushed shoreline environments landward, a process termed **marine transgression**. Seawater invaded and inundated the mouths of stream valleys, widening them into our modern estuaries.

An **estuary** is a partially enclosed body of water that develops in places where freshwater from rivers and streams flows into the ocean, mixing with the salty seawater. Estuaries, and the lands surrounding them, mark the transition from terrestrial/freshwater systems to marine/saltwater systems. They are protected from the full force of marine waves and storms by reefs, barrier islands, or fingers of land (*spits*) that absorb the ocean's energy. Generally thought of as relatively calm bodies of water, estuaries are, nonetheless, influenced by the tides, although their restricted openings alters the timing and height of tides compared to those in the nearby open ocean.

Estuaries—also called bays, lagoons, harbors, inlets, or sounds—come in many different sizes and shapes. But all, regardless of name or size, are marked by a single defining feature: the mix of freshwater and saltwater. Familiar examples of estuaries include San Francisco Bay, Puget Sound, Chesapeake Bay, Boston Harbor, and Tampa Bay (**FIGURE 20.30**).

Estuarine environments are among the most biologically productive on Earth, producing more organic matter each year than similar-size areas of forest, grassland, or agriculture. Their sheltered waters

Skyscan/Photo Researchers

© William Helsel/Age Fotostock America, Inc.

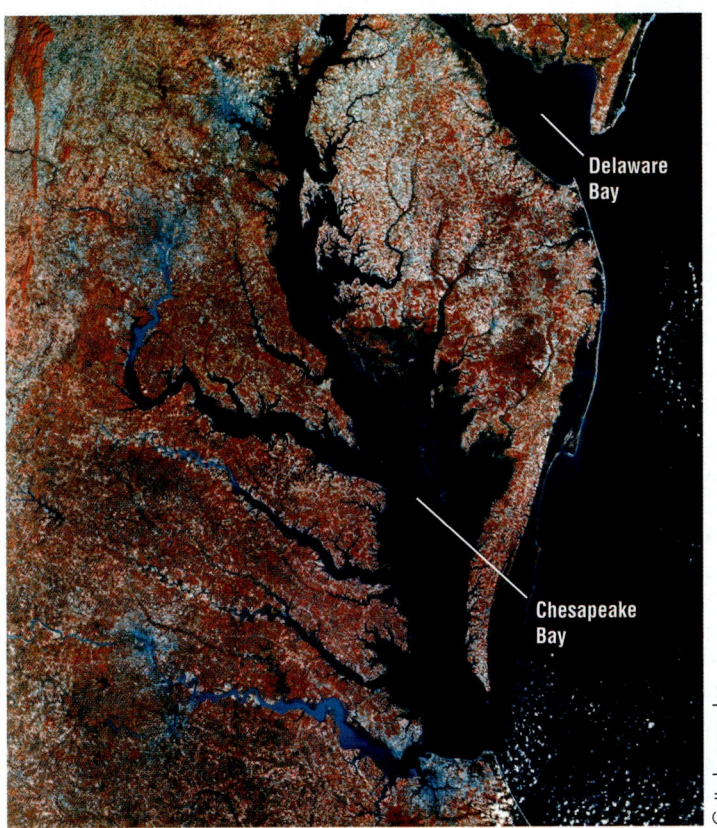

FIGURE 20.30 The Chesapeake and Delaware bays on the Atlantic coast were formed by marine transgression of the mouths of the Susquehanna and Delaware rivers.

Ⓠ **Explain the role of sea level rise in the development of the Chesapeake and Delaware bays.**

the marsh surface. The stalks and leaves of salt-tolerant plants assist this process by capturing grains of silt and clay while they are still suspended in floodwaters; rain later washes these onto the marsh surface. As sediment accumulates, the coastal basins and embayments holding marshes are filled with mud.

The history of estuaries and wetlands can be clearly linked to sea level rise. Globally, sea level rise slowed about 5000 to 3000 years ago; by 2000 years ago, the rate of global sea level rise was only about 1 millimeter per year or less (less than one-third its present rate). Under this new environment of slow marine transgression on the U.S. Atlantic coast, a single species of tidal marsh grass, smooth cord-grass (*Spartina alterniflora*), established the first permanent foothold in the low-energy embayments of drowned stream valleys. Grass shoots slowed the movement of the water, trapping and accumulating even more sediment. This enhanced sedimentation, coupled with the increasing volume of root systems, allowed the newly developing marsh surface to keep up with rising sea level, and even expand seaward into the open estuary. As a result, tidal wetlands have become important, widespread estuarine environments that play key roles in coastal sedimentation and biological diversity.

are home to unique communities of plants and animals specially adapted for life at the margin of the sea. These floras and faunas occupy various habitats in and around estuaries, including shallow open waters, tidal wetland, sandy beaches, mud and sand flats, oyster reefs, mangrove forests, river deltas, coral lagoons, and sea-grass beds.

Tidal Wetlands

Tidal wetlands, also called *salt marshes*, develop along the fringe of many estuaries. Water draining from the uplands carries sediments, nutrients, and pollutants that are deposited in mud-rich wetlands but prevented from entering estuarine waters.

Wetland plants and soils act as a natural buffer between the land and ocean, absorbing floodwaters and dissipating storm surges. Salt-marsh grasses and other plants help stabilize the shoreline and counteract coastal erosion (**FIGURE 20.31**).

Salt marshes are formed in quiet, low-energy environments that are protected from direct wave action yet still experience storm surge, weather, and daily tide changes. Marshes typically are flooded daily by high tide, but it is the spring flood tide that reaches farthest inland and covers the entire wetland surface. During flood tide, fine-grained suspended silts and clays settle out of the water and accumulate on

FIGURE 20.31 Tidal wetlands are characterized by broad expanses of salt-tolerant grasses and numerous meandering tide channels.

Ⓠ **EXPAND YOUR THINKING**

Refer back to Figure 20.30: What will happen to the Delaware and Chesapeake estuaries as sea level continues to rise due to global warming? What major impacts can be expected?

20-10 Coasts May Be Submergent or Emergent, Depositional or Erosional, or Exhibit Aspects of All Four of These Characteristics

LO 20-10 Distinguish between submergent and emergent coasts and between depositional and erosional coasts.

The strong influence of sea level rise since the end of the last ice age can be classified according to either uplift rate or net physical processes. Coasts where the land is tectonically uplifting at a rate that exceeds sea level rise are known as emergent coasts; those where sea level rise exceeds uplift are known as submergent coasts. Erosional coasts are those dominated by erosional sedimentary processes, and depositional coasts by depositional sedimentary processes.

Emergent and Submergent Coasts

Emergent coasts are characterized by fossil wave-cut platforms, or *beach ridges,* that are uplifted above the present-day sea level. Consider the coast of California (**FIGURE 20.32**), which is being uplifted by tectonic forces associated with the San Andreas Fault System marking the western boundary of the North American Plate. Uplift tends to be sudden and is associated with large earthquakes—the origin of the term *coseismic uplift.* Uplift occurs at the same time as an earthquake and is propelled by the same seismic event. This type of uplift produces a cliff at the coastline that is eroded by waves during the interval of time between coseismic events.

Another type of emergent coast is one that is uplifting because the weight of a glacier has been removed by climate warming since the end of the last ice age. In these areas, usually located in extreme northern and southern latitudes along continental shores (e.g., Norway and Canada), the lithosphere slowly rebounds in response to the removal of the weight of a past glacier, a process that is called **glacioisostatic uplift.** It occurs because the lithosphere has elastic properties that enable it to recover from the former weight (**FIGURE 20.33**). Often, sequences of emerged beach ridges mark the history of the uplift.

Submergent coasts, in contrast, experience slow flooding by the ocean as a result of sea level rise, lithospheric subsidence, or both. A submerging coast is any place where the net result of sea level change and vertical land movement is marine drowning. In the current era of accelerated sea level rise, many coastal systems are submerging. These include areas where estuaries form in the flooded mouths of river systems, such as along the eastern coast of the United States (e.g., Chesapeake Bay, shown in Figure 20.30).

In places where great volumes of sediment collect, such as major delta regions, the enormous weight of the sediment causes downward flexing of the lithosphere. When a rise in sea level is added to this subsidence, rapid flooding occurs across broad expanses of low-lying shoreline. The Ganges–Brahmaputra river delta region is an example of a submerging coast. This region, on the coast of Bangladesh, one of the world's poorest countries, is expected to undergo a rise of 1.5 meters in sea level in response to the combination of global warming and submergence. It is estimated that the nation will lose 16% of its land area and that 15% of its population will be displaced (**FIGURE 20.34**).

Erosional and Depositional Coasts

Another way to classify coastal systems is by the net result of physical processes, either erosional or depositional. **Erosional coasts** are those that are experiencing net landward migration under the force of

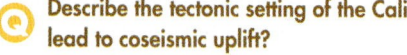

FIGURE 20.32 The coast of central California is an emergent coast because tectonic forces cause uplift faster than the rate of sea level rise. In this photo, a broad fossil marine terrace marks the top of the coastal cliff.

Describe the tectonic setting of the California coast. Why does it lead to coseismic uplift?

FIGURE 20.33 When the weight of a glacier is removed, the lithosphere elastically rebounds. A series of gravel beach ridges may develop, with each ridge marking a former shoreline that has since been lifted above the wave zone.

What is the evidence that this coast is emergent?

Today
Total population: 150 million
Total land area: 134,000 km²

Sea-level rise: 1.5 m
Total population affected: 17 million (15%)
Total land area affected: 22,000 km² (16%)

(a)

(b)

FIGURE 20.34 Continued sea level rise will displace over 17 million people in Bangladesh and submerge 16% of the nation's land.

Q Why are deltas especially vulnerable to sea level rise?

erosion. Most often they include rocky shorelines undergoing landward retreat. Sea stacks, wave-cut beaches, and undercut cliffs characterize erosional coasts.

Erosional coasts typically have steep topography associated with tectonically active margins, with a narrow continental shelf, steep coastal plains, and "youthful" mountain systems—those that have not existed long enough to shed high volumes of sediments. Rivers tend to cut into steep mountainous watersheds with high gradients. Sediments on these coasts are coarse grained, and wetlands and estuaries on them tend to be narrow and poorly developed. All these features are common to shorelines that generally act as sediment sources rather than as sites of sediment accumulation.

Depositional coasts, in contrast, often are associated with old mountain systems that are no longer tectonically active. With long periods of geologic time during which to shed sediments into the coastal zone, a broad continental shelf to diffuse wave energy, and wide coastal plains to collect and store sediments, these coastal systems are characterized by abundant sediment that forms barrier island chains, deltas, broad lagoons, and extensive tidal wetlands. River valleys in them tend to be wide and flat. In general, depositional coasts are found in midplate settings where little active mountain building takes place (**FIGURE 20.35**).

Test your understanding of these complex environments by completing the Critical Thinking exercise, "Coastal Systems."

Marco Simoni/Getty Images, Inc.

(a)

Ingram Publishing/SuperStock

(b)

FIGURE 20.35 (a) Erosional coasts are characterized by eroding sea cliffs, sea stacks, and short beaches in narrow embayments. (b) Depositional coasts collect large quantities of sediment. Coastal wetlands tend to be found on depositional coasts.

Q How is the tectonic setting likely to differ between erosional and depositional coasts?

Q EXPAND YOUR THINKING

Considering that global sea level rise has accelerated, which of these four coastal types is most vulnerable to coastal hazards, and which is least vulnerable? Explain your reasoning.

CRITICAL THINKING

Coastal Systems

Coastal environments are complex geologic systems that receive sediments, dissolved compounds, nutrients, and a mixture of fresh and salty water from terrestrial and marine environments. The growth of human communities along coasts has led to declines in environmental resources, increased levels of pollution, heightened vulnerability to coastal hazards, and other negative impacts.

Keeping these facts in mind, and referring to **FIGURE 20.36**, please work with a partner to complete the following exercise:

1. Label the coastal environments.

2. Make a table with five columns, labeled (left to right): Environment, Sediment Type, Rock Type, Coastal Processes, and Sea Level Rise. List each coastal environment in the left column. In the next four columns, do the following:

   a. Describe the sediments that collect in each environment.

   b. Name the type of sedimentary rock that will result from each environment.

   c. Describe how waves and tides influence each environment.

   d. Consider sea level rise: How will it affect each environment?

3. Label the diagram with the coastal hazards that will affect each environment.

4. You are the mayor of a small coastal town with a population of 5000. In the fall of 2016, a hurricane hits the barrier island on which your town is located. Seventy-five percent of the buildings in town are completely destroyed by high winds and storm surge. The other buildings are damaged to some degree. Roads are torn up by waves; electricity and sewage infrastructure are destroyed. The damage to the community totals $7 billion. The Federal Emergency Management Agency (FEMA) asks you to consider abandoning the town and offers you land to rebuild on the adjacent mainland. What should you do? How would you involve the whole community in making this decision? Write a scenario in which you lead the community to a consensus.

FIGURE 20.36 Coastal environments.

(20-11) ## Coral Reefs Are Home to One-Fourth of All Marine Species

LO 20-11 Define the main types of coral reefs and describe their formation.

A large and interdependent ecosystem relies on coral reefs for survival, yet they occupy a mere 0.2% of the world's oceans. Home to 25% of all marine species, these underwater habitats are crucial to the ocean's biodiversity.

FIGURE 20.37 Home to diverse marine organisms, the world's coral reefs have been called the rain forests of the sea.

David Nardini/Getty Images, Inc.

Reefs around the world are threatened by ocean warming and acidification, pollution, and overfishing. What steps would you take to manage them, to ensure their survival?

Coral Reefs

Coral reefs are shallow coastal environments where numerous types of *coral* (an animal) and *algae* (a plant) build massive solid structures of calcium carbonate ($CaCO_3$—limestone). Corals secrete this material, which composes their skeletons, as part of their metabolic activity (**FIGURE 20.37**).

Individual coral animals are called *polyps*. A polyp has a single opening that functions both to take in food and to expel waste; this opening is surrounded by a fringe of tentacles. Each polyp grows within its own hard cup, called a *calyx*, depositing a solid skeleton of $CaCO_3$ as it grows.

As layer upon layer of calcium carbonate builds up over time, the reef framework is constructed, with individual coral animals, numbering in the hundreds of thousands, living like a skin on the surface. Under ideal conditions, some corals can grow several centimeters per year. By secreting calcium carbonate in enormous groups, the corals can expand into huge colonies. Each colony is made up of individual polyps, all of which are linked to their neighbors by connective tissue, including their stomachs; so when one polyp eats, they all eat!

Living among the reef builders are myriad fish, mollusks, echinoderms, and other types of organisms in the world's most biologically diverse marine ecosystem. Up to 3000 different species of plants and animals may be found living together on a single coral reef. It is believed that one-quarter of all known fish species live on or around coral reefs. Therefore, besides being remarkable ecological environments, reefs also are valuable natural assets; they provide food, jobs, protection from large waves and storm surge, and billions of dollars in revenues each year to local communities and national economies.

Inside the corals live tiny *symbiotic algae*, a beneficial parasite that converts sunlight and nutrients into food for coral growth and calcium carbonate production. These algae depend on light for photosynthesis, so corals require direct access to sunlight for growth and can be damaged by sediment or other factors that muddy the clarity or quality of the water around them.

Coral reefs are generally found only between 30° N and 30° S of the equator (**FIGURE 20.38**); but within this belt, they do not occur in areas where freshwater and heavy coastal sedimentation prevent coral growth. Over geologic time, reefs have survived fluctuating sea levels, uplifting landmasses, periods of widespread warming and re-

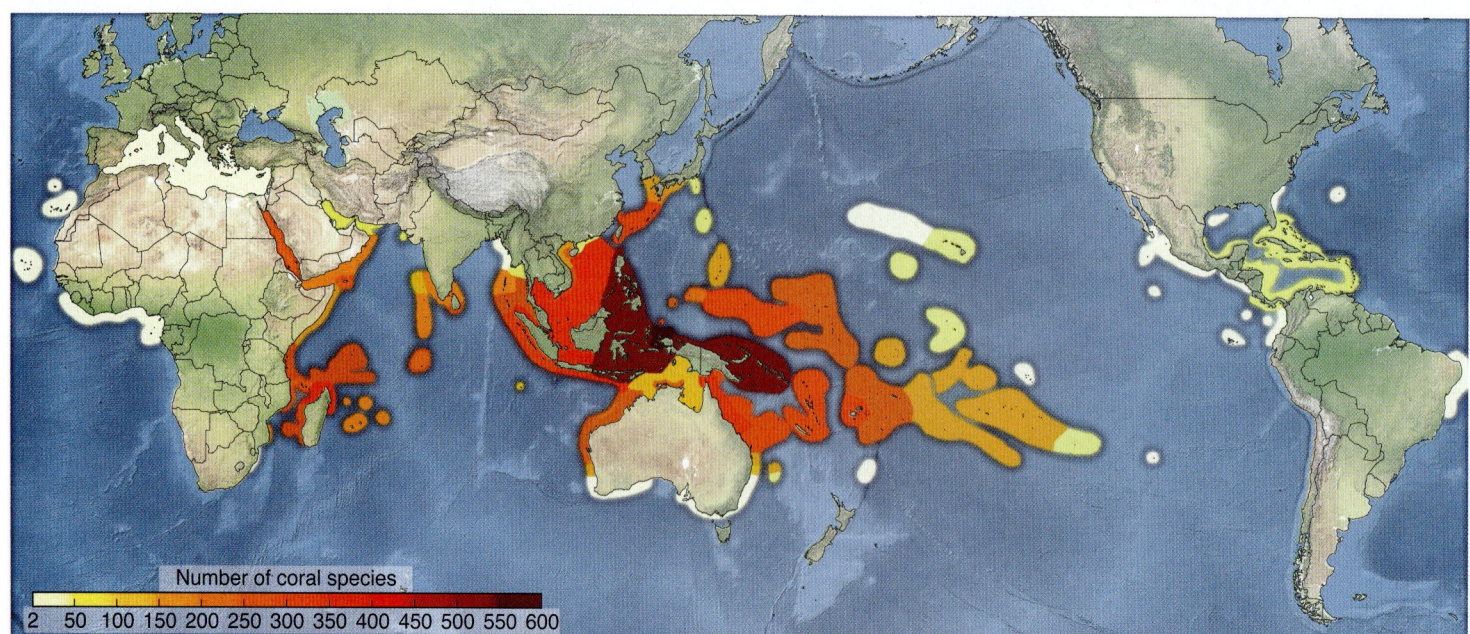

Number of coral species
2 50 100 150 200 250 300 350 400 450 500 550 600

FIGURE 20.38 Located between 30° N and 30° S of the equator, coral reefs require shallow, warm, sunlit seas.

What types of environmental conditions do corals prefer?

(a)

Courtesy of Chip Fletcher

© Reuters/CORBIS

(b)

FIGURE 20.39 (a) Fringing reefs are attached to a mainland coast. (b) Barrier reefs have a deep lagoon that separates the reef from the coast. (c) Atolls develop when the island to which reefs are attached subsides below sea level. Small sandy islets often emerge on the atoll's rim. (d) Patch reefs are small, isolated reefs that grow in lagoons or on reef flats.

Explain how atolls form.

Douglas Faulkner/Photo Researchers

(c)

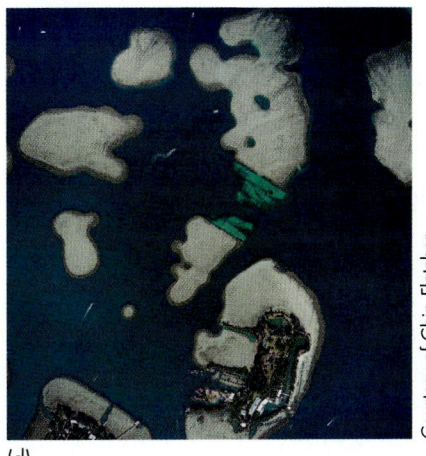

Courtesy of Chip Fletcher

(d)

peated glaciations, as well as recurrent short-term natural disasters, such as cyclones and hurricanes. In short, coral reefs have a remarkable capability to adapt and survive.

Unfortunately however, coral reefs are threatened by a host of troubles. Climate change is causing the oceans to warm and this disrupts the ecosystem built by reefs. Ocean acidification is also a threat as acidic waters interfere with the ability of coral to secrete the calcium carbonate that builds their exoskeleton. Reefs also suffer from pollution out of nearby watersheds where humans live (see the discussion "Coastal Pollution" in section 20-12). Because corals are animals, they are subject to viruses, diseases, poisoning by pollution, and other ills coming from *polluted runoff* discharged from land. Another major impact is the problem of *overfishing*. Coastal communities frequently take fish from reefs for human consumption but too often the amount of fish that are hunted exceeds the capacity of the reef to replace them. Many of these fish are herbivorous, that is, they eat algae on the reef. With the loss of these fish, algae succeed in out-competing corals for space on the seafloor and many reefs are being destroyed by the spread of noxious algae where herbivorous fish are absent.

Types of Reefs

There are four basic kinds of coral reefs: *fringing reefs*, *barrier reefs*, *atolls*, and *patch reefs* (**FIGURE 20.39**).

Fringing reefs grow in shallow water and are directly attached to a coastline. They consist of several ecological zones (**FIGURE 20.40**)

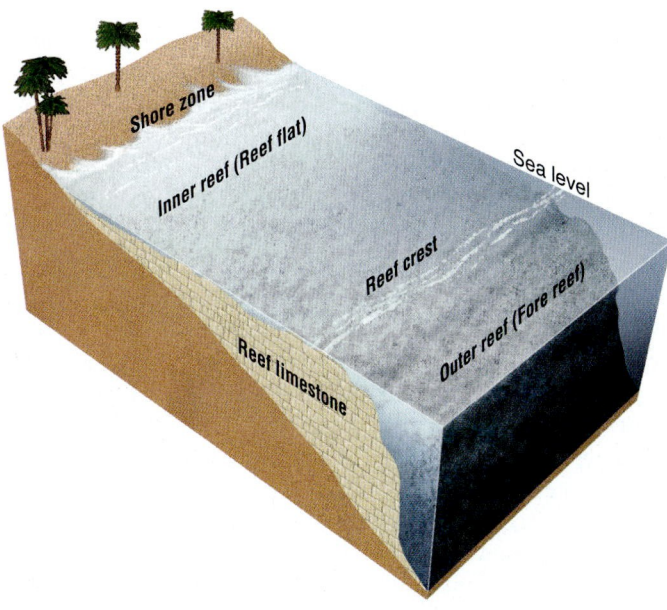

FIGURE 20.40 Reefs have four ecological zones: shore zone, reef flat, reef crest, and fore reef.

What types of environmental conditions do corals prefer?

that are defined by their depth, the shape of the reef's surface, and the plant and animal community it supports. These zones include the *shore zone*; the *inner reef* or *reef flat,* located in shallow water with little wave energy; the *reef crest,* the shallowest seaward portion of the reef, which causes waves to break; and the *outer reef* or *fore reef,* the seaward-sloping front below the reef crest. Each zone is characterized by a unique assemblage of coral, algae, and other organisms. All four types of reefs have similar zones.

Barrier reefs are separated from land by a lagoon. Lagoons can be very deep, exceeding 30 meters in depth, and are characterized by sandy floors, prolific coral growth, and strong tidal currents. Barrier reefs can be large, regional-scale features marked by separate islands, inlets, and patch reefs. The zones found on a fringing reef also appear on a barrier reef.

Atolls are shaped like irregular doughnuts with a broad, deep lagoon in the middle. They originate from fringing reefs attached to volcanic islands that subside over time. As the island subsides (usually, as a lithospheric plate cools when it moves away from a hotspot) and erodes (in the wake of stream and wave erosion), the reef grows upward continuously in order to remain in shallow sunlit waters. Eventually, a barrier reef may develop when the volcanic island subsides below sea level. By the time the last of the island sinks below the waves, the reef has become nearly circular in shape, with a deep central lagoon.

Patch reefs are small, isolated reefs typically several tens of meters in diameter; they are located in lagoons or on fringing reef flats.

Q EXPAND YOUR THINKING

Explain the various ways in which global warming could affect the world's reefs.

20-12 Coastal Problems Are Growing as Populations Increase

LO 20-12 Discuss the effects of human activities on coastal environments.

The coastal zone has become the focus of dramatic population growth, development and investment, thereby exposing more people to coastal hazards and causing more damage to coastal resources and ecosystems.

Coastal Erosion

Coastal erosion is now a worldwide problem, threatening beaches and coastal lands, as well as the economic and environmental systems that rely on them. According to a study released by Federal Emergency Management Agency (FEMA), approximately 25% of homes and other structures within 150 meters of the U.S. coastline and the shorelines of the Great Lakes will fall victim to the effects of coastal erosion within the next 60 years.

The areas expected to be worst-hit include the Atlantic and Gulf coastlines, which are estimated to account for 60% of nationwide losses from erosion. Costs to U.S. homeowners are estimated to average more than $500 million per year, calculated in terms of higher insurance premiums, replacement of damaged buildings, and engineering measures necessary to counteract further erosion (**FIGURE 20.41**).

This study has prompted FEMA to establish an *avoidance policy,* recommended steps for coastal communities to prevent erosion. It is designed to circumvent what experts predict will happen when erosion and human land use run afoul of each other—high nationwide costs, hardships, and financial burdens on homeowners, and negative impacts on shoreline environments that suffer.

In point of fact, there is really only one good response when coastal erosion strikes: Get out of the way. Seawalls and other types of barriers may protect a house or road temporarily, but by setting an artificial boundary on the ecosystem, thus preventing it from migrating with changing sea level and other shifting conditions, they usually backfire, in the form of beach loss, wetland loss, and other types of environmental repercussions. Seawalls also separate beaches from sand dunes, an important source of sand. Given the inevitability of continued sea level rise and fiercer coastal storms, moving to, building on, or remaining on an eroding coast are unwise choices to make.

Coastal Pollution

Among the major threats to coastal resources is the higher levels of **polluted runoff** in coastal waters adjacent to urban, suburban, industrialized, and agricultural lands. Polluted runoff is the contaminated water that runs off city streets, building sites, parking lots, and farmlands. It now accounts for 75% of the pollution in coastal waters.

Brett R. Henry/Alamy

FIGURE 20.41 Over the next 60 years, approximately 25% of homes within 150 meters of the U.S. shoreline will be lost to erosion.

 **How will this scene change as sea level rise continues?**

© J. Emilio Flores/Corbis/Corbis

FIGURE 20.42 Eutrophication occurs when algae populations in coastal waters increase rapidly because of high nutrient content. Their growth robs the water column of oxygen that is necessary to other forms of life, often leading to massive die-offs of fish in coastal waters and estuaries.

Polluted runoff comes from many sources: used engine oil poured down storm drains; pesticides and fertilizers applied to farms and yards; dirt that washes to the coastal zone from logging, agriculture, or mining sites; industrial solvents and chemicals; and manure from livestock operations. The problem of polluted runoff is especially serious along U.S. coasts where more people live in coastal counties than anywhere else in the nation. These millions of people produce huge quantities of pollution that is ultimately released into coastal waters.

Agriculture is the largest contributor of polluted runoff. Fertilizer and manure, along with more than a billion tons of eroded soil, flow from farm fields into coastal waters every year. Over 13,000 metric tons of farming chemicals are applied annually in areas that drain into the nation's coasts. Some of these compounds kill native marine ecosystems while others fuel the growth of noxious species of algae that use all the oxygen in the water, thereby suffocating other forms of life, a process called **eutrophication** (**FIGURE 20.42**).

Coastal Hazards

Sea level rise is bringing coastal hazards literally to the doorsteps of our buildings, an especially worrisome trend, according to researchers who have been documenting changes in the behavior of hurricanes in reaction to higher sea surface temperatures. Storms have been battering our coasts for millennia, but in recent decades they have become stronger and more destructive. Mounting scientific evidence links this trend to global warming, because hurricanes get their power from warm surface waters underlying the storm and warmer air along its path.

Throughout geologic history, high waves, coastal erosion, tsunamis, high winds, hurricanes, and changing sea levels have come and gone without causing harm. Natural coastal processes raise concerns only when humans are affected by them. When they occur near man-made developments—buildings and roadways—these natural processes become **coastal hazards**, a term that refers to any coastal process that may cause loss of life or property.

Today, with sea temperatures rising with global warming, fiercer storms are brewing around the globe. In the United States, annual losses attributable to natural disasters have climbed from and average $20 billion per year in the 1980s to $85 billion per year today. In 2005, Hurricane Katrina alone was responsible for over $100 billion in losses to the City of New Orleans. Superstorm Sandy, the second most costly event in United States history, caused a total of $68 billion in damages and killed 286 people. Figures such as these have raised a red flag for many government agencies and private insurance companies, which are tracking the escalating costs nationwide from natural hazards.

Underlying these statistics is population growth in hazard-prone locations, notably coastal areas; more people translates into more homes, businesses, schools, highways, and hospitals being built on sites that are particularly vulnerable to chronic or catastrophic coastal hazards—hurricanes, severe storms, sea level rise, coastal erosion, and tsunamis (**FIGURE 20.43**).

FIGURE 20.43 As more people move into the coastal zone, buildings, roads, homes, and hospitals are built in locations vulnerable to natural coastal hazards. Sea level rise associated with global warming will make these locations even more unsafe in the future.

Guy Vanderelst/Getty Images, Inc.

The future promises more of the same. Studies show a global trend toward higher wind speeds of the strongest hurricanes over the past two or three decades, especially in the North Atlantic Ocean and the Indian Ocean. According to the Intergovernmental Panel on Climate Change (IPCC), it is "more likely than not" that humans are a contributing factor to hurricane intensification since the 1970s. Furthermore, reports the IPCC, "It is likely that future hurricanes will become more intense, with larger peak wind speeds and heavier precipitation."

Researchers have concluded that global warming will intensify the strongest storms and that there may be fewer weak ones. More specifically, they are estimating that the overall strength of storms, as measured in wind speed, will rise by 2% to 11%, but that the number of storms will decline, by between 6% and 34%. In sum, there would be fewer weak and moderate storms and more of the powerful, damaging ones, which, as just noted, are projected to be stronger in response to warming.

The size and scope of these highly damaging events have already had a profound effect on public perceptions of, and therefore policies relating to, hazards, especially regarding what can, or should, be done to minimize their impact on U.S. populations. In response to escalating costs of catastrophic natural disasters in recent years, agencies in charge of emergency management have broadened their focus beyond disaster preparedness to include **hazard mitigation**, an effort to minimize individual and community vulnerability to natural disasters. The primary purpose of mitigation is to ensure that fewer individuals and their property become victims of such disasters, in particular to reduce the number of injuries and deaths, as well as lower the overall costs of and relieve the economic burden imposed on communities by natural hazards.

Q EXPAND YOUR THINKING

What options do coastal communities have for dealing with coastal erosion, coastal pollution, and coastal hazards?

LET'S REVIEW

Coastal problems grow along with the numbers of people moving near or into coastal regions. Development and investment in these areas has dramatically escalated coastal pollution, thereby speeding degradation of the natural ecosystem; heightened human exposure to coastal hazards; and magnified damage to natural coastal resources. Sea level rise, storm surge, coastal erosion, and many other coastal hazards now actively threaten the lives of residents in these areas, as well as the billions of dollars of infrastructure built along the edges of oceans.

Government agencies are working to mitigate these threats, but in the end, the simplest, though perhaps not the easiest, solution is for people to move away from the water's edge and avoid the areas of greatest vulnerability. To achieve this goal in an era of global warming will require a citizenry educated in the science of coastal geology and committed to protecting the environment for future generations.

STUDY GUIDE

20-1 Change occurs constantly on the global shoreline.

- Coastlines are unique and complex geologic systems. Influenced by processes and materials from the land, sea, and atmosphere, they are everchanging environments. Their temperate climate, beautiful vistas and abundant resources have made coastal lands the most crowded and developed areas in the world. Fifty percent of the population of the industrialized world lives within 100 kilometers of a coast, a population that is predicted to grow by 15% during the next two decades.

- Winds, waves, tides, sea level variations, and currents act as agents of change, collectively called **coastal processes**. They sculpt the shoreline through erosion and deposition of sediment and by flooding low areas. Understanding how coastal processes and sedimentary materials interact is crucial to interpreting the geologic history of coastal systems and finding ways for people to live in the coastal zone in a sustainable manner.

20-2 Wave energy is the dominant force driving natural coastal change.

- Waves on the ocean surface are generated by the friction of air blowing across the water, the reason they are called wind waves. A wave travels through water without making a current. Energy radiates outward from the source but water does not. The size of a wave is governed by wind velocity, the fetch or distance over which the wind blows, and the length of time the wind disturbs the water.

- As a wave enters shallow water, it slows before those following it. The next wave, still moving at its original speed, tends to catch up with the wave in front of it; the wavelength decreases and the wave height increases. The crest becomes narrow and pointed with a steep face while the trough becomes a wide, shallow curve. Because the wavelength decreases and the speed slows, the wave period remains unchanged.

- The strongest winds (greater than 54 kilometers per hour) and highest waves (greater than 6 meters) are found in the Southern Ocean and are associated with stormy regions in the North Atlantic and the North Pacific.

20-3 Wave refraction and wave-generated currents occur in shallow water.

- Wave **refraction** occurs when the wave crest bends so that it aligns parallel to the seafloor. For straight coasts with parallel contours on the shallow seafloor, refraction tends to orient waves parallel to the coastline.

- Water piles up against the shore, raising the water level and creating sea level setup, which leads to the formation of **wave-generated currents**. Sea level setup is released periodically in the form of currents that set off three-dimensional circulation within the surf zone—longshore currents and rip currents.

20-4 Longshore currents and rip currents transport sediment in the surf zone.

- When waves approach a shoreline from an angle, they generate a **longshore current** that moves along the shoreline in the direction of wave movement. This current develops in part due to the periodic release of sea level setup and in part due to wave momentum directed obliquely to the shoreline.

- An important consequence of sea level setup along a shoreline is the emergence of **rip currents**. As water piles up along a beach, it is released periodically in the form of offshore-directed currents that surge through the surf zone. These currents carry suspended sand and floating debris, and so appear to an observer as dirty or dark water that extends through a surf zone beyond the breakers.

- Wave-generated currents move sand that has been eroded from a rocky headland into the adjoining embayment, thereby building a beach. Over time, the headland is worn down and the bay fills with sand. Eventually, the shoreline becomes a long, straight beach.

20-5 Gravity and inertia spawn two tides every day.

- The oceans rise and fall twice every day in response to **tide-raising** forces related to the gravitational attraction among the Moon, the Sun, and Earth.

- Two tide-raising forces are responsible for the lunar ocean tide: the gravitational attractive force and the inertial force. The balance between these two opposing forces defines the tide-raising force. On the side of Earth facing the Moon, the gravitational attraction force is greater than the inertial force, and a tidal bulge is created on the ocean surface. On the side of Earth facing away from the Moon, the inertial force is greater and produces a tidal bulge. Earth rotates beneath its watery envelope and encounters these two bulges (high tide) twice every day.

- When the Sun, the Moon, and Earth are aligned, the spring tide occurs. When the Moon is at 90° to the Sun–Earth alignment, the neap tide occurs.

20-6 Hurricanes and tropical storms cause enormous damage to coastal areas.

- Called the greatest storm on Earth, a **hurricane** is capable of annihilating coastal areas with sustained winds of over 200 kilometers per hour, intense rainfall, and flooding ocean waters pushing huge waves. During its life cycle, a hurricane is capable of expending as much energy as 10,000 nuclear bombs.

- **Storm surge** is a combined effect of low atmospheric pressure above the ocean surface, causing a bulge of water to travel beneath the hurricane, and the effect of wind shearing the water surface in the direction of the hurricane's forward movement. Sea level setup due to wave momentum also contributes to storm surge. The combined processes of wind shear, low pressure, and setup can raise the water level by several meters. Add to this the waves, 6 to 10 meters high or more, propelled by the winds of a hurricane, and it is easy to understand why hurricanes can be so destructive when they intersect a coastline, especially one with low topography. They cause massive damage and flooding of low-lying coastlines.

20-7 Sea level rise since the last ice age has shaped most coastlines, and continues to do so.

- One of the most influential geological processes shaping coastal environments around the world is the rise in sea level since the end of the last ice age, which reached its peak between 20,000 to 30,000 years ago. Rapid global warming began shortly thereafter. Sea level at that time was over 120 meters below the present level, so much of the world's shallow seafloor and continental shelves were exposed.

- As glaciers around the world melted, the meltwater flowed to the sea and raised sea level. The rising sea flooded river mouths and continental shelves on every coastline in the world. This deluge shaped the look and character of modern coastal environments. Barrier islands emerged, estuaries expanded, tidal wetlands accreted, and coral reefs flourished in warm seas.

- Scientists have projected that by the end of this century, sea level will rise between 0.75 and 1.8 meters. The physical effects of sea level rise along the coast will be dramatic. They can be grouped into five categories: inundation of low-lying areas, erosion of beaches and bluffs, salt intrusion into aquifers and surface waters, higher groundwater, and increased flooding and storm damage.

20-8 Barrier islands migrate with rising sea level.

- Low-elevation coastlines with an abundance of sand tend to form long, narrow **barrier islands** characterized by sandy beaches, tidal inlets, lagoons, and tidal marshes.

- Barrier islands are fragile, constantly changing geologic environments that shift their positions when overrun by storm surge. When a storm such as a hurricane encounters a low barrier island, the floodwaters carry sand from the ocean side of the island and deposit it on the lagoon side.

- Like a giant treadmill, the island rolls landward by eroding on the ocean side and building on the lagoon side while at the same time losing some sand offshore. This process is referred to as barrier rollover. Barrier rollover, driven by storm surge and sand transport through tidal inlets into lagoons, allows an island to migrate landward under conditions of sea level rise, and thereby avoid drowning in place.

20-9 Rocky shorelines, estuaries, and tidal wetlands are important coastal environments.

- **Rocky shorelines** are found around the world in places where steep hills or mountains descend directly into the sea. Rocks that more easily succumb to erosion will retreat while those more resistant protrude into the surf. Sea stacks, composed of resistant rock capable of withstanding the pounding of waves and bioerosion by intertidal organisms, are shaped in this manner. Sea stacks mark the former position of the shoreline before erosion forced the coast to retreat landward.

- As ocean waters rose following the last ice age, they flooded coastal uplands and moved the shoreline inland, a process termed **marine transgression**. Seawater invaded and flooded the mouths of stream valleys, forming estuaries. An **estuary** is a partially enclosed body of water that develops where freshwater from rivers and streams flows into the ocean, mixing with the salty seawater.

- Tidal wetlands, also called salt marshes, develop along the fringe of many estuaries. Water draining from the uplands carries sediments, nutrients, and pollutants that are deposited in mud-rich wetlands and prevented from entering estuarine waters. Wetland plants and soils act as a natural buffer between the land and ocean, absorbing floodwaters and dissipating storm surges. Salt marsh grasses and other plants help stabilize the shoreline and counteract coastal erosion.

20-10 Coasts may be submergent or emergent, depositional or erosional, or exhibit aspects of all four of these characteristics.

- Coasts where the land is tectonically uplifting at a rate that matches or exceeds sea level rise are known as emergent coasts. Those where sea level rise exceeds uplift are known as submergent coasts.

- Coastal systems also may be classified by the relative magnitude of erosional versus depositional processes; thus, erosional coasts are dominated by erosional processes, and depositional coasts are dominated by depositional processes.

20-11 Coral reefs are home to one-fourth of all marine species.

- Reefs are crucial to supporting the ocean's biodiversity, as they are home to 25% of all marine species. Coral reefs are shallow coastal environments where numerous types of coral (an animal) and algae (a plant) build massive solid structures of calcium carbonate ($CaCO_3$). There are four basic kinds of coral reefs: fringing reefs, barrier reefs, atolls, and patch reefs.

20-12 Coastal problems are growing as populations increase.

- As the coastal zone became the site of dramatic population growth, these areas became heavily developed, heightening their exposure to coastal hazards and magnifying damage to natural coastal resources. Sea level rise brings coastal hazards literally to the doorsteps of buildings in these areas.

- Coastal erosion is, today, a worldwide problem that threatens beaches, coastal lands, and the economic and environmental systems that rely on them. Among the major threats to coastal resources is the more frequent occurrence of polluted runoff in coastal waters adjacent to urban, suburban, industrialized, and agricultural coastal lands. When natural coastal processes encounter buildings and roadways, they become coastal hazards, a term that refers to any coastal process that may cause loss of life or property.

KEY TERMS

barrier islands (p. 608)
beach (p. 597)
beach profile (p. 609)
coastal erosion (p. 620)
coastal geology (p. 596)
coastal hazards (p. 621)
coastal processes (p. 596)
coral reefs (p. 618)
depositional coasts (p. 615)
emergent coasts (p. 614)
erosional coasts (p. 614)

estuary (p. 612)
eutrophication (p. 621)
glacioisostatic uplift (p. 614)
hazard mitigation (p. 622)
hurricane (p. 605)
longshore current (p. 601)
longshore drift (p. 601)
marine transgression (p. 612)
polluted runoff (p. 609)
refraction (p. 599)
rip currents (p. 601)

rocky shorelines (p. 612)
sand bars (p. 602)
storm surge (p. 605)
submergent coasts (p. 614)
tidal wetlands (p. 609)
tides (p. 603)
washover (p. 611)
wave-generated currents (p. 600)

ASSESSING YOUR KNOWLEDGE

Please complete this exercise before coming to class. Identify the best answer to each question.

1. The coastal zone is constantly changing because:
   a. Waves and currents are constantly shifting the position of sand and other sediments.
   b. Storms frequently strike the shore.
   c. Sea level rises and falls through time, changing the position of the shoreline.
   d. Humans influence coastal environments.
   e. All of the above.

2. As waves move from deep to shallow water, they:
   a. Grow smaller.
   b. Become more energetic.
   c. Travel faster.
   d. Grow steeper.
   e. Do not change significantly.

3. Waves in shallow water do which of the following?
   a. Erode and deposit sediment.
   b. Cause the tide to change.
   c. Create currents.
   d. The wavelength decreases and height increases.
   e. Develop flat troughs and sharp peaks.

4. Wave-generated currents include:
   a. Rip currents and longshore currents.
   b. Hurricanes and storm surge.
   c. Refraction and diffraction.
   d. Shoaling and water orbitals.
   e. Stokes drift and setup.

5. Longshore currents and rip currents develop:
   a. In deep water.
   b. On the subaerial beach.
   c. In nearshore circulation.
   d. Because of global warming.
   e. All of the above.

6. Spring tide occurs when:
   a. March and April tides reach their highest point.
   b. The lunar and solar tides are aligned.
   c. The lunar and solar tides are opposed.
   d. The Moon is farthest from Earth.
   e. Winter turns to spring.

7. Storm surge is:
   a. The product of high pressure under a storm.
   b. A result of high winds occurring at high tide.
   c. Caused by high winds and waves and low atmospheric pressure.
   d. Severe only at high tide.
   e. None of the above.

8. Hurricanes:
   a. Do not occur outside of the tropics.
   b. Have decreased their impact on U.S. shores.
   c. Cause coastal damage by high winds and flooding.
   d. Are not strong enough to inflict serious damage on houses and roads.
   e. None of the above.

Cotton Coulson/NG Image Collection

9. Sea level rise:
   a. Has exceeded 120 meters since the last ice age.
   b. Has led to the formation of estuaries in the mouths of major rivers.
   c. Is a growing problem on all coastlines.
   d. Is a major cause of coastal erosion.
   e. All of the above.

Brett R. Henry/Alamy

10. What is the ratio of coastal erosion to sea level rise as sea level goes up?
    a. 1 to 1.
    b. The rate of erosion is typically two orders of magnitude greater than the rate of sea level rise.
    c. 50 to 1.
    d. Coastal erosion usually occurs at 1,000 times the rate of sea level rise.
    e. None of the above.

11. Do barrier islands migrate with rising sea level?
    a. Yes, this process is called "rollover."
    b. No, barrier islands cannot migrate with rising sea level.
    c. Yes, but rollover only occurs because wind can shift sand dunes on the island.
    d. No, because barrier islands cannot migrate if they are damaged by a hurricane.
    e. Yes, barrier islands migrate in a seaward direction as sea level rises.

12. Tidal deltas develop in places where:
    a. Sediments collect in the mouths of rivers.
    b. Barrier islands roll over.
    c. Storm surge is especially frequent.
    d. Tides enter and exit lagoons at inlets.
    e. Rocky shorelines accumulate beaches.

13. Chesapeake Bay and San Francisco Bay are:
    a. Not true estuaries.
    b. The result of marine transgression.
    c. Formed by sediment erosion in the mouths of streams.
    d. Major river deltas.
    e. Examples of emergent coastlines.

14. A barrier island:
    a. Is a depositional coast.
    b. Is an emergent coast.
    c. Migrates with rising sea level.
    d. Is a type of delta.
    e. Has a barrier lagoon on its landward side.

15. Shorelines on tectonically stable lands that collect sediment are termed:
    a. Emergent coasts.
    b. Submergent coasts.
    c. Erosional coasts.
    d. Depositional coasts.
    e. Estuaries.

16. Corals are:
    a. Animals that build huge colonies of calcium carbonate.
    b. Organisms known as polyps.
    c. Builders of reefs, along with plants (algae).
    d. Dependent on sunlight.
    e. All of the above.

17. The main types of reefs are:
    a. Coral, algal, and oyster.
    b. Barrier, atoll, fringing, and patch.
    c. Submergent, emergent, stable, and unstable.
    d. Tropical, arctic, and temperate.
    e. None of the above.

18. Humans and coastal environments make for uneasy neighbors because of:
    a. Pollution, erosion, and exposure to hazards.
    b. Building too close to the ocean.
    c. Rising sea level, placing human development at risk.
    d. Polluted runoff, which may cause eutrophication.
    e. All of the above.

19. In time, the shoreline in this photo will:
    a. Develop deeper embayments.
    b. Grow larger headlands due to beach erosion.
    c. Not change significantly.
    d. Tend to straighten.
    e. None of the above.

G.R. Roberts/The Natural Sciences Image Library

20. Coastal hazards are costing more in recent years because:
    a. Hurricanes last longer and are stronger.
    b. There is more vulnerable development on the coastline.
    c. Sea level is higher.
    d. The coastal population is growing.
    e. All of the above.

FURTHER RESEARCH

1. What will be the future impact of rising sea level on barrier islands on which towns and highways have been constructed?

2. How will future sea level rise influence coral reefs? (Explain in terms of all four types of reefs.)

3. How do barrier islands migrate landward? Why do tidal wetland deposits become exposed in the ocean beach?

4. Make a chart with two columns labeled "Emergent Coasts" and "Submergent Coasts" and two rows labeled "Erosional Coasts" and "Depositional Coasts." Fill in geographic examples of each of the four types.

5. Name five coastal processes.

6. How do waves change as they approach and finally intersect a shoreline?

ONLINE RESOURCES

Explore more about coasts on the following Web sites:

National Oceanic and Atmospheric Administration (NOAA) Ocean and Coastal Resource Management:
http://coastalmanagement.noaa.gov/welcome.html

National Oceanic and Atmospheric Administration (NOAA) Hurricane page:
http://hurricanes.noaa.gov

Environmental Protection Agency sea level rise reports:
http://www.epa.gov/climatechange/effects/coastal/slrreports.html

The U.S. Geological Survey Marine and Coastal Geology Program monthly newsletter:
http://soundwaves.usgs.gov

Additional animations, videos, and other online resources are available at this book's companion Web site:
www.wiley.com/college/fletcher

This companion Web site also has more information about WileyPLUS and other Wiley teaching and learning resources.

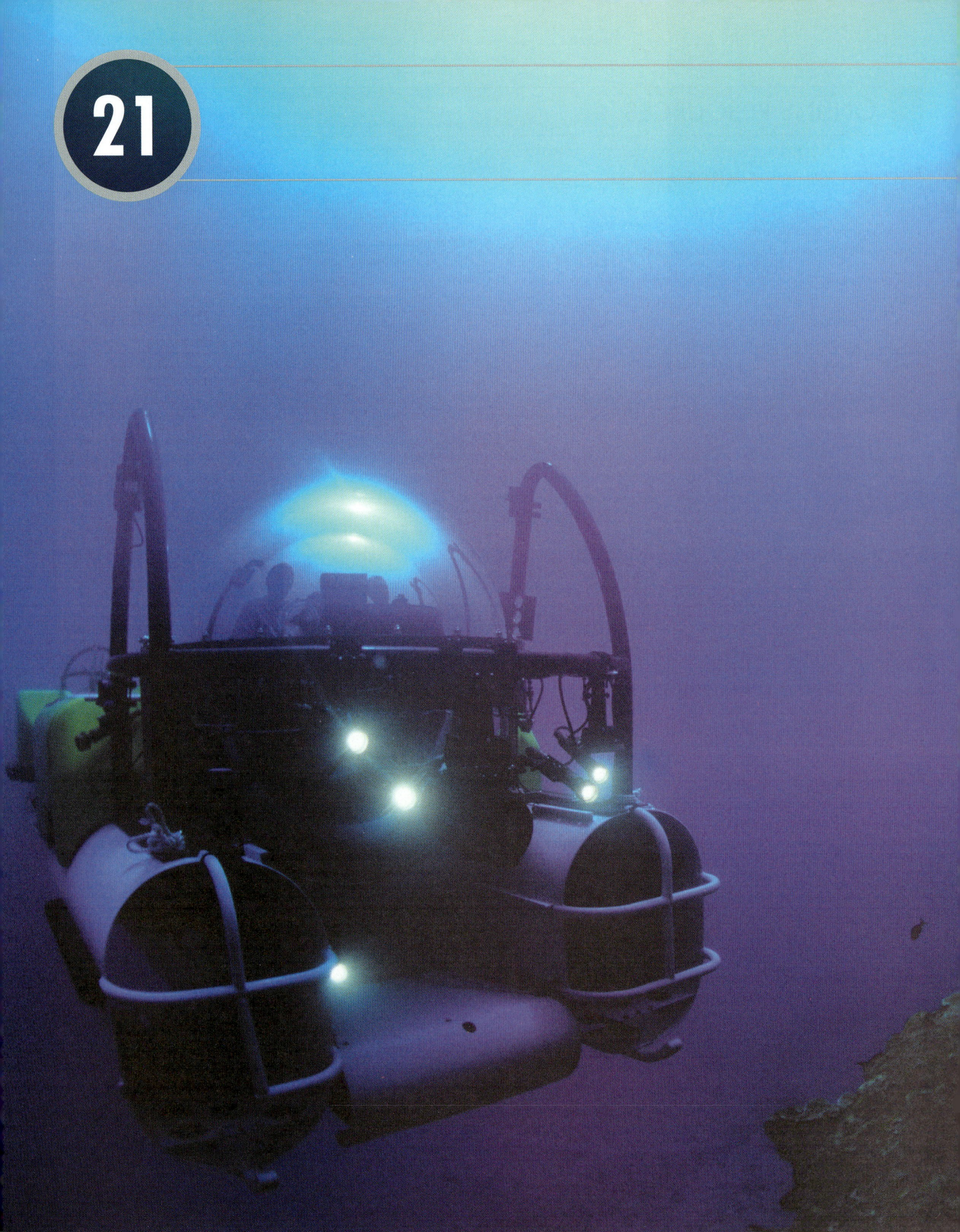

MARINE GEOLOGY

CHAPTER CONTENTS & LEARNING OBJECTIVES

GEOLOGY IN OUR LIVES

Oceans cover more than 70% of Earth's surface, and as such are the planet's largest and most important environment. Numerous geological processes are essential to the character of the oceans. Consider, for one, the currents in the five major oceanic circulation systems, which transfer heat from the tropics to the poles. Currents also carry heat and dissolved compounds (such as carbon dioxide) to the deepest parts of the ocean, where they can be stored for centuries. Heat exchange between the ocean and the atmosphere controls the climate, weather, winds, precipitation, and other fundamental aspects of Earth's environment. Plankton communities are among the many ecosystems that are dependent on ocean circulation. In turn, plankton governs the type of sediment that accumulates on the seafloor and the compounds that are buried with these sediments. These processes and others regulate and stabilize the chemistry and temperature of the atmosphere and ocean water. Unfortunately, human activities are negatively impacting the oceans in very serious ways, and to a global extent.

Q **What are the most important geologic processes governing the character of the oceans?**

21-1 Marine Geology Is the Study of Geologic Processes within Ocean Basins

LO 21-1 List and describe the world's five oceans.

Viewed from outer space, Earth appears as a sphere painted in swirling whites and deep blues. What these colors depict, in fact, is water in its many forms: circulating clouds of water vapor, deep oceans, and seasonally shifting snow and ice. Dominating this scene, however, are the oceans.

The Five Oceans of the World

With an average depth of 3,800 meters, the oceans of the world (**FIGURE 21.1**) contain roughly 97% of the water on Earth. (The other 3% is found in the atmosphere, on the land surface, or locked within the lithosphere.) The largest single feature on the planet surface, covering more than 70% of it, our oceans are unique in the solar system. As far as we know, no other planet has liquid water, although the discovery of ice on Mars and the well-developed channels across Martian landscapes indicate there may have been liquid water there in the past.

It is highly likely that life on Earth originated in the oceans, which are still home today to an amazingly diverse web of ecosystems. Yet despite the dominant presence of the oceans among Earth's environments, we know more about the surface of the Moon than we do about the floor of these great bodies of water.

Until the year 2000, there were four recognized oceans: the *Pacific, Atlantic, Indian,* and *Arctic;* then, in the spring of that year, the International Hydrographic Organization identified a new ocean, the *Southern Ocean,*

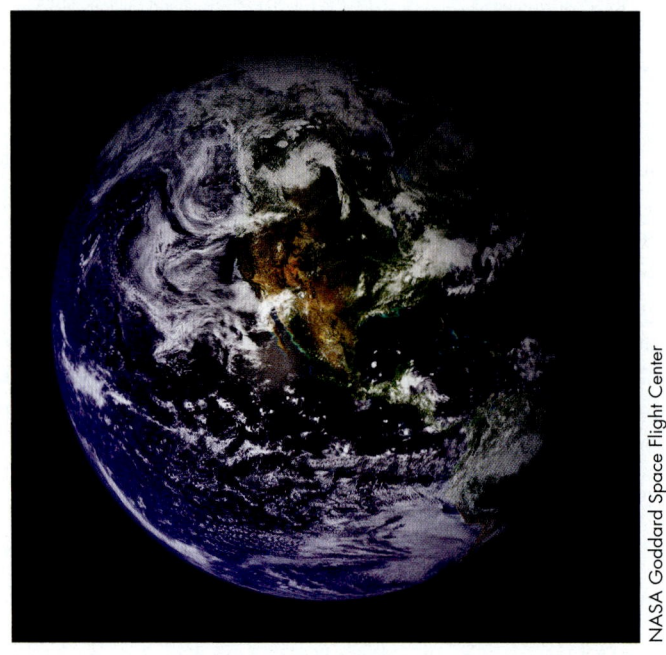

FIGURE 21.1 Earth is the water planet, and its oceans are the largest single feature on its surface.

 Why do oceans dominate Earth's climate?

TABLE 21.1 Oceans

Ocean	Area (Square kilometers)	Average Depth (meters)	Greatest Depth (meters)
Pacific Ocean	166,240,977	4637	Mariana Trench: 11,033
Atlantic Ocean	86,557,402	3926	Puerto Rico Trench: 8605
Indian Ocean	73,426,163	3963	Java Trench: 7725
Southern Ocean	20,327,000	4000 to 5000	Southern end of the South Sandwich Trench: 7235
Arctic Ocean	13,224,479	1204	Eurasia Basin: 5450

which surrounds Antarctica and extends to 60° latitude. The oceans are listed in **TABLE 21.1**, along with their sizes, and illustrated in **FIGURE 21.2**. There are also many *seas* (smaller branches of an ocean); seas often are partly enclosed by land. The largest of these are the South China Sea, the Caribbean Sea, and the Mediterranean Sea.

Given that these oceans influence the entire planet, it is understandable that many fields of science are now devoted to their study. The field that is the topic of our discussion in this chapter, **marine geology**, involves geophysical, geochemical, sedimentological, and paleontological investigations of the ocean floor and coastal margins. Marine geology has strong ties to *physical oceanography* (study of ocean currents and waves), *chemical oceanography* (study of chemical processes in the ocean), *biological oceanography* (study of living organisms in the ocean), and plate tectonics.

Marine geologists assess the ocean in terms of the geological processes that take place within ocean basins. They study and observe sedimentary and geochemical processes on the seafloor and along the coasts, collect and analyze cores of fossilized plankton from the seafloor, interpret how seafloor topography relates to plate tectonic processes, and examine aspects of water circulation and climate that are central to the way oceans interact with Earth's other geological systems.

Why the Ocean Is Salty

When it comes to the oceans, perhaps the first question on everyone's mind is, "Why is the water salty?" The answer is because it contains dissolved ions and compounds that come from the chemical weathering of continental rocks and from hydrothermal processes at spreading centers and subduction zones where seawater interacts with hot crust. Over 90% of the dissolved ions in seawater are chloride (Cl) and sodium (Na), which form NaCl, halite. In addition, marine saltiness derives from dissolved sulfur (S), calcium (Ca), magnesium (Mg), bromine (Br), potassium (K), and bicarbonate (HCO_3).

All together, the concentration of these dissolved ions, known as the **salinity**, has a global average of about 35 parts per 1000. In other words, approximately 35 of every 1000 (3.5%) parts of the weight of seawater come from dissolved ions; thus, in 4 cubic kilometers of seawater, the weight of the sodium chloride (NaCl) would be about 120 million tons. The same 4 cubic kilometers would also contain up to 25 tons of gold and 45 tons of silver. But before you plan to get rich by extracting gold and silver from seawater, think first about how big 4 cubic kilometers actually is: 1 cubic mile. It has been estimated that if all the dissolved salt in the ocean were removed and spread evenly across the

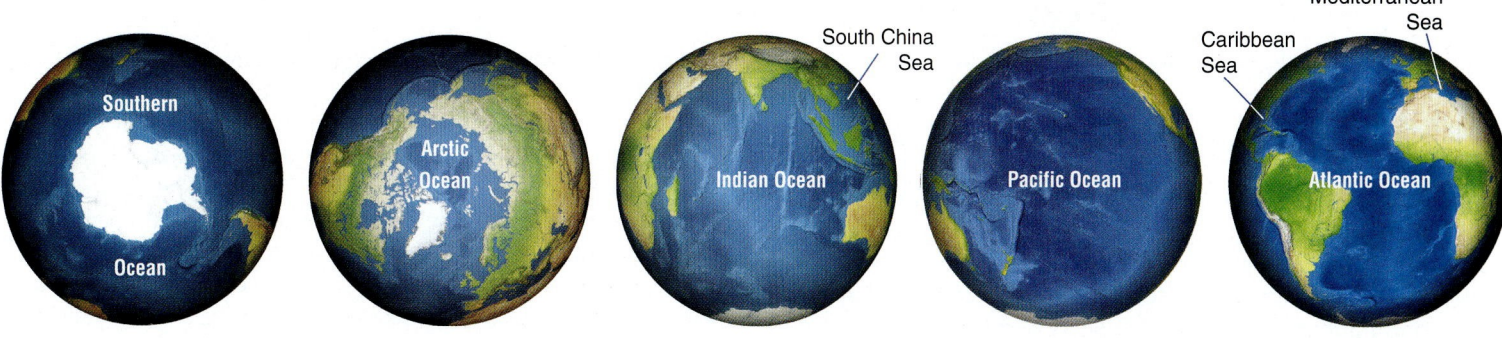

FIGURE 21.2 Earth has five oceans: Southern, Arctic, Indian, Pacific, and Atlantic.

What aspects of the ocean do scientists study?

land surface, it would form a layer more than 166 meters thick, about equivalent to the height of a 40-story office tower.

Although the average salinity of the ocean is relatively stable through time (with some variation due to environmental processes that change through geologic time, such as climate), it does vary from place to place. Salinity is higher in midlatitude oceans, where evaporation exceeds precipitation, and in relatively enclosed areas, such as the Mediterranean and Red Seas. Salinity is lower near the equator, because of high rainfall, and near major rivers where freshwater enters the ocean (**FIGURE 21.3**).

Ocean water is, clearly, a complex solution of dissolved ions and compounds, as well as decaying organic matter. Most of the dissolved components are derived from chemical weathering of the crust. Others (e.g., Cl, S) come from hydrothermal venting at spreading centers and other sites where seawater leaches ions from the crust.

There is geologic evidence that the oceans have been as salty as they are today for over 1 billion years. Hence, if dissolved ions and compounds are being perpetually delivered to the oceans by runoff and hydrothermal activity, something must be regulating the mixture such that both the amount of water and the amount of dissolved material are stable.

The stable salinity of the oceans is attributable to geologic processes that regulate the concentration of the dissolved components; that is, inputs of water and dissolved ions and compounds must be matched by outputs. The amount of water in the oceans is regulated by the hydrologic cycle. Over time, the input of water by runoff and precipitation is evenly matched by evaporation. This balance keeps the volume of water in the oceans approximately stable, except when it is influenced by climate changes and volcanism.

Likewise, the volume of dissolved ions and compounds is held approximately stable over time. Sedimentary processes that remove the dissolved components at nearly the same rate that they are introduced accomplish this stability. Processes that remove dissolved components include:

1. *Inorganic precipitation* (mostly along the ocean margins), where evaporation concentrates the dissolved components, to form crystalline evaporite deposits;

2. *Salt storage* in the pore water among marine sediments as they are deposited;

3. *Chemical interactions* between seawater and the basalt seafloor that create new minerals, which are stored in the crust; and

4. *Organic precipitation* of dissolved components by marine plants and animals to build solid tissue and skeletal parts.

These components are eventually buried as sediment when an organism (usually, plankton) dies and falls to the seafloor. All of these processes regulate the volume of dissolved ions and compounds. In sum, geological processes are the reason ocean salinity remains essentially stable through time.

EXPAND YOUR THINKING

Are the dissolved ions and compounds removed from seawater stored permanently in the lithosphere, or do they eventually return to the surface? Explain.

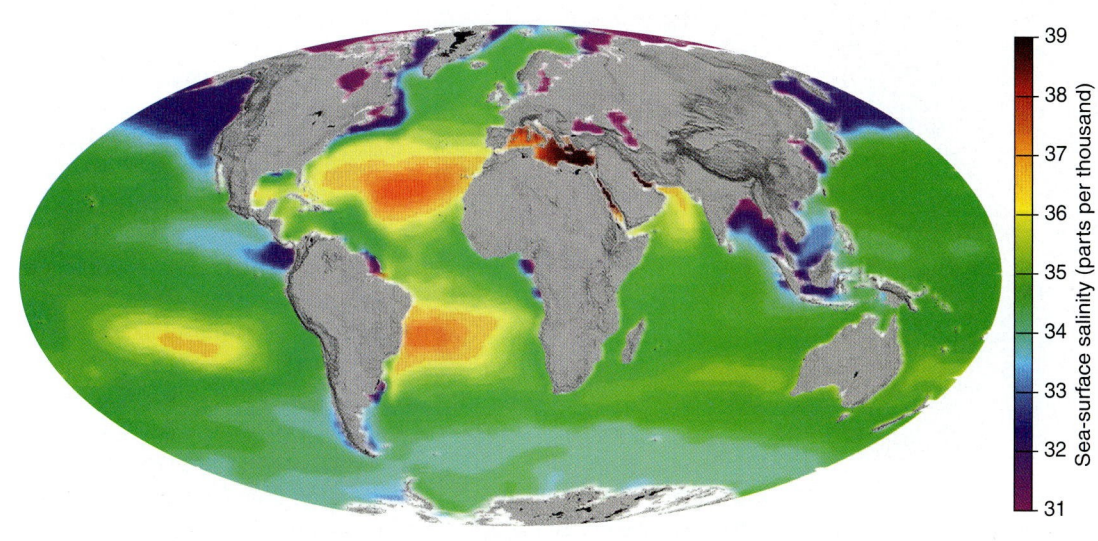

FIGURE 21.3 Ocean salinity is high in midlatitude oceans and restricted areas, such as the Mediterranean and Red Seas. Salinity is low near the equator and major rivers.

What are the primary factors that influence ocean salinity?

21-2 Ocean Waters Are Mixed by a Global System of Currents

LO **21-2** Draw the patterns of ocean circulation.

The oceans influence many other natural systems on the planet, in particular the weather and climate. Ocean water moderates surface temperatures by absorbing heat from the Sun and transporting that heat toward the poles and the seafloor.

Ocean Circulation

Restless ocean currents (**FIGURE 21.4**) distribute heat energy from the Sun around the globe, warming the land and air during winter and cooling it in summer. *Ocean circulation* is the large-scale movement of water by **currents** (a large mass of continuously moving water).

There are basically two types of oceanic circulation: 1) winds drive *surface circulation*, and 2) cool water at the poles that sinks and moves through the lower ocean drives *deep circulation*. The general pattern of circulation consists of surface currents carrying warm water away from the tropics toward the poles while releasing heat to the atmosphere during the journey.

Winter at the poles further cools this surface water; now cooler, the water sinks to the deep ocean, creating currents along the seafloor and at mid-depths in the water column. This process is especially pronounced in the North Atlantic and in the Southern Ocean in the coastal waters of Antarctica, where *deep water* production is the strongest.

Deep ocean water gradually returns to the surface nearly everywhere in the ocean. Once at the surface, it is carried back to the tropics by surface currents, where it is warmed again and the cycle begins anew. The more efficient the cycle, the more heat is transferred from the tropics to the poles, and the more this heat warms the climate.

Gyres

In reaction to Earth's rotation, currents are deflected to the right in the northern hemisphere and to the left in the southern hemisphere. This is known as the **Coriolis effect**, named after the French scientist *Gaspard–Gustave Coriolis* (1792–1843) who described the transfer of energy in rotating systems.

These currents eventually come into contact with the continents, which redirect them (again, to the right and left, depending on which hemisphere they are in), creating large-scale circulation systems called **gyres** that sweep the major ocean basins. There are five major basinwide gyres: the North Atlantic, South Atlantic, North Pacific, South Pacific, and Indian Ocean. Each gyre is composed of a strong and narrow *western boundary current* and a weak and broad *eastern boundary current*.

Let us examine the surface circulation of the *North Pacific Gyre* as a typical example of how winds and the Coriolis effect combine with continental deflection to generate circulation. In the North Pacific atmosphere, a descending column of dry air that originated at the equator, the northern end of the Hadley cell (see Chapters 14 and 19), blows toward the equator but is deflected to the west by the Coriolis effect.

This southwest-flowing wind is known as the **trade wind**. All the major ocean basins in both the northern and southern hemispheres have trade winds. The Pacific trade wind drives the *North Equatorial Current* to the west just north of the equator at about 15° N latitude. This current is deflected north near the Philippines to establish the warm western boundary current known as the Japan or *Kuroshio Current*.

The Kuroshio Current carries warm water away from the tropics until it turns to the east at approximately 45° N latitude, where it becomes the *North Pacific Current,* which moves across the basin toward North America. As it approaches the North American continent, the North Pacific Current branches, with one arm moving north to circulate through the Gulf of Alaska and the Bering Sea, as the *Alaska Current*. A southern arm becomes the cool, slow-moving eastern boundary current called the *California Current*.

The California Current moves from about 60° N to 15° N latitude and merges with the North Equatorial Current. From there it travels, once again, thousands of miles across the basin to Asia. Each of the five major gyres in the oceans has similar systems of currents.

Vertical Circulation

In the North Atlantic basin, the western boundary current is known as the *Gulf Stream*. The Gulf Stream carries warm tropical water from the Caribbean to the cold waters of the North Atlantic. As it moves, the Gulf Stream cools and evaporates, thus greatly increasing its density. By the time it arrives in the North Atlantic as a cold, salty body of water, it can no longer stay afloat and begins a long descent of 2 to 4 kilometers, where it becomes a deep current known as the *North Atlantic Deep Water*.

FIGURE 21.4 Currents flow in complex patterns, affected by wind, Earth's rotation, and water density.

 **Explain the general pattern of ocean circulation.**

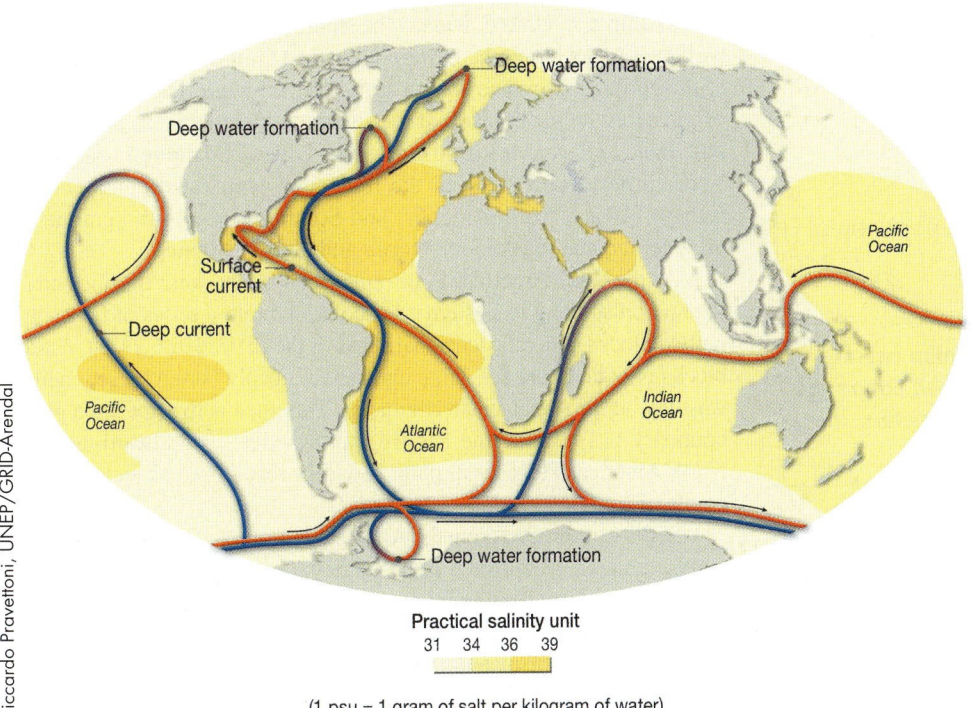

Riccardo Pravettoni, UNEP/GRID-Arendal

Deep water formation

Deep water formation

Surface current

Deep current

Pacific Ocean

Pacific Ocean

Atlantic Ocean

Indian Ocean

Deep water formation

Practical salinity unit
31 34 36 39

(1 psu = 1 gram of salt per kilogram of water)

FIGURE 21.5 The thermohaline circulation is a global pattern of currents that carries heat, dissolved gas, and other compounds on a trip that can take up to 1300 years.

What drives the oceanic conveyor belt?

The Deep Water travels south through the Atlantic and eventually joins similar deep water that is forming in the Southern Ocean; there it becomes the *Circumpolar Deep Water*, which journeys throughout the Southern Ocean. An arm of the Circumpolar Deep Water migrates into the North Pacific, and there, after a voyage of approximately 35,000 kilometers, water that originated in the North Atlantic Gulf Stream eventually surfaces into the sunshine.

It has been estimated that up to 1300 years may pass before this water returns to its place of origin. This **thermohaline circulation**

(**FIGURE 21.5**), also called the *oceanic conveyor belt* (recall the discussion of thermohaline circulation in Chapter 14), travels through all the world's oceans, connecting them in a truly global system that transports both energy (heat) and matter (solids, dissolved compounds, and gases), and thereby influences global climate.

Ekman Transport

Ocean currents may cause the water beneath them to flow in a different direction. This unexpected result, called *Ekman transport* after Swedish oceanographer *Vagn Walfird Ekman* (1874–1954), who first investigated the phenomenon, is a consequence of the Coriolis effect interacting with a column of water.

Recall that the Coriolis effect causes objects in the northern hemisphere to deflect to the right and objects in the southern hemisphere to deflect to the left. As wind blows across the water and creates a current, the Coriolis effect will deflect it slightly. Water in lower layers deflects as well, but not as fast as the surface water. This effect translates downward through the water column in the shape of a weakening spiral. Theoretically, this spiral results in water transport that is 90° to the right of the wind direction in the northern hemisphere.

Ekman transport is especially notable in coastal zones, where wind blowing parallel to the shore can generate a deeper current that is perpendicular to the shoreline. For instance, along the Pacific coast of Canada, the wind generally blows to the south in the summer. Ekman transport stimulated by this wind drives coastal water to flow offshore, 90° to the right of the wind. To replace this water, a deep, nutrient-rich current rises along the seafloor into the coastal zone in a phenomenon called *upwelling*. (This upwelling brings nutrient-rich water that sustains a seasonal fishery that is crucial to the economy of western Canada.) In the winter, the wind reverses direction and blows along the shoreline to the north, stimulating Ekman transport that is directed onshore, forcing shallow waters into the coastal zone (**FIGURE 21.6**). As these waters pile up against the coast, they drive a current along the seafloor that is directed offshore, a process called *downwelling*.

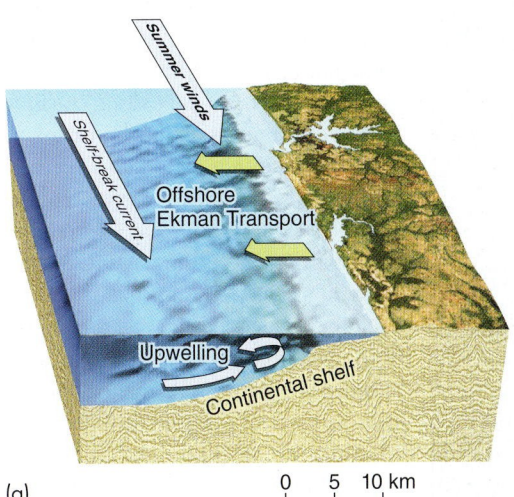

(a)

Summer winds

Shelf-break current

Offshore Ekman Transport

Upwelling

Continental shelf

0 5 10 km

(b)

Winter winds

NE Pacific Coastal current

Onshore Ekman Transport

Continental shelf

0 5 10 km

FIGURE 21.6 (a) Along the Pacific coast of Canada, summer winds from the north generate Ekman transport to the west and stimulate upwelling of nutrient-rich deep water. (b) In the winter, winds flow to the north, compelling Ekman transport to the east, onto the shoreline.

How does Ekman flow influence coastal waters?

EXPAND YOUR THINKING

Deep water pumped from the seafloor off Hawaii is desalinated and sold as the cleanest drinking water on Earth. Explain the reasoning behind this claim.

21-3 A Continental Shelf Is the Submerged Border of a Continent

LO 21-3 Describe the main features of a continental shelf.

Continental shelves make up about 8% of the entire ocean area. An undersea extension of a continent, a shelf can range from over 100 kilometers wide to only a few kilometers, depending on its geologic history.

The U.S. Continental Shelf

If you look closely at a map of the seafloor around most continents (FIGURE 21.7), you will notice a shallow platform that extends offshore from the coastline. This is the **continental shelf,** an area of the ocean floor that is generally less than 200 meters deep surrounding each continental landmass. The fact that the shelf happens to be underwater can be traced to its geologic origin as the outer margin of the continent.

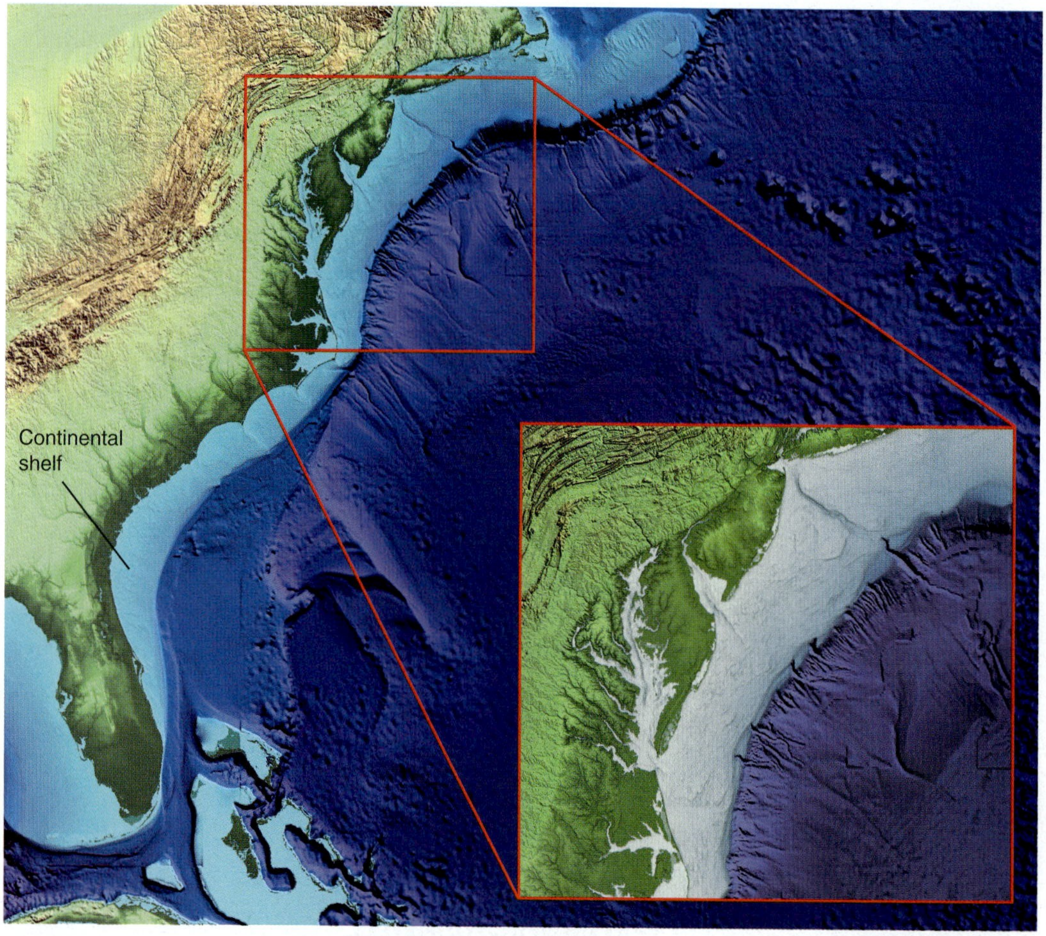

Within the United States, the continental shelf is particularly well developed along the eastern coast and Gulf of Mexico. It is part of the geologic province known as the *coastal plain*. The coastal plain originated with the breakup of the supercontinent Pangaea in the Late Paleozoic to Early Mesozoic eras, some 280 to 230 million years ago. A basin developed between the four newly separated continents—Europe, Africa, North America, and South America—and gradually grew to become the Atlantic Ocean and Gulf of Mexico.

As the North American Plate pulled away to the west, the thick continental crust along its eastern edge collapsed into a series of normally faulted blocks under the tensional stress of the rift (FIGURE 21.8). These blocks now lie under the continental shelf, deeply buried by sediments washing off the Appalachian Highlands. This once-active divergent plate boundary has become the passive trailing edge of westward-moving North America. The coastal plain is widely recognized as a classic example of a *passive continental margin*.

Sediments eroded from the Appalachians moved to the continental margin and were deposited in broad deltas, estuaries, barrier islands, and other coastal and marine sedimentary environments. They gradually covered the faulted continental margin, burying it under sedimentary layers thousands of meters thick. These units are Mesozoic to Cenozoic in age and include thick sequences of Cretaceous rocks.

In the northern region of the Coastal Plain Province, boreholes have been drilled to a depth of 5 kilometers without encountering the igneous or metamorphic *basement*. The sediment sequence is even thicker along the Texas portion of the province, with up to 10 kilometers of Jurassic to Quaternary sedimentary beds. These layers are well-known as a rich source of petroleum.

As sediments accumulated along the U.S. coast, the level of the sea rose and fell with changes in global climate (warm climate causes a rise in sea level, cool climate causes a fall in sea level) and vertical shifts in the elevation of the continent due to changing heat flow associated with the rifting process. As a result, the seashore migrated back and forth across the newly forming coastal plain, depositing various types of sediments on the landscape.

Over many tens of millions of years, the margin of North America cooled and subsided as it pulled away from the hot environment of the Mid–Atlantic spreading center. During this time, the rate of sediment delivery and deposition from the Appalachians approximately matched the rate of continental subsidence, and a great wedge of sediment, 5 to 15 kilometers thick, developed into the flat continental shelf we see today.

FIGURE 21.7 A continental shelf is the submarine extension of a continent. It may range in width from over 100 kilometers to only a few kilometers, depending on its geologic history.

Continental shelf

Q **What is the origin of the eastern U.S. continental shelf?**

Shelf Areas

Continental shelves extend from the shore (at an average slope of about

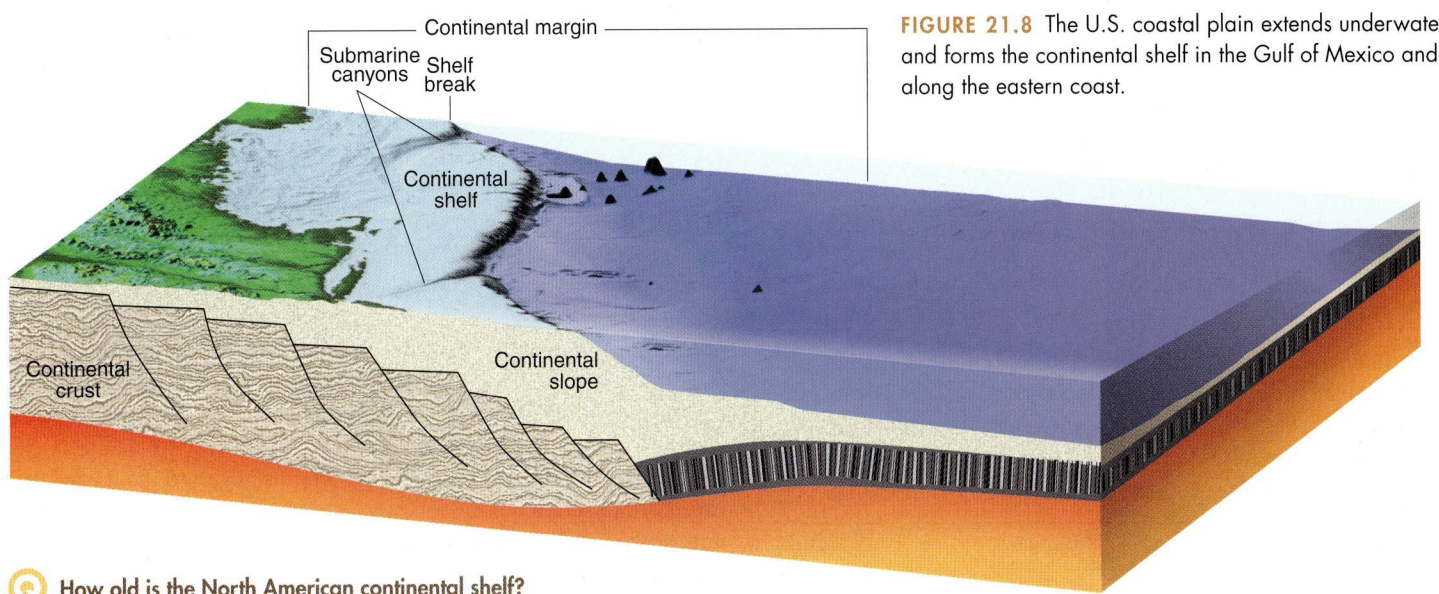

Continental margin

Submarine canyons

Shelf break

Continental shelf

Continental crust

Continental slope

FIGURE 21.8 The U.S. coastal plain extends underwater and forms the continental shelf in the Gulf of Mexico and along the eastern coast.

How old is the North American continental shelf?

(0.1) to the *shelf break*, the point at which the seafloor steepens sharply, marking the boundary between the continental shelf and the **continental slope**. The break is found at a depth ranging from 100 to 200 meters. It has been theorized that this depth may mark a former coastline, and that global variations in the depth of the continental break are the result of differences in the vertical movement of continents.

The shelf surface is swept by waves, tidal currents, and storms, which in places continually move a layer of loose sediment consisting of sand, silt, and silty mud. In other locations, the shelf is starved of sediment, and the seafloor consists of outcropping sedimentary rocks of (usually) Cenozoic age. The shelf surface exhibits some relief, featuring small hills and ridges that alternate with shallow depressions and troughs. In some cases, steep-walled V-shape **submarine canyons** cut deeply into both the shelf and the slope below.

Many of the world's continental shelves are rich in natural resources, such as marine life, sand and gravel deposits, natural gas and oil reserves (**FIGURE 21.9**), phosphorite beds (sedimentary rock that contains high amounts of phosphate often produced by biological activity in the overlying water column), and others.

Most nations have claimed *mineral and land rights* to their shelves and actively manage these resources through systems of laws and rules. In the United States, mineral (including oil and gas) exploration and production from our continental shelves are managed through the Department of Interior according to the Outer Continental Shelf Lands Act of 1953 (amended in 1978). The act directs the Secretary of the Interior to conserve the nation's natural resources; develop natural gas and oil reserves in an orderly and timely manner; meet the energy needs of the country; protect the human, marine, and coastal environments; and receive a fair and equitable return on the resources of the continental shelf.

The federal government's role in managing the shelf for mineral resources is largely confined to the outer portions of the shelf; coastal states assume jurisdiction on the inner portion. Researching and mapping shelf areas and resources has been an important task of marine geologists over the years.

EXPAND YOUR THINKING

If you were to make a map of a continent, where would you draw the border? Explain.

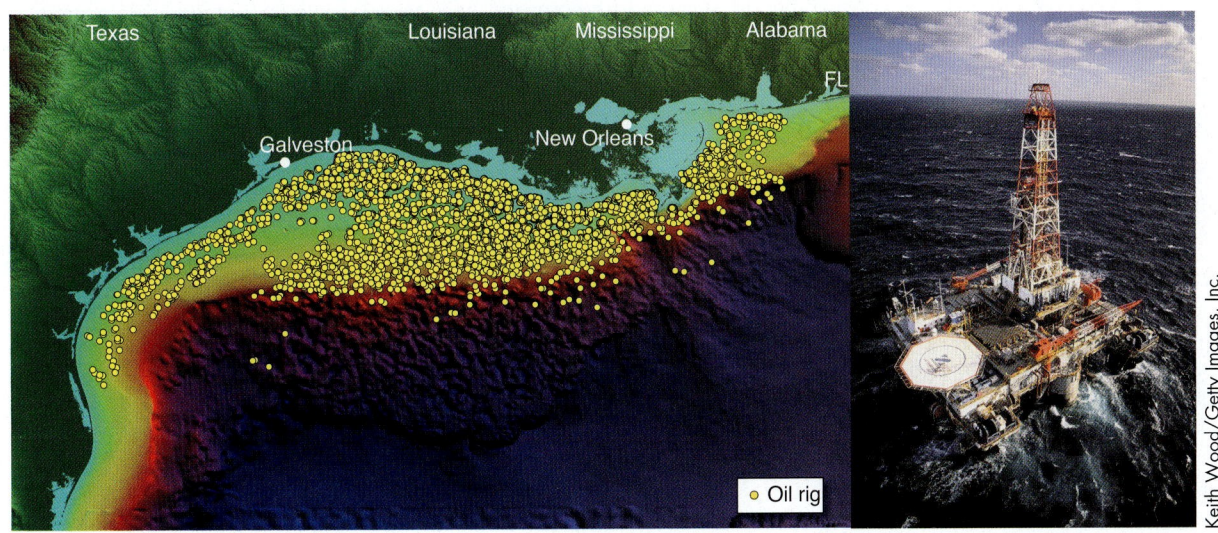

Texas   Louisiana   Mississippi   Alabama   FL

Galveston   New Orleans

Oil rig

FIGURE 21.9 The U.S. continental shelf in the Gulf of Mexico is rich with oil and gas deposits, which are drilled by almost 4000 active oil rigs.

Keith Wood/Getty Images, Inc.

Who manages the natural energy resources of the continental shelf?

21-4 The Continental Margin Consists of the Shelf, the Slope, and the Rise

LO 21-4 Identify the major processes shaping the geologic history of a continental margin.

The continental margin is the broad zone that includes the continental shelf, the continental slope, and the continental rise. The margin separates the thin oceanic crust of greater density from the thick continental crust of lesser density and constitutes about 28% of the total oceanic area.

Active and Passive Margins

The transition from the continental to oceanic crust typically occurs at the outer part of the margin under the **continental rise**. Beyond the seaward edge of the rise lies the **abyssal plain** on fully oceanic crust (discussed in section 21–6).

The tectonic setting of a **continental margin** establishes the basis for interpreting its history and structure. Margins are classified as "active" (also called *leading edge*) or "passive" (also called *trailing edge*), each of which has certain fundamental characteristics (**FIGURE 21.10**). Active continental margins, the site of a plate boundary (convergent, divergent, or transform), are tectonically active, generally narrower, and with less sediment input than passive margins.

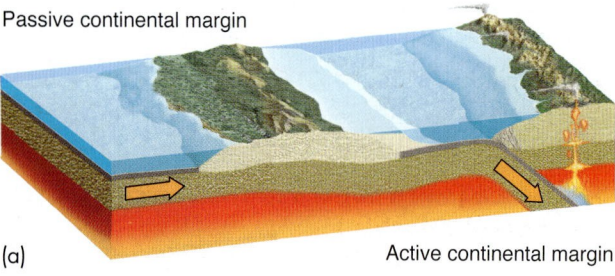

Passive continental margin

(a)   Active continental margin

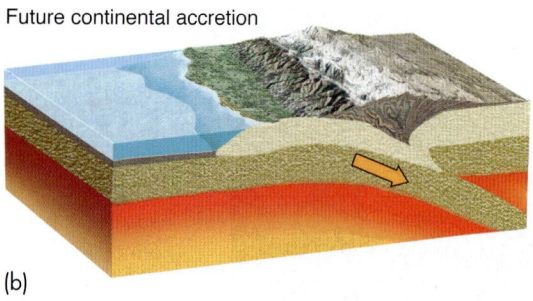

Future continental accretion

(b)

FIGURE 21.10 (a) A passive margin is not located at a plate boundary, whereas an active margin is. (b) As a passive margin, on a moving plate, approaches and collides with an active margin, an accreted terrane develops.

Where have accreted terranes been influential in shaping the margins of North America?

Active margins located at convergent boundaries where there is a subduction zone may have **accreted terrane**. Also known as *exotic terrane*, accreted terrane develops where foreign blocks of crust, carried on a subducting plate toward the margin, collide with the overriding plate. These terranes add to the volume and topography of a continental margin and complicate the structure of the crust.

Accreted terrane may include island arcs, small bits of continental-type crust (think New Zealand or Japan), or hotspot islands that collide with the active margin. Since the active margin is on the overriding plate, it is also likely to have a volcanic arc fed by magma generated by water released from the subducting plate. Each new accreted terrane that collides with the volcanic arc may cause the subduction site to close down and a new one to open on the seaward side of the collision where there is oceanic crust. Subsequently, the subduction zone "jumps" back to the advancing front of the active margin, and the accreted terrane becomes part of the active continental margin.

The coast-parallel ranges and valleys of the Cascadia margin on the Oregon coast are a series of accreted terranes that have formed as the North American Plate migrated to the west with the breakup of Pangaea. These terranes were island arcs and continental fragments in the Pacific that were swept up in the westward movement of the plate and added to its active western margin.

Passive continental margins are located within plates. The passive margin of the U.S. East Coast and Gulf Coast, for example, is separated from the Mid–Atlantic Ridge by an expanse of oceanic crust that was generated after Pangaea rifted. Oceanic and continental crusts meet in a region of low tectonic activity beneath the continental rise off the East Coast of the United States. Passive margins are generally wide and may receive a large influx of sediment coming from large and mature watersheds (such as the Mississippi River) and intrabasinal carbonate sedimentation associated with plankton deposition or coral reef construction.

Submarine Canyons

Cutting into the shelf and slope of both active and passive continental margins are submarine canyons, steep-sided valleys in the shape of a meandering channel. Many submarine canyons end with a thick sedimentary deposit on the continental rise called a **submarine fan.**

Complex features, submarine canyons doubtless develop in several different ways, each based on its own unique set of circumstances. Here we describe a common history that is found on margins around the world.

Submarine canyons have winding valleys with V-shape cross sections and a central axis that slopes outward and downward (typical of a stream channel) across the continental slope. They may have numerous tributaries entering from both sides, and relief that is equivalent to major canyons on land. Two good examples are the Monterey Submarine Canyon in central California (**FIGURE 21.11**) and the Grand Canyon of the Colorado River, which are comparable in size.

In light of the facts that submarine canyons resemble terrestrial stream channels, and that many canyons align with continental watersheds, it is believed that the heads of submarine canyons were originally cut by rivers during periods of low sea level (such as during Pleistocene glaciations). Dendritic drainage patterns (typical of stream-carved watersheds) are well developed in many shallow

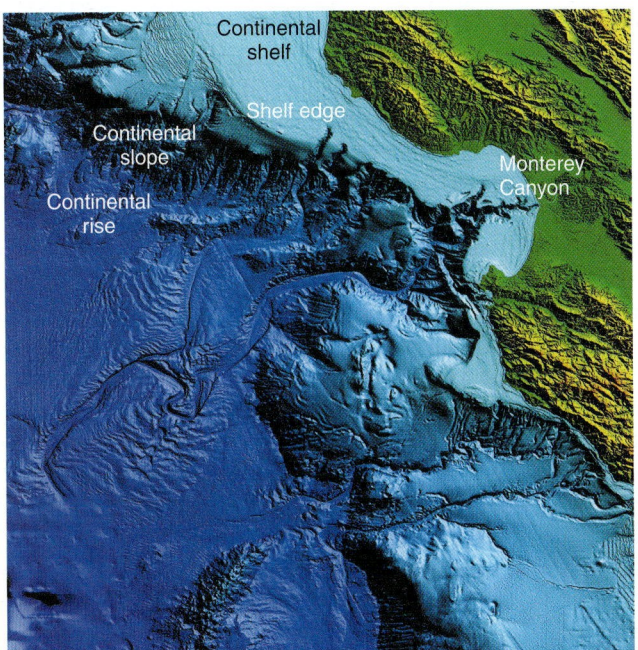

FIGURE 21.11 The origin of Monterey Submarine Canyon is unknown, but one hypothesis suggests that the shallow portion of the canyon was cut millions of years ago by a large river that no longer flows in the region.

What is the evidence that the shallow portion of submarine canyons was originally carved by a stream?

canyon heads but become less prominent in deeper water. Also, various continental margins are known to have undergone subsidence over millions of years, which would have submerged upper canyons originally eroded by streams. This evidence has convinced many

marine geologists that the upper portion of submarine canyons were carved by stream erosion during low sea levels on continental margins that have subsided over recent geologic history.

That said, even if fluvial erosion processes are important in defining the shallow extent of canyons, they cannot have influenced deep canyon sections at depths of thousands of meters. From this evidence, a *composite origin* to submarine canyons has been hypothesized, involving the development of erosive submarine landslides known as **turbidity currents** (**FIGURE 21.12**).

Turbidity currents arise following the collapse of a sedimentary deposit on the slope or shelf edge. The collapse may be caused by an earthquake, a storm, or internal waves and currents within the ocean that travel along thermal and salinity layers in the water column. Sediments thrown into suspension by the collapse develop a dense, sediment-laden current that travels down the axis of the canyon at high speed, carving the floor and sides.

The density and speed of a turbidity current carries it into deep water (where it scours the lower sections of the canyon) and out onto the continental rise (where it feeds sediment to the submarine fan). Turbidity currents generally are credited with both excavating submarine canyons and transporting great quantities of sediment down the canyons to shape the fans that build the continental rise at the base of the continental slope. Continual marine deposition on the continental slope outside of a canyon can raise the level of the surrounding seafloor over time, compelling the canyon to gain greater relief.

Turbidity currents carry sand and mud, which are then deposited on the rise when the currents spread out laterally and decelerate at the base of the canyons. These deposits, called *turbidites*, normally have graded bedding—their grains become progressively finer upward, reflecting sedimentation as the current velocity gradually decreases.

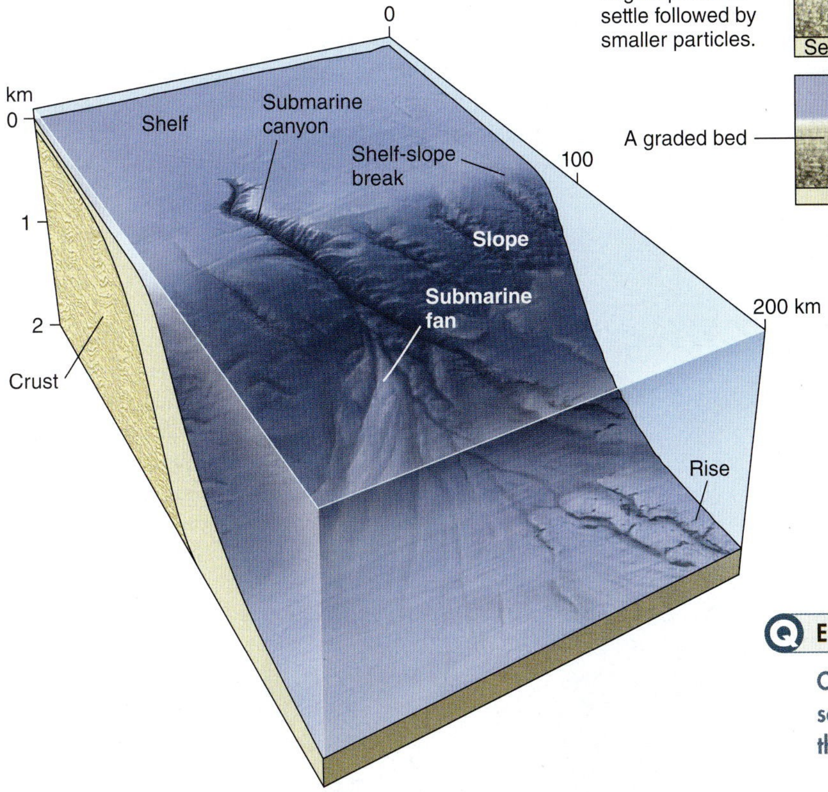

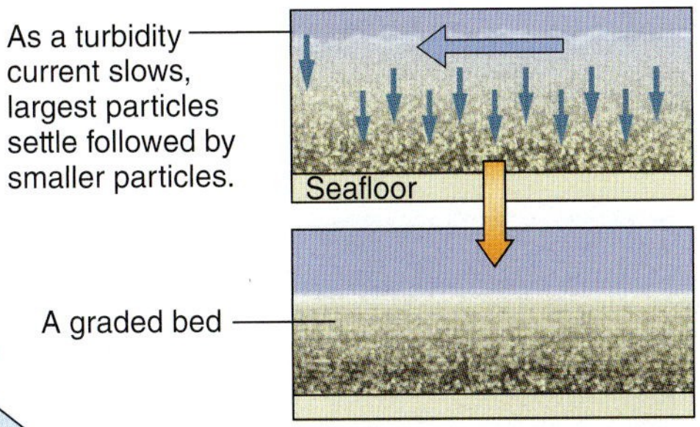

As a turbidity current slows, largest particles settle followed by smaller particles.

Seafloor

A graded bed

FIGURE 21.12 Turbidity currents carry sediment from the continental shelf and slope onto the continental rise. As the current slows and the sediment is deposited in a submarine fan, the largest/heaviest grains settle first, the fine grains last. This makes a sedimentary deposit known as a turbidite.

What kind of rock is made by turbidites? In what type of tectonic setting would you find such deposits on land?

EXPAND YOUR THINKING

Considered simply, a submarine canyon consists of two types of sedimentary deposits. Turbidity currents deliver one kind. What is the other kind?

21-5 Most Ocean Sediment Is Deposited on the Continental Margin

LO 21-5 Describe sediment that collects on a continental margin.

Most of the sediment (about 75% by volume) from Earth's five oceans is deposited on the continental shelf, slope, and rise, where accumulations can exceed 10 kilometers in thickness.

Marine Sediments

The many types of sediment found in the ocean at all depths reveal characteristics of its environments, history, physical and chemical processes, and ecosystems. But there are distinctions: Sediments in the deep differ from sediments in the coastal zone and the continental shelf (**FIGURE 21.13**), for all marine sediment reflects the chemical and physical processes that formed them, as well as the environmental characteristics that govern their deposition. Marine geologists collect and interpret these sediments to deepen their understanding of the biogeochemical aspects of the oceans' basins and their margins.

Sediment of the Coastal Zone

Water characteristics (such as temperature, sediment load, and nutrient content), climate, tectonic setting, depth, history of sea level change, and other factors govern the type of sediment found on a continental margin. Some margins are glaciated; volcanic arcs dominate others; and arid deserts or humid rain forests surround still others.

These environments, the lithology of the continental crust, and the processes that weather the crust all determine the nature of sediments that enter the watersheds and travel onto the margin.

The coastal zone of most margins receives *detrital sediment* via watersheds that drain neighboring uplands. Recall that coastal environments (**FIGURE 21.14**) include tidal wetlands and mudflats, which collect silt and clay delivered by daily tides; lagoons and estuaries, which collect muds and sands transported by streams and ocean tides; sandy barrier islands, beaches, and dunes; tidal inlets, which are typically sand rich; and deltas and coral reefs. (To refresh your memory on the geologic character of these environments, refer to Chapters 8 and 20.)

During the last ice age, many rivers flowed directly across the exposed continental shelf and deposited their sediments at the outer shelf or onto the continental slope. But when the ice age ended and sea level rose around the world, the coast retreated, over 100 kilometers in many cases.

The rising seawater flooded into river channels, driving them back into the watersheds, and formed estuaries that today trap modern sediments. This history, coupled with the fact that most large rivers have human-made hydroelectric dams and their sediment-trapping reservoirs, results in modern shelves that tend to be "sediment starved." Chesapeake and Delaware bays on the U.S. East Coast, and San Francisco Bay and Puget Sound on the West Coast are examples of such sediment-storing estuaries. Thus, many shelves do not receive large amounts of modern sediment; instead, they hold *relict sediments* deposited as sea level migrated across the shelf 15,000 to 3000 years ago, when global climate and sea level were changing. Relict sediment, rather than modern sediment, composes the seafloor on many shelves.

Sediment Character of the Continental Shelf

Across the shelf, away from the coastal zone where many watershed sediments are trapped, the sediment character reflects a different set of influences. Sediment that collects on the seafloor between the shoreline and the edge of the shelf is termed **neritic sediment** (from *nerita*, Latin for "sea mussel").

On some margins, neritic sediment is composed of the skeletal remains of plankton that live in the water column; these remains collect on the seafloor after the plankton die, forming *biogenic sediment*. Some shelves get their sediment from winds that blow off the land (**FIGURE 21.15**) or by weathering of the rocks exposed on the seafloor

FIGURE 21.14 Coastal environments that trap sediments include tidal wetlands, lagoons, estuaries, inlets, barrier islands, dune fields, and beaches.

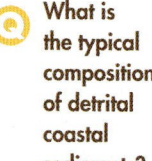
What is the typical composition of detrital coastal sediments?

Bill Bachman/Alamy

FIGURE 21.13 Fine-grained sediments in suspension along the continental margin of the Gulf of Mexico.

MODIS/NASA

FIGURE 21.15 Off the western coast of Africa, strong winds blow desert silt into the ocean.

What other types of sediment can collect on continental shelves?

NASA

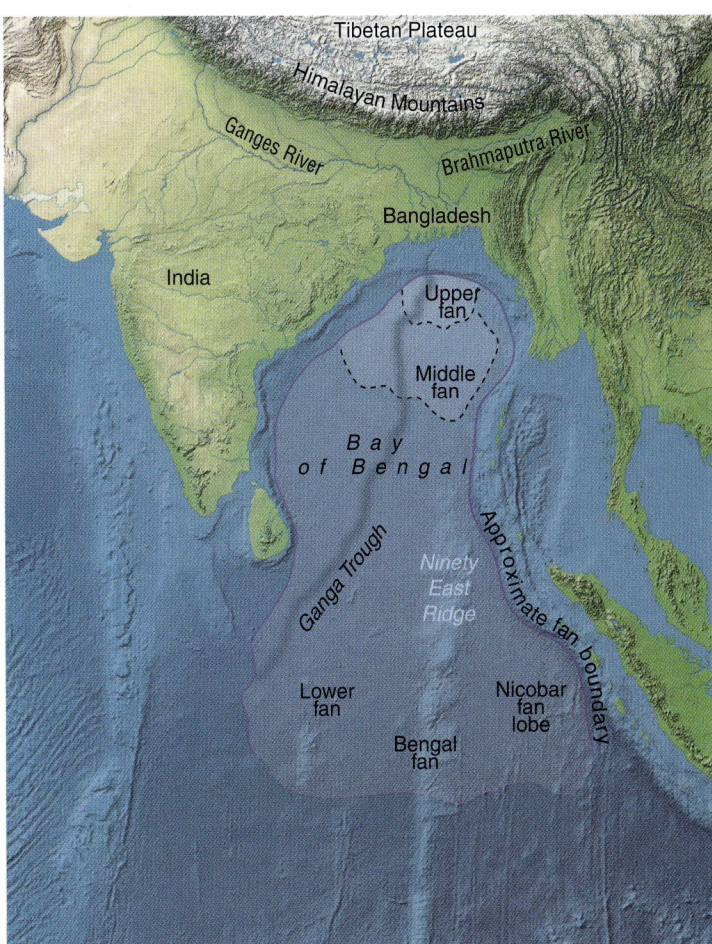

FIGURE 21.17 The Ganges and Brahmaputra rivers transport sediment from the highest mountains on Earth (the Himalayas) to the Bengal Fan, the world's largest submarine accumulation of sediment.

Why is so much sediment available to build the Bengal Fan?

(termed *residual sediment*). Still others collect *authigenic sediment*, a type that inorganically precipitates from the water column or whose mineralogy is changed by interactions with seawater.

Authigenic sediment comprises *glauconite*, clay formed when anoxic (oxygen-starved) seawater interacts with sedimentary deposits on the shelf, or *phosphorite*, a rock of phosphates derived from marine invertebrates that secrete shells of calcium phosphate and the bones and excrement of vertebrates. On low-latitude shelves, coral reefs may be widespread, producing carbonate sediment.

On many shelves, neritic sediment is relict, not modern; it is left over from the time when the shoreline was transgressing across the shelf during rising sea level. The resulting sedimentary record of these variations can be highly complex, often containing combinations of many sediment sources (**FIGURE 21.16**).

But there are also shelves that receive massive amounts of modern sediment because they have large watersheds that drain broad areas of land. Deltas formed by rivers such as the Mississippi (United States), Amazon (Brazil), Ganges–Brahmaputra (India), and the Yangtze (China) contain so much detrital sediment that they bury the topography of the margin.

The Ganges and Brahmaputra rivers, for example, empty into the Bay of Bengal on the north coast of the Indian Ocean (**FIGURE 21.17**). This, the *Ganges–Brahmaputra Delta*, is the world's largest delta and one of the most fertile regions on the planet. A large part of the nation of Bangladesh lies on the delta, with between 125 and 143 million of its people living on it. Many of them depend on the delta for survival,

despite risks from geologic hazards—floods caused by seasonal monsoon rains, heavy runoff from melting glaciers in the Himalayas, and the threat of storm surge rolling across the low-lying delta in the path of frequent tropical cyclones sweeping through the bay. Sea level rise caused by global climate change also threatens communities living on the low-lying delta plain.

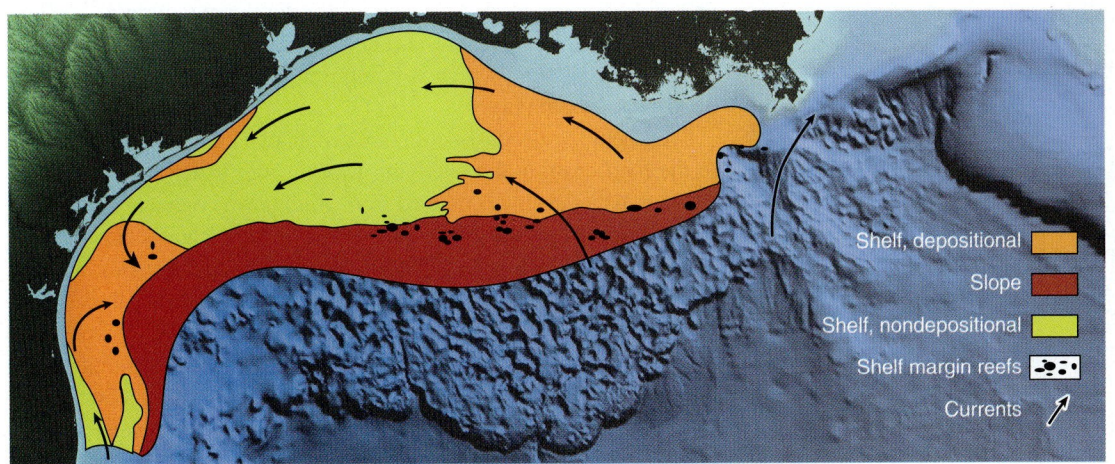

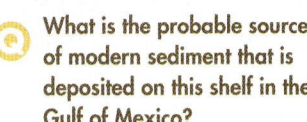

FIGURE 21.16 In the Gulf of Mexico, the U.S. continental shelf has large areas of relict sediment (yellow), left over from the rise of sea level after the last ice age.

What is the probable source of modern sediment that is deposited on this shelf in the Gulf of Mexico?

Offshore of the Ganges–Brahmaputra Delta is the *Bengal Fan*, the largest submarine accumulation of sediment on Earth. River sediment is carried from the mouth of the Ganges and Brahmaputra rivers through several submarine canyons, some of which are over 2000 kilometers long, and across the seafloor of the Bay of Bengal, finally coming to rest up to 30° latitude (over 3000 kilometers) from where it originated.

Hemipelagic Sediments

Along many margins, currents and waves redistribute sediment laterally on the shelf. Some sediment, however, is moved across the shelf edge to collect on the continental slope and rise. Sediments that drape upper and middle continental slopes around the world are known as **hemipelagic sediments**. These grade from predominantly continental muds (closer to watershed sources) into biogenic sediments (farther from watershed sources). Even where biogenic constituents are dominant, hemipelagic sediments typically have a dark color, reflecting the detrital mud component.

Variations in the composition of terrigenous muds (derived directly from the weathering and erosion of rocks on Earth's surface) reflect weathering intensity on the adjacent land. Terrigenous sediments are delivered to the ocean by rivers and remain in suspension as they are carried beyond the shelf edge by surface currents or by downwelling currents related to Ekman transport.

Ⓠ EXPAND YOUR THINKING

Describe factors that control the types of sediment that collect on continental margins.

Ⓔ 21-6 Pelagic Sediment Covers the Abyssal Plains

Ⓛ 21-6 Identify the processes that control the composition of pelagic sediments.

Beyond the continental margin lies the abyssal plain. Untouched by sunlight, the abyssal plain is the horizontal or very gently sloping, sediment-draped region of the deep ocean floor that lies between the continental rise and a mid-ocean ridge. Abyssal plains are among Earth's largest and least explored geologic regions.

Abyssal Plain Configurations

Generally located at depths of 3000 to 6000 meters, abyssal plains (**FIGURE 21.18**) are hundreds of kilometers wide and thousands of kilometers long, and they cover approximately 40% of the ocean floor. They are, in fact, one of the major topographic provinces on the planet. The *Sohm Abyssal Plain*, for example, in the North Atlantic basin between New England and the Mid–Atlantic Ridge, is massive, comprising an area of approximately 900,000 square kilometers.

Abyssal plains are globally distributed, but the largest are located in the Atlantic Ocean. In the topographically complex Indian and Pacific oceans, with their numerous mid-oceanic ridges, island chains, seamounts, and wide spreading centers, abyssal plains occur mainly as the small, flat floors of marginal seas (such as the floor of the *Philippine Sea*) or as the narrow, elongate bottoms of deep sea trenches at subduction zones.

A prominent aspect of abyssal plains is their sediment, known as **pelagic sediment** (pelagic means "in the water column"). Much of the reason the abyssal seafloor lacks dramatic relief is that the rugged topography created by faulting and fracturing at spreading centers is buried under thick blankets of marine sediment.

Abyssal sediment deposits thicken with the age of the oceanic crust, generally increasing away from the spreading centers, with sediment typically from 100 to 500 meters thick in the central abyssal plains (**FIGURE 21.19**). Sediment accumulation increases dramatically beneath the continental rise and slope, where turbidity currents and other sources of clastic sediment (such as river deltas) lead to sediment thicknesses of 3,000 meters and more.

The abyssal environment is very calm, far removed from storms that disturb shallow ocean waters, rivers that drain the continents, and thick deep sea fans at the base of the continental margin. The low energy of the abyssal plain is reflected in the fine-grained character of pelagic sediment, which is composed predominantly of **biogenic ooze** and **abyssal clay**.

Biogenic ooze consists of microscopic plankton remains produced in the food chain in the overlying waters, from which they fall to the seafloor. Abyssal clay is derived from windblown continental sediment that falls from the atmosphere into the sea. Other pelagic sediment components include *volcanogenic particles*, *cosmogenic particles*, and *authigenic sediments*, each of which is discussed in the next subsection.

Pelagic sediment reflects the nature of the marine environment: water temperature, depth, and proximity to a continent. In equatorial to temperate regions where the water is shallower than 4 to 5 kilometers, pelagic sediment is composed primarily of calcareous shells of microscopic plankton consisting of *foraminifera* (an animal) and

FIGURE 21.18

A tripod fish momentarily comes to rest on the sediment-covered abyssal plain of the Pacific Ocean near Hawaii.

Jason Midnight-4 AM team, D. Weis (UBC) team leader and photographer, Kilo-Moana September 2007, University of Hawaii.

Ⓠ What is the source of sediment on the abyssal plains?

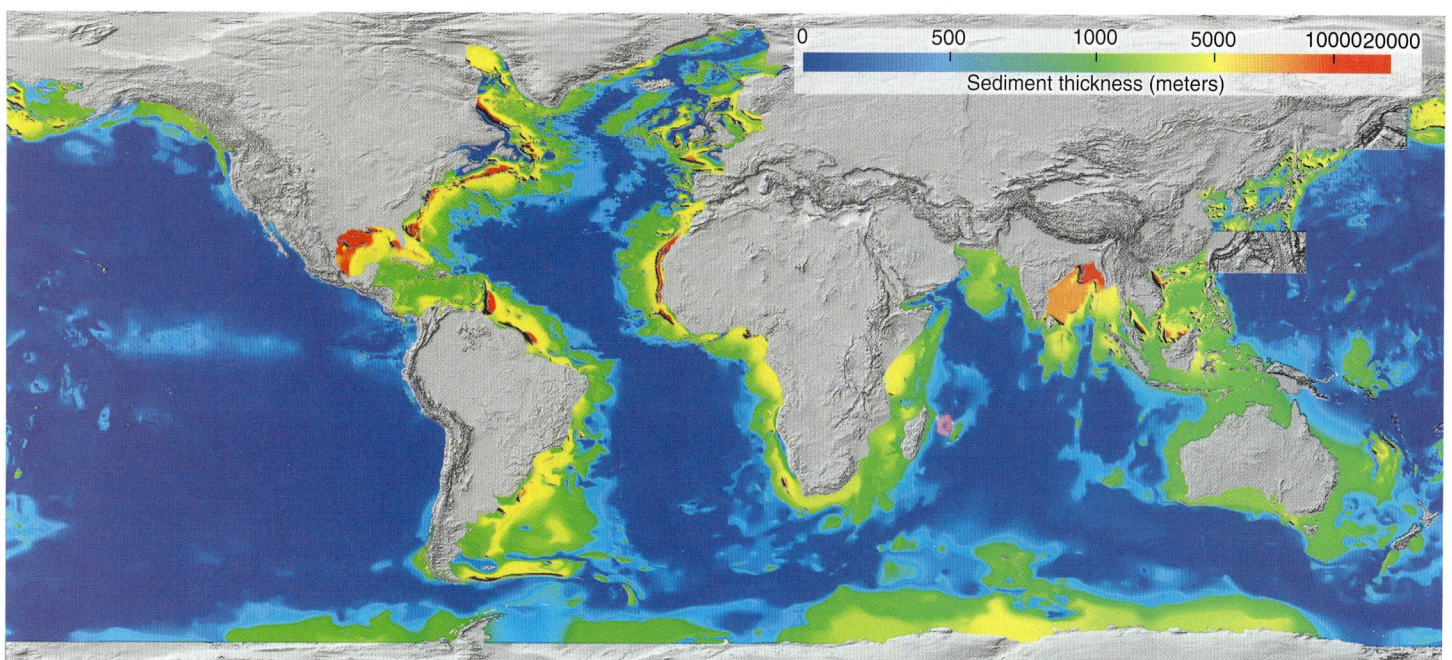

FIGURE 21.19 Sediment thickness on the seafloor varies from a few meters on young oceanic crust to thousands of meters along continental margins.

⊙ **What is the composition of abyssal sediment?**

(a)   Thierry Berrod, Mona Lisa Production/ Science Source

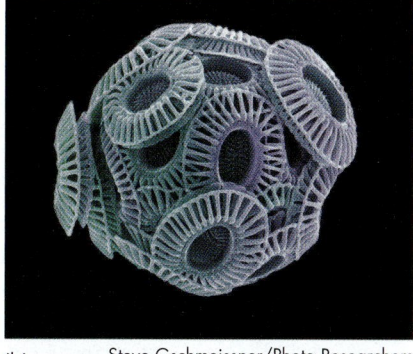

(b)   Steve Gschmeissner/Photo Researchers

(c)   Eye of Science/Photo Researchers

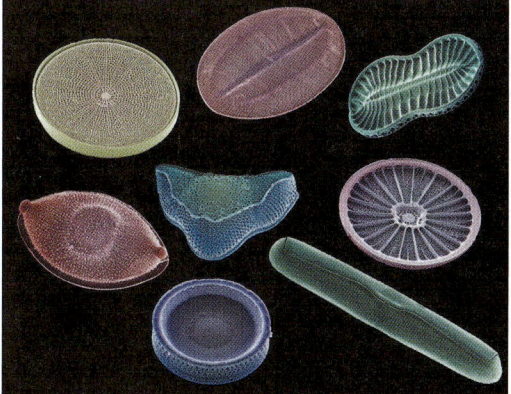

(d)   Steve Gschmeissner/Science Source

FIGURE 21.20 Pelagic sediment usually has a high percentage of microscopic skeletal debris consisting of calcareous ooze, such as (a) foraminifera and (b) coccolithophores, or siliceous ooze, such as (c) radiolarians and (d) diatoms.

⊙ **What factors determine the composition of biogenic ooze?**

coccolithophores (a plant). These sediments (**FIGURE 21.20**) make *calcareous ooze,* any carbonate sediment composed of more than 30% microscopic skeletal debris. Below a depth of 4 to 5 kilometers, calcareous shells dissolve, and the principal sediment is abyssal clay mixed with the remains of plankton called *radiolarians* (an animal) and *diatoms* (a plant) that are made of silica.

Carbonate Compensation Depth

The depth at which calcium carbonate ($CaCO_3$) completely dissolves is known as the **carbonate compensation depth**, or CCD. Whether the seafloor is deeper or shallower than the CCD is a major factor governing the composition of pelagic sediment. In the absence of cold water and abundant terrigenous sediment, seafloor that lies above the CCD tends to be rich in calcareous ooze. Seafloor that lies below the CCD tends to be rich in *siliceous ooze* and abyssal clay.

Calcareous sediment dissolves at the CCD, in interaction with a weak acid that is formed by dissolved carbon dioxide in seawater. Cold seawater under high pressure (deep water) holds more dissolved carbon dioxide than warmer water at low pressure (shallow water). The dissolved carbon dioxide (CO_2) combines with water (H_2O) to form *carbonic acid* (H_2CO_3), a reaction we discussed in Chapter 7 in terms of rock weathering. Carbonic acid reacts with calcareous particles falling through the water column and they dissolve.

Below the CCD, calcareous particles will be completely dissolved. The exact depth of the CCD varies according to the temperature, pressure, and chemical composition of seawater around the world (and through geologic time), caused by variations in water circulation, salinity, and other factors. The CCD can also be driven downward by high rates of calcareous sediment production in the water above, because waters become saturated with respect to carbonate.

Sediments Below the CCD

Below the CCD, pelagic sediment consists largely of abyssal clay (also termed *red clay*) and siliceous ooze. Abyssal clay is blown off the continents into the ocean, particularly on the western coasts of continents adjacent to major deserts. This aeolian dust is deposited everywhere, but it is abundant only in abyssal regions because low biological productivity and the dissolution of calcareous skeletal debris prevent other, normally dominant, types of sediment from accumulating. Of concern today is that aeolian dust is accumulat-

ing in greater quantities, as humans continue to degrade the landscape and set large brushfires, and as desertification intensifies.

Although less common, abyssal sediments also comprise volcanogenic particles, cosmogenic particles, and authigenic sediments, mentioned earlier. Volcanogenic particles, typically consisting of fine ash that falls from the atmosphere, can be a rich source of pelagic sediment in localities near volcanically active regions.

- Cosmogenic sediments accumulate in very low amounts. They derive from dust created by asteroid impacts, from comet debris that falls to Earth from space, and from meteorites that release particles as they encounter friction in the atmosphere.

- Authigenic sediments are made up of minerals that precipitate from seawater. These include *metallic sulfides,* which form where hydrothermal processes at spreading centers precipitate minerals, and *manganese nodules* and *phosphate deposits,* which accumulate from precipitated ions dissolved in seawater.

ⓠ EXPAND YOUR THINKING

Make a table listing the types of sediment that accumulate on the abyssal plain, and identify the major controls on their abundance.

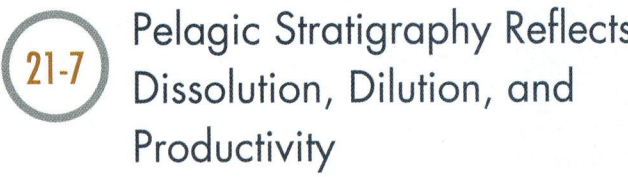

21-7 Pelagic Stratigraphy Reflects Dissolution, Dilution, and Productivity

LO 21-7 Explain the stratigraphy of pelagic sediments.

The rate at which pelagic sediment accumulates depends on three processes: *dissolution* (of biogenous particles), *dilution* (mixing with other types of sediment), and *productivity* (production of new organic matter by living organisms in the water).

Dissolution

Typical ocean waters are *undersaturated* with respect to silica. Seawater that is undersaturated with respect to a substance will tend to continuously dissolve that substance until the water becomes saturated, at which point no more will dissolve. Seawater, therefore, tends to dissolve silica plankton shells.

However, conditions favoring deposition of silica or calcium carbonate are opposite from one another. As seawater warms and pressure is lowered, silica solubility (the tendency to dissolve) increases. Consequently, silica is undersaturated in the oceans; but it is *less* undersaturated in the cold, pressurized water of the deep ocean. This means: 1) that silica tends to favor cold surface water; and 2) that as it falls through cold deep water its chances of arriving at the seafloor and forming a sedimentary deposit are enhanced. On the other hand, carbonate solubility increases with depth, and bottom waters become *more* undersaturated in calcium carbonate.

The patterns of carbonate and silica deposits on the seafloor reflect the different processes of formation and preservation, resulting in calcareous oozes that have little biogenic silica and silica oozes that have little biogenic carbonate.

The solubility of calcium carbonate varies from one ocean to another because of variations in temperature and carbon dioxide content. Near the surface of all oceans, where the temperature is relatively high, seawater generally is saturated with respect to calcium carbonate (hence, calcium carbonate sediment tends not to dissolve). But at average depths of 4 to 5 kilometers, the dissolved carbon dioxide content in seawater is sufficiently high to dissolve calcium carbonate. Consequently, calcareous shells are rarely found on the ocean floor at depths below 5 kilometers; however, this may vary. For instance, the CCD occurs at 6 kilometers in the Atlantic and 3.5 kilometers in parts of the Pacific.

Dilution and Productivity

Generally speaking, two types of sediment cover most of the seafloor away from continents: abyssal clay and biogenic ooze (**FIGURE 21.21**). Abyssal clay accumulates at the very slow rate of about 1 millimeter per 1000 years and covers most of the deep ocean floor below the CCD. As mentioned in section 21–6, abyssal clay is deposited everywhere in the oceans, but it is found in high concentrations only where other types of sediment (such as biogenic oozes) are absent and, hence, do not dilute the clay accumulation.

As discussed earlier, there are two main groups of ooze: siliceous ooze and calcareous ooze. Siliceous ooze is composed mostly of the tiny remains of diatoms and radiolarians. Both diatoms and radiolarians secrete skeletal hard parts made up of *opaline silica* ("opaline" simply means that there is water inside the silica molecule, which makes it appear iridescent).

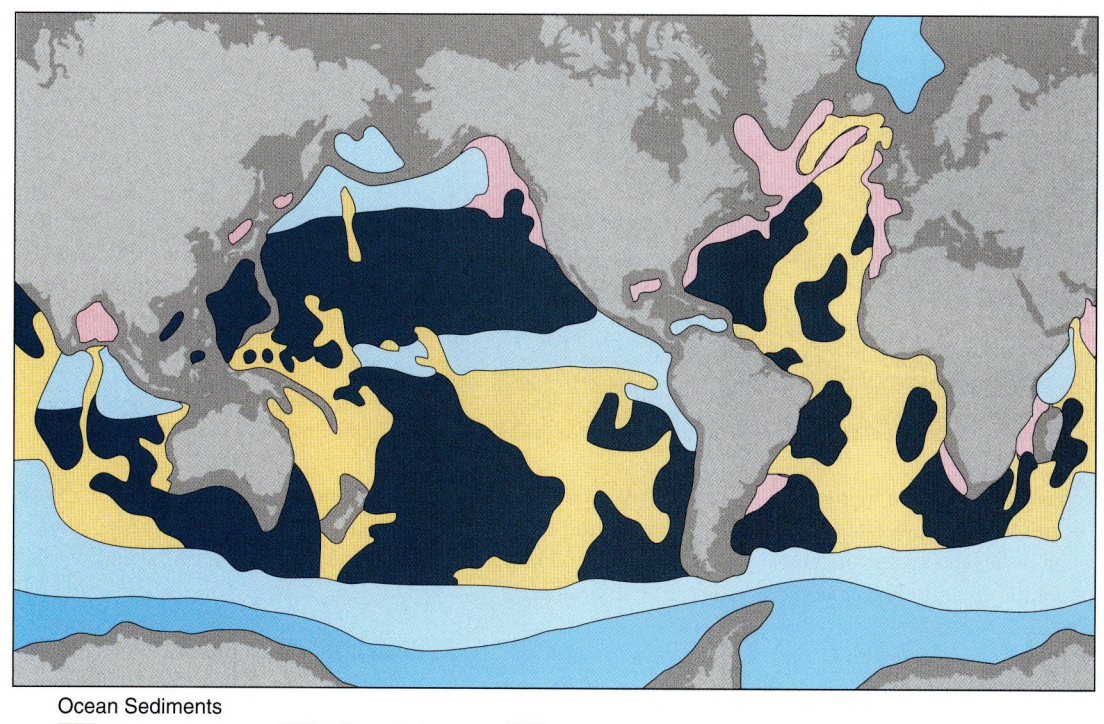

FIGURE 21.21 The modern ocean floor is covered with various sediments. Near Antarctica and to the east of Greenland, the sediment is detrital and comes from ice-rafting and glacial deposition. In the cold water of the Southern Ocean and in the north Pacific, the seafloor tends to collect siliceous ooze. Siliceous ooze also collects in the equatorial Pacific and Indian oceans. Below the CCD, the sediment is mostly abyssal clay. Above the CCD, away from the continents, most of the seafloor is covered in calcareous ooze. Along continental margins, watersheds on the land deliver detrital (terrigenous) sediment in great quantities.

Ocean Sediments
- ⬛ Calcareous ooze
- ⬛ Siliceous ooze
- ⬛ Abyssal clay
- ⬛ Glacial sediments
- ⬛ Terrigenous sediments
- ⬛ Continental margin sediments

Point to one area on this map and explain the composition of the pelagic sediment there.

Calcareous oozes are composed of the remains of coccolithophores and foraminifera. They secrete hard parts composed of calcium carbonate. Coccoliths are a type of algae; foraminifera may live on the seafloor as well as float in the water column, and they, like radiolaria, feed mainly on algae.

Marine Stratigraphy

The depth of the Mid–Atlantic Ridge is about 2500 meters at its crest and 7500 meters at its base. The CCD intersects the ridge midway down its flanks and controls the nature of pelagic sedimentation.

Calcareous ooze collects on the seafloor on the shallow portions of the ridge and forms a thick layer where the slowly spreading seafloor approaches the CCD. But once the seafloor passes through the CCD, the deposition of calcareous ooze halts, and abyssal clay (no longer diluted by biogenous sediment) becomes the dominant component of the pelagic sediment. This pattern produces a relatively simple two-layer stratigraphy consisting of a layer of calcareous ooze underlying a layer of abyssal clay (FIGURE 21.22).

In the Pacific Ocean, the stratigraphy is more complex (FIGURE 21.23). The East Pacific Rise, the mid-ocean ridge that is the birthplace of the Pacific Plate, is located in the southeast portion of the Pacific basin. From there the plate moves to the northwest through the southern hemisphere, across the equator, and eventually to subduction zones in the northwestern Pacific.

With a typical depth of approximately 2500 meters, the East Pacific Rise collects calcareous ooze along its crest and upper flanks. However, much of the Pacific seafloor lies below the CCD, and abyssal clay collects there as the plate moves toward the equator.

At the equator, something interesting happens. Because of high biological productivity in the warm, nutrient-rich water, abundant siliceous and calcareous skeletal debris falls through the water column. Dissolution of this large volume of material results in seawater that is only slightly undersaturated with respect to calcium carbonate and silica. Hence, there is little dissolution of biogenic particles, so the seafloor collects a mixture of siliceous and calcareous ooze—essentially, the CCD is depressed to the depth of the seafloor.

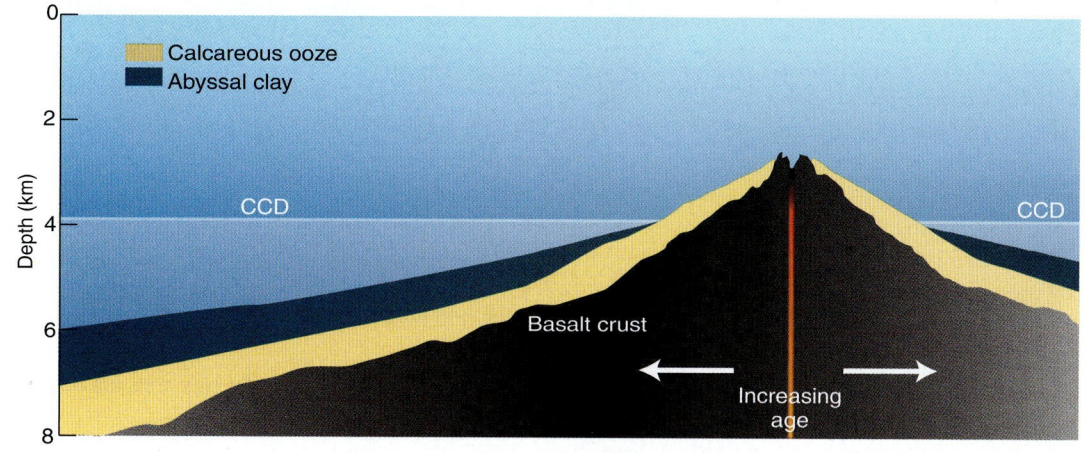

FIGURE 21.22 The shallow Mid–Atlantic Ridge collects calcareous ooze until the seafloor passes through the CCD. As the seafloor moves into deeper water, it collects predominantly abyssal clay. The resulting stratigraphy consists of two layers, ooze overlain by clay.

Describe the stratigraphy of the sediments if you were to drill a core into the Atlantic seafloor.

Equator

Calcareous ooze
Abyssal clay
Mixed siliceous and
calcareous ooze

High productivity

East
Pacific
Rise

CCD

CCD

Basalt crust

Northward movement
of Pacific Plate

FIGURE 21.23 The stratigraphy of abyssal sediments in the Pacific Ocean is more complex than in the Atlantic Ocean.

Ⓠ Explain what happens to the CCD in the Pacific Ocean at the equator.

Once the constantly moving plate passes beyond the highly productive equatorial waters, the CCD re-forms at a shallower depth, and a second layer of abyssal clay collects and buries the mixed siliceous/calcareous sediments that developed under the equator. The result is a four-layer model of stratigraphy produced in the Pacific Ocean. From the base to the top, this model consists of calcareous ooze, abyssal clay, a layer of mixed siliceous and calcareous ooze, and another layer of abyssal clay.

Ⓠ **EXPAND YOUR THINKING**

If abyssal clay is prominent only when ooze is absent, describe in relative terms the rates of accumulation of the two types of sediment. Why do they accumulate at different rates?

21-8 The Mid-Ocean Ridge Is the Site of Seafloor Spreading

LO 21-8 Specify the principal components of oceanic crust.

The world's longest mountain range is, in fact, the global mid-ocean ridge—actually, a series of ridges of the world's oceans connected to form a single global ridge system. This system runs across the seafloor of all of Earth's five oceans, to a total length of about 60,000 kilometers.

The Mid-Ocean Ridge

The **mid-ocean ridge** (**FIGURE 21.24**) appears where lithospheric plates separate; as they gradually move apart, magma rises to fill the gap. This magma, interacting with cold seawater that descends through fractures in the seafloor, drives the eruption of *hydrothermal vents* along the ridges (**FIGURE 21.25**). Seawater that contacts hot rock or magma heats up to 400°C and dissolves elements such as sulfur (S), lead (Pb), copper (Cu), and zinc (Zn) from the hot crust. This superheated water rises through the seafloor and gushes from vents, carrying with it these newly acquired elements. Valuable metallic sulfide minerals precipitate from the water. (Recall the sulfide mineral group from Chapter 4.)

Researchers descending in deep-sea submersibles to study mid-ocean ridges may be greeted with the astonishing site of tall, thin "chimneys" of precipitated metallic sulfide minerals spewing clouds of black smoke. Known as *black smokers*, these mark the location of hydrothermal vents. Unique *hydrothermal vent communities* consisting of giant tube worms, clams, crabs, and shrimp have evolved to live around these seafloor geysers.

FIGURE 21.24 The mid-ocean ridge is the world's longest mountain range. Here, it marks the spreading center between the Nubian Plate and the South American Plate.

Ⓠ In addition to the minerals in basalt, what other minerals are added to the seafloor at spreading centers?

FIGURE 21.25 The hot springs at the surface of hydrothermal vents are called "black smokers" in reference to their billowing clouds of precipitating metals.

NOAA PMEL Vents Program/SCIENCE PHOTO LIBRARY

What is the basis for the food chain at hydrothermal vent communities?

How do living things survive in such an inhospitable environment? Most life on Earth depends on sunlight as the ultimate source of energy. Green plants use sunlight to make food by the process of photosynthesis, and other organisms then feed on plants. But in the darkness of the ocean depths there is no sunlight, so instead of photosynthesis, a special class of bacteria use dissolved sulfur compounds in hydrothermal fluids as a source of energy. These bacteria are the base of vent food webs, and all other animals ultimately depend on them for nourishment. Thus, hydrothermal vent communities are forever tied to plate tectonics for their living environment.

Volcanism at Mid-Ocean Ridges

Spreading center volcanism occurs at the site of mid-ocean ridges where two lithospheric plates diverge from one another. A rift valley at the crest of the ridge is characterized by extension of oceanic crust, normal faulting, magma intrusion, seafloor volcanic activity, and the creation of a rugged terrain made of faulted blocks of rock.

As two plates separate, magma from the asthenosphere rises to fill fissures and fractures, and in the process produces new oceanic crust. The rise of this hot mantle gives thermal buoyancy to the seafloor; this is the reason that ridges stand high above the surrounding abyssal plain. Researchers estimate that, every year, approximately 20 seafloor eruptions occur along mid-ocean rift valleys and that 2.5 square kilometers of new seafloor is laid. About 4 cubic kilometers of new ocean crust is formed every year, with a thickness averaging 1 to 2 kilometers along the ridge axis.

The lower two-thirds (5 kilometers) of the newly forming oceanic crust consists of slow-cooling intrusive magma composed of phaneritic gabbro and layered ultramafic rocks. Above this layer is, on average, a 1.5-kilometer-thick unit of vertical basalt dikes, often called *sheeted dikes* in reference to their dense spacing. This layer is buried by a 0.5-kilometer-thick unit of volcanic pillow basalt interbedded with rich metallic sulfide minerals produced by the hydrothermal process.

Within rift valleys, the seafloor is composed of pillow basalts and metallic sulfide deposits; after the basalts and deposits form, volcanism and hydrothermal action ceases, and these rocks accumulate a blanket of pelagic sediment. The sedimentary layer thickens and begins to bury the rugged topography of the rift valley as the crust moves away from the spreading center (**FIGURE 21.26**).

Abyssal Hills Arise on the Flanks of Mid-Ocean Ridges

Abyssal hills may be among the lesser known of Earth's topographic features, even though some scientists think they are the most abundant

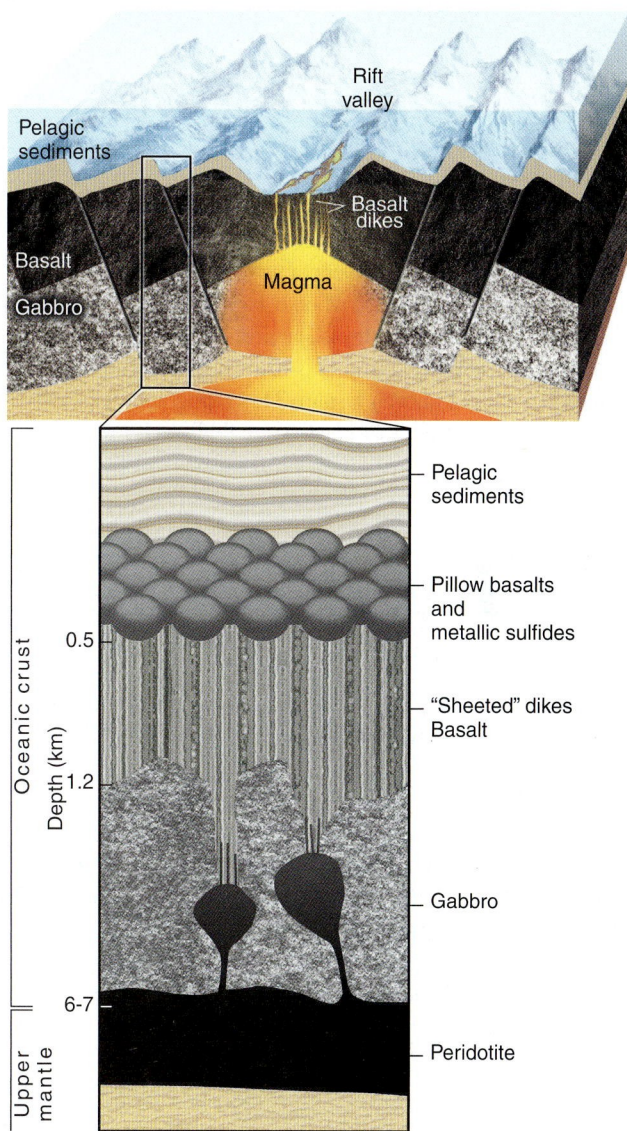

FIGURE 21.26 Oceanic crust consists of a layer of phaneritic gabbro and ultramafic rocks. This layer is overlain by densely spaced basalt dikes that carry lava to the seafloor. Extrusive basalts are interbedded with deposits of metallic sulfide minerals. As the seafloor is carried away from the rift valley, it is slowly buried in pelagic sediment.

How do the igneous rock composition and texture change from the upper mantle to the seafloor?

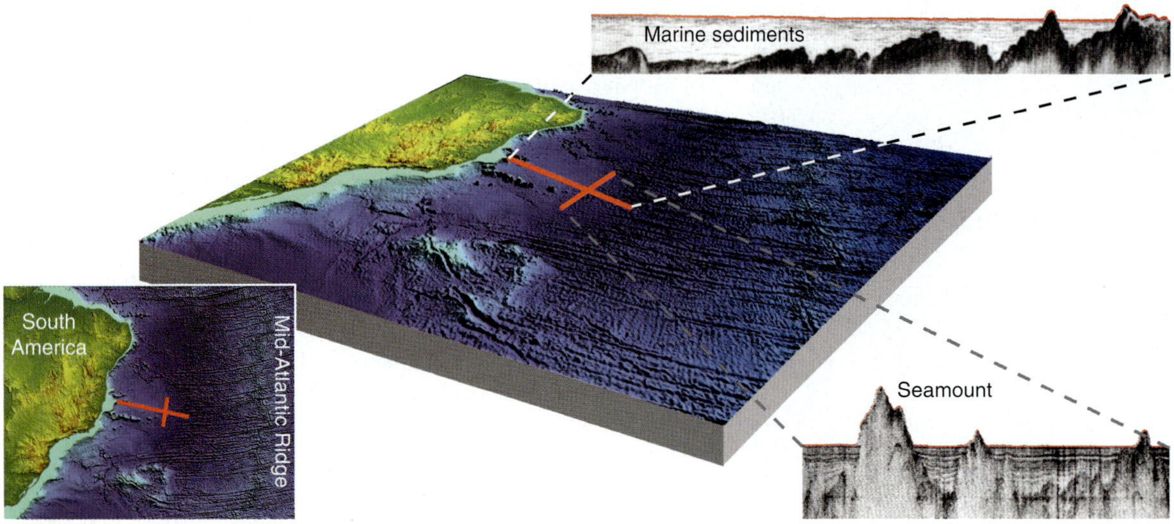

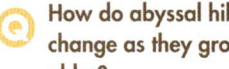

FIGURE 21.27 Abyssal hills have been described as Earth's most abundant topographic feature. They are sediment-covered terrain produced by rifting at mid-ocean ridges.

How do abyssal hills change as they grow older?

(**FIGURE 21.27**). These hills are small, well-defined elevations, ridges, and peaks that rise a few meters to several hundred meters above the abyssal seafloor. Typically, they are most abundant near mid-ocean ridges; fewer appear farther away from ridges, because they eventually disappear under layers of sediment.

Abyssal hills are produced when pelagic sediment buries the rugged topography formed in rift valleys. The normal faulting and volcanism at rift zones makes new seafloor that is marked by high topographic relief and very jagged and uneven terrain. This topography is slowly buried under layers of sediment that express the highs and lows of the relief, though not its craggy nature, and thus produce abyssal hills. As the hills age and move away from the rift valley, they are buried in pelagic sediment, which yield subdued but abundant topographic features.

To expand your knowledge of the seafloor, complete the Critical Thinking exercise titled "Structure of the Seafloor."

Q EXPAND YOUR THINKING

What principal factors control the distribution and relief of abyssal hills?

21-9 Oceanic Trenches Occur at Subduction Zones

LO 21-9 Describe the characteristics of oceanic trenches.

Plate subduction results in the formation of one of the most dramatic features in the ocean basins: **oceanic trenches**. These trenches range in depth from 6 to 11 kilometers and, as such, help to define plate boundaries.

Oceanic Trenches

In our study of plate tectonics, you learned that convergent margins are characterized by the meeting of two plates, with the denser plate subducting beneath the shallower plate. Typically, the older oceanic lithosphere subducts beneath the younger oceanic lithosphere; or, the oceanic lithosphere of any age subducts when it converges with the continental lithosphere.

In the case of ocean-ocean convergence, a volcanic *island arc* (**FIGURE 21.28**) will be created. In the case of ocean-continent convergence, a *volcanic arc* will develop. Both these volcanically active ranges are the product of partial melting beneath the overriding plate, where water released from the subducting slab lowers the melting point of rock in the upper mantle. This process generates magma that intrudes the overlying crust and produces volcanism.

The oceanic trenches are the deepest parts of the ocean floor. In one of them, the *Challenger Deep* of the *Mariana Trench* in the west Pacific, the seafloor is 11 kilometers deep and sits 3 to 4 kilometers deeper than the surrounding ocean floor. The Mariana Trench is the world's deepest, the site where two plates, both of oceanic lithosphere, converge. The older Pacific Plate subducts beneath the younger Philippine Plate.

Other major trenches around the world are (**FIGURE 21.29**):

- The *Aleutian Trench*, 7.6 kilometers deep, where the Pacific Plate subducts beneath an oceanic portion of the North American Plate.

- The *Kuril Trench*, 10.5 kilometers deep, which links the Aleutian Trench with the Japan Trench in the northwest Pacific basin.

- The *Japan Trench*, 9 kilometers deep, where the Pacific Plate subducts beneath the Eurasian Plate.

- The *Tonga Trench*, 10.8 kilometers deep, where the Pacific Plate subducts beneath the Tonga Plate and the Indo–Australian Plate.

- The *Kermadec Trench*, 10 kilometers deep, where the Pacific Plate subducts beneath the Indo–Australian Plate.

- The *Middle America Trench*, 6.6 kilometers deep, where the Pacific Plate subducts beneath the Caribbean Plate.

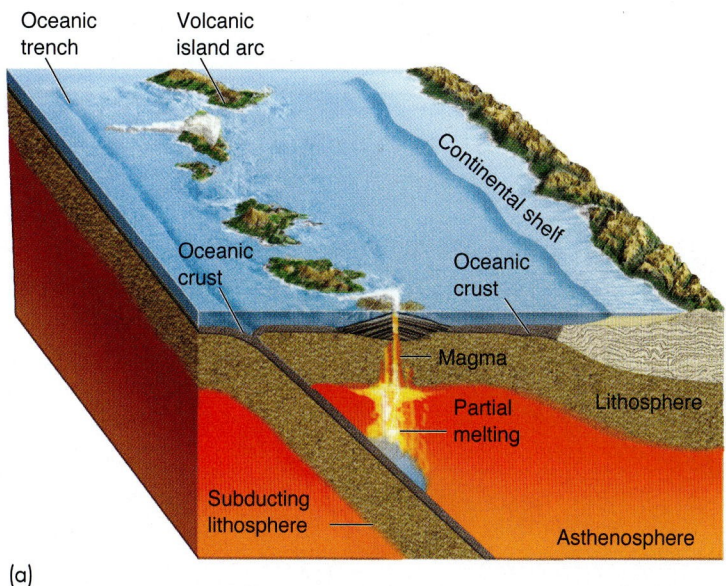

(a)

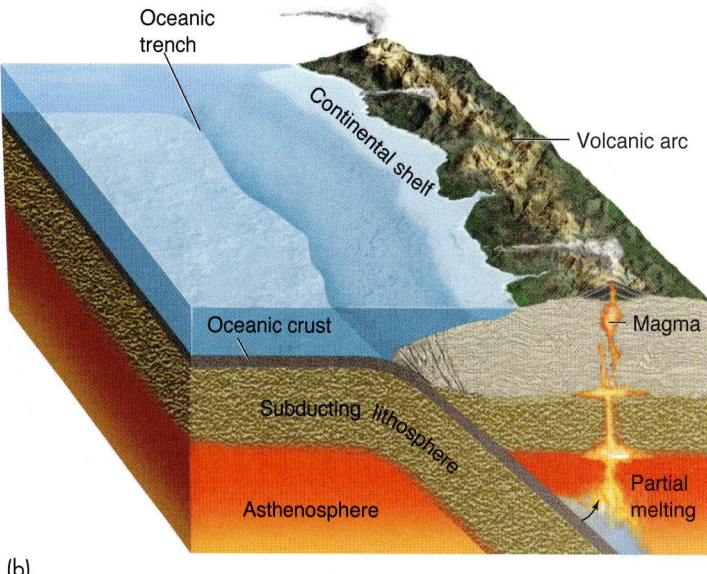

(b)

FIGURE 21.28 Oceanic trenches develop where one plate subducts beneath another. (a) An island arc is shaped by ocean-ocean convergence, and (b) a volcanic arc is formed by ocean-continent convergence.

Q **What is the source of magma that generates volcanism on the overlying plate?**

- The *Peru–Chile Trench*, also known as the *Atacama Trench*, 8 kilometers deep, where the Nazca Plate subducts beneath the South America Plate.

- The *Cayman Trough*, 7 kilometers deep, which is not a proper trench but is nonetheless one of the deepest locations in the oceans as a result of plate to plate interaction, is formed by a transform fault between the North American Plate and the Caribbean Plate.

- The *Puerto Rico Trench*, 8.6 kilometers deep, which marks the boundary of the North American Plate and the Caribbean Plate. Read more about this trench in the Geology in Action feature.

Oceanic trenches generally have a V-shape cross section, similar to a stream-carved valley, except that the side on the overriding plate tends to be steeper than the side on the subducting plate. Trench walls can be as steep as 45° (the Tonga Trench), although walls dipping between 4° and 16° are more typical.

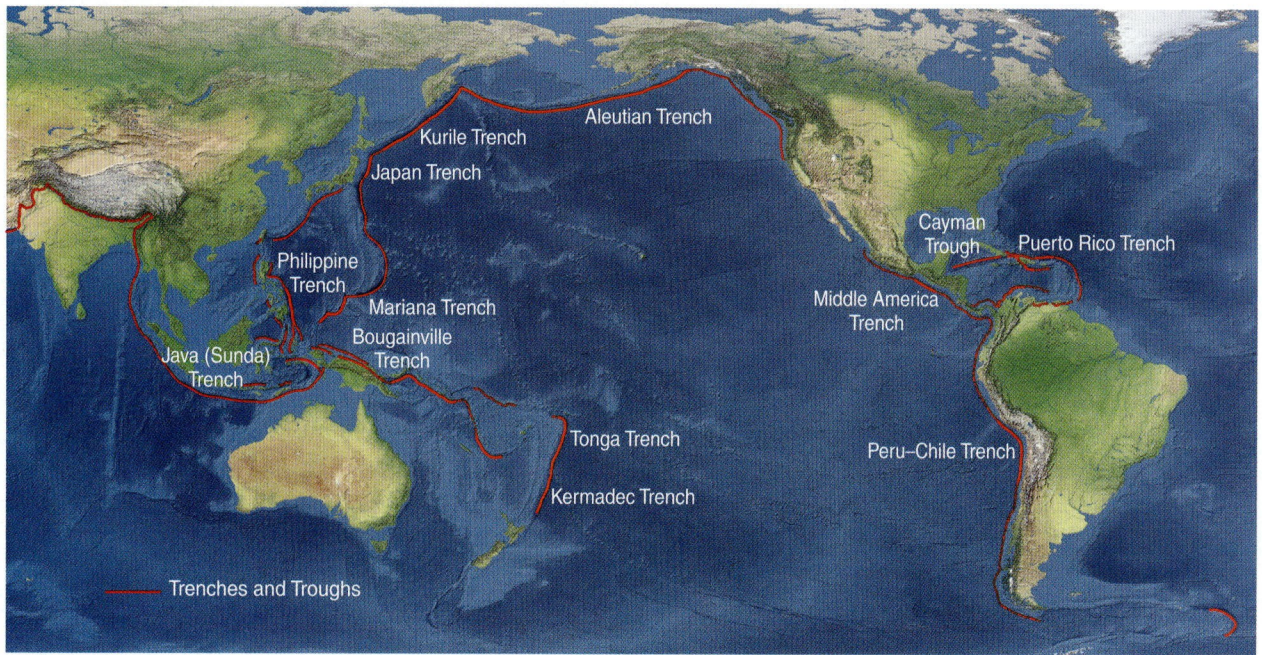

FIGURE 21.29 Oceanic trenches occur at convergent plate margins.

CRITICAL THINKING

Structure of the Seafloor

On the North American Plate, the Atlantic seafloor is created at the Mid-Atlantic Ridge and moves to the west.
Working with a partner, and referring to **FIGURE 21.30**, please complete the following exercise:

1. Draw an expanded cross section of the rock and sediment types that form the oceanic crust. Be sure to include igneous as well as sedimentary rock types and label the important features.

2. The figure shows several important features. Label as many as you can.

3. Circle the abyssal hills. What is the origin of these features?

4. Describe the processes that form submarine canyons. How are they related to submarine fans?

5. Why and how do the source and composition of sediment change between the coast, the continental shelf and rise, an abyssal plain, and the oceanic ridge?

6. Classify the types of rocks and minerals that develop in the rift valley of the Mid–Atlantic Ridge.

7. As a marine scientist, you have a research submersible available for four dives, an expensive and, so, rare opportunity. Where will you dive? Why? What research will you conduct during each dive? State a hypothesis motivating each dive and specify the data you will collect to test your hypotheses.

FIGURE 21.30 The Atlantic seafloor.

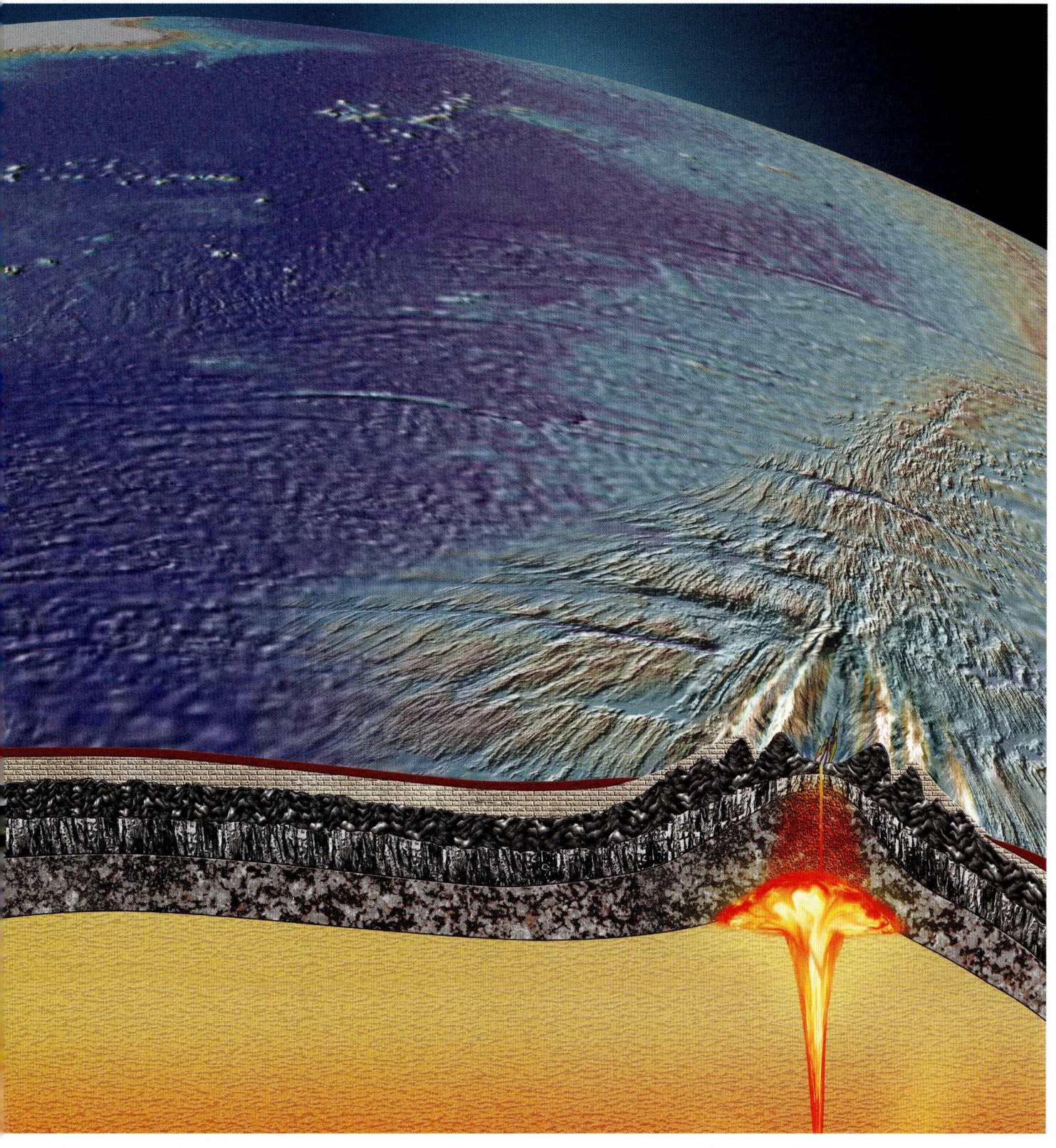

The Puerto Rico Trench

The Puerto Rico Trench (**FIGURE 21.31**) marks the convergence of the North American Plate and the Caribbean Plate. South and east of Puerto Rico, the boundary is convergent, but at Puerto Rico and Hispaniola, the plates define a transform boundary, one with a significant seismic and tsunami risk. This risk, long recognized by geologists, was made known to the world on January 12, 2010, when a violent earthquake rocked nearby Haiti, with fatal consequences.

Because Puerto Rico lies on an active plate boundary, earthquakes are a constant threat to the more than 4 million U.S. citizens who live there, and tsunamis threaten the densely populated coastal region, where most communities on the island are located. No surprise, then, that the region has a history of large earthquakes. At least seven quakes measuring Richter magnitudes of 6.9 or greater have occurred since 1787. In 1946, a magnitude 8.1 earthquake occurred north of Hispaniola, a large island to the east of Puerto Rico, where Haiti is located. This quake generated a tsunami that drowned nearly 1800 people in northeastern Hispaniola. Other earthquakes in the region are known to have produced tsunamis as well.

But the site of the most damaging earthquake in the history of the Caribbean region—7.0 magnitude—was Haiti, on that terrible January

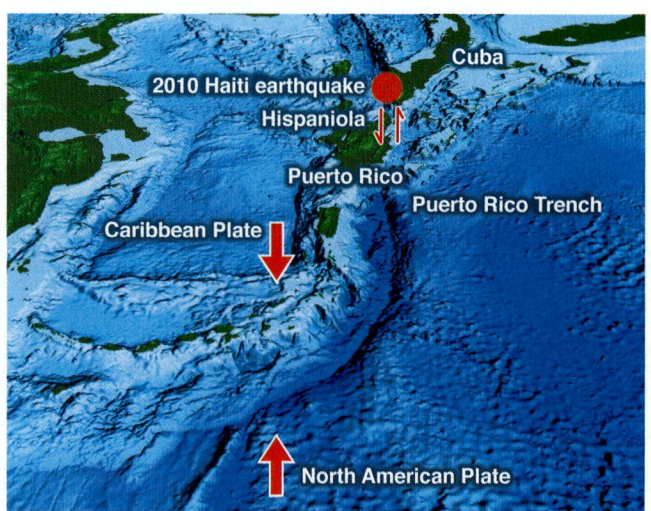

FIGURE 21.31 The Puerto Rico Trench marks the boundary between the North American Plate and the Caribbean Plate.

day in 2010. Over 200,000 people died as a result—an unusually high death toll, attributed to poor construction methods used in buildings in the impoverished nation.

The floor of a trench is usually a flat abyssal plain, well below the CCD, blanketed in pelagic sediment, although not in the thicknesses found in mid-ocean areas, because oceanic lithosphere is being subducted at trenches, and there is less time to accumulate sediment than at more stable settings. Abyssal clay, siliceous ooze, volcanic ash and lapilli, layers of turbidites (coarse sediments), and disturbed layers of sediment that have slumped from steep trench walls are the typical material that layer a trench floor.

Q EXPAND YOUR THINKING

Why would oceanic trenches be a source of tsunamis?

21-10 Human Impacts on Earth's Oceans Are Global in Extent

LO 21-10 Identify the human impacts on the oceans.

Human impacts on the marine environment are now known to be global in scale.

Studying Human Impacts

The science of marine geology is concerned primarily with the history of ocean basins, the nature of sediments that collect there and what they can tell us about past events, and the processes related to

plate tectonics that influence the oceanic lithosphere and adjoining continents. At the same time, many geologists are deeply involved in helping to manage marine resources and environments associated with coral reefs, beaches, estuaries, barrier islands, the shelf, and the larger ocean in an effort to alleviate the pressure imposed on these environments by people and their activities.

In February 2008, researchers at the University of Santa Barbara, California, released a map showing areas of the world where ocean environments had been negatively impacted by human activities. One fact stood out: Across the entire 360 million square kilometers covered by the oceans, 70% of Earth's surface, less than 4% of the ocean remains unaffected by humans (**FIGURE 21.32**).

Nineteen scientists collaborating from a number of universities and government agencies compiled the map using past studies of human impacts on marine ecosystems, such as coral reefs, sea-grass beds, fisheries, continental shelves, and the deep ocean. Whereas most previous work had focused largely on localized impacts, the

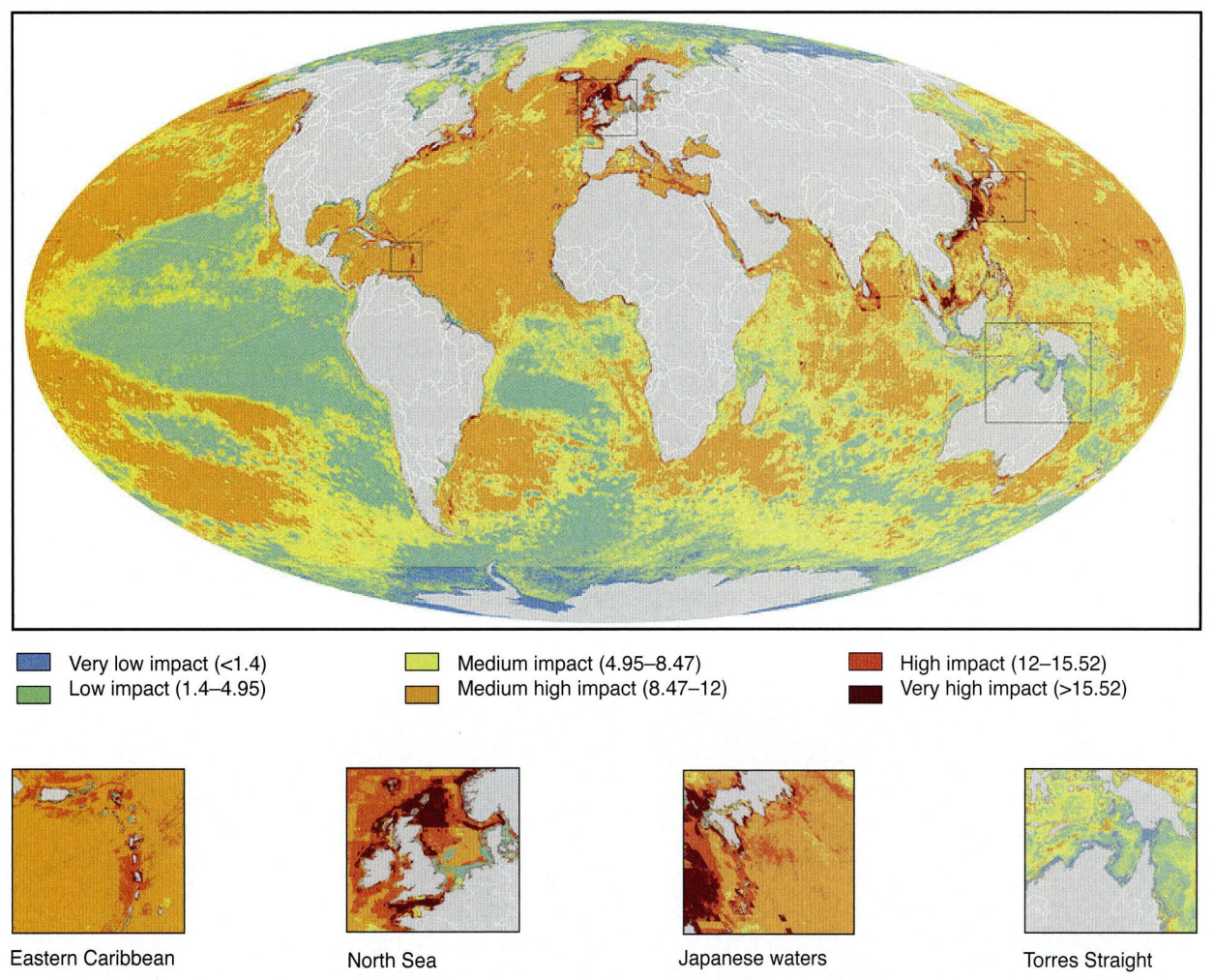

Very low impact (<1.4)
Low impact (1.4–4.95)
Medium impact (4.95–8.47)
Medium high impact (8.47–12)
High impact (12–15.52)
Very high impact (>15.52)

Eastern Caribbean

North Sea

Japanese waters

Torres Straight

FIGURE 21.32 Map of human impacts on the global ocean. Color coding is used to depict the degree of impact: darker orange and red colors indicate greatest impact.

Q Where is the degree of human impact on the oceans the highest? Where is it the lowest?

new study expanded the scope to encompass human impacts on marine ecosystems on a global scale.

The study investigated a total of 17 ways humans and their activities affect the oceans— among them, fishing, fertilizer runoff, pollution, shipping, and climate change. All were mapped as individual "geographic information layers" and overlain on one another to produce a composite map of global impacts. The resulting data revealed that, in fact, the degree of human impact over large areas of the ocean is still poorly understood. But what was clearly understood—an important outcome of this study—is that the past practice of managing resources by addressing one type of impact at a time fails to acknowledge the combined sum of multiple impacts.

Researchers concluded that managers can no longer afford to focus just on, say, fishing or coastal wetland loss or pollution, as if they each produce only separate effects. The reality is that human impacts overlap in space and time, and in too many cases their combined magnitude is frighteningly high. Conservation action that cuts across the entire range of human impacts is what's needed—now—around the globe if the oceans are to recover.

We examine in greater detail the human impact on one marine environment in the Earth Citizenship feature, "Loss of Coral Reefs."

Q EXPAND YOUR THINKING

Managers of marine environments often work in different agencies and so do not communicate with one another. How would you correct this situation? What barriers might you encounter?

EARTH CITIZENSHIP

Loss of Coral Reefs

Although reefs (**FIGURE 21.33**) cover less than 1% of the ocean floor, they play an integral role in coastal communities. They buffer coastal areas from large waves, produce sand for beaches, offer a shallow protective environment for numerous marine species, and provide humans with economic benefits through fisheries and tourism. The economies of many coastal communities rely on fish caught on coral reefs. When corals die, this income quickly disappears. Coral disease, predators, rising ocean temperatures and ocean acidification due to climate change, nutrient pollution, destructive fishing practices, and sediment runoff from coastal development all can jeopardize reef communities.

A study of corals in the central and western Pacific Ocean and the eastern Indian Ocean has revealed that coral is dying faster than was previously thought. Coral communities are disappearing at a rate of 1% a year, and nearly 1554 square kilometers of living coral have disappeared every year since the late 1960s. Researchers analyzed 6000 quantitative surveys of reef communities performed between 1968 and 2004 from more than 2600 coral reefs. They found that living coral declined from 40% in the early 1980s to approximately 20% by 2003.

Perhaps more alarming is that researchers also concluded that half of the world's reef-building corals already have been lost, and that this is a global phenomenon probably attributable in part to large-scale impacts such as climate change. Continued decline of Indo–Pacific reefs could have dire consequences for marine ecology and diversity, as well as humans, who stand to lose millions of dollars of income from fisheries and tourism.

(a)
Darryl Leniuk/Getty Images, Inc.

(b)
Michael Patrick O'Neill/Photo Researchers

FIGURE 21.33 (a) Healthy coral reefs are some of the most ecologically diverse communities in the ocean; they provide important environmental as well as economic benefits. (b) Humans negatively impact reefs through polluting, overfishing, and accelerating climate change. Shown here is a dead coral reef, the victim of such impacts.

LET'S REVIEW

Sediment accumulation and plate tectonics are connected processes. Mid-ocean ridges and oceanic trenches mark the location of spreading centers and subduction zones. Plates move between these sites beneath waters of varying temperatures, depths, and productivity. These conditions govern the stratigraphy of pelagic sediment. Continental margins contain most of the sediment. Detrital sediment is trapped in the coastal zone, and the shelves collect biogenic and authigenic sediment. Some margins pour sediment into the ocean, forming massive deltas and fans that become dominant topographic features. Sediment moving across the continental shelf is funneled through submarine canyons to deep-sea fans composed of turbidite sequences. In sum, the oceans are an integrated geological system where living organisms, plate tectonics, the continents and atmosphere, and the water column all influence one another.

STUDY GUIDE

21-1 Marine geology is the study of geologic processes within ocean basins.

- There are five recognized oceans on Earth: the Pacific, Atlantic, Indian, Southern, and Arctic. There are also many seas, smaller branches of an ocean that are partly enclosed by land. The largest seas are the South China Sea, the Caribbean Sea, and the Mediterranean Sea.

- **Marine geology** is a branch of science focused on geophysical, geochemical, sedimentological, and paleontological investigations of the ocean floor and coastal margins.

- The reason the ocean is salty is because it contains dissolved ions and compounds that come from chemical weathering and hydrothermal processes whereby seawater interacts with hot crust. Chloride and sodium make up over 90% of all dissolved ions in seawater. The concentration of dissolved ions in seawater, known as the **salinity**, has a global average of about 35 parts per 1,000. Ocean salinity is a result of geologic processes that regulate the concentration of dissolved components; inputs of water and dissolved ions must be matched by outputs.

21-2 Ocean waters are mixed by a global system of currents.

- Ocean circulation is the large-scale movement of water by **currents**, large masses of continuously moving ocean water. There are two types of oceanic circulation: surface circulation, which is driven by winds, and deep circulation, which is driven by cool water at the poles that sinks and moves through the lower ocean.

- The general pattern of oceanic circulation is that surface currents carry warm water away from the tropics toward the poles while releasing heat to the atmosphere during the journey. At the poles, surface water is cooled during winter and sinks to the deep ocean, generating currents along the seafloor and at mid-depths in the water column.

- Due to Earth's rotation, currents are deflected to the right in the northern hemisphere and to the left in the southern hemisphere. This deflection is known as the **Coriolis effect**. Continents redirect currents and produce large-scale circulation systems called **gyres**. There are five major gyres: the North Atlantic, South Atlantic, North Pacific, South Pacific, and Indian Ocean gyres.

- Ekman transport occurs when the Coriolis effect interacts with a column of water or air. As the Coriolis effect deflects a current, water in lower layers deflects more slowly than surface water. This effect translates down as a weakening spiral resulting in water transport that is 90° to the right of the wind direction in the Northern Hemisphere.

21-3 A continental shelf is the submerged border of a continent.

- A **continental shelf** is the area generally less than 200 meters deep surrounding a continental landmass. It is the undersea extension of a continent, ranging from over 100 kilometers wide to a few kilometers, depending on its geologic history. Shelves make up about 8% of the oceanic area.

- The shelf surface is swept by waves, tidal currents, and storms, which, in places, continually transport a layer of sediment to the seafloor consisting of sand, silt, and silty mud. In other locations, the shelf is starved of sediment, and the seafloor consists of outcropping sedimentary rocks of (usually) Cenozoic age.

- Many continental shelves are rich in natural resources, such as marine life, sand and gravel deposits, natural gas and oil reserves, phosphorite beds, and others. Most nations claim mineral and land rights to their shelves and actively manage the resources found on and in them.

21-4 The continental margin consists of the shelf, the slope, and the rise.

- The **continental margin** is made up of the continental shelf, the **continental slope**, and the continental rise. The margin, which separates thin oceanic crust of greater density from thick continental crust of lesser density, constitutes about 28% of the oceanic area. Beyond the rise lies the **abyssal plain** on fully oceanic crust.

- Active margins mark plate boundaries (convergent, divergent, or transform) and typically have narrower width and less sediment input than passive margins. Active margins located at convergent plate boundaries where there is a subduction zone may have **accreted terrane**. Passive margins are located within plates. For instance, the passive margin of the United States is separated from the Mid–Atlantic Ridge by an expanse of oceanic crust.

- A **submarine canyon** is a valley in the shape of a meandering channel that cuts into the shelf and slope. Many canyons end with a thick sedimentary deposit on the continental rise called a **submarine fan**. The upper portions of submarine canyons likely were carved by stream erosion during low sea levels on margins that have subsided. Lower portions of canyons are believed to have been carved by **turbidity currents**, dense, sediment-laden flows that travel down the axis of a canyon at high speed, sculpting the floor and sides. Turbidity currents carry sand and mud that is deposited on the rise when the currents spread out laterally and decelerate at the base of the canyon. These deposits are called turbidites.

21-5 Most ocean sediment is deposited on the continental margin.

- Most of the sediment of Earth's oceans (about 75% by volume) is deposited on the shelf, slope, and rise, where accumulations can exceed 10 kilometers in thickness. When it settles between the shoreline and the edge of the shelf, it is termed **neritic sediment**, which may be detrital, biogenic, or eolian. Sediments that drape upper and middle continental slopes around the world are known as **hemipelagic sediments**, and they grade from terrigenous mud to biogenic sediments.

- The coastal zone receives detrital sediment via watersheds. When the last ice age ended and sea level rose, rising water flooded river channels, driving them back into the watersheds. This process resulted in the development of estuaries, which today trap modern sediments, thereby starving shelves of sediment. Shelf sediment tends to be relict, from the period of sea level rise.

- Deltas formed by rivers such as the Mississippi (United States), Amazon (Brazil), Ganges–Brahmaputra (India), and the Yangtze (China) contain so much detrital sediment that they bury the topography of the margin.

21-6 Pelagic sediment covers the abyssal plains.

• The abyssal plain lies between the continental rise and a mid-ocean ridge. Abyssal plains are among Earth's least explored geologic features. Abyssal sediment is composed predominantly of **biogenic ooze** and **abyssal clay**. Biogenic ooze consists of plankton remains, and abyssal clay is derived from continents where sediment is eroded by the wind and falls into the sea. Other abyssal sediments are volcanogenic particles, cosmogenic particles, and authigenic sediments.

• Calcareous shells come from foraminiferans (animals) and coccolithophores (plants). These make calcareous ooze (more than 30% skeletal debris). Siliceous ooze comes from plankton with silica shells, such as radiolarians (animals) and diatoms (plants).

• The **carbonate compensation** depth (CCD) is the depth at which calcium carbonate ($CaCO_3$) completely dissolves, where the water is acidic. Whether the seafloor is deeper or shallower than the CCD governs the composition of pelagic sediment. Seafloor that lies below the CCD tends to be rich in abyssal clay.

21-7 Pelagic stratigraphy reflects dissolution, dilution, and productivity.

• **Pelagic sediment** accumulation depends on dissolution of biogenous particles, mixing with other types of sediment, and production of new organic matter.

• Calcareous ooze collects on the shallow portions of a mid-ocean ridge. Below the CCD, abyssal clay is the dominant sediment. This pattern produces a two-layer stratigraphy consisting of calcareous ooze underlying abyssal clay.

• The East Pacific Rise collects calcareous ooze on its crest and upper slope. Below the CCD, abyssal clay collects as the seafloor moves toward the equator. At the equator, high biological productivity suppresses the CCD, so mixed calcareous/siliceous ooze collects. Beyond equatorial waters, the CCD re-forms and abyssal clay collects and then buries the biogenous sediment from the equator. The result is a four-layer model of stratigraphy in the Pacific Ocean.

21-8 The mid-ocean ridge is the site of seafloor spreading.

• The world's longest mountain range, about 60,000 kilometers, is the global mid-ocean ridge, which is actually made up of all the mid-ocean ridges of the world. Valuable metallic sulfide minerals precipitate in the rift valley at mid-ocean ridges from hot water expelled at black smokers that mark the location of hydrothermal vents.

• Rock in the upper mantle is ultramafic with the composition of peridotite. This rock partially melts to produce mafic magma, which intrudes the crust to form gabbro. Crystallization yields layered ultramafic rocks among the gabbros. Basaltic magma rising rapidly to the seafloor forms sheeted dikes and fissure eruptions, generating submarine lava fields. Nestled among pillow lava in the rift valley are metallic sulfides formed by hydrothermal activity.

• **Abyssal hills** are small, well-defined hills, ridges, and peaks that rise a few meters to several hundred meters above the abyssal seafloor. They are produced when pelagic sediment blankets the rugged topography formed in rift valleys.

21-9 Oceanic trenches occur at subduction zones.

• Plate convergence results in **oceanic trenches** that range in depth from 6 to 11 kilometers. The Mariana Trench, the world's deepest, appears where the older Pacific Plate subducts beneath the younger Philippine Plate. The floor of a trench is usually a flat abyssal plain, well below the CCD, blanketed in pelagic sediment—although not in the thicknesses found in mid-ocean settings, because oceanic lithosphere is subducted at trenches and so there is less time to accumulate sediment than at more stable settings.

• Ocean-ocean convergence produces a volcanic island arc, and ocean-continent convergence produces a volcanic arc. Convergent boundaries are marked by seismic activity. Earthquake foci outline the position of a subducting oceanic plate.

21-10 Human impacts on Earth's oceans are global in extent.

• Marine geologists play important roles in managing marine resources, such as coral reefs, coastal environments, and others.

• Researchers have produced a map showing areas of the world where ocean environments have been negatively impacted by human activities. Across the entire 360 million square kilometers covered by the oceans (70% of Earth's surface), less than 4% remains unaffected by humans.

• Managing resources by addressing one type of negative impact at a time ignores the combined force of multiple impacts. Managers cannot, therefore, afford to focus only on, for example, overfishing or coastal wetland loss or pollution as if they are separate problems. All these issues overlap in space and time; and in many cases, the combined magnitude is high. If the oceans are to recover, conservation action that cuts across the full range of impacts is needed around the globe.

KEY TERMS

abyssal clay (p. 640)

abyssal hills (p. 645)

abyssal plain (p. 636)

accreted terrane (p. 636)

biogenic ooze (p. 640)

carbonate compensation depth (p. 641)

continental margin (p. 636)

continental rise (p. 636)

continental shelf (p. 634)

continental slope (p. 635)

Coriolis effect (p. 632)

currents (p. 632)

gyres (p. 632)

hemipelagic sediments (p. 640)

marine geology (p. 630)

mid-ocean ridge (p. 644)

neritic sediment (p. 638)

oceanic trenches (p. 646)

pelagic sediment (p. 640)

salinity (p. 630)

submarine canyons (p. 635)

submarine fan (p. 636)

thermohaline circulation (p. 633)

trade wind (p. 632)

turbidity currents (p. 637)

ASSESSING YOUR KNOWLEDGE

Please complete this exercise before coming to class. Identify the best answer to each question.

1. The five oceans are the:
   a. Mediterranean, Caribbean, Indian, Atlantic, and Pacific.
   b. Antarctic, Arctic, Pacific, Atlantic, and Mediterranean.
   c. Southern, Arctic, Pacific, Atlantic, and Indian.
   d. Pacific, Atlantic, Arctic, and Indian.
   e. Indian, Red Sea, Arctic, Pacific, and Antarctic.

2. Gyres in the northern hemisphere rotate (counterclockwise; clockwise) because of the (Coriolis; Eckman) effect.

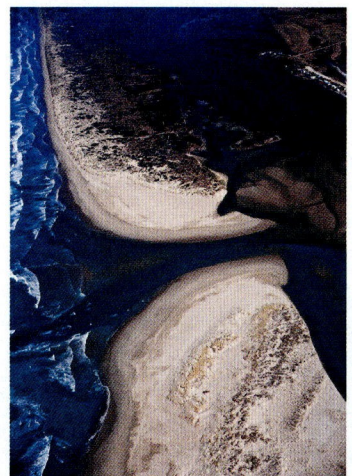

Bill Bachman/Alamy

3. Surface circulation is propelled principally by:
   a. Thermohaline circulation.
   b. Earth's rotation.
   c. Plate tectonics.
   d. The wind.
   e. The tides.

4. Ekman transport occurs when:
   a. The Coriolis effect causes surface water to travel to the left in the northern hemisphere.
   b. Water below the surface lags behind surface water in responding to the Coriolis effect.
   c. Winds blow water toward the coast, causing upwelling.
   d. Surface water and water that is deeper travel in the same direction due to the Coriolis effect.
   e. Currents sink due to density and evaporation.

5. The continental margin consists of the:
   a. Abyssal plain, slope, rise, and mid-ocean ridge.
   b. Shelf, slope, rise, and abyssal plain.
   c. Shelf, slope, and rise.
   d. Watershed, shelf, and slope.
   e. Abyssal hills, submarine fans, and submarine ridges.

6. Submarine canyons are:
   a. Formed by erosion of the continental slope by streams and turbidity currents.
   b. Drowned watersheds.
   c. Inactive drainage systems shaped by past tectonic processes.

d. Caused by sediment migration across the continental rise.
e. The result of hurricane storm surge.

7. Active margins are characterized by:
   a. Wide, sediment-rich shelves.
   b. Narrow shelves along tectonic boundaries.
   c. High sediment accumulation due to large watersheds and their deltas.
   d. Abyssal clay accumulation because of a shallow CCD.
   e. Slow subsidence.

8. Modern shelves tend to be sediment starved because:
   a. Coastal environments trap detrital sediments.
   b. Actually, shelves are not sediment starved.
   c. Biogenic sediment overwhelms detrital sediment accumulation.
   d. Strong currents prevent sediment from accumulating.
   e. Neritic sediments are buried by pelagic sediments on shelves.

9. The majority of sediment moving down watersheds to the ocean will:
   a. Eventually end up on the abyssal plain.
   b. Be deposited on the continental shelf.
   c. Develop into a turbidity current.
   d. Be trapped in the coastal zone.
   e. None of the above.

10. Pelagic sediment consists predominantly of:
    a. Relict, authigenic, and detrital sediments.
    b. Biogenic ooze and abyssal clay.
    c. Hemipelagic sediment and turbidites.
    d. Metallic sulfide deposits.
    e. Abyssal hills and turbidites.

11. What is the chemistry of the following organisms? (calcareous; siliceous)

    Diatoms _____
    Coccolithophores _____
    Foraminifera _____
    Radiolarians _____

Thierry Berrod, Mona Lisa Production/Science Source

Steve Gschmeissner/Photo Researchers

(a)   (b)

Eye of Science/Photo Researchers

Steve Gschmeissner/Science Source

(c)   (d)

12. Calcareous ooze is composed mostly of:
    a. Foraminifera and coccolithophores.
    b. Abyssal clay and diatoms.
    c. Diatoms and radiolarians.
    d. Partially dissolved abyssal clay.
    e. Cosmogenic and volcanogenic sediments.

13. The carbonate compensation depth is:
    a. The depth at which calcium carbonate precipitates in the water.
    b. Generally at 2 kilometers in most oceans.
    c. Controlled by the salinity.
    d. The level of silicate precipitation.
    e. None of the above.

14. Pelagic sediment stratigraphy is influenced by:
    a. The CCD.
    b. Productivity in overlying waters.
    c. Proximity to continents.
    d. Dissolution and dilution of sediments.
    e. All of the above.

15. Oceanic crust is composed of:
    a. Gabbro, andesite, and basalt.
    b. Gabbro, basalt, hydrothermal deposits, and pelagic sediment.
    c. Basalt, pillow lava, peridotite, andesite, and marine rhyolite.
    d. Rhyolite lava flows, gabbro, basalt, and hemipelagic sediment.
    e. None of the above.

16. The stratigraphy of oceanic crust includes a rock unit produced by hydrothermal venting. This layer is composed of:
    a. Manganese nodules.
    b. Biogenic ooze and abyssal clay.
    c. Hemipelagic sediment and turbidites.
    d. Metallic sulfide deposits.
    e. Abyssal hills and turbidites.

17. Most oceanic trenches are characterized by:
    a. Great depth.
    b. Mixed types of sediments.
    c. Earthquake activity.
    d. Nearby active volcanism.
    e. All of the above.

18. Managing marine environments requires:
    a. Solving one problem at a time.
    b. Attending to the coastal zone because that is where most of the problems are.
    c. Managing overlapping human impacts simultaneously.
    d. In reality, most marine environments do not need management.
    e. Focusing mostly on the surface waters.

19. The ocean is known as "Earth's heat engine" because:
    a. Surface circulation carries heat from the equator toward the poles and releases it to the atmosphere during the journey, warming the air.
    b. Thermohaline circulation carries heat to the deep ocean, releasing it around the globe as the water circulates through the deep sea.
    c. Ocean water stores heat during the day and releases it at night, warming the air when the Sun is absent.
    d. Water stores heat and moderates global warming of the atmosphere.
    e. All of the above.

20. Coral reefs are threatened by:
    a. Ocean acidification.
    b. Rising ocean temperatures.
    c. Sediment and nutrient pollution.
    d. Coral disease and destructive fishing practices.
    e. All of the above.

Darryl Leniuk/Getty Images, Inc.

FURTHER RESEARCH

1. What type of sediment is most likely to accumulate in the Arctic Ocean?

2. Why does the mid-ocean ridge rise above the surrounding seafloor?

3. How do abyssal hills form?

4. Predict the type of sediment you would find at these locations:

   a. At 60 meters of water depth offshore New Jersey

   b. On the abyssal plain around Hawaii

   c. In the Puerto Rico Trench

   d. South of Greenland on the Atlantic seafloor

5. Describe the major topographic provinces of the seafloor.

6. Why are many modern continental shelves "starved" of sediment?

7. Draw the stratigraphy of the Pacific seafloor where it descends into the Aleutian Trench.

8. What is the origin of submarine canyons?

ONLINE RESOURCES

Explore more about oceans on the following Web sites:

Find maps of the seafloor and various types of marine geology at: information:www.ngdc.noaa.gov/mgg/mggd.html

Learn about marine geology research being conducted by the U.S. Geological Survey at: http://marine.usgs.gov

To discover how the federal government manages oceans and coasts and how the National Ocean Service makes maps of the seafloor, go to: http://oceanservice.noaa.gov

Additional animations, videos, and other online resources are available at this book's companion Web site: www.wiley.com/college/fletcher

This companion Web site also has more information about WileyPLUS and other Wiley teaching and learning resources.

Common Rock-Forming Silicate Minerals

Silicate Mineral	Composition	Physical Properties
Quartz	Silicon dioxide (silica, SiO_2)	Hardness of 7 (on scale of 1 to 10)*; will not cleave (fractures unevenly); specific gravity: 2.65
Potassium feldspar group	Aluminosilicates of potassium	Hardness of 6.0–6.5; cleaves well in two directions; pink or white; specific gravity: 2.5–2.6
Plagioclase feldspar group	Aluminosilicates of sodium and calcium	Hardness of 6.0–6.5; cleaves well in two directions; white or gray; may show striations on cleavage planes; specific gravity: 2.6–2.7
Muscovite mica	Aluminosilicates of potassium with water	Hardness of 2–3; cleaves perfectly in one direction, yielding flexible, thin plates; colorless; transparent in thin sheets; specific gravity: 2.8–3.0
Biotite mica	Aluminosilicates of magnesium, iron, potassium, with water	Hardness of 2.5–3.0; cleaves perfectly in one direction, yielding flexible, thin plates; black to dark brown; specific gravity: 2.7–3.2
Pyroxene group	Silicates of aluminum, calcium, magnesium, and iron	Hardness of 5–6; cleaves in two directions at 87° and 93°; black to dark green; specific gravity: 3.1–3.5
Amphibole group	Silicates of aluminum, calcium, magnesium, and iron	Hardness of 5–6; cleaves in two directions at 56° and 124°; black to dark green; specific gravity: 3.0–3.3
Olivine	Silicates of magnesium and iron	Hardness of 6.5–7.0; light green; transparent to translucent; specific gravity: 3.2–3.6
Garnet group	Aluminosilicates of iron, calcium, magnesium, and manganese	Hardness of 6.5–7.5; uneven fracture; red, brown, or yellow; specific gravity: 3.5–4.3

(Biotite mica, Pyroxene group, Amphibole group, and Olivine are bracketed as Ferromagnesian minerals.)

*The scale of hardness used by geologists was formulated in 1822 by Frederich Mohs. Beginning with diamond as the hardest mineral, he arranged the following table:

10 Diamond	8 Topaz	6 Feldspar	4 Fluorite	2 Gypsum
9 Corundum	7 Quartz	5 Apatite	3 Calcite	1 Talc

The Periodic Table

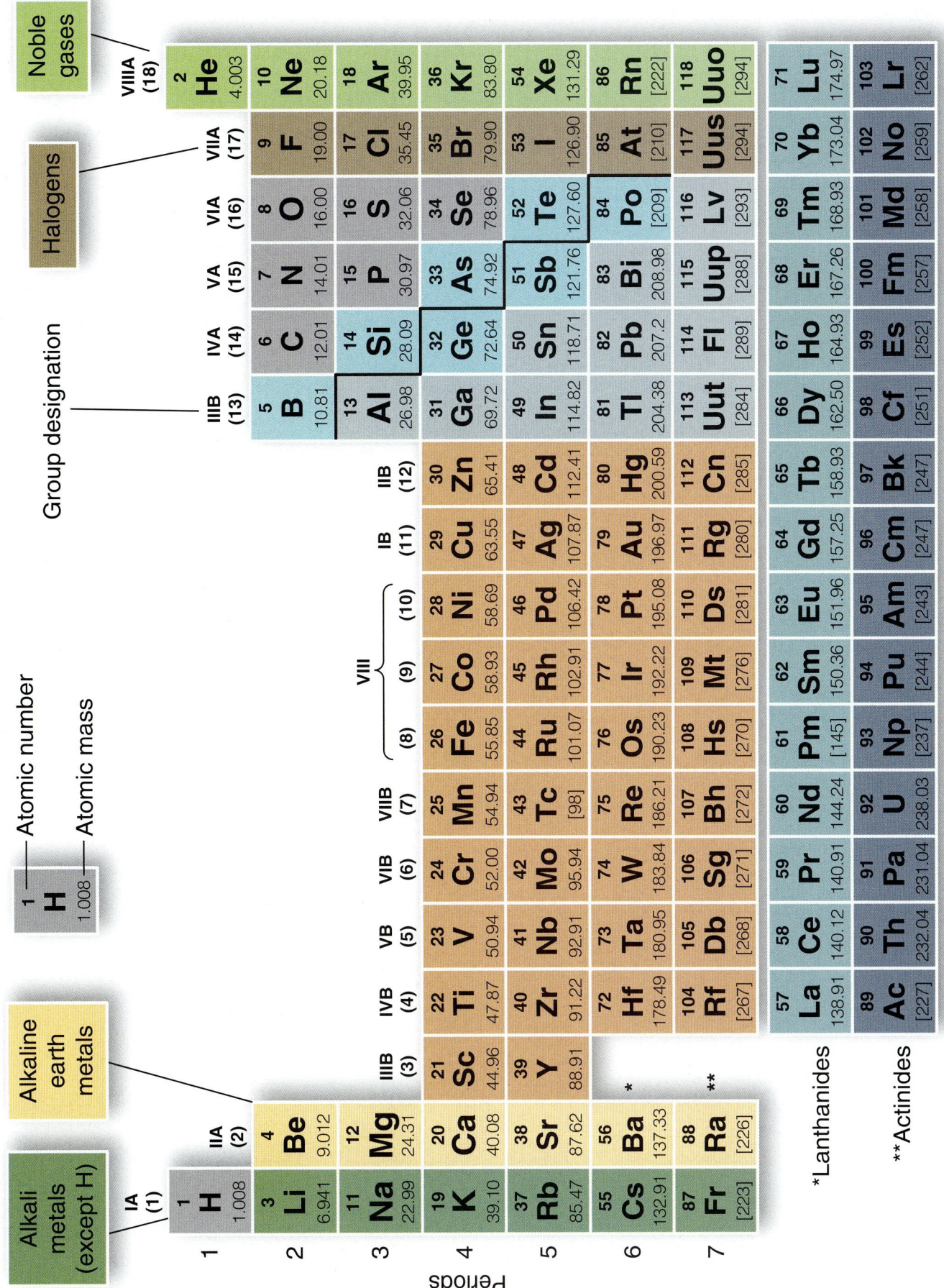

Atomic number
Atomic mass

1
H
1.008

Group designation

Alkali metals (except H)

Alkaline earth metals

Noble gases

Halogens

Periods

Elements and Their Chemical Symbols

Actinium	Ac	Einsteinium	Es	Mendelevium	Md	Samarium	Sm
Aluminum	Al	Erbium	Er	Mercury	Hg	Scandium	Sc
Americium	Am	Europium	Eu	Molybdenum	Mo	Seaborgium	Sg
Antimony	Sb	Fermium	Fm	Neodymium	Nd	Selenium	Se
Argon	Ar	Fluorine	F	Neon	Ne	Silicon	Si
Arsenic	As	Francium	Fe	Neptunium	Np	Silver	Ag
Astatine	At	Gadolinium	Gd	Nickel	Ni	Sodium	Na
Barium	Ba	Gallium	Ga	Niobium	Nb	Strontium	Sr
Berkelium	Bk	Germanium	Ge	Nitrogen	N	Sulfur	S
Beryllium	Be	Gold	Au	Nobelium	No	Tantalum	Ta
Bismuth	Bi	Hafnium	Hf	Osmium	Os	Technetium	Tc
Bohrium	Bh	Hassium	Hs	Oxygen	O	Tellurium	Te
Boron	B	Helium	He	Palladium	Pd	Terbium	Tb
Bromine	Br	Holmium	Ho	Phosphorus	P	Thallium	Tl
Cadmium	Cd	Hydrogen	H	Platinum	Pt	Thorium	Th
Calcium	Ca	Indium	In	Plutonium	Pu	Thulium	Tm
Californium	Cf	Iodine	I	Polonium	Po	Tin	Sn
Carbon	C	Iridium	Ir	Potassium	K	Titanium	Ti
Cerium	Ce	Iron	Fe	Praseodymium	Pr	Tungsten	W
Cesium	Cs	Krypton	Kr	Promethium	Pm	Ununnilium	Uun
Chlorine	Cl	Lanthanum	La	Protactinium	Pa	Unununium	Uuu
Chromium	Cr	Lawrencium	Lr	Radium	Ra	Uranium	U
Cobalt	Co	Lead	Pb	Radon	Rn	Vanadium	V
Copernicium	Cn	Lithium	Li	Rhenium	Re	Xenon	Xe
Copper	Cu	Lutetium	Lu	Rhodium	Rh	Ytterbium	Yb
Curium	Cm	Magnesium	Mg	Rubidium	Rb	Yttrium	Y
Dubnium	Db	Manganese	Mn	Ruthenium	Ru	Zinc	Zn
Dysprosium	Dy	Meitnerium	Mt	Rutherfordium	Rf	Zirconium	Zr

GLOSSARY

A

Aa   Rubbly, rough-looking form of lava, basaltic in composition.

Abrasion   Process wherein solid rock is worn away by impact; grains become smaller and more spherical the longer they are transported.

Abyssal Clay   Component of pelagic sediment (silt and clay) that is typically windblown from a continent.

Abyssal Hill   Well-defined hills, ridges, and peaks that rise a few meters to several hundred meters above the abyssal seafloor.

Abyssal Plain   Large flat area of the deep seafloor having slopes less than about 1 m/km and ranging in depth below sea level from 3 km to 6 km.

Accreted Terrane   Landmass that originated as an island arc or a micro-continent that was later added onto a continent, typically by convergent tectonics.

Accretionary Prism   Wedge-shaped formation of rock along the front of an overriding plate; formed as a plate subducts and some of the sediment is scraped off and collects as a series of angular rock slabs.

Acid Mine Drainage (AMD)   Acidic, metal-rich, polluted runoff from a mine that can contaminate the surrounding ecosystem.

Active Margin   Evolving margin of a continent on an overriding plate, typically containing a volcanic arc and accretionary prism.

Aftershocks   Earthquakes that occur after a large earthquake, on the same fault or nearby fault, related to the stress environment of the mainshock.

Agassiz, Louis (1807–1873)   Influential paleontologist, glaciologist, geologist, and originator of the "Theory of the Ice Ages."

Alluvial Channel   Channel located in its own alluvium, giving it a tendency to erode laterally and create a wide, flat-bottomed valley.

Alluvial Fan   Fan-shaped depositional feature of stream sediment where a channel in a steep watershed discharges into a flat valley, typically in arid, tectonically active environment.

Alpine Glacier   Glacier confined to a mountain valley.

Andesitic Magma   Magma that is andesitic (intermediate) in composition.

Anion   Atom that gains one or more electrons and acquires an overall negative charge.

Anthracite   Hard coal formed by metamorphosis.

Anticline   Convex upward arch (a fold) with an axial plane that is vertical, having the oldest rocks in the center.

Appalachian Highlands   Located inland of the Coastal Plain, reaching from Maine to Alabama, a rugged region composed largely of Paleozoic rocks.

B

Aquiclude   Body of impermeable, or distinctly less permeable, rock that prevents groundwater movement.

Aquifer   Underground formation of saturated crust that can produce useful quantities of water when tapped by a well.

Archean Eon   Period of geologic time that followed the Hadean Eon, between 3.8 to 2.5 billion years ago, during which Earth formed a solid volcanic crust able to support life.

Artesian Spring   Natural spring that rises out of the ground under its own pressure without the need for a pump and draws its supply of water from an artesian aquifer.

Atoll   Coral reef, often roughly circular in plan, which has no mainland and which encloses a shallow lagoon and forms on a subsiding shield volcano.

Atom   Smallest component in nature that has the properties of a given substance.

Atomic Number   Number of protons in an atom's nucleus by which each element is defined.

Avalanche   Fast-flowing, fluidized mass of snow, ice, air, and occasionally some regolith that cascades down a mountainside due to the collapse of a snowfield.

Avoidance   Learning to recognize the presence and characteristics of a geologic hazard and not building or otherwise using the land there.

Banded Iron Formation (BIF)   Formation with fine layering of cherty silica and iron, generally in the form of hematite, magnetite, or siderite; dating from late Archean time and a major source of iron ore.

Barrier Islands   Long, low, narrow sandy islands with one shoreline that faces the open ocean and another that faces a saltwater tidal lagoon protected by the barrier.

Basaltic Magma   Magma that is basaltic (mafic) in composition.

Base Flow   Water that enters a stream system from the water table through the channel bed.

Base Level   Theoretical lowest level to which the land will erode, ultimately equal to sea level.

Basin and Range Province   Area that extends from California across Nevada and Utah, and from Idaho into Mexico, with normally faulted blocks of crust; formed by extension.

Bauxite   Ore of aluminum; formed by intense weathering and found as a residual soil; largely gibbsite [$Al(OH)_3$].

Beach   Unconsolidated accumulation of sand and gravel along the shoreline.

Beach Profile   Topographic cross-section of a beach.

Biogenic Ooze   Component of pelagic sediment consisting of microscopic plankton remains

produced in the food chain in the overlying waters, from which they fall to the seafloor.

Biogenic Sediments   Types of sediment produced by organic precipitation of the remains of living organisms.

Biological Weathering   Type of weathering that occurs when rock disintegrates due to the chemical and/or physical activity of a living organism.

Bituminous   Soft coal.

Black Smoker   Hydrothermal spring underwater that releases dark, billowing clouds of metal-rich particles.

Blocks   Pyroclastic fragments that are ejected in a solid state, measuring more than 63 mm across.

Body Waves   Seismic waves that travel into and through Earth's interior.

Bomb   Pyroclastic fragments that are ejected in a semi-solid or plastic condition, measuring more than 63 mm across.

Bowen's Reaction Series   Process in which magma evolves to a new chemical state that is relatively enriched in silica as a result of fractionally crystallizing iron, magnesium, and calcium-rich compounds.

Braided Channel   Channel with more sediment than it can transport, which accumulates in the channel and forms bars and islets to give it a "braided" appearance.

C

Calcium Carbonate ($CaCO_3$)   Found in rock in all parts of the world as the mineral calcite and forming the rock limestone and the main component of shells of marine organisms, such as coral, certain plankton, gastropods, mollusks, and others.

Caldera   Roughly circular, steep-walled volcanic basin usually marking the summit of a volcano and one type of site from which eruptions occur.

Canadian Shield   Stable continental region not subjected to orogenesis since its origin; composed of Precambrian igneous and metamorphic rocks that have been exposed by glacial erosion during recent ice ages.

Cap Rock   Rock usually composed of shale or evaporite layers that seals a hydrocarbon trap.

Carbon Cycle   Production, storage, and movement of carbon on Earth.

Carbon Dioxide (CO_2)   Colorless, odorless, incombustible gas consisting of one carbon and two oxygen atoms; greenhouse gas produced by human activities such as burning carbon-based fuels.

Carbonate Platform   Large, flat, shallow marine environment where carbonate sediments collect, typically tropical or subtropical.

Cation   Atom that loses electrons and acquires an overall positive charge.

Cave   Natural opening in the crust that extends beyond the range of light and is large enough to permit the entry of humans.

Cenozoic Era   Youngest era of the Phanerozoic Eon.

Central Vent Volcano   Volcano built around a central vent.

Channel   Physical confine of flowing water consisting of a bed and banks.

Channeled Scablands   Area of eroded channels, giant ripple marks and gravel bars, scour depressions, and remnants of eroded basalt interpreted to represent past episodes of megafloods due to ice dam outbursts in the Pacific Northwest.

Chemical Differentiation   Process of forming chemically distinct magma by the incomplete melting of rock or incomplete crystallization of magma.

Chemical Sediments   Types of sediment produced by inorganic precipitation of dissolved compounds.

Chemical Weathering   Chemical decomposition of minerals in rock.

Chondrite   Primitive meteorite containing the four lead isotopes but no uranium or thorium parents; considered most primitive material in the Solar System.

Cinder Cone   Built-up ash, lapilli, blocks, and bombs of congealed lava ejected from a single vent.

Clastic Sediments   Broken pieces of crust deposited by water, wind, ice, or some other physical process.

Clay   Most abundant sediment, created by the decomposition of silicate rocks, usually during hydrolysis; composed of fine-grained minerals (phyllosilicates).

Climate Feedback   Secondary change that occurs within the climate system in response to a primary change, either positive or negative.

Climate Proxy   Indicator of past climate.

Coal Seam   Layer of coal deposited by "coal swamp" and lithified to compressed and concentrated carbon; typically 1 m to 10 m in thickness.

Coastal Erosion   Erosion of coastlines due to sea-level rise and/or decreased sediment availability to the shoreline.

Coastal Geology   Study concerned with understanding coastal processes, the geologic framework of coastal areas, and how humans live in coastal environments.

Coastal Hazard   Coastal process that may cause loss of life or property.

Coastal Plain   Flattest of all the provinces; stretches over 3500 km, from Cape Cod in New England south along the Atlantic coast into the Gulf of Mexico and to the Mexican border, which contains some of the youngest rocks in the continent.

Coastal Processes   Agents of change on coastal areas such as winds, waves, storms, tides, sea-level change, and currents.

Colorado Plateau Province   Centered around the "Four Corners" region where the states of Utah, Colorado, Arizona, and New Mexico come together; area of arid conditions, steep relief, sparsely vegetated mesas, and deep canyons.

Columbia Plateau Province   Located in the northwest corner of the nation in the states of Oregon, Nevada, Idaho, and Wyoming; constitutes the world's youngest large igneous province and was inundated by the largest documented flood in geologic history.

Compound   Combination of atoms of different elements bonded together.

Compressional Stress   Type of directed stress that squeezes rock.

Cone of Depression   Conical depression in the water table immediately surrounding a well.

Contact Metamorphism   Metamorphism adjacent to an intrusive igneous rock.

Contaminant Plume   Combination of toxic liquids that flows at the same hydraulic gradient that drives groundwater flow in aquifer systems.

Continental Accretion   Process of adding exotic terranes to the active margins of continents.

Continental Crust   Part of Earth's crust that comprises the continents.

Continental Glacier   Glacier covering a large area and not confined by topography.

Continental Margin   Broad zone that includes the continental shelf, the continental slope, and the continental rise.

Continental Rise   Region of gently changing slope where the abyssal plain meets the continental margin.

Continental Shelf   Area of the ocean floor that is generally less than 200 m deep surrounding each continental landmass.

Continental Slope   Pronounced slope beyond the seaward margin of the continental shelf and ending offshore at the continental rise.

Continent–Continent Convergent Boundary   Location where two plates composed of continental crust collide, building mountains in the process.

Convection   Transfer of heat by movement from areas of high heat to areas of low heat.

Convergent Boundaries   Locations within the crust where two or more plates push together.

Coral Reef   Shallow coastal environment where numerous types of coral and algae build massive solid structures of calcium carbonate.

Cordillera   Complicated mosaic of ranges, valleys, fault zones, and volcanic regions that runs from the western tip of Alaska to the southern tip of South America.

Core   Mass, largely metallic iron, at the center of Earth; composed of inner and outer regions.

Coriolis Effect   Effect that, due to the rotation of Earth, causes air and water currents to be deflected to the right in the Northern Hemisphere and to the left in the Southern Hemisphere.

Country Rock   Phrase for intruded crust used by miners and geologists.

Crater   Smaller, circular depression created primarily by the explosive excavation of rock during an eruption.

Craton   Core of ancient rock in the continental crust that has attained tectonic and isostatic stability.

Creep   Slow, down-slope migration of soil under the influence of gravity.

Critical Thinking   Use of reasoning to develop hypotheses and theories that can be rigorously tested.

Crust   Outermost and thinnest of Earth's compositional layers, consisting of brittle rock that is less dense than the mantle.

Crystalline Structure   Geometric pattern that atoms assume in a solid.

Crystallization   Process through which atoms or compounds that are in a liquid state are arranged into an orderly solid state.

Cumulate   Layer of dense minerals forming a concentrated deposit on the floor of a magma chamber.

Current   Mass of continuously moving water.

D

Darcy's Law   Relationship among discharge, permeability, and hydraulic gradient in groundwater movement.

Darwin, Charles (1809–1882)   English natural scientist who formulated a theory of evolution by natural selection.

Debris Fall   Relatively free fall or collapse of regolith from a cliff or steep slope.

Debris Flow   Downslope movement of a mass of unconsolidated, water-saturated regolith, more than half of which is coarser than sand.

Debris Slide   Slow to rapid downslope movement of regolith across an inclined surface.

Decay Chain   Steps of radioactive decay involving multiple radioactive daughters.

Decompression Melting   Melting that occurs when pressure is lowered; *also* Pressure Release Melting.

Deforestation   Removal of trees for human use of the land or the timber that results.

Deformation   Process of solid rock of the crust experiencing stress.

Delta   Sedimentary deposit formed where a channel reaches base level and deposits its load of sediment.

Deposition   Material, such as sediment, being added to a landform.

Depositional Coast   Coast dominated by depositional processes.

Desert   Arid region that receives less than 25 cm of precipitation per year, or an area where the rate of evapotranspiration exceeds the rate of precipitation.

Desert Pavement   Surface layer of coarse particles concentrated chiefly by wind deflation.

Desertification   Expansion of desert into nondesert areas.

Dikes Arms of intrusive rock that cut vertically (or nearly so) across other rocks in the crust.

Dip Angle in degrees between a horizontal plane and an inclined plane, measured down from horizontal in a plane perpendicular to the strike.

Dip-Slip Fault Fault in which displacement occurs vertically along the dip of the fault plane.

Direct Mitigation Practices that reduce vulnerability to mass wasting hazards when it is too late for avoidance.

Directed Stress Pressure acting on the crust that is greatest in one direction.

Discharge Volume of water passing any point on a stream over time.

Discontinuity Change in rock density, causing seismic wave velocity to change suddenly.

Disseminated Ore Deposit Mineral ore dispersed throughout a rock in sparse quantities.

Dissolution Chemical weathering reaction in which a mineral dissolves in water.

Divergent Plate Boundaries Locations within the crust where two or more plates pull away from each other.

Drainage System Network of channels that pass runoff downhill.

Dune Mound or ridge of sand deposited by wind.

E

Earthquake Sudden shaking of the crust.

Effusive Eruption Volcanic eruption that pours out fluid lava.

Elastic Rebound Theory Theory that earthquakes result from the release of stored elastic energy by slippage on faults.

Emergent Coast Coast where the land is tectonically uplifting at a rate that exceeds sea-level rise.

Energy Resources Nonrenewable (oil, coal, natural gas, and uranium) and renewable (solar, wind, bioenergy, geothermal, fuel cells, hydroelectric, and ocean energy) resources that create energy to power machinery.

Environments of Deposition Various sedimentary basins where sediment comes to rest.

Eons Hundreds of millions to billions of years.

Ephemeral Stream Stream where flowing water is present only after specific rainfall events or seasonally.

Epicenter Geographic spot on Earth's surface that is located directly above an earthquake focus.

Eras Tens of millions to hundreds of millions of years.

Erosion Complex group of related processes by which rock particles are physically moved.

Erosional Coast Coast dominated by erosional processes.

Eruption Column Massive, high-velocity, billowing cloud of gas, molten rock, and solid particles that is blasted into the air with tremendous force.

Estuary Partially enclosed body of water that is formed in places where freshwater from rivers and streams flows into the ocean, mixing with salty seawater.

Eutrophication Increase of noxious forms of algae, due to nutrient contamination, which use up oxygen in the water when decaying and cause other forms of life to suffocate.

Evaporites Minerals formed when evaporation leads to inorganic precipitation of dissolved minerals; a form of chemical sediment.

Evolution Changing of plant and animal community through time.

Exotic Terranes New pieces of continental crust added by accretion at convergent margins.

Explosive Eruption Dramatic volcanic eruption that throws volcanic products kilometers into the air; characterized by high gas discharge, an eruption column, ash, and extreme violence.

Extrapolation Form of critical thinking used to make estimations of phenomena that cannot be directly observed, consisting of inferring or estimating an answer by projecting or extending a known value.

Extrusive Igneous Rock Rock formed by the solidification of magma erupted onto Earth's surface.

F

Fault Zone or plane of breakage across which layers of rock are displaced relative to one another.

Fault-Block Mountain Mountain characterized by tensional stress, crustal thinning, and normal faulting.

Felsic Igneous rocks composed of minerals with the lowest melting temperatures, light in color, and relatively enriched in sodium, potassium, oxygen, and silicon.

Field Geology Obtaining data directly from the rocks exposed at Earth's surface.

Fissure Eruption Eruption that originates from an elongated fracture on the side of a volcano.

Flank Eruption Eruption from the side or slope of a volcano.

Flood Discharge great enough to cause a stream to overflow its banks.

Flood Basalt Location where so much basaltic lava has issued from the ground that it forms a high plateau.

Floodplain Part of any stream valley that is regularly inundated during floods and whose geologic framework is formed by floods.

Fluvial Of, or pertaining to, streams or rivers, especially erosional and depositional processes of streams and the sediments and landforms resulting from them.

Fluvial Erosion Active movement of sediment particles by running water.

Focus Point where an earthquake originates due to fracture of the crust.

Fold Individual bend or warp in layered rock.

Fold-and-Thrust Mountain Mountain that develops at a continental collision site along a convergent margin due to orogenesis.

Foliation Texture in metamorphic rock due to directed stress, caused by alignment of recrystallizing minerals.

Foreshocks Small earthquakes that precede a larger earthquake on the same fault or a nearby fault; related to the same general stress environment.

Fossil Remains of animals and plants, or traces of their presence, that have been preserved in the crust.

Fossil Fuels Concentrated forms of burnable organic carbon contained in rock originating from fossil accumulations.

Fracture Breakage of brittle material.

G

Gas Giants Outermost planets of the Solar System (Jupiter, Saturn, Uranus, and Neptune) that are characterized by great mass, low density, and thick atmospheres consisting primarily of hydrogen and helium.

Genetic Mutations Random changes to genetic material, either RNA or DNA.

Genetic Variation Differences in inherited traits.

Geologic Events Notable occurrences of common geologic processes: deposition, deformation, faulting, intrusion, and erosion.

Geologic Hazards Dangerous natural processes such as landslides, floods, erosion, volcanic eruptions, tsunamis, storms, earthquakes, and hurricanes.

Geologic Provinces Regions with a distinctive common geologic history.

Geologic Resources Nonliving materials that are mined to maintain our system of industry and quality of life.

Geologic Time Scale Calendar of events that covers Earth's entire history.

Geological Engineering Study of geological processes and the role they play in engineering projects and safety.

Geological Map Map that shows the topography of an area, as well as the lithology and age of rock units exposed at the surface, or shallowly buried under loose sediment and soil, and their strike and dip.

Geology Study of Earth and other objects.

Geomagnetic Field Magnetic field that surrounds Earth.

Geothermal Energy Heat that originates within Earth through decay of radioactive isotopes.

Geyser Thermal spring equipped with a system of groundwater recharge and heating that causes intermittent eruptions of water and steam.

Glacial Landforms Landforms created due to glaciation.

Glacial Moraine Deposit of till that marks a former position of the ice.

Glacial Outwash Poorly sorted stratified sediment deposited in an outwash plain.

Glacial Till Unstratified and unsorted sediment that has been deposited directly by melting ice.

Glacier Semi-permanent body of ice, consisting largely of recrystallized snow, that shows evidence of downslope or outward movement, due to the stress of its own weight.

Glacier Mass Balance Process in which the terminus of an active glacier advances or retreats, depending on the relative rates of ice delivery and wastage.

Glacioisostatic Uplift Process in which crust is uplifting because the weight of a glacier has been removed.

Global Circulation Models (GCMs) Computer-based mathematical programs that simulate the physics of Earth's oceans, land, and atmosphere.

Global Climate Change Large-scale change in climate over time, whether natural or as a result of human activity.

Global Warming Increase in the average temperature of Earth's surface, including the oceans.

Graded Stream Stream that has apparently achieved, throughout long reaches, a state of dynamic equilibrium between the rate of sediment transport and the rate of sediment supply.

Greenhouse Effect Atmosphere's natural ability to store heat radiated from Earth.

Greenhouse Gas Gas that is in the atmosphere, mainly H_2O, CO_2, CFCs, N_2O, and CH_4, that cause the greenhouse effect.

Groundwater All the water contained in the pore spaces within bedrock and regolith.

Groundwater Remediation Process by which water pollution is extracted and filtered.

Gyre Large-scale circulation system that sweeps the major ocean basins.

H

Half-Life Time required for half of a radioisotope species in a sample to decay.

Hawaiian Islands Exposed tips of the Hawaiian Ridge, a long volcanic mountain range on the seafloor in the central North Pacific, originating at the Hawaiian hotspot.

Hazard Mitigation Effort to minimize individual and community vulnerability to disasters.

Hemipelagic Sediment Sediment that drapes upper and middle continental slopes around the world.

High-Grade Metamorphism Metamorphism under conditions of high temperature and high pressure.

Historical Geology Study of Earth's history.

Hominidae Family of organisms in which humans are a member.

Homologous Structures Similarity in bone structure.

Hotspot Single, stationary sources of magma in the mantle that periodically erupt onto the crust, forming active volcanoes.

Humus Layer Soil layer rich in partially decomposed organic plant debris.

Hurricane Tropical cyclonic storm having winds that exceed 119 km/h.

Hutton, James (1726–1797) Scottish farmer and naturalist, known as the founder of modern geology, originator of uniformitarianism and deep time (extended Earth history).

Hydraulic Gradient Slope of the water table.

Hydraulic Pressure Fact that pressure on one part of a water body will be transmitted throughout the body; movement of water from regions of high pressure to regions of low pressure.

Hydrogeologist Geologist who studies groundwater.

Hydrologic Cycle Processes that drive water movement among the atmosphere, the ocean, and the crust.

Hydrolysis Weathering process involving silicate minerals and their decomposition by acidic water; chemical reaction in which the H^+ or OH^- ions of water replace ions of a mineral.

Hydrosphere Liquid water on Earth, including atmosphere, ocean, rivers, and lakes.

Hydrothermal Fluid Chemically enriched hot fluid in the crust when groundwater comes into contact with hot rock.

Hydrothermal Vein Fillings Newly crystallized minerals in cracks and fractures, precipitated from hot fluid enriched in dissolved minerals leached from hot rock associated with intrusions.

Hydrothermal Vent Hot water springs in the seafloor that occur when cold seawater seeps downward through the cracks in the crust and meets the hot intrusive rock below.

Hypothesis Testable educated guess that attempts to explain a phenomenon.

I

Ice Ages Cool climate periods, also called glacial periods, during which as much as 30% of Earth's surface is covered by glaciers, sea level lowers, and other global phenomena occur.

Ice Wedging Process in which the growth of ice crystals, due to the increased volume of water when it turns to ice, forces a joint to split open.

Igneous Evolution Process in which igneous magma fractionally crystallizes and partially melts to produce myriad igneous products, which enter the rock cycle, and thus produce diversity throughout all rock families.

Igneous Mineral Mineral created by crystallization from cooling magma.

Igneous Rock Rock produced by crystallization of magma.

Index Minerals Metamorphic mineral whose first appearance marks a specific zone of metamorphism.

Induced Fission Inducing a material to radioactive decay.

Industrial Revolution Period beginning in the mid-1800s in which there was rapid population growth; activities such as deforestation and burning fossil fuels increased, affecting the mixture of gases in the atmosphere.

Inner Core Rock body at Earth's center thought to be composed of solid metal alloy consisting mostly of iron and nickel (about 5000°C).

Insolation Amount of energy Earth receives from the Sun.

Insoluble Residue Mineral product of weathering.

Interglacials Warm intervals in global climate.

Intergovernmental Panel on Climate Change (IPCC) International panel created by the United Nations and the World Meteorological Organization to address the complex issue of global warming.

Interior Highlands Province Encompassing parts of Texas, Oklahoma, Arkansas, and Missouri, it is composed of sequences of folded Paleozoic sedimentary rocks underlain by rocks of Precambrian age.

Interior Plains Province Broad, generally flat area of the central United States between the Rocky Mountains and the Appalachians that is composed of flat-lying or gently dipping sedimentary strata.

Intermediate Igneous rock whose composition is between mafic and felsic.

Intrusion Movement of magma through the crust.

Intrusive Igneous Rock Igneous rock formed by crystallization of magma below Earth's surface.

Ion Atom with a net charge from the gain or loss of electrons.

Iron Catastrophe Event in Earth's history in which the planet's temperature passed the melting point of iron (1538°C), which resulted in the internal layers that characterize Earth today.

Iron Oxide Compound composed of oxidized iron (e.g., hematite; Fe_2O_3).

Island Arc Chain of insular andesitic stratovolcanoes on oceanic crust, parallel to a seafloor trench, produced by partial melting of upper mantle at a subduction zone.

Isostasy Process that restores equilibrium when mass of crust changes.

Isotope Atoms of an element having the same atomic number but differing mass numbers.

J

Joints   Openings or "partings" in rock where no lateral displacement has occurred.

K

Karst Topography   Assemblage of topographic forms resulting from dissolution of carbonate bedrock and consisting primarily of closely spaced sinkholes.

L

Lahar   Mudflow on a volcano typically composed of fluidized ash.

Landfill   Location used for the mass disposal of municipal and industrial waste.

Lapilli   Larger pyroclastic fragments (tephra), measuring 2.5 mm to 63 mm across.

Large Igneous Province   Voluminous emplacements of predominantly mafic extrusive and intrusive rock from fissure eruptions.

Large-Scale Volcanic Terrain   Large-volume volcanically derived crust with complex origin, typically not originating with "classic" central vent volcano.

Laterite   Iron oxide-rich soil produced by weathering.

Laurentian Upland Province   Small corner of the Canadian Shield that reaches into Wisconsin and Minnesota, consisting of the shield and the stable platform.

Lava   Magma that reaches Earth's surface through a volcanic vent.

Lithification   Process that converts sediment into sedimentary rock.

Lithosphere   Consists of the upper mantle and the crust.

Little Ice Age   Two- to four-century-long cold climate interval ending in the nineteenth century.

Loess   Wind-deposited silt, sometimes accompanied by some clay and fine sand.

Longshore Current   Wave-generated current, within the surf zone, that flows parallel to the coast.

Longshore Drift   Sediment transport that moves in a longshore current.

Love Wave   Type of seismic surface wave that moves the ground from side to side.

Low-Grade Metamorphism   Metamorphism under conditions of low temperature and low pressure.

Lyell, Charles (1797–1875)   Author of *Principles of Geology*, which popularized uniformitarianism—the idea that Earth is shaped by slow-moving forces still in operation today. Lyell was a close and influential friend of Charles Darwin.

M

Maar Vent   Low-relief, broad volcanic crater formed by shallow explosive eruptions typically where groundwater comes in contact with magma.

Mafic   Igneous rocks composed of minerals with the highest melting temperatures, mostly dark in color, and relatively enriched in iron, magnesium, and calcium.

Magma   Molten rock, together with any minerals, rock fragments, grain and dissolved gases, that forms when temperatures rise and melting occurs in the mantle or crust.

Magma Differentiation   Changing of the composition of magma, by several types of processes, as it migrates into and through the crust.

Mainshock   Largest earthquake in a cluster.

Mantle   Thick shell of dense, hot, rock that surrounds Earth's core.

Marine Geology   Geophysical, geochemical, sedimentological, and paleontological investigations of the ocean floor and coastal margins.

Marine Isotopic Stages   Numbering system developed to name the alternating warm periods and ice ages of recent geologic history based on isotopic climate proxies.

Marine Transgression   Shoreline movement toward the land; process in which ocean waters rose after the last ice age, flooding the coast and moving the shoreline inland.

Mass Extinctions   Events in which large numbers of species permanently die out within a very short period.

Mass Number   Number of neutrons plus protons in the nucleus of an atom.

Mass Wasting   Process in which gravity pulls soil, debris, sediment, and broken rock down a hillside or cliff.

Meandering Channel   Channel with loop-like bends that evolves over time.

Mélange   Chaotic mixture of broken, jumbled, and thrust-faulted rock above a subduction zone.

Meltwater   Layer of water below a glacier, formed when increased pressure toward the base of a glacier lowers the melting point of ice and by the accumulation of percolating meltwater from above.

Mesozoic Era   Middle era of the Phanerozoic Eon.

Metamorphic Facies   Set of metamorphic mineral assemblages repeatedly found together and designating certain pressure and temperature conditions.

Metamorphic Mineral   Mineral created by recrystallization of rock under conditions of high heat and pressure within the crust.

Metamorphic Rock   Rock whose original compounds, textures, or both, have been transformed by reactions in the solid state as a result of high temperature, high pressure, or both.

Metamorphism   All changes in mineral assemblage, rock texture, or both, that take place in rocks in the solid state within Earth's crust as a result of changes in temperature and pressure.

Methane   Colorless, odorless, clean-burning fossil fuel.

Mid-Ocean Ridge   Ridge that develops along a rift zone in the ocean due to high heat flow and magma upwelling.

Milky Way Galaxy   Galaxy that contains billions of stars, including the Sun, held together by mutual gravitation, in which Earth is located.

Mineral   Naturally occurring, inorganic, crystalline solid with a definite, but sometimes variable, chemical composition.

Mineral Resources   Useful rock and mineral materials that have value in their natural state or in a processed state.

Mohorovicic Discontinuity   Boundary between crust and mantle.

Mohs Hardness Scale   Scale that uses numerical values from 1 to 10 as a relative measure of hardness based on comparison with specific minerals.

Monocline   Local steepening in an otherwise uniformly dipping layer of strata.

Monogenetic Field   Collections of volcanic vents and flows, sometimes numbering in the hundreds or thousands typically from a single magma source.

Mountain Belt   Group of several mountain ranges, usually with related histories extending across a broad range of time.

Mud   Unconsolidated, water-saturated sediment of silt and clay grains.

Mud Flow   Flowing mass of water-saturated silt and clay grains.

N

Natural Attenuation   Technique for groundwater remediation involving shutting down any source of ongoing pollution, defining and isolating a known contamination plume, and allowing natural processes to reduce the toxic potential of contaminants.

Natural Law   Successful theory that over time has been shown to be always true for a broad array of natural processes.

Natural Resources   Materials that occur in nature and are essential or useful to humans, such as water, air, building stone, topsoil, and minerals.

Natural Selection   Process by which favorable traits that are heritable become more common in successive generations of a population of reproducing organisms.

Neritic Sediment   Sediment that collects on the seafloor between the shoreline and the edge of the continental shelf.

Nonfoliated Texture   Texture found in metamorphic rocks that are undergoing contact metamorphism or in rocks that, when viewed with the unaided eye, show no evidence of foliated texture.

Nonrenewable   Resource whose total amount available on Earth is reduced every time it is used, and what is used is not replenished within a human generation.

Nonrenewable Energy Resources Oil, coal, natural gas, and uranium used for energy, whose total amount available on Earth is reduced every time it is used, and what is used is not replenished within a human generation.

Normal Fault Fault, generally steeply inclined, along which the hanging-wall block has moved relatively downward.

Nuclear Fusion Thermonuclear reaction in which the nuclei of light atoms join to form nuclei of heavier atoms, releasing energy and creating new heavier elements.

Nucleation Initial grouping of a few atoms that starts the process of crystal growth.

O

Ocean Acidification Increasing acidity, in the form of carbonic acid, of oceanic waters.

Ocean–Continent Convergent Boundary Formed where oceanic crust has collided with continental crust, which causes the denser oceanic crust to recycle below the continental crust.

Ocean–Ocean Convergent Boundary Formed where two plates composed of oceanic crust converge; typically the denser (older) will subduct.

Oceanic Crust Crust beneath the abyssal oceans, generally basaltic in composition.

Oceanic Fracture Zone Type of transform boundary consisting of inactive and active portions.

Oceanic Trenches Arc-linear seafloor depressions at plate boundaries where an oceanic plate subducts into the mantle beneath a converging plate.

Orbital Parameters There are six Keplerian parameters derived from the "Laws of Planetary Motion," but in paleoclimatology, typically there are three: axial tilt, eccentricity, and precession.

Ore Aggregate of minerals from which one or more minerals can be extracted profitably.

Ore Minerals Minerals that are smelted more readily than others.

Orogenesis Mountain building resulting from continent–continent plate convergence.

Orographic Effect Topographic effect in the water cycle that leads to the rate of condensation exceeding the rate of evaporation; causes the windward side of mountains to be humid and the leeward side to be arid or semi-arid.

Outcrop Place where layers of rock are exposed.

Outer Core Surrounding the inner core, it is extremely hot, but is melted because it is under less pressure than the inner core.

Oversteepening Slope failure due to the addition of material to the top of a slope, or due to the removal at the base of a slope, increasing the slope angle so that it is unstable.

Oxbow Lake Crescent-shaped, shallow lake or marsh occupying the abandoned channel of a meandering stream.

Oxidation Process in which a chemical element loses electrons by bonding with oxygen.

P

P Wave Meaning *primary wave*, a type of seismic body wave that compresses and stretches the material through which it moves.

Pacific Mountain System Geologically young, tectonically active highland that forms a rugged, mountainous landscape along the Pacific coast.

Pahoehoe Smooth, ropy-surfaced basaltic lava flow.

Paleoclimatology Study of past climate.

Paleomagnetism Remnant magnetism in ancient rock recording past conditions of the geomagnetic field.

Paleozoic Era Oldest era of the Phanerozoic Eon.

Pangaea Historical "supercontinent" in which Europe, North America, Africa, Antarctica, South America, and all continental lands were formerly joined together.

Partial Melting Melting of rock that occurs when some compounds melt while others remain solid.

Passive Margin Tectonically quiet margin of a continent.

Pegmatite Exceptionally coarse-grained intrusive igneous rock, commonly granitic in composition.

Pelagic Sediment Sediment consisting of the remains of marine organisms in the open ocean.

Periods Millions to tens of millions of years.

Permafrost Soil whose moisture is frozen for more than two years.

Permeability Degree to which a fluid is able to flow through the crust; the interconnectedness of pore space.

Petroleum Gaseous, liquid, and semisolid substances consisting of compounds of carbon and hydrogen.

Phreatomagmatic Eruption Explosions caused by rapid expansion of steam when magma comes in contact with groundwater.

Phylogeny Evolutionary lineage of an organism.

Physical Geology Study of materials that compose Earth; the chemical, biological, and physical processes that create them and the ways in which they are organized and distributed throughout the planet.

Physical Weathering Type of weathering that occurs when rock is fragmented by physical processes that do not change its chemical composition.

Pillow Lava Submarine deposits of talus and bulbous rock created when basalt erupting onto the seafloor is quickly frozen by the cold seawater.

Placer Deposit Deposit formed when weathered soil erodes into a channel and dense mineral grains are concentrated by the moving water.

Planetesimal Accretion Largely collisional process by which matter gathered to form planets.

Plate Boundaries Locations within the crust where lithospheric plates meet.

Plate Tectonics Special branch of geophysics that deals with the processes by which the lithosphere is moved laterally over the asthenosphere.

Playa Lake Shallow lake held by the floor of a basin after a rain event, the hydrology of which is controlled solely by precipitation and evaporation.

Plinian Eruption Large volcanic explosions that push thick, dark columns of pyroclasts and gas high into the stratosphere.

Plutonic Igneous Rock Intrusive rock that occurs at depths within the crust.

Polluted Runoff Contaminated runoff that comes from a variety of sources, including roads, gutters, lawns, construction sites, and contains contaminants.

Porosity Measure (percentage) of the amount of empty space contained in a rock.

Porphyry Copper Type of hydrothermal deposit containing disseminated minerals in country rock.

Pressure Release Melting Melting that occurs when pressure is lowered.

Primary Sedimentary Structure Feature that reflects physical processes that act on sediments.

Principle of Uniformitarianism Principle stating that Earth is very old, natural processes have been essentially uniform through time, and that study of modern geologic processes is useful in understanding past geologic events.

Pumice Volcanic rock with high gas content (floats in water).

Pyroclastic Debris Erupted bits of rock.

Pyroclastic Flow Hot, highly mobile avalanche of tephra that rushes down the flank of a volcano during eruption.

Q

Quaternary Period Most recent period of geologic time (2.6 mya to present).

R

Radioactive Decay Natural process of transformation by which the nucleus of one kind of atom transforms into the nucleus of another kind of atom and emits energy in the process.

Radioisotopes Radioactive atoms, each of which decay at a unique, fixed rate.

Radiometric Dating Uses radioactive decay to estimate the age of geologic samples.

Rayleigh Wave Seismic surface wave that rolls along the ground, moving up and down and forward to backward in the direction of movement.

Reasoning by Analogy Scientific thinking in which one thing is inferred to be similar to another, usually the more simplified thing, in order to promote understanding.

Recharge   Process through which precipitation adds new water to the groundwater system.

Reclamation   Actions such as stream preservation, protecting groundwater, preserving topsoil, handling waste rock, controlling erosion, sediment, and dust, and restoring the shape of the land after a mine operation has ceased.

Recrystallization   Growth of new minerals in a rock by recycling old minerals in the solid state.

Refraction   Change in direction when a wave passes from one medium to another.

Regional Metamorphism   Metamorphism affecting large volumes of crust typically involving high pressure conditions.

Regolith   Soil, debris, sediment, and broken rock.

Relative Dating   Determines the order of formation among geologic events based on the relationship of layers of rock to one another.

Renewable Energy   Energy taken from inexhaustible sources.

Reverse Fault   Fault, generally steeply inclined, along which the hanging-wall block has moved upward relative to the foot wall.

Rhyolitic Lava   Lava that is felsic in composition.

Richter Scale   Measurement of the largest amplitude on a seismogram, describing an earthquake's magnitude.

Rift Valley   Linear, fault-bounded valley along a divergent plate boundary or spreading center.

Ring of Fire   Outer boundary of the Pacific Plate where it meets the surrounding plates, creating the most seismically and volcanically active zone in the world.

Rip Current   High-velocity, wave-generated current flowing seaward from the shore.

Ripple   Migrating, small wave of sand deposited by wind or water.

Rock   Solid aggregation of minerals.

Rock Cleavage   Property by which rock breaks into plate-like fragments along flat planes.

Rock Cycle   Recycling of rock material, in the course of which rock is created, destroyed, and altered through the operation of internal and external Earth processes; powered by gravity, geothermal energy, and solar energy.

Rockfall   Free fall of detached bodies of bedrock from a cliff or steep slope.

Rock Record   History of geologic events preserved in rocks, fossils, and sediments of the crust.

Rock Slide   Sudden and rapid downslope movement of detached masses of bedrock across an inclined surface.

Rocky Shoreline   Shoreline where solid rock rather than unconsolidated sediment characterizes the water's edge.

Rocky Mountains   Series of ranges that stretch from Canada to New Mexico.

Root Wedging   Fracture of rocks created by the pressure put forth by growing roots; physical weathering.

S

S Wave   Meaning *secondary wave*, a type of seismic body wave that can only move through solid materials, moving up and down or side to side.

Sahel   Region of North Africa that is the border between the shifting sands of the Sahara in the north and the moist tropical rainforest regions of central Africa.

Salinity   Concentration of dissolved sulfur, calcium, magnesium, bromine, potassium, and bicarbonate in water.

Sand   Rock and mineral particles 0.0625 mm to 2 mm in diameter.

Sand Bar   Shallow areas along a coast due to sand accumulation.

Scientific Method   Use of reasoning to interpret evidence in order to construct a testable hypothesis as an explanation of a natural phenomenon; testing of hypotheses.

Seafloor Spreading   Hypothesis proposed during the early 1960s in which lateral movement of oceanic crust away from mid-ocean ridges was postulated and explained by mantle convection.

Sea-Level Rise   Increase in the level of the world's oceans, typically at a rate and with variability that is unique to a location.

Seamount   Extinct shield volcano capped by fossil reef.

Sediment   Particles of mineral, broken rock, and organic debris that are unconsolidated, formed by weathering.

Sedimentary Mineral   Mineral created by crystallization among Earth surface environments including weathering.

Sedimentary Quartz   Weathering product that results from crystallization of silica.

Sedimentary Rock   Rock formed by organic or inorganic precipitation or by deposition and cementation of sediment.

Seismic Shaking Hazards   Damaging, dangerous processes related to seismicity, including building damage and collapse.

Seismic Tomography   Method of imaging Earth structure using differences in seismic wave behavior.

Seismic Waves   Energy produced by an earthquake.

Seismometer   Instrument that detects earthquakes.

Shear Stress   Directed stress that shears rock side to side.

Shield Volcano   Volcano that emits fluid lava, typically basaltic, and builds up a broad dome-shaped edifice with a low surface slope.

Silicate   Type of mineral composed of silica.

Silicate Structure   Mineral structure of silica tetrahedra.

Silt   Particle between 0.002 mm and 0.063 mm.

Sinkhole   Depression created when an underground cavern grows, coalesces, and undermines an area until the roof eventually collapses.

Slope Stability   Degree to which a slope or cliff is stable from mass wasting.

Slump   Landform that occurs when regolith suddenly drops (or rotates) a short distance down a slope.

Smelting   Heating or melting a mineral to obtain the metal it contains.

Soil Classification   Classifying soils on the basis of several related factors including mineralogy, texture, organic composition, moisture, and others.

Soil Profile   Set of regolith layers formed when percolation forms and moves dissolved ions that may recombine to form clay minerals.

Solar Energy   Radiant light and heat from the Sun.

Solar Nebula Hypothesis   Hypothesis stating that the Sun, planets, and other objects orbiting the Sun originated at the same time from the same source through the collapse and condensation of a planetary nebula, and have evolved in varying ways since that time.

Solar System   Sun, planets, and other objects trapped by the Sun's gravity.

Solar Wind   Subatomic particles and radiation originating from the Sun.

Solifluction   Soil creep that occurs in water-saturated regolith on permafrost.

Sorting   Separation of grains by density and size.

Spheroidal Weathering   Exfoliation characterized by preferential weathering of edges and corners.

Spring   Groundwater emerging naturally at the ground surface.

Stalactites   Downward-hanging inorganic precipitates in a cave.

Stalagmites   Upward-oriented inorganic precipitates on a cave floor.

Steno, Nicholas (1638–1686)   Considered the father of geology and stratigraphy, credited with "Law of Superposition," "Principle of Original Horizontality," "Principle of Lateral Continuity," and other fundamental geological principles.

Storm Surge   Coastal flooding resulting from high sea level under a storm.

Storm Water Runoff   Water that flows overland during a rainstorm.

Straight Channel   Channel whose underlying topography and structure of the bedrock force the channel to be straight.

Strain   Change in shape and volume of a rock caused by stress.

Stratigraphic Principles   Seven basic but important concepts that provide the foundation for relative dating, including the principles of superposition, original horizontality, original lateral continuity, cross-cutting relationships, inclusions, unconformities, and fossil succession.

Stratovolcano   High, steep-sided volcano that is formed typically from intermediate magma, often above convergent plate margins.

Stream   Channel that carries water.

Stream Piracy   Headward erosion that breaches a drainage divide, intersects another channel, and captures its flow.

Stress Magnitude and direction of a deforming force exerted on a surface.

Strike Compass direction of a horizontal line that marks the intersection of an inclined plane with Earth's surface.

Strike-Slip Fault Fault on which displacement has been horizontal and parallel to the strike of the fault.

Stromatolites Microscopic bacteria that form dome-like structures by trapping sediments between fine strands.

Structural Geology Branch of geology concerned with determining the architecture of the crust and inferring from it the nature of folds and faults in the rock that are usually below ground and therefore not in full view.

Subduction Zones Plate margin where one plate is recycled into the mantle beneath another.

Submarine Canyon Steep-sided valley on the continental shelf or slope resembling a river-cut canyon on land.

Submarine Fan Thick, sedimentary deposit on the continental rise at the end of a submarine canyon.

Submergent Coast Coast where sea-level rise exceeds the rate of land uplift.

Summit Crater Crater situated atop a volcanic mountain formed from erupted volcanic deposits, such as lava and tephra.

Superfund Program U.S. government program whose goal is to locate, investigate, and clean up the worst toxic-waste sites nationwide.

Surface Waves Seismic waves that travel along Earth's surface.

Sustainable Resource Any natural product used by humans that meets the needs of the present without compromising the ability of future generations to meet their own needs.

Syncline Concave upward fold or arch with an axial plane that is vertical, having the youngest rocks in the center of the fold.

T

Talus Slope of fallen rock that collects at the base of cliffs and steep hillsides.

Tensional Stress Directed stress that stretches rock.

Tephra Loose assemblage of airborne pyroclasts.

Terrestrial Planets Innermost planets of the solar system (Mercury, Venus, Earth, and Mars) that have high densities and generally silicate and metallic compositions.

Thalweg Line running along a channel connecting the deepest parts of the stream, marking the natural direction of the watercourse.

Theory Hypothesis that passes repeated and rigorous tests and is applicable to a broad phenomenon.

Theory of Evolution Theory which states that changes occur in the inherited traits of a biological population from one generation to the next.

Thermohaline Circulation Global pattern of ocean circulation propelled by the sinking of dense cold and salty water.

Thrust Fault Low-angle reverse fault with dip less than 20°.

Tidal Wetland Marsh flooded by high tide on a daily basis and characterized by growth of salt-tolerant plants in the muddy substrate.

Tides Regular and predictable oscillations in the sea surface due to gravitational forces.

Trade Winds Flow of air from subtropical regions toward the equator that ships carrying goods relied on for several centuries.

Transform Boundaries Plate boundary characterized by lateral slip past one another.

Trap Closed layer of porous and permeable reservoir rock that is sealed by an impermeable cap rock usually composed of shale or evaporite layers.

Tributary Stream that joins a larger stream.

Tsunami Series of waves in the ocean caused by rapid movement of the seafloor.

Tuff Rock formed when pyroclastic debris accumulates on the ground and solidifies.

Turbidity Current Gravity-driven current consisting of a dilute mixture of sediment and water, which has a density greater than the surrounding water and travels down the continental slope and rise.

U

Ultramafic Igneous rock enriched in iron and magnesium.

Unconformity Removal of rock or sediment by erosion, creating a gap in the temporal record of Earth history.

Undercutting Process in which rock or regolith is removed from the base of a hillside, such as by stream erosion, thereby increasing the slope.

Uniformitarianism Concept stating that Earth's history is best explained by observations of modern processes, suggesting that geologic principles have been uniform over time.

Uplift Raising of crust by geologic processes.

V

Vent Opening in the crust through which lava flows.

Ventifacts Rocks with unusual shapes and flat faces that have been abraded by wind-blown sediment.

Vestigial Structures Vestiges of body parts that were used by ancestral forms but are now nonessential.

Volcanic Arc Line of explosive volcanoes.

Volcanic Mountain Mountain formed at a volcanic arc above a subduction zone.

Volcanic Outgassing Process in which moisture trapped in minerals is released by melting within Earth's interior.

Volcano Landform that releases lava, gas, or ashes, or has done so in the past.

Volcanologist Scientist who studies volcanoes.

W

Wadati-Benioff Zone Steeply dipping plane defined by earthquake foci that corresponds to a subducting oceanic slab.

Washover Process of marine inundation carrying sand from the ocean side of a barrier island to the lagoon side.

Water Table Upper surface of the saturated zone of groundwater.

Watershed Total area feeding water to a stream.

Wave Refraction Process by which the direction of waves, moving into shallow water at an angle to the shoreline, is changed by interaction with the seafloor.

Wave-Generated Currents Currents created by the movement of waves.

Welded Tuff Welded mass of glass shards solidified when pyroclastic flows lose their energy and come to rest as thick beds of ash and lapilli.

Wetland Land whose soil is saturated with moisture either permanently or seasonally.

Z

Zone of Accumulation Upper zone on a glacier, covered by snowfall, and representing an area of net gain in mass.

Zone of Wastage Area of a glacier where ice is lost not only by melting, but also by sublimation and calving.

INDEX